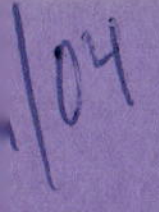

3000 800055 83289
St. Louis Community College

D0211939

WITHDRAWN

The Museum of
Broadcast Communications

Encyclopedia of Radio

The Museum of
Broadcast Communications

Encyclopedia of Radio

VOLUME 3
O–Z

Editor
CHRISTOPHER H. STERLING

Consulting Editor
MICHAEL C. KEITH

FITZROY DEARBORN
AN IMPRINT OF THE TAYLOR & FRANCIS GROUP

NEW YORK • LONDON

Published in 2004 by
Fitzroy Dearborn
An imprint of the Taylor and Francis Group
29 West 35th Street
New York, NY 10001

Published in Great Britain by
Fitzroy Dearborn
An imprint of the Taylor and Francis Group
11 New Fetter Lane
London EC4P 4EE

10 9 8 7 6 5 4 3 2 1

Library of Congress Cataloging-in-Publication Data

The Museum of Broadcast Communications encyclopedia of radio / editor,
Christopher H. Sterling ; consulting editor, Michael Keith.
p. cm.
ISBN 1-57958-249-4 (set : alk. paper) -- ISBN 1-57958-431-4 (vol. 1 :
alk. paper) -- ISBN 1-57958-432-2 (vol. 2 : alk. paper) -- ISBN
1-57958-452-7 (vol. 3 : alk. paper)
1. Radio--Encyclopedias. 2. Radio programs--Encyclopedias. I.
Sterling, Christopher H., 1943- II. Museum of Broadcast Communications.
III. Title.
TK6544.M84 2004
384.54'03--dc22

2003015683

First published in the USA and UK 2004

Typeset by Andrea Rosenberg
Printed by Edwards Brothers
Cover design by Peter Aristedes, Chicago Advertising and Design, Chicago, Illinois

Cover photos: (from upper left): Edward R. Murrow; *National Barn Dance;* Eve Arden of *Our Miss Brooks;* Irna Phillips; Earl Nightingale; Ed Wynn; *The Shadow;* Groucho Marx, *You Bet Your Life;* Eleanor Roosevelt; William S. Paley

CONTENTS

Advisers and Contributors vii

List of Entries xi

Volume 1, entries A–E 1–560

Volume 2, entries F–N 561–1032

Volume 3, entries O–Z 1033–1595

Contributor Affiliations 1597

Index 1607

ADVISERS

Stanley R. Alten
Syracuse University

Frank J. Chorba
Washburn University

Lynn A. Christian
International Radio Consultant

Ed Cohen
Clear Channel Broadcasters

Norman Corwin
Radio Playwright

Susan J. Douglas
University of Michigan

James E. Fletcher
University of Georgia

Robert S. Fortner
Calvin College

Stan Freberg
Radio Producer

Donald G. Godfrey
Arizona State University

Marty Halperin
Pacific Pioneer Broadcasters

Gordon H. Hastings
Broadcasters' Foundation

Robert Henabery
Radio Producer

Robert L. Hilliard
Emerson College

Michele Hilmes
University of Wisconsin, Madison

Chuck Howell
Library of American Broadcasting
University of Maryland

Stanley Hubbard
Hubbard Communications

John D. Jackson
Concordia University

Jack Mitchell
University of Wisconsin, Madison

Graham Mytton
BBC World Service (ret.)

Horace Newcomb
Director, Peabody Awards
University of Georgia

Peter B. Orlik
Central Michigan University

Ed Shane
Shane Media Services

Marlin R. Taylor
XM Satellite Radio

CONTRIBUTORS

Michael H. Adams
Alan B. Albarran
Pierre Albert
Craig Allen
Steven D. Anderson
Larry Appelbaum
Edd Applegate
Sousan Arafeh
John S. Armstrong
Philip J. Auter
Robert K. Avery
Glenda R. Balas
Mary Christine Banwart
Warren Bareiss
Ray Barfield
Kyle S. Barnett
Douglas L. Battema
Mary E. Beadle
Christine Becker
Johnny Beerling
Alan Bell
Louise Benjamin
ElDean Bennett
Marvin Bensman
Jerome S. Berg
Rosemary Bergeron
William L. Bird, Jr.
Howard Blue
A. Joseph Borrell
Douglas A. Boyd
John Bradford
L. Clare Bratten
Mark Braun
Jack Brown
Michael Brown
Robert J. Brown
Donald R. Browne
John H. Bryant
Joseph G. Buchman
Karen S. Buzzard
Paul Brian Campbell
Dom Caristi
Ginger Rudeseal Carter
Dixon H. Chandler II
Frank J. Chorba
Lynn A. Christian
Claudia Clark
Kathleen Collins
Jerry Condra
Harold N. Cones
Bryan Cornell
Elizabeth Cox
Steve Craig
Tim Crook
Marie Cusson
Keri Davies
E. Alvin Davis
J.M. Dempsey
Corley Dennison
Neil Denslow
Steven Dick
John D.H. Downing
Pamela K. Doyle
Christina S. Drale
Susan Tyler Eastman
Bob Edwards
Kathryn Smoot Egan
Lyombe Eko
Sandra L. Ellis
Ralph Engelman
Erika Engstrom
Stuart L. Esrock
Charles Feldman
Michel Filion
Howard Fink
Seth Finn
Robert G. Finney
Margaret Finucane
James E. Fletcher
Corey Flintoff
Joe S. Foote
Robert C. Fordan
Robert S. Fortner
James C. Foust
Ralph Frasca
James A. Freeman
Elfriede Fürsich
Charles F. Ganzert
Ronald Garay
Philipp Gassert
Judith Gerber
Norman Gilliland

Donald G. Godfrey
Douglas Gomery
Jim Grubbs
Joanne Gula
Paul F. Gullifor
Linwood A. Hagin
Donna L. Halper
Tona J. Hangen
Margot Hardenbergh
Jeffrey D. Harman
Dorinda Hartmann
Gordon H. Hastings
Joy Elizabeth Hayes
John Allen Hendricks
Alexandra Hendriks
Ariana Hernandez-Reguant
Robert L. Hilliard
Jim Hilliker
Michele Hilmes
John Hochheimer
Jack Holgate
Herbert H. Howard
Chuck Howell
Kevin Howley
W.A. Kelly Huff
Peter E. Hunn
John D. Jackson
Randy Jacobs
Glen M. Johnson
Phylis Johnson
Sara Jones
Lynda Lee Kaid
Stephen A. Kallis, Jr.
Steve Kang
Michael C. Keith
Ari Kelman
Colum Kenny
John Michael Kittross
Frederica P. Kushner
Philip J. Lane
Matthew Lasar
Laurie Thomas Lee
Renée Legris
Frederic A. Leigh
Lawrence W. Lichty
Lucy A. Liggett
Val E. Limburg
Robert Henry Lochte
Jason Loviglio
Gregory Ferrell Lowe
Christopher Lucas
Mike Mashon
Marilyn J. Matelski
Peter E. Mayeux
Dennis W. Mazzocco
Thomas A. McCain
Jeffrey M. McCall
David McCartney
Tom McCourt
Brad McCoy
Allison McCracken
Drew O. McDaniel
Michael A. McGregor
Robert McKenzie
Elizabeth McLeod
Mike Meeske
Fritz Messere
Colin Miller
Toby Miller
Bruce Mims
Jack Minkow
Jack Mitchell
Jason Mittell
Barbara Moore
Matthew Murray
Graham Mytton
Gregory D. Newton
Greg Nielsen
D'Arcy John Oaks
William F. O'Connor
Cary O'Dell
Robert M. Ogles
Ryota Ono
Peter B. Orlik
Pierre-C. Pagé
Brian T. Pauling
Manjunath Pendakur
Douglas K. Penisten
Stephen D. Perry
Patricia Phalen
Steven Phipps
Joseph R. Piasek
Gregory G. Pitts
Mark Poindexter
Tim Pollard
Robert F. Potter
Alf Pratte
Patricia Joyner Priest
Dennis Randolph
Lawrence N. Redd
David E. Reese
Patton B. Reighard

Andre Richte
Edward A. Riedinger
Terry A. Robertson
Melinda B. Robins
América Rodríguez
Eric W. Rothenbuhler
Richard Rudin
Joseph A. Russomanno
Anne Sanderlin
Erica Scharrer
Steven R. Scherer
Karl Schmid
Clair Schulz
Ed Shane
Pam Shane
Mitchell Shapiro
Jason T. Siegel
Ron Simon
B.R. Smith
Ruth Bayard Smith
Lynn Spangler
David R. Spencer
David Spiceland
Laurie R. Squire
Michael Stamm
Christopher H. Sterling
Will Straw
Michael Streissguth
Mary Kay Switzer
Rick Sykes
Marlin R. Taylor
Matt Taylor
Herbert A. Terry
Richard Tiner
Regis Tucci
David E. Tucker
Don Rodney Vaughan
Mary Vipond
Randall Vogt
Ira Wagman
Andrew Walker
Peter Wallace
Jennifer Hyland Wang
Richard Ward
Mary Ann Watson
Brian West
Gilbert A. Williams
Sonja Williams
Wenmouth Williams, Jr.
Roger Wilmut
Stephen M. Winzenburg
Richard Wolff
Roosevelt "Rick" Wright, Jr.
Edgar B. Wycoff
Thimios Zaharopoulos

LIST OF ENTRIES

Volume 1
A.C. Nielsen Company
Adult Contemporary Format
Adventures in Good Music
Advertising
Advertising Agencies
Affirmative Action
Africa
Africa No. 1
African-Americans in Radio
Album-Oriented Rock Format
Aldrich Family
Alexanderson, E.F.W.
All India Radio
All News Format
All Night Radio
All Things Considered
Allen, Fred
Allen, Mel
Alternative Format
Amalgamated Broadcasting System
Ameche, Don
American Broadcasting Company
American Broadcasting Station in Europe
American Family Robinson
American Federation of Musicians
American Federation of Television and Radio Artists
American School of the Air
American Society of Composers, Authors, and Publishers
American Telephone and Telegraph
American Top 40
American Women in Radio and Television
America's Town Meeting of the Air
Amos 'n' Andy
AM Radio
Antenna
Arab World Radio
Arbitron
Archers
Armed Forces Radio Service
Armstrong, Edwin Howard
Asia
Association for Women in Communications
Audience
Audience Research Methods
Audimeter
Audio Processing
Audio Streaming
Audiotape
Auditorium Testing
Australia
Australian Aboriginal Radio
Automation
Automobile Radios
Autry, Gene
Awards and Prizes
Axis Sally

Barber, Red
Barnouw, Erik
Bell Telephone Laboratories
Benny, Jack
Beulah Show
Beville, Hugh Malcolm
Big D Jamboree
Biondi, Dick
Birch Scarborough Research
Blacklisting
Black-Oriented Radio
Black Radio Networks

Block, Martin
Blore, Chuck
Blue Book
Blue Network
Blues Format
Board for International Broadcasting
Bob and Ray
Border Radio
Bose, Amar G.
Brazil
Brice, Fanny
Brinkley, John R.
British Broadcasting Corporation (BBC)
British Broadcasting Corporation: BBC Broadcasting House
British Broadcasting Corporation: BBC Local Radio
British Broadcasting Corporation: BBC Monitoring
British Broadcasting Corporation: BBC Orchestras
British Broadcasting Corporation: BBC Radio Programming
British Broadcasting Corporation: BBC World Service
British Commercial Radio
British Disk Jockeys
British Forces Broadcasting Service
British Pirate Radio
British Radio Journalism
Broadcast Education Association
Broadcast Music Incorporated
Broadcasting Board of Governors
Brokerage in Radio
Brown, Himan

Cable Radio
Call Letters
Can You Top This?
Canada
Canadian Radio Archives
Canadian Radio Policy
Canadian Radio Programming
Canadian Broadcasting Corporation
Canadian Radio and Multiculturalism
Canadian Radio and the Music Industry
Canadian Radio Drama
Canadian Radio Satire
Canadian News and Sports Broadcasting
Canadian Talk Radio
Cantor, Eddie
Capehart Corporation
Capital Radio
Captain Midnight
Car Talk
Cavalcade of America
Censorship
CFCF
CHED
Chenault, Gene
Children's Novels and Radio
Children's Programs
CHUM
Churchill, Winston Spencer
Citizen's Band Radio
CKAC
CKLW
Clandestine Radio
Clark, Dick
Classical Music Format
Classic Rock Format
Clear Channel Communications Inc.
Clear Channel Stations
Cold War Radio
College Radio
Collingwood, Charles
Columbia Broadcasting System
Columnists
Combo
Comedy
Commentators
Commercial Load
Commercial Tests
Communications Act of 1934
Community Radio
CONELRAD
Conrad, Frank
Conrad, William
Consultants
Contemporary Christian Music Format
Contemporary Hit Radio/Top 40 Format
Control Board/Audio Mixer
Controversial Issues
Cooke, Alistair
Cooper, Giles
Cooper, Jack L.
Cooperative Analysis of Broadcasting
Copyright
Corporation for Public Broadcasting
Corwin, Norman
Coughlin, Father Charles
Country Music Format

Cowan, Louis
Critics
Crosby, Bing
Crosley, Powel
Cross, Milton J.
Crutchfield, Charles H.
Crystal Receivers
Cuba

Daly, John Charles
Davis, Elmer
Dees, Rick
De Forest, Lee
Delmar, Kenneth
Demographics
Deregulation of Radio
Desert Island Discs
Developing Nations
Diary
Digital Audio Broadcasting
Digital Recording
Digital Satellite Radio
Dill, Clarence Cleveland
Disk Jockeys
Documentary Programs on U.S. Radio
Dolby Noise Reduction
Donahue, Tom
Don Lee Broadcasting System
Dr. Demento
Drake, Bill
Drake, Galen
Drama on U.S. Radio
Drama Worldwide
Drew, Paul
Duffy's Tavern
Duhamel, Helen
Dunbar, Jim
Dunlap, Orrin E.
Durante, Jimmy
Durham, Richard
DXers/DXing

Early Wireless
Earplay
Easy Aces
Easy Listening/Beautiful Music Format
Edgar Bergen and Charlie McCarthy
Editorializing
Education about Radio
Educational Radio to 1967
Edwards, Bob
Edwards, Ralph
Election Coverage
Ellis, Elmo
Emergencies, Radio's Role in
Emergency Broadcast System
Emerson Radio
"Equal Time" Rule
European Broadcasting Union
Evangelists/Evangelical Radio
Everett, Kenny

Volume 2
Fadiman, Clifton
Fairness Doctrine
Family Theater
Fan Magazines
Far East Broadcasting Company
Farm/Agricultural Radio
Faulk, John Henry
Federal Communications Commission
Federal Radio Commission
Female Radio Personalities and Disk Jockeys
Fessenden, Reginald
Fibber McGee and Molly
Film Depictions of Radio
Fireside Chats
First Amendment and Radio
Fleming, Sir John Ambrose
Flywheel, Shyster, and Flywheel
FM Radio
FM Trade Associations
Foreign Broadcast Information Service
Formats
France
Freberg, Stan
Frederick, Pauline
Freed, Alan
Freed, Paul
Free Form Format
Frequency Allocation
Fresh Air
Friendly, Fred
Fritzsche, Hans

Gabel, Martin
Gambling, John
Gangbusters
Gay and Lesbian Radio
General Electric
George Burns and Gracie Allen Show
German Wireless Pioneers

Germany
Gillard, Frank
Godfrey, Arthur
Goldbergs
Goldsmith, Alfred Norton
Goon Show
Gordon, Gale
Gospel Music Format
Grand Ole Opry
Great Gildersleeve
Greece
Green Hornet
Greenwald, James L.
Ground Wave
Group W
Gunsmoke

Ham Radio
Harvey, Paul
Hate Radio
Hear It Now
Heavy Metal/Active Rock Format
Herrold, Charles D.
Hertz, Heinrich
Hicks, Tom
High Fidelity
Hill, George Washington
Hill, Lewis
Hindenburg Disaster
Hispanic Radio
Hoaxes
Hogan, John V.L. "Jack"
Hollywood and Radio
Hooperatings
Hoover, Herbert
Hope, Bob
Horror Programs
Hottelet, Richard C.
Howe, Quincy
Hulbert, Maurice "Hot Rod" Jr.
Hummert, Frank and Anne

I Love a Mystery
Imus, Don
Infinity Broadcasting
Inner Sanctum Mysteries
Intercollegiate Broadcasting System
International Radio Broadcasting
International Telecommunication Union
Internet Radio
Ireland
Isay, David
Israel
Italy

Jack Armstrong, the All-American Boy
Jamming
Japan
Jazz Format
Jehovah's Witnesses and Radio
Jepko, Herb
Jewish Radio Programs in the United States
Jingles
Joyner, Tom
Judis, Bernice

Kaltenborn, H.V.
Karmazin, Mel
Kasem, Casey
KCBS/KQW
KCMO
KDKA
Keillor, Garrison
Kent, A. Atwater
Kesten, Paul
KFFA
KFI
KGO
KHJ
King Biscuit Flower Hour
King, Larry
King, Nelson
Kirby, Edward M.
Klauber, Edward A.
Kling, William
KMOX
KNX
KOA
KOB
KPFA
KRLA
KSL
KTRH
Kuralt, Charles
KWKH
Kyser, Kay
KYW

Landell de Moura, Father Roberto
Lazarsfeld, Paul F.
Let's Pretend
Lewis, William B.

Liberty Broadcasting System
Licensing
Lights Out
Limbaugh, Rush
Little Orphan Annie
Localism in Radio
Local Marketing Agreements
Lodge, Oliver J.
London Broadcasting Company
Lone Ranger
Long, Huey
Lord Haw-Haw
Lord, Phillips H.
Low-Power Radio/Microradio
Lum 'n' Abner
Lux Radio Theater

Mankiewicz, Frank
Ma Perkins
March of Time
Marconi, Guglielmo
Market
Marketplace
Markle, Fletcher
Maxwell, James Clerk
Mayflower Decision
McBride, Mary Margaret
McDonald, Eugene F., Jr.
McGannon, Don
McLaughlin, Edward F.
McLendon, Gordon
McNamee, Graham
McNeill, Don
McPherson, Aimee Semple
Media Rating Council
Mercury Theater of the Air
Metromedia
Metropolitan Opera Broadcasts
Mexico
Middle of the Road Format
Milam, Lorenzo
Minnesota Public Radio
Monitor
Morgan, Henry
Morgan, Robert W.
Mormon Tabernacle Choir
Morning Edition
Morning Programs
Morrow, Bruce "Cousin Brucie"
Motorola
Murray, Lyn
Murray the K
Murrow, Edward R.
Museums and Archives of Radio
Music on Radio
Music Testing
Mutual Broadcasting System

Narrowcasting
National Association of Broadcasters
National Association of Educational Broadcasters
National Association of Educational Broadcasters Tape Network
National Barn Dance
National Broadcasting Company
National Federation of Community Broadcasters
National Public Radio
National Radio Systems Committee
National Religious Broadcasters
National Telecommunications and Information Administration
Native American Radio
Netherlands
Network Monopoly Probe
New Zealand
News
News Agencies
Nightingale, Earl
Non-English-Language Radio in the United States
North American Regional Broadcasting Agreement
Nostalgia Radio

Volume 3
Oboler, Arch
Obscenity and Indecency on Radio
Office of Radio Research
Office of War Information
Oldies Format
One Man's Family
Osgood, Charles
Our Miss Brooks
Owens, Gary
Ownership, Mergers, and Acquisitions

Pacifica Foundation
Paley, William S.
Payola
Pay Radio
Peabody Awards and Archive

Perryman, Tom
Philco Radio
Phillips, Irna
Playwrights on Radio
Poetry on Radio
Politics and Radio
Polk, George
Popov, Alexander
Portable Radio Stations
Prairie Home Companion
Premiums
Press-Radio War
Production for Radio
Programming Research
Programming Strategies and Processes
Progressive Rock Format
Promax
Promenade Concerts
Promotion on Radio
Propaganda by Radio
Psychographics
Public Affairs Programming
Public Broadcasting Act of 1967
"Public Interest, Convenience, or Necessity"
Public Radio International
Public Radio Since 1967
Public Service Radio
Pulse Inc.
Pyne, Joe

Quaal, Ward L.
Quiz and Audience Participation Programs

RADAR
Radio Academy
Radio Advertising Bureau
Radio in the American Sector (Berlin)
Radio Authority
Radio City
Radio Corporation of America
Radio Data System
Radio Disney
Radio Free Asia
Radio Free Europe/Radio Liberty
Radio Hall of Fame
Radio Luxembourg
Radio Martí
Radio Monte Carlo
Radio Moscow
Radio Sawa/Middle East Radio Network
Receivers
Recording and Studio Equipment
Recordings and the Radio Industry
Re-Creations of Events
Red Channels
Red Lion Case
Regulation
Reith, John C.W.
Religion on Radio
Renfro Valley Barn Dance
Retro Formats
Rock and Roll Format
Rogers, Will
Rome, Jim
Roosevelt, Eleanor
Russia and Soviet Union

Sarnoff, David
Scandinavia
Schechter, A.A.
Schulke, James
Science Fiction Programs
Sevareid, Eric
"Seven Dirty Words" Case
Shadow
Shaw, Allen
Shepard, John
Shepherd, Jean
Shirer, William L.
Shock Jocks
Shortwave Radio
Siemering, William
Siepmann, Charles A.
Silent Nights
Simon, Scott
Simulcasting
Singers on Radio
Situation Comedy
Skelton, Red
Sklar, Rick
Smith, Kate
Smulyan, Jeffrey H.
Soap Opera
Social Class and Radio
Social and Political Movements
Soft Rock Format
Sound Effects
Soundprint
South America
South Pacific Islands
Sportscasters
Sports on Radio

Stamberg, Susan
Stanton, Frank N.
Star Wars
Station Rep Firms
Stereo
Stereotypes on Radio
Stern, Howard
Storer, George
Storz, Todd
Striker, Fran
Subsidiary Communications Authorization
Suspense
Sustaining Programs
Swing, Raymond Gram
Syndication

Taishoff, Sol
Talent Raids
Talent Shows
Talk of the Nation
Talk Radio
Taylor, Deems
Taylor, Marlin R.
Technical Organizations
Telecommunications Act of 1996
Television Depictions of Radio
Ten-Watt Stations
Terkel, Louis "Studs"
Tesla, Nikola
Theater Guild on the Air
This American Life
Thomas, Lowell
Thomas, Rufus
Tokyo Rose
Topless Radio
Totenberg, Nina
Tracht, Doug "Greaseman"
Trade Associations
Trade Press
Transistor Radios
Tremayne, Les
Trout, Robert

Underground Radio
United Fruit Company
United Nations Radio
United States
United States Congress and Radio
United States Influences on British Radio
United States Navy and Radio
United States Presidency and Radio
United States Supreme Court and Radio
Urban Contemporary Format

Vallee, Rudy
Variety Shows
Vatican Radio
Vaudeville
Vic and Sade
Violence and Radio
Virtual Radio
Voice of America
Voice of the Listener and Viewer
Vox Pop

WABC
Walkman
War of the Worlds
WBAI
WBAP
WBBM
WBT
WBZ
WCBS
WCCO
WCFL
WDIA
WEAF
Weaver, Sylvester (Pat)
Wertheimer, Linda
Westerns
Westinghouse
Westwood One
WEVD
WGI
WGN
WHA and Wisconsin Public Radio
WHER
White, Paul
White, Wallace H.
Williams, Bruce
Williams, Jerry
Williams, Nat D.
Wilson, Don
Winchell, Walter
WINS
Wireless Acts of 1910 and 1912; Radio Acts of 1912 and 1927
Wire Recording
WJR
WLAC
WLS

WLW
WMAQ
WNBC
WNEW
WNYC
Wolfman Jack
Women in Radio
WOR
World Radiocommunication Conference
World War II and U.S. Radio
WQXR
Wright, Early
WRR
WSB
WSM
WTOP
WWJ
WWL
WWVA
WXYZ
Wynn, Ed

Yankee Network
You Bet Your Life
Your Hit Parade
Yours Truly, Johnny Dollar

Zenith Radio Corporation

O

Oboler, Arch 1909–1987

U.S. Writer, Producer, Director

Arch Oboler was one of the great auteurs in radio history, using the medium as his personal means of expression. During his career as one of radio's premier and prodigious dramatists, Oboler estimated that he wrote more than 850 plays, many of which represent the highest achievement of aural composition.

Early Years

Born in one of the nation's most creative broadcasting centers, Chicago, on 7 December 1909, Oboler became obsessed with the medium after building his first crystal set. After graduating from the University of Chicago, he wrote more than 50 plays before any station showed an interest in his fledgling ability. During the early 1930s radio was, according to Oboler, "an imitation of motion pictures and an echo of the stage." Oboler was searching for a new way to realize the potential of the medium. In 1933 he sold his first script, *Futuristics,* to the National Broadcasting Company (NBC) for national broadcast to inaugurate the new Radio City facility in New York.

In 1934 Oboler established a name for himself as a contributor of original playlets for Don Ameche in the anthology series *Grand Hotel.* A year later, Rudy Vallee was so impressed with Oboler's work for his radio series that he repeated several scripts. Oboler continued his association with Ameche on the *Chase and Sanborn Hour,* which also starred Edgar Bergen and Charlie McCarthy. His satirical takeoff on the Adam and Eve myth, written for Ameche and guest star Mae West, created a national uproar because of the Hollywood siren's salacious reading of Oboler's suggestive lines. Oboler's career continued to soar.

Network Success

In May 1936 NBC gave creative control of the popular series *Lights Out* to Oboler. The original creator and producer of this "ultimate in horror," Wyllis Cooper, had left to become a Los Angeles screenwriter, and Oboler seized the opportunity to formulate *Lights Out* as a "theater of the mind." He experimented with narrative techniques, developing his patented stream-of-consciousness mode to delve into the minds of his characters. Oboler also used sound effects to construct bizarre worlds that were often analyses of contemporary social situations.

Oboler's own macabre voice was heard opening each episode of *Lights Out* with the signature warning, "It . . . is . . . later . . . than . . . you . . . think." What usually followed was an examination of the dark side of Oboler's own imagination. In "Chicken Heart," a scientific experiment goes awry, until the expanding organ overwhelms the world with an incessant thumping. (The program inspired one of Bill Cosby's classic routines about his childhood, heard on his *Wonderfulness* album.) In another critique on science, "Revolt of the Worms," the modest earthworm grows to absurd proportions and suffocates the lead characters. The tense situations were abetted by inventive sound effects—bacon frying signified the body being electrocuted; the chewing of Lifesavers candy simulated the crushing of bones; and the manipulation of warm spaghetti meant that something ungodly was happening to human flesh. Hollywood stars, such as Boris Karloff, were intrigued by Oboler's fantasies and journeyed to Chicago to contribute their services. By 1938 Oboler felt confined by the horror genre and wanted instead to take on the evils of the times, Hitler and the rise of Nazism.

A progressive NBC executive, Lewis Titterton, suggested the title for Oboler's new series upon hearing a recording of the pilot, "The Ugliest Man in the World." Titterton thought the future of radio depended on the vision of the writer, and the new series was crowned *Arch Oboler's Plays,* debuting in March 1939. Oboler was now among radio's elite, the first writer accorded name-in-the-title status. The new program was sustained by NBC, presented without advertising, and Oboler's imagination was now unfettered.

Oboler, writing about "the terrors and monsters within each of us," used his stream-of-consciousness technique to shattering effect and made radio a viable new art form. He revealed the inner psyche of a soldier who returned home a vegetable, played by the tenacious James Cagney, in his controversial adaptation of Dalton Trumbo's antiwar novel, *Johnny Got His Gun.* Joan Crawford joined forces with Oboler and played an anxious mother awaiting birth in *Baby.* The acclaimed Russian actress Alla Nazimova offered her services, and Oboler wrote the antifascist drama *Ivory Tower* for her. He also pushed radio drama into new directions: he was the first to integrate a full orchestra into a radio play, his dramatization of the life of Russian composer Piotr Ilich Tchaikovsky.

After his series left the air in March 1940, Oboler lent his talents to the propaganda effort to rally American morale during World War II. Oboler's plays explicitly warned about the atrocities of "the Jap-Nazi world." In "Chicago, Germany," which was presented on the *Treasury Star Parade,* he imagined what would happen if Hitler conquered the United States. Urging that radio needed "an injection of hatred and passionate feeling," Oboler created several patriotic series for NBC: *Everyman's Theatre,* which he directed and wrote, using previous scripts from *Lights Out* and *Arch Oboler's Plays; Plays for Americans,* which featured such Hollywood stars as Bette Davis, James Stewart, and Dick Powell; and *Everything for the Boys,* a unique collaboration with actor Ronald Colman to provide entertainment for servicemen.

By the end of the war, Oboler was exhausted by the nerve-wracking demands of his chosen medium. In the mid-1940s he brought back *Lights Out* for the Columbia Broadcasting System (CBS) and *Arch Oboler's Plays* for Mutual. He lamented that "radio, for the dramatist, is a huge, insatiable sausage grinder into which he feeds his creative life to be converted into neatly packaged detergents."

Arch Oboler
Courtesy Library of American Broadcasting

Other Media

He took his dark vision almost exclusively to the motion pictures, wanting to control as much of the production as possible. His first directorial effort, *Bewitched,* was based on an award-winning radio script he had written for Bette Davis. Other films he wrote and produced had overtly political themes: *Strange Holiday,* about a fascist invasion of America, and *Five,* the story of survivors of a nuclear holocaust. Always fascinated by technology, he wrote, directed, and produced the first commercially successful film in 3-D, *Bwana Devil.* He spent millions of his own money to gain further credibility for 3-D movies, but to little avail. His only foray into television was, of all things, a comedy anthology series, which critics acclaimed for its imagination and craftsmanship. Oboler's experimental *Comedy Theatre* was discontinued by the American Broadcasting Companies (ABC) after only six episodes.

When Oboler died in 1987, a syndication firm was releasing much of his radio work that had not been heard for more than 40 years. Throughout his career, Oboler gave the horror genre legitimacy and seriousness, leading author Stephen King to call Oboler "the genre's prime auteur." Inheritor of Edgar Allan Poe's dark, fantastical vision, Oboler paved the way for such future masters of the grotesque as Rod Serling and Ray Bradbury.

RON SIMON

See also Lights Out; Playwrights on Radio

Arch Oboler. Born in Chicago, Illinois, 7 December 1909. Attended University of Chicago; sold first radio play, *Futuristics,* to NBC, 1934; wrote short plays for *Grand Hotel,*

1934–35; also wrote for *Rudy Vallee Show* and *Chase and Sanborn Hour,* mid-1930s; oversaw own series, *Arch Oboler's Plays*, 1939–40; wrote and directed various films, 1940–72; wrote and directed Hans Conreid satire on television, *The Twonky*, 1953; wrote theatrical play *Night of the Auk*, 1958; created new 3-D process, Space Vision, for film *The Bubble*, 1966; negotiated deal with Metacom to syndicate and release on audiocassette *Lights Out* and compilation of other plays, *Yesterday, Today, and Tomorrow*, 1986. Died in Westlake Village, California, 19 March 1987.

Radio Series

1936–38, 1942–43	*Lights Out*
1939–40, 1945	*Arch Oboler's Plays*
1940–41	*Everyman's Theater*
1942	*Plays for Americans;* writer, *Treasury Star Parade*
1944	*Everything for the Boys*

Television Series

Comedy Theatre, 1949

Films

Escape, 1940; *Gangway for Tomorrow*, 1943; *Bewitched*, 1945; *Strange Holiday*, 1945; *The Arnelo Affair*, 1947; *Five*, 1951; *Bwana Devil*, 1952; *The Twonky*, 1953; *One Plus One: Exploring the Kinsey Reports*, 1961; *The Bubble*, 1966 (later reissused as *Fantastic Invasion of Planet Earth*); *Domo Arigato* (unreleased), 1972

Selected Publications

Fourteen Radio Plays, 1940
New Radio Plays, 1940
This Freedom: Thirteen Radio Plays, 1942
Oboler Omnibus: Radio Plays and Personalities, 1945
Night of the Auk: A Free Prose Play, 1958
House on Fire, 1969

Further Reading

Skutch, Ira, editor, *Five Directors: The Golden Years of Radio: Based on Interviews with Himan Brown, Axel Gruenberg, Fletcher Markle, Arch Oboler, Robert Lewis Shayon*, Lanham, Maryland: Scarecrow Press, 1998

Obscenity and Indecency on Radio

Radio has long been considered a guest in the home or car, and so the medium has been constrained in the kinds of language broadcast. The original language of the Radio Act of 1927 (Sec. 29) indicated that "No person within the jurisdiction of the United States shall utter any obscene, indecent, or profane language by means of radio transmission." However, it was not until 1948 that Congress put teeth into this provision by incorporating this prohibition into the criminal code (18 U.S.C.A. 1464). But does such a constraint violate First Amendment rights of free expression? And has the cultural climate of language usage changed since 1948? Do we know what constitutes *obscene* or *indecent* for everyone, or do these meanings differ from one person to another? Is there any good way for the Federal Communications Commission (FCC) or the courts to enforce this measure without violating a provision in the Communication Act (Section 326), let alone the First Amendment, that prohibits the censorship of broadcast communication? These questions complicate any clear answer to or brief discussion of this issue. However, some court decisions and activities of the FCC give us a history upon which we can base an informed discussion.

Developing Concern

Early in the history of broadcasting, those who announced or otherwise spoke on radio did so with a great deal of decorum and civility. Often they would wear formal attire, even though no one could see them. The language was precise, enunciation was as perfect as possible, and certainly there was neither slang nor profanity. "Radio speakers" were guests in listeners' homes, and they spoke with careful politeness.

The FCC recognized in 1975 that

> broadcasting requires special treatment because of four important considerations: (1) children have access to radios and in many cases are unsupervised by parents; (2) radio receivers are in the home, a place where people's privacy interest is entitled to extra deference; (3) non-consenting adults may tune in a station without any warning that offensive language is being or will be broadcast; and (4) there is a scarcity of spectrum space, the use of which the government must therefore license in the public interest. (56 FCC 2nd 97)

However, radio evolved to the point that speakers became less formal and more conversational in their radio dialogues with the listener. American culture was changing. Nudity in over-the-counter magazines such as *Playboy* appeared. Frank discussions about sex were no longer as taboo as they had been previously. It was during this time, in the late 1950s and 1960s, that the U.S. Supreme Court tried yet again to define obscenity issues in the media.

In *Roth v United States* (1957), the U.S. Supreme Court tried for the first time to establish some definite measure to define obscenity, after noting that it was not protected by the First Amendment. It was a matter of "whether to the average person, applying contemporary community standards, the dominant theme of the material, taken as a whole, appeals to prurient interest." There was still no certainty as to what *prurient* meant. A definition was left unarticulated in the case of *Jacobellis v Ohio* (1964): Justice Stewart stated that although he had a hard time defining pornography, "I know it when I see it."

In *Memoirs v Massachusetts* (1966), the new element added to the definition was that material is patently offensive when it affronts contemporary community standards relating to the description or representation of sexual matters. "Pandering material," that which openly advertises and appeals to erotic interests, was yet another element added to the definition in 1966 (*Ginzburg v United States*).

But by 1967, the courts had become flooded with obscenity cases, resulting in confusion. In *Redrup v New York,* the court articulated a kind of reverse definition: nothing was obscene except when it fell under the specific circumstances of (1) "pandering," (2) failure to uphold specific statutes designed to protect juveniles, or (3) an assault upon individual privacy by publication in a manner so obtrusive as to make it impossible for the unwilling individual to avoid exposure to it. It is perhaps this third provision that prevented obscene language in broadcasting in an era in which such language was becoming common in other media.

By 1973, in one of the last attempts by the U.S. Supreme Court to define obscenity, in *Miller v California,* the Court fell back to the elements of the "Roth rule" from the 1957 case. It also added the notion of the "SLAPS rule," which takes into consideration whether the work in question lacks "Serious Literary, Artistic, Political or Scientific value"; the Court thereby rejected the previous obscenity standard of being "utterly without redeeming social value."

It was during this era that American mass culture continually presented the courts (and, where broadcasting was concerned, the FCC) with dilemmas of staying with the traditional or liberalizing policies to accommodate new language and attitudes about sex.

As American culture changed, how was radio to reflect this shift? Slang expressions, double entendres, dirty jokes, and derogatory terms became popular. Indeed, during that era, some talk radio programs, known as "topless radio," discussed matters of sex in a frank manner. Here, talk show hosts, disc jockeys, and phone-in callers engaged in sexually explicit dialogue, apparently for the express purpose of titillating listeners. Some stations were found to be broadcasting indecent material, however, and fined by the FCC in 1973.

But the pivotal case as it relates to obscenity or indecency on radio came in the case of comedian George Carlin. In his comedy routine "Seven Dirty Words You Can't Say on Radio or Television," Carlin expressed thoughts about the nature of some taboo words and how nonsensical their expressions were in many colloquialisms. His descriptions poked fun at society's view of such words. His humorous satire examined the "language of ordinary people . . . and our silly attitudes toward those words." Carlin's routine on the subject was recorded and released with the provision "not for broadcast." However, a New York City radio station owned by the Pacifica Foundation aired the dialogue one afternoon. A man heard the broadcast of the dialogue while driving with his young son and wrote a letter of complaint to the FCC. He stated that he could not understand why the recording had been broadcast over the air that the FCC was supposed to control (*FCC v Pacifica,* 1978).

The challenge worked its way through the courts for nearly five years, from 1973, when it was first aired, until 1978, when the U.S. Supreme Court decided on it. Because of the dilemma between the FCC's need to prohibit indecent broadcasts and the constitutional rights guaranteed by the First Amendment, the U.S. Supreme Court agreed to hear the case. The Court in considering this dilemma had to take into consideration both the changing social climate of increasing latitude and also the fact that the sketch had been broadcast in mid-afternoon, at a time when any child might have been listening. The Court looked at the FCC's enforcement role and its mandate for enforcement from the U.S. Code on indecency and reflected on the cases over the previous decade, which had maintained no First Amendment protection for obscene material. The Court also looked at the careful definitions of *obscenity* and *indecency.*

However, the need for more stringent definitions seemed necessary for broadcasting. One distinction was that *indecency,* unlike *obscenity,* may have First Amendment protection. The concept of *obscenity* uses a more serious standard than does *indecency,* which was defined by the court in the *Pacifica* (1978) case as "intimately connected with the exposure of children to language that describes, in terms patently offensive as measured by contemporary community standards for the broadcast medium, sexual or excretory activities and organs, at times of the day when there is a reasonable risk that children may be in the audience."

The Court went on to suggest that the material might not have been indecent under some circumstances, such as when children would not be present. This notion led to the concept of *safe harbor,* a time, for example late at night, when the number of children listening is minimal and when it might be safe to consider a different standard.

For many succeeding years, the FCC applied this ruling to indecent programming airing before 10:00 P.M. By 1987 the FCC changed its definition of *safe harbor* and came to consider a more general definition of indecency. Time of day became less of a factor, because children or youth could often be found listening at all hours. Context became the important factor in determining indecency. Later, the Commission returned to the safe harbor idea, changing the start of the harbor to midnight, provided that the questionable materials began with appropriate warnings. It was assumed by the Commission that parents would maintain some control over their children's listening after midnight. Thus, the focus shifted from the FCC's policing to setting up zones in which parents were responsible.

Broadcasters, together with other interested parties, challenged this new post-midnight safe harbor, because there didn't seem to be any data on such a safe zone. However, Congress intervened, and in the 1989 appropriations bill signed by President Reagan, the FCC was required to enforce a ban on indecent and obscene speech 24 hours a day. This was Congressional grandstanding—they knew such a full-time ban would never survive Court review, but they could look virtuous. Courts eventually did overturn the ban.

Further litigation continued well into the next decade with continually varying definitions of *safe harbor.* The FCC decided in 1993 to look at indecency from the other side; rather than addressing the issue of a safe harbor for questionable language, the FCC declared that it would enforce a safe harbor from 6:00 A.M. to 8:00 P.M., during which time indecency would not be aired. This left the later evening hours up for grabs, but the Commission still maintained a watch on complaints about indecency.

Elements of indecency evolved from topless radio to "shock jocks," disc jockeys and hosts who used shocking and titillating language to enhance their popularity, language and descriptions to which listeners had become more accustomed. Don Imus and Howard Stern were examples of this type of rawness, and unpredictability became popular. Stern, who proclaimed himself "King of all Media," often asked his guests about their sexual habits. References to women were generally about their sexual attributes. His female co-host was the object of frank discussions about sexual habits, and sometimes he would spank bikini-clad females. Radio leaves everything to the imagination, so it was difficult to determine whether it was the act on radio or the imagination of the listener that made things unacceptable.

Stern, who wrote books about his misadventures, and about whom a movie was made, remained unabashed about his boldness. It was this brash style that took him from one market to the many markets that carried him. As long as there was an audience to listen to his material, it appeared to be socially acceptable, even though most of his audience members were adolescent white males who chuckled at the suggestive material. He spoke of masturbation, incest, and the breast size of famous women, and he included segments on "bestiality dial-a-date." His antics caused him to be fined by the FCC on the basis of violating the federal statute prohibiting the use of obscene, indecent, or profane language on the air. His parent company, Infinity Broadcasting, fought the fines, and they remained in litigation for several years until Infinity moved for a deal to acquire other broadcast stations, a proposal that had to have FCC approval. An agreement was struck, and Infinity paid some of the fines incurred by Stern.

But the shock jock's language and antics continued to trouble some segments of listeners. One church group called for a widespread boycott of products advertised on Stern's shows. The advertisers found that their sales dropped; in turn, they dropped their advertising support of the radio personality. Occasionally, this infuriated Stern, who lashed out against such groups. His infamy was noted in the national press in the spring of 1999, when, while speaking of the shooting tragedy at Columbine High School, he indicated that the shooters were kind of stupid—they should have had sex with those girls before they took them out. This statement put Stern in a questionable light in the minds of many, but his program continued.

Specific challenges regarding the definition of indecency remained problematic. In the early 1990s, an album by the rap group 2 Live Crew, *As Nasty as They Wanna Be,* proved to be troublesome for many. It contained hundreds of sexual references and obscenities. Live performances of material from the album caused arrests of the group members. There was word that some stations tried to air the work, but no complaints ever reached the FCC, and no action was taken against 2 Live Crew by the FCC.

In Santa Barbara, a station licensed to the Regents of the University of California played sexually explicit lyrics in the recorded music "Making Bacon" by the Pork Dudes. A Pacifica station in Los Angeles played excerpts of the play *The Jerker,* about a homosexual dying of AIDS. Although the story and theme were not found to be objectionable by the FCC, the extensive use of patently offensive language referring to sexual and excretory organs and functions caused the FCC to rule that the broadcast was indecent.

There is an ever-growing list of broadcasts that either have been or could be cited for the use of indecent language. The examples given previously illustrate the problems with definition and enforcement by the FCC.

Profanity can be equally offensive for some listeners, but judgments regarding profanity are highly subjective, and profanity is considered to have different levels of offensiveness as part of the artistic expression. Such complications make it difficult for the FCC to enforce the prohibition of "profane language." *Profanity* is "the use of irreverent or irreligious words, including cursing by invoking deity." In earlier generations, profanity was common but was usually confined to specific groups that would be tolerant of such words—not in "polite, mixed company" and certainly not in broadcasting.

Recently, profanity has become more common, perhaps because of its widespread use in movies. The FCC has viewed profanity as being judged from the perspective of the listener or viewer. Generally, the Commission will not bring action unless a speaker uses profanity so repeatedly in invoking a curse as to cause a public nuisance. So radio announcers and talk show hosts who occasionally invoke deity in a profane manner would not likely incur any action from the FCC. More recently, the use of profanity in everything from the language of some musical artists to prime-time television shows would make any enforcement of the profanity prohibition difficult. Indeed, one letter from a conservative watchdog group in the fall of 1999 about indecent language on network television received the reply from the FCC that as long as language was used "in context," there were no words that could be forbidden. It was a trend that concerned the more conservative observers.

A newer problem is the wide diversity of the media and the invisible nature of those media that are not really "broadcast" and are therefore not punishable under the indecency rule that applies only to "broadcast" language. For example, when one watches premium cable channels and sees R-rated movies with a plethora of obscene, indecent, and profane language, it is easy to get the idea that such language has become acceptable in broadcasting on radio and television. However, cable dissemination is another medium, not a broadcast entity. Although the distinction remains clear in the minds of regulators and enforcers, it may be less clear in the minds of consumers—listeners and viewers. Such popularizing of language once found taboo makes for mixed signals about the implementation of unlawful language in broadcasting.

Was there a legislative cure? In 1996 Congress passed the Communications Decency Act, seeking to protect minors from harmful material, specifically on the internet, but the Act included the term *transmission,* which could be construed to include broadcasting. The American Civil Liberties Union challenged the law on the basis that it violated First Amendment rights. The U.S. Supreme Court in 1997 ruled that the Act's provisions of "indecent transmission" and "patently offensive display" abridged the freedom of speech protected by the First Amendment; the ruling thus struck down the Communications Decency Act.

The convergence of broadcast, cable, and the internet has further confused the picture of exactly what constitutes "indecent, profane or obscene" language and when and under what circumstances such language may be broadcast or prohibited. As social mores change, so too will the laws and policies affecting materials seen in the media or heard on radio.

VAL E. LIMBURG

See also Censorship; First Amendment and Radio; Pacifica Foundation; Seven Dirty Words Case; Shock Jocks; Stern, Howard; Topless Radio; United States Supreme Court and Radio

Further Reading

Action for Children's Television v FCC, 852 F2d 1332 (1988)
Broadcast Regulation (annual; published by the National Association of Broadcasters Legal Department)
Clark, Anne L., "As Nasty As They Wanna Be: Popular Music on Trial," *New York University Law Review* 65 (1990)
"FCC Takes Strong Stance on Enforcement of Prohibition against Obscene and Indecent Broadcasts," *FCC Fact Sheet* (November 1987)
FCC v Pacifica Foundation, 438 US 726 (1978)
Federal Communications Commission, "Policy Statement: In the Matter of Industry Guidance on the Commission's Case Law Interpreting . . . Policies Regarding Broadcast Indecency," Washington, D.C.: FCC (Release 01-90), 6 April 2001 (available on www.fcc.gov)
Ginsburg, Douglas H., *Regulation of Broadcasting: Law and Policy towards Radio, Television, and Cable Communications,* St. Paul, Minnesota: West, 1979; 3rd edition, as *Regulation of the Electronic Mass Media: Law and Policy for Radio, Television, Cable, and the New Video Technologies,* by Michael Botein, 1998
Ginzburg v U.S., 383 U.S. 463 (1966)
Hilliard, Robert L., and Michael C. Keith, *Dirty Discourse: Sex and Indecency in American Radio,* Ames: Iowa State University Press, 2003
Infinity Broadcasting Corporation of Pennsylvania, 3 FCC Rcd 930 (1987)
Jacobellis v Ohio, 378 U.S. 184 (1964)
LaFayette, Jon, "A Stern Challenge," *Electronic Media* (23 August 1993)
Limburg, Val E., "Obscenity: The Struggle for Definition," in *Mass Media and Society,* compiled by Alan Wells, Palo Alto, California: National Press Books, 1972; 4th edition, edited by Wells, Lexington, Massachusetts: Lexington Books, 1987
Lipschultz, Jeremy Harris, *Broadcast Indecency: FCC Regulation and the First Amendment,* Boston: Focal Press, 1996
Memoirs v Massachusetts, 383 U.S. 413 (1966)
Miller v California, 413 U.S. 15 (1973)
Redrup v New York, 386 U.S. 767 (1967)
The Regents of the University of California, 2 FCC Rcd 2708 (1987)

Reno v ACLU, 117 S. Ct. 2329 (1997)
Rivera-Sanchez, Milagros, "The Origins of the Ban on 'Obscene, Indecent, or Profane' Language of the Radio Act of 1927," *Journalism and Mass Communication Monographs* 149 (February 1995)
Roth v U.S., 384 U.S. 476 (1957)
Smith, F. Leslie, Milan D. Meeske, and John W. Wright, II, *Electronic Media and Government: The Regulation of Wireless and Wired Mass Communication in the United States*, White Plains, New York: Longman, 1995
Zoglin, Richard, "Shock Jock," *Time* (30 November 1992)

Office of Radio Research

Applied Studies of Audiences

The Office of Radio Research was established in 1937 after the Rockefeller Foundation extended a two-year grant to Hadley Cantril, a psychologist at Princeton University (best remembered for his studies about the effects of Orson Welles' radio broadcast *Invasion from Mars*), and Frank Stanton, a recent psychology Ph.D. from Ohio State University who headed up research for the Columbia Broadcasting System (CBS) and ultimately directed the network as president from 1948 to 1971. The two researchers had proposed a series of studies to broaden the methods for assessing the impact of radio on the public, including the motivations for listening, radio's psychological value to the audience, and the role it played in their lives. Although they had envisioned extensive use of experimental methods, they chose as their research director an Austrian-trained psychologist, Paul Lazarsfeld, who proved to have a deep interest in applied research and methodological innovation. Lazarsfeld had come to the United States on a Rockefeller fellowship in 1933 and remained there after a protofascist government took power in Austria in 1934.

As associate directors, Cantril and Stanton oversaw the work of the office, but Lazarsfeld was in charge of its day-to-day operations, which were initially handled at an office in Newark where Lazarsfeld had already set up an applied social research center in conjunction with the University of Newark. As an intellectual alliance became firmly established between Stanton and Lazarsfeld, Cantril began to withdraw from the project, and the offices, still officially known as the Princeton Office of Radio Research, migrated in 1938 to New York City's Union Square. In 1939, Princeton severed its ties and the project was moved to Columbia University. Given the relatively limited amount of funding available, Lazarsfeld was forced to pursue a wide variety of audience studies, depending often on archival data previously gathered for Gallup polls and radio program audits as well as soliciting applied research projects from commercial organizations. Lazarsfeld was especially adept at secondary analysis of program ratings, with a particular focus on the social differences between the audiences of various programs, an approach that was nurtured by his European concern for social stratification.

In a successful attempt to win additional Rockefeller Foundation support, Lazarsfeld initiated plans to publish the research studies generated by his small staff. Twice, in 1939 and in 1940, he arranged for an entire issue of the *Journal of Applied Psychology* to be wholly devoted to radio research completed by the office; he was also responsible for a thematic compilation of studies in book form, titled *Radio and the Printed Page*, published in 1940. This tradition continued with three volumes of essays titled *Radio Research* that Lazarsfeld and Stanton jointly published in 1941, 1944, and 1949. The major proposal that Lazarsfeld put forth to ensure a three-year renewal of Rockefeller funding to begin in March 1940 was the idea of creating a study involving two large panels of radio listeners to be interviewed at intervals to assess their response to a sequence of radio broadcasts.

Initially, the research was to be centered on a radio program sponsored by the Department of Agriculture, but ultimately the panel design was implemented for a study of the impact of mass media on voting during the 1940 presidential campaign between Democrat Franklin Roosevelt, running for an unprecedented third term, and Wendell Willkie, a Republican utility executive. The unexpected results—only 54 out of 600 panel members shifted from their initial voting preferences between May and November—led Lazarsfeld and his colleagues, Bernard Berelson and Hazel Gaudet, to devise their theory of the two-step flow of communication effects, in which opinion leaders provide a critical interpersonal link between mass media messages and their intended audiences (*The People's Choice*, 1944, 1948).

In the decade of the 1940s, the office supported a number of studies that marked their authors as prominent innovators in communication research, including sociologist Robert Merton's study of Kate Smith's radio war bond campaign; Bernard Berelson's study of the effects of the 1945 New York City newspaper strike; and Herta Herzog's study of the motivations of women soap opera listeners using in-depth interviews, a precursor of focus group methodology. Joseph Klapper, who later headed audience research at CBS, completed his dissertation under Lazarsfeld's direction in 1949. A revised version published in 1960, *The Effects of Mass Communication,* forcefully argued against the notion of powerful mass media. Lazarsfeld also invited fellow European émigrés to the office. Film theorist Rudolf Arnheim analyzed the dramatic content of soap operas, and critical theorist Theodor Adorno studied the role of popular and classical music, although Adorno was uncomfortable with Lazarsfeld's quantitative techniques.

Key to the success of the office may have been the relationship between Lazarsfeld and Stanton. Lazarsfeld was ever mindful that he needed the confidence of the broadcast industry to acquire access to the data it gathered as well as political support, but he also realized that he had to persuade executives that studies with negative or unorthodox findings might nevertheless help them better understand and run their industry. Stanton not only had financial resources at CBS to fill in funding gaps at the office, but he was an ally who shared Lazarsfeld's commitment to innovative research techniques. Their joint development of what was called the Lazarsfeld–Stanton Program Analyzer reflected their shared zeal. This device, used both at the office and at CBS studios, allowed the simultaneous recording of the individual opinions of a test audience. Ten listeners at a time could indicate when they liked or disliked the program they were hearing. Such an instant audience analysis machine would be replicated 40 years later using networked personal computers.

Curiously, the Office of Radio Research benefited as well from the social and technological disruptions occasioned by the outbreak of World War II in Europe and by the United States' eventual entry into the conflict. The Rockefeller Foundation's support of radio research was motivated in no small part by a concern about the uses of radio programming for propaganda in wartime. And when America found itself at war, social psychologist Samuel Stouffer, who headed research for the U.S. Army, pursued some of his projects using Lazarsfeld's personnel and even the office's program analyzer. Finally, the exigencies of the wartime economy interrupted the development of commercial television in the United States, artificially extending radio's dominant role in American life.

The vitality of broadcast radio and the importance of this wartime research no doubt aided Lazarsfeld in his quest to fully integrate the office into the structure of Columbia University as a research unit in the graduate school. This he achieved in 1945, shortly after renaming his hybrid organization for academic, governmental, and commercial research the Bureau of Applied Social Research. It is unclear whether the name change was an attempt to achieve academic legitimacy, to broaden its research mission, or both, but by the time the bureau celebrated its 20th anniversary in 1957, it had effectively abandoned commercial broadcasting as a focus of study.

SETH FINN

See also Audience Research Methods; Education about Radio; Lazarsfeld, Paul F.; Stanton, Frank

Further Reading

Delia, Jesse G., "Communication Research: A History," in *Handbook of Communication Science,* edited by Charles R. Berger and Steven H. Chaffee, Beverly Hills, California, and London: Sage, 1987

Lazarsfeld, Paul F., editor, "Radio Research and Applied Psychology," *The Journal of Applied Psychology* 23:1 (special issue, February 1939)

Lazarsfeld, Paul F., *Radio and the Printed Page: An Introduction to the Study of Radio and Its Role in the Communication of Ideas,* New York: Duell, Sloan, and Pearce, 1940; reprint, New York: Arno Press, 1971

Lazarsfeld, Paul F., "An Episode in the History of Social Research: A Memoir," in *The Intellectual Migration: Europe and America, 1930–1960,* edited by Donald Fleming and Bernard Bailyn, Cambridge, Massachusetts: Harvard University Press, 1968; expanded edition, 1969

Lazarsfeld, Paul F., and Frank N. Stanton, editors, *Radio Research, 1941,* New York: Duell, Sloan, and Pearce, 1941

Lazarsfeld, Paul F., and Frank N. Stanton, editors, *Radio Research, 1942–43,* New York: Duell, Sloan, and Pearce, 1944

Lazarsfeld, Paul F., and Frank N. Stanton, editors, *Communication Research, 1948–49,* New York: Harper, 1949

Merton, Robert K., *Mass Persuasion: The Social Psychology of a War Bond Drive,* New York and London: Harper, 1946

Schramm, Wilbur, *The Beginnings of Communication Study in America: A Personal Memoir,* edited by Steven H. Chaffee and Everett M. Rogers, Thousand Oaks, California, and London: Sage, 1997

Sills, David L., "Stanton, Lazarsfeld, and Merton—Pioneers in Communication Research," in *American Communication Research: The Remembered History,* edited by Everette E. Dennis and Ellen Wartella, Mahwah, New Jersey: Erlbaum, 1996

Office of War Information

World War II U.S. Government Agency

The Office of War Information (OWI) was intended to be the primary voice of the United States government during World War II. It cleared government radio programs and provided background information on the war for use by broadcasters and periodical publishers.

Origins

The Office of War Information began its life as a response by President Franklin D. Roosevelt to the petty sniping and conflicting reports released by rival government information agencies. Established on 13 June 1942, the mission of the Office of War Information was described in Executive Order 9182 as "the facilitation of the development of an informed and intelligent understanding, at home and abroad, of the status and progress of the war effort, and of the war policies, activities, and aims of the Government." The OWI's domestic branch was formed by combining all or parts of three existing entities: the Office of Facts and Figures, the Office of Government Reports, and the Information Division of the Office of Economic Stabilization, while its foreign or overseas branch was created from the existing Foreign Information Service of the Office of the Coordinator of Information. The OWI soon expanded its scope beyond the duties of the agencies it replaced, moving into production and policy along with information dissemination.

Although similar to its predecessors in function, the OWI was envisioned initially as having greater authority than the agencies it replaced. But Roosevelt refrained from giving the OWI the necessary teeth to accomplish the task at hand. Many federal departments still had their own information units, and the OWI's authority allowed it only to coordinate their various activities. It could suggest whether a given piece of information should be released or withheld, but it had no recourse if other agencies ignored their advice. Thus in practice the OWI had no more real authority than the agencies it replaced had exercised.

Structure of the OWI

Appointed to head the OWI was journalist and former CBS radio commentator Elmer Davis, who as director was to be involved in larger policy issues both at home and abroad. The choice was hailed almost universally in all quarters, for his radio work had given Davis a reputation for honesty and common sense. His associate director was Milton Eisenhower, a former official of the Department of Agriculture. Eisenhower was appointed specifically to handle administrative matters, as Davis had no experience in that area.

Former Office of Facts and Figures head Archibald MacLeish assumed the post of Assistant Director for Policy Development and reported directly to Davis' office. MacLeish's office served as a think tank of sorts for policy questions, which were then referred (with recommendations) to Davis for final action.

The two branch heads for domestic and foreign operations were next in the hierarchy. Robert Sherwood, noted playwright (*Abe Lincoln in Illinois*) and one-time Republican who became speechwriter and advisor for Franklin Roosevelt, was chosen as Administrative Director for Overseas Operations. Moderate Midwestern publisher and radio station owner Gardner "Mike" Cowles, Jr., was appointed Assistant Director for Domestic Operations.

Foreign Branch Activities: The Voice of America

The difference in the titles of Sherwood and Cowles was no accident. Sherwood, the former head of the Foreign Information Service of the Office of the Coordinator of Information, had centered his burgeoning operation in New York, not Washington, and much of its later work was actually based in England. Davis and Eisenhower had no experience in international affairs, and much of their time would later be spent trying to smooth the ruffled feathers of "a new and often cantankerous staff on the one hand, and dissatisfied executive departments on the other." From the outset Sherwood and the overseas branch maintained a level of autonomy and distance from its superiors (and from congressional criticism) that the domestic branch could not duplicate.

The Voice of America (VOA), the radio service of the overseas branch, was created in the belief that the U.S. was lagging behind both the Germans and the British in the use of radio as a tool of war. Originally part of the Foreign Information Service, its formation predates that of the OWI by several months. John Houseman, a theatrical and radio producer best known for his collaboration with Orson Welles, was appointed by Sherwood to head the Voice of America. Under his leadership, the Voice, as it was known, was a bastion of liberalism. Its sound was unique, making use of agit-prop and experimental radio and theater techniques to communicate its views. Organized on the principal of language desks (much like the shortwave operations of the BBC), the VOA became a haven for expatriates from every nation under Nazi occupation. As a part of the foreign branch, the VOA exercised direct control not only over its programming but eventually over its broadcast facilities, acquiring the use of all 14 of the existing shortwave outlets in the U.S. by early 1943.

Under Houseman, the VOA played an important role in defending America against criticism for the lack of a second front during the dark days before the tide of war had turned. By the summer of 1943, however, as victory began to appear likely for the Allies, the liberal propaganda style of Houseman and his staff gave way to a less dramatic, more journalistic approach, one meant not to rouse people to action but to inform. This trend accelerated after Houseman and many of his adherents left the VOA. Military news took precedence over the political, and anything smacking overtly of propaganda was dropped (on the principle that those living under Nazi occupation had heard enough of it already). This approach continues to be effective for the VOA to this day.

Domestic Branch Activities

At its peak, the OWI's domestic branch had its finger in nearly every aspect of the war effort, to the dismay of its critics, and scores of federal agencies cleared information through it daily. This broad reach was not to last, however.

In a democracy, a domestic government propaganda operation cannot help but be slightly suspect, even in wartime, so it is surprising that OWI homefront activities went unchallenged as long as they did. Elmer Davis had been on the job a little more than a year before the dam broke. During the spring and summer of 1943 a congressional coalition of Republicans and conservative Democrats attacked the agency, accusing it and its director of using its resources to work for the re-election of Roosevelt in 1944. The fact that its director had been a frequent critic of congressional Republicans before leaving CBS for the OWI made the situation even worse. Davis fought a losing battle with Congress over the organization's domestic activities and budget, even threatening to resign. By the time the situation had eased somewhat in late June, domestic operating expenditures had been trimmed to $2.75 million, a meager amount compared to the $24-million budget approved for the overseas branch. The appropriation was just enough, Davis mused, to avoid "the odium of having put us out of business, and carefully not enough to let us accomplish much." As it was, the cuts forced closings of regional offices around the country. The motion picture production and publishing arms ceased to exist, and much of the responsibility formerly held by the OWI was lost to other government agencies or returned to private-sector entities such as the Advertising Council. As a result, the OWI's domestic operations became precisely those of the "coordinating super-agency" that had been decried.

One of the few domestic activities not drastically reduced in scope was OWI placement of government messages on radio programs and stations, both national and local. As all time was "donated" (participation was mandatory), expenses were minimal. The system put in place to organize this effort was devised and managed by the staff of the OWI Radio Bureau.

The Radio Bureau

According to a preliminary inventory of the files of the OWI, published by the U.S. Government Printing Office in 1952, the Radio Bureau "reviewed, cleared and approved all proposed radio programs sponsored by Government agencies and served as a central point of contact and clearance for other agencies in other relationships with the radio broadcasting and advertising industries." It also "obtained the use of radio programs with a known audience," and "kept the radio industry informed of the relative importance of the many requests for contribution of free time for Government programs.

Within the bureau were various divisions, each with its own specific area of concentration. These included the Government Liaison Division, the Program Services Division, and the Industry Relations Division. The most important, however, was the Allocation Division. Its primary function was the management of the seven "facilities plans," four providing for the orderly inclusion of "action"-oriented messages (those requiring activity on the part of the listening audience, such as "Don't Buy Black Market Meat") and three plans devoted to communicating war-related "background" information (messages which were meant to educate the American people about why they were fighting and what they were fighting for). The seven facilities plans administered by the Allocation Division were developed to "better insure effective, well planned dissemination of all war information (exclusive of war news) via radio."

First and most extensive among the more specific "action" plans was the Network Allocation Plan. The primary goals of this plan included a "determination as to what needs were paramount and deserved priority," an "orderly allocation and distribution of needs over the radio network structure," and a "wise distribution of war messages which would not surfeit audience and harm established listening habits." To accomplish these tasks, an elaborate bureaucracy was developed linking the four national networks, the various national sponsors, and the program-producing advertising agencies to the OWI Radio Bureau's Allocation Division in Washington, D.C. Under this plan every network program classified as entertainment would carry such information as the Allocation Division provided. Shows broadcasting weekly were required to include a message once every month, while those airing more frequently, such as the many soap operas of the day, were expected to air two such messages in the same period. Schedules were designed to prevent audience overexposure to any one campaign, and many radio campaigns could be run simultaneously owing to the large number of programs available for use.

Copies of all OWI allocation directives concerning a specific program were sent to each of the parties involved. The networks appointed OWI liaisons for both sponsored and sustaining programming, while sponsors usually delegated the job to someone in their advertising department (often the manager himself). These communications to sponsor and the network sponsored program liaison were just courtesies, however, as the bulk of radio's creative work at this time was carried on in the radio departments of advertising agencies around the country (though mostly in New York, Chicago, and Hollywood). These agencies were hired by the sponsor to handle every facet of a show, from conception to production. It was through the like of Kenyon & Eckhardt, Lord & Thomas, BBD&O, and J. Walter Thompson that the government's information truly flowed to the ears of the American people.

OWI information was distributed in the form of a fact sheet that contained the information to be stressed in a program's scheduled allocation. This, along with a cover letter listing the name of the program it was for and the projected date of broadcast, was sent to the ad agency's OWI liaison, usually a staffer in the agency's radio division. Once the information reached the hands of the program's writers, its treatment was completely up to them. Participation was viewed as mandatory, and Radio Bureau staffers monitored broadcasts to verify that assigned allocations had been carried, judging them according to perceived effectiveness at the same time.

The Station Announcement Plan and National Spot Plan were similar in form and function to the Network Allocation Plan but targeted independent stations, syndicated and local sports programming, and other non-network fare.

The final plan in the action group was the Station Live Program Plan, which was under the control of the various OWI Regional Radio Directors who, working with local stations, would try to produce one-shot war programs with more of a hometown flavor. The OWI promoted the importance of a local character coming through over the air so that people would feel more connected to the war effort.

The first of the background information plans was the Feature Series plan. This plan, proposed jointly by the four national networks, called for a "series of network programs embracing background issues of the war which cannot be fully delineated by radio in any other way." The focus was "not on the progress of the war, and not on the things the citizen must do to help win the war, but on the things the citizen must know and understand about the war effort in order wholeheartedly to play his part during the war and in the establishment of a just and lasting peace."

The next background plan made use of existing network programming in much the same way as the Network Allocation Plan did, but shows on the Special Assignment Plan were those that volunteered their time and talent "over and above" the requirements already set by the Network Allocation Plan. These programs developed a stronger relationship with OWI and were given more personal treatment by the Radio Bureau, along with tremendous access to government facilities and resources.

The final background plan was the Station Transcription Plan, which consisted of "transcribed war programs produced for station use by government agencies." Stations could receive programs upon request. The two main programs offered were *The Treasury Star Parade* (in cooperation with the Treasury Department), which was a three-a-week series, and the OWI's own show, *Uncle Sam*, a five-a-week "strip."

It is not possible to truly gauge the overall importance of radio as a tool of government information dissemination during World War II, even from the vantage point of more than a half-century later. Still, despite the difficulties inherent in its task, the OWI was correct to let radio do the job at home with a minimum of government interference, and in many ways it was a success story of amazing proportions. Some expressed alarm at the use of advertising techniques in the service of the war effort, but a great deal of important government information was communicated effectively to the nation's civilian population using the familiar forms of radio advertising. Though it is hard to isolate radio's impact from other media, its role as a common focal point of American life undeniably helped foster a unity of purpose among those on the home front that may not otherwise have existed. That the work of the former radio arm of the overseas branch, the Voice of America, goes on to this day is an indication of its perceived effectiveness. Though World War II and the Cold War have ended, the VOA continues to serve the foreign policy goals of the United States, broadcasting 900 hours of programming each week in 53 languages.

CHUCK HOWELL

See also BBC World Service; Cold War Radio; Davis, Elmer; Politics and Radio; Propaganda by Radio; Radio Free Europe/Radio Liberty; Shortwave Radio; Voice of America; World War II and U.S. Radio

Further Reading

The records of the Office of War Information comprise Record Group 208 of the National Archives.

Burlingame, Roger, *Don't Let Them Scare You: The Life and Times of Elmer Davis*, Philadelphia, Pennsylvania: Lipincott, 1961

Dryer, Sherman Harvard, *Radio in Wartime,* New York: Greenberg, 1942

Kirby, Edward Montague, and Jack W. Harris, *Star Spangled Radio,* Chicago: Ziff-Davis, 1948

Koppes, Clayton R., and Gregory D. Black, *Hollywood Goes to War: How Politics, Profits, and Propaganda Shaped World War II Movies,* New York: Free Press, and London: Collier-Macmillan, 1987

Shulman, Holly Cowan, *The Voice of America: Propaganda and Democracy, 1941–1945,* Madison: University of Wisconsin Press, 1990
Steele, Richard, "Preparing the Public for War: Efforts to Establish a National Propaganda Agency, 1940–1941," *American Historical Review* 75 (1970)
Weinberg, Sydney, "What to Tell America: The Writers Quarrel in the OWI," *Journal of American History* 55 (1968)
Winkler, Allan M., *The Politics of Propaganda: The Office of War Information, 1942–1945,* New Haven, Connecticut: Yale University Press, 1978

Oldies Format

The Oldies radio format features the greatest pop music hits of the 1950s and 1960s. The format was created in the early 1970s and initially targeted the musical tastes of the 18- to 34-year-old "baby boomers." The format still targets baby boomers today, although they are now predominantly 35 to 54. This audience was the first generation to grow up with rock and roll. The format's core artists include such 1950s standouts as Elvis Presley, Chuck Berry, Buddy Holly, and the Everly Brothers and 1960s superstars such as the Beatles, Supremes, Beach Boys, and Four Seasons.

The Oldies format was based on a segmentation strategy that targeted older Top 40 listeners. Top 40 was the first radio format to target a young audience, and from its inception in the late 1950s through the late 1960s, it was virtually the only younger-appeal format. When musical styles changed in the late 1960s and early 1970s, a certain portion of the Top 40 audience ceased to relate to the current style of popular music. This disenfranchisement created a format void that was ultimately filled by the creation of the Oldies format.

Top 40 stations play mostly current hits, with a few oldies thrown in each hour. For Oldies stations, the entire focus is on 1950s and 1960s oldies, although some first-generation Oldies stations occasionally played current hits, calling them "Future Gold." This practice was dropped when Oldies stations discovered that their audience didn't want to hear current music, not even one song an hour. Research showed the audience wanted Oldies music exclusively, and this exclusivity became the format's most powerful listener benefit. The success of this appeal was the genesis for the format's most popular positioning statement, "All Oldies. All The Time."

The Oldies format was a natural, evolutionary outgrowth of Top 40 radio. Consequently, it was shaped by the same cultural factors that affected our society at that time. In the 1950s and 1960s America was a "mass market" society. There were few media choices. Most markets had one or two daily newspapers; three network TV stations (American Broadcasting Company [ABC], Columbia Broadcasting System [CBS], and National Broadcasting Company [NBC]); and a handful of AM radio stations. (FM stations didn't gain significant listening levels until the mid-1970s.)

Because of the limited media choices, media (and virtually all other consumer offerings) were targeted to the mass audience. Most radio stations were programmed to appeal to a broad, adult audience. When the Top 40 format was created, it too was programmed to a wide demographic target audience that ranged in age from subteens to people in their 30s and 40s. Therefore, Top 40 music of the day was an amalgam of many tastes and styles and represented a variety of music, including such divergent artists as Fats Domino; Percy Faith; Peter, Paul, and Mary; The Singing Nun; Roger Miller; Herman's Hermits; Dean Martin; James Brown; Cream; and others. This diversity of musical styles is reflected in the playlists of today's Oldies stations.

The first stations involved in the development of the Oldies format included Westinghouse's WIND-AM in Chicago; CBS FM's WCAU-FM in Philadelphia; and the station that became the format's standard bearer, CBS FM's flagship station, WCBS-FM in New York City. Other early Oldies stations included KRTH-FM in Los Angeles, WFYR-FM in Chicago, and KOOL-FM in Phoenix.

The early success of WCBS-FM was notable for two reasons. First, the station operated in the spotlight of the nation's largest market, and anything of consequence that happened in the number-one market made news. Second, it was an FM station that was beginning to get noticeable ratings at a time when the overwhelming majority of all listening still occurred on the AM band.

WCBS-FM's success had a major impact on the format. Radio operators surmised, "If WCBS-FM can pull a three-share in New York City, with all of the stations in that market, imagine what we can pull in ours." What they discovered, in most cases, was that their Oldies station also was able to pull a three-share, which was indicative of the format's appeal at that time.

Oldies was viewed as a niche format. It had a small, loyal audience but was not considered a format that would make a station a market leader. Most Oldies stations were operated on that premise. Generally, they tended to be on the AM dial with inferior signals (at a time when radio listening was increasingly focused on the FM dial). Most ran network or syndicated programming that was not local, they weren't promoted or marketed, and they were low-cost operations with modest profit goals. There were some notable exceptions, but for the most part the radio industry perceived Oldies to be a second-tier format. Like many beliefs, this had the possibility of becoming a self-fulfilling prophecy. Since the radio industry did not believe Oldies could deliver big ratings, it did not invest resources in the format, and thus the format did not grow.

This perception lasted until the mid-1980s, when two major broadcasting companies began achieving big ratings with their Oldies properties. Shamrock Broadcasting's WWSW-AM-FM in Pittsburgh, KXKL-FM in Denver, and WFOX-FM in Atlanta and Sconnix Broadcasting's WQSR-FM in Baltimore and WMXJ-FM in Miami served as pivotal success stories that forced the industry to reevaluate Oldies as a format. These two companies asked the question, "What would happen if you treated an Oldies station like you would any other station? What would happen if you offered local programming, hired programming consultants, did local research, invested heavily in programming and marketing, and—most important—offered the format on FM?" From the late 1980s to the mid-1990s, market after market saw the emergence of a new FM Oldies competitor. Today, Oldies is considered a major radio format, and in virtually every large and medium market, and in most smaller markets, listeners can find at least one station that specializes in it.

E. ALVIN DAVIS

See also Classic Rock Format; Rock and Roll Format

Further Reading

Jack, Wolfman, *Have Mercy: Confessions of the Original Rock 'n' Roll Animal,* New York: Warner Bros., 1995
Sklar, Rick, *Rocking America: How the All-Hit Radio Stations Took Over,* New York: St. Martin's Press, 1984

Old-Time Radio. *See* Nostalgia Radio

One Man's Family

Serial Drama

Created by one of broadcasting's neglected auteurs, Carlton E. Morse, *One Man's Family* was radio's most acclaimed and popular primetime serial drama. Relating the multi-generational saga of the Barbours, an upper middle class family in San Francisco, *One Man's Family* aired 3,256 episodes from 1932 to 1959, making it the longest uninterrupted narrative in the history of American radio. With an opening dedication to "the mothers and fathers of the younger generation and to their bewildering offspring," *One Man's Family* reflected the aspirations and tensions of the American family over three decades.

Morse imbued *One Man's Family* with a novelistic aura. He modeled his series on *The Forsyte Saga,* John Galsworthy's sprawling study of an aristocratic family in late Victorian and Edwardian England. To underscore the literary parallel, Morse divided his series into "books" and "chapters," which were announced at the beginning of each show. The final program closed the run at Chapter 30 of Book 134. The artistic trappings help give *One's Man's Family* a critical legitimacy, making it seem more than soap opera, closer to literature. Much of the critical discourse about the program echoed Gerald Nachman's assessment that the serial "was emblematic of America's ongoing faith in the home as the savoir of the nation and the wellspring of its spiritual strength."

A Morse Creation

Louisiana-born Carlton Morse studied drama at the University of California at San Francisco, where he first became intrigued with the city that would take hold of his imagination. After struggling as a journalist, he joined the staff of KGO, the San Francisco radio affiliate of NBC two weeks before the stock market crash of 1929. He began writing scripts for the series *House of Myths* and later wrote for such mystery programs as *Chinatown Tales* and *Split-Second Tales*. He gained a reputation for his "blood and thunder" scripts, including "Dead Man Prowl" and "City of Dead," first heard on NBC *Mystery Serial* and later revived for the syndication series *Adventures* by Morse in the 1940s. Morse also crafted plays based on the files of the San Francisco Police Department for *Barbary Coasts Nights*.

Concerned about increasing juvenile delinquency after World War I, Morse turned his back on the action genre and developed a series that would affirm family bonds. Serving as both producer and director, he chose many young actors with whom he worked at the university and at *Mystery Serial* for a new program that would emphasize relationships over plot. *One Man's Family* debuted on 29 April 1932 as a 13-week trial on NBC's San Francisco, Los Angeles, and Seattle stations. Broadcast weekly from the NBC radio studios at 111 Sutter Street in San Francisco, the program was aired in May to the full West Coast lineup of stations and one year later was carried nationwide, becoming the first program based in the West to be heard regularly in the East.

One Man's Family quickly attained commercial viability. In 1934 Penn Tobacco became a regular sponsor for a year, soon replaced by Standard Brands for the next 14 years. With a half-hour Wednesday evening slot and an increasingly devoted audience, the production moved to Hollywood in 1937. *One Man's Family* achieved its highest rating during the 1939–40 season when it was on Sunday evenings following the Jack Benny and Edgar Bergen/Charlie McCarthy comedy programs, and ranked among the top five programs in the nation. In September 1949 there was such a public outcry when Standard Brands dropped the series that NBC sustained it until another advertiser could be found. Miles Laboratories became the sponsor in June 1950 and reorganized *One Man's Family* as a daily quarter-hour program. When the series ended on 8 May 1959, Morse mourned its passing with a note to the *Los Angeles Times*: "The signposts for sound family life are now few, and I feel the loss of *One Man's Family* is just another abandoned lighthouse."

The Ensemble

One Man's Family was immediately recognized for its unprecedented realism. Morse's relationship with his cast was partly responsible for the program's authenticity. Morse wrote the scripts with the personal quirks of each actor in mind, and the ensemble responded by sticking with the program for years. When it ended in 1959, many actors were still playing roles that they had originated decades before. J. Anthony Smythe starred as the patriarch Henry Barbour, the crusty, conservative stockbroker, for the entire run of the series. Minetta Lane continued as Fanny Barbour, the patient mother of five children, until 1955 as did Michael Raffetto who held forth as the eldest son Paul, a battle scarred veteran who became the trusted moral center of the family. Page Gilman, the son of NBC vice president Don Gilman, joined the cast as a 15-year old playing the youngest son Jack and stayed to the cancellation. When Barton Yarborough, who played the mercurial son Clifford, died suddenly in 1951, his character was written out of the script.

The inaugural episode began with a household of seven; by the end of the run the extended family totaled more than 90. The story lines unfolded slowly and realistically, dealing with such traditional serial subjects as romance, marriage, children, and divorce. Morse tried to capture the rhythms of daily life by listening to his actors and the audience. When the actress Bernice Berwin, who played the eldest daughter Hazel, was pregnant, so was her character. Although much of the domestic action centered on the large family home in the wealthy community of Sea Cliff near the Golden State Bridge and their weekend retreat at Sky Ranch, the well-off Barbours were also affected by national events, especially World War II, which threatened the lives of several leading characters. The tragic heroine of the program, Claudia, was lost at sea when Germans torpedoed her ship, a situation constructed by Morse to allow actress Kathleen Wilson to leave the series and raise her own family. Listeners became so upset that their impetuous and star-crossed Claudia was presumably dead, that Morse revived the role with another actress, Barbara Fuller, who stayed until the end of the series.

Impact

One Man's Family became a national ritual as millions of Americans embraced the Barbour family as their own. Many collectibles were marketed to the ardent fans, including cookbooks, diaries, and family albums. Radio satirists Bob and Ray parodied this sentimental side of the series with their look at the "Butcher family" in *One Feller's Family*. Standard Brands and its advertising company J. Walter Thompson were so pleased with the success of *One Man's Family* that they asked Morse to create another radio series. Morse returned to his action roots and conceived *I Love a Mystery*, which employed many of *One Man's Family* regulars. Morse adapted his family serial to television several times. *One Man's Family* was presented first in 1949 as a weekly prime-time series on NBC. Lasting three years, the live, half-hour television version started the Barbour story practically from the beginning, almost where it began decades before on radio. Consequently, none of the radio stars were asked to participate. This first video adaptation is notable because it featured a young Eva Marie Saint as the adventurous Claudia with Tony Randall and Mercedes McCambridge in lesser roles. Again starting at the beginning, *One's Man Family* was brought to television as an afternoon quarter-hour soap opera in 1954 but lasted only a year. This time Anne Whitfield played Claudia, while she was also playing Claudia' s daughter Penelope on radio. Finally in 1958 Morse tried again with a pilot produced for *The Loretta Young Show,* focusing on the relationship between Claudia (Jean Allison) and Johnny Roberts (Keefe Brasselle), an early radio story line.

Carlton Morse estimated that he wrote more than 10 million words to bring to life his vision of an American dynasty. Like the rest of America, his radio family, although prosperous WASPs, persevered through the Great Depression, World War II, and the beginnings of the atomic age. Morse truly believed that if the integrity and moral strength of the family could stay intact, then no great harm could come to the nation. *One Man's Family* was the embodiment of his patriotic and patriarchal philosophy. If *One Man's Family* is still warmly remembered by radio partisans, Morse's immense talents as a producer/writer/director have largely been forgotten, and references to his creative contribution are missing from several radio histories. *Variety* even misspelled his first name when he died on 24 May 1993 in Sacramento, California. But Morse was diligent to save and copyright his scripts, donating them to Stanford University for serious scholarship.

For those critics who fell under the sway of *One Man's Family,* the collection at Stanford will prove that Morse was not only a pioneer of the serial narrative, but also one of broadcasting's most gifted and compelling storytellers. The program series is one of those honored by the Radio Hall of Fame.

RON SIMON

See also I Love a Mystery; Radio Hall of Fame

Cast

Henry Barbour	J. Anthony Smythe
Fanny Barbour	Minetta Ellen (1932–55)
	Mary Adams (1955–59)
Paul Barbour	Michael Raffetto (1933–55)
	Russell Thorson (1955–59)
Hazel Barbour	Bernice Berwin (1932–58)
Claudia Barbour	Kathleen Wilson (1932–43)
	Barbara Fuller (1945–59)
Clifford Barbour	Barton Yarborough (1932–51)
Jake Barbour	Page Gilman (1932–59)

Creator
Carlton E. Morse

Writers
Carlton E. Morse, Harlen Ware, Michael Raffetto, Clinton Buddy Twiss, Charles Buck

Directors
Carlton E. Morse, George Fogle, Michael Raffetto

Announcers
William Andrews, Ken Carpenter, Frank Barton

Organists
Paul Carson (1932–51), Sybil Chism (1951–54), Martha Green (1954)

Programming History
NBC (NBC Red until 1943) 29 April 1932–8 May 1959

Further Reading

Cox, Jim "One Man's Family," Chapter 15 of *The Great Radio Soap Operas,* Jefferson, North Carolina: McFarland, 1999

Herman, James, One Man's Family website, <http://www.geocities.com/californiajamesh/OMF/>

Morse, Carlton, *One Man's Family Album: An Inside Look at Radio's Longest Running Show,* Woodside, California: Seven Stones Press, 1988

Nachman, Gerald, "The Royal Family," Chapter 19 of *Raised on Radio,* New York: Pantheon Books, 1998

One Man's Family Tree website, <http://kinnexions.com/reunion/oneman.htm>

Steadman, Raymond William, "Family Saga," Chapter 20 of *The Serials: Suspense and Drama by Installment,* 2nd edition, Norman: University of Oklahoma Press, 1977

Osgood, Charles 1933–

U.S. News Reporter

Radio news personality Charles Osgood has received many accolades for his unique style. One of his colleagues, Charles Kuralt, referred to him as "one of the last great broadcast writers."

His ability to take a simple off-beat news story or little-known fact and develop a commentary attracts many radio listeners to his daily morning Columbia Broadcasting System (CBS) Radio News program, *The Osgood Files,* which runs in four installments. CBS News reports that his program is noted for its large audience.

Osgood worked as general manager of WHCT (the first pay television station in the United States) in Hartford, Connecticut, and as program director and manager for WGMS in Washington, D.C. Before going to WCBS, Osgood spent almost four years on general assignment for American Broadcasting Companies (ABC) Radio News, where he did a show called *The Flair Reports,* a radio program similar to *The Osgood Files.*

Prior to joining CBS News, Osgood was a morning anchor and reporter for WCBS Newsradio 88, the CBS flagship AM station in New York City, from 1967–91. On 28 August 1997, WCBS Newsradio 88 celebrated 30 years of all-news format by highlighting Charles Osgood as one of the legends of radio in a program called "Let's Find Out."

In addition to his work on CBS Radio Network, Osgood is also a correspondent for CBS Television Network. On 1 April 1994, he was named anchor of CBS News *Sunday Morning.* Before being named to this post, he had provided commentary for *CBS This Morning* and had been a regular contributor to *Up to the Minute* and *Sunday Morning.* Until June of 1992, he was co-anchor of the *CBS Morning News* and a contributor to the *CBS Evening News with Dan Rather.* From 1981–87, Osgood anchored the *CBS Sunday Night News.*

His features for *The Osgood Files,* which he writes and delivers, often display a wit that is blended with a childlike sense of wonder, as exemplified in his frequent sign-off, "I'll see you on the radio." To Osgood, radio permits the listener to build his or her own pictures, which are not confined by the "limits of a television screen" but are constructed by limitless imagination.

Early-morning commuters appreciate Osgood's keen perspective on life, which sheds new light on a fast-paced society. In fact, the Society of Silurians, a distinguished group of New York journalists, have honored him for his "fresh approach to news and its background."

Often called CBS News' "poet in residence," Osgood uses the Dr. Seuss approach to poetry to deliver some of his commentaries in verse. As he says, "The news of the day is so goofy at times/It just seems to fit into couplets and rhymes." Consequently, it seemed only fitting for President Richard Kneedler of Franklin and Marshall College to present Osgood with an honorary Doctor of Human Letters Citation (1998) that was written in verse. Osgood acknowledged the honor by quoting from Theodore Geisel (a.k.a. Dr. Seuss).

Osgood's commentaries have earned him some of the most prestigious awards ever given to a broadcast journalist. For five consecutive years, he received the *Washington Journalism Review*'s "Best in the Business" Award as "Best Radio Reporter" (1988–92). He also won two George Foster Peabody awards (1985–86) for *Newsmark,* a weekly CBS Radio public-affairs broadcast.

In 1990 Osgood was inducted into the National Association of Broadcasters Hall of Fame. In 1993 he was presented with the Marconi Radio Award for Syndicated Network Personality of the Year. In 1995 he received the Lowell Thomas

Charles Osgood
Courtesy of CBS News

Electronic Journalism Award from the International Platform Association and the John Connor Humanitarian Service Award from Operation Smile.

Osgood, a native New Yorker, received a B.S. in economics from Fordham University. He also holds an honorary doctorate from Fordham, an honorary law degree from St. John's University School of Law, and honorary degrees from St. Bonaventure University, Stonehill College, College of St. Rose, LeMoyne College, St. Peter's College, and the College of Mount St. Vincent.

MARY KAY SWITZER

Charles Osgood. Born in New York City, 8 January 1933. Attended Fordham University, B.S. in Economics, 1954; reporter, ABC, 1963–67, WCBS, 1967–91, CBS News, 1991–present; news anchor, *CBS Sunday Night News,* 1981–87; wrote and delivered features on *The Osgood Files.* Inducted into National Association of Broadcasters' Hall of Fame, 1990; received honorary degrees from various U.S. universities; received George Foster Peabody Award, 1985, 1986, *Washington Journalism Review*'s "Best in the Business" Award as "Best Radio Reporter", 1988–92, Marconi Radio Award, 1993, Radio Mercury Award, 1999.

Radio Series

The Osgood Files

Selected Publications

Nothing Could Be Finer Than a Crisis That Is Minor in the Morning, 1979
There Is Nothing I Wouldn't Do If You Would Be My POSSLQ, 1979
Osgood on Speaking: How to Think on Your Feet without Falling on Your Face, 1988
The Osgood Files, 1991
See You on the Radio, 1999

Further Reading

The Osgood File website, <http://www.cbsradio.com/news_osgood.asp>
Risley, Ford, "Charles Osgood," in *Encyclopedia of Television News,* edited by Michael D. Murray, Phoenix, Arizona: Oryx Press, 1999

Our Miss Brooks

Situation Comedy

One of the most beloved female characters in radio comedy was Eve Arden's Connie Brooks, an English teacher at a Midwestern high school. Miss Brooks was good-humored, witty, and sardonic, a change from the scatterbrained or merely sarcastic female characters who had been plentiful in supporting or co-starring roles in radio comedy up to that time. Miss Brooks was radio's first single woman as lead character in prime time, and Eve Arden brought to the role the character and intelligence that she had become famous for in films such as *Mildred Pierce.* In films, however, Arden had usually played supporting roles, such as the heroine's best friend. *Our Miss Brooks* allowed Arden to shine on her own, and her series became one of the most popular of the postwar period.

Arden was not the first choice for the role of Connie Brooks. Producer Harry Ackerman had originally wanted Lucille Ball, who turned it down, and he had then asked Shirley Booth. Booth refused the series because, Ackerman recalled, "all she could see was the downside of being an underpaid teacher. She couldn't make any fun of it" (quoted in Nachman, 1998). Meanwhile, Columbia Broadcasting System (CBS) President William Paley had become acquainted with Arden and proposed that she do the series. She passed on the original script, but a rewrite by Joe Quillan and Al Lewis proved more to her taste (Lewis would serve throughout the show's run as writer-director). There was still the matter of the broadcast schedule to negotiate, however. The program was scheduled to premiere at the beginning of the summer, which Arden had planned to spend with her children in Connecticut. She asked if the programs could be transcribed, allowing her to do all 15 in a short time and then depart. The network agreed, and *Our Miss Brooks* became one of the first radio programs to be broadcast by transcription. It quickly became the number-one program of the 1948 summer season.

Arden's warm and witty delivery anchored the program, but she was surrounded by an equally talented supporting cast, many of whom went on to fame in film and television. Gale Gordon, famous for his portrayal of Mayor La Trivia on *Fibber McGee and Molly* (and later as Lucille Ball's television

Eve Arden, *Our Miss Brooks*
Courtesy Radio Hall of Fame

nemesis), played Madison High's blustering Principal Conklin. Future film stars Jeff Chandler and Richard Crenna both had prominent roles—Chandler as Miss Brooks' love interest, biology teacher Mr. Boynton, and Crenna as her adoring, mischief-making student Walter Denton. The other women in the cast were Jane Morgan as Brooks' addled landlady Mrs. Davis; Gloria McMillan as the principal's daughter Harriet; and Mary Jane Croft as Miss Enright, a possible competitor for Mr. Boynton's affections.

Stories often revolve around some scheme of Denton's going comically awry (such as his plan to make Miss Brooks teacher of the year); he inevitably involves Miss Brooks in some way, which ends up getting her into hot water with Mr. Conklin. Conklin constantly suspects Brooks of being at the root of whatever problem is at hand (with some justification), but she is also adept at mollifying him and helps to protect him from the dreaded school board. As Gerald Nachman (1998) has noted, part of the uniqueness of Brooks' character is that "she treated men with refreshing suspicion—but as undeserving equals." Miss Brooks is always the smartest, most rational, most sophisticated person in any room. In her romance with Mr. Boynton, she is clearly the aggressor in the relationship, trying to figure out ways to manipulate Boynton into asking her out on dates. As she explains to Mrs. Conklin in one episode, "In a moment of weakness, I promised Mr. Boynton the entire weekend." Mrs. Conklin asks, "When did you do that, Miss Brooks?" and she replies, "At lunch in about an hour from now. That is, I'm sure he'll accept—er—invite me." Although Mr. Boynton remains friendly and admiring throughout the series, he is perpetually dim with regard to romance and often seems more enamored of his frog, McDougall, than of Miss Brooks.

Connie Brooks' spirit, however, remains undaunted by the lack of progress in her romance or the shortcomings of her profession. Although the program addressed some of the problems of being a schoolteacher—especially low wages and lack of appreciation—Brooks' obvious sense of her own worth helped her rise above her circumstances and served as an inspiration to teachers around the country. Eve Arden received thousands of approving letters from teachers, was often asked to address Parent-Teacher Association (PTA) meetings, and was even offered jobs teaching English at various high schools. Most important, perhaps, Brooks suggested that marriage and children were not the only road to fulfillment for young women in the 1950s. The popularity of the program attests to the audience's desire for more alternatives than those that the narrowly defined culture of the postwar period offered them, and Miss Brooks made such an alternative seem not only possible, but thoroughly enjoyable as well.

In 1952 *Our Miss Brooks* moved to television, where it ran successfully until 1956, earning Eve Arden an Emmy as Best Actress. Although she continued to appear on stage and in films, Arden's low, throaty voice was instantly recognizable to fans of the show, and she always remained best known as Miss Brooks. Fortunately, most of the program's eight-year radio run is available in recordings, providing a remarkable testament to one of radio's comedy heroines.

ALLISON MCCRACKEN

See also Comedy; Gordon, Gale

Cast

Miss Connie Brooks	Eve Arden
Principal Osgood Conklin	Gale Gordon
Philip Boynton	Jeff Chandler (1948–53), Robert Rockwell (1953–57)
Walter Denton	Richard Crenna
Mrs. Margaret Davis	Jane Morgan
Harriet Conklin	Gloria McMillan
Stretch Snodgrass	Leonard Smith
Miss Enright	Mary Jane Croft
The French Teacher	Maurice Marsac
Jacque Monet	Gerald Mohr
Announcer	Verne Smith, Bob Lamond

Producer/Creator

Larry Berns

Director/Writer

Al Lewis

Programming History

CBS 1948–57

Further Reading

Arden, Eve, *Three Phases of Eve: An Autobiography*, New York: St. Martin's Press, 1985

Nachman, Gerald, *Raised on Radio: In Quest of The Lone Ranger, Jack Benny . . .*, New York: Pantheon, 1998

Owens, Gary 1936–

U.S. Radio Personality

In his more than 40-year career, Gary Owens has hosted more than 12,000 radio shows. Owens began working in radio in the 1950s, paying his way through college by working at KORN, in Mitchell, South Dakota. As a boy, Owens listened to such radio programs and personalities as *Superman, The Shadow,* Jack Benny, and Fred Allen. From KORN, Owens went on to KMA in Shenandoah, Iowa; to KOIL in Omaha, Nebraska; and to KIMN in Denver, Colorado. Gordon McLendon hired him to increase ratings at KILT in Houston, KTSA in San Antonio, WNOE in New Orleans, and WIL in St. Louis. Owens said in a *Billboard* interview: "I was a troubleshooter for McLendon, doing unusual things to gain ratings. I worked in three markets in one year" (Rusk, 1996). In 1959 McLendon moved Owens to the morning drive slot at KEWB in San Francisco, which was within arm's length of Owens' ultimate goal of Hollywood.

Owens moved to Hollywood's KFWB in 1961. In 1962 he began a 20-year stint at KMPC in Hollywood. TV producers and directors hired him for countless parts and voiceovers in movies, TV shows, cartoons, and commercials. Owens, broadcasting direct "from beautiful downtown Burbank," became one of the more recognizable voices in entertainment. Owens is most famous for his trademark hand-over-the-ear delivery on NBC-TV's *Rowan and Martin's Laugh-In* from 1968 to 1973. He was one of many regulars on the one-hour television show. Endless gags were crammed into the hour-long program,

Gary Owens
Courtesy Radio Hall of Fame

because the producers believed the fast pace would not require all attempts at humor to work so long as the audience was not given time to be bored. From 29 September to 26 December 1969, Owens hosted the daytime game show *Letters to Laugh-In,* a spin-off of the regular show in which four guest celebrity panelists read jokes sent in by *Laugh-In*'s viewers. The studio audience chose jokes to be used on the prime-time program.

In 1982 Owens left KMPC for several other Los Angeles stations. He first went to KPRZ, then to KKGO-FM in 1985, KFI in 1987, KLAC in 1992, and KJQI in 1993. Since 15 June 1996, Owens has hosted *Music of Your Life,* a radio program syndicated by Jones Satellite Networks. While being seen or heard on TV, Owens has also worked regularly in radio. He told *Billboard* magazine: "I occasionally will get off the air and strictly do television, cartoons, and commercials, but I eat, drink, and sleep radio. I love radio. I always have" (Rusk, 1996). Owens has won 50 Clio Awards and earned a star in the Hollywood Walk of Fame. He has been hailed as a "Legend" by *Billboard* magazine and (along with Gordon McLendon) was named to the Museum of Broadcast Communications Radio Hall of Fame in 1994.

W.A. KELLY HUFF

See also McLendon, Gordon

Gary Owens. Born Gary Altman in Mitchell, South Dakota, 10 May 1936. Began radio career at KORN, Mitchell, South Dakota, 1950s; hosted morning drive slot on KEWB, San Francisco, California, 1959; host at KMPC, Hollywood, California, 1962–82; regular on *Rowan and Martin's Laugh-In,* 1968–73; Worked at various Los Angeles radio stations1982–96; host of syndicated radio show *Music of Your Life,* 1996–present; recipient: star on Hollywood Walk of Fame, 1994; more than 50 Clio Awards; inducted into Museum of Broadcast Communications Hall of Fame, 1994.

Radio Series

Watermark's Soundtrack of the 60s, 1981–84
Gary Owens' Soundtracks, 1984–88
Music of Your Life, 1996–

Television

Rowan and Martin's Laugh-In, 1968–73; *The Gong Show,* 1976–77

Films

Narrator, *The Naked Witch,* 1960 (a.k.a. *The Naked Temptress*); *The Love Bug,* 1969; narrator, *Dr. Phibes Rises Again,* 1972; narrator, *Loose Shoes,* 1980 (a.k.a. *Coming Attractions* and *Quackers*); *Hysterical,* 1983; narrator, *Aliens, Dragons, Monsters and Me,* 1986; *I'm Gonna Git You Sucka,* 1988; *Destroyer,* 1988 (a.k.a. *The Edison Effect* and *Shadow of Death*); *How I Got Into College,* 1989; *Kill Crazy,* 1990; *Ed Wood: Look Back In Angora,* 1994; *Diggin' Up Business,* 1990; *Spy Hard,* 1996; *Border to Border,* 1998

Further Reading

Elias, Thomas D., "'Befuddled' Gary Owens Keeps 'Laugh-In' along Scripps Howard News Service," *Chicago Tribune* (24 August 1988)
"The Next Voice You Hear . . . Is Probably Gary Owens," *Broadcasting* (8 September 1986)
Nidetz, Steve, "Golden Tones: Gary Owens Honors Radio As Its Museum Honors Him," *Chicago Tribune* (4 November 1994)
Rusk, Bob, "The Airwaves of Gary Owens' Life: Radio Is Still Tops for 'Laugh-In' Regular," *Billboard* (20 July 1996)

Ownership, Mergers, and Acquisitions

During radio's first seven decades, through the mid-1990s, ownership of radio stations was limited by Federal Communications Commission (FCC) rules. Since the passage of the 1996 Telecommunications Act, owners have been allowed to own not just a few stations, but hundreds. Through multiple mergers and acquisitions, a handful of new radio owners—led by Clear Channel and Infinity/Columbia Broadcasting System (CBS)—have consolidated hundreds of stations under a single corporate umbrella.

The First 70 Years

During radio's development stage in the 1920s, stations were most often owned and operated as secondary sidelines to other businesses, such as hotels, retail stores, or radio-related businesses such as receiver manufacturing or sales. Newspapers acquired stations out of fear of a new competitor for local advertising dollars. Department stores and hotels bought stations to promote their sales. But with the development of the national networks, the National Broadcasting Company (NBC, owned by radio manufacturer Radio Corporation of America) and then CBS, ownership began to consolidate.

By 1936 about half of all stations were affiliated with a network, but the FCC frowned on networks' directly owning too many outlets. NBC and CBS acquired stations in major cities, reaching most of the population, and these owned and operated units accounted for a quarter of the networks' income. In 1943 the FCC issued a rule limiting owners to no more than one AM and one FM station per market. Thus groups of stations were developed across multiple markets by both networks and by other firms. Later in the 1940s, the Commission set as seven the number of AM stations that could be owned in common. In 1953 the FCC raised the FM limit to seven and formalized the limit of seven AM and seven FM stations that would define radio ownership for three decades.

The New World of Radio Ownership

The dramatic growth in the number of radio stations led the FCC (under pressure from the broadcast industry) to allow several increases in ownership limits after 1980—to 20 AM and 20 FM stations, for a potential national total of 40 outlets—by the early 1990s. Also gone, as of 1992, was the "duopoly rule" that had limited a single group or individual to no more than a single AM and FM outlet in a single market. On 8 February 1996, President Clinton signed into law the Telecommunications Act of 1996, which, among its many provisions, directed the FCC to eliminate the national multiple radio ownership rule and to relax the local ownership rule. In an Order adopted 7 March 1996, the FCC implemented these provisions, and soon the former ownership limits of seven AM and seven FM stations seemed quaint.

The Telecommunications Act of 1996 loosened but did not eliminate ownership restrictions. For example, in markets with 45 or more commercial stations, a single company may own up to 8 stations with no more than 5 as either AM or FM. If the market has 30–44 commercial radio stations, the total number one owner can acquire drops to 7, with a maximum of 4 in the same class (AM or FM). For smaller markets, those with 15–29 radio stations, the total "cap" (permitted absolute amount) drops to 6, with 4 of any one modulation. Finally, in markets with fewer than 14 commercial radio stations, the total that any one company can own is 5, with no more than 3 as AM or FM (in each case, up to half the stations in the market).

Passage of the 1996 Act set off the greatest merger wave in radio history. CBS merged with Infinity Broadcasting, and in a telling metaphor Infinity's founder, Mel Karmazin, noted, "it's like combining two ocean front properties." He meant that the new empire would not be some "mom and pop" collection of stations, as had often been the case in the past, but would own seven outlets in New York City, six in Los Angeles, ten in Chicago, eight in San Francisco, and four in Washington, D.C. By 2000, the new CBS/Infinity combination commanded nearly a third of all radio advertising revenues in the top ten markets.

During just the first year after passage of the 1996 law, the FCC calculated that some 2,066 radio stations (about 20 percent of the total) changed owners. As a result of this trading activity, the Commission observed that there were a score of new owners and a significant increase in the number of large group owners—and therefore of concentration of station ownership overall. There were also considerable changes in the composition of the top 50 radio group owners, reflecting mergers between companies on that list as radio groups began to develop into vast empires. And the top media companies continued consolidating: the top ten of 1996 shrank to six by the end of 1997, to a top three by the fall of 1998, and to a Big Two by 2000—Clear Channel and CBS/Infinity after its merger with Viacom. Disney's American Broadcasting Companies (ABC) and Cox's radio division followed as part of major media conglomerates and were the only powers that could offer Clear Channel or Viacom's CBS/Infinity a true challenge.

Clear Channel

Clear Channel Communications, based in San Antonio, Texas, owned more than 900 radio stations by early 2000, the largest radio group in history. It remained the largest in mid-2003, with more than 1,200 stations nationwide. Clear Channel owned stations in the top markets of the United States, yet it was still best thought of as a force in small and medium markets in communities such as Grand Rapids, Michigan (media market 66), and El Paso, Texas (media market 69). Clear Channel had also expanded abroad, acquiring radio stations in Australia, New Zealand, and the Czech Republic.

But Clear Channel's October 1999 purchase of Hicks Muse's collection of radio stations pushed it into an ownership category never fathomed by the creators of the Communications Act of 1934. Hicks Muse, an investment company unknown to the radio business before 1995, owned in excess of 400 stations when it sold to Clear Channel. The deals had come rapidly. In February 1997, in two deals worth more than $1.6 billion, the largest radio group was formed. The combined company, to be called Chancellor Media Corporation, was put together by Hicks Muse; later that month, Hicks Muse took over Evergreen, and the rush was on. Next came the takeover of ten radio properties owned by Viacom for $1.075 billion, bringing Hicks Muse to widespread notice with a then combined enterprise of 103 radio stations in 21 markets, aggregate net revenues of more than $700 million, and an enterprise value of about $5 billion. With its late 1999 takeover of Hicks Muse, Clear Channel had acquired stations in nearly every radio market in the United States—certainly in every major one. Clear Channel had become the greatest owner of radio in the medium's 80-year history.

CBS/Infinity

CBS/Infinity owned far fewer radio stations than Clear Channel, but its 161 stations by early 2000 placed it in a strong second place. Despite growing to 185 stations by mid-2003, CBS/Infinity had slipped to fourth place in number of stations but remained second in terms of revenue and audience size because the outlets were all located in major markets. With stations in the top ten markets, CBS/Infinity was a media conglomerate, with a very profitable radio division functioning at the heart of the corporation's strategy for the future. Karmazin argued that radio offered an advertising vehicle that even television could not match, because radio could target listeners far more efficiently.

Under Mel Karmazin, founder and head of Infinity Broadcasting (which Westinghouse CBS acquired in July 1996 for $4.9 billion in the costliest merger to that date in radio history), CBS/Infinity radio ownership in the top ten markets had become impressive by 2000. In the next ten markets, CBS/Infinity owned strong radio positions in Atlanta, Minneapolis, St. Louis, Baltimore, and Pittsburgh. All this added up to equal status with Clear Channel, but one that was accomplished in a far different manner.

Disney/ABC

Disney—although far more famous for its other media operations—took sizable revenues from its radio division, although its radio holdings paled in comparison to CBS, let alone Clear Channel. Yet Disney offered a significant presence because overall it ranked as the largest media conglomerate in the world before the America Online/Time Warner merger early in 2000. Chief Executive Officer Michael Eisner and his management team kept a significant position in radio; Disney did not sell off these assets as it did with the newspapers it acquired from ABC/Capital Cities, but it also chose not to try to match the merger frenzy of its larger radio rivals. Disney surely had the resources to grow into a larger radio power, but as the 1990s ended the company had chosen not to expand. It rested on its ownership of stations in the top media markets and so should best be thought of as a smaller version of CBS/Infinity. Yet radio did not rank high on Michael Eisner's radar, because Disney management looked to expand the television side of ABC with the potential of more sizable synergies. Disney expanded only the AM penetration of ABC radio to provide outlets for its Radio Disney children's network. By mid-2003, Disney/ABC had slipped to ninth place, with 74 stations, though still ranking third in total audience, and fifth in revenue.

Radio Station Ownership in Top 10 Markets, by Major Groups: 2000

	Number of Stations Owned by:		
Market	*Clear Channel*	*Infinity/ CBS*	*ABC/ Disney*
1. New York	5	6	3
2. Los Angeles	15*	8	3
3. Chicago	6	8	5
4. San Francisco	9	7	3
5. Philadelphia	6	5	1
6. Detroit	7	6	4
7. Dallas/Ft. Worth	8	8	4
8. Boston	0	5	0
9. Washington	8	5	3
10. Houston	16*	4	10
Top-10 Market Stations Owned	80	62	36
Total Stations Owned	959	161	43
Stations Rank	1	4	11
Revenue Rank	1	2	3
Audience Reach Rank	1	2	3

*Some of these will be spun off to meet FCC ownership limits–this process in a number of markets will bring Clear Channel to just under 900 stations.

Source: Who Owns What (10 January 2000)

Cox and Cumulus

Like Disney/ABC, Cox by 2000 represented a diverse media corporation far more famous for other operations. Its radio division was sizable, but the public focused on its newspaper operations, and the company was best known for its newspaper, the *Atlanta Constitution,* and for its move into new media. Although this Sunbelt company had substantial interests in broadcast and cable television, Cox also had a large number of radio stations, with four in Los Angeles, and clusters in Houston, Atlanta, Tampa, and Orlando. Cox merged with New City Communications to add markets such as Tulsa, Oklahoma, and New Haven, Bridgeport, and Norwalk-Stamford, Connecticut. Then the company traded its Los Angeles cluster of outlets to Clear Channel to acquire Houston properties as well as to add stations to their existing clusters. By mid-2003, Cox had more than 75 stations—third in revenue, fourth in audience, but seventh in number of outlets.

Cumulus Media aggregated more than 260 stations by 2003 (second only to Infinity in number of outlets, though ninth or tenth in terms of audience or revenues), buying and consolidating in markets smaller than the top 100. The company's strategy paralleled in the smallest audience-rated markets what Clear Channel had accomplished in markets of all sizes, and what CBS/Infinity achieved in the largest cities.

Minor Mergers and Public Policy

There were holdouts to the radio merger frenzy of the late 1990s. Stubborn single station owners did hang on. WRNR-FM in the metro Washington, D.C., market illustrates the frustrations of operating as a single independent company in a world of radio consolidation. Jack Einstein is a throwback to the days when the FCC restricted ownership of radio. His Annapolis, Maryland–based WRNR-FM sought simply to survive as the bigger consolidated companies took over the Washington, D.C., market. He had no advantages of scale economies to reduce costs, nor could he sell a whole set of stations and formats to big advertisers. The temptation was to cash out, but Einstein—as of early 2000—could not resist the lure of running a radio station programming vintage and progressive rock, and so with a "group" of three stations all in the Washington, D.C., market, the eighth largest U.S. radio market, he sought narrow formats and held on with his son as featured disc jockey.

The U.S. Department of Justice remains an important player as well. Since the FCC has lifted its ownership limits, it has been up to the antitrust division to determine if a merger violates antitrust laws. During the late 1990s, the Department of Justice negotiated a number of consent decrees, such as one in Cincinnati in which Jacor agreed to reduce its share of the advertising dollars from 53 percent to 46 percent, and another whereby CBS (as a result of its Infinity takeover) had to divest itself of stations in nine separate markets.

Faced with growing complaints about radio's continuing consolidation, however, in mid-2003 the FCC adjusted its radio market rules. The commission replaced its "signal contour" method of defining local radio markets with a geographic market approach used by the Arbitron rating service. The commission concluded that its signal contour method created anomalies in radio ownership that congress could not have intended in the 1996 act. The FCC's 2003 decision closed a seeming loophole by applying Arbitron's definitions of geographic radio markets to better reflect radio industry practice.

Radio Ownership Outlook

The aforementioned top radio groups are among the biggest companies in an ever-consolidating radio industry. There are many others, but they are all far smaller than the dominant companies. Clear Channel, Infinity, Cox, and Cumulus point to a continuing trend toward greater consolidation during the first years of the 21st century as the radio industry continues to adapt to the far looser FCC caps on ownership set in 1996 and modified in 2003.

Although no one knows how far this consolidation will go—it was considerably slowed by the poor economic conditions of the early 2000s—media conglomerates such as McGraw Hill Companies, the New York Times Company, and the Tribune Company continued to own and operate radio station groups awaiting the radio industry's shakeout from mergers and acquisitions. All had acquired radio years earlier, and by the early 2000s they chose simply to sit on their relatively small holdings as management determined how far a Clear Channel or Cumulus would grow and what effects of that consolidation would spill over to the relatively small holders of radio stations. Most observers agreed, however, that the first decade of the new century would likely see radio consolidate even further.

DOUGLAS GOMERY

See also Clear Channel Communications; Deregulation; Federal Communications Commission; Infinity Broadcasting; Karmazin, Mel; Licensing; Localism in Radio; Radio Disney; Regulation; Telecommunications Act of 1996

Further Reading

Albarran, Alan B., *Media Economics: Understanding Markets, Industries, and Concepts,* Ames: Iowa State University Press, 1996; 2nd edition, 2002
Chan-Olmsted, Sylvia M., "A Chance for Survival or Status Quo? The Economic Implications of Radio Duopoly Ownership Rules," *Journal of Radio Studies* 3 (1995–96)
Compaine, Benjamin M., et al., *Who Owns the Media?* White Plains, New York: Knowledge Industry, 1979; 3rd edition, by Compaine and Douglas Gomery, Mahwah, New Jersey: Erlbaum, 2000
Ditingo, Vincent M., *The Remaking of Radio,* Boston and Oxford: Focal Press, 1995
Federal Communications Commission, Mass Media Bureau, Policy and Rules Division, *Review of the Radio Industry, 1997,* MM Docket No. 98–35 (13 March 1998) (issued as part of 1998 Biennial Regulatory Review)
Sterling, Christopher H., "U.S. Communications Industry Ownership and the 1996 Telecommunications Act: Watershed or Unintended Consequences?" in *Media Power, Professionals, and Policies,* edited by Howard Tumber, London and New York: Routledge, 2000
Who Owns What (M Street Publications, weekly)
Williams, Wenmouth, Jr., "The Impact of Ownership Rules and the Telecommunications Act of 1996 on a Small Radio Market," *Journal of Radio Studies* 5, no. 2 (1998)

P

Pacifica Foundation

U.S. Noncommercial Radio Network

The Pacifica Foundation inaugurated the first listener-supported, noncommercial radio network in the United States in the two decades after World War II. By the 1990s it owned five noncommercial FM stations throughout the country: KPFA-FM in Berkeley, California (acquired in 1949); KPFK-FM in Los Angeles (1959); WBAI-FM in New York City (1960); KPFT-FM in Houston (1970); and WPFW-FM in Washington, D.C. (1977). Characterized by unconventional and dissent-oriented programming, Pacifica also provides news and public-affairs material for about 60 affiliated community radio stations.

Pacifist Origins

The Pacifica Foundation was created in 1946 by a small group of World War II–era conscientious objectors (COs) who had participated in the pacifist student club movement of the 1930s. Most of them belonged to the War Resister's League or the American Friends Service Committee. Lewis Hill is generally credited as the guiding force behind the creation of Pacifica. A CO himself, Hill sought the development of institutions that would foster what he called a "pacific world in our time" through the encouragement of public dialogue. In pursuit of nonprofit status, Hill filed Articles of Incorporation with the state of California in the summer of 1946.

The Pacifica Foundation's first project—listener-supported KPFA-FM in Berkeley—received a license in 1948 and went on the air on 15 April 1949. KPFA is recognized as the world's oldest listener-supported noncommercial FM station, but it barely made it through its first five years. Equipped with a 250-watt transmitter during a period when hardly anyone owned FM receivers, KPFA was saved by the Ford Foundation from an early death with a $150,000 grant in 1951. This windfall allowed Hill to test his "2 percent theory," first articulated in that year. His theory stated that any listener-sponsored radio station could function effectively with the regular monetary support of 2 percent of a given metropolitan area's radio listeners.

Having obtained a larger transmitter by the mid-1950s, KPFA's early programs reached an audience of approximately 4,000 subscribers, about one-quarter of whom held advanced degrees. The audience regularly tuned in for the commentaries of movie critic Pauline Kael, poet Kenneth Rexroth, and Zen scholar Alan Watts. The station became a mecca for the leading lights of what scholars generally call the San Francisco Literary Renaissance, especially Jack Spicer, Robert Duncan, and Lawrence Ferlinghetti. In 1957 KPFA broadcast the first radio airing of Allen Ginsberg's poem *Howl,* copies of which were subsequently seized at Ferlinghetti's City Lights Books by the San Francisco police department on charges of obscenity.

Transition to Dissent Radio

By 1959 the Pacifica Foundation had inaugurated its second radio station, KPFK-FM in Los Angeles; this acquisition was followed in 1960 by WBAI-FM in New York City, a gift to Pacifica by a philanthropist.

Although Pacifica came into existence during the early years of the Cold War, it had its ideological roots in 1930s pacifism. In its public-affairs programs Pacifica encouraged town-hall style discussions revolving around pacifist/anarchist questions. A KPFA panel entitled "Does atomic power threaten our civil liberties?" included participants on the political left, center, and right. On its three stations, the organization endeavored to include the commentaries of conservatives such as William Rusher, Russell Kirk, and Caspar Weinberger.

But the government took far more interest in the handful of communists and communist sympathizers who regularly appeared on Pacifica's three frequencies, such as historian Herbert Aptheker and Sovietologist William Mandel. In 1960 KPFA broadcast Mandel excoriating the House Un-American Activities Committee (HUAC) during his subpoenaed appearance in San Francisco. When, in October 1962, WBAI in New York broadcast the comments of a disgruntled former Federal Bureau of Investigation (FBI) agent, it triggered a complete bureau investigation of the entire Pacifica network, personally ordered by an irritated J. Edgar Hoover. In early 1963 the principals of Pacifica's national board were subpoenaed to appear before the Senate Internal Security Subcommittee, dominated by Senator James Eastland, a personal friend of Hoover.

Seeking public sympathy for the network, Pacifica's leaders gradually revised their ideology, adapting the Federal Communications Commission's (FCC) fairness doctrine to the organization's mission. Rather than attempting to encourage left/right dialogue, the network would serve the dissenter and his or her audience. "Just as I feel little obligation to spend time on my broadcasts saying what is wrong with communist governments," declared Pacifica president Hallock Hoffman in 1963, "since everyone hears what is wrong with communist governments from every side, I think Pacifica serves the ideal of balance if it spends little time reinforcing popular beliefs." KPFA public-affairs director Elsa Knight Thompson made a similar appeal in 1970: "Pacifica Foundation was created to implement the 'Fairness' doctrine on the air rather than on paper, but implementing this policy of balanced programming is not achieved by having someone say yes for five minutes and then finding someone to say no for five minutes." "Pacifica is high-risk radio," concluded a 1975 brochure. "When the theater is burning, our microphones are available to shout fire."

Pacifica's tactical response to McCarthyism gave birth to what a later generation would describe as "alternative media." By the early 1960s the growing availability of FM receivers enabled Pacifica to broaden its audience. More than 27,000 people subscribed to a Pacifica station by 1964. By the late 1960s WBAI staff estimated that approximately 600,000 people tuned in to the station for its mélange of music, commentary, and live coverage of the Vietnam antiwar movement, an audience level probably never matched by any Pacifica station since. The network became "free speech, First Amendment" radio, famous for its broadcasts of the remarks of Ernesto "Che" Guevara and for its news dispatches directly from Hanoi.

By 1970 the network's reputation had spread to the point that it caused a backlash. Pacifica station KPFT-FM in Houston went on the air that year, only to have its transmitter bombed twice by the Ku Klux Klan. Ultimately a Klansman was apprehended while en route to California to continue his sabotage at the Los Angeles and Berkeley stations. In 1971 a WBAI manager spent time in prison for refusing to turn over to the police taped statements of men incarcerated in New York's notorious "Tombs" city jail. Three years later KPFK's manager was incarcerated for refusing to surrender to the FBI taped statements of the Symbionese Liberation Army. So far did Pacifica push the envelope of free speech that the Supreme Court in 1978 ultimately ruled as indecent a 1973 WBAI broadcast of comedian George Carlin's "Seven Words You Can't Say on Television" routine.

Community Radio

As the Pacifica network grew, its public statements became increasingly populist. After the Berkeley Free Speech movement of 1964 and the Columbia student uprising of 1968, the organization's principals spoke in the language of grassroots democracy. "We have been too academic in the past," declared a KPFA news director in 1975, "and now we want to go to the people and get their feelings." Such rhetoric drew to the organization an unprecedented wave of feminists and minority activists who sought the chance to express their feelings without the assistance of white middle-class mediators. These programmers, often adherents of "Third World" ideologies such as Maoism, coexisted uneasily with an earlier generation of Pacifica activists. Ultimately this generational tension resulted in difficult and lengthy staff strikes at KPFA in 1974 and at WBAI in 1977.

Out of these conflicts came the basic precepts of "community radio": first, that decisions at a Pacifica station ought to be made collectively, and second, that the network should dedicate its efforts to giving the "voiceless" a forum. The principles of Pacifica's fifth acquisition, WPFW-FM in Washington, D.C., most clearly articulated this philosophy. When WPFW was inaugurated in 1977 after seven years of bureaucratic wrangling with the FCC, the station's news director issued a directive to WPFW's news staff that exemplified the idea of community broadcasting: "We are here to tell people what PARTICIPANTS (the perpetrators and those affected) are saying, doing, planning and thinking—NOT what WE THINK they stand for or really mean. Their actions speak louder than your adjectives." Many noncommercial radio stations across the United States adopted this stance and created the National Federation of Community Broadcasters in 1975.

The community radio philosophy allowed the Pacifica network to accommodate an unprecedented new wave of staff who provided distinct programming. At KPFT came a bevy of programs that served the Gulf Coast in no less than 11 different languages. From WPFW in Washington, D.C., a disk jockey broke the news of the 1983 U.S. invasion of Grenada through live telephone interviews with musicians in Jamaica. At all the Pacifica stations, women's and Third World departments sprang up, with refurbished broom closets often functioning as their offices. The organization's self-conscious decision to operate on the ground floor of American life allowed it to billboard the talents of artists such as Whoopi Goldberg, Alice Walker, and Bobby McFerrin long before they became enshrined in American culture. It also enabled Pacifica stations to provide unique coverage of ongoing stories, such as the 1970s campaign to stop the closing of the International Hotel in San Francisco, a refuge for Filipino workers, and the efforts of solidarity groups to challenge U.S. policies in Central America.

Toward a Centralized Network

Although the Pacifica Foundation has crossed swords with the FBI, local police departments, and Senate investigation committees, its most turbulent battles have usually been with itself. After 1980, the most difficult question the organization faced was the extent to which it should centralize its operations and programming schedule. Pacifica began its first experiments with national programming in the late 1970s with the creation of a national news program. These efforts were followed by live "gavel-to-gavel" coverage of the 1986 Senate Iran-Contra hearings and subsequent confirmation hearings on the candidacy of Robert Bork for the Supreme Court. Emboldened by these successful ventures, the network inaugurated "Democracy Now" in 1996, a one-hour public-affairs program taken by all the Pacifica stations and 65 affiliated community radio stations. By 1997 Pacifica estimated that 700,000 people a week listened to its programming either direct via a Pacifica station or through an affiliate.

But in the 1990s these gains were accompanied by painful purges of volunteers, especially at KPFA and KPFT. Managers who had lost patience with the democratic process initiated these staff reorganizations. They perceived station program schedules as fragmented, broken down into too many individually controlled shows of poor quality. The problem of managing Pacifica frequencies was exacerbated by government policies that, beginning in the late 1970s, limited the amount of local-access, noncommercial airtime available in most metropolitan regions by eliminating most of the educational low-power FM outlets still on the air. Faced with scores of programmers who had hardly anywhere else to go, Pacifica stations found that personnel and programming decisions had become perilous—guaranteed to provoke trouble. In addition, the emergence of the internet in the early 1990s enabled dismissed programmers throughout the network to create effective dissident organizations, complete with discussion lists and websites.

In February 1999, under pressure from the Corporation for Public Broadcasting to increase audience share, the national board of the Pacifica Foundation voted to centralize its operations, removing all station Local Advisory Board (LAB) members from its body. LAB members had sat on the national board since the late 1970s. This reform was followed by the dismissal of KPFA's general manager two weeks before that radio station's 50th anniversary celebration and by the cancellation of programs whose hosts discussed the controversy surrounding that dismissal over the airwaves. These actions exacerbated long-standing frustrations at KPFA and throughout the Pacifica network. On 13 July 1999, after media activists discovered a memorandum from a member of the national board proposing the sale of KPFA or WBAI, the Pacifica Foundation hired a security firm to expel KPFA's staff from the building. More than 50 programmers were arrested. In response, some 10,000 Bay Area residents staged a demonstration demanding the restoration of the station and the resignation of the national board. The crisis became an international cause, receiving press coverage throughout the United States and Europe. KPFA drew expressions of support from as far away as the staff of Serbia's banned Radio B92. KPFA's staff returned to work on 5 August and the board, at a meeting in Houston in late October, pledged not to sell or transfer the license of any Pacifica station.

Unfortunately, the network's leadership continued to try to solve its problems via personnel purges. Later in 1999 Pacifica's executive director removed the news bureau's program director from his position shortly after the Pacifica Network News broadcast a brief story about community radio affiliate dissatisfaction with the service. Then in November of 2000 the general manager of WBAI was removed. Pacifica once again fired staff who protested the dismissal over the station's airwaves. These actions sparked a nationwide listener-subscriber boycott of the network, which, in tandem with three lawsuits filed against the board and a public pressure campaign, forced Pacifica's leadership to sue for peace in the winter of 2001. A settlement resulted in an interim governing board, which set itself the task of creating a more democratic structure for the organization.

MATTHEW LASAR

See also Community Radio; Controversial Issues; Educational Radio to 1967; Fairness Doctrine; Free Form Format; Hill, Lewis; KPFA; Public Affairs Programming; Public Radio Since 1967; Seven Dirty Words Case; United States Supreme Court and Radio; WBAI

Further Reading

Engelman, Ralph, *Public Radio and Television in America: A Political History,* Thousand Oaks, California: Sage, 1996

Hill, Lewis, *Voluntary Listener-Sponsorship: A Report to Educational Broadcasters on the Experiment at KPFA,* Berkeley, California: Pacifica Foundation, 1958

Land, Jeff, *Active Radio: Pacifica's Brash Experiment,* Minneapolis: University of Minnesota Press, 1999

Lasar, Matthew, *Pacifica Radio: The Rise of an Alternative Network*, Philadelphia, Pennsylvania: Temple University Press, 1999
Lewis, Peter M., and Jerry Booth, *The Invisible Medium: Public, Commercial, and Community Radio*, London: Macmillan, 1989; Washington, D.C.: Howard University Press, 1990
McKinney, Eleanor, editor, *The Exacting Ear: The Story of Listener-Sponsored Radio, and an Anthology of Programs from KPFA, KPFK, and WBAI*, New York: Pantheon, 1966
Post, Steve, *Playing in the FM Band: A Personal Account of Free Radio*, New York: Viking Press, 1974
Walker, Jesse, *Rebels in the Air: An Alternative History of Radio in America*, New York: New York University Press, 2001

Paley, William S. 1901–1990

U.S. Broadcast Executive

William S. Paley developed and long exercised control over the Columbia Broadcasting System (CBS), one of the most successful networks in broadcasting. His contributions and influence make him one of the giants of radio and television history.

The son of Jewish Russian immigrants who prospered with a cigar manufacturing business, first in Chicago, and later in Philadelphia, Paley became involved with radio when his sister married into the Levy family, part owners of the fledgling Columbia radio network. His father, Sam Paley, agreed to invest in the network, partly to help rescue Columbia from financial instability and also because of the success he had advertising his La Palina cigars on WCAU, the Philadelphia affiliate. After his first exposure to radio, supervising *The La Palina Smoker*, a program sponsored by the family cigar company, young William Paley became enthralled with the medium.

In 1928 Paley and members of his family purchased controlling interest in the struggling radio network that Paley would continue to nurture and oversee for more than 60 years. Although his number of shares and percentage of ownership dwindled over time, Paley ruled the network as if it were his alone. Serving as chairman of CBS, Paley developed the network by buying additional stations, adding to the number of affiliated stations carrying CBS programming, and conducting a number of savvy business negotiations to enhance the network's competitive position. Under Paley's aggressive leadership, CBS rapidly advanced to challenge its older and larger rival, the National Broadcasting Company (NBC), in audience size and profitability.

One of the strategies Paley used to entice new stations to carry CBS programming was to increase the number of hours of sustaining programming (programming without advertising) provided at no cost to the affiliates, but with the proviso that CBS would have exclusive rights to furnish programs to the stations. Along with his business acumen, Paley was especially adept at gauging public taste and programmed his network to appeal to large audiences while at the same time providing news, cultural, and public-affairs programming widely acknowledged to represent outstanding quality. CBS was a leader, especially in the area of news programming, where the broadcasts of Edward R. Murrow and the staff of correspondents assembled during World War II became legendary. By providing quality programming and also carrying popular and entertaining shows for the mass audience, Paley was able to maximize both prestige and profits for CBS.

Paley's extravagant personal lifestyle and involvement in high New York social circles were also demanding of his time. A man of considerable personal charm, he thoroughly enjoyed the company of well-known and important people, including royalty, entertainers, and other socialites. He was married twice, and both women, Dorothy and Barbara (Babe), were symbols of beauty, style, and elegance. Frequently an absentee administrator, Paley leaned heavily on such notable CBS executives as Edward Klauber, Paul Kesten, and Frank Stanton. During World War II, Paley served as a consultant to the Psychological Warfare Branch of the Office of War Information, spending time in Africa and Europe attending to organizational matters and supervising broadcasts. CBS was run in his absence by Paul Kesten, who resigned in 1946. Following the war, Frank Stanton became president of CBS, supervising its daily operation. Stanton became a leading spokesman for the entire broadcasting industry and an articulate defender of the First Amendment. Stanton served as president of CBS for many years under Paley's chairmanship and played an especially crucial role in guiding the network, but it was Paley who held and exercised final authority.

George Burns, William Paley, Gracie Allen, Mary Livingston, and Jack Benny
Courtesy CBS Photo Archive

Unlike NBC's David Sarnoff, Paley frequently involved himself directly in programming matters, personally courting and encouraging well-known personalities to work for CBS. In the late 1940s, in what became known as "talent raids," Paley successfully lured several popular NBC stars to work for CBS, including Jack Benny, Bing Crosby, George Burns and Gracie Allen, Edgar Bergen, Red Skelton and the program *Amos 'n' Andy.* These big names drew large audiences and boosted the ratings for CBS, allowing the network to charge higher rates for advertising time and resulting in greater profits, some of which were invested in television, making CBS more competitive with RCA and NBC.

By the 1950s, networks were concentrating far more on television than on radio. Many of the stars and some entire programs were moved from one medium to the other. Edward R. Murrow's *Hear It Now* became *See It Now* on television. Programming on network radio became little more than news summaries, sports, and coverage of special events.

Along with Frank Stanton, Paley continued to guide CBS through the development of its television network and the building of new headquarters, known as "Black Rock," completed in 1964. In 1976, Paley was instrumental in creating the Museum of Broadcasting in New York City. After Stanton retired, a number of executives were brought in, each dismissed in turn until 1983, when Paley was succeeded as CBS chairman by Thomas Wyman. After an unsuccessful hostile attempt by Ted Turner to buy CBS, Lawrence Tisch gained control in 1987 and reinstated Paley, although by that time his influence had been greatly reduced. Paley continued as a figurehead chairman until his death in 1990 at age 89.

B.R. SMITH

See also Columbia Broadcasting System; Keston, Paul; Klauber, Edward; Stanton, Frank N.; Talent Raids

William S. Paley. Born in Chicago, Illinois, 28 September 1901. Bachelor's degree from the University of Pennsylvania's Wharton School of Finance, 1922. Served as a civilian overseas, Office of War Information; commissioned as a U.S.

Army colonel; served as deputy chief of the Psychological Warfare Division, 1943–45. Named in charge of production, Congress Cigar Company, Philadelphia, Pennsylvania, 1922–28; promoted to vice president, 1923–28; also named secretary, 1925–28; family purchased United Independent Broadcasters and its Columbia Network, 1928; company was renamed the Columbia Broadcasting System (now CBS, Inc.), New York City, 1928; president, 1928–46, chief executive officer, 1928–77, chairman of board of directors, 1946–83, executive committee chairman, 1983–90, CBS, Inc., New York City. Museum of Modern Art trustee,1937–90, president, 1968–72, chairman, 1972–90, New York City. Recipient: Medallion of Honor, City of New York, 1965; Gold Achievement medal, Poor Richard Club; Keynote Award, National Association of Broadcasters; special award, Broadcast Pioneers; Concert Artist Guild Award. Died in New York City, 26 October 1990.

Selected Publications

Radio as a Cultural Force, 1934
In Honor of a Man and an Ideal . . . Three Talks on Freedom (with Archibald MacLeish and Edward R. Murrow), 1942
As It Happened: A Memoir, 1979

Further Reading

"First 100 Fifth Estaters," *Broadcasting and Cable* (20 December 1999)
Metz, Robert, *CBS: Reflections in a Bloodshot Eye,* Chicago: Playboy Press, 1975
Paper, Lewis J., *Empire: William S. Paley and the Making of CBS,* New York: St. Martin's Press, 1987
Slater, Robert, *This is CBS: A Chronicle of 60 Years,* Englewood Cliffs, New Jersey: Prentice-Hall, 1988
Smith, Sally Bedell, *In All His Glory: The Life of William S. Paley,* New York: Simon and Schuster, 1990

Payola

Illegal Payments to Disc Jockeys

Payola is an undisclosed payment by a music promoter to a broadcaster for the purpose of influencing the airplay of a particular song. This business practice of paying to play has been known to be associated with radio broadcasting since a national scandal rocked the radio industry in the late 1950s, but sheet music publishers were paying popular artists to perform specific songs even before the start of radio broadcasting.

The term *payola* was coined by *Variety* in 1938 in the wake of numerous stories covering music "pluggers" who promoted their songs to big-name orchestra leaders with network radio shows. It was common for as many as a dozen pluggers to attend a remote broadcast begging a popular bandleader to perform their music. In 1935, the National Broadcasting Company (NBC) barred pluggers from entering the Radio Corporation of America (RCA) building in an attempt to insulate directors from the relentless promotion men.

As postwar television viewing grew, many network radio programs disappeared. Local radio stations, forced to develop low-cost local programming, embraced recorded music formats. Local disc jockeys were soon selecting the music of a new generation—rock and roll.

From 1945 until 1959, the number of recording companies grew from about a half dozen to nearly 2,000. By the end of the 1950s, large-market radio stations were receiving as many as 250 new record releases each week. With many more records being produced than could be broadcast, record companies followed the industry formula of compensating the most popular platter spinners. Disc jockeys justified payola as "consulting fees" to "audition" new releases. In most cases, payment was legal as long as it was reported on income tax forms.

In addition to cash, disc jockeys were offered gifts, such as liquor, TV sets, clothes, and sometimes prostitutes. In a 1959 *Life* magazine expose, former WXYZ Detroit disc jockey Ed McKenzie remembered that a "record plugger once offered to install a bar in my basement" ("A Deejay's Expose—and Views," 23 November 1959). That same year, covering the Second International Radio Programming Seminar and Pop Music Disk Jockey Convention in Miami Beach, *Time* magazine described the 2,500 attendees as members of "one of the most pampered trades in the U.S." ("Disk Jockeys: The Big Payola," 8 June 1959). Record companies flocked to Florida to backslap the jocks. *Variety* described the convention as a drunken orgy. A headline in the *Miami Herald* exclaimed, "Booze, Broads, and Bribes."

Elected officials noticed the attention payola was receiving in the popular press. It was also a wake-up call for naive station owners and managers who were unaware of payola or who did not mind their popular personalities earning a little extra on the side. When several government investigations were announced, many paranoid station executives conducted their own internal investigations. Fearing eventual action by the Federal Communications Commission (FCC), owners suspended or in some cases fired disc jockeys suspected of accepting payola. Many disc jockeys resigned before they were asked questions. Detroit WJBK-TV news director Jack LeGoff was let go after he defended payola on the air as "a part of American business." His fellow employee, disc jockey Tom Clay, was fired for taking payola, and two other WJBK-AM staffers resigned. The Federal Trade Commission (FTC) began issuing complaints against record labels, and record giant RCA Victor later agreed to a consent judgment to end payola practices. The New York district attorney's office began a payola inquiry and subpoenaed the financial records of a dozen record companies.

The House Legislative Oversight Subcommittee conducted the most highly publicized investigation. In late 1959 it sent investigators to Boston, Chicago, Los Angeles, Milwaukee, New York, and Detroit to examine the payola racket. An aide said the subcommittee "had been receiving complaints from all parts of the country about disk jockeys and music programs." The subcommittee claimed it had received many letters "from irate parents complaining about particular types of music—specifically certain types of rock and roll and the music aimed at the teenage market" ("Hogan Starting 'Payola' Inquiry in Radio and TV," *New York Times,* 20 November 1959). Claiming the moral high ground, the American Society of Composers, Authors, and Publishers (ASCAP) and other publishing interests lobbied Capitol Hill, asserting that payola was the reason rock and roll music existed. The subcommittee began hearings on "Payola and Other Deceptive Practices in the Broadcasting Field" in February 1960. However, the focus was clearly on rock and roll. Critics argued that the hearings were merely a witch hunt. Broadcasting trade papers believed it was election-year grandstanding by the same politicians who had produced the recent quiz shows scandal. Nevertheless, more than 50 witnesses testified before the committee, including disc jockeys, program managers, radio station owners, record distributors, music surveyors, and FCC officials.

The consequences of the payola hearings and scandal varied. Several popular East Coast disc jockeys who lost their jobs found employment in West Coast markets, including former New York rock and roll pioneer Alan Freed. Freed was later indicted by a Manhattan grand jury on charges of commercial bribery and was penalized by the Internal Revenue Service. In 1960 the Communications Act of 1934 was amended to discourage payola. Stations were required to announce on the air any "promotional considerations" they accepted to broadcast specific programming. This made the licensee responsible for disclosing gifts accepted by its employees. Violators faced fines of up to $10,000 and jail terms of up to one year. As a result of new anti-payola regulations and the growing popularity of the Top 40 format, most disc jockeys no longer wielded the power to pick the hits. Managers (program directors) made the final decisions.

Payola did not end with new regulations. After 1960 it simply became more sophisticated. A 1977 FCC investigator explained, "It's not simply a matter of someone handing a disk jockey a $100 bill and a record he is expected to play" ("Say It Again Sam, But You'll Have to Tell the FCC Why," 14 February 1977). Trade lists compiled from the reports of a few select radio stations determined hit songs. Manipulating radio playlists was more important than frequent airplay. Labels hired independent promoters to do the dirty work of bribery, making it more difficult to trace payola back to record companies. This introduced underground connections to organized crime. Marijuana and cocaine became payoffs in some markets ("drugola"). Rumors and occasional investigations continued to surface as the 20th century ended. However, the actual extent of present-day payola is difficult to determine.

Another industry-coined term for a different behind-the-scenes scheme in broadcasting is *plugola.* Plugola involves a payoff in exchange for hiding a "plug"—in essence, an advertisement within a radio program. These disguised commercials have their roots in the film industry, in which advertisers paid to have brand names strategically placed in film scenes. The practice eventually infested the golden era of radio as promoters provided under-the-table incentives to program producers and writers willing to plant plugs in radio scripts. For the right price, big-name talent would even adlib a subtle plug during a coast-to-coast show. By the 1950s a "complimentary" plug for a product or service planted on a top television or radio show cost $250 with neither the network nor program sponsor receiving compensation. Plugola was not examined as closely as payola during the 1959–60 Congressional probe. By that time plug planting was disappearing as network radio eroded and television networks began clamping down on the practice.

STEVEN R. SCHERER

See also Clark, Dick; Contemporary Hit Radio Format/Top 40; Disk Jockeys; Freed, Alan; Recordings and the Radio Industry; Rock and Roll Format; United States Congress and Radio

Further Reading

Committee on Interstate and Foreign Commerce, *Responsibilities of Broadcasting Licensees and Station Personnel: Hearings before a Subcommittee of the Committee on Interstate and Foreign Commerce, House of Representatives, 86th Congress, 2nd Session, on Payola and Other Deceptive Practices in the Broadcasting Field,* Washington D.C.: GPO, 1960

Committee on Interstate and Foreign Commerce, Special Subcommittee on Legislative Oversight, *Songplugging and the Airwaves: A Functional Outline of the Popular Music Business: Staff Study for the Committee on Interstate and Foreign Commerce, House of Representatives, 86th Congress, 2nd Session,* Washington D.C.: GPO, 1960
"A Deejay's Expose—and Views," *Life* (23 November 1959)
"Disk Jockeys: The Big Payola," *Time* (8 June 1959)
"Hogan Starting 'Payola' Inquiry in Radio and TV," *New York Times* (20 November 1959)
"Payola Blues," *Newsweek* (30 November 1959)
"Say It Again Sam, But You'll Have to Tell the FCC Why," *Broadcasting* (14 February 1977)
Segrave, Kerry, *Payola in the Music Industry: A History, 1880–1991,* Jefferson, North Carolina, and London: McFarland, 1994
Smith, Wes, *The Pied Pipers of Rock 'n' Roll: Radio Deejays of the 50s and 60s,* Marietta, Georgia: Longstreet Press, 1989

Pay Radio

Although at first an odd proposition to American radio listeners, the notion of pay radio—paying to listen to one or more radio program services—is neither new nor rare. And with developing digital services, pay radio is rapidly becoming an accepted part of radio's landscape.

Early Notions

As radio struggled to find a means of financial support in the early 1920s, several suggested options included some means of direct payment by listeners. Station WHB in Kansas City, Missouri, for example, sold tickets in 1924 to an "invisible theater of the air." For a set price, listeners would obtain a literal ticket—but then, of course, those who did not pay could also "attend" the broadcasts, so the idea rapidly died. This was an early example of listener "donations" to help cover the cost of radio broadcasting—an idea still common in public radio and television today.

Outside the U.S., many other nations adopted a system of radio licensing wherein for each radio or household, an annual fee (effectively a service tax) was set by the government to be paid (usually to the postal authorities), with the proceeds being turned over to the national broadcasting service to meet its operational expenses. The British Broadcasting Corporation (BBC) was perhaps the prime example of this type of funding, which continued into the 21st century despite advertising's inroads in commercial systems in most other countries.

Although advertising was widely adopted in the United States by the late 1920s as radio's chief means of support, the beginning of the Depression placed heavy financial pressure on emerging stations and networks. Therefore, additional means of seeking radio revenue were occasionally offered. For example, in 1930, at the tenth anniversary dinner celebration of KDKA's (Pittsburgh, Pennsylvania) first broadcast, one politician suggested developing a radio with a key device, with said key to be sold for a set price (he suggested a dollar) each year.

With the competition of television looming after World War II, the pay radio idea was again briefly touted. Before becoming a U.S. senator, William Benton backed the idea of subscription or pay radio. Listeners would pay $18 for a year (essentially a nickel a day) to receive radio programs without advertising. At about the same time, in 1947–48, several stations expressed interest in a home music service based on patents of the Muzak corporation.

Neither idea came to fruition—in part because of controversy over even introducing the notion of paying for something that was then free to the audience. And with that, any discussion about pay radio disappeared for several decades. Indeed, the heated debate over possible introduction of a pay television service, which raged from 1948 to 1968, made any consideration of pay systems for radio even more remote.

Modern Pay Radio

By the 1990s, however, some radio listeners were paying to receive programs. Many of the nation's cable television systems offered audio channels, usually for an additional monthly subscription fee.

The 1990s also saw development of two new radio technologies, both of which revived notions of pay radio. The first was satellite delivery of many radio channels, some without advertising, in return for an annual subscription fee by listeners—

essentially the same idea that Benton had proposed a half-century earlier, updated with a new means of transmission.

Satellite pay radio got a boost when, in early March 1997, the Federal Communications Commission (FCC) approved frequencies for digital satellite radio transmissions to be made available for payment of a monthly subscription fee. Four companies sought to provide the new service—American Mobile Satellite Corporation (Reston, Virginia), CD Radio of Washington, D.C., Digital Satellite Broadcasting (Seattle, Washington), and Primosphere (New York), though in the end only two (operating as XM Radio and Sirius) actually took to the air. The broadcast industry fought satellite radio in a doomed effort to forestall the competition.

Subscription fees for the compact disc (CD) quality digital satellite service ranged from $10 to $13 per month when they were introduced in 2001–02. Available channels—some carrying advertising—included weather, sports, opera, talk, and a variety of musical formats. Service providers focused marketing of the service especially to those spending large amounts of time in their cars, such as urban commuters.

The second technological innovation, finally approved by the FCC in late 2002, was terrestrial CD-quality digital transmission, or digital audio radio (DAR) or broadcasting (DAB). This technology was already available in Japan and parts of Europe, and some American industry figures suggested that listeners might pay a subscription fee for the service, which would mark a vast improvement over AM and FM analog transmissions.

In neither the satellite nor the terrestrial radio systems, however, is the term *pay radio* very widely used, for it carries negative connotations in an advertiser-supported business.

CHRISTOPHER H. STERLING

See also Cable Radio; Digital Audio Broadcasting; Digital Satellite Radio

---

Peabody Awards and Archive

Now one of the more prestigious prizes granted in broadcasting, the Peabody Awards have since 1941 recognized the best people and programs in both radio and (since 1948) television. From 1,200 entries annually, only about 25 to 35 programs are selected, making these among the most competitive of broadcast awards.

Origins

In 1938 the National Association of Broadcasters (NAB) formed a committee to establish a prize to recognize distinguished achievement and meritorious service in radio programming. One of the committee members was Lambdin Kay, longtime manager of WSB Radio in Atlanta. Kay became a champion of the awards program and made it his special project. One day, WSB's continuity editor, Mrs. Lessie Smithgall, overheard Kay talking about setting up a fair and impartial system for administering the awards. She suggested that he contact her former professor, John E. Drewry, then dean of the Henry Grady School of Journalism at the University of Georgia, about university sponsorship of the award.

Basing his concept on the Pulitzer program administered at Columbia University, Kay approached Dean Drewry and received his enthusiastic support. Drewry contacted the university's president, Eugene Sanford—who had initiated the first journalism classes at the university in 1913—for his endorsement. Sanford approved the plan to house the broadcasting award at the University of Georgia, and during 1939, the board of regents of the University of Georgia authorized the award to be named in honor of George Foster Peabody, a native Georgian and major benefactor of the university. At the 1939 meeting of the NAB in San Francisco, Kay and Drewry presented the plan to the association and received unanimous support for establishing the Peabody Award.

Peabody, born in 1852 in Columbus, Georgia, moved to New York with his family after the Civil War. Largely self-educated, Peabody became a successful industrialist and financier. He supported humanitarian causes, especially education, and helped finance a library, a forestry school, and a classroom building at the University of Georgia. He was also the university's first nonresident trustee. In appreciation, the university awarded him an honorary degree. After his death in 1938, the university named the broadcasting award for him. Today, the name George Foster Peabody has become synonymous with excellence in electronic media.

The first awards were presented at a banquet at the Commodore Hotel in New York on 31 March 1941 for programs broadcast in 1940 and were jointly sponsored by NAB and the

University of Georgia's School of Journalism. The Columbia Broadcasting System (CBS) carried the ceremony live, and the broadcast carried addresses by CBS Chairman William S. Paley and noted reporter Elmer Davis, the recipient of the first personal Peabody Award. After the program, the NAB bowed out of future sponsorship of the awards to avoid any conflict of interest. In 1990, at the 50th anniversary ceremony, the NAB once again sponsored the event.

The Peabody Awards recognized television programs for the first time in 1948. Among early television winners were *Disneyland,* Ed Sullivan, and Edward R. Murrow for his *See It Now* series. Cable television was first recognized in 1981 when Home Box Office (HBO) and *Ms. Magazine* won for "She's Nobody's Baby: A History of American Women in the 20th Century." Over the years, Peabody Award winners for meritorious programming read like a "Who's Who" in broadcasting and cable. A complete list of winners is available on the Peabody website (address listed below).

The Peabody Awards are the most competitive of the many broadcasting honors. The awards are given without regard for program genre, and recipients represent programs in news, documentaries, education, entertainment, public service, and children's shows. Personal and organizational awards are also presented for outstanding achievement in broadcasting and cable. A national advisory board selects the Peabody Award recipients, and its members are practitioners, educators, critics, and other leaders in the broadcast and cable industries.

One by-product of the awards is the formation of one of the country's richest broadcast archives. Housed at the University of Georgia Library, the Peabody Collection dates to 1940 and preserves the best of 20th-century radio and television for today's students and scholars and for future generations. The collection has grown to over 40,000 radio and television programs and related print materials. In breadth and program diversity, the collection contains significant news, documentary, educational, and entertainment programming from radio, broadcast television, and cable.

The collection also reflects the full range of independent, local, and network programming and contains many of the finest, most significant moments in broadcast and cable history. Virtually all major news events, from World War II to today, may be found. In addition to spot news, documentary, and public-affairs programs, the Peabody Award also recognizes excellence in entertainment, educational, children's, and sports programming. One definite advantage of the collection is its accumulation of local programming.

The collection's entry books list submissions and provide information on programs submitted in a given year, with radio and television entries inventoried separately. Entries are listed alphabetically by state and are grouped according to submission categories: news, entertainment, documentary, education, youth/children, public service, and individual/institutional. Each volume has a table of contents, and entries are described in brief paragraphs provided by the submitters.

Because copyright is retained by copyright holders, not the Peabody Collection, no programs are loaned, but researchers may use the collection on site. Generally, shows from 1976 onward may be found in the card catalog of the Media Center of the University of Georgia's Main Library. These programs are listed alphabetically by title within categories. In most cases, screening copies exist for on-site viewing. Many older programs are not readily accessible because screening copies have not been dubbed owing to lack of funding.

LOUISE BENJAMIN

See also Museums and Archives of Radio

Further Reading

Peabody Awards, <www.peabody.uga.edu/index2.html>

Priest, Patricia J., "Mining the Peabody Archive," *Journal of Radio Studies* 1 (1992)

Reid, John Edgar, "A Half Century of Peabody Radio: Tracking Public Service Entries 1940–1989," *Journal of Radio Studies* 1 (1992)

Sherman, Barry L., "The Peabody Collection: A Treasure Trove For Radio Research," *Journal of Radio Studies* 1 (1992)

Ware, Louise, *George Foster Peabody: Banker, Philanthropist, Publicist,* Athens: University of Georgia Press, 1951

Perryman, Tom 1927–

U.S. Country Music Promoter and Disc Jockey

As a disc jockey, promoter, and radio station owner, Tom Perryman played an important role in the dissemination of country music from the 1940s into the 1970s. From radio stations in Gladewater, Texas, and later Nashville, Tennessee, Perryman enthusiastically introduced new country music artists and their records to a wide audience. In addition, acting as a sort of southern Alan Freed, he booked and promoted concerts for aspiring artists, most notably Elvis Presley, whose career was in its infancy when Perryman worked with him. He belongs to a fraternity of influential country music disc jockeys who exemplified the widening role of DJs in the commercialization of country music during the post–World War II era. Perryman's handprints on the commercialization of country music are as plain as those of DJs Nelson King (WCKY, Cincinnati, Ohio), Hugh Cherry (WKDA, Nashville, Tennessee), Squeakin' Deacon Moore (KXLA, Pasadena, California), and Biff Collie (KLEE, Houston, Texas).

Before Perryman joined radio station WSM in Nashville, Tennessee (where he rose to the apex of his power as a disc jockey), he was active in booking and promoting country artists in the music-mad northeast Texas area. He performed these tasks in addition to his disc jockeying at radio station KSIJ in Gladewater, Texas. It was common for disc jockeys to book artists during Perryman's record-spinning stint in Texas during the late 1940s and early 1950s, but Perryman is distinguished for giving soon-to-be popular musical acts some of their earliest breaks. When a 19-year-old Elvis Presley was stuck in Shreveport, Louisiana, without money or bookings, after playing KWKH's *Louisiana Hayride* in the fall of 1954, Perryman put Presley and his band to work in a Gladewater club. It was one of Presley's first non-radio gigs outside his home turf of Memphis, Tennessee; Perryman would continue to book Presley in Texas over the next year, until the rocker rose to national prominence in 1956.

Perryman's efforts as a disc jockey and concert promoter also helped to generate a Texas following for developing country music acts such as Johnny Cash, Johnny Horton, and Jim Reeves, artists who appeared weekly on KWKH's *Louisiana Hayride*. These artists relied on regional bookers and promoters to get work for them between their weekly *Hayride* appearances, and Tom Perryman worked tirelessly for them in the nearby northeast Texas region. Perryman formed a particularly close bond with Jim Reeves, allying with him in numerous ventures including the purchase of KGRI radio in Henderson, Texas.

As Perryman consolidated his influence in Texas, his profile on the national country music scene grew with his frequent letters to the music trade magazine *Billboard*, which recorded his impressions of country music artists and songs. His published plugs often helped to sustain an artist's or song's momentum. "Tom Perryman, KSIJ, Gladewater, Texas," one *Billboard* item read in 1952, "reports that Slim Whitman, who has 'Indian Love Call' coming up on Imperial, is running neck and neck with Hank Williams for top station popularity." Record company representatives realized that Perryman was an important figure on the Texas music scene partly because they saw his name in *Billboard* virtually every week during the early 1950s; as a result, recording executives often consulted him as they searched for new artists.

In 1953 Perryman's work on behalf of the country music industry broadened when he joined the first board of the Country Music Disc Jockey Association (CMDJA), which would evolve into the Country Music Association (CMA), country music's most influential trade group. Perryman's impact on the distribution of country music increased dramatically with his move in 1956 to radio station WSM in Nashville. He became the first host of the station's influential all-night country music disc jockey show *Opry Star Spotlight*. (Others such as Eddie Hill and T. Tommy Cutrer had hosted WSM's overnight country programming before him, but Perryman was the first to handle the show after it was named *Opry Star Spotlight*.) Over the 50,000-watt clear channel station, Perryman brought country music to the station's vast overnight audience that was spread over most of the contiguous United States. During his two years at WSM (1956–1958), he also served as the talent coordinator of the station's powerful talent agency and the assistant manager of WSM's hallmark country music show, the *Grand Ole Opry*.

Perryman continued to help open markets for country music after leaving WSM. In 1959, after a brief return to KSIJ in Texas, he became general manager and co-owner (with country music star Jim Reeves) of radio station KGRI in Henderson, Texas. Under Perryman and Reeves, KGRI became one of the nation's first all-country-music stations. In 1967, three years after the death of Jim Reeves in an airplane crash, Perryman and Reeves' widow, Mary, became co-owners of WMTS-AM and FM in Murfeesboro, Tennessee, another all-country station. He served as general manager of KGRI and WMTS concurrently. In 1969, the Country Music Association named WMTS "Station of the Year."

Perryman retired from regular radio work in 1978 upon the sale of WMTS. At the beginning of the 21st century he served as president of the Reunion of Professional Entertainers of Country Music, a group that acts as an advocate for veteran country music performers. He was inducted into the Country Music Disc Jockey Hall of Fame in 1988 and the Texas Country Music Hall of Fame in 1999.

MICHAEL STREISSGUTH

See also Country Music Format; Disk Jockeys; Grand Ole Opry; King, Nelson; KWKH; WSM

Thomas Eugene ("Tom") Perryman. Born in Dallas, Texas, 16 July 1927. One of three children born to W.C. Perryman and Margaret Morris Perryman; attended Tyler Commercial College, Tyler, Texas, 1945–47; received first class radio telephone license; disc jockey, KEBE, Jacksonville, Texas, 1947–49; KSIJ, Gladewater, Texas, 1949–56; host, *Opry Star Spotlight,* and talent coordinator, *Grand Ole Opry,* WSM, Nashville, Tennessee, 1956–58; returned to KSIJ, 1958–59; co-owner and general manager, KGRI, Henderson, Texas, 1959–74; co-owner and general manager, WMTS, Murfreesboro, Tennessee, 1967–78; joined Country Music Disc Jockey Association, 1953; president of the Reunion of Professional Entertainers of Country Music, 2000; inducted into Country Music Disc Jockey Hall of Fame, 1988, and Texas Country Music Hall of Fame, 1999.

Radio Series

1956–58 *Grand Ole Opry; Opry Star Spotlight*

Further Reading

Guralnick, Peter, *Last Train to Memphis: The Rise of Elvis Presley,* Boston: Little Brown, 1994

Guralnick, Peter, and Ernst Jorgensen, *Elvis Day by Day,* New York: Ballantine Books, 1999; London: Hi Marketing, 2000

Malone, Bill C., *Country Music, U.S.A.,* Austin: University of Texas Press, 1968; revised edition, 1985

Streissguth, Michael, *Like a Moth to a Flame: The Jim Reeves Story,* Nashville, Tennessee: Rutledge Hill Press, 1998

Philco Radio

Radio Receiver Manufacturer

Philco radios were first introduced by a small Philadelphia electrical manufacturer in 1928. Within two years, they became the top-selling radios in the United States, and they continued to lead the market for more than a decade.

Early Years

In the initial years of radio development, radios operated from electricity provided by batteries. The Philadelphia Battery Storage Company (the Philco trademark was registered in 1919) began producing batteries for radios in the early 1920s, an offshoot of its earlier production of automobile and truck batteries. In 1925 Philco produced an innovation that eliminated the need for batteries and allowed radio owners to operate the set with electricity from a light socket—the "Socket-Power" unit. The company grew quickly with its two radio-related products and soon began sponsoring a national radio broadcast known as the *Philco Hour,* which appeared on National Broadcasting Company (NBC) Blue from 1927 to 1929 and on the Columbia Broadcasting System (CBS) from 1929 to 1931.

The Radio Corporation of America (RCA) introduced a set powered by alternating current (AC) in 1927. Not long after, RCA announced that it would license other manufacturers to use the new AC technology, effectively wiping out the market for radio batteries.

Philco management recognized the challenge to its survival and completely changed the focus of its business. The company chose to design and build its own radios and set mid-1928 as a target date to introduce the new product. Most manufacturers were producing radios that all looked very much alike. Philco chose to change that look. Instead of the sturdy, sensible wooden box that most table-model radios resembled, the company produced a metal case radio in five bright colors. To attract female consumers, Philco offered four of the models decorated with hand-painted floral designs. Each had color-coordinated matching speakers (early radios did not contain internal speakers).

Recognizing the importance of a strong advertising campaign, Philco introduced its radios in expensive double-page color ads exclaiming, "COLOR! VIVID COLOR!" The copy described them as "enhanced with color effects by Mlle. Messaros, one of the foremost colorists in the decorative arts. The colors are applied *by hand* under her personal direction."

In addition to the unusual color sets, Philco also offered three console models, radios that were cabinet-style pieces of furniture based on the Louis XVI period; these were created by an internationally known furniture designer. The striking line of radios would be the first in a long line of innovations from the company. Philco introduced its first eight models at an industry trade show in June 1928.

It was a stunning accomplishment. Many had thought that RCA's new AC-powered radios spelled the end for Philadelphia Battery Storage. Philco's response was an indication of an innovative ability that would stand it in good stead for the next 30 years and help maintain its reputation for superior-quality products.

At the end of 1928, Philco was in 26th place among radio manufacturers. The company's management believed the best way to compete in 1929 was to improve quality and lower prices. At a time when most manufacturers used the labor-intensive method of assembling each radio individually by hand, Philco borrowed $7 million to convert and expand its facilities for mass production.

Philco advanced to third place in the industry by selling 408,000 radios in 1929. Despite the October 1929 stock market crash, Philco continued to thrive. The $7 million debt was paid off early in 1930. Later that year, Philco increased its employee base from 1,500 in May to 4,000 by September. Orders for radios continued to increase, and the company announced it was hiring 75 men a day. Soon Philco announced the formation of a subsidiary to manufacture Philco-Transitone automobile radios.

The radio manufacturer continued to recognize the value of advertising, and, in an early version of a common promotional technique today, a Philco radio was prominently featured in the hotel room of movie star Bing Crosby in Paramount's 1932 film *The Big Broadcast*. At the same time that the company was proving such a success in radio manufacturing, it was also looking toward the future and diversifying. Noted inventor Philo T. Farnsworth came to Philco in 1931 to join a research team devoted to television development. In 1932 the Federal Radio Commission licensed Philco's experimental television station W3XE. The company also began to produce home appliances. During the 1930s Philco offered a variety of products, including automobile radios, phonographs, radio-phonograph combinations, air conditioners, and refrigerators.

Near the end of the decade, Philco began a shift from the large tabletop "tombstone" and "cathedral" radios (so called because of their shape) and produced smaller and less experience radios. Industry leaders Atwater Kent, Crosley, Majestic, and Zenith initially ignored the shift to undersized table models. Consumers, however, responded quickly to the new, space-saving radios. In 1938 Philco's 10 millionth radio came off the assembly line.

World War II and After

With the United States' entry into World War II, Philco shifted to production of military items, including radios for tanks and planes. The company also trained military personnel in the installation, operation, and maintenance of electronic aircraft equipment. Only after the war's end in 1945 were new radios, phonographs, refrigerators, and air conditioners produced again by Philco.

Philco sponsored two national network programs: *Philco Radio Time*, starring Bing Crosby, and a radio anthology of plays adapted for radio, the *Philco Radio Playhouse*, hosted by actor Joseph Cotten. *Philco Radio Time* was the first prerecorded program. Prior to its introduction in 1946, all radio programs were broadcast live.

Philco was active in transistor research in the 1950s, producing a number of transistor radio models. The 1960s began a troublesome time for Philco. The company was losing money—more than $4 million in 1961. Ford Motor Company purchased Philco late in 1961. The Philco-Ford division, producing television sets, computers, and satellite communication equipment, once again became profitable.

During the 1970s Americans became more interested in antiques and collectibles, and among the items drawing their interest were cathedral radios from 40 years earlier. Philco-Ford introduced a miniature, transistorized AM-FM replica of its model 90 Baby Grand (originally produced in 1931).

General Telephone and Electronics purchased the Philco division from Ford in 1974 and then sold Philco to North American Philips Corporation, where the division now produces televisions.

SANDRA L. ELLIS

See also Automobile Radios; Receivers; Transistor Radios

Further Reading

"1,250,000 out of 4,200,000 U.S. Radios Sold Last Year Have the Philco Trademark," *Fortune* (February 1935)

Douglas, Alan, *Radio Manufacturers of the 1920s*, 3 vols., Vestal, New York: Vestal Press, 1988–91; see especially vol. 2, 1989

Johnson, David, and Betty Johnson, *Guide to Old Radios: Pointers, Pictures, and Price,* Radnor, Pennsylvania: Wallace-Homestead Book, 1989; 2nd edition, Iola, Wisconsin: Krause, 1995

McMahon, Morgan E., *A Flick of the Switch: 1930–1950,* Palos Verdes Peninsula, California: Vintage Radio, 1975

Ramirez, Ron, and Michael Prosise, *Philco Radio, 1928–1942: A Pictorial History of the World's Most Popular Radio,* Atglen, Pennsylvania: Schiffer, 1993

Wolkonowicz, John P., "The Philco Corporation: Historical Review and Strategic Analysis, 1892–1961," Master's thesis, MIT, Alfred P. Sloan School of Management, 1981

Phillips, Irna 1901–1973

U.S. Creator of Radio Serial Dramas

Every woman's life is a soap opera.
—Irna Phillips

Irna Phillips, one of the most prolific and successful creators of daytime serials, has long been regarded by contemporaries and historians as the "Queen of Soap Opera." Indeed, Dan Wakefield (1976), in his study of soap operas, concluded that Irna Phillips "is to soap opera what Edison is to the light bulb and Fulton to the steamboat." Although competitors such as Frank and Anne Hummert oversaw the development of dozens of serials, Phillips was the individual creator, writer, and producer of the largest number of serials on daytime radio (9) and television (8). Not only was Phillips responsible for inventing the first daytime serial in 1930, she influenced the rise of daytime dramas in radio and developed the programming form over more than 40 years. In her radio career, Irna Phillips helped to establish many of the distinguishing characteristics of the soap opera genre: a continuous and never-ending narrative, multiple plot lines, dialogue-based drama, complex characterization, slow-paced story lines, a focus on female characters and emotional intimacy, suspenseful cliff-hangers, and the use of dramatic organ music to bridge scenes. However, her most notable programming achievement may be the creation of the *Guiding Light.* Broadcast—first on radio and then on television—for more than six decades, Irna Phillips' *Guiding Light* is the longest continuous story ever told in American broadcasting.

Irna Phillips was born 1 July 1901 in Chicago, Illinois, the youngest of ten children of a German-Jewish family. After her father died when Irna was only eight, she maintained a close relationship with her mother. After attending Northwestern University for one year (1918–19), Phillips transferred to the University of Illinois, became a drama student, and graduated with a degree in education in 1922. After a brief stint teaching at a junior college in Fulton, Missouri, Phillips earned a master's degree at the University of Wisconsin in 1924. After her graduate work, Phillips spent four years teaching speech and drama at Dayton's Teacher College in Dayton, Ohio. On a visit to her hometown at the age of 28, Irna Phillips toured the radio station WGN. According to legend, while she toured the facility, Irna was mistaken for an actress looking for an audition. After reading Eugene Field's poem, "The Bowleg Boy," Phillips was offered a job as an intern. Phillips initially turned down the offer, but the visit rekindled her interest in drama. By May 1930 Phillips was a paid employee of WGN, acting in small roles and reading inspirational verse on a program entitled *Thought for a Day.*

Accounts differ as to how Phillips began writing for radio. Her work starring in and writing for a 15-minute drama, *Sue and Irene,* and designing a Memorial Day program, among other projects, apparently caught the attention of WGN station manager Henry Selinger. Selinger, aware of the untapped selling potential of daytime radio, had approached the Lord and Thomas advertising agency to develop programming for female audiences. Selinger offered Phillips $50 per week to write a daily drama about the "Sudds" family, which could be sponsored by a soap company. Although various sponsors remained unconvinced of the potential of a daytime serial, the program, renamed *Painted Dreams,* aired on WGN as a sustaining local program on 20 October 1930. *Painted Dreams,* broadcast six days a week, was radio's first soap opera, the first daytime serial narrative aimed at a female audience. The drama about an Irish-American household centered upon three characters: Mother Moynihan (voiced by Phillips); her daughter Irene; and a boarder, Sue Morton; the show dealt with the conflict between the traditional Mother Moynihan and her modern, career-minded daughter.

After two years and 520 scripts, Phillips wanted to sell her successful serial to a network. WGN, claiming ownership of the program, refused to allow Phillips to take *Painted Dreams* national. In protest, Irna Phillips quit her job and filed suit against the company. After an eight-year court battle over the program's copyright, Phillips lost the case. After leaving WGN, Phillips was determined to establish the rights to her own material. For the rest of her career, Phillips acted as an independent producer, creating programs and subsidizing all their

Irna Phillips
Courtesy CBS Photo Archive

production costs before selling the package to a network. In 1932 Phillips moved to WGN's rival in Chicago, WMAQ. Collaborating with Walter Wicker, she developed another serial, a virtual replica of *Painted Dreams. Today's Children* (1932–37; 1943–50) featured a widowed Irish-American mother, Mother Moran; her daughters, Eileen and Frances; and a boarder, Kay Norton, living on Chicago's Hester Street.

Phillips developed seven other original serial dramas for daytime radio. After the birth of *Today's Children,* Phillips developed a short-lived drama about a painter involved with many glamorous women, entitled *Masquerade* (1934–35; 1946–47). This serial, designed to sell cosmetics, lasted only a few months, although it would be revived and redesigned nearly a decade later. Phillips' next and more successful creation was *Road of Life* (1937–59). This serial was the first major soap opera about doctors and nurses. Jim Brent, a young intern at City Hospital, was the serial's hero, equally adept at mending the "broken legs and broken hearts" of his patients (cited in "Queen of Soaps," 1964). That same year, Phillips' seminal daytime drama, *Guiding Light* (1937–56 on radio; 1952–present on television) was first broadcast. Dr. Preston Bradley, a Chicago clergyman admired by Phillips, was the inspiration for the serial's central character, Dr. John Ruthledge. The assimilationist drama centered upon Dr. Ruthledge and his attempts to minister to the Italian, Irish, and Polish families in the slums of Five Points. After the death of Phillips' mother (who was the model for the mothers in *Painted Dreams* and *Today's Children*) in 1937, Phillips replaced her popular and highly-rated serial *Today's Children* with the drama *The Woman in White.* In tribute to the nurses who cared for her ailing mother, Phillips developed a daytime drama about a young nurse named Karen Adams and her struggle to save the souls of her patients.

Phillips' next two contributions were more focused on the personal lives of their heroines. *The Right to Happiness* (1939–60), initially a spin-off of *Guiding Light,* became the story of an oft-married heroine, Carolyn Allen, and her fight for personal happiness. Influenced by Irna Phillips' experience as a single career woman in 1940s New York City, *Lonely Women* (1942–46) detailed the life of a group of women, separated from husbands and boyfriends during World War II, living at the Towers, an all-female hotel. In 1944 Phillips played with the traditional 15-minute format of the genre. She combined three of her serials in an hour-long block entitled the *General Mills Hour* and experimented with different program lengths, integrated story lines, and characters that traveled from serial to serial. Her final contribution to daytime radio was *Brighter Day* (1948–56), a drama about a widowed clergyman and his children in the town of Three Rivers.

Not only was Phillips prolific, her dramas were incredibly popular. Phillips and her sponsors regularly received over a quarter of a million requests for each promotion advertised on her serials. By 1940 Phillips' most enduring product, *Guiding Light,* was broadcast to 12 million listeners over 300 stations, and Phillips' salary was estimated at over $200,000 a year.

Irna Phillips believed there were three main objectives of a successful daytime radio drama: to entertain, to teach, and to sell. Although Phillips, a never-married career woman, had little in common with the 37 million homemakers targeted by daytime radio serials, she knew this audience would connect with complex female characters and domestic story lines. In her serials, Phillips threatened the security of her main female characters with illnesses, problem children, other women, and any other obstacles that might destroy their domestic bliss. Her tension-filled formula—"small-town-woman-with-an-emotional-problem-in-physical-danger"—shaped the development of many of her dramas. Irna Phillips was also a staunch advocate of the educational potential of her daytime dramas. Serials, she believed, taught women valuable lessons about how to address "real-life" problems. Phillips was quite progressive in integrating social issues into her plot lines, a practice that would become a hallmark of the genre.

To make her serials as realistic as possible, Phillips consulted physicians, lawyers, and government experts on how to incorporate social issues in her serials and how to advise her audience to respond to these matters. For example, in 1945 Phillips worked with the Office of War Information to produce a story line to instruct women on how to rehabilitate disabled war veterans. Phillips—often the target of critics such as James Thurber and Dr. Louis Berg, who feared the dangerous influence of these programs on American women—was so convinced of the public service her serials offered that she spent $45,000 of her own money to combat these critical perceptions in the press. But despite her lofty ideals, Irna Phillips also understood the commercial realities of daytime programming. From her earliest days in radio, Phillips realized that the daytime serial was a "selling drama," a program intended to attract a lucrative female audience and to sell goods to that audience. Phillips was quite careful throughout her career to cultivate a reputation as a producer dedicated to her sponsors and responsible for making her serials effective advertising vehicles. For example, to charm a potential sponsor, Montgomery Ward, Phillips planned an engagement story line for one of her main characters to sell the home goods carried by that department store. Her success entertaining, teaching, and selling goods to female audiences made Phillips (with her longtime sponsor and business partner, Procter and Gamble) one of the most powerful women in soap opera production.

Phillips' unusual writing technique was strained as her serials became more successful. Phillips dictated up to six scripts per day to her longtime secretary, acting out each line of dialogue and changing her voice to indicate different characters. By the early 1940s, Phillips had five daytime serials on network radio and over 60 characters to write for. In her most prolific period, Phillips wrote between 2 and 3 million words a year for radio, the equivalent of 27 novels each year. To manage the workload, she eventually developed a system similar to that of the Hummerts, plotting story lines and delegating the work of writing dialogue to a stable of writers.

Irna Phillips' success in developing and producing daytime serials continued even after network radio began to decline. Two of her serials, *Guiding Light* and *Road of Life,* were among the few programs to survive the transition to television. She also produced the first television soap opera in 1949, the short-lived *These Are My Children.* Just as she was influential in developing the soap opera genre on radio, Phillips pioneered the adaptation of radio serials to television, inventing an intimate visual style for the genre. She also created enduring television soap operas such as *Another World* (1964–99), *As the World Turns* (1956–present), and *Days of Our Lives* (1965–present). Not only did Phillips develop a record number of radio and television soap operas, she also trained the next generation of soap opera creators on television. Agnes Nixon (creator of *All My Children, Loving, The City,* and *One Life to Live*); William J. Bell (creator of *The Young and the Restless* and *The Bold and the Beautiful*); and Ted Corday (co-creator of *Days of Our Lives*) were all protégés of Irna Phillips in radio and early television. In all, seven of the ten soap operas on daytime television in 2003 can be traced to Irna Phillips. Although traditional broadcast histories have failed to acknowledge the extent of her contribution to the development of radio and television, recent historians, her sponsor, Procter and Gamble, and the daytime community have publicly recognized Phillips as perhaps the most important force in the creation and development of the soap opera.

JENNIFER HYLAND WANG

See also Soap Opera; Women in Radio

Irna Phillips. Born Erna Phillips in Chicago, Illinois, 1 July 1901. Youngest of ten children to German-Jewish parents, William and Betty Phillips; adopted two children, Thomas Dirk, 1941 and Katherine Louise, 1943; attended Northwestern University, 1918–19; studied drama at University of Illinois, B.A. in education, 1922; attended University of Wisconsin, M.A., 1924; taught dramatics and storytelling at junior college in Missouri, 1923; taught dramatics and speech at Teacher College, Dayton, Ohio, 1925–29; employee of WGN, 1930–32; creator, writer, and actress of radio series, 1930–60; consultant on daytime programs for J. Walter Thompson Advertising Agency, 1940s; taught classes on writing radio serials at Northwestern University, 1945–46; script consultant, Radio Scripts Incorporated, 1946–47; developed first daytime network soap opera on television, *These Are My Children*, 1949; adapted several of her own radio serials for television in the 1950s; consultant for ABC's prime-time soap *Peyton Place*, 1964; co-creator of *Days of Our Lives* with Ted Corday, Betty Corday, and Allan Chase, 1965. Died in Chicago, Illinois, 23 December 1973.

Radio Series

1930	*Thought for a Day*
1930–34	*Painted Dreams*
1932–37, 1943–50	*Today's Children*
1934–35, 1946–47	*Masquerade*
1937–56	*Guiding Light*
1937–59	*Road of Life*
1938–48	*Woman in White*
1939–60	*Right To Happiness*
1942–46	*Lonely Women*
1944	*General Mills Hour*
1948–56	*Brighter Day*

Television Series

These Are My Children, 1949; *Guiding Light,* 1952–present; *Road of Life,* 1954–55; *Brighter Day,* 1954–62; *As the World Turns,* 1956–present; *Another World,* 1964–99; *Days of Our Lives,* 1965–present; *Our Private World,* 1965; *Love Is a Many Splendored Thing,* 1967–73

Further Reading

The Irna Phillips Collection is housed at the State Historical Society of Wisconsin, Madison.

Allen, Robert Clyde, *Speaking of Soap Operas,* Chapel Hill: University of North Carolina Press, 1985

Cox, Jim, *The Great Radio Soap Operas,* Jefferson, North Carolina: McFarland, 1999

Horton, Gerd, *Radio Goes to War,* Berkeley: University of California Press, 2002

Lavin, Marilyn, "Creating Consumers in the 1930s: Irna Phillips and the Radio Soap Opera," *Journal of Consumer Research* 22 (June 1995)

MacDonald, J. Fred, *Don't Touch That Dial! Radio Programming in American Life, 1920–1960,* Chicago: Nelson-Hall, 1979

Museum of Television and Radio, *Worlds without End: The Art and History of the Soap Opera,* New York: Abrams, 1997

O'Dell, Cary, *Women Pioneers in Television: Biographies of Fifteen Industry Leaders,* Jefferson, North Carolina: McFarland, 1996

Procter and Gamble Soap Opera Collection, Bowling Green, Ohio: Bowling Green Popular Culture Library

"Queen of Soaps," *Newsweek* (11 May 1964)

Russell, Norton, "Expert on Happiness," *Radio-Television Mirror* 13, no. 1 (November 1939)
Seiter, Ellen, "To Teach and to Sell: Irna Phillips and Her Sponsors, 1930–1954," *Journal of Film and Video* 41 (Spring 1989)
Wakefield, Dan, *All Her Children*, Garden City, New York: Doubleday, 1976
Wyden, Peter, "Madame Soap Opera," *The Saturday Evening Post* (25 June 1960)

Playwrights on Radio

Writing drama for radio was a challenging task, for, unlike on the stage or screen, a radio playwright had to depend solely on sound, dialogue, and description to tell a story. The inability to use any visual devices derailed more than a few early would-be radio authors. Those who succeeded, however, created a wonderful art form that flourished for nearly two decades of radio's golden age. A half century later it is understandable that few people under the age of 60 remember the names of Norman Corwin or any of the other major radio dramatists. Corwin was the dean of America's radio playwrights during the golden age that began in 1935 and lasted until the late 1940s. Poets, novelists, and mystery writers joined Corwin and others who wrote mainly for radio in having their works aired to a vast listening public. A number of writers for the stage also sought out the broadcasting studios. A 1945 anthology of radio plays featured works by a priest, two army sergeants, a Noble prize–winning novelist, a musicologist, and even a blind man. Novelist John Steinbeck, TV personality Steve Allen, and TV newscaster Chet Huntley took brief stabs at radio play writing.

Golden Age Radio Drama

The first golden age radio playwrights were heard on two dramatic anthologies, the *Cavalcade of America* and the *Columbia Workshop*. The former was a showcase for largely patriotic shows; the latter, an experimental dramatic anthology, offered more artistic productions. Soon after the *Columbia Workshop* was launched in 1936, the show received a manuscript for *The Fall of the City*, a play in verse, from Pulitzer prize winner Archibald MacLeish. It dealt allegorically with the growth of fascist dictatorships in Europe. Its broadcast struck a chord and inspired a number of other talented and experienced writers. Because poetry lends itself particularly to aural expression, Stephen Vincent Benét, Edna St. Vincent Millay, and other poets joined MacLeish's ranks on radio. So did William Saroyan, Pare Lorentz, Dorothy Parker, and a number of other prose writers.

Howard Koch and John Houseman

On the eve of Halloween 1938, CBS broadcast what is probably the medum's best-remembered show, an adaptation of H.G. Wells' novel *The War Of the Worlds*. Orson Welles produced and acted in the program. However, despite the popular notion that Orson Welles also wrote the radio script, the primary authors were his two collaborators, John Houseman and Howard Koch.

John Houseman, a Romanian born Jew, was originally a grain dealer until the Depression put him out of business. By 1938, he was a stage producer, director, and writer, working in two undertakings with Welles, the Mercury Theatre (a stage troupe) and the *Mercury Theatre of the Air*. For the latter, a show on CBS, Houseman initially did the writing until he and Welles took on Howard Koch, a tall, shy Columbia Law School graduate, to relieve Houseman. It was Koch who wrote the original script for *War of the Worlds*. Houseman collaborated on two rewrites and Welles applied the finishing touches.

Even after the show created panic nationwide, Koch, its principal author, still remained unknown to the public, for Welles never attempted to set the record straight. After a shift in the Welles-Houseman relationship, Houseman left. Subsequently, Campbell's Soup offered Welles sponsorship of another radio series, and Koch continued as writer. Not long afterwards, Koch became a film writer, most notably co-author of the famed *Casablanca*.

Among the many talented writers who took to radio was Brooklyn born Lucille Fletcher. After receiving her degree from Vassar College in 1933, Fletcher took a job as a typist in CBS's publicity department. After Norman Corwin produced a play based on a story that she wrote, Fletcher decided to try radio writing herself. One of her first plays was performed on *Mercury Theatre of the Air*. Her most successful one, *Sorry, Wrong Number*, first broadcast in 1943, was translated into 15 languages, made into two films, and served as the basis for two operas. Other radio plays by Fletcher were produced on the *Columbia Workshop* and the *Suspense* series.

In the four or five years before Pearl Harbor, in response to the threat of fascism that had materialized in Europe, a group of writers began to produce a "social consciousness" body of radio drama that displayed a strong concern for human freedom. Thus Arch Oboler inserted political themes into about a third of the shows he wrote for the horror series *Lights Out.* Like virtually all of these writers in the pre-Pearl Harbor days, however, he did so in an allegorical manner. Also, a year before the Japanese attack, a government official had taken an initiative that led to the creation of *The Free Company,* an allegorical, antifascist series that featured plays by Stephen Vincent Benét, MacLeish, Maxwell Anderson, Robert Sherwood, Marc Connelly, and Sherwood Anderson.

World War II

After Pearl Harbor, both the networks and the government promoted programs intended to boost morale and otherwise assist in the pursuit of victory. Norman Corwin directed two four-network efforts at the request of the government. The first, only a week after Pearl Harbor, celebrated the Bill of Rights. The second, which began two months later, was a 13-part series about the war, for which he wrote about half of the shows. Its broadcast roughly coincided with production of Arch Oboler's first post–Pearl Harbor series, *Plays for Americans* on NBC. Oboler produced at least 70 "beat the Axis" radio plays in 1942. He played a particularly prominent role in the propaganda campaign against Germany, arguing forcefully that Americans needed to hate the enemy in order to conduct a successful war effort.

Twenty-seven year old Ranald MacDougall was one of the writers for the Corwin series. Like many others of his generation, as a boy he had been fascinated by the technology of radio and had wound copper wire around an oatmeal box to make his first set. MacDougall began to write seriously at the age of 12. In his late teens he was working as an usher at New York City's Radio City Music Hall when a conversation with an NBC executive landed him an office job with the network. Eventually MacDougall worked his way onto the script staff. In 1942 he began working for a new war series, *The Man Behind the Gun,* one of the best dramas depicting the war era. Other notable and prolific writers of wartime radio drama included Allan Sloane, Peter Lyon, and poets Stephen Vincent Benét, Langston Hughes, and Norman Rosten. New Jersey–born Millard Lampell and playwright Arthur Miller, both close friends of Rosten, wrote for radio as well.

Allan Sloane, originally a newspaper journalist, broke into radio in 1943 when, temporarily jobless, he walked unannounced into the office of the producer of *The Man Behind the Gun* and handed in a script he had written at home. He was hired within a few days. Peter Lyon, another journalist, got started writing for *The March of Time,* a dramatized news documentary. Later, he also wrote for the *Cavalcade of America.* Like many of his radio colleagues, Lyon was a progressive. Among other interests, he was a strong trade unionist. He and Millard Lampell both wrote for a wartime series entitled *Labor for Victory.* Lyon also served in 1944 as president of the Radio Writers Guild.

In his earlier years Pulitzer prize winner Stephen Vincent Benét was rather indifferent to politics. But during the 1930s, he grew increasingly interested in national and international events and developed wide friendships among European refugees. By the time of Pearl Harbor, he was driven to assist the war effort, attending meetings and giving radio readings of his poetry. His best-known contribution to the war effort was a radio drama in verse, "They Burned the Books."

Langston Hughes had many fewer opportunities with radio than did his white peers. He first wrote for radio in 1940 when CBS asked him to prepare some scripts for a Norman Corwin series. Soon, Hughes was lending his talent to the war effort even though he found that he was radio's "token" black writer, defending a democracy whose fruits he could not fully share.

Norman Rosten, a protégé of Stephen Vincent Benét, was one of the most prolific writers of radio plays of the 1930s and 1940s. Rosten received part of his preparation for writing radio drama on a playwriting fellowship at the University of Michigan. The broadcast of MacLeish's *The Fall of the City* made him realize radio's potential as a vehicle for poetry. After completing his studies, Rosten returned to New York where, with an introduction from Benét, he began to write patriotic radio plays for the *Cavalcade of America.*

Millard Lampell grew up in the same New Jersey hometown as Allan Sloane. Son of an immigrant garment worker, Lampell, a short, athletic, ebullient, man, worked in his youth as a fruit picker and coal miner. He sold his first piece of writing, an article about fascist groups on campus, while he was in college in West Virginia. After college, Lampell moved to New York City and sang for a while with folk singers Woody Guthrie and Pete Seeger. One Sunday, after they finished a gig, a radio producer who had been in the audience approached Lampell and asked him if he would be interested in writing for radio. Lampell's writing career was off and running.

In 1942 and 1943 Lampell wrote scripts for *It's the Navy* and several other war-related radio programs. In 1943, he was drafted. After training, he was assigned to the Air Force radio section in New Haven, Connecticut. There he wrote, produced, and directed programs. In 1944, Lampell was released from the service to visit veterans hospitals around the U.S. and gather material for radio scripts about returning soldiers. Afterwards, he produced *First in the Air,* a series for the Army Air Forces program. Subsequently he lectured on radio writing at several New York area colleges. In 1946, Lampell went to Hollywood as a contract writer at Warner Brothers.

The fact that Arthur Miller wrote most of his early plays for radio has not been well known. But he wrote perhaps 25 radio plays between 1939 and1946, most of them war related. New York–born Miller attended the University of Michigan, where he befriended Norman Rosten and began writing plays. After graduation, Miller also moved back East. Within a few months, a film studio offered him $250 a week to work in Hollywood. Miller rejected the offer, opting instead for a job at $22.77 per week with the Federal Theatre, a government work program for writers. At around the same time, the *Columbia Workshop* accepted his first play for radio. Miller also wrote for a series entitled *The Doctor Fights* and, after Rosten recommended him, for the *Cavalcade of America.*

During the war, Miller was rejected twice for military service because of a knee injury. As a substitute, he took a job in the Brooklyn Navy Yard where he helped recondition ships for service. He also threw himself into writing patriotic shows for radio. He worked quickly, completing a half-hour play in less than a day and spending only three months per year writing for radio. During the other nine he wrote for the stage.

Arthur Laurents graduated from Cornell University in 1940. At the urging of a friend, he enrolled in a radio writing course at New York University. His teacher, a CBS director, was so impressed with a play that Laurents wrote for the course that he sold it to the Columbia Workshop. After that Laurents wrote for numerous commercial shows before moving away from radio in favor of musicals and films.

Morton Wishengrad was born in New York's Lower East Side in 1913 to Russian-Jewish immigrant parents. A tall, thin, and reserved man, he shared many of the concerns of the progressive-minded writers of his generation. Wishengrad's first job was as the educational director, editor, and researcher for the International Ladies Garment Workers Union. During the war he worked for the American Federation of Labor as director of a joint Labor Short Wave Bureau that broadcast to organized labor in Europe. He also wrote scripts for NBC's *Labor for Victory,* the *Cavalcade of America,* and several other shows.

Writing for Minorities

Wishengrad stands out among a number of Jewish radio writers as perhaps the only one to clearly bring his Jewish consciousness to radio. Although he wrote for mainstream shows, Wishengrad is best remembered for his writing for programs that primarily addressed a Jewish audience. In 1943, he produced a script about the battle of the Warsaw Ghetto for the American Jewish Committee. It was one of a very few wartime shows that touched on the Nazis' genocide policy. The following year, Wishengrad began to write for *The Eternal Light,* a new Jewish religious drama series, sponsored by the Jewish Theological Seminary.

Mitchell Grayson and Richard Durham, two African American radio dramatists, also appealed to special audiences. Grayson, a New York writer, wrote and directed *New World A'Coming.* Durham wrote *Destination Freedom* in Chicago. Both were provocative collections of half-hour black history dramas about prominent African Americans that helped pave the way for the civil rights movement.

Postwar Era

Despite radio drama's great success from the mid 1930s, by the 1950s the genre was in decline, a consequence of the ascent of television and the postwar increase in commercialization in broadcasting. The latter helped make radio vulnerable to a destructive broadcasting industry blacklist carried out by anticommunist vigilantes. Corwin, Grayson, Hughes, Lampell, Rosten, and virtually all of the other writers discussed here were its targets. By 1957, most radio dramatists also had to write for television and film to make a living.

With the departure of many of the war-era radio dramatists, the American radio audience lost a steady source of progressive ideas. The public also lost some fine entertainment. For a short time, during television's Golden Age, the new medium filled the gap. But then it too faded from its glory days.

HOWARD BLUE

See also Blacklisting; Cavalcade of America; Corwin, Norman; Drama, U.S.; Drama, Worldwide; Durham, Richard; Lights Out; March of Time; Mercury Theater of the Air; Oboler, Arch; War of the Worlds

Further Reading

Barnouw, Erik, editor, *Radio Drama in Action: Twenty-Five Plays of a Changing World,* New York: Farrar and Rinehart, 1945

Benet, Stephen Vincent, *We Stand United and Other Radio Scripts,* New York: Farrar, 1944

Blue, Howard, *Words at War: World War II Era Radio Drama and the Postwar Broadcasting Industry Blacklist,* Lanham, Maryland: Scarecrow Press, 2002

Boyd, James, editor, *The Free Company Presents . . . A Collection of Plays about the Meaning of America,* New York: Dodd Mead, 1941

Brady, Frank, *Citizen Welles: A Biography of Orson Welles,* New York: Scribner, 1989

Fenton, Charles A., *Stephen Vincent Benét: The Life and Times of an American Man of Letters, 1898–1943,* New Haven, Connecticut: Yale University Press, 1958

Lampell, Millard, *The Long Way Home,* New York: Messner, 1946

Laurents, Arthur, *Original Story By: A Memoir of Broadway and Hollywood,* New York: Knopf, 2000

Liss, Joseph, editor, *Radio's Best Plays,* New York: Greenberg, 1947

MacDonald, J. Fred, *Don't Touch That Dial! Radio Programming in American Life, 1920–1960,* Chicago: Nelson Hall, 1979

Miller, Arthur, *Timebends: A Life,* New York: Grove Press, and London: Methuen, 1987

Rampersad, Arnold, *The Life of Langston Hughes,* 2 vols., New York: Oxford University Press, 1986–88

Poetry on Radio

Poetry has played many roles on radio, from filling dead airtime to giving voice to the grief of mourning, as well as many of the places in between.

Introduction and Themes

Poetry on radio was initially occasional and spontaneous, with an announcer reading a poem before or after a music segment or the farm reports. Ted Malone, who was later to be one of the best-known hosts of poetry on radio, began his career this way. In the late 1920s, quickly substituting for an act that failed to show, the program director thrust an anthology of poetry and a mike at Malone and left him to his fate. His reading was a success, and Malone went on to a radio career of more than 25 years with *Between the Bookends.* Because of the ease of production and minimal expense involved, this tradition expanded in the 1930s and 1940s and continues to this day.

The poetry heard on radio has tended to fall into several categories. Early on and into the golden days of radio it has mostly been of the sentimental or "light" variety, especially on shows that were commercially sponsored. Poetry was often taken from popular magazines, such as *Good Housekeeping* or *Redbook.* Many times it was read over music, sometimes accompanied by philosophizing or "gentle wisdom." When more serious or "highbrow" poetry found its way onto the airwaves, it was usually in a sustaining program, and later in noncommercial radio. Some shows included music underneath, or before or after the poetry, while others were straight readings of poems.

By the late 1930s poetry found a new form with the appearance of radio verse plays. The originators of radio verse plays in the U.S. were Archibald MacLeish and Norman Corwin. It was here that poetry reached its greatest number of listeners, as illustrated by the estimated audience of 60 million for Corwin's "We Hold These Truths" in 1941. The paragraphs that follow, by no means inclusive, highlight selected examples and events throughout the nearly 100 years of radio.

Origins

In the early part of the 20th century, radio was primarily a medium of individuals and amateurs some of whom would occasionally read poems, sometimes just to fill dead air, or sometimes in loosely organized "shows." Unfortunately, because the technology was not easily available to record these "programs," and scripts were rarely kept, few names or titles have survived.

In the early 1920s poetry was often read within other programs, such as variety shows hosted by people such as Major Bowes, Rudy Vallee, and Fred Allen. The earliest poetry appeared on local radio shows, such as a program in Yankton, South Dakota, broadcast in 1921 on WNAX and supported by the Guerney Seed and Nursery Company. One listener won a "poetry" contest with, "the Guerney's are the farmer's friend / They always will be to the end."

Later, when radio became more structured, stations broadcast identifiable announcers, and programs with titles appeared, such as *Cheerio, Tony's Scrapbook, Between the Bookends,* and *Poet's Gold.* One of the earliest shows featured Edward Godfrey reading poetry on a children's program titled *Stories and Poems Read by Uncle Ed over KDKA* (Pittsburgh, Pennsylvania). Beginning 26 October 1923, it was a variety program, usually lasting 15 minutes, that ranged from imitations of birds and animals to music on the mouth organ and guitar to the reading of poems, some original.

One the first shows to feature poetry was *Poems,* with Beatrice Meisler reciting poetry each week on WGBS in New York City in 1925. Another was *Poet's Corner,* sponsored by Hewitt's Bookstore, with an unidentified performer reading poetry on KFON in Long Beach, California, in 1926. A third was *Poetry Club,* in which Mrs. David Hugh read and discussed poetry on KHJ, Los Angeles, California, in 1929.

While most of this early poetry on the radio was of the sentimental type, more serious poetry did appear. As radio became more regulated in the 1920s, one of its roles was to provide a public service, as defined in the Radio Act of 1927. With this in mind, "highbrow" material, including some poetry, also appeared in sustaining programs (programs funded by the station or network without commercial advertising). David Sarnoff, vice president and general manager of RCA, suggested that the masses needed to be uplifted by culture, including poetry. While these early poetry broadcasts sometimes were on commercial radio, they were not necessarily commercially sponsored and often served as filler between commercially sponsored programs.

Network Radio

Poetry began on the networks in 1929 when CBS asked David Ross, with his fine voice, to read poetry in a half-hour Sunday afternoon sustaining show titled *Poet's Gold.* However, like many radio shows and personalities, Ross began locally, first airing in 1926 on WMAQ, Chicago. *Poet's Gold* continued on CBS into the late 1930s, mostly on Sunday afternoons. Ross read classic and contemporary verse by poets such as Ben Jonson, Samuel Taylor Coleridge, Percy Shelley, Robert Frost, Edna St. Vincent Millay, and Stephen Vincent Benet, usually over music. *Variety* commented that it was one of the best of the many programs of the 1930s featuring poetry and quiet music. Ross participated in another trend of poetry on radio programs by publishing a related book named after the show, *Poet's Gold.*

Ted Malone (Frank Alden Russell) and Tony Wons (Anthony Snow) were the two major personalities of poetry shows on network radio. On *Tony's Scrapbook* Wons gave down-home wisdom and commentary mixed with sentimental verse from sources such as *Good Housekeeping* and *Redbook,* as well as drawing occasionally from literary giants. Wons' gentle, intimate sounding delivery made him a favorite with female listeners. Popular or "light" poetry was more likely to be commercially sponsored, and Wons was able to find companies to foot the bill, including International Silver and Johnson Wax in the 1930s and Hallmark Cards in the early 1940s. Before that he had supported his program by selling yearly collections of poems read on the air, thus the show's name. CBS participated in the publishing of the scrapbooks and later had a deal whereby the network got a cut of sales from the books in exchange for air time.

Wons also appeared on other shows. In 1931 he contributed to the *Camel Quarter Hour,* a variety program directed by Erik Barnouw and sponsored by R.J. Reynolds, where he read a poem on each show. In the commercial spirit, he also read two poetically phrased messages about Camel cigarettes, as part of an attempt to market cigarettes to women. From September 1934 to August 1935, Wons appeared on *The House by the Side of the Road,* a half-hour Sunday afternoon dramatic program, sponsored by Johnson's Wax and Allied Products and broadcast on NBC.

Ted Malone hosted *Between the Bookends,* a book review program that also featured conversation and poetry, for more than 25 years. It included Malone's poetry as well as that of others and was usually amusing and almost always uplifting. Malone sat in a studio with the lights dimmed and read poetry to organ accompaniment played by Rosa Rio. To Malone, radio was an intimate medium that fit the simple and sentimental poetry he read. Malone was poetry editor at *Good Housekeeping,* and the poems he read were mostly taken from it and similar magazines. He published several *Between the Bookends* anthologies, which were widely available.

Malone also hosted a 30-minute *Pilgrimage of Poetry* program for one season on NBC Blue that featured 32 of the most famous poets in America. Malone traveled to the homes of writers such as Henry Wadsworth Longfellow, Stephen Crane, Edgar Allan Poe, and Walt Whitman, among others, as a way of sharing their lives with his radio audience. He followed this with a season of *American Pilgrimage,* broadcast from 1940 to 1941, where he visited the homes of such literary stalwarts as Mark Twain and Herman Melville.

Sentimental/Cheerful Poetry

Readings of sentimental/cheerful poetry over a background of organ music or in between announcers' talking and philosophizing had the widest audience for poetry on the radio. *Cheerio,* a program of good cheer that included poetry, was also the broadcast name used by the popular broadcaster and host of the show, Charles K. Field. He read poetry and gave inspirational talks starting in 1925 on local radio in the San Francisco Bay area. He moved to the networks in March of 1927 until he was off the air in April of 1940. Field claimed never to have received a penny from his broadcasting, but he wrote many best-selling books. They contained the poetry and wisdom used on his shows and provided his income. In a program called *Cheerio Exchange,* Field's staff maintained a fund to purchase radios at a discount to be lent to shut-ins.

Moon River was broadcast on WLW (Cincinnati, Ohio) in the 1930s and 1940s. It was one of the best-known and best-loved local shows of the network radio era. It began when WLW's owner, Powel Crosley, Jr., told Ed Bryan to create a poetry show to make use of the organ he had just purchased. Narrated by Bob Brown among others, *Moon River* occasionally found its way to the networks. One reference mentions it being broadcast on NBC Red at 12:30 A.M. on 15 March 1942.

George Work hosted *Melody and Rhyme* from 1927 until his death in 1947. It was broadcast on WNYC on Sunday mornings from 8:00 to 8:45 A.M. During those 45 minutes, he read four to six sentimental poems over music, or between musical selections.

Edgar Guest's poems were a regular feature of many music-poetry programs. He also had his own radio career as a poet/host. It began in Boston in the 1920s, although Guest was most well known in the Midwest. He hosted his own show in the 1930s on WASH (Grand Rapids, Michigan) and later on a show in which he read poems between selections by the Detroit Symphony on CBS. Guest's career with poetry on radio continued into the 1940s as he provided poetry for an NBC radio show in 1941 along with Eddie Howard.

Highbrow/Serious Poetry

Although most of the poetry on the radio was of the sentimental variety, there has long been a presence of "highbrow" or serious poetry allied with the more elite poets, poetry establishments, and the academy. One show that included more serious poetry was the previously mentioned *Poet's Gold*. Another was *Poetry Hour* hosted by A.M. (Aloysius Michael) Sullivan, an officer of the Poetry Society of America. Sullivan began broadcasting *Poetry Hour* on WOR/Mutual Network in 1930 and was still on the air into the 1940s. *Poetry Hour* presented the poets reading their own work and discussing techniques and trends in poetry. He presented more than 300 poets on the air including Stephen Vincent Benet, Edgar Lee Masters, Padraic Colum, Sara Teasdale, Mark Van Doren, and Kimball Flaccus. In 1937 this program helped provide an entrée onto the radio airwaves for Norman Corwin, who filled in for Sullivan on several occasions. In 1948 Sullivan created *The Poet Speaks*, which combined the reading of poetry with discussion about it.

Among the highest of the highbrows, Harriet Monroe, the founding editor of *Poetry: A Magazine of Verse*, appeared on *Here, There and Everywhere*, a 15-minute Chicago area program between 1930 and 1936. She also was on CBS's *American School of the Air* in a 10-minute segment on modern poetry.

Radio Verse Plays

The late 1930s brought a new development to the history of poetry on radio with the beginning of radio verse plays. The legendary first broadcast was Archibald MacLeish's verse play for radio *The Fall of the City*, which was broadcast 11 April 1937 from 7:00 to 7:30 P.M. on CBS. *The Fall of the City*, said critics, represented the American broadcasting industry's discovery of great radio in poetry and at the same time an American poet's discovery of great poetry in radio. In MacLeish's own words, "I realized at the time how much 'The Fall of the City' owed to not being seen, how much it owed to the fact that the imagination conceives it."

The Fall of the City raised the bar for the level of poetry on radio. First, its author was a writer of high repute—MacLeish won the Pulitzer Prize for Poetry in 1933 and later became the Librarian of Congress. Second, the play recognized radio as a literary medium, well suited to verse and poetry. This furthered the view that radio was highbrow and artistic. *Fall of the City* was referred to frequently by critics in later years as an eminent first in literary broadcasting.

It was, however, Norman Corwin who was to become known as "radio's poet laureate." The *New York Times* said that "Corwin writes with a poet's vision, a good reporter's clarity and a technician's precise knowledge of his craft—three attributes that have made him preeminent in radio literature." While Norman Corwin's name eventually overshadowed all others in the world of radio verse plays, he went on to become much more as he made his reputation as a writer, producer, and director.

Corwin began his radio career at WBZA (Springfield, Massachusetts) reading the news and then later hosting a show called *Rhymes and Cadences* in which he read poems aloud. He moved to New York in 1936, where he hosted *Poetic License* beginning in late 1937. It was broadcast from 9:45 to 10:00 P.M. on WQXR, a New York station then known for its high-quality ("for the discriminating listener") innovative programming. *Poetic License* featured some of the leading poets of the day in conversational poetry, or what was to become known as "talking verse." On one show Corwin presented an adaptation of *Spoon River Anthology* that caught the attention of W.B. Lewis, a CBS vice-president, and led to his becoming a major fixture there.

Soon after moving to CBS, Corwin started *Words without Music, a* sustaining show broadcast on Sundays from late 1938 to June 1939. Here Corwin appealed to a variety of listeners by reading poetry from Mother Goose to Walt Whitman, Carl Sandburg, Robert Frost, and occasionally African-American poets such as James Weldon Johnson and Sterling Brown. *Words without Music* used a form called "vitalized" or "orchestrated" poetry that consisted of dramatizing a poem, giving its lines to several voices, and sometimes adding lines and repetition for dramatic effect. By dramatizing poems and/or adding a sense of humor, Corwin opened up the audience for "good" poetry beyond the highbrow, to include those who enjoyed drama and/or comedy.

Words without Music inspired an NBC poetry program Fables without Music, featuring Alfred Kreymborg. It was, however, much more serious in tone and subject. Broadcast in the spring of 1939, it consisted of 10 radio verse plays of 15 minutes each.

Among the hundreds of verse plays written for radio in the 1940s are Stephen Vincent Benet's *A Child is Born*, W.H. Auden's *The Dark Valley*, Pearl S. Buck's *Will This Earth*

Hold?, Norman Rosten's *Concerning the Red Army*, and Edna St. Vincent Millay's *The Murder of Lidice*.

Postwar Years

After World War II, poetry played a lesser role on network radio. Some earlier shows, such as *Between the Bookends*, survived into the 1950s. Others suffered as the networks became more commercialized and reduced the number of sustaining programs. Norman Corwin left CBS in 1948 because of a contract dispute and moved to Hollywood. In addition, the networks were less interested in promoting these shows. One example was their reduction in support for the publication of books to support and advertise their on-air poetry shows.

Shows containing poetry or of interest to poetry fans did, however, appear, some locally and some on the networks. One was hosted by Mary Margaret McBride, who broadcast from 1935 to 1955, first as Martha Deane, then under her own name. Her guests included Langston Hughes, Mark and Carl Van Doren, Carl Sandburg, and William Carlos Williams.

Like so many artists, poets of the 1950s suffered under the pressures of McCarthyism. Unofficial lists of "communists" included many prominent poets such as W.H. Auden, Archibald MacLeish, Carl and Mark Van Doren, and Stephen Vincent Benet.

New poetry shows that began in the 1950s included *Poetry of Our Time*, featuring author and poet Katherine Anne Porter reading her poetry. Another was *Anthology*, which mixed music, poetry, and other literature in a series presented in cooperation with the Poetry Center of the Young Men's-Young Women's Hebrew Association. It was produced by Steve White, directed by Draper Lewis, and hosted by Harry Fleetwood. Guests included W.H. Auden, Edna St. Vincent Millay, William Carlos Williams, and Wallace Stevens.

David Ross' long career of reading poetry on the air continued in the 1950s with *Words in the Night*. Here Ross read poetry with music provided by guitarist Tony Mattola and vocalist Sally Sweetland. *The Poet Speaks* was heard on WGBH (Boston) featuring Wallace Stevens, e.e. cummings, and Adrienne Rich, among many others.

Non-Commercial and Other Venues

With the decline in network support for poetry, non-commercial and educational radio filled some of the gap. University and other stations serving cultured and academic populations have supported numerous poetry programs that have lasted as long as their mostly unpaid hosts could manage. Broadcasting since the late 1920s, WOSO at Ohio State University did some of the most progressive work in poetry including *Lyric Ohio*, which presented the work of Ohio poets. On the air since 1930, *Invitation to Reading* on WHA at the University of Wisconsin included poetry programs for high school students.

In the late 1950s, music and literature shows made up 70 percent of the schedule for KPFA (Berkeley, California), flagship station of the Pacifica Foundation. In 1959 KPFA broadcast *The Poetry of Lawrence Ferlinghetti* as well as numerous other shows of beat poets from the thriving poetry scene in the San Francisco Bay Area at the time. Chicago's fine arts station, WFMT, has throughout its history featured live and recorded poetry readings. One notable example is Ken Nordine's *Word Jazz*, which was broadcast in the late 1950s and early 1960s.

The Library of Congress' presence on the radio began in the early 1940s with the Radio Research Project headed by Archibald MacLeish. Although poetry was not its main emphasis, the Research Project featured some of it in a series titled *Books and the News*. In 1950, consultant in poetry Conrad Aiken inaugurated the broadcasting of readings over a local radio station, WCFM, under the Library of Congress' sponsorship. These broadcasts continued in the 1950s and included Katherine Garrison Chapin reading the poems of Emily Dickinson as well as the works of other women poets. The programs were broadcast from the Coolidge Auditorium at the Library of Congress and were often aired only locally in Washington, D.C., although some reached a national audience as well.

Since the early 1950s the Library of Congress has continued to have an irregular but important presence on the airwaves. It has often featured the current consultant in poetry (now called Poet Laureate Consultant in Poetry) as well as a long list of distinguished poets of note. The Library of Congress continues to host poetry broadcasts to this day. Recently these have primarily been on WETA-FM in Washington with national distribution via satellite. One example is *The Poet and the Poem from the Library of Congress*, hosted by Grace Cavalieri, which began broadcasting in 1989 and has been annual since 1997. Robert Pinsky provided the first live poetry webcast on the internet from the Library of Congress on 8 October 1998.

The Library of Congress' Archive of Recorded Poetry and Literature contains many broadcasts by known and not-so-known poets. Many recordings were made at local stations and given to the Library for inclusion in the archive. The Archive of Hispanic Literature, also at the Library of Congress, contains the recordings of hundreds of poets, including six or more Nobel Prize winners from Latin America and Spain. It is difficult to know how many of these were broadcast.

The decade of the 1960s opened with a major American poet, Robert Frost, reading a poem he had written at the inauguration of John F. Kennedy. It was broadcast live on radio and began

> *A Golden Age of poetry and power,*
> *Of which the noonday's the beginning hour*

At that point he was blinded by the sun and snow and could hardly read his words. Quickly he moved into a poem he knew well, "The Gift Outright," holding his head high as the words reached out. Frost also read on WAMF (Amherst, Massachusetts) many times between 1948 and 1962. Most of those recordings are in the Archive of Recorded Poetry and Literature at the Library of Congress.

In the 1970s, poetry's access to the airwaves increased with the advent of National Public Radio (NPR). Since its beginning in 1971, NPR has broadcast a number of shows that have featured poets and poetry. *Voices in the Wind* broadcast beginning in 1974 with Oscar Brand as host and included Nikki Giovanni, Lucille Clifton, and Allen Ginsberg. *Poet Speaks,* a 30-minute show that originated at WGBH, broadcast with Herbert Kenny as host from April to June 1972. Guests included Richard Eberhart, John Updike, May Sarton, and Allen Tate. Some shows had very limited runs, perhaps one or only a few broadcasts. *Spoon River Anthology* was broadcast in March 1973 in four weekly installments, originating at WGBH. *The Archibald MacLeish Tribute* was broadcast 15 April 1981 with MacLeish and John Ciardi. MacLeish was also interviewed in *Book Beat* on 31 October 1971. *Voice of the Poet* was broadcast 15 January 1975 with Jerome Rothenberg and Marge Piercy among the guests. Some of the other shows that have featured or included poetry are *Talk of the Nation, Fresh Air,* and *Children's Radio Theatre.*

Poetry on Radio Today

At the beginning of the 21st century, Garrison Keillor hosted a daily five-minute radio program called *The Writer's Almanac,* in which he notes milestones of the day and closes with a poem or two. It is heard each day on public radio stations throughout the country. Occasionally Keillor also includes poetry in his weekly program *A Prairie Home Companion.* Other short shows similar to *The Writer's Almanac* include *The Osgood File* with Charles Osgood, heard daily on the CBS radio network, and *Bookbeat,* a daily report on new books and authors with Don Swain as host on WCBS in New York.

New Letters on the Air, the radio companion to the printed publication *New Letters,* was first broadcast locally in Kansas City, Missouri, beginning in 1977. *New Letters on the Air* is a half-hour weekly show designed primarily to introduce the author with a short interview and then a number of poems. Typically, the program has about 10 minutes of poetry, 15 minutes of interview, and 3 minutes of introductions, credits, and musical bridges. *New Letters on the Air* has featured four Nobel Prize Winners, as well as 50 winners of various other literary awards, including the Pulitzer Prize. Approximately one-third of the featured writers are members of ethnic minorities. It has been syndicated over the NPR satellite and broadcast in more than 60 cities. From 1984 to 1995 it was hosted and produced by Rebekah (Presson) Mosby.

The Poet and the Poem is a one-hour show broadcast locally on WPFW in Washington, D.C., nationally by Pacifica Radio, and internationally by the Voice of America. From 1977 to 1997 it was broadcast weekly, first on Thursday, then on Sunday evenings. It has featured more than 2,000 poets ranging from United States Poet Laureates and Pulitzer Prize winners to unpublished and/or fledgling poets of consequence. In *The Poet and the Poem,* host Grace Cavalieri first provides biographical information about the poet who then reads several poems. A poet herself, Cavalieri asks probing and insightful questions that draw the poets out in revealing and informative ways. From 1978 to 1993, the program also hosted an all-day poetry broadcast once a year featuring 35 performers present. Titled *Ribbon of Song,* it featured Sterling Brown among others and the archival works of Paul Laurence Dunbar and Langston Hughes among others. In May (National Poetry Month) of 2000, in celebration of the bicentennial of the Library of Congress, *The Poet and the Poem from the Library of Congress* was a one-hour weekly show with W.S. Merwin, Louise Gluck, Robert Pinsky, and Rita Dove.

Other contemporary shows of note include *Poems to a Listener* with host Henry Lyman on WFCR-FM, Amherst, Massachusetts; *A Moveable Feast* with Tom Vitale and guests such as Allan Ginsberg, Charles Bukowski, and Joyce Carole Oates; *Soundings* with Wayne Pond; *Bookworm* with Michael Silverblatt from KCRW, Santa Monica, California; *Enjoyment of Poetry* with Florence Becker Lennon on WEVD, New York; *Booktalk* with Rus Morgan on WYPL, and *The Book Show* with Douglas Glover.

Poetry continues to be heard on radio in much the way it was in the early parts of the 20th century. It is heard on stations big and small, to inform and entertain, as filler, inspiration, and as something to soothe. As the internet becomes more of a force in radio, more programs will be available live over the web. Recorded poetry as well is becoming available on line, being broadcast, or "webcast."

Perhaps the close and continuing association of poetry and radio should come as no surprise. The development of the radio restored the power of the spoken word. Both are reflections of the original oral traditions that gave birth to our literary heritage. In the words of Archibald MacLeish, "The ear is the poet's perfect audience, his only true audience. And it is radio and only radio which can give him public access to this perfect friend."

BRAD MCCOY

See also Canadian Radio Drama; Corwin, Norman; Drama, U.S.; Drama, Worldwide; Playwrights on Radio

Further Reading

Cifelli, Edward M., *John Ciardi: A Biography*, Fayetteville: University of Arkansas Press, 1997

Donaldson, Scott, and R.H. Winnick, *Archibald MacLeish: An American Life*, Boston and London: Houghton Mifflin, 1992

Everett, Michael, editor, *The Radio Book of Verse*, New York: Poetry House, 1939

Godfrey, Edward, *Stories and Poems Read by Uncle Ed over KDKA*, Pittsburgh, Pennsylvania: Godfrey, 1926

Kaplan, Milton Allen, *Radio and Poetry*, New York: Columbia University Press, 1949

MacLeish, Archibald, *The Fall of the City: A Verse Play for Radio*, New York and Toronto: Farrar and Rinehart, and London: Boriswood, 1937

Malone, Ted, *A Listener's Aid to "Pilgrimage of Poetry": Ted Malone's Album of Poetic Shrines*, New York: Columbia University Press, 1939

McGuire, William, *Poetry's Catbird Seat: The Consultantship in Poetry in the English Language at the Library of Congress, 1937–1987*, Washington, D.C.: Library of Congress, 1988

Poetry Broadcast: An Anthology Compiled for Radio Programs, 4 vols., New York: Exposition Press, 1946

Poetry House, editors, *Poems for Radio*, New York: Poetry House, 1945

Ross, David, compiler, *Poet's Gold: An Anthology of Poems to Be Read Aloud*, New York: Macaulay, 1933

Selch, Andrea, "Engineering Democracy: Commercial Radio's Use of Poetry, 1920–1960," Ph.D. diss., Duke University, 1999

Tate, Allen, compiler, *Sixty American Poets, 1896–1944*, Washington, D.C., 1945; revised edition, edited by Kenton Kilmer, Detroit, Michigan: Gale Research, 1969

Wons, Anthony, compiler, *Your Dog and My Dog: From Tony Wons' Famous Radio Scrap Book*, Chicago: Reilly and Lee, 1935

Politics and Radio

The late 1920s witnessed the dramatic expansion of radio as it became a powerful new form of mass communication. The 1930 census reported that nearly 70 percent of all homes in the United States had at least one radio. As the 1930s came to a close, more families owned radios than owned telephones and automobiles or subscribed to newspapers. Nowhere was the impact of radio more widespread than in the political arena. Radio had several effects on political discourse and campaigns. It made campaigning more expensive, as the cost of radio air time added millions to campaign budgets, and it brought the advertising agency into politics. Perhaps even more important, radio was the first technological medium that allowed presidents to "go public"—that is, to go over the heads of Congress and directly to the people, thereby changing the method of governance. This essay, detailing radio's impact on the political process, chronicles the major eras of political radio in the U.S., which include 1) politics and radio in the early years—from Harding to Hoover; 2) Roosevelt and radio; 3) the postwar years and the rise of television; and 4) the rebirth of political radio.

Politics and Radio in the Early Years—From Harding to Hoover

The first president to speak on radio was President Woodrow Wilson in 1919. However, only a few people heard Wilson's address, and those listening could distinguish no more than a few clear words. Historians theorize that if radio had evolved ten years earlier and provided him the opportunity to speak directly to the people, Woodrow Wilson might have been more successful in his appeal for League of Nations membership.

It was not until after the landslide Harding election of 1920 that both the public and politicians realized that radio could be a pragmatic and efficient communication medium. After the election, Harding periodically spoke on the radio concerning national matters. His 1923 State of the Union speech was carried widely, and later that same year, while on a Western tour, Harding energized the populace's acceptance of political radio with a series of commentaries entitled "Stewardship of the Administration." The effects of these radio announcements were swiftly imprinted upon the American political landscape.

For example, development of radio during the Harding administration was evident in the broadcasting arrangements made for his Western tour. The railroad car in which he traveled was equipped with a radio transmitter in order to broadcast speeches to a large portion of the nation.

There is little doubt that the novelty of these presidential addresses made at least some impression on the populace. Nonetheless, Harding, it seems, was not altogether comfortable with the new medium. A *New York Times* observer reported, "He is dominated by the restraining influence of the radio-telephone amplifiers, into which he has talked while making these addresses. The mechanical contrivance worries him . . . and he is tempted at times to revert to the old style oratory" ("Cordial to Harding, Cold to Speeches," *New York Times*, 25 June 1923).

As troubling as the mechanics of the radio may have been to Harding, his successor, Calvin Coolidge, found the medium to be suitable to him rhetorically, even as he recognized radio's political possibilities. "Silent Cal" was anything but silent when it came to radio broadcasts. In 1924 it is estimated that Coolidge spoke more than 9,000 words per month over radio and that more than 50,000 people heard his voice during the first eight months of 1927, more than had heard any previous president. That radio broadcasts benefited Coolidge is rarely disputed. Both writers and politicians who assessed Coolidge's radio abilities gave him enthusiastic endorsements. Coolidge himself once made the observation, "I am very fortunate that I came with the radio. I can't make an engaging, rousing, or oratorical speech, but I have a good radio voice, and now I can get my message across to them without acquainting them with my lack of oratorical ability" (Chester, 1969). Coolidge tended, however, to refrain from utilizing radio as a tool for practical political or party gain. Even though radio played a large role in establishing Coolidge as president and in getting him re-elected, he did not feel the need to speak habitually to the nation.

Certainly presidential addresses made up the largest number of political programs on radio in these early years, but the public also seemed to pay attention to other political events carried on the airwaves. Starting in 1924, political party conventions were covered by radio and heard by a large audience, and that same election witnessed the beginning of paid broadcast advertising for political parties and candidates. The early years of radio also saw the beginning of radio's use as a medium for advocating political viewpoints. The most famous of these advocates was the "Radio Priest," Father Charles Coughlin, who used radio to promote his views on social and political issues of the day during the 1920s and 1930s.

If Coolidge was the harbinger of radio as beneficial to the democratic ideal, Hoover was the forerunner of radio as an integral part of a campaign. Hoover pushed the Republican party toward spending the major portion of its publicity budget on radio. Indeed, in May 1928 Hoover indicated that he would engineer his campaign largely through radio and films. The use of radio in the 1928 campaign by both Hoover and his Democratic challenger, Alfred Smith, was remarkable on a number of levels. First, despite his interest in drama as a younger man, Smith did not use the radio airwaves well. Problematic was his East Side accent, which may have endeared him to the immigrant population in New York City but which hurt him in the South, where he desperately needed votes. Second, the spending for airtime in this campaign, by both parties, reached nearly $2 million. This represented only about 15 percent of the total publicity budgets of both parties for a medium that was able to reach nearly 40 million voters.

Radio as a political medium was growing rapidly. For example, the League of Women Voters developed a bipartisan series of programs designed to inform voters, claiming to provide background information, differing points of view on issues, and information on the political and voting process. Campaigning notwithstanding, political scientists argue that radio may have had an even more powerful influence upon American presidential politics than simply the obvious effect of helping candidates attract votes. The power of the president rose with the ability to go "over the head of Congress" directly to the American public. The potential power of the presidency was strengthened during the Harding, Coolidge, and Hoover administrations, and radio played an important role in this development.

Roosevelt and Radio

The harshness of the Great Depression, which had shrouded the country since 1929, almost ensured a victory for the Democratic party in the 1932 election. One of the more notable aspects of his nomination was Franklin Delano Roosevelt's outstanding radio speaking ability. He began his campaign with a radio address at Albany, New York, and he accepted the nomination of his party at the convention over radio, breaking the precedent of waiting for a time lapse of one month. Long before his nomination, Roosevelt suspected that the power of radio would be important to his political livelihood. For instance, in an address to the Tammany Speaker's Bureau in 1929, Roosevelt argued that American politics "had passed from an era in which silver tongues had swayed many votes through a period of newspaper domination to the present age in which radio was king" (Chester, 1969).

The election of 1932 witnessed a remarkable juxtaposition of quality versus quantity. Herbert Hoover had, for example, launched his campaign for re-election over the largest political radio hookup in history, nearly 160 stations. The Republican party used 73 hours of network time to boost its candidate, compared to 51 hours for the Democrats, and ultimately Republicans outspent Democrats on the radio. This did little to enhance the possibility of Hoover's being re-elected, however, as he was not a particularly effective performer on the radio.

The 1936 election was one of the more remarkable campaigns in the history of presidential politics. The Republicans chose a notoriously poor public speaker in Kansas governor Alfred Landon. Furthermore, 1936 witnessed the arrival of a third-party candidate, William Lemke of the Union party, whose candidacy was driven by Father Coughlin. Faced with a strong Democratic candidate as well as Father Coughlin's demagoguery, Republican strategists developed and employed several innovative radio strategies to assist their candidate. The innovations included spot radio advertisements for their candidate as well as the radio drama *Liberty at the Crossroads,* which played on WGN in Chicago. One of the more creative techniques was a one-sided "debate" in which Republican linchpin Arthur Vandenburg "debated" a phonographically recorded Franklin Roosevelt. The debate was designed to illustrate Roosevelt's failed campaign promises from 1932. The "debate" caused consternation among the 66 stations scheduled to carry it: 23 broadcast it, 21 cut it off, and the remainder vacillated back and forth. Also during 1936, both parties placed a great deal of emphasis on foreign language broadcasts, creating about 2,000 political broadcasts altogether. The Democrats had foreign-language transmissions in over a dozen cities, and the Republicans employed 29 languages in everything from 100-word spot advertisements to 30-minute talks.

Roosevelt's overwhelming victories in 1932 and 1936 were due, in part, to his use of radio. However, these victories were not so much caused by the use of radio in the campaign itself; instead, it was the cultivating of the electorate during his first term in office that ensured Roosevelt's success. Republican innovation aside, the introduction of Roosevelt's "fireside chats" was one of the most effective uses of mediated political communication in the 20th century.

The Postwar Years and the Rise of Television

With the advent of television, Harry Truman as well as other politicians had to adapt to a new broadcast medium. The future of politics in radio also evolved. Significant during the 1948 campaign was a debate that served as the antecedent to the famous televised Kennedy and Nixon debates 12 years later. The leading contender for the Republican nomination in 1948 was Thomas Dewey, but Minnesota Governor Harold Stasson was a strong challenger. The two men faced off at a Portland radio station in May 1948, and Dewey profited from the confrontation. A poll taken by the *Oregonian* showed that Stassen's popularity dropped following the debate, although by the eve of the election his numbers reflected his pre debate strength.

Harry Truman's most significant innovations in the political use of radio in 1948 were to record press conferences to assist the White House in checking sources and notes. The Truman White House then began to allow radio broadcasters to transmit portions of the recordings to the general public. Truman's administration will probably be best remembered in broadcast history for the ascent of television. Though radio continued to be used, television quickly became the medium that affected the electorate. Truman became the first president to participate in a television broadcast from the White House when on 5 October 1947 he asked people to cooperate in the President's Food Conservation Program.

Although radio would not again dominate the American political scene, radio remained an important medium for political progress in many less developed countries. The Voice of America continues to broadcast political and informational programming around the world, and many nations, divided by political subcultures and a myriad of differing languages, still find radio superior to television as a way of communicating political messages.

Rebirth of Political Radio

Television made a rather subdued entry into politics at the 1948 Democratic convention. Only a handful of cameras provided live coverage for people who owned sets between Boston and Washington. Even so, television was about to create dramatic changes in American politics. Radio had already laid the foundation for these changes. One of the changes that had developed with radio was that it was a more politically neutral medium than the print medium of the press. Barnard (1924) pointed out, "The listener can form his own opinion for the candidate's utterance before the press or the parties can instruct him." Television continued, to a degree, that aura of neutrality.

Although the days of radio *dominance* of presidential politics were over, radio continues to this day to play an important role in political campaigns for state and local elections. Thousands of elections take place in the United States below the presidential level, and in each election cycle radio serves not only to cover candidates and issues in these races, but also to provide the only affordable and viable broadcast medium for campaign advertising. Even in presidential campaigns and many statewide races, radio remains a viable advertising medium because it provides an avenue for targeting much more specific subgroups of the population than television's more generic audience.

The dichotomy between television and radio grew, and by 1956 television had become a more important source of information than radio. However, radio retained an important role in politics. For example, scholars still argue about the controversial finding from the 1960 Kennedy/Nixon debate that

Kennedy was judged the winner by those who watched the debate on television, whereas Nixon was thought to be the superior debater by those who heard the debate on the radio. This finding fueled a continuing debate of its own—why some candidates apparently are more successful on television while others excel on radio.

Radio also played a major role in several aspects of the 1968 presidential campaign. The candidacy of Senator Eugene McCarthy of Minnesota was strengthened when he devoted a large portion of his budget for the New Hampshire primary to radio (McCarthy's campaign team created some 7,200 spots for 23 New Hampshire radio stations to run within a three-week period). The impact of McCarthy on the 1968 primaries was due in large part to his use of radio in New Hampshire. Richard Nixon also relied on radio in his 1968 campaign: believing the studies that touted his superiority on radio in the 1960 debates, Nixon delivered radio addresses on 14 straight nights leading up to the general election. Not only did Nixon use radio extensively again in his 1972 re-election campaign, but he regularly devoted time to major radio addresses during his presidency.

Not until Ronald Reagan would a president give such attention to radio as a method of communicating with the American public. Unhappy with press representation of his policies, Ronald Reagan initiated regular Saturday afternoon radio broadcasts in order to talk directly with the people. George Bush occasionally delivered radio addresses as well, though less frequently than Reagan. President Bill Clinton returned to Reagan's routine of Saturday radio conferences. The general consensus has been that the effects of these messages are limited and may be of more importance in creating news for other media to disseminate. For instance, even though barely half of all radio stations broadcast Reagan's radio addresses, coverage of the talks by the television networks was extensive, and thus Reagan may have succeeded in putting many of his ideas on the table by using the media to emphasize his own agenda.

Another political radio phenomenon developed in the 1980s and 1990s. Political talk radio erupted in the 1980s with impressive audience demographics. For example, in 1995 Rush Limbaugh attracted nearly 20 million listeners, 92 percent of whom were registered voters, 39 percent of whom had college degrees, and 30 percent of whom had a family income of over $60,000. Industry officials argue that talk radio affects politics and elections by reaching a small target group of active citizens. The decline in the popularity of political talk radio in the late 1990s, however, suggests that the dramatic impact attributed to political talk radio in the 1992 and 1996 elections may not be repeated in the next millennium.

The next development in radio's marriage with politics will undoubtedly evolve from its melding with yet another new medium, the internet. Increasingly, radio stations are finding outlets for their programming through internet broadcast, and political talk radio as well as campaign advertising and airing of political issue positions are all sure to provide increased venues for political impact.

Finally, radio set the stage for a new type of political communication. Radio and television are more intimate in nature, bringing political leaders and candidates into the home, where families watch and listen to the candidate in informal settings. Roosevelt's fireside chats introduced a new model of communication, that is, one leader or candidate sitting in his or her living room speaking with millions of people also sitting in the privacy of their living rooms (Jamieson, 1988). These intimate settings allow a politician to educate, to remind, and in large part to garner support for his or her programs. Radio was the building block upon which politically intimate communication developed and the springboard for the success of television in the political arena.

Although radio may not have revolutionized politics, it did help to change the atmosphere in which the political system operated. Radio may have limited the old-style political oratory and led to the new genres of intimate political address represented well by Ronald Reagan and Bill Clinton. Radio also increased the president's ability to "go public" with issues, forever altering the political dynamics of interaction between the executive and legislative branches.

It is significant that only a few years after the advent of radio in the political arena, one of the greatest radio politicians, Roosevelt, came to the forefront. Very early in the Roosevelt years it was clear that he demonstrated in his low-key fireside chats a mastery of intimate personal delivery. But by 1932, this was altogether natural. Radio was, by then, clearly the way a president spoke to the nation's people.

LYNDA LEE KAID AND TERRY A. ROBERTSON

See also British Radio Journalism; Canadian Radio News; Churchill, Winston S.; Commentators; Controversial Issues; Coughlin, Father Charles; Editorializing; Election Coverage; Equal Time Rule; Fairness Doctrine; Fireside Chats; First Amendment and Radio; Hoover, Herbert; Limbaugh, Rush; News; Roosevelt, Eleanor; Social and Political Movements; Talk Radio; United States Congress and Radio; United States Presidents and Radio

Further Reading

Archer, Gleason Leonard, *History of Radio to 1926*, New York: American Historical Society, 1938; reprint, New York: Arno Press, 1971

Barber, James David, *The Pulse of Politics: Electing Presidents in the Media Age*, New York: Norton, 1980

Barnard, Eunice F., " Radio Politics," *New Republic* 38 (19 March 1924)

Becker, Samuel L., "Presidential Power: The Influence of Broadcasting," *Quarterly Journal of Speech* 67 (1961)
Braden, Waldo W., and Earnest Brandenburg, "Roosevelt's Fireside Chats," *Speech Monographs* 22, no. 5 (November 1955)
Casey, Robert D., "Republican Propaganda in the 1936–1937 Campaign," *Public Opinion Quarterly* 1 (1937)
Chester, Edward W., *Radio, Television, and American Politics,* New York: Sheed and Ward, 1969
"Cordial to Harding, Cold to Speeches," *New York Times* (25 June 1923)
Cornwell, Elmer E., Jr., *Presidential Leadership of Public Opinion,* Bloomington: Indiana University Press, 1965
Dryer, Sherman H., "Air Power," *Colliers* 106 (14 September 1940)
Freidel, Frank Burt, *Franklin D. Roosevelt: The Triumph,* Boston: Little Brown, 1956
Hollander, Barry A., "Political Talk Radio in the '90s: A Panel Study," *Journal of Radio Studies* 6, no. 2 (Autumn 1999)
Jamieson, Kathleen Hall, *Eloquence in an Electronic Age: The Transformation of Political Speechmaking,* New York and Oxford: Oxford University Press, 1988
Kernell, Samuel, *Going Public: New Strategies of Presidential Leadership,* Washington, D.C.: CQ Press, 1986; 3rd edition, 1997
Martin, Howard H., "President Reagan's Return to Radio," *Journalism Quarterly* 61 (1984)
Tulis, Jeffrey K., *The Rhetorical Presidency,* Princeton, New Jersey: Princeton University Press, 1987
Willis, Edgar E., "Radio and Presidential Campaigning," *Central States Speech Journal* 20 (1969)
Wolfe, G. Joseph, "Some Reactions to the Advent of Campaigning by Radio," *Journal of Broadcasting* 13, no. 3 (Summer 1969)

Polk, George 1913–1948

U.S. Radio Correspondent

Columbia Broadcasting System (CBS) radio reporter George Polk was murdered in Greece in May 1948 at the age of 34, and the identity of his murderer was never discovered. It is a great irony that a prestigious award named for him is given annually to journalists whose tenacious investigative reporting demonstrates the importance of a free press, for Polk was discouraged—not only by the government, but by well-respected members of journalism's elite—from the type of in-depth digging that would later garner other reporters the coveted Polk Award. The evidence indicates that, in order to further U.S. Cold War policies, CBS and a committee of Washington journalists were willing to accept the unlikely conclusion that Communists were behind Polk's murder.

Shortly after Polk arrived in Europe in 1945 as a freelancer for the *Los Angeles Daily News* and United Daily Features Syndicate, he met CBS reporter Edward R. Murrow. Impressed by the former Navy pilot, who soon became a respected member of the Middle East press corps, Murrow urged Polk to consider reporting for CBS. Polk was writing for *Newsweek* when CBS hired him in 1946 to be their Cairo correspondent. Within months, the network promoted him to chief Middle East correspondent.

He was tireless in his efforts to report on the Greek civil war that was punishing a population that had already suffered under Nazi occupation during World War II. The United States government was backing the Royalist regime in power. In 1947 President Harry Truman convinced Congress to provide a controversial $300 million aid package to Greece that included funding, military personnel, supplies, equipment, and civilian advisors. In its battle against Communism, America's Cold War policy of supporting the Greek government had come under fire from Polk, who wrote critical stories of "corruption and venality," charging that money was being siphoned from the millions of dollars in American funding. In the days prior to his death, Polk had received information confirming his suspicions of malfeasance and planned to report it on his return to the United States.

Polk's last broadcast report aired from Greece on 6 May 1948. He told a CBS colleague that he was taking a few days of vacation before heading home to the United States and a Nieman fellowship. Eight days later, his body washed ashore in Salonika Bay. He had been shot, execution style, in the back and his hands and feet were bound. Members of the American press were quick to respond, and when the New York Newspa-

George Polk
Courtesy CBS Photo Archive

per Guild set out to conduct an independent investigation of the correspondent's death, a group of nationally known media representatives stepped in.

Political columnist Walter Lippman headed the self-appointed committee looking into the murder. Included in the group were CBS head William Paley, James Reston of the *New York Times,* Ernest Lindley of *Newsweek,* and *Washington Post* publisher Eugene Meyer, among others. Lindley, an editor at *Newsweek,* in later years described the committee's purpose: to do "everything within their power to see that the murderers of George Polk were arrested, brought to trial, and convicted." Lippman chose attorney and former Office of Strategic Services (OSS) head General William "Wild Bill" Donovan to represent the committee. But Donovan's investigator, Lieutenant Colonel James Killis, was recalled by the State Department after he discovered evidence pointing toward the Greek government and away from the Communists.

Who killed George Polk? A number of theories have been advanced in the intervening years. The one accepted without skepticism by the U. S. government and rubber-stamped by the Lippman Committee, despite weaknesses and inconsistencies in the evidence, blames Communist guerrillas who were supposedly hoping the right-wing Greek government would be blamed for Polk's death. Thus, the theory postulates, the American support for the Greek government would diminish.

There is compelling evidence that upper-level Greek officials were trying to prevent Polk from reporting on government corruption. Shortly before his death, Polk met with Greek Foreign Minister Constantine Tsaldaris. Polk told the powerful politician that he knew about U.S. assistance money that had found its way into Tsaldaris' personal bank account. He threatened to "blow this story sky high." Polk had also been reporting on the brutality of the repressive Royalist regime. In his broadcast on the day he disappeared, Polk described the effects of martial law in Greece and reported the more than 200 executions recently conducted by military firing squads.

Another theory suggests that agents from Great Britain, the leading power in the Middle East before being replaced by the United States, murdered Polk to sabotage U.S.-Greek relations. In the mid-1990s, an alternative theory was suggested. In this scenario, Polk was tracking down information about drug smuggling and black marketeering. The criminals supposedly got rid of Polk before he could reveal what he knew about their activities.

During the summer of 1948, as the Greek authorities showed few results from their investigation, Donovan began to pressure for an arrest. In August police brought in journalist Gregory Staktopoulos and tortured him during the next six weeks. Eventually Staktopoulos "confessed" to assisting the Communists, who he said were directed by the Kremlin. He was convicted in a show trial the following April. There were two supposed conspirators who were tried in absentia. Later it was determined that one was dead at the time Polk was killed. Staktopoulos was sentenced to life in prison but the sentence was later reduced and he was released in 1961. He maintained his innocence until his death in 1998. Polk's Greek widow described pressure from officials to sign a document implying that her husband had been killed by a jealous lover. She left the country with a vow never to return.

Donovan's report, described by some as a whitewash, was three years in coming. It appears that his goal throughout the investigation was to deflect attention and embarrassment from the Greek government. Criticism of Lippman and the committee members focuses on their willingness to support the U.S. government's version of events rather than the truth. Iconoclast I.F. Stone appears to be the only media representative at the time to report on the conflict of interest evidenced by the relationship between the committee and the U.S. government. Stone, who called Polk "the first casualty of the Cold War," published a 1952 series of articles challenging the Lippman Committee's report.

The George Polk Awards are presented annually by Long Island University for special journalistic achievement. They go to famous as well as small town reporters.

SANDRA L. ELLIS

See also Murrow, Edward R.

George Polk. Born in Fort Worth, Texas, 17 October 1913. One of five children of George Washington Polk, Sr., and Adelaide Roe; attended Virginia Military Institute three years; graduated University of Alaska, 1938; hired by Shanghai *Evening Post,* 1938; hired by Paris bureau of *Herald Tribune,* 1949; transferred to *Herald Tribune* New York City office, 1940; took leave of absence for graduate study at New York University, 1941; joined U.S. Navy, 1941; assigned as fighter pilot to Marine unit at Henderson Field, Guadalcanal; awarded Purple Heart and presidential citation after being wounded; discharged 1944; assigned to Washington, D.C. bureau of *Herald Tribune*; hired as foreign correspondent for Los Angeles *Daily News,* 1945; feature writer, United Features Syndicate, 1946; named CBS News correspondent, 1946. Died (murder) in Salonika, Greece, 9 May 1948.

Further Reading

Bernhard, Nancy E., *U.S. Television News and Cold War Propaganda, 1947–1960,* Cambridge and New York: Cambridge University Press, 1999

Keeley, Edmund, *The Salonika Bay Murder: Cold War Politics and the Polk Affair,* Princeton, New Jersey: Princeton University Press, 1989

Marton, Kati, *The Polk Conspiracy: Murder and Cover-Up in the Case of CBS News Correspondent George Polk,* New York: Farrar Straus and Giroux, 1990

Unger, Sanford, "The Case of the Inconvenient Correspondent," *Columbia Journalism Review* 29 (November/December 1990)

Vlanton, Elias, and Zak Mettger, *Who Killed George Polk? The Press Covers Up a Death in the Family,* Philadelphia, Pennsylvania: Temple University Press, 1996

Popov, Alexander 1859–1906

Russian Wireless Pioneer

Although claimed by the Soviet Union for many decades as the inventor of radio, Alexander Popov was in reality one of several important early experimenters with wireless apparatus. Popov was an academic interested in research results more than commercial applications. A serious problem for modern researchers is that Popov left few laboratory notes; most of what is known of his work comes from his few published papers and the recollections of contemporaries.

Origins

Popov, born in a small town in a Ural Mountains mining region, grew up surrounded by various applications of technology. These fascinated him from a young age. After two years of elementary schooling, in 1873 he entered the seminary (secondary school) in the city of Perm and studied mathematics, natural science, and theology, and he began to develop his great interest in physics.

In 1877 he went on to university studies of mathematics and physics at St. Petersburg University, where he excelled at both experimental work and building the equipment necessary for such physics research. At the same time he worked with the local power company, which was developing an arc light system for the city. After his graduation in 1882, Popov stayed on as a laboratory assistant, but needing more support for his growing family, Popov accepted a teaching position at the Russian Torpedo School located at the naval base at Kronstadt, on the Gulf of Finland, beginning in 1883. The home of the country's Baltic Fleet, the base offered fine laboratory and library facilities, indeed some of the best in the country. Popov's teaching duties and original research work there focused on applications of electricity on board ships. Because of the pressure of work, however, he published few scientific papers.

In 1893 Popov represented the Torpedo School on a visit to the Chicago World's Fair and the Third International Electrical Congress, where he got a chance to interact with other electrical researchers on visits to companies and laboratories. Such exposure helped him refine his own work at home and sharpened his interest in the application of what were then called Hertzian waves.

Wireless Telegraphy

By 1894 Popov was increasingly focused on wireless transmission and reception. He read of Lodge's improvements in the "coherer" device then used to detect wireless signals and worked to improve on them. His initial application of this technology concerned atmospheric electricity. This was a kind of crude electrical weather forecasting, specifically in the detection of nearby lightning flashes that could be picked up because of their electrical discharge. In a sense Popov found a use for signals that would later be scorned as static noise to be overcome.

On 7 May (April 25 on the old-style calendar then still in use in Russia) 1895, Popov's paper "On the Relation of Metal Powders to Electric Oscillations" provided a demonstration of his lightning detecting (and warning) wireless apparatus to members of the physics department of the Russian Physical and Chemical Society, noting he had achieved success at distances up to 600 yards. A continuing operation of lightning detection was established later that summer with the Institute of Forestry and was soon detecting lightning discharges up to 20 miles distant. A local newspaper report on Popov's demonstration and the potential it represented appeared a few days later. A published description of his apparatus first appeared in the initial 1896 issue of a respected Russian scientific quarterly. Beginning in the 1920s, May 7 was celebrated in the Soviet Union as "radio day."

In a December 1895 note Popov added that he entertained the hope "that when my apparatus is perfected it will be applicable to the transmission of signals to a distance by means of rapid electric vibrations—as soon as a sufficiently powerful generator of these vibrations is discovered." He was among the first to foresee the practical potential of experimental wireless work.

On 24 March 1896 Popov may have conducted a wireless demonstration before the same society, on the campus of St. Petersburg University, where he transmitted the name "Heinrich Hertz." Unfortunately no one present recorded their recollections until some three decades later, by which time the Soviet Union was already touting the inventor as *the* inventor of radio. This demonstration is potentially important, as Marconi only conducted his first public demonstration in Britain in July 1896, four months later. Both Marconi and Popov were among the first experimenters to achieve wireless transmission and detection over a distance of several miles.

Popov recognized the many limitations in his system and sought to improve it (amidst his many other duties and interests). He focused on increasing transmission power but paid little attention to improving his aerials (antennas) or the quality of the signal detectors (receivers) used and thus achieved only limited gains in coverage. Still, by 1897 he was installing wireless apparatus on Russian naval vessels for short-range tests, and within two years he was conducting demonstrations 20 miles at sea. In 1899 Popov traveled to Germany and France to meet with other wireless pioneers, including Adolph Slaby, and compared notes on what each was accomplishing and how. At about the same time, Popov wireless devices were used in the complex rescue of a stranded Russian battleship and to save nearly 50 fishermen who had floated out to sea on an iceberg.

Final Years

While there will always be controversy over who accomplished what and when in early wireless, Popov remains an important pioneer. He helped to establish the first Russian manufacture of wireless equipment (at the Torpedo School), helped to train others, and worked closely with, among others, French engineer Eugene Ducretet, who manufactured equipment to Popov's specifications. Popov earned the Grand Gold Medal for his research at the Paris International Exposition (World's Fair) of 1900. In 1901 he was named professor and in 1905 director of the prestigious Electrotechnical Institute in St. Petersburg.

Popov's early death (he was 46) was due to his generally poor health brought on by a lifetime of overwork, but it was hastened by Czarist government pressures placed on him to discipline institute students with whose political protests he largely agreed. He thus died too soon to witness the growing exploitation of improved wireless systems in a host of different ways. He never claimed to be the inventor of wireless—that came from others long after his death.

CHRISTOPHER H. STERLING

See also Early Wireless; German Wireless Pioneers; Hertz, Heinrich; Lodge, Oliver J.; Marconi, Guglielmo

Alexander Stepanovitch Popov. Born in Turyinskiye Rudniki, Russia, 16 March 1859, the fourth of seven children of a priest. Graduated with a degree in physics from St. Petersburg University, 1882. Appointed an instructor at the Navy Torpedo School, Kronstadt, 1883. Professor (1901) and director (1905) of Electrotechnical Institute, St. Petersburg. Died in St. Petersburg, Russia, 13 January 1906.

Further Reading

Howe, G.W.O., "Alexander S. Popov," *Wireless Engineer* 25 (January 1948)

Ioffe, K., "Popov: Russia's Marconi?" *Electronics World and Wireless World* (July 1992)

Kryzhanovsky, Leonid, and James P. Rybak, "Recognizing Some of the Many Contributors to the Early Development of Wireless Telegraphy," *Popular Electronics* (1993)

Radovskii, M.I., *Alexander Popov: Inventor of Radio,* Moscow: Foreign Language, 1957

Rybak, James P., "Alexander Popov: Russia's Radio Pioneer," *Popular Electronics* (August 1992)

Susskind, Charles, "Popov and the Beginnings of Radiotelegraphy," *Proceedings of the Institute of Radio Engineers* 50 (October 1962)

Portable Radio Stations

In the early 1920s, hundreds of entrepreneurs were bitten by the radio bug and decided to set up their own stations. In cities all over the United States, local businessmen (and several women) put radio studios in their stores, their houses, their garages, or their factories. But some station owners had a different plan: to put a radio station in a truck and drive it to cities that had no station of their own. Such mobile stations were called "portables," and from about 1923 to 1928, they were often invited to county fairs, expositions, and amusement parks. The owner would remove the equipment from the truck and set up an actual broadcasting station on the grounds. Most of the portables were not very powerful—between 10 and 50 watts—but their purpose was to introduce the new technology to people who lived far from the big cities. Portables also served as a good gimmick to get more people to attend a local sales event: attendees could not only shop but also watch a live radio broadcast. One portable, WTAT (later renamed WATT), was owned by the Edison Electric Illuminating Company, which first put it on the air (and on the road) in the summer of 1923. WTAT was usually driven to the hall in which a home show or electronics exhibition was taking place; Edison personnel would first entertain and then demonstrate the wonders of the company's various products.

Another successful portable operated in Rhode Island and throughout New England. Owner Charles Messter was written up in the *Providence Journal* on 7 January 1925: the reporter discussed some of the cities where the station had been and then explained how the portable worked.

> [Mr. Messter's station, WCBR] consists of a 50-watt standard Western Electric transmitter using 600 volts on the plate. He carries storage batteries and a charger so that he will not be caught without power. His three-wire outside antenna is 200 feet long and is usually erected on top of the building in which the outfit is being used . . . The entire outfit can be easily set up and taken down, and this makes practicable its shipment from place to place on short notice.

Perhaps the best-known owner of a portable station operated in the Midwest throughout the mid-1920s. Charles L. Carrell, formerly a theater impresario, operated five portables based in Chicago, and he took them wherever he was hired to broadcast. One of Carrell's portables, WHBM, appeared in East St. Louis in December 1927, having been invited there by the Chamber of Commerce. The station remained for three months of broadcasts, giving many local performers an opportunity to be heard.

Of course, the novelty of portables wore off, but they might have continued their work were it not for the increasing number of stations on the AM band. In November 1925, at the Fourth National Radio Conference, Herbert Hoover, then Secretary of Commerce, stated that the airwaves had become too crowded. He warned that soon, not everyone who wanted to put a new station on the air would be able to do so. This and other problems (such as wave jumping—in which a station operated from a different frequency than the one it had been assigned) would lead to the creation of the Federal Radio Commission (FRC), an agency that had the authority to license and supervise radio broadcasting, in an attempt to bring order to a chaotic situation. The FRC regarded the portable transmitters as part of the chaos. Portables interfered with an already crowded radio spectrum, and the agency decided that eliminating them would be a positive step. It might not solve the problem of crowding—by 1927, there were not that many portables left—but it would at least keep the airwaves free of sudden interference that might be caused when a portable came to town.

The FRC General Order 6 of 26 April 1927 warned that portable broadcasting stations would only be relicensed for a limited period of time—120 days. (Originally, portables tended to operate mainly in the summer, when fairs and outdoor shows were taking place, but some portables had become nearly year-round operations.) It wasn't long before the FRC began to strongly suggest that any portable that wanted to select a specific city of license could become a permanent part of that city, but that licenses to operate portables as portables would not be permitted for much longer. The end came in mid-1928, when the FRC issued General Order 30, officially terminating the portables. Some of the owners, anticipating this, had found homes for themselves and their stations—Charles Messter, for example, joined his friend Harold Dewing in Illinois, where they anchored a portable to the city of Springfield. By 1928 Edison Electric had long since put a full-time station on the air (WEEI) and no longer needed the promotional value of a station in a truck, so WATT was shut down, as were several portables in other cities.

But the person who owned the most portables was the one who didn't want to see them taken off the air. Charles Carrell demanded a hearing from the FRC, and several months later, he went to Washington to plead his case. Unfortunately for him, the FRC seemed to have its mind made up. After reading Carrell's materials and considering his argument, the Commission decided there was just no room for portables any longer. In fact, the commissioners did not mince words: they called the portables "a menace" and went on to say that permitting the portables would not be in the public interest, because the frequencies they chose were usually already occupied by permanent stations, and the closer together on the dial two stations were, the worse their signals would be received. Thus, the renewal of the portable station licenses could not be allowed.

Carrell took his case to the U.S. District Court on appeal, but the court would not overturn the FRC's ruling. Having lost most of his stations, he moved one of the Chicago portables (WBBZ) to Ponca City, Oklahoma, and made it a permanent station; he moved his family there, too. Only four years after losing his legal battle, he died in 1933 at the age of 58. The station in Ponca City still exists, but its early days as a portable are seldom if ever mentioned. In fact, few people realize how innovative portables were and how, for a brief period of time, they delighted radio fans who had never seen a live radio broadcast before.

DONNA L. HALPER

See also Federal Radio Commission; Frequency Allocation; Licensing

Further Reading

"Criticizes Roving Radio: Commission Replies to Appeal from Revocation of License," *New York Times* (27 August 1933)

"Discontinuance of Portable Stations" and "General Order No. 30," in *Second Annual Report of the Federal Radio Commission,* Washington, D.C.: GPO, 1928

"Licenses for Portable Stations," in *First Annual Report of the Federal Radio Commission,* Washington, D.C.: GPO, 1927

"Six Appeal Radio Board Orders," *New York Times* (6 January 1929)

A Prairie Home Companion

Public Radio Comedy Variety Program

A Prairie Home Companion (*PHC*) is one of the most successful programs produced on public radio in the United States. The show received the prestigious George Foster Peabody award in 1980 and its creator and host, Garrison Keillor, is considered an American cultural treasure. He was awarded a National Arts and Humanities medal by the Clinton White House in 1999. Nearly 4 million listeners tune in weekly to more than 500 public radio stations across the United States for the live two-hour broadcast. The show also airs abroad on America One and the Armed Forces Networks in Europe and the Far East.

Often compared to humorist Mark Twain, Keillor writes the script for each week's show. It includes comedy sketches with recurring characters ("The Lives of the Cowboys," "Guy Noir Private Eye"), mock commercials ("Ketchup Advisory Board," "Bebopareebop Rhubarb Pie," and "Café Boeuf" and the occasional competition ("Talent from Towns under 2,000" contest). Musicians from around the globe provide a diverse mix of live folk, jazz, rock and roll, classical, gospel, and ethnic tunes.

A program host on Minnesota Educational Radio in the early 1970s (it became Minnesota Public Radio in September 1974), Keillor was inspired to try an old-fashioned radio variety show back home in Minnesota after a leave of absence to research the *Grand Ole Opry* for a *New Yorker* magazine article. The show first played in a nearly empty auditorium at Macalester College in St. Paul on 6 July 1974. Twelve people (who paid one dollar for adult admission—50 cents for children) made up the audience. The show's popularity slowly grew, and national broadcasts began in 1980.

By 1987, 13 years after the initial performance, 4 million listeners were tuning in to hear Keillor open the show with its signature tune, Hank Snow's "Hello Love." In the same year, however, Keillor announced that *PHC* was coming to an end; he was heading off to Denmark to devote himself to writing. There was a farewell broadcast in June 1987 in St. Paul. One year later, there was a second farewell show from Radio City Music Hall in New York. Keillor told the crowd, "It was so much fun leaving that we're coming back to say goodbye again," to enthusiastic applause. The following year, the cast crisscrossed the United States, performing in 13 cities, for the "Third Annual Farewell Tour."

In 1989 Keillor started a new variety show, *The American Radio Company*, broadcast from the Brooklyn Academy of Music. Eventually more than 200 public radio stations carried the program. In 1993 the show moved to Minnesota and its name was changed back to *A Prairie Home Companion*. The show was still being broadcast at the dawn of the 21st century.

About half the programs are produced in the Fitzgerald Theater in downtown St. Paul. The remainder are broadcast from a tour of cities scattered across the U.S., as well as Europe. In large civic centers and college auditoriums, fans of all ages gather to see the stage set with its worn Oriental rugs, musical instruments, and microphones. Behind it all there's a clapboard house-front with a light in the upstairs window and several lucky audience members sitting on the front porch.

Dressed in his signature black suit, white shirt, and red tie, Keillor sings tunes he has written during the previous week, accompanied by the house musicians, the Guys' All-Star Shoe Band led by pianist Rich Dworsky. Keillor plugs the fictitious sponsor of the show, Powdermilk Biscuits, "with that whole-wheat goodness that gives shy persons the strength to get up and do what needs to be done."

The audience waits in anticipation as the ensemble cast, actors Tim Russell and Sue Scott, and sound effects wizards Tom Keith and Fred Newman, step up to their microphones. Russell, a master of impersonation, brings roars of laughter as he converses with Keillor in the voices of Presidents Bill Clinton or George W. Bush, Julia Child, Bob Dylan, Ted Koppel, and many other celebrated personalities. Keith is famous for his ability to produce sound effects with his voice (the sounds of animals, automobiles, motorcycles, missiles, helicopters, and explosions) and a variety of props (including a box of gravel, cellophane, and a miniature door).

The musical segments are eclectic and could include a gospel group, a rhythm and blues singer, or a classical pianist. Rockabilly band Jack Knife and the Sharps, guitarist Leo Kottke, mandolin player Peter Ostroushko and singers Suzy Bogguss and Iris DeMent have performed for *PHC* audiences. Special guests may include writers, actors, comedians, or poets (humorists Al Franken and Roy Blount, Jr., authors Studs Terkel and Frank McCourt, U.S. Poet Laureate Billy Collins, comedienne Paula Poundstone, and actress Sarah Jessica Parker have made appearances).

Halfway through the show Keillor reads messages scribbled by members of the audience on small pieces of paper. Birthday wishes, parental advice, and words of affection go out to friends and family across the country. Throughout the show the audience responds with delighted applause to the mix of songs, comedy routines, brief interviews with guests, and commercials for old familiar products and services.

Near the end of *PHC*, 15 to 25 minutes are reserved for Keillor's weekly monologue. His opening words, "It's been a quiet week in Lake Wobegon," are greeted with a surge of applause. In Keillor's melodious baritone, the latest tales about the imaginary town's residents (including the Tolleruds, Krebsbachs, and Pastor Ingqvist) lull the audience into a sense of community in the darkened theater. The words flow steadily until finally it all comes to a tidy close, "And that's the news from Lake Wobegon, where all the women are strong, all the men are good-looking, and all the children are above average."

A Prairie Home Companion stands at the end of a tradition that stretches back to the earliest days of radio. Its predecessors include such variety programs as *The Eveready Hour, The Rudy Vallee Show* (also called *The Fleischmann Hour*), *The Maxwell House Show Boat* and *The Collier Hour.* With the arrival of television in the 1940s, variety shows disappeared or shifted to the new medium. The old-fashioned variety owes its survival almost completely to the appearance in the 1970s of *A Prairie Home Companion.*

SANDRA L. ELLIS

See also Comedy; Keillor, Garrison; Kling, William; Minnesota Public Radio; Public Radio International

Cast

Host	Garrison Keillor
Actors	Tim Russell, Sue Scott
Sound Effects	Tom Keith, Fred Newman

Producer/Creator

Garrison Keillor

Programming History

Minnesota Public Radio	1974–80
American Public Radio/ Public Radio International	1980–present

Further Reading

Barol, Bill, "A Shy Person Says So Long," *Newsweek* (15 June 1987)

Larson, Charles, and Christine Oravec, "A Prairie Home Companion and the Fabrication of Community," *Critical Studies in Mass Communication* 4 (September 1987)

Lee, Judith Yaross, *Garrison Keillor: A Voice of America,* Jackson: University Press of Mississippi, 1991

Nelson, Michael, "Church on Saturday Night: Garrison Keillor's *Prairie Home Companion,*" Virginia Quarterly Review 77 (Winter 2001)

Prairie Home Companion website, <www.prairiehome.org>

Scholl, Peter A., "Garrison Keillor and the News from Lake Wobegon," *Studies in American Humor* (Winter 1985–86)

Scholl, Peter A., *Garrison Keillor,* New York: Twayne, and Toronto: Macmillan Canada, 1993

Selix, Casey, "At Home on the Prairie," *Houston Chronicle* (1 July 1999)

Premiums

Toys and Gifts Offered over the Air

The radio premium became a significant device for measuring listeners' product and program loyalties, and it confirmed sponsors' identifications with admired personalities and attractive fictional characters. Rushing strongly through the 1940s, the avalanche of radio premiums—recipe or inspirational booklets, club badges and membership manuals, costume jewelry, character rings and pins, whistles, and other paraphernalia—finally abated with the advent of television.

In radio's early years, premiums were introduced obliquely in unsponsored "talks." Initially forbidden to describe specific products on the air, many companies offered staff representatives to discuss consumer-interest subjects, and afterward the program host might suggest that the speaker's employer had *permitted* him to offer a token of appreciation for the listener's interest—a 1923 recipe booklet from a Chicago meatpacker, for instance. Thus, through the Trojan horse of ostensibly objective information giving, companies could gain listeners' gratitude by dropping logo-marked "gifts" into their mailboxes.

In the 1930s, the postman's premium-bearing burden increased. Premiums were compellingly described in often-lengthy commercials, and they were sometimes integrated into program content. "Missing a commercial proved almost as much a disappointment as missing a moment of the action itself," recalls a veteran children's serial listener; "I enjoyed sending for the advertised products, especially those, like the decoder ring, which became part of the story." To get the required labels or panels, children spooned through boxes of breakfast food, and their mothers baked and fried their way through packages of flour and shortening. Each week thousands of labels, dimes, and postcards reached premium fulfillment addresses in Chicago and St. Louis, and the listener's eager wait began for the mail carrier to bring the Jack Armstrong Hike-o-Meter or the brooch "just like the one our heroine wears in today's episode."

The premium's role in building audience loyalty may be seen in a handwritten note reproduced inside the front cover of Standard Brands' 1938 souvenir script *One Man's Family Looks at Life,* where Paul Barbour, the radio clan's philosopher-son, says, "As you know, this book comes to you not only from *One Man's Family,* but also from the makers of Tender Leaf Tea. When you think of one, think of the other. For Tender Leaf Tea makes it possible for *One Man's Family* to meet *your* family over the radio." Five years earlier, the pioneering serial *Clara, Lu, and Em* had courted such consumer loyalty by offering a 1933 Chicago World's Fair spoon for a Super Suds box top. For a time *David Harum,* the story of a small-town banker and horse trader, gave away a horse each week; later, when the protagonist took up photography, the program promised a working camera for a quarter and a Bab-O cleanser label. Pepsodent toothpaste's sponsorship of *Amos 'n' Andy* produced such premiums as sheet music, scripts of key episodes, and maps of the characters' adopted hometown.

General interest programs of the 1930s also devised apt premiums. *Captain Tim Healy's Stamp Club of the Air* explored the historical or geographical backgrounds of postal designs and would send a starter stamp collection for an Ivory soap wrapper. In an age fascinated by aviation, a shoe store chain's program *Friendly Five Footnotes* encouraged listeners to pick up copies of the booklet *It's Easy to Learn to Fly* at the local outlet. *The Court of Missing Heirs,* dramatizing stories of unclaimed fortunes, published a bulletin listing such cases, and *The University of Chicago Round Table* furnished transcripts of its radio discussions.

Many children's adventure serial premiums were tied to annual memberships in clubs and secret societies, and announcers energetically persuaded the listener to be "the first on your block" to obtain "your very own" message decoder badge, club manual, glow-in-the-dark ring, or other "swell" object for play and display. Don Gordon tutored children to claim their free *Captain Midnight* Flight Patrol membership cards "on the spot" when their parents next filled the gas tank at the Skelly service station, and Ovaltine spokesman Pierre Andre elicited many a "thin dime" for send-away *Orphan Annie* and later *Captain Midnight* premiums. *Jack Armstrong,* Wheaties' serial of a high school athlete turned adventurer, featured devices useful in hiking and camping. In *Tom Mix,* the cowboy hero was able to escape being tied up by nudging a magnifying glass into position so that the sun would burn into the rope, and soon requests for the magnifying glass premium filled Ralston-Purina's redemption offices. Quaker Oats offered each *Sergeant Preston of the Yukon* listener a certificate representing "actual" ownership of a square inch of Klondike land; the property was later forfeited for nonpayment of taxes, but the deed certificate continues to rise in value as an artifact of 1940s childhood.

In the 1950s, radio premiums faded with the single-sponsor programs that had offered them. Some radio shows attempted TV adaptations or simulcasts and continued to feature premiums for a time, but the success of the radio premium, like that of the host program, had depended on the enlarging power of the imagination. The camera would show too clearly that the giveaway periscope was a flimsy thing of plastic or heavy paper and that the soap opera premium jewelry gave off a glassy glare in the black-and-white TV picture. Television would develop its own lures, and the radio premium was put away in closets and memories.

RAY BARFIELD

See also Nostalgia Radio; Promotion on Radio

Further Reading

Hake, Theodore L., *Overstreet Presents Hake's Price Guide to Character Toy Premiums: Including Comic, Cereal, TV, Movies, Radio, and Related Store-Bought Items,* Timonium, Maryland: Gemstone, 1996; 2nd edition, York, Pennsylvania: Gemstone, 1998

Harmon, Jim, *Radio and TV Premiums: A Guide to the History and Value of Radio and TV Premiums,* Iola, Wisconsin: Krause, 1997

Heide, Robert, and John Gilman, *Dime-Store Dream Parade: Popular Culture, 1925–1955,* New York: Dutton, 1979

Marchand, Roland, *Advertising the American Dream: Making Way for Modernity, 1920–1940,* Berkeley: University of California Press, 1985

Stedman, Raymond William, *The Serials: Suspense and Drama by Installment,* Norman: University of Oklahoma Press, 1971; 2nd edition, 1977

Tumbusch, Tom, *Tomart's Price Guide to Radio Premium and Cereal Box Collectibles, Including Comic Character, Pulp Hero, TV, and Other Premiums,* Dayton, Ohio: Tomart, 1991

Press-Radio War

Newspapers' Attempt to Stifle Radio News

The Press-Radio War proved to be an early example of the new medium of radio broadcasting competing against established newspapers to define roles and control the flow of information to the public. The radio industry emerged from this so-called war as a formidable medium that could not be restrained by newspaper publishers.

Origins of the "War"

Radio stations broadcast virtually no news in the early days of the medium. Most of what could be considered news broadcasts were actually commentaries delivered perhaps no more than once a week by broadcast pioneers such as H.V. Kaltenborn and Frederic Wile. The Columbia Broadcasting System (CBS) radio network broadcast its first regular daily news summary beginning in early 1929. By that time, the nation's radio listeners had grown to appreciate this broadcast medium's ability to inform them in a timely manner. News broadcasts of election results and the sensational kidnapping of the Lindbergh baby in 1932 had whetted the public appetite for information delivered via radio. An estimated 63 million listeners tuned in to radio broadcasts the day Herbert Hoover was inaugurated in 1929.

The newspaper industry became alarmed at what was clearly becoming a threat to print. Radio threatened to take away the "breaking news" role of newspapers. The newspaper "extra" was becoming a thing of the past by the early 1930s. Perhaps more important, newspaper executives feared that the growth of radio news would continue an erosion of advertising revenue from print to broadcast. The newspaper industry was ready to wage "war" against radio.

A committee of the American Newspaper Publishers Association (ANPA) brought two key recommendations to the 1933 ANPA convention. One was for the wire services to stop supplying news material to radio. The other was for newspapers to publish radio program listings only if the radio stations paid for the space as advertising. The wire services (Associated Press [AP], United Press [UP], and International News Service [INS]) were largely controlled by newspaper publishers. Pressure from the publishers led the wire services to stop providing news to radio broadcasters. The newspaper industry mistakenly believed that the removal of wire service access from the radio industry would force broadcasters out of the news business.

The radio industry fought back, largely through the efforts of CBS. CBS president William Paley directed Paul White to establish the Columbia News Service, a news organization that could supply the news needs of CBS network without the help of the wire services. White, a former UP executive, quickly established news bureaus for CBS in major U.S. cities and arranged for part-time stringers (temporary, on-call reporters) in other news centers. The Columbia News Service bought international news reports from foreign news agencies around the world. CBS network continued its broadcasting of news, further angering the newspaper industry. The National Broadcasting Company (NBC) also continued broadcasting the news, largely through the reporting efforts of Abe Schechter, but on a more limited scale than CBS. The newspaper industry retaliated by dropping CBS program listings from many newspapers around the nation.

It was clear that neither broadcasters nor newspaper publishers were happy with how this war was developing. Representatives of Publishers' National Radio Committee, the wire services, and CBS and NBC met in December 1933 in New York City to discuss ways in which they might work out their differences. The two-day meeting, held at the Hotel Biltmore, resulted in a list of understandings that became known as the Biltmore Agreement.

The agreement called for the establishment of a Press-Radio Bureau that would provide news from the wire services twice a day for unsponsored five-minute newscasts on the radio networks. The morning newscasts, however, could only be broadcast after 9:30 A.M. so as to protect sales of morning newspapers, and evening newscasts were to be broadcast after 9:00 P.M. to protect evening newspaper sales. CBS agreed to break up its own news service, and NBC agreed not to begin one. News bulletins of "transcendental" significance would be provided by the Press-Radio Bureau in a timely fashion between newscasts when circumstances dictated.

The terms of the agreement clearly favored the newspaper industry. The radio industry suffered little, however, because the agreement quickly began to unravel. Several factors led to the quick failure of the Biltmore Agreement. First, independent radio stations and even network affiliates not owned by the networks were not included and thus did not feel compelled to adhere to the terms of the agreement. They scheduled newscasts whenever they chose, with whatever information they could put together. Next, news material was provided to radio stations from several new news-gathering services, including the Transradio Press, which had jumped in to provide information to broadcasters at the time when wire services refused to provide news to radio. These new services were referred to in the newspaper industry as "outlaw" press associations, but there was clearly nothing in the Biltmore Agreement that could restrict them. The Press-Radio Bureau then began sending the networks more and more news updates during the day under the agreement's provision that allowed for timely release of "bulletin" information. Finally, in early 1935, UP and INS announced that they would renew selling news service to radio broadcasters as a way to maintain a competitive position against Transradio and the other "outlaws." AP soon followed. In a little over one year's time, the agreement had fallen apart. The Press-Radio Bureau executive committee would not meet again after May 1935.

The Biltmore Agreement failed owing to both practical and philosophical weaknesses. Practically, the power of radio broadcasting was beyond the point at which arbitrary restrictions imposed by the newspaper industry could be effective. From a philosophic standpoint, the agreement was clearly a narrow-sighted effort to stifle information flow in a democracy. The effort to suppress news in a legitimate, although relatively new, channel of expression was a violation of the principles of press freedom in the United States. Had the effort to restrict news from the radio airwaves been more successful, the clear losers in the matter would have been America's consumers of radio broadcasting.

JEFFREY M. MCCALL

See also News; News Agencies; Schechter, A.A.; White, Paul W.

Further Reading

"The Biltmore Agreement," in *Documents of American Broadcasting,* edited by Frank Kahn, New York: Appleton-Century-Crofts, 1968; 4th edition, Englewood Cliffs, New Jersey: Prentice Hall, 1984
Bliss, Edward, Jr., *Now the News: The Story of Broadcast Journalism,* New York: Columbia University Press, 1991
Chester, Giraud, "The Press-Radio War: 1933–1935," *Public Opinion Quarterly* 13 (Summer 1949)
Jackaway, Gwenyth L., *Media at War: Radio's Challenge to the Newspapers, 1924–1939,* Westport, Connecticut: Praeger, 1995
Lott, George E., Jr., "The Press-Radio War of the 1930s," *Journal of Broadcasting* 14 (Summer 1970)
Paley, William S., *As It Happened: A Memoir,* Garden City, New York: Doubleday, 1979
White, Paul W., *News on the Air,* New York: Harcourt Brace, 1947

Prizes. *See* Awards and Prizes

Production for Radio

Creating Radio Programs

Production is an important, if not the most important, function at a radio station. Without production there is no sound from the radio speaker. As used in radio, the term *production* refers to the assembly of various sources of sound to achieve a purpose related to radio programming. Production is the intermediate step that translates ideas into audible content. The preceding steps are planning and writing, and transmission is the last step to deliver the program to the listener.

The production process in radio has changed significantly since radio first became a broadcast medium. All program production was live in the early days: actors, musicians, and announcers gathered around the microphone at the scheduled air time and created the radio show, commercials and all. However, turntables, records, and recorders soon gave producers the ability to reproduce and enhance production efforts. Today, computers and other digital equipment play major roles in the creation of production for radio.

Early Radio Production

Production for radio during the time of the experimental broadcasts before 1920 was rudimentary: the radio equipment operator spoke into the microphone himself. As the technology of radio improved during the 1920s, interest in radio grew. By the late 1920s, as radio programming had become more complex, the production of those programs required more people and equipment.

At the National Broadcasting Company (NBC) Red network, the person responsible for the complete supervision of a program, including conducting rehearsals, was called the "production director." At NBC Blue, the Columbia Broadcasting System (CBS), and Mutual, he was called the "producer" or "production man." In a 1944 radio production text, Albert R. Crews defined the production director as "a painter who uses a loudspeaker for his canvas; actors, speakers, music, sound effects for his colors; and a mixing panel for his palette. He must consider himself a conductor as well as a partner-creator of a symphony in sound." From the time the completed script was delivered until the program aired, the production director was the final authority on all matters relating to the broadcast. The production director was responsible for devising the best arrangement of musicians, vocalists, speakers, actors, and sound effects technicians in order to create the program.

Radio programming of that era could be grouped into two major categories: spots and programs. Spots varied from ten seconds to five minutes and included commercial announcements, news broadcasts, and weather reports. Programs featuring speech (serials and dramas, speeches, instruction, news commentary, audience participation, sports, and religion), music/song, and novelty/variety ran up to one hour.

Most of these programs were created live in the studio. The typical studio was a large room, usually acoustically isolated in some way from the rest of the building, hung with heavy drapes or acoustical wall treatments. Here the actors, musicians, announcers, and sound effects people were arranged around one or more microphones.

Many programs were broadcast live from remote locations. Stations used what were called "pick-up locations" connected with telephone lines. Theaters, churches, baseball fields, hotels, and dance halls were some of the locations from which radio stations originated live productions of special events or regular programs.

Many of the programs heard on radio throughout these early decades of radio were uncomplicated from a production viewpoint, especially those created at smaller independent stations. One or two people were heard talking or reading, occasionally with musical accompaniment. More elaborate programs were produced by larger stations and the networks. Sponsor messages became elaborate mini-productions within a program. Humor, melodrama, and jingle singers helped create memorable messages. Musicians, singers, actors, and complex live sound effects all contributed to the production of these live commercials.

Drama programs made regular and frequent use of one of the most fascinating aspects of live radio production, the assembly of sound effects to create the illusions that were so important in making radio theater of the mind. Often, a sound effect suggested time and location or created exposition in these dramas. Coconut shells "stomped" in a tray of sand and a few stones created the sound of horses. Crash boxes filled with broken glass were ready to create the sound of glass breaking. The manipulation of uncut broom corn created the sound illusion of walking through brush. Windows and doors mounted in portable frames, a splash tank, and a walking platform were just some of the various mechanical devices used to create realistic, and sometimes unrealistic, sound effects.

From Live to Recorded Production

As productions became more sophisticated, producers also began to use sound effects recorded on discs. These were ten-inch, double-sided discs revolving at 78 rpm and were used to produce effects that could not be created realistically in a studio, such as train and airplane sounds, machinery of various sorts, and sounds of warfare. Often, effects were used in combination, played at variable speeds on multiple turntables that were specifically designed for sound effect reproduction. Changing the speed on an effect often created a realistic sound of something other than the original. For many programs, the sound effects staff and their equipment were the most important part of the production team assembled for a broadcast.

Even before the introduction of the tape recorder, transcription discs were used to record programs for rebroadcast, archiving, or distribution to other stations, but they were also used to create libraries of music, sound effects, and commercials. Thanks to transcription discs, many popular radio programs and commercials from the 1920s through the 1940s have been preserved. A 16-inch transcription disc, revolving at 33 1/3 rpm, could hold 15 minutes of program material. By the late 1940s, CBS had introduced the vinyl long-playing record, the 33 1/3 LP. Although radio also adopted this format as a means of recording and distribution, nothing would match the tape recorder for production purposes.

The Germans developed the magnetic tape recorder during World War II. The "Magnetophon" design was brought back to the United States at the end of the war. In 1946 singer Bing Crosby's program *Philco Radio Time* was the first program to make use of the tape recorder to record and edit the program in advance of airing. The Ampex Corporation produced its first recorder in 1948, and broadcasters quickly began to use the stationary tape recorder in production. Now producers were released from the bonds of time and place. Networks used recorders to delay programs for the different time zones. Advertising agencies were able to put produced commercials on disc and tape, allowing for continuous re-airing by stations and networks. Audiotape also introduced the ability to edit program content, correcting mistakes or making changes by cutting out or inserting additional tape. In 1959 the endless-loop tape cartridge recorder was introduced to radio. This led to a major improvement in the way commercials were aired on radio. A continuous loop of audio tape housed in a cartridge allowed an individual commercial to be selected by the engineer, inserted into the player, and played back immediately without cueing. Broadcast "carts," reel-to-reel recorders, turntables, and microphones were the basic pieces of equipment used in almost every radio station through the early 1990s.

How Production Works: A Hypothetical Case Study

To illustrate the various components of modern radio production and how they interrelate, the first part of a typical broadcast morning at a hypothetical radio station using a digital recording and playback system is described. All of the activities in the following descriptions represent production tasks, either live or recorded.

The morning announcer turns on a computer monitor and loads a playlist containing most of the day's commercials and messages into the computer-based digital audio delivery system, which was installed a year ago to replace the station's aging cart machines and reel-to-reel recorders. The computer used by the announcer in the radio control room, along with those in the production studio, the newsroom, the music director's office, and the traffic director's office, are all connected to a fileserver, so that as soon as a recording is created at one location, it is available for playback anywhere on the network.

The announcer begins the broadcast day by playing a sign-on announcement (although increasingly, especially in larger markets, stations remain on the air at all times). This daily message was recorded in the station's production studio a month ago by the production director after the new ownership of the station was approved. After sign-on, the announcer turns on the network feed potentiometer or "pot" in order to air a network newscast, which is coming to the station via satellite. After the newscast, a 60-second commercial for a local furniture store is aired from the digital audio system; this commercial was produced by the afternoon announcer last week. The commercial uses a track from the station's music production library and the announcer's voice, which has been run through a microphone patched through a microphone processor in order to make the voice sound more powerful.

Following this commercial, the announcer turns on the newsroom microphone, and the news director begins the first of several live morning newscasts. The first two stories each require the announcer to insert an actuality or sound bite at the appropriate place in the story, using the minidisc player mounted just above the compact disc (CD) players. These actualities were extracted from an interview the news director recorded over the telephone yesterday afternoon, edited on the newsroom computer, and transferred to minidisc. The newscast concludes with the weather, which the announcer plays from a broadcast cart recorded over the telephone earlier this morning. After the weather, the announcer plays a station jingle, a short recorded musical promotion for the station. This jingle was recorded as part of an image campaign package recorded by a Dallas production company specifically for this station about three months ago. The announcer quickly looks at the computer monitor to make sure that the music playlist for the day is loaded on the other side of the screen and starts the first song of the morning. This selection, along with six other songs, was recorded into the digital audio delivery system yesterday by the music director.

After a second piece of music, the announcer finally opens her own microphone to say good morning. While talking, she adds the sound of a bugle to punctuate her remarks. The bugle, like many other standard sound effects, comes from a library of digitally recorded sounds licensed for broadcast use and is recorded, along with about 20 other sound effects, directly on a stand-alone hard disk recorder. Each sound effect has its own selector button. The first traffic report of the day is scheduled next. The announcer checks the console to make sure that the remote feed is in cue. Right on schedule, the line comes alive. Today the traffic reporter is downtown at the scene of an accident and is using a mobile transmitter. After a short traffic report, the announcer performs the well-practiced routine of reading advertising copy for the sponsor of the traffic report over the traffic music theme played back from the computer. At the end of the theme music, two more recorded commercials are triggered by the computer to start automatically. These came preproduced from the clients' advertising agencies. The first spot was produced at a Los Angeles production house and features four different actor voices and customized music. The second spot was produced for the client through a Chicago advertising agency. The recording itself was made in Chicago, while the voice talent was in New York, connected to the Chicago recording studio by a digital telephone line. The first spot was mailed to the station, and the second was sent as a sound file over the internet. An intern dubbed the first spot using a reel-to-reel recorder in the production studio and placed the second spot in the playlist by using the cut/paste function of the computer.

After several more music cuts, the announcer plays the new morning show contest open. Over the past several days, she has worked in the production studio recording this rather complex opening for the contest using a multitrack digital editor. She used six different segments from previous contest winners, a music bed, sound effects, and several electronic production elements to create the background. Then she voiced her copy through the digital effects processor to completely change the sound of her voice. After the open plays, she reads the trivia question and asks for the sixth caller. As music plays, she answers the phone until she gets to the sixth caller. After receiving permission to record the call, she feeds the phone input of the console into audition (so that it doesn't go out over the air) and records the contest winner onto the hard disk recorder. With just enough time to electronically edit out the beginning of the phone call, she adjusts the edit markers on the waveform editing screen, presses the cut button, and saves her work. The song ends, and she plays back the contest winner recording on the air, adjusting the output level on the audio console. On to the next commercial break.

This hypothetical excerpt of daily activity in a radio control room effectively illustrates the fact that production is a multilevel activity necessary to create a radio program. Segments of the assembled program might have been produced at different places, different times, and using different equipment. Or the production of a segment might be happening live, concurrent with the program's airing. In some respects, digital recording simplifies the production process. Digital recording into a networked system allows instantaneous delivery of the completed spot, program, or sound element throughout the system. However, there is always the potential for computer or other equipment failure. Radio stations that depend on computer-based audio systems usually implement a redundant backup approach as part of the system or retain some analog equipment for emergency use.

Production Personnel

Most stations have at least one person whose primary function is to record and manage the station's production of commercials, promotional announcements, public service messages, identifiers, and the many other sound elements used hourly and daily on the air. The person primarily responsible for this work is the production director. There are more people involved in a station's production than just the production director, however. As the previous illustration suggests, a large number of people, both outside and inside the station, have responsibilities that are related to production: a reporter editing an interview to create an excerpt to be used in an upcoming newscast; an on-air announcer reading the weather while mixing in a music bed underneath; an account executive dubbing some new spots supplied by a client's advertising agency; a station's disc jockeys on the air live; jingle singers in Dallas recording a new set of jingles for a station in Detroit; an engineer recording or feeding a live airing of an orchestra concert syndicated via satellite; and a sports producer mixing multiple announcer and field microphones with recorded features to create a live baseball broadcast—all are involved in production.

The Recording Process

Since the 1940s recording has become one of the most important parts of the production process. Radio today depends on quality recordings to create the bulk of the program schedule. The recording and production process actually starts outside the radio station for much of the program content. The station modifies most of this material very little. Most music aired by stations today comes to the station already recorded in some format: CD, hard disk, or digitally via the internet. Some commercials come to the station prerecorded; these elements are simply dubbed or rerecorded to an appropriate format for use in the on-air playback equipment.

Much of the material heard on a radio station is, however, recorded in its own production studio. The recording process can be illustrated by following the recording of a typical commercial: (1) a copy writer at an agency or the station writes commercial copy; (2) the producer working for the agency or at the local station reviews the copy and selects appropriate music and sound effects for the spot if not already specified; (3a) the spot can be recorded in real time by mixing all three elements (voice, music bed, sound effects) at the console and routing it to a recorder (reel, cart, or computer); (3b) alternatively, the spot can be mixed as a multitrack production in three successive recordings, recording the music at full volume on track one, voice on track two, and sound effects on track three. The levels can be adjusted during final mixdown and recording to the format to be used for on-air playback. If the copy changes, the producer can go back to the master multitrack recording and either rerecord the entire voice track or edit, cutting and pasting the changes from another audio file.

As automation becomes more prevalent and refined in radio, the recording process becomes even more central to the production function. Precision recordings, timed perfectly and recorded digitally, allow customized voice tracking by announcers from remote locations. An announcer working virtually anywhere in the country can function as a shift announcer for multiple stations anywhere, providing individualized current information for each separate station, all during the same shift. Digital distribution, digital recording, and digital automation create the illusion that the announcer is physically present at the station. Radio has truly become a virtual medium.

The Editing Process

A big part of production for radio is the editing of radio program material. Editing of recorded audio material for radio is undertaken for one or more of the following three reasons: (1) to correct mistakes; (2) to shorten or lengthen the running time of an element; or (3) to creatively enhance or change the content. In the days of analog recording, the quickest way to add, delete, or reorder material in a recording was to splice the tape, physically cutting the tape with a razor blade or scissors; then removing, adding, or replacing tape; and finally rejoining the segments with splicing tape. Done well, a splice edit is imperceptible. The editor finds the beginning point of the edit, marking the tape over the playback head on the recorder. The tape is then advanced until the end point is found and marked over the playback head. Using a splicing block, an angled cut is made over each mark, followed by the insertion of a similarly marked and cut audiotape segment or the joining of the two ends of tape. Assured that there is no gap or overlap between the two segments of tape, the editor places a piece of specially formulated splicing tape over the splice and then closely trims the excess splicing tape.

The splice block has been all but replaced by the digital editor. There are many brands, types, and approaches to digital editing. Some units are stand-alone units, a combination of hardware and software. Others are software packages for a computer with sound card. Some are basic two-channel editors; others are multitrack recorder/editors that allow almost unlimited additions of sound layers with no degradation of audio quality. What they all have in common is the ability to allow the editor to visually and audibly determine precise edit points, cut and paste audio from one file to another or to the same file, and perform non-destructive modification of the original audio. Much of the recording and editing work that takes place in radio production studios is focused on commercials.

Producing Creative Commercials

Producing the radio station's commercials is a creative challenge. The production for each spot has to accurately interpret the details of the written copy, capture the mood of the spot as intended by the writer, attract attention, sound different from all the other spots running on the station, and yet be consistent with the overall format and sound of the radio station.

There are many ways to meet the challenges of producing compelling commercials. Turning the process over to an advertising agency is one way. As reviewed earlier, ad agencies were once in almost total control of the radio networks' program and commercial production. Although agencies no longer exercise such a stranglehold on the programming decisions of local stations and networks, much of the advertising content heard on radio today is produced through the efforts of ad agencies representing clients, especially at the network and syndicated program level. These larger clients can afford to pay for the creative writing, production, and celebrity talent used to create memorable advertising. Stations merely schedule and dub or pass through these commercials from the network or syndicator.

Networks are not the only place to find creative spots. There are many techniques and resources available to producers at local stations as well. Radio stations still use music and sound effects production libraries to enhance production. First available on 78-rpm records, then on vinyl LPs, these libraries are now digitally recorded on CD. The production music libraries offer precisely timed versions of instrumental music beds that can be licensed and used as the backgrounds for commercials and other announcements. These music tracks are usually recorded in 10-, 15-, 30-, and 60-second lengths of the same theme. These creatively titled compositions are available in a variety of music styles, tempos, and instrumentation. Most of the music production libraries are buyout libraries: the station pays a flat fee for the right to use the entire library indefinitely. Some libraries use "needle drop" fees (the term goes back to the days of turntables and discs): the station pays a fee for use of a specific music track for a specified length of time for a specific commercial. Some companies specialize in the composition and recording of customized, personalized jingles or music beds for station clients.

Sound effects are usually sold as buyout libraries containing a comprehensive array of digitally recorded sounds of every possible situation, activity, or device. Babies crying, rocket launches, train whistles, wind, rain, and a computer modem are examples of sounds that have been digitally recorded for inclusion in a sound effects library. Continuing a tradition from the early days of radio, the use of sound effects adds realism and interest to radio content.

The Role of Production in Creating a Station's "Sound"

Since the advent of television, radio has gone from being a general interest entertainment medium to programming using a specialized, formatted approach. Radio stations rely on a "format," or the creation of a consistent mix of programming elements, to attract and maintain a target audience. The station can then maximize its listenership within specific demographic characteristics, carving out a specific niche among all the stations in a competitive market. All the format elements are carefully selected and positioned in order to maximize and maintain listeners. Most formats are music based, but some formats are based on music alternatives, such as talk, sports, and all-news radio.

The basic components of format are often the same from station to station. Two competing stations could be programming exactly the same music lists. The differences listeners would hear in the stations would stem from the different approaches taken with production and related elements like promotion. Beyond format, production is the key element influencing the sound of a radio station. The music beds used in the commercials, the promotional announcements for the station itself, the station identifiers, the jingles, the voices on the air, and how they deliver content are just some of the many production elements that contribute to the overall sound of the station. Production is what ties the different programming elements together and makes the whole package seamless. Listeners are not necessarily aware of good production, but they certainly notice the lack of it.

JEFFREY D. HARMAN

See also Audio Processing; Audiotape; Automation in Radio; Control Board/Audio Mixer; Recording and Studio Equipment; Sound Effects

Further Reading

Adams, Michael H., and Kimberly K. Massey, *Introduction to Radio: Production and Programming,* Madison, Wisconsin: Brown and Benchmark, 1995
Alten, Stanley R., *Audio in Media,* Belmont, California: Wadsworth, 1981; 6th edition, Belmont, California, and London: Wadsworth, 2002
Carlile, John Snyder, *Production and Direction of Radio Programs,* New York: Prentice-Hall, 1939
Crews, Albert R., *Radio Production Directing,* Boston: Houghton Mifflin, 1944
Keith, Michael C., *Radio Production: Art and Science,* Boston and London: Focal Press, 1990
Keith, Michael C., and Joseph M. Krause, *The Radio Station,* Boston: Focal Press, 1986; 4th edition, by Keith, Boston and London: Focal Press, 1997
MacDonald, J. Fred, *Don't Touch That Dial! Radio Programming in American Life, 1920–1960,* Chicago: Nelson-Hall, 1979
McLeish, Robert, *The Technique of Radio Production: A Manual for Local Broadcasters,* London: Focal Press, and New York: Focal/Hastings House, 1978; 4th edition, as *Radio Production: A Manual for Broadcasters,* Oxford and Boston: Focal Press, 1999
O'Donnell, Lewis B., Philip Benoit, and Carl Hausman, *Modern Radio Production,* Belmont, California: Wadsworth, 1986; 5th edition, 2000
Oringel, Robert S., *Audio Control Handbook: For Radio and Television Broadcasting,* New York: Hastings House, 1956; 6th edition, Boston and London: Focal Press, 1989
Reese, David E., and Lynne S. Gross, *Radio Production Worktext: Studio and Equipment,* Boston and London: Focal Press, 1990; 3rd edition, Boston and Oxford: Focal Press, 1998
Siegel, Bruce H., *Creative Radio Production,* Boston: Focal Press, 1992

Programming Research

Most programming research conducted today is done to measure motives and habits of a radio station's target audience. The programmer's goal is to deliver audience to advertisers to generate revenue from the sale of advertising time. By 1985 most programming decisions were based on attracting audiences rather than on providing "necessary" information to them. Advertisers tend to trust research conducted by parties outside the station, for example, by ratings companies such as Arbitron. Research data gathered by commercial research organizations and paid for by the radio station owner is considered less reliable, and research conducted in house is the least reliable, according to advertisers. Station programmers use all three types of research to learn listener motivations that will inspire loyalists to listen longer and that will attract new listeners to sample the station.

In-house research is usually more valid to the specific station; it is also much less expensive than vendor research. In-house research begins before the station goes on the air for the first time, with tests of signal strength within the broadcast reach. Dial testing of all signals within the Area of Dominant Influence (ADI) that will be competing for the same audience indicates missing formats that audiences might tune in to if they were available. Once a format is selected and the station is on the air, station telephone call-outs to audience loyalists and potential listeners can help determine which tunes need to be dropped from the playlist because they are too familiar (boring), and which need to be added to make the sound more current. The station programmer uses ethnographic techniques to study lifestyles of the target audience—observing them in everyday settings, reading the magazines and newspapers they read, and noting when they are tuned in to radio.

Psychographic research also provides lifestyle and buying information about the station's audience. The research is survey based, and the populations are sampled by zip code. The assumption is that people who share a zip code also share values, lifestyles, and consumer motivation. Advertising time is sold based on a match in psychographics between the advertiser's target and the station's audience.

Outside research vendors are contracted to provide a more objective view of audience perceptions of the station's format, programming elements, promotions, and even the call letter colors. Auditorium testing is a quantitative method used to measure music perceptions. As many as 500 subjects, screened for age and other demographics and loyalty to the station or its closest competitor, are gathered in an auditorium (a hotel convention room, for example) to listen to "hooks" of 10 to 20 seconds of a tune, enough for audiences familiar with the tune to identify it. Between 200 and 500 tunes might be tested for responses of "like" to "dislike," "tired of it," or "unfamiliar." If a tune is familiar and liked, but the audience is tired of it, it is played less frequently on the station despite its national popularity. In addition to music testing, auditorium testing is used to measure response to advertising and marketing campaigns, disc jockeys, talk show hosts, and other talent, including news, contests, and marketing. Focus groups are the most common qualitative method used to understand *why* audience members (about 12 are chosen to participate) respond as they do. The moderator assesses motivation by using psychological projective techniques, brainstorming, laddering, role playing, role reversal, and others. Information from focus groups is usually not generalizable unless a great number of groups are conducted with subjects chosen randomly from a population.

The task for the radio station program director in assessing the value of research data and findings is to ask a number of questions. First, is this a quantitative study? If so, how was the sample drawn? Were there enough subjects to analyze the data statistically? What is the margin of error? Can the findings be generalized to the station's listening population? Second, were the questions asked unbiased? Were they valid? (i.e., did they test what the station programmer wanted to know?). Finally, are the findings reliable? If we did a similar study, with subjects drawn from the same population, would we get similar results?

Responsible researchers address these issues and explain their conclusions in language that is clear to the programmer. The goal is to provide information to enable the programmer to select programming that will "deliver" audiences profitably to advertisers but that will also respond to listeners' "convenience, interest, or necessity." Audience members are to be considered fellow community constituents, not just ears delivered to advertisers.

KATHRYN SMOOT EGAN

See also Audience; Audience Research Methods; Auditorium Testing; Consultants; Demographics; Psychographics

Further Reading

Carroll, Raymond L., and Donald M. Davis, *Electronic Media Programming: Strategies and Decision Making,* New York and London: McGraw Hill, 1993

Eastman, Susan Tyler, Sydney W. Head, and Lewis Klein, *Broadcast Programming, Strategies for Winning Television and Radio Audiences,* Belmont, California: Wadsworth, 1981; 5th edition, as *Broadcast/Cable Programming: Strategies and Practices,* by Eastman and Douglas A. Ferguson, 1996

Vane, Edwin T., and Lynne S. Gross, *Programming for TV, Radio, and Cable,* Boston and London: Focal Press, 1994

Programming Strategies and Processes

Programming was born of the combination of scheduling segments and tabulating appearances, yet it has an elusive definition. As Les Brown wrote in the foreword to *Broadcast Programming: Strategies for Winning Television and Radio Audiences* (Eastman), there is "a vast lore of programming wisdom, much of it self-contradictory because what works at one time or place may not work at all at another time or place." In the same textbook, Sydney W. Head worked at a definition:

> Programming is strategy. It deals with the advance planning of the program schedule as a whole. It involves searching out and acquiring program materials and planning a coherent sequence, a program service. Production is tactics. It deals with arranging and maneuvering the people and things needed to put programming plans into action. It selects and deploys the means for achieving program plans on the air.
>
> A program service is much more than the sum of its parts. Decisions about how to combine programs, or program elements, into an effective whole are just as important as decisions about which program items to accept or reject.

Head cautioned that a "seemingly obvious" distinction between programming and production is often overlooked.

> This oversight arises for understandable reasons: in the first place, production is much easier to define, teach, and practice than is programming. The production end product is visible, audible, observable, assessable. Programming, however, is far more elusive. It cannot be

practiced unless one has on-air access to an actual station and perhaps a year to await results. Production, on the other hand, can be practiced with modest facilities, and the results can be recorded for instant analysis and evaluation.

Finally, the programming function varies so much in the scope and nature of its operations from one programming situation to another that it is difficult to discern what, if anything, all these situations have in common.

Radio's Golden Age

Radio's best-known programming strategies were introduced during the 1930s and early 1940s when network programs emerged featuring former vaudeville stars who, through radio, became a part of everyday life in America. That was a unique period. Radio was the only free entertainment medium (once you owned a receiver) for a nation emerging from a disastrous depression. Radio changed listeners' attitudes about themselves, about their world, and especially about their leisure time. The radio occupied the same central place in U.S. life that television would achieve in the latter half of the century.

The key strategy was to entertain with words and sounds that stimulated listener imagination. The Golden Age of radio has also been called the "theater of the mind" days when listeners turned words and sounds into mental pictures. A man named Raymond opened a squeaking door on *Inner Sanctum* and the stories told behind that door made spines tingle for half an hour. That age also brought *The Shadow*, a Gothic thriller whose main character was a mental projection against a foggy night full of smoke from coal-burning furnaces.

On the lighter side of that period was Fibber McGee's closet, a packed-to-the-gills jumble that fell with a crash to the floor once per episode in an avalanche that usually included samples of the sponsor's product. After the cacophony there was a pause. Finally, a dinner bell crash-tinkled to the floor as punctuation. Each member of the radio audience "saw" each scene exactly as he or she wanted to see it. Each listener "saw" a different show, yet each show was perfect because it was all a product of the mind, stimulated by the spoken word in conjunction with musical interludes and sound effects. (When attempting a television revival of radio's *Fibber McGee and Molly*, the National Broadcasting Company [NBC] left the famous closet out of sight—in the hands of sound-effects experts and the imagination of the viewer.)

Programming strategy quickly evolved. The orchestras, sopranos, and baritones who were radio's first performers were on the air to give receiver manufacturers live demonstrations of the new audio medium. Once radio took hold, the strategy was to amass large audiences for advertisers who longed to have their products associated with those performers who became household names, such as Jack Benny, George Burns and Gracie Allen, Fred Allen, and W.C. Fields.

Surviving the Challenge of Television

The arrival of television abruptly changed radio strategy. As the 1950s began, the radio networks were collapsing under the impact of the new visual medium. Jack Benny's deadpan face could now be seen, not just imagined. Radio performers moved into brightly lit video studios and added scenery, building sets of the houses and neighborhoods that previously had been part of listener imagination.

To owners, radio stations became liabilities, not the assets they had been just a few years before. Many broadcasting companies sold their radio stations to invest in television. Radio entrepreneurs sifted through the wreckage of big-time radio and improvised new ideas to attract audiences. What saved radio during the encroachment of television was what appeared to be a new innovation in programming—the disc jockey.

Programmer Rick Sklar offered this scenario in *Rockin' America: How the All-Hit Radio Stations Took Over* (1984), his story of New York's legendary Top 40 station, WABC:

> Imagine the dilemma of the first person who proposed playing records instead of broadcasting live bands over the radio. Records? Who will listen to records played over the radio? People play records on phonographs. They'll think we're putting one over on them if we play records on the radio. But as early as 1948 the first bands were being laid off.

Playing records wasn't actually new in the mid 1950s. In New York, WNEW's Martin Block had created his *Make Believe Ballroom* program 20 years earlier. Beginning in 1935, Block made music using turntables and "platters," as the 78-rpm records were called, not with a baton or a piano as his contemporaries did.

When radio was disrupted by the introduction of television, broadcasters stretched individual record programs into 24-hour formats. Thus were born Top 40, Middle of the Road, Beautiful Music, and other full-time formats. Each station became something distinct. The days of radio stations' attempts to be "all things to all people" virtually disappeared. Fortunately, however, disc jockeys such as Todd Storz and Gordon McLendon did not view the future of radio in terms of its past. The change to format radio stimulated another golden age as teenagers discovered Top 40.

The Top 40 format was an all-new strategy of playing the best-selling 45-rpm records over and over, so the listener was never more than a few minutes away from hearing a favorite song. On Top 40 radio, Elvis Presley was king and so was the disc jockey, who spun the sound track to the lives of 1950s teenagers. The Top 40 format began the resurgence of public interest in radio. It also spawned other music-based program services based on the strategy of repetition of songs: the Adult Contemporary, Country, and Urban formats. The all-music strategy was to dominate radio for half a century.

Programming as Science

Michael C. Keith calls programming a radio station during this period "an increasingly complex task." In his book *The Radio Station*, Keith writes "The basic idea, of course, is to air the type of format that will attract a sizeable enough piece of the audience demographic to satisfy the advertiser." The ability to attract and hold an audience requires science as well as art in programming strategy. The science involves studying what motivates the audience to stay with long-form 24-hour music programs.

Examination of the audience was distilled into nine essential questions by Sydney Head in *Broadcast Programming: Strategies for Winning Television and Radio Audiences* (Eastman, Head, and Klein, 1981):

1. How much time does the average person listen to my station?
2. Am I doing a good job of reaching my target audience?
3. How many different groups of people contribute to my station's average audience?
4. What percentage of the listeners in one of my time periods also listen to my station in another time period?
5. During which hours of the day does my station do the best job of reaching listeners?
6. How much of my audience listens only to me and to no other stations?
7. Is my station ahead of or behind the market average of away-from-home listening?
8. Which are the most available audiences during certain times of day?
9. How often do my listeners hear the same record?

The questions are best answered in ratings reports from Arbitron and similar companies. Ratings questions about size and quality of the audience often distract programmers from the issue of usefulness or innovation of the programming itself. Amassing an audience requires constant feedback on how well the audience is satisfied. Yet none of these questions concerns the quality or the content of programming.

Radio is measured not only in terms of cumulative audience (the number of people who tune into a given station in a week) but also by time spent listening (the number of quarter hour segments heard, or "average quarter hours"). The ratings process influences programming strategy, and programmers find themselves aiming their efforts at the ratings methodology rather than the audience by attempting to extend time spent listening in order to increase average quarter hour shares.

A sure way to effect longer listening is to combine programming elements—hit records, for example—with positions on the clock. If a familiar and popular classic song is played at the top of the hour, then followed in the next music position with a current hit, then by an up-and-coming record, and so on around the hour, the station's programmer sets up a sound that includes a constant change of era or year of origin for each song. The result is a "Hot Clock" or "Music Wheel" that when drawn resembles the face of a clock with lines extending from the center like bicycle spokes separating songs, commercials, and other programming elements. (Hot clocks are not exclusive to music stations. News and news-talk stations use them as well—often calling the visual version a "News Wheel"—to designate places on the hourly clock for certain types of news stories or talk show segments.)

The music programmer further attempts to mix upbeat songs with slow songs, male vocals with female vocals, large production sounds with solo instruments, etc., so that a sense of musical variety is achieved within the hour. This definition of category is from a client memo from the consulting firm Shane Media Services:

> What is a category? It's a group of songs organized by a primary characteristic. On the current/recurrent side of the library, categorization is determined by age and amount of play. For example, new songs are put in an "add" or "light" category; these titles receive the least amount of play and are protected by categories made of more familiar songs. This category's function is to introduce new songs in a palatable manner. It's placed on the clock so announcers can sell the songs it includes.
>
> As songs build up play, they begin to be tested in audience research. Songs listeners show interest in are worthy of more play. They move up to medium current—enough rotation for average listeners to develop affection for the songs.
>
> Medium songs that excel in research move to hot current. Hot current is a category determined by value—current songs listeners care about most. Recurrent categories are made of proven hits and age again becomes the primary organizational criterion: Hot recurrents are the strongest hits with high play. Medium and bulk recurrents are seven to fourteen months old and represent the best of the recent hits.
>
> Gold categories are organized by value: the Power category is made up of the best testing titles, songs listeners want to hear every day. These songs have high "love"

scores, high "play more" scores and little "tired of" scores.

The medium or secondary gold category is made of songs listeners like but aren't involved with enough to want to hear daily. Sometimes records are put into secondary gold because high "love" scores are combined with high "tired of" scores. Resting songs by playing them every two to three days instead of daily usually results in higher scores the next time songs are tested.

If a station uses a third level of gold, it's usually made of songs that have sufficient "like" scores but no "love." These are "okay" songs that won't make listeners tune out but also don't cause them to turn up the station or tell their friends to stop working and listen.

Music rotation software such as Selector, PowerGold, and MusicMaster helps stations achieve balance through elaborate sound codes and type codes. Use of the popular software also allows efficient management and diagnosis of play histories in a music library once criteria are established.

Formats

A brief overview of the major formats will shed some light on programming strategy.

Adult Contemporary (AC) The most familiar musical selections found on radio are from Adult Contemporary stations. In terms of the number of listeners, AC was the most popular format of the 1980s and 1990s. It has many permutations owing to the broad age range of listeners (25–54) and their diversity of tastes. AC's territory covers the softest jazz instrumental and light rock to modern hits. Collectively, the varieties of AC are radio's most listened-to formats.

Top 40 or Contemporary Hit Radio (CHR) "Contemporary Hit Radio" was the euphemism used by radio people who thought "Top 40" was a term used by kids. Listening to their public, they discovered that "Top 40" was what people called radio that played the hits. The strategy remains the same as it was during the earliest years of the format: play the best-selling songs and repeat them often.

Rock Rock music (formerly called "rock and roll") takes many different forms, most of which have resulted in radio formats. For example, Classic Rock is an offshoot of Mainstream Rock. Offering the comfort of familiarity, the Classic Rock format presents a mix of well-known bands, primarily from the 1970s. Then there's the easy-to-identify rift between Mainstream and Modern (or Alternative) rock. Alternative rock music begins as a splinter of some other style. If it grows, it is embraced by the mainstream and is no longer "alternative," thus losing its cachet.

Country Since the 1970s, Country has been adopted by more stations than any other format. In its earliest days, the Country format was considered to be aimed at a blue-collar audience. As the audience grew and young artists revitalized the music in the early 1990s, the blue-collar image became mainstream. The format scores very well among adults aged 25 to 54 and is the least prone to audience fragmentation. Its strategy is rooted in the Top 40 tradition: play the most popular songs often. Country stations tend to offer a mix of songs from a broad spectrum of years.

News-Talk News-Talk radio could be called "personality radio," even though the caller interaction and tendency toward political topics often obscures the impact of on-air personalities. Rush Limbaugh and Dr. Laura Schlessinger were the reigning superstars at the beginning of the 21st century, with hundreds of newcomers trying to establish themselves in second and third place. Programming News-Talk requires a talent-driven strategy. Talented program hosts who give the listeners access by telephone to highly interactive, relevant discussions are the big winners.

Marketing as Radio's Product

In the late 1970s and early 1980s, radio station operators changed their focus from product and programming to positioning and perception. The shift was in large part a result of the efforts of consultant George A. Burns, whose books postulated that the way a radio station sounded was secondary to what listeners imagined or perceived about the station. Burns urged his readers to leave "the product dimension" and to enter "the marketing dimension."

Programming had long been radio's primary product concern. However, in the competitive environment after the rise of FM music formats, product alone was not enough to achieve differentiation and success. As Burns wrote in *Radio Imagery: Strategies in Station Positioning* (1980):

> Standard wisdom originally held that fragmentation would provide a wider spectrum of listening opportunities for the public. Indeed, this seemed to be the case at first. Formats such as "classical" and "progressive rock" supplied fuel for the work of spreading FM. AM operators first became aware of FM in a serious way when they noticed "beautiful music" or "progressive rock" beginning to hurt them. Fragmentation began as alternative programming.

As audiences fragmented, more stations became viable listening options even though each station's audience was smaller. For instance, more than 70 stations could be heard in most parts of the Los Angeles metro area in 1999, according to the *M Street Radio Directory.* In smaller cities, too, the number of stations proliferated. Tulsa, Oklahoma, had more than 25 stations serving some segment of its population during that same period. Large numbers of stations usually meant several stations in the same format with little differentiation. To program effectively against direct competition, stations enlisted research companies to test music libraries and to probe audience perceptions. The research allowed stations to play proven, safe songs that received the highest scores and to reduce the risk of unfamiliar or overplayed songs.

Research became part of the marketing loop. With so many stations vying for a diminishing segment of the audience, the paramount concern became keeping a station's name in the forefront of listeners' minds. Naming stations was as much an art as programming them. Mnemonics like Z-93, Star 104, K-FROG, and FROGGY 98 caught attention because they were memorable. Advertising campaigns urged listeners to remember one station for country music, another for news, and so on.

Station management concerned with marketing did not ignore programming, but they did take programming for granted. Effective programming was accepted as a given, and emphasis shifted to marketing, promotion, and advertising to influence audience recall.

Programming after Consolidation

The Telecommunications Act of 1996 eliminated many radio ownership limits and simultaneously shifted programming strategy. Consolidated ownership now makes it possible for one company to own a cluster of stations in the same market. An operator company that can do so chooses its cluster by format in order to control all or most stations in a format or to minimize competition. Thus a company may acquire several varieties of rock stations, or both of a city's country stations, or a combination that includes an all-news station, an all-talk station, and an all-sports station, just to cite three possibilities.

Consolidation offers opportunities for cost savings in programming by combining air talents to perform on several stations in a cluster and operating one news department to serve many stations. For instance, Capster Broadcasting Partners assembled a staff of air talents in Austin, Texas, to feed voice tracks to more than 100 stations in the Capstar and AM/FM groups via a wide area network of linked computers. Another approach to cost savings in programming is the "hub and spoke" system employed by Clear Channel Communications, in which a centrally located station in a large or medium market feeds programming to be simulcast in nearby smaller cities.

Consolidation also has given rise to syndicated programs, both national and regional. Many morning shows are syndicated: for example, radio personalities Bob and Tom from Indianapolis, John Boy and Billy from Charlotte, Mark and Brian from Los Angeles, and Rick Dees also from Los Angeles. Such shows seem to be a return to the network style of early radio, but without the national attention that network programs received in the pre-television era. Syndicated programming allows local stations to air shows they could not (or would not) produce themselves.

ED SHANE

See also individual radio formats (Adult Contemporary, etc.) as discussed above; Arbitron; Audience; Audience Research Methods; Auditorium Testing; Demographics; Disk Jockeys; Music Testing; Office of Radio Research; Programming Research; Promotion on Radio

Further Reading

Burns, George A., *Radio Imagery: Strategies in Station Positioning*, Studio City, California: Burns Media, 1980

Burns, George A., *Playing the Positioning Game: Aiming at the Core*, Studio City, California: Burns Media, 1981

Eastman, Susan Tyler, Sydney W. Head, and Lewis Klein, *Broadcast Programming, Strategies for Winning Television and Radio Audiences*, Belmont, California: Wadsworth, 1981; 5th edition, as *Broadcast/Cable Programming: Strategies and Practices*, by Eastman and Douglas A. Ferguson, 1996

Keith, Michael C., *The Radio Station*, Boston: Focal Press, 1986; 5th edition, 2000

Shane, Ed, *Cutting Through: Strategies and Tactics for Radio*, Houston, Texas: Shane Media, 1991

Shane, Ed, "The State of the Industry: Radio's Shifting Paradigm," *Journal of Radio Studies* 5, no. 2 (Summer 1998)

Sklar, Rick, *Rocking America: An Insider's Story: How the All-Hit Radio Stations Took Over*, New York: St. Martin's Press, 1984

Progressive Rock Format

Progressive rock is a radio format designed to appeal to rock music fans who were initially represented by the counterculture of the late 1960s. At times referred to as *progressive radio, underground, free-form, album-oriented rock* (AOR), *alternative,* and *classic rock,* the progressive rock format has its roots in the underground rock movement in the years leading up to the Woodstock Music Festival in 1969.

As an alternative to the repetitive hit music of Top 40, which was pervasive on AM radio at the time, progressive radio began on undeveloped FM stations as so-called free-form programming, an ephemeral concept that encouraged each disc jockey to program his or her own show with a newer and broader brand of rock and roll than merely its pop progeny, interspersing music with commentary that referenced an emerging cultural change in the United States. Progressive radio drew an audience previously unserved by radio, and it is credited with establishing FM as a dominant force in music radio.

By the mid-1970s, FM progressive rock in a variety of permutations spearheaded audience migration to the new frequency band. Although the radio industry paid little attention to FM, FM was handily supported by the recording industry and the electronics industry, both of which had high stakes in high-fidelity products. By the end of the decade "progressive" stations had seized a majority of the mainstream radio audience owing to the widespread popularity of progressive rock music and the ability of FM to broadcast high fidelity in stereo. As progressive music hit the top of the charts, Top 40 stations reacted by retooling their playlists and moving their operations to the FM band, forcing AM to all but abandon music as a viable format in most competitive markets. To distinguish themselves from Top 40's encroachment, many progressive rock stations began adopting the AOR format designation. Although the progressive rock format has not disappeared, defining it has become contentious and subjective. Some prefer to tie the format to the music of its breakthrough, programmed as classic rock, whereas others consider it a timeless concept based on a musical alternative to mainstream pop, peppered with a left-of-center banter. In either case, progressive rock radio is connected to at least three distinct breeds of broadcaster: rock and roll radio pioneers such as Alan "Moondog" Freed, community or pirate radio operators, and shock jocks.

The rock and roll of Elvis Presley, Jackie Wilson, Buddy Holly, The Shirelles, Chuck Berry, and countless others successfully altered radio's musical landscape in the 1950s as disc jockeys developed a relationship with their youthful audiences and an ear for an alternative sound. Similarly, early progressive rock disc jockeys such as Tom "Big Daddy" Donahue brought a new sensibility and attitude to the medium, playing a variety of alternative music styles, including folk music; a more sophisticated brand of rock and roll; and cross-genre hybrids that combined blues, country, soul, funk, or jazz with increasingly amplified guitar rock, creating new subgenres such as folk-rock, progressive country, country rock, jazz fusion, latin rock, acid rock, hard rock, and heavy metal. Progressive stations did not play hit singles. They played albums, sometimes an entire LP at a time. The format was built around artists who broke new ground musically and lyrically. Artists such as The Doors, Jimi Hendrix, Joni Mitchell, Vanilla Fudge, Creedence Clearwater Revival, Crosby Stills and Nash, Led Zeppelin, Jefferson Airplane, Janis Joplin, Allman Brothers, Santana, Beatles, Rolling Stones, and others certainly had pop hits, but it was their other album tracks that made progressive playlists, songs often distinguished by lyrics that questioned authority, embraced social protest, and celebrated the sexual revolution.

This music did not fit neatly into the under-three-minutes song length common for airplay on Top 40. There were also rare recordings, live performances, and rock star interviews, which added spontaneity and primed the syndication efforts of new radio networks such as Westwood One. And rather than announce the time and temperature, progressive disc jockeys slipped in pithy social commentary and politically charged comedy bits—often in the same breath. The unique combination of music, structure, and editorial point of view conspired to spawn an identity all its own. Not since the seminal rock and roll disc jockeys of the 1950s had radio responded so directly to popular culture. In both eras, too, popularity brought increased scrutiny, claims of amorality, and scandal. And both saw the execution of their respective preformats move from the individual programmer to station owners and consultants who went on to develop successfully marketable formats.

As playlists shrank and advertising increased, purist progressive announcers became disenchanted with an apparent squandering of the airwaves by strictly commercial interests. Some became radio pirates in the spirit of the underground movement, broadcasting without a license from hideout locations. Others joined or started licensed, noncommercial, low-power community radio. Still others abandoned the need to control the music by developing talk show personalities that led to the advent of FM talk and the so-called shock jock.

Early progressive stations were instrumental in introducing and "breaking" new artists and music. They also provided a voice for "lifestyle" news services such as Earth News, the National Broadcasting Company's (NBC) The Source, and the American Broadcasting Companies' ABC FM network, as well as forging opportunities for targeted advertising. Progressive announcers would categorically reject ad copy with hype or that promoted products antithetical to the desires of their perceived audience. Traditionally, progressive stations did not yell at their audiences, sensationalize in the purely promotional sense, or hard sell, except in parody or satire. They adopted a reduced commercial load of 9 to 15 minutes of advertising per hour as opposed to the 18 to 30 minutes of Top 40, partly because of the fact that advertisers were not yet sold on the value of the progressive rock audience.

The format's original structural characteristics included commercials scheduled in "spot sets" and music programmed in "song sets," which established the music mix "segue" as an art form, setting a mood that organically led to increased time spent listening. Many progressive rock formatics were so successful in holding audience that they have long been incorporated into other music formats.

Early adapters of commercially successful progressive rock formats include San Francisco's KMPX and KSAN; Boston's WBCN; Los Angeles' KMET and KLOS; Chicago's WXRT; and New York's WOR-FM, WNEW-FM, and WABC-FM/WPLJ. Perhaps the most consistent preservation of the original progressive rock spirit has been on student-run college radio, as well as on some of the Pacifica stations.

JOSEPH R. PIASEK

See also Album-Oriented Rock Format; Alternative Format; Classic Rock Format; Community Radio; Free Form Format; Freed, Alan; Pacifica Foundation; Underground Radio; WABC; WNEW; WOR

Further Reading

Ditingo, Vincent, *The Remaking of Radio,* Boston and Oxford: Focal Press, 1995
Fornatale, Peter, and Joshua Mills, *Radio in the Television Age,* Woodstock, New York: Overlook Press, 1980
Keith, Michael C., *Voices in the Purple Haze: Underground Radio and the Sixties,* Westport, Connecticut: Praeger, 1997
Routt, Edd, James McGrath, and Frederick Weiss, *The Radio Format Conundrum,* New York: Hastings House, 1978

Promax

Industry Trade Association

The primary trade organization in electronic media focusing on promotion and marketing, Promax is a nonprofit, mutual-benefit association for promotion and marketing executives in the electronic media. It conducts annual trade conventions, distributes videotapes of award-winning on-air promotions and print materials, and serves as a clearinghouse for ideas and projects related to electronic media promotion via its resource center and weekly fax memos.

Although broadcast and cable television has been its most salient focus, Promax also serves the commercial and noncommercial radio industry. About 10 percent of its 2,000 member companies are in the radio business, mostly from major-market stations and large groups and from production companies offering creative concepts and marketing services to radio. Among the best known in the radio area have been such companies as Columbia Broadcasting System (CBS) Radio, American Broadcasting Companies (ABC) Radio Networks, and the Westwood One networks. Nearly all large and midsized market stations hold membership and send representatives to the annual Promax conventions.

The association's avowed goals are to share promotion and marketing strategies, to share research techniques and information, and to spread the word on new creative concepts and technologies. Like other trade associations, Promax provides the opportunity for the marketing executives of networks and program syndicators to meet with their counterparts at affiliated stations and for production companies and consultants to show off their wares and attract the attention of potential clients.

On a daily basis, a president/chief executive officer (CEO) and a staff of 20 full-time personnel run the association. The president is selected by an executive committee of the 27-person board of directors overseeing Promax. A portion of the board turns over annually, and new directors are elected by the entire association membership. Board members are nominated to represent a mix of networks, studios, stations, cable systems, production companies, distributors, agencies, and consulting companies.

Founded in 1956 as the Broadcasters Promotion Association (BPA) and operating in tandem with the Broadcast

Designers Association (BDA) since the late 1970s, the association grew from a few hundred company members to well over 1,000 by the mid-1980s and to nearly 2,000 companies by the year 2000. Administered in the 1970s and 1980s out of Lancaster, Pennsylvania, by Executive Director Lance Webster, the organization's growth in the late 1980s resulted in a move to Los Angeles. Webster is credited with increasing the association's national visibility, stabilizing its procedures and structures, and creating active ties with other trade and educational organizations.

In 1984 the association's name was changed to Broadcast Promotion and Marketing Executives to acknowledge participation by highest station and network management and to recognize two factors: the rising importance of promotional activities in an increasingly competitive situation and the industry's increased focus on marketing strategies—which involve a broader conception than sales promotion and audience program promotion. But by the early 1990s, the incongruity of a large cable membership had led to still another evolution in the association's name, this time to Promax in 1993. (Although the name is frequently spelled in all capitals, the word is not an acronym; it draws on the association's traditional connection with the field of promotion, adds an *x* for executives, and hints at "maximum something" to users, an acceptable bit of hyperbole suiting this profession.)

Another major change was the expansion of worldwide promotion conventions, such as the association's first international conference in Leeds, England, in 1990. This became the annual Promax–United Kingdom meeting, and it has been followed by Promax Asia, Promax Europe, Promax Latin America, and, in 1999, the first Promax Australia/New Zealand convention. As of 2000, the association had member companies in 35 different countries and was dealing with rapid membership growth outside the United States.

Two individuals clearly stand out as the biggest contributors to the evolution of the association: Lance Webster, publications coordinator in 1979 and executive director for much of the 1980s, and Jim Chabin, president and CEO during most of the 1990s (now president of the Academy of Television Arts and Sciences in Hollywood), who implemented the association's international vision. In 1999 Glynn Brailsford stepped into the joint position of president and CEO of both Promax and BDA and has continued building and consolidating the international partner associations and melding the administration of Promax and BDA.

Association board president in 1980, Tom Dawson, a CBS Radio vice president, was a key individual in building the radio membership in the organization from 1976 through 1988. Elected to the association's Hall of Fame in 1992, Dawson is building an archive and writing a history of the association. Erica Farber, executive vice president at Interep and president of the Board in 1991–92, took on the role of fostering the growth of radio membership. Farber is now publisher and CEO of *Radio and Records.*

The annual North American convention, usually held in the United States, is the centerpiece of Promax's offerings. Called the Promax Conference and Exposition, by 1999 the association reported that nearly 7,000 industry executives attended. The convention offers about 65 sessions and workshops with some 200 expert speakers and presenters; about one-half of those panels directly or indirectly address radio interests. Topics have included branding and copyright and music, sports, and performance rights, in addition to radio-only panels about contesting and games, image marketing, segmentation research, audience measurement, and audio technologies. By the late 1990s, the internet had become the conference subject of the most riveting interest to radio executives.

At the annual convention, Promax makes the International Gold Medallion Awards for excellence in marketing and promotion, recognizing achievement in 280 categories for local television, networks, cable, radio, and program syndication with gold Muse statuettes or silver certificates. The categories recognize image and program promotion in on-air and print media as well as multimedia campaigns and contests. Special achievement awards have gone to such celebrities as Casey Kasem, Dick Clark, Stan Freberg, and Chuck Blore.

The BDA was formed in 1978 with the support of what was then BPA, and it has since met annually conjointly with Promax in North America, Europe, and Australia. It spotlights the needs and interests of the creative staffs of stations and advertising agencies and gives its own awards for outstanding artwork at a separate meeting held at the same time and in the same conference site as Promax. It also conducts panel sessions and workshops focusing on cutting-edge design concepts for program guides, magazines and newspapers, outdoor billboards, posters, and transit signs, as well as on-air spots for both television and radio. In recent years, considerable attention has gone to the implementation of creative ideas via digital technology.

In the 1980s, Promax began issuing weekly faxed communications (*PromofaX*), replete with practical ideas for station promotion, many of which are innovative ideas for radio contests or image promotion. Promax also publishes an annual *Image* magazine at the time of the North American convention, which incorporates examples of promotion and marketing in both television and radio from around the world. In addition, Promax conducts periodic surveys that track changes in salaries, status, and backgrounds of commercial promotion managers and changes in the technologies and methods used in daily promotion and image promotional campaigns. It summarizes these results in *PromofaX* and *Image.* Central to member relations, the Promax Resource Center is a repository for materials about promotion and marketing. It sells audio- and videotapes of panels, workshops, and keynote speakers as well as copies of award-winning print materials.

SUSAN TYLER EASTMAN

See also Promotion on Radio

Further Reading

Eastman, Susan Tyler, and Robert A. Klein, editors, *Strategies in Broadcast and Cable Promotion: Commercial Television, Radio, Cable, Pay-Television, Public Television,* Belmont, California: Wadsworth, 1982; 4th edition, as *Promotion and Marketing for Broadcasting, Cable, and the Web,* edited by Eastman, Klein, and Douglas A. Ferguson, Boston and Oxford: Focal Press, 2002

McDowell, Walter, and Alan Batten, *Branding TV: Principles and Practices,* Washington, D.C.: National Association of Broadcasters, 1999

Promax website, <www.promax.tv/main.asp>

PROMAX Image (1993–)

Promenade Concerts

BBC Classical Music Series

They are a classic example of the old meeting the new, the young meeting the aged, and the traditional meeting the contemporary. They are at once casual and formal, simultaneously hip and sophisticated, sometimes playful and at other times serious. They are the Henry Wood Promenade Concerts, the summer music series of the British Broadcasting Corporation (BBC), known simply as "the Proms."

Origins

Begun in 1895, the Proms were created in order to expose the widest possible audience to the classical music. To serve this purpose, the Proms were presented as less formal than traditional classical concerts, and with at least some very inexpensive ticket prices. Those strategies remain in place to this day. Young and old alike attend, some seated in luxury boxes and others lounging on the promenade floor. Some wear formal attire while others sport cutoffs and sandals. The result is that a very diverse audience continues to attend the annual series.

Although the name Henry Wood has long been associated with the Proms, the series was actually the brainchild of Robert Newman. In fact, the original name for the concert series was "Mr. Robert Newman's Promenade Concerts." Newman was perhaps more businessperson than musician, but during his tenure as the manager of the Queen's Hall in London (the original home of the Proms), he made known his desire to reach a wider audience with finer music. He envisioned a type of educational environment in which audiences might be attracted by more popular music, and then—once attracted—exposed to the more serious higher forms of music. He envisioned a less formal environment in which audience members could enjoy performances as they stood or sat in a more relaxed promenade arrangement. He also desired a more affordable environment that would allow everyone, even those of limited means, to enjoy fine music. Robert Newman had the vision; Henry Wood was the man he selected to give life to that vision as the first conductor of the Queen's Hall Orchestra and the Promenade Concerts.

Henry Wood's contributions to the Proms were legendary. He introduced audiences to new music, new performers, and new composers. Each season featured a certain amount of standard repertoire, but Wood insisted upon including new works—works that came to be known as the "novelties." The Proms orchestra consisted of the finest established musicians available, but Wood was also known for being a champion of aspiring young players. Wood also introduced his audiences to many of the leading composers of the day, including Richard Strauss and Sergei Rachmaninoff.

The Proms on the Air

The BBC became involved with the Proms in 1927. Because of financial trouble, Newman had earlier relinquished the management of Queen's Hall and its orchestra, as well as the concert series, to Chappell and Company, a publishing house. But the Proms continued to operate at a deficit, and in 1927 Chappell and Company's management decided that the company could no longer sustain the concerts financially. As a result, the BBC took charge and a new era began. Not only did the BBC bring financial support, but now the Proms could reach an even wider audience than ever, as people throughout England could listen to every performance on their radios.

World War II brought about several changes in the Proms. Temporary withdrawal of the BBC's financial support (necessitated by the war) forced Wood to seek and secure private sponsorship for a few years. The BBC Orchestra, which had been formed in 1930 (three years after the BBC took over the Proms) was temporarily replaced by the London Symphony. Perhaps most significant, however, was a war-induced change in venue for the concert festival. In 1941, Queen's Hall was severely damaged by German bombing, forcing the Proms to be moved to the Royal Albert Hall. This historical and elegant auditorium, which has a capacity of 6,000, remains the home of the Proms to this day.

Henry Wood conducted 50 years of Promenade Concerts. He died in August 1944, only three weeks after his last Proms concert. In the years that followed his death, several conductors and music directors helped to define the future direction of the Proms. Two primary trends of the post-Wood era were an increase in the number of participating orchestras and an increase in the variety of musical styles represented. During the 1950s, orchestras from places other than London were included in the Proms lineup for the first time. As other orchestras participated, the Proms took on a more international flavor. Through the 1960s this international flavor began to be noticed more in the musical repertoire as well.

A very popular feature of the Proms is its "Last Night" celebration that closes the season each year. It combines a festive party atmosphere with the Proms tradition of great classical music and a sense of camaraderie on the part of "Prommers" who participate in a robust audience sing-along of traditional Proms favorites. The Last Night is so popular, and tickets are so hard to come by, that London's Hyde Park now hosts the BBC Proms in the Park, an event that allows an additional 40,000 people to view the Last Night of the Proms telecast live on a giant video screen.

More than 100 years after their beginning, the BBC Henry Wood Promenade Concerts remain true to their original purpose: to be enjoyable and accessible to a wide audience. Their popularity continues to increase, and BBC Radio 3 still airs the entire 72-concert series each summer. Additionally, some performances are broadcast by BBC Television and the BBC World Service, further extending the Proms experience to audiences around the world. The concert series attracts some of the world's finest musicians, who perform some of the world's best music.

RICHARD TINER

See also BBC Orchestras; Classical Music Format

Further Reading

BBC Online: BBC Proms, <www.bbc.co.uk/radio3/proms/>
Briggs, Asa, *The BBC: The First Fifty Years,* Oxford and New York: Oxford University Press, 1985
Cain, John, *The BBC: 70 Years of Broadcasting,* London: British Broadcasting Corporation, 1992
Cox, David Vassall, *The Henry Wood Proms,* London: British Broadcasting Corporation, 1980
Hall, Barrie, *The Proms and the Men Who Made Them,* Boston and London: Allen and Unwin, 1981
Miall, Leonard, *Inside the BBC: British Broadcasting Characters,* London: Weidenfeld and Nicolson, 1994

Promotion on Radio

During the early years of commercial radio in the United States, most stations had little reason to promote themselves to listeners because of the relative lack of competition in the local marketplace. Moreover, since the homogeneous programming provided by national networks resulted in every station's sounding basically alike, there was little to promote about a station's "distinctive" qualities. By the middle 1950s, however, local radio programmers began experimenting with new entertainment and musical formats in an attempt to regain audiences lost to the upstart medium of television.

Once stations began to differentiate their programming and audiences found a choice of offerings, the need to attract listeners to specific stations intensified. Since then, the variety of station formats and programming has increased, and along with it the value of promotion in the eyes of radio professionals. One reason promotion is vital in today's radio marketplace is the current Arbitron ratings system, which depends on listeners' ability to recall a station's call letters or name when filling out the ratings diary. Arguably, this methodology means that the station with the highest top-of-mind awareness, not the greatest number of actual listeners, may win the ratings battle.

In an effort to increase this station awareness in the local community, stations depend on promotion to help accomplish four major goals: (1) to increase the number of people who sample the station; (2) to give the current audience a reason to listen for a longer time; (3) to provide listeners who must tune out incentives to tune in later; and (4) to create and reinforce the station's image.

Audience acquisition promotions are designed to encourage station sampling by people who don't regularly listen. By necessity, this is accomplished through promotional campaigns designed for and delivered through other media. Television commercials on local stations and cable systems, direct mail pieces, roadside billboards, bumper stickers, T-shirts, key chains, refrigerator magnets, and a variety of other types of promotional merchandise are used to introduce the public to the station's call letters, frequency, format, personalities, and contests. The strategy behind audience acquisition promotions is to inform or remind potential audience members about the programming a station delivers, hoping to match the station's product with listener wants and needs.

Promotions designed to increase the amount of time current listeners spend tuned to the station are called audience maintenance promotions. Often, audience maintenance goals are accomplished through the design and implementation of on-air contests. For example, a music-oriented station might implement a contest in which listeners must hear a specific song, or any song by a specific artist, before phoning the station to try and win a prize. By strategically working with the programming or music departments to determine when the designated song or artist will play, stations can increase the time listeners spend tuned to the station waiting for a chance to win. Other popular variations on this audience maintenance contest include the scavenger hunt (in which listeners wait to hear different items they must collect in order to have a chance to win a prize) and the treasure hunt (in which clues are periodically given for the hidden location of a valuable gift certificate).

Contests may also be designed to provide audience members who must stop listening with incentives to tune back sometime later. These contests are referred to as recyclers, and they take one of two forms. Horizontal recyclers are designed to entice listeners to tune in again at the same time the following day. For example, a common horizontal promotion on music stations is to encourage listeners to fax the midday host a list of their favorite songs. The listener whose list is chosen wins prizes plus gets to hear his or her favorites played during the lunch hour. When stations promote this as a programming element done every weekday at the same time, it recycles listeners to the lunch hour of the following day to see if they have won. Vertical recyclers, on the other hand, are designed to entice listeners to tune in later during the same day. For example, having a morning announcer promote that the next chance to win concert tickets is during the afternoon show recycles those who may have to tune out during the workday to the station for their drive home.

Regardless of the strategic goals of a station contest, promotion directors must keep in mind applicable federal regulations. For example, the FCC prohibits the broadcast of information about most lotteries other than those sponsored by nonprofit organizations and state governments. In order for a radio promotional contest to constitute a lottery, it must contain three components: a prize, chance, and consideration. Consideration is something of value paid by the contest participant, such as an entry fee. For example, if a station decided to give away a new car (a prize) to someone chosen at random (chance) from those who bought a ticket to the station's annual concert/birthday party (consideration), that station would violate lottery laws when promoting the contest on air. Stations often alleviate this problem by instructing listeners how to enter without consideration in the official contest rules—which the FCC requires be fully and accurately disclosed. Other laws that may impact radio promotions prohibit the broadcast of obscene content and limit the broadcast of indecent material to certain hours.

Another goal of radio promotion is to establish and reinforce the station's image in the minds of the audience. It is important that every contest, billboard, bumper sticker, and website associated with the station is consistent with the desired image. Often, program directors and promotion departments work together to establish the target image for their station. This way the program director's vision of the station image is known by the promotion department, and they can be sure that all aspects of the promotion mix are developed with the image in mind.

In order to identify a station's target image, a list is sometimes created of words or phrases that would be desirable for listeners to associate with the station. For example, a news station would want to have the image of being dependable, honest, timely, and involved with the community. A rock station, on the other hand, would be more interested in being known as the rock concert station, as being knowledgeable about entertainment news, and as the station that gives away music-related prizes. Creating lists such as these can help ensure that the content of radio promotion is consistent with the desired image.

When trying to reinforce station image, however, content may not be the only variable that promotion directors can use. Radio consultant Lee Abrams argues that the sound of on-air promotions can go a long way to improving a station's image. Abrams advises that when creating on-air promotion, stations should "hit the production room and come out with great stuff" (Lynch and Gillispie, 1998). Recent research suggests that Abrams may be correct. Potter and Callison (2000) have experimented with different types of production techniques in on-air promotions and found that promotions containing sound effects, music, and multiple announcers create more positive images in listeners than comparatively simple promotions do.

Promotional activities—from contests and giveaways to billboards and bumper stickers—cost money. Many stations regularly budget for ongoing promotional campaigns, funding them through management's commitment to marketing the station. Sometimes operating capital for promotions at a current-based music station is enhanced by money from an independent music promoter who pays the station for the right to discuss music decisions with the program director on a weekly basis.

In all formats, however, there is increasing pressure to develop more sales promotions than ever before. Sales promotions are campaigns that accomplish marketing and image goals for the station while simultaneously providing the sales department with tie-ins that can be sold to advertisers. Sales promotions take many forms; for example, co-sponsorship of a contest or event by both the station and the client, remote broadcasts from a client's location where listeners enter to win prizes, coupons for the client's business as the "removable" backs of station bumper stickers, and advertiser logos screen printed on station t-shirts. Anything that the station's promotion can do to help generate advertising revenue can be viewed as a way to offset the cost of the promotion itself.

ROBERT F. POTTER

See also Arbitron; Consultants; Programming Strategies and Processes; Promax

Further Reading

Buchman, J.G., "Commercial Radio Promotion," in *Promotion and Marketing for Broadcasting and Cable,* edited by Susan Tyler Eastman, Douglas A. Ferguson, and Robert A. Klein, 4th edition, Boston: Focal Press, 2002

Lynch, Joanna R., and Greg Gillispie, "Creating an Image," in *Process and Practice of Radio Programming,* by Lynch and Gillispie, Lanham, Maryland: University Press of America, 1998

Newton, G.D., and R.F. Potter, "Promotion of Radio," in *Research in Media Promotion,* edited by Susan Tyler Eastman, Mahwah, New Jersey: Erlbaum, 2000

Norberg, Eric G., "Promoting Your Station," in *Radio Programming: Tactics and Strategy,* by Norberg, Boston and Oxford: Focal Press, 1996

Potter, R.F., and C. Callison, "Sounds Exciting! The Effects of Audio Complexity on Listeners' Attitudes and Memory for Radio Promotional Announcements," *Journal of Radio Studies* 7 (2000)

Propaganda by Radio

Propaganda has had many definitions. A basic definition is that it is selective and biased information aimed at indoctrinating, converting, and influencing people. From its source in Roman Catholicism, when in 1622 Pope Gregory XV created the *Sacra Congregatio de Propaganda Fide* (an agency charged with spreading the Christian faith in foreign missions), to efforts to use radio for propaganda, particularly by Nazi Germany in the 1930s and 1940s, *propaganda* has taken on increasingly negative connotations and is currently associated with lies, deception, and disinformation designed to control populations or dishearten adversaries.

Origins

It is difficult to say with any accuracy when radio was first used as a medium of propaganda. Several countries had radio services on the air prior to World War II, including Belgium (1934), France (1931), Japan (1935), the Netherlands (1929), the Soviet Union (1927), and the United Kingdom (1932). Arguably some of these broadcasts aimed to present these nations in the most positive light (especially those of Radio Moscow in 1925 and 1927, celebrating the October Revolution and the Bolshevik victory). Some historians consider the Soviet broadcast in 1917 (featuring Vladimir Lenin and the announcement of the beginning of a "new age" on 30 October) to be the first recorded propaganda broadcast using radio. It was aimed at potential revolutionaries in Europe. However, Nazi Germany is often credited with initiating the earliest sustained effort to broadcast propaganda via radio, using radio as an instrument of conquest. There is some dispute on this score, as there was intermittent use of radio even during World War I, although this was more accurately wireless telephony and consisted mostly of news reports. One famous "broadcast" of this type carried Woodrow Wilson's "Fourteen Points" plan to end the war and to assure that it was the "war to end wars." During the Versailles peace conference in 1919, Germany was prohibited by the Allied powers from using its transmitting stations in Nauen, Hanover, or Berlin, and from constructing any new transmitting facilities for a period of three months after the signing of the peace treaty.

One reason for the rapid development of radio as an instrument of propaganda after World War I was that age-old tensions continued to exist in Europe as radio developed there under state ownership or control. With nation-states having broadcast monopolies, it was a short step to considering the medium as a vehicle for pursuing national interests, for putting a national spin on events—in short, for propaganda. In 1925 the Soviet Union put the world's first shortwave station on the air, greatly extending the reach of the Bolshevik political agenda. The Soviet Union then used radio as a weapon in a dispute with Romania (provoking a Romanian radio response) over Bessasrabia in 1926. Japan also used radio as a weapon in its invasion of Manchuria shortly after putting its first station on the air in 1931.

By the early 1930s international radio services distinguished programming in their native language from that in other languages, with the latter considered propaganda. By this definition the Third Reich's inaugural English-language broadcast to North America, on 1 April 1933, was propaganda regardless of its actual content. This also applied to its creation of radio services in several other foreign languages the following year. The earliest broadcasts were really aimed at Germans who had emigrated, providing them with a continuing link to the "Fatherland," but gradually the broadcasts became more overtly propagandistic, with the *Reichsender* encouraging listener parties to hear the latest speeches by Adolf Hitler and other Nazi Party leaders. Germany also used radio as a weapon to influence politics, to gain sympathy for the Austrian Nazi Party in 1933 and 1934, and prior to its intervention in Austrian politics in 1938. In the latter campaign, the Nazis distributed 100,000 radio sets in Austria to assure an audience for its broadcasts. (This tactic was also used by the Italians for broadcasts from Radio *Bari* to the Middle East and North Africa beginning in 1935. By 1937 the Italians were broadcasting in 16 languages to countries in the Mediterranean basin.) The Nazi goal was to create what was called a "fifth column," a group of convinced believers in the Nazi cause who would assist it by working within their own countries to promote it.

Some efforts were made to stem the use of radio for propaganda purposes. The League of Nations, which had been advocated by U.S. President Wilson as part of his "Fourteen Points," adopted a convention (agreement) in 1936 to outlaw aggressive radio propaganda and incitements to war or insurrection. This agreement also called upon its signatories to use radio to promote peace, better knowledge of other civilizations and conditions of life, and the true nature of international relations between neighbors. Although this convention was signed by 28 countries, it was eventually ratified by only 19 and had little effect on actual uses of radio in Europe. It did not come into effect until 1938 and by then it was too late to stem the tide of radio use to promote national aims. (Neither Germany nor Italy signed the convention.)

Fascist propaganda (especially from Italy to the Middle East, where the British had extensive colonial and commercial interests) led the British Broadcasting Corporation (BBC) to initiate its first foreign-language broadcasts. Its request to Parliament to do so was premised on the need to respond to such propaganda, provoking a debate about whether the United Kingdom should engage in propaganda, defined as foreign-language broadcasting. Approval was secured, however, and on 3 January 1938 the BBC began broadcasting in Arabic to the Middle East and North Africa. Its first broadcast caused an immediate stir because it reported the execution of an Arab accused of carrying a concealed weapon. The British Foreign Office feared that the report would result in widespread unrest. This broadcast solidified the BBC's independence from government interference and became the foundation of its reputation for truthful reporting, however, as it demonstrated that the service would broadcast the truth even when the British government disapproved.

It was not merely the use of foreign languages that was considered propaganda, however. Prior to World War II, both Nazi Germany and Fascist Italy used propaganda as a vital part of their peacetime diplomacy. Some nations also used radio for domestic propaganda, seeking to enlist their citizens' support for proposed national efforts or to convince them of the correctness of government policies, even when individual citizens' fortunes didn't seem to be improving. Germany, Italy, and Japan all became masters of such strategies, employing not just radio but a variety of other media, including print and film, to engage their own citizens in preparation for conflict with other countries. In Japan, the emperor's enthronement was broadcast throughout the country in 1928, and in 1936 the Japanese government created an Information Committee under cabinet control to coordinate propaganda activities. All Japanese broadcasts (domestic and international) became subject to censorship. Transmissions that were prohibited included those that "impaired the dignity" of the imperial house (the emperor), those that impaired the honor of the government or of the military services, and those that were determined to be political speeches or discussions. Similar measures were taken in Germany through the *Propagandaministerium* under Joseph Goebbels, whose opinion was that all read broadcasting is true propaganda.

World War II

With the outbreak of World War II in 1939, radio became a full-fledged tool of war. Germany's government declared it illegal to listen to foreign broadcasts, even including broadcasts from Nazi-occupied areas in the ban. Radio sets were confiscated to prevent unauthorized listening and it became a capital offense to listen to certain broadcasts, particularly those from the BBC. It engaged non-German citizens to broadcast in foreign languages to its adversaries, with the most famous broadcaster being the Irishman William Joyce, the second broadcaster dubbed "Lord Haw Haw" in Britain. He was executed by the British at the conclusion of the war for treason. The Italians likewise employed foreigners for its broadcasts, including the American poet Ezra Pound, who was also tried for treason after the war. (Pound was determined to be insane and committed to a mental institution.) Other infamous broadcasters who broadcast specifically to troops included those dubbed "Axis Sally" and "Tokyo Rose" by the Allied forces. Their broadcasts were designed to demoralize Allied troops by exaggerating both Allied losses and naval or battlefield successes of the Axis powers (Germany, Italy, and Japan). They also attempted to make the soldiers homesick by suggesting that their wives and girlfriends were lonely or being unfaithful to them or suggesting that Allied battlefield commanders had little or no concern for the welfare of their troops. Stations carrying such broadcasts also played music popular with the troops in an effort to attract them to listen, and they often were heard by the troops because of that music.

The British became adept at "black propaganda" or clandestine broadcasting, putting signals on the air that purported to be aired from German-occupied territory. The first such British broadcast occurred on 26 May 1940, when *Deutscher Freiheitssender* (German Freedom Station) went on the air, featuring Carl Speiker. The French, too, sponsored such broadcasts until the Nazi *Blitzkrieg* overran their forces, leading to German occupation of Northern France and the French collaborationist *Vichy* government in the south. The Soviet Union and the United States engaged in similar activities. For instance, Radio 1212 purported to be a German radio station broadcasting after D-day but was in fact operated by the U.S. military's psychological warfare unit using the facilities of Radio Luxembourg, which were captured and put back on the air on 22 September 1944. Radio 1212 operated as a clandestine radio station between 2:00 and 6:00 A.M., calling attention to mistakes made by the German high command and other authorities. During the daytime and evening hours the station operated as Radio Free Luxembourg, broadcasting portions of letters captured by the Allies in a program called "Letters That Didn't Reach Them" and greetings to those back home from a parade of carefully selected German prisoners of war, along with other programs. The U.S. serviceman's magazine *Yank* eventually claimed that this station broadcast the first on-air execution, when two German soldiers were shot after being captured while engaged in espionage.

Another strategy employed during World War II was called "ghost-voicing." This involved surreptitiously sliding a signal onto a broadcast from another station at a higher power so that the broadcast sounded as seamless as possible. The ghost signal would then be heard by the listener as if from the original station. Both the British and the Soviets became adept at interrupting Hitler's radio speeches with broadcasts designed to make him appear mentally unbalanced or foolish, or to foment unrest by putting different words in his mouth that were designed to upset his listeners. A similar technique involved recording propaganda broadcasts and then waiting for the right time to replay them to discredit the original source. For instance, when the German army was pushing into the Soviet Union, Nazi broadcasts carried pronouncements from Adolf Hitler concerning the timing of the Soviet state's total collapse. When the Soviet Red Army held Moscow and Stalingrad and began to push the German army back, the BBC rebroadcast Hitler's original claims to demonstrate his fallibility.

The United States entered the "radio war" in February 1942, when the Voice of America (VOA) was first broadcast to Europe. In June the Office of War Information was created for the purpose of disseminating both information and propaganda. It contracted with privately owned U.S. shortwave stations to carry VOA programs. By November the federal government began supervising all privately owned international broadcasting stations. The VOA, like the BBC, saw its role as being that of telling the truth or spreading the gospel of democracy.

Cold War Radio

Propaganda by radio continued after the conclusion of World War II as a result of increasing tensions between the Soviet Union and its former Western wartime allies. The tensions created by this use of international radio made it difficult to resolve differences in various international assemblies, including the International Telecommunication Union and the United Nations, over regulation of the airwaves, issues of national sovereignty, freedom of information, and the right to communicate. Countries using radio for propaganda tended to demonize each other, with radio a major weapon in the Cold War that emerged in the aftermath of World War II. The two "superpowers"—the United States and the Soviet Union (so called due to their size, population, and early development of nuclear arsenals)—engaged in a titanic struggle to win the hearts and minds of the world's peoples. As the old colonial empires began to be replaced by newly independent states, particularly after 1960, both of these countries pressed their propaganda war into new arenas in Africa and Asia. Of particular significance in this postwar propaganda world were the issues of arms control and disarmament, aggressive or military intent—especially in Europe where two military alliances, NATO and the Warsaw Pact, faced off across the Iron Curtain—and the extension or containment of world revolution in response to the ideological rhetorics of democracy versus dictatorship (the Western version) and equality versus hegemony (the Soviet version).

Even though the Voice of America had been dismantled by President Harry Truman in 1945, a new National Security Act passed in 1947 re-established U.S. intelligence operations and created the Central Intelligence Agency (CIA). The CIA secretly funded the creation and maintenance of both Radio Liberty (RL) broadcasting into the Soviet Union itself and Radio Free Europe broadcasting to the Central and Eastern European countries that had fallen under Soviet hegemony at the end of the war. These stations, called "surrogate radio services," were designed to provide domestic news for each of the target countries in order to inform the occupied populations of the truth about what was happening in their own countries. Both Radio Free Europe and Radio Liberty depended on émigré staff from the countries to which they broadcast to lend an aura of authenticity to their broadcasts. All of the countries to which the two services broadcast were considered to be behind the Iron Curtain.

In 1948, after most U.S. information services abroad had been shut down, the U.S. Congress changed its opinion on the value of information and of international broadcasting and passed the Smith-Mundt Act, creating a permanent international information agency and providing operating funds for a revitalized Voice of America. In 1950, with the outbreak of the Korean conflict, President Truman declared a "campaign of truth" against Communist distortions of American actions, to be carried out by U.S. government-funded organizations, including the VOA.

On the Soviet side, Radio Moscow continued operations at the conclusion of World War II. The Soviet Union also saw to it that countries in the Warsaw Pact (a military alliance to offset the NATO alliance of the Western European powers) established radio services to supplement its own activities, which led to the creation of Radio Prague (Czechoslovakia), Radio Berlin (East Germany), Radio Budapest (Hungary), Radio Bucharest (Romania), and Radio Sofia (Bulgaria), among others. Although these stations put their own spin on broadcasts, they were also subject to strict control through national communist parties subject to Moscow. So there was little difference in their opinions on issues that were considered important by the Soviet Union's Communist Party.

The Soviet Union also funded the creation of radio stations by revolutionary movements that it supported throughout the world and engaged in clandestine radio broadcasts such as the Voice of the Turkish Communist Party and Our Radio into Turkey from Romania and East Germany, the National Voice of Iran from the Soviet Union itself, and the Voice of the Iranian Toilers from Afghanistan during the Soviet occupation and war in Afghanistan. In similar fashion, the CIA funded clandestine radio stations such as Radio Swan, directed to Cuba; Voice of Liberation, directed to Guatemala; and Radio *Quince de Septiembre,* directed to Nicaragua, as well as stations in southeast Asia (Voice of the National United Front of Kampuchea) and Iran (Free Voice of Iran). And in the Allied-occupied portion of Berlin, the military established Radio In the American Sector (RIAS), ostensibly to broadcast to Allied troops stationed in the city but with the knowledge that most of the citizens of East Germany could easily tune it in as well.

The nature of the propaganda used by the Cold War powers shifted over the period from 1946 to 1985, too. In the early days, beginning with Truman's campaign of truth, the role of American propaganda radio was to respond to what were seen as Soviet lies and provocations by using radio to correct the record. It was radio of reaction. Gradually, however, the Voice of America began to concentrate instead on presenting the face of the United States to the world by providing more ongoing features about life there, including its culture and its economic and political systems. It became more pro-active, leaving some of the more onerous "corrective" work to be accomplished by Radio Free Europe and Radio Liberty, and later by Radio Martí (directed to Cuba) and Radio Free Asia.

To a degree the VOA followed the BBC's lead. The BBC had been freed from its reactive role to Nazi and Fascist radio by the victory of the Allied powers in World War II. On the Soviet side, too, the nature of the propaganda changed under Premiers Kosygin and Brezhnev, becoming less strident in its complaints about Western portrayals of its intentions and concentrating more on the achievements of Soviet science and technology. By 1981 a publication of Progress Publishers (the English-language press of the Soviet Union), written by Vladimir Artemov, recognized the shift, claiming that radio propaganda, although still providing what he called "ideological interference," had become more versatile and sophisticated.

These changes reduced the amount of explicit propaganda; that is, propaganda using restrictive themes and often demonizing other countries, their leaders, or activities, or relying on catch words, sloganeering, or stock ideological interpretations to focus listeners' attention on the particular aspect of a dispute or event that would provide the most positive spin for the broadcaster's government. The emotional temper of the competitive broadcasts also decreased, even though control of the information flow continued in an attempt to have a particular point of view accepted as the most accurate or legitimate one.

With the inauguration of *perestroika* (restructuring) and *glasnost* (openness) in the Soviet Union by Mikhail Gorbachev in the mid-1980s, the role of propaganda between the USSR and the U.S. (and their respective allies) began to shift again. Gorbachev began to adopt the public relations techniques of the West in an effort to achieve new respectability for the Soviet posture in international relations, again moving away from the most obvious forms of propaganda to those that were more subtle and harder to characterize. The United States, for its part, began to focus its attention even more on "civilized persuasion" conducted through the public vehicles of information distribution via radio and satellite television. It contrasted this approach with what it considered the continuing propaganda of the Soviet state, regardless of the new media strategies that Gorbachev adopted.

With the "velvet revolution" (non-violent political change) that occurred in Central and Eastern Europe beginning in 1989, not only did the role of the former Eastern bloc radio services change but so did the role of Radio Free Europe and Radio Liberty. The most controversial broadcasts by either of these services occurred between 1953 and 1956, when Radio Free Europe was accused of encouraging the Polish and Hungarian uprisings that were put down by Soviet tanks. Apparently many central Europeans had interpreted Radio Free Europe broadcasts as promising U.S. intervention if they rose up against Soviet power. Ironically, the early broadcasts of 1953, based on the defection of a colonel in the Polish secret police, were considered the most effective U.S. political broadcasts since 1945. But when the uprisings occurred in 1956, based on extensive use of the Polish colonel's revelations about secret police files as well as reports on Khrushchev's attack on Stalin for excesses in February 1956 and adoption of an agreement between Khrushchev and Tito on different paths to socialism, they also became the most controversial. After the velvet revolution there was some talk in Washington circles of shutting these services down altogether. But when leaders such as Vaclav Havel of the Czech Republic indicated the need for the stations to continue broadcasting and to assist these countries in the transition to democracy, their broadcasts continued. They began to set up in-country bureaus to facilitate news flow and began to function as truly domestic radio stations, often finding local facilities from which to broadcast their programs. Radio Free Europe even moved most of its operational functions from Munich to Prague and took up residence in the former parliamentary building.

After the Cold War

When Soviet generals attempted a coup against Mikhail Gorbachev in 1991, locking him in his country house until they were stopped in Moscow by its mayor (Boris Yeltsin) and thousands of protesters, Gorbachev later credited international shortwave radio, and particularly the broadcasts of the BBC World Service, with keeping him apprised of unfolding events. Although the Soviet Union collapsed as a unitary state shortly thereafter, Gorbachev's dependence on Western broadcasts underscored the earlier comments of many Soviet dissidents, including Aleksandr Solzhenitsyn, regarding their value.

More old-style propaganda continued to thrive during this period of redefinition, however. A new surrogate station, Radio Free Afghanistan, operated during the Soviet occupation of that country, Radio Martí's operations were supplemented by a new Television Martí and a new Radio Free Asia, all with the operational principles that had guided Radio Free Europe and Radio Liberty during the Cold War largely intact. During the Gulf War of 1991, American fighter-bombers attacked the transmitting complex and studios of Radio Baghdad on several occasions, and the Voice of America found itself subject to review over its suspected role as a propaganda agency.

In continuing regional conflicts, radio also continued to be used for propaganda purposes. In Rwanda, for instance, the Hutu government used radio to adroitly fan the flames of tribal rivalries, inspiring many Hutus to go on killing rampages, although they formerly lived at peace with their Tutsi neighbors. In the Balkans, too, the warring factions used radio to whip up ethnic hatred and to justify policies of genocide. During the Balkan air campaign, U.S. bombers targeted Yugoslavian radio and television operations with the justification that the Milosovik regime in Belgrade was using the media to spread disinformation and hatred toward Bosnians and Kosova Muslims among ethnic Serbs.

In 2000 several clandestine radio operations continued to function throughout the world, with stations operating in Afghanistan, Burma (Myanmar), Colombia, Congo, Eritrea, Ethiopia, Georgia, Iran, Iraq, Kurdistan, Somalia, Sri Lanka, Sudan, Tibet, Vietnam, and West Sahara. Virtually all of these countries either had active liberation movements operating (such as the Tamil Tigers in Sri Lanka and the Kurds in Iraq), had continuing tribal or ethnic rivalries (such as in Congo), or were engaged in cross-border warfare (such as Eritrea and Ethiopia). In other words, all types of military conflicts have included the use of radio as a vehicle of propaganda or psychological warfare.

The arrival of new technologies for international information distribution have meant some reduction in the significance of radio as a medium of propaganda. The Ayatollah Khomeini, for instance, smuggled audiotapes into Iran prior to the Iranian revolution that overthrew the Shah, and Chinese dissidents attempting to get their message to the West have used both facsimile machines and the internet. Even in the depths of the Cold War, Soviet dissidents distributed information clandestinely through printed publications and videotapes of surreptitiously performed dissident plays, distributed by hand from one person to another. But radio, as an inexpensive and widely available medium of communication, will probably remain the medium of choice for protest, propaganda, insurrection, and revolution for many years to come.

ROBERT S. FORTNER

See also Axis Sally; BBC World Service; Board for International Broadcasting; Clandestine Radio; Cold War Radio; International Radio; Lord Haw-Haw; Radio Free Europe/Radio Liberty; Radio Martí; Radio Moscow; Radio Sawa/Middle East Radio Network; Shortwave Radio; Tokyo Rose; Voice of America; World War II and U.S. Radio

Further Reading

Artemov, Vladimir L'vovich, *Information Abused: Critical Essays,* translated by Dmitry Belyavsky, Moscow: Progress, 1981

Browne, Donald R., *International Radio Broadcasting: The Limits of the Limitless Medium,* New York: Praeger, 1982

Elliston, J., Propaganda Pages <www.parascope.com/articles/0797/propaganda.htm>

Hale, Julian Anthony Stuart, *Radio Power: Propaganda and International Broadcasting,* Philadelphia, Pennsylvania: Temple University Press, and London: Elek, 1975

Nelson, Michael, *War of the Black Heavens: The Battles of Western Broadcasting in the Cold War,* Syracuse, New York: Syracuse University Press, and London: Brassey, 1997

Rolo, Charles J., *Radio Goes to War,* New York: Putnam, 1942; London: Faber, 1943

Soley, Lawrence C., *Radio Warfare: OSS and CIA Subversive Propaganda,* New York: Praeger, 1989

Taylor, Philip M., "Propaganda in International Politics, 1919–1939," in *Film and Radio Propaganda in World War II,* edited by Kenneth R.M. Short, Knoxville: University of Tennessee Press, and London: Croom Helm, 1983

Whitton, John B., and John H. Herz, "Radio in International Politics," in *Propaganda by Short Wave,* edited by Harwood L. Childs and John B. Whitton, Princeton, New Jersey: Princeton University Press, and London: Milford and Oxford University Press, 1942

Wood, James, *History of International Broadcasting,* vol. 1, London: Peregrinus, 1992; vol. 2, London: Institute of Electrical Engineers, 2000

Psychographics

Grouping Radio Listeners by Psychological Characteristics

Psychographics is the term for a method of market segmentation that groups consumers on the basis of their psychological characteristics. Unlike demographics, which describes consumer or audience attributes such as sex, age, income, or occupation, psychographics is concerned with unobservable personality traits, such as confidence, aggressiveness, extroversion, curiosity, conscientiousness, agreeableness, and so forth. Psychographics draws inspiration from an array of conceptual perspectives, including theories such as trait-factor, motivation, self-concept, psychoanalytic, and social psychology. Lifestyle characteristics—activities, interests, and opinions—are generally considered a conceptual framework distinct from psychographics. In practice, however, blending personalities and lifestyles is key to producing useful marketing information, and lifestyle characteristics are routinely considered part of a psychographic profile.

Development of Psychographics

Although media and market research about consumer psychology was common as early as the 1920s, the term *psychographics* did not appear until the late 1960s and early 1970s, when target marketing emerged as a predominant business and communication strategy. As more and more companies focused product development and communication efforts on narrowly defined consumer groups, advertisers and marketers called for more sophisticated market segmentation techniques. It had become clear that demographic classifications were insufficient because they lacked the detail necessary for crafting the style of persuasive messages advertisers now preferred. Also inadequate were the prevailing methods for collecting psychological data. Researchers often chose in-depth interviews to uncover psychological and lifestyle dimensions about their subjects. Although rich in detail, the qualitative data were unwieldy for marketers. Interviews were time consuming, which realistically limited the total number of conversations, and therefore, the sample sizes of studies. Interviewing also generated a vast amount of material that was slow to code and cumbersome to analyze.

The emergence of psychographic research paralleled the rapid increase of computer accessibility. Psychographics emphasized easily administered survey instruments with objective questions and precoded responses. Computer data analysis helped psychographic studies include large numbers of subjects, which in turn gave them more general results that could be processed in less time.

Psychographics and Radio

Advertisers' increased emphasis on psychographics also coincided with, and contributed to, the resurgence of radio as a marketing medium in the 1960s and 1970s. Radio was moving from a mass-appeal medium, the something-for-everyone sound and style, to a format-driven medium focused on listener niches and format specialization. At the same time, the proliferation of FM was increasing the number of radio stations and, consequently, competition for advertising dollars. Advertisers were looking beyond standard demographic groupings of target audience; they wanted more tightly focused audience profiles. Differentiating station formats and delivering the audiences for which advertisers were asking became an economic necessity. And it also became necessary to back up claims about number and types of listeners with acceptable cumulative (also known as "cume") audience figures and other ratings details. Not only did the fusion of these various factors stimulate format specialization in radio, it also spurred the creation of hyperspecializations—finely tuned variations on the already flourishing number of general format-types.

The division of the daily radio schedule into dayparts also enhanced radio's attractiveness to advertisers intent on applying psychographic research to their media buys. The accent on lifestyle characteristics in psychographics found a perfect complement in the radio programming day. Advertisers could not only narrow the type of listener to whom they were speaking but could also isolate message sending to the time of day most likely to match target consumers' listening habits.

Although some radio stations used psychographic research to profile their own audiences, most commercial stations continued to market their audiences using demographic descriptions. This practice continues today for several reasons, not the least of which is the prohibitive cost of psychographic research. The fact that radio stations do not typically provide psychographic data about their audiences is not, however, a significant barrier to advertiser purchases of radio spots. Advertisers buy airtime based on both demographic and psychographic data and generally have explicit knowledge of what type of listeners they want.

Public radio is an exception. Public stations regularly use audience personality and lifestyle profiles to entice program sponsors and fortify fund-raising efforts. Understanding listeners' motivations for donating to public broadcasting helps stations to construct persuasive messages and, as a result, to boost financial support from listeners.

Psychographic Measures

Many researchers customize their own segmentation studies as they attempt to predict consumer behavior based on psychographic profiles. Instruments designed to measure various constructs, such as learning style, locus of control, sensation seeking, or general personality traits, illustrate potential tools for gathering psychographic data.

A variety of research firms offer proprietary psychographic research models and syndicated research services. Among the best known are the *Yankelovich Monitor* and the Values and Lifestyles Systems (VALS) from SRI International (formerly the Stanford Research Institute). Since 1970 the *Yankelovich Monitor* has published an annual report on the changing attitudes of adults aged 16 and older based on 2,500 two-hour in-home interviews, combined with written questionnaires. The *Monitor* study is designed to identify broad consumer trends and to build in-depth profiles of target segments.

SRI International created the original VALS in 1978. It offered a psychographic typology that categorized American adults into nine mutually exclusive groups based on consumer responses to questions about lifestyles and social values. The VALS segments were revised and renamed in 1989 in an effort to make VALS more useful to SRI's business customers. Rather than classifying consumers by responses to topical, attitude-oriented issues, the new VALS2 system uses eight profile groups that cluster consumers based on fixed psychological qualities.

CLAUDIA CLARK

See also Audience; Audience Research Methods; Demographics

Further Reading

Gunter, Barrie, and Adrian Furnham, *Consumer Profiles: An Introduction to Psychographics,* London: Routledge, 1992
Heath, Rebecca Piirto, "The Frontiers of Psychographics," *American Demographics* 18, no.7 (July 1996)
Keith, Michael, *The Radio Station,* Boston: Focal Press, 1986; 4th edition, 1997
Riche, Martha Farnsworth, "Psychographics for the 1990s," *American Demographics* 11, no. 7 (July 1989)
Schulberg, Bob, *Radio Advertising: The Authoritative Handbook,* Lincolnwood, Illinois: NTC Business Books, 1989

Public Affairs Programming

From the earliest days of broadcasting, public affairs has been a vital part of the program service of most radio stations and networks. Although it has come in various forms through the years, and despite its decline in recent years in commercial radio, this type of programming has been a resilient, integral player in the public's efforts to understand the vital issues of the day.

"Public affairs" is a broadly construed program type in which current issues of public concern are discussed, analyzed, and debated. The issues may be of broad public interest (such as a presidential election) or designed to appeal to a more narrowly based set of interests (such as the building of an overpass by a public school). The "public" may be defined as the general population of listeners or a more narrowly defined segment (gay men, farmers, housewives, etc.). Yet despite the wide variety of approaches to this program type, there have been two basic approaches to its conceptualization: in the more common one, the public affairs program is designed to have an expert or group of experts discuss the matter at hand; in the other, a more widely drawn segment of opinion and analysis is tapped.

Origins

From the inception of regular broadcasting, radio stations (and later the networks) had an interest in maintaining a public affairs presence. Such programs were inexpensive and helped to build radio's public reputation. Among the earliest were speeches given by prominent people—local and national. In 1923 President Warren G. Harding spoke about the World Court in St. Louis. His speech was carried by local station KSD and by AT&T stations in New York and Washington, D.C., producing the largest audience ever to hear a presidential address at one time. Listeners took these programs quite seriously. Later that year, former President Woodrow Wilson's speech on Armistice Day was broadcast by radio stations in New York City and Schenectady, New York; Washington, D.C.; and Providence, Rhode Island, despite his having been in ill health and out of the public eye for some time. (More than 20,000 people showed up at his house the next day to wish him well.) The following summer, 18 stations linked to WEAF, New York, carried coverage and commentary of the 1924 Democratic National Convention. Listeners heard arguments and violent debates between members of the Ku Klux Klan, and also between New York Governor Alfred E. Smith, former Treasury Secretary William Gibbs MacAdoo, and William Jennings Bryan. Heated arguments went on for hours, complete with cheering and booing from the assembled galleries. Fistfights broke out on the air; they were so tumultuous that the Democratic Party stationed an official censor to stand on the platform in order to shut off the microphone when speeches became too heated.

Despite the popularity of the format, broadcasters found early on that any serious discussion of public affairs was bound to be contentious. Because the espousal of any particular position on a disputed issue was bound to receive favorable comments from those who agreed with it and criticism from those who did not, broadcasters feared alienating any part of their audience, or the current political powers, or (worst of all, from their perspective) existing or potential advertisers. From this concern would come a firm broadcast business stand against allowing purchase of advertising time for expression of views on controversial issues—a ban that lasted well into the 1980s. On 4 April 1922, Hans von (later H.V.) Kaltenborn, then associate editor of the *Brooklyn Eagle,* began a series of half-hour reviews of current world affairs on station WEAF, New York, the radio station of AT&T. These talks were something new for broadcasting, especially his editorial commentary on the affairs of the day. As Kaltenborn notes in his autobiography, *Fifty Fabulous Years* (1950), radio management was reluctant to air such discussions because they feared "the expression of opinion on the air might have dangerous repercussions and might even jeopardize the future of broadcasting." From its beginnings, public affairs programming demonstrated the conflicting pressures on broadcasters of informing the public and protecting the bottom line. Program producers often were caught in a struggle between the public and the commercial interests of station management.

One way to mitigate this "problem" was to air public affairs program series that offered a variety of viewpoints. Among the first public affairs program series were regular broadcasts of *Meetings of the Foreign Policy Association,* which ran on National Broadcasting Company (NBC) Blue from November 1926 to 1940, *Meetings of the Government Club* (NBC, 1926 to 1930), and *Our Government,* a series hosted by journalist David Lawrence discussing the relationships between the federal government, business, and various professions (NBC, 1927 to 1933). In 1929, 227 officials of the U.S. Department of Agriculture gave more than 500 addresses on various issues concerning agricultural issues and policies on NBC stations.

Major Network Series

In the tightening grip of the Depression and the coming of the Roosevelt years, NBC broadcast two public affairs series focusing on economics: *The Economic World Today* (November 1932 to June 1933) and *Economics in a Changing World*

(October 1934 to March 1935). During the mid- to late 1930s, NBC broadcast a number of public affairs series explaining the roles of various New Deal programs such as the National Recovery Administration (1933), Federal Housing Administration (three series from 1934 to 1939), and the Social Security Act (1936 to 1940).

By the mid 1930s public affairs programs had become a regular part of network program schedules. Their primary format was either to present debates or discussions between experts on particular issues or to broadcast interviews of prominent individuals by journalists. While none expected nor achieved large audiences, the relatively small number of listeners were generally those with strong social and political ties, and thus of importance greater than their number. Among the most well known were the *University of Chicago Round Table,* broadcast on WMAQ, Chicago (1931 to 1933), and then on the NBC Red Network (1933 to 1955); *American Forum of the Air,* hosted by Theodore Granik on the Mutual and later NBC networks (1937 to 1956); and *America's Town Meeting of the Air,* hosted by George V. Denny on the NBC Blue/American Broadcasting Companies (ABC) network (1935 to 1956). All shared the format of a panel presenting various viewpoints on issues of the day to the audience. Based on the notion that somehow "scholarly objectivity" would remove any fear that the program could be controversial, *The University of Chicago Round Table* was aimed at an elite, educated audience (panelists were intellectuals, primarily college professors).

The real breakthrough program was *American Forum of the Air.* Sponsored by Gimbel's Department Store in New York, *American Forum* was initially hosted by store employee and law student Theodore Granik. His idea was to provide legal advice and a weekly discussion of legal issues over the air in a panel discussion format. When the program moved to WOR (Newark, New Jersey), Granik started to move the panel toward more controversial questions in a more adversarial format. Guests included members of Congress, Cabinet secretaries, journalists, and other prominent citizens. The topics discussed included the New Deal, labor unrest, civil rights isolationism, fascism and Communism. (It should be noted that no communists were ever allowed to speak on the program even when the subject was the nature of Communism itself.) The program was considered important enough to be printed verbatim in the Congressional Record, resulting in many floor debates initiated by the program. Fireworks erupted when a heated debate (virtually unknown in radio up to that point) broke out on the subject of prohibition, between New York Congressman Emmanuel Celler and Emma Boole of the Women's Christian Temperance Union. Boole charged members of Congress with being illegal drunkards, arguing that there were "underground passages" running directly from Washington speakeasies to congressional offices. The charges caused a national uproar drawing widespread attention, and a large audience, to the program.

The best-remembered series of public affairs speeches were the Fireside Chats of President Franklin D. Roosevelt. The first of these, on the banking crisis, was broadcast 12 March 1933, just a few days after his inauguration. Speaking to an audience estimated at more than 60 million radio listeners, President Roosevelt explained banking practices, his reasons for instituting a "bank holiday," and a call for people to have confidence in the government's ability to carry out his plans. In 104 radio addresses between 1933 and 1936, Roosevelt drew large audiences and an array of support for his New Deal policies. The series lasted until 1944.

But it was in reaction to a neighbor who refused to listen to anything Roosevelt had to say that George V. Denny, Jr. created the best-known public affairs debate program: *America's Town Meeting of the Air.* For more than two decades, *Town Meeting* was the public affairs program of choice for millions of listeners. The program received more than 4,000 pieces of fan mail per week. More than 1,000 *Town Meeting* debate and discussion clubs were formed in libraries, churches, schools, community organizations, and local homes for people to gather, listen, and then continue the debate long into the night after a program was over. The National Women's Radio Committee named *Town Meeting* the best educational program in the country in 1936. High school students in New York City listened to the programs and then participated in similar classroom discussion the next day. In 1938 and 1939, listeners purchased more than 250,000 copies of program transcripts so that they could have a permanent record of what had been said.

Town Meeting's popularity stemmed largely from the range of program debate and the volatility of its format. From the beginning it hosted debates that easily led to heated argument. The initial broadcast on 30 May 1935 was "Which Way America—Communism, Fascism, Socialism or Democracy?" In other broadcasts, Eleanor Roosevelt debated Mrs. Eugene Meyer on the benefits of the New Deal, and Langston Hughes discussed "Let's Face the Race Question" (at a time when the voices of African-Americans were seldom heard). Other speakers included justices of the Supreme Court, Norman Thomas, William Randolph Hearst, Jr., Cabinet secretaries, members of Congress, leading educators, and noted authors. Whereas other programs eschewed contentious feedback from the audience, *Town Meeting* promoted it; audience condemnation and heckling of speakers was expected. In some programs, speakers came close to physical violence on the air.

An October 1931 talk by British playwright George Bernard Shaw on the Columbia Broadcasting System (CBS) created a major stir. The network had wanted Shaw to come to its London studios to give a talk on the current situation in Europe. Network management's elation at his appearance was deflated when he focused the majority of his remarks on praising the Communist system in the Soviet Union: "Hello, all my friends in America! How are all you dear old boobs who have been telling one another for a month that I have gone dotty about Russia? . . . Russia has a laugh on us. She has us fooled, beaten, shamed, shown up, outpointed, and all but knocked out." A resulting widespread public outcry led CBS executives quickly to broadcast a rebuttal from a clergyman in order to counteract what they feared was Shaw's inference that the communist system was divinely favored.

Among the more notable public affairs programs of the 1940s were *Life Begins at 80* (on Mutual and ABC from 1948 to 1953), a discussion of world affairs by a group of senior citizens that was reportedly so frank in its discussions that programs had to be taped and edited before broadcast; *Juvenile Jury* (carried by Mutual and NBC from 1946 to 1953), a program featuring young people giving their perspectives on current issues of the day; and *Leave It to the Girls* (on Mutual, 1945 to 1949), which started as a discussion program featuring career women talking about problems submitted by their listeners, before becoming more comedic.

By the 1950s much of radio's major programming was migrating to television; public affairs programs either followed this trend or met their demise with the growing popularity of the new medium. In December 1950, CBS began a weekly series, *Hear It Now*, hosted by highly regarded correspondent Edward R. Murrow. The program was short-lived, moving to television in September 1951 as *See It Now*. Another radio program, *Meet the Press*, began in the late 1940s on NBC Radio and also moved to television in the 1950s. (It remains a Sunday morning fixture to this day.) The long-running *America's Town Meeting of the Air* left the air in 1956.

Many local stations continued to produce public affairs programs targeted at specific immigrant groups such as Poles, Basques, Japanese, Haitians, and Mexicans. Typical was the *Hellenic Radio Hour* hosted by Penelope Apostolides at stations around Washington, D.C., between 1950 and 1995. The program was a one-hour broadcast featuring news and discussion by and for the Greek community, as well as aspects of Greek culture.

Non-Commercial Programs

While public affairs radio was declining on commercial radio in the 1950s, a newer, more robust format was taking its place: listener-sponsored radio typified by the broadcasts from stations of the Pacifica Foundation, originally of Berkeley, California. Public affairs was one of the four primary areas of station programming. As Eleanor McKinney notes in *The Exacting Ear* (1966), Pacifica Radio's intention was to provide a program service different from that provided by commercial broadcasters, because "(w)e were all convinced that the commercial notion of 'all us bright people in here broadcasting to all you sheep-like masses out there' was completely false."

McKinney cites Lewis Hill, Pacifica's founder, who held that the problem was how to provide listeners with truly provocative programming that addressed significant alternative viewpoints, analyses, and proposals for fixing the major problems of the day. "Radio which aims to do that," Hill argues, "must express what its practitioners believe to be real, good, beautiful and so forth, and what they believe is truly at stake in the assertion of such values." Hill went on to claim that "either some particular person makes up his mind about these things and learns to express them for himself, or we have no values or no significant expression of them."

A cross-section of Pacifica's programs published in 1966 shows that the public affairs commitment of the three stations in Berkeley, Los Angeles, and New York spanned a wide range: a 1953 broadcast of a talk on the "First Amendment: Core of Constitution" delivered before a congressional committee by legal scholar Alexander Meiklejohn; a 1958 hour-long interview with Ammon Hennacy, editor of the *Catholic Worker*, in which he discussed conscientious objection to war and the benefits of a decentralized state; the 1960 broadcast and subsequent documentary productions reporting on House Un-American Activities Committee hearings in San Francisco; a much requested interview by Irish poet and author Ella Young, discussing environmental issues; Supreme Court Justice William O. Douglas discussing racial discrimination, on Independence Day 1962; a documentary, "Freedom Now!" produced from field recordings of blacks and whites during the racial struggles of Birmingham, Alabama, in 1963; and regularly scheduled series of commentaries by William Rusher, editor of the *National Review*, and noted author Ayn Rand. These outlets for a wide range of public affairs remain on the air to this day.

In the 1970s, several former staff and volunteers at the five Pacifica stations (Pacifica had added stations in Houston and Washington, D.C.) were among the first reporters and producers at National Public Radio (NPR). This non-commercial network has been producing two daily news programs (*Morning Edition* and *All Things Considered*) that regularly feature documentaries on a wide range of subjects, such as health care, poverty, environmental concerns, electoral campaign financing, war and peace, famine, and many other subjects.

While commercial radio largely abandoned its public affairs commitment in the wake of Reagan era deregulation, it remains a vital component of non-commercial and community stations around the country. Public affairs programming is one of the hallmarks of NPR. It provides three daily public affairs talk programs: *Talk of the Nation*, a national call-in program that runs for two hours Monday through Thursday afternoons; *Fresh Air*, a daily hour-long interview program that focuses on the arts and culture, and the ways they are imbedded within current events; and *The Dianne Rehm Show*, a daily two-hour call-in program with many distinguished guests, offering listeners opportunities to hear and participate in lively, thoughtful dialogues on a variety of topics.

NPR also provides stations with three weekly public affairs programs. *Latino USA* and host Maria Hinojosa provide public radio audiences with information about the issues and events affecting the lives of the nation's growing and increasingly diverse Latino communities. News round-ups and acclaimed cultural segments promote cross-cultural understanding and develop a forum for Latino cultural and artistic expression. *Living on Earth*, which has won a number of awards, is hosted by Steve Curwood. The program explores the environment—what people are doing to it and what it's doing to us. In-depth coverage, features, interviews, and commentary examine how the environment affects medicine, politics, technology, economics, transportation, agriculture, and more. The third series, *The Merrow Report*, focuses on education, youth, and learning, hosted by John Merrow.

In 1983 a second public radio programming source, American Public Radio (since renamed Public Radio International or PRI), began operations from the Twin Cities of Minnesota. Its stated mission was "to develop distinctive radio programs and to diversify the public radio offerings available to American listeners. Among APR's first program offerings was a two-hour weekly talk, essay, interview, and listener call-in program, *Modern Times with Larry Josephson*. The program (first aired by local station KCRW in Santa Monica, California) was about the basic moral and philosophical questions posed by current issues such as abortion; Supreme Court decisions; the Joel Steinberg/Hedda Nussbaum tragedy (hosted by Susan Brownmiller); the atomic age (hosted by McGeorge Bundy); and the end of the Cold War (hosted by Arthur Schlesinger, Jr.).

During the Persian Gulf War in 1990, APR carried a half-hour nightly program, *Gulf War: Special Edition*, consisting of reports from more than 20 BBC reporters in the Middle East, combined with CBC coverage. The series, modeled after the *Nightly Vietnam Report* of Pacifica's WBAI (New York), provided international perspectives on the war that were unavailable from any single producer or network.

In 1994 APR changed its name to Public Radio International to focus more of its efforts, in part, on globally relevant programming. *The World*, public radio's first global news program, was begun in 1996.

More recently PRI has provided public radio stations with daily public affairs programming from the British Broadcasting Corporation (BBC) and the Canadian Broadcasting Corporation (CBC). Among the BBC offerings are *Newshour*, a 60-minute thrice daily program of news reporting, commentary, and analysis; *The World Today*, a 15-minute program that looks into one international issue each day; and *Outlook*, a 25-minute magazine-style program on international issues. The weekly program *Dialogue* is produced by the Woodrow Wilson International Center for Scholars in association with Radio Smithsonian. This program focuses on topics of national, international, historical and cultural affairs.

Alternative Public Affairs Formats

There have also been efforts to broaden the ways in which stations reach out to their audiences to engage them in discussions and actions concerning issues of current interest. A brief experiment in an alternative form of public affairs radio was *America's Hour*, which ran on CBS from July to September 1935. Its aim was to "boost America" while decrying public dissatisfaction at the height of the Depression. The format was an hour-long melodrama on such issues as railroads, hospitals, mining, and aviation; the intent was to praise mutual management/worker relationships, denouncing "radicals who breed discontent." This program was also noteworthy as being the early breeding ground for radio dramatic actors of later prominence: Orson Welles, Joseph Cotten, Ray Collins, Betty Garde, and Agnes Moorehead.

Some of the more interesting public affairs programs were experiments in using the station to initiate dialogue within the community, with the station seen as a forum for the active engagement of various segments of the local community. Two of the earliest such efforts were produced at commercial FM stations. One of these was the work of Danny Schechter at WBCN-FM, Boston, between 1970 and 1977. WBCN was a major pioneer underground or progressive commercial music station in the country. Calling himself "The News Dissector," Schechter created a public affairs format to match the diverse interests of the station's listeners who, he believed, were interested in public affairs not slanted in the traditional way. Writing in *The More You Watch the Less You Know* (1997), Schechter says his approach was to dissect the news; that is, to

> break it down into elements that explained what was going on, rather than just report the familiar surfaces. . . . [He] wanted to present news that looked at the world from the point of views of people who were trying to change it, rather than those who would keep it the way it was.

Schecter provided in-depth analyses and discussions of such issues as the anti-Vietnam movements, racism and apartheid in South Africa, and the needs and interests of workers in Boston-area manufacturing industries. He sought to bring a more inclusive format to public affairs programs by inviting community activists to participate in discussions of major issues of the day.

Wes "Scoop" Nisker produced public affairs programs in the same vein at KSAN, San Francisco, in the mid 1960s to mid 1970s. Nisker created person-on-the-street packages in which he incorporated a variety of voices, music, and sounds to create programs on subjects such as the annual Gay Pride Day parade, political campaigns, and sex and violence on television. Nisker concluded each broadcast by urging his listeners to become involved directly in the issues of the day, saying, "If you don't like the news you hear on the radio, go out and create some of your own."

Another experiment in engaging the community actively in public affairs was *The Drum* on WBUR, Boston. Begun in 1968 by a consortium composed of staff from WBUR, the Boston mayor's office, Action for Boston Community Development, and local commercial radio and television stations, *The Drum* was an effort to provide both a radio forum for public affairs discussions and a job training site for young adult members of Boston's minority communities. At the time, the only station that had programming targeted at minority communities in Boston signed off at sundown each day. To reach out to these segments of the population, *The Drum* provided a nightly program of news, music, and public affairs features (on issues such as health care, housing, employment, violence, drug abuse, and education) aimed at those communities. Program staff was composed primarily of young men and women recruited from the communities covered. The recruits were contracted to work for the program for a year, during which they were given rigorous training in news and public affairs reporting, writing and program production, announcing, publicity, and community outreach. At the completion of the training year, commercial radio and television stations provided them with jobs as a means of increasing minority staff presence and, for the stations, as a way to gain larger audiences in inner city communities.

Thus *The Drum* provided a model, demonstrating ways that local broadcasters could more closely cater to the needs and interests of under-represented communities in their areas while also expanding their audience base. *The Drum* project was terminated in August 1971. However, many former *Drum* trainees are still working in the broadcasting industries as on-air talent and station management.

The 1970s also saw the founding of radio stations produced by and for Native American communities. More than two dozen Native American stations now have their own national satellite network and have garnered a large audience in indigenous communities. On the reservations the stations produce public affairs news and cultural features of interest to the communities. The sole source of Indian news, these stations act as preservers of Native American languages and culture. They have become the new *eyapaha* (a Lakota word for "town crier"). Typical is station KBRW in Barrow, Alaska. The station's management considers its most important product to be programs about Native American issues and interests, local and state news and discussions, broadcasts of local governmental meetings, personal messages, and public service announcements.

Among the longest running of public affairs commentary programs was *Uncommon Sense: The Radio News Essays of Charles Morgan,* which ran on KPFK, Los Angeles, from 1974 to 1991. This was a series of twice-weekly 15-minute essays on current events covering such topics as the power of the multinational oil companies (which Morgan decried as "The Dictatorship of the Petroletariat"), the Rockefeller-sponsored Trilateral Commission, political extremism, and the increasing disconnection between official news and politics and the growing underclass of people of color in south central and east Los Angeles. From the mid 1980s, Morgan added a listener call-in program, *Talk to Me,* that allowed listeners to respond to what he was saying on the air. *Talk to Me*'s significance was that Morgan encouraged listeners to debate him and each other in a lively exchange of ideas. This was very different from the "question-the-expert/hang-up-for-the-answer" format that dominated talk radio.

Alternative Radio is a public affairs program service in Boulder, Colorado, that attempts to breach the near-monopoly of corporate control over commercial radio outlets. Founded in 1986 by producer David Barsamian, *Alternative Radio* provides lectures by and interviews with outstanding analysts (and individuals usually shunned by mainstream media sources) on a variety of topics. Barsamian offers the hour-long program free to stations and then sells copies of the programs to listeners. Among recent national program bestsellers are historian Howard Zinn on "The Use and Abuse of History," Vermont independent congressman Bernie Sanders on "Single Payer Health Care," and journalists Molly Ivins on "American Political Culture and Other Jokes" and Barbara Ehrenreich on "Trash Media: The Tabloidization of the News." Other regular speakers are Michael Parenti, Noam Chomsky, Ralph Nader, and Dr. Helen Caldicott.

The Women's International News Gathering Service (WINGS) is an all-woman independent radio production company that produces and distributes news and current affairs programs by and about women around the world. WINGS programs are used by non-commercial radio stations, women's studies, and individuals. Programs can be heard on local radio stations, on shortwave, on the internet, and on cassettes. The WINGS mailing list provides updates on stories and new information about women's media. Headquartered in Austin, Texas, WINGS collects programs and news stories to distribute to public radio stations around the country.

A more recent radio public affairs program with an alternative format is the daily *Democracy Now!* This is a national, listener-sponsored public radio and TV show, pioneering the largest community media collaboration in the country. The program started in 1997 as the only daily election show in public broadcasting, and has since broadened its focus to national and international public issues. In 1998, *Democracy Now!* went to Nigeria, Africa's most populous country, to document the activities of U.S. oil companies in the Niger Delta. The program won the 1998 George Polk Award for the radio documentary "Drilling and Killing: Chevron and Nigeria's Military Dictatorship." In November 1999 *Democracy Now!* produced an eight-day series of special reports on the demonstrations against the World Trade Organization meetings in Seattle, Washington. Following these programs, in 2000 *Democracy Now!* pioneered a unique multi-media collaboration involving non-profit community radio, the internet, and satellite and cable television through the Free Speech TV satellite channel. This is the first radio public affairs program to utilize, and to engage, voices from around the world via converging communications technologies.

JOHN HOCHHEIMER

See also All Things Considered; America's Town Meeting of the Air; Commentators; Controversial Issues; Documentary Programs; Editorializing; Educational Radio to 1967; Fairness Doctrine; Fireside Chats; Fresh Air; Hear It Now; Hill, Lewis; Kaltenborn, H.V.; Morning Edition; Murrow, Edward R.; National Public Radio; Native American Radio; News; Pacifica Foundation; Politics and Radio; Public Radio International; Public Radio Since 1967; Public Service Radio; Talk of the Nation; Talk Radio

Further Reading

Barnouw, Erik, *A History of Broadcasting in the United States,* 3 vols., New York: Oxford University Press, 1966–70; see especially vol. 1, *A Tower in Babel: To 1933,* 1966

Chase, Francis Seabury, Jr., *Sound and Fury: An Informal History of Broadcasting,* New York and London: Harper, 1942

Keith, Michael C., *Signals in the Air: Native Broadcasting in America,* Westport, Connecticut.: Praeger, 1995

McKinney, Eleanor, editor, *The Exacting Ear: The Story of Listener-Sponsored Radio and an Anthology of Programs from KPFA, KPFK, and WBAI,* New York: Pantheon, 1966

Schechter, Danny, *The More You Watch, the Less You Know: News Wars/(Sub)Merged Hopes/Media Adventures,* New York: Seven Stories Press, 1997

Public Broadcasting Act of 1967

U.S. Legislation Creating a Support Mechanism for Public Broadcasting

The Public Broadcasting Act of 1967 was the first federal legislation that enabled Congressional support for a national public radio and television system for the American people. As a direct result of the Act signed into Law on 7 November, the Corporation for Public Broadcasting was created, and subsequently National Public Radio (NPR) and the Public Broadcasting Service (PBS) were established as radio and television distribution networks, respectively.

Origins

Historians are fond of recalling that federal support of public radio in the United States was largely an afterthought. The impetus that led to the creation of the Public Broadcasting Act of 1967 is rooted in efforts to gain public awareness and funding for what was then known as educational television. That the Act was rewritten to explicitly include the medium of radio is a testament to the enormous commitment of a handful of radio enthusiasts.

Noncommercial educational radio frequencies were first set aside by the Federal Communications Commission (FCC) in 1938 (for special high-frequency AM stations to broadcast to school classes) and in 1941 for the new FM service. When FM's frequency band shifted in 1945, the educational reservation was shifted as well. In each case, however, the building of stations to use the allocated frequencies was slow in coming. Educational institutions found it difficult to gather funds to put stations on the air and then sustain their on-going operation. Many of the educational radio stations that were built during the 1950s and 1960s failed to achieve the professional standards of their commercial counterparts, and hence the audiences for such outlets were relatively small. So little attention was being given to this "hidden medium" that radio representatives had little influence in the power circles of Washington, D.C.

By contrast, the FCC had created noncommercial educational television channels in April of 1952. The medium of television had caught the public's imagination as an educational resource, although except for Ford Foundation grants its early funding picture was comparable to that of educational radio. Concerns about American education during the late 1950s led to the availability of limited monies to support educational television via the National Defense Education Act of 1958, but educational television advocates saw this modest infusion merely as an important first step. A major political offensive was launched by groups such as the National Association of Educational Broadcasters (NAEB) to take advantage of the pro-educational television campaign rhetoric of President John F. Kennedy. This effort reached fruition on 1 May 1962 when President Kennedy signed into law the Educational Television Facilities Act of 1962. The legislation authorized $32 million over a five-year period to construct new stations or improve the coverage of existing stations. During this period the number of educational television stations nearly doubled.

A New Legislative Initiative

The passage of the Educational Television Act of 1962 not only generated new funding for the construction of educational television stations but also created a new awareness and support base in both houses of Congress. Representatives and senators alike who had been actively involved in the passage of the facilities legislation remained openly impressed with the promise of this new educational medium. One of the senators who had helped mount the charge for educational television funding in the 1950s was Lyndon B. Johnson. When Johnson became president of the United States after the Kennedy assassination, educational television appeared to fit well with his Great Society programs.

The Educational Television Stations (ETS) division of the NAEB held a conference on the long-range financing of educational television in December of 1964. That gathering, and the national survey of station needs associated with it, served as the launching pad for creating a blue ribbon commission to study the future of this important educational resource and to make recommendations to the president. Days after the conference concluded, C. Scott Fletcher of ETS had secured funding from the Carnegie Foundation to form the Carnegie Commission on Educational Television, to be chaired by Ralph Lowell. After a year of study, the Commission issued its report—*Public Television: A Program for Action.* President Johnson received and endorsed the recommendations and then called for legislation that would give life to the vision outlined. Educational broadcasters also applauded the report at a second funding conference held in March 1967 that was designed to encourage prompt congressional action. That action came quickly, as Senate hearings on S.1160 began 11 April 1967. But for advocates of educational radio, or public radio as it was now being called, the legislation had a definite weakness: there was no explicit provision for the radio medium. The proposed law circulating in both houses of Congress (S.1160) was for a Public Television Act.

If there is a single individual who deserves credit for changing the course of this legislation it is Jerrold Sandler, executive director of NAEB's National Educational Radio (NER) division. Sandler was well aware of the ETS division's intention to play down the role of educational radio because of its uneven track record. Without fanfare, Sandler began a campaign to have public radio included in the language of the Act. Among his initiatives were a conference at the Johnson Foundation's Wingspread Center in Racine, Wisconsin, and the commissioning of a national fact-finding study to demonstrate that public radio was indeed alive and well. The resulting report, *The Hidden Medium: A Status Report on Educational Radio in the United States,* was distributed to Washington policy makers after the Senate and House bills had already been scheduled for hearings. But even at this late date, the report had a significant impact. Jerrold Sandler's impassioned testimony during the Senate hearings prompted Senator Griffin of Michigan to propose that the bill be broadened to include radio and the name of the forthcoming legislation be retitled the Public Broadcasting Act of 1967. In addition, the name of the oversight agency to be created by the Act was changed from the Corporation of Public Television to the Corporation for Public Broadcasting. Public radio had scored a major policy victory.

This landmark legislation became Section 396 of the Communications Act of 1934. Congress mandated the FCC to uphold the law that was designed to "encourage the growth and development of public radio and television broadcasting" in the United States. Yet with all its public-interest language and the creation of a new nonprofit organization to ensure that public radio and television would develop and prosper, the Act failed to provide the insulated long-range funding mechanism recommended in the Carnegie Commission report. That failure would consume the energies of the public broadcasting community for decades to come.

ROBERT K. AVERY

See also Communications Act of 1934; Corporation for Public Broadcasting; Educational Radio to 1967; National Association of Educational Broadcasters; National Public Radio; Public Radio International; Public Radio Since 1967; Public Service Radio

Further Reading

Avery, Robert K., and Robert Pepper, "Balancing the Equation: Public Radio Comes of Age," *Public Telecommunications Review* 7, no. 2 (March/April 1979)

Burke, John Edward, *An Historical-Analytical Study of the Legislative and Political Origins of the Public Broadcasting Act of 1967,* New York: Arno Press, 1979

Carnegie Commission on Educational Television, *Public Television: A Program for Action,* New York: Harper and Row, 1967

Herman W. Land Associates, *The Hidden Medium: A Status Report on Educational Radio in the United States,* New York: Herman W. Land Associates, 1967

Witherspoon, John, Roselle Kovitz, Robert K. Avery, and Alan G. Stavitsky, *A History of Public Broadcasting,* Washington, D.C.: Current, 2000

"Public Interest, Convenience or Necessity"

"Public interest, convenience or necessity" is perhaps the most significant phrase in the Communications Act of 1934. Through this durable but flexible set of words, first employed in the Radio Act of 1927, Congress guides (but also allows vital discretion to) the Federal Communications Commission (FCC), radio's most important federal regulator. The agency must act in the public interest. In turn, the rules and policies it creates exist, in part, to prod radio licensees to serve the public interest.

Statutory Origins

The earliest federal radio statutes (the Wireless Ship Acts of 1910 and 1912) were short-lived, although the subsequent Radio Act of 1912 lasted for 15 years. Initially written primarily for maritime wireless telephony, these laws proved inadequate when radio expanded to include broadcasting. Courts ruled that the secretary of commerce (the major regulator under the early acts) lacked the discretion or flexibility to adopt new rules or regulations as radio changed. The secretary's actions, courts said, were limited to the specifics of the Act. Broadcasting required new legislation.

After ignoring this problem for several years, Congress finally adopted the Radio Act of 1927. The new statute created a Federal Radio Commission (FRC) and charged it with keeping up with the rapidly changing field of radio. Unlike the secretary of commerce, the FRC could adopt rules and regulations with the force of law. Such discretion, however, required the statutory limitation that Congress provided by mandating that the FRC regulate radio in the "public interest, convenience or necessity." When the Radio Act of 1927 was replaced by the Communications Act of 1934, Congress re-enacted the public interest standard.

The phrase was derived in part from earlier statutes regulating usage of scarce public resources such as public lands and establishing federal agencies to manage natural monopolies such as railways. By creating the FRC and the FCC and giving them this general statutory charge, Congress could step back from the day-to-day details of regulating rapidly changing radio, but it could also always rein in those agencies by saying, formally and informally, that they had not acted in the public interest. In theory, courts could do the same, if it can be argued that an FCC action is not in the public interest. In practice, however, the Communications Act has granted the FCC wide and rarely challenged discretion. For seven decades, the FCC has justified various rules, regulations, and policies under the standard. As radio (and its social and business contexts) has changed, the FCC has repeatedly altered its understanding of what the public interest requires.

Who Determines the Public Interest?

It is not flippant to say that the public interest in radio is whatever a majority of FCC commissioners believe it to be at any given time. There are two limits to this statement, however. First, the FCC runs the risk of being overturned in court if it cannot justify a rule, regulation, or action as being in the public interest. Courts have historically been reluctant to make this finding, however. More often, they either rule that the FCC has not compiled an adequate record to support its decision—and give the FCC a second chance on remand—or conclude that the commission simply lacked statutory authority to act in the area. FCC actions, of course, must also comply with the Constitution, especially the First Amendment. On rare occasions, courts have ruled that an FCC policy thought to promote the public interest cannot stand because it violates the First Amendment.

Second, the FCC's opinions on the public interest can be undone by Congress, the commission's ultimate source of both budget and policy authority. If FCC rules, policies, or actions substantially distress Congress, legislators can seek to substitute their view of the public interest for the commission's by simply overriding an FCC decision or amending the Communications Act. Such steps are rarely taken, however. It is more common for Congress, through budgetary and oversight hearings, to telegraph warnings to the FCC about its expectations. The FCC usually heeds these warnings and rarely offends Congress even if commissioners believe an offending action to be in the public interest. A mid–2003 package of FCC decisions concerning media ownership sparked considerable congressional concern and disagreement, and an attempt in the Senate to roll back the rules to those existing before the FCC change. But House (and White House) support for the FCC action doomed the Senate initiative.

The FCC is typically a light-handed regulator of broadcasters. In many areas, the FCC leaves them great discretion as to how to fulfill general FCC mandates. Radio broadcasters, for example, must make "reasonable efforts" to provide "reasonable access" to their stations to candidates for federal elective office. In determining what is reasonable, the FCC expects radio broadcasters to consider what would best serve the public interest.

Public Interest Standard and Radio

In the early 1920s, attempts to regulate radio by the secretary of commerce collapsed as courts ruled that the secretary had limited authority to deal with broadcasting. When the Federal Radio Commission was formed in 1927, it had to deal with the consequences of this breakdown. There were more radio stations on the air than the technology of the day could handle, with resulting interference reducing service quality for all. The FRC had to clear the air and reduce the number of licensees.

Some urged that this be done on technical grounds alone, removing from the air stations that caused interference or could not maintain a reliable transmission schedule. Advocates of this approach often pointed to Section 29 of the Radio Act of 1927 prohibiting the FRC from censoring the uses of radio. This limitation, they argued, meant the FRC could not consider the content of the service a broadcaster was providing.

Others concluded, however, that the public interest standard compelled the FRC to consider content, despite the no-censorship clause. The FRC adopted this position and denied licenses and license renewal to radio broadcasters whose content was not at least generally in the public interest. The FRC ruled that the interests of the public, in good service as the commission defined it, were superior to the interests of broadcasters or advertisers. The public interest mandate led the FCC to regard broadcasters as "proxies" or "trustees" for the public. Reasoning that broadcasting was a scarce public resource and that there were more who wanted to broadcast than frequencies to accommodate them, the FCC developed the trusteeship model and assumed the ability to oversee and define the duties of the trustees—radio broadcasters.

The criteria for being a good trustee were not especially burdensome. Licensees were not to run stations solely to serve their own interests or the interests of advertisers. They were not to air programs or hoaxes (such as Orson Welles' 1938 *War of the Worlds* broadcast) that would scare or disrupt the community. They were not to carry programming, such as on-air diagnoses of disease and prescription of remedies, that might cause harm to others. In the very earliest days of the FRC, being a good trustee meant not playing recorded music, on the assumption that people who wanted to hear records could buy them—a public interest perspective the FRC quickly abandoned.

These programming policies eventually evolved into a general FCC expectation that every radio broadcaster would offer a "balanced" or "well-rounded" program service—a something-for-everyone-at-some-time approach that required every radio broadcaster to offer both paid and sustaining (unpaid) programs during the broadcast day that would be of some interest to everyone. These general expectations persisted until the 1960s when radio, as a result of the ascendance of TV, began to develop specialized formats serving narrowly targeted audiences. The FCC acquiesced in this specialization.

Overall FCC regulation of radio content switched from expecting balanced or well-rounded programming to anticipating that broadcasters would offer minimum amounts of non-commercial, non-entertainment programming and refrain from over-commercialization. Although there were never any specific FCC rules setting quantitative news and public affairs expectations, until the early 1980s radio broadcasters ran the risk of having license renewals designated for review by the full FCC if they failed to offer minimum amounts of news or public affairs shows (8 percent for AM, 6 percent for FM), or if they ran too many commercials. Under the public interest standard, the FCC also expected that radio broadcasters would regularly and formally, through surveys of community leaders and the general public, ascertain the problems, needs, and interests of their communities and use their findings to formulate non-commercial, non-entertainment programming.

For many years radio broadcasters were also required to comply with the Fairness Doctrine, another policy the FCC promulgated under the public interest standard. This doctrine imposed two obligations on broadcasters: to devote "reasonable" attention to the coverage of controversial issues of public importance in their community, and to provide a "reasonable opportunity" for opposing views on those issues to be heard. The doctrine was never codified in the Communications Act of 1934; rather, it was another example of a policy created by the FCC under the public interest standard.

Public Interest and the Marketplace

During the 1970s, winds of deregulation swept through the regulatory world, including the FCC. Under both Democratic (Carter) and Republican (Reagan) administrations, and in many areas beyond communications (banking, transportation, etc.), the theory was advanced that marketplace forces, rather than regulation, should be relied upon whenever possible. Regulation, including regulation of radio, should be a last resort, used only when the marketplace produced clearly dysfunctional results. Public interest regulation traced its origins to New Deal responses to the Depression, the greatest marketplace collapse in U.S. history. Fifty years later, with a more robust, capitalistic economy in place, economists, industry leaders, and regulators argued that FCC behavioral regulations, such as the Fairness Doctrine and news programming guidelines, were no longer appropriate. The commission, it was argued, should only rarely substitute its assessment of the public interest for what consumers and the radio industry, responding to marketplace forces, wanted and chose to do.

By the 1980s the number of radio stations was also much greater than in the 1930s and 1940s when many FCC public interest regulations began. The few hundred AM stations of 1934 had grown into thousands of AM and FM stations. With so many more stations on the air, it was argued, members of the public could choose those fitting their own standards or preferences. Radio deregulation orders in 1981 and 1984 eliminated the FCC's expectations about minimal amounts of news and public affairs programs. The commission reasoned that stations should not be compelled to provide specific amounts of such content in the public interest if they did not want to do so and if consumers were uninterested in such content. Other radio stations or perhaps other media would step in and fulfill any need for news if stations decided to cut back. Similarly, the FCC dropped its limits on the amount of time stations could devote to airing commercials. A station running too many commercials would presumably suffer in the marketplace. That marketplace, rather than the FCC, would henceforth protect the public interest from over-commercialization. Finally, the FCC decided that the public interest no longer required broadcasters to formally ascertain the problems, needs, and interests of their communities on a regular basis. Broadcasters out of touch with their communities, it reasoned, would be held in check by marketplace forces, so the rules mandating ascertainment were dropped.

Three years later, again responding to marketplace-based theories of deregulation, the FCC dropped the Fairness Doctrine. It concluded that the doctrine might be counter-productive, as it could push broadcasters to play it safe in order to avoid Fairness Doctrine complaints. But more significantly, the commission believed that a multiplicity of voices in the electronic media marketplace of the late 1980s would, if deregulated, better serve the public's interest in receiving diverse and antagonistic information than regulation by the FCC. Some scholars and many radio industry leaders trace the growth of highly opinionated talk radio in the 1990s to the elimination of the Fairness Doctrine, as the FCC no longer believed that it was contrary to the public interest for a broadcaster to be unfair or unbalanced in the treatment of public issues.

Regulation under the Public Interest Standard Today

At the start of the 21st century, little remains of the public interest regulation of radio as practiced during most of the 20th century. All radio broadcasters are expected by the FCC to offer some "issue responsive" programming, but the commission has almost never questioned broadcasters' interpretations of this vague standard. If licensees prepare and properly place in local public files quarterly "issues/programs" lists identifying at least five issues that have received treatment on the station during the previous quarter, the FCC assumes that the issue-responsive programming obligation has been met. It is considered contrary to the public interest to broadcast false information concerning a crime or catastrophe if it is foreseeable that the broadcast will cause "substantial public harm"—a prohibition aimed mostly at shock-jock hoaxes. The terms and conditions of station contests and promotions must be fully disclosed and, except in extraordinary circumstances, adhered to. Misleading the public about such contests is considered contrary to the public interest.

As the FCC reduced or eliminated its content- and/or conduct-related public interest rules, it sharpened some structural regulations rooted in the public interest standard. The goal has been to promote diversity of station ownership and employment, as a complement to marketplace based deregulation. If the commission is to rely on the marketplace, then that marketplace must be diverse and competitive. Although Congress, through the Telecommunications Act of 1996, eliminated a decades-old FCC public-interest-based cap on the number of radio stations a single owner could own and set off massive consolidation of radio ownership, the FCC adopted limits on the number of stations that could be owned by a single owner within markets. The commission rooted its limits in the theory that dominance of a market by a single voice was contrary to the public interest.

And, since the late 1960s, the FCC has attempted to promote diversity in the broadcast employment marketplace by adopting policies attempting to enhance the employment of minorities and women by broadcasters and the ownership of stations by members of ethnic groups. In the late 1990s, with national standards on equal employment opportunity, affirmative action, and "minority set-asides" shifting, courts questioned and in some instances overturned these FCC policies. Believing that ethnic and gender diversity in employment and station ownership was in the public interest, however, the

FCC adopted revised policies that it hoped would survive judicial scrutiny.

At the start of the 21st century, radio and TV broadcasting are in the process of converting from analog to digital transmission. Digital transmission may dramatically alter the services broadcasters can deliver. Digital radio broadcasting will surely continue to provide listeners with entertainment, information, and advertising. It may also permit additional services to piggyback onto the aural services broadcasters have provided since the 1920s. Digital radio, for example, may have a greater capacity to transmit emergency information than analog broadcasting. Public policy debates are likely to emerge about whether digital radio should have new and different public interest obligations beyond the minimal obligations currently imposed on analog radio. Public interest advocates are likely to argue for such regulations, whereas the radio industry will surely argue that continued reliance on competition and the marketplace is the course of action most in the public interest.

HERBERT A. TERRY

See also Blue Book; Communications Act of 1934; Controversial Issues; Deregulation; Editorializing; Fairness Doctrine; Federal Communications Commission; Federal Radio Commission; Hoaxes on Radio; Telecommunications Act of 1996; United States Congress and Radio; United States Supreme Court and Radio; Wireless Acts of 1910 and 1912/ Radio Acts of 1912 and 1927

Further Reading

The FCC's broadcasting rules, in part reflecting its implementation of the public interest standard, are codified in Title 47 of the *Code of Federal Regulations, Telecommunication.*

Aufderheide, Patricia, *Communications Policy and the Public Interest: The Telecommunications Act of 1996,* New York: Guilford Press, 1999

Benjamin, Louise M., *Freedom of the Air and the Public Interest: First Amendment Rights in Broadcasting to 1935,* Carbondale: Southern Illinois University Press, 2001

Fowler, Marc, and Daniel Brenner, "A Marketplace Approach to Broadcast Regulation," *Texas Law Review* 60 (1982)

Huber, Peter William, *Law and Disorder in Cyberspace: Abolish the FCC and Let Common Law Rule the Telecosm,* Oxford and New York: Oxford University Press, 1997

Krattenmaker, Thomas G., and Lucas A. Powe, Jr., *Regulating Broadcast Programming,* Cambridge, Massachusetts: MIT Press, and Washington, D.C.: American Enterprise Institute Press, 1994

Rowland, Willard, "The Meaning of 'The Public Interest' in Communications Policy, Part I: Its Origins in State and Federal Regulation," *Communication Law and Policy* 2, no. 3 (1997)

Rowland, Willard, "The Meaning of 'The Public Interest' in Communications Policy, Part II: Its Implementation in Early Broadcast Law and Regulation," *Communication Law and Policy* 2, no. 4 (1997)

Public Radio International

Although many people view National Public Radio (NPR) as synonymous with public radio broadcasting in the United States, its rival network, Public Radio International (PRI), actually distributes more programs to public radio stations.

PRI originated as American Public Radio Associates (APR), which grew out of concerns by large-market stations that NPR's control over programming distribution led it to favor its own programs over those produced by member stations. Several of these stations, including Minnesota Public Radio, New York's WNYC, Cincinnati's WGUC, San Francisco's KQED, and KUSC in Los Angeles, founded APR in January 1982. Minnesota Public Radio president William Kling chaired the new organization. From the outset, APR followed an explicitly entrepreneurial model of organization that was frequently at odds with NPR's slow-moving membership model. Whereas NPR was governed by an elected board of station managers, APR operated under an independent board of directors. Whereas NPR developed and produced the majority of its programs, using staff and facilities subsidized by member stations, APR distributed already completed shows from stations and independent producers. Finally, NPR offered an entire program service for a single price to member stations, whereas APR provided individual programs to stations on an exclusive basis.

APR's initial program offering was *A Prairie Home Companion,* which Minnesota Public Radio had syndicated since 1980, after NPR president Frank Mankiewicz had rejected the program as "too parochial." *A Prairie Home Companion* skyrocketed in popularity, ranking second to NPR's *All Things Considered* as an audience (and station fundraising) draw. In

1983 APR incorporated itself as a fully independent organization. The following year, APR began to distribute *Monitorradio,* a news and public-affairs program produced by the *Christian Science Monitor.* In 1985 APR surpassed NPR as the largest supplier of cultural programs in public radio.

APR also benefited from changes in public radio funding in the mid–1980s. Beginning with fiscal year 1987, nearly all federal dollars went directly to stations. Public radio stations could then purchase programs from NPR or from other organizations and stations. By directing federal funds to stations instead of sending money to stations through NPR, local stations gained more control over programming and NPR was buffered from unstable federal funding.

Emboldened by the increase in direct funding, and claiming that NPR's distribution policies (which required member stations to purchase a full schedule rather than individual programs) posed a significant barrier to entry, APR threatened to bring an antitrust suit against NPR. In late 1987 NPR responded by "unbundling" its program service by offering groups of programs, rather than an entire schedule, to stations. However, APR offered producers higher fees, and popular programs such as *Fresh Air, Mountain Stage,* and *Whad'ya Know* began to jump from NPR to APR. NPR responded by further paring back its cultural programming in favor of news and public affairs.

Reflecting its global designs on public broadcasting, American Public Radio changed its name to Public Radio International (PRI) in July 1994. Two years later, PRI made its first venture into program production with *The World,* an ambitious news and public-affairs program designed to compete directly with *All Things Considered.* PRI also distributes the highly popular *Marketplace* financial program, a show that appeals to corporations as well as listeners: by the mid-1990s, *Marketplace* drew 4 percent of all corporate sponsorship money for public radio and brought in the highest sponsor income of any public radio program.

In 1997 PRI counted 591 affiliates; NPR had 635 member stations. The two networks had long been considered bitter rivals; therefore, many observers were stunned in late 1997 when NPR President Delano Lewis approached PRI president Steven Salyer to discuss the possibilities of merging the organizations, believing that a merger would attract more corporate sponsors to public radio, help position public radio against commercial competitors, and also allow public radio to act quickly on entrepreneurial ventures. The plan died quickly, however. Whereas NPR was controlled by stations, PRI was not interested in station representation. An NPR/PRI merger also would have limited the number of opportunities for program distribution and would have reduced diversity. Yet given potential economies of scale and declining federal funding for public radio, such a union may someday prove irresistible. In the meantime, many public radio listeners remain understandably confused about the two services, and their relationship. At the time of writing, the two are totally separate program services for noncommercial radio stations, competing for corporate underwriting dollars.

PRI's leaders have stated that their service offers a competitive alternative to NPR, creating more diversity for the public radio system. Indeed, PRI has distributed many of public radio's outstanding programs throughout its history. Yet much of PRI's success stems from the fact that (unlike NPR) it primarily distributes completed programs, therefore avoiding the costs incurred through production. PRI's attempts at producing programs, such as *The World,* have proved to be highly problematic. Competition may lead to more pluralistic programming, yet competition has its pitfalls when applied to public goods and services. Critics have charged that PRI has focused on reaching upscale audiences from the outset, emphasizing classical music and business-oriented news and public affairs programming while relegating the less popular "conscience" items to NPR. More than any other organization, PRI has played an instrumental role in introducing marketplace economics into the public radio system. Although PRI's financial success is incontestable, its overall contribution to public radio remains subject to debate.

TOM MCCOURT

See also Fresh Air; Keillor, Garrison; Kling, William; Marketplace; Minnesota Public Radio; National Public Radio; Public Affairs Programming; Public Radio Since 1967

Further Reading

Engelman, Ralph, *Public Radio and Television in America: A Political History,* Thousand Oaks, California: Sage, 1996

Ledbetter, James, *Made Possible By . . . : The Death of Public Broadcasting in the United States,* London: Verso, 1997

Looker, Thomas, *The Sound and the Story: NPR and the Art of Radio,* Boston: Houghton Mifflin, 1995

McCourt, Tom, *Conflicting Communication Interests in America: The Case of National Public Radio,* Westport, Connecticut: Praeger, 1999

Witherspoon, John, Roselle Kovitz, Robert K. Avery, and Alan G. Stavitsky, *A History of Public Broadcasting,* Washington, D.C.: Current, 2000

Public Radio Since 1967

Although educational radio had existed for 50 years—longer than commercial radio—a study published in 1967 aptly described it as "the hidden medium." Commercial radio overwhelmed its noncommercial alternative on the AM band through the 1940s. Then, just when the reservation of 20 channels for noncommercial use on the new FM band gave educational radio a new start after World War II, the attention of educational broadcasters switched to television. Radio continued to languish.

The Beginning

A handful of professionally staffed educational stations served mostly rural areas of the country from state universities in 1967. New York City's WNYC and Boston's WGBH were urban exceptions and accounted for much of educational radio's very limited total national listenership. The Pacifica Foundation radio stations in Berkeley, Los Angeles, and New York attracted more notoriety and more listeners than did most educational stations. Voices of political and social dissent, these stations stood somewhat apart from mainstream educational radio stations. They would continue that independent course as leaders of the "community" radio movement separate from, and sometimes in conflict with, the "public" radio discussed in this article.

When educational broadcasters organized to seek federal funding in the mid-1960s, they sought support only for television. Radio, they believed, had no future. Funding educational radio would divert precious resources from television. The clandestine efforts of a small band of maverick educational radio managers, however, quietly slipped the words "and radio" into President Johnson's Public Television Act of 1967. Deputy Undersecretary of Health, Education, and Welfare Dean Costen added those crucial words as he drafted the legislation for the administration. Costen was an old friend and former employee of Ed Burrows, manager of the University of Michigan radio station, WUOM, and a friend of another former WUOM employee, Jerrold Sandler, the Washington lobbyist for educational radio. At the behest of Burrows, Michigan's Senator Robert Griffin sealed radio's victory by amending the name of the 1967 Public Television Act to "Public Broadcasting" and creating a Corporation for Public Broadcasting rather than a Corporation for Public Television. Thus began the modern history of educational—now called public—radio.

Required by law to create a national public radio system out of virtually nothing, the new Corporation for Public Broadcasting (CPB) appointed as director of radio activities Al Hulsen, the manager of the university station at Amherst, Massachusetts, whose soft-spoken style belied a ferocious determination. Hulsen pursued a two-pronged strategy. He offered financial aid to noncommercial stations that reached minimal professional standards: maintaining a staff of at least three full-time members, broadcasting six days per week for 48 weeks a year, and providing a program service of cultural and informational programming aimed at the general public rather than student training, instructional, or religious programming. The minimum requirements were to increase gradually to five full-time staff and 18 hours a day operations for 365 days a year. Though hardly rigorous, these standards excluded all but 72 of the 400 noncommercial radio stations operating in 1969.

The second part of Hulsen's strategy would create a national entity to produce, acquire, and distribute quality programming to those stations. For this part of the strategy, the weakness of the public radio stations proved advantageous. Whereas public television boasted a handful of relatively strong local stations with national programming ambitions that might be thwarted by a single strong national production center, no public radio station felt capable of producing a significant amount of national-quality programming. Moreover, its weakness and obscurity allowed public radio to avoid the scrutiny of the Nixon administration when it put an end to any dreams of a strong "fourth network" for public television. No one objected if the anemic public radio system created a single independent production and distribution entity that could give strong direction and a clear identity.

The argument for a national radio production center went beyond politics, however. Radio programming differed from television programming. Whether commercial or public, television built schedules of unrelated programs from a variety of producers. Contemporary radio stations, commercial or public, built their services around integrated *formats* rather than a series of programs, and such coherence would best be provided by a single production center. An initial planning board incorporated that production center in Washington, D.C. on 3 March 1970 as National Public Radio (NPR).

National Public Radio defined the national identity for public radio. Its initial board of directors, elected by and largely composed of station managers, in turn articulated that identity. Board member William Siemering of WBFO in Buffalo, New York, captured the spirit of the board deliberations and the anti-authoritarian political climate on college campuses in the late 1960s. His "National Public Radio Purposes" set out a series of expectations that significantly modified the formal, elitist quality of traditional educational radio and its model, Britain's British Broadcasting Corporation (BBC). To Siemering and his fellow board members, public radio should be an instrument of direct democracy. It would listen to the nation as much as it would talk to it. Yes, public radio would pursue the highest standards of journalism. Yes, public radio would tap the academic resources of the nation as never before. Yes, public radio would preserve and foster the cultural life of the nation. All of this might have been expected of traditional educational broadcasting. *Public* radio, however, would also reflect the diversity of the nation, giving voice to the unheard, establishing dialogue among those who seldom speak with one another, and seeking wisdom in ordinary people as well as those with credentials. In Siemering's memorable phrase, public radio would "celebrate the human experience."

The chief celebrant would be Siemering himself. NPR's first president, Donald Quayle, hired Siemering as his program director and told him to bring into reality the ideals he had so eloquently enunciated. The implementation proved to be more difficult than the promises, all of which were expected to be realized in NPR's first program offering, *All Things Considered,* a 90-minute daily magazine that debuted at 5:00 P.M., Monday, 3 May 1971. As its title suggested, *All Things Considered* was intended to be something more than a "news" program. It would "contain some news," Siemering said, but *All Things Considered* would also reflect public radio's egalitarian values and a commitment to "quality" in a whole range of topics. Any subject might be considered, as long as it was approached in a considered manner. Quayle and the board intended *All Things Considered* to be public radio's *Sesame Street,* a defining program that would break into the public's consciousness and call attention to this newly defined medium of public radio. The program did indeed come to define public radio, but only slowly and incrementally over the course of the next decade. *All Things Considered* won its first Peabody Award in 1973 and its first Dupont Award in 1976.

All Things Considered contributed to the gradual growth in listenership for and awareness of public radio. The other strand of Hulsen's strategy, building and strengthening local stations, proved to be even more important. The incentive of federal money and the opportunity to bolster their image in their local communities by carrying *All Things Considered* caused universities and other local licensees throughout the country to upgrade their small stations to meet Hulsen's standards or, in many cases, to start new stations from scratch. Hulsen put particular emphasis and resources into upgrading several small college stations with virtually no listenership in the Los Angeles basin into significant enterprises. He focused on turning the Chicago Board of Education's instructional station into a large, powerful public station covering all of "Chicagoland" from the top of the John Hancock Building. He invested significant federal funds in several production centers, most notably Minnesota Public Radio (MPR), which burgeoned from a student radio station at St. John's College to two statewide networks and the largest locally based operation in public radio.

Listenership doubled in the five years after public radio audiences were first measured in 1973, from roughly 2 million listeners a week to a little more than 4 million in 1978. The largest part of that growth came from new and upgraded stations added to the system, and a smaller part came from increased listening to the initial core stations. Federal, state, and local taxes, primarily through state universities, provided more than 80 percent of the funding for these stations. Listeners provided only about 10 percent.

Public radio grew from almost nothing to something real in the 1970s. Nonetheless, it remained in the shadow of public television and found itself responding to the often troubling developments in the visual medium. Although hostile to all media, the administration of Richard Nixon was particularly unhappy about a television system created by Lyndon Johnson's Great Society, supported by tax money, and presumed to be liberal and hostile to Nixon and his policies. The administration was able to express its displeasure through its control of federal appropriations for public media, which the president vetoed in 1972. To strengthen itself politically, public television folded its lobbying activities into its programming organization, the Public Broadcasting Service (PBS), and recruited a politically influential board and board chairman to provide effective leadership.

Public radio might have responded with a parallel structure at NPR. Quayle objected, however, that political activities should be kept separate from an organization like NPR that produced programming, particularly news programming. As a result, public radio established a separate organization in 1973 to handle its lobbying activities, the Association of Public Radio Stations (APRS), with its own board of directors that consisted, as did the NPR board, primarily of elected station managers. The heads of the two largest local public radio organizations, Minnesota Public Radio and Wisconsin Public Radio, assumed the leadership of APRS, and NPR found itself facing a strong rival. William Kling of Minnesota and Ronald Bornstein of Wisconsin criticized the leadership of NPR. More fundamentally, they rejected the concept of a single national production center. They preferred the television model of program production by the larger stations in the system, the largest of which they happened to manage.

The rivalries within public radio weakened further its already weak position relative to public television, particularly when the two media fought one another over the division of CPB funds between radio and television. Television's victory in formulating CPB's 1975 budget forced the radio system to conclude that the division between NPR and APRS needed to end, and the two organizations merged. In reality, the merger in 1977 constituted a takeover of NPR by the APRS leadership, which vowed to give public radio the dynamism they felt it had heretofore lacked.

The Second Beginning

To provide that dynamism, the board of the "new" National Public Radio chose as president Frank Mankiewicz, son and nephew of Hollywood producers and writers Herman and Joseph. Mankiewicz was a lawyer and a reporter, and he had been press secretary to the late Senator Robert Kennedy and manager of George McGovern's 1972 campaign for president. A showman, a journalist, a politician, and a natural promoter, Mankiewicz had all the qualities that might put public radio on the map. He would provide public radio with a second beginning.

Public radio made important strides in four areas during the Mankiewicz years, 1977–83.

Politics. Mankiewicz resolved the continuing conflict with public television over the division of CPB money by convincing Congress to earmark 25 percent of CPB funding for radio, leaving 75 percent for television, a more favorable division for radio than it had ever had—or even hoped for—in the past.

Visibility. Mankiewicz was everywhere. Suddenly national media paid attention to the hidden medium, in part because of Mankiewicz's perseverance and in part because he gave them programs they could write about. NPR gained particular notoriety with its live broadcast of the Senate debate over the Panama Canal treaty and an exclusive call-in program from the White House with President Jimmy Carter. Public radio gained similar publicity for a radio adaptation of the movie hit *Star Wars.*

Programs. Less important in attracting attention, but ultimately far more important in attracting listeners, NPR in 1981 added a two-hour *Morning Edition* complement to its showcase afternoon program, *All Things Considered.* Morning, of course, is radio prime time, and Mankiewicz made public radio competitive where it mattered most, lifting listenership not only in the morning but throughout the day. The birth of *Morning Edition* brought with it a more basic change. With newsmagazines in the morning and the late afternoon, NPR became an around-the-clock news organization and moved firmly into coverage of breaking news. What had been perceived as an "alternative" medium focusing on the offbeat, the whimsical, the arts, ideas, and the lives and opinions of ordinary people in addition to carrying "some news," *All Things Considered* joined *Morning Edition* in focusing more heavily on the news of the day as reported by a greatly expanded system of reporters, most notably the female trio of Nina Totenberg, Linda Wertheimer, and Cokie Roberts. Mankiewicz redefined "alternative" as doing what they other guy does but doing it better. "In depth" replaced "alternative" as public radio's raison d'être.

Satellite Distribution. Congress agreed to fund satellite distribution systems for public radio and television in 1979. The radio satellite dramatically improved the technical quality of public radio's national programming. More important, it allowed multiple programs to be distributed at the same time and allowed live program origination from various places in the country other than Washington, D.C. Multiple origination points gave individual public radio stations the ability to send programming to the system without going through NPR in Washington. It made feasible a television model of multiple program producers, long advocated by Minnesota's Kling, the principle architect of the new satellite system.

Though Bornstein and Kling had been largely responsible for his elevation to the NPR presidency, Frank Mankiewicz could not bring himself to accept a diluted role for NPR as leader and sole programmer for public radio nationally. He would not agree to Kling's proposal that NPR fund, distribute, and promote a live weekly variety show that Minnesota Public Radio produced in St. Paul. Perhaps Kling never really expected—or even wanted—Mankiewicz to accept his proposal, for he used the rejection as a rationale to establish a second network in competition with NPR. American Public Radio (APR) would resemble PBS more than NPR. Like PBS, American Public Radio would not produce its own programs. Rather, it would schedule and promote national programs produced by individual stations, particularly those of Minnesota and four other founding stations that Kling brought into his enterprise: WNYC, New York; WGBH, Boston; KUSC, Los Angeles; and WGUC, Cincinnati. Bill Kling served as president of both American Public Radio and Minnesota Public Radio. Headquarters of the new network started in the MPR building before moving to an office building a few blocks away in downtown St. Paul. Ultimately, American Public Radio signaled its independence from its parent by moving out of St. Paul—all the way to Minneapolis. In 1995 APR changed its name to Public Radio International.

APR burst on the public radio scene in 1980 with the St. Paul–based variety show *A Prairie Home Companion.* Veteran MPR announcer and freelance writer Garrison Keillor hosted the show in an intimate style that conveyed listeners to the mythical town of Lake Wobegon, Minnesota, "where all the women are strong, all the men are good looking, and all the children above average." In addition to the News from Lake Wobegon, *A Prairie Home Companion* included a wide range of musical styles, skits, and commercial parodies. When it declined to fund and distribute the program, NPR had claimed that *A Prairie Home Companion* was too regional, but listeners across the country proved otherwise as they turned on their radios Saturday evenings to enjoy the latest from Lake Wobegon.

Beginning national distribution less than a year after the launch of *Morning Edition* by NPR, *A Prairie Home Companion* joined it and *All Things Considered* as the three programs that defined public radio and drew listeners to it. Garrison Keillor became public radio's most recognized personality, his face reaching the cover of *Time* magazine in 1985. *Morning Edition* and *A Prairie Home Companion*, plus continued growth in the number of public radio stations, caused public radio's audience to double again, to over 8 million by 1983.

The best of times turned into the worst of times in 1983. The 25 percent budget cuts imposed on public broadcasting by Ronald Reagan launched the crisis, but Frank Mankiewicz's reaction to the cuts almost turned it fatal. In every crisis lies an opportunity, and Bill Kling saw the impending budget cuts at public radio stations as a chance for APR to sell those stations large quantities of low-cost, high-quality programming that would allow stations to reduce staffs and costs. Not willing to let Kling steal this market, Mankiewicz responded with an even better package at an even lower cost—so low, in fact, that it could not support itself. Kling wisely ended his project. Mankiewicz went full speed ahead with his, as part of a burgeoning concept of entrepreneurial activities that would allow NPR to "get off the federal fix by '86." Whatever federal money Reagan continued to provide should go to support the stations, Mankiewicz said. NPR would support itself by selling programs to stations and selling a wide variety of services to business partners.

Whatever the merits of these projects in the long run, they required substantial cash investments in the short run, and the need for those investments came in the same year that NPR's federal support, which made up most of its budget, dropped by 25 percent. In March 1982 NPR staff realized they would run a $3 million deficit in that fiscal year. Despite drastic cuts imposed by the NPR board, the deficit projection doubled to $6 million a month later, and to $9 million by June: NPR was insolvent. Mankiewicz, other senior officials, and about a quarter of NPR's 500 employees lost their jobs, and the same Ron Bornstein from Wisconsin who had been instrumental in placing Mankiewicz in the NPR presidency took over as interim president. Bornstein arranged a series of loans from the Corporation for Public Broadcasting. NPR's member stations guaranteed repayment of the loans, and NPR continued to operate.

The Third Beginning

The public radio system that emerged from the NPR crisis looked quite different from the system that went into it. Though NPR was still the most important single organization, individual stations asserted their independence and leadership. The stations, after all, had deposed Mankiewicz and guaranteed the loans that saved NPR. Their responsibility for NPR became permanent in 1985, when they and CPB agreed that federal money that had formerly gone directly to NPR would now go to the stations, who would purchase programs from NPR, APR, or other sources. The larger stations in the system banded together as the Station Resources Group (SRG), expressly designed to exert leadership within the system in place of NPR. The SRG contended that the public radio system brought into existence by federal money through the Corporation for Public Broadcasting should no longer look to government as its primary source of funding. Public radio's future rested not with government support, but with the ability of public radio to raise private funds from corporations, foundations, and, above all, directly from listeners. Reagan's 25 percent cut in federal support, and the anti-government, free-market philosophy it represented, suggested to these stations that government funding was less certain than private money. Moreover, some of them, most particularly Bill Kling of Minnesota, concluded that private funding was preferable to government funding and set out aggressively to seek "underwriting," the euphemism for soft advertising, listener memberships, and major gifts. Again led by Kling, the station managers of the SRG decided that the academic institutions and state and local governments that held the licenses to most public radio stations hampered the new entrepreneurial spirit. Some actually separated from their institutional licensees to become freestanding, not-for-profit corporations; others found ways to operate more independently within their institutional structures.

The results of these efforts were dramatic. By 1998, public radio was a half billion dollar-a-year industry, with the private sector providing more than half of its income. Listeners provided 30 percent of public radio's revenue; business 17 percent; and foundations 10 percent. Only 13 percent came from the federal government, and state and local governments generated 30 percent, primarily through the budgets of state universities. By 1998, public radio was reaching more than 20 million listeners each week, ten times its 1973 listenership.

The ability of public radio to raise money through memberships and underwriting depended directly on its ability to attract, hold, and satisfy listeners. Even those most committed to the more traditional mission of public radio recognized that they fulfilled their mission best when their programs reached the most people. Hence, public radio's priorities after 1983 emphasized audience growth through research. The strategy developed by a national Audience Building Task Force in 1986 set the tone for much of the subsequent development of public radio. Recognizing that most public radio listeners listened to commercial radio more than they listened to public radio, the task force determined that the most direct way to increase public radio listening was to get current listeners to spend more time with public radio and to attract more people like the current listeners. Rather than increasing the diversity of programs in order to appeal to more people, the most successful public radio stations focused on programs that appealed most to existing listeners.

Stations set out to eliminate those program elements that they believed caused listeners to tune away to other stations.

Out went some of the more esoteric and self-indulgent offerings. A similar fate awaited programs too blatantly academic in tone. The hour or two a week of programs aimed at targeted groups turned off those who were not members of those groups; these programs disappeared from most stations. Music in general assumed a reduced role, replaced by news and information, the primary appeal of public radio. The music that remained tended to be confined to classical and jazz, and particularly to the more mainstream selections within those genres. Any hopes for a revival of radio drama evaporated in the quest for consistent appeal.

To replace programming that no longer fit, NPR added programs with qualities that echoed the appeal of *Morning Edition* and *All Things Considered*—programs such as *Weekend Edition, Fresh Air,* and *Talk of the Nation.* Public Radio International offered *Marketplace,* a daily news program focused on economics that was created by two former producers of *All Things Considered* and that appealed to NPR news listeners. Nonetheless, the biggest public radio "hit" of the 1990s turned out to be a quirky program that followed no known formula. *Car Talk,* a less-than-serious advice program, became the most listened to hour on public radio. It was like no other public radio program, yet it appealed to most of the same people who liked *Morning Edition, All Things Considered,* and *A Prairie Home Companion.* It gave them yet another reason to spend more time with public radio.

Research in the late 1980s demonstrated the wisdom of such a strategy as public radio sought to raise money from listeners. The research found that the propensity to make donations related directly to the loyalty of listeners to the station. Those who spent the most time with public radio were the most likely to give. Those who tuned in only occasionally were less likely to give. Survival and growth, then, depended on each public radio station's becoming extremely important in the lives of some people rather than marginally important to many people. Whereas Siemering's philosophy urged public radio to bring together people of all backgrounds, races, regions, ages, and educational levels, the new imperative suggested focusing on a particular subset of the potential audience.

A subsequent study called *Audience 88* identified the people most attracted to public radio programming. It identified educational attainment as the primary predictor of an interest in public radio programming—not surprising, perhaps, for a medium that began in universities as "educational" radio. The more years of education an individual had, the more likely he or she would be to listen to public radio. A substantial part of the audience had earned advanced graduate and professional degrees. Since education correlates directly with income, well-educated public radio listeners tended to be very comfortable financially, but education level, not wealth or social class, predicted loyalty to public radio. Indeed, the most likely of all to love public radio was the individual with a lot of education and a more modest income, the teacher rather than the doctor, the social worker rather than the investment banker. The ultimate public radio listener turned out to be the Ph.D. who drives a cab. In the values and lifestyle terminology of the time, public radio listeners came largely from the psychographic group called "Inner Directed and Societally Conscious."

Public radio had by the end of the 1980s identified its audience and committed itself to serving that audience well, much as commercial media identify target audiences and attempt to give them what they want. Unlike commercial radio, however, public radio formulated its mission and values initially without consideration of an intended audience. It produced programs that reflected the democratic purposes enunciated by Bill Siemering in 1970. It produced programs that sought to be thoughtful, fair, open-minded, and in-depth; programs of substance, not hype; programs that represented the best traditions of the universities that gave birth to public radio, but in an accessible, non-academic style. It produced programs driven not by commercial values but by the desire to "celebrate the human experience."

Listeners who heard this programming, and liked it, self-selected. They chose to listen to programs that resonated with them. This self-selection happened before public radio professionals learned—or cared—who these people were. When public radio decided its life depended on keeping, pleasing, and deepening the loyalty of those listeners, the most effective strategy was clear: public radio needed to commit itself ever more firmly to the original values that attracted those listeners in the first place. Public radio gives its target listeners what they want when it presents programs that reflect the initial academic values of universities, a commitment to depth and quality, and Bill Siemering's faith in the intelligence and openness of ordinary people.

JACK MITCHELL

See also All Things Considered; Car Talk; Community Radio; Corporation for Public Broadcasting; Earplay; Educational Radio to 1967; Edwards, Bob; Fresh Air; Keillor, Garrison; Kling, William; Mankiewicz, Frank; Minnesota Public Radio; Morning Edition; National Public Radio; Pacifica Foundation; Prairie Home Companion; Public Affairs Programming; Public Broadcasting Act; Public Radio International; Public Service Radio; Siemering, William; Simon, Scott; Soundprint; Stamberg, Susan; Star Wars; This American Life; Totenberg, Nina; Wertheimer, Linda; WHA and Wisconsin Public Radio

Further Reading

Collins, Mary, *National Public Radio: The Cast of Characters,* Arlington, Virginia: Seven Locks Press, 1993

Engelman, Ralph, *Public Radio and Television in America: A Political History,* Thousand Oaks, California: Sage, 1996

Fedo, Michael, *The Man from Lake Wobegon,* New York: St. Martin's Press, 1987
Harden, Blaine, "Religious and Public Stations Battle for Share of Radio Dial," *New York Times* (15 September 2002)
Herman W. Land Associates, *The Hidden Medium: A Status Report on Educational Radio in the United States,* Washington, D.C.: The National Association of Educational Broadcasters, 1967
Ledbetter, James, *Made Possible By . . . : The Death of Public Broadcasting in the United States,* London: Verso, 1997
Looker, Thomas, *The Sound and the Story: NPR and the Art of Radio,* Boston: Houghton Mifflin, 1995
McDougal, Dennis, "The Public Radio Wars," four part series, *The Los Angeles Times* (October 8, 9, 10, 11, 1985)
Porter, Bruce, "Has Success Spoiled NPR?" *Columbia Journalism Review* 29, no. 3 (September–October 1990)
Salyer, Stephen, "Monopoly to Marketplace-Competition Comes to Public Radio," *Media Studies Journal* 7, no. 3 (1993)
Savitsky, A.G., "Guys in Suits and Charts, Audience Research in U.S. Public Radio," *Journal of Broadcasting and Electronic Media* 39, no. 2 (1995)
Stamberg, Susan, *Every Night at Five: Susan Stamberg's All Things Considered Book,* New York: Pantheon, 1982
Wertheimer, Linda, *Listening to America: Twenty-five Years in the Life of a Nation, As Heard on National Public Radio,* Boston: Houghton Mifflin, 1995
Witherspoon, John, Roselle Kovitz, Robert K. Avery, and Alan G. Stavitsky, *A History of Public Broadcasting,* Washington, D.C.: Current, 2000
Zuckerman, Lawrence, "Has Success Spoiled NPR?" *Mother Jones* 12, no. 5 (June–July 1987)

Public Service Radio

Public service radio has its roots in Great Britain and is based upon civic principles that envision radio broadcasting as contributing to the betterment of society and the promotion of democratic ideals. The concept of public service radio values public welfare and social good over competitive market forces.

Origins

Public service radio is based on the principles of universality of service, diversity of programming, provision for minority audiences and the disadvantaged, support of an informed electorate, and cultural and educational enrichment. The concept was conceived and fostered within an overarching ideal of cultural and intellectual enlightenment of society. The roots of public service radio are generally traced to documents prepared in support of the establishment of the British Broadcasting Corporation (BBC) by Royal Charter on 1 January 1927. As public trustee, the BBC was to emphasize serious, educational, and cultural programming that would elevate the level of intellectual and aesthetic tastes of its audience. In turn the BBC would be insulated from both political and commercial influence. Therefore, the corporation was a creation of the crown rather than parliament, and funding to support the venture was determined to be derived from license fees on radio (and later television) receivers rather than advertising. Under the skillful leadership of the BBC's first director general, John Reith, this institution of public service radio embarked on an ethical mission of high moral responsibility to utilize the electromagnetic spectrum—a scarce public resource—to enhance the quality of life of all British citizens. Critical inspection of the performance record of public service radio in England since the 1930s would suggest that although there has been a consistent effort to adhere to this lofty idealism, actual practice has never been totally exempt from the political and economic imperatives of modern society. However, in contrast to the performance of the profit-driven commercial radio stations of the United States and elsewhere, the BBC has been repeatedly singled out as a standard bearer of some of the highest quality radio programming available anywhere in the world.

Development of Principles

The notions underlying public service radio undoubtedly grew out of the belief that since the airwaves are an invaluable public resource, the use of this resource must always be driven by a sense of ethical purpose. The medium of radio was seen as being especially well suited to the exploration of society's educational and cultural potential. Fundamental to John Reith's philosophy was a commitment to universality—the idea that services of radio should be made available to the greatest number of citizens possible, thereby elevating the quality of life of an entire society. In defiance of basic capitalistic principles,

Reith saw the radio audience as a set of people needing to be served and uplifted rather than exploited for financial gain. Instead of seeking the largest audiences possible in order to maximize profits, public service radio was supposed to awaken tastes in serious literature, challenge an awareness of the human condition, and stretch the minds of listeners to explore new cultural horizons.

Critics of the principle of universality as applied to public service radio argue that broadcasting the highest quality programming available is of little value if no one is listening. Advocates of public service radio maintain that quality programming is not necessarily dull or boring. Indeed, one of the principles of public service radio is that it offers a wide range of program fare, including entertainment. A commitment to being comprehensive in its approach to diverse service suggests that, within the range of programming offered listeners during a typical broadcast day, some segments do entertain as well as educate and inform. The original conception of multiple services (e.g., Radio 1, Radio 2, and Radio 3), was to offer a range of choices simultaneously that were, by design, distinctively different, as opposed to the virtually identical commercial programs that were all vying for the same mass audience.

Of concern to the original founders of public service radio was that news and public affairs programs would be free from any partisan influence, whether from government or the commercial marketplace. The insulation of revenue from the direct control of any agency that could bias the programming content was seen as a fundamental principle for assuring that the radio service would truly be in service to the entire citizenry. However, as suggested above, this ideal has always existed more in theory than practice.

Global Growth

Within the governance of national authorities, public service radio was recreated across Western European democracies and beyond in various forms. At the core of each was a commitment to operating radio services in the public good. The principal paradigm adopted to accomplish this mission was the establishment of a state-owned broadcasting system that either functioned as a monopoly or at least as the dominant broadcasting institution. Funding came in the form of license fees, taxes, or similar noncommercial options. Examples of these organizations include the Netherlands Broadcasting Foundation, Danish Broadcasting Corporation, Radiodiffusion Télévision Française, Swedish Television Company, Radiotelevisione Italiana, Canadian Broadcasting Corporation, and Australian Broadcasting Corporation. While the ideals on which these and other systems were based suggested services that were characterized by universality and diversity, there were notable violations to these ideals, especially in Germany, France, and Italy. In some cases the state-owned broadcasting system became the political mouthpiece for whomever was in power. Such abuse of the broadcasting institutions' mandate made public service broadcasting the subject of frequent political debates.

Contemporary accounts of public service radio worldwide often include the United States' National Public Radio (NPR) and Public Radio International (PRI) as American examples. However, unlike the British model which was adopted across Europe, the U.S. system came into being as an alternative to the commercially financed and market driven system that has dominated U.S. broadcasting from its inception. Whereas 1927 marked the beginning of public service radio in Britain, the United States Radio Act of 1927 created the communication policy framework that enabled advertiser-supported radio to flourish. Language contained within this act explicitly mandated radio stations to operate "in the public interest, convenience, or necessity," but the public service ideals of raising the educational and cultural standards of the citizenry were marginalized in favor of capitalistic incentives. When the Radio Act was replaced by the Communications Act of 1934, the Federal Communications Commission (FCC) recommended to Congress that "no fixed percentages of radio broadcast facilities be allocated by statute to particular types or kinds of non-profit radio programs or to persons identified with particular types or kinds of non-profit activities." It was not until 1938 that the FCC created an experimental license for "noncommercial educational" radio stations. In 1941 the commission reserved some of the new FM channels for non-commercial licensees. But even though these stations were envisioned to be the United States' answer to the ideals of public service radio, the government's failure to provide any funding mechanism for noncommercial educational stations for decades resulted in a weak and undernourished broadcasting service. Educational radio in the United States was referred to as the "hidden medium." Until passage of the Public Broadcasting Act of 1967, there was no formal government-supported mechanism for ensuring that public (service) radio could develop into a viable national communications medium.

Growing Criticism

During the 1970s and 1980s, public service broadcasting worldwide came under attack, as the underlying principles on which it was based were called into question. Much of the dissatisfaction came as a response to public service radio's television counterpart. The arrival of new modes of television delivery (cable television, satellites, video cassettes) had created new means of access to broadcast services and thus changed the public's perception about the importance and even legitimacy of a broadcasting service founded on the principle of spectrum scarcity. From an ideological perspective, conservative critics were raising questions about the very notion of a public culture and about public service broadcasting as a closed, elitist, inbred, white male institution. Movement toward a global economy was having an ever increasing impact on the way policy-makers saw the products of radio and television. The free market viability of some educational and cultural programming as successful commercial commodities seemed to support the arguments of critics that public service broadcasting on reserved channels was no longer justified.

Deregulation of communication industries was a necessary prerequisite to the breakdown of international trade barriers, and the shift toward increased privatization brought new players into what had been a closed system. The growing appeal of economic directives derived from consumer preferences favored the substitution of the U.S. market forces model for the long-standing public trustee model that had been the backbone of public service radio. Adding to this appeal was the growing realization that program production and distribution costs would continue to mount within an economic climate of flat or decreasing public funding.

By the early 1990s, the groundswell of political and public dissatisfaction with the privileged position of public service broadcasting entities worldwide had reached major proportions. Studies were revealing bureaucratic bungling, cost overruns, and the misuse of funds. One commission after another was recommending at least the partial dismantling or reorganization of existing institutions. New measures of accountability demanded more than idealistic rhetoric, and telecommunication policymakers were turning a deaf ear to public service broadcasting advocates.

Prospects for the Future

Communication scholars in the U.S. who had been reticent on these issues began to mount an intellectual counterattack based largely on the experiences of public broadcasting in the United States. Critiques of U.S. communications policy underscored concerns about the evils of commercialization and the open marketplace. Studies pointed to the loss of minority voices, a steady decline in programs for segmented populations, and a demystification of the illusion of unlimited program choices introduced by the new communications delivery systems. Content analyses revealed program duplication, not diversity, and the question of just how far commercial broadcasters would venture away from the well-proven formulas and formats was getting public attention.

A concerned electorate was beginning to ask whether the wide-scale transformation of telecommunications was not without considerable risk, and whether turning over the electronic sources of culture, education, and political discourse to the ever-shifting forces of the commercial marketplace might have profound negative consequences. Many of the studies that expressed concern about the ever-accelerating growth of market forces pragmatism at the expense of public trust idealism were drawing their interpretive power from theoretical writings about the vital importance of the public sphere to the future of democracy worldwide. The central argument advanced was that within contemporary society individuals pursue their own private self-interests, whereas it is within the public sphere that individuals function in their role as citizens. Public service radio and television were characterized as being essential to the preservation of the public sphere.

By the end of the 20th century, the environment of electronic communications was in a state of flux as newer technologies vied for a piece of a quickly expanding and constantly evolving marketplace. Both commercial and public service radio stations were adding audio streaming on the worldwide web to their traditional modes of distribution. Public service radio institutions were reassessing their missions and were building new alliances with book publishers, computer software manufacturers, and commercial production houses. In the United States, public radio stations were experimenting with enhanced underwriting messages that were sounding more and more like conventional advertising. The relative success of these and other new ventures worldwide was still unknown. Whether public service radio would survive the enormous media transformation that was taking place around the globe had become a frequent topic of both academic and political debate.

ROBERT K. AVERY

See also Africa; Arab World Radio; Australia; British Broadcasting Corporation; Canadian Broadcasting Corporation; European Broadcasting Union; France; Germany; India; Ireland; Israel; Italy; Japan; National Public Radio; Netherlands; Public Affairs Programming; Public Radio International; Public Radio Since 1967; Reith, John C.W.; Scandinavia

Further Reading

Avery, Robert K., editor, *Public Service Broadcasting in a Multichannel Environment: The History and Survival of an Ideal*, New York: Longman, 1993

Blumler, Jay G., and T.J. Nossiter, editors, *Broadcasting Finance in Transition: A Comparative Handbook*, New York: Oxford University Press, 1991

Burns, Tom, *The BBC: Public Institution and Private World*, London: Macmillan, 1977

Emery, Walter B., *National and International Systems of Broadcasting: Their History, Operation, and Control*, East Lansing: Michigan State University Press, 1969

McChesney, Robert Waterman, *Telecommunications, Mass Media, and Democracy: The Battle for the Control of U.S. Broadcasting, 1928–1935*, New York: Oxford University Press, 1993

Raboy, Marc, *Missed Opportunities: The Story of Canada's Broadcasting Policy,* Montreal and Buffalo, New York: McGill-Queen's University Press, 1990
Reith, John Charles Walsham, *Broadcast over Britain,* London: Hodder and Stoughton, 1924; reprint, 1971
Tracey, Michael, *The Decline and Fall of Public Service Broadcasting,* Oxford and New York: Oxford University Press, 1998

Pulse, Inc., The

Audience Research Firm

The Pulse, Inc., provided audience research reports for up to 250 markets from 1941 to 1976. Throughout its history, the company was associated with its founder, Dr. Sydney Roslow.

Origins

In 1940 and 1941 Roslow worked as a psychologist for the Psychological Corporation, a company composed of academic psychologists. In 1939 the corporation had experimented with a roster personal interview technique in an audience study commissioned by station WBEM (Buffalo, New York).

Encouraged by noted social scientist Paul Lazarsfeld, Roslow published the first official *New York Pulse* for October/November 1941. Stations WABC, WEAF, WNEW, and WOR, as well as advertising agency N.W. Ayer, were subscribers to the report, which summarized interviews with 300 respondents per day (2,100 per week). Equal numbers of respondents were assigned to each of the three dayparts included in the report.

World War II slowed the young company. As Roslow went into government service during the war (in the Department of Agriculture in Washington, D.C.), his wife carried on the business. Two large studies were conducted for WCAU (Philadelphia) in 1944 and 1945. Shortly after the war, Pulse added surveys in Boston, Chicago, Cincinnati, Ohio, and Richmond, Virginia. Eventually Pulse became the dominant local radio audience research supplier. By early 1963 Roslow reported that Pulse was publishing reports on 250 markets. Its clients included 150 advertising agencies and 650 radio stations.

At the peak of its popularity, Pulse referred to its research method as "personal house-to-house interviews":

> The emphasis in this survey is that the interview is made at the home.... The roster as used in PULSE surveys is a schedule of radio stations and programs by day-part periods. . . . After the introduction the interviewer is instructed to elicit from the respondents an estimate of when they listened to the radio as the first step. To obtain this information, the interviewer tries to proceed hour by hour through the day, beginning with the time the respondent gets up. Then, the respondents are invited to look at the roster and report their listening.

Pulse respondents reported listening at home and away from home, a unique feature of its audience reports during the company's heyday. Listening preferences, demographics, and other measures were also collected.

The sampling procedure for the Pulse surveys was a sample of "sampling points" (geographical locations distributed at random through the survey areas of the market being studied). This process may be clearer by reviewing the particulars from the October–December 1973 Pulse survey of Atlanta, Georgia. According to the report, Metromail (a market research firm under contract to Pulse) selected at random the addresses of telephone households from the counties surveyed. Each of these addresses became the center of an interviewing "cluster" of approximately 15 interviews. In all, the data for Atlanta came from 146 sampling points, with 2,791 persons interviewed (19 persons per sampling point). As the survey period was 7 October through 28 December 1973 (82 days), it can be seen that the survey could have involved as few as two interviewers, each interviewing persons at one sampling point per day.

The counties surveyed were divided into two parts: the central zone, representing the counties where market radio stations delivered stronger signals; and a larger area, the radio station area, where Atlanta stations were not as dominant and listeners lived farther from the business areas of Atlanta.

In most other respects, the Pulse reports present data similar to that in contemporary radio market reports: listening estimates, cumulative estimates, daypart estimates, in-home and out-of-home estimates, etc. Special sampling provisions were instituted for Hispanic and African-American listeners.

Decline

During the 1970s Pulse was gradually supplanted by Arbitron (formerly the American Research Bureau [ARB]) as the dominant radio audience research company. According to Beville (1988), Roslow attributed this trend to several factors. One was the acceptance by advertising agencies of the diary technique for the measurement of television audiences. It was only a small step further to consider personal diary technique (the Arbitron method) acceptable for measuring radio audiences.

A second factor in Roslow's view was that radio advertising was not profitable enough for agencies to buy more than one audience measurement service; that is, they could not afford to buy reports from both Arbitron and Pulse. A third factor was the technology of the Arbitron parent company at the time, Control Data Corporation (CDC). CDC's large computers could produce more elaborate reports in shorter periods of time and sooner after completion of data collection than the computers available to Pulse.

Another factor in Pulse's loss of business may have been the broadcast industry's All-Radio Methodology Study (ARMS) of 1965. This survey compared the various techniques for collecting radio listening data. ARMS gave the Pulse technique good marks but also provided validation of the personal diary method used by ARB (later Arbitron). In addition, the ARB methods included larger survey areas, making it possible for stations to justify advertising sales on the strength of larger areas included within the surveys of their markets.

During this downward trend in Pulse's business, a number of management changes were also taking place. In 1975, Roslow (who had remained the company's sole owner) retired and moved to Florida after appointing his son Richard president. The following year, two other long-time key officers of the company departed: Laurence Roslow, Sydney's nephew, who had directed all research operations; and sales manager George Sternberg (Sydney's brother-in-law). Left to manage the company, Richard Roslow and his younger brother Peter failed to reverse its fortunes, and in April 1978, Pulse, Inc., closed its doors.

There will always be a lingering nostalgia for Pulse reports among radio managers who remember them, as no other radio audience research methods interviewed listeners face to face. Radio and sales managers had come to feel that this feature alone justified their confidence in the service, as they valued face to face contacts very highly. In addition, the disappearance of Pulse meant that future ratings services for radio had much higher overhead expenses, involving not only huge computers but centralized calling centers and large numbers of workers to collect the necessary data. Prices for rating services inevitably edged upward as well.

JAMES E. FLETCHER

See also Arbitron; Audience Research Methods; Lazarsfeld, Paul F.

Further Reading

Beville, Hugh Malcolm, Jr., *Audience Ratings: Radio, Television, and Cable,* Hillsdale, New Jersey: Erlbaum, 1985; revised edition, 1988

Fletcher, James E., "Commercial Sources of Audience Information—The Rating Report," in *Handbook of Radio and TV Broadcasting: Research Procedures in Audience, Program, and Revenues,* edited by James E. Fletcher, New York: Van Nostrand Reinhold, 1981

Hartshorn, Gerald Gregory, editor, *Audience Research Sourcebook,* Washington, D.C.: National Association of Broadcasters, 1991

The Pulse, *Radio Pulse Atlanta, Georgia, October–November, 1973,* New York: The Pulse, 1973

Webster, James G., and Lawrence W. Lichty, *Ratings Analysis: Theory and Practice,* Hillsdale, New Jersey: Erlbaum, 1991; 2nd edition, as *Ratings Analysis: The Theory and Practice of Audience Research,* by Webster, Lichty, and Patricia F. Phalen, Mahwah, New Jersey: Erlbaum, 2000

Pyne, Joe 1925–1970

U.S. Talk Show Host

In the 1960s, Joe Pyne pioneered the "controversial" talk show format, first on radio and then on television, in which political and economic discussion became a vehicle for entertainment. He was combative with guests, using sharp barbs to insult them and at the same time to please his audience.

Pyne was born in Chester, Pennsylvania, in 1925. A Marine during World War II, he won three battle stars while fighting in the Pacific, and he lost his left leg. After the war, Pyne's broadcasting career began in Chester, at WVCH, in 1948. A year later he hosted a call-in radio talk show, "It's Your Nickel," at WILM in Wilmington, Delaware. He later worked at seven different radio stations in the United States and Canada, including WDEL-TV in Wilmington, before eventually moving to Los Angeles in 1957 to work in radio and television, including KLAC-TV.

As the public became increasingly concerned about the direction the country was headed, listeners wanted not only to hear those and other issues discussed, but also to express their own thoughts on them. Controversy surrounding the conclusions reached in the Warren Commission's 1964 report on the assassination of President Kennedy, as well as the growth of three political movements—the Berkeley Free Speech Movement, opposition to the Vietnam War, and the American civil rights struggle—all provided issues of concern to radio listeners in the mid-1960s. With the telephone, two-way radio was possible, and listeners developed loyalties to radio talk show hosts who reflected their own opinions and viewpoints.

As political and economic issues increasingly seemed to divide Americans, Pyne's style of confrontation with guests, in which he baited and insulted them rather than engaging in intellectual discussion, grew to be extremely popular with audiences. He routinely invited, and then lambasted, guests who held extreme positions, such as black Muslims, American Nazis, and Ku Klux Klansmen. Sometimes the guests self-destructed under intense questioning and ridicule by Pyne, as was the case on one program with a self-described religious bishop who admitted to having had sexual affairs, or on another show when the advocate of the free distribution of the drug LSD revealed he had once spent time confined in a mental institution.

Pyne seemed to relish shocking his guests with his rude behavior, and the more he displayed bad manners, the more his audience seemed to enjoy his performance. Pyne occasionally used airtime to advocate extreme political policies, including bombing Communist China and sentencing those who smoked marijuana to life in prison. Vietnam War protesters were called "peace creeps," and liberals and homosexuals were told they were "stupid" or "jerks." Those Pyne did not agree with were told to go home and "gargle with razor blades." Members of the studio audience were invited into the "Beef Box," where they were permitted to sound off about issues until Pyne grew weary and dismissed them, often after first deriding them.

Metromedia placed Pyne's program in syndication, and in 1966 it was heard over 250 radio stations. In 1966 Pyne expanded his radio show to television, airing first on Los Angeles station KTTV. Metromedia syndicated the program, and *The Joe Pyne Show* eventually was carried in over 80 cities throughout the United States and Canada. Pyne was also tapped to become the host of *Showdown,* a daytime TV quiz show on the National Broadcasting Company (NBC) network.

At its peak, Pyne's morning radio show was ranked number one in its time slot in the Los Angeles market and was syndicated in more than 400 cities and towns throughout the country. His shows were on the air for a total of 27 hours every week.

Pyne's critics called him "Killer Joe," believing he displayed a lack of fair play with his guests. They chided him for behaving cruelly by inviting people onto the show in order to make fools of them, purely for the entertainment of his audience. Some critics regarded Pyne as the host of nothing more than a tasteless "electronic peepshow." *New York Times* critic Jack Gould referred to Pyne as "the ranking nuisance of broadcasting," but he added that if Pyne were to exercise self-restraint, he had the potential to air a show of "vigor and value."

However, Pyne responded to his critics that he did not want to be known as a nice person, and that he required his programs to be "visceral" and not to involve intellectual discussions. He believed his role was to expose extremists, hucksters, and kooks. Pyne also believed his guests were masochistic, looking for him to punish them for their polemics or con games.

Pyne's show was not the first of its kind. In 1958 David Susskind aired a late-night talk show, *Open End,* in which guests argued over political and economic issues, often raising their voices and interrupting each other. However, *The Joe Pyne Show* was one of the first to mold the program around the host's ability to exploit the guests' views for entertainment value. In 1966 Pyne's show spawned similar syndicated programs, such as *The Alan Burke Show,* hosted by another abrasive radio talk announcer, and *Firing Line,* moderated by conservative political columnist and editor William F. Buckley.

Pyne, who chain-smoked on the air, contracted lung cancer and died on 23 March 1970. He was 44.

ROBERT C. FORDAN

See also Controversial Issues; Metromedia; Talk Radio

Joe Pyne. Born in Chester, Pennsylvania, 1925. Served in U.S. Marine Corps during World War II; began broadcasting career in Chester at WVCH, in 1948; worked at several radio stations in United States and Canada; radio show host, KLAC-AM, Los Angeles; host, *Joe Pyne Show,* 1964–70, eventually syndicated by Metromedia; known for confrontational interview style. Died in Hollywood, California, 23 March 1970.

Radio Series
1964–70 *Joe Pyne Show*

Television Series
Joe Pyne Show, 1966–69; *Showdown,* 1966

Films
Unkissed Bride, 1966; *Love-Ins,* 1967

Further Reading
Gould, Jack, "Joe Pyne's Electronic Peepshow," *New York Times* (5 June 1966)
Himmelstein, Hal, *Television Myth and the American Mind,* New York: Praeger, 1984
Hirsch, Alan, *Talking Heads: Political Talk Shows and Their Star Pundits,* New York: St. Martin's Press, 1991
"Killer Joe," *Time* (29 July 1966)
Rose, Brian Geoffrey, and Robert S. Alley, editors, *TV Genres: A Handbook and Reference Guide,* Westport, Connecticut: Greenwood Press, 1985
Williamson, Bruce, "Joe Pyne: TV's New Fun Game: Savagery," *Life* (7 April 1967)

Q

Quaal, Ward L. 1919–

U.S. Broadcasting Executive and Consultant

Although most radio listeners would not recognize his name, Ward L. Quaal is probably one of the most influential figures in U.S. broadcasting. In a career spanning nearly six decades, Quaal has worked in numerous on-air, management, lobbying, and consulting positions.

Quaal began his career in broadcasting while a student at the University of Michigan, working as an announcer, writer, and producer at WBEO in Marquette, Michigan, and later at WJR in Detroit. In 1941, after earning a degree in speech and radio, he began working as an announcer and producer at WGN in Chicago, where he broke into a Chicago Bears football broadcast to announce the Japanese bombing of Pearl Harbor on 7 December 1941.

Quaal entered the service in 1942 and worked as a communications officer in the navy until 1945. After the war, he returned to WGN as special assistant to general manager Frank P. Schreiber. In this capacity, he oversaw the station's farm and public service programming; represented the station in Washington, D.C.; and helped plan the development of WGN-TV, which became Chicago's first full-time television station in April 1948. The contacts Quaal made in Washington, coupled with his effectiveness in communicating his station's interests in a variety of political arenas, would make him a valuable industry spokesperson in the ensuing decades.

In 1949 Quaal took a leave of absence from WGN to become executive director of the Clear Channel Broadcasting Service (CCBS), at the time perhaps the most influential trade group in broadcasting. The CCBS represented independent (non-network-owned) clear channel radio stations seeking to maintain and enhance their status as the only stations broadcasting on their respective frequencies at night. Clear channel stations, through the CCBS, also sought power increases of up to 750,000 watts. One of the clear channel stations' chief political opponents, Senator Edwin Johnson of Colorado, called the CCBS "a well-entrenched, well financed, well staffed group who are determined to have radio control in the United States."

To bolster the argument for higher power, Quaal touted the clear channel stations' efforts to provide farm and rural service programming. He backed up the argument by encouraging member stations to improve such programming, simultaneously building close relationships with leading farm groups. Quaal, in fact, helped orchestrate support for clear channels from influential national farm lobbies such as the National Grange and American Farm Bureau Federation, both of which went on the record in support of clear channel broadcasting.

As the clear channel debate dragged on into the 1950s, Quaal left the CCBS to join Cincinnati's Crosley Broadcasting Corporation in 1952. He continued to be intimately involved with the clear channel debate, however, making frequent trips to Washington to lobby on behalf of the CCBS. After four years at Crosley, Quaal returned once again to WGN and WGN-TV as vice president and general manager in 1956. Following the death of founder Robert R. McCormick, the management at the Chicago *Tribune,* which owned WGN, neglected the radio property in favor of television. "Radio has had it," Quaal was told by one *Tribune* official before he took the job. Quaal proceeded to revamp the station's programming, pulling numerous paid religious programs off the air and concentrating on increased local programming; in fact, he was able to rebuild WGN radio to national prominence. During this time, Quaal also oversaw the expansion of the *Tribune's* radio and television holdings and developed WGN-TV as the model for the modern independent television station. He became president of WGN Continental Broadcasting Company (now Tribune Broadcasting Company) in 1963. In 1975 he retired from WGN and began his own consulting firm, The Ward L. Quaal Company, the following year.

His political clout reached its zenith during the 1980s, because Quaal had been close friends with Ronald Reagan since the 1940s, when the two met at a Chicago radio concert Quaal was announcing. Reagan sought Quaal's input on Federal Communications Commission (FCC) appointments, and in many ways Quaal became *the* voice of the broadcasting industry, as far as the president was concerned. Quaal took the industry's case for blocking reinstatement of the fairness doctrine directly to Reagan, who in 1987 vetoed a government funding bill because it also included the fairness doctrine. Similarly, Quaal supported other measures designed to deregulate the broadcasting industry during the Reagan years. "I take special pride in my efforts to bring full First Amendment rights to broadcasting," Quaal said.

Ironically, by the 1990s Quaal's dissenting voice was drowned out by the industry's quest for raising limits on ownership of radio and television stations. Quaal opposed the provisions in the Telecommunications Act of 1996 that would soon allow companies to own hundreds of radio stations. Quaal maintained that such acquisitions have ruined radio's local service. "They've wrecked radio," Quaal said. "How can you own more than one hundred stations and keep track of their local programming?"

Quaal continues to run his company from offices in Chicago and Los Angeles. He works with clients on management and personnel issues, acquisitions, and lobbying activities. Quaal says that the main things he brings to his clients are experience and "a lot of contacts." Many in the industry agree. "In the trenches of Washington's regulatory battlefields," *Electronic Media* wrote in 1988, "few industry lobbyists can boast the accomplishments of broadcasting's Ward Quaal."

JAMES C. FOUST

See also Clear Channel Stations; Consultants; WGN

Ward L. Quaal. Born in Ishpeming, Michigan, 7 April 1919. Attended University of Michigan, A.B. in Speech and Radio, 1941; announcer/writer, WBEO, Marquette, Michigan, 1936–37; announcer, writer, producer, WJR, Detroit, Michigan, 1937–41; special events announcer, producer, WGN, Chicago, Illinois, 1941–42; served in U.S. Navy, lieutenant, 1942–45; special assistant to general manager, WGN, 1945–49; executive director, Clear Channel Broadcasting Service, 1949–52; assistant general manager, Crosley Broadcasting Corporation, Cincinnati, Ohio, 1952, vice president/general manager, Crosley Broadcasting Corporation, 1953–56; vice president/general manager/member of the board, WGN Continental Broadcasting, 1956; executive vice president, then president, WGN Continental Broadcasting (now Tribune Broadcasting Company), 1960–75; president Ward L. Quaal Company, 1975–; director and member of executive committee, U.S. Satellite Broadcasting Company, 1982–2000; served on numerous boards of organizations. Received Distinguished Service Award, National Association of Broadcasters, 1973; "Blue Ribbon" advisory panel on Advanced Television, FCC, 1987; Hall of Fame, *Broadcasting* magazine, 1991; Silver Circle Award, Chicago Chapter of National Academy of Television Arts and Sciences, 1993.

Selected Publications

Radio-Television-Cable Management (with James A. Brown), 3rd edition, 1998

Further Reading

Fink, John, and Francis Coughlin, *WGN: A Pictorial History,* Chicago: WGN, 1961

Foust, James C., *Big Voices of the Air: The Battle over Clear Channel Radio,* Ames: Iowa State University Press, 2000

Halonen, Doug, "Lobbyist Succeeds with Friends in High Places," *Electronic Media* (8 February 1988)

Quiz and Audience Participation Programs

Few genres used radio's strengths of live broadcasting, spontaneity, and listener involvement more effectively than audience participation programs and their most successful incarnation, the quiz show. Few genres created as much of a sensation as quiz shows at the height of their popularity, or as much of a backlash when condemned by the institutions of broadcasting. And few genres demonstrated the radical transformation of radio in the television era by so quickly abandoning the very medium that gave birth to the quiz show itself. Although quiz shows are best remembered at the center of the infamous TV quiz show scandals of the late 1950s, the genre had a rich history in the early days of radio.

The quiz show is one of the only genres that could truly be called "native" to broadcasting, not stemming from the common sources of vaudeville, theater, film, or literature. Certainly the program format drew upon a number of important antecedents, including newspaper puzzles, parlor games, spelling bees, gambling, carnival contests, and movie-house games such as "Screeno." Despite these varied sources, the specific incarnation of the quiz show on radio was unique, combining the informational content of an educational program, the competitive thrill of sports spectatorship, the humorous patter of comedy and variety shows, and the musical performance featured on much radio programming. The quiz show was the most successful type of audience participation program, which was a general term for any program that incorporated the audience—whether in the studio or at home—into the program's proceedings. Other forms of audience participation programs included "stunt" programs, amateur hours, and "sob shows."

Origins

The radio quiz show had its roots in the earliest days of the medium's commercialization in the mid-1920s, albeit in a form quite different from the way the genre would thrive in the 1940s. Question-and-answer quizzes were a common feature of local radio broadcasts on shows such as WJZ-New York's *The Pop Question Game.* On this program and on other segments on local stations, announcers would ask questions and provide the correct answers after a pause; there were no contestants or prizes, and the audience was expected to try to guess the correct answer at home. These early question-and-answer shows were not terribly popular and were mostly used to fill time and provide an educational diversion within a fairly barren radio schedule. As radio programming became more sophisticated, these early local quizzes were relegated to the sidelines of the radio schedule.

The first major breakthrough of the audience participation genre occurred in the mid-1930s. As national networks came to dominate the airwaves with their high-priced stars, local unaffiliated stations such as New York's WHN needed to devise innovative formats to compete. In 1934 WHN hired theatrical manager Major Edward Bowes to create an inexpensive program to boost their ratings; Bowes drew upon a vaudeville tradition in airing an amateur talent contest. The program was a huge hit—within a year, the National Broadcasting Company (NBC) had purchased the show, and Major Bowes' *The Original Amateur Hour* was voted the most popular show of 1935. The formula was simple—parade a succession of unpolished performers in front of the microphone and reward talented individuals with prizes. An additional gimmick caught on, allowing audiences at home to vote for their favorite amateur via telephone, making the show a truly participatory endeavor. As with most successful radio innovations, a number of imitations followed, leading to an all-out amateur craze in the mid-1930s. Although the amateur hour itself faded in popularity by the end of the decade, its effects lasted far longer through its popularization of the audience participation format.

Although amateur hours showcased ordinary people as the "talent" featured on the radio, they presented traditional forms of entertainment, such as music, comedy, and dancing. It took another program to shift the focus of audience participation to everyday people doing everyday activities. *Vox Pop* emerged on NBC in 1935 with an unusual format—the hosts asked ordinary people questions, broadcasting the resulting dialogue as representative of everyday life. Additionally, the program pioneered the practice of giving prizes to its participants, offering both cash and merchandise from sponsors to interesting guests. Other programs followed in *Vox Pop*'s footsteps, most notably *We, the People* (1936), as the audience participation format developed into a successful alternative to comedies, musical programs, and dramas.

The quiz show itself developed out of these precedents, taking the contest form from amateur hours and the everyday guests from *Vox Pop.* In 1936 the show hailed as the first radio quiz debuted on Washington, D.C., station WJSV; *Professor Quiz* offered ten silver dollars to the contestant who answered the most questions during each program and invited listeners to send in questions for additional cash prizes if their questions were used on the air. The program was picked up by the Columbia Broadcasting System (CBS) in 1937 to great success, leading to numerous clones, including *Uncle Jim's Question Bee, Dr. I.Q., True or False,* and *Ask-It-Basket.* By the late 1930s, quiz shows were established as a popular genre for prime-time radio, with many programs among the top network offerings. The genre established its formula quickly—prizes were modest, questions were intellectual but not too difficult for average listeners, and audience members were invited to participate by sending in their own questions for additional prizes.

Golden Age

Though certainly a fad, quiz shows were able to survive far longer than other radio fads, such as the amateur hour. One reason for the genre's continued success was the creativity of quiz producers in devising variations on the basic formula. One of the first and most successful innovations was NBC's *Information Please.* Producer Dan Golenpaul thought that most quiz show listeners might want to turn the tables on the "know-it-alls" who ran quiz shows and try to "stump the experts." His program, debuting in 1938, offered the chance to do just that—*Information Please* featured a panel of experts on a number of topics, including columnist Franklin Pierce Adams, sportswriter John Kiernan, pianist Oscar Levant, and one rotating guest panelist, all presided over by erudite host Clifton Fadiman. Listeners were invited to submit questions designed to stump the panel, and listeners were rewarded with $10 and an encyclopedia set if they offered a question that could not be answered correctly. The program incorporated a great deal of wit and sophisticated patter among the panel, gave audience members a chance to show up the alleged experts, and created the celebrity panel format that has since become a staple of game shows. The program was hailed not only as fine entertainment but as legitimate education as well, receiving awards from literary magazines and becoming a favorite on college campuses. *Information Please* demonstrated that the quiz show was not dependent on featuring everyday people but could also captivate audiences with celebrities participating in the quizzes.

Whereas *Information Please* succeeded on its intellect and wits, another innovation pushed the genre in the opposite direction. Since the emergence of broadcasting, music has been at the center of radio programming. Quiz shows were quick to incorporate this radio staple to appeal to listeners less interested in intellectual and informational questions. Struggling bandleader Kay Kyser teamed with up-and-coming quiz producer Louis Cowan to devise the *College of Musical Knowledge*, a quiz focused on musical questions interspersed with numbers played by Kyser's band. The program was a hit for Chicago's WGN in 1936 and transferred to NBC in 1938, where it would run for ten years. Similar programs followed *College*'s mixing of quizzes and music in the late 1930s and early 1940s, including *Melody Puzzles, Beat the Band,* and *So You Think You Know Music.* But a more controversial example of the musical audience participation program debuted in 1939—NBC's *Pot o' Gold.*

It was not until *Pot o' Gold* that a radio program took full advantage of the widespread availability of the telephone to allow listeners to participate more fully in broadcasting. The program was not a quiz show by most definitions—most of each show consisted of typical band numbers from Horace Heidt and His Musical Knights. But *Pot o' Gold* had a gimmick that made it a national sensation for two years—during each episode, the hosts would spin a large wheel to randomly select a phone number from a collection of telephone books spanning the country. Heidt would then call the number and award $1,000—then a vast sum for most families still suffering the effects of the Great Depression—to whomever answered the phone. The gimmick was a huge success, creating a new type of audience participation program termed the *giveaway.* For two years, the show was enormously popular, leading to reported drops in movie attendance and phone calling during its Tuesday night time slot, as all of America awaited Heidt's lucrative call. The show even spawned a 1941 movie musical (also entitled *Pot o' Gold*), starring Heidt and Jimmy Stewart, that fictionalized the show's origins. Despite the program's success, it was off the air after two years, because it generated as much controversy as popularity.

Some of the controversy surrounding *Pot o' Gold* stemmed from accusations that the mechanism for choosing telephone numbers was biased, not representing some locations and discriminating against people who had moved since phone books had been issued. Other people objected to the telephone system's inability to guarantee that calls would be put through effectively, worrying that they would miss out on the jackpot. And in the 1940s, not everyone owned a telephone. But the biggest controversy involved the Federal Communications Commission's (FCC) accusation that the program was a lottery and thus violated a provision of the Communication Act of 1934. Although the FCC could not censor programming, it was empowered to prevent stations from broadcasting illegal material, including programs that were deemed to violate federal lottery laws. In 1940 the FCC decided that *Pot o' Gold,* along with a number of local programs, violated the lottery section of the Communication Act and recommended that the responsible broadcasters be prosecuted by the Department of Justice. NBC and other broadcasters denied that giveaway programs were lotteries, because no listener needed to provide any money or other "consideration" (except owning a telephone) to be eligible to win. The Department of Justice refused to prosecute, and the FCC dropped its case, yet broadcasters took the action as a warning; rather than risk being denied license renewal by the FCC, broadcasters canceled or retooled most giveaway programs to make sure they did not violate lottery laws. The link between quiz shows, the FCC, and lottery laws did not disappear, however, and the issue would become even more controversial in the late 1940s.

Although *Pot o' Gold* and other giveaways created a brief sensation, other innovations in the quiz show format proved to have more long-lasting success. Just as *College of Musical Knowledge* thrived by mixing quiz shows with music programs, another genre mixture provided a number of hits: blending children's programming with quiz shows. Cowan, again the crucial innovator, decided to combine the format of *Information Please*'s panel of experts with the widespread appeal of precocious children. The resulting hybrid was *Quiz Kids,* which debuted on NBC in 1940 and lasted for 13 seasons. The program featured a panel of erudite children who amazed audiences both with the extent of their knowledge and their more typically childlike personalities. The program was a hit among both adult and child audiences, and many teachers praised the show's ability to make learning and education seem fun and entertaining. Again the quiz show was held up as both entertaining and educational, although that balance would start to shift throughout the 1940s.

Funny Stunts

As the quiz show entered the 1940s, most of the programs were viewed as respectable, entertaining, and even educational. But one innovation would drastically change the tone of the audience participation format, pushing the genre away from the intellectual pursuits of *Information Please* and toward more outlandish and comic pleasures. Ralph Edwards, a radio announcer, decided to capitalize on the audience participation boom. He felt that many potential audience members would enjoy the participatory aspect of quiz shows but were put off by the intellectual nature of shows such as *Information Please* and *Professor Quiz*; thus he set out to devise a quiz show that focused more on humor and participation than on knowledge and education. His inspiration came from a parlor game he remembered playing as a youth—one person would ask another a question, and if the answer were wrong, the person who answered incorrectly would have to pay for his or her mistake by being forced to do some humiliating "consequence." Edwards named his show after the game, and in 1940 *Truth or Consequences* debuted on CBS. The prizes were small—$15 for a correct answer, $5 for performing a consequence—but audiences were enthralled by the show for other reasons. The consequences became the centerpiece of the program, leading contestants purposely to answer questions wrong to perform comic stunts.

Initially the consequences were quick and modest—one contestant had to spell words while sucking on a lollipop, another had to be a one-man band with pots and pans, and a construction worker had to imitate a bawling baby. As the show grew in popularity, sponsor Ivory Soap upped the production budget to devise more elaborate stunts. The show added remote broadcasts, putting contestants out on the streets to interview strangers or to lie in bed with a seal on a New York street corner. The program also upped the ante for cash prizes, often offering large rewards for completing a stunt—one contestant was promised $1,000 if he could fall asleep during the course of the program. In one of the more notorious stunts, Edwards told a contestant that a cash prize was buried on a street corner in Holyoke, Massachusetts; the man immediately boarded a train to dig up his loot, but he was beaten to the punch by hundreds of local residents who intercepted his $1,000 bounty. Another famed stunt resulted after Edwards told audience members to send a contestant pennies to buy war bonds in 1943; the woman received over 300,000 coins as a result, requiring Edwards to provide helpers to open her mail. As a result of such excessive and outrageous participatory stunts, *Truth or Consequences* became a radio sensation in the 1940s, leading the way for other "stunt" shows to reach the air (and for a town in New Mexico to rename itself after the show).

The most notable clone was NBC's *People Are Funny*, starring Art Linkletter. The show took the basic format of *Truth or Consequences*, adding different stunts and Linkletter's comic personality to generate a large fan base beginning in 1942. One of Linkletter's stunts pitted two contestants against each other to see who could hitchhike across the country faster; the winner was given a new car. Another stunt gave a family their own airplane just for answering the question, "What is your name?" The success of these shows proved the importance of humor to the audience participation format—producers saw that quiz shows did not need to rely on the question-and-answer format to entertain an audience and draw high ratings. *Can You Top This?* was one successful comic quiz (1940–54)—listeners sent in jokes, which were read on the air; then, a panel of comedians tried to "top" each joke with another on the same topic, and the studio audience judged the results. Another program was *It Pays to Be Ignorant* (1942–51), an outright satire of *Information Please* in which panelists humorously failed to answer questions such as "What animal do you get goat's milk from?" The program was fully scripted with no audience participation, but the parody mined the same terrain as quiz shows. By far the most successful comedy quiz was 1947's *You Bet Your Life*, starring the well-known comedian Groucho Marx; the format was that of a typical quiz show, made distinctive only by Marx's comic ad libs and friendly harassment of contestants. Soon the show became more focused on Groucho's quips, with the quiz providing only a basic structure for the comedy.

Besides leading to comedy quizzes, the stunt programs popularized another variation on the audience participation format—the ongoing telephone contest. Although *Pot o' Gold* had made the random telephone giveaway an important feature of audience participation programs, it took the well-established success of *Truth or Consequences* to bring the giveaway back after the FCC's concerns in the early 1940s. Edwards started a contest called "Mr. Hush" in 1946—each week a mystery voice read a riddle and a series of clues. Edwards would then call a random phone number, asking whomever answered to identify the mysterious Mr. Hush; after weeks of failed attempts by other listeners, eventually the listener who gave the correct answer of Jack Dempsey won an enormous jackpot of sponsor-provided merchandise. Subsequent contests, such as "Walking Man" and "Mrs. Hush," were expanded to allow listeners to submit their phone numbers along with contributions for health-related charities; Edward's contests raised millions of dollars for organizations such as the March of Dimes and the American Heart Association. These telephone contests became a national sensation, with winners making headlines and boosting ratings to record levels and inspiring another Jimmy Stewart film, *The Jackpot* (1950). Edwards' contests reinvigorated the giveaway format, which would reappear to greater controversy in the late 1940s.

Quiz shows saw little innovation beyond the stunt programs during the war years. Shows like *Information Please*, *Quiz Kids*, and *Dr. IQ* all continued their success, with few new programs competing against their formula. The new shows that did emerge followed the basic formulas set up by the genre's forerunners, with a few added twists. Two long-lasting programs added a gambling element to the quiz format—

both *Take It or Leave It* (1940–52) and *Double or Nothing* (1940–54) allowed contestants either to take their winnings or to risk them on another question for double the amount. The prizes were modest—*Take It or Leave It*'s grand prize was $64, leading to a new catchphrase, "the $64 question." This basic format would be revisited on television in the late 1950s by the higher-priced quiz show *The $64,000 Question* and in 1999 by *Who Wants to Be a Millionaire?* Although quizzes went mostly unchanged during World War II, there were concerns about the open microphone featured in all audience participation shows—the U.S. government worried about foreign agents using these programs to communicate coded messages. The U.S. Office of Censorship issued guidelines for programs to avoid "man-on-the-street" interviews like those on *Vox Pop* and to be careful in selecting audience members to participate. Quiz shows joined the war effort, donating prizes to war relief and encouraging listeners to participate in war bond drives.

Sob Shows

Another variant in the audience participation show emerged in the mid-1940s. Many daytime programs had established solid audiences, especially among women, by focusing on human-interest stories. The audience participation format was compatible with this type of program, and thus producers created what were often deemed "sob shows." The most famous and long running of these programs was *Queen for a Day*, debuting in 1945. The program featured a panel of women who testified to the hardships of their lives and told listeners their greatest wish. The studio audience would then judge which woman was most "worthy" of rewards, naming her "Queen for a Day." The Queen would be awarded her wish as well as a package of sponsor-provided merchandise. Although the program capitalized on women's poverty and desperation to garner ratings, it also provided both material and emotional uplift for thousands of women over its 20-year run. Other programs succeeded in the daytime schedule, including both more traditional quizzes like *Double or Nothing* and *Give and Take* and other human-interest contests like *Bride and Groom* and *Second Honeymoon*. Audience participation shows had taken root in daytime schedules, a position they continue to inhabit on television to this day.

If *Queen for a Day* established the emotional potential of audience participation programs, it took *Strike It Rich* to fulfill that potential. CBS brought the show to the air in 1947, almost immediately creating controversy. The program featured down-and-out contestants who competed in a short quiz to win up to $800; the real drama followed, as audience members called in on a "heartline" to offer help, in the form of money, jobs, goods, or services, to the needy contestants. People highlighted their hardships to capture the pity of the enthralled home audience, who listened in high numbers, but controversy followed. Critics decried the program as exploiting human misery for profit. One contestant was successful in garnering pity from the audience and was given a good deal of charity, but it was soon discovered that he was an escaped convict from Texas. The New York Department of Welfare complained that people traveled to the city to appear on the program, only to be refused and end up on the welfare rolls; the Department of Welfare demanded that the program be required to get official licensure for providing public welfare aid to contestants. All of this negative publicity merely boosted the show's ratings, and it transferred to television and ran for over a decade. But the quiz show was still in for its most dire round of negative publicity.

Before discussing the last wave of radio quizzes in the late 1940s, it is important to consider not only the programs that composed the genre on radio, but also the cultural values associated with the genre. Although certain formats were celebrated as "quality" radio (such as prime-time drama) and others were derided as inappropriate (such as soap operas), quiz shows and audience participation programs were seen as mostly harmless entertainment with little controversial content. Nevertheless, many critics felt that the genre promoted the "un-American" value of receiving "something for nothing," because people could receive lavish prizes for answering simple questions (or sometimes just for answering the telephone). Fans of the genre, however, felt that the quiz show offered hope during tough times, giving people the promise that their dire straits might be turned around with a simple phone call. Most people thought the genre was "simple entertainment," although some held up the educational possibilities of quizzes to provide knowledge to the masses and to popularize education among listeners. Some audience members became die-hard fans, looking to participate in the programs enough to call themselves "professional contestants" because they frequented the studio broadcasts in New York. Though audience participation shows ran the gamut of cultural legitimacy—from *Information Please*'s highbrow appeals to *Queen for a Day*'s often shameless exploitation of human misery—the quiz show was generally accepted as a valuable part of the radio schedule.

The radio industry saw the genre in more stark economic terms—quiz shows were an inexpensive programming form, simple to produce, with proven popularity. Sponsors liked the programs because they were a highly profitable format—they required little money for "talent" (only hosts and announcers), needed small writing staffs, and could be produced quickly without many rehearsals. Although prizes were often lavish, especially in the late 1940s, producers usually persuaded companies to contribute products to the prize packages in exchange for on-air mentions; this practice was eventually discontinued by the National Association of Broadcasters (NAB) in 1948, because they felt that sponsors were "freeloading" on their programs. Network censors were a bit less enthusiastic about the genre because the ad-libbed format often led to comments that were viewed as inappropriate and hard to control. Despite general industrial support for the audience participation format, the late 1940s would see quiz shows gain a powerful enemy: the FCC.

Stop the Quiz Show!

Following the success of the phone contests on *Truth or Consequences,* a number of programs emerged to capitalize on the giveaway format in the late 1940s. Shows such as *Get Rich Quick* and *Everybody Wins* used the telephone call as a mechanism to draw in listeners by giving away large jackpots, but the most successful and notorious giveaway was ABC's *Stop the Music!* Premiering in early 1948, the show had a simple premise—a band played songs until the announcer yelled, "Stop the Music!" Then host Bert Parks called a random phone number and asked the listener to identify the song. If the listener got the correct title, he or she would win a prize and a chance at the huge jackpot—usually over $20,000 in merchandise—for identifying the "Mystery Melody." The Cowan-produced show became a huge sensation, with high ratings and widespread press coverage. The program's success sparked other giveaways, such as *Sing It Again,* creating the biggest boom in prime-time quiz shows in radio history. Yet the rise of the giveaways prompted numerous protests by various players within the broadcasting industry.

Fred Allen, whose program had been the perennial ratings champion on Sunday nights, found his show sliding when *Stop the Music!* aired in his time slot. He launched a high-profile anti-giveaway campaign in the press, even offering $5,000 to any listener who, if called by *Stop the Music!*, would claim to be listening to Allen instead, an offer Allen never had the opportunity to fulfill. The NAB also felt that the giveaway trend was potentially a detriment to radio; in 1948 the NAB issued a policy statement positioning itself firmly against "buying an audience" instead of offering solid entertainment programs. But the biggest, and most powerful, enemy to the giveaways was the FCC. In the name of the public interest, the commission issued a policy in August 1948 claiming that giveaways violated lottery laws and threatening to revoke licenses of stations that continued to broadcast the programs. Although the order was immediately enjoined by the courts when ABC filed a lawsuit against the FCC's orders, the policy became a lightning rod for the variety of opinions swirling around the genre in the late 1940s and helped put an end to the radio quiz show.

The FCC's ban received overwhelming press coverage, with critics and commentators taking sides on the matter. Thousands of audience letters poured in to the FCC, expressing divergent opinions on the value of both the programs and the FCC's actions. Critics of the genre condemned it as gambling, pandering to base instincts, and not offering wholesome entertainment. Defenders of quiz shows claimed that they were real-life dramas, that they were more entertaining than scripted programs, and that people did not listen simply for a chance to win. The issue dragged on in the courts for years; it was finally resolved in 1954, when the U.S. Supreme Court ruled *(FCC v ABC,* 347 US 284 [1954]) that the FCC had misinterpreted the lottery laws—giveaways were legitimate because contestants were not required to provide "consideration" to be eligible to win. Although successful programs like *Stop the Music!* continued while the courts deliberated, new programs were careful to avoid the giveaway format so they would not suffer from negative publicity. By the time the format was cleared, the giveaway had mostly disappeared as a brief fad. Yet shows like *Stop the Music!* had upped the ante for quiz shows, leading to the huge jackpots that would become staples of the prime-time television quizzes of the late 1950s—and of early 2000.

Although the radio quiz show had moments of intense popularity and publicized controversy, its decline was quiet and swift. As television began to spread into more homes in the postwar era, quiz shows were quick to make the transition to the visual format. Unlike dramatic programs, quizzes did not have to create elaborate sets or visuals to appear on television—cameras could easily capture the inexpensive live proceedings that studio audiences had been witnessing for years. Thus programs like *Stop the Music!, You Bet Your Life, Quiz Kids,* and *Truth or Consequences* all made the transition to television in the early 1950s. Although the networks initially aired these shows on both television and radio, they soon realized that the audiences for radio were dwindling; by removing the shows from the radio, they encouraged fans to purchase televisions when the newer medium became the primary entertainment form of the 1950s. When the TV quiz shows of the late 1950s were exposed as being scripted and "fixed," radio quizzes had long been off the air and thus remained immune from the scandals.

The quiz show was an important part of radio's "golden age," captivating audiences with high-minded questions and emotional appeals, precocious youth and outrageous stunts. Although the format did not last beyond this era of radio broadcasting, its impact is still seen today on television game shows—the inflating jackpots, engaging contestant personalities, amusing celebrity panels, and quick-witted hosts all were conventions established by radio quiz shows. Occasionally radio refers back to these traditions—public radio's *Whad'ya Know* and local stations' call-in giveaways are both updates of classic radio audience participation techniques. Yet these programs were once a broadcasting staple, equal in popularity to better-known genres such as suspense dramas, musical performances, news reports, and celebrity comedy shows. The quiz show was an important, though often ignored, component of radio history, one that warrants a greater examination and appreciation by media historians.

JASON MITTELL

See also Kyser, Kay; Vox Pop; You Bet Your Life

Further Reading

Adams, Franklin P., "Inside 'Information, Please!'" *Harper's* 184 (February 1942)

Beatty, Jerome, "Have You a $100,000 Idea?" *American Magazine* 143, no. 3 (March 1947)

Beatty, Jerome, "Backstage at the Give-Aways," *American Magazine* 148, no. 1 (July 1949)

Cox, Jim, *The Great Radio Audience Participation Shows: Seventeen Programs from the 1940s and 1950s,* Jefferson, North Carolina: McFarland, 2001

DeLong, Thomas A., *Quiz Craze: America's Infatuation with Game Shows,* New York: Praeger, 1991

Eddy, Don, "Daffy Dollars," *American Magazine* 142, no. 6 (December 1946)

Gould, Jack, "Jack Benny or Jackpot?" *New York Times Magazine* (15 August 1948)

James, Ed, "Radio Give-Aways," *American Mercury* 67 (October 1948)

Lear, John, "Magnificent Ignoramus," *Saturday Evening Post* 216, no. 2 (8 July 1944)

Lear, John, "Part-Time Lunatic," *Saturday Evening Post* 218, no. 5 (4 August 1945)

Marks, Leonard H., "Legality of Radio Giveaway Programs," *Georgetown Law Journal* 37 (1949)

Mittell, Jason, "Before the Scandals: The Radio Precedents of the Quiz Show Genre," in *The Radio Reader: Essays in the Cultural History of Radio,* edited by Michele Hilmes and Jason Loviglio, New York: Routledge, 2002

Pringle, Henry F., "Wise Guys of the Air," *Saturday Evening Post* 218, no. 45 (11 May 1946)

Robinson, Henry Morton, "Information Please," *Reader's Digest* 34 (January 1939)

R

RADAR

National Radio Ratings Service

Because radio is woven into the patterns of consumers' day-to-day lives, reaching them at home, in the office, and on the road, research is needed to understand how radio's role in today's media mix may be changing. As with the more commonly known television ratings conducted by Nielsen Media Research, Statistical Research Inc. (SRI) uses a sample of radio listeners to determine listener patterns and characteristics. Radio's All Dimension Audience Research (RADAR) was first produced in 1967 and has been produced by SRI since 1972. Using nearly 30 measured networks, RADAR can provide data on national and network radio audiences in a variety of formats to SRI clients.

For its RADAR product, SRI uses an 8-day telephone interview methodology, which it claims establishes a rapport with respondents that allows them to create an accurate and complete picture of radio exposure over the course of one week. SRI also has the ability to merge its respondent data with some 3 million "clearances" (records of carriage), thus allowing RADAR to provide ratings for specific programs and commercials. Since 1972, SRI has increased the RADAR sample size (from 4,000 to 12,000 respondents annually), the number of measurement weeks (from two to 48 annually), and the frequency of reporting (from annually to quarterly).

One component of the RADAR product allows users to process data and conduct analyses, including profiles of national radio audiences, profiles of network radio audiences, custom electronic ratings books, estimates of the reach and frequency for rotation plans, estimates of the reach and frequency for broadcast schedules, and optimal network radio advertising plans. Another component allows users to combine RADAR data with information from other sources and generate overall reach-and-frequency estimates.

Data from RADAR reach and frequency applications can be used as a base for combinations with other information—about radio, the internet, print, or any medium—that the user has obtained and entered. Up to 12 different other-media properties or sources can be included and collectively weighed at the user's discretion.

There are also two RADAR software applications that enable users to estimate audiences for local markets and programs not measured directly by RADAR. With these tools, users can approximate how much a plan's reach may be increased by scheduling units on non-RADAR programming. In addition, users with access to local data can distribute RADAR's national audience data for a particular schedule to individual markets.

The founders and principals of SRI are Gale Metzger, president (formerly with A.C. Nielson, and the first research supplier to be elected chair of the Advertising Research Foundation's Board of Directors), and Gerald J. Glasser, a former professor of business statistics at New York University. In 1990 Metzger and Glasser were joint recipients of the Hugh Malcolm Beville, Jr., Award of the National Association of Broadcasters and the Broadcast Education Association. The SRI founders were cited for "integrating audience research into the broadcast managerial process" and for "superior leadership in the development of the audience measurement field," among other contributions. The SRI staff includes more than 70 full-time employees. The company is based in Westfield, New Jersey.

RADAR studies are based on probability sampling, high response rates, in-depth interviewing by trained personnel, and multiple checks to ensure accuracy. The key component is the sampling process.

Representative Samples

If every member of the population has an equally good chance of being in the sample, it is a representative sample. Through statistical theory, we know that fairly drawn (or random) samples usually vary in small ways from the population. Over time, these small differences tend to average out. A representative sample does not have to be very large to represent the population from which it is drawn, but it does need to be selected in a way that gives all members of the population the same chance of being chosen. Although it is impossible to determine the exact number of listeners spread over a particular area, well-conducted samples generally provide a good estimate.

In sampling, it is common to select a small portion of the entire population to test. If a more accurate estimate is desired, a larger sample is taken. For example, suppose we select 10,000 people to question regarding their radio listening habits. It would be very unlikely that we would get exactly 5,000 that listened to a particular radio station. Likewise, it would be very unlikely to get 0 or 10,000. However, the percent of the listeners we did find in our sample group would be close to the percent that existed in the entire population. In fact, according to sampling theory, the larger the sample we used, the more confidence we would have in our estimated answer. With a sample of 10,000 we might reasonably conclude that the actual percentage would be between 48 and 52 percent of the listeners. This would suggest that possibly one out of every twenty times we would estimate an answer outside this range. With a larger sample, this range might decrease to 49 to 51 percent, and thus decrease this error probability to more than one out of thirty times.

The RADAR product uses a structured respondent recruiting process. This process includes incentives, advance contacts with potential respondents, and flexibility in the week for which the radio listening data are compiled. This process results in a tabulated sample of about 50 percent.

In addition to the RADAR product, SRI has conducted numerous proprietary studies on radio for the major broadcast networks and programming suppliers. Studies resulting from this work include: the ability of radio commercials to evoke images from familiar TV ads; the way people relate to radio, including the role it plays in their daily lives; awareness of and attention to public radio, as well as reactions of public radio audiences to programming and on-air fund raising; perceptions of local radio stations—reasons for listening and non-listening, station image, and evaluations of station personalities; and the accuracy ascribed to radio news broadcasts by listeners.

Tracking Trends

Because the methodology of RADAR research has remained consistent, there exists a capability to track trends across more than 25 years of data. For example, trend comparisons have shown that as women have moved increasingly into the labor market, their radio usage has become more similar to men's. In addition, listening itself has also moved outside the home. In 1998, 38 percent of radio usage occurred in homes, as opposed to 61 percent in 1980, and car radio listening has nearly doubled over the same time, from 17 percent to 33 percent. Finally, the FM and AM bands have essentially switched places in terms of listenership during the past 25 years; AM's audience share has dropped from 75 percent to 18 percent, and FM's share has risen from 25 percent to 82 percent.

A significant share of SRI's work has been in media audience measurement. SRI's clients have included all major television and radio networks, professional sports leagues, Fortune 500 manufacturers, and foundations.

DENNIS RANDOLPH

See also Audience Research Methods

Further Reading

Beville, Hugh Malcolm, Jr., "RADAR," in *Audience Ratings: Radio, Television, and Cable,* Hillsdale, New Jersey: Erlbaum, 1985; 2nd edition, 1988

Webster, James G., and Lawrence W. Lichty, *Ratings Analysis: Theory and Practice,* Mahwah, New Jersey: Erlbaum, 1991; 2nd edition, as *Ratings Analysis: The Theory and Practice of Audience Research,* by Webster, Lichty, and Patricia F. Phalen, Mahwah, New Jersey, and London: Erlbaum, 2000

Radio Academy

Promoting and Celebrating British Radio

The Radio Academy in the United Kingdom was established in 1983 "to encourage the pursuit of excellence in all aspects of the radio broadcasting industry, and to foster a greater understanding of the medium" (from the mission statement). It includes members from the British Broadcasting Corporation (BBC) and British commercial radio spheres.

Origins

In 1980, the idea of a professional body offering neutral ground for all of those interested in the development of radio broadcasting in the U.K. was proposed at the Radio Festival in Edinburgh by Dick (later Lord Richard) Francis, the BBC's managing director of radio. His companion on the platform at the time was the chairman of public affairs for the commercial radio companies, John Bradford, who endorsed the idea on behalf of the commercial sector.

Francis acknowledged in his initial remarks that many people believed that a radio academy had existed for over 50 years and that it was known as the BBC. He now acknowledged, some six years after the launch of independent (i.e., commercial) radio in Britain, that there was indeed an alternative source of radio production with demonstrated merit.

A working committee was formed under the chairmanship of Caroline Millington from the BBC. It included representatives of the BBC, commercial radio, and those with unaligned interests. Their first task was to assume responsibility for the organization and administration of an annual celebration of radio, a Radio Festival.

Operations

From extremely small beginnings with a handful of members and an occasional public event such as the yearly festival, the Radio Academy has grown to become the most important voice for individual practitioners of the craft and business of radio in the U.K. Since 1980, the Academy has been a registered nonprofit organization operating under the direction of a council elected by its membership. Although it offers members a wide range of services, its central purpose remains as clear in the 21st century as it was in the original concept Francis proposed: the provision of neutral ground on which all of those who care about radio can debate and celebrate the power of the medium.

The key event in the Academy's calendar continues to be the Radio Festival. This has now been supplemented with an event for the music industry, Music Radio, as well as events for radio technicians, promotions departments, news staffs, and the web community. The Council of the Radio Academy also confers an annual Fellowship of the Radio Academy, the highest honor given in recognition of service to the U.K. radio industry, to an individual who has made an outstanding and sustained contribution to the industry. In addition, since 1998 the Academy has hosted the annual Sony Radio Awards ceremony. Its members nominate candidates to be considered for the Sony Gold Award.

The Radio Academy fosters close working relationships with student broadcasters under the aegis of the Student Radio Association and also with the academic community and its association, the Radio Studies Network. The Academy also provides administrative support to both of these organizations.

The Sony Radio Awards

The Sony Radio Awards are the Oscars of British radio, an annual award program that recognizes excellence in all aspects of radio. Established under the auspices of the Japanese electronics company Sony in the early 1980s, the Sony Awards evolved from a similar program supported by the U.K.-based Pye electronics company. The Radio Academy now administers the program and hosts the awards ceremony, although Sony continues to support the awards program financially.

The governance of the awards scheme is entrusted to an invited committee who report through their chairman to the Council of the Radio Academy. It is the responsibility of this group to determine individual categories and criteria to be applied by the judges, a small team of professionals in or closely allied to the radio industry.

For the Silver and Bronze awards, nominations are open to programs and broadcasters from any station that has operated continuously through the preceding 12 months. Submissions usually consist of unedited recordings of the programs and individuals being nominated. Only members of the Radio Academy may submit nominations for the Gold Award, which recognizes either an individual or program judged to have made the greatest contribution to the industry during the previous year. Nominees submitted by the members are considered by the awards committee, which makes the final decision on a yearly recipient. A short list of five nominations for each award is published about six weeks before the awards ceremony, and the final results remain confidential until their announcement at the ceremony. In exceptional circumstances, the committee may also give a special award for outstanding service over an extended period of time.

JOHN BRADFORD

See also British Commercial Radio

Further Reading

Radio Academy Yearbook and Directory Online: The Radio Academy Annual, <www.radioacademy.org/>

Radio Acts. *See* Wireless Acts of 1910 and 1912; Radio Acts of 1912 and 1927

Radio Advertising Bureau

U.S. Radio Trade Association

The Radio Advertising Bureau (RAB) is the sales and marketing arm of the U.S. radio industry. The RAB promotes the effectiveness of radio advertising to potential national advertisers, helps its members effectively market radio advertising to station clients, provides sales training for station employees, and serves as an information resource for station members.

Origins

A Broadcast Advertising Bureau (BAB) was established by the National Association of Broadcasters in 1950, but it quickly failed for lack of support. A second attempt a year later was more successful, and by 1954 the bureau had all four U.S. networks, more than 835 stations, and 11 station representatives as dues-paying members. At the beginning of 1955, BAB became the Radio Advertising Bureau with a continued aim to provide sales information, especially radio advertising success stories, to prospective advertisers and their agencies. RAB had more than 1,000 station members and a million dollar annual budget by 1959.

The RAB had arrived at a crucial time in radio history. Audiences and advertisers were concentrating on television, and radio was in the midst of its transition from traditional or middle-of-the-road programming aimed at a broad audience to increasing specialization based on various popular music formats, especially variations on Top 40. RAB opened branch offices in Chicago, Los Angeles, and Detroit in the early 1960s. In 1964, RAB initiated the All-Radio Methodology Study (ARMS) audience research program to determine the best ways to measure and describe radio's listeners. On a lighter note, RAB retained comic Stan Freberg to develop a series of commercials touting radio's benefits over television. One of them applied sound effects to create the image of turning Lake Michigan into a huge cherry sundae. It was effectively used for many years.

RAB expanded its efforts and output in the 1970s and 1980s, catering to both large and small radio outlets. Monthly publications, sales meetings, and demonstrations spread the word on how best to utilize radio for advertising. RAB also targeted such large advertisers as Sears and Procter and Gamble—neither of which then advertised on radio—and turned both companies into major users of the medium. A 1989 campaign used brief moments of silence to explore what the world would be like without radio. RAB began to service FM outlets and soon moved to a "radio is radio" campaign, arguing that both AM and FM provided valuable services to advertisers. In 1994 much of the RAB moved to Dallas in a cost-cutting move, though headquarters remained in New York.

RAB Objective

The RAB promotes the effectiveness of radio advertising, helps its members effectively market radio advertising to station clients, provides sales training for station employees, and serves as an information resource for station members. The annual *Radio Marketing Guide and Factbook for Advertisers*, published by RAB, compiles the most recent data on radio audiences, provides information on the top radio advertisers, and includes comparative media information and radio listener facts.

As the primary sales association for the industry, the RAB also tracks the performance and financial health of the radio industry. As many large radio group owners have begun to sell stock through initial public offerings, the RAB has become an important spokesperson for the health and prosperity of the radio industry. More than 5,000 stations, networks, and sales organizations in the United States and abroad are members of RAB.

Member Services

The RAB's Member Services Helpline provides members with access to the radio industry's largest database of marketing, media, and consumer behavior information. The database includes more than a half-million individual reports on some 3,000 different marketing, media, and consumer topics. RAB uses the internet to supply station members with information (its website is www.rab.com). Available on the site is information to help radio account executives prospect for clients, prepare client proposals, make client presentations, and become a marketing resource for advertising clients.

RAB members can find RAB Instant Backgrounds on 160 distinct business categories, products, or services. A radio account executive needing information to prepare presentations for clients as diverse as accountants, a women's clothing retailer, an air conditioning repair service, or warehouse shopping service could obtain specific information about the customers who typically use these services, including competitive characteristics of each business category and the times customers prefer to shop. RAB Research also provides seasonal promotional and sales ideas, consumer information, and media information—including not only facts on radio usage but information to help account executives sell against other media such as newspapers, television, yellow pages, and the internet. An audio library of 1,000 MP3 format commercials, a database of commercial scripts, and a co-op advertising directory are also available online. RAB PROposal Wizard can be downloaded to assist account executives in creating attractive, organized, and problem-solving sales proposals.

Sales Training

A continuing theme for the RAB has been to promote the effectiveness of the radio industry as an advertising medium against other competing media. Although radio broadcasts have entertained and informed listeners since the 1920s, the radio industry receives less than ten cents of every dollar spent on advertising. Newspapers and television receive the greatest percentages of revenue. RAB's efforts are intended to assist local stations in getting a larger share of the local advertising revenue and to see that advertisers nationwide are aware of radio's effectiveness. RAB's awareness campaign, entitled "Radio Gets Results," focuses on how local stations have provided marketing solutions for their clients. Gary Fries, president and chief executive officer of the RAB, described the "Radio Gets Results" campaign as a way to provide the radio industry with documented proof of radio's unique ability to deliver outstanding results for its advertisers.

Professional development of station account executives is another role of the RAB. Radio consolidation, one result of the passage of the Telecommunications Act of 1996, has decreased the number of radio station owners. Large radio groups of several hundred radio stations are now possible. The larger radio ownership groups have done two things to the industry. First, they have put increased pressure on station managers and sales managers to increase revenue. Second, stations are increasingly aware of the need to invest in sales training for their employees.

RAB station members receive daily sales and marketing emails to help sales managers conduct successful sales meetings and to highlight new sales opportunities for account executives. The RAB offers sales training and accreditation through the Academy Certified Radio Marketing Professional. RAB began offering sales training courses in 1973; they estimate that only 5 percent of all radio salespeople have ever qualified for accreditation. Once the Radio Marketing Professional status is reached, persons wishing to receive advanced designations must combine knowledge gained from studying RAB materials with what they know from their day-to-day experience as radio account executives.

Because the Radio Advertising Bureau is a member-supported trade group, much of RAB's information is available only to members. The RAB's website includes free information, including the "Radio Gets Results" station testimonials, media statistics, links to other sites, and the latest press releases from RAB, which often highlight industry trends. Instant Backgrounds, audio files, the co-op database, and other features are available to members only.

GREGORY G. PITTS

See also Advertising; Advertising Agencies; Consultants; FM Trade Associations; Promax; Promotion on Radio; Station Rep Firms; Trade Associations

Further Reading

Albarran, Alan B., and Gregory G. Pitts, *The Radio Broadcasting Industry,* Boston: Allyn and Bacon, 2000

Arbitron Radio Market Report Reference Guide: A Guide to Understanding and Using Radio Audience Estimates, New York: Arbitron, 1996

Marx, Steve, and Pierre Bouvard, *Radio Advertising's Missing Ingredient: The Optimum Effective Scheduling System,* Washington, D.C.: National Association of Broadcasters, 1990; 2nd edition, 1993

Radio Advertising Bureau website, <www.rab.com>

Shane, Ed, *Selling Electronic Media,* Boston: Focal Press, 1999

Streeter, Thomas, *Selling the Air: A Critique of the Policy of Commercial Broadcasting in the United States,* Chicago: University of Chicago Press, 1996

Warner, Charlie, *Broadcast and Cable Selling,* Belmont, California: Wadsworth, 1986; 2nd edition, by Warner and Joseph Buchman, 1993

Radio in the American Sector (Berlin)

U.S. International Radio Station

Radio in the American Sector (RIAS) of Berlin was one of the longer-lasting international broadcasting services in the U.S. "arsenal" of Cold War propaganda weaponry. Aside from its longevity, it is also significant as the earliest example of a "surrogate" (what audiences presumably want but do not get from their own domestic stations) international radio service for an entire nation. The founders of Radio Free Europe and Radio Liberty modeled their services on RIAS.

Origins

The U.S. military occupation force founded RIAS in 1946 to reach Germans living in the American-occupied sector of Berlin with a few hours per day of German-language material. The schedule included news programs and some entertainment, but there were also programs about the evils of the former Nazi regime and the steps being taken to make Berlin and Germany into democratic entities. At first it operated as a wired radio service through what remained of the city's telephone system. Soon, however, tensions between the Western Allies (comprised of the United States, Great Britain, and France) and the Soviet Union increased, and the station acquired a low-power AM transmitter. It also began to employ more and more refugees from the Soviet-occupied zone of Berlin and East Germany.

The new transmitter enabled RIAS to reach the entire Berlin area, which was a decided advantage when in 1948 the Soviet Union imposed a blockade of all roads, rails, and canals leading into the Allied-occupied western half of the city. The Berlin Airlift mounted by the Allies brought coal, medicine, potatoes, and other vital supplies; RIAS brought messages of support for the West Berliners and continuing news of the successful defiance of the blockade for the Soviets and East Berliners. The station also seemed by its very presence to symbolize Western determination to remain in West Berlin.

By the time the blockade ended in 1949, it was clear that the division between East and West Germany was hardening into something more permanent and that RIAS should be employed to reach *all* of Soviet-occupied East Germany. At that time, it was placed under the control of the U.S. High Commission for Germany. The station already had developed a well-rounded program service, with a wide range of information and entertainment that included full coverage of activities in East and West Berlin as well as radio dramas, quiz shows, jazz, pop, classical music (RIAS had its own band, orchestra, and chorus), cabaret, and satire. Much of what it broadcast was most unwelcome to Soviet and East German authorities, who did not want East German listeners to hear of the growing economic strength of West Germany or to be exposed to such "decadent" Western material as jazz, church services, or music by avant-garde composers such as Stravinsky and Schoenberg.

Cold War Operations

Most of all, those authorities deeply resented and feared RIAS's daily coverage of events that showed Soviet and East German administrators in a bad light. They jammed (electronically blocked) the incoming broadcasts, but RIAS employed an increasing number of frequencies and more powerful transmitters in the AM, FM, longwave and shortwave bands and added a second 24-hour-a-day program service. Although some of the jamming was successful, much was not—and even the successes may simply have aroused the curiosity of listeners to learn what it was that the government wanted to keep them from hearing.

Intensification of the Cold War during the 1950s drove RIAS to try even harder to gather every bit of information possible on life in what now bore the title of German Democratic Republic (GDR). National and (brief) local newscasts were intensely monitored and transcribed; national and local newspapers were examined minutely; and intelligence reports were studied from every angle. All of that information was filed on cards, which were cross-cataloged under numerous headings. This system enabled RIAS to link many developments, no matter how small, with past developments. In many instances, East Germans would learn that life actually was getting worse for them—or it would be if they didn't use every legal resource at their disposal to block or modify deleterious changes to labor contracts, educational content, and the like. Yet the station was most careful to refrain from anything that might encourage its East German listeners to openly rebel against their communist rulers—a step that would almost certainly bring the large and well-armed Soviet military forces stationed throughout the country into what would be a very uneven battle.

When East German factory workers, distressed over contractual changes that forced them to work even harder for no increase in already low wages, took to the streets of Berlin and several other cities in March 1953, RIAS knew that it had to cover the demonstrations. As it did so, the station also reminded the workers and their supporters time and again to refrain from destroying property or openly calling for removal of the communist government. When the demonstrations dispersed, there was no retaliation by the government, which quietly improved the contract. It also was in 1953 that RIAS was brought under the newly created United States Information Agency (USIA).

The Hungarian Uprising of 1956 had a very different outcome, with workers and supporters—including the U.S. government–supported Radio Free Europe—at first seeming to win a change in government, but soon facing Soviet tanks and brutal repression from a new communist administration. This event may have underscored the wisdom of RIAS's decision to caution East Germans against rash actions, but it also underscored the limits to which the Western Allies would go in supporting such insurrections. RIAS now had to reappraise its overall policy of encouraging its East German audience to think of the communist government as temporary. That policy was replaced by the concept of gradual evolution, through which the communist government would be encouraged to move away from doctrinaire Marxism and toward a more citizen-friendly form of socialism, or at least "communism with a human face."

Gradual evolution was based on an interesting assumption: if the people *and* their rulers were aware of the ever-growing prosperity, political stability, and military strength of Western Europe and North America, there would be great pressure on communist governments to change or be left hopelessly behind. It was particularly easy for RIAS to contrast the West with the East because the Federal Republic of Germany (more widely known as West Germany) clearly represented the former, and the German Democratic Republic the latter. The two nations shared a common history and language but were worlds apart in their present economic strength and degree of personal freedom (to travel, to change jobs, or to vote in open elections). Since RIAS had access to a large amount of information about East Germany that its rulers would not share with their people, it could present detailed and generally accurate comparisons on a daily basis.

Much of that advantage disappeared overnight on 13 August 1961, when the East German government erected a wall that literally divided Berlin in two. It became much more difficult to obtain daily newspapers from around the GDR; the heavy flow through West Berlin of East German refugees, who often had provided valuable and detailed information on life in their former homeland, ground to a virtual halt. Nevertheless, RIAS persevered, now emphasizing Western progress and changes taking place in other communist nations in Eastern Europe. The hope in the latter case was that East Germans might be able to persuade their government to at least follow the example of other "fraternal" communist partners. The station also developed special programs such as a 5-minute daily broadcast for East German military personnel stationed along the Berlin Wall and along the heavily guarded border with West Germany, in which those on duty were reminded that would-be escapees were Germans, too, and certainly not criminals in any of the usual senses; therefore, the program urged guards to consider shooting to miss, and not to kill.

It was more difficult than ever to measure the effectiveness of any RIAS broadcasts to East Germany, since refugees had been far and away the most important source of survey-gathered data in pre-Wall days, and now very few escaped. In 1972 the two Germanys signed an interstate treaty permitting a limited number of personal visits in both directions. That seemed to signal the possibility that there might someday be some form of confederation. West German politicians and media, RIAS among them, referred less and less often to "the so-called German Democratic Republic." The West German government was now covering most of RIAS's annual budget, although a few U.S. administrators remained nominally in charge and the station remained within USIA. Yet jamming of the station continued, even as it was becoming easier for East Germans to watch West German television without interference from the East German authorities. (RIAS itself began a regularly scheduled TV service, but not until August 1998.)

Demise

By the 1980s, there were further signs of relaxation, accommodation, and even cooperation by both Germanies, and RIAS broadcasts were careful to take note of the changing climate. However, jazz and rock music more or less indigenous to East Germany (but often sounding suspiciously like current Western jazz and rock stars) began to appear on East German radio and TV. Mild forms of criticism of the government (but not of communism itself) appeared in East German cabarets and novels. Such moves made RIAS less unique, but since the GDR remained one of the most rigid communist-governed nations in Eastern Europe, RIAS broadcasts about the greater freedoms developing in some of the GDR's neighbors continued to find ready ears. Still, very few Germans on either side of the Berlin Wall were prepared for its entry points to open freely on 9 November 1989, and even fewer to see large numbers of East Germans walking through to mingle with West Berliners. Within a year, the two German governments had signed a treaty of unification and become a single nation, but hardly as equal partners: the "new" Federal Republic was a larger version of the old one, and very much a part of the West.

The unification ended RIAS's raison d'être. In 1992 RIAS TV was incorporated with West Germany's international radio service, Deutsche Welle, and became DW Auslandfernsehen ("international TV"). In that same year, the station's second radio service, RIAS 2, became a commercial radio station ("r.s.2") operating on the same frequency. In 1994 the West

German longwave radio service Deutschlandfunk incorporated parts of the East German radio services and parts of RIAS 1 as a new Deutschlandfunk "Berlin-programm." Thus, two erstwhile Cold War enemy services now found themselves partners, even as the release of hitherto secret data from East Germany showed that RIAS Berlin indeed had been a major thorn in the side of the East German government and an important factor in bringing about its demise.

DONALD R. BROWNE

See also Cold War Radio; Germany; Jamming; Propaganda by Radio; Radio Free Europe/Radio Liberty

Further Reading

Browne, Donald, "History and Programming Policies of RIAS: Radio in the American Sector of Berlin," Ph.D. diss., University of Michigan, 1961

Browne, Donald, "RIAS Berlin: A Case Study of a Cold War Broadcast Operation," *Journal of Broadcasting* 10 (Spring 1966)

Browne, Donald, "Radio in the American Sector, RIAS Berlin," in *Western Broadcasting Over the Iron Curtain,* edited by K.R.M. Short, London: Croom Helm, 1986

Ostermann, Christian F., "The United States, the East German Uprising of 1953, and the Limits of Rollback," <cwihp.si.edu/pdf/Wp11.pdf>

Radio Authority

Regulating British Commercial Radio

The Radio Authority is at the center of the United Kingdom's regulatory system for radio. The Broadcasting Acts of 1990 and 1996 gave it responsibility for advertising and awarding licences and subsequently for overseeing the performance of all commercial, community, and other radio services in the U.K., whether they be national, local, cable, satellite, or other broadcast services. Its reach covers everything in the British radio business in the country with the exception of the publicly owned and non-commercial British Broadcasting Corporation (BBC). The radio services for which it is responsible range from national radio networks such as Classic FM and Talk Sport to hospital and student radio stations and special short-term radio services covering, for example, special events and trial services.

The Radio Authority also launched a project known as Access Radio, a not-for-profit pilot scheme, in 2002. This may lead to a series of small-scale local community radio services licensed on a permanent basis in the future. The Access Radio stations under the pilot scheme are intended to serve a particular neighborhood, or community of interest, and to have clear social gain aims. The Radio Authority took over the responsibility for all non-BBC radio from the Independent Broadcasting Authority (IBA) in 1991. That body had been responsible for licensing and supervising British commercial television until the first commercial radio license in Britain in 1973, when it added radio to its responsibilities. The 1990 Broadcasting Act separated the functions of commercial radio and television, giving the former to the Radio Authority.

The Authority carries out three main tasks and responsibilities: it plans the use of frequencies allocated to radio broadcasting, awards licences to bidders with a view to broadening listener choice, and regulates content, in both advertising and programming. It publishes codes that broadcasters must adhere to, covering the use of frequencies, the content of programs, and advertising and sponsorship. It also supervises the radio station ownership landscape.

Although it may appear that Britain's radio industry remains highly regulated, in general regulations are designed to ensure that those who are awarded licenses abide by the undertakings that they have made in their applications. If, for example, a station bids for a license to program talk radio, it cannot then play a lot of adult contemporary rock music. If one bids for a license to be a local mixed-programming station, it cannot then broadcast only religious evangelism.

The Authority has the power to demand that stations make apologies for breaches in Authority codes. Offending stations may be asked to make corrections, may be fined, or may, in extreme cases, have their licenses shortened or revoked entirely.

Concerning advertising, the Authority code sets out standards to ensure that advertising is legal, decent, honest, and truthful. The Authority code on news and current affairs requires all coverage of these to be both accurate and impartial. Unbalanced politically partisan broadcasting is not permitted. The code covers discussion and phone-in programs, personal view programs, documentaries and features, and programs at election times. There are also codes on general programs and on engineering and the way in which allocated frequencies are used.

The Radio Authority advertises when new radio licenses become available and is open to anyone to apply. There are several levels at which licenses are advertised and awarded. National licenses are for services covering the entire country on AM or FM, of which there are three. They are advertised nationally and in open competition. Applicants have to make a cash bid; provided that they meet the requirements laid down in the Broadcasting Acts, licenses are awarded for eight years' duration to the highest bidder.

Local licenses are available for both AM and FM services and are advertised in the respective locality. The Authority has to decide whether a suitable frequency is available and whether the local market has the commercial resources to support another service. When deciding on the applications, it has to consider whether the proposed service will cater to local interests and tastes and broaden listener choice. There is public consultation during the decision making process. The Authority also has to decide whether the applicant has the financial resources to sustain the service for the license period. As with the national licenses, local licenses are awarded for eight years.

Cable and satellite licenses are subject to fewer requirements and obligations. All services intended for general reception (i.e., those without any coding or restricted access provision) need to be licensed by the Authority. Such licences are awarded for five years. Like many other European countries, Britain now has digital radio using the Eureka DAB system. The Radio Authority is required to license all commercial services using the platform. The Radio Authority had awarded 42 digital licences by mid–2003, and together these provide more than 250 program services.

The Authority is funded by the license fees paid to it by each of the licensees and by the fees payable by all applicants. The chair, deputy chair, and all other members are appointed by the government's secretary of state for Culture, Media, and Sport; the organization consists of 47 full time and part time staff.

A new communications regulator, Ofcom, was to be established by the end of 2003. The new Communications Bill that went through Parliament in 2003 is expected to bring about the transfer of the functions of the five existing regulatory bodies, the Radio Authority, Independent Television Commission, Broadcasting Standards Commission, Oftel, and the Radiocommunications Agency, to Ofcom. The existing bodies will be disbanded as a result.

GRAHAM MYTTON

See also British Commercial Radio

Further Reading

Radio Authority website, <www.radioauthority.org.uk/>
Ofcom website, <www.ofcom.org.uk>

Radio City

New York City Headquarters of NBC

Few station or network headquarters are well known as tourist spots. However, Radio City, a central part of Rockefeller Center in midtown Manhattan, has been an exception virtually since it opened in 1933 as the operational headquarters of the National Broadcasting Company (NBC). Public tours are offered, and Radio City has even been the subject of at least three novels.

Origins

In 1928, John D. Rockefeller, Jr., leased 12 acres in midtown Manhattan from Columbia University. Called the "Upper Estate" by the landowner, the plot was bounded by Fifth and Sixth Avenues and 49th Street to 52nd Street, and was then occupied by low-rise brownstones, tenements, and theaters. Rockefeller planned to revitalize the area with three large office buildings and a new Metropolitan Opera House, but the stock market crash of 1929 forced him to scrap the original plans. Still wanting to develop a commercial district on the property, however, Rockefeller hired three architectural firms and a consultant to refine his plans.

In 1930, a $250-million, 11-building project for the area was announced. Raymond Hood had overall architectural control of what would become the largest privately owned prewar business and amusement complex in the world. The first art-deco-style building to open was the Radio-Keith-Orpheum

(RKO) Theater, which seated about 3,500 moviegoers (it was torn down in 1954 to make room for an office building). Delayed somewhat by the Depression, the last of the original buildings was not completed until 1940.

In the 1950s what had become known as Rockefeller Center was extended west of Sixth Avenue to incorporate several high-rise office buildings; the Center now comprises 18 buildings. Control of the Rockefeller Center was sold by the family in 1985 in a complex deal. By 1989 the Japanese Mitsubishi company had become the majority owner, but after a real estate turndown, ownership went to Tishman Speyer Properties in 1997.

The centerpiece of Rockefeller Center, however, was the 70-story building at 30 Rockefeller Plaza, soon known for its chief occupant as the "RCA Building" ("GE Building" after 1986), but often referred to simply as "30 Rock." It opened in 1933.

NBC Headquarters

The Radio Corporation of America (RCA) moved its headquarters into many floors of the skyscraper in late 1933, bringing the NBC network along with it. NBC occupied its new studios early in November and dedicated them on 18 November 1933.

Studios for the network and its New York flagship station, then WEAF (later WNBC), would eventually occupy 11 floors, of which only the second and fifth had windows. This was part of an unprecedented effort to keep outside sounds isolated. Extensive sound filtration and insulation systems made the many studios among the finest anywhere. Their entire space (4.5 million square feet) was air conditioned, a rarity in that early period. The studios ranged in size from 14 feet by 23 feet to the world's largest studio—78 feet by 135 feet with ceilings 30 feet high. This huge room, Studio 8-H, could accommodate an audience of 1,300 persons.

Not all of the NBC space was immediately used; postwar auditorium studios (Studios 6, 7, and 8) were developed on the sixth and seventh floors as needed. Most studio equipment—master control and the like—was on the fifth floor with 50 tons of backup batteries in case of a power loss. No fewer than 275 synchronized clocks appeared throughout the NBC and WEAF floors.

The new NBC space had also been designed for eventual television operations, although that medium was technically very crude in the early 1930s. Later in the decade, however, NBC began studio television experiments with a much improved all-electronic system that soon extended to two large mobile vehicles, often seen parked outside 30 Rockefeller Center. Right from the start, NBC provided studio tours to the general public, which proved to be highly popular and grew even more so when extended to the new television spaces. By the end of World War II, more than half a million people took either the basic or extended studio tour of NBC every year.

The Radio City Music Hall

Perhaps the most famous single part of the Rockefeller Center complex is the Radio City Music Hall on Sixth Avenue. Built at a cost of about $8 million, it was the largest theater in the world with almost 6,000 seats under a huge arching ceiling. Its handsome art deco interiors were designed by Donald Deskey. It opened on 27 December 1932, and the first of its now-famous Christmas stage shows was offered a year later.

For decades the Music Hall offered movie and stage show combinations that were hugely successful. But as television developed and films began to draw smaller audiences, so did the Music Hall. In 1978 some discussions were held with regard to tearing the place down and erecting another office building; however, the building was granted city landmark status and saved. Its presentations after 1979 focused on what people could not see in their home towns or on television—spectacular stage shows.

In the late 1990s, the Hall underwent a massive $77 million renovation paid for by Cablevision, which currently operates the facility. The complete renovation even included changing each of the nearly 6,000 seats by the company that had made the originals. Much of the original lighting was repaired and rejuvenated, as were all wall surfaces. Computers now control lighting and sound effects, yet much of the Music Hall's original stage equipment remains in place and in use.

CHRISTOPHER H. STERLING

See also National Broadcasting Company; Radio Corporation of America; WEAF; WNBC

Further Reading

Balfour, Alan H., *Rockefeller Center: Architecture As Theatre,* New York: McGraw-Hill, 1978

Barbour, David, "The Showplace of the Nation Reborn," *Entertainment Design* (April 2000)

Brown, Henry Collins, *From Alley Pond to Rockefeller Center,* New York: Dutton, 1936

Chamberlain, Samuel, editor, *Rockefeller Center: A Photographic Narrative,* New York: Hastings House, 1947; revised edition, 1952

Francisco, Charles, *The Radio City Music Hall: An Affectionate History of the World's Greatest Theater,* New York: Dutton, 1979

Karp, Walter, *The Center: A History and Guide to Rockefeller Center,* New York: American Heritage, 1982

Loth, David Goldsmith, *The City within a City: The Romance of Rockefeller Center,* New York: Morrow, 1966

NBC's Air Castles, New York: NBC, 1947

Novels about Radio City
Morland, Nigel, *Murder at Radio City: A Mrs. Pym Story,* New York and Toronto, Ontario: Farrar and Rinehart, 1939
Spence, Hartzell, *Radio City,* New York: Dial Press, 1941
Wheeler, Ruthe S., *Janet Hardy in Radio City,* Chicago: Goldsmith, 1935

Radio Corporation of America

No single U.S. communications company has had a more fundamental and important association with the worldwide development of radio broadcasting than the Radio Corporation of America (RCA). Although RCA no longer exists as a separate corporate entity today, having been acquired by General Electric (GE) in 1986, its brand-name products continue to be marketed by Thomson S.A. of France.

Origins

In the early 20th century, the American Marconi Company was a wholly owned subsidiary of Marconi Company, a British corporation. Marconi had a virtual monopoly on maritime ship-to-shore wireless communication when England entered World War I against Germany in 1914. Because of the important role of wireless in maritime operations, President Woodrow Wilson directed the U.S. Navy to assume control of the American Marconi stations and all German-owned stations in 1917 when the United States entered the war. The Navy operated these until the end of the war in 1918, when the U.S. government was reluctant to return control of the American Marconi stations to the British parent company.

Franklin D. Roosevelt was then Assistant Secretary of the Navy, and the experience gained from overseeing operation of these wireless stations during the war convinced him that all radio patents and operations in the United States should be kept under U.S. control. After World War I, British Marconi and General Electric began negotiations for the GE Alexanderson Alternator, which was the state-of-the-art hardware for long-distance wireless transmission. The U.S. Congress expressed concern that Marconi's acquisition of this equipment would result in a foreign company (and, by implication, a foreign country, albeit a friendly one) gaining complete control over maritime communication, which would not be in the national interest. Congress and the Navy pressured General Electric to buy American Marconi, which it did in October 1919. RCA was established to operate all the American Marconi wireless stations that General Electric had acquired. RCA's charter mandated that all board members were to be U.S. citizens and that stock interest by foreign companies or individuals could not exceed 20 percent. The chief of GE's legal department, Owen Young, was appointed RCA board chairman, and two former American Marconi executives, Edward McNally and David Sarnoff, were appointed its president and commercial manager respectively.

GE and RCA developed cross-licensing agreements that permitted each firm to use the other's radio patents. Over the next three years (through 1922), cross-licensing agreements and RCA stock purchases were made by Westinghouse, American Telephone and Telegraph (AT&T), and United Fruit Company, which had patents on crystal radio, as radio was United Fruit's primary means of communication with its Central and South American plantations. The radio business at that time consisted primarily of international, maritime, and amateur radio services, and just a handful of radio broadcasting stations.

Radio broadcasting grew rapidly during the early 1920s and fostered corporate competition, much of it bitter, over whether the Federal Government or private corporations should control radio broadcasting and how. The corporations, for that matter, where engaged in vicious competition for control and experimented with various ways to shoulder the costs of radio broadcasting and the need for its further development. During this fractious period, AT&T and its manufacturing subsidiary Western Electric were referred to as the "telephone group." WEAF in New York was its first AM station and was in the vanguard for many experiments and innovations during the 1920s. An important experiment was known as "toll broadcasting." The telephone group introduced the first commercial announcements on radio (WEAF) in 1922. RCA, GE, and Westinghouse were referred to as the "radio group" and pooled their development efforts, including operation of the GE and Westinghouse stations. The two groups had very different ideas about how radio and broadcasting should develop.

The Federal Trade Commission, following a congressional mandate, began an investigation into alleged monopolistic practices brought about as a result of the more than 2,000 pooled patents and cross-licensing agreements. The results of this 1923 investigation led to a binding arbitration agreement between the telephone and broadcasting groups to resolve their differences. Eventually, in 1926 they resolved the disputes when AT&T agreed to leave the broadcasting field, selling the popular WEAF station to RCA. RCA then created the National Broadcasting Company (NBC) as a wholly owned subsidiary of RCA. NBC began a regular network service using WEAF as its flagship program source. In time, 25 stations in different markets were affiliated with the NBC network. Shortly thereafter NBC established a second radio network using Westinghouse's WJZ, also in New York, as its anchor for programming. The WEAF operation was to be known as the Red Network and the WJZ operation was known as the Blue Network, ostensibly because an RCA engineer drew connecting lines in red and blue on a map showing the locations of the stations served by each. NBC became the base for RCA's expansion into commercial radio broadcasting on a national scale.

RCA became a giant in the radio set manufacturing business initially by marketing GE and Westinghouse radios. In 1929 the three companies consolidated their research, manufacturing, and marketing operations. RCA then bought the Victor Talking Machine Company and its famous logo and slogan "His Master's Voice"—the Francis Barraud painting of "Nipper" the Fox Terrier looking into the bell of a Victrola. The $154 million paid for Victor enabled RCA Victor to manufacture its own radios and phonographs at a new plant in Camden, New Jersey. The new RCA Victor then established the RCA Radiotron Company to manufacture radio tubes, and by 1930 RCA controlled a substantial portion of the several markets in which it was active. This raised eyebrows in Washington, and investigations soon followed.

The Justice Department instituted antitrust proceedings that lasted almost three years. Finally, in 1932 the companies signed a compromise consent decree that resulted in GE and Westinghouse divesting themselves of RCA stock, relinquishing their positions on the RCA board of directors, and making their license agreements (dating back to the early 1920s) nonexclusive. RCA was now an autonomous and independent corporation. Its new president, David Sarnoff, introduced the RCA Photophone in 1932, a device that allowed moviegoers to hear Al Jolson on screen in the first "talkie."

Expansion

The newly independent Radio Corporation of America grew rapidly, expanding some operations to Hollywood, and growing both its manufacturing base and NBC networks. A research and development center opened near Princeton, New Jersey. Research on television focused at RCA's plant in Camden, New Jersey. RCA joined with a chain of vaudeville theaters to start Radio-Keith-Orpheum (RKO) Movie Studios and produced many successful feature films. The NBC radio networks flourished in the 1930s. Rockefeller Center was built in New York City during the early 1930s, and the high-rise building in that complex was named the RCA Building as part of the agreement by RCA to occupy multiple floors, including the first eight (which housed both WEAF and the NBC network facilities), and the top floors (which housed RCA's corporate offices).

RCA's involvement in television began in 1930. Sarnoff hired Vladimir Zworykin, the inventor of the "Iconoscope" (forerunner of today's camera tube) and "kinescope" (forerunner of today's picture tube) away from Westinghouse at the end of the 1920s to establish a laboratory in Camden to develop television. Ten years and $50 million later, Sarnoff introduced RCA's electronic television at the 1939 World's Fair in New York City. Commercial television, using largely RCA technology, began operation in mid 1941.

When the United States entered the war in 1941, RCA plants were converted to war production, making tubes, sound equipment, sonar bomb fuses, mine detectors, and ultimately radar, which was introduced later in the war by the British. Commercial radio and television production resumed less than two months following the end of World War II in 1945. Many new television stations sprang up, resulting in a freeze on authorizations (licensing) by the Federal Communications Commission (FCC) for almost four years, in order to assure equitable allocation of stations and non-interference of signals throughout the country.

Following a round of controversial Congressional hearings, RCA's all-electronic color TV system was adopted as the national standard in 1953. RCA held the patents on that technology at the time, so virtually every color television set produced until the mid 1960s contained RCA parts. All color kinescopes were manufactured by RCA. When a finished tube came off the assembly line it was given one of several different brand labels, packed into a matching brand-labeled box, and forwarded to an RCA competitor such as General Electric, Sylvania, Philco, or Motorola, if not labeled for sale as an RCA set.

Decline

The 1950s and early 1960s marked the peak of RCA's role in the broadcast and electronics industry. Sarnoff had actively supported the research underlying the company's success in black-and-white and later color television. The firm had developed a huge collection of patents, further strengthening its position. RCA was active in virtually all parts of the electronics field, including military and space communications, and it held major market positions in all those industry segments. At the same time RCA had become a major military equipment manufacturer and had entered the computer business, among others, spreading its resources across new fields.

Sarnoff remained in charge until his 1969 retirement, but other firms made the breakthroughs that would dominate the business in years to come. After Sarnoff's retirement, massive investments in technology that rapidly became obsolete (such as a type of video disc recording and mainframe computers) led to huge losses, weakening RCA in the 1970s (then under the leadership of Sarnoff's son Robert). As manufacturing lagged, soon the NBC television network subsidiary was providing RCA's margin of profit. The younger Sarnoff was soon replaced by a quick succession of other presidents and chairmen as RCA sought to regain its former electronics preeminence within a far more competitive and deregulated marketplace. Much of its consumer electronics business (as with other U.S. firms), including radio and television manufacturing, faded in the face of new competition from abroad. The company's attempt to break into the computer manufacturing business resulted in a huge loss—more than $500 million when RCA finally pulled out. When a strong hand was most needed, severe management infighting broke out that would prove fatal to the company's survival as an independent entity.

Finally secret negotiations (initially RCA's board and other leaders were not included) were begun between RCA chairman Thornton Bradshaw and General Electric, which was now rich with profits from manufacturing and takeovers of other companies. The situation was ironic, as GE had created RCA so many years before. Bradshaw said later that some kind of takeover was the only way RCA's many parts might be kept together. But this was not to be.

In 1986 GE took over all of RCA for $6 billion (a huge deal at the time) and shortly thereafter began to dismantle the empire David Sarnoff and others had created. Some parts were incorporated into GE's operations. The research labs that had helped to create television and other products were sold because GE had its own research labs. In 1987 the RCA trade name was also sold for use on consumer products that would now be sold by the French company, Thomson S.A. The NBC radio network that had pioneered national radio programming in the 1920s was sold, as were all of the NBC-owned radio stations.

RCA exists today merely as a product trade name, the giant company that once dominated U.S. radio having disappeared less than two decades after the retirement of its longtime leader.

ROBERT G. FINNEY

See also American Telephone and Telegraph; Armstrong, Edwin Howard; Blue Network; General Electric; National Broadcasting Company; Sarnoff, David; Westinghouse

Further Reading

Archer, Gleason L., *History of Radio to 1926,* New York: New York Historical Society, 1938; reprint, New York: Arno Press, 1971

Archer, Gleason L., *Big Business and Radio, New York: American Historical Company,* 1939; reprint, New York: Arno Press, 1971

Barnum, Frederick O. III, *His Master's Voice: Ninety Years of Communications Pioneering and Progress: Victor Talking Machine Company, Radio Corporation of America, General Electric Company,* Camden, New Jersey: General Electric, 1991

Bilby, Kenneth W., *The General: David Sarnoff and the Rise of the Communications Industry,* New York: Harper and Row, 1986

Demaree, Allan T., "RCA After the Bath," *Fortune* (September 1972)

Douglas, Alan, *Radio Manufacturers of the 1920s,* vol. 3: *RCA to Zenith,* Vestal, New York: Vestal Press, 1991

Douglas, Susan Jeanne, *Inventing American Broadcasting, 1899–1922,* Baltimore, Maryland: Johns Hopkins University Press, 1987

Federal Trade Commission, *Report of the Federal Trade Commission on the Radio Industry,* Washington, D.C.: Government Printing Office, 1924; reprint, New York: Arno Press 1971

The First 25 Years of RCA: A Quarter-Century of Radio Progress, New York: RCA, 1944

Howeth, L.N., *History of Communications-Electronics in the United States Navy,* Washington, D.C.: Government Printing Office, 1963

Lewis, Tom, *Empire of the Air: The Men Who Made Radio,* New York: Burlingame Books, 1991

Sobel, Robert, *RCA,* Briarcliff Manor, New York: Stein and Day, 1986

Warner, J.C. et al., *RCA: An Historical Perspective,* New York: RCA, 1957, 1958, 1963, 1967, 1978

Radio Data System

Transmitting Additional Information

Radio Data System (RDS), or Radio Broadcast Data System (RBDS) as it is called in the United States, is a transitional technology for FM radio, important parts of which will be incorporated into the developing digital radio systems. RDS technology allows a station to transmit an eight character digital message (e.g., station call letters, identification of music being played) to suitably equipped receivers. A small digital readout tells a listener what station is tuned, what music or talk is being provided at the time, as well as other types of information.

Development

Because of the line-of-sight limit to analog FM radio transmission, many transmitters are required to cover a large geographical area. Adjacent transmitters cannot broadcast on the same frequency because they would interfere with each other. In order to stay tuned to the same radio network or program service, a listener driving long distances would have to constantly seek out a new signal as he moved. Unless the listener knew which transmitter served which area, he would not know the optimal frequency for his favorite network or music or talk program service. This was the problem that engineers sought to solve with the development of RDS beginning in the late 1960s.

Swedish engineers began development of what became RDS in 1976. They sought a means of sending data to radio pagers. Soon a group of broadcast engineers working under the auspices of the European Broadcasting Union (EBU) developed the RDS through the 1970s to meet the requirements of European countries, and subsequently it became a European standard under the umbrella of the Comité Européen de Normalisation Electronique. Initial field tests began in 1980. A large-scale operational trial took place in Germany five years later. Regular service began in Ireland, France, and Sweden, in addition to Germany, in 1987, the same year that Volvo made available the first car radio featuring RDS capability. More European countries—and radio manufacturers—followed over the next two years. By the early 2000s, the number of RDS sets in use totaled over 60 million, most of them in Europe.

Radio Data System Features

RDS technology uses a separate and inaudible digital signal that is a subcarrier (an additional signal) of an FM transmission. An RDS receiver can decode this information to enable digital display of station or program information including the following features.

RDS allows automatic retuning to alternative frequencies. When the radio detects that the signal for a particular program service is becoming poor and hard to hear (due to distance from the transmitter), it seeks another one with the same program identification, and if that station provides better quality, the radio switches over so quickly that the listener is not aware of it. More expensive and sophisticated radios have two tuners at the front of the set, and these are constantly searching for a better service, making for even more efficient switching. Not having to look at the radio to retune while driving has obvious implications for safety on congested roads.

Indicating the type of program provided by the station is one popular RDS feature, while providing additional information such as useful telephone numbers, record titles, and so on via the display is another. This information can be up to 64 characters long and is displayed by scrolling through successive eight character screens. This feature is obviously of more use in home tuners than in moving cars.

On an RDS-equipped radio, the listener is offered an 8-character alphanumeric display of the call letters of the station. With the use of abbreviated indicators, this display can also inform the listener whether that station will provide any sort of traffic or travel program. The radio can interrupt listening when these traffic announcements are being broadcast.

A feature that has made the traffic and travel service work efficiently is the enhanced other networks (EON) feature. This is used to update information stored in a receiver about other program services than the one currently tuned. In other words, a listener can be listening to one radio station or even to a cassette or CD, and if a different local news station is about to broadcast travel information, the receiver can switch away from the primary source of entertainment to that travel information and back at the end of it. If manufacturers were convinced of a demand, they could build sets that would be able to vector onto any particular program type. EON availability is demonstrated by the logo "RDS-EON" which is normally displayed on a radio's front panel or on receiver packing.

In Europe, many broadcasters are currently implementing another feature using the RDS traffic message channel (TMC). Through this, it is possible to broadcast encoded travel announcements, and by means of a voice synthesizer, these can be heard or printed out by the listener in his or her own language regardless of the country through which he or she is driving. Because of the complexity of the process, in order to make the best use of the service, receivers should have two tuners at the front, one listening for conventional RDS services and one tuned to the service carrying the TMC information. To date, probably because of cost, the manufacturing industry is dragging its feet over the development of these more sophisticated radios, though there are several simple TMC-capable sets on the market.

RDS can also supply the smart radio with the current date and time, which adjusts automatically for time zone changes. Among the more subtle features especially useful in Europe is the extended country code, which provides supplementary information to tell the radio in which country (and thus language) it is operating. To keep the RDS system flexible and to adapt to new developments, an "open data application" retains unallocated data groups for control of potential new tasks.

The system can also control the relative volumes of speech and music via a music-speech switch. One feature extensively used in some European countries (and indeed, the feature that began RDS development), is radio paging, which enables broadcasters to use existing networks in a cost-effective way to deliver messages to personal receivers. Up to 40,000 subscribers can take advantage of the service on one program service. There is a related emergency warning system so that those broadcasters who wish to do so can transmit confidential warning messages in the event of a national emergency.

Radio Data System in the United States

Initial RDS demonstrations took place in the United States in 1984 in Detroit. Ford began development work on an RDS-equipped automobile radio. Research and further demonstrations continued in various locations for several years. RDS was demonstrated at the 1986 NAB convention. The National Association of Broadcasters and the National Radio Standards Committee (NRSC) formed a subcommittee to develop an American technical standard recommendation for the Federal Communications Commission. The United States adopted a Radio Broadcast Data System (RBDS) standard in 1993 that added functionality to the basic RDS offering. Further developments in the 1990s sought to retain basic RDS and RBDS compatibility.

As stations are all identified by call letters, a unique set of program identification codes was devised for the transmitter RDS encoders in North America, and a new set of program types was needed to meet the specific needs of the American market. The 31 numerical program type (PTY) codes thus vary in the European RDS and American RBDS systems.

Future

RDS was designed for use with analog FM broadcasting. As such, it achieved some success in several European broadcasting systems and in a few other countries. But as RBDS, it never took off in the U.S. American station managers saw little value in the system, deciding not to invest in a technical patch for a medium—FM—threatened with obsolescence. They foresaw that most of the RDS/RBDS features would be provided in the more revolutionary digital services coming on line in the new century.

JOHNNY BEERLING

See also Digital Audio Broadcasting

Further Reading

Kopitz, Dietmar, and Bev Marks, *RDS: The Radio Data System,* Boston: Artech House, 1999
RDS Forum website, <www.rds.org.uk/>
Wright, Scott, *The Broadcaster's Guide to RDS,* Boston: Focal Press, 1997

Radio Disney

Radio Network for Children

Radio Disney is a radio network for children and a marketing unit of The Walt Disney Company. The network is distributed nationally by Disney's American Broadcasting Companies (ABC) Radio Networks to an affiliate base of AM radio stations and is also streamed on the internet.

Radio Disney is a 24-hour synthesis of contemporary hit music, oldies, radio theater, and game shows, all programmed for "kids and moms." Its live feed originates from an ABC production facility in Dallas, and some programming occurs remotely from locations such as Disneyland in Anaheim, California. Disc jockeys play music, talk to children on the telephone, and interview celebrities. Signature Disney cartoon characters (Goofy, Donald Duck, Minnie and Mickey Mouse) pop in regularly to keep the broadcast environment "fun."

Commercial breaks include avails (available commercial spot positions in a program) for local affiliate sponsors, although advertising and promotion for other Disney businesses (theme parks, live events, websites, movies, television, and retail) is an inherent aspect of program content and contest prizing.

Radio had scores of shows for children in the pre-television age, but as those comedies, adventure series, and westerns migrated to the tube, so did their audiences. In fact, Disney's *Mickey Mouse Club* was long responsible for television's ability to keep kids entertained. Consequently, radio essentially abandoned children, reacting to the onslaught of TV with increasingly sophisticated strategies to identify and sell its listeners to advertisers, supported by ratings methodologies that were unable to effectively quantify the under-12 demographic. Although pop music radio has attracted kids since the early days of rock and roll, children had not been directly targeted, nor could they be accurately counted.

Nonetheless, experiments in format radio for children began in earnest in the early 1980s on both commercial and public stations, laying the groundwork for an ongoing children's radio network, which led to Disney's entering the field. A number of concepts for children's radio were tested by the *Children's Radio Network,* established in 1982 by William C. Osewalt; these were broadcast on a handful of underutilized commercial AM stations from Florida to Oregon and eventually found a full-time home at WWTC in Minneapolis. The station's owner, Christopher Dahl, designated WWTC the flagship station for the first full-time kids' radio network, adopting the moniker "Radio Aahs." The network was programmed with specific segments for younger and older children; featured music, contests, and stories; and encouraged audience participation. Because its survival depended on increased distribution and advertising revenues, Radio Aahs entered a marketing relationship with ABC Radio.

It seemed that radio for children would flourish in the 1980s, as others, too, explored opportunities, some supported by Peter Yarrow of the singing group Peter Paul and Mary and by Peggy Charren of Action for Children's Television. KPAL in Little Rock, Arkansas, had a full-time children's format, and WGN in Chicago was among the affiliates of the short-lived weekly music and news series for children, *New Waves,* which was cohosted by Fred Newman of Nickelodeon and funded by the Markle Foundation.

Public radio's significant children's radio venture, *Kids America* (which started locally in 1984 on New York's WNYC as *Small Things Considered*), was a live, 90-minute daily program distributed by American Public Radio that featured music, wordplay, call-ins, jokes, celebrity interviews, and problem-solving advice for a national audience until 1988.

Discussions at ABC about a children's network began as early as 1989, but it was not until 1996 that Radio Disney began broadcasting. Although Radio Disney's format is derivative of its predecessors (ABC Radio's marketing relationship with Radio Aahs resulted in a $30 million judgment against Disney in 1998), its uniqueness resides in the use of Disney characters—the first time Disney cartoon stars have had a regular radio presence—and in Radio Disney's internet site (www.pcs.disney.go.com/disneyradio/), which promotes the music it plays on the air, conducts listener polls, and attempts to engage kids in an array of participatory content. Disney's relationship with Infoseek, which established go.com, and distribution via satellite (both XM and Sirius Satellite Radio offer the channel to subscribers), has propelled Radio Disney beyond its progenitors and the limitations of AM. Because Arbitron (the radio ratings company) does not provide listening statistics on children, Radio Disney relies on a marketing model to determine its media value rather than the more typical advertising model based on cume audience (or CPM, cost per thousand), common with other radio networks. Radio Disney uses the might of its brand to cross-promote, creating a children's media environment that is as much aural amusement park as it is radio network.

JOSEPH R. PIASEK

See also Children's Programs

Radio Free Asia

U.S. International Radio Service

The collapse of the Soviet empire in the so-called velvet revolution that began in 1989 led many to credit the work of international radio services, particularly Radio Free Europe (RFE) and Radio Liberty (RL), for breaking the information monopoly in Eastern and Central Europe. Because of the Tiananmen Square massacre of June 1989, opinions developed that perhaps a similar broadcasting effort might be attempted in Asia. Yet there was also a feeling that the work of the surrogate stations was over and that they should be phased out.

During the administration of President George H.W. Bush (1989–93), two ad hoc committees were established to examine this issue. In 1991—the year of the Soviet Union's own collapse—one of these groups, the President's Task Force on U.S. Broadcasting, recommended that the United States increase its efforts to broadcast into China and other communist states in Asia. The following year, the second group mandated by the U.S. Congress, the Commission on Broadcasting to the People's Republic of China, made a similar recommendation. Both groups suggested that the United States follow the RFE/RL model of setting up a surrogate radio service, although a long-standing group, the U.S. Advisory Commission on Public Diplomacy, disagreed with these recommendations in 1992, endorsing a minority opinion that the United States should augment its current broadcasting activities through the Voice of America (VOA). No action was taken, however, during the remaining years of the Bush administration.

In 1993 the new Clinton administration proposed shutting down the RFE/RL operation to save $210 million per year. A fight to save the services was mounted in Congress, led by Senator Joseph Biden, and the Clinton administration gave way. Daniel A. Mica, Chairman of the Board for International Broadcasting, parent organization of RFE/RL, despite the U.S. Advisory Commission's reservations, argued that surrogate radio stations had been effective in the past. As he put it, "I have talked with Yeltsin, Havel, Walesa, people in the streets who would come up and tell me that they huddled in closets for 40 years to listen to our broadcasts. It helped make possible, I think, what we see today . . . the fall of the Berlin Wall, the new and emerging democracies. Is surrogate radio, VOA, necessary any longer? Do you believe that everything is going to be fine in Russia, in some of the bloc countries that are going through elections, that are electing some of the very people who were just toppled a few years ago? Any knowledgeable person knows that is not the case. As a baby being born, you cannot abandon it. As a new nation being structured, we cannot abandon it."

In 1994 Congress passed the Radio Free Asia Act to establish a new surrogate radio service for Asia. It was to be housed, along with RFE/RL, under the Board for International Broadcasting. Although funded by Congress, it was to have—like its predecessor services—a quasi-independent status. Radio Free Asia began broadcasts in March 1996. It uses 12 leased transmitters in Asia, the Pacific, Europe, and the United States to broadcast into China, Laos, North Korea, Burma, Vietnam, Tibet, and Cambodia. The service functions as a "home" or surrogate radio station, with its programming aimed at providing internal news to populations living within its target areas. It also promotes democratization and the establishment of a market economy in its broadcasts. Its stated mission is

> [t]o broadcast domestic news and information in nine languages to listeners in Asia who do not have access to full and free news media. The purpose of RFA [Radio Free Asia] is to deliver accurate and timely news, information and commentary and to provide a forum for a variety of opinions and voices from within Asian countries. RFA seeks to promote the rights of freedom of opinion and expression—including the freedom to seek, receive and impart information and ideas through any medium regardless of frontiers.

In 1997 the Radio Free Asia Act passed by Congress declared that the government of the People's Republic of China was systematically controlling the flow of information to the Chinese people and that the Chinese government was more interested in maintaining its political monopoly than in economic development. The act called for increasing the hours of broadcasting in Mandarin, Cantonese, and Tibetan to 24 hours a day and for increasing efforts to add other dialects to Radio Free Asia broadcasts; the act also endorsed the idea of adding Mandarin television broadcasts through Worldnet seven days a week.

Like the VOA, Radio Free Asia broadcasts in Mandarin have been jammed with limited success. There have also been reports that its programs in Korean and Vietnamese have been jammed from time to time.

In March 2000 Radio Free Asia, with a staff of 248 people in eight locations (Washington, D.C.; Hong Kong; Tokyo; Taipei; Phnom Penh; Dharamsala; Bangkok; and Seoul), was broadcasting 12 hours of Mandarin per day, 8 hours of Tibetan, 2 hours of Burmese, 2 hours of Vietnamese, 2 hours of Korean, 2 hours of Lao, 2 hours of Khmer, 3 hours of Cantonese, and 1 hour of Uygher. All services were aired seven days a week.

ROBERT S. FORTNER

See also Board for International Broadcasting; Cold War Radio; International Radio Broadcasting; Jamming; Propaganda by Radio; Shortwave Radio; Voice of America

Further Reading

Hennes, David A., *China/Asia Broadcasting Proposals for U.S. Surrogate Services,* Washington, D.C.: Congressional Research Service, Library of Congress, 1992

Radio Free Asia, <www.rfa.org>

Snyder, Alvin A., *U.S. Foreign Affairs in the New Information Age: Charting a Course for the 21st Century,* Washington, D.C.: Annenberg Washington Program, Communications Policy Studies, Northwestern University, 1994

Radio Free Europe/Radio Liberty

U.S. International Radio Services

Radio Free Europe (RFE) and Radio Liberty (RL) represent the most ambitious of all Western international broadcast operations developed especially for the Cold War. Directed at the Soviet Union and the communist-governed nations of Eastern and Central Europe, by the 1980s they broadcast over 1,000 hours per week in 21 languages. As the century closed, those figures stood at over 800 and 22, respectively. RFE and RL have proven to be remarkably adaptable and resilient over time, escaping extinction on more than one occasion, and both continue to operate years after the end of the Cold War.

Origins

A clear majority of international radio services are financed openly by national governments, but several governments also covertly finance clandestine (concealed identity, often unlicensed) radio stations, most of them small-scale operations using one or two languages and with active lives of weeks or months. From their origins in the early 1950s until the early 1970s, RFE/RL functioned as very large-scale operations using many languages, and with a highly unusual form of concealed identity: they masqueraded as nongovernmental services financed by contributions from citizens and private corporations throughout the United States. Even after the U.S. journal of opinion *Ramparts* published an article in March 1967 that provided a detailed account of the U.S. Central Intelligence Agency's (CIA) many financial conduits used to support such operations as RFE/RL, the now not-so-covert system of financing continued.

That the stations were provided with such an elaborate disguise in the first place has much to do with a modus operandi that the U.S. government first established during World War II. The Voice of America (VOA) began operation in February 1942 as the official international radio "voice" of the government. Very soon, however, the predecessor to the CIA—the Office of Strategic Services (OSS)—began to create clandestine stations to broadcast to Germany, Italy, Japan, and the territories they occupied. Since there were no official links between the government and those stations, government officials could deny knowledge of them and of anything they broadcast, which they clearly could not do in the case of VOA. The CIA was created in 1947, and within two years it had established the National Committee for a Free Europe (NCFE) as a private corporation. The NCFE supported unofficial (i.e., nongovernmental) services that would follow an independent policy line where Eastern Europe was concerned. Diplomatic custom prevented the U.S. government from calling for the liberation of Eastern European nations from communist rule, because they were sovereign states; NCFE, as an independent body, could and did advocate such a position.

RFE came on air in 1950 with a service to Czechoslovakia. It was staffed mainly with Eastern Europeans who had fled to the West when the communist governments came to power, and it was administered by some of the same individuals who had worked for the OSS during World War II. It identified itself as "the sort of station that Czechoslovaks [then, in turn, Romanians, Hungarians, Poles, Bulgarians, and from 1951 to 1953, Albanians] would want if they had a real choice." In other words, the various RFE language services, following the lead of Radio in the American Sector (RIAS) Berlin, functioned as surrogate domestic stations. Radio Liberation from Bolshevism, as the eventual Radio Liberty was called at its inception in 1953, served the entire Soviet Union in much the same manner, except that its surrogate services were often vehicles for encouraging separatist sentiments in Soviet republics such as Georgia, Ukraine, and Uzbek, using the languages of those republics as well as Russian to do so.

Strategies and Cooperation

Although RFE and RL were separate organizations until 1976, they did cooperate in a number of ways, particularly in sharing audience research findings. Each had its own office of research and relied heavily on data gathered from refugee centers, as well as from visitors to Western nations. (When the Cold War ended, unpublished research studies that had been conducted by the communist governments revealed that the RFE/RL data were remarkably accurate.) Also, both were administered by U.S. citizens covertly appointed and paid by the CIA, and both received CIA intelligence reports concerning RFE/RL target nations.

The stations did follow somewhat different pathways where the theme of "liberation" was concerned: The Radio Free Europe appeal was more likely to urge individuals to take small-scale actions, such as work slowdowns or minor acts of sabotage, that would weaken the communist governments to the point where large numbers of citizens might rise up and force them out of power, perhaps with the military support of the Western democracies. The Radio Liberty appeal regarded the liberation of the Soviet Union as much more of a long-term undertaking, and exposure of governmental corruption, inefficiency, brutality, and unequal treatment of many of the republics was seen as an important contribution to the eventual downfall of communism. However, hints at overt Western military intervention were exceedingly rare. To the communist governments of the target nations, distinctions between the stations did not matter: both were jammed (their transmissions blocked by electronic interference) from the time they first came on air.

When the Hungarian Revolution of October 1956 failed to draw Western intervention and brought an even tougher communist government into power, both RFE and RL were forced to reappraise their strategies. Anything suggesting such intervention now was forbidden. One overt symbol of that change was the renaming of Radio Liberation, which became Radio Liberty in 1958. The stations now began to emphasize the growing economic strength of the West, particularly Europe. The hope was that listeners in the Eastern European nations and the Soviet Union would compare their own circumstances with those of their Western counterparts, eventually conclude that they were falling farther and farther behind, and then put increasing pressure on their own governments to liberalize economic policies (with possible effects on political policies) so that they would more closely resemble the obviously successful Western policies. The message now was one of evolution, not revolution. This message was often supported by news reports and other current events programs that informed all of the target nations about reforms undertaken in any one of them—in RFE parlance, "cross-reporting."

Whether that policy shift impressed RFE/RL audiences is difficult to say. The communist media publicized allegations that RFE's Hungarian service had in effect caused the deaths of thousands of Hungarians in the 1956 revolution by leading them on with false promises of Western intervention. Research conducted in the years immediately following the revolution revealed that, although there were no unequivocal promises, there were strong suggestions of intervention in several broadcasts, usually added spontaneously by RFE announcers to reports from Western sources. That caused both RFE and RL to tighten their systems of internal supervision, which until then had given the heads of the various language services (many of them not U.S. citizens and usually driven by strong personal hatred of communist rule in their old homelands) great freedom to interpret and apply overall policy directives as they wished. Stricter supervision proved easier to manage with RFE than with RL, in part because of the scarcity of individuals with sufficient fluency in the many non-European languages, such as Uzbek, in which RL broadcast. What is more, any supervisor would have to be sensitive to the emotional nuances of those languages yet possess sufficient emotional detachment from the Soviet political culture that affected life in the Uzbek Republic to be able to render valid independent judgment on the style and content of RL's Uzbek Service broadcasts—no easy task.

Challenges in the 1960s

In the 1960s, RFE/RL faced three major challenges. The first was the Cuban Missile Crisis of 1962, in which nuclear war at times appeared imminent. RL increased the broadcast time of its Russian-language service and added a special Russian-language service for Soviet military personnel stationed in Cuba. Its message was one of restraint, although it underlined in the clearest possible terms the determination of the U.S. government to keep Soviet nuclear missiles out of Cuba. Within days, the Soviet ships carrying the missiles turned around in mid-ocean and headed back home. The second crisis—the move by Czechoslovakia in 1968 to become more independent of Soviet influence—was in certain ways more difficult for RFE/RL. The stations were aware that what they broadcast to Czechoslovakia itself and what they broadcast about the Czech situation to the other Eastern European countries and the Soviet Union would be compared with RFE's performance during the Hungarian Revolution in 1956. Supervision of broadcast content was tight enough this time that the basic message of U.S. and overall Western support for Czechoslovakia's seeming shift toward democracy was accompanied by messages urging Czech listeners to proceed with caution. There were no messages even suggesting the possibility of Western military intervention. The Soviet military already stationed in Czechoslovakia did intervene and in effect installed a more loyal communist government, but bloodshed was minimal.

The third crisis—disclosure of CIA funding for RFE/RL in 1967—sent many station staff members into shock. There were already rumors that the CIA was losing interest in the stations, chiefly because the Cold War seemed to be easing off, but also because the stations were costly and difficult to administer. (Power struggles among and within the various RL language services over "correct" policies for "their" republics, religious denominations, and so forth were legendary and sometimes vicious.) The stations' handling of the Czech situation during the following year earned them some praise, but the seeming shift in U.S. policy toward greater accommodation of the Soviet Union and the People's Republic of China at the end of the decade and on into the early 1970s gave the anti-RFE/RL forces within the CIA fresh hope and encouraged some members of Congress (including the highly influential Democratic senator from Arkansas, J. William Fulbright) to argue that what Fulbright labeled "this relic of the Cold War" should be put to rest.

The stations lobbied Congress on their own behalf and, with the support of Eastern European and Soviet Republic exile groups around the United States—many of them well connected with U.S. senators and representatives—managed to avoid extinction, albeit under a new administrative structure that was far more open to public scrutiny than that under the CIA had ever been. Congress was also anxious to see some visible signs of increased efficiency in the stations and pushed for a merger of their respective staffs into one building, with one administrative board, a consolidated management, and joint policy guidance. All of these changes took place in 1976.

The Era of Détente

The spirit of détente (relaxation of Cold War tensions) introduced by President Richard Nixon as he visited China and the Soviet Union in the early 1970s held up reasonably well in succeeding years, and by the early 1980s there were signs of relaxation in some of the more doctrinaire Marxist economic and cultural policies. Those signs became far more evident after Mikhail Gorbachev became president of the Soviet Union in 1985. Artistic and economic freedoms multiplied and were joined by a degree of political freedom, particularly in the form of the increasing freedom accorded to the Soviet media to criticize graft and corruption even when committed by high-ranking officials. The government's failure to promptly disclose the magnitude of the Chernobyl nuclear plant disaster in 1986 infuriated Gorbachev, who urged the media to redouble their investigative efforts while virtually ordering Soviet officials to cooperate with the media in those efforts. By the end of the 1980s, the Soviet Union and most of the Eastern European countries had shut down their jamming transmitters, and RFE/RL broadcasts could be received clearly, but there was less reason to listen to them now that the communist media were able to report more fully, frankly, and immediately. It looked very much as if the old RFE/RL mantra—"We're in this business to work ourselves out of a job"—might finally be coming true.

Finding a New Mission

But if the RFE/RL cat had already used up some of its nine lives, it had others in reserve. There were numerous calls in Congress during the early 1990s for abolition. What purpose could those services have, opponents argued, now that the Cold War was over and the former communist nations were establishing media systems that operated along democratic and even free enterprise lines? RFE/RL and its supporters argued that it was premature to assume an immediate, or even swift, conversion to democracy and free enterprise on the part of radio and television in Eastern Europe and the former Soviet Union. Furthermore, RFE/RL could be a key player in effecting such a conversion, both by the example of its own programming and by consultation services that it was well suited to provide, since its staff spoke the languages of those countries and understood democratic broadcasting. The service also agreed to shift its operational base from Munich to Prague—a move that would save a great deal of money because of the lower cost of living in Prague and freedom from high German labor contract costs.

The stations also could point out the support they continued to receive from prominent political figures in their target regions. Russian President Boris Yeltsin helped RFE/RL establish a bureau in Moscow following the unsuccessful coup attempt of August 1991, during which RL had played an important role in keeping Russian listeners informed of developments. Czech President Vaclav Havel told station staff that Czech Republic stations needed the professional example of RFE's Czech service as a reference point and that Czech listeners could profit from the broad perspective on regional and world events that it supplied. Even so, RFE's Czech service was cut back, and it entered into a cooperative relationship with Czech public radio to set up a new public-affairs program. The RFE Polish and Hungarian services ceased. But both RFE and RL found new outlets by making arrangements with more than 100 radio stations throughout Eastern Europe and the former Soviet Union to rebroadcast RFE/RL programs.

As Yugoslavia began to break apart in the early 1990s, RFE and its supporters raised the possibility of a Yugoslav service, and in 1994 RFE began broadcasting in Serbian, Croatian, and Bosnian. Disorder in Albania in the mid-1990s brought a resurrection of RFE's former Albanian service. Continuing crises in the Balkans during the late 1990s, the war in Kosovo in particular, served as fresh reminders of the fragility of that part of the world and provided one more reason for maintaining the services. RL acquired responsibility for two tactical radio operations originally established by the CIA. Radio Free Afghanistan and Radio Free Iraq both came under RL in the 1990s; the former has survived the disappearance of a harsh Islamic government in Afghanistan in 2002, while instability in Iraq virtually guarantees continuation of the latter. RL also added a Persian (Farsi) service, Radio Azad, in 1998, then renamed and reoriented it in December 2002 as Radio Farda, for Iranians under 30. (Farda is a joint venture between RFE/RL and VOA.) Finally, RL initiated a web-based Tajik service in 2001.

Thus, RFE/RL has been able to establish for itself what appears to be a permanent place in the overall structure of international broadcasting activity financed by the U.S. government. The International Broadcasting Act of 1994 (Public Law 103-236) brought together all U.S. government nonmilitary international services under a Broadcasting Board of Governors. The nine presidential appointees who serve on the board also function as the RFE/RL board of directors. They also provide oversight for the International Broadcasting Bureau (IBB), which includes the VOA, Radio-TV Martí (to Cuba), Radio Free Asia (RFA), and the WORLDNET Television and Film Service. As of 1 October 1999, the IBB became independent from the U.S. Information Agency and is now financed through a separate annual appropriation by Congress. RFE/RL also receives such an appropriation. So does the new (broadcasts began in 1996) RFA, which serves listeners mainly in the People's Republic of China, Laos, Cambodia, and North Korea.

This situation throws into sharper relief than ever one of the longer-running questions about RFE/RL and RFA vis-à-vis VOA: why does the U.S. government need two broadcast services to reach the same parts of the world in the same languages? There is little evidence to suggest that listeners see RFE/RL or RFA as anything other than voices of the U.S. government, even though official publicity from them stresses their "private" identities. There is one notable difference in program strategies: RFE/RL and RFA both place far more emphasis on coverage of events taking place within the nations they serve, whereas VOA stresses political, economic, and cultural life in the United States. But all of them provide coverage of international events, and there is duplication in that coverage. It seems likely that there will be increased pressure by Congress on all of the services to coordinate that and possibly other aspects of their programming, with a full merger into a single service as a not-unlikely prospect at some future date.

DONALD R. BROWNE

See also Board for International Broadcasting; Cold War Radio; International Radio Broadcasting; Jamming; Propaganda by Radio; Radio Free Asia; Radio in the American Sector; Shortwave Radio; Voice of America

Further Reading

Alekseeva, Liudmila, *U.S. Broadcasting to the Soviet Union,* New York: U.S. Helsinki Watch Committee, 1986
Critchlow, James, *Radio Hole-in-the-Head/Radio Liberty: An Insider's Story of Cold War Broadcasting,* Washington, D.C.: American University Press, 1995
Holt, Robert, *Radio Free Europe,* Minneapolis: University of Minnesota Press, 1958
Lisann, Maury, *Broadcasting to the Soviet Union: International Politics and Radio,* New York: Praeger, 1975
Michie, Allan Andrew, *Voices through the Iron Curtain: The Radio Free Europe Story,* New York: Dodd Mead, 1963
Mickelson, Sig, *America's Other Voice: The Story of Radio Free Europe and Radio Liberty,* New York: Praeger, 1983
Nelson, Michael, *War in the Black Heavens: The Battles of Western Broadcasting in the Cold War,* Syracuse, New York: Syracuse University Press, 1997
Panfilov, Artyom F., and Yuri Karchevsky, *Subversion by Radio: Radio Free Europe and Radio Liberty,* Moscow: Novosti Press Agency, 1974
Picaper, Jean-Paul, *Le pont invisible: Ces radios et televisions que l'est veut reduire au silence,* Paris: Plon, 1986
Shanor, Donald, *The New Voice of Radio Free Europe,* New York: Columbia University, 1968
Short, K.R.M., editor, *Western Broadcasting Over the Iron Curtain,* London: Croom Helm, 1986
Urban, George, *Radio Free Europe and the Pursuit of Democracy: My War within the Cold War,* New Haven, Connecticut: Yale University Press, 1997

Radio Hall of Fame

Recognizing Important Contributions to American Radio

Operated by the Museum of Broadcast Communications (MBC), the Radio Hall of Fame honors those who have done the finest work in U.S. radio broadcasting, past or present. Each year several new winners are added to the growing list of honorees.

How It Operates

The idea of creating a hall of fame for radio broadcasting originated with the Emerson Radio Corporation in New York. The first awards were given in 1988. The MBC in Chicago took over administration of the awards in 1991.

Each year the Radio Hall of Fame steering committee (a group of about 30 radio professionals and others named by the president of the MBC) nominates worthy individuals in each of four categories: pioneer network or syndicated; pioneer local or regional ("pioneer" meaning at least 20 years of service); active network or syndicated; and active local or regional ("active" meaning at least 10 years of service). Up to four nominations may be made in each category.

Nominations are sent to members of the Radio Hall of Fame and the MBC in May and (to new members) July of each year. Ballots are tabulated by a Chicago accounting firm. Nominees who receive at least 50 votes are eligible for consideration for induction for up to five subsequent years. New winners are announced in August and presented at an annual Chicago dinner/broadcast in November.

Inductees: 1988–2001

Each of the inductees (person or program) is listed below alphabetically by program category.

Comedy
Fred Allen
Amos 'n' Andy
Eddie Anderson
Jack Benny
Bob and Ray
Burns and Allen
Eddie Cantor
Can You Top This?
Car Talk
Easy Aces
Edgar Bergen and Charlie McCarthy Show
Fibber McGee and Molly
Stan Freberg
Bob Hope
Our Miss Brooks
Red Skelton
You Bet Your Life

Disc Jockeys
Dick Biondi
Martin Block
Dick Clark
Yvonne Daniels
Rick Dees
Alan Freed
Karl Haas
Hal Jackson
Tom Joyner
Casey Kasem
Murry "the K" Kaufman
Herb Kent
Robert W. Morgan
"Cousin Brucie" Morrow
Gary Owens
Wolfman Jack

Drama and Adventure
Don Ameche
Himan Brown
William Conrad
Norman Corwin
The Goldbergs
Jack Armstrong, The All-American Boy
Little Orphan Annie
The Lone Ranger
The Lux Radio Theater
Ma Perkins
The March of Time
The Mercury Theater of the Air
One Man's Family
The Romance of Helen Trent
The Shadow
Les Tremayne

Music and Variety
Bing Crosby
Tommy Dorsey
Ralph Edwards
Arthur Godfrey
Benny Goodman
Grand Ole Opry
Garrison Keillor
Kay Kyser
Don McNeill
Chuck Shaden
Kate Smith
Take It or Leave It
Your Hit Parade

News and Talk
All Things Considered
Paul Harvey Aurandt
Jack Carney
CBS World News Roundup
Jim Dunbar
Paul Harvey
Gordon Hinkley
Don Imus
Larry King
Rush Limbaugh
J.P. McCarthy
Edward R. Murrow
Charles Osgood
Wally Phillips
Susan Stamberg
Bob Steele
Lowell Thomas
Bruce Williams
Jerry Williams

Sports
Mel Allen
Red Barber
Jack Brickhouse

Jack Buck
Harry Caray
Don Dunphy
Ernie Harwell
Vin Scully
Bill Stern
Bob Uecker

"Emerson" Winners
(presented "for distinguished lifetime achievement in production, management, or technology")
Edwin Howard Armstrong
Jesse B. Blayton, Sr.
Andrew Carter and Edward Pate, Jr.
Lee de Forest
Lynne "Angel" Harvey
Leonard Goldenson
Ralph Guild
Edward F. McLaughlin
Gordon McLendon
Guglielmo Marconi
William S. Paley
James H. Quello
David Sarnoff
Rick Sklar
Frank Stanton

CHRISTOPHER H. STERLING

See also Awards and Prizes; Museums and Archives of Radio

Further Reading

Radio Hall of Fame website, <www.radiohof.org>

Radio Liberty. *See* Radio Free Europe/Radio Liberty

Radio Luxembourg

English Language Service, 1933–1992

Radio Luxembourg has a unique place in radio history. It broadcast to three generations of United Kingdom listeners, and it is the best-remembered—though not the first—station broadcasting from Europe. Some of the most famous personalities in broadcasting and light entertainment were heard on the station.

Origins

Before World War II, the British Broadcasting Corporation (BBC), which had an official monopoly on U.K. radio services until 1973, broadcast mainly serious and high cultural programming, especially on Sundays—often the only rest day for working people. This left a clear gap in demand for lighter, entertainment-led fare.

The tiny country of Luxembourg had been allocated only a single low-power, medium wave transmitter by the International Broadcasting Union, but in order to gain lucrative advertising from much bigger audiences elsewhere in Europe, the Compagnie Luxembourgeoise de Radiodiffusion (CLR), which had been incorporated in 1931, appropriated a long wave frequency. It began broadcasting services in several languages over the most powerful transmitter in Europe from studios at the Villa Louvigny in the city of Luxembourg. The English service, initially broadcast only on Sundays to counter the limited BBC programming, began on 4 June 1933. Most of the programs were produced by London advertising agencies, principally J. Walter Thompson and the London Press Exchange; most programs were variety shows, dance-band half hours, and personality showcases, with the artists closely associated with the products they advertised. The most fondly remembered show of this period was a children's program, *The Ovaltiney's Concert Party,* sponsored by the makers of the malted drink Ovaltine. In an inspired marketing ploy, a club called the League of Ovaltineys was formed in 1935; by 1939 the club claimed some 5 million members.

In 1936 the BBC's audience research indicated that most listening in the United Kingdom on Sundays was to continent-based stations, with Luxembourg being the most successful. The post office in the United Kingdom, under pressure from the BBC, refused to carry programs from England by landline to the Luxembourg transmitter, so most were recorded in London on either 78 rpm discs or film celluloid and shipped to the Grand Duchy.

Indeed, the BBC, the U.K. government, and other national and international bodies made strenuous and persistent efforts to thwart the Luxembourg broadcasts, but the broadcasts continued until World War II and Luxembourg's 1940 occupation by Germany. Even then, the station continued broadcasts in English to the United Kingdom—this time for Nazi propaganda. The most notorious broadcaster on this "service" was William Joyce, whose outrageous claims about the progress of the war and the Allied leaders—all delivered in an exaggerated upper-class accent—led the British public to mock him as "Lord Haw Haw." Joyce was executed for treason in 1946.

Postwar Operations

Toward the end of 1944, U.S. forces took over the station, initially to broadcast morale-boosting programs for Allied prisoners of war and, after peace in 1945, as an entertainment station for troops remaining in Europe. The U.S., French, and British governments were all involved in negotiations to take over the transmitters, but in September 1946, Radio Luxembourg announced that sponsored programs in English would begin again at the end of the year. The U.K. audience quickly returned, along with the Ovaltineys, and in 1948 British listeners heard their first record chart show. However, the range of programs also included comedy shows and quiz and talent shows; several of the latter, including *Take Your Pick* and *Opportunity Knocks,* made a successful transition to the new commercial television service in the United Kingdom.

The rapid success of television's claim on U.K. audiences led CLR to believe they could make more money by using the long wave frequency for their French programs. So in 1951 the English language service was moved to medium wave—the soon-to-be-famous 208 (meters) frequency—and confined to evening broadcasts. In 1954 CLR began a television service and became CLT—the Compagnie Luxembourgeoise de Télédiffusion. In the mid- to late-1950s Radio Luxembourg's content became more concentrated on record programs, sponsored by the major record companies, which, naturally, insisted that only their own releases be played in their slots. Nevertheless, this was the first time U.K. listeners had the opportunity to hear rock and roll records—the BBC broadcast only a couple of hours of "pop" music every week and for several years frowned on the new music genre, as did many parents. This, of course, only added to the illicit thrill for the young generation of listening to Luxembourg—especially with a "secret" transistor radio under the bedclothes.

The limitations of the content and presentation of the sponsored record programs became increasingly unsatisfactory after "pirate" radio stations began operations in 1964 off the coast of Britain, mainly broadcasting "live" programs and based on a much wider mix of popular music. Crucially, in this era Luxembourg lost its place as the buccaneer of broadcasting: the offshore stations were now the ones pushing the legal and musical boundaries and defying authority—Luxembourg now almost sounded like the establishment.

Demise

In 1968, after a new law had forced most of the pirate stations off the air, Luxembourg introduced a "live" disc jockey–presented Top 40 format, broadcast directly from the Grand Duchy. The BBC had introduced the officially sanctioned replacement for the pirates, Radio 1, in September 1967. Broadcasting on this station was mostly restricted to daytime, so for the pop-hungry audience, Luxembourg had evening hours virtually to itself until the start of authorized local commercial radio services in October 1973.

In the 1970s and 1980s, in order to stave off increasing competition from BBC and newer commercial stations, Radio Luxembourg made several further attempts to redefine itself and appeal to a more tightly defined niche audience; formats included disco and album rock. However, U.K. listeners had an increasingly sophisticated range and quality of media on which to listen to music, and the famous "Luxembourg fade"—the signal struggling against nighttime AM interference—lost much of its charm. All of this led to a dramatic loss of audiences and, with them, advertising revenues.

The decision was made to reallocate the transmitter and frequency used for the English language service: programs on 1440 kilohertz ended on 30 December 1991. An English service continued on the ASTRA satellite—using an audio channel on SKY TV—and on shortwave, but even this restricted service was deemed not to be viable and was closed just 12 months later. In 1997 CLT merged with the Hamburg-based Film-und-Fernseh-GmbH (UFA) to form CLT–UFA.

Richard Rudin

See also British Broadcasting Corporation; British Pirate Radio; British Radio Formats; Lord Haw Haw

Further Reading

Barnard, Stephen, *On the Radio: Music Radio in Britain,* Milton Keynes, Buckinghamshire, and Philadelphia, Pennsylvania: Open University Press, 1989

Chapman, Robert, *Selling the Sixties: The Pirates and Pop Music Radio,* London and New York: Routledge, 1992

Crisell, Andrew, *An Introductory History of British Broadcasting,* London and New York: Routledge, 1997
Crook, Tim, *International Radio Journalism: History, Theory, and Practice,* London and New York: Routledge, 1998
Nichols, Richard, *Radio Luxembourg, The Station of the Stars: An Affectionate History of 50 Years of Broadcasting,* London: Allen, 1983
Shingler, Martin, and Cindy Wieringa, *On Air: Methods and Meanings of Radio,* London and New York: Arnold, 1987

Radio Martí

U.S. Service Directed at Cuba

Radio Martí is a U.S. government operated news, public affairs, and music radio service designed for listeners in Cuba. Named after an early 20th century Cuban independence hero José Martí (who, ironically, often fought against the U.S.), Radio Martí was modeled after other government-sponsored propaganda radio services such as Radio Free Europe, Radio Liberty, and the Voice of America (VOA). Radio Martí began operating in 1985 and generated considerable controversy, much of which continues today.

Origins

The U.S. government began to oppose Fidel Castro's regime shortly after it came to power at the beginning of 1959. Opposition escalated in the early 1980s with Cuba's support of rebellions in El Salvador and Guatemala, and Cuba's influence on Nicaragua's Sandinista regime. A 1981 Reagan Administration initiative, "Radio Broadcasting to Cuba" became the seed for what would become Radio Martí, and was a revision of a proposal by Senator Jesse Helms (R-NC). In June 1981 he asked that existing Cuban broadcasts by VOA be designated "Radio Free Cuba" and be transmitted over an AM frequency as there were few shortwave receivers available in Cuba.

American AM broadcasters were alarmed as they were already experiencing escalating radio interference from domestic Cuban transmitters. Noted first by stations in southern Florida, AM stations across the Southeast were suffering growing interference from often high-powered (50 kilowatts and more) Cuban transmitters. U.S. radio broadcasters charged Cuba with violating the North American Broadcast Agreement, which both nations had signed in 1950 (and from which Cuba withdrew in 1981). In response to station pleas for action, the Federal Communications Commission began granting "temporary" power increases to the U.S. outlets most strongly affected.

After Cuba announced plans to launch almost 200 new radio stations, many of them also using high power, a Western hemisphere AM broadcasting conference was convened in late 1981 in Rio de Janeiro. The American proposal to develop a Cuban-focused radio service angered Cuban representatives who walked out in the middle of the six-week conference. Despite this growing domestic and international tension, however, the Reagan Administration continued to push for establishment of the service.

A Presidential Commission on Broadcasting to Cuba was established in September 1981 to recommend the operating principles and details of the new station. Its report a year later suggested what became Radio Martí's charter, stating that the purpose of the station was to "tell the truth to the Cuban people about their government's domestic mismanagement and promotion of subversion and international terrorism in this hemisphere and elsewhere." In the meantime, congressional funding to initiate Radio Martí development was delayed for two years as debate dragged on.

Use of AM Frequencies

At the heart of the debate was the proposal to use an AM frequency for the Cuban service due to the dearth of shortwave receivers on the island. This provoked a conflict between American broadcasters fearing greater interference and the government's anti-Castro policies. U.S. AM stations also feared Cuban retaliation in the form of high-powered signal jamming that could wreak havoc with American stations and not just Radio Martí. The National Association of Broadcasters promoted several alternatives for Radio Martí, including use of a shortwave frequency and making the service a function of VOA rather than a free-standing service. Radio Martí supporters argued that VOA was prohibited by its charter from performing this function.

In its final report, the presidential commission acknowledged that government use of a civilian AM frequency was a violation of principle, but was not without precedent. VOA already used 1580 kHz for Caribbean broadcasts from the island of Antigua. Radio Free Europe also used an AM transmitter to broadcast to Eastern Europe from Germany. In the late 1960s, a government-sponsored anti-Cuban AM station was established on Swan Island, about 400 miles southwest of Cuba. Named Radio Americas, the clandestine station (supported by the Central Intelligence Agency) was successfully jammed by Cuba and operations were quickly terminated. Since none of these stations were located in the U.S., they posed little threat to American broadcasters.

The most important precedent invoked by the commission, however, was an AM frequency VOA had been using since the Cuban missile crisis in October 1962. Located in Marathon Key, Florida, it broadcast to Cuba on 1180 kHz for three decades even though the military crisis originally used to justify the operation had quickly faded away. Radio Marathon, the first government operated station within the continental United States to use an AM frequency, shares the AM channel used by radio station WHAM (Rochester, New York) and other outlets.

In 1982 lawyers for station WHO submitted testimony showing that it would not be a violation of the VOA charter for it to assume responsibility for Radio Martí. This new evidence allowed the existing Radio Marathon operation to evolve into a new Radio Martí. Senate Bill 602 was signed into law by President Reagan in September 1983, formally creating Radio Martí as a part of VOA. Controversies continued over who would actually control station content—professional VOA administrators and broadcasters, or the anti-Castro Cubans based in Miami. The latter wanted a far stronger political tone to the operation and sought to make it a voice for the Miami-based opposition movement. This battle would extend over the next two decades.

Operations and Controversy

Radio Martí signed on the air on 20 May 1985 from Marathon in the Florida Keys, transmitting on 1180 kHz. Initial reports from listeners suggested that while the station was less strident in its message than many had feared, it was also dull. Music programming was popular, but provided little that was not already available on Cuban stations (or those in Miami often picked up in Cuba). One Radio Martí soap opera had been produced a decade earlier (and sounded that way) while a Spanish-language comedy show was older still. Lack of personnel was cited by station programmers as the reason for what was planned as temporary use of the old material.

Initially programming 14 hours per day, Radio Martí expanded to 17 hours in 1986, and eventually into a 24-hour service. Audience research was difficult given the understandable refusal of Cuban authorities to allow on-site surveys. Instead, Radio Martí officials relied largely upon interviews with recent Cuban emigrants for a sense of what was being listened to and listener reactions.

Radio Martí operation and an appreciation of its growing audience led to the creation of a television counterpart, TV Martí, and an Office of Cuba Broadcasting (OCB) to supervise both radio and television stations, in March 1990. Thanks to consistent jamming of its signal, however, the late afternoon telecasts were seen by few on the island. Its relative lack of success prompted calls for a cut in funding to both services, though political support assured that funding continued into the 21st century, with most of the money going into the radio service. The U.S. government experimented in May 2002 with delivery of both radio and television signals using transmitters in circling airplanes as well as direct broadcast satellite links in an attempt to overcome Cuba's jamming efforts. By early 2003, broadcasts also encouraged tuning the services over the internet, though internet access is limited (and tightly controlled) within Cuba.

In 1994 the International Broadcasting Act required that all American international broadcasts, including Radio Martí, be consistent with broad American foreign policy objectives. It also reorganized all U.S. international radio services and later made them part of the Department of State. New management for Radio and Television Martí was installed in 1997 and fairly regularly thereafter. By late 1998, 80 percent of Radio Martí programs were live, replacing pre-recorded material and making the service more appealing to younger listeners. Five years later, the service was operating on a budget of about $15 million annually, employed just over 100 people full-time (both in Miami and at the transmitter site in Marathon Key) and made use of many part-time broadcasters. Its two AM transmitters (upgraded to 100 kilowatts in 1999) utilize four antenna towers that focus signals toward Cuba. Cuban jamming efforts varied depending on relations between the two countries.

Despite strong congressional and administration political support, problems persisted with station administration and internal controls of Radio Martí program balance and objectivity. News commentary and public affairs programs were often found to be too shrill and more reflective of Miami Cuban exile thinking than broader American foreign policy as was required by law. A State Department study in mid–1999 confirmed these concerns and recommended more internal program reviews and effective logging of what exactly was being broadcast as well as external oversight. As just one high-profile example, debate erupted over Radio Martí coverage of the Elián Gonzalez controversy in early 2000. A six year-old boy had escaped the island with others, and was living in Miami with a relative. After much political wrangling over several months, he was taken by U.S. police (resulting in some inflammatory news coverage and photos) and returned to Cuba and his father. Raging controversy over this "surrender" pitted the Miami exile community against the federal government and the radio service was caught in the middle. Radio Martí delayed the news of what many Cuban exiles perceived as a political abduction for four hours. The station director was soon replaced.

Well into the early 2000s, continuing controversy enveloped direction of both the OCB and Radio Martí, and a succession of directors of both operations came and went, reflecting continued infighting over the content and tone of the service.

PAUL F. GULLIFOR AND CHRISTOPHER H. STERLING

See also Board for International Broadcasting; Broadcasting Board of Governors; Cuba; International Radio Broadcasting; Jamming; North American Regional Broadcasting Agreement; Propaganda by Radio; Voice of America

Further Reading

Cody, Edward, "Marti Wafts Time Warp to Listeners in Havana," *Washington Post* (3 June 1985)
"The Continuing Saga of Radio Marti," *Broadcasting* (26 July 1982)
Frederick, Howard H., *Cuban-American Radio Wars: Ideology in International Telecommunication*, Norwood, New Jersey: Ablex, 1986
Goshko, John M., "Radio Marti Broadcasts Soft-Sell Propaganda," *Washington Post* (3 June 1985)
Parker, Laura, "TV Marti: Igniting War of Airwaves," *Washington Post* (27 March 1990)
The Presidential Commission on Broadcasting to Cuba: Final Report, Washington, D.C.: Government Printing Office, 1982
Radio Broadcasting to Cuba: Hearings. U.S. Senate, Committee on Foreign Relations, 98th Cong., 1st Sess., July–August 1982, April 1983
"Radio Marti Gets a Second Year Report Card," *Broadcasting* (18 May 1987)
Radio Martí Observer website, <www.cubapolidata.com/rmo/>
Radio Marti: Program Review Processes Need Strengthening, Washington, D.C.: U.S. General Accounting Office, September 1994
"U.S. Considering Special Radio Broadcasts to Cuba," *New York Times* (27 August 1981)
U.S. Information Agency: Issues Related to Reinvention Planning in the Office of Cuba Broadcasting, Washington, D.C.: U.S. General Accounting Office, May 1996

Radio Monte Carlo

Commercial Radio in France

Nominally the radio service of the principality of Monaco, Radio Monte Carlo was in fact controlled by France until 1998, providing news and entertainment programs in a commercial format for both French and European audiences. A later Middle East service had more of an international flavor.

Origins

Prior to World War II, France experimented with both commercial and state-operated radio, with some of the most popular radio heard in Britain coming from French-established radio stations such as Radio Luxembourg, Radio Normandie, and Radio Paris. During the war, Vichy Prime Minister Pierre Laval created an organization to operate what were eventually known as *radios périphériques*, commercial stations that could broadcast into France from the fringe. Radio Monte Carlo was the first of these, established in 1942. The Vichy (collaborationist) government in the southern part of France purchased the property in 1943. The German government began to install high-powered transmitters in an underground bunker located in the principality of Monaco, for propaganda purposes. Allied forces captured the station before it could be put on the air by the Nazis, however, and in 1945 at the conclusion of the war, all radio stations in France itself were nationalized through the organization Société Financière de Radiodiffusion (SOFIRAD).

Between 1945 and 1958, SOFIRAD controlled more than 80 percent of the interest in Radio Monte Carlo. In 1964 the French state took over indirect control of the *radios périphériques*, creating a quasi-state corporation. This was part of what was called France's "audiovisual foreign policy," coordinating all schools, technical and scientific cooperation, and cultural activities (including radio, television, and film) as part of the state's cultural diplomacy. Through this activity it subsidized the operations of public broadcasting in France; the operations of TV5; Canal France International; Radio-France Internationale; the *périphériques;* Radio Monte Carlo Moyen-Orient (Middle East), an AM (medium waveband) station broadcasting in Arabic and French from Cyprus; and Médi 1, broadcasting to the Maghreb countries from Morocco.

Radio Monte Carlo's programs were generally livelier, more informal, and more commercial than those of the state monopoly radio stations. In 1964 it began FM broadcasting and in 1965 it started experimental broadcasts using long waves at 1,250,000 watts, making it the most powerful long wave radio station in the world. It also developed a reputation as a more reliable source of news in France than the state monopoly broadcasters, even though the French government itself had such a large (over 80 percent) interest in it and appointed ten of the 12 members of its board of governors. This reputation for news coverage became especially noteworthy in its reporting of the student riots in Paris in 1968.

Transition to Private Ownership

Although it was originally a profitable venture, as it and the other *radios périphériques* attracted more than half the French radio audience, by the end of the 1980s Radio Monte Carlo's audience had dropped significantly, partly due to the rise of pirate radio stations in 1977 and 1978. These pirate stations were operated not only by individual entrepreneurs but also by French opposition political parties themselves so that by 1981 the French state radio monopoly was effectively broken. Radio Monte Carlo attempted to attract people back by initiating talk radio and purchasing the Radio-Montmartre.

In 1986 the government granted domestic FM licenses to the *radios périphériques* and in 1989 Radio Monte Carlo purchased the Nostalgie FM network. By 1992 there were 1,700 radio stations operating in France, 1,200 of which were commercial. Radio Monte Carlo operated on a variety of wavelengths including AM or medium wave frequencies and shortwave, from both Monaco and Paris. This prompted questions about the long-term viability of the *radios périphériques* in the newly competitive and privatized environment of French media, as well as question about the French government's participation in them. In 1994 an agreement was reached whereby Radio Monte Carlo would be privatized, thus ending the participation of the French state in its finances and operations. This occurred by mutual agreement in 1998.

Middle East Service

When Radio Monte Carlo Middle East was first created, it relayed its programs from the studios of Radio Monte Carlo in Monaco, but in the mid 1970s the production headquarters was shifted to Paris. (Although it broadcast on medium wave, it also began a short-lived experiment in shortwave delivery after 1985.) Increasing in popularity in the Middle East, particularly for its pro-Arab reporting during the 1973 Middle East war, Radio Monte Carlo played a prominent role in reporting the 1970 Mecca Mosque incident, the 1981 assassination of Egyptian President Anwar Sadat, Lebanon hostage-takings, and the 1981 *Achille Lauro* cruise liner hijacking. It broadcast a mix of French, American, and Arabic music, in addition to news and commentary. As Boyd puts it, Radio Monte Carlo Middle East was an especially profitable station from the mid 1970s until the early 1980s, when the oil slowdown occurred in the region. Its profits were further reduced in 1986 when Saudi television began accepting advertisements.

ROBERT S. FORTNER

See also France; International Radio Broadcasting; Shortwave Radio

Further Reading

Boyd, Douglas A., *Broadcasting in the Arab World: A Survey of Radio and Television in the Middle East,* Philadelphia, Pennsylvania: Temple University Press, 1982; 3rd edition, as *Broadcasting in the Arab World: A Survey of the Electronic Media in the Middle East,* Ames: Iowa State University Press, 2000

Kuhn, Raymond, *The Media in France,* New York and London: Routledge, 1995

Wood, James, "Radio Monte Carlo," in *History of International Broadcasting,* by Wood, vol. 1, London: Peregrinus, 1992

Radio Moscow

International Radio Service of the Soviet Union

The Soviet Union was one of the first countries in the world to engage in international broadcasting. Its first official international broadcasts aired in 1927 during the tenth-anniversary celebration of the Bolshevik Revolution. Listeners outside the Soviet Union had been able to hear Soviet shortwave broadcasts as early as 1922 from Radio Comintern, although these broadcasts were officially designated as domestic. In 1925 broadcasts about the anniversary of the October Revolution were broadcast in English, French, and German (although also ostensibly to internal audiences), and the 1927 broadcasts from Moscow were also aired in foreign languages. All these efforts, however, were sporadic, ending as quickly as they had begun when their official purpose concluded. Radio Moscow was officially inaugurated as a continuous service in 1929, with broadcasts in Chinese, Korean, and English from Khabarovsk, Siberia, near Manchuria, and in German, French, and English from Moscow itself.

Radio Moscow was the Soviet Union's entry into the radio war waged against Nazi Germany by the Allied powers in World War II. Although the Germans had used radio against the Soviet state in the mid-1930s, the battle between these two powers ceased once the Russo-German non-aggression pact was signed in 1939, a situation that continued until Hitler attacked the eastern front in June 1941, at which time Radio Moscow resumed its anti-Nazi broadcasts. Radio Moscow continued to be heard throughout Europe during the war, including in Nazi Germany, and it was a prime target for Luftwaffe bombing runs over the Soviet Union.

Cold War Transmissions

Following the end of the war in 1945, Radio Moscow became the principal adversary of the radio services emanating from North Atlantic Treaty Organization (NATO) countries in the Cold War that began in 1946. It broadcast about 48 hours a day (or about 335 hours per week) in various languages. Like the Western international radio stations, it broadcast a mix of news, features, music, commentaries, mailbag (comments from listeners or answers to their questions), and interview programs. By 1950 Radio Moscow was broadcasting 533 hours a week, by 1960 1,015 hours, by 1970 1,908 hours, and by 1980 2,094 hours. In 1986 Radio Moscow, Radio Peace and Progress, and other Soviet regional stations were broadcasting 2,229 hours per week (second only to the American stations, Voice of America, Radio Free Europe, Radio Liberty, and Radio Martí, in total hours per week). In 1985 Radio Moscow was broadcasting in 82 languages to Europe, Asia, Africa, the Americas, and the Pacific Island nations (including Australia).

From its earliest years, Radio Moscow was under the control of various state ministries and committees, with the State Committee for Radio and Television in charge from 1957 until the demise of the Soviet Union. Radio Moscow was thus subject to the same ideological expectations as those that applied to the Soviet Union's domestic services. These expectations were enforced by the Communist Party itself, exercised through the Council of Ministers. The principal divisions of the State Committee—mirrored in the organization of each of the broadcasting services—were information, propaganda, children's programs, youth programs, literary programs, music, audience research, and program exchange.

Voice of Russia

A new policy of *glasnost* (openness) began to filter into the broadcasts of Radio Moscow in 1987, based in the avowed commitments of the Soviet leader Mikhail Gorbachev. Radio Moscow became less adversarial toward the West, and its broadcasts began to take on more of the tone of the Western stations, particularly the British Broadcasting Corporation (BBC) World Service. But with the fall of the Soviet Union and the financial difficulties of the new Russian state, Radio Moscow also fell on hard times. It began to cut back its broadcast hours and to reduce its programming languages. In 1995 it laid off 30 percent of its staff. By 1999 the "Voice of Russia," Radio Moscow's new name, was down to 31 languages, and its weekly airtime was down to 504 hours. The Voice of Russia claims to broadcast 340 feature programs, providing its listeners with insights into the various "fields of life" in Russia, and to have listeners in 160 countries. It uses 50 transmitters in Russia, 30 in former states of the Soviet Union, one in Germany, and two in China to reach its audiences.

The change in Radio Moscow's programming philosophy could be seen between 1987 and 1990 in its coverage of international news, particularly in events that included the two superpowers. Its ideological edge was softened over this period, and it began to grope for ways to report the news using different linguistic frames than those that had influenced its broadcasts from early in its history. It shifted away from a perspective grounded in rivalry or enmity between the Soviet

Union and the West, to different linguistic categories that were more dynamic and neutral in their treatment of the American president and life in the United States.

Despite the changes in Russia, the Voice of Russia continues to be a state broadcasting company. It is financed by the Russian government and is governed by a chairman and a board of directors. It is also a member of the Federal Teleradio/Broadcasting Service of Russia, which also includes the state-financed domestic radio and television services in Russia. Although it is expected to provide multiple views in its programming, it continues to give priority to views of the state in its broadcasts. Its mission continues to be that of telling the world about Russia, portraying the country realistically, and explaining it to its international audiences.

ROBERT S. FORTNER

See also Cold War Radio; International Radio Broadcasting; Jamming; Propaganda by Radio; Shortwave Radio

Further Reading

Aster, Howard, and Elzbieta Olechowska, editors, *Challenges for International Broadcasting: The Audience First?* Oakville, Ontario: Mosaic Press, 1998
Bumpus, Bernard, and Barbara Skelt, *Seventy Years of International Broadcasting,* Paris: Unesco, 1984
Fortner, Robert S., *International Communication: History, Conflict, and Control of the Global Metropolis,* Belmont, California: Wadsworth, 1993
Fortner, Robert S., *Public Diplomacy and International Politics: The Symbolic Constructs of Summits and International Radio News,* Westport, Connecticut: Praeger, 1994
Nelson, Michael, *War of the Black Heavens: The Battles of Western Broadcasting in the Cold War,* Syracuse, New York: Syracuse University Press, 1997
Wood, James, *History of International Broadcasting,* vol. 1, London: Peregrinus, 1992; vol. 2, London: Institute of Electrical Engineers, 2000

Radio Sawa/Middle East Radio Network

U.S. International Radio Service

Early in 2002 the United States initiated a new 24-hour Arabic language radio broadcasting service in the Middle East. The service was a product of Voice of America (VOA) rethinking about the United States' image in the region. Implementation was speeded up after the 11 September 2001 terrorist attacks on New York and Washington, D.C., and Radio Sawa ("together" in Arabic) began service in March of 2002.

Origins and Mission

For several decades the Voice of America provided an Arab-language service by shortwave and medium wave. These broadcasts were usually for only a few hours a day, and there was often considerable criticism of this single service for such a varied area. More specifically, the service was often criticized for what was or was not broadcast. With the expansion of FM service throughout the Middle East, shortwave listening dropped off. United States radio "presence" across the region was thus increasingly left behind by the domestic and international radio efforts of other countries. Consideration of a dedicated Arabic service was initiated at the end of the 20th century.

The idea for what would become Radio Sawa originated with the U.S. Broadcasting Board of Governors (BBG), which supervises all U.S. international radio services. Norman J. Pattiz, chairman of Westwood One, and a Clinton administration appointee to the BBG (and the only radio professional member), spearheaded the process to get the new service on the air. He led a study mission to the Middle East in early 2001, becoming even more convinced of the need for a full-time Arab-language service to counteract the many area government-supported media services that provide less-than-positive views of the United States. Pattiz's idea took on more urgency in the aftermath of the September 2001 terrorist attacks on the United States and soon became a "pilot project" of the Voice of America. Congress provided $35 million to launch the project in fiscal year 2002, half of which was to be used for one-time investments in transmitters. The second year (fiscal year 2003) financial request to Congress was for nearly $22 million.

The intent of the Middle East Radio Network (MERN), as it was originally dubbed, was expressed in an early press release: "to broadcast accurate, timely and relevant news and information about the region, the world and the United States, and, thereby, to advance long-term U.S. national interests." MERN's primary audience was defined as those in the Middle East under the age of 30—roughly 60 percent of the population. A secondary audience was defined as anyone seeking news.

United States delegations visited various Arab nations in late 2001 and early 2002, seeking agreements to establish both medium wave (AM) and FM transmitter locations for the new service. Operations were established in Washington and in Dubai, and Arab-speaking staff was hired in both locations. Broadcasts would be provided in five regional Arabic dialects aimed primarily at listeners in Jordan and the Palestinian areas, Egypt, Iraq, Sudan, and the Gulf States.

Initiation

MERN first went on the air as Radio Sawa on Friday, 22 March 2002, with popular Western and Arabic musical programming and promotions for news broadcasts that began in early April. Initial broadcasts came from FM transmitters based in Amman, Jordan, and in Kuwait City. The service was also carried on the digital audio channels of the satellite services Nilesat, Arabsat, and Eutelsat's "Hotbird." Termed "phase one" of the service's rollout, these first transmitters were joined in mid-April with two more from both Abu Dhabi and Dubai in the United Arab Emirates. Medium wave or AM transmissions were provided from both the Eastern Mediterranean and the Gulf. (The VOA's Arabic language service was terminated in favor of Radio Sawa shortly after the latter went on the air.)

Programming was planned to expand to include news analysis, interviews, opinion pieces, sports, weather, and features on a variety of political and social issues. Various regional programs were also projected. By early 2003, FM transmitters were operating 24 hours a day from Amman (Jordan); Kuwait; Dubai; Abu Dhabi; Doha; Djibouti; and Manama (Bahrein), along with the two medium-wave (AM) transmissions—and the service could also be tuned in numerous short-wave frequencies. Plans included six different program streams for different parts of the Middle East. Initial Radio Sawa audiences were difficult to ascertain, although by mid-2002 audiences seemed to be growing, especially in Jordan. Original expectations were that 4 million would listen during the service's first year, with closer to 7 million a year later.

CHRISTOPHER H. STERLING

See also Arab World Radio; Broadcasting Board of Governors; Voice of America; Westwood One

Further Reading

Radio Sawa webpage, <www.ibb.gov/radiosawa/index.html>

Schneider, Howard, "A Little Pop-aganda for Arabs: At New VOA Radio Station, Music Is Light and So Is the News," *Washington Post* (26 July 2002)

"US-funded Radio Sawa Makes Its Debut on Local Airwaves," *Jordan Times* (26 March 2002)

Zacharia, Janine, "Tuning into the Voice of Freedom," *Jerusalem Post* (25 April 2002)

Receivers

Over the course of some eight decades of radio broadcasting, receivers for tuning into broadcasts have evolved and become generally lighter, more efficient, and less expensive. They have changed radically in both design and internal features. While radio receivers are now virtually ubiquitous in homes, cars, and offices, such was not always the case. This entry focuses primarily on U.S. commercial development of consumer radios, with some discussion of developments outside the U.S. as well.

Early Receivers

The earliest consumer radio receivers were relatively crude handmade devices created by amateurs. Throughout the 1920s radio enthusiasts built a variety of sets: sometimes simple crystal tuners, often more complex sets featuring several vacuum tubes (these were sometimes sold as ready-to-make kits). Except for the crystal sets, early radios required power provided by bulky wet-cell batteries and later by rechargeable storage batteries. Users listened using earphones or, by mid-decade, separate horn-shaped acoustic speakers. These early sets were limited in sensitivity (ability to pick up weaker signals) and selectivity (ability to distinguish between signals and to tune sharply) and were not very handsome to look at, consisting of a mass of wires, tubes, batteries (often leaking), and controls. They required careful tuning with several dials, and delivered (by present day standards) poor audio quality.

The first commercial receivers became available in 1920–21, but initial store-purchased sets were expensive and thus limited in appeal. Manufactured initially by Westinghouse and General Electric (for sale under the RCA "Radiola" label) and soon thereafter by other manufacturers, they came with key components simply mounted on boards (termed *breadboards* in the trade) or built into wooden cases (sometimes referred to as *breadboxes*), and they required the same complex battery and antenna rigs used by the home-built sets. As the radio craze peaked in 1922, hundreds of other manufacturers—including Atwater Kent, Crosley, Grebe, and many smaller firms—offered receivers from inexpensive single-tube sets (selling for up to $10) to far more sophisticated multi-tube receivers selling for well over $100. By 1924 more than a million commercial sets a year were being sold to consumers. Most were hand-wired, thanks to relatively cheap labor (this industry was one of the first modern manufacturing businesses to employ women) and the complexity of their designs.

The first receivers able to operate on AC power, thus eliminating the need for messy batteries, appeared from Atwater Kent, Grigsby-Grunow ("Majestic" radios), and then RCA in 1927–28. From 1924 until 1931, only RCA offered the superior superheterodyne circuit, for the company had purchased Edwin Armstrong's patent. The "superhet" provided far better selectivity and sensitivity than did radios made by other firms. As part of the settlement of an antitrust proceeding, RCA had to make the circuit available to other manufacturers. Superheterodyne circuits dominated radio receivers for decades.

By the late 1920s receiver design was becoming more sophisticated, and the first console (free-standing) "hi-boy" floor models became available as radios became an item of furniture for the home. Indeed, the design of the wooden cabinetry became an important radio selling point, as did annual model changes (sometimes models changed more frequently than annually), somewhat paralleling automobile sales techniques. Radios were widely sold in various historical "period" styles as well as utilitarian furniture—desks and end tables, for example—all intended to fit with varied home decor. Some table models appeared in all-metal cases, which added to their weight but protected the delicate tubes and circuits. Again attesting to inexpensive labor costs, some were even offered with hand-painted floral designs in multiple colors, made to order. Most radios offered circular dials for selecting stations, leading to users referring to a given frequency "on the dial."

On the eve of the Great Depression, some 60 manufacturers of radios were operating, although four of them held two-thirds of the market and two controlled a third. Fancy console models dominated the radio market. Overproduction, however, helped to force prices down (from an average of $136 in 1929 to about $90 in 1930 and down to $47 by 1932). A host of manufacturers were forced out of the market and then out of business. Depression realities forced a return to cheaper small table model radio receivers, and by the mid-1930s Philco was making a third of them. Crosley and Emerson also specialized in small and inexpensive radios (around $15).

Mass Market Receivers

In part as a move toward greater efficiency, radio circuits, vacuum tubes, and designs all became more standardized. Most table radios (by the mid-1930s, these combined the receiver and loudspeaker in a single enclosure and made up 75 percent of the market) were of the "cathedral," "gothic" (with a rounded or pointed top), or later "tombstone" (with a flat top) design. As economic conditions improved, console radios resumed their former popularity, and better models featuring shortwave as well as medium wave (AM) bands, pushbutton station selection, better speakers, and larger lighted radio dials for easier tuning became available. The first car radios appeared around 1930, although only the more expensive models included radios as factory-installed equipment. By 1938 more than half of the world's radio receivers were in the United States. That radio had become central to daily life is clear as more homes owned radios than owned telephones, vacuum cleaners, or electric irons.

Radios came in an increasing variety of shapes and sizes. Smaller shelf or "mantle" radios became popular because they were easy to move from one room to another. Compact or "midget" radios by Emerson and other firms were made possible by smaller vacuum tubes. Some were manufactured in novelty designs tied to popular radio shows (e.g., Charlie McCarthy or Hopalong Cassidy) or reflecting popular culture (e.g., the Dione Sextuplets) themes. Prices dropped below $10 for some models. In the late 1930s and early 1940s, tabletop sets manufactured from special Bakelite and Catalin plastic resins became popular owing to their reasonable prices and huge decorative appeal. Most sets featured two to three colors and fit in well as bright additions to kitchens and family rooms. Some models were purchased because of their value as both colorful and practical accessories in bedrooms and even bathrooms. (Today many of these same sets are expensive collector's items, costing thousands of dollars.)

In April 1942 manufacture of all civilian radios was halted as part of the U.S. war effort. Finding spare parts grew harder, and thus radio repairs became increasingly difficult as the war continued. Pent up demand led to massive numbers of AM radios being made again starting in late 1945. More than 50 million were sold from 1946 to 1948 alone. The availability of more receivers meant more multi-receiver households, a phenomenon that had first appeared in the 1930s. Plastic was now the radio cabinet material of choice, available in many different colors and shapes and less expensive to manufacture than the former wooden casings.

Radios in the 1950s continued the trend to smaller and lighter formats. Few consoles were sold (households devoted available console space to television) as sales concentrated on table radios, most of which were AM only despite the appearance of FM stations. Clock radios with "wake-up" features became popular. Radio technology had changed very little over two decades because designs were still based on the use of vacuum tubes. Only in 1954 did the first tiny transistor models become available, their ready portability a trade-off for very poor sound quality. Only in the 1960s did transistor radio prices drop sufficiently to drive tube models off the market.

FM Receivers

The first FM radio receivers became available in 1941 just as the first FM stations took to the air. Perhaps 400,000 were manufactured before wartime consumer product restrictions came into force. All of the pre-war sets were designed for the 42–50 megahertz FM band, so when the Federal Communications Commission (FCC) shifted the service up to 88–108 megahertz in 1945, the older sets were made obsolete. A few receivers were sold with both old and new FM bands in the 1945–48 period, when dual station operation was allowed. Only very slowly did sets on the new band become available, and then only at prices far higher than comparable AM models. Sales were slow—a tiny fraction of AM levels—and declined into the 1950s.

As FM service began slowly to expand in the late 1950s, however, so did the manufacture and sale of FM receivers. The addition of stereo service in 1961 was a major factor in rising FM receiver sales, as was the growing high fidelity movement (which originated in Britain) among audiophiles. FM reception was a core feature of component stereo systems. Companies such as Fisher, Marantz, and Scott built high-quality tuners that required separate amplifiers and speakers but delivered superb sound.

By the mid-1960s FM radios were found in more than half of the nation's households, but FM radios remained rare in automobiles into the 1970s. FM proponents sought to persuade Congress to require that all radios receive both AM and FM bands (parallel to television legislation that mandated UHF as well as VHF reception capability), but slowly rising FM sales made the attempt unnecessary. As prices came down and more FM stations took to the air, more radios featured both AM and FM tuning capability.

Portable Receivers

Initial portable radios were portable in name only. Handmade "portables" were fragile, bulky, and heavy, even without the required horn speaker and batteries. Yet the ability to take a radio with you was a strong lure, as is evident even in radio advertising of the early 1920s. Some homemade models were tiny and featured earphones. Still, the first commercial portables resembled midsize suitcases, and not light ones at that. One knew they were intended to be portables simply because they had handles. All these difficulties drove the portable radio off the market by 1926–27 while consumers focused on the new plug-in home receivers.

While many small radios were sold with handles, they still had to be plugged in (they did not operate with batteries), making them what Schiffer (1991) terms *pseudo-portables*. They looked the part, but they were not. Almost no true U.S. portable radios were manufactured in the 1930s save by some smaller companies operating on the fringe of the business. The market was simply too small.

The availability of smaller and more efficient tubes that drew battery power more sparingly helped spark a revival in manufacturer interest in portable radios in 1938–39. Led by Philco and other firms, battery-powered portables began to flood the U.S. market: more than 150 models were available by the end of 1939. Many still looked like cloth-covered suitcases, but now they were small, light, and clearly intended to be carried about. Others made use of the then-new plastic cases and weighed only five or six pounds even with their batteries. Some were even touted as being pocket sized. Most could also double as plug-in table radios. One of the best was Zenith's Trans-Oceanic, which went on sale early in 1942 featuring several shortwave bands with AM reception, a radio log, special antenna, and a large battery pack. Despite its high price ($100), it was hugely popular in the few months it was available, and it resumed production after the war, lasting in improved models well into the 1960s.

Postwar portables were central to the AM radio boom. Although based on prewar technology, they appeared in bright plastic cases that emphasized modern styling, many of them dubbed "lunchbox" radios because of their size. Manufactured by many companies, these were hugely popular in the late 1940s and into the 1950s, selling more than a million units a year. Smaller "miniature" or "shirt-pocket" models appeared in the early 1950s. No portables included FM tuning, in part because of the limited distance FM stations could transmit.

Announcement of the transistor in 1948 and its first appearance in a small radio (the Regency TR-1 in 1954) sparked a new generation of portables in the late 1950s that emphasized their tiny size, light weight (11 ounces in the Regency's case), long-lasting batteries, and portability. Combined with radio's development of Top 40, the transistor portable helped to transform radio's audience from an at-home family image to an on-the-road active youth image. Yet tube-based (or tube-transistor combination) portables lasted into the early 1960s, in part because of the initial high cost of transistors.

Receivers Outside the U.S.

Radio receivers in other countries generally followed the same trends as in the United States. Tight patent control encouraged domestic manufacturing in most industrial nations before World War II, limiting imports from the United States or elsewhere. What most set different nations apart was the design of the radios in each era.

Early British wireless sets were battery powered; sets designed for "mains" (AC) power appeared in 1926 but took longer to catch on as electrical service was spotty. Annual (1926–49) 8- to 10-day "Radiolympia" shows near London highlighted the annual model changes of the dozens of receiver and radio equipment manufacturers. There was little standardization of components or design, although regulations required the use of largely British materials and vacuum tubes. Receiver cabinets were made largely of wood until the inception of Bakelite products in the mid-1930s. Many receivers were rented rather than sold. Thanks to the lingering lack of rural electricity, the battery-powered portable radio was very popular in the 1930s, well before that occurred in the United States. Although civil production continued well into 1940, the war's demand for rapid construction of military equipment brought about far more standardization of parts and processes with a dramatic reduction in the different numbers of devices manufactured. Postwar radios were made under conditions of rationing yet achieved prewar levels very quickly. Promotion centered on homes purchasing a "second set" to expand radio listening. Radios appeared in more modern designs. Although BBC broadcasting in FM began in 1955, few FM radios were sold until the 1960s. The inception of FM and availability of improved recordings sparked the high-fidelity movement that began in Britain. High quality audio products became a chief British export. At the other end of the radio scale, the first British transistor radio was sold in 1956 but by 1964 (as in the United States), the market was dominated by Asian imports.

In Europe, Germany had scores of radio manufacturing firms before World War II. Cheap labor and solid patent control (especially by Telefunken) allowed manufacture of good radios at low cost, resulting in few imports. During the war, many receivers were made to receive only local German stations in an attempt to reduce foreign listening. Saba made some of the best German radios from the mid-1920s into the 1960s. They offered advanced features such as motor drives for more rapid tuning, multiple bands, and remote controls. Many German table radios of the 1950s, such as those by Grundig, Telefunken, and Blaupunkt, became high-end imports to the U.S. market. They featured multiple bands (including FM well before most American sets), good speakers, push-button tuning, and often had rounded and polished wooden cabinetry. The German industry finally succumbed to Far Eastern imports, as had U.S. firms.

In France, prewar radios were nearly all of domestic manufacture, with a fair number being exported to European and South American markets. As in many nations receivers were expensive at first, leading to considerable community listening to radios in centrally located public places. French postwar radios featured some of the most dramatic styling of receivers anywhere. A combination of art deco and moderne, their cabinets combined wood with metal and plastic highlights.

A few nations emphasized the manufacture of receivers for wired radio systems. Designed to keep receiver costs down while preventing listening to foreign stations, receivers could be very cheaply designed (essentially they were loudspeakers) as they were connected by wire to only one or two nearby transmitters. Some 12 million of these existed in 1950 and 40 million by 1960, nearly all in Russia (70 percent of all sets) and Eastern Europe (25 to 40 percent depending on the country). China made considerable use of wired radio systems as well.

In Japan most receivers into the late 1920s were of the crystal type. Battery-powered tube receivers (with from one to three vacuum tubes) cost ten times as much, and the best models were imported from the U.S. Set makers were usually small companies with limited output, which partially explains why early Japanese radios did not achieve the technical prowess of those in other nations. AC-powered receivers appeared in 1926 and by late in the decade were becoming more affordable and thus common. By 1930 loudspeakers were built into radio sets and were soon of the vastly improved dynamic type providing better quality sound.

Superheterodyne receivers were on the market by 1933 but did not become commonly available until after World War II due to cost. By 1938 the NHK network developed a "standard receiver" type with three or four tubes and magnetic speakers, and it "approved' though did not license the 15 companies chosen of the more than 60 operating set makers. During the war, government-encouraged consolidation resulted in about 30 firms being allowed to manufacture approved types of receivers, none of which included shortwave bands in order to better control listener choices. Price controls were imposed that continued under the occupation. By the late 1940s rationing and price controls were lifted, and Japanese radios began to match the best of those made abroad, often sold on the installment plan to spread out their high prices. The number and quality of receivers both expanded dramatically in the 1950s. All-wave receivers became standard, high-fidelity sound was available by 1955, FM reception in 1957, and stereo receivers after 1960. Transistor radios, manufactured in small numbers in the 1950s, dominated the industry by the 1960s, when many were exported to the United States.

Modern Receivers

The primary change in the U.S. radio market after 1960 (other than the addition of FM stereo) was less in their technology than their source. Japan began producing small tube portable radios for sale in the United States in the 1950s and sold its first transistor radio (a Sony TR-63) in 1957. Soon thousands of radios were being imported, combining solid (and often better) engineering with low prices. Some U.S. manufacturers shifted set making (or licensed their designs) to companies in the Far East. The revolution happened quickly; by 1963 there was no U.S. small transistor radio, for they all came from Japan. Cheaper foreign labor soon wiped out U.S. radio (and later television) manufacturing of all types; the final Zenith Trans-Oceanic model of 1973 was made in Taiwan.

In the 1990s some sets (called "smart receivers") featured RDS (Radio Data Systems) technology that enabled AM and FM stations to transmit data to receivers, thus allowing them to perform several automatic functions. Users of such sets could interface with stations and access programming information from them via a built-in light-emitting diode (LED) screen. Radio thus became a visual medium. At the turn of the 21st century, the first digital radio receivers were becoming available in the United States, having already appeared in Britain and several European countries. These were high-end products designed for use with satellite radio services.

CHRISTOPHER H. STERLING

See also Automobile Radios; Capehart; CONELRAD; Crystal Receivers; Developing Nations; Digital Satellite Radio; Dolby Noise Reduction; Early Wireless; Emerson Radio; General Electric; Ham Radio; Japan; Kent, Atwater; Motorola; Philco Radio; Radio Corporation of America; Radio Data Systems; Shortwave Radio; Stereo; Transistor Radios; Walkman; Westinghouse; Zenith

Further Reading

Collins, Philip, *Radios: The Golden Age*, San Francisco: Chronicle Books, 1987

Douglas, Alan, *Radio Manufacturers of the 1920s*, 3 vols., Vestal, New York: Vestal Press, 1988–91

Geddes, Keith, and Gordon Bussey, *The Setmakers: A History of the Radio and Television Industry*, London: British Radio and Electronic Equipment Manufacturers' Association, 1991

Grinder, Robert E., and George H. Fathauer, *The Radio Collector's Directory and Price Guide*, Scottsdale, Arizona: Ironwood Associates, 1986

Hill, Jonathan, *The Cat's Whisker: 50 Years of Wireless Design*, London: Oresko Books, 1977

Hill, Jonathan. *Radio! Radio!* Bampton, England: Sunrise Press, 1986; 3rd edition, 1996

McMahon, Morgan E., *Vintage Radio: 1887-1929*, Palos Verdes Estates, California: Vintage Radio, 1973

McMahon, Morgan E., *Flick of the Switch, 1930-1950*, Palos Verdes Estates, California: Vintage Radio, 1975

The Old Timers' Bulletin, Breesport, New York: Antique Wireless Association, 1960– (quarterly)

"Radio Broadcasting Receivers," in *Statistics on Radio and Telvision, 1950–1960*, Paris: Unesco, 1963

Schiffer, Michael Brian, *The Portable Radio in American Life*, Tucson: University of Arizona Press, 1991

Sideli, John, *Classic Plastic Radios of the 1930s and 1940s*, New York: Dutton, 1990

Stokes, John W., *70 Years of Radio Tubes and Valves: A Guide for Electronic Engineers, Historians and Collectors*, Vestal, New York: Vestal Press, 1982

Tyne, Gerald F.J., *Saga of the Vacuum Tube*, Indianapolis: Howard W. Sams, 1977

Recording and Studio Equipment

From the carbon microphones used by the early radio experimenters to the virtual studio of today, radio equipment has evolved to meet the demands of changing programming strategies. In the earliest days of radio, the equipment used to create programs was very basic. The radio operator spoke into a microphone connected to a transmitter. Soon, radio control rooms were equipped with mixing consoles to mix and route microphones, remote lines, network feeds, and transcription players. In the 1950s, tape recorders were added to the equipment inventory of the best-equipped radio studios. In the 1960s, broadcast cartridge recorders and players were adopted by broadcasters as the industry standard for recording and playback of commercials and other short-form production.

In the 21st century, the basic functions of recording, editing, and playback are still central to the production process at all radio stations, but the choice of type and brand of equipment is extensive. At many radio stations, most of the recording and studio equipment is now digital, and the computer hard disk and other digital alternatives are the storage media of choice. The typical production and on-air studios of today are equipped with a variety of microphones, mixers, and consoles, analog and/or digital recording and playback equipment, audio processing equipment, and monitors. This essay describes the typical recording and audio equipment that either has been used or is being used in radio studios to meet production needs.

Microphones

The microphone is the most fundamental of all the recording and studio equipment in use today. Most radio production usually starts with a microphone. A microphone is a transducer: it changes the sound energy of an announcer's voice, musical instrument, or other sound into an electrical signal that can then be mixed with other microphones and audio sources to create the radio program. Depending on production requirements and budget, a number of different types of microphones have been used in radio broadcasting.

Microphones can be classified by the method of creating the electrical signal, or the means of transducing the sound. Adapted from their applications in telephones, carbon microphones were the first microphones used in radio. As the carbon microphone was prone to distortion and easily damaged, it was soon replaced by more durable and accurate microphones. Dynamic, or moving-coil, microphones use a diaphragm attached to a moving coil in a magnetic field to generate electrical energy. The dynamic microphone is accurate and fairly inexpensive, making it popular for both studio and remote situations. The Electro-Voice RE-20 and Shure SM-7 are two of the most commonly used dynamic microphones in radio studios today.

Condenser (or more accurately, capacitor) microphones use a built-in battery or phantom power supply from a console or mixer to charge a conductive diaphragm and backplate, creating a changing capacitance that generates the electrical signal. Condenser microphones, because of their wide frequency response, are the microphones of choice in many radio studios.

A third type, the velocity (also known as ribbon) microphone, suspends a strip of corrugated aluminum ribbon in a magnetic field. When the ribbon is vibrated by the sound pressure, an electric current is generated. While this type of microphone may still be found in some radio studios, the dynamic and condenser microphones are the microphones typically used in today's radio operations.

Microphones can also be classified by pickup patterns. A microphone's pickup pattern describes how the microphone responds to sound coming from different directions. There are three major types of pickup patterns: omnidirectional, bidirectional, and unidirectional. All three types have been or are being used extensively in radio production. An omnidirectional microphone picks up sound uniformly from all directions. The bidirectional pattern picks up sound best from the front and rear, rejecting sound from the sides. A unidirectional pattern picks up sound from the front.

A specific type of unidirectional pattern is the cardioid. Its heart-shaped pattern picks up sound best from the front, with more rejection as the sound source moves to the back. Subcategories of the cardioid pattern are the supercardioid, hypercardioid, and ultracardioid. Each of these successively rejects more sound from the sides, focusing more narrowly on the area in front of the microphone pattern. The pickup pattern of a microphone is determined by the type of pickup element it uses together with the number, size, and positioning of ports in the microphone housing used to direct the pickup of sound. Specific production situations dictate the type of pattern to be used. An omnidirectional microphone works well to pick up a group of people gathered around a microphone. Bidirectional microphones were useful in producing radio dramas because the actors could face each other. Cardioid microphones are especially popular in modern radio production because of their ability to reject unwanted studio noises behind the microphone.

Audio Mixers and Consoles

The audio mixer or console is the focal point of operations in a radio or recording studio. A mixer or console is a device that selects, amplifies, routes, mixes, processes, and monitors input signals, sending the resulting output(s) to the transmitter, recorder, or other destination. A mixer is distinguished from a console in that the mixer is smaller and sometimes portable, and is used for basic production such as mixing the three or four audio sources typically used in a newscast or sportscast. An audio console or audio board is larger and more complex, sometimes with 20 or more channels providing space for the numerous audio selection and processing options needed regularly in on-air and production situations. The console used for on-air operations is designed to be easy to use and efficient in selecting, controlling, and mixing the sources typically needed during a live radio program. The production console, because it is used in recording situations, often has a completely different design and layout from an on-air console. Flexibility, in terms of assigning and processing inputs and outputs, is the key to production console design.

On-Air Console

The electronics and design of on-air consoles have changed significantly from the early days when consoles were large, custom-designed, vacuum-tube units with large rotary controls that dominated the studio. Today's solid state, often digital, fader-control consoles meet the same needs as those first audio consoles. The on-air console is used to select, mix, and control the audio signals used to create the on-air programming of the station. This console's main purpose is to facilitate the simultaneous playback of several microphones, compact disc (CD) players, broadcast cartridge players, and other audio playback equipment. The primary output of this console is typically fed to the final audio processing and on to the transmitter.

Most on-air consoles have similar standard features and layouts. Variation comes from the number of channels and sources that can be connected to the console and special features in processing and monitoring the inputs and outputs. The number of channels on the on-air consoles varies greatly. In small or mostly automated stations, a console with six channels may handle all required operations; stations with complex live programming may have an on-air console with 20 or more channels. Each of these channels has a linear fader control that increases the output of the channel as it is pushed up, allowing the operator to visually monitor the settings of the channels. Each channel usually has two, three, or more selectable inputs to allow alternative sources to be selected depending on specific requirements. For example, channel five on a console may have CD player one as the A input, mini-disc player one as the B input, and reel one as the C input. Most on-air consoles have two or three outputs, often labeled program, audition, and utility. Each channel can be assigned individually to program, audition, or utility. The program output is often fed directly to the on-air audio processing equipment and then on to the transmitter. The audition output is often fed to recording equipment, to record telephone calls off-air, for example. Some facilities use the audition and utility feeds to send program material to a second radio station or other location. These outputs usually have corresponding VU (Volume Unit) or LED (Light Emitting Diode) meters to allow for visual monitoring of the audio outputs. The console will also have switchable monitoring of these outputs using loudspeakers and/or headphones. On-air consoles also have a completely separate monitoring circuit called "cue" to allow the operator to preview or cue the audio before it is added to the mix. Each channel, where appropriate, has remote start/stop capability to control an audio source such as a tape recorder or CD player. Other standard console features typically include a digital clock/timer, built-in connection circuits for adding telephones as audio sources, and intercom circuits to facilitate communication with announcers in other studios.

Production Console

Until the 1970s, equipment in most radio production studios was not significantly different from equipment in on-air studios. Each studio was equipped with a console, microphone(s), open-reel audiotape recorder, broadcast cartridge (cart) player, and turntables. A broadcast cart recorder and perhaps additional open-reel audio recorders and audio processing distinguished the studio used for production from the on-air studio. As the popularity of stereo FM radio increased after 1961 and affordable multitrack tape recorders were introduced, the motivation and the means existed to create more elaborate commercial and program productions. The contemporary radio production console reflects this more complex approach. The console is designed to facilitate concurrent selection of more audio sources, compound audio processing of the various signals, and provide flexibility in routing inputs and outputs depending on the specific needs of the production project.

Much of contemporary radio production, especially in larger markets, begins as a multitrack recording project. Generally, multitrack recording involves a two-stage process. In the first stage, the different elements of the production are each recorded individually, at full volume, on separate tracks. The second stage involves playing back all of the separate tracks concurrently, adjusting the relative outputs to their appropriate levels, and mixing down to two-track stereo. This procedure provides a more efficient method of changing one or any combination of elements without re-recording all the elements. Experiments with the mix can be conducted without affecting the final recording. The production console needed to support this method provides the ability to route each of the tracks from a multitrack recorder to a separate fader for adjustment and processing during mixdown.

Digital Console

One of the most recent innovations in recording and studio equipment is the digital console. Digital consoles accept the digital output of CD players, hard disk recorders, and other digital sources without conversion to analog for the purpose of routing and mixing. Maintaining the signal in digital form minimizes the possibility of noise and other artifacts introduced during conversion to analog and reconversion back to digital. The continuity of the digital signal can be maintained throughout the audio chain. A digital console is different from an analog console in that the audio controls on the surface of the digital console are not physically connected to the audio circuitry. The controls are actuators that send digital control signals to the circuitry to carry out the console functions. These digital signals are commands that can be stored, grouped, recalled, and assigned as needed to various channels.

This process creates simplicity and flexibility, allowing console size to remain compact and efficient to operate.

The digital console also provides sample-rate standards conversion and synchronization, converting differing standards from different equipment to one station standard. The CD uses a sampling rate of 44.1 kilohertz, or 44.1 thousand times per second, to measure the height of the analog signal. Hard disk audio systems usually sample at 32 kilohertz while digital audiotape uses a 48 kilohertz rate. The console can generate one standard and all devices can be converted to it, along with providing a reference signal for synchronization.

Most digital consoles also accept analog inputs (open-reel audiotape recorder, microphones, etc.) and provide analog to digital conversion for insertion of these sources into the digital air chain. The console has been the last piece of major radio production equipment to complete its evolution to digital. At the same time, digital developments have, in some respects, eliminated the need for the hard-wired console.

Virtual Console

As more production is created in the digital domain, there becomes less need for an audio console to be a separate piece of hardware, manipulated by an operator. Audio sources can be connected directly to a computer and the functions previously controlled by an operator working a console (such as source selection, routing, mixing, and processing) can be programmed on the computer screen as a software function. In essence, the same process described above for digital console operation can take place entirely within the computer. The virtual console is usually a component of many hard disk recording systems. The console may simply be used to select and route audio sources for two track recording and editing or the console may be the starting point for a complex multitrack recording and editing session. Radio has come a long way from the early days of recording programs on transcription disks.

Transcription Disk Recorder

The earliest recording method used in radio was the transcription disk recorder. Modified from the early phonograph technologies, radio stations began using the transcription recorders and players in the 1920s. By the 1930s, most of the larger radio stations had transcription disk recorders. The recording process used a 16-inch flat disk of aluminum or glass covered with cellulose nitrate in which a lateral or vertical groove was cut. The transcription was cut at a speed of 33 1/3-rpm and provided a transcription time of 15 minutes. These disks were used by affiliate stations to record network programs and play them back at a later time, by program producers to distribute non-network programs to stations, and by local stations to record their own programs for rebroadcast or air check purposes. Even after the introduction of the magnetic tape recorder, the use of transcription recordings continued into the late 1950s.

Open-Reel Audiotape Recorder

The analog open-reel audiotape recorder was introduced to radio in the United States after World War II. Both Rangertone and Ampex Corporation manufactured professional open-reel audiotape recorders based on the designs brought back from Germany. The analog open-reel audiotape recorder has been an integral part of every radio studio since then. At first, recorders were monaural, recording and playing one track. The two track recorder was then introduced and, as FM radio and stereo transmission developed, soon became the recording standard.

The fundamentals of tape recorder technology have changed very little since the tape recorder's introduction. The audiotape used today is of better quality, resulting in better fidelity and lower noise, but the electronic principles and mechanics of operation are largely unchanged.

The main components of the open-reel audio tape recorder are the magnetic heads, the tape transport mechanics, the recording and playback electronics, and the tape itself. The magnetic heads of the recorder are the focal point in the process of recording audio on a tape and reading audio from a tape in order to recreate it. There are three heads in a professional recorder, mounted left to right in the following order: erase, record, playback. The tape goes past the erase head first, where any previous signal recorded on the tape is removed during the record function. At the record head, if the tape recorder has been placed in the record mode and a signal is being sent to input of the recorder, this signal will be deposited on the tape, magnetizing the metal particles present on the tape to create an analog of the original sound waves. This signal is now stored on the tape, and as the tape passes the playback head, the arranged magnetized particles create an electrical signal representing the original sound, which is then sent to the output section of the tape recorder electronics.

A tape recorder can be described by the number of tracks it can record or play. A recorder with one head each for erase, record, and play is a one (full) track monaural recorder. Its record and playback electronics are configured using one channel. A recorder with two heads is a two channel, two (or half) track stereo recorder. A recorder with four heads is a four channel, four track recorder. Larger radio production facilities will often have four or eight track recorders available for multitrack radio production. Because multitrack recordings are typically mixed down to two track stereo for on-air playback, the most common audiotape recorder used in radio is the two track stereo recorder.

The open-reel tape transport mechanism is designed primarily to move the tape past the heads at a consistent, exact speed. Tape speeds used in radio are seven and a half inches per second (ips) and 15 ips. Tape reels of seven inches, holding 1200 feet of one and a half mil (thousandths of an inch) tape will record and play for 30 minutes at seven and a half ips. Tape reels measuring 10.5 inches will hold 2400 feet of one and a half mil tape for 60 minutes at seven and a half ips. The tape transport system consists of the motors and tensioning systems to move the tape efficiently from supply side to take-up side. The key component in this process is the capstan and pinch roller. The capstan turns at a precise speed while the pinch roller holds the tape against the capstan, pulling the tape past the heads at the selected speed. Any variation in speed or interference in the transport process will result in inaccurate recording or reproduction. Most recorders also have some type of tape counter to display elapsed time or the amount of tape used.

The open-reel audiotape recorder also has electronic circuitry to support the recording and playback processes. Professional recorders typically have both microphone and line level inputs with individual level adjustments for each channel. The line level input accepts outputs from the console or other equipment, including other recorders. The record mode can be engaged separately for each channel as well. This feature, combined with a feature often called Sel Sync (Selective Synchronization) or Sel Rep (Selective Reproduction) allows the user to listen to material recorded on one channel while recording on the other channel in synchronization with the playback channel. Professional recorders also have monitor selector switches, one for each channel, that allow listening to the source audio at the input or the audio coming from the playback head. These monitor select switches also control the signal sent to the VU meter for visual monitoring of the audio. Output level controls for each channel are also part of the electronics of a professional recorder.

The audiotape itself is an important component of the recording process. Better quality tape costs more but provides better reproduction and long-term storage. Inexpensive tapes will deteriorate faster, causing audio dropouts and flaking of the magnetic oxide coating. Tape thickness is either one or one and a half mils (thousandths of an inch). Open-reel audiotape is available in quarter-, half-, one-, or two-inch widths. The quarter-inch tape is used with full track, two track, and some four track recorders. Multitrack recorders of four or more tracks use half-inch or wider tape. The two inch tape is used on recorders that record and play 16 or 24 tracks used in professional recording applications.

Broadcast Cartridge Recorder/Player

Shortly after the open-reel audiotape recorder became a mainstay of the radio production process, the development of the broadcast tape cartridge and recorder/player created another major refinement in the ability of radio to efficiently record and reproduce content. The cartridge recorder and player were introduced to radio in the late 1950s. Using a quarter-inch tape traveling at seven and a half ips, the tape is spliced into an endless loop wound on a single hub so that it comes out from the center of the reel, travels past the record/playback heads, and is rewound on the outside of the hub. A cue tone recorded on the tape when the start button is pressed in the record mode is sensed by the player during playback mode, stopping the tape at exactly the start of the message. A standard sized cartridge allows recording times of five seconds to ten minutes. Multiple messages can be recorded on one cart; each has its own cue (stop) tone. At one time, recording and dubbing carts was the primary activity in any radio production facility. Commercials, jingles, music, news actualities, and any program or message shorter than ten minutes was recorded on a cart. Almost every radio station used carts until the late 1990s, when digital media rapidly began to replace them.

Digital Recording, Playback, and Editing Equipment

The development of digital audio recording and playback equipment has introduced major changes in radio recording and production. The changeover from analog to digital started with the compact disc player. The once ubiquitous turntable is now almost nonexistent in radio studios. Radio quickly adopted the CD as a playback medium for music and other programs. A read-only medium until recently, CD recorders and re-writeable CDs are now being used in radio for a variety of purposes, including program production and archival storage.

Digital versions of the audiotape recorder are also being used in radio production. The rotary-head digital audiotape recorder (R-DAT) uses a helical scanning process very similar to that used in a video cassette recorder to record the digital information necessary to encode CD-quality two-channel stereo audio on a small cassette tape. The open-reel stationary head digital tape recorder manufactured today is a multitrack recorder with 24 or 48 tracks, and because of its complexity it is generally found in professional recording studios rather than in radio production studios.

Introduced as a consumer application, the digital minidisc combines the laser optical technology of the CD with the magnetic recording process of tape. The minidisc format has been adopted by many radio stations because it provides digital-quality recording, random access, and portability for field recording, all at a more modest cost than many other digital recording and playback systems. In these facilities, the minidisc recorder/player has replaced the cart recorder/player for recording and playing back short-form programs, announcements, and messages.

Digital options for recording, storing, and playing back radio programming have increased dramatically in recent years. Another option is now the hard disk of a computer. When a sound card is added to a computer with sufficient speed and a large enough hard drive, the computer can effectively replace a tape recorder. As the cost of hard drives has decreased and disk storage space has increased, recording audio on hard drives has rapidly become a viable alternative to tape and disk-based audio recording systems in radio stations. The tapeless digital radio studio exists in many different configurations and formats.

Hard disk recording systems used in radio include self-contained units such as the 360 Systems Shortcut, a hard disk two track recorder/editor, and the 360 Systems Instant Replay, a hard disk audio player, capable of playing back up to 1,000 different audio cuts when a button is pushed. Another approach to the hard disk system is the software-based integrated system installed on a personal computer. The Enco Digital Audio Delivery system and Arrakis Digilink are two of the many examples of this approach to digital audio. These systems integrate, in varying combinations, the mixing and routing functions of a console, audio editing and processing, playback, and automation functions into a single unit. While hard disk recording allows efficiency, quality, and creativity not possible with tape-based systems, broadcasters must contend with a new set of interface and technical support issues as they adopt these technologies. Hard disks can fail without warning. Where these systems completely replace tape and cartridge-based playback systems, some type of drive redundancy or removable tape or disk-based back-up is a necessary component of the system.

Once cost-prohibitive for many radio stations, digital multitrack recording and editing is now a key feature of much of the digital recording software and equipment. Many of the products are software-based so that any computer with a sound card and a large hard drive can become a digital recorder/editor. Many are self-contained units capable of recording eight, sixteen, or more tracks onto a built-in hard disk, Hi-8 videotape, or removable disk or tape system. Many of these devices are capable of being synchronized together using MIDI (Musical Instrument Digital Interface) or SMPTE (Society for Motion Picture and Television Engineers) time code so that the number of recording tracks can be expanded.

Digital recording has also changed the methods of editing audio. The process of recording over material to be replaced, dubbing to a new tape, or physically marking, cutting, and splicing the audio tape to remove or insert material has largely been replaced by a virtual editing process that takes place on the computer screen. A visual representation of the waveform of the audio to be edited is displayed on the computer screen or LCD (Liquid Crystal Diode) panel of the hard disk editor. Various editing functions can be carried out by marking and highlighting segments of the waveform. Material can be cut from one file and pasted into another. Adjustments can be made to the beginning and ending points of the audio file. The whole file or just a portion can be programmed to loop continuously. Files can be combined in a virtual over-dubbing mode.

Audio Processing

In addition to the audio sources, console, and recording devices, most radio recording studios have additional equipment to process the audio for creative and technical reasons. There are four general categories of audio processing: frequency, amplitude, time, and noise. An equalizer is a type of frequency processor, increasing or decreasing selected parts of a sound's frequency response. Equalizer controls are included on many production consoles and in digital audio software programs and are used to refine and enhance the sounds during the recording or production. An amplitude processor is used to control the volume of the audio. Compressors and limiters are commonly used to even out the level of audio. Reverberation and echo units are time processing units that give a sound a distinctive characteristic. Noise reduction processors such as Dolby and DBX units help lower noise caused by analog recording and electronics. Audio processing today is often created digitally, allowing multiple functions to be manipulated. Some processing and effects equipment uses MIDI as an interface system to connect to synthesizers, keyboards, samplers, and other electronic equipment for creative control purposes in radio production.

Monitors

Every studio needs at least one pair of monitor speakers in order to hear the production being created. A speaker performs the exact opposite function from a dynamic microphone; electrical energy is transduced into sound energy when the electrical energy representing the sound is sent through two wires into a magnetic field at the back of the speaker. A moving coil and speaker cone suspended in this field are induced to vibrate by this electrical energy, causing surrounding air molecules to vibrate, creating sound analogous to the original sound. In critical production and recording situations, the monitors must be capable of accurately reproducing the sound from the audio that has been mixed and recorded. Most speakers used as radio studio monitors are dynamic speakers. A separate woofer is used to reproduce the lower frequencies and a tweeter reproduces the higher frequencies. A crossover is an electronic circuit that sends lower frequencies to the woofer and higher frequencies to the tweeter. An acoustic suspension design has speakers and crossover mounted in a sealed box enclosure. The bass reflex design uses a tuned port to release some of the lower frequency sound from the rear of the speaker to combine with the main sound of the speakers, resulting in stronger bass sound. Typically, radio studio speakers are hung from the ceiling or mounted on the walls behind the audio console. This placement, with the speakers slightly angled in toward the operating position, enables the operator to accurately hear the stereo imaging in the mixed sound. Since an open microphone near a monitor speaker will cause feedback, in radio studios a mute circuit is usually used so that when a microphone is turned on, the monitor speaker will be muted. Headphones are worn so that the operator or other personnel may continue to hear the mixed audio while the microphone is on.

Studios

At the beginning of radio broadcasting, studios did not exist; the announcer used a microphone connected directly to the transmitter, housed in whatever space was available, often in a garage or shack. As the process of creating live radio programs grew more sophisticated, so did the studios. Soon the engineers and equipment were housed in a control room, separated from the performers (and often studio audiences) in the studio. A variety of materials including burlap, plush velour curtains, and solid surface baffles were used to control the acoustics of the studios. Larger stations and networks constructed very elaborate and large studio complexes, rich in architectural design and isolated from the rest of the building through special suspensions. Some studios were large enough to seat full orchestras and large studio audiences.

As radio began to rely on local programs featuring recorded music, radio stations began to build smaller, more specialized studios. Initially, most stations relied on an engineer to operate the audio console and adjust the transmitter from the control room while the announcer or program host remained in the announcer booth, receiving cues through the glass separating the two rooms. As transmitters became more stable and stations looked for ways to cut costs, it soon became common for the announcer to also engineer his or her own program. This operational change resulted in a smaller radio studio where the announcer stands or sits at a microphone and console, surrounded by other necessary equipment within easy reach. While proper acoustic design and isolation is still important in these contemporary studios, audio processing helps overcome acoustical problems. Digital remote equipment also minimizes the need for radio studios. A telephone line and a digital encoder/decoder set make it possible for a radio station to be able to have a studio virtually anywhere.

JEFFREY D. HARMAN

See also Audio Processing; Audiotape; Automation in Radio; Control Board/Audio Mixer; Digital Recording; Dolby Noise Reduction; High Fidelity; Production for Radio; Stereo

Further Reading

Adams, Michael H., and Kimberly K. Massey, *Introduction to Radio: Production and Programming,* Madison, Wisconsin: Brown and Benchmark, 1995

Alten, Stanley R., *Audio in Media,* Belmont, California: Wadsworth, 1981; 5th edition, 1999

Burroughs, Lou, *Microphones: Design and Application,* Plainview, New York: Sagamore, 1974

Crews, Albert R., *Radio Production Directing,* Boston and New York: Houghton Mifflin, 1944

Hausman, Carl, Lewis B. O'Donnell, and Philip Benoit, *Modern Radio Production,* Belmont, California: Wadsworth, 1986; 5th edition, Belmont, California, and London: Wadsworth, 2000

Kefauver, Alan P., *Fundamentals of Digital Audio,* Madison, Wisconsin: A-R Editions, 1998

Keith, Michael C., *Radio Production: Art and Science,* Boston: Focal Press, 1990

Keith, Michael C., and Joseph M. Krause, *The Radio Station,* Boston: Focal Press, 1986; 5th edition, by Keith, Boston and Oxford: Focal Press, 2000

Oringel, Robert S., *Audio Control Handbook,* New York: Hastings House, 1956; 6th edition, Boston: Focal Press, 1989

Reese, David E., and Lynne S. Gross, *Radio Production Worktext: Studio and Equipment,* Boston: Focal Press, 1990; 3rd edition, 1998

Rumsey, Francis, and Tim McCormick, *Sound and Recording: An Introduction,* Oxford and Boston: Focal Press, 1992; 3rd edition, 1997

Siegel, Bruce H., *Creative Radio Production,* Boston: Focal Press, 1992

Talbot-Smith, Michael, *Broadcast Sound Technology,* London and Boston: Butterworths, 1990; 2nd edition, Oxford and Boston: Focal Press, 1995

Talbot-Smith, Michael, editor, *Audio Engineer's Reference Book,* Oxford: Focal Press, 1994; 2nd edition, 1999

Watkinson, John, *An Introduction to Digital Audio,* Boston and Oxford: Focal Press, 1994

Watkinson, John, *The Art of Sound Reproduction,* Woburn, Massachusetts: Focal Press, 1998

Recordings and the Radio Industry

The radio business and music recording industry have been symbiotically related from the inception of broadcasting in the 1920s. While live music performed on radio initially devastated the record business, radio's improved technology later helped to revitalize the manufacture of recordings. As radio shifted in the 1950s from programs to formats, popular recorded rock music became a staple on the air. Music played on radio promoted record sales—indeed, radio and records (the title of one trade periodical) became virtually inseparable.

Origins

However, radio's relationship with recordings was not so close at first; the two were separated by their differing technologies. While records were played on some early experimental broadcasts, and on smaller market stations in the 1920s, generally radio avoided the use of recorded music. This was due in part to the poor quality of the largely mechanical recordings of the time, and in part to a widely held feeling that radio should not merely play recordings people could easily buy on their own. This tradition of live radio carried over to the networks as they began forming in the late 1920s. Stations and networks provided hours of live music programming of all types, popular to classical, and the best techniques of studio design and microphone placement became issues of scientific analysis.

Live music on radio nearly doomed the record industry. The poor quality of mechanically recorded records could not hold up to the higher quality of live singing and orchestral music on the air. As record sales declined sharply, many firms left the business. Radio's technology came to the rescue in the form of the electrical transcription (ET). Developed in the late 1920s to allow longer recording times (up to 15 minutes on a side), the 33 1/3 rpm discs featured better frequency response through all-electronic means of recording. They were typically used by stations (WOR in New York was one of the first) to record and sometimes to archive programs. ETs encouraged the development of program syndication as well. Initial attempts failed to sell the better quality 33 1/3 format as commercial records, probably because requiring consumers to buy a new record player in the depths of the Depression was doomed from the start.

The big band music of the 1930s and 1940s was programmed widely on radio and helped to revive record sales. Many programs featured top singers and orchestras whose records were often promoted on the shows. Musicians had plenty of live venues at which to play, and radio carried many of them. At the same time, however, worries about recordings concerned the American Federation of Musicians, which instituted two strikes in the 1940s to preserve live radio in the face of some pressure to allow greater use of recordings.

Even as recording methods improved, the standard commercial recording, a 78-rpm disc, could still play only a few minutes on a side. World War II delayed further progress on all consumer products, but wartime research would contribute to a post-war revolution.

Postwar Revolution

When popular singer Bing Crosby heard about audio tape recording, he offered his top-rated program to a new network (ABC) on the condition that they allow him to record the show to avoid having to perform it twice each time for the East and West Coast time zones. Although networks had stoutly resisted the use of recordings on the air save in emergency situations, ABC was desperate for a popular star to build its competitive position. Crosby got his wish, and the ban on use of recordings began to waste away in the face of this new means of making high quality recordings that sounded almost live.

In 1947–48, engineers at CBS Laboratories, working under Peter C. Goldmark, developed a considerably enhanced 33 1/3 rpm record that squeezed in more grooves on each side of a 12-inch disc. Other parts of the system were also upgraded, including the vinyl from which the record was made, the stylus, microphones, amplifiers, and record players themselves. The "microgroove" long-playing (LP) records, with 20 to 25 minutes of music on each side, were a sensation when introduced on the market. Now consumers could hear opera or symphony concerts without annoying record changes every few minutes. Shortly thereafter, RCA introduced its own version of the improved records, a 45-rpm 7-inch disc, called "extended play" (EP) designed particularly for popular songs. Soon "the 45" became the standard form for selling popular music, while "the 33" was the standard album for longer-form music. Record players were designed to switch among the three speeds, although the 78 quickly disappeared. The most expensive console models combined radios (and soon televisions) with the record player.

As new radio stations flooded onto the air in the late 1940s and the 1950s, they sought programs to attract listeners and advertisers. Live programming was expensive while the use of records was cheaper and thus far more profitable. The result was more playing of recordings on the air—unheard of a mere few years earlier. At the same time a quest for high-fidelity (hi-fi) began to spread from Britain, encouraging an interest in better sound. The new FM stations catered to this interest as did a market for often expensive sound systems. But more attention—and far more money—was focused on the mid-1950s rise of rock-and-roll music.

Rock music saved radio as the networks gave way to television competition. As the top-40 music system developed in the mid-1950s, record sales grew modestly. After the rise of Elvis Presley in 1956, the growth in sales of 45 and 33 rpm records through the end of the decade was more than 125 percent over previous years. The payola scandal late in the decade served to underline what everyone in the business now understood—that radio's on-air "plays" were essential to popular record sales just as the existence of those records was vital to radio's success. Though more closely controlled after 1960, music companies continued to provide free or heavily discounted records to radio stations to encourage their inclusion on program playlists, and thus to promote record sales.

Record sales continued to skyrocket into the 1960s. Capitalizing on the popularity of rock, radio programmed more of it—and record companies produced more recordings. Radio and records were both aimed at a youthful audience with expendable income, listeners of increasing interest to advertisers. By the 1970s the splintering of music formats as well as the increasing separation of programs between AM and FM stations provided more outlets for recorded music of all types.

Radio's continuing popularity clearly affected music types and formats. "Crossover" music became more common: a specialized recording (say, a country song) might begin to sell well more broadly in the pop music market. Or a specialized song might sell well in another niche market. A crossover can be accidental, but most often it is the result of a deliberate attempt by a music company to bring a song or performer to the attention of more buyers to increase sales. Crossovers have become more common since the 1950s. The constant changing of their music formats by many radio stations has encouraged this constant mixing and changing of musical types.

Modern Era

The arrival of both television and especially cable TV channels changed radio's role. By the late 1980s, radio was no longer the sole broadcast outlet for recorded music as listeners flocked to music concerts and channels offering video with sound. Radio remained the most mobile means of listening to music.

Development of the relatively short-lived cartridge and then cassette audio recording devices aided not only radio stations, but also consumers who could now more readily travel with their recordings. The appearance of compact disc (CD) recordings in 1984 soon displaced the LP as the standard consumer recording format, and the even higher capacity of DVDs may do the same to CDs in the early 21st century. Through all these recording format changes, however, radio (along with some cable music channels) continued largely to define what music America liked.

In the late 1990s, internet streaming of music added yet another channel of music delivery to consumers. While the internet offered radio stations far greater reach, this development directly threatened music companies as listeners could download digital copies of desired music without buying it. A host of legal cases, especially involving an internet service called Napster, tested what could and could not be offered online, generally having the effect of limiting downloads only to music that customers paid for. Even here, however, the relationship of music delivery and sales was somewhat symbiotic as recordings (CDs or tapes) were increasingly being sold online as well as through retail outlets.

CHRISTOPHER H. STERLING

See also American Federation of Musicians; Contemporary Hit Radio Format/Top 40; Internet Radio; Payola

Further Reading

Channan, Michael, *Repeated Takes: A Short History of Recording and Its Effects on Music,* London: Verso, 1995

Compaine, Benjamin M., and Christopher Sterling, et al., *Who Owns the Media? Concentration of Ownership in the Mass Communications Industry,* New York: Harmony Books, 1979; 3rd edition, with subtitle *Competition and Concentration in the Mass Media Industry,* by Compaine and Douglas Gomery, Mahwah, New Jersey: Erlbaum, 2000

Hickerson, Jay, *The Ultimate History of Network Radio Programming and Guide to All Circulating Shows,* Hamden, Connecticut: Hickerson, 1992; 3rd edition, as *The New, Revised, Ultimate History of Network Radio Programming and Guide to All Circulating Shows,* 1996

Hull, Geoffrey P., *The Recording Industry,* New York: Allyn and Bacon, 1998

Kenney, William Howland, *Recorded Music in American Life: The Phonograph and Popular Memory 1890–1945,* New York: Oxford University Press, 1999

Millard, Andre, *America on Record: A History of Recorded Sound,* Cambridge and New York: Cambridge University Press, 1995

Morton, David, *Off the Record: The Technology and Culture of Sound,* New Brunswick, New Jersey: Rutgers University Press, 2000

The M Street Journal (1990–)

Radio and Records, <www.rronline.com>

Reinsch, John Leonard, *Radio Station Management,* New York: Harper, 1948; 2nd edition, by Reinsch and Elmo Israel Ellis, New York: Harper and Row, 1960

Sanjek, Russell, and David Sanjek, *American Popular Music Business in the 20th Century,* New York: Oxford University Press, 1991

Re-Creations of Events

Creative Use of Sound Effects

The term *re-creations* refers to the creative use of sound effects and other inputs to provide a program that is not what it appears to be. To an extent, any radio drama that uses sound effects to create in listeners a mental image involves re-creations. But the most famous examples in radio history involved the use of wire service reports (for the facts) and recorded sound effects (for the color, especially crowd noises such as applause and shouts) to encourage listeners to think they were listening to sports play-by-plays from an observer right on the field rather than, in fact, as visualized by an announcer often hundreds of miles from the scene.

In the early days of radio, news played a relatively small role in radio broadcasting. With entertainment programming serving as the main precedent for new programs, some early radio newscasters re-created events in the news. Given the available technology at the time, radio broadcasters found it difficult to easily produce actualities of people in the news, so the dramatizations of events in the news became a well-accepted substitute.

The best-known example of a news re-creation was *The March of Time,* a program produced by *Time* magazine. *Time* general manager Rob Edward Larsen and Fred Smith of WLW in Cincinnati originally produced a program in 1928 called *NewsCasting,* a summary of the news syndicated to about 60 stations. (Larsen receives credit in some sources for coining the term "newscasting.") In late 1929, Smith hit upon the idea of dramatizing the news, and the Larsen-Smith team began producing a program called *NewsActing.*

On 6 March 1931, *The March of Time* debuted on CBS, produced by Larsen and Smith and using actors, sound effects and music. The program took its name from its theme song *The March of Time,* composed by Ted Koehler and Harold Arlen. The script of the first program began: "Tonight, the editors of *Time*, the weekly newsmagazine, attempt a new kind of reporting of the news, the re-enacting as clearly and dramatically as the medium of radio will permit, some themes from the news of the week."

Many of the "historical" recordings of early radio broadcasting are, in fact, re-creations. Especially in the 1920s, most radio stations generally did not have the capability to record their programs. The legacy of *The March of Time* made such re-creations acceptable in the late 1940s and early 1950s when a number of records celebrating radio-station anniversaries were released. For example, the recording of "the first broadcast" of KDKA in Pittsburgh on election night, 1920, is actually a re-creation made in the mid–1930s.

Sports, especially baseball games, became popular occasions for radio re-creations. Until the late 1950s, major league baseball franchises were concentrated in the Northeast. No team was located farther west than St. Louis or farther south than Cincinnati. As a way of bringing major league games to vast areas of the nation without big-league teams of their own, radio stations began to re-create games that were taking place in ballparks around the country.

"Dutch" Reagan

Certainly the most famous of the re-creators, now if not then, is Ronald Reagan. "Dutch" Reagan, as he was then known, broadcast more than 600 re-creations of Chicago Cubs games on WOC in Davenport, Iowa, and WHO in Des Moines, Iowa, between 1933 and 1936. The future president's system was typical of the play-by-play re-creators. Reagan sat in a radio studio in Des Moines, 300 miles away from Wrigley Field, and received a pitch-by-pitch description of the game in progress via telegraph. As Reagan described it many years later:

> Looking through the window I could see [the producer] "Curly" (complete with headphones) start typing. This was my cue to start talking. It would go something like this: "The pitcher (whatever his name happened to be) has the sign, he's coming out of the windup, there's the pitch," and at that moment Curly would slip me the blank. It might contain the information "S2C," and without a pause I would translate this into "It's a called strike breaking over the inside corner, making it two strikes on the batter."

Occasionally the telegraph line would fail and the announcer would have to improvise until it was restored. In those situations, the announcer had to use his wits, as Reagan once was forced to do in a game tied in the ninth inning:

> I knew of only one thing that wouldn't get in the score column and betray me—a foul ball. So I had Augie [Galan] foul this pitch down the left field foul line. I looked expectantly at Curly. He just shrugged helplessly, so I had Augie foul another one and still another. . . . I described in detail the red-headed kid who had scrambled and gotten the souvenir ball. . . . He fouled for six minutes and forty-five seconds until I lost count.

Gordon McLendon

Whereas Reagan's colorful exploits as a play-by-play re-creation announcer are well known because of his later fame as an actor and political figure, the person best known as a practitioner of the skill was Gordon McLendon, "the old Scotchman, 83-years-old this very day" (quoted in Garay, 1992). Re-creations had been around since the earliest days of radio and were considered an "honorable practice of the time," according to famed sports broadcaster Lindsey Nelson. But McLendon elevated re-creation to an art with his productions of baseball on the Liberty Broadcasting System (LBS) in the late 1940s and early 1950s.

McLendon was the young owner of fledgling radio station KLIF in Dallas, Texas. One of his early programming ideas at KLIF was to re-create sporting events. In fact, his first re-creation was of a professional football game between the Detroit Lions and the Chicago Cardinals on the day that KLIF took the air, 7 November 1947. But baseball was McLendon's true love.

The leisurely pace of baseball has always allowed play-by-play announcers great freedom, and it was said that in re-creating baseball games, there were no limitations—the broadcaster's imagination could run wild. McLendon started re-creating major league baseball games on KLIF in 1948. This soon led to the establishment of LBS. By 1951, 458 affiliates had joined LBS, and the network was second in size only to the Mutual Broadcasting System.

An important aspect of re-creations was the use of sound effects. Author Ronald Garay described McLendon's technique:

> Three or four turntables were kept spinning throughout the re-created games with disks containing the various crowd noises always cued on one or more of the turntables. . . . What would be distinguishable above the crowd noise every so often would be a vendor hollering out the name of a sponsor's product. . . . Gordon's passion for realism and accuracy led him to send an engineer to record every ballpark sound that a radio listener might expect to hear. [Engineer] Glenn Callison recorded many of the sounds at Burnett Field in Dallas. . . . Gordon sent [technician] Craig La Taste to every major league baseball stadium in the country to record sounds identified with each particular stadium.

The success of a re-creation also depended, of course, on the skill of the individual announcer. Author Willie Morris, in his autobiography *North Toward Home* (1967), wrote of McLendon:

> His games were rare and remarkable entities; casual pop flies had the flow of history behind them, double plays resembled the stark clashes of old armies, and home runs deserved acknowledgement on earthen urns. Later, when I came across Thomas Wolfe, I felt I had heard him before, from Shibe Park, Crosley Field, or the Yankee Stadium.

McLendon himself described his approach this way:

> No picture that is shown on television could be possibly as vivid as the picture I painted in my own mind of a baseball game. To me . . . those players were far bigger than life, and Ebbetts Field, even though there were 3,000 people there if you actually were broadcasting from the field, was always in my mind's eye crowded with 35,000 people. The walls were a thousand feet tall that those home runs were hit [over]. . . . I could come out with a far more vivid picture than any that I could have ever painted from the baseball park itself.

The end of this creative but somewhat misleading use of the airwaves came in the early 1950s as both major- and minor-league audience attendance began to decline, in part owing to growing television coverage of the majors. A glimmer of the potential profits from controlling television and radio rights to their franchises pushed baseball owners to forbid unauthorized use of the activities in their ballparks. They sharply increased the fees charged to LBS for game rights (from $1,000 for a whole season in 1949 to $225,000 just two years later) and then turned the matter over to individual baseball clubs for renegotiation and higher fees. Thirteen of the teams refused to grant LBS rights at all. This fee increase, in conjunction with the departure of key advertisers, spelled the end of re-created baseball games.

J.M. DEMPSEY

See also Liberty Broadcasting System; March of Time; McLendon, Gordon; Sound Effects; Sportscasters

Further Reading

Fielding, Raymond, *The March of Time, 1935–1951,* New York: Oxford University Press, 1978

Garay, Ronald, *Gordon McLendon: The Maverick of Radio,* New York: Greenwood Press, 1992

Harper, Jim, "Gordon McLendon: Pioneer Baseball Broadcaster," *Baseball History* 1 (Spring 1986)

Lichty, Lawrence W., and Thomas W. Bohn, "Radio's March of Time: Dramatized News," *Journalism Quarterly* 51, no. 3 (Autumn 1974)

Morris, Edmund, *Dutch: A Memoir of Ronald Reagan,* New York: Random House, and London: HarperCollins, 1999

Morris, Willie, *North Toward Home,* Boston: Houghton Mifflin, 1967

Nelson, Lindsey, *Hello Everybody, I'm Lindsay Nelson,* New York: Beech Tree Books, 1985

Nelson, Lindsey, and Al Hirschberg, *Backstage at the Mets,* New York: Viking Press, 1966

Smith, Curt, *Voices of the Game*, South Bend, Indiana: Diamond, 1987; updated and revised edition, New York: Simon and Schuster, 1992

Wills, Garry, *Reagan's America: Innocents at Home*, Garden City, New York: Doubleday, and London: Heinemann, 1987; revised edition, New York: Penguin, 2000

Red Channels

1950 Blacklist

In 1950, as Julius and Ethel Rosenberg were being arrested for atomic spying and the Korean War was commencing, the American Business Consultants published *Red Channels, The Report of Communist Influence in Radio and Television,* which formalized the practice of blacklisting in the broadcasting industry. The publication listed 151 performers and artists who were deemed to be communist sympathizers ("fellow travelers" in 1950s parlance). Many of the personalities who were charged and others who were whispered about were denied employment in show business.

The publication of *Red Channels* was an outgrowth of the investigation of communist infiltration of the motion pictures by the House on Un-American Activities (HUAC), chaired by Parnell Thomas in 1947. After World War II, as the Cold War developed between the United States and the Soviet Union, someone suspected of being a supporter of communist and, in many instances, liberal causes could be branded a traitor and forced to reveal his or her political ideology. There was a great concern among certain members of Congress and other patriotic organizations that writers, directors, and actors were using popular culture to spread their nefarious beliefs. Many studio and broadcasting executives, under pressure by the HUAC hearings and by advertiser concerns, agreed never to hire a known communist on staff.

Founded in 1947, the American Business Consultants was a private organization headed by three former Federal Bureau of Investigation agents: John G. Keenan, Kenneth Bierly, and Theodore Kirkpatrick. Proclaiming that the government's efforts had failed to combat the communist message, the group published a newsletter, *Counterattack: The Newsletter of Facts on Communism,* to "obtain, file, and index factual information on communists, communists fronts and other subversive organizations." The group canvassed volumes of the *Daily Worker,* pamphlets of leftist rallies, and unpublished findings of the HUAC committee to uncover names of potential traitors. In June 1950 they published their special report, *Red Channels*, a formal list of 151 people whom the Communist Party used as "'belts' to transmit pro-Sovietism to the American public." The report claimed that the Russians were using radio and television as a means of indoctrinating U.S. citizens even more than press or film. Many of radio's most influential and talented artists were cited, including producers/directors Himan Brown, Norman Corwin, and William Robson; personalities Ben Grauer, Henry Morgan, and Irene Wicker; and commentators Robert St. John, William L. Shirer, and Howard K. Smith. Some retribution was immediate: Wicker, the "Singing Lady" who had entertained children for years on radio, was dropped by her sponsor Kellogg, and Robson, who had directed many of *Columbia Workshop*'s innovative dramas, was mysteriously dismissed from his Columbia Broadcasting System (CBS) assignments, receiving payments until his contract expired.

Many of the 151 industry members were listed because they were social activists, from New Deal supporters to civil rights demonstrators. The writers of *Red Channels* (who were not credited by name) admitted that not all those named were political radicals; in fact, some "dupes" advanced "communist objectives with complete unconsciousness." Nevertheless, everyone was under suspicion, and *Red Channels* became one of several blacklists circulating on Madison Avenue (though one of the only ones formally published), consulted by broadcasting executives, advertising agencies, and sponsors. Some performers were given the opportunity to recant their previous beliefs or risk being barred from the industry. For a fee, the American Business Consultants also advised radio and television producers as they were casting on which performers had problematic backgrounds.

With the publication of the 215-page *Red Channels,* many companies began to institutionalize blacklisting. CBS, considered the most progressive of the networks, demanded loyalty oaths from its employees and hired a vice president in charge of "security." The other networks quickly followed suit. Batten, Barton, Durstine, and Osborn legitimized the hiring of "security officers" to clear names for the advertising agencies. Both networks and agencies, working as quietly as possible behind the scenes, abided by one principle: don't hire controversial personnel, so that you won't have to fire them.

The Report of
COMMUNIST INFLUENCE IN RADIO AND TELEVISION

Published June, 1950

By AMERICAN BUSINESS CONSULTANTS
Publishers of
COUNTERATTACK
THE NEWSLETTER OF FACTS TO COMBAT COMMUNISM
55 West 42 Street, New York 18, N. Y.

Title page of *Red Channels*
Courtesy Wisconsin Center for Film and Theater Research

For many, the *Red Channels* list was the culmination of many years' worth of whispered accusations. Radio director William Sweets had been under investigation since the late 1940s. Sweets directed two hit series, *Gangbusters* and *Counterspy,* for the radio production company Phillips H. Lord, Inc. A charge had been made to the series' sponsors that Sweets, also national president of the Radio Directors Guild, had mandated that only communists could work for him. He was forced to resign and, following the distribution of *Red Channels,* had difficulty finding any employment. Sweets' difficulties were later attributed to Vince Harnett, who had worked in the Lord office and who became a specialist in communist infiltration. As a freelancer, he wrote the introduction to *Red Channels* and worked with Lawrence Johnson's supermarket chain in Syracuse, New York, to boycott products of sponsors that allegedly advertised on shows employing subversives.

The institutional pressure also affected radio news departments. William Shirer had been a member of "Murrow's boys" and was one of radio's most respected commentators. In 1947 he resigned from CBS when his news program lost its sponsor. The *New Republic* (13 January 1947, cited in Cloud and Olson, 1996) charged that after World War II, 24 liberal analysts, such as Shirer, were dropped by the radio networks because of objections from sponsors. Shirer did work for other stations, but after his name appeared in *Red Channels,* no major network regularly employed him again. Considering himself a victim of the blacklist paranoia, Shirer contended "that if the major networks had taken a firm stand in the beginning . . . [by] making a fair determination of individual cases, this thing would never have gotten off the ground."

Before the dissemination of *Red Channels,* discrimination based on political reasons was informal and subjective. The publications of the American Business Consultants commenced systematic ideological screenings, although always behind closed doors. Such blacklisting also coincided with corporate pressures in broadcasting to make a profit in postwar America by pleasing the largest audience possible. Any controversy was strictly to be avoided. Even after the 1954 downfall of one of *Red Channels'* most ardent supporters, Senator Joseph McCarthy, the purging of subversive elements in the entertainment industry persisted quietly into the 1960s. In 1962, when a radio raconteur, John Henry Faulk, won a libel suit against *Counterattack* and other accusers, the mechanics of blacklisting finally became part of the public record just as the process died out.

RON SIMON

See also Blacklisting; Corwin, Norman; Faulk, John Henry; Shirer, William L.

Further Reading

Cloud, Stanley, and Lynne Olson, *The Murrow Boys: Pioneers on the Front Lines of Broadcast Journalism,* Boston: Houghton Mifflin, 1996

Cogley, John, *Blacklisting: Radio-Television,* New York: Fund for the Republic, 1956

Faulk, John Henry, *Fear on Trial,* New York: Simon and Schuster, 1964

Foley, Karen Sue, *The Political Blacklist in the Broadcast Industry: The Decade of the 1950s,* New York: Arno Press, 1979

Navasky, Victor S., *Naming Names,* New York: Viking Press, 1980

Vaughn, Robert, *Only Victims: A Study of Show Business Blacklisting,* New York: Putnam, 1972

Red Lion Case

Landmark Supreme Court Decision

In this 1969 decision, the U.S. Supreme Court upheld the constitutionality of the Federal Communications Commission's (FCC) fairness doctrine and of related personal attack rules. These FCC regulations in some instances required broadcasters to air viewpoints with which they disagreed or to provide airtime to persons criticized by or on the stations. The case is important because it reaffirmed the notion that radio stations' use of a scarce public resource—the electromagnetic spectrum—justified a different First Amendment standard for broadcasting than for print media.

Although the case is commonly referred to as simply *Red Lion,* the court's decision involved a second lower court case as well, *United States v Radio Television News Directors Association.* Because the two cases involved similar issues, the Supreme Court consolidated them and issued one opinion. The cases involved the FCC's fairness doctrine and a specific application of the doctrine known as the personal attack rules. Under the general fairness doctrine, all radio broadcasters had an affirmative duty to cover controversial issues of public importance in their programming and to provide a reasonable opportunity for all sides of the controversy to be aired. The personal attack rule stated that if a person's character, integrity, or honesty was attacked during the discussion of a controversial issue of public importance, the station airing the attack had to notify the person of the attack; provide a tape, transcript, or summary of the program; and afford the attacked person an opportunity to reply to the attack on the air.

The *Red Lion* litigation arose when WGCB, a radio station licensed to the Red Lion Broadcasting Company in southeastern Pennsylvania, aired a 15-minute syndicated program by the Reverend Billy James Hargis. During the program Hargis discussed a book, *Goldwater—Extremist on the Right,* written by Fred J. Cook. Hargis attacked Cook as being a communist sympathizer, a newspaper reporter who was fired for writing false charges against public officials, and a critic of J. Edgar Hoover and the FBI and of the Central Intelligence Agency. When Cook heard of the broadcast he demanded time to reply on the station pursuant to FCC policy. Red Lion refused, and the FCC ordered the company to afford Cook an opportunity to reply. Red Lion appealed the order, and the U.S. Court of Appeals for the District of Columbia Circuit upheld the FCC's decision. (Years later, respected journalist Fred Friendly alleged that Cook may have been working with the Democratic National Committee, using the fairness doctrine and the personal attack rule to harass stations that carried ultraconservative programming. In response, both Cook and the Committee insisted that Cook had acted alone.)

After the *Red Lion* litigation had begun, the FCC adopted specific regulations clarifying its personal attack rules. The Radio Television News Directors Association challenged the FCC's action in court, and the U.S. Court of Appeals for the Seventh Circuit ruled that the FCC's regulations violated the First Amendment free speech and free press rights of broadcast stations.

In its ruling on the consolidated cases, the Supreme Court confronted two primary legal issues: whether the FCC had jurisdiction under the Communications Act to adopt the fairness doctrine and its related rules, and whether the policies abridged the free speech and free press rights of broadcasters. Justice Byron White wrote for a unanimous court. (Eight justices voted in the case. Because he had not participated in the oral arguments, Justice Douglas took no part in the Court's decision.)

With respect to the jurisdictional issue, the Court noted that the FCC had been given extensive powers to regulate in the public interest, convenience, and necessity. In previous cases, the power to regulate had been described by the Court as "not niggardly but expansive." The Court also cited various congressional actions that seemingly approved of the FCC's actions in adopting and enforcing the fairness doctrine. Given the Court's own precedents and the implied congressional approval of the FCC's actions, the Court determined that the FCC did in fact have the necessary authority under the act to establish the fairness doctrine and personal attack rules.

Regarding the constitutional issue, the Court first stated that broadcasting was clearly a medium protected by the First Amendment, but that differences in the characteristics of various media justified different treatment under the First Amendment. The Court then addressed the issue of spectrum scarcity. Because of the interference that would result, not everyone who wants to broadcast can do so; there is simply not enough spectrum for all would-be broadcasters. For this reason, Congress enacted legislation giving the FCC the authority to license and regulate broadcasting to serve the public interest. Justice White wrote that, given this scarcity of frequencies, "it is idle to posit an unabridgeable First Amendment right to broadcast comparable to the right of every individual to speak, write, or publish."

The Court determined the public has a right to hear the voices of those who, because of spectrum scarcity, cannot speak through their own broadcast station. According to Justice White,

> Because of the scarcity of radio frequencies, the government is permitted to put restraints on licensees in favor of

> others whose views should be expressed on this unique medium. But the people as a whole retain their interest in free speech by radio and their collective right to have the medium function consistently with the ends and purposes of the First Amendment. It is the right of the viewers and listeners, not the right of the broadcasters, which is paramount. It is the purpose of the First Amendment to preserve an uninhibited marketplace of ideas in which truth will ultimately prevail, rather than to countenance monopolization of that market, whether it be by the government itself or a private licensee. . . . It is the right of the public to receive suitable access to social, political, esthetic, moral, and other ideas and experiences which is crucial here. That right may not constitutionally be abridged either by Congress or the FCC.

Having found that the First Amendment rights of viewers and listeners to receive diverse viewpoints were paramount over the First Amendment rights of broadcasters to control the speech over their stations, the Court concluded that the fairness doctrine and its related regulations passed constitutional muster.

The *Red Lion* case is important not so much because of the individual disputes it settled but because of its lasting contribution to our understanding of the First Amendment rights of broadcasters. Even though the FCC stopped enforcing the fairness doctrine in 1987 and the personal attack rule in 2001, the case still stands for the proposition that, because of spectrum scarcity, the government can apply different First Amendment standards to broadcasting than it does to other media.

MICHAEL A. MCGREGOR

See also Controversial Issues; Fairness Doctrine; Federal Communications Commission; First Amendment and Radio; United States Supreme Court and Radio

Further Reading

Friendly, Fred, *The Good Guys, the Bad Guys, and the First Amendment: Free Speech vs. Fairness in Broadcasting,* New York: Random House, 1976

Red Lion Broadcasting Company v Federal Communications Commission, 395 US 367 (1969)

Regulation

Regulation of radio in the United States began in 1910 when Congress passed modest legislation to control the use of wireless at sea, a law updated in 1912. A separate and more substantial Radio Act of 1912 could not foresee broadcasting. Following the beginning and surge of radio broadcasting in the early 1920s, pressure rose for a new law. The Radio Act of 1927 created an independent commission to determine regulatory policy for radio and broadcasting in the United States. The venerable Communications Act of 1934 expanded the powers of what became the Federal Communications Commission (FCC) to determine regulatory policy, subject to congressional oversight.

Over the past eight decades, government regulation of American radio has varied from minimal oversight, to pervasive control, and more recently to substantial deregulation. Beginning in the late 1970s the commission began adopting less stringent regulatory policies for broadcasters, replacing specific requirements with market-based competition. The Telecommunications Act of 1996 introduced significant relaxation in broadcasting ownership requirements. The 1934 Act, though amended many times, still provides the overarching schema for American radio regulatory policy.

Early Radio Regulation

The "commerce" clause of the U.S. Constitution (art. 1, sec. 8) assigns to Congress the option of regulating interstate and foreign commerce. Early radio stations served as basic communication systems, transmitters of messages that were meant to facilitate commerce and to protect the health and well-being of U.S. citizens. The Wireless Ship Act of 1910 (PL 262, 61st Cong.) reflected congressional intent to institute modest regulatory requirements on the nascent wireless communications industry. Oceangoing ships traveling to or from the United States were required to have transmitting equipment if carrying more than 60 passengers. The secretary of the Department of Commerce and Labor was given the authority to make additional regulations to secure the execution of the 1910 Act.

The 1912 sinking of the *Titanic,* with the tragic loss of 1,500 lives, forced congressional action to meet American international treaty obligations in wireless communication. The 1910 law was expanded, and was followed a few months later by the more comprehensive Radio Act of 1912. The law provided for licensing all transmitting apparatus for interstate or foreign commerce by the secretary of commerce and required that each station operator be licensed and that the government prescribe regulations to minimize interference. Other sections of the act provided for the licensing of experimental stations, regulation over the type of modulation, prohibition against divulging the content of private messages, and a requirement to give preference to distress signals.

The Radio Act of 1912 did not foresee and thus did not mention radio broadcasting. Public interest and service obligations were not discussed except as they pertained to point-to-point communication. However, the emergence of broadcasting after 1920, with the rapid proliferation of new stations seeking licenses and vying for airtime on an extremely limited allocation of frequencies, created administrative problems for Secretary of Commerce Herbert Hoover. The 1912 act allowed the secretary no discretion to develop and modify administrative regulation as needed, depending on changes in the radio business.

In 1922 Hoover convened the first of what became four annual National Radio Conferences designed to elicit voluntary self-regulation by broadcasters and other interested parties. Attendees realized the inadequacy of the 1912 Act and many called for better government oversight through more comprehensive legislation. But interference among broadcasting channels dramatically increased as the number of stations proliferated and as operations increased power and moved transmitters; early broadcasting entered a period of chaos without any significant government oversight.

Between 1922 and 1923, Hoover expanded the number of frequencies assigned to radio broadcasting in an attempt to relieve interference conditions. The secretary, who strongly endorsed the notion of self-regulation, had some success persuading stations to share frequencies, limit power, and split up the broadcast day. However, despite his attempt to facilitate solutions, a growing dissatisfaction with time allotments and frequency sharing created problems for Hoover's policy of "associationalism." It was becoming apparent to Hoover and the industry that self-regulation could not solve increasing interference and allocation problems. At the fourth and final radio conference in November 1925, all agreed that "public interest" should be the basis for broadcasting policy. Attendees also convinced the secretary to stop issuing new radio licenses. Thus, the licensing provisions of the 1912 Act were suspended under an ad hoc regulatory policy that was agreed to by government and by the large radio manufacturing and broadcasting interests.

On 16 April 1926, however, a federal appeals court dealt the final blow to the 1912 Act when it ruled that the secretary had overstepped his authority. As a result, Hoover was powerless to enforce any operating requirements on licensees (*United States v Zenith Radio Corp. et al.*, 12 F. 2nd 614 [N. D. Ill. 1926]). Immediately after the *Zenith* decision, stations began switching frequencies, increasing power, and ignoring previously agreed-upon time-sharing arrangements. Interference levels grew dramatically, particularly at night, when signals were prone to long-distance skipping.

Under growing pressure that was fueled by dissatisfaction among broadcasters and the listening public, Congress passed the Radio Act of 1927. The legislation specifically regulated broadcasting for the first time by establishing a framework to regulate the industry and investing decision-making powers in an independent agency. Seven years later, with the passage of the Communications Act of 1934, Congress merged oversight of wired and wireless communication under the new FCC.

The Radio Act of 1927—Real Beginning of Broadcast Regulation

The Radio Act of 1927 (PL 632, 69th Cong.) conceived that a newly constituted Federal Radio Commission (FRC) would be able to resolve numerous interference problems that had emerged during radio's development. Drawing on a combination of earlier legislative efforts introduced by Representative Wallace White of Maine and Senator Clarence C. Dill of Montana, the 1927 legislation provided for continued but more effective licensing and for the assigning of station frequencies, power, and fixed terms for all radio licenses. The legislation also provided for the creation of a temporary commission with authority to designate licensees and to regulate stations' operating conditions. The act asserted a public interest in broadcasting and public ownership of the airwaves, extended considerable rulemaking discretion to the commission, and provided commissioners with considerable discretion to decide questions of law and policy.

One of the significant outcomes of the 1927 Act, still debated today, was that broadcasters were accorded more limited rights under the First Amendment than was traditional for the press. The legislation clearly designated the electromagnetic spectrum as part of the public domain, allowing the commission the power to grant rights to users of the spectrum but forbidding private ownership over communication channels. In addition, extreme interference problems encountered with the breakdown of the 1912 Act suggested that a real scarcity of available channels existed. This complicated the task of the commission to devise a permanent allocation scheme that would suit all political and business constituents. Because the known radio spectrum and limited engineering capabilities could not afford all who wanted to speak an opportunity to do so, the FRC was empowered to impose rules and regulations limiting the number of entities actually using the airwaves. Legislators provided the commission with broad discretionary powers, subject to adjudication by the federal courts.

Many of today's expectations for regulatory policy emanate from the 1927 legislation. The Radio Act called for a commission comprising five members, each appointed from and responsible for representing a specific geographical zone of the United States. Congress initially conceived that the agency would dispense with the interference problems within the first year, after which the commission would become a consultative, quasi-judicial body meeting only when necessary. Sections 4 and 9 of the act invoked an undefined public-interest standard and gave commissioners the power to license and regulate wireless stations; federal radio stations were exempt from regulatory oversight. Licensing decisions made by the commission were subject to adjudication by the court of appeals, essentially as a *de novo* review. Legal scholars point to the fact that oversight was essentially a limited review of specific issues within a narrow class of petitioners.

Although the act did not contain specific language to regulate broadcasting "chains" or networks, legislators gave commissioners the ability to "make special regulations applicable to radio stations engaged in chain broadcasting" (sec. 4 [h]). Therefore, whatever control the commission could impose over radio networks had to be accomplished at the station level. Similarly, the act dealt with advertising in a minimal fashion. Some historians point to the fact that advertising was not widely accepted in 1926, when the bill was written, as one possible reason to explain the apparent oversight in the legislation.

Scholars are divided over the effectiveness of the FRC, but they generally give the organization little credit for effecting consistent and strong regulatory policy. During its six-year tenure, relations with Congress were stormy, sometimes to the point of hostility. By the end of its first year, congressional members who wrote provisions of the act called FRC commissioners "cowards" for their lack of regulatory action. Other critics pointed to flawed decision making based on poor information collection. The FRC's inability to resolve interference problems and redistribute licenses caused Congress to impose the Davis Amendment, which called for equality of service standards, in the 1928 reauthorization bill.

The broadcasting industry, led by the Radio Corporation of America (RCA), Westinghouse, and General Electric, succeeded in convincing the FRC that the general framework of broadcasting developed under the secretary of commerce should be retained. Robert McChesney (1993) points out that reauthorizing the existing commercial stations without redistributing licenses caused many noncommercial licensees to have their allocations and times of operation reduced. At the end of the FRC's tenure in 1934, commercial broadcasting was well established in the United States.

Passage of radio legislation clearly reflected the congressional view that the electromagnetic spectrum represented a valuable natural resource that was to be carefully cultivated and conserved for the general population. *United States v Zenith* had opened the floodgates to far too many licenses, creating substantial interference and chaos for listeners and broadcasters alike. It is not surprising, therefore, that resolution of licensing-related controversies became the first priority of the FRC. Regulatory decisions of the early commission were frequently politically motivated. Client politics stifled regulatory efforts by pitting interests that favored policies to support the growth of a nascent broadcasting industry on one hand against the desires of congressional members who wanted a solution that redistributed licenses along geographical regions on the other. Thus, partisan politics made the FRC sensitive to criticism from both large industry players and the regional constituents of various members of Congress. Furthermore, because only two of the FRC commissioners were actually confirmed by the Senate during the FRC's first term, the initial action of the agency was tentative, depriving the commission of an opportunity to regulate boldly. With passage of the Davis Amendment in 1928, Congress specifically directed the FRC to solve interference problems that plagued the AM band and to provide equalization of services to the different geographical regions of the country. With newly confirmed commissioners and a better sense of purpose, the FRC redistributed the broadcast band in the fall of that year. General Order 40 put into place a structure that allocated certain broadcast frequencies for long-distance, regional, and local services. The outcome of this new engineering calculation was the development of clear channel stations, which came to dominate radio during the 1930s, 1940s and 1950s.

The basic regulatory structure embodied in the Radio Act of 1927 became the basis for the permanent body designated under the Communications Act of 1934. By 1934 broadcasting had evolved into a highly profitable business. The structural components of the network radio system, almost wholly outside the purview of the FRC, had developed into a series of highly successful operations. The breakup of the RCA trust had created powerful forces within the communications industry vying for different segments of the industry. Many of the most powerful broadcasting stations, designated as "clear channels," were licensed to the large broadcasting or radio manufacturing companies, and the FRC's adoption of a system of clear (national), regional, and local AM channels (General Order 40) solidified the interest of stations already on the air, but has lasted in large part to the present as the chief means of allocating channels and reducing interference.

The Federal Communications Commission

The passage of the Communications Act of 1934 (PL 416, 73rd Cong.) established a permanent commission to oversee and regulate the broadcasting and telecommunications industries. In creating the FCC, Congress invested the permanent agency with the same broad regulatory powers that had been given to the FRC. These powers were extended to include wired telecommunications services, which had previously been under the jurisdiction of the Interstate Commerce Commission. Most provisions of the 1927 Act were incorporated word for word into Title III of the more comprehensive Communications Act. The language of both the 1927 and 1934 acts allowed the agency to employ a wide variety of sanctions, incentives, and other tools to fulfill regulatory or policy mandates. Over the years, court rulings concerning the Act and appealed FCC decisions have helped to delineate the boundaries of permissible government action.

The new agency did not intend to upset the broadcasting systems that had developed under the Secretary of Commerce and the FRC. However, Congress provided the new agency with some regulatory flexibility by repealing the specific requirements of the Davis Amendment. The general themes of the 1934 Act exemplified the principles of the New Deal by consolidating federal powers under one agency and centralizing the decision-making power for all communications industries. Both the 1927 and 1934 acts are significant because they invested regulatory powers in an independent "expert" agency. The realization that broadcast regulation should not be limited to supervision of interference and other technical aspects was fully apparent to legislators.

The newly formed FCC was confronted with the need to develop both an immediate and a long-term agenda. Immediate tasks included identifying and defining what constituted service in the "public interest, convenience, and necessity." The FRC had developed some regulatory policies, but clarification would be needed for long-term administrative policy development. As a corollary to this process, the FCC would have to develop reflective criteria to determine whether stations were doing an acceptable job of meeting their public service obligations. Consistent with this goal, the agency would be required to articulate and give meaning to broad phrases such as *public interest.*

Secondly, the FCC was now charged with developing a plan for utilizing the expanding electromagnetic spectrum. The years between creation of the FRC and creation of the FCC yielded important discoveries regarding the extent and usage of the spectrum. The commission, charged with "the larger, more effective use of radio" (sec. 303 [g]), needed to develop a more complex mechanism for determining which users should be allowed to use what radio band and for what purposes. Different uses would require differing amounts of spectrum space, and conflicting requests for spectrum utilization would require the FCC to make determinations for which services and how many users the spectrum could provide for.

FCC *Annual Reports* through 1939 illustrate that the commission undertook much more sophisticated record collection than the FRC. Between 1936 and 1937, the commission required all broadcasters to file comprehensive information regarding income, property investment, number of employees, and nature and types of programs. The FCC reported much of the statistical data to Congress during 1938. In that same year, the Commission began the first full-fledged review of the practices of chain (network) broadcasting. The initial years of the FCC reflected the need to collect and collate sufficient data to implement a long-term broadcast regulatory policy.

Localism and Trusteeship—A Framework for Broadcast Agenda Setting

The broad nature of the language used in the 1927 and 1934 acts did not prescribe specific tests or mandates for users of the radio spectrum. Consequently, it became necessary for the FRC and FCC to articulate policies that it could use as touchstones for measuring the service of the licensee. Over time, these pronouncements, coupled with rules and regulations, allowed the commission to establish a baseline regulatory policy.

Early decisions of the FRC and the FCC generally illustrate the importance placed on local operation and on a "trusteeship" model in broadcasting. Local outlets were seen as "trustees" of the public interest, despite the growing power and programming of the radio networks throughout the 1930s and 1940s. These two principles became the bedrock of federal radio policy making for decades.

The commission realized that defining what constituted acceptable service for a licensee required developing a set of standards that a broadcast licensee could aspire to or be measured against. In *Great Lakes Broadcasting Company et al v FRC* (37 F. 2nd 993 [D.C. Cir.]), the FRC devised an important set of principles to delineate what constituted public service and to inform licensees as to what their obligations would be as trustees using a natural public resource. The principle of trusteeship was based on a rationale of spectrum scarcity and required broadcasters to provide that

> the tastes, needs, and desires of all substantial groups among the listening public should be met, in some fair proportion, by a well-rounded program, in which entertainment, consisting of music both classical and lighter grades, religion, education, and instruction, important public events, discussion of public questions, weather, market reports, and news, and matters of interest to all members of the family find a place.

In asserting that stations were trustees, the FCC attempted to develop a policy that scrutinized the economic impact that proposed stations would have on current station trustees. Thus, the FCC denied some licenses when it feared that an applicant had inadequate resources. At other times, the FCC refused to issue a license when an applicant was financially secure but would provide harmful competition to an existing licensee. Taken as a whole, FCC decisions published between 1934 and 1940 do not illustrate how the agency evaluated the merits of potential economic injury, nor do they demonstrate a uniform record of policy making. Robert Horwitz notes that the FCC's decision to protect the broadcast system resulted in de facto protection of existing broadcasting facilities. In apparent ad hoc fashion, the FCC sometimes approved license applications where there were existing stations, and other times it refused to grant construction permits in cities that had no primary radio service at all. Engineering factors were not critical in many decisions.

Encouraging localism (the idea that stations should reflect their own communities) became the second fundamental principle and proved useful for several reasons. Both network and local programs were being provided to listeners via a local licensee assigned to serve a particular community or via a clear channel station meant to serve a wide geographical area. Although the FCC had very limited control over national networks, it discerned that its power to regulate was essentially the power to control local stations, the stations' relationships with network program suppliers, and the stations' relationship with the community of license. Thus, policy evaluation based on serving the interests of the city of license provided the FCC with sufficient leverage over the whole of the broadcast industry through station regulation.

Regulatory Policies in the 1940s

As the 1930s ended, the FCC expressed concern that radio networks held too much power over licensees through affiliation agreements that prevented stations from programming more independently. The commission's actions during this period illustrate a desire to increase the responsiveness of local licensees to their listening public, reduce the anticompetitive behavior of the powerful radio networks, and effectively increase competition in local broadcasting. In addition, the commission wanted to end the competitive advantage the National Broadcasting Company (NBC) held over the Columbia Broadcasting System (CBS) and the smaller Mutual network as a result of its ability to program both the NBC Red and Blue networks. The Chain Broadcasting Regulations issued in early 1941 were challenged by NBC in *National Broadcasting Company et al v United States* (319 U.S. 190 [1943]). The Supreme Court decision, written by Justice Frankfurter, upheld the FCC's authority to regulate the business arrangements between networks and licensees. More important, the decision upheld the constitutionality of the Communications Act and reaffirmed the commission's presumption that it had substantive discretionary power to regulate broadcasting.

But hoped-for changes in the relationship between networks and affiliates failed to emerge. Even though the commission faced increased criticism and oversight hearings between 1942 and 1944 (resulting in the resignation or non-reappointment of several commissioners), FCC staff investigated what licensees proposed to program when they filed applications compared to what they actually programmed. The results of the investigation were summarized in the 1946 "Blue Book." The Blue Book restated the commission's interpretation of what constituted public-interest obligations and articulated four primary concerns: the quantity of sustaining programming aired by the licensee, the broadcast of live local programs, the creation of programming devoted to the discussion of public issues, and elimination of advertising abuse.

Although the Blue Book reflected the first attempt by the FCC to articulate a fully developed policy statement regarding what constituted good service, the FCC never fully enforced application of the service statements it advocated. Still, the Blue Book was an attempt to make broadcasters more responsive to their listeners. First, it articulated the commission's view as to what constituted good service. Second, it started a debate within the industry as broadcasters objected to a perceived governmental attack on their ability to program without censorship or government interference. Third, in an attempt to forestall the promulgation of formal content-based rules, the National Association of Broadcasters strengthened its own self-regulatory radio code. Finally, though perhaps unintentionally, the commission increased record-keeping requirements for broadcasters.

Following the end of World War II (a period of unparalleled prosperity in broadcasting), the commission encouraged local competition by allowing a rapid and dramatic increase in the number of licenses in the standard (AM) broadcast band. Some experts note that the pressure to expand broadcasting may be seen less as a regulatory initiative and more as a result of renewed interest in entertainment due to the end of the Depression of the 1930s and the repeal of war priority restrictions. The expansion of radio broadcasting was further enhanced with the FCC's creation of a new expanded FM band after 1945. As the decade drew to a close, the FCC revoked its former prohibition on station editorializing, thus laying the groundwork for what would become known as the fairness doctrine.

Postwar Radio Policy

The shift in revenue from network to local radio and the increased competition in the AM band forced significant changes in radio programming. Morning and afternoon "drive" times became major sources of revenue. Stations abandoned the block programming structure typical of network affiliation in favor of generic formats that were stripped across the broadcast week. Because concepts developed in the Blue Book held little significance in the context of this new local competition, the FCC issued a *Programming Policy Statement* in 1960 (25 Fed. Reg. 7291; 44 FCC 2303) to restate a licensee's broadcast obligations. Reflecting the changes in radio with the rise of television, requirements for sustaining programs (those without advertising support) were dropped and other program guidelines were updated.

The 1960 *Programming Policy Statement* required broadcasters to discover the "tastes, needs, and desires" of the people through local area surveys known as "community ascertainment." With the 1960 Policy Statement, the commission delineated a 14-point list of major program elements that broadcasters were supposed to provide for the service area. Broadcasters criticized the laundry-list approach to programming requirements and the agency itself for using specific "quotas" of programs enumerated in guidelines as a litmus test for automatic license renewal.

A decade later, the FCC issued a "primer" that placed significance on programs that were responsive to community problems rather than serving the "tastes, needs, and desires" of the community. In 1976 the commission reaffirmed a commitment to the ascertainment process by making it a continuous requirement. Many felt that with this requirement, combined with other filing requirements, the FCC was imposing a significant record-keeping and filing burden on licensees.

Between 1964 and 1980, the fairness doctrine was fairly rigorously enforced by the FCC, and critics of the doctrine claimed that the specter of a fairness complaint, with its potential for legal entanglements, frequently prevented ("chilled") broadcasters from airing more discussion of public controversies. Supporters of the doctrine claimed that broadcasters merely used the threat of a fairness complaint as an excuse for not airing more controversial material. The fairness doctrine was a source of discomfort for broadcasters and First Amendment advocates alike. By the 1980s, because of the increasing competition among stations and with radio deregulation under way, the commission began looking for a way to eliminate its own doctrine.

Until 1980 the FCC set policy through implementation of a series of behavioral rules meant to provide guidelines for broadcasters as to what was or was not acceptable. For example, the ban on indecent language was proscriptive, telling stations what was the boundary of acceptable speech. Rules banning some types of cross-media ownership and simulcasting of AM and FM programs were examples of agency rulemaking designed to *structurally* organize the industry. Taken as a whole, the accretion of rules and record keeping led broadcasters and policy makers alike to question whether the time was ripe for regulatory reform.

Radio Deregulation

Serious thinking about deregulation began at the FCC in the mid-1970s, and a move to lift some rules on radio had been outlined in 1979. The 1980 election of Ronald Reagan increased the pace of thinking about radio deregulation. Radio had grown from a few hundred stations in the 1920s to thousands of outlets. Under the lead of FCC Chairman Mark Fowler, the agency pushed to deregulate four required station activities, leaving more up to licensees. The eventual report and order (84 FCC2d 968) eliminated minimal advertising and non-entertainment program guidelines, program log requirements, and rigid approaches to ascertaining community needs. Instead, the commission planned to rely more on "marketplace forces" to provide checks against program or advertising abuses. Many filing requirements were abolished. In the deregulatory process, radio license renewal literally became a pro forma postcard process, increasing most licensees' expectations of renewal. Deregulation, combined with an easing of ownership restrictions in the early 1990s, reflected the agency's attempt to make market economies define broadcasters, who now faced increasing operating costs and stagnant advertising revenues.

Commission Regulation as Ad Hoc Policy Making and a Diminishing Public Trustee Model

Robert Horwitz notes that FCC deregulatory policies championed under Mark Fowler (1981–87) mirrored the 50-year-old demands of the broadcasting industry to make the commission a neutral technical oversight agency. The liberal interpretation of the First Amendment, characterized by the equal-time requirements of the FCC's political broadcast rules, faded with the commission's deregulatory efforts. With fewer content and structural controls left in place, radio broadcasting illustrated erosion of the public trusteeship model that had characterized radio since 1927. The dismantling of the trustee concept, combined with the reduction in structural requirements in regulation, left radio open to freer market competition. Passage of the Telecommunications Act of 1996 eased radio station ownership limitations, although the maximum number of stations allowed in any specific market is still restricted.

Even though deregulation and ownership consolidation have increased competition among the top radio stations in most radio markets, critics of deregulation note that the FCC has failed to create a diversity of ownership to mirror the demographic characteristics of America itself. Despite agency attempts to encourage diversity, the number of minority-held licenses has actually decreased, whereas the overall number of station licenses has increased. Other critics point to the failure of AM stereo and the delayed introduction of a digital radio broadcasting standard as indications that the FCC has rarely been successful in inducing the industry to embrace technological innovations. Pirate radio broadcasters and growing disenchantment with increasingly stratified radio programming led the FCC to create a new low-powered FM broadcasting service, which faced strongly negative reactions from the industry and Congress.

The commission's decision making can be viewed from a number of useful perspectives. Commission policy can usually be traced on a track parallel to congressional initiatives. For example, with the growth of radio into a large, mature industry, constituent pressure on the agency from members of Congress became less significant, and as a result social regulatory or structural policies were relaxed. Congressional intent, as manifested in legislative efforts such as the Telecommunications Act of 1996, illustrates Congress' desire to treat radio broadcasters less like public trustees and more like a price- and entry-controlled industry segment.

Early critics of the FRC and later the FCC complained that forced social regulation created artificially close ties between the regulatory agency and the broadcasting industry. Under "capture theory" analysis, such a regulatory agency becomes overly concerned with maintaining the economic well-being of the public trustees it licenses. The result is the creation of an oligopoly with limited, managed competition. During the 1980s, both conservatives and liberals promoted broadcast deregulation as a way to deconstruct the relationship that had developed between the regulators and the maturing broadcasting industry. Because competition creates long-term economic uncertainty, deregulation undermines the agency-client relationship.

The success of U.S. radio regulation can be measured in a number of ways. Radio is a vibrant industry with several large competitive players owning hundreds of radio stations each. Competition for listeners within specific demographic segments in most medium and large radio markets is fierce. Large radio markets support many different formats. Critics of the FCC liberalization policies complain that such policies may serve large listening segments but tend to marginalize smaller populations that seek more diversity in programming or increased access to the media to express divergent viewpoints. Whether one believes that the government should be worried about First Amendment issues largely depends on one's view of whether social regulation should mandate public access to the airwaves. However, a consequence of the expansion of the number of radio outlets and deregulation as a governmental policy is that both factors undermine the public trustee argument. In the long term, radio broadcasting will continue to be more concerned with economics and less concerned with the tastes, needs, and desires of the community of license.

FRITZ MESSERE

See also Blue Book; Censorship; Clear Channel Stations; Communications Act of 1934; Controversial Issues; Copyright; Deregulation; Editorializing; Equal Time Rule; Fairness Doctrine; Federal Communications Commission; Federal Radio Commission; First Amendment and Radio; Frequency Allocation; Licensing; Localism in Radio; Mayflower Decision; Network Monopoly Probe; Obscenity/Indecency on Radio; Payola; Public Interest, Convenience or Necessity; Red Lion Case; Seven Dirty Words Case; Telecommunications Act of 1996; Topless Radio; United States Congress and Radio; United States Supreme Court and Radio; Wireless Acts of 1910 and 1912/Radio Acts of 1912 and 1927

Further Reading

Benjamin, Louise M., *Freedom of the Air and the Public Interest: First Amendment Rights in Broadcasting to 1935*, Carbondale: Southern Illinois University Press, 2001

Bensman, Marvin R., *The Beginning of Broadcast Regulation in the Twentieth Century,* Jefferson, North Carolina: McFarland, 2000

Corn-Revere, Robert, *Rationales and Rationalizations: Regulating the Electronic Media,* Washington, D.C.: Media Institute, 1997

Davis, Stephen, *The Law of Radio Communication,* New York: McGraw-Hill, 1927

Dill, Clarence C., *Radio Law: Practice, Procedure,* Washington, D.C.: National Law Book Company, 1938

Dominick, Joseph R., Barry L. Sherman, and Fritz Messere, *Broadcasting, Cable, the Internet, and Beyond: An Introduction to Modern Electronic Media,* Boston: McGraw Hill, 2000

Federal Communications Commission, *Annual Report,* Washington, D.C.: Government Printing Office, 1935–97 (issues for 1935–55 reprinted by Arno Press, New York, 1971)

Federal Radio Commission, *Annual Report,* Washington, D.C.: Government Printing Office, 1927–33; reprint, New York: Arno Press, 1971

Horwitz, Robert Britt, *The Irony of Regulatory Reform: The Deregulation of American Telecommunications,* Oxford: Oxford University Press, 1989

Lindblom, Charles E., and Edward J. Woodhouse, *The Policy-Making Process,* Upper Saddle River, New Jersey: Prentice Hall, 1968; 3rd edition, 1993

McChesney, Robert W., *Telecommunications, Mass Media, and Democracy: The Battle for the Control of U.S. Broadcasting, 1928–1935*, New York and Oxford: Oxford University Press, 1993

Rose, Cornelia B., *National Policy for Radio Broadcasting,* New York: Harper, 1940; reprint, New York: Arno Press, 1971

Rosen, Philip, *The Modern Stentors: Radio Broadcasters and the Federal Government, 1920–1934*, Westport, Connecticut: Greenwood Press, 1980

Warner, Harry P., *Radio and Television Law,* Albany, New York: Matthew Bender, 1948

Reith, John C.W. 1889–1971

First Director-General of the British Broadcasting Corporation

In its 75-year history, only two directors-general of the British Broadcasting Corporation (BBC) have seen their names used as descriptions for a style and philosophy of management. One was the most recent, John Birt, and the other was the first man to hold the post, indeed, the man credited with giving the BBC its unique character among broadcasting organizations, John Reith.

Reith was a unique character, a towering presence both physically (he stood 6 feet, 6 inches tall) and in the force of his personality. He was a man driven by the belief that it was his destiny to perform a great service for his country and constantly tortured by the fear that he was failing to fulfill that destiny. He was capable of the breadth of vision required to see the potential for the infant broadcasting industry, but at the same time he harbored petty grudges and resentments against those he felt were hindering him in his task. And just as his personality was complex, so was the institution he fashioned in his own image. The term *Reithian* has come to embody the first principles of British public service broadcasting: to provide the best in order to enlighten and educate. The term also carries overtones of paternalism and elitism.

Early Years

Reith was born in 1889, the seventh child after a gap of ten years, to the Reverend Doctor George Reith and his wife Mary. Dr. Reith was a minister of the Presbyterian Church in the fashionable part of Glasgow. He ran an austere and joyless home based on strict Christian principles. The influence of his father's puritanism ran through the whole of John Reith's life and work.

The fact that his father refused to send him to university after leaving school and instead apprenticed him to an engineering firm was a perpetual source of resentment. Although Reith never felt satisfied with engineering, he discovered a talent for organization and found a good job with another engineering firm in London.

After being invalided out of the army in 1915 with a World War I injury that left him with a huge scar down his left cheek, Reith again turned to engineering. But, in common with many after the war, he had a period out of work until he was given a temporary post in London as the personal assistant to a Conservative politician, Sir William Bull.

A few months later, he answered an advertisement for the general manager of the BBC. Reith applied although, as he wrote in his diary, he knew next to nothing about broadcasting. Probably in part because of the influence of Sir William Bull, Reith was appointed, and in December 1922 he began to organize British broadcasting.

The BBC began as a private company formed from a group of the most successful wireless set manufacturers. It was answerable to the postmaster general, who issued broadcasting licenses to the BBC to be distributed regionally. The conditions of the licenses were that the service should not carry advertising and that it should not broadcast news that had not previously been published. This last restriction was a sop to the newspaper industry, which feared the BBC would put newspapers out of business.

It did not take Reith long to find out what broadcasting was about. Along with the small group of talented and enthusiastic men who had been recruited to help launch the BBC, he quickly saw its potential. He also decided from the very early days that the BBC should perform a public service regardless of the effect on the company's profits.

Regional stations were quickly established in the North in Manchester and in the Midlands in Birmingham. But Reith strongly believed that a national broadcaster operating in the public interest should be able to serve everyone. A number of smaller relay stations were quickly opened, and by 1925, 80 percent of the population could receive the BBC. Reith quickly introduced a networking system that would allow the regional stations to broadcast simultaneously important programs from London and that would also allow the regional stations to supply the network. By 1930 the BBC was transmitting two services, the National Programme, from London, and the Regional Programme, which offered a more local schedule.

Reith's declared aim was to bring the best of Britain to all the British. To this end he set up advisory committees to guide the content of the broadcasts. The first of these was the Religious Advisory Committee, which was formed in March 1923. It was the prototype of others set up not only in London but also in the regional centers. The Spoken English Advisory Committee had as members George Bernard Shaw, Rudyard Kipling, and the Poet Laureate Robert Bridges, among others. He wanted the committees to bring the best minds to consider the quality of the programs, but his was a very highbrow view of quality. The Music Committee, for example, set up in 1925, comprised only classical musicians.

In 1925 the government appointed Lord Crawford to head an enquiry into the future of the BBC. Reith took it upon himself to deliver a memorandum to the committee, not on behalf of his board but in a personal capacity. In it he urged the committee to turn the BBC into a public corporation along the lines of the utilities. In March 1926 the committee agreed, and the BBC ceased to be a private company on 1 January 1927 and became the British Broadcasting Corporation.

BBC and the General Strike

Reith had once again gotten his own way, but he was soon plunged into one of the most serious crises of the BBC's history, one that would establish the character of the BBC forever. Britain's coal miners had been protesting about their working conditions and pay for many months, and the Conservative government, under Prime Minister Stanley Baldwin, had been stalling. The storm finally broke in May 1926, when the Trades Union Congress called for a general strike. The broadcast of dance music was interrupted so that Reith himself could broadcast the news of the start of the strike. Because the newspapers were not being printed, restrictions on the broadcasting of news were relaxed, and the BBC put out five bulletins a day. Reith insisted that only announcers who spoke English with an Oxford accent should be allowed to read the news, so as to "build up in the public mind a sense of the BBC's collective personality" (Briggs, 1965).

Immediately the BBC clashed with the government. Chancellor of the Exchequer Winston Churchill, who was putting out a government propaganda sheet, saw the potential of the BBC and wanted to commandeer it. Reith got wind of this and was determined to stop him. He persuaded Baldwin that if the BBC were to retain any credibility with the public, it must be seen as impartial. At the same time, in the national interest, he promised that the BBC would not do anything that might prolong or be seen as justifying the strike.

To this end, the BBC broadcast news of demonstrations and read out statements from the Labour Party leader, Ramsay MacDonald, and various prominent Trades Unionists. But neither the representatives of the strikers nor the parliamentary opposition were allowed to appear in front of the microphone in person. Eight days after the strike began, Baldwin went on air to tell the British people that the strike was over and that they should forgive and forget. When Baldwin finished speaking, Reith the showman paused and then started to declaim the words of the hymn "Jerusalem," which begins, "And did those feet in ancient times, / Walk upon England's mountains green?" A massed choir and orchestra he had set up in the studio picked up the hymn, and it ended in a great climax.

This event marked the future of relations between the BBC and successive governments. The BBC has remained impartial and often a thorn in the flesh of government, but it has always supported the concept of parliamentary democracy and has always been vulnerable to government pressure on such things as the renewal of its charter or the amount of the license fee. Reith defined the BBC as "an institution within a constitution."

Sir John C.W. Reith (left) and Henry Adams Bellows of Columbia Broadcasting System
Courtesy CBS Photo Archive

The events surrounding the strike have led some commentators to conclude that Reith failed the BBC at this crucial time, that because of his desire to be in with the people at the center of power and his desire for a knighthood, he allowed the BBC to be used by the government. Others say that he could have done no more. With Churchill breathing down his neck, the indecisive Baldwin could easily have allowed the BBC to be taken over if Reith had not fought so hard to keep the company unaligned.

Reith's whole vision for the BBC was as a force for social cohesion. One of the ways he sought to achieve this was through the live broadcasting of national events. The first of these was the speech by George V to open the British Empire Exhibition on 23 April 1924, heard by an audience of 10 million people. From 1927 onward, many more public events were covered on a regular basis, such as The Trooping the Colour ceremony, and also major sporting events, such as the Oxford and Cambridge boat race, international rugby matches, the Grand National horse race, and the Wimbledon tennis tournament.

Final BBC Years

Directors-general of the BBC have to report to the board of governors. Reith's first four years after incorporation were a time of great tension between himself and the board. The chairman was Lord Clarendon, who was ineffectual in the role and easily bullied by more assertive board members, among them Mrs. Philip Snowden, the wife of the former chancellor in the Labour government. She took issue with almost everything Reith did. She felt he was an overbearing autocrat and spoke in letters of his "overwhelming egoism." Although Mrs. Snowden, or "the Scarlet Woman," as Reith called her in his diaries, was undoubtedly out of line with her approach to her work on the board, she made some telling points. Reith certainly was an autocrat. He wrote in his diary about his frustration at having to refer to committees and work with others when he could do the job so much more quickly if he were left to do it alone. His conviction that he alone could run the BBC properly was mirrored by his conviction that the BBC alone could provide the best service to the British people. He was adamant that the BBC should remain a monopoly, but he believed equally that it must remain in the hands of the state and not become a commercial operation.

Mrs. Snowden was particularly bothered by the lack of any staff associations at the BBC (all staff complaints went directly to the director-general) and by Reith's unorthodox recruitment procedures. Very few senior jobs were advertised; they mostly went to friends of friends. All senior appointments were made by the director-general himself, and every new person taken on by the BBC had an interview with him soon after starting. Reith also sacked people himself. One of the more controversial firings was of the chief engineer, Peter Eckersley, who was reputed to know more about broadcasting than anyone else in the country. Reith fired him in 1929 for being involved with a woman before her divorce had come through. Reith later explained that it was not a simple moral case. He would not fire everyone who had an extramarital affair. A conductor or a variety show producer could behave differently from a manager, an announcer, or a senior talks producer. But he did believe the BBC was the keeper of the nation's conscience, and as such, those who worked for it had to be seen as above reproach themselves.

But Reith started to get bored. He announced loudly that there was not enough to do at the BBC now he had gotten it running and that he was ready to take on a really big job and so be of more service to his country. Unfortunately, his character and his oft-professed contempt for politics if not politicians told against him. Nothing was forthcoming. He set about restructuring the BBC, but his style of management was becoming rather outdated and very cumbersome. He created far too many committees, which acted as a dead weight on the program makers, who, in the national center at least, became dutifully conformist.

In the regions, however, where Reith was less able to interfere, there was room for more creativity. The northern region in particular saw a burgeoning of innovative programs, using regional accents, live broadcasts, and the voices of ordinary people talking about their work.

By the late 1930s it was clear that Reith had to go. What is not clear is how his departure was engineered. One theory is that the chairman and various board members plotted to oust him, but there is little concrete evidence to support this. In 1938 Reith was summoned by the prime minister's office to be told that he was to leave the BBC forthwith and take over Imperial Airways.

After his last day at Broadcasting House, he was driven, with his wife, to the Droitwich transmitter, and at the stroke of midnight he closed down all the engines and switched it off, without, he was pleased to be able to tell anyone who would listen, any engineer telling him what to do. As he left he signed the visitors' book: "JCW Reith, late BBC."

For a while he lobbied to be allowed to become chairman of the board or to be otherwise involved, ex officio. When these pleas fell on deaf ears (he would undoubtedly have made his successor's life very difficult), he sent back all his radio sets and turned his back on the organization to which he had devoted the previous 16 years.

Although he had another 33 years in public life, and although his subsequent posts were in prestigious government departments and public companies, Reith frequently succumbed to bouts of depression over his own failure to achieve anything of what he deemed real greatness. To others, however, his legacy is still in evidence whenever anybody turns on a radio to listen to the BBC.

SARA JONES

See also British Broadcasting Corporation; Public Service Radio

John C.W. Reith. Born in Stonehaven, Scotland, 20 July 1889. Educated at Glasgow Technical College; served in Scottish Rifles in World War I; served in Officers' Training Corps, 1911–14; lieutenant, Royal Navy, 1942–44; engineering apprenticeship, Hydepark Engineering Works, Glasgow, 1906–14; junior engineer, Pearsons, London, 1914; general manager, Beardmore's engineering works, Coatbridge, 1920–22; general manager, British Broadcasting Company (BBC), 1922–26; first director general, BBC, 1927–38; started first regular schedule of public television broadcasts in the world, 1936; chairman, Imperial Airways, 1938; Minister of Information, 1940; member of Parliament for Southampton, 1940–41; Minister of Transport, 1940; Minister of Works, 1940–42; director, Cable and Wireless, 1944; chairman, Commonwealth Communications Council (later the Commonwealth Telecommunications Board), 1944–50; chairman, National Film Finance Board, 1949; director, Tube Investments, 1953. Received GBE, 1934; knighted, 1927; GCVO Doctor of Laws, Manchester University, 1933; honorary doctorate, Oxford

University, 1935; member of Royal Company of Archers, 1937; GCVO, 1939; titled Baron Reith of Stonehaven, 1940; Commander of the Bath, 1944; Lord High Commissioner of Church of Scotland, 1967–69; Lord Rector, Glasgow University, 1966; Knight of the Thistle, 1969. Died in Edinburgh, Scotland, 16 June 1971.

Publications

Broadcast over Britain, 1924
Into the Wind, 1949
Wearing Spurs, 1966
The Reith Diaries, edited by Charles Stuart, 1975

Further Reading

Allighan, Garry, *Sir John Reith,* London: Stanley Paul, 1938
Boyle, Andrew, *Only the Wind Will Listen: Reith of the BBC,* London: Hutchinson, 1972
Bridson, D.G., *Prospero and Ariel: The Rise and Fall of Radio,* London: Gollancz, 1971
Briggs, Asa, *The History of Broadcasting in the United Kingdom,* London and New York: Oxford University Press, 1961– ; see especially vols. 1–2, *The Birth of Broadcasting,* 1961, and *The Golden Age of Wireless,* 1965
MacCabe, Colin, and Olivia Stewart, editors, *The BBC and Public Service Broadcasting,* Manchester: Manchester University Press, 1986
Matheson, Hilda, *Broadcasting,* London: Thornton Butterworth, 1933
McIntyre, Ian, *The Expense of Glory: A Life of John Reith,* London: Harper Collins, 1994
Milner, Roger, *Reith: The BBC Years,* Edinburgh, Scotland: Mainstream, 1983
Scannell, Paddy, and David Cardiff, *A Social History of British Broadcasting,* vol. 1, *1922–1939: Serving the Nation,* Oxford and Cambridge, Massachusetts: Blackwell, 1991
Shapley, Olive, and Christina Hart, *Broadcasting a Life: The Autobiography of Olive Shapley,* London: Scarlet Press, 1996

Religion on Radio

The message, "What hath God wrought?" sent over Morse's telegraph in 1837 indicated that religious topics might well be prominent in electronic communications. This proved to be the case when, shortly after Marconi succeeded in transmitting messages by wireless telegraphy in 1895, hymns and prayers were included in the earliest test broadcasts. As stations were granted licenses, religious programming instituted itself as a vibrant and often controversial element in radio.

Beginnings

One of the country's first radio broadcast stations, Westinghouse's KDKA in Pittsburgh, aired *Sunday Vespers* from the nearby Calvary Episcopal Church on 2 January 1921. The pastor was not very interested in the event and so asked his junior associate, Rev. Lewis Whittemore, to conduct the service. In order not to distract those attending, the two KDKA engineers (one Jewish, the other Roman Catholic) donned choir robes. This broadcast was so well received by the listening audience that it soon became a recurring feature of KDKA's Sunday schedule and was presided over by the senior pastor.

In November 1921 the first continuous religious program was broadcast as the *Radio Church of the Air,* and in the following month the Church of the Covenant in Washington, D.C., obtained a broadcast license in order to set up WDM, the nation's first religious radio station. In 1922 Chicago mayor William Hale Thompson invited Paul Rader, who had recently founded the Gospel Tabernacle, to broadcast from a radio station set up in City Hall. Rader quickly grasped the potential reach of radio and negotiated the use of WBBM's studios and airtime on Sundays to run his own once-a-week station, WJBT ("Where Jesus Blesses Thousands"). Besides broadcasting services, Rader presented talks and aired performances by his Gospel Tabernacle Musicians.

Rush to Radio

By 1923 religious organizations held 12 broadcasting licenses and other church groups offered programs to nonreligious channels. That year, fundamentalist preacher R.R. Brown launched the first weekly non-denominational program over Omaha's WOW. Later known as the *Radio Chapel Service,* the program continued until Brown's death in 1964. In 1924 Walter Maier, an Old Testament professor at Concordia Seminary, was the force behind the establishment of KFUO, which still broadcasts from St. Louis under the auspices of the Lutheran Church, Missouri Synod. Also in 1924, Aimee Semple McPherson's International Church of the Foursquare Gospel in Los Angeles set up KFSG. By 1925 the number of religious organizations holding broadcasting licenses had increased to 63. Faced with the explosion of interest in all types of radio, the Department of Commerce passed a rule limiting any new religious or public service stations to broadcasting at 83.3 kilocycles, forcing the sharing of this one frequency by several stations.

Controlling the Flow

On the East Coast in 1923, S. Parkes Cadman started broadcasting services from the Brooklyn YMCA over WEAF and, through a telephone hookup, to a few New England stations. WEAF liked his interdenominational and low-key approach, but other religious groups quickly began pressing for access to the airwaves, so WEAF approached the Greater New York Federation of Churches (GNYFC) for help. It agreed to provide the station with three program streams: Protestantism was represented by the *National Radio Pulpit* (which remained on the air until the early 1970s), while Roman Catholics and Jews were given their own airtime. When WEAF became part of NBC, its religious programs became available coast-to-coast.

Sustaining Time

Looking at the varieties of religious broadcasting available and conscious of the diversity of their vast audiences, the networks were eager to keep radical and controversial religious broadcasts off their airwaves. The Federal Council of Churches (which, paradoxically, was subservient to the GNYFC) was consulted by NBC's Religious Advisory Council, which subsequently decided that mainline religious groups would receive sustaining (free) time on the network provided the groups paid their own production costs and avoided proselytizing. All programs would be non-denominational in nature and would be presented by a single speaker so that a preaching format could be maintained.

Under the system of sustaining time, not only were minority groups such as Muslims and Buddhists excluded, but so were Christian Fundamentalists, Evangelicals, Pentecostals and, for many years, Southern Baptists, Lutherans, and other sizable denominations. Mormonism was represented (on CBS) by *Music and the Spoken Word,* featuring the Mormon Tabernacle Choir. In effect, religious broadcasting became segregated—mainline Protestantism, Catholicism, and Judaism were welcomed by the networks at both the local and national levels, while Fundamentalists, Evangelicals, and other independent groups were forced to rely on paid broadcasting. Furthermore, NBC refused to sell them any network airtime.

The Federal Radio Commission's (FRC) attempt in 1927 to end the chaos caused by too many stations trying to broadcast on too few available frequencies made matters even more difficult for non-mainstream religious groups. In tightening up its regulations, within six years the FRC had forced about half of the religious radio stations to close because they could not provide the required equipment and personnel.

In 1934 the GNYFC ceded responsibility for religious programming on NBC and CBS to a new committee of the Federal Council of Churches. The mainstream groups accepted the idea of sustaining-time religious broadcasting because it guaranteed them access to large audiences through programs such as *National Radio Pulpit, Catholic Hour,* and *Message of Israel.* Cooperation with the networks lessened any threat that they would seek to ban all religious broadcasting because of the often provocative views of conservative groups. Protestant conservatives, however, believed that their voice was being silenced by federal intervention and collusion between mainline religions and liberal network bosses.

Paid-Time Programs

Despite their inability to purchase airtime on the networks and the increased technical regulations on their facilities, many Protestant conservative broadcasters prospered. In 1926, for instance, WMBI, a station sponsored (then and now—it is the nation's oldest audience supported station) by the Moody Bible Institute, went on the air with up-to-date equipment and a professional staff who avoided both direct financial appeals and demagogic attacks on other religious viewpoints.

Charles E. Fuller was perhaps the quintessential fundamentalist preacher of the 1930s and 1940s. He made his first radio broadcast in 1923, but did not begin a full-time career in religious broadcasting until 1933. The Mutual Broadcasting System (MBS) was happy to sell airtime to Fuller and to other non-mainline religious broadcasters in order to get enough revenue to establish itself in the marketplace. The success of Fuller's *Old Fashioned Revival Hour (OFRH)* may be tracked by the number of Mutual stations that carried the program: 66 in 1937, 117 in 1938, and 550 by 1942.

As its name indicates, Fuller's *OFRH* largely followed the format of a revival meeting; its message was simple and represented an amalgamation of conservative religious and cultural values, such as a certain anti-intellectualism and concern with apocalyptic themes. Fuller was given to addressing the audience as his "friends in radioland" and would encourage them to take out their Bibles and gather round the radio set. The *OFRH* was in many ways a model for later broadcasts of a similar nature: its charismatic presenter was at the center of the show, the program was positioned within larger church activities, fundraising activities were emphasized, and production values were kept extremely high.

Walter Maier, who had been instrumental in the founding of KFUO, started *The Lutheran Hour* in 1930. It aired on CBS until the network changed its policy on paid time broadcasting

Billy Graham
Courtesy Billy Graham Evangelistic Association

in 1935 and so had to move to the Mutual network. Underwritten by General Motors, *The Lutheran Hour* became the most popular religious program of its day, broadcast in 36 languages over 1200 stations worldwide, and receiving more mail from listeners than *Amos 'n' Andy.* Unlike many other radio preachers, Maier avoided any sort of star status, insisting that the message rather than the messenger was the only thing that mattered.

Other significant programming was supplied by Paul Rader, the founder of WJBT in Chicago, who by 1930 had an hour-long network show on CBS, the *Breakfast Brigade,* which featured his Tabernacle musicians. J. Harold Smith's *Radio Bible Hour* and Theodore Epp's *Back to the Bible Hour* also had many devoted listeners.

Anti-Bias Stance Strengthened

In 1939 issues concerning sustaining time versus paid time once more came into sharp relief with the decision by the National Association of Broadcasters (NAB) to restrict its member stations from editorializing about "controversial" matters, such as anti-Nazi or anti-communist statements. The NAB was responding, at least in part, to the mounting anti-Semitic and anti-Roosevelt rhetoric in broadcasts on CBS by Father Charles Coughlin, a Roman Catholic priest.

One result of the NAB's actions was that the Federal Communications Commission (FCC) began to deny license renewals to stations judged to have neglected warnings against bias and controversial themes. Further, the FCC insisted that religious broadcasters and other such groups could no longer seek new members or beg for financial contributions over the air. In reaction to the latter ruling, religious broadcasters devised the "free will offering" pitch which is still used today.

Although by 1941 Coughlin had been removed from the airwaves, the FCC took its anti-bias stance even further with the "Mayflower Decision" which, in commenting upon a Boston radio station, held that "the broadcaster cannot be an advocate." This ruling was relaxed somewhat during World War II when paid time religious programming was allowed to return to evangelization activities as long as national politics were not commented upon.

Mutual Changes Its Course

By 1940 about a quarter of the Mutual network's revenues came from Fuller's *OFRH* and other paid-time religious broadcasting. Two things, however, caused it to change its position: it was becoming more financially stable and was looking to further diversify its revenue stream, and in 1942 it was attacked by a liberal interest group, the Institute of Education by Radio, which was opposed to all forms of paid religious broadcasting.

Responding to this onslaught, conservative broadcasters met in St. Louis and set up the National Association of Evangelicals in an attempt to prevent further restrictions on paid religious broadcasts. Despite this move, Mutual announced in 1943 that it intended to make deep cuts in the time it made available for such programs in its 1944 season. The NAE reacted with consternation and worried that individual stations might also try to limit the time sold to religious broadcasters. In 1944, therefore, the members of the NAE decided that they needed to organize an effective pressure group and so set up the National Religious Broadcasters (NRB).

Fuller, forced by Mutual to accept only one half-hour program a week, chose to put a shortened version of *Pilgrim's Hour* in that slot and moved the *OFRH* to a hastily assembled group of independent radio stations, which gave him most of the coverage of the Mutual network. Fuller's solution of non-network syndication through independent stations soon became, and remains, a favorite method for distributing paid religious programming. Mainline religious groups, in fact, also decided in 1944 to start their own syndication campaign through the Joint Religious Radio Committee of the Congregational Christian Church, the Presbyterian Church USA, and the United Church of Canada. Today, in the same way that the NRB supports the conservative or independent wing of Protestantism, mainline Protestantism is represented by the Communications Commission of the National Council of Churches.

After World War II

During World War II, the government froze construction of new stations and the FCC granted few new licenses. By the end of 1945, however, the FCC had already issued more than a thousand licenses for stations and, for the first time since the 1920s, new noncommercial stations began to appear. This provided new outlets for religious programming. Choices also expanded in commercial operations. In 1946, for instance, KDRU in Dinuba, California, began as the first ever Christian radio station run as a commercial enterprise.

In the 1940s and 1950s new voices and new programs found their way onto the airwaves. ABC, the new network founded after the split-up of NBC's Red and Blue networks, aired *The Greatest Story Ever Told* between 1947 and 1956. The Southern Baptists finally got sporadic access to sustaining time for *The Southern Baptist Hour,* and the Christian Reformed Church's *Back to God Hour* started in 1947. The Church of the Nazarene was represented by *Showers of Blessings,* the Seventh Day Adventists had the *Voice of Prophecy,* the Free Methodists produced the *Light and Life Hour,* the Mennonites had their *Mennonite Hour,* and the Assemblies of God sponsored *Sermons in Song.*

Under the direction of the National Council of Catholic Men, the Catholic Church produced four programs: a drama,

The Ave Maria Hour, The Catholic Hour with Fulton Sheen (who presented the program from 1930 until he transferred to television in 1952), *Faith in Our Times,* which was broadcast on the Mutual network, and *The Hour of Faith.* Judaism was represented by one new program, *The Eternal Light,* which ran on NBC from 1946 until 1955. Noted for its excellence, it was funded by NBC and produced in conjunction with Moishe Davis, director of the Jewish Theological Seminary.

Just as many secular shows moved from radio to television, so also did several well-known religious programs. Nevertheless, radio continued to host a wide variety of religious shows. In 1950 *The Hour of Decision* carried Billy Graham's Atlanta Crusade over more than 150 ABC affiliates, and the next year Norman Vincent Peale and his wife Ruth became the first husband-and-wife team to host a religious program on radio.

Decline of Radio Networks

The rise of television caused a huge shakeout in the radio industry. At the same time that conservative and independent religious broadcasters seemed to find new confidence and dynamism, some mainline churches (including some that were having major financial upheavals) appeared to be losing interest in maintaining a vital presence on the radio.

The mainline churches had bought into the notion of sustaining time. As network revenues fell, however, and many local stations wished to maximize their profitability, there was a marked decrease in the amount of time given to sustaining programs and a trend to place these programs into fringe timeslots, either very late at night or early in the morning. The NCC, noting that sustaining time had declined from 47 to 8 percent of religious broadcasting, charged that this trend moved against the public interest, but it did not get very far with its complaints.

In 1960 the NRB and the NCC reached a compromise between their two positions, declaring that broadcasters' public service obligations could be fulfilled by either sustaining or paid time. Their announcement banned program-length fundraising, declaring it unconstitutional, and further stated that religious programs were exempt from the FCC's Fairness Doctrine. The FCC's "hands off" stance facilitated this posture, but it was challenged in 1964 when an author claimed that he had been slandered by a conservative preacher and demanded equal time to reply on WCCB, a Pennsylvania Station of the Red Lion Broadcast Group. This demand was upheld by the FCC and, ultimately, by the U.S. Supreme Court.

Toward the Current Era

In the mid-1970s, word started to spread that Madelyn Murray O'Hair, a leading atheist, was asking the FCC to put an end to all religious broadcasting. Although a total fabrication, it was taken very seriously by conservatives and independents who had already been riled at an unsuccessful attempt by Jeremy Lanzman and Lorenzo Milam to limit the number of licenses given to religious broadcasters. A result of this was a significant increase in donations to conservative and independent religious broadcasters.

In 1977 James Dobson, a lay psychologist whose "traditionalist parenting" ideas pitted him against the likes of Benjamin Spock, began broadcasting *Focus on the Family* on local radio. Within 15 years his show had grown into the most popular Christian program ever, being broadcast on nearly 1,500 radio stations in the United States and abroad. Unlike many others who started in radio, Dobson insists that he remains happy working in radio and has no desire to move into television.

The NRB became explicitly involved in the political process during the late 1970s. Alarmed at what it considered the "liberal drift" of the Carter Administration, the NRB convened a meeting of religious broadcasters to urge them to become more involved in "educating" Christians about the political process. As a result of this gathering, Jerry Falwell formed the Moral Majority and Pat Robertson not only organized the Freedom Council, but seven years later launched an unsuccessful campaign to win the Republican presidential nomination. Since the early 1980s, Christian fundamentalists have been active and loyal supporters of Republican candidates for political office.

In 1978 WYIS in Philadelphia, the first religious station owned by African-Americans, went on the air. Although it is conservative in nature, there are other preachers of the "Black Gospel" who are not quite as traditional. Frederick J. Eikerenkoetter II, known to his followers as the "Reverend Ike," began to make a name for himself by preaching a "gospel of prosperity," mailing out "miracle prayer cloths" and pamphlets about how to become rich and stay that way. At one point in the late 1970s, the Reverend Ike had a program in 56 radio markets, but he has since faded into obscurity. The American Muslim community has been served by programs featuring the Honorable Elijah, Malcolm X and, more recently, Minister Louis Farrakhan. Mainstream Muslim radio is now available via the internet.

Jewish radio is largely served by syndicated programs such as *Israel Today Radio* and programs made by the Jewish Federation and the Union of American Hebrew Congregations. There are a few Jewish radio stations, notably in Florida and Boston, but as with the Muslim community, the biggest source of Jewish religious programming can be found through streaming audio on the internet.

Catholic radio appears to have fallen on hard times since the days when it was the responsibility of the National Council of Catholic Men. Certain dioceses, mostly in the Northeast, the states bordering Mexico, and on the West coast, have their own radio stations, and some syndicated programs are produced by groups such as Franciscan Communications, the

Christophers, and the Paulists. But given the hierarchical structure of the Roman Catholic church, the lack of a consistent radio strategy is puzzling. In 1999 a group of Roman Catholic entrepreneurs funded Catholic Family Radio, a network with an avowedly conservative viewpoint, but it almost immediately fell into financial difficulties and faced bankruptcy within a year. Mother Angelica's EWTN network, although principally a television operation, also broadcasts on shortwave radio, provides syndicated programming to local stations, and can also be heard over the internet.

Entering the 21st Century

In 2003, www.radio-locator.com listed 1184 religious radio stations in the United States and Canada, about equally divided between AM and FM. Most carry inspirational and spiritual talk and music, and many of them continued to air sermons. Another 411 stations were listed as "Christian Contemporary" and there were 495 "Gospel" stations, most of these in the South.

Clearly, religious broadcasting by radio continues to flourish, although media consolidation and population shifts have meant that some forms of Christian radio (local stations in Appalachia, for example), are in decline. Conversely, the continuing influx of Hispanics into the United States has meant that considerably more resources are being dedicated to the religious needs of Spanish-speaking peoples.

The excitement that greeted the first religious broadcasts 80 or more years ago has, of course, been considerably tempered. Radio was the first medium that could communicate the sense of a speaker's presence to a mass audience in distant locations. Christians especially hoped that radio broadcasting would allow them to obey the command of Jesus to "Go ye therefore and teach all nations." Experience has shown that radio is not a very efficient tool for gaining converts, but that it can be very useful in providing comfort and support to those already committed to the broadcaster's viewpoint.

Future developments in religious broadcasting by radio will depend on at least three major factors: the creedal and liturgical orientations of the people who will want to listen to the programming, the structures of ownership and control of the media, and ongoing technological developments.

PAUL BRIAN CAMPBELL

See also Blue Book; Contemporary Christian Music Format; Controversial Issues; Coughlin, Father Charles; Evangelists/Evangelical Radio; Fairness Doctrine; Far East Broadcasting Company; Gospel Music Format; Jehovah's Witnesses; Jewish Radio Programs in the United States; McPherson, Aimee Semple; Mormon Tabernacle Choir; National Religious Broadcasters; Red Lion Case

Further Reading

Alexander, Bobby Chris, *Televangelism Reconsidered: Ritual in the Search for Human Community,* Atlanta, Georgia: Scholars Press, 1994

Apostolidis, Paul, *Stations of the Cross: Adorno and Christian Right Radio,* Durham, North Carolina: Duke University Press, 2000

Armstrong, Ben, *The Electric Church,* Nashville, Tennessee: Nelson, 1979

Bachman, John W., *The Church in the World of Radio-Television,* New York: Association Press, 1960

Dinwiddie, Melville, *Religion by Radio: Its Place in British Broadcasting,* London: Allen and Unwin, 1968

Dorgan, Howard, *The Airwaves of Zion: Radio and Religion in Appalachia,* Knoxville: University of Tennessee Press, 1993

Erickson, Hal, *Religious Radio and Television in the United States, 1921–1991: The Programs and Personalities,* Jefferson, North Carolina: McFarland, 1992

Harden, Blaine, "Religious and Public Stations Battle for Share of Radio Dial," *New York Times* (15 September 2002)

Hill, George H., *Airwaves to the Soul: The Influence and Growth of Religious Broadcasting in America,* Saratoga, California: R and E, 1983

Hoover, Stewart M., *Mass Media Religion: The Social Sources of the Electronic Church,* Newbury Park, California: Sage, 1988

Melton, J. Gordon, et al., *Prime-Time Religion: An Encyclopedia of Religious Broadcasting,* Phoenix: Oryx Press, 1997

Morris, James, *The Preachers,* New York: St. Martin's Press, 1973

Oberdorfer, Donald N., *Electronic Christianity: Myth or Ministry,* Taylors Falls, Minnesota: Brekke, 1982

Parker, Everett C., Elinor Inman, and Ross Snyder, *Religious Radio: What to Do and How,* New York: Harper, 1948

Schultze, Quentin James, editor, *American Evangelicals and the Mass Media: Perspectives on the Relationship between American Evangelicals and the Mass Media,* Grand Rapids, Michigan: Academie Books/Zondervan, 1990

Ward, Mark, Sr., *Air of Salvation: The Story of Christian Broadcasting,* Grand Rapids, Michigan: Baker Books, 1994

Renfro Valley Barn Dance

Country Music Program

One of country music's important radio stage shows, the *Renfro Valley Barn Dance* could be heard in the U.S. South and Midwest from the late 1930s until the late 1950s. Along with the *Grand Ole Opry* on WSM (Nashville, Tennessee), the *Jamboree* on WWVA (Wheeling, West Virginia), the *National Barn Dance* on WLS (Chicago, Illinois), and other similar country music shows, the *Renfro Valley Barn Dance* provided a widely heard forum for country music as it grew commercially in the 1940s and 1950s.

The Saturday night showcase for country music talent debuted on 9 October 1937 over the 500,000-watt radio station WLW in Cincinnati, Ohio. Initially, *Renfro Valley Barn Dance* broadcast from the Cincinnati Music Hall and then from the Memorial Auditorium in Dayton, Ohio, but in 1939 John Lair, the program's originator, moved operations to Renfro Valley, Kentucky, some 60 miles from Lexington and not far from Lair's birthplace in Rockcastle County. In Renfro Valley, Lair stationed his show in a converted barn and built around it a rustic pioneer village for tourists to visit. This idea of building a tourist destination around the Renfro Valley stage show predated the *Grand Ole Opry*'s Opryland megaplex by some 30 years.

With the show's physical move to Kentucky came also a move to a new radio station home. In 1941 the 50,000-watt WHAS in Louisville began airing the *Renfro Valley Barn Dance*, propelling to the South and Midwest the sounds of the show's country singers and comedians. Over the span of the show's run on radio, it would also be carried by the National Broadcasting Company (NBC), the Columbia Broadcasting System (CBS), and the Mutual Broadcasting System.

Many of the early performers on the *Renfro Valley Barn Dance* had come to the show from WLS radio in Chicago, where John Lair had organized the popular Cumberland Ridge Runners—a musical act on the *National Barn Dance*—and had worked as a music librarian in the 1930s. Former *National Barn Dance* acts on WLS who followed Lair when he headed South were the musical acts Red Foley, Lily May Ledford's Coon Creek Girls, and Karl and Harty. Foley, who was an original investor in the Renfro Valley complex, would go on to be the best-known graduate of the *Renfro Valley Barn Dance*, garnering many hit country songs on the Decca recording label and a prominent spot on the nation's most popular barn dance, the *Grand Ole Opry*. Other nationally known talent who appeared regularly on the *Barn Dance* included comedian Whitey "The Duke of Paducah" Ford (another original investor in the Renfro Valley complex), comedians and song parodists Homer and Jethro, and steel guitar legend Jerry Byrd.

Although the *Barn Dance* gave valuable exposure to country musicians and comics and, in general, helped to establish country music as a commercial force, the show was also important in preserving many of the pre–World War II elements of country music. Founder John Lair was an avid collector of folk songs and ensured that those songs continued to be performed by *Renfro Valley Barn Dance* performers, even as other country music radio shows and performers were forgetting such songs. In addition, as electric instruments and drums became increasingly common in country music during the 1940s, Lair maintained an emphasis on traditional acoustic music, such as that performed by Renfro Valley acts Manuel "Old Joe" Clark, the Callaway Sisters, the Mountain Rangers, and the Laurel County Boys.

As the popularity of the *Renfro Valley Barn Dance* grew in the 1940s, it spun off other musical showcases that brought country music to various audiences. Tent shows featuring *Renfro Valley* talent played one-nighters throughout the East, Northeast, and South. In addition, the Renfro Valley troupe performed on daily shows broadcast over WHAS. For fans of gospel music, Lair and his "Renfro Valley folks" produced the *Renfro Valley Gatherin'*, a program that aired on the CBS network in the 1950s and that still airs today in syndication over more than 150 radio stations in the United States and Canada.

In 1958 WHAS and CBS dropped the *Renfro Valley Barn Dance* from their schedules, marking the end of the show's wide distribution. The program was a victim of the rise of rock and roll music and the decline of network radio. Virtually the only broadcast outlet for Renfro Valley talent would be the tiny radio station WRVK, which Lair established in 1957. As a live performance, however, the *Renfro Valley Barn Dance* continues to be staged every Saturday night, drawing tourists to Lair's pioneer village and often featuring major country music artists. Lair died in 1985 at the age of 91, but his mission to bring country music to the people continues to be fulfilled.

MICHAEL STREISSGUTH

See also Country Music Format; Grand Ole Opry; National Barn Dance

Programming History

Mutual	1938; 1946–47
NBC	1940–41
NBC Blue	1941
CBS	1941–49; 1951 (as *The Renfro Valley Country Store*)

Further Reading

Daniel, Wayne W., "A Voice Like a Friendly Handshake," *Journal of Country Music* 16, no. 1 (1993)
Hall, Wade, *Hell-Bent for Music: The Life of Pee Wee King*, Lexington: University Press of Kentucky, 1996
Kingsbury, Paul, editor, *The Encyclopedia of Country Music: The Ultimate Guide to the Music*, New York: Oxford University Press, 1998
Malone, Bill C., *Country Music U.S.A.: A Fifty-Year History*, Austin: University of Texas Press, 1968; revised edition, 1985
McCloud, Barry, editor, *Definitive Country: The Ultimate Encyclopedia of Country Music and Its Performers*, New York: Berkley, 1995
Rice, Harry S., "Renfro Valley on the Radio, 1937–1941," *Journal of Country Music* 19, no. 2 (1997)
Stamper, Pete, *It All Happened in Renfro Valley*, Lexington: University Press of Kentucky, 1999

Retro Formats

Oldies/Nostalgia/Classic

Although these programming formats are not identical, they all derive the music they air from years gone by. Whereas the nostalgia station, sometimes referred to as *Big Band*, builds its playlist around tunes popular as far back as the 1940s and 1950s, the oldies outlet focuses its attention on the pop tunes of the 1950s and 1960s. A typical oldies quarter hour might consist of songs by Elvis Presley, the Beatles, Brian Hyland, Three Dog Night, and the Ronettes. In contrast, a nostalgia quarter hour might consist of tunes from the pre-rock era performed by Duke Ellington, Benny Goodman, Frank Sinatra, the Mills Brothers, Tommy Dorsey, and popular ballad singers of the past few decades.

Nostalgia radio caught on in the late 1970s, the concept of programmer Al Ham. Nostalgia is a highly syndicated format, and most stations go out of house for program material. Because much of the music predates stereo processing (1958), AM outlets are frequently the purveyors of this brand of radio, although in recent years more and more nostalgia programming has appeared on FM because recordings have been remixed in stereo. Music is invariably presented in sweeps, and for the most part disc jockeys maintain a low profile. Similar to easy listening, nostalgia emphasizes its music and keeps other program elements at an unobtrusive distance. In the 1980s, easy listening stations lost some listeners to this format, which claimed a viable share of the radio audience.

The oldies format was first introduced in the 1960s by programmers Bill Drake and Chuck Blore. Whereas nostalgia's audience tends to be over the age of 55, most oldies listeners are somewhat younger. Unlike nostalgia, many oldies outlets originate their own programming, and very few employ syndicator services. In contrast with its vintage music cousin, the oldies format allows greater disc jockey presence. At many oldies stations, air personalities play a key role. Music is rarely broadcast in sweeps, and commercials, rather than being clustered, are inserted in a random fashion between songs.

In the 1990s, oldies stations attracted a broader age demographic than they had in previous years because of a continuing resurgence in the popularity of early rock music. At the same time, nostalgia listener numbers remained fairly static but substantial enough to keep the format on the air in several markets. As of 2002, some 700 radio stations featured one or the other retro sound. A more dance/contemporary approach, called "jammin' oldies," has attracted additional listeners in recent years.

Another variety of vintage radio, classic rock/classic hit (also called boomer rock and adult hits), rose to prominence in the late 1980s. Stations employing this music schematic draw their playlists from the chart toppers (primarily in the rock area) of the 1970s through the early 1990s and often appear among the top-ranked stations in their respective markets.

Whereas classic rock concentrates on tunes essentially featured by album-oriented rock stations over the past quarter century, classic hit stations fill the gap between oldies and Contemporary Hit Radio (CHR) outlets with playlists that draw from Top 40 charts of the same period, although there may be an emphasis on more recent tunes at some classic hit stations.

MICHAEL C. KEITH

See also Album-Oriented Rock Format; Blore, Chuck; Classic Rock Format; Drake, Bill; Formats; Oldies Format; Rock and Roll Format

Further Reading

Hall, Claude, and Barbara Hall, *This Business of Radio Programming: A Comprehensive Look at Modern Programming Techniques Used Throughout the Radio World,* New York: Billboard, 1977

Keith, Michael C., *Radio Programming: Consultancy and Formatics,* Boston: Focal Press, 1987

Rock and Roll Format

Radio's 1950s Transition

Rock and roll was a hybrid musical form that grew out of rhythm and blues and country boogie, adapting the adult themes of the lyrics found there to the concerns of teenagers. Electric guitars and saxophones were predominant. The rhythm was usually marked by a strong backbeat, though shuffle, swing, straight-eight, rumba, and other rhythms were used. Harmonically, rock and roll adopted the blues chord changes and the standard song structures of the music that preceded it.

Background

Rock and roll radio in the United States was part of a massive set of changes in the industry beginning in the late 1940s and leading to modern formatted radio. After World War II and through the 1950s, the radio industry in the United States underwent fundamental changes, including an increase in the number of AM stations from less than 1,000 in 1945 to about 3,600 in 1960. This radically increased the competition for advertising income, on-air talent, programming materials, and audiences. At the same time, broadcasting networks were shifting their advertising finances, talent, and programming to television, leaving many of the older, established stations in need of programming, income, and management ideas. As television began to dominate the prime-time evening audience, radio increasingly depended on daytime audiences and on audience segments outside the urban, middle- and upper-class living rooms where television was adopted early, audiences such as African-Americans, teenagers, rural dwellers, and the less affluent. The displacement of the living room radio by the television, the postwar increase in the prevalence of car radios, and the later transistor revolution led to a dispersion and segmentation of the audience. People listened outside the family group, as individuals in different rooms of the home and outside the home.

Record shows served the need for cheap programming that appealed to audiences who tended to be listening secondary to other activities. The shows were usually built around a disk jockey personality, who often chose the music and might also work with local record stores and other sponsors. The personality and the music became a programming package with special appeal to targeted audience segments—as opposed to the old network model of wholesome entertainment for the whole family. Rock and roll reflected the shift from mainstream homogeneity to diversity and special-appeal programming.

Origins

In the late 1940s, commercial necessity began to overcome racist habits among radio station owners, managers, and advertisers, who began to program and advertise for African-American audiences and to hire African-Americans as on-air talent and program advisers. The first experiments were so successful that they led to a revolution in what was called "Negro appeal radio," featuring rhythm and blues music and disc jockeys who used the argot of working-class black folks. Disc jockeys in this format quickly became local celebrities, with their personal styles growing correspondingly more flamboyant. Attention-getting nicknames, rapping, rhyming, signifying, and characteristics of older verbal insult games and of the later rap and hip-hop were present in the style of African-American personality disc jockeys. As this became the hot new trend in radio, white disc jockeys learned to talk like hep cats too, and sometimes African-American voice coaches and programming consultants were hired at otherwise segregated radio stations. Later white proponents of the style, such as Dewey Phillips and Alan Freed, dropped the rhyming and much of the stylized wit, replacing them with a kind of wildness that may have reflected the liberty of white release into black style—as well as the booze and pills they were famous for consuming.

Negro appeal radio was not only a boon to the African-American community but also led to the discovery and development of white audiences for what had been conceived of as race music. Two important contingents were white entrepreneurs, often with working-class roots and rebellious attitudes, and white teenagers with spare time and disposable income. Disc jockeys such as John Richbourg (WLAC, Nashville), Dewey Phillips (WHBQ, Memphis), and Alan Freed (WJW, Cleveland); record producers such as Leonard Chess (Chess Records, Chicago), Sam Phillips (Sun Records, Memphis), and Randy Woods (Dot Records, Nashville); and record store owners and mail-order entrepreneurs such as Randy Woods (who turned mail-order business for Randy's Record Shop in Gallatin, Tennessee, into financing for Dot Records) and Leo Mintz in Cleveland were key players. They were white people with more than casual contact with African-American culture and with their own complicated mix of motives. Although some were primarily exploiting business opportunities, others were responding to a genuine affinity for African-American people and culture, and for others rebelliousness appears to have been the primary motive. The disc jockeys became the spoken voice bringing rhythm and blues music to white teenagers, and thus their rebellion—explicit in their loud, rude, on-air style and implicit in their love for forbidden black culture—became an essential component of rock and roll.

Heyday and Controversy

Beginning in 1955, rock and roll records became an ever-larger presence on music sales charts. These were songs by both white and black artists, mostly produced by independent record companies, bought by both white and black audiences. Eventually these songs dominated both the pop and the rhythm and blues charts, with notable presence on the country chart as well. In 1956 Elvis Presley made his first release after his contract was bought from Sun records by the major Radio Corporation of America (RCA); "Heartbreak Hotel"/"I Was the One" was in the top 10 of all three charts simultaneously. In the following years, the pop charts became completely dominated by rock and roll and soul, and so many songs crossed over from the rhythm and blues chart to the pop chart that *Billboard* actually suspended a separate listing for a short period in the early 1960s.

Across the same years that rock and roll came to dominate the pop charts, Top 40 radio became the new standard model for popular music radio programming. In this model, the popularity charts were used as a guide to radio programming, with the most popular songs played the most often. Complications about the validity of the charts or about the necessity that radio stations chose songs to play before they could appear on the charts, were ignored. The programming logic was hailed as a dispassionate, even scientific advance. Questions of taste were irrelevant, it was said; the new Top 40 programmers gave the public what it wanted. The result was that the charts and the radio were locked into a positive feedback system, so that some popularity led to more popularity—and rock and roll took over.

Rock and roll was controversial, and not only because of its associations with rebellion and forbidden fun. Though it took a while to catch on in the white middle class, by the late 1950s it swept up teenage interest in a manner that disconcerted adults. Reports of conflict between police and crowds at a few concerts were widely publicized. Exploitation movies capitalized on the association of rock and roll and delinquency. Racists objected to the mixing of black and white musicians and audience members. Rock and roll was predominantly produced by small, independent record companies that quickly came to dominate the older major companies in the popular music market. It was pioneered on independent radio stations, and when it crossed to more established stations, the flamboyant, independent character of the personality disc jockeys came with it. More established interests in the music industry, white backlash groups in the South, conservative ministers, parent-teacher associations, and politicians found a convergence of interests in their suspicion that rock and roll was a conspiracy led by the disc jockeys and damaging to (white) youth.

The payola scandals of 1959–60 were the most prominent component of the anti–rock and roll backlash. The practice of record companies' plying radio and other industry personnel with money and favors was decades old and not illegal; as early as the 1890s, song publishers had aided sheet music sales by paying prominent band leaders to perform their songs. What was new was the power of individual disc jockeys and the success of new, small record companies outside the New York music industry establishment. The disc jockeys were the primary target of the scandal, and station owners used the opportunity to wrest control of programming away from them. The model of management-controlled Top 40 programming spread throughout the industry, and by the early 1960s few disc jockeys anywhere selected their own music to play. The free-spirited and entrepreneurial era of rock and roll radio in the United States was over.

ERIC W. ROTHENBUHLER

See also African Americans in Radio; Black-Oriented Radio; Disk Jockeys; Freed, Alan; Music; Payola; Recordings and the Radio Industry; Social Class and Radio; Contemporary Hit Radio Format/Top 40

Further Reading

Barlow, William, *Voice Over: The Making of Black Radio*, Philadelphia, Pennsylvania: Temple University Press, 1999
Eberly, Phillip K., *Music in the Air: America's Changing Tastes in Popular Music, 1920–1980*, New York: Hastings House, 1982

Ennis, Philip H., *The Seventh Stream: The Emergence of Rocknroll in American Popular Music,* Hanover, New Hampshire: Wesleyan University Press, 1992
Fornatale, Peter, and Joshua E. Mills, *Radio in the Television Age,* Woodstock, New York: Overlook Press, 1980
Gillett, Charlie, *The Sound of the City: The Rise of Rock and Roll,* New York: Pantheon, 1983
Jackson, John A., *Big Beat Heat: Alan Freed and the Early Years of Rock and Roll,* London: Macmillan, 1991
Shaw, Arnold, *The Rockin' 50s: The Decade That Transformed the Pop Music Scene,* New York: Hawthorne Books, 1974
Smith, Wes, *The Pied Pipers of Rock 'n' Roll: Radio Deejays of the 50s and 60s,* Marietta, Georgia: Longstreet Press, 1989

Rogers, Will 1879–1935

U.S. Radio Humorist

Will Rogers became one of the United States' most popular entertainers of the 1920s and early 1930s. His appeal was enormous through his newspaper columns, movies, lectures, books, and radio shows.

A photo at the Will Rogers Memorial Museum in Claremore, Oklahoma, shows him making his first radio appearance in 1922 over pioneer station KDKA in Pittsburgh. A 1924 photo shows him with members of the Eveready Orchestra at WEAF in New York City on election day. He did a series titled "Fifteen Minutes with a Diplomat" on the *Eveready Hour,* the first commercially sponsored pre-network hookup.

Rogers was initially uncomfortable with radio. He had difficulty with time restrictions and performing in studios without audience laughter to give him clues as to whether his remarks were funny. He wrote after an early broadcast, "Well that little microphone that you are talking into, it's not going to laugh, so you don't know when you tell anything whether to wait for the laugh, or just go right on." As radio studios became more elaborate, producers included live audiences as part of his programs.

Departing from his wealthy upbringing, Will Rogers approached his radio audiences in the guise of a simple cowboy from the plains of Oklahoma. Always an enthusiastic reader of newspapers, Rogers delivered a style of humor encompassing current social and political issues while gently roasting the key personalities or offering advice on various matters.

During the 1920s his numerous professional endeavors included radio appearances and participation in various national radio hookups. He also wrote humorous radio ads for Bull Durham tobacco. On 17 August 1926, while in Europe, Rogers did a broadcast in London for the largest fee ever paid a radio personality in Great Britain. Known for his support of relief organizations, his fee went to a hospital charity.

On 4 January 1928 Rogers hosted an ambitious nationwide show connecting 45 stations. Sponsored by Dodge, it was the first broadcast featuring performances in four different locations. After introducing the main guests, Fred and Dorothy Stone, Al Jolson, and Paul Whiteman's orchestra, Rogers announced that he had a surprise. Then imitating the high-pitched voice of Calvin Coolidge, he delivered his own version of the president's state of the union message. Listeners actually believed that Coolidge was talking. Although Rogers sent an apologetic telegram to the White House, biographer Ben Yagoda claims that Rogers never got back into the president's good graces.

By the early 1930s Rogers' broadcasts and daily newspaper pieces reached 40 million people. His income from his various ventures amounted to $600,000 annually. Beginning in 1930 he had his own regular program, sponsored by E.R. Squibb and Sons, a drug company. From that date until his death, he made regular appearances over radio. He did 12 radio broadcasts for Squibb in 1930 and 53 programs called *The Good Gulf Show* for the Gulf Oil Company between 1933 and 1935. According to the *New York Times,* Squibb paid Rogers $77,000, almost as much as Babe Ruth's annual salary. The Squibb shows were monologues on Charles Lindbergh, President Hoover, Alfred E. Smith, the Prince of Wales, Henry Ford, and other popular persons and topics.

In 1932 Rogers agreed to be a regular on *Ziegfeld's Follies of the Air.* The show's tight format allotted just four minutes for Rogers' monologue. Rogers' segment, done from Los Angeles, was hooked into the rest of the program originating in New York. When cued to close his segment, Rogers went on talking. After finding out that he was cut off, Rogers sent the sponsor, Chrysler, a three-word telegram: "Get Another Boy."

Rogers' difficulty with time constrictions can be attributed to the freewheeling style that he had perfected over years of ad-libbing on the Ziegfeld Follies and as a vaudeville entertainer. He said whatever came into his head while rambling from one anecdote to another without any concern for time. He took advantage of the situation to get laughs by bringing an alarm clock to the studio during a *Good Gulf Show* on 7 May 1933. "When the clock's alarm sounds," he promised, "I don't care whether I am in the middle of reciting Gunga Din or the Declaration of Independence, I am going to stop." Each Sunday thereafter the announcer opened the program with, "Here is Will Rogers and his famous alarm clock." The alarm became the signature for his time to sign off.

Rogers appeared weekly on *The Good Gulf Show* beginning on 30 April 1933. He was paid $50,000 for the series and donated all the money to unemployment relief. The broadcast normally originated from Los Angeles, but his commentaries were transmitted from wherever he was on tour. After the commercials and orchestral segments, Rogers' presentations lasted 15 minutes. The show aired on Sunday evenings at 9:00 P.M. and was followed by President Franklin D. Roosevelt's "Fireside Chats."

Rogers drew from the headlines and combined materials from his newspaper column with extemporaneous horseplay. The programs were live and unedited, offering a less mediated, rawer version of Rogers than do his writings.

On a Gulf program from Chicago, he employed his vocal version of the voices of *Amos 'n' Andy.* On another program, Rogers discussed radio's invisible audience while asking the listeners if his humor was too political. One Gulf program opens with a tribute to Will Rogers from the Senate read by Colonel Edwin A. Halsey describing Rogers as "the poet laureate of wisecracks." Another memorable Gulf broadcast features him talking about gold; however, he discusses it as a sportscaster announcing a football game between the United States and the rest of the world.

The *Good Gulf Show* lasted until June 1935. Rogers signed an agreement to continue the programs, but he and his friend Wiley Post died in a plane crash in Alaska on 15 August 1935. Both NBC and CBS went off the air for 30 minutes as a tribute to his life.

Few recordings of Rogers' air work have survived. Printed transcriptions of the Gulf programs are available at the Will Rogers Memorial. Unfortunately, only 16 sound recordings of the Gulf series exist. Verbatim transcripts of the 12 Squibb shows can be found in *Radio Broadcasts of Will Rogers* (1983) along with an important single broadcast made in 1931 for the Organization of Unemployment Relief. An audio cassette tape collection of his radio talks, mostly from the Gulf show, is available at the memorial.

FRANK J. CHORBA

William Adair Rogers. Born 4 November 1879, near Oologah, Indian Territory (later Oklahoma), one of eight children of a successful rancher and banker. After an extensive rodeo career, came to New York as vaudeville performer, eventually with Ziegfeld Follies, 1904–14; moved to Hollywood in 1918 and appearance in dozens of films, 1918–35; syndicated newspaper column, "Will Rogers Says," 1922–35. Killed in a plane crash with Wiley Post near Point Barrow, Alaska, 15 August 1935.

Selected Publications

The Illiterate Digest, 1924
There's Not a Bathing Suit in Russia and Other Bare Facts, 1927
Letters of a Self-Made Diplomat to His President, 1926
Ether and Me or "Just Relax," 1929
Twelve Radio Talks, 1930

Further Reading

Gragert, Steven K., editor, *Radio Broadcasts of Will Rogers,* Stillwater: Oklahoma State University Press, 1983
Ketchum, Richard M., *Will Rogers: His Life and Times,* New York: American Heritage, 1973
O'Brien, Patrick Joseph, *Will Rogers: Ambassador of Good Will, Prince of Wit and Wisdom,* Chicago and Philadelphia, Pennsylvania: Winston, 1935
Rollins, Peter C., *Will Rogers: A Bio-Bibliography,* Westport, Connecticut: Greenwood Press, 1984
Yagoda, Ben, *Will Rogers: A Biography,* New York: Knopf, 1993

Rome, Jim

U.S. Sports Radio Personality

Jim Rome is one of the most popular and respected sports radio personalities of the day, with a following so rabid that it approaches cult status. Known for his aggressive, "in-your-face" style, Rome asks only one thing of the callers to his Los Angeles-based, nationally syndicated program: "Have a take and don't suck."

Rome's radio career began while attending college at the University of California, Santa Barbara (UCSB) and working for campus station KCSB. After graduation in 1987 with a degree in Communications, he worked for a Santa Barbara radio station KTMS in a number of capacities, including sports director and play-by-play announcer for UCSB sports. The essence of his radio program today can be traced to its roots in that period, when he tried to make the program he hosted both different and appealing.

In late 1990 Rome received a big break when he began working for XTRA Sports 690 in San Diego, a 50,000-watt station whose signal could be heard up and down the Pacific coast. His show ran from 7:00 P.M. to midnight, and this was when the program really developed. Off-the-wall callers were commonplace, prompting Rome to characterize the climate as a jungle where only the strong survive. The label stuck. The phenomenon grew through mid-1995, when the Nobel Broadcast Group began syndicating Rome's program, initially to four affiliates. The program is now syndicated by the Premiere Radio network and has more than 200 affiliates with 2.5 million listeners in the United States and Canada.

The Jim Rome Show is much more than telephone calls and interviews. When listeners tune in, they are transported into a unique radio environment—the Jungle. To survive in this jungle, the visitor is advised to understand its unique language—its "gloss" (short for glossary). A few examples include *crib* (a team's home field or arena, or one's home territory); *blowing up* (becoming popular, thriving); *props* (credit, praise); fishwrap (*newspaper*); and *monkeys* (program directors of stations that air *The Jim Rome Show*). That gloss, along with the overall texture of the program, can make it difficult to follow initially. Rome characterizes his program as an acquired taste and asks new listeners to give it a two-week trial before rendering a verdict. The three-hour program consists largely of running "smack"—the host and his callers lining up with highly critical takes on one or more topical issues or personalities. It is not for the faint of heart. Only the thick-skinned need apply—or call. Failure to have a take that does not "suck" will generally lead Rome to "running" the caller—sounding a buzzer and hanging up. Especially good takes, however, often punctuated with "Out!!!" by the caller, will be followed by the host's exclaiming, "Rack him!" (or, on occasion, her). This means that the call is now eligible for the "Huge Call of the Day" and may be replayed at the end of the program. For those unwilling or unable to call, they may fax or e-mail their takes, and Rome ("Romey" to Jungle veterans—the "clones," as they are called) will read them.

Although the Jungle can be a rough place, it is a principled place, where the host's takes—though delivered with his typical edginess—are replete with moral integrity. Rome's contributions to radio include bringing intelligence, knowledge, and insight to a genre—sports talk radio—that is often lacking in those attributes. An expert interviewer, Rome frequently elicits information that others would not. Moreover, because he is so widely respected, people who typically refuse interview requests often agree to talk with Romey.

Rome's respect from the public and sports figures alike stems from his knowledge of his subjects plus a penchant for asking the tough questions. Although his style remains hard-hitting and no holds barred, Rome is highly respected by many people for the insight and preparedness he brings to his interviews. Much of his day, in fact, is devoted to preparation. He claims that for every hour he is on the air, he spends at least two doing his homework.

Linked to *The Jim Rome Show* is the phenomenon of the "Tour Stop." Tour Stops began with Rome's occasionally originating his radio program from a location in one of his affiliate cities. His clones attend en masse, some traveling long distances. In addition to listeners, several sports figures who live in the Tour Stop city also drop by. Like the program itself, the Tour Stop happening has grown exponentially since its inception. The first was in Omaha in 1996. Because of escalating crowd sizes and a sort of "Jungle fever" that ensued as Rome attempted to conduct his program, Tour Stops are now conducted off the air on Saturdays.

Although Rome's résumé is dominated by his work in radio, he also has utilized his talents in television. This includes two years hosting *Talk2* on ESPN2 in the mid-1990s. In that venue, an incident occurred that continues to follow Rome. Then-Los Angeles Rams quarterback Jim Everett was a guest on the show, and Rome called him "Chris" three times, a reference to female tennis player Chris Evert, in questioning his toughness. Once Everett had enough, he flipped over the table and nearly did the same to Rome. Rome now acknowledges he mishandled the situation and calls it "regrettable."

Rome's television programs also include *The Last Word* on Fox Sports Net, which he hosted for five years nightly. In May 2003, he began a weekly program on ESPN, *Rome is Burning*.

Rome was named California's sportscaster of the year in 2000. He lives in Los Angeles with wife Janet and son Jake.

JOSEPH A. RUSSOMANNO

See also Sports on Radio; Sportscasters

Jim Rome. Received degree in Communications, University of California, Santa Barbara, 1987. Worked for station XTRA, San Diego, beginning in 1990. *The Jim Rome Show* became nationally syndicated in 1995, eventually gaining more than 200 affiliates.

Further Reading

Campbell, Barre, "The Rise of the Rome Empire," *Ottawa Sun* (21 August 1999)

Fay, John, "King of the Jungle," *Cincinnati Enquirer* (18 December 1999)

Magenheimer, Lisa, "Off Air, Controversial Rome's Your Average Jim," *Tampa Tribune* (26 March 1999)

Jim Rome website, <www.jimrome.com>

"Rome Tour Stop Rocks 'C-Town,'" *Cincinnati Enquirer* (24 January 1999)

Ruelas, Richard, "Rome's Legion Hail Their Voice," *Arizona Republic* (15 November 1998)

"Welcome to the Jungle," *Ottawa Sun* (21 August 1999)

Roosevelt, Eleanor 1884–1962

U.S. First Lady and Broadcaster

Among her many roles as first lady, Eleanor Roosevelt delivered commercially sponsored broadcasts. She was the first to do so. Eleanor Roosevelt was on the air more often than any other Roosevelt and for a greater variety of reasons. She used radio to sway public opinion on issues she was concerned about, to make money, and to help her friends and family. And like her husband, she had a personal, conversational approach to the medium that worked well for radio.

Roosevelt's radio career included a few political talks before 1927, a number of sponsored series from 1930 through the 1950s, and hundreds of speeches and interviews. From 1921 to 1924, while Franklin D. Roosevelt was struggling with his polio, Eleanor Roosevelt was reaching out into political activism and giving radio speeches to keep the Roosevelt name before the public. She gave a very well organized speech on WRNY about an upcoming referendum in New York. It was clearly a political speech supporting the Democratic platform, but it started off on a personal note—"On one of the registration days here in NYC a friend of mine tells me that she had the following experience . . . ." Other speeches were on topics that reflected her own interests in helping others: charities such as Children's Aid Society, Salvation Army, Emergency Unemployment Relief Fund, and later the United Nations. She also gave messages that reflected her set of values; they could be about parenting, the duties of a wife, or the life of a first lady. She was aware of her prominence and desired to keep that stature to continue to be able to influence social change. She worked at maintaining a platform and a channel to her audience, one that she assumed to be mostly women.

Her first series of sponsored radio programs in 1932–33—*Pond's Radio Program Speeches,* for Pond's, a cold cream manufacturer—was clearly directed to women during the daytime. A sampling of her subjects included "Keeping Your Husband Happy" and "Official and Social Life in Washington." The very fact that Roosevelt, soon to be first lady, was being paid for her broadcasts made the news, but she replied that the monies were going to charity via a special fund set up with the American Friends Service Committee, and she reluctantly held off making more commitments for sponsored programming until 1934.

The sponsored series that aired from 9:30 to 10:00 P.M. in 1934 for the Simmons Mattress company was intended to be a blend of feature material and current events. Roosevelt actually referred to herself as a news commentator. She had other sponsored series, but the series that had the most controversial programming was her 25-program Sunday night series for the Pan American Coffee Association—an organization that represented eight coffee-exporting countries—while she was co-director of the Office of Civil Defense. Her topics included rumors of the president's dictatorial powers, torpedoing of ships, German propaganda, U.S. civilian defense, anti-Semitism, and others. The programs were remarkable for their

Eleanor Roosevelt at CBS's WABC Studios
Courtesy CBS Photo Archive

incorporation of propaganda for the president's policies, but research indicates that only the socialist Norman Thomas asked for reply time to give the isolationists' perspective—which he was denied.

After the war and out of the White House, Roosevelt's selection of broadcast topics continued to vary. Her daughter Anna produced a 15-minute weekday radio series with her from October 1948 to August 1949; Eleanor Roosevelt participated two or three times a week. Samples of Eleanor's topics for this included the value of Wiltwyck School, a small school for delinquent children across the Hudson River from Hyde Park that she had supported throughout its years; radio's responsibility, which was heavily drawn from Edward R. Murrow's acceptance speech for an award; and spring at Hyde Park. But apart from this series, Roosevelt also gave radio broadcasts on very serious issues such as communism and the value of freedom of expression, and the significance of the Declaration of Human Rights. She also gave partisan speeches for Democratic candidates, and she was always ready to be interviewed.

The lack of continuity of broadcast series and the variety of topics indicate to some that Roosevelt did not take radio especially seriously. She was not a broadcaster first and foremost, but she was clearly aware of the business aspects of broadcasting. Her correspondence with her agents, who demanded high fees for her, make apparent her awareness of the need for exclusivity with agents or with networks, and illustrates her desire to protect her output so that it was always valued. One agent, Myles Lasker, was adamant about trying to get high fees. He found a sponsor, the J. Walter Thompson advertising agents for The Johns-Manville Company, willing to pay a $3,000 honorarium for a 5-minute talk on "The American Home—Its Traditions and Hopes." Another agent noted: "I've chased down several radio inquiries and find they're phony. I guess it's someone trying to pull some cheap stunt." Roosevelt was careful to protect any content that had been sponsored. She notified the National Broadcasting Company (NBC) that they could list her speeches in a catalog intended to sell their programming but that she would not allow for any of her commercially sponsored broadcasts to be listed. She was also economical in her use of content. Just as she had a selection of four or five topics for her lecture tours, she often repeated subjects over the years. There are countless programs whose topic is "A Day in the Life of the First Lady," "Role of the Wife," or the value of "Freedom of Expression."

She also used the medium as a means to work with and help friends. At the request of her friend Esther Lape, for example, Roosevelt broadcast a speech in January 1935 to counter the Hearst-Coughlin onslaught against the World Court. Then, in 1941, she was more than willing to speak on behalf of FDR's policies, and she was sure to have him, or his aides, go over the scripts. To help her children make money she worked with her daughter Anna (1948–49) on a radio series, and then with son Elliott (1950–52) on a television series.

Roosevelt appreciated the power and influence the medium afforded her. In her later years she gave up her lecture circuit in favor of broadcasting, so that more people could hear her. In 1951 she was asked to be a charter member of the women's broadcasting organization, AWRT (American Women in Radio and Television).

MARGOT HARDENBERGH

See also American Women in Radio and Television; Fireside Chats; United States Presidency and Radio

Eleanor Roosevelt. Born Anna Eleanor Roosevelt (fifth cousin to Franklin Delano Roosevelt) in New York City, 11 October 1884. Educated privately and at Allenswood School, England; married Franklin Delano Roosevelt, 1905; active in Women's Trade Union League, League of Women Voters, and women's division of Democratic Party; helped form furniture factory, Val-Kill Industries, 1926–38; co-owner with Marion Dickman, Todhunter School, New York City, 1927–32; as first lady she initiated weekly press conferences with women reporters, lectured throughout country, delivered sponsored radio broadcasts with proceeds going to American Friends Service Committee, co-director, Office of Civilian Defense, 1941–42; during WWII visited troops throughout world; named "The First Lady of Radio" by WNBC, 1939; appointed representative to United Nations, 1945–51, 1961–62; elected chairman of Commission on Human Rights; helped draft Universal Declaration of Human Rights, 1946; wrote books, magazine articles, and syndicated daily newspaper column "My Day," 1936–62; hosted radio and television programs and traveled widely as lecturer. Died in New York City, 7 November 1962.

Radio Series

1932–51 (with some hiatuses)	*Eleanor Roosevelt* or *Eleanor Roosevelt's Chat Show*

Television Series

Today with Mrs. Roosevelt (packaged by Elliott Roosevelt), 1950–51; *Prospects of Mankind*, 1960–62

Selected Publications

This Is My Story, 1937
This I Remember, 1940
India and the Awakening East, 1953
On My Own, 1958
Autobiography of Eleanor Roosevelt, 1961
Tomorrow Is Now, 1963

What I Hope to Leave Behind, 1995
Courage in a Dangerous World, 1999

Further Reading

Beasley, Maurine Hoffman, *Eleanor Roosevelt and the Media: A Public Quest for Self-Fulfillment,* Urbana: University of Illinois Press, 1987

Beasley, Maurine Hoffman, and Paul Belgrade, "Eleanor Roosevelt: First Lady As Radio Pioneer," *Journalism History* 11, nos. 3–4 (Autumn/Winter 1985)

Cook, Blanche Wiesen, *Eleanor Roosevelt,* 2 vols., New York: Viking, 1992–2000

Lash, Joseph P., *Love, Eleanor: Eleanor Roosevelt and Her Friends,* New York: Doubleday, 1964

Roosevelt, Franklin D. *See* Fireside Chats

Russia and Soviet Union

Prerevolutionary Russia, although in many ways technologically undeveloped, was a leader in the development of wireless. Despite the Russian Revolution and the civil wars that followed it, the country's deployment of radio in European (western) Russia in the earliest decades of the 20th century was comparable to that occurring in other industrialized nations. The Union of Soviet Socialist Republics (USSR) subsequently developed extensive wired and wireless radio systems that, in content and structure, were consistent with communist ideology and the aims of communist states. The collapse of the USSR in 1991 put radio, like many aspects of life in the Russian Federation, on a chaotic course. Commercialism came rapidly to some areas, but state-run systems struggle on. Despite exploiting some new technologies, such as the internet, for global as well as national distribution, the overall preparedness of Russian radio broadcasters for the radio technologies of the 21st century seems inadequate.

Prerevolutionary Russian Radio (to 1918)

The Soviet Union long hailed St. Petersburg academician Alexander Stepanovitch Popov as the "father of radio," and Russia continues to celebrate 7 May as "Radio Day" in recognition of his demonstration of the detection of electromagnetic discharges from lightning on that day in 1895. It is possible that Popov transmitted the Morse-coded words *Heinrich Hertz* at St. Petersburg University on 24 March 1896, but documentation of this achievement has always been hazy. If one counts the 7 May 1895 demonstration as a true demonstration of radio telegraphy, then Popov invented radio more than a year before Marconi's 2 June 1896 patent application. However, Marconi's July 1896 demonstration of sending and receiving coded messages is better documented than Popov's alleged March transmission. Unlike Marconi, Popov did not capitalize on his invention. He seems to have had less interest in radio than Marconi did, and in any event, the czarist navy took over radio's development by 1899. Popov worked independently of Marconi, although they both built on the common work of scientists and inventors such as Maxwell, Hertz, and Lodge, and Popov certainly counts as an important inventor of early radio. Others were also at work on radio in St. Petersburg at the time, including Vladimir Zworykin, who immigrated to the United States just two years following the 1917 Russian Revolution, joined Westinghouse Electric Corporation, and later became known in the West as a father of electronic television. Although the West benefited from the flight from revolutionary Russia, many scientists and technicians remained in the country after 1917 and contributed to the development of radio in the Soviet Union.

Radio in the USSR (1918–90)

Vladimir Lenin appreciated the potential for radio to advance the causes of the Bolsheviks and to build support for the party. He considered radio sufficiently important as a potential means of overcoming the huge distances in the country to justify devoting a substantial portion of the country's scarce gold reserves to its development. Newspapers, Lenin's preferred means of communication, could not reach the masses of illiterate workers and peasants as radio might. Therefore, as early as December 1918, Lenin set up an experimental radio laboratory in Nizhniy Novgorod (Gorki in the Soviet era). Activity at the lab was halted during the civil wars that followed the 1917 revolution, but it resumed in 1924. A 12-kilowatt station (among the most powerful in the world at the time) went on the air from Moscow beginning 17 September 1922, irregularly transmitting music and an occasional speech by Lenin. In October 1924 the Russian government's Council of People's Commissars established a Joint-Stock Company for Radio Broadcasting, *Radioperedacha,* with the stock held by trade unions and teachers. Control of the station by such a joint-stock company reflected a level of freedom from Communist Party control that would later disappear in the USSR. Under this company, the Moscow station on resumed systematic, regular broadcasts on 12 October 1924. Within a year Moscow also had the first Soviet wired radio system (*radiotranslyatsionni uzel*), an interconnection of 50 speakers. Although Russia may not have invented the radio, a good case can be made that it invented the wired nation, because up until the mid-1960s, the number of wired receivers—*radiotochki*—in the country generally exceeded the number of over-the-air receivers.

With Stalin's ascent to power, a movement began to increase central government and party control of wired and wireless radio. *Radioperedacha* was dissolved in July 1928, and the Commissariat of Posts and Telegraphs, which already regulated radiotelegraphs, unsuccessfully assumed control. On 31 January 1933 the Council of People's Commissars of the USSR established the All-Union Committee for Radiobroadcasting and Radiofication (Russian acronym, VRK) and put it broadly in charge of wired and wireless radio throughout the USSR. Ultimately, each USSR republic (except, notably, Russia) established its own radio committee to oversee services to the republics, but these committees worked under close supervision from Moscow, which continued to determine nationwide programming. There were local committees for cities and even for factories, collective farms, and the like.

Wired and wireless radio grew slowly prior to World War II. In 1928 there were about 20 over-the-air stations, and by the time Hitler invaded Russia in 1941, that number had grown to about 90 stations and an estimated 760,000 over-the-air receivers. But VRK estimated that in 1940 there were 11,000 "radio exchanges" powering about 5 million speakers.

The "Great Patriotic War," as Russians call World War II, devastated Soviet radio. Nearly all over-the-air sets were confiscated by Soviet authorities and, it is claimed, returned at the war's end. In European (western) Russia, German invaders either captured wired radio exchanges or destroyed them. When the war turned for the Soviet Union, it discovered that it was unable to broadcast pro-Soviet propaganda to the western part of the nation because very few individuals had wireless receivers. At the war's end, only 5,500 exchanges were operable, most of them east of Moscow. The wired systems, however, were quickly rebuilt, and it is estimated that by 1947 9,250 exchanges were operable.

Wired radio was attempted in other nations, both in the East and in the West, but no nation took the idea as far as the Soviet Union did. The addiction to wired radio is sometimes traced to the desires to limit information access of Soviet citizens, and that may have been part of the motive. But there were other reasons for the adoption and expansion of the system.

Until they grow to gargantuan size (which eventually they did), wired exchanges are cost efficient. Because the home or work receivers consist of little more than a case, a transformer, a switch, and a speaker, they are something that the chronically underdeveloped Soviet consumer electronics industry could produce in quantity. Over-the-air radios were more complex and expensive; in 1936, for example, there were just 650,000 such sets in the country. Around 200,000 of these were regarded as outmoded, and 270,000 were crystal sets.

Wired systems had some other advantages. They did not require batteries—chronically unreliable and in short supply—as inexpensive receivers often did. They could even operate in households without electricity. Through wind-powered exchanges, some Soviet villages got wired radio before they were electrified. During the Nazi siege of Leningrad (September 1941–January 1944), the wired radio system continued to function despite the collapse of the electrical system in the city. Throughout the 900-day siege, a metronome continued to beat as the heart of Leningrad, varying in pace, some said, with the level of threat to the city. Speeches, war information, and even live symphony concerts were transmitted to the beleaguered city.

Wired exchanges could also be operated to provide some time for "local" broadcasts. At the very least, services from Moscow could be interrupted for short periods of time to accommodate programming specific to a republic (including programming in languages other than Russian) or city, or, in some instances, programming intended for a specific factory, collective farm, or workplace. Given the low signal capacity of the systems, however, such programming displaced Moscow programming and therefore never amounted to a large percentage of transmission time.

Wired radio systems were repaired and expanded after the war. By 1946 Moscow (and later other parts of the country) had two-channel service. A third channel was added in 1947, when it was estimated that only 18 percent of radio listening came from over-the-air services. A fourth channel was added in 1965, and a five-channel system was the never-realized Soviet goal.

Slowly after the war, however, over-the-air stations were begun in earnest. Because AM had not become as entrenched in the USSR as in other countries (notably in the United States), FM was introduced relatively early. In 1963 the country was thought to have 170 long-, medium- (AM), and shortwave stations but also 86 FM stations. Whereas wired speakers accounted for 75 percent of receiving equipment in the 1950s, by about 1965 half of all receivers in the country were wireless. The country settled into a pattern, still true today, of hybridized wired and wireless transmission.

Following Stalin's death in March 1953, VRK evolved into the USSR State Committee for Television and Radio (*Gosteleradio*), which, until the collapse of the USSR in 1991, monopolized domestic radio and television as well as international broadcasting. Choices expanded somewhat as *Gosteleradio* added services. At the end of the Soviet era, the "First Program," a mixture of news, commentary, classical music, and folk music, used wired and wireless services to reach 97 percent of the population. The second program, *Mayak* (lighthouse or beacon), a highly popular, slightly less political or stuffy service, reached 85 percent of the country. In 1964 *Mayak* was the first service to go full-time in the USSR. Its lighter, sometimes even deliberately entertaining, service and round-the-clock schedule were intended to make it, in part, a competitor to Western shortwave international radio services, such as Radio Free Europe. The Western services attracted audiences through their brighter blend of music and talk, and, until *Mayak* was introduced, they exploited the fact that Soviet radio channels shut down at midnight. The third program service, emphasizing classical music and education, reached only 40 percent of the nation by the 1980s, and the fourth channel could be heard by only 7 percent of listeners, mostly in Moscow.

Although not as political as in Stalinist times, Soviet radio services remained generally faithful to the communist ideal of using the mass media primarily to raise the cultural level of the populace. Entertainment was secondary. There was some popular music and jazz by the 1960s but also, compared to Western standards, heavy doses of opera, drama, and literary programs. Many programs were specifically targeted at children. Sporting matches were also frequent fare. The national service emphasized Russian, but services in the republics produced programming in many of the Soviet Union's more than 80 major languages. About 65 percent of receivers were over-the-air, picking up predominantly shortwave and FM signals. Although the Soviet AM band was similar to the band used in Europe and in the United States (570–1479 kilohertz, but on 9-kilohertz spacing rather than the 10-kilohertz spacing used in the United States), the Soviet FM band used a different range than that used in Europe and the United States. In part to discourage reception of Western FM signals in border states, the Soviet Union and its Warsaw Pact allies used a lower (64–74 megahertz) FM band, commonly known as the UKW or ultrashortwave band, than was used in the West (88–108 megahertz). Unlike FM in the United States, Soviet FM stations transmitted on even as well as odd (e.g., 88.1 and also 88.2) frequencies in the FM band. Soviet radio also used a longwave band of 150–350 kilohertz and a shortwave band of 5.9–17.6 megahertz. Wired systems, by then usually carrying three channels, except in one-channel rural areas, continued to serve both cities and the countryside. In rural areas, they were often the only services available. Some urban wired systems used stereo. The *Gosteleradio* empire, on the eve of the Soviet collapse, had grown colossal, employing an estimated 82,000 people, second in size only to the Ministry of Defense.

Radio in the Russian Federation (Since 1991)

Some liberalization of Soviet radio occurred early under General Secretary Mikhail Gorbachev's policies of *glasnost* (openness) and *perestroika* (restructuring) once he succeeded Konstantin Chernenko in March 1985. Reporting of the failings of Soviet government (obvious to many Soviet citizens anyway) was allowed, although reporting of disasters such as Chernobyl in 1986 still did not take place immediately. Although a reformer, Gorbachev sought to preserve the USSR and, at least initially, the Communist Party. But *glasnost* empowered critics of the Soviet system. Boris Yeltsin, a well-known political figure, resigned from the party in June 1990, only to become the first freely elected president of the Russian Republic in July 1991. Independence movements in the non-Slavic republics grew. Often, anti-Soviet forces seized radio and television facilities. On 13 January 1991, 13 Lithuanians were killed in battles over studio and transmission facilities in Vilnius. The shocking bloodshed, many believed, hastened the end of Soviet rule, already undercut by dissent and rot at its heart. Significant squabbling began between Yeltsin—advocate of an independent Russia—and Gorbachev, defender of the Union.

The end really began on 19 August 1991 when, with Gorbachev out of Moscow, hardliners led a short-lived putsch aimed at removing him. After heroic defense at the Russian White House—headquarters for Yeltsin's Parliament of the Russian Republic—the ill-planned coup collapsed and Gorbachev returned to Moscow, weakened and disgraced. In short order, the USSR had to accept the independence of the Baltic and, later, other republics. On 25 December 1991—after a 74-year run—the curtain fell on the USSR. Gorbachev's goal of an open, but still communist, society had proved unsustainable.

Yeltsin realized that Gorbachev and *Gosteleradio* directly controlled electronic mass media in Russia. As noted earlier, under the USSR Russia had been the only republic without its own radio and television committee: the interests of Russia in electronic media were supposedly represented by *Gosteleradio*, but this situation was unacceptable to Yeltsin. In August 1990 the Russian Parliament created the All-Russian State Radio and Television Company (*Vserossyjskaya Gosudarstvennaya Teleradiokompaniya* [*VGTRK*, also known as RTR]), which launched Radio Russia (*Radio Rossii*) on 6 December 1990. During the Lithuanian revolt in January 1991, Radio Russia supported the rebels. Gorbachev officially acknowledged RTR on 13 May 1991, and it assumed control of the national radio channel 2. When, in August 1991, coup leaders gained control of the main transmitter in northeast Moscow, RTR continued to transmit from the White House via shortwave and, in effect, through CNN.

As Yeltsin consolidated power following the coup, Radio Russia became the "first button" on wired radios throughout the Russian Federation, although the by then withering *Gosteleradio* held on to other national radio services. During an attempted coup against Yeltsin in fall 1993—following the collapse of the Soviet Union—anti-Yeltsin forces attempted to seize the shared RTR/Ostankino production facilities, but RTR had downtown studios and somehow maintained its link to the transmitter.

Today, Russia is served by four interregional, theoretically state-funded, radio services. Radio Russia, run by RTR, is the second most widely attended to service and is generally regarded as the official voice of the Putin government that succeeded Yeltsin in 1999. It broadcasts 18 hours per day on channel 1. Radio *Mayak,* a subsidiary of VGTRK, is still the most popular single service in the nation. Radio *Yunost* (youth), now associated with *Mayak,* broadcasts programs for audiences from 14 to 25 years old. Together, they now share channel 2. Radio *Orfey* (Orpheus) broadcasts 18 hours daily of classical music, educational programs, and newscasts about cultural events as a noncommercial state institution, typically on channel 3. Regional governments continue to run regional services, sometimes in languages other than Russian. There is also a state-run international service, the radio broadcasting company *Golor Rossii* (Voice of Russia), formerly *Radio Moscow,* which transmits programs in Russian and 31 foreign languages. As with many Russian state institutions, however, the economics of all state-run services are unsettled. Salaries are low and irregular. The state often fails to pay agreed-upon sums, and as a result, state-run services accept some advertising and are often behind in payments to other state-run agencies that still control towers and transmitters.

Given the continuing dependence on state-controlled over-the-air radio transmitters and wired radio systems, it should not be surprising that "independent" (nonstate) radio had a difficult birth. The first nonstate radio broadcast station (since Lenin's experiments just after the revolution) began in August 1990, when radio journalists at the faculty of journalism at Moscow State University founded *Echo Moskvy* (Moscow's Echo). It stayed on the air even when, during the 19 August 1991 putsch, KGB officials attempted to close it. Today it is one of about 20 nonstate FM stations available in Moscow. Nonstate over-the-air radio broadcasting has blossomed in many other larger Russian cities, but in nonurban areas, state broadcasting (by wire and over the air) is often all that is available. Regional governors often have fairly tight control over state, and even nonstate, radio in their regions.

Many independent over-the-air radio stations operate in multiple cities. In U.S. terms, they might be described as para-networks or certainly as closely controlled by a group owner. Many independent radio stations function as Russian-Western joint ventures of some sort. *Europa plus,* for example, is a joint French-Russian company that operates through about 60 "partner stations." American media entrepreneur John Kluge (after selling Metromedia's television properties to Rupert Murdoch to form the heart of Fox broadcasting) turned Metromedia International Group into a major player in Eastern Europe. It broadcasts on two frequencies in Moscow and on one in St. Petersburg and has other properties in Hungary, Estonia, Georgia, Latvia, and the Czech Republic.

According to the Licensing Administration of the Ministry for Press, Television and Radio Broadcasting, and Mass Communications of Russia, 946 licenses for the operation of radio transmitters were in use in Russia as of 1 September 2000. These include 671 licenses for over-the-air radio broadcasting, 216 for wired transmission, 10 for satellite radio, 7 for over-the-air and wired broadcasting, and 2 for multiprogram broadcasting.

Programming of independent radio generally does not stress politics. Rather, the stations often target the small and continually endangered Russian middle class, especially its youth. Programming is often an eclectic mix of Russian and Western rock and pop and can include weather and helicopter traffic reports and disc jockey patter that would be familiar to Western ears, although Russia has yet to develop an indigenous Howard Stern. Hard hit by the Russian economic collapse of 1998, which wiped out much of the advertising marketplace, the economics of independent stations are often problematic, although Western investors seem committed for the long haul. Independent stations are often in arrears to state-run transmission facilities, and continual government threats to collect the debt compromise their independence. Some often generate revenue by accepting payment for retransmitting the British Broadcasting Corporation (BBC), Deutsche Welle, or Radio Liberty. Others retransmit paid religious programming from Western (often fundamentalist Christian) sources. Weak economics tempt playola on a scale considered unlawful in the West, and because of lax copyright law enforcement (and no effective organization such as the American Society of Composers, Authors, and Publishers [ASCAP] or Broadcast Music Incorporated [BMI]), Russian popular musicians are eternally complaining that they are undercompensated for their music. One response is frequent co-promotion of raves and concerts between artists and radio stations.

Many of the independent stations found frequency space in the upper (Western) FM band (88–108 megahertz), which was not used in Soviet times. Since 1991, the Russian consumer electronics industry has largely collapsed. Today, radios in Russia are often imported from Europe or Asia and (at least in the early to mid-1990s) often did not even receive the "Russian" (64–74 megahertz) band. Wideband FM radios are now available but are more expensive than sets without the old Soviet frequencies. The result is that those who can afford new radios sometimes cannot receive the Soviet frequencies at all but have easy access to the Western band, which may not be heard by Russians who have not replaced their radio sets. Car radio listening (almost insignificant in Soviet times because of the lack of private cars) is important in major cities. Despite tough times, urban Russians increasingly own cars. Imported used and new vehicles inevitably come with Western FM radios, and whenever they can, Russian youth mount Japanese radios in even the most decrepit of Russian cars. A response to this frequency chaos, by both independent and state-run over-the-air broadcasters, has been, in effect, cluster broadcasting. The same "station" may transmit on a variety of frequencies—from shortwave to Soviet FM to Western FM.

The Future of Russian Radio

It is as risky to predict the future of Russian radio as it is to predict the future of the Russian Federation itself. Political pressures continue on independent Russian television, and although less on radio, they are still present there. The economics of independent broadcasting, given the sad state of the Russian economy, remain less than promising. Transmission towers and transmitters themselves, still overwhelmingly state owned (and leased by independent radio), have received few technical upgrades and often not even basic maintenance for almost 20 years. A fire, allegedly started by a short circuit, at the Ostankino television tower in August 2000 knocked out most radio and television broadcasts in Moscow for several days. Other transmission facilities are likely to be in even more decayed condition. Russia lags behind Europe (and even the United States) in making a transition to digital radio.

Compared to Soviet times, however, some things are brighter. There are more radio services and more diverse content. There is political debate. There is more popular entertainment, and somehow, high culture hangs on. Perhaps because of their familiarity with wired radio (or conceivably out of fear of the state of their transmitters), Russian radio has embraced the internet, and those of the international Russian-speaking diaspora, at least, can access a wide range of state and nonstate audio services. Indeed, if the money could be found for nationwide wireless broadband services (a not entirely fanciful idea in a country with decaying traditional broadcasting and wired telephone systems), "radio" in Russia's future might someday be widely delivered by the internet.

HERBERT A. TERRY AND ANDREI RICHTE

See also Metromedia; Popov, Alexander; Propaganda by Radio; Radio Moscow

Further Reading

Androunas, Elena, *Soviet Media in Transition: Structural and Economic Alternatives,* Westport, Connecticut: Praeger, 1993
Browne, Donald R., *Electronic Media and Industrialized Nations: A Comparative Study,* Ames: Iowa State University Press, 1999
Ellis, Frank, *From Glasnost to the Internet: Russia's New Infosphere,* New York: St. Martin's Press, 1999
Emery, Walter B., *National and International Systems of Broadcasting: Their History, Operation, and Control,* East Lansing: Michigan State University Press, 1969 (see especially chapter 22)
Ganley, Gladys D., *Unglued Empire: The Soviet Experience with Communications Technologies,* Norwood, New Jersey: Ablex, 1996
Mickiewicz, Ellen Propper, *Changing Channels: Television and the Struggle for Power in Russia,* New York: Oxford University Press, 1997; revised and expanded edition, Durham, North Carolina: Duke University Press, 1999
Paulu, Burton, *Radio and Television Broadcasting in Eastern Europe,* Minneapolis: University of Minnesota Press, 1974
Radio and Television Systems in Central and Eastern Europe, Strasbourg, France: European Audiovisual Observatory, 1998

S

Sarnoff, David 1891–1971

U.S. Broadcast Executive

Though not an engineer and lacking much formal education, David Sarnoff played an important role supporting the technical development of American radio and television through his leadership of the Radio Corporation of America (RCA), a key manufacturer of receivers and broadcast equipment. He was also instrumental in the formation of the first permanent radio network, the National Broadcasting Company (NBC). The originator of many myths about his own background, Sarnoff's actual accomplishments needed little embellishment.

Early Years

Sarnoff's father, Abraham, left Russia for America when David was four, leaving behind his pregnant wife, David, and a younger son. When David was five, his mother sent him to Korme to begin training as a rabbi with his granduncle. He remained there until he was nine, when his father sent for the family to move to America. They left Russia in 1900 in the midst of considerable political confusion, traveling by ship to their new country. Soon after their arrival in New York City, David's father became an invalid. The boy took jobs delivering meat and newspapers, and he entered school to learn English. Within two years 11-year-old David bought a newsstand for $200, and he took a job singing soprano in a synagogue choir at $1.50 per week. Sarnoff's voice changed when he was 15, so he lost the choir job, and in that same year his father died. Sarnoff decided to pursue a career in newspapers and set his sights on the *New York Herald*. The first of many legends sprang from his visit to the paper. It has been said fate took Sarnoff the wrong way in the *Herald* building, causing him to end up at a Commercial Cable Company branch office. In reality, Sarnoff merely stopped at the first desk he saw and he became an office boy for $5 a week. He discovered operators made more money, so he studied on his own to learn telegraphy and Morse code.

Sarnoff moved to a job with the Marconi Wireless and Telegraph Company, where he first worked as office boy and then served as wireless telegraph operator wherever needed, be it on land or sea. At 17 he was assigned to Marconi's Siasconset station on Nantucket Island, where he read every technical book in the small station library and in Nantucket. In 1911 he volunteered for an Arctic sealing exposition that some considered to be so dangerous that the crews might not make it back alive. Sarnoff installed and operated the Marconi Company's wireless equipment onboard the *Beothic*. After the journey, the Marconi Company wanted him to continue at Nantucket, but he wanted to transfer to Sea Gate station in Brooklyn. Sarnoff took a pay cut to work at the busiest American wireless station, and within months, before he turned 20, he became manager.

Sarnoff parlayed the Sea Gate position into a better one at the Marconi facility located in the Wanamaker department store in New York City. Now on duty only during store hours, Sarnoff had time to pursue technical training at the Pratt Institute in Brooklyn. Although the story was widely repeated in later years, Sarnoff was not directly involved in the wireless reporting of the 1912 Titanic disaster. By that time he had moved into the lower rungs of Marconi management and was no longer a regular operator.

Sarnoff's rise was rapid. He was appointed as an operator instructor, then inspector of wireless equipment installed onboard ships, and then inspector of wireless stations. Soon he was appointed assistant traffic manager, assistant chief engineer (1913), and commercial manager (1917). At some point during this period, he may have composed a memo proposing the American Marconi manufacture of a "radio music box"—what today we would call a radio receiver—as a consumer device. The idea was initially ignored.

David Sarnoff
Courtesy Radio Hall of Fame

Radio Corporation of America (RCA)

Sarnoff's future shifted immediately after World War I when the U.S. Navy became concerned about British control of American Marconi. Given how central the company's stations were to government and commercial wireless operations, the navy pressed for elimination of the foreign (albeit Allied) control of its facilities. General Electric was prevailed upon to buy American Marconi and spin it off into a subsidiary to be named the Radio Corporation of America (RCA). Sarnoff continued in his role as commercial manager with the new firm and reframed his "radio music box" memo in 1920 or 1921, this time to broader acceptance given the initial appearance of radio broadcast stations.

RCA played a central role in the development of radio in the 1920s, with Sarnoff often directly involved in important innovations. He was a key figure in the formation of radio's first permanent network, the National Broadcasting Company, in 1926. He was named to RCA's board of directors in 1927, and he became acting president of the company and president in his own right on 3 January 1930. A government anti-trust suit forced General Electric and Westinghouse to give up their stock in RCA by 1932, and the corporation emerged for the first time as a fully independent entity.

In the early 1930s, Sarnoff pushed RCA into the front line of radio research. The company began active television research in 1929. Sarnoff was intrigued when his old friend Edwin Armstrong proposed a wholly new system of radio that could successfully eliminate the static. After receiving his initial patents on FM in 1933, Armstrong was invited by Sarnoff to test the system in RCA facilities on top of the Empire State Building. However, Armstrong's claim that his new FM would make AM obsolete and eventually replace it—despite the fact that RCA and NBC were based on the existing AM technology—soured Sarnoff on the FM idea, and RCA turned to a major research push for television. Armstrong was asked to remove his apparatus, and he soon went his own way. The former friends soon became enemies as Armstrong pushed for FM development while RCA concentrated on television, which it introduced to American audiences at the 1939 World's Fair. RCA research was central to the system of black-and-white television that began regular operation in mid-1941.

In 1942 Sarnoff consolidated RCA's many research efforts in a new laboratory facility in Princeton, New Jersey (in 1951 it took his name). For much of 1944, Sarnoff was on active duty with the U.S. Army, assisting in development of communication systems for the invasion of Europe. He returned to RCA at the beginning of 1945 as a brigadier general (and was known as General Sarnoff for the remainder of his life).

Sarnoff's final 25 years (until his 1969 retirement) at RCA were increasingly dominated by two circumstances: competition with CBS and William S. Paley (a very different kind of industry leader), and the development of color television. Paley's network lacked a manufacturing arm and so concentrated more fully on programming, leading to his "raid" on NBC's chief stars and programs in 1948–49, just as network television was taking hold (and a year after Sarnoff moved up to become RCA's chairman). CBS led in the ratings war for the next two decades, but the rivalry extended to technology as the two firms fought over improved records (CBS developed the 33 1/3-rmp LP, while RCA played catch-up with the 45 rpm disc) and development of a system of color television. RCA worked feverishly on an all-electronic system compatible with existing black-and-white receivers. CBS pursued a different route with a partially mechanical system that was not compatible with existing sets and won initial approval from the Federal Communications Commission (FCC) in 1950. RCA research to improve their system continued during the Korean War (1950–53) when manufacture of color sets was suspended because of wartime needs. Late in 1953 RCA persuaded the FCC to reverse its decision and accept the RCA all-electronic system as the basis for American color television. NBC began color telecasts early in 1954.

After a series of RCA presidents passed through the executive suite, Sarnoff turned to his son Robert, naming him president of NBC in 1956 and of RCA itself in 1965. Sarnoff remained CEO and fully in charge through the 1960s, years that were filled with public homage and honors to his long career. He retired in 1970 and died in 1971.

W.A. KELLY HUFF

See also Armstrong, Edwin H.; Columbia Broadcasting System; FM Radio; National Broadcasting Company; Paley, William S.; Radio Corporation of America; Talent Raids

David Sarnoff. Born in Uzlian, Russia, 27 February 1891; immigrated to United States, 1900. Attended Pratt Institute, New York City; Army War College, 1926. Took delivery jobs and at age 11 bought newsstand; office boy for Commercial Cable Company; taught himself telegraphy and Morse code; office boy and then wireless operator with American Marconi Wireless and Telegraph Company; manager, Sea Gate Station, Brooklyn, 1909; installed and operated Marconi wireless equipment on *Beothic* for Arctic exposition, 1911; employed by Marconi, 1912–19; allegedly first to propose "Radio Music Box," 1915 or 1916; married Lizette Hermant, 1917; RCA commercial manager, 1919–21; RCA general manager, 1921–22; proposed network radio, 1922; RCA vice president and general manager, 1922; wrote memo to RCA board predicting television, 1923; helped form NBC, 1926; elected to RCA's board, 1927; acting RCA president, 1928; created Radio-Keith Orpheum (RKO) motion pictures company with Joseph P. Kennedy, 1928; became chairman of board, RKO, 1928; RCA executive vice president, 1929; served as chairman of board, RCA, 1947–70; forged RCA purchase of Victor Talking Machine Company, 1929; RCA president, 1930–65; gained FCC approval of RCA's color TV system, 1953. Died in New York City, 12 December 1971.

Film

Empire of the Air: The Men Who Made Radio, 1991

Selected Publications

Principles and Practices of Network Radio Broadcasting: Testimony of David Sarnoff, President, Radio Corporation of America, Chairman of the Board, National Broadcasting Company, before the Federal Communications Commission, Washington D.C., November 14, 1938, and May 17, 1939, 1939

Pioneering in Television: Prophecy and Fulfillment, 1945; 4th edition, 1956

Looking Ahead: The Papers of David Sarnoff, 1968

Further Reading

Benjamin, Louise, "In Search of the 'Sarnoff Radio Music Box' Memo," *Journal of Broadcasting and Electronic Media* 37, no. 3 (1993)

Bilby, Kenneth W., *The General: David Sarnoff and the Rise of the Communications Industry*, New York: Harper and Row, 1986

"The David Sarnoff Era Ends at RCA," *Broadcasting* (12 January 1970)

Dreher, Carl, *Sarnoff: An American Success*, New York: Quadrangle Books, 1977

Lewis, Tom, *Empire of the Air: The Men Who Made Radio*, New York: Burlingame Books, 1991

Lyons, Eugene, *David Sarnoff: A Biography*, New York: Harper and Row, 1966

Sobel, Robert, *RCA*, New York: Stein and Day, 1986

Tebbel, John William, *David Sarnoff: Putting Electrons to Work*, Chicago: Encyclopaedia Britannica Press, 1963

Wisdom: The Magazine of Knowledge and Education 22 (1958) (special issue entitled "The Universe of David Sarnoff")

Satellite Radio. *See* Digital Satellite Radio

Scandinavia

Broadcasting in the Nordic Countries

Radio in the five Scandinavian countries typifies the public service tradition in Europe, as private commercial stations appeared only in the mid 1980s. Although now competing with commercial ventures for listeners, the public service systems remain the foundation of radio in this region.

Scandinavian radio is heavily influenced by the geography of the region and by language traits. Variations in geographic size and population density explain comparative distinctions, but there is also a shared cultural and social heritage (broadly construed) that is responsible for its similarities. Thus geographic and linguistic traits influence how radio is organized as well as explain its comparative importance.

It costs a great deal to broadcast in Norway due to its mountainous terrain, which requires a large number of transmitters and relay stations. In Finland the per capita costs for serving the Swedish-language and Laplander minorities are high. Iceland has a small and comparatively isolated population, and not surprisingly the fewest number of radio channels. Denmark is densely populated and has the highest number of channels per capita, although Norway actually has more total channels.

None of these countries speak a language that is widely spoken elsewhere, although Swedish is the best known due to Sweden's legacy as a former colonial power. Linguistic distinctiveness supports the popularity and importance of radio broadcasting in Scandinavia, whereas shared cultural and historical experience explains the cooperative framework between regional public service broadcasters.

Radio's Potential

History explains much about the organization and importance of Scandinavian radio. Radio was used as a propaganda tool by neighboring countries during World War II and the Cold War, and more generally as a way of strengthening national resolve and consciousness throughout Europe. The potential for abusing its capabilities for social influence combined with a desire to strengthen democracy and facilitate postwar reconstruction proved decisive in confirming the public service monopoly approach in the mid 1900s.

With the collapse of the Soviet Union and growing American influence, Scandinavia adopted processes and practices characterizing the European Union (EU) and Common Market in the 1990s. Deregulation targeted a range of monopolies in the early 1980s, especially electronic media. As a result, contemporary Scandinavian radio is characterized by dual systems, with private commercial and public service channels competing in an increasingly open and international media marketplace.

Early History of Scandinavian Radio

There are broad similarities in the early history of radio broadcasting in the Nordic region. Wireless transmission was initially restricted to the military. Transmission typically depended on connections provided by a public trust telegraph company, and radio amateurs helped to develop and popularize radio. The economic depression of the late 1920s and early 1930s contributed to the establishment of the public service approach and monopoly organization, coinciding with concerns about the growth of fascism and communism in mainland Europe. The early history of radio was characterized by rivalry with newspaper companies and, paradoxically, benefited from newspaper innovations in developing program services.

Denmark

In the early 1920s radio amateurs (army engineers, etc.) were illegally experimenting with radio transmission, which was restricted to military control. Public interest grew when Svend Carstensen from the Copenhagen daily *Politiken* began broadcasting news bulletins in 1923 (via *Radioavis*). There were three transmitters in Denmark from 1923 to 1925. Radio manufacturers were also involved, holding seats in the first radio council of the Danish State Broadcasting Company (Statsradiofonien). Neither manufacturers nor amateurs had strong preferences about funding. Early Danish radio did not have advertising or sponsorship, although newspapers did produce radio shows during the experimental period between 1922 and 1925.

Radio became an issue of public debate in the mid 1920s. Engineers, industrialists, and the political right were in favor of organizing radio as a private industry. Union representatives, educators, and the political left favored public ownership, voicing concerns about the cultural impact of radio. The business interests were weak and disorganized, and in 1925 a one-year experiment for nationwide broadcasting by the Statsradiofonien was legislated.

Statsradiofonien became a permanent institution in 1926, and the name was changed to Danmarks Radio (DR) in 1959. DR is regarded as an independent institution of the state: it is financed totally by license fees, regulated by public service obligations, and administered by the radio council (*Radioraadet*).

Iceland

The Icelandic Parliament (Althingi) passed the first radio law in 1925, which authorized a private company for a seven-year period. That company, Utvarp HF, only lasted two years because its listeners never exceeded 500 and the company was denied government financial support. In 1927 a government committee investigated possibilities for creating a public broadcasting company. Icelandic Radio or Útvarpsstöð Íslands í Reykjavík (later known as RUV) was established in 1928 in the capital city of Reykjavík. The broadcasting board had its first meeting in late 1929, and broadcasting began in December 1930 with a 16-kilowatt station at Vatnesendi, five miles from the capital city. (The studio was located in Reykjavík.) Early broadcasting was restricted to a couple of hours each evening.

RUV was financed with a license fee and with supplemental advertising. The birth of radio broadcasting was not a conflicted event in Iceland because it began later here than elsewhere in Scandinavia, and also because characteristic problems had already been addressed in the 1920s when telephony was an issue.

Norway

Early Norwegian broadcasting was technically illegal because the Telegraph Law of 1914 prohibited radio listening. Wireless announcements were restricted to military and commercial correspondence under telegraph monopoly. Legal broadcasting began in October 1924 when the radio ban ended and license fee funding was established. Some advertising was also allowed. Advertising revenue peaked in 1929 at a bit less than 5 percent of total income. There was also a 10 percent "turnover tax" levied on all broadcasting equipment.

Experimental transmissions actually began in 1923 with Norway's first radio station, Kristiania Broadcasting, which broadcast using a 500-watt transmitter located at Tryvannshoegda, near Oslo. The station was moved to the center of Oslo when radio made its official broadcasting debut in 1925. The Norwegian Broadcasting Company (Kringkastingselskapet AS) was operated as a private company from 1925 to 1933. Other private broadcasting companies were established in Bergen (1925), Aalesund (1926) and Tromsoe (1927). In 1933 the Norwegian government acquired the stock and created The Norwegian Broadcasting Corporation (NRK) that exists today. Despite opposition, advertising continued until 1939 (all NRK today are completely without advertising). The turnover tax ended in 1988.

Sweden

In the 1920s most Swedes supported a public monopoly because of concerns about radio's propaganda potential and out of a desire for "order." Broadcasting was, however, first organized as a private company under strict government control. News content was prohibited until 1956 because of pressure from newspaper companies; this was supported by political parties. Radiotjänst AB (Svergies Radio) was then established in 1925 by a group of publishers and businesses. Manufacturers also applied for licensing. License fee funding was the preferred model; advertising was legally possible, but never accepted in practice. Radiotjänst AB was dependent on Televerket (the Telecommunications Administration). Televerket collected the license fees and distributed the monies. Until 1951 the state received about 30 percent of the annual levy. Televerket was responsible for the terrestrial infrastructure and also signed the annual operating agreement with the government.

As was typical in the Nordic region, most Swedes of the day lived in the countryside. (Heavy urbanization occurred in the postwar era.) Radio was assigned obligations to provide distance education and cultural services, and was mandated with an "enlightenment mission." The license fee was inexpensive to encourage rapid diffusion and to avoid discrimination against lower-income families, and a broadcasting commission was set up in 1927 to insure that Radiotjänst fulfilled its legal obligations. Since 1935 the commission has focused on reviewing programs.

Finland

Early Finnish radio is linked with the first decade of Finland's independence in the 1920s (it had been a Russian Grand Duchy until 1918). Clandestine radio had begun in 1917 under Czarist rule. The Radio Act was first enacted by the new Finnish government in 1919, specifying that equipment use be an exclusive right of the state.

The Finnish Radio Amateur League (Nuoren voiman liito) was formed in 1921 by young officers in the army's Signals Battalion. The Finnish government granted them a license to transmit radio signals and within a few years they had established 75 radio stations in Finland. The Finnish Radio Association (Suomen radioyhdistys r.y.) was independently formed in 1923 to represent the "interests of listeners." The association began offering evening programs on radio stations in 1924.

Most activity in early Finnish radio was situated in and around Helsinki and Tampere, the two biggest cities. The league and the association agreed on a public service approach. The Finnish government supported this as well because of a shared understanding that radio was a useful educational tool for people in remote regions. Most Finns lived in rural locations, and radio was also perceived as an appropriate tool for "sharing the cultural wealth." Early independence was marked by social turmoil between Finnish Finns and Swedish Finns, so radio was given a national unity mandate. In 1926 the Finnish government approved a private corporation financed by stockholders involved with radio in the period, as well as the Finnish State, and called Oy Ylesiradio—AB Finlands Rundradio (YLE), the first part of the name in Finnish and the second in Swedish. An administrative council represented stockholder interests. After 1927, however, only members of Parliament were nominated to the council.

In the early years YLE was situated as a production company that could lend programs to local private stations. The transmission network was owned by the Finnish military. Programming requirements specified non-political content, no advertising, license fee revenue, bilingual services (Finnish and Swedish), and an emphasis on rural populations in need of distance education and cultural literacy services. In 1934 YLE became a complete public service radio company. Thereafter the state owned 90 percent of all stock, private sector involvement ended, and program content as well as organizational management became increasingly linked with the Finnish State via Parliament. This coincided with the depression era when money was tight and there was increasing apprehension over the rise of fascism and communism in other parts of Europe.

Broadcasting Principles

Unlike the situation in the United States, where public broadcasting operates at the fringes of a dominant commercial media market, in Scandinavia public broadcasting has been the foundation of the broadcast media. That foundation rests on five principles that position broadcasting as a service for citizens: (1) public broadcasters are required to provide a comprehensive range of program services for everyone who pays an annual license fee, typically referred to as "universal service"; (2) public broadcasting is a non-commercial venture. Annual license fees are collected as a form of taxation, the payment of which confers rights on the receiver and obligations on the source; (3) public broadcasting is accountable to parliamentary oversight; (4) broadcasting must serve cultural and educational intentions (frequently summarized as an "enlightenment mission," this principle gives entertainment a lower priority); and (5) radio should have a tight domestic focus premised on language, culture, and unity.

The public service approach to broadcasting was created, in part at least, as a reaction against U.S. commercialism. Analyses of U.S. radio convinced many in Scandinavia that the commercial approach did not provide sufficient minority services and placed educational radio in a marginal context. Because U.S. radio focused on popular culture programs, to some observers it appeared that network and station owners demonstrated little interest in or potential for developing "high culture" services. Finally, although the commercial approach can address audiences as citizens, marketplace imperatives tend to favor a consumer-oriented emphasis. These value judgments, in turn, were used to legitimate the public service ethic, and by extension a monopoly organization. The monopoly preference was also linked to limitations in available frequencies before FM radio was developed as well as the large investment required to create nationwide broadcasting systems for such small populations.

Competition and Deregulation

The pirate radio phenomenon of the 1960s and 1970s directly challenged the public monopoly in radio. These unlicensed and unregulated private channels broadcasting from offshore locations in international waters successfully targeted teenagers and young adults in the largest metropolitan areas, particularly Stockholm and Copenhagen. They took advantage of the underutilized VHF bandwidths in this part of the world. Although clandestine political activists were involved, most pirate stations were advertising-supported ventures. They capitalized on the lack of popular culture programs that were preferred by young people. Radio Luxembourg also filled this need, although it was a licensed commercial channel.

Public broadcasting monopolies fought back by launching regional channels in the 1970s and by offering limited and select popular music targeting young people. Although the regional channels were popular with adult and elderly listeners, youngsters were not satisfied. The popular success of the pirate radio channels strengthened the move to liberalize broadcasting policies.

With the exception of Sweden, private local radio was introduced in the Scandinavian countries in the mid to late 1980s. In light of the pirate channels, it was clear that there were sufficient frequencies available for new commercial stations. The public monopolies were now challenged as being contrary to democratic virtues. The 1990s saw growing competition between public and private channels, the latter becoming increasingly nationwide and network-oriented by the end of the decade.

In Finland, for example, the government approved legislation in 1993 called the Act on YLE that provides increased security for that company and its public service character but in exchange requires much higher productivity, efficiency, and openness to public scrutiny. This followed a major internal restructuring initiative launched in 1990 when Yleisradio reformed its radio channels. Radiomafia was launched to compete directly with the private commercial channels for the ears and hearts of young Finns, whereas Radio Suomi was structured as a network of regional channels featuring news and current affairs programming. In 2003 YLE launched a second radio reform in response to increasing competition, especially for young adults. Radiomafia was replaced with YLEX, for example. The situation in Finland is paralleled elsewhere in the Nordic region. In Norway, the combined private channels account for about 10 percent of gross advertising, and P4 is a commercially financed national channel that recently purchased a substantial share of the company that owns Radio Nova, Finland's first nationwide commercial channel. The Disney corporation is also involved in Finnish radio, sharing ownership in Kiss FM.

Increases in competition coincide with and are increasingly fueled by European Union initiatives that decry protectionist policies and pursue European integration on the basis of commerce and competitive open markets in the European Economic Community (EEC). Member states find it increasingly difficult to regulate and control broadcasting as a purely domestic matter.

Even more pervasive is the increasing influence of concepts and tools developed in U.S. commercial radio. Today's public radio broadcasters are increasingly aware of rotation clocks, target audiences, formats, and profiling. Such concepts and tools were anathema a dozen years ago, but today one commonly hears programs referred to as products, audiences as customers, and listeners as markets. Media consultants, researchers, and scholars from the U.S. are no longer rare, but quite commonplace. For example, the European Broadcasting Union offered a 1999 seminar for public broadcasters on "Marketing Public Service Values." A radio training curriculum developed at Yleisradio in Finland features such coursework as "Competition Analysis" and "Program Soundscape." Broadcasters in the private sector are similarly inclined to learn about and rely on U.S. concepts and tools, partly because those channels are also commercial and thus fully compatible with the approach, and partly because U.S. media companies are investors and partners with various private channels and media concerns in Europe today.

The Future

The future of Nordic radio broadcasting depends on at least four factors. The first is the extent to which public broadcasting companies will be able to maintain the public service ethic, which will be increasingly difficult in light of growing competition with the private sector. Private competition is the second factor—the intensity and amount of private competition that will be permitted by the respective governments.

Third is technology, specifically digital audio broadcasting (DAB). In the late 1990s public broadcasting companies and governments in Scandinavia have pursued DAB as a priority, although the relative ranking of that priority varies from country to country (e.g., high in Finland but subject to recent setbacks in Sweden). DAB provides the only viable option for expanding the range of services radio can provide because the bandwidth required for one VHF/FM channel is sufficient for six DAB channels (and possibly more, depending on the type of content). To date sales of DAB receivers have been sluggish because costs are comparatively high. Public broadcasters are carrying much of the responsibility for DAB development; the private sector has limited investment risks until a market develops.

A fourth and final factor is ultimately catalytic in determining how all of the others unfold in practice. The dynamics of radio broadcasting in the Nordic region depend on the balance between European Union policies emphasizing a business-oriented approach to just about everything, and a popular backlash that appeared in the late 1990s against such a strong economic orientation. EU policy is increasingly castigated for lacking sensitivity and balance with regard to regional concerns and cultural distinctions.

There are at least two major challenges facing radio broadcasting in the Nordic region. First, as public broadcasting has formed radio's foundation, the commercial sector has been supplemental. If this changes under increasing economic pressure, Nordic countries might create a public service ghetto wherein only the commercially unattractive programs and audiences are left to public service entities. That would mark the end of universal service and weaken the democratic orientation in Nordic radio. Second, there is the danger of an ethical dilution of the public service approach caused by the necessity of mastering commercial logic, concepts, and tools in order to cope with increasing competition. Ironically, the competitive dynamic that has fueled development could be weakened.

GREGORY FERRELL LOWE

See also Digital Audio Broadcasting; European Broadcasting Union

Further Reading

DR (Danish Broadcasting Corporation), <www.dr.dk>

Emery, Walter B., "The Norden [sic] Countries," chapters 10–13 in *National and International Systems of Broadcasting: Their History, Operation, and Control,* East Lansing: Michigan State University Press, 1969

European Broadcasting Union, <www.ebu.ch>

Hans-Bredow-Institut für Medienforschung an der Universität Hamburg, <www.rrz.uni-hamburg.de/hans-bredow-institut/english/index.html>

Hujanen, Taisto, and Per Jauert, editors, "Symposium: Radio Broadcasting in Scandinavia," *Journal of Radio Studies* 5, no 1 (Winter 1998) and 5, no. 2 (Summer 1998)

Inetmedia: The Broadcasting Link, <www.markovits.com/broadcasting>

Lowe, Gregory Ferrell, and Taisto Hujanen, editors, *Broadcasting and Convergence: New Articulations of the Public Service Remit,* Göteborg, Sweden: Nordicom, 2003

Nordicom Review: Nordic Research on Media and Communication, <www.nordicom.gu.se/review.html>

NRK: Forsida, <www.nrk.no>
Ríkisútvarpið (Icelandic National Broadcasting Service), <www.ruv.is>
Sveriges Radio (Swedish Broadcasting Corporation), <www.sr.se>
YLE (Yleisradio), <www.yle.fi>

Schechter, A.A. 1907–1989

U.S. Reporter, Broadcast Executive, Creator of NBC News

With ingenuity but few resources, Abe Schechter built a news operation at the National Broadcasting Company (NBC) in the 1930s that could compete with print journalists.

He was born in Central Falls, Rhode Island, in 1907, the son of George Schechter and Celia Riven. In high school, he began as a reporter for the *Providence Journal* and continued there until he received a bachelor's degree from Boston University with a major in journalism. Then he worked at the *Newark Star-Eagle* and moved onto the *New York World*. He went to work for the wire services, first the Associated Press (AP) and then International News Service, where he became the youngest city editor in New York City.

In 1932, he joined NBC in the publicity department and then in the news and special events department. When he took the job, the network did little to cover news on a daily basis with the exception of a 15-minute, five-day-a-week newscast that had been launched in 1930 by Lowell Thomas. There were also a few commentaries, some with objective content and some with predictable political bias. But neither NBC nor the Columbia Broadcasting System (CBS) had a team of reporters and editors to collect, write, and present the news. Instead, the networks relied on the wire services for their information.

As the Depression deepened during the 1930s, the newspaper industry began to feel that broadcasters were becoming too competitive in the fight for the advertising dollar. To solve the problem, wire services were told to stop providing material to broadcasters, and because the print industry controlled AP and was the dominant client of the other services, the wire services agreed.

Schechter was forced to find his own news. Story ideas came from affiliates and, of course, from newspapers. To gather information for the Lowell Thomas newscast, Schechter relied on telephone interviews, because he had no staff and worked out of a storage room. He found that the words *NBC* and *Lowell Thomas* were effective tools for getting his calls through to important people, and the promise of tickets to live broadcasts of their programs was helpful, too. Sometimes, he even managed to scoop the print reporters. But for the most part, he had to settle for broadcasting features, rather than news.

Schechter headed NBC's news and special events department from 1938 to 1941. He described the early years of NBC news in his autobiography, *I Live on Air*, published in 1941. Despite the success in gathering news during the press-radio war, NBC returned to using Schechter's department primarily for features, such as a singing mouse contest or the voice of a deep sea diver fathoms under the ocean.

Then, as the tensions in Europe tightened, NBC hired more correspondents and allowed more news on the air. Perhaps the network's motivation was to promote itself in the competition with CBS for listeners, but the result was more and better coverage of international events and an increase in advertising revenues as the audience for news grew. Schechter hired a news team in Europe to cover the coming war (although some criticized his choice of reporter for Munich, Max Jordan, who seemed to favor the Germans and was often able to get information out first).

Schechter's most famous scoop occurred in 1939. The German battleship *Graf Spee* was attacked by British cruisers and sought refuge in Montevideo, Uruguay. When the South American country asked her to leave, the captain decided to scuttle his ship rather than surrender her to the British. An NBC reporter described live on air the explosions and the sinking as millions listened. In 1940 Schechter scored another coup when he got a statement from the minister of The Netherlands as his homeland was being invaded by the Germans; Schechter followed this up with an on-the-scene report of Nazi paratroopers landing on Dutch soil.

When the United States joined World War II, Schechter left NBC to work with the Office of War Information but then moved to the War Department's Bureau of Public Relations as a civilian adviser; there he oversaw radio communications for journalists covering the war in the Pacific Islands. He received the Legion of Merit for his work during the Philippine campaign.

After the war, he became vice president in charge of news at the Mutual network, and then for a short time he was an executive with Crowell Collier Publishing Company. He returned to NBC in 1950 and became executive producer for a new television program, a morning news and talk show called *The Today Show,* but he left soon after the program went on the air to start up his own public relations firm, A.A. Schechter Associates, in 1952. The company was purchased by Hill and Knowlton in 1973. Schechter and his wife were killed in an automobile accident on Long Island in 1989.

BARBARA MOORE

See also National Broadcasting Company; News; Office of War Information

Abel Alan Schechter. Born in Central Falls, Rhode Island, 10 August 1907. Attended Boston University, B.A. in journalism, 1928; reporter, *Providence Journal,* 1924–28; reporter, *New York World,* 1928–31; reporter, Associated Press; editor, International News Service, New York, 1931–32; publicity specialist, NBC, 1932–38; head of news and special events, NBC, 1938–41; Office of War Information, 1942; civilian adviser, War Department Bureau of Public Relations, 1942–45; vice president in charge of news, Mutual network, 1945–50; creator and executive producer, *Today Show,* NBC-TV, 1950–52; owner of public relations firm, A.A. Schechter Associates, 1952–73. Died in Southampton, New York, 24 May 1989.

Publications

"Go Ahead, Garrison!" A Story of News Broadcasting, 1940
I Live on Air (with Edward Anthony), 1941

Further Reading

Bliss, Edward, Jr., *Now the News: The Story of Broadcast Journalism,* New York: Columbia University Press, 1991
The Fourth Chime, New York: NBC, 1944
McAndrew, William R., "Newscasts Grow Up—With Schechter," *Broadcasting* (3 March 1941)

Schulke, James 1922–1999

U.S. FM Radio Pioneer

James A. Schulke is recognized by many as the father of FM musical programming. Schulke's uniqueness in radio broadcasting is directly related to the legendary discipline that he instilled in radio management, programming, engineering, talent, and marketing. At the time Schulke embarked on a career in radio, less than 5 percent of U.S. radio listening was to the FM band.

An Ohio native, Schulke received his B.A. from Denison University and his M.B.A. from Harvard. He served in the U.S. Marine Corps during World War II.

Schulke began his career far from the established radio industry. Following an early New York City career with talent agent J.L. Saphier and advertising agency Young and Rubicam, Schulke was hired as vice president of the Paramount Sunset Corporation in Hollywood. He was charged with the responsibility of revitalizing the Paramount Lot and brought the production of such popular television shows as *Gunsmoke, Have Gun Will Travel,* and *Bonanza* to the facility. Having built a reputation as a turn-around specialist, Schulke was transferred by Paramount to become general manager of KTLA-TV, Los Angeles, one of three independent TV stations operating in the competitive Los Angeles market.

Schulke's television career flourished with creativity. He established the first dual-anchor newscast in the market, was first to use a helicopter for covering television news, and built the first videotape newsroom. Under his leadership, KTLA became the first independent television station to rank number one in late news.

Schulke's career in radio began in 1962 when he joined the Magnavox Company, which was searching for methods to promote the sale of FM radio receivers. Schulke helped convince Magnavox to underwrite the creation of the National Association of FM Broadcasters (NAFMB). Schulke was the organization's first president. Under his leadership, the NAFMB lobbied the media research firm Arbitron to begin measuring FM listening separate from AM. The organization was also instrumental in pushing the Federal Communications Commission (FCC) in the mid-1960s into establishing a rule requiring separate programming for AM and FM stations. The original rule required 50 percent separation. Up until that time nearly all FM stations had been simulcast with an AM station.

Upon leaving Magnavox, Schulke teamed with Robert Richer and Marlin Taylor in 1965 to form Quality Media Incorporated (QMI), an organization designed exclusively to market FM radio to advertisers. QMI's specialty was selling "good music" radio stations, including those with classical music formats. It was at QMI that Schulke first examined the potential of a tightly controlled "beautiful music" format. He saw the potential of matched-flow music and the need to appeal to the female listeners who controlled at-home listening. Schulke left QMI in 1970 to form Schulke Radio Productions (SRP). Marlin Taylor later left QMI to form a competitive beautiful music programming company for the Bonneville Corporation.

SRP's original operation was located in a brownstone apartment on New York's upper east side, where Schulke resided. Schulke recruited Phil Stout as his music director, and together they developed a matched-flow all-instrumental beautiful music format. The programming was based on long-term quarter-hour audience maintenance to drive average quarter-hour audience ratings. Schulke's control was so complete that he personally approved every music selection and every matched-flow quarter hour syndicated by SRP. When the supply of original instrumental and cover recordings recorded in the United States was exhausted, he was the first U.S. radio programmer to enter into an exclusive agreement with the British Broadcasting Corporation (BBC) in London for original arrangements.

SRP was very selective in choosing its clients because of Schulke's extremely rigid contract requirements. He insisted upon maintaining what was tantamount to total creative control. Commercial load was limited to no more than six minutes of advertising per hour, and the commercial acceptance policy rigidly forbade personal hygiene products or any loud multiple-voice pitches. News was kept to the bare minimum, so as not to distract the listener from the music flow. Schulke's contracts forbade subsidiary use of a station's subcarrier and demanded circularly polarized antennae, which would improve FM reception in the home and in the automobile. He even retained the right to veto the selection of any announcer.

Among the early Schulke believers was Woody Sudbrink, the founder of Sudbrink Broadcasting and the owner of the greatest number of independent FM stations in America from 1970 to 1978. Other original clients were Mike Lareux of WOOD-FM, Grand Rapids, Michigan, and J.D. McArthur, owner of WEAT-FM, Palm Beach, Florida. Each was the first top-rated FM station in its market.

Innovative marketing was also a Schulke signature. He is credited with creating phonetic call letters such as WLIF, "Life in Baltimore"; WLAK, "Lake in Chicago"; and WPCH, "Peach in Atlanta." He was also an early believer in using television to promote FM listening, with on-air ads noted for serene scenes and beautiful music that always culminated with an exact focus on the station's numeric dial position and phonetic call letter reference.

When the business grew, Schulke moved the operation to a self-contained facility in New Jersey where a state-of-the-art stereo duplication facility was constructed and the famous "black box" was located. The "black box" was developed there and was later installed in the audio-processing chain at every SRP station. The technical details of this engineering development were never disclosed (legend has it that the box was empty), leading some to regard it as a marketing ploy. It is factual that Schulke would not allow stereo commercials on any of his stations because SRP was not in a position to control the quality of the stereo duplication of commercials.

By 1974, two years after forming SRP, Schulke stations ranked either number one or two in the top 20 U.S. radio markets. SRP, which had been renamed Stereo Radio Productions, was sold to Cox Radio in 1979. At that time, SRP stations held major audience positions in the top 150 U.S. radio markets. James A. Schulke died on 6 August 1999 in Fort Lauderdale, Florida, at the age of 77. His papers are archived at the Library of American Broadcasting at the University of Maryland, College Park, Maryland.

GORDON H. HASTINGS

See also FM Trade Organizations; Taylor, Marlin R.

James Schulke. Born in 1922. Received B.A. from Denision University and M.B.A. from Harvard; served in United States Marine Corps in World War II; worked at Magnavox Company, 1962–65; first president of National Association of FM Broadcasters; formed Quality Media Inc., 1965; formed Schulke Radio Productions, 1970; sold operation to Cox Radio, 1979. Died in Fort Lauderdale, Florida, 6 August 1999.

Science Fiction Programs

Science fiction programming takes full advantage of radio's ability to transport us through time and space—at a fraction of the cost of a bus ticket. With a well-written story, good voice actors, a few inexpensive sound effects devices, and a willingness to suspend disbelief, we can easily find ourselves lost in *Dimension X,* refugees in a *War of the Worlds,* or leaping tall buildings with our pal *Superman.*

The genre, which traces its roots to the pulp magazines and comic strips of the 1920s, has most often been labeled "thriller drama," but it has actually infused almost every type of fiction, from action-adventure to comedy. There have been sci-fi detective programs, sci-fi adventure shows, sci-fi comedies, sci-fi kids' shows, even sci-fi soap operas. Many programs such as *The Shadow* dealt, at least periodically, with science fiction themes. Regardless of the other elements of a program (or episode), to be science fiction, a work should integrate the relationship between humans and "futuristic themes" such as new technology or alien races.

Science fiction radio dates back to the earliest days of commercial radio. *Ultra Violet,* a program few people remember, was first syndicated as early as 1930. More famous, however, were programs such as *Buck Rogers in the 25th Century,* which was first broadcast in 1932 and is commonly credited as being the first science fiction radio program.

Based on a popular comic strip, *Buck Rogers* was a 15-minute serial that aired five times a week at 7:15 P.M. on the Columbia Broadcasting System (CBS). Aimed predominantly at children, the series focused on Buck, a man from the present (the 1930s) who finds himself transported to the 25th century. The cast of characters included a very strong female character, Wilma Dearing, and the amazing scientist Dr. Huer. Interestingly, many of the fanciful technological devices invented by Dr. Huer in the show became commonplace technologies in the late 20th century. Good and evil were very clearly defined in *Buck Rogers,* and good always prevailed, but there were no truly memorable villains such as *Flash Gordon*'s Ming the Merciless. Like all good serials, most episodes of *Buck Rogers* closed with a "cliff-hanger" ending that left many questions unanswered. Listeners had to "tune in tomorrow" for the next exciting installment. The series and sponsor also held the attention of their audience by allowing them to become "Solar Scouts" and to receive items such as "planetary maps" by responding to Kellogg's premiums.

Superman first arrived from Krypton on the Mutual Broadcasting System in 1940 and is a good example of how science fiction merged with other genres. In this show, also aimed predominantly at children, our superhero fought crime both as Superman and as his alter ego, mild-mannered reporter Clark Kent. "Girl reporter" Lois Lane and "kid photographer" Jimmy Olsen, along with gruff editor Perry White, made up the rest of the regular cast. Often categorized as an action-adventure, crime, or thriller-drama program, the show's central character was an alien with superhuman powers.

Although most 1930s sci-fi radio programming was aimed at children, a few shows were designed for adults. Most of these were episodes of anthology programs such as *Mercury Theater of the Air,* which premiered on CBS in the fall of 1938. On 30 October, just a few short weeks after the premiere, this prestigious drama program, hosted by Orson Welles, pulled off the greatest hoax in radio history—the radio adaptation of H.G. Wells' "War of the Worlds." The pre-Halloween dramatization of Martians landing in Grover's Mill, New Jersey, led some listeners to panic—and many to leave their homes.

Orson Welles' program played on the fears of an audience worried about war in Europe. When actual fighting broke out in 1939, more adult science fiction programs were broadcast as episodes of anthology shows. Series such as *Lights Out, Radio City Playhouse,* and *Escape* featured science fiction entries concerning time travel, alien invasion, and world conquest. America, including its radio audience, was becoming more technologically savvy, and more world-weary. Consequently, adult science fiction programs were becoming less reliant on horror and fantasy and more focused on actual science and technology.

By the 1950s, television was pulling a significant number of listeners away from radio. In order to hang on to adult audiences, the radio networks experimented with science fiction series aimed at adult audiences. Several adult anthology series devoted exclusively to science fiction premiered in the early 1950s, including *Year 2000 Plus* on Mutual and *Dimension X* on the National Broadcasting Company (NBC). Later known as *X Minus One, Dimension X* was one of the first radio drama series to be recorded on tape rather than broadcast live. As a result, programs could be more involved and could be post-produced to clean up mistakes. As an anthology, stories changed from week to week. Some shows were quite serious, but one of the most famous is an ironic comedy titled "A Logic Named Joe." Originally broadcast on 1 July 1950, this humorous tale is about a world where futuristic computers, or "logics," can do "everything for you." The logics are interconnected in a worldwide web of computers that exchange information. "A Logic Named Joe" takes a comic look at a common theme in science fiction: humanity's fear that technology will take over and corrupt society. Unlike the adult-oriented anthology programs, most series science fiction of the era was limited to such children's shows as *Tom Corbett, Space Cadet,* and *Space Patrol.*

As U.S. radio comedy and drama moved to television, science fiction on the radio declined but did not disappear. Later series such as *CBS Radio Mystery Theater* often included science fiction, as well as fantasy and horror themes. National Public Radio stations also imported programs from the British Broadcasting Corporation (BBC) and the Canadian Broadcasting Corporation (CBC). One of the more famous imports, originally aired in England in 1978, was *The Hitchhiker's Guide to the Galaxy* series, which later spawned a BBC television series, several novels, and an interactive computer game.

Radio science fiction programs are actually more plentiful today than they have ever been, thanks to cassette sales, the internet, and a variety of interest groups. Not only are episodes of many classic programs such as "A Logic Named Joe" available for audio streaming, but original programming is being produced, such as the Sci-Fi Channel's web "radio" program, *Seeing Ear Theater.* Thanks to continued interest in the form and some very fantastical technological advances, science fiction radio is not only alive and well, but its future is very exciting.

PHILIP J. AUTER

See also Children's Programs; Hoaxes on Radio; Shadow; Sound Effects; Star Wars; War of the Worlds

Further Reading

Buxton, F., and Bill Owen, *The Big Broadcast: 1920–1950,* New York: Viking Press, 1972

Science Fiction on Radio, <www.mtn.org/~jstearns/sf-otr.html>

Widner, James F., "To Boldly Go . . . ", <www.otr.com/sf.html>

Widner, James F., and Meade Frierson III, *Science Fiction—On Radio: A Revised Look at 1950–1975,* Birmingham, Alabama: A.F.A.B., 1996

Sevareid, Eric 1912–1992

U.S. Journalist and Radio Commentator

One of the first "Murrow Boys" at Columbia Broadcasting System (CBS) news, Arnold Eric Sevareid reported many aspects of World War II and then became well known for his radio and television commentaries that surveyed the several sides of various controversies, often leaving the final conclusion up to the listener. He spent his entire broadcast career of 38 years with CBS.

Origins

Sevareid grew up first in North Dakota and then in Minnesota, graduating from high school in 1930. That same year he undertook an arduous 2,200 mile canoe trip with a friend, Walter Port, traveling from Manitoba all the way to the Atlantic Coast. Five years later he wrote his first book, a children's tale about the trip called *Canoeing with the Cree.* He spent a few months in the summer of 1931 out in the California gold fields, gaining useful experience, though little gold.

While in college at the University of Minnesota, he first entered journalism, working as a reporter on the student paper, the *Minnesota Daily.* He was active in and reported on the university's move to drop compulsory military training. But his radical views and activities probably cost him a chance for the paper's editorship. Before graduating in 1935, he had begun working as a reporter at the downtown *Minneapolis Journal.* After graduation he helped organize the local chapter of the Newspaper Guild at the paper, which led to his being fired soon thereafter, supposedly for a minor error in a story.

In 1937 Sevareid and his wife sailed to Europe on a freighter, seeking a change of scene and new challenges. He studied for a time at the London School of Economics and at the Alliance Française in Paris. At this point he began using his middle rather than his first name and again became a reporter (and eventually night city editor) for the *Paris Herald.* He also worked part time for United Press in Paris. His writing soon attracted attention.

Radio Years

Sevareid's broadcast career began with a telephone call from London. Edward R. Murrow called him in mid-1939 to offer him a CBS radio job. Murrow noted that while he knew little of Sevareid's background, he liked his writing and ideas. The radio career nearly ended at its beginning when, nervous and with a halting delivery, Sevareid underwhelmed CBS officials in New York on his first broadcast in August 1939 just before

Eric Sevareid
Courtesy CBS Photo Archive

World War II. A subsequent broadcast was only marginally better, but Murrow stuck by his new discovery and urged CBS to do the same.

As with other members of the "Murrow Boys" band, Sevareid covered many different aspects of the war. Based in Paris, he covered the fall of the city (getting his wife and new-born twin sons out just in time) in one of the last broadcasts before German troops entered. He was the first to report the surrender of the country a few weeks later. He reported from London with Murrow during the Blitz in 1940 and was then called back to the United States to report the war from New York and Washington. In 1943 he moved to the complex and remote China-Burma-India theater, where he at one point was lost for several weeks in the Burmese jungle after having to bail out of an airplane. By 1944–45, Sevareid was covering Allied advances in the Mediterranean theater and into Germany itself.

After the war he returned to the United States and headed the CBS news bureau in Washington, D.C., from 1946 to 1959. He published *Not So Wild a Dream,* a memoir of the Great Depression and the war years that was widely praised and long remained in print. CBS moved him to London for two years (1959–61) as a roving correspondent for the network. The new decade saw him briefly back in New York (until 1964) before a permanent return to Washington, where he remained for the remainder of his career. Not initially fond of the nation's capital (he felt it was too provincial), he soon found it an exciting place to work, later terming it the news center of the world.

Later Career

Never entirely comfortable on the air even after years of radio experience, Sevareid was even less comfortable on television. As with Murrow and his other contemporaries, he thought little of television at first, considering video merely a means of entertainment and not a means of covering serious news.

Nevertheless he undertook CBS assignments on such early Sunday "news ghetto" programs as *Capitol Cloakroom* and *The American Week* and was both host and science reporter on the CBS series *Conquest* in the mid-1950s. Sevareid reported eight presidential elections from 1948 to 1976, fulfilling different roles in CBS coverage of those campaigns. His important 1965 interview with former candidate Adlai Stevenson (for which he won a New York Newspaper Guild "Page One" award), however, was not broadcast over CBS but instead appeared as an article in *Look* magazine.

Sevareid became best known to millions of Americans as the avuncular senior commentator on *The CBS Evening News with Walter Cronkite* from 1963 to his retirement in late 1977. His two-minute commentaries ranged over many subjects and were criticized by some observers as covering many sides of a controversy without clearly coming to any firm conclusions. In return Sevareid argued that his role was to enlighten, not to lead. Indeed, he would often readily admit he did not know the answer to some controversial issue. His last CBS television commentary appeared on the Cronkite program 30 November 1977, just four days after his 65th birthday, then the mandatory retirement age at the network.

Sevareid defined himself as a cultural conservative and political liberal, later refining the latter to conservative on foreign affairs and liberal on domestic matters. He would always consider himself first a writer, even during his decades with CBS. For many years (until 1966) he wrote a widely syndicated weekly newspaper column and also produced many magazine pieces and several books, the latter usually based on his radio or television commentaries. He also lectured widely.

Sevareid's radio and television journalism won prestigious Peabody awards in 1950 (for his radio work), 1964, and 1967, and television Emmys, two in 1973 and one each in 1974 and 1977.

CHRISTOPHER H. STERLING

See also Commentators; Murrow, Edward R.; News; World War II and U.S. Radio

Arnold Eric Sevareid. Born in Velva, North Dakota, 26 November 1912, one of four siblings, to Alfred Sevareid, a local banker, and Clara Hougen. Moved to Minnesota and graduated from high school, 1930; graduated from University of Minnesota (political science, 1935). Reporter, *Minneapolis Journal*, 1933–36. Studied at London School of Economics and Alliance Française in Paris, late 1930s. Reporter, Paris *Herald*, 1936–39. CBS News, 1939–1977. Commentator, *CBS Evening News* 1963–1977. Died in Washington, D.C., 9 July 1992.

Selected Publications

Not So Wild a Dream, 1946
In One Ear, 1952
Small Sounds in the Night: A Collection of Capsule Commentaries on the American Scene, 1956
This Is Eric Sevareid, 1964
Conversations with Eric Sevareid: Interviews with Notable Americans, 1976

Further Reading

Auster, Albert, "Eric Sevareid," in *Encyclopedia of Television*, edited by Horace Newcomb, vol. 3, Chicago and London: Fitzroy Dearborn, 1997
Cloud, Stanley, and Lynne Olson, *The Murrow Boys: Pioneers on the Front Lines of Broadcast Journalism*, Boston: Houghton Mifflin, 1996
Nimmo, Dan, and Chevelle Newsome, "(Arnold) Eric Sevareid," in *Political Commentators in the United States in the 20th Century: A Bio-Critical Sourcebook*, by Nimmo and Newsome, Westport, Connecticut: Greenwood Press, 1997
Schroth, Raymond A., *The American Journey of Eric Sevareid*, South Royalton, Vermont: Steerforth Press, 1995

"Seven Dirty Words" Case

Supreme Court Decision on Broadcast Obscenity

For five years in the 1970s, the case of the "seven dirty words," in which a complaint by one listener brought the issue of electronic free speech all the way to the Supreme Court, was a direct challenge to the whole underpinning of Federal Communications Commission (FCC) regulation of the airwaves. The "seven dirty words" trial originated when a George Carlin record was played on a New York City radio station and became one of the landmark cases concerning indecency on the public airwaves.

Following in the tradition of his mentor Lenny Bruce, stand-up comedian George Carlin revolutionized comedy with a hip, irreverent attitude and an uninhibited social commentary. By the early 1970s Carlin turned his back on an established, middle-class audience and played almost exclusively to the counterculture. In 1972 he was arrested at the Milwaukee Summerfest while performing one of his satirical routines on language, entitled "Filthy Words" (sometimes called the "Seven Dirty Words You Can Never Use on Television" because of the monologue's central joke). A local judge threw out the charges, but the same routine became the linchpin of a Supreme Court case involving the FCC.

In 1973 Carlin recorded "Filthy Words" for his album *Occupation: Foole,* which was distributed by Little David Records. *New York Times* critic Peter Schjeldahl noted that "Carlin's playing with words, through meaning and emotive permutations, is so mild and earnest it's almost wholesome" and that his "subliminal puerilities are often dissipated by lovely flashes of pure sensitivity." In the earthy monologue, Carlin claims that there are 400,000 words and only seven you can't say on television. On 30 October 1973 a recording of "Filthy Words" was played on the Pacifica Foundation's New York FM station, WBAI. An announcer for the early afternoon program, *Lunchpail,* preceded the album cut with a warning that some listeners might deem Carlin's language offensive.

One month later, the FCC received a complaint from a man who was listening to the station while driving into the city with his 15-year-old son. This was the only objection about the Carlin broadcast that was forwarded to either the commission or WBAI, a listener-supported public radio station.

After writing to WBAI to confirm the broadcast, the FCC used the outraged letter to define the nature of indecency. Ruling that Carlin's recording was not obscene, the commissioners argued that the language was indeed indecent, depicting "sexual or excretory activity and organs, at a time of day when there is a reasonable risk that children may be in the audience." The FCC did not want to censor this material, which it cited as "patently offensive as measured by contemporary community standards for media broadcast," but wished to develop a principle of "channeling," or finding a time of day when the fewest children would be in the audience. Several commissioners thought their statement did not go far enough and wanted indecent language prohibited from the airwaves at any time, but the majority felt that Carlin's material would be appropriate sometime after midnight. The Pacifica Foundation appealed the decision, contending that the definition of indecent did not take into account any serious literary, artistic, political, or scientific value.

In March 1977 the U.S. Court of Appeals of the District of Columbia overturned the FCC ruling in a two-to-one vote because it violated Section 326 of the Communications Act, which forbids the FCC from censoring any work. Calling the FCC's position "overbroad and vague," Judge Edward A. Tamm of the Appeals Court also deemed the attempt to channel offensive material into the late hours a form of censorship. The FCC appealed.

In July 1978 the U.S. Supreme Court, in a five-to-four split decision, reversed the appeals court and affirmed the FCC's right to limit the use of profane language. In the majority opinion Justice John Paul Stevens wrote that "of all forms of communication, it is broadcasting that has received the most limited First Amendment protection" (*Federal Communications Commission v Pacifica Foundation* [438 US 726]). No distinction was made between radio and television or between AM and FM radio. Stevens concluded that offensive satire is available elsewhere but should not be broadcast in the afternoon when it is accessible to children. The Carlin broadcast was not labeled obscene, but it was considered by Justice Lewis Powell as "a sort of verbal shock treatment." Justice William Brennan dissented, finding that by its definition of indecency the FCC would deem works by Shakespeare, Joyce, Hemingway, and Chaucer to be inappropriate. He concluded that it is "only an acute ethnocentric myopia that enables the Court to approve censorship of communications solely because of the words they contain."

The Supreme Court's decision did not fully settle the question of what is free expression on the public airwaves. Many writers and producers considered the ruling a major setback to the First Amendment rights of broadcasters. Some newspapers, including the *Washington Post* and the *New York Times,* included editorials in favor of the limited ruling but nevertheless did not print the notorious seven words. FCC chairman Charles Ferris worried that broadcasters would not tackle controversial subjects for fear of the ruling and reassured the industry that so-called indecent language would not be barred in news programs.

But by the end of the 1970s, the growth of cable television transformed the entire world of communications. Home Box Office (HBO) presented the entire "Filthy Words" routine in the special *George Carlin Again* in 1978. This time, Carlin was seen on premium cable, airing in the evening; there were no complaints to the FCC. Given that one has to pay a separate fee to receive HBO programs, the FCC or the courts would likely see such a presentation in a different light than the Carlin broadcast over WBAI. Over the next two decades, competition from cable would largely obliterate the definition of indecency for broadcasters and the FCC. But Carlin's seven words would still rarely be heard on over-the-air radio or television. A clear legal distinction remained between cable service, to which one subscribes, and free over-the-air broadcasting. And the FCC maintained a partial ban on such material, channeling it to the 10 P.M. to 6 A.M. period.

In 2001 the FCC announced new decency guidelines for radio and television broadcasters, making it simpler for stations to determine unacceptable material. The guidelines include instructions about intent and context as well as warnings against repeated swearing and explicit language. Under the new ruling, Carlin's routine would still be problematic if played during morning, afternoon, or evening hours.

RON SIMON

See also Obscenity and Indecency on Radio; WBAI

Further Reading

Brenner, Daniel L., and William L. Rivers, *Free but Regulated: Conflicting Traditions in Media Law,* Ames: Iowa State University Press, 1982
Carlin, George, *Brain Droppings,* New York: Hyperion, 1997
Douglas, Susan J., *Listening In: Radio and the American Imagination: From Amos 'n' Andy and Edward R. Murrow to Wolfman Jack and Howard Stern,* New York: Times Books, 1999
FCC v Pacifica Foundation, 438 US 726 (1978)
Kahn, Frank, "Indecency in Broadcasting," in *Documents of American Broadcasting,* by Kahn, Upper Saddle River, New Jersey: Prentice Hall, 1968
Lipschultz, Jeremy Harris, *Broadcast Indecency: FCC Regulation and the First Amendment,* Boston: Focal Press, 1996

The Shadow

Crime Drama

In radio drama's heyday, from the 1930s to the mid-1950s, few program openings matched—and perhaps none surpassed—the recognizability of *The Shadow's* aural calling card. The first straining phrases of Camille Saint-Saëns' tone poem *Le Rouet d'Omphale* (Omphale's Spinning Wheel) are established and begin to fade as the filtered voice cuts through in measured intensity: "*Who* knows *what e*-vil *lurks* in the *hearts* of *men*? The *Shad*-ow *knows.*" A decidedly unfunny laugh follows, and the rushing music swells again. Here the listener can imagine the unblinking stare of the mysterious figure who has gazed into the dark side of human nature. Thus begun, the late Sunday afternoon broadcasts of *The Shadow* prompted many a schoolchild's nightmares and not a few adults' nervous glances into the evening darkness.

Origins

Although The Shadow became one of the best-known characters in radio drama, he first held the more modest role of introducer of stories about other protagonists, dramatized from pulp magazines published by Street and Smith, whose fiction factory had been churning out nickel and dime novels well before the end of the 19th century. Street and Smith's *Detective Story Magazine* and *Western Story Magazine*, both begun during World War I, were well established by the end of the 1920s, but competition tightened as rival mystery and Western publications as well as new aviation, sports, college humor, and romance titles crowded the market. Street and Smith decided to seek new readers through the rapidly evolving rival popular medium, radio.

For *The Detective Story Hour,* first heard on CBS at 9:30 P.M. on 31 July 1930, Harry Engman Charlot wrote scripts based on stories soon to be published in *Detective Story Magazine.* Charlot (whose poisoning death in a seedy Bowery hotel five years later was never solved) also suggested that the host-narrator might be a mysterious figure who told the story from the shadows, his identity disguised. James La Curto was the first to impersonate The Shadow, but after several weeks he accepted a Broadway role and was replaced by Frank Readick, Jr., whose sidelit publicity photographs shielded his face behind a visor mask worn under a broad-brimmed fedora. Stimulated as much by the novelty of the program's mysterious host as by interest in the dramatized samples of Street and Smith fictional wares, listeners rushed to newsstands asking for "The Shadow's magazine." In April 1931 the publisher obliged with the first issue of *The Shadow, A Detective Magazine,* announced as a quarterly publication. Increased demand quickly turned it into *The Shadow Monthly* and then into simply *The Shadow,* published twice a month.

The Detective Story Hour was discontinued after a year, but audience interest led to the revival of The Shadow character as a part of *The Blue Coal Review,* a 40-week Sunday variety series that debuted on 6 September 1931 at 5:30 P.M. on CBS. Half of the hour was devoted to Street and Smith story dramatizations introduced by Readick in The Shadow persona. A few weeks later, Street and Smith reclaimed its Thursday evening

9:30 CBS spot with *Love Story Drama* (retitled *Love Story Hour* later in the season), in which The Shadow recounted tales from *Love Story Magazine*. These hosting duties kept The Shadow at work for a year.

In January 1932 Frank Readick was also heard in a Tuesday evening CBS series that used *The Shadow* as the program title for the first time. Other runs featuring Readick and the briefly returning James La Curto were aired in various time slots on NBC and CBS until 27 March 1935. During this period Street and Smith wanted the broadcast tales to reflect the magazine Shadow's protagonist role, while Blue Coal, by then a well-entrenched sponsor, did not want to alter the successful format that restricted the on-air Shadow to host-narrator duties. In 1937 the impasse was broken when Blue Coal agreed to a trial run in which The Shadow would serve as the main character, on the understanding that the program would return to its Shadow-as-host-only format if the public did not welcome the change.

Star Series

With Orson Welles newly cast as The Shadow, the program had a sensational re-opening on the Mutual network at 5:30 P.M. on Sunday, 26 September 1937, and it would remain a late Sunday afternoon fixture for many years. Ironically, just as Street and Smith had achieved a radio Shadow mirroring the print character's function as an active avenger against lawbreakers, the freshened broadcast figure pursued a new direction different from that which the magazine hero had taken from the beginning of the decade.

Having seen a market for The Shadow's own magazine in early 1931, Street and Smith offered the writing task to Walter B. Gibson, an experienced hand at turning out detective stories and a magician skilled enough to be the ghost writer of "how to" manuals published under the names of Harry Houdini, Harry Blackstone, Howard Thurston, and others. Gibson's interest in the occult would help in defining The Shadow's extraordinary powers. Borrowing names from two literary agents, Gibson began producing The Shadow novels under the pseudonym Maxwell Grant, and when the publisher increased and then redoubled the magazine's frequency, he found himself typing 5,000 to 10,000 words a day while his cabin in rural Maine was literally being built around him. The publisher arranged one meeting between Gibson and an early scripter of the radio series, Edward Hale Bierstadt, but otherwise the handling of the broadcast *Shadow* was left to Street and Smith and to the Ruthrauff and Ryan Advertising Agency. Turning out 28 novels of approximately 60,000 words per issue in the first year and 24 novels in each of the next six years, Gibson wrote of an international network of agents, lookouts, and operatives who reported to The Shadow, who himself took a number of guises, aviator Kent Allard being the chief of these. The well-heeled and well-traveled Lamont Cranston, employer of a chauffeur named Stanley, was simply one identity that The Shadow assumed as needed when the "real" Cranston was abroad. The Shadow's print tales took him to the Caribbean, Asia, the Pacific islands, and other places where strange religions, philosophies, and practices abounded.

When Ruthrauff and Ryan, the program's packager, planned *The Shadow*'s return to the airwaves in 1937, the agency decided that Lamont Cranston would be the sole alter ego of the crimefighter. Although the pulp Shadow had been a loner, Orson Welles' Shadow gained the companionship of "the lovely Margo Lane," first played by Agnes Moorehead, who shared Welles' background in the Mercury Theater and in *The March of Time* dramatizations of news events. According to the announcer's weekly spiel, Miss Lane (named after Margot Stevenson, whom producer Clark Andrews had been dating) was "the only person who knows to whom the voice of the invisible Shadow belongs." While the pulp Shadow had merely concealed himself in the shadows, the radio Shadow had learned "the power to cloud men's minds so they cannot see him." In the fifth episode of the revived series ("The Temple Bells of Neban," broadcast 24 October 1937), Cranston explained to Margo Lane how an Indian yogi had tutored him in "the mesmeric trick that the underworld calls invisibility."

The scripts gave Lamont and Margo a breezy sophistication reminiscent of the marital chat of Nick and Nora Charles in Dashiell Hammett's *The Thin Man* (novel, 1934; film, 1934; radio series, 1941–50) and anticipated the "darling" this and "darling" that conversations of amateur sleuths Pam and Jerry North in CBS's 1942–54 series *Mr. and Mrs. North*. Always on the go, Lamont and Margo might find danger in Haiti, at the opera, on the road to a ski resort, or in a carnival fun house. While Margo sometimes saved Lamont from danger or stumbled onto a key clue, she gradually became a convenient bait for psychopaths and a vulnerable "lady in distress" on whom to hang ten minutes' worth of plot tension. The pulp novels' supporting network of crimefighting agents was gone, and Lamont and Margo were generally left to their own resources. Crusty, snapping, and demanding Police Commissioner Weston, representing the civic establishment in contrast to The Shadow's benevolent vigilantism, dealt testily with Cranston on many occasions, and the taxi driver Shrevvy provided comic relief with his self-mocking repetitions of phrase. Even these secondary characters gradually disappeared.

When Orson Welles took the radio lead in *The Shadow*, he was promised that he would not have to attend rehearsals; he arrived at the WOR studios by taxi just before airtime, and the on-air performance was his *only* read-through of each script. His Mercury Theater and other commitments became so demanding, however, that he relinquished The Shadow's role after the regular 1937–38 Blue Coal season and a 1938 transcribed summer season (repeated in 1939) for B.F. Goodrich.

The Shadow played by Jimmy La Curto
Courtesy CBS Photo Archive

(The Shadow participated in the tiremaker's commercials, delivering a stern warning about worn tires on slick roads.) Ironically, although Welles was the best-known actor to portray The Shadow on radio, his sardonic laugh did not match the scariness of Frank Readick's, whose recorded "Crime does not pay" warning continued to be used at the end of each episode during Welles' tenure.

Bill Johnstone undertook the Cranston/Shadow role on 25 September 1938 and held it through five seasons, while Agnes Moorehead later yielded Margo Lane's part to Marjorie Anderson, and Ken Roberts became something of an institution as the announcer. Bret Morrison inherited the lead for the 1943–44 season but left over salary differences. John Archer and Steve Cortleigh had brief runs in playing The Shadow in the usual Sunday afternoon time slot as well as in occasional appearances on the mid-1940s quiz program *Quick as a Flash,* where a rotating series of radio detectives dramatized crimes for the panel to solve. Bret Morrison returned to the lead role in 1945, held it until the program ended in 1954, and later recorded a few new episodes for non-broadcast tape distribution. In those last years Margo Lane was played by Lesley Woods, Grace Matthews, and Gertrude Warner.

During the late 1930s and early 1940s, most episodes of *The Shadow* began *in medias res,* and the typical Sunday found Lamont, Margo, and their antagonists speeding from place to place by car, motorboat, or private airplane. To make an inquiring visit to a crime scene or to deliver a warning to a miscreant ("The Shadow already knows enough to hang you, Joseph Hart!"), Lamont Cranston often assumed The Shadow's identity shortly before the mid-program commercial break, and his second appearance, at the end of the episode, brought the crack of a handgun and the echoing cries of the guilty as they crumbled, sank, burned, or fell to their deaths. In later years the mad-scientist villains (often lisping and spitting in middle European accents) largely gave way to more commonplace thugs and hoodlums (grumbling in accents of Brooklyn and the Bronx). Early 1950s listeners often felt that they could predict, to the minute, the plot turn when a loudly protesting criminal's death would illustrate The Shadow's view that "The *weed* of *crime* bears *bitter fruit.* Crime does *not pay.*"

Neither Walter Gibson's pulp novels nor *The Shadow*'s radio scripts gave much attention to subtleties of characterization or incident, and listeners have concluded that scarcely any story in all those broadcast years fully made sense. Villains, often nursing misconceptions of being wronged by society or stinging from previous encounters with The Shadow, were often megalomaniacs with ambitions for dominating large populations through the risky deployment of flimsy devices. One antagonist strove to block the sun's rays and leave the city in darkness, while in another tale a gang of robbers placed exploding light bulbs in fixtures throughout town so that people would become fearful of using any lights, even car headlights or hospital operating room lamps. In "The Ghost of Captain Baylor" (15 January 1939), The Shadow freed 50 innocent sailors from the island dungeon of a group that was attempting to control naval traffic by rising in a submarine, popping open the hatch, and shooting a machine gun at passing vessels. The Shadow locked the thugs in their own dungeon and left them "to suffer, as long as their lives last, the terrible fate they designed for others," and while this typically articulated bit of poetic justice unfolded, Lamont Cranston's yacht proved capacious enough to return the victims to shore.

The Shadow's own powers and degree of vulnerability sometimes differed from one episode to the next. While he often seemed able to slip into any cell, cave, locked basement, vault, or ship's cabin, he was nearly burned alive in an ordinary room when its sole window was locked from the outside. In "Appointment with Death" (12 March 1939), a vengeful ex-con tricked The Shadow into swimming toward an island hideout so that the villain might shoot at the gap which the invisible Shadow's body would make in the rippling water. Other antagonists noticed the impressions of The Shadow's feet in deep-pile carpets or sought to trap his image in an early television receiver.

The last episode of *The Shadow* was broadcast on 26 December 1954, five years after Blue Coal had dropped its sponsorship and the same year that the Street and Smith *The Shadow* magazine had ceased publication. Communism and McCarthyism were the shadows then looming over the U.S. horizon, and television had begun to push drama programs from radio. In a 1960s reminiscence, Walter Gibson noted that The Shadow had become a creature of "camp," although he felt that those who derided the series also privately enjoyed it. In 1963 transcriptions of classic episodes were syndicated to WGN and other stations.

In the early 1940s *The Shadow* had been the highest-rated dramatic program on radio, and it prompted a 1940 movie serial starring Victor Jory, a short-lived comic strip drawn by Vernon Greene, a comic book series, three "Big Little Books," and a number of Blue Coal premiums, including ink blotters and glow-in-the-dark rings. The Shadow had been showcased in several "B" features in the 1930s and 1940s, and in 1994 Alec Baldwin starred in a large-budget film adaptation in which lavish costuming and set decoration adorned a typically loose-jointed plot.

The Shadow programs are reasonably well represented in the circulating libraries of radio clubs today, and many episodes starring Welles, Johnstone, and Morrison have been released on commercial records, cassettes, and CDs. Several internet sites offer easy access to program episodes, pulp novel texts, and splashy magazine covers. *The Shadow* persists, too, as a shuddering delight in the memories of those who knew radio before they knew television, when Sunday evening meant sharing Lamont Cranston and Margo Lane's adventures among the obsessively evil and the picturesquely insane. With *The Lone Ranger, Amos 'n' Andy,* and a very few others, *The Shadow* remains an essential figure of golden age radio.

RAY BARFIELD

See also Drama, U.S.; Mercury Theater of the Air; Mutual Broadcasting System; Quiz and Audience Participation Programs; Violence and Radio

Cast

The Shadow/ Lamont Cranston	James La Curto (1930; 1934–35), Frank Readick, Jr. (1930–35), Robert Hardy Andrews (1932), Orson Welles (1937–38; 1939 repeats); Bill Johnstone (1938–43), Bret Morrison (1943–44; 1945–54), John Archer (1944–45), Steve Courtleigh (1945)
Margo Lane	Agnes Moorehead (1937–40), Marjorie Anderson (1940–44), Marion Sharkley (1944), Laura Mae Carpenter (1945), Leslie Woods (1945–46), Grace Matthews (1946–49), Gertrude Warner (1949–54)
Announcer	Ken Roberts (1931–32; 1935; 1937–44), Dell Sharbutt (1934), Don Hancock (1945–47), Andre Baruch (1947–49), Carl Caruso (1949–51), Sandy Becker (1951–53), Ted Mallie (1953–54)

Programming History

CBS	1930–32
NBC	1932–33
Mutual	1937–54

Further Reading

Brower, Brock, "A Lament for Old-Time Radio," *Esquire* 53 (April 1960)
Gibson, Walter Brown, and Anthony Tollin, editors, *The Shadow Scrapbook,* New York: Harcourt Brace Jovanovich, 1979
Harmon, Jim, *The Great Radio Heroes,* Garden City, New York: Doubleday, 1967
Harmon, Jim, *Radio Mystery and Adventure: And Its Appearances in Film, Television, and Other Mediums,* Jefferson, North Carolina: McFarland, 1992
Settel, Irving, *A Pictorial History of Radio,* New York: Citadel Press, 1960
Stedman, Raymond William, *The Serials: Suspense and Drama by Installment,* Norman: University of Oklahoma Press, 1971; 2nd edition, 1977

Shaw, Allen 1943–

U.S. Radio Programmer and Station Owner

Known first for the "Love" format on American Broadcasting Companies (ABC)–owned FM stations in the late 1960s and early 1970s, Allen Shaw was instrumental in the success of several major broadcast groups and became a station owner. During his career, Shaw was responsible for the operations of 43 radio stations in 31 markets for five different companies.

Origins

Radio attracted Shaw early. At age 11 he experimented with electronics, building transmitters and communicating via Morse code. A chance visit with his father to a local radio station turned young Shaw from aspiring engineer to aspiring air talent and programmer. The elder Shaw owned a building in Haines City, Florida, where 50-kilowatt WGTO-AM located its studios in 1956, and one weekend he asked his son if he would like to see the facility. "It was the first time I had ever seen a disc jockey at work," Shaw remembered, "and it looked magical." He recalled his comments to his father about being able to hear the disc jockey in the car and in the kitchen. The performance side of radio hooked the 13-year-old.

WGTO was the first station licensed to Shaw's hometown, but it was not to be his first radio job. "They were a professional station; they hired good people," he said self-effacingly. In 1959 an "owner operating on a shoestring" established 500-watt WHAW-AM in Haines City, and, to use Shaw's words, "I was available for a buck an hour." An after-school job at WHAW launched Shaw's career at age 15.

In college he pursued radio, earning a degree in radio-television from Northwestern University in 1965. The proximity of Evanston to Chicago introduced Shaw to the legendary WCFL-AM, one of the two battling Top 40 stations in the Windy City at the time.

Radio Career

After college, Shaw moved to Albany, New York, for a brief stint as a disc jockey at WPTR, working the 7 to 11 P.M. shift. As Shaw told author Michael Keith in *Voices in the Purple Haze,* "I was supposed to be playing the top-selling hits of the day, such as 'Everybody Loves a Clown' by Gary Lewis and the Playboys and 'Lover's Concerto' by the Toys. I found myself slipping in unapproved cuts off the Rolling Stones, Beatles, and Bob Dylan albums. This was, of course, breaking the format rules of the station, a clear act of revolutionary protest." WPTR's program director told Shaw—"in no uncertain terms," as Shaw put it—that the public would decide through the purchase of 45-rpm singles what the station would play. Shaw, however, was hooked on the new music being released on long-playing albums.

A year later, in 1966, he was asked back to Chicago to become assistant program director of WCFL-AM. That station, like WPTR, was true to the top 40 singles. But Shaw listened at home to promotional copies of new rock albums by bands unfamiliar to him—Vanilla Fudge, Velvet Underground, and the Jimi Hendrix Experience.

"I found these sounds very interesting and exciting," he said, and he asked WCFL management if he could play some of them on the air. They agreed to one hour on Sunday nights. As Shaw described it, "Underground radio had come to Chicago, if only an hour a week and in monophonic sound."

During 1967 two events deepened Shaw's interest in the new music, called both "progressive rock" and "underground" at the time. The first was a radio conference featuring Tom Donahue, a former Top 40 disc jockey who became known as the "father of underground radio." Donahue's KMPX-FM, San Francisco, played an eclectic mix of blues, folk, and rock without structure or a playlist. The other event was a concert in Chicago where Lou Reed's band, Velvet Underground, opened for Jimi Hendrix. Shaw talked to Reed about the growing interest in the new music in New York and other cities. "[Reed] assured me that I'd be certain to succeed if I put this music on FM full-time," Shaw said.

That encouragement set Shaw working nights and weekends with WCFL colleague and former college friend George Yahraes on sample music tapes and graphics for what they decided to call the "Love" format. The two presented their ideas to owners of Chicago FM stations to no avail. Then they went to New York to meet with major radio networks. ABC Radio gave them not only a hearing, but also an invitation to make their presentation at the annual ABC managers' meeting in February 1968. The new format was so well received by ABC managers that by June of that year Shaw was director of FM Special Projects at ABC.

On 28 February 1969, ABC launched the Love format on FM stations it owned in New York, Los Angeles, San Francisco, Detroit, Houston, and Pittsburgh. Tapes were produced in New York and shipped to the other stations for playback. Of the period, Shaw said, "We knew at ABC-FM that we were, as unlikely as it was, the largest corporate entity broadcasting the drumbeat of the flower-children tribe over powerful FM stations. . . . There was a certain amount of headiness in our attitude." The success of the Love format prompted ABC to name Shaw—at age 26—president of ABC-Owned FM Stations, a post he held from 1970 until 1980.

The executive experience he gained at ABC made Shaw valuable to other broadcast companies, too. In 1981 he was named executive vice president of radio for Summit Communications, based in Winston-Salem, North Carolina. His move to that city would turn out to be a long-term affair, as he remained in Winston-Salem when he was appointed executive vice president and chief operating officer of the North Carolina–based Beasley Broadcast Group.

Shaw moved into station ownership as president and chief executive officer (CEO) of Crescent Communications, which acquired KYLD-FM in San Francisco in November 1993. The company expanded with two additional stations in the San Francisco–San Jose market and acquisitions in Albuquerque, New Mexico and Las Vegas, Nevada. The aggregate purchase price for all the stations was $63 million, and they were sold three years later to four separate buyers for a total of $135 million.

The experience at Crescent allowed Shaw to build yet another company, Centennial Broadcasting, of which he was owner, president, and CEO. He understood the increasing value of scarce radio properties and in 1997 built clusters with three FM stations in Las Vegas and two FMs and an AM in New Orleans. A programmer at heart, Shaw developed yet another format by modernizing the adult standards format to make it attractive to the older baby boom audience. His standards format became the audience leader in Las Vegas on KJUL-FM. In early 2001 Centennial was sold to the Beasley Broadcast Group, and Shaw rejoined Beasley's corporate team.

ED SHANE

See also Underground Radio

Allen Shaw. Born in Haines City, Florida, 24 September 1943. Attended Northwestern University, Evanston, Illinois, degree in radio/television, 1965. Began radio career at WHAW-AM, Haines City, while in high school; after college, worked as a disc jockey, WPTR-AM, Albany, New York, 1965–66;

assistant program director, WCFL-AM, Chicago, 1966–68; developed FM Rock "Love" format on ABC Radio's FM stations, 1968; program director, ABC Owned FM Stations, 1968–70; president, 1970–80; executive vice president, Radio Division, Summit Communications, Winston-Salem, North Carolina, 1981–85; executive vice president and chief operating officer, Beasley Broadcast Group, 1985–93; president and chief executive officer, Crescent Communications, 1993–96; president and chief executive officer, Centennial Broadcasting, 1997–2002; rejoined Beasley Broadcast Group, 2002. Active on the Boards of Directors of the Radio Advertising Bureau, the Broadcast Pioneers Library, and Goodwill Industries of N.W. North Carolina.

Further Reading

Keith, Michael C., *Voices in the Purple Haze: Underground Radio and the Sixties,* Westport, Connecticut: Praeger, 1997

Keith, Michael C., compiler, *Talking Radio: An Oral History of American Radio in the Television Age,* Armonk, New York: Sharpe, 2000

Ladd, Jim, *Radio Waves: Life and Revolution on the FM Dial,* New York: St. Martin's Press, 1991

Shepard, John 1886–1950

U.S. Radio Pioneer

John Shepard was an important New England radio broadcaster who helped pioneer many of the medium's program and technology innovations. Little remembered today, his was a name to be reckoned with in the first three decades of radio broadcasting.

Origins

Born into a family of Boston merchants, Shepard did not set out to work in radio. His grandfather had founded a department store called Shepard, Norwell & Company in the 1860s; it later expanded and was re-named the Shepard Stores. John Shepard III followed in his grandfather's and father's footsteps by joining the business and soon became manager of the store in Boston while his brother Robert managed the Providence store.

Early Boston Radio

With his ability to spot a sales trend, Shepard noticed during the 1920s the growing popularity of amateur radio and the first broadcast stations, and made sure to carry equipment and receiving sets in his store. He became friendly with several of the amateurs, and when they began working with newly emerging broadcast operations, Shepard became even more intrigued with the medium's potential. Convinced that radio was no passing fad, he planned a station in the Providence store—WEAN, which was licensed in June 1922; and by the end of July, a Boston station, WNAC, was up and running from the 4th floor of the Boston store. WNAC became so identified with the department store that in its first few months, virtually nobody knew its call letters; even local newspapers referred to it only as the "Shepard Stores station." That was fine with John Shepard, who believed the novelty of a radio station would bring people into his store to watch a broadcast under way, and then, he hoped, do some shopping. In WNAC's first few months, he even helped with some of the announcing, calling himself merely "J.S." after the under-spoken fashion of the day.

All of the initial announcers were Shepard store employees, and the entertainment was derived mainly from music schools or local entertainers who wanted some radio exposure. But as the station continued to gain new fans, Shepard did something unusual for the time: he hired some well-known performers, and he paid them to be on the air. In the early 1920s, most stations had no budget for talent, so they depended on volunteers. Shepard was determined to run his station as a business, however, and that meant paying performers. While good for the performers, this move soon proved unfortunate for smaller stations like Boston competitor WGI, which, with few dollars for talent, soon lost many of its most popular musicians.

Shepard quickly became one of radio's most passionate fans. When the National Association of Broadcasters was founded in 1923, he became the group's first vice president and a member of their board of directors. Later that year, he was one of a small group of owners and executives who met with President Coolidge to discuss the future of radio, and in subsequent years, he was part of delegations that spoke with congressmen about up-coming legislation which might affect radio.

Shepard had great respect for radio's ability to spread ideas. He invited members of the clergy from all the major faiths to give inspirational talks at a time when diversity was not commonly accepted; his station was the first in Boston to air synagogue services. His interest in politics led to broadcasting some very contentious political debates, and also led to accusations that he was allowing the candidates he liked more air time than the ones he did not. Of course, Shepard was also devoted to radio because of its ability to sell products; his staff became known for creative copywriting and memorable commercials.

Innovations

Shepard's station was the first in Massachusetts to put an African-American musical, "Shuffle Along," on the air, in November 1922. WNAC was also the first to use "house names." When "Jean Sargent" or "Nancy Howe" left his station, her replacement took the same name as the woman she succeeded. There is also some evidence that he aired the first play-by-play broadcasts of Boston Red Sox baseball.

In early 1927 Shepard experimented with what today would be called a home shopping operation. His new station, WASN (*A*ll *S*hopping *N*ews), was run almost entirely by women, several of whom went on to long careers in broadcasting. Though it turned out that home shopping's time had not yet come, it was only one of many innovations that came from Shepard. Although he never built a receiver nor invented anything, he knew how to spot a trend and hire the right people.

Shepard was among the first broadcasters to link two stations (he undertook an early version of networking with WEAF in New York City in January 1923). Periodically during the 1920s, he also linked WNAC with sister station WEAN to share programming. (In fact, a friendly rivalry ensued between the Boston and Providence stores, beginning with a road race between the two cities in the early 1930s, and culminating in the '40s with a weekly show called "Quiz of Two Cities.") From the successful linking of WNAC and WEAN, the Yankee Network was born; a local network that offered smaller New England stations the resources of both WNAC and WEAN, the Yankee Network first began signing up affiliates in the spring of 1930 and quickly became very popular.

Shepard helped to create a wider acceptance of broadcast journalism. He threw his support behind his own local newsgathering organization, in direct opposition to the Associated Press, which wanted to restrict radio news. Established in 1934, the Yankee Network News Service won the right for local radio journalists to obtain press credentials. It was soon in direct competition with the region's newspapers, using the slogan "News while it IS news," a direct slap at the inability of newspapers to react instantly as radio could.

Yankee Network technical director Paul DeMars had become an early supporter of Edwin Howard Armstrong's FM research. When in 1936 Shepard saw what FM could accomplish, he too was won over. He invested more than a quarter million dollars on his own experimental FM outlets, placing W1XOJ on the air in 1939 near Worcester as the first FM station in Massachusetts and one of the first half-dozen such experimental outlets anywhere. Its 50,000-watt signal nearly blanketed all of New England. Programs were transmitted to the site 42 miles west of Boston by another FM transmitter near the main Yankee Network studios in downtown Boston. This was followed in late 1940 by a second FM transmitter, W1XER atop Mt. Washington in New Hampshire, thereby creating the world's first (occasional) FM network by early in 1941.

The Yankee Network applied, unsuccessfully, for an FM license in New York City. But despite that setback, Shepard was the first broadcaster to create daily programming for FM, and in 1941 he was the first to get FM programs sponsored. He continued to promote the benefits of the new technology by sponsoring demonstrations, as well as selling FM receivers in his stores.

A landmark legal ruling came about because of a Shepard station practice. A disgruntled ex-employee, Lawrence Flynn, contested Shepard's Boston license by demonstrating how Shepard allowed politicians he favored to receive extra air time and favorable editorial mention. Shepard received a reprimand but retained his station when the Federal Communications Commission's *Mayflower* Decision of 1941 forbid radio stations from editorializing.

Nor did Shepard focus only on radio. When Charles Francis Jenkins' innovations in mechanical television in Washington, D.C., showed promise, Shepard supported development of a station for Boston. W1XAY, with studios in Lexington, Massachusetts, was on the air from the summer of 1928 until it ran out of money in early 1930.

Shepard served in numerous executive positions with the National Association of Broadcasters throughout the 1920s and 1930s, and he was actively involved in Broadcast Music Incorporated; he also chaired the first organization for FM owners (FM Broadcasters Inc.), and served on the board of the Mutual Broadcasting System. In the mid 1940s, in rapidly declining health, he sold controlling interest in the Yankee Network to General Tire Company, staying with the company for a while before retiring in 1949. He died in June 1950.

DONNA L. HALPER

See also Editorializing; FM Radio; FM Trade Associations; Mayflower Decision; National Association of Broadcasters; Yankee Network

John Shepard III. Born in Boston, 19 March 1886. One of two sons of John Shepard Jr., owner of several department stores, and Flora Martin Shepard. Attended Brookline (Massachusetts) High School and the Wertz Naval Academy Preparatory School and then joined the family business. Founder and President of WNAC, Boston (on air 31 July 1922). Formed Yankee Network, 1930. One of the original members of the National Association of Broadcasters, serving as Vice President and later Treasurer. Helped pioneer FM radio, 1939–40. Helped create Broadcast Music Inc., was chairman of Board of Providence Shepard Store, on Board of Trustees at Suffolk College (Boston). Retired 1949. Died in Boston, 11 June 1950.

Further Reading

"Boston Stores Adopt Radio to Send Out Shopping News," *New York Times* (30 January 1927)
"Dedicate Yankee Exclusive AM-FM Studios; Galaxy of Celebrities to Attend," *Boston Post* (15 March 1942)
"Description of Station W1XOJ," *FM* 1, no. 1 (November 1940)
"Shepard Fights Press: Yankee Network Goes on the Air," *Variety* (6 March 1934)
"Shepard Obituary," *Boston Post* (12 June 1950)
"Station WASN Will Broadcast Latest Shopping News," *Boston Globe* (29 January 1927)
"We Pay Our Respects to John Shepard III," *Broadcasting* (August 1932)
"WNAC-Shepard Stores in the Hub City," *Radio Digest* (January 1925)
"Yankee Frequency Modulation about Ready," *Broadcasting* (June 1939)
"Yankee Sets Up Its Own 24 Hour News Service," *Broadcasting* (March 1934)

Shepherd, Jean 1921–1999

U.S. Radio Humorist and Monologist

His fans called him "Shep." Media guru Marshall McLuhan hailed him as the "first radio novelist." Like a modern-day Scheherazade, Jean Shepherd was a master storyteller who, with wit, tempered irreverence, and a gimlet eye for the minutiae of growing up, spun an inexhaustible supply of tales to a loyal following of late-night radio listeners.

He was born Jean Parker Shepherd 26 July 1921 in Hammond, Indiana, a steel-mill town just outside Chicago. During World War II, Shepherd served in the Army Signal Corps (an experience that provided fodder for a number of his stories) and briefly attended Indiana University. He began his career in entertainment as a performer at Chicago's Goodman Theatre.

Between 1950 and 1954, Shepherd was a disc jockey at WSAI in Cincinnati, doing live remotes from a restaurant called Shuller's Wigwam and hosting a nightly comedy show, *Rear Bumpers,* on WLW. In 1956 he moved to New York's WOR, where the *Jean Shepherd Show* broadcast for the next 21 years to an audience that swelled to as many as 100,000 listeners all along the eastern seaboard, courtesy of WOR's 50,000-watt clear channel signal.

The 45-minute show opened with the familiar racetrack bugle call that heralded his theme song (the "Bahn Frei Polka" by Edouard Strauss). Working without a script, Shepherd embellished tales of his boyhood years hanging out with pals "Flick," "Schwartz," and "Brunner" and of time spent in the army. Sometimes an entire show would be built around an absurd news story, and on occasion he read selections from favorite literary figures, such as poet Robert Service. No Shepherd tale ever proceeded in linear fashion: there were detours everywhere. He would go off on a tangent, digress, interrupt himself with an overlapping story, and then, even as his closing theme started to play, easily and logically tie all the loose ends together.

His narratives evoked nostalgia without cloying sentiment and were cynical without being destructive. Shepherd spoke of the ordinary, the remembered things—his mother standing at the kitchen sink in her stained chenille bathrobe making a meal that was always red cabbage, meatloaf, and Jell-O; the perpetual whining of a younger brother; his father's Blatz Beer burp. His delivery was conversational, punctuated by an occasional staccato burst of laughter; a chortle; a conspiratorial whisper; or a musical interlude, which could include anything from a kazoo solo to *kopfspielen* (musical sounds created by tapping on one's head) to recorded selections (most requested was "The Bear Missed the Train," a parody to the tune of "Bei Mir Bist Du Schön"). And as expression of the very apex of human triumph, he would utter the word "Excelsior!" ("Excellence!").

Shepherd's stories were richly detailed. Walking to school during an Indiana winter meant wearing "a 16-foot scarf wound spirally from left to right until only the faint glint of two eyes peering out of a mound of moving clothing told you that a kid was in the neighborhood." The exaggerated anticipation of waiting for the mailman to deliver a coveted radio premium (the *Little Orphan Annie* "Secret Decoder") was described as, "At last, after at least 200 years of constant vigil, there was delivered to me a big fat lumpy letter. There are few things more thrilling in life than lumpy letters. . . . Even to this day I feel a wild surge of exultation when I run my hands over an envelope that is thick, fat and pregnant with mystery." Many of the tales were written later as short stories for a variety of magazines, and several were collected into books.

Shepherd described himself as a humorist rather than a comic. "A humorist looks outward and sees the world," he said, "a comic looks inward and sees himself." His familiar manner combined with the intimacy of the medium encouraged almost cultlike devotion, prompting Shepherd to observe, "I had 5 million listeners and each thought he was the only one." Fans felt like they had a secret pact with him, as if he and they were the only ones in on a big joke—and sometimes they were. More than once he would tell his audience to crank up the volume on the radio and shout along with him, "Drop the tools, we've got you covered!" One evening, Shepherd encouraged listeners to leave their radios, go to a street corner in Manhattan, and just mill around. Thousands showed up—and so did the police, who had gotten reports of a mob gathering. But the WOR listeners had been advised to simply and quietly mill—and then go home.

The greatest prank ever played by Shepherd and his devotees was the celebrated *I, Libertine* hoax, in which he told listeners to go to their bookstores and ask for a nonexistent book called *I, Libertine.* Prompted by the sudden demand, booksellers frantically tried to locate the book. Articles began appearing about the publishing sensation—the *New York Times Book Review* even included the book in its list of newly published works. "Friends would call to tell me that they'd met people at cocktail parties who claimed to have read it," Shepherd recalled. When the hoax was finally revealed, one of the publishers who had been pursuing paperback rights to the "sensation" persuaded Theodore Sturgeon to actually write it (under the pseudonym of Frederick R. Ewing) based on Shepherd's idea. It is now considered a collectible volume.

A WOR staffer once commented, "nobody at the station worked with Shepherd, instead they tried to work around him." His working relationship with WOR tended to be scornful, even antagonistic, and Shepherd made little attempt to soften this contempt: when giving the station identification, he would say, "speaking of relics, this is WOR Radio." He was annoyed by the necessary interruptions imposed by the commercial break and would instruct the engineer to "hit the money button." His 21-year run on WOR Radio ended in April 1977.

Jean Shepherd
Courtesy AP/Wide World Photos

In the 1970s, the *Jean Shepherd Show* was syndicated nationwide to public radio and college campus stations, and for the next two decades Shepherd made a series of personal appearances, including Carnegie Hall and an annual Princeton University show. He also began a longtime collaboration with the Public Broadcasting Service and eventually became involved in feature films. The film *A Christmas Story,* which Shepherd cowrote and narrated, has become a holiday classic.

LAURIE R. SQUIRE

See also WOR

Jean Shepherd. Born in Hammond, Indiana, 26 July 1921. Attended University of Maryland, 1948, Indiana University, 1949–50, and Goodman Theater School in Chicago; served in U.S. Army Signal Corps during World War II; host, *Rear Bumpers*, WLW, Cincinnati, Ohio, 1950–54; host, *The Jean Shepherd Show*, WOR, New York City, 1956–77, nationwide syndication, 1975–77; worked in off-Broadway stage productions in the 1950s and 1960s; writer/narrator, *American Playhouse* series, PBS/WGBH, 1976, 1982–83, 1989; co-writer/performer, *New Faces of 1962;* one-man show, live broadcast, Limelight Café, New York City, 1962–68; one-man shows, Carnegie Hall, New York City, 1971–75, Princeton University, New Jersey, 1966–96; author, co-screenwriter and cameo appearance in film *A Christmas Story*, 1983. Received *Playboy Magazine* Humor/Satire Award; Mark Twain Award, International Platform Speakers Association, 1976; honorary doctorate, Purdue University, 1995. Died in Sanibel Island, Florida, 16 October 1999.

Radio Series

1950–54 *Rear Bumpers*
1956–77 *The Jean Shepherd Show*

Films

Silver Darlings, 1947; *Fame Is the Spur,* 1947; *Light Fantastic,* 1964; *Tiki Tiki,* 1971; *Lenny Bruce without Tears,* 1971; *A Christmas Story*, 1983; *My Summer Story* (aka *It Runs in the Family),* 1994

Stage

Voice of the Turtle, 1961; *New Faces of 1962,* 1962

Selected Publications

America of George Ade, 1866–1944: Fables, Short Stories, Essays, 1960
Night People's Guide to New York: A Darien House Project, 1965
In God We Trust All Others Pay Cash, 1966
Wanda Hickey's Night of Golden Memories and Other Disasters, 1971
Ferrari in the Bedroom, 1972
Phantom of the Open Hearth: A Film for Television Co-ordinated by Leigh Brown, 1978
Living Thoughts, 1980
Fistful of Fig Newtons, 1981

Further Reading

Jaker, Bill, Frank Sulek, and Peter Kanze, *The Airwaves of New York: Illustrated Histories of 156 AM Stations in the Metropolitan Area, 1921–1996,* Jefferson, North Carolina: McFarland, 1998
Miller, Mark K., "Humorist Jean Shepherd Dies at 78," *Broadcasting & Cable* (25 October 1999)
Ramirez, Anthony, "Jean Shepherd, a Raconteur and a Wit of Radio, Is Dead," *New York Times* (18 October 1999)
Sadur, Jim, "FLICK LIVES! Jean Shepherd Web Site," <www.keyflux.com/shep/shepmain.htm>

Shirer, William L. 1904–1993

U.S. Radio Journalist and Author

William L. Shirer was an accomplished newspaper foreign correspondent when in 1937 he became an important member of the Columbia Broadcasting System (CBS) radio news team assembled by Edward R. Murrow. From the late 1930s into the early years of World War II, his was the voice people often heard from Berlin, Vienna, and other European cities. Shirer continued his career as a radio commentator until 1949 and later published a number of influential books growing out of his reporting experience.

Early Years

Shirer was born in Chicago, son of a crusading U.S. attorney who died when the boy was only nine. His mother moved William and his brother and sister to Cedar Rapids, Iowa, where in 1921 Shirer entered the denominational Coe College and graduated in 1925, having worked on the school paper. He worked his way to Europe on a cattle boat, arriving in Paris, which was then at its peak as an arts and literary hot spot.

Shirer's journalistic career began with a stint as reporter on the *Paris Tribune,* a European arm of the *Chicago Tribune,* from 1925 to 1927. His reporting of Lindbergh's flight to Paris in 1927 landed him a spot as a foreign correspondent for the parent *Chicago Tribune* that year but he was laid off in a 1932 downsizing. He was able to land a job, albeit one at a lower status, as a copy editor for the *New York Herald* Paris office that lasted for a couple of years. His next post was as the Berlin-based correspondent for Hearst's Universal News Service beginning in 1934. His bad luck continued, however, and in 1937 he was laid off from that post as well.

Radio Reporting

Now Shirer's luck turned around. On the very day he lost the Universal position, he received a cable from Edward R. Murrow, then CBS director of "talks" based in London. Murrow, seeking seasoned reportorial help as he developed a European news team, took Shirer to dinner at the fashionable Adlon Hotel in Berlin and asked him to join CBS at the same salary Shirer had been earning (Murrow did not know Shirer had lost his job). With few prospects and a pregnant wife, Shirer accepted.

While he had extensive journalism and European experience and was multi-lingual, Shirer did not have an obvious "radio voice," especially when compared with Murrow. Nonetheless, after a rather shaky initial few reports for CBS from Berlin (CBS news director Paul White was less than overwhelmed), Shirer became the first member of what years later would be known as "Murrow's Boys," a team of reporters that would all but dominate radio's coverage of rising European tensions and eventually war.

After several frustrating months of fairly minor reporting chores, Shirer hit his stride with an eyewitness account of Hitler's takeover of Austria in early 1938. Unable to broadcast from Vienna because of street barricades, Shirer flew to London (not an easy thing to do at the time) and broadcast to CBS in New York by means of shortwave. On 13 March 1938 Shirer led off the first multi-city CBS "World News Roundup" (although the program title was created later) with reporters in several European cities. Months later he reported on the growing Czech crisis from both Prague and Berlin, again using shortwave links. Throughout this period, Shirer and Murrow were close personal as well as professional colleagues, working virtually as equals.

Perhaps his finest radio news coup came in June 1940 when, by carefully reading what was going on and ignoring German news officials, he was on hand to report the French surrender at Compiègne, north of Paris. His report, carried on both CBS and the National Broadcasting Company (NBC), beat the official news release by some six hours. As censorship became tighter in Berlin, Shirer found himself unable to report the full story from the German capital city, and at the end of the year, he traveled home to New York to arrange for publication of his diary.

Shirer's *Berlin Diary* was a huge best-seller in 1941, eventually selling half a million copies. He went on the lecture circuit and did well. For the rest of the war, he reported from the United States, with two brief trips, one to London and one to Paris. He and Murrow, the latter angry that Shirer had never returned to his European news haunts, began to grow apart. Beginning in 1944 he had a regular, sponsored, Sunday evening news commentary program on CBS and developed a considerable following.

William L. Shirer
Courtesy Library of American Broadcasting

Murrow became Shirer's boss when the former became a CBS vice president in charge of the network's news operations. The tension between the two men grew as Shirer often rejected or ignored Murrow's suggestions for reports or news approaches. In March 1947 Shirer lost his program's sponsorship, weakening his position within the CBS news hierarchy, and costing him about $1,000 a week in lost sponsor fees. Although the reasons were never made clear, Shirer became convinced he was dropped for his liberal views in a time of increasing concern about communism in the United States. Others argued that he had become lazy and insufferable, making the rupture inevitable. The end of Shirer's program was widely reported and caused a storm of protest against (but not within) CBS, some of it aimed at Murrow. Shirer resigned from the network, well aware he no longer had a future there. Although he offered a weekly radio commentary on the Mutual network for two years, that stopped in 1949 and Shirer never worked full-time in broadcast news again.

Later Years

Shirer felt betrayed by Murrow and CBS and was bitter about it for the rest of his life. He wrote two nonfiction books on postwar Europe and two novels, neither especially successful, but one of them, *Stranger Come Home* (1954), was felt by many who knew both men to be starkly critical of Murrow.

The last period of Shirer's life began with the publication and huge success of his *Rise and Fall of the Third Reich* (1960), a history of the German Third Reich. Drawing on his direct experience (and personal acquaintance with many of the German leaders) and research, it became a Book of the Month selection, went through 20 reprintings in the first year alone, sold millions of copies, and has remained in print for decades. Once again Shirer was successful, respected, and financially well off.

For the last quarter century of his life, Shirer lived in the Berkshires of western Massachusetts and devoted himself to his writing, regularly turning out nonfiction books, several of which did quite well. Among them was a three-volume autobiography. His last book, a biography of Leo Tolstoy and his wife, appeared just a year before his death at age 89. Seven years later, transcripts of many of his Berlin broadcasts from 1934–1940 were published, allowing a new generation to appreciate his commentary and insight.

CHRISTOPHER H. STERLING

See also Columbia Broadcasting System; Murrow, Edward R.; News; White, Paul

William L. Shirer. Born in Chicago, Illinois, 23 February 1904. Attended Coe College, Cedar Rapids, Iowa, B.A. 1925, and Collège de France, Paris, 1925–27; reporter, *Paris Tribune*, 1925–27; foreign correspondent, *Chicago Tribune*, 1927–29; chief of Central European Bureau, *Chicago Tribune*, Vienna, 1929–32; European correspondent, *New York Herald*, Paris office, 1932–34; Berlin correspondent, Universal News Service, 1935–37; CBS radio continental representative, news correspondent, and commentator, 1937–47; columnist, *New York Herald Tribune*, 1942–48; broadcast commentator, Mutual Broadcasting system, 1947–49; full-time writer, 1950–93. Listed in *Red Channels*, 1950, blacklisted for five years; president, Authors Guild, 1956–57. Received Headliners Club Award, 1938, 1941; honorary doctorate, Coe College, 1941; George Foster Peabody award, 1947; National Book Award, 1961; Sidney Hillman Foundation Award, 1961. Died in Boston, Massachusetts, 28 December 1993.

Television Series

Nightmare Years, 1989

Films

Magic Face, 1951; *Rise and Fall of the Third Reich*, 1968

Selected Publications

Berlin Speaking, 1940
Berlin Diary: The Journal of a Foreign Correspondent, 1934–1941, 1941
Poison Pen, 1942
End of a Berlin Diary, 1947
Traitor, 1950
Midcentury Journey: The Western World through Its Years of Conflict, 1952
Stranger Come Home, 1954
Challenge of Scandinavia, 1955
Consul's Wife, 1956
Rise and Fall of the Third Reich: A History of Nazi Germany, 1960
Collapse of the Third Republic: An Inquiry into the Fall of France in 1940, 1969
Twentieth Century Journey (autobiography), 3 vols., 1976–90
Gandhi: A Memoir, 1979
Love and Hatred: The Troubled Marriage of Leo and Sonya Tolstoy, 1994
"This Is Berlin:" Radio Broadcasts from Nazi Germany, 1999

Further Reading

Cloud, Stanley, and Lynne Olson, *The Murrow Boys: Pioneers on the Front Lines of Broadcast Journalism*, Boston: Houghton Mifflin, 1996
Rosenfeld, G.C., "The Reception of William L. Shirer, Rise and Fall of the Third Reich," *Journal of Contemporary History* 29, no. 1 (January 1994)

Shock Jocks

During a blizzard in January 1982, an Air Florida passenger jet lifted off from National Airport in Washington, D.C., and almost immediately crashed into the Fourteenth Street Bridge, close to the end of the runway. The resulting tragic loss of life led to a program segment by radio shock jock Howard Stern. Stern called Air Florida personnel and asked for the price of a ticket from National Airport to the bridge.

Radio shock jocks purvey a radio format known as "shock radio," "raunchy radio," or "topless radio." Shock jocks such as New York's Howard Stern and Chicago's Erich Mancow are aptly labeled, for the goal of their program content is to appeal to a predominantly male audience using a panoply of sexual and scatological references and stunts. Listeners are barraged by sexually explicit references, cultural and ethnic attacks, off-color listener telephone calls, and sexually based interviews and antics. It is not unusual for Howard Stern and his on-air staff to cajole a female guest into disrobing as Stern details her anatomy for his listening audience. Other radio jocks associated with this program style have included Steve Dahl, Gary Meier, Jonathon Brandmeier, Danny Bonaduce, and Don Imus.

Although it is difficult to specifically date the onset of the shock jock format, the trend can be traced to the early 1970s. Prior to that date, the Federal Communications Commission (FCC) heard very few cases dealing with obscenity and indecency, and most objectionable programs were isolated in nature; for example, a station in North Carolina lost its license in 1962 for broadcasting blue material by country and western singer Charlie Walker. In 1972 a Detroit radio station was fined $2,000 for a sexual parody of a popular song.

The origin of the shock radio issue, however, might rest with WGLD-FM, Oak Park, Illinois, for airing a syndicated program entitled *Femme Forum.* This early morning broadcast contained an array of explicit sexual references, including a female listener's technique for oral sex. Popularly labeled "topless radio," the format subsequently led to the FCC's imposition of a $2,000 fine in 1973 for obscene broadcasting. The issue of indecency arose again in 1975 with the airing of George Carlin's comedy album *Occupation: Foole,* from which WBAI-FM, New York, broadcast the "seven dirty words" segment at 2 P.M. The FCC fined the station only a small amount but issued a ruling on the scheduling of adult-oriented and indecent broadcast material. Shock jock Howard Stern created the benchmark for such fines in September 1995, when Infinity Broadcasting agreed to pay almost $2 million to settle a number of FCC fines against Stern's programming.

Although *Newsweek* has termed New York's Don Imus a "*former* shock jock," many instead consider him to be the "*father* of shock jocks." The nature of his early broadcasting career confirms his iconoclastic approach. More recently, Imus has assumed the position of watchdog over the political and social arena, and his barrages of criticism have an acerbic edge. Imus' attacks on President Clinton and his verbal caricatures of Dan Rather and CNN's Wolf Blitzer, among others, have been widely reported. This may be a new arena for shock jocks.

Today's shock jocks continue to sustain a large youthful listening audience with programming challenging the limits of taste and legal suitability. The most notable is New York's Howard Stern, self-appointed "King of All Media." His program is a pastiche of sexual patter and risqué material. Although Stern's syndicated program has been canceled in several markets, he continues to command a large listening audience. Fans of Stern will find a myriad of webpages discussing his programming and personal life. On one website, "Howard Stern News Desk," Stern is quoted as remarking, "I always resented the term 'shock jock' that the press came up with for me . . . because I never intentionally set out to shock anybody. What I intentionally set out to do was to talk as I talk off the air, to talk the way guys talk sitting around a bar." Stern's rise as a shock jock is detailed in both his book and film, both entitled *Private Parts.*

Another popular shock jock, Erich "Mancow" Muller, has been successful in Chicago. His irreverent barrage of ethnic and cultural slurs, coupled with references to bodily functions, has led to a wide following in his markets. His broadcast sidekick, "Turd," provides on-the-street antics and interviews.

The longevity of shock jocks will depend on many factors, including continued audience acceptance and advertising support; intervening variables may include the role of self-regulation and the FCC's interpretation of regulations and statutes relating to obscenity and indecency. Historically, radio programming formats and personalities have changed with the whims of the radio audience. For a period in the late 20th and early 21st centuries, shock jocks occupied a popular position on the radio dial.

CHARLES FELDMAN

See also Imus, Don; Obscenity and Indecency on Radio; Seven Dirty Words Case; Stern, Howard; Topless Radio

Further Reading

Feldman, Charles, and Stanley Tickton, "Obscene/Indecent Programming: Regulation of Ambiguity," *Journal of Broadcasting* 20 (Spring 1976)

Holloway, Lynette, "A Contrite and Gentler Brand of Shock Radio, For Now," *New York Times* (16 September 2002)

Lipschultz, Jeremy, *Broadcast Indecency: F.C.C. Regulation and the First Amendment,* Boston: Focal Press, 1997

"Morning Mouth Mancow Muller's Animal House of the Air," *Chicago Reader* (31 March 1995)
Reed, Jim, *Everything Imus: All You Ever Wanted to Know about Don Imus,* Secaucus, New Jersey: Carol, 1999
Stern, Howard, *Private Parts,* New York: Pocket Books, 1994
Stern, Howard, *Miss America,* New York: Harper Paperbacks, 1995

Shortwave Radio

Most radio broadcasting takes place in the AM and FM bands (and, outside the United States, in the long wave band—148.5–283.5 kHz). However, the shortwave bands are home to domestic, regional, and international broadcasting as well.

Many different types of radio service use the shortwave frequencies, including maritime, aeronautical, data transmissions, and amateur radio. Within the shortwave spectrum of 1.7–30 MHz (often referred to as high frequency or "HF"), 14 bands have been allocated to shortwave broadcasting (ranges vary slightly in different parts of the world):

2.3–2.495 megahertz (120-meter band)
3.2–3.4 MHz (90-meter band)
3.9–4.0 MHz (75-meter band)
4.75–5.06 MHz (60-meter band)
5.9–6.2 MHz (49-meter band)
7.1–7.35 MHz (41-meter band)
9.4–9.9 MHz (31-meter band)
11.6–12.1 MHz (25-meter band)
13.57–13.87 MHz (22-meter band)
15.1–15.8 MHz (19-meter band)
17.48–17.9 MHz (16-meter band)
18.9–19.02 MHz (15-meter band)
21.45–21.85 MHz (13-meter band)
25.67–26.1 MHz (11-meter band)

The first three of these bands are referred to as the tropical bands and are generally reserved for broadcasting from countries in equatorial regions. The remaining "international" bands are open to everyone. A limited amount of shortwave broadcasting also takes place outside the designated bands.

Origins

The development of shortwave broadcasting traces its roots to early radio experimenters. Known as "hams," these amateur radio operators would utilize their radio equipment to communicate with each other, transmitting on any frequencies they chose.

Exploration of the shortwave bands began as a result of the Radio Act of 1912, which limited amateurs to operation above 1.5 MHz, bands then unexplored and thought to be of little value. All radio transmissions to that point (including those by the hams) had taken place *below* 1.5 MHz. What was widely thought to be ham radio's banishment to useless frequencies above 1.5 MHz proved to be its greatest asset, however, for the hams soon discovered that the reflective properties of the ionosphere made reliable radio transmission over great distances possible on the shortwave frequencies. Moreover, such transmissions could be accomplished with less power and smaller transmitting facilities than had theretofore been required for long-distance communication. (Ham radio operators continue to be an important element of the modern radio scene.)

Although Marconi had experimented with shortwave spark transmitters as early as 1901, the earliest shortwave broadcaster was Westinghouse engineer Frank Conrad, who experimented with the shortwave rebroadcast of Westinghouse station KDKA in 1921–22. These transmissions were heard in other countries, some of which rebroadcast the programs on their local standard broadcast stations. The long-distance capability of shortwave was a result of the reflective properties of a portion of the ionosphere known then as the Kennelly-Heaviside layer, so named after Britain's Sir Oliver Heaviside and Harvard professor Arthur Kennelly, who had independently suggested the existence of such a phenomenon.

By 1925 some large U.S. stations, including General Electric's WGY, Crosley's WLW, and RCA's WJZ, were simulcasting their regular AM programming on shortwave for experimental purposes. In the United States, it was hoped that shortwave could substitute for the long-distance cables needed to connect AM stations for network broadcasting. The cables were owned by the American Telephone and Telegraph Company (AT&T), which, as a competing broadcaster, was reluctant to lease the cables to others. When AT&T settled its disputes with the radio industry in 1926, ceased its own broadcasting, and agreed to lease its long-distance lines, a commercial rationale for reliance on shortwave as an adjunct to domestic broadcasting was lost.

Several objectives supported the use of shortwave for broadcasting purposes in radio's early days. Most important was basic technical experimentation: determining how far and how reliably signals could be transmitted, and at what times and frequencies. Another objective was to cover remote areas not easily reached by limited-range AM ("medium wave") signals, and thus provide news and entertainment to those not otherwise served by radio.

A third objective was to serve as a unifying national force. In the United States, broadcasting was essentially a private function. The government set technical standards and provided regulation on matters such as frequency, power, and hours of operation, but the broadcasters themselves were private enterprises. In most other countries, however, broadcasting was a government monopoly, and the government the principal broadcaster. (A modified American model was in force in Central and South America, where broadcasting was in mainly private hands, but with government broadcasting permitted as well.) Where the government controlled broadcasting, it was hoped that shortwave would be a useful tool for nation building. In some cases, geography dictated even loftier goals: Britain's "Empire Service," as its early international shortwave broadcasts were known, served as a means of communication with subjects in distant colonies.

The relaying of programs from one country to another by shortwave for rebroadcasting on local AM frequencies was another objective of early shortwave broadcasting. Although in theory such relays would provide a means of enriching local programming, the relays were usually small in scale until World War II, when rebroadcasting took on a propaganda objective. Both Germany and the United States entered into a large number of arrangements with South American stations for the rebroadcasting on local AM frequencies of programs delivered by shortwave.

The use of direct shortwave broadcasting—that is, shortwave programs intended for direct reception by listeners in another country, without local rebroadcasting—also proved a valuable propaganda technique for both sides during World War II. Although the reception of shortwave signals still required special equipment and some technical skill, it was not as complex as in the experimental days. In addition, the directional properties of shortwave—the ability to beam transmissions so as to maximize reception in specified geographic areas—lent itself to shortwave broadcasts specially targeted for particular parts of the globe. Thus Germany had its "U.S.A. Zone," and Italy and Britain had their North American services. The broadcasters made their programs more attractive by using the native language of the target audience.

After the war, the growth of shortwave broadcasting was largely a product of the propaganda needs of Cold War antagonists and the desire of many countries to have a place at the international broadcasting table and a voice that served national pride. Shortwave broadcasting maintained its usual shape, however: international services presented by government broadcasters in the listener's own language, at convenient times, and on multiple frequencies so as to provide the best reception in the target zone.

Although never as prevalent on the shortwave bands as the government broadcasters, private stations use shortwave as well. In some cases these are private AM stations simulcasting on shortwave simply to increase their range. Private religious organizations have also used shortwave as a means of transmitting their message worldwide. This has been a growing phenomenon, with many religious organizations boasting modern, high-power transmitting plants. First in this category was HCJB in Quito, Ecuador, which has been broadcasting religious programs worldwide since 1931. Most private shortwave broadcasting in the United States is by religious stations.

Although its roots can be traced as far back as the 1936–39 Spanish Civil War, jamming became a serious problem in international shortwave broadcasting after World War II. The intentional transmission of noise on or near the frequency of an offending station by a transmitter located in the listener's area was routine in communist countries and was effective in preventing the reception of unwanted broadcasts. One response to jamming was to broadcast on multiple frequencies in hopes that one would get through. This led to more jamming and rendered significant parts of the shortwave broadcasting bands useless. Jamming largely ceased in 1988–89.

Modern Era

A number of modern trends have affected shortwave broadcasting both positively and negatively. The solid-state revolution greatly simplified the reception of shortwave signals. Frequency drift, a problem for many years in vacuum tube receivers, has disappeared; direct dial tuning has eliminated guesswork in finding the desired frequency; and synchronous detection has improved fidelity. In addition, miniaturization has made possible even portable shortwave receivers with sufficient sensitivity to give good reception. Higher transmitter power has meant stronger shortwave signals, the elimination of jamming has reopened previously unusable band space, and increased international coordination has led to better frequency allocation and less interference.

However, shortwave broadcasting is often underfunded by the parent authorities, and this, along with the absence of a marketplace ethic, has often resulted in unexceptional programming. And despite improvements in the quality of shortwave receivers, several factors, including the superior fidelity of local AM and FM reception, the "second nature" operation of regular radios, the absence of shortwave on car radios, and the scarcity of widely available information on station schedules, have frustrated broad acceptance of shortwave. In addition, communication with faraway places is no longer a novelty. The ubiquity of information media such as cable TV and the internet makes shortwave broadcasting look quaint.

The number of shortwave broadcasters is on the decline. In addition to the loss of many local stations, some of the major broadcasters, including Radio Moscow, Radio Canada International, and the British Broadcasting Corporation (BBC), have reduced their output, sometimes drastically, and some countries have left shortwave broadcasting altogether. In some places, shortwave broadcast time has become a commodity, with high-power stations selling transmitter time to unrelated program producers (often in other nations). Although reliance on someone else's broadcasting facility is not without risk, purchase of airtime on a transmitter closer to the target zone can improve reception and eliminate the producer's need to maintain an expensive transmitting plant, while providing the transmitting station with additional revenue. With the demise of the Soviet Bloc and the introduction of market forces, Western stations have been able to purchase airtime in places that were previously off-limits. Thus, a religious station such as Trans World Radio can be heard broadcasting from Albania and the former Soviet Union. "Freedom" programs also rent time on transmitters in countries not directly related to the broadcaster's underlying message; for example, the Democratic Voice of Burma purchases time on transmitters in Norway and Germany, and the Voice of Tibet has broadcast from the Seychelles.

Notwithstanding the hurdles that shortwave broadcasting has faced, its oft-predicted demise does not appear imminent. Experiments in the use of the single-sideband transmitting mode and digital shortwave broadcasting continue the effort to improve signal quality, lessen interference, and reduce the power necessary to push broadcast signals around the globe on shortwave.

JEROME S. BERG

See also BBC World Service; Cold War Radio; Ham Radio; International Radio Broadcasting; Jamming; Propaganda by Radio; Radio in the American Sector; Radio Free Asia; Radio Free Europe/Radio Liberty; Radio Luxembourg; Radio Martí; Radio Monte Carlo; Radio Moscow; Radio Sawa/Middle East Radio Network; Religion on Radio; Vatican Radio; World War II and U.S. Radio

Further Reading

Berg, Jerome S., *On the Short Waves, 1923–1945: Broadcast Listening in the Pioneer Days of Radio,* Jefferson, North Carolina: McFarland, 1999
Browne, Donald R., *International Radio Broadcasting: The Limits of the Limitless Medium,* New York: Praeger, 1982
Hale, Julian, *Radio Power: Propaganda and International Broadcasting,* Philadelphia, Pennsylvania: Temple University Press, 1975
Magne, Lawrence, editor, *Passport to World Band Radio,* Penn's Park, Pennsylvania: International Broadcasting Services (annual; 1984–)
Sidel, Michael Kent, "A Historical Analysis of American Short Wave Broadcasting, 1916–1942," Ph.D. diss., Northwestern University, 1976
Wood, James, *History of International Broadcasting,* vol. 1, London: Peregrinus, 1992; vol. 2, London: Institution of Electrical Engineers, 2000
World Radio TV Handbook: The Directory of International Broadcasting, Oxford: WRTH Publications (annual; 1947–)

Siemering, William 1934–

U.S. Public Radio Executive and Producer

Author of the Statement of Purpose of National Public Radio (NPR) and NPR's first program director, Bill Siemering is widely regarded as the philosopher and conscience of contemporary public radio in the United States. He defined a future for public radio in the late 1960s based on a firm foundation in educational radio's traditions.

Origins

Siemering's parents had toured on the Chautauqua circuit in the late 1920s until radio killed the movement in the early 1930s. The Chautauqua shows toured small towns in the United States, bringing entertainment, culture, inspiration, and education to the largely isolated residents. When radio supplanted the touring shows, Siemering later believed, educational radio picked up their mission. When his parents settled in Wisconsin in 1934, rural Americans were receiving much the same sort of stimulation and diversion from the radio stations operated by state universities in the Midwest. Indeed, the family moved into a stone farmhouse literally in the shadow of the transmitting towers of the University of Wisconsin radio station, WHA, just south of Madison.

In his student days at the university, Siemering joined WHA's part-time staff and was immersed in the "Wisconsin Idea" of university outreach and public service. (He reminded public radio aficionados many years later that the first dictionary listing for "broadcast" refers to farmers throwing seeds in a wide circle, confident that some will take root but never certain about which ones or where.) This romantic rural vision of broadcasting met a gritty urban reality in 1962 when Siemering became manager of WBFO, the FM radio station of the State University of New York (SUNY) at Buffalo. He set out to do for a contemporary urban population what the traditional land-grant university stations had done for rural residents. Siemering saw the radio station as a potential link between the involved university and the community it served. "Public broadcasting can no longer be content as a refreshing cultural oasis," he said. "The emphasis must now be on solutions; how the individual can become politically involved to effect change."

To foster understanding between the races and to provide a voice for Buffalo's minority community, Siemering set up a storefront studio and provided airtime for volunteer broadcasters from poor neighborhoods. When the inevitable student strike to protest the Vietnam War hit the SUNY campus and 300 police moved in, he made the station a forum for discussion of the issues surrounding the strike. To the traditional authority of academics and journalists, Siemering added the unfiltered views of participants from all sides of the controversy. He saw a United States in which people were becoming compartmentalized, selecting those sources for their information that reinforced their existing beliefs. Public broadcasting, he said, should be an information source in which diverse groups could put their faith. He chastised both sides in the Vietnam War debate for avoiding meaningful discussion. Both sides, he wrote, feared that "discourse would modify, or in the eyes of some, weaken or compromise a position which has all the righteousness of a fundamentalist religion." The station's attitude was that all sides represented perspectives worth listening to: "We assumed they had a respectable point of view; our attitude was more like that of a counselor trying to have an individual share his perception of reality than an interrogating journalist."

Elected by fellow managers to the organizing board of National Public Radio in 1969, Siemering became the Thomas Jefferson of the new enterprise, capturing in eloquent words the thoughts of his colleagues, which he had done much to shape by his example in Buffalo:

> National Public Radio . . . will regard the individual differences among men with respect and joy rather than derision and hate; it will celebrate the human experience as infinitely varied rather than vacuous and banal; it will encourage a sense of active constructive participation, rather than apathetic helplessness. . . .
>
> The programs will enable the individual to better understand himself, his government, his institutions, and his national and social environment so he can intelligently participate in effecting the process of change. . . .
>
> The total service should be trustworthy, enhance intellectual development, expand knowledge, deepen aural aesthetic enjoyment, increase the pleasure of living in a pluralistic society and result in a service to listeners which makes them more responsive, informed human beings and intelligent, responsible citizens of their communities and the world.

Within a year, Siemering had moved to Washington as NPR's first program director, assigned the task of implementing his ideals, particularly in the form of NPR's initial offering, *All Things Considered*. Almost from the beginning, his philosophical approach clashed with bureaucratic necessities. Management seeking concrete plans were not reassured by phrases like, "Let's all hold hands and run a race." Yet it was exactly his humane philosophy that inspired much of the initial NPR staff and the emerging public radio family. He hired staff more on the basis of personal chemistry than professional experience, and his participatory management style relied on inspiration rather than direction. The results were both exciting and chaotic.

All Things Considered got off to a shaky start: occasionally brilliant, often less than competent. Ultimately the program found its stride, and it was at about that point when, to the shock of most staff and certainly Siemering himself, he was fired in December of 1972. The timing of the dismissal confounded him, although the underlying reason did not. From his first day on the job, his poetic operating style had bewildered and frustrated the NPR president and top management, who concluded after two years that his style would not change. Despite his rather brief tenure as head of programming, however, Siemering's influence on the institution endured. Many were inspired by his ideas and personality long after he departed. Even those who did not agree with his vision had to accept it as the point of departure from which NPR would evolve.

Siemering continued to work in public radio for 20 years after leaving NPR, holding leadership positions at Minnesota Public Radio and in Philadelphia, where he nurtured *Fresh Air* into a national program. Ultimately he was given what seemed a perfect assignment for him when he was made executive producer of the documentary series *Soundprint,* which was designed to showcase the best work of the most creative producers in the public radio system. The father of *All Things Considered* and *Fresh Air* had a third child, one in which quality and creativity would prevail, where his ability to inspire would find an appreciative audience. Even *Soundprint* faced practical realities, however, and after five years more pragmatic leadership replaced him.

In 1993, the MacArthur Foundation selected Siemering as recipient of one of its coveted Genius fellowships, providing him with a stipend to pursue his interests. He used the opportunity to work with the emerging democratic radio systems in South Africa, Asia, Eastern Europe, and the former Soviet Union. He continued his international work after the MacArthur fellowship ended and well into his "retirement."

JACK MITCHELL

See also All Things Considered; Fresh Air; Minnesota Public Radio; National Public Radio; Public Radio since 1967; Soundprint; Stamberg, Susan; Wertheimer, Linda; WHA and Wisconsin Public Radio

William Siemering. Born in Sun Prairie, Wisconsin, 26 October 1934. Attended University of Wisconsin, Madison, B.S. 1956, M.S. 1960; station manager, WBFO-FM, Buffalo, New York, 1962–70; became first director of programming, National Public Radio (NPR), 1970–72; noted for writing NPR's mission statement; founding member of NPR's board of directors; creator of NPR's *All Things Considered;* producer/reporter/manager, KCCM, Morehead, Minnesota, 1973–77; vice president of programming, Minnesota Public Radio, St. Paul, 1977–78; manager and vice president, WHYY-FM, Philadelphia, 1978–87; executive producer, *Soundprint,* WJHU-FM, Baltimore, Maryland, 1987–92; president, International Center for Journalists. Received Edward R. Murrow Award, 1986; MacArthur Foundation "Genius" Fellowship, 1993–98.

Further Reading

Collins, Mary, *National Public Radio: The Cast of Characters,* Washington, D.C.: Seven Locks Press, 1993
Engelman, Ralph, *Public Radio and Television in America: A Political History,* Thousand Oaks, California: Sage, 1996
Looker, Thomas, *The Sound and the Story: NPR and the Art of Radio,* Boston: Houghton Mifflin, 1995
Stamberg, Susan, *Every Night at Five: Susan Stamberg's All Things Considered Book,* New York: Pantheon, 1982
Wertheimer, Linda, editor, *Listening to America: Twenty-Five Years in the Life of a Nation, As Heard on National Public Radio,* Boston: Houghton Mifflin, 1995

Siepmann, Charles A. 1899–1985

U.S. (British-Born) Radio Critic, Author, and Educator

Co-author of the Federal Communications Commission's (FCC) controversial "Blue Book" program policy statement and author of two important postwar books on American radio broadcasting, Siepmann was a strong advocate for exploiting the educational and public service potential in radio and later in television.

Charles A. Siepmann was born and educated in England and spent most of his first four decades in that country. After an education in the classics (interrupted by service with the British Army's Royal Field Artillery on the Italian front in 1917–1918), he graduated from Oxford's Keble College in 1922. While in college and soon thereafter, he worked with Brown Shipley near London. For three further years, he was a housemaster at a British Prisons reform school for delinquent boys.

The important turning point to the second phase in Siepmann's career—practical experience in radio broadcasting—came in the fall of 1927, when he joined the British Broadcasting Corporation (BBC), then just five years old and in only its second year of operating as a state-supported service. Under the firm direction of its director general, Sir John Reith, the BBC was already making an important impression in Britain and elsewhere in defining what public service radio could provide. Siepmann's next dozen years with the BBC, in a series of progressively more responsible roles, firmly defined his mindset concerning the positive benefits of public service radio broadcasting.

He began as a deputy to the director of the BBC's adult education section, rising to director in 1929. After the two sections were amalgamated, he became director of talks in 1932.

As Asa Briggs writes, "Under Siepmann's general direction, the reorganized Talks Branch settled down to plan some of the liveliest talks in the BBC's history" (see Briggs, vol. 2). After rising tensions on political and organizational matters in the Talks Branch, Siepmann was shifted to a newly created post, director of regional relations, in mid-1935. Six months later, after a careful study, he produced an important assessment of regional transmissions "in which for the first time an official of the BBC fully explored the social and cultural aspects of regional broadcasting" (Briggs, vol. 2). This marked the beginning of his own understanding of and appreciation for local, rather than only centralized, program planning. For his last three years at the BBC, until 1939, Siepmann was director of BBC program planning and a member of the BBC's Control Board. His final role was to assist the BBC in planning for its informational and related roles in the event of what seemed an increasingly likely European war.

The remainder of Siepmann's life was spent largely in the United States. He had first visited the U.S. in 1937 with the support of a Rockefeller Foundation grant to study how U.S. universities utilized radio. He returned two years later as a lecturer and assistant to President James Conant at Harvard University, serving from November 1939 until he entered government service in 1942, the same year he became a naturalized American citizen. Siepmann's wartime government work applied his knowledge of radio to special wartime needs. He worked first with the radio division of the short-lived Office of Facts and Figures and then in various posts with the Office of War Information (OWI), concluding the war as director of OWI's Oriental Broadcasting Section, based in San Francisco.

Siepmann is perhaps best remembered today for what came next—several months' work as a consultant for the FCC in 1945. Working closely with FCC staff members, he authored some of the text of a landmark FCC report issued in March 1946: *Public Service Responsibilities of Broadcast Licensees,* which was quickly dubbed the "Blue Book" after its cover color. The Blue Book was a scathing critique of the program and advertising practices of selected commercial radio stations in the 1939–44 period and became the center of a long-lasting controversy between the industry and its regulators.

He paralleled his government work with an extended and more personal argument in *Radio's Second Chance* (1946), his first full-length book. Siepmann concluded that although the American system of radio was "basically sound" and had much to offer, it could do much more by offering a broader spectrum of content in addition to music and entertainment. He suggested the developing FM service as one means of reinventing American radio to serve a wider public interest, an image drawn in considerable part from his public-affairs program experience with the prewar BBC. He called for more talks and discussions, more local programs, and a clearer separation of advertising from programming.

In 1946 Siepmann moved into the third, longest, and final phase of his career, this time as a university academic. Appointed a professor of education, he soon became chairman of New York University's department of communications, a post he held for more than two decades. In addition to teaching and other university duties, he published widely in both research journals and magazines of opinion, writing about public policy concerns in both radio and television. He also consulted briefly with the Canadian government's Massey Commission, undertaking a content analysis of the country's network radio programming in 1949 that appeared as an appendix in the commission's final report two years later.

Siepmann's *Radio, Television, and American Society* (1950) became an influential college textbook that remained in widespread use for many years. Perhaps his definitive statement of what broadcasting was and might be within the U.S. context, the book's chapters were divided into two parts. The first several chapters were devoted to a description of the United States and other systems of broadcasting, and those in the second part dealt at considerable length with the "social implications" of broadcasting. Writing at the dawn of commercial television, he emphasized the likely impact of video on radio and other media and argued again that although the U.S. system was good, it could be made still better by providing a broader choice of programs, especially more public service content.

In the early 1950s, Siepmann's attention turned almost totally to television's role in education, which Siepmann made the subject of several articles in *The Nation* and elsewhere; of a UNESCO-published study (1952) describing early American efforts to develop television's educational potential; and of his last book, *TV and Our School Crisis* (1958), on how television might help the educational crisis, which won a Stanton award. On his retirement from New York University, he taught for a few additional years at nearby Sarah Lawrence College before his final retirement in 1971.

CHRISTOPHER H. STERLING

See also Blue Book; Critics; Education about Radio; Office of War Information; Public Service Broadcasting

Charles Arthur Siepmann. Born in Bristol, England, 10 March 1899. Attended Oxford University, B.A., 1922; served in British Royal Field Artillery, 1917–18; worked for Brown Shipley and Co., 1920–24; housemaster and education officer, British Prisons Service, 1924–27; worked for British Broadcasting Corporation, 1927–39; lecturer, Harvard University, 1939–42; became naturalized U.S. citizen, 1942; consultant and deputy director, Office of War Information, 1942–45; consultant, Federal Communications Commission (FCC), 1945–46; drafted FCC publication *Public Services Responsibilities of Broadcast Licensees,* 1946; chairman, Department of Communications, New York University, 1946–

68; chairman, New York Civil Liberties Union, 1956–60; professor, Sarah Lawrence College, 1968–71. Received Rockefeller Foundation Fellowship, 1937; first Frank Stanton Award for Meritorious Research on the Media of Mass Communications, 1958. Died in London, 19 March 1985.

Selected Publications

Radio in Wartime, 1942
Radio's Second Chance, 1946
Radio Listener's Bill of Rights: Democracy, Radio, and You, 1948
Radio, Television, and Society, 1950
Television and Education in the United States, 1952
TV and Our School Crisis, 1958

Further Reading

Briggs, Asa, *The History of Broadcasting in the United Kingdom,* London and New York: Oxford University Press, 1961– ; see especially vol. 2, *The Golden Age of Wireless,* 1965
Meyer, Richard J., "Charles A. Siepmann," *The NAEB Journal* (May–June 1963)
Richards, Robert K., "Siepmann Finds Flaws in U.S. Radio," *Broadcasting* (6 August 1945)

Silent Nights

Enabling Distant Listening

The concept of *silent nights* is unique to American radio broadcasting in the early 1920s. Imagine a time when radio stations went off the air for two days because a president had died: it happened in August 1923, when, out of respect for the late President Harding, numerous stations voluntarily left the air until after his funeral. In the early 1920s, radio stations might also take a day off for a major holiday. A station could shut down because normal hours of operation were still very limited. No stations broadcast 24 hours a day yet—in fact, few were on for more than 4 hours, usually just in the evening. Even if a local station were broadcasting, static and interference from other stations might keep listeners from enjoying the entertainment.

In 1920, when regular broadcasting began in the United States, there were only a handful of stations on the air, and they shared a common frequency—360 meters (about 833 kilohertz). Sharing time was not yet an issue. By mid-1922, however, there were over 150 stations, and to have them all share one frequency was impossible. Trying to offer a solution, the government opened another frequency—400 meters (750 kilohertz). It was left to government radio inspectors to seek compromise with the stations to determine the times during which each would broadcast. Broadcast historian Erik Barnouw notes that in Los Angeles there were as many as 23 stations sharing one frequency at any given time.

And then came the idea of silent nights. It probably seemed like a good idea at the time. Stations in a city would voluntarily remain silent for several hours or for an entire evening, thus enabling radio fans to listen in to other cities and receive distant stations. Some cities also used the silent night concept to reduce interference, because with so many stations occupying only two frequencies, reception was getting worse and worse. From late 1922 to the mid-1925, stations grappled with the problem and tried various versions of silent nights. Some cities (such as Boston) couldn't seem to agree on it, so stations would suspend operation for each other if one station had a special broadcast planned; individual stations also agreed to stay silent one night a week to allow other stations more time on the air, but gradually this plan was abandoned. In Minneapolis, according to the *Morning Tribune* (6 October 1923), a "quiet hour" was chosen rather than an entire evening, but this idea also ran into trouble when a major event occurred during the time that was supposed to be the quiet hour.

Silent nights seem to have received the most support in Chicago. According to Barnouw, by 1923 Chicago stations voted to have Monday evenings after 7 P.M. as their silent night. In fact, Chicago was quite organized in its efforts. The Chicago Broadcasters Association sent out a very detailed press release to the major newspapers in early October of 1925 explaining why member stations had voted to continue with their silent night even though many other cities had tried it and given up. The Chicago Broadcasters Association believed that local listeners wanted to hear stations in other cities, and they felt they could bring about goodwill by continuing to make this possible. Atlanta, Kansas City, San Francisco, and Dallas were other cities that experimented with a silent night.

Unfortunately, silent nights didn't solve the problem. Although having stations voluntarily go off the air pleased those people listening for distant stations (called "DXers") the overall problems of crowding and interference were not alleviated by one city's stations going silent for several hours once a week. Also, even in cities where one night was agreed upon, it didn't take long for one or more stations to refuse to cooperate, and soon things were back to where they had begun. The Department of Commerce opened more of what became the AM band in 1923 and 1924, but the result was that even more stations came on the air, and radio columnists were once again noting complaints from listeners regarding poor reception.

By 1926–27 network broadcasting and paid advertising were becoming more common, and stations could no longer afford a night without revenue. The idea of silent nights, as noble as it may have been, became increasingly rare. In late 1927 even Chicago finally abandoned it, with station owners admitting that as much as they wanted to allow fans to hear distant stations, they also wanted to make a living, and they couldn't do that by staying silent. The days of shutting down for a holiday or going off the air to benefit the DXers were over, for radio had become a business and would be run like one.

DONNA L. HALPER

See also DXers/DXing

Further Reading

Barnouw, Eric K., *A History of Broadcasting in the United States,* 3 vols., Oxford: Oxford University Press, 1966–70; see especially vol. 1, *A Tower in Babel: To 1933,* 1966
"Chicago Ends 'Silent Nights'," *New York Times* (16 November 1927)
"New York Studios Not Likely to Adopt Silent Night Plan," *New York Times* (14 June 1925)

Simon, Scott 1952–

U.S. Public Radio Journalist and Host

Scott Simon has since 1985 served as host of National Public Radio's (NPR) *Weekend Edition Saturday,* a two-hour morning show known for its offbeat take on the past week's news events intermixed with aurally distinctive long-form feature pieces.

Simon credits the Canadian Broadcasting Corporation (CBC), Edward R. Murrow's broadcasts, and his family's show business background as early influences on his career in radio journalism. While working for Chicago public television station WTTW, Simon began filing stories as a freelancer for NPR in the wake of the death of Chicago Mayor Richard J. Daley in 1976; he eventually became the full-time Chicago bureau chief for the radio network. Simon first came to national attention for his coverage of an American Nazi Party rally. Simon's report deftly edited the sounds coming from the rally organizers, the counterdemonstrators, and even the apathetic, creating for listeners a three-dimensional account of the 1978 Chicago event unrivaled by any other media report. Later assignments moved Simon to NPR headquarters in Washington, D.C., where he specialized in covering wars and human rights.

NPR created *Weekend Edition* in November 1985 as a showcase for Simon's talents and as a symbol of NPR's reemergence after a period of financial difficulty. The Saturday morning show quickly earned critical approval from adults dissatisfied with commercial television's competing lineup of programming designed for children. In an interview, Simon names three main factors that have given his show staying power over the years. First, the show is not news-driven but rather news-intensive. For example, the show is flexible enough to cover breaking news, but it is also more reflective, with longer features than NPR's daily radio news programs, such as *Morning Edition,* typically have. Second, Simon credits the regular cast of characters who contribute to the show, such as entertainment critic Elvis Mitchell and political news veteran Daniel Schorr. Finally, Simon says that staff of *Weekend Edition Saturday* do not run story ideas by focus groups but instead try to find interesting stories that are often overlooked by other media.

Independent observers also credit Simon's signature style as the key to the longevity of *Weekend Edition Saturday.* Like Garrison Keillor, Scott Simon's style of presentation is distinctive. His soft-spoken voice, somewhat high-pitched by commercial radio standards, is capable of conveying subtle emotional emphasis, especially when powered by his skillful writing, which colleagues have described as being among the best in broadcasting. Indeed, the show gives Simon a weekly slot for commentary, which is known within NPR as the "music cue" because of the music used as a bridge after the feature. Simon's commentaries often speak to overnight news developments, because he frequently works on his scripts until very close to airtime. In 1989 Simon won a George Foster Peabody Award for his radio essays and commentaries, which are often the subject of listener mail.

Simon's writing kindles audience members' attention in part because of his willingness to discuss his personal views. This attitude also fuels Simon's critics, who praise journalistic objectivity and see Simon's injection of his personality into the broadcast as being at odds with that aim. An example is Simon's Quaker faith: though raised in a Catholic–Jewish household, Simon is affiliated with the Society of Friends. His strong views on the death penalty, alcoholic beverages, and human rights are well known to listeners. While respecting the notion of objectivity, Simon believes it is important for journalists to have opinions and to be willing to stand publicly for issues of conscience. Addressing critics, Simon points out that he was a respected war correspondent and covered military engagements fairly despite the conflict with his religious beliefs.

In August 1992, Simon went on hiatus from NPR and hosted National Broadcasting Company (NBC) Television's revamped weekend *Today* morning programs. By all accounts, Simon's first shows were awkward and riddled with technical mistakes. Contributing to the problem were miscasting with his cohost and Simon's discomfort with the soft news features that the television show's producers wanted. After disagreements with management about the direction of the show, Simon returned within a year to host NPR's *Weekend Edition Saturday.* Since NBC, Simon's most frequent television appearances have been on the Public Broadcasting Service (PBS). In 1997 he narrated a documentary, *Affluenza,* reviews of which again focused on Simon's personality. One reviewer described *Affluenza* as Simon's sermon about Americans' wasteful consumption habits. To less controversy, Simon co-anchored segments of PBS's millennium celebration coverage.

Simon's fascination with Chicago's sports teams is a frequent topic of conversation on *Weekend Edition Saturday* and the subject of Simon's autobiographical *Home and Away: Memoir of a Fan.*

Despite his long association with public radio, Simon maintains he is not a "radiohead." He sees himself primarily as a communicator and remains loyal to public broadcasting because of his conviction that good journalism, like the best work of Edward R. Murrow, should challenge the audience. Reflecting on the state of contemporary commercial radio, Simon says NPR alone offers him that opportunity.

A. JOSEPH BORRELL

See also National Public Radio

Scott Simon. Born in Chicago, Illinois, 16 March 1952. Attended the University of Chicago; joined National Public Radio (NPR) as chief of Chicago Bureau, 1977; became host of NPR's *Weekend Edition Saturday,* 1985; took special leave from *Weekend Edition Saturday* to cover Kosovo War for NPR, 1999, and the war in Iraq, 2003. Received Unity Award in Media, 1978; Major Armstrong Award, 1979; Emmy for television documentary, 1982; Silver Cindy Award, 1986; Robert F. Kennedy Journalism Award, 1986; George Foster Peabody Award, 1989; DuPont-Columbia University Award, 1992; Presidential End Hunger Award.

Radio Series

1985– *Weekend Edition,* later known as *Weekend Edition Saturday*

Television Series

Weekend Today, 1992–93; *Coming and Going,* 1994; *Life on the Internet,* 1996

Selected Publications

Home and Away: Memoir of a Fan, 2000
Jackie Robinson and the Integration of Baseball, 2002

Further Reading

Austin, April, "Listening to What Simon Says," *Christian Science Monitor* (2 February 1989)
Collins, Mary, *National Public Radio: The Cast of Characters,* Washington, D.C.: Seven Locks, 1993
Fager, Charles, "Scott Simon: Friend on the Air," *The Christian Century* (10 December 1986)
Looker, Thomas, *The Sound and the Story: NPR and the Art of Radio,* Boston: Houghton Mifflin, 1995
Weinstein, Steven, "NPR's 'Weekend Edition': An Alternative to the Saturday Blahs," *Los Angeles Times* (2 July 1987)

Simulcasting

As television replaced radio in the 1950s, many previously successful network radio programs shifted to television. For a time, some television programs simultaneously aired their audio portion on network radio stations. Although radio listeners were sometimes annoyed by performers' references to images seen by the television viewer but invisible to the radio listener, the simulcast broadcast, with visual references, often provided a program incentive for the radio listener to purchase a television set. Music and variety programs were especially favored for radio simulcast because there was little or no loss of content for the radio listener. As television grew in popularity, simulcasting of television audio on radio declined.

A far more common form of simulcasting that went on for years involved duplication of AM programming on FM stations. As FM stations began going on the air in the 1940s, FM stations often provided 100 percent duplication of all network programs supplied to an AM sister station. Most new FM outlets were owned by and co-located with AM stations. The former nearly always carried the latter's programs. Broadcasters claimed they were trying to assist the new medium with popular programs from the old, but in fact the chief reason was that because of FM's small audiences, extra program expense made little sense. In reality, of course, simulcasting created little incentive for listeners to purchase an FM receiver. The public saw little advantage in buying an FM receiver to pick up programs they were already receiving from an AM station. About 80 percent of the FM stations signing on the air in the late 1940s were co-owned with an AM station—and most simulcast.

Duplication of programming by FM stations began to decline only in July 1964 when the first rules limiting such practice went into effect. The Federal Communications Commission specified that in markets of 100,000 residents or more, at least half the programming aired on an FM station had to be original. Full implementation of the nonduplication rule took place through the late 1960s, and the rule was extended to small-market stations by the 1970s. Although duplication did not always mean simulcasting, simultaneous delivery of programs from an AM station on an FM station was the most common form of duplication. The nonduplication rule also ended the practice of recording AM programs for playback in a nonsimulcast manner. Although widely criticized by broadcasters at the time (who would have to provide separate programming at considerable expense), the end of simulcasting soon provided a huge boost to FM popularity and a concomitant increase in demand for FM receivers.

In the late 1990s, simulcasting took on a very different meaning. Passage of the Telecommunications Act of 1996 allowed operation of multiple radio stations by a single owner in individual markets. Some group owners have used simulcasting to extend the reach of successful urban stations to additional stations in outlying suburban areas. This strategy has also been used to introduce news or talk programs found on AM stations to listeners accustomed to FM stations. Rather than hoping to convince listeners to change their car radio programming to the AM band, a simulcast allows the listener to find the same programming on FM.

A closely related use of simulcasting has been to create a local "network" by purchasing two or more stations, usually FM outlets licensed to communities outside a major metropolitan area, and to simulcast programs on both. When the stations simulcasting the signal are not of sufficient power to cover the market, the simulcast enables listeners to change to a sister "network" station as the listener travels from the coverage area of one station into the coverage area of another.

Simulcasting has also returned as a promotional vehicle for radio stations and music video channels. Radio stations have simulcast the audio of concerts or music videos appearing on cable program services such as Music Televison (MTV), Video Hits One (VH1), and Country Music Television (CMT). The simulcast enables the listener to enjoy the visual aspect of the program through cable while hearing a stereo broadcast of the same program via a radio station.

Simulcasting has also become a revenue source for some radio stations. These stations air the audio portion of a local television station's newscast either as paid programming or in exchange for station promotional mention by the television station. In some markets, group owners of radio and television stations in the same market regard simulcasting as a mutually beneficial promotion for both operations.

GREGORY G. PITTS

See also FM Radio; Programming Strategies and Processes

Further Reading

Albarran, Alan, and Gregory G. Pitts, *The Radio Broadcast Industry*, Boston: Allyn and Bacon, 2001

Eastman, Susan Tyler, Sydney W. Head, and Lewis Klein, *Broadcast Programming, Strategies for Winning Television and Radio Audiences*, Belmont, California: Wadsworth, 1981; 5th edition, as *Broadcast/Cable Programming: Strategies and Practices*, by Eastman and Douglas A. Ferguson, 1996

Smulyan, Susan, *Selling Radio: Commercialization of American Broadcasting, 1920–1934*, Washington, D.C.: Smithsonian Institution Press, 1994

Sterling, Christopher H., and John M. Kittross, *Stay Tuned: A Concise History of American Broadcasting*, Belmont, California: Wadsworth, 1978; 2nd edition, 1990

Streeter, Thomas, *Selling the Air: A Critique of the Policy of Commercial Broadcasting in the United States*, Chicago: University of Chicago Press, 1996

Singers on Radio

From the beginning of radio broadcasting, the singing voice has been a staple. While some early microphones had trouble adequately carrying powerful voices (the carbon powder would congeal, cutting off sound transmission), as technology improved, vocal music and musical variety became—and have remained—radio's program norm.

Origins

Many pioneer stations used local singers who craved audiences and would gladly perform free for the honor of singing on radio. Early announcers were often selected for their singing ability, as they could be called upon to fill unused air time at a moment's notice. (There were pianos in many early studios for just this purpose.) When WEAF dedicated its New York City studio on Broadway in 1923, the broadcast featured an assortment of singers ranging from opera stars to popular songsters. Indeed, well over 60 percent of radio air time in the 1920s was devoted to some form of music, often singing.

Reginald Fessenden engineered the first broadcast of a human voice—as distinguished from transmissions of Morse code—in December 1906. His broadcast included a recording of Handel's "Largo," a tenor aria from the opera *Serse*. Fessenden himself sang a Christmas song, thus becoming the first person to sing live on radio. Fessenden's audience was made up mostly of radio operators at sea in the North Atlantic off the coast of Brant Rock, Massachusetts. About two months later, in February 1907, vaudeville performer Eugenia Farrar became the first woman to sing live on radio when she performed "I Love You Truly" as part of a similar broadcasting experiment by radio pioneer Lee de Forest.

It would be another 13 years before radio broadcasting as it is known today had its true beginning. But from the earliest days of radio, listeners left little doubt about what they wanted to hear. For example, in 1922 listeners responding to a poll by WBAY (later WEAF) in New York City said that music was what they most enjoyed hearing on radio. But their tastes in music were widely divergent, just as they are today. Various factions wanted to hear dance music, symphony concerts, old-time ballads, religious hymns, and brass band selections, among other styles. One decision that played a large role in determining the type of music and singers to be heard on the radio came following World War I when Congress decided that radio broadcasting would be a commercial enterprise. Secretary of Commerce Herbert Hoover, who guided the fledgling industry in its earliest days, was aghast: "It is inconceivable that we should allow so great a possibility for service, for news and entertainment, for education and vital commercial purposes, to be drowned in advertising chatter." However, the decision to support the operation of radio broadcasting through advertising helped to ensure that most of the music broadcast on radio eventually would be popular in nature.

Early radio broadcasters were image-conscious, and operas played a major role in the content of early radio programs. Broadcasts by operatic vocalists and orchestras generated a wider appreciation for "fine music," often among those who previously had little interest in music of any kind. As classical music impresario Sol Hurok said:

> People who own sets look up programs to find out what is being broadcast. They read that an aria from *La Boheme* will be sung that night. They become interested and ask themselves, 'What is *La Boheme*?' . . . In this way an interest in music is created which is beneficial because all of the listeners are prospective attendants.

Evidence of this came from the sale of phonograph recordings. Early on, phonograph record manufacturers were hostile to radio, as the sale of records initially dropped noticeably following the emergence of radio. In response, the Victor Talking Machine Company (which would later become part of RCA Victor) kept almost all of its major artists off the air, reasoning that if listeners could hear singers free over the radio, no one would pay for their records.

But by late 1924, more visionary ideas had prevailed. Victor announced "the beginning of a new era in radio broadcasting." The company would feature its greatest recording artists in a series of radio programs. Every selection was, or soon would be, available on Victor's prestigious Red Seal label. On New Year's Day 1925, two of Victor's most popular singers,

Lucrezia Bori and John McCormack, performed an hour-long program on WEAF. The station estimated that 6 million listeners tuned in the broadcast, and within a week, listeners purchased more than 200,000 Bori and McCormack disks. The alliance between the recording industry and radio had been established.

Crooners

The phenomenon of the national pop music star had its beginnings on network radio. On 29 October 1929, only two days after the Wall Street crash known as "Black Tuesday," Rudy Vallee sang "My Time Is Your Time" for the first time as the host of *The Fleischmann Hour* on the National Broadcasting Company (NBC) network. Vallee's greeting, "Heigh-ho, everybody," had already become familiar to New York listeners on WABC and WOR. Soon fan mail was pouring in to NBC, so much so that the network bought the rights to "The Maine Stein Song" and gave it to Vallee to record. The song became a huge hit and earned what the *New York Times* described as "a small fortune" for NBC.

Vallee remained a popular entertainer well into the television era. He was the first well-known exponent of the singing style known as "crooning," which employed a soft, sensual style in contrast to the booming, straightforward manner of singing that had existed before electronic amplification. Crooning elicited strong attacks from traditionalists such as Cardinal William Henry O'Connell of the Boston Roman Catholic Archdiocese, who strongly denounced crooning in the January 1932 edition of *Literary Digest,* calling it

> Immoral and imbecile slush. A degenerate, low-down sort of interpretation of love. A love song is a beautiful thing in itself . . . But listen again with this new idea in your head and see if you do not get a sensation of revolting disgust at a man whining a degenerate song, which is unworthy of any American man. . . . It is a sensuous, effeminate luxurious sort of paganism. . . . Think of the boys and girls who are brought up with that idea of music.

But as author Thomas DeLong explains in *The Mighty Music Box* (1980), crooning was a natural consequence of technology: "In essence, it was an adaptation to the techniques of radio broadcasting. The highly sensitive mike demanded a different mode of vocal production. Singing into the delicate carbon microphones compelled artists to use soft, almost caressing tones lest a loud or high note shatter a transmitter tube."

The Columbia Broadcasting System (CBS) soon had a large stable of crooners, including Morton Downey, Will Osborne, Kate Smith, Ruth Etting, the Boswell Sisters, and Art Jarrett. The network featured one or two 15-minute song programs during every hour of evening programming. NBC had its own programs with Russ Columbo, Little Jack Little, Jack Fulton, Jane Froman, and the Pickens Sisters.

But crooning found its ultimate expression in Bing Crosby's vocal stylings. Crosby had already found some measure of success as a vocalist with the Paul Whiteman orchestra, singing in a group called the Rhythm Boys. As part of Whiteman's band, they had sung on a CBS program, but Crosby came to the attention of CBS's young president, William Paley, when he repeatedly heard a recording of Crosby and the Rhythm Boys performing "I Surrender, Dear" while he was on board a cruise ship. He personally signed Crosby to a CBS contract and gave him a nightly program opposite NBC's *Amos 'n' Andy* in the fall of 1931.

Singers Eclipse Dance Bands

Until this time, instrumental music performed by dance bands had enjoyed at least as much popularity on radio as vocal music. If a bandleader wished to include a vocal within an arrangement, one of the band members, often the leader himself, would step forward to sing a verse. The quality of the singing was of little concern. But after Bing Crosby shot to stardom, the singer of popular songs on the radio became as important as the dance band, and bands began to feature star vocalists. The big bands remained extremely popular, but they were fronted by popular singers such as Jo Stafford, Rosemary Clooney, Doris Day, Dick Haymes, Tex Beneke, and the young Frank Sinatra. Like Sinatra, who sang with the Tommy Dorsey band, many of the singers came to eclipse the popularity of their bands and went on to become stars in their own right. Crosby's interpretive ability also brought a new appreciation for the words of the song, and as time went on, a musical performance without the singing of a song's lyrics became increasingly rare.

Sinatra's popularity as a singer blossomed in 1943 and 1944 on *Your Hit Parade,* an NBC program that featured an ensemble of male and female singers performing the 15 most popular songs of the week. His ardent following soon surpassed those of Vallee and Crosby at their peaks. Sinatra returned for a second stint on *Your Hit Parade* from 1947 to 1949, by which time song stylists had come to dominate popular music on the radio. In a time when several recorded versions of a popular song might compete for popularity on the airwaves, the *Your Hit Parade* stable of singers, popular in their own right, would present their own versions of the songs.

Royalty Fight Brings Diversity

A struggle over the payment of royalties to composers of music broadcast on radio had the unintended consequence of bringing new musical styles to the medium. The American

Society of Composers, Authors, and Publishers (ASCAP) and the National Association of Broadcasters (NAB) clashed over a new contract that was to take effect in 1941. The broadcasters felt that the fees being demanded were far too high. When discussions over a new contract reached an impasse, the NAB created its own performing rights society, Broadcast Music Incorporated (BMI). Stations stopped playing ASCAP songs and played only BMI or public-domain tunes. BMI's creation had opened the airwaves to new styles of music. ASCAP represented the traditional popular-music composers of Broadway, Tin Pan Alley, and Hollywood. BMI opened the door to the composers of regional music—rhythm and blues, country and western, and eventually rock and roll. As a result, the talents of songwriters and singers such as Fats Domino, Chuck Berry, and Hank Williams eventually found a place on the air.

New Song Styles

Even before BMI opened the airwaves to new, sometimes earthier styles of music, black singers had been prominent on radio. Although civil rights leaders criticized the creators of radio comedies and dramas for allowing blacks to perform only in limited and often demeaning roles, numerous black singers performed on sponsored programs. As early as the 1930s, the Mills Brothers, the Ink Spots, and Louis Armstrong had network programs. Although many advertisers shied away from sponsoring programs featuring African-Americans, Ethel Waters had a program sponsored by the American Oil Company, and Paul Robeson was the featured singer on General Electric and Eastman Kodak programs. At the beginning of World War II in Europe, when Nazi oppression focused attention on the broader implications of racism and intolerance, Robeson sang the "Ballad for Americans" on the CBS series *Pursuit of Happiness.* He received a 15-minute ovation from the studio audience that continued after the program left the air. In his commanding bass voice, Robeson performed the song, which proclaimed the values of freedom and human rights, on radio programs several times during World War II. Gospel music sung by black choirs and smaller vocal groups such as the Southernaires also received a significant amount of airtime during this period.

But radio also perpetuated unflattering racial stereotypes with offerings such as the minstrel program *Plantation Nights* on KFI in Los Angeles, in which "the imaginary locale is an old Southern plantation where darkies come to serenade the owner." As with music sung by African-Americans, country and western music had a place on radio prior to the dispute with ASCAP. Network radio carried many Western programs that were especially popular with children. Many of these programs featured singing cowboys. Two of the most popular, Gene Autry and Roy Rogers, had long-running network radio programs. *The Grand Ole Opry,* a live, Saturday night barn dance style program from Nashville, debuted on WSM in 1925. *The National Barn Dance* on WLS in Chicago became the first such program to be broadcast to a national audience. Both featured a host of new singers on radio.

Postwar Radio

By the end of the 1940s, radio had a powerful new competitor—television. As network radio lost more and more of its audience to the new medium in the early 1950s, it also lost many of its biggest stars, such as Bing Crosby, to network television. Obviously, a new localized programming format was needed if radio was to survive. The answer was "Top 40," a format that first emerged in 1949 at KOHW in Omaha, Nebraska, and at KTIX in New Orleans in 1953 under the guidance of Todd Storz. The fast-paced format was based on repetitious playing of the most popular hit records of the moment, the "Top 40." Gordon McLendon is credited with giving the Top 40 format much of its brash, flashy quality at about the same time at KLIF in Dallas.

The foundation for the Top 40 format had been laid by WNEW in New York. Under the leadership of station manager Bernice Judis, WNEW had eschewed network programming, building its success on local news and the playing of popular records since going on the air in 1934. On one of its programs, *Make Believe Ballroom,* Martin Block created the illusion of a live performance in a dance hall with a "revolving bandstand," as he played records by various groups and singers.

In New Orleans, Storz's original Top 40 format originally was a broad-based, adult-oriented format featuring hits by traditional pop singers such as Perry Como and Patti Page. Gradually, as the 1950s unfolded, the new, energetic, sexually charged music called "rock and roll" began to replace such conventional performers. Soon the growing population of young people who were its natural constituency found a messiah in Elvis Presley. When Presley hit it big in 1956 with "Heartbreak Hotel" and "Don't Be Cruel," the rock and roll landslide soon hit Top 40 radio, which quickly became synonymous with the new music. The denigration of crooning 25 years earlier paled in comparison with the criticism of rock and roll singers. But Storz defended Top 40 radio's right to play rock and roll music in the May 1958 edition of *U.S. Radio*: "Our desire is that our stations shall please the majority of the people the majority of the time. . . . Our format was built on the premise that it is not within our province to dictate by censorship, programming tastes to the American public."

Many changes have occurred in radio programming since the Top 40 format emerged. FM radio grew in acceptance throughout the 1960s and 1970s. It eventually surpassed the popularity of AM radio, greatly increasing the number of stations on the air and encouraging the development of specialized music formats such as urban contemporary and alternative rock. But the voices of singers heard on radio, as during the Top 40 era, continue to emanate from popular recordings.

J. M. DEMPSEY

See also, in addition to individual formats discussed, Crosby, Bing; Formats; Grand Ole Opry; McLendon, Gordon; National Barn Dance; Recordings and the Radio Industry; Smith, Kate; Storz, Todd; Vallee, Rudy; Your Hit Parade

Further Reading

De Lerma, Dominique-Rene, editor, *Black Music in Our Culture,* Kent, Ohio: Kent State University Press, 1970
DeLong, Thomas A., *The Mighty Music Box,* Los Angeles: Amber Crest Books, 1980
Eberly, Philip K., *Music in the Air: America's Changing Tastes in Popular Music, 1920–1980,* New York: Hastings House, 1982
Ewen, David, editor, *American Popular Songs: From the Revolutionary War to the Present,* New York: Random House, 1966
Fong-Torres, Ben, *The Hits Just Keep on Coming: The History of Top 40 Radio,* San Francisco: Miller-Freeman Books, 1998
Kinkle, Roger D., *The Complete Encyclopedia of Popular Music and Jazz, 1900–1950,* 4 vols., New Rochelle, New York: Arlington House, 1974
Lonstein, Albert I., and Vito Marino, *The Compleat Sinatra,* Ellensville, New York: Cameron, 1970
MacDonald, J. Fred, *Don't Touch That Dial! Radio Programming in American Life,* Chicago: Nelson-Hall, 1979
Pleasants, Henry, *The Great American Popular Singers,* New York: Simon and Schuster, 1974
Routt, Edd, James B. McGrath, and Fredric A. Weiss, *The Radio Format Conundrum,* New York: Hastings House, 1978
Thompson, Charles, *Bing: The Authorized Biography,* New York: McKay, 1976
Tosches, Nick, *Country: The Biggest Music in America,* Briarcliff Manor, New York: Stein and Day, 1977
Vallee, Rudy, and Gil McKean, *My Time Is Your Time: The Story of Rudy Vallee,* New York: Obolensky, 1962
Williams, John R., *This Was "Your Hit Parade,"* Camden, Maine: Williams, 1973

Situation Comedy

Fibber McGee trying to get something out of his junk-filled hall closet without starting an avalanche; Amos 'n' Andy caught up in the Kingfish's latest scheme; Jack Benny considering his options when confronted by a mugger with the classic question, "Your money or your life"—these are but a few of the vivid memories from the "golden era" of radio situation comedy. With their offbeat personality flaws, idiosyncratic neighbors, and disrespectful domestic help, these characters were not just friends to their millions of listeners—they were "family."

Defining a Format

"Family" is, in fact, the linchpin of radio situation comedy. Unlike its comedy/variety relative, the "sitcom" retained the recurring cast of the dramatic serial. In fact, historians once labeled programs such as *The Goldbergs, Henry Aldrich,* and *The Life of Riley,* which we call situation comedies today, as "comedy dramas," thus emphasizing their dramatic story line. Each character in the situation comedy is often a two-dimensional parody of one or two human foibles. Listen to any classic radio sitcom and you often find the "drunk," the "tightwad," the "know-it-all," the "dumbbell," and many other stereotypes. These exaggerated personality flaws define each "family member," and determine how that character interacts with the rest of the show's family, and how he or she will deal with this week's adventure. Radio sitcoms are very consistent in basic structure, but they do vary in length. Although many radio sitcoms ran for 15 minutes, most eventually settled into the more popular 30-minute length. A few even stretch to 45 or 60 minutes, but these are rare.

In the simplistic world of the radio sitcom, with its recurring characters, settings, and themes, stories focus on the main character's adventures—be they big or small. Although most stories were about the central personality, episodes occasionally spotlighted secondary characters. Unlike the radio drama, though, the situation comedy played story lines for laughs.

The basic structure of a radio situation comedy is very consistent. The show's regular cast of characters is (re)introduced to the audience. At the same time, the "comfortable" environment of their sitcom world is made clear. Then someone or something upsets the routine, adding instability to this self-contained world. The story line takes the characters through a series of dramatic yet comic adventures, each one building until the climax of the show. Along the way the audience is exposed to "running gags" and a comedy of character that transcends the week's episode. The audience also hears commercials, sometimes performed by the characters and "subtly" embedded into the story. Although not a variety show, the radio sitcom would sometimes rely on such variety staples as musical numbers and celebrity guests. At the end of the comedy drama, the adventure is resolved, and the characters are back to where they started. Change is rarely permanent in the radio sitcom world.

The term *family* is used broadly when describing the sitcom cast of characters. It identifies traditional family members but also friends and coworkers. Any group of people that the main character spends significant amounts of time with and cares a great deal for make up his or her sitcom family. Because of this liberal definition of *family,* the situation comedy might be primarily centered on the home but might just as often gravitate to a social gathering place (such as a bar) or a work environment. Any time a small group of characters could gather together, interact, and share adventures, a situation comedy was born. Radio sitcoms have often appeared in the form of soap operas, adventure programs, science fiction, even as variety shows. In fact, among the earliest sitcoms were the fictional adventures of performers such as Jack Benny and Fred Allen as they went about the day-to-day tasks of putting on their variety shows!

An important characteristic of radio comedy was that the home audience had to imagine certain elements. Radio's lack of a visual element created "theater of the mind," allowing listeners to imagine Jack Benny's clunky old Maxwell car (played by veteran voice actor Mel Blanc) and to assume that the many characters in *Amos 'n' Andy* were actually African-American (when in fact they were initially all portrayed by two white actors, Freeman Gosden and Charles Correll). But in its early days, this radio format proved to be a challenge for its stars. Coming primarily from the vaudeville circuit, radio's comedians were accustomed to interacting directly with their audience, and they often relied on visual as well as verbal humor. The former problem was solved by adding an in-studio audience.

Origins

Both radio comedy/variety and situation comedy programs trace their roots to the days of the touring circuses, burlesque shows, medicine shows, musical reviews, and vaudeville companies that thrived from the late 19th century into the early 20th. Troupes of actors, singers, dancers, poets, and comics—plus an almost infinite variety of more esoteric acts (such as sword swallowers, jugglers, and animal acts)—would take their show on the road, playing in various towns and cities on a predetermined "circuit." The makeup of these troupes may have differed, but their basic components tended to be similar. An introduction by a master of ceremonies, emcee, or troupe manager would be followed by a wide variety of acts strung together with interim commentary by the emcee. This would often build up to a grand finish featuring a more extravagant sketch or a featured humorist or singer of the day. And in most of these forms, the salesman hawking his products—an early example of a program "sponsor"—was one of the more important parts of the show.

The typical nine-act vaudeville bill would usually include as its seventh act a full-stage comedy or drama playlet as a preliminary act to the bill's climactic eighth act—often a famous comedian or vocalist. Not every vaudeville house could afford playlets featuring well-known stars. Consequently, another sort of playlet, one that relied more upon action than upon stars, was developed. Most of these were comedies, and the vaudeville comic playlet became a well-recognized model for stage comedy. These comedies of situation structure are the ancestors of the modern sitcom.

When the radio networks were first looking for talent in the late 1920s, they turned to the vaudeville circuits for acts that might make the transition to an "audio-only" medium. Radio variety was born of this siphoning of vaudeville talent for use on radio. The radio programs usually included one or two hosts, whose presence provided a skeletal structure for the program, which would showcase a variety of acts by both new and established performers. Radio adopted many vaudeville program types. The situation comedy, or "comic playlet of situation," was one of the last formats borrowed from vaudeville, possibly because it did not promote star value as other formats did.

Sitcoms in Radio's Golden Era

Situation comedy premiered nationally during the 1929–30 radio season with *Amos 'n' Andy.* Soon, situation comedies such as *Our Miss Brooks, Beulah, Leave It to Joan, My Favorite Husband, The Goldbergs,* and *My Friend Irma* filled the airwaves, and a new genre for a new medium was born.

Many of the earliest radio sitcoms were not much more than a showcase for vaudeville and film comedians who cobbled together bits from their existing bag of tricks. The Marx Brothers' situation comedy *Flywheel, Shyster, and Flywheel*—a Monday night installment in the *Standard Oil Five Star Theater* series on the National Broadcasting Company (NBC) in the early 1930s—is a prime example. Ostensibly a sitcom about the mishaps of three "shyster" lawyers, scripts were mostly a rehash of gags from the brothers' vaudeville and film performances.

Gradually, though, more and more of the comedy in radio sitcoms was based on character, plot, and story line. A large number of shows in the 1930s straddled the fence between sitcom and variety show. Stars such as Jack Benny "played themselves," and stories were set around their fictional adventures with their equally fictional friends, family, and coworkers. ("Real person" and radio sitcom star Fred Allen maintained a fake feud with Benny for years, although the two admired each other very much in real life.) In the work environment, stories often involved putting on the star's radio variety show; thus the situation comedy was able to sneak in many of the conventions of the variety format. Although Jack Benny's fictional variety show was never actually heard during the sitcom, audiences were treated to performances by guest acts during "rehearsals" that Jack and other characters were involved in "at the studio."

As radio and its audience evolved, so did the quantity and quality of its programs. The situation comedy became one of the staples of 1930s and 1940s radio entertainment. But sitcoms about "real people" were supplanted by the adventures of fictional characters. Stories about *Fibber McGee and Molly, Blondie,* and *Our Miss Brooks* soon dominated the airwaves. In another indicator of radio's impact, *Lum 'n' Abner* was set in the fictional town of Pine Ridge, Arkansas, and in 1936 the real Arkansas town of Waters changed its name to Pine Ridge in honor of the show.

Demise of the Radio Situation Comedy

As radio had borrowed from vaudeville, so television borrowed from radio—for both talent and program formats. Television's first situation comedies were "inherited" from radio, beginning with *The Goldbergs* and *The Life of Riley* in 1949. Network television turned to successful formats on radio, partly as a quick fix to find programming and partly to save money. Three-quarters of early television station owners were already radio station owners.

The direct ancestry of radio to television allowed radio to contribute format styles and even entire programs to the new medium. Many programs, such as *The Chesterfield Supper Club* (hosted by Perry Como), were simulcast in an effort to save money and provide programs for the new medium. Popular radio shows were not necessarily picked up by their respective networks' fledgling television franchises. The big-three radio networks soon foresaw that their future was in television, and bidding wars erupted for the most popular radio programs. Columbia Broadcasting System (CBS) TV "stole" many popular radio shows from rival NBC's radio programs. NBC retaliated, and the American Broadcasting Companies (ABC) participated—but on a smaller scale. Radio networks became a less important part of the national media picture. Most of their familiar program formats shifted to television, as did advertiser dollars. Radio eventually evolved into a provider of music, talk, and news.

Many established radio stars, such as Jack Benny, Red Skelton, Bob Hope, and Fred Allen, attempted to make the transition to television. Some were successful, but others were not. There were many advantages for the situation comedy in the new medium. Viewers could now see how characters fit in with their surroundings. More important, thanks to the television camera, gestures and mannerisms assumed a role impossible on radio. However, there were quite a few problems to overcome in the transition. George Burns and Gracie Allen had to throw out their scripts and learn to memorize their complex verbal comedy routines, and cameras had to be placed so they did not block the live audience that Burns and Allen and other performers needed. Ironically, performers who, several decades earlier, had had to learn how to entertain through sound alone now had to relearn how to *appear* before an audience's very eyes and still stay in character.

When *The Jack Benny Show* first aired on television, Benny had several things to overcome. At first he could not decide between an hour format or a half hour, so he settled on 45 minutes for his debut program. Future programs settled into the increasingly popular half-hour mold. Sets had to be designed and built to portray what had been left to the imagination on radio. One-time scenes and elaborate sets, such as Benny's famous vault, had to be deleted because of cost or the inability to create them effectively. But now viewers could *see,* not just hear, Benny's slow burn and his look of malaise. Visuals added a wealth of information for the viewer, but producers had to spend a lot on props, costumes, and set pieces to show us all how cheap Jack was.

One of the biggest changes to a transitioning sitcom occurred on the new television version of *Amos 'n' Andy.* Because the entire cast of characters was black, but many of the roles had been played on radio by white series creators Freeman Gosden and Charles Correll, CBS decided to do a four-year talent search for experienced black comedy actors to portray the roles. Only African-American actors Ernestine Wade and Amanda Randolph were retained from the original radio cast. Like its radio ancestor, the television version of *Amos 'n' Andy* relied on many stereotypical sitcom personalities, including ignorant, naive, and conniving characters. During its run and afterward, many groups, such as the National Association for the Advancement of Colored People (NAACP), protested the wildly popular series because of its negative representation of blacks. In spite of these protests, CBS moved the popular radio show to television, and in 1951 *Amos 'n' Andy* became America's first television sitcom with an all-black cast (it ran for two seasons). The radio version continued but evolved into a quasi-variety show called *The Amos 'n' Andy Music Hall,* which ended in 1955.

By the mid-1950s radio sitcoms—like most network radio formats—had migrated almost completely to television. One strange "reverse crossover" was *My Little Margie,* a sitcom about a well-to-do widower and his 21-year-old daughter, who was intent on "protecting him" from various female suitors. The show premiered on CBS television in June 1952 for a three-month run. NBC ran the series for a few months before it resumed broadcast on CBS in January 1953. At about the same time, the series began producing new episodes for CBS network radio. The television series returned to NBC in September 1953 and stayed there until August 1955. The radio version remained on CBS, but it also ended in 1955.

Only three radio sitcoms, *Our Miss Brooks, The Great Gildersleeve,* and *Fibber McGee and Molly,* were still broadcast during the 1955–56 season. *Our Miss Brooks* began on CBS radio in 1948, but it began running on television as well in 1952 with almost the same cast. Both versions of the show ended in 1956. *Gildersleeve,* a character on the *Fibber McGee and Molly* show, spun off into his own radio series in 1941. Although it had a 15-year run on radio, *Gildersleeve* was not popular enough to make the transition to television. *Fibber McGee and Molly*—which aired for 22 years—left NBC radio in 1957, permanently closing the door on network radio's situation comedy closet. The show reappeared on NBC television for a very short 6-month run in 1959. The characters and situations in the McGee household did not transfer well to the new television neighborhood.

Although radio sitcoms ceased to air nationally in the United States in 1957, the format has not entirely disappeared. Occasionally, comedy dramas have been produced for American public radio. Imports from Canada, the British Broadcasting Corporation (BBC), and other international markets have also made their way to American airwaves. One of the more popular of these was a BBC radio sitcom disguised as a science fiction episodic serial, *The Hitchhiker's Guide to the Galaxy.* It premiered in Great Britain in 1978 and traveled to American public radio in the early 1980s.

PHILIP J. AUTER

See also, in addition to performers and programs mentioned in this essay, Comedy; Variety Shows; Vaudeville

Further Reading

Barson, Michael, editor, *Flywheel, Shyster, and Flywheel: The Marx Brothers' Lost Radio Show,* New York: Pantheon Books, 1988; London: Chatto and Windus, 1989

Buxton, Frank, and William Hugh Owen, *The Big Broadcast, 1920–1950: A New, Revised, and Greatly Expanded Edition of Radio's Golden Age,* New York: Viking Press, 1972

Dunning, John, *On the Air: The Encyclopedia of Old Time Radio,* New York: Oxford University Press, 1998

Firestone, Ross, *The Big Radio Comedy Program,* Chicago: Contemporary Books, 1978

Grote, David, *The End of Comedy: The Sit-Com and the Comedic Tradition,* Hamden, Connecticut: Archon Books, 1983

Harmon, Jim, *The Great Radio Comedians,* Garden City, New York: Doubleday, 1970

Josefsberg, Milt, *The Jack Benny Show,* New Rochelle, New York: Arlington House, 1977

Poole, Gary, *Radio Comedy Diary: A Researcher's Guide to the Actual Jokes and Quotes of the Top Comedy Programs of 1947–50,* Jefferson, North Carolina: McFarland, 2001

Settel, Irving, *A Pictorial History of Radio,* New York: Grossett and Dunlap, 1967; 2nd editon, New York: Ungar, 1983

Summers, Harrison Boyd, editor, *A Thirty-Yyear History of Programs Carried on National Radio Networks in the United States, 1926–1956,* Columbus: Ohio State University, 1958

Wertheim, Arthur Frank, *Radio Comedy,* New York: Oxford University Press, 1979

Widner, James F., "Comedy Central," <www.otr.com/comedy.html>

Skelton, Red 1913–1997

U.S. Comedian and Actor

An entertainer whose career extended from vaudeville to television, Red Skelton first gained national prominence on U.S. network radio during the late 1930s and 1940s.

Origins

Richard Bernard Skelton was born in Vincennes, Indiana, in 1913, two months after the death of his father. His mother worked as a cleaning woman and elevator operator while raising her four sons. In 1923, at age ten, young Skelton enjoyed his first taste of the entertainment business when he went backstage with actor Ed Wynn following a performance in Vincennes. Five years later, at age 15, he joined a traveling show and began to develop his many trademark characters as a vaudeville and circus clown. He appeared in the famed Haggen and Wallenbach Circus, the same circus his father had joined during the 1890s. Throughout his teens, his varied jobs included newsboy, traveling medicine show pitchman, and showboat entertainer.

While working at a Kansas City theater in 1930 he befriended an usher, Edna Marie Stilwell, and married her the following year. She became his manager and sidekick, writing some of the original material for their acts even after their marriage ended in 1943.

Radio Years

By 1937 Skelton's stage appearances were gaining national attention. Following a Toronto performance that year, he returned to the United States and appeared as a guest comedian on *The Red Foley Show* on WLW radio in Cincinnati, Ohio (his first radio appearance), and later that year on *The Rudy Vallee Show,* where he continued to appear during 1938. In 1939 he was the headline act on *Avalon Time,* a half-hour network variety show, and in October 1941 was given his own program, *The Red Skelton Show,* also known as *Red Skelton's Scrapbook,* on NBC radio.

Sponsored by Raleigh Cigarettes, the Tuesday night show featured comedy skits and musical numbers by Harriet Hilliard, Wonderful Smith, and the Ozzie Nelson Orchestra. Usually at the center of attention was one of Skelton's many zany characters, including Junior the Mean Widdle Kid ("I dood it," Junior's favorite expression, became a popular national catch phrase in the early 1940s), San Fernando Red, Willie Lump Lump, Clem Kadiddlehopper, Bolivar Shagnasty, J. Newton Numbskull, and Sheriff Deadeye, "the fastest gun in the west." By the end of its second year, *The Red Skelton Show* was a smashing success, rated second among the national radio audience.

During World War II, Skelton performed at several military bases and munitions plants, and in March 1944 was drafted into the U.S. Army, where he served as a private. Following his discharge in December 1945, he resumed his weekly NBC broadcast with a new supporting cast, including vocalist Anita Ellis and the David Forrester Orchestra. (Harriet Hilliard and the Ozzie Nelson Orchestra had their own NBC show by that time.)

Like many popular NBC radio programs following the war, *The Red Skelton Show* attracted the attention of rival CBS chief William Paley, who lured Skelton to his network in 1949 during a talent raid that also included renowned comedians Jack Benny and Edgar Bergen. Skelton's radio variety show continued on CBS until 1952. With the advent of television, he returned to NBC to host a weekly TV variety show. After two seasons he moved to CBS television. For nearly two decades, from 1951 to 1970, *The Red Skelton Show* enjoyed consistently high audience ratings, ranking among the top 15 prime time network television series for 16 of its 19 seasons on the air.

Typical of Skelton's humor were his exaggerated jokes. "My electric toaster broke down," he once said. "So I repaired it with parts from an airplane. Now when the toast pops out it circles the table twice before coming in for a landing." In addition to radio and television, Skelton appeared in over 40 films, many of them MGM comedies in the 1940s and 1950s. They include *A Southern Yankee, Watch the Birdie, Whistling in Dixie, Whistling in the Dark,* and *Whistling in Brooklyn,* in which Skelton pitches in a baseball game against the Brooklyn Dodgers.

In 1986, 15 years after the cancellation of his television variety program (CBS cited "rising production costs" as the reason for canceling), Skelton accepted the Governors Award of the Academy of Television Arts and Sciences at the Emmy Awards show. His lasting distress over his program's cancellation was evident after the audience had given him a standing ovation for his award. "I want to thank you for sitting down," he said. "I thought you were pulling a CBS and walking out on me." Two years later, he was inducted into the Academy's Television Hall of Fame. He also received the Screen Actors' Guild's Golden Globe Award.

An accomplished artist, Skelton made and sold paintings of clown faces after his television career ended, fetching thousands of dollars. He also supported children's charities, including the Shriner's Crippled Children's Hospital. A foundation bearing his name is based in Vincennes, Indiana, and aids children in need.

Following a lengthy illness Skelton died at a hospital in Rancho Mirage, California, on 17 September 1997, at age 84.

DAVID MCCARTNEY

Richard Bernard "Red" Skelton. Born in Vincennes, Indiana, 18 July 1913. Performed in burlesque and vaudeville from age 15; worked in the circus as a clown; made Broadway debut, 1937; debuted on radio, 1937; debuted on film, 1938; made television debut, 1951. Recipient: 3 Emmy Awards; Screen Actors Guild Achievement Award; Governor's Award, Academy of Television Arts and Sciences; Freedom Foundation Award, 1970; National Commanders Award, American Legion, 1970; Golden Globe Award, 1978. Died in Rancho Mirage, California, 17 September 1997.

Radio Series

1937	*The Red Foley Show*
1937–38	*The Rudy Vallee Show*
1939	*Avalon Time*
1941–49	*The Raleigh Cigarette Program* aka *The Red Skelton Show* and *Red Skelton's Scrapbook*
1949–53	*The Red Skelton Show*

Television Series

1951–70	*The Red Skelton Show*

Films

Having Wonderful Time, 1938; *Seeing Red,* 1939; *Flight Command,* 1940; *Whistling in the Dark,* 1941; *The People vs. Dr. Kildare,* 1941; *Lady Be Good,* 1941; *Dr. Kildare's Wedding Day,* 1941; *Whistling in Dixie,* 1942; *Ship Ahoy,* 1942; *Panama Hattie,* 1942; *Maisie Gets Her Man,* 1942; *DuBarry was a Lady,* 1943; *Thousand Cheers,* 1943; *Whistling in Brooklyn,* 1943; *I Dood It,* 1943; *Radio Bugs,* 1944; *Bathing Beauty,* 1944; *Ziegfeld Follies,* 1946; *The Show-Off,* 1946; *The Luckiest Guy in the World,* 1946; *Merton of the Movies,* 1947; *The Fuller Brush Man,* 1948; *Southern Yankee,* 1948; *Neptune's Daughter,* 1949; *Three Little Words,* 1950; *Duchess of Idaho,* 1950; *Moments in Music,* 1950; *The Yellow Cab Man,* 1950; *The Fuller Brush Girl,* 1950; *Watch the Birdie,* 1951; *Texas Carnival,* 1951; *Excuse My Dust,* 1951; *Lovely To Look At,* 1952; *The Clown,* 1952; *Half a Hero,* 1953; *The Great Diamond Robbery,* 1953; *Susan Slept Here,* 1954; *Hollywood Goes to War,* 1954; *Around the World in Eighty Days,* 1956; *Public Pigeon No. One,* 1957; *Ocean's Eleven,* 1960; *MGM's Big Parade of Comedy,* 1964; *Those Magnificent Men in Their Flying Machines, or How I Flew from London to Paris in 25 Hours 11 minutes,* 1965; *That's Dancing!,* 1985

Stage

Red Skelton in Concert, 1977

Selected Publications

A Red Skeleton in Your Closet, 1965
Red Skelton's Gertrude and Heathcliffe, 1971, also known as *Gertrude and Heathcliffe,* 1974

Futher Reading

Harmon, Jim, *The Great Radio Comedians,* Garden City, New York: Doubleday, 1970
Marx, Arthur, *Red Skelton,* New York: Dutton, 1979
Wertheim, Arthur Frank, *Radio Comedy,* New York: Oxford University Press, 1979

Sklar, Rick 1930–1992

U.S. Program Innovator

Rick Sklar was an important and widely imitated figure in the development and promotion of Top 40 radio, especially for ABC in New York.

Early Years

Sklar was born in New York and grew up in the Brighton Beach area of Brooklyn. He attended New York University and while there volunteered as a writer at the city-owned WNYC radio station. After graduation he got his first commercial job at WPAC in Patchogue, on Long Island, where, as was typical with such a post, he undertook everything from writing copy to announcing news. Seeking more income, in 1954 he answered a want ad that read "copy/contact-Radio" and went to work for WINS in New York City. There he worked with legendary disc jockeys Alan Freed, Murray the K, and Al

"Jazzbo" Collins. While at WINS, Sklar wrote station jingles, created original contests and promotions, and was instrumental in the station's dramatic rise in the ratings.

By 1960 he was at WMGM (soon changed to WHN) as program director, but within a couple of years he departed for WABC, the station that would make him a legend in the industry.

ABC Years

Sklar started at WABC in June 1962 as director of community affairs, although he played a role in most program operations from the start, including the production of promotional announcements for the station. He became the station program director late in 1963, and in a very few years he had moved the station from mediocre to spectacular ratings based on both his program and promotional ideas.

Despite the station's miniscule promotion budget, Sklar devised brilliant ways to promote the station. His first big promotion required listeners to paint a likeness of the Mona Lisa. Sklar scored a major victory when he got surrealist painter Salvador Dali to serve as the judge. Listeners who painted the best, worst, biggest, and smallest rendition of the Mona Lisa were each awarded $100. The promotion resulted in the station being overwhelmed with entries, but it significantly raised the public's awareness of the low-rated outlet. By the early 1970s, nearly 6 million people a week were listening to WABC, which had jumped to the top two or three New York stations in ratings reports.

Ironically, this success got somewhat in the way of further promotional ideas. No longer could the station offer simple call-in contests for fear of wiping out the entire phone system. Taking a different tack, Sklar devised the "$25,000 button" contest in 1974. Rather than have thousands of calls jamming the New York phone system, button spotters were hired to comb the New York area streets and award prizes to people wearing WABC promotional buttons. Sklar's last big WABC promotion was "The Big Ticket," in which local newspapers carried an insert that contained a ticket from WABC with a number on it. Those who heard their number announced over the air won prizes. Naturally such stunts encouraged people to listen in the hope of winning something.

Among Sklar's many special talents was his ability to create a music playlist that attracted a young audience. Record companies courted Sklar's imprimatur for their new artists, because exposure on the nation's flagship Top 40 station meant other pop music outlets around the country would add the artists to their playlists. Many top performers, among them Barbra Streisand, John Lennon, Neil Sedaka, and Stevie Wonder, became admirers of Sklar and sought his support for their own recording efforts. No contemporary music station in the country was as influential as WABC during its heyday. While at WABC Sklar was also instrumental in the development of the careers of several of the nation's most popular disc jockeys, most notably Bruce "Cousin Brucie" Morrow, Dan Ingram, Ron Lundy, Herb Oscar Anderson, and Harry Harrison.

In March 1977 Sklar shifted focus from station to network operations when he was promoted to vice president of programming for ABC's radio division. He helped to further develop the various ABC networks, including developing a talk-radio format. Within a couple years of Sklar's departure from WABC, the station began a precipitous slide in the ratings, which led it to abandon its music programming in favor of talk.

Later Life

In 1984 Sklar left ABC to start his own consulting firm known as Sklar Communications, and he became a consultant to many stations around the country, including some with which he had once competed. In addition to his consulting, Sklar found great satisfaction in sharing his knowledge with students at New York area colleges. A marathon runner in his spare time, in 1992 Sklar entered the hospital for minor foot surgery to repair a torn tendon in his left ankle. An anesthesia complication took his life on 22 June 1992.

CHRISTOPHER H. STERLING AND MICHAEL C. KEITH

See also American Broadcasting Company; Contemporary Hit Radio Format/Top 40; Disk Jockeys; Freed, Alan; WABC

Richard Sklar. Born in New York City, 1930. Graduated from New York University. Volunteered as a writer, WNYC, New York. Hired by WPAC in Patchogue, New York; hired by WINS, New York, 1954. Program director for WMGM, New York, 1960–62. Director of community affairs, WABC, New York, 1962–63, and program director, 1963–77. Vice president of programming, ABC's radio division, 1977–84. Started Sklar Communications and became a radio consultant, 1984–92. Died in New York City, 22 June 1992.

Selected Publications

Rocking America: An Insider's Story: How the All-Hit Radio Stations Took Over, 1984

Further Reading

Morrow, Cousin Bruce, and Laura Baudo, *Cousin Brucie! My Life in Rock 'n' Roll Radio*, New York: Beech Tree Books, 1987

Musicradio WABC Rick Sklar Page, <www.musicradio77.com/Sklar.html>

Smith, Kate 1907–1986

U.S. Radio Singer and Personality

Often referred to as "the First Lady of Radio," Kate Smith was one of the most popular radio singers and personalities of the 1930s and 1940s. Her radio variety show rivaled that of Rudy Vallee in its popularity and impact; her daily commentary at "high noon" was part of the routine of millions of homemakers; and she set sales records in her marathon radio war-bond drives during World War II. She was known especially for her patriotism; Irving Berlin's song "God Bless America," which she debuted on her program in 1938, became her trademark. A generously proportioned woman with a motherly attitude toward her listeners, Smith projected a warm friendliness that suited the intimacy of the radio medium and caused many listeners to identify with her.

According to her birth certificate, Kate Smith was born in Washington, D.C., on 1 May 1907, although she later claimed to have been born in Greenville, Virginia, which more appropriately represented her small-town persona. Her father was a newspaper agent, and both her parents were amateur singers. Kate first sang publicly at a war-bond rally in 1917 at the age of nine. She was a born performer and proved a popular local attraction, winning many amateur vaudeville competitions in the D.C. area. After she graduated from high school, her father insisted that she attend George Washington University's School of Nursing rather than go on the stage. After nine months, Smith quit and began performing in vaudeville in Washington. *Variety* reviewed her in February 1926, noting that she had "not only a good voice, but one of much volume." Smith was a contralto and was described most often in early reviews as a "blues singer" or as a "coon-shouter" in the tradition of Sophie Tucker. Her renditions of songs were "straight" rather than jazzy; she sang in a clear, strong voice that reflected the melody and lyrics as written.

Broadway producers soon took note of her, and she appeared in three successful Broadway productions between 1926 and 1931: *Honeymoon Lane, Hit the Deck,* and *Flying High*. She also enjoyed several vaudeville successes, including one at the famed Palace Theatre. Smith was unhappy with stage work, however, because she felt that she did not fit in with the Broadway crowd. She neither drank nor smoked, and her weight (topping 200 pounds) often made her the object of ridicule both on stage and off. But a 1930 backstage meeting with Columbia Records talent scout Ted Collins changed her life. Collins was impressed by the power of her voice and persuaded her to make some records with Columbia. Eventually, Collins offered to become her manager, splitting the profits 50-50. Smith agreed, and their relationship, which began on a handshake, became one of the most successful partnerships in radio history. In addition to managing her career, Collins also served as the announcer on all her programs, carefully framing and protecting her public image.

Collins decided the growing medium of radio would be an appropriate venue for Smith. He arranged a 15-minute music segment on the National Broadcasting Company (NBC) in the spring of 1931, *Kate Smith Sings,* but he switched networks when he was able to arrange a better deal with the Columbia Broadcasting System (CBS). *Kate Smith Sings,* as well as a few well-placed appearances on Rudy Vallee's popular *Fleischmann Hour,* made Kate Smith a radio star by mid-1931. In July 1931, she was crowned "Queen of the Air" by New York mayor Jimmy Walker on the city hall steps. By the fall, her program had found a sponsor in La Palina cigars, and the program *Kate Smith and Her Swanee Music* was a big hit for CBS from 1931 through 1934. Listeners responded so quickly to Smith not only because of her vocal talent but also because she used the radio medium in particularly successful ways. From the beginning of her broadcasts, she used the same theme song, "When the Moon Comes Over the Mountain," and the same greeting ("Hello everybody") and closing ("Thanks for listenin', and good-bye folks"). The repetition of these elements, as well as Smith's simplicity and uncondescending manner, greatly appealed to her domestic audience.

When the La Palina program ended, Collins decided it was time for Smith to tour America so that her radio fans could see her. The tour, *Kate Smith's Swanee Revue,* was meant to last several weeks, but it proved so popular that it ran for eight months. Smith also used this time to appear in a couple of films: *The Big Broadcast of 1932,* in which she had a cameo, and *Hello Everybody!* (1933), a feature film for Paramount in which she starred. Although she always liked performing live and enjoyed her tour, she did not like Hollywood and was happy to return to radio. After trying a few different formats, CBS, Collins, and Smith struck gold with *The Kate Smith Hour,* a variety program that ran from 1936 to 1945 (1936–37 as *Kate Smith A&P Bandwagon*). This enormously popular program combined song with comedy sketches, dramatic scenes from movies and Broadway, and a talk segment called "Women of America" in which Smith spotlighted and interviewed women she found noteworthy. This program launched the careers of comedians Henny Youngman and Abbott and Costello, among others, as well as the popular 1940s comedy *The Aldrich Family.*

Smith's "talk" segments proved so successful that Collins and CBS arranged a second program for her, *Kate Smith Speaks,* a daily 15-minute segment addressed primarily to

The Kate Smith Show, 1932
Courtesy CBS Photo Archive

housewives. "It's high noon in New York, and time for Kate Smith" began Collins' famous introduction to the series, which ran from 1939 to 1951 and was the number-one daytime program. It began with a news segment by Collins, and then Smith commented informally on various timely topics she thought would interest her readers, from current movies to recipes to social welfare issues. She often "chatted" with guest stars who "dropped by." Smith framed these programs as "heart-to-heart" talks with her listeners and did not present herself as an authority on any subject.

Smith's patriotism and her homey persona made her an especially effective morale builder during World War II, and she devoted herself to this work with uncommon energy. She expanded her broadcast schedule for Armed Forces Radio, sang countless patriotic songs, entertained the troops whenever she could, and conducted the most profitable war-bond drives of the war. She conducted four bond-drive marathons over the radio, in which she promised to stay awake as long as listeners donated money. The most famous of these drives took place on 21 September 1943; Smith stayed on the air from 8:00 A.M. to 1:00 A.M., raising over $36 million. The success of this particular broadcast was so remarkable that sociologist Robert Merton of Columbia University's Department of Psychology published an analysis of it in 1946. Merton interviewed several hundred of Smith's listeners and found that people responded to her appeals more than those of other stars because they believed so completely in her sincerity and felt a personal relationship with her.

Smith's popularity declined after the war, although she hosted a fairly successful television variety program early in the 1950s. She remained a popular and respected guest star on television, however, until her health began to fail in the late 1970s. She died in Raleigh, North Carolina, on 17 June 1986.

ALLISON MCCRACKEN

See also Singers on Radio

Kate Smith. Born in Washington, D.C., 1 May 1907. Attended George Washington School of Nursing, 1924–25; started singing career as young child; various vaudeville appearances, 1924–26; Broadway debut in *Honeymoon Lane,* 1926; starring role in Broadway shows *Hit the Deck* (revival), 1927, and *Flying High,* 1930; radio debut with *Freddy Rich's Rhythm Kings,* Winter, 1930; notable film appearance in *The Big Broadcast of 1932;* starring film role in *Hello Everybody!* 1933; cross-country vaudeville tour, *Kate Smith and the Swanee Revue,* 1933–34; own radio variety series, *The Kate Smith Hour* (with some variations in title), 1936–51; debuted Irving Berlin's song "God Bless America," 10 November 1938; had own radio commentary series, *Kate Smith Speaks,* 1939–51; set records for selling war bonds during World War II; had own television series, 1950–60; guest roles in numerous radio and television shows. Recipient: Women's International Center Living Legacy Award, 1985. Died in Raleigh, North Carolina, 17 June 1986.

Radio Series

1930–31	*Freddy Rich's Rhythm Kings*
1931, 1947–49, 1951	*Kate Smith Sings*
1931–34	*Kate Smith and Her Swanee Music*
1934–35	*The Kate Smith Matinee; The Kate Smith New Star Revue*
1936–37	*The Kate Smith A&P Bandwagon*
1937–45	*The Kate Smith Hour*
1938	*Kate Smith's Column*
1938–39	*Speaking Her Mind*
1939–51	*Kate Smith Speaks*
1945–47, 1951–52, 1958	*The Kate Smith Show*
1947	*Kate Smith's Serenade*
1949–50	*Kate Smith Calls*

Television Series

The Kate Smith Hour, 1950–54; *The Kate Smith Evening Hour,* 1951–52; *The Kate Smith Show,* 1960

Films

Kate Smith—Songbird of the South (short), 1930; *Newsreel* (1 minute), 1931; *Rambling Round Radio Row #1* (short), 1932; *Paramount Pictorial* (short), 1932; *The Big Broadcast,* 1932; *Hello Everybody!* 1933; *Hollywood on Parade* (short), 1933; *America Sings with Kate Smith* (short), 1942; *This Is the Army* (cameo), 1943

Stage

Honeymoon Lane, 1926–27; *Hit the Deck,* 1927; *Honeymoon Lane,* 1929; *Flying High,* 1930–31; *Kate Smith and the Swanee Revue,* 1933–34

Selected Publications

Living in a Great Big Way, 1938
Kate Smith Stories of Annabelle (with Bill Martin and Bernard Herman Martin), 1951
Upon My Lips a Song, 1960

Further Reading

Hayes, Richard K., *Kate Smith: A Biography, with a Discography, Filmography, and List of Stage Appearances,* Jefferson, North Carolina: McFarland, 1995
Merton, Robert K., *Mass Persuasion: The Social Psychology of a War Bond Drive,* New York: Harper, 1946; Westport, Connecticut: Greenwood, 1971

Smulyan, Jeffrey H. 1947–

U.S. Radio Executive and Entrepreneur

As a young boy, Jeffrey H. Smulyan had three things in mind: sports, radio, and the desire to start a business. A native of Indiana, Smulyan was born in Indianapolis in 1947 and grew up in a family of entrepreneurs, so such a vision came naturally and was in his blood.

He attended the University of Southern California and graduated cum laude in 1969 with a bachelor's degree in history and telecommunications; he went on to receive his J.D. degree from the USC School of Law in 1972. He experienced his first taste of radio at USC's student radio station and developed leadership skills as senior class president. Smulyan's love of radio was greater than his experience, yet his inborn understanding and leadership abilities served him well. He returned to his native state in 1973 and officially entered broadcasting by taking on the management of WNTS, a small AM station in Indianapolis. Three years later Smulyan became manager of another small AM station, KCRO in Omaha, Nebraska. In each case, Smulyan's father, Samuel W. Smulyan, purchased the station after his son became its manager. WNTS was acquired in 1974 and KCRO in 1979. Today both are still owned by Samuel Smulyan's estate, but they have no affiliation with Emmis.

By 1979 Smulyan was ready to begin making an impact on radio on a grander scale. He organized and became principal shareholder of the Emmis Broadcasting Corporation (*emmis* means "truth" in the Hebrew language), which purchased a small FM station licensed to Shelbyville, Indiana, located some 25 miles from the center of Indianapolis. Smulyan's intent was to build a tower close to the city and compete in the larger market. This was accomplished, and WENS went on the air on 4 July 1981 as the Emmis flagship station.

With WENS quickly becoming competitive and profitable, Smulyan and Emmis began their expansion. During the next five years, six stations were acquired, and Emmis entered the nation's number-one market, New York, in 1986 with the purchase of WHN. Smulyan renamed it WFAN (the "Fan") and launched the nation's first 24-hour all-sports radio station. He later purchased the National Broadcasting Company (NBC)–owned AM and FM stations in New York, moving WFAN to WNBC's frequency of 660 kilohertz, which provided better coverage of the market.

The all-sports concept rapidly took hold and has been copied in many other cities. Although not achieving high audience ratings at WFAN or anywhere else, it is extremely popular with advertisers, as the prime listeners are young adult males. This factor has helped WFAN become the highest revenue-generating radio station in America for several years running.

By 1989 Emmis was the owner of ten stations, and Smulyan began to diversify his interests by leading a group of investors in purchasing the Seattle Mariners baseball team. However, shortly after the acquisition, the combination of a heavy debt load, a change in bank credit rules, the oncoming national recession, and low baseball revenues forced Smulyan to sell some of Emmis' prized station holdings, most notably WFAN. The team was sold two and a half years later. Just prior to the baseball team purchase, Smulyan had moved the company into the publishing arena, acquiring *Indianapolis Monthly,* the first of several regional magazines Emmis would buy. This action would lead to the eventual name change to Emmis Communications.

After returning the company to a solid financial footing, Smulyan again began to move forward, although more prudently than during the 1980s, on the acquisition front. In 1994 it became a publicly owned company traded on the Nasdaq exchange. In 1998 Smulyan took Emmis into television ownership with the purchase of six medium-market stations. Others have since been added. It became an international company when in 1997 Emmis was awarded a license to operate a new national radio network in Hungary, which was named Slager Radio. *Slager* is a Hungarian word, adopted from German, that means "hit." This was followed in 1999 by the purchase of FM News and Radio 10 in Buenos Aires, Argentina. The year 1999 also brought a new commitment to Indianapolis—which has been Emmis' corporate home since the company was founded—with the unveiling of a state-of-the-art building for headquarters and for housing its several area radio stations. At the time of completion, the $35 million structure was considered by some to be the best radio facility in the nation.

Some two decades after its beginnings, the names Jeffrey H. Smulyan and Emmis are still virtually synonymous. Smulyan continues as the driving force behind the organization and has built a corporate environment known for its highly focused commitment to excellence and for its willingness to invest in its people, creating a culture that keeps employees on board for a long time. Many of the Emmis stations are among the best-performing stations, in terms of both audience ratings surveys and advertising revenues, in the radio industry.

To formalize his different approach to traditional business philosophy, some years ago Smulyan created the 11 commandments of Emmis, which are prominently displayed in all facilities. They are: admit your mistakes; be flexible and keep an open mind; be rational and look at all opinions; have fun and don't take this too seriously; never get smug; don't underprice yourself or your medium, and don't attack the industry—build it up; believe in yourself—if you think you can make it happen, you will; never jeopardize your integrity—we win the right way or we don't win at all; be good to your people—get them into the game and give them a piece of the pie; be passionate about what you do and compassionate about how you do it; and take care of your audience and advertisers. Smulyan holds or has held leadership positions within industry organizations such as the National Association of Broadcasters and the Radio Advertising Bureau, as well as with the U.S. Olympic Committee.

By 2003 Emmis owned 27 radio stations, including three FM outlets in New York, two in Los Angeles and one in Chicago, plus four stations in its hometown of Indianapolis. It also owned 16 television stations, with the announced intention of stepping up its acquisition pace by adding several more in the coming years.

MARLIN R. TAYLOR

Jeffrey H. Smulyan. Born in Indianapolis, Indiana, 6 April 1947. Attended University of Southern California, B.A. in History and Telecommunications, 1969; University of Southern California School of Law, J.D., 1972; manager, WNTS-AM, Indianapolis, Indiana, 1973–76; manager, KCRO-AM, Omaha, Nebraska, 1976–79; formed Emmis Communications Corporation, 1979; director, National Association of Broadcasters; member of American, Indiana and Federal Communications bar associations; chairman,

Board of Radio Advertising Bureau; board of trustees, Ball State University, from 1996. Received American Women in Radio and Television's Star Award, 1994; named by White House to lead U.S. delegation to Plenipotentiary Conference of International Telecommunications Union, 1994; Entrepreneur of the Year Award, Ernst and Young, 1995; Entrepreneur of the Year Award, *Indianapolis Business Journal,* 1995; ranked one of 10 most influential radio executives of past two decades, *Radio and Records*, 1995; ranked one of 40 most powerful people in radio, *Radio Ink*, 1998 and 1999; honored by National Association of Broadcasters with its Year 2002 National Radio Award.

Selected Publications

Power to Some People: The FCC's Clear Channel Allocation Policy, 1971

Further Reading

Jones, Tim, "Emmis Plans to Get Radio Active," *Chicago Tribune* (6 February 2000)

Peers, Martin, "Emmis Stops Watching Radio Deals from the Sidelines," *Wall Street Journal* (9 November 1999)

Rathbun, Elizabeth A., "Indiana Pacer: Jeff Smulyan Is Building a Media Empire with All Deliberate Speed," *Broadcasting and Cable* (21 June 1999)

Soap Opera

Daytime Radio Drama

Although many critics disagree on when the first soap opera was actually broadcast, most would concede that the earliest prototype for serial drama appeared on Chicago radio in the 1920s: Irna Phillips' *Painted Dreams,* a mosaic of fanciful stories about heroes, villains, and helpless victims. *Painted Dreams* did not fare very well on local radio initially, but Phillips was never discouraged by her perceived failure. She (along with Frank Hummert, an advertising executive, and his wife Anne) was convinced that a successful serial format in newspapers and magazines could translate well into radio. Within a short time, Phillips and both of the Hummerts were proven right; *The Smith Family,* premiering nationally in 1925, became an instant hit. The program was built around two vaudevillians, Jim and Marion Jordan (who later became Fibber McGee and Molly). Later, *The Smith Family* was joined by *Clara; Lu 'n Em; Vic and Sade; Just Plain Bill; The Romance of Helen Trent; Ma Perkins;* and *Betty and Bob.*

The meteoric rise of daytime serial drama was a phenomenon that had not been foreseen by most programmers, and certainly not by any advertisers. In fact, the networks' first impulse was to reject the notion of any type of series targeted toward women. They thought it foolhardy and unprofitable because of the seemingly unattractive listening population of unpaid workers (housewives—unattractive because of their perceived lack of impact on revenues generated for sponsors) during the afternoon time block, and also because of the questionable cost efficiency of providing serious drama in continuous segments.

Despite these reservations, however, the networks decided to experiment with several 15-minute "episodes," provided at discounted prices, to interested sponsors in the early 1930s. Most advertising support for these daytime dramas came from corporations such as the Colgate Palmolive Peet Company and Procter and Gamble, who sold household products to interested female listeners. Thus, the term *soap opera* was coined to describe the melodramatic plotlines sold by detergent companies.

In retrospect, those who gambled on the success of radio soap operas need not have worried; the format seemed to be a perfect complement to the medium. Relying completely on sound, radio producers spread news, information, musical entertainment, and folktales. In minstrel tradition, narrators could easily set the stage for radio drama, providing descriptions of characters and settings for the stories. Within minutes, listeners (mostly women) were ushered into an imaginary world (guided by the narrator) with friends and enemies they might never encounter otherwise. In short, they became participants in a place more exciting, dramatic, and compelling than the home from which they listened.

Thus, the introduction of daytime drama met with as immediate a success as the evening serial counterpart had enjoyed. Devoted listeners faithfully followed the lives and loves of their favorite soap opera characters. And, much to the networks' surprise, housewives were not an unattractive listening demographic to possess. In fact, programmers soon discovered that homemakers, though not directly in the labor force, often controlled the purse strings of the household economy. By 1939 advertising revenue for the popular serials had exceeded $26 million. Less than ten years later, Procter and Gamble was spending over $20 million each year on radio serials. Housewives had indeed found an alluring substitute for previous programming fare (such as hygienic information, recipe readings, and household tips) and were demonstrating their consumer power as well. Network programmers and advertisers had inadvertently stumbled onto an undiscovered gold mine. However, creative programming was not the only reason for the immediate popularity of radio soap operas. To better understand the success of daytime drama in the 1930s, it is important to look at two additional factors: the story formula and its relationship to Depression-era America

Serial Drama and the Cultural Landscape of 1930s America

Irna Phillips was a major contributor to early soap opera formula and content. She concentrated on characterization more than plotline fantasy and later became noted for introducing "working professionals" (doctors and lawyers) to daytime serials. To her, the events were far less important than how they were interpreted or acted upon by her characters. Unlike Phillips, Frank and Anne Hummert believed strongly in plot-driven stories, developing an "assembly-line" or formulaic approach to soap operas that has continued to be successful in today's media. Together, these radio pioneers created a solid genre for future generations.

The Hummerts originated many of the popular early daytime dramas such as *Just Plain Bill, The Romance of Helen Trent,* and *Ma Perkins.* They based most of their stories in the Midwest—an ideal setting for several reasons. First of all, the Hummerts' ad agency was located in Chicago. Practically speaking, they felt their soap operas should be produced there to cut expenses and to enable them to exert more creative control. Further, since most of the Hummerts' life experience came from the Midwest, they were more confident having their ideas and plots set there. Finally, the Hummerts felt that the Midwest carried with it an accurate reflection of American values, attitudes, and lifestyles. It seemed to be an ideal part of the country for audiences to associate with the familiar themes of daytime drama, known as the "Hummert formula."

The Hummerts' story formula was really quite simple: they combined fantasies of exotic romance, pathos, and suspense with a familiar environment of everyday life in a small-town or rural setting. Combined with an identifiable hero or heroine, this formula produced an overwhelming audience response: people everywhere shared common needs, common values, and common problems.

This broadcast unity of beliefs and attitudes was especially important during the Depression era, when poverty, unemployment, and general political pessimism threatened the very fiber of American family life. Women, in particular, felt threatened. Although most were not laid off from jobs themselves, they found themselves demoralized as those around them, one by one, lost work. Household incomes declined markedly, and women were forced to feed, clothe, and shelter their families with far fewer resources than before. Amid their discouragement, listeners relied on soap opera characters such as Ma Perkins and *Just Plain Bill* Davidson—common folks who could survive despite overwhelming odds. Their victories over the trials and tribulations of daily living gave many Americans the feeling that they, too, could and would survive.

By 1936 soap operas began to dominate the daytime radio dials. *The Goldbergs* moved from its prime-time perch to afternoons (followed in 1937 by *Myrt and Marge*); several Hummert dramas premiered (including *David Harum, Rich Man's Darling, Love Song,* and *John's Other Wife*); and a soon-to-be-famous soap writer, Elaine Carrington, debuted her first work, *Pepper Young's Family.* In 1937 more daytime drama appeared, some worthy of note (such as *The Guiding Light,* the longest-running soap opera in radio/television history), and some better forgotten. However, the total impact of radio serials had finally been realized—both negatively and positively—and as such, the serials became open to criticism from women's groups such as the "Auntie Septics," who argued that story lines with suggestive sex, faulty marriages, and subsequent divorces threatened the survival of the American family unit; or followers of New York psychiatrist Louis Berg, who argued through his "hypodermic theory" that messages from soap operas, when "injected" into American listeners' heads, directly precipitated all sorts of psychosomatic traumas, including blood pressure problems, heart arrhythmias, and gastrointestinal disorders. These political action groups were often supported by male doctors who resented the implied superiority of serial female protagonists. The moral proselytizers ultimately faded away, in large part because of network and advertiser resistance as well as public admonishment of people like Berg, who was found to have based his research solely on his *own* blood pressure and pulse. Soap operas had survived not only the effects of the Great Depression, but the potential ruin caused by their detractors as well.

The 1940s: A Golden Age for Radio Soap Operas

As daytime drama entered the 1940s, several characteristics of serial writing emerged. First, characterization was simple, straightforward, and easily recognizable. Since most daytime radio listeners were women, listeners could identify with a woman who led a simple life yet was also a solid citizen and model for others in her mythical community.

Second, characters found themselves in predicaments that were easily identifiable by their listeners, with settings easily imaginable to those who had never traveled far beyond their home environment. As Rudolf Arnheim discovered in his study "The World of the Daytime Serial," soap opera characters seemingly preferred commonplace occurrences in their own hometown, as opposed to problems in an unknown environment. And, when circumstances necessitated travel, the new setting invariably was in the United States. Arnheim surmised that soap opera producers refrained from international travel because they felt listeners would not enjoy a foreign setting that would demand that they imagine a place outside their own realm of experience.

Third, most of the action revolved around strong, stable female characters, who were not necessarily professionals, but who were community cornerstones nonetheless. Men were very definitely the weaker sex in soap opera life—a direct reflection on the primary listening audience during the daytime hours.

Finally, daytime drama was often used as a vehicle for moral discussions or a rededication to American beliefs and values. Soap opera heroines often voiced the platitudes of the Golden Rule as well as the rewards that would come to those who could endure the trials and tribulations of living in a troubled society.

After World War II, economic "happy days" returned and soap operas reflected this boom with more career-oriented characters (especially women). But negative postwar elements also emerged, such as postwar mental stress and alcoholism. All these and more were discussed on Ma Perkins' doorstep—with often easy solutions—keeping the "painted dream" of America alive and well.

Career women became more numerous in the 1940s because of writers like Irna Phillips. Phillips also introduced mental problems and amnesia to daytime drama, to reflect America's postwar interest in psychology. Usually a central character suffered some type of emotional malady such as memory loss, a nervous breakdown, alcoholism, or shell shock as a result of wartime stress. Also, psychosomatic paralysis was a common affliction of the long-suffering soap opera heroes and heroines.

Toward the end of the 1940s, crime emerged as an important plotline theme, especially in the area of juvenile delinquency. This direction was also reflective of the times, for Americans were becoming increasingly concerned about youth crime. Criminal story lines continued throughout the early 1950s and continue to be an important theme in daytime drama on television (although the situations have been updated considerably).

The 1950s: The Move from Radio to Television

In the early 1950s, most soap operas moved from radio to television, and the resulting change in technology was felt at all levels, including scriptwriting, acting, and production. The visual medium of television allowed for a wider choice in soap opera settings, because writers were not forced to limit themselves to the experiential world of radio listeners. Rather, they could take their characters anywhere, as long as they visually established the appropriate setting. However, the visual element in television also had distinct limits, for soap writers could no longer rely on "imagination" to set a scene.

Despite many writers' strong preference for radio, most writers, like Irna Phillips, readjusted themselves to the new medium. Phillips' way to explore television's strengths was to use more reality-based themes in established soap operas like *The Guiding Light* (a Phillips creation, moving from radio to television in 1952) as well as to write new soaps, such as *As the World Turns* (with Agnes Nixon), *Another World* (with Bill Bell), *Days of Our Lives*, and *Love Is a Many Splendored Thing*. Phillips' stories, along with those of other serial writers, such as Roy Winsor (*Search for Tomorrow, Love of Life*, and *The Secret Storm*) and Irving Vendig (*Edge of Night*), were lauded by both viewers and critics: ratings skyrocketed and scholars now asserted that daytime TV drama was both entertaining and informative.

In the mid-1950s, some soap operas expanded to 30 minutes, as compared to the 15-minute capsules of the 1930s and 1940s. Because viewers could now see their characters, plotlines became more slowly paced to capitalize on all the advantages of the visual medium such as character reactions and new locales. In fact, a common plotline such as a marriage proposal could last for weeks in a 1950s television soap. After the male character "popped the question" in *Secret Storm*, for example, several days of programming would be spent learning the reactions of both principal and supporting characters for this event: the bride-to-be, her mother, her old boyfriend, his old girlfriend or ex-wife, his secret admirer, her secret admirer, and so on. The possibilities were endless. Thus, one major plotline could sustain itself for weeks longer on television than would be possible on radio despite the added 15 minutes of programming each day. But was it any more effective? Some would argue yes; others would disagree.

The fact remains that many television soap opera plots in the 1980s and 1990s seemed to have changed little since their birth in the late 1920s. Stories still revolve around issues of love, family, health, and security within a cultural context—much like early radio serial drama. And even today, according to author J. Fred MacDonald (in *Don't Touch That Dial*), TV soap opera characters respond to these issues with the same philosophy as their radio predecessors used, which was best expressed by Kay Fairchild, a central character in a 1940s radio serial called *Stepmother:*

> All we can be sure of is that nothing is sure. And that tomorrow won't be like today. Our lives move in cycles—sometimes that's a good thing to remember,

sometimes bad. We're in a dark valley that allows us to hope, and to be almost sure that we'll come out after awhile on top of a hill. But, we have to remember, too, that beyond every hill, there's another valley.

One of the greatest frustrations of serial drama is that the storylines are neverending—though the last seven daytime serials left the Columbia Broadcasting System (CBS) altogether in late 1960. Ironically, this is also one of its greatest attractions. As a result, soap operas are one of most recognized genres of broadcasting today, enjoying consistent audience devotion and popularity in a world where most success is as ephemeral as the last ratings period. And it all began on radio.

MARILYN J. MATELSKI

See also The Goldbergs; Hummert, Anne and Frank; Ma Perkins; Phillips, Irna

Further Reading

Allen, Robert Clyde, *Speaking of Soap Operas,* Chapel Hill: University of North Carolina Press, 1985

Arnheim, Rudolph, "The World of the Daytime Serial," in *Radio Research, 1942–1943,* edited by Paul F. Lazarsfeld and Frank Nicholas Stanton, New York: Essential Books, 1944

Buckman, Peter, *All for Love: A Study in Soap Opera,* Salem, New Hampshire: Salem House, 1985

Cantor, Muriel G., and Suzanne Pingree, *The Soap Opera,* Beverly Hills, California: Sage, 1983

Cox, Jim, *The Great Radio Soap Operas,* Jefferson, North Carolina: McFarland, 1999

"Era Ends as Soaps Leave Radio," *Broadcasting* (22 August 1960)

LaGuardia, Robert, *Soap World,* New York: Arbor House, 1983

MacDonald, J. Fred, *Don't Touch that Dial! Radio Programming in American Life, 1920–1960,* Chicago: Nelson-Hall, 1979

Matelski, Marilyn J., *The Soap Opera Evolution: America's Enduring Romance with Daytime Drama,* Jefferson, North Carolina: McFarland, 1988

Matelski, Marilyn J., *Soap Operas Worldwide: Cultural and Serial Realities,* Jefferson, North Carolina: McFarland, 1999

Rouverol, Jean, *Writing for Daytime Drama,* Boston: Focal Press, 1992

Schemering, Christopher, *The Soap Opera Encyclopedia,* New York: Ballantine, 1985; newly updated and expanded edition, 1988

Stedman, Raymond William, *The Serials: Suspense and Drama by Installment,* 2nd edition, Norman: University of Oklahoma Press, 1977

Wilder, Frances Farmer, *Radio's Daytime Serial,* New York: Columbia Broadcasting System, 1945

Social Class and Radio

Research in social class attempts to explain how and why societies are divided into hierarchies of power, prestige, or wealth. Studies of social class and media examine how mass media reinforce or reproduce this persistent "stratification" of society. Of all the mass media, radio is an especially rich subject area for the study of social class because radio itself has had several definitions over its history. Each of its several incarnations—wireless telegraph, popular hobby, mass medium, and music utility—presents different opportunities for the exploration of social class.

Radio was a conspicuous newcomer in both U.S. and British culture between 1920 and 1950, the subject of constant debate among scholars, journalists, and politicians. During this period, radio's prominence may be likened to the role of television in today's culture. As a result, nearly all in-depth studies of radio take this golden age as their subject. Very few studies focus on social class in their research; however, from the many cultural, social, and oral histories of radio as well as ethnographic accounts that make up our understanding of radio's audience, it is possible to arrive at some generalizations about radio and social class. Foremost of these is radio's image as an essentially middle-class medium and the creation of a new middle-class culture of consumption and home-centered leisure in the advanced industrial world.

What Is Social Class?

Journalists, critics, and academics often use the concept of class rather casually, referring to the lower, middle, and upper classes, without explaining what they mean by these terms. Typically class is used as a synonym for level of income or wealth. Class is considerably more complex, however, than this casual use suggests. Most media researchers adopt one of two approaches to class: either the approach in the United States that emerged from a sociological tradition in the 1940s and 1950s, or the British Cultural Studies (BCS) approach that developed out of a Marxist literary studies tradition in the 1960s.

American media studies rarely use social class as an object of study or a unit of analysis. Social critic Benjamin DeMott has criticized U.S. culture and other cultural critics for accepting a "myth of classlessness," the assumption that nearly everyone in U.S. culture is (or aspires to be) within the bounds of middle-class taste and values. U.S. media, according to DeMott, reinforce this myth (DeMott, *The Imperial Middle: Why Americans Can't Think Straight about Class,* 1990). On the rare occasion that social class is defined in the sociological tradition, it is often considered a product of four factors: occupation, education, income, and/or self-identity. As an example of this approach, Melvin Kohn's 1969 book *Class and Conformity: A Study in Values* uses the factors listed above, coupled with statistical analysis, to arrive at useful insights into the workings of social class within families and society.

BCS has grappled with the meanings and definition of social class in a much more critical manner than has sociological analysis in the United States. Accordingly, there are many interpretations of class within BCS. However, most cultural critics would agree that the BCS conception of class fuses the thinking of three important figures: Karl Marx, Antonio Gramsci, and Pierre Bourdieu.

In Marx's conception of class, the divisions between the "three great classes" (capitalist, bourgeoisie, and proletariat) are essentially economic. The capitalist class uses its wealth and property ownership to dominate and organize society according to its own interests.

Antonio Gramsci, an Italian intellectual writing in the 1920s and 1930s, highlighted the role of conflict and negotiation in social structures. Gramsci's theory of "hegemony" states that maintaining control in a social order requires the upper class to constantly *reinvent* its social dominance. Because the upper classes cannot simply dominate all classes at all times, Gramsci argues, class power must operate through the power to define the nature of "prestige" in a society as well as control of political and military force. Gramsci's ideas are significant in their recognition that power relations between social classes are inherently unstable. He also acknowledges schools, families, and media as important battlegrounds in the ongoing struggle for class hegemony. These ideas opened the door to thinking about social class as more than a product of economics, but also a cultural process.

Pierre Bourdieu, a French sociologist, coined the term "cultural capital" to describe how cultural products (books, media) and cultural knowledge (languages, rituals) serve as markers to reinforce social class as it passes from generation to generation. Cultural capital not only helps reproduce social class, but can also translate into an economic advantage and thus a better economic class position for individuals. Using Bourdieu's ideas, media researchers are able to link cultural experiences, such as listening to the radio, accessing the internet, or attending "high art" events such as the opera, directly to the formation and reproduction of social stratification.

Social class can be a powerful analytic tool, a way to better understand group or individual behavior and choices within a historical and social context. However, social class is also difficult to define; the stratification it describes is subject to change and multiple interpretations. Social class may be defined as a product of economic position, cultural position, psychology, or values, yet any particular set of class labels—lower class, working class, middle class, upper class, or elite—will always be inadequate, a simplification of a complex reality. Furthermore, the boundaries between classes are often unclear, and individuals may belong to multiple classes.

Radio: The Middle-Class Medium?

The prevailing attitude toward social class and radio in most historical or institutional research can be summed up by a quote from Fred MacDonald's 1979 book *Don't Touch That Dial:* "Seeking to please an audience of millions of relatively free-and-equal, middle-class citizens, radio inevitably reflected the democratic environment which it served." In other words, radio is widely perceived as a medium successfully catering to all people, serving up a stew of politics, news, sports, music, and drama that represents a perfect mix of the tastes and values of the population as a whole. Some see radio as the beginning of "homogenizing" the culture of the United States. National networks standardized the nation's news and entertainment as it pulled together the first mass audience in history. It ironed out sectional differences and gave the nation a common culture. However, more recent studies have begun to question whether radio is truly a "middle-class medium."

Radio's association with the middle class began early in its history. In her 1987 book, Susan Douglas describes radio's most prominent inventors and entrepreneurs, most of whom emerged from middle-class or upper-middle-class backgrounds. After the turn of the century, thousands of hobbyists converged on the new technology, building transmitters and receivers to experiment and socialize in "the ether." This subculture, Douglas says, was "primarily white, middle-class boys and men who built their own stations in their bedrooms, attics, or garages" (see Douglas, 1987). The amateur operator credited with making the first professional broadcasts, Frank Conrad, was a white, middle-class engineer employed by Westinghouse.

The popular press and news media supported the image of resourceful, average, middle-class boys and men mastering this "new frontier." However, there were female and working-class radio buffs as well. Michele Hilmes (1997) has countered Douglas' description of the amateur operator subculture, arguing that the popular press ignored amateurs outside the mainstream of male, middle-class culture. From its beginnings, Hilmes argues, radio was seen as a threat to established social hierarchies such as class, race, and gender—that is, a threat to the hegemony of white, middle-class values. According to Hilmes, the popular press chose to portray white men and boys as icons of wireless because they were less threatening to the social order than women or working-class male operators.

The opposition between "working-class culture" and "middle-class culture" animates many of the inquiries researchers have made into radio's development as a mass medium. Cultural critics, including Raymond Williams, Jacques Donzelet, and Simon Frith, have noted how radio's success shifted entertainment into the domestic sphere, away from collective spaces such as taverns, dance halls, or simply the front porch or street. Such a change was related to class because outdoor amusements were associated with the lower classes, the "unruly masses," while home-based entertainment was considered more genteel and middle-class. In an oral history of early radio, "The Box on the Dresser," Shaun Moores writes that the radio could only become a fixture in the home if it could be accommodated within an existing structure of family relations, routines, and patterns. The radio schedule was quickly shaped to fit the patterns of middle-class family life, intertwining with mother's housework, father's job, and the children's schoolwork and bedtimes. Radio's reorganization of leisure time is one example of how radio interacted with large-scale social transformations in the early 20th century, such as industrialization, urbanization, and the rise of consumer culture. If radio seemed to represent the interests of the middle class, it was in part because middle-class interests were beginning to occupy the culture as a whole.

Radio's threat to established hierarchies extended into its golden age. Hilmes locates class awareness in the advertising men who produced radio's most popular programs, most of whom were well-educated, upper-middle-class men. As the group responsible for filling the radio schedule, they reinforced existing class, race, and gender distinctions through their programming choices. For example, the dialects heard on *Amos 'n' Andy* or *The Goldbergs* and the ethnic stereotypes used by Fred Allen in his "Allen's Alley" sketches reinforced long-standing social differences. Meanwhile, the stories on dramatic programs and soap operas and ubiquitous home economics programs presented middle-class family life as the social norm.

Economic reasoning lay behind this appeal to the middle class. In the United States, radio programs were built with a powerful imperative: to appeal to an identified market of listening consumers. As Eileen Meehan has written, ratings systems developed to measure this market were deeply influenced by class. For instance, the first two widely used ratings methods, the Cooperative Analysis of Broadcasting (CAB) and C.E. Hooper's "Hooperatings," relied on telephone interviews to gather data. However, the telephone was still a relative luxury in the early 1930s. Thus, CAB and Hooperatings did not measure the listening habits of all homes with radios but rather only the habits of homes with radios *and* telephones—homes that represented what Meehan calls the "thoroughly modern, consumer-oriented middle-class" (see Meehan, 1990). Subsequent ratings systems for radio and television, including the Arbitron ratings, still rely on this conception of the audience, literally defining *audience* as only those people that most appeal to advertisers.

Unlike the American broadcasting model, which served predominantly commercial interests, the British Broadcasting Corporation (BBC) was formed to serve the principle of public service. Perhaps as a result, social class is more prominent in British studies of mass media. By most historical accounts, assumptions about class differences and tensions deeply influenced the BBC's programming. "Serious" programs, such as classical music and public affairs, targeted the bourgeoisie, while popular "light entertainment," such as dance music, was included for the working-class audience. In a 1983 article, Simon Frith challenged this interpretation of the BBC's popular programming. Frith argued that light entertainment on the BBC was a "middlebrow" form that both entertained a mass audience and fulfilled the BBC's public service charge, but at the cost of producing programs that were wholesome, yet bland and repetitive. David Cardiff revisited Frith's argument in a 1988 article to suggest that comedy on the BBC was also a middlebrow creation, but in no way bland or repetitive. Rather, by simultaneously addressing a mass audience and representing the culture of the elite through comedy, the programs resolved class tensions by deflating perceived differences with laughter.

The construction of middle-class audiences does not tell the entire story of social class and radio. The power of radio (or any form of media) to reshape the identity of its audience is an open debate. Lizabeth Cohen (1989) has questioned the ability of institutions of mass culture—chain stores, movies, and radio—to erase the differences between classes at the grassroots. In her studies of Chicago's working class, Cohen has pointed out that broadcasting was an intensely local enterprise through the 1920s and into the Great Depression. It featured talks by local personages, ethnic/nationality hours, labor news, church services, and vaudeville acts familiar to local communities. Furthermore, working-class audiences were likely to build their own radios (avoiding high-priced, mass-produced "parlour sets") and engaged in communal listening in stores, social clubs, and neighbors' homes. In other words, rather than blanketing communities with a unified mass culture, early radio promoted affiliations within existing groups through their ethnic, religious, or working-class identities. Similarly, radio had an appeal to rural populations distinct from its urban, middle-class image. As late as the 1940s, the differences between rural and urban life in the United States were stark. To label these families as representative of either the middle class or the working class would be inaccurate, for rural America was a culture apart in many ways. Nonetheless, as Richard Butsch (2000) describes, radio broadcasts as early as the 1920s began to address the particular needs and tastes of rural audiences. Although radios spread slowly into rural areas because of the prohibitive cost of owning and operating the radio sets, they were among the most coveted and prized possessions for rural families. Like Cohen's working-class audiences, rural families built and maintained their own radios and practiced communal listening well after most urban audiences had retreated into isolated domesticity.

The emergence of talk radio as a potent political force over the last 20 years has raised questions about the potential for radio to represent social classes traditionally cut off from political discourse. For instance, the populist uprising that derailed President Bill Clinton's nomination of Zoe Baird for Attorney General in 1993 has been attributed to a "spontaneous combustion" of working- and lower-class voices. Benjamin Page and Jason Tannenbaum (1996) have described how mainstream media communicators, influenced perhaps by their own class position, initially treated Baird's tax violations as a trivial matter (she failed to pay Social Security taxes for illegal aliens she had employed). However, an intense reaction materialized through the medium of participatory talk radio, influencing members of Congress and the media and leading to the withdrawal of Baird's nomination. The Baird case may be an exception. In other instances, talk radio commentators have been critiqued for the intensity of their political rhetoric and attraction to scandal. Whether talk radio has inspired a kind of "direct democracy" for the working-class, as the Baird example suggests, or simply given voice to the more angry and intolerant voices in society is a matter for debate.

The case of WCFL, the only radio station to be owned and operated by a labor union, presents another example of working-class interest in radio. The Chicago Federation of Labor founded WCFL in 1926 as a platform to broadcast entertainment and information of interest to the labor unions and the working class. WCFL struggled to survive as a listener-supported station, undermined by a hostile business and regulatory environment, high operating costs, and stiff competition from better-financed commercial stations. By the late 1930s, WCFL had adopted the commercial model of broadcasting. However, the station remained a symbol of resistance to corporate-controlled media and continued to provide news from the perspective of organized labor. WCFL's organizers also hoped to preserve working-class culture, including the music, theater, and art of local ethnic groups and pro-union artists, much of which was being replaced by the avalanche of popular culture coming from Hollywood and New York radio studios. WCFL stayed on the air into the 1970s, although it had lost nearly all pretensions of being the "Voice of Labor." The barriers faced by WCFL are a good illustration of the problems faced by all alternative media. Today only community radio, low-power radio, and, to a lesser extent, public radio provide media outlets for social classes whose art and issues are rarely seen or heard, yet all face perennial problems securing adequate funding and political support, and proving their "relevance" in an industry driven by audience ratings.

Although radio technology and programming have changed dramatically since the 1930s, the economic realities of commercial radio have changed very little. Nonetheless, there are few studies of modern radio that address social class in even a peripheral way. There is some irony here, because the modern radio environment may be more class-inflected than ever. Contemporary radio is marketed to narrow, well-defined niches of the population as ratings companies track the age, gender, ethnicity, and education of audiences in fine detail. Formats are built with a particular audience's tastes, politics, and languages in mind. Meanwhile, the ideal listener is still conceived of in terms of his or her ability to buy advertised products. Understanding how society is organized economically and culturally is central to the task of understanding social class, and radio remains a powerful expression of both economics and culture in the 21st century.

CHRISTOPHER LUCAS

See also Critics; Educational Radio to 1967; Pacifica Foundation; Playwrights on Radio; Poetry and Radio; Public Radio Since 1967; Public Service Radio; Stereotypes and Radio; WCFL

Further Reading

Bolce, Louis, Gerald De Maio, and Douglas Muzzio, "Dial-In Democracy: Talk Radio and the 1994 Election," *Political Science Quarterly* 111, no. 3 (1996)

Butsch, Richard, *The Making of American Audiences: From Stage to Television, 1750–1990*, Cambridge and New York: Cambridge University Press, 2000

Cardiff, David, "Mass Middlebrow Laughter: The Origins of BBC Comedy," *Media, Culture, and Society* 10 (1988)

Cohen, Lizabeth, "Encountering Mass Culture at the Grassroots: The Experience of Chicago Workers in the 1920s," *American Quarterly* 41, no. 1 (1989)

Douglas, Susan Jeanne, *Inventing American Broadcasting, 1899–1922*, Baltimore, Maryland: Johns Hopkins University Press, 1987

Frith, Simon, "The Pleasures of the Hearth: The Making of BBC Light Entertainment," in *Formations of Pleasure*, edited by Tony Bennett, et al., London and Boston: Routledge and Kegan Paul, 1983

Garnham, Nicholas, and Raymond Williams, "Pierre Bourdieu and the Sociology of Culture: An Introduction," in *Media, Culture, and Society: A Critical Reader,* edited by Richard Collins, et al., London and Beverly Hills, California: Sage, 1986
Godfried, Nathan, *WCFL: Chicago's Voice of Labor, 1926–78,* Urbana: University of Illinois Press, 1997
Hilmes, Michele, *Radio Voices: American Broadcasting, 1922–1952,* Minneapolis: University of Minnesota Press, 1997
Kohn, Melvin L., *Class and Conformity: A Study in Values,* Homewood, Illinois: Dorsey Press, 1969; 2nd edition, Chicago: University of Chicago Press, 1977
MacDonald, J. Fred, *Don't Touch that Dial! Radio Programming in American Life, 1920–1960,* Chicago: Nelson-Hall, 1979
Meehan, Eileen, "Why We Don't Count: The Commodity Audience," in *Logics of Television: Essays in Cultural Criticism,* edited by Patricia Mellencamp, Bloomington: Indiana University Press, and London: BFI Books, 1990
Moores, Shaun, "The Box on the Dresser: Memories of Early Radio and Everyday Life," *Media, Culture, and Society* 10 (1988)
Page, Benjamin, and Jason Tannenbaum, "Populistic Deliberation and Talk Radio," *Journal of Communication* 46, no. 2 (1996)
Scannell, Paddy, and David Cardiff, *A Social History of British Broadcasting: Serving the Nation, 1922–1939,* Oxford and Cambridge, Massachusetts: Blackwell, 1991
Thomson, David Cleghorn, *Radio Is Changing Us:* A Survey of Radio Development and Its Problems in Our Changing World, London: Watts, 1937

Social and Political Movements

Radio has often played a very significant role in social and political movements. Its relative cheapness and portability (especially in the age of transistors), and even its ability to be concealed, have rendered radio particularly suitable for waves of public protest and movements of dissent. The focus here will be not the rendition of social movements by mainstream radio stations, but rather three examples of the direct use of radio by such movements: the anti-colonial revolution in Algeria in the years 1956–62; the role of two "movement" radio stations in the defense of Portugal's democracy following the overthrow of its fascist dictatorship in 1974; and the free radio movement in Italy in the mid-1970s.

Algeria, 1956–62

The Algerians' anti-colonial revolt was a supreme litmus test for France. Defeated militarily in its colony of Vietnam just two years previously, the French elite officially viewed Algeria not as a colony at all, but as "overseas" France. Seemingly confirming this was the presence of a million French settlers, some there for generations. To lose Algeria in battle meant a second, unthinkable humiliation, striking at the heart of the French elite's sense of their nation's world power. Even the French Communist Party sharply distanced itself from the Algerian rebel.

Frantz Fanon, a psychiatrist from Martinique working in Algeria who subsequently joined the rebels, has provided an extraordinary account of the role of radio in the uprising. Until the revolutionaries began to use radio to communicate their message, radio had been widely seen among ordinary Algerians as a technology to support French cultural domination and the settler colony. For language reasons in part, but also for reasons of cultural propriety, a radio set in the house would have symbolized the welcome of French authority and an alien culture.

However, when the Voice of Fighting Algeria went on the air, there was an almost immediate change in Algerians' attitude to the medium. Receiving sets began to sell in droves. The organizers of the station were members of the Front de Libération Nationale (FLN), a guerilla organization. The station began to broadcast more frequently from 1954 and announced schedules from the end of 1956. It broadcast from Rabat (Morocco), Tunis (Tunisia), Cairo (Egypt), and Damascus (Syria). The broadcasts were made in Arabic, French, and Berber in an attempt to emphasize the unity of all anti-colonial forces in the country.

The French occupation forces' propaganda now had an interlocutor who could challenge such claims as that the rebels were responsible for invading a village, blowing up its houses and massacring its inhabitants. The rebels' radio could also bring news of United Nations votes condemning the French occupation forces, and of support from Egypt and other voices in the Arab world. Not least, it could set out its proposals for the future when colonial rule had been overthrown.

The French authorities' response was to ban the sale of battery chargers (few Arab homes had piped electricity) and to take the presence of a radio set in any house they invaded as direct evidence of support for the rebels, with corresponding violent reprisals. They also jammed the station, forcing it to relocate its signal several times a day. People would spend hours roaming the dial to find the new frequency, but even when they could not, the mere sound of the continued jamming static signified that the struggle was still continuing.

The Algerians' story has global implications. Algeria, Cameroon, Cyprus, Vietnam, and Zimbabwe were among the relatively few colonial territories that fought their colonial masters militarily. They were tremendously important in convincing Britain and France that they could not hold on to their empires forever, and in persuading them to negotiate with independence movements peacefully elsewhere. The role of radio in helping to galvanize the Algerian revolt thus had implications far beyond Algeria's borders.

Radio and the End of Portuguese Fascism and Colonialism

In April 1974, the Portuguese dictatorship (in place since 1926) was overthrown by a group of army captains. They were of different minds, but two desires united them: their determination to be rid of the regime and its brutal secret police, and their resolve to end Portugal's centuries-long but hopelessly costly colonial rule in Africa. There were several attempts to restore the former regime in the 18 months that followed. During this period, two stations, Rádio Renascença and Rádio Clube Português, steadfastly operated in the capital of Lisbon as a forum for the public to unite around the country's new direction.

It is important in evaluating these stations' role to realize that not only Portugal had a stake in the outcome. The new independence of Angola and Mozambique left apartheid South Africa and the white racist regime in former Rhodesia geographically isolated as the last holdouts of European rule on the African continent. Spain, in the last years of dictator Francisco Franco (1939–75), no longer was a haven for fascists as it transitioned to democracy, which in turn was a beacon for many Latin Americans in the 1970s as they struggled with their own military regimes.

Until April 1974, Rádio Renascença had been nominally a Catholic station, but with minimal religious programming. Its political slant was extremely conservative, with its news staff forbidden to mention the reform-oriented social doctrines of the Second Vatican Council. Rádio Clube Português had had owners close to the Franco regime. With the long-suppressed voices of the Portuguese public suddenly unleashed in the streets (in wall-posters and murals and graffiti) and in parts of the press, these two stations opened up the airwaves to the throng of those delirious with the opportunity to be heard at long last. It was over Rádio Clube's transmitter that the popular song "Grandola" was broadcast, serving as the signal for the army captains' bloodless overthrow of the dictatorship.

Of the two, Renascença was the more free-form and anarchic, Rádio Clube the more recognizable in conventional professional terms. Renascença made its microphones available to virtually every group involved in the movement of change. It also tried to broadcast regular religious services (which its previous management had not done), but ran into a wall of opposition from the Catholic hierarchy, itself deeply compromised with the former regime. Some Rádio Clube newspeople were close to the Communist Party, but they were a small minority. Basically its structure and format, although hospitable to the new political climate, strove to provide balanced news.

Both stations were popular in different ways. When it looked as though Renascença might be closed down because of Vatican and Western governments' pressure, a huge demonstration spontaneously erupted in its support. Rádio Clube always had solid finances because advertisers accepted that its determination to be fair to all sides was popular with its listeners. However, both stations, but particularly Renascença, became international political footballs among the member nations of the North Atlantic Treaty Organization (NATO), whose elites were terrified that Portugal might become a Soviet neo-colony. Even Pope Paul VI publicly denounced Renascença. The actual prospect of a Soviet takeover was highly implausible, but was energetically touted by other political parties in Portugal who saw this as a means to garner foreign support and funds, as Portugal was the poorest nation in Western Europe at that time.

In November 1995, as a major gesture to the stations' symbolic status for NATO governments, a new government first dynamited Renascença's transmitter and then took over Rádio Clube. It was a sad and messy ending, in the name of democracy, for two stations that had sought to serve the public rather than the prevailing power structure.

Free Radio Explosion in Italy

In 1976, faced with a series of unauthorized broadcasts that challenged the government's airwaves monopoly, the Italian Constitutional Court ruled that local broadcasting need not be under that monopoly. There followed an immediate and dramatic expansion of such stations, such that two years later there were more radio stations listed in Italy (a nation then of about 55 million people) than in the United States. Admittedly, these stations varied tremendously in type. Some were proto-commercial stations (indeed, it was one such, a conservative operation called Radio Parma, that had mounted the main challenge), but others were voices of the revolutionary left, of specific local communities, or simply of a group of teenagers with a pile of discs on the floor and a mini-transmitter.

One important aspect of this sudden transformation of Italy's radio scene was its relationship to the labor and student political movements that roiled Italy from 1969 onward. Unlike the political earthquake in France in May–June 1968, which seemed to subside after a year or two, the turbulence in Italy continued for at least a decade, beginning in 1969. The use of radio by the labor and student political movements was extensive, and the fact that their broadcasts could not be national in scope was not a huge limitation. Such stations were concentrated in cities in northern and central Italy, as southern cities were generally less hospitable to the political upsurge. The distances between many of the northern and central cities were not large, and the local orientation made for an immediacy that might have been difficult to achieve on a national plane.

The stations of the far left varied in style. Some, such as Radio Popolare in Milan, set out from the beginning to be forums for a variety of voices and interests within the general public in a city. These also included the concerns of migrant workers. Others, such as Controradio in Florence, after brief initial periods of 100 percent support for one particular far left group, set their sights on serving a city's youth, who were often overlooked in the bustling tourist and visitor trade. Some stations positioned themselves on the far left, but with no particular affiliation to any political group. And still others adopted the Leninist role of transmitting the official propaganda of the moment for their chosen political sects.

One fairly short-lived radio station in this movement—and more than a few were short-lived—was Radio Alice in Bologna, which was named after the character in Lewis Carroll's *Alice in Wonderland.* The programming here was an intense mixture of poetry, far left politics, performance art, innovative music, call-ins (the far left pioneered call-ins on Italian radio) and other dramatic moments. The most dramatic broadcast occurred in March 1977, when the station was the epicenter of a major student revolt at the ancient University of Bologna. At one point, armed city police broke the door down and listeners heard the commotion and the station's young staff shouting, "We have our hands up! Don't shoot!"

Interestingly, many of these stations also had a long-term impact on the musical realm. For all their fiery political questioning and deep involvement in demonstrations and other challenges to the status quo, they also provided a new generation with access to some of the finest international popular music available, some of which was dedicated to challenging injustice and poverty. Thus their work in the zone of the imagination continued long after the demonstrations and the eventual dissolution of the far left movements in the wake of Red Brigades terrorism in the late 1970s. They also served to fuel the 1980s ecological and antinuclear movements in Italy, attracting some of the most sincere movement activists, who were disillusioned with sectarian politics. And finally, these stations sparked the ensuing free radio movement in France during the late 1970s.

Many other instances of radio's role in social and political movements could be cited. The medium's directness, flexibility, and low cost have made it a favorite instrument of horizontal and interactive communication.

JOHN D.H. DOWNING

See also Developing Nations; Hate Radio; Italy; Pacifica Foundation; Politics and Radio; Propaganda by Radio

Further Reading

Downing, John Derek Hall, *Radical Media: The Political Experience of Alternative Communication,* Boston: South End Press, 1984; as *Radical Media: Rebellious Communication and Social Movements,* Thousand Oaks, California, and London: Sage, 2000

Fanon, Frantz, *L'an cinq de la Révolution Algérienne,* Paris: Maspero, 1959; as *Studies in a Dying Colonialism,* translated by Haakon Chavalier, New York: Monthly Review Press, 1965

Girard, Bruce, editor, *A Passion for Radio: Radio Waves and Community,* Montreal, Quebec: Black Rose Books, 1992

Huesca, Robert, "A Procedural View of Participatory Communication: Lessons from Bolivian Tin Miners' Radio," *Media, Culture, and Society* 17 (1995)

Jankowski, Nick, Ole Prehn, and James Stappers, editors, *The People's Voice: Local Radio and Television in Europe,* London: Libbey, 1992

Keith, Michael C., *Voices in the Purple Haze: Underground Radio and the Sixties,* Westport, Connecticut: Praeger, 1997

Land, Jeffrey Richard, *Active Radio: Pacifica's Brash Experiment,* Minneapolis: University of Minnesota Press, 1999

López Vigil, José Ignacio, *Las mil y una historias de Radio Venceremos,* San Salvador, El Salvador: UCAEditores, 1991; as *Rebel Radio: The Story of El Salvador's Radio Venceremos,* translated by Mark Fried, Willimantic, Connecticut: Curbstone Press, 1994; London: Latin America Bureau, 1995

Mata, Marita, "Being Women in the Popular Radio," in *Women in Grassroots Communication: Furthering Social Change,* edited by Pilar Riaño, Thousand Oaks, California: Sage, 1994

Shanor, Donald R., *Behind the Lines: The Private War against Soviet Censorship,* New York: St. Martin's Press, 1985

Soley, Lawrence C., *Free Radio: Electronic Civil Disobedience,* Boulder, Colorado: Westview Press, 1999

Soft Rock Format

The soft rock format of radio programming features songs containing elements of both rock and roll and pop music. This musical subgenre originated in the early days of rock and roll, developed into a separate radio format on FM in the early 1970s, and eventually evolved into the adult contemporary format of the 1980s and 1990s.

The term *soft rock* refers to that side of rock and roll music characterized by, naturally, a softer, less raucous style than classic rock and roll. Stuessy and Lipscomb (1999), in *Rock and Roll: Its History and Stylistic Development,* trace the beginnings of soft rock as a musical trend back to the 1950s and the sock hops popular in U.S. high schools. Danceable, upbeat music needed a counterpart musical style, one to which teenagers could "slow dance." Soft rock also filled a need "to balance the harder mainstream rock with a softer, less raucous alternative, while still maintaining some essential elements of the rock style."

In terms of composition and musical style, Stuessy and Lipscomb describe soft rock songs in general as possessing the following characteristics: slow to moderate beat, a soft backbeat, triple division of the rhythmic pattern, strong emphasis on "beautiful" melodies, harmonies that follow a "major tonic—minor subdominant—major subdominant—dominant" chord progression, chords held for two to eight beats, musically conjunct tunes, love-oriented lyrics, and a lead singer with backup vocals. Regarding the singing found in soft rock music, vocalists might use falsetto, blue notes, and variations of rhythm and blues and gospel style. Specific components aside, the label "soft rock" applies to music that includes either a bass line or a rhythmic pattern derived from rock and roll, or both. Pure pop songs lack these basic rock elements.

Popular artists with soft rock hits during the early days of the genre included "white soft rock" singers Elvis Presley and Pat Boone and "black soft rock" acts such as Frankie Lymon and the Teenagers, the Five Satins, and the Monotones. The songs "Crying in the Chapel" by the Orioles and "Sh-Boom" by the Chords epitomize the softer style of rock music during that era. Other artists whose hit songs exemplified the soft rock style included Paul Anka, Bobby Darin, Frankie Avalon, Johnny Mathis, Connie Francis, and Brenda Lee. Soft rock as a trend continued into the 1960s, with Neil Sedaka, Bobby Vee, Bobby Vinton, and Frankie Valli and the Four Seasons among the more successful teen idols whose songs fit the genre (Stuessy and Lipscomb, 1999).

Although soft rock as a popular music style had already coexisted with harder-edged rock and roll since the 1950s, it became a separate and distinct format during the 1960s, when the radio industry began to experience widespread fragmentation resulting from diversification in the music business itself. Keith (1997) notes that the wide array of popular artists' styles, such as those of the Beatles and Glen Campbell, resulted in myriad format variations: "the 1960s saw the advent of the radio formats of Soft Rock and Acid and Psychedelic hard rock." During this decade of change, there emerged the "chicken rock" stations, those that only "flirted" with rock and roll by airing softer tunes of the popular rock music genre: "While a Chicken Rocker would air 'Michelle' and 'Yesterday' by the Beatles, it would avoid 'Strawberry Fields Forever' and 'Yellow Submarine'" (Keith, 1987).

The soft rock format truly came into its own during the 1970s, when, as Keith (1987) outlines, chicken rock, which appealed to the younger end of the 24- to 39-year-old adult demographic, evolved into mellow rock, which found its core audience among 18- to 24-year-olds. Mellow rock, an FM specialty, at first featured playlists containing soft rock tunes both popular and somewhat unknown. Eventually the word *mellow,* a throwback to the 1960s drug culture vocabulary, became outdated. With the change in terminology to soft rock came a narrowing of the soft rock playlist, which included more popular hits of the time. By the mid-1970s, soft rock found its target audience—"the young adult who had grown weary of the hard-driving rock sound but who still preferred Elton John over Robert Goulet" (Keith, 1984).

Soft rockers came from a variety of musical backgrounds, including folk, country and western, and soul. Stuessy and Lipscomb (1999) point to the Carpenters, Barry Manilow, Neil Diamond, America, the Osmonds, John Denver, and Roberta Flack as examples of important artists in soft rock music of the 1970s. Gregory (1998) credits the success of the country-rock band The Eagles with the proliferation of "an entire sub-genre of Soft Rock bands," such as Fleetwood Mac.

As the 1970s came to an end, so, too, did soft rock as a distinct format. Keith (1987, 1997) attributes its demise to the emergence of disco as a station format, the rise in popularity of hit music stations, and the updating of playlists featuring "easy listening" music. Soft rock the format might have disappeared, but its style lived on in the 1980s through the music of groups such as Air Supply, the Alan Parsons Project, the Police (and their number-one hit single "Every Breath You Take"), and Huey Lewis and the News (Stuessy and Lipscomb, 1999).

Soft rock eventually evolved into the adult contemporary format popular during the 1980s and 1990s. In the late 1990s and at the turn of the century, some adult contemporary stations still use the term *soft rock* to describe themselves and the softer rock style of the artists and music they feature.

ERIKA ENGSTROM

See also Adult Contemporary Format; Middle of the Road Format

Further Reading

Gregory, Hugh, *A Century of Pop,* Chicago: A Cappella, and London: Hamlyn, 1998
Howard, Herbert H., and Michael S. Kievman, *Radio and TV Programming,* Ames: Iowa State University Press, 1986; 2nd edition, as *Radio, TV, and Cable Programming,* by Howard, Kievman, and Barbara A. Moore, 1994
Keith, Michael C., *Production in Format Radio Handbook,* Lanham, Maryland: University Press of America, 1984
Keith, Michael C., *Radio Programming: Consultancy and Formatics,* Boston: Focal Press, 1987
Keith, Michael C., and Joseph M. Krause, *The Radio Station,* Boston: Focal Press, 1986; 5th edition, by Keith, 2000
Stuessy, Joe, *Rock and Roll: Its History and Stylistic Development,* Englewood Cliffs, New Jersey: Prentice Hall, 1990; 3rd edition, by Stuessy and Scott Lipscomb, Upper Saddle River, New Jersey: Prentice Hall, 1999

Sound Effects

As vaudeville faded from popularity and silent films grew, many of the musicians who created music and sound moved to providing accompaniment for silent films. The dexterity required of these musicians, especially the drummers, allowed them to expand their repertoires to include additional sounds that enhanced the movie for viewers. Robert L. Mott, a sound effects artist from the early days of radio, wrote that the earliest actors who moved from live theater and vaudeville to radio brought with them their props, costumes, and gags. These comedians, who relied on the visual gags for laughter, were often lost when they tried to make it in radio. Radio presented a completely new set of opportunities and problems for actors and producers.

The need for people able to produce sound effects to complement radio performances was recognized early. The earliest use of sound effects in a dramatic radio program was, according to Mott (1993), an unknown radio show in the early 1920s in Schenectady, New York. It wasn't until the late 1920s that the CBS network hired sound effects personnel to work with their dramatic programs. Arthur and Ora Nichols, a husband and wife team who were the first sound effects artists in radio, had been musicians for vaudeville acts. The couple perfected their sound effects skills in movie houses, learning additional instruments and ways to perform sounds so the silent films were more realistic to audience members. Arthur Nichols learned to use drums to provide a more diverse range of sounds. Many of the early sound effects artists were drummers, who found the dexterity instilled by their instrument was useful when producing multiple sound effects at one time. These musicians were also experienced in using a variety of contraptions ("traps") to produce different kinds of sounds.

Before the Nicholses introduced the concept of sound effects to network radio, writers had to script cues that allowed listeners to understand what was happening. The dialogue was often confusing but necessary to let listeners know, for example, that someone was supposed to have knocked on the door. The actors found it as awkward as did the listeners. With the introduction of sound effects, the listening experience was enhanced for the radio audience.

As if by magic, sound effects aided listeners in more fully visualizing the stories they heard on the radio. Mott (1993) writes that sound effects relied on the art of deception. Nisbett (1962) emphasizes that the responsibility of the sound effects artist is not to reproduce sounds exactly as they are but to *suggest* the sound—that to be too realistic may even be detrimental to the desired effect. Most programs had one or two sound effects artists working on any production, though a network record of eight sound effects people were used for CBS Radio's production of *Moby Dick* in 1977.

A sound effects artist needed a variety of skills and talents to be successful. The first requirement, according to Mott, was timing. The second was to be ambidextrous; there were numerous occasions that required independent movement of each hand simultaneously. Additionally, sound effects artists had to be aware of pitch, timbre, harmonics, loudness, attack, sustain, and decay. These nine components, according to Mott, were integral to producing realistic sound effects. Despite the tremendous amount of skill and work involved in producing sound effects, the artists rarely received mention in the credits of a show. Management believed that to acknowledge that the sounds of radio were merely people creating the noises rather than the actual sound would damage the credibility of the programs.

Types of Sound Effects

Sound effects can be categorized into two types: spot (live) effects and those that are pre-recorded. Spot effects are those sounds that are produced live in the studio, whereas recorded effects are produced independently and inserted into the program at the appropriate time. In the earliest days of radio, spot effects were the only type of sound effect used. The technology for recorded effects was developed for sounds that could not be conveniently created within the studio such as cars, airplanes, weather, large crowds, etc.

Sound effects artists frequently needed to create sounds for ordinary sounds and for extraordinary sounds. Ordinary sounds (those we hear all the time) were often the most difficult to create in a realistic fashion. For example, Mott noted that visually-oriented producers were the most difficult to work with, as they often believed that if a telephone bell is to ring, there must be a telephone with the bell to "make it sound like a telephone." The artists were skeptical because experience had shown them that producers often thought it was a telephone ringer until they saw it wasn't attached to a telephone, and only then did they insist that the ringing "Just didn't sound like a real telephone."

In the early days of radio, sound effects artists were often called on to produce sounds that no one had ever heard before. For example, the early science fiction programs often included an invasion from outer space. The sound effects artists needed to produce sounds of a space ship entering Earth's orbit for an invasion. The artists relied on imagination, creativity, and the odds and ends that comprised the sound effects prop room to create sounds that would induce an audience to believe a space ship was fast approaching Earth.

Important Sound Effects People

Ora Nichols was one of the first sound effects artists and the first female in the industry. Ora and her husband, Arthur, spent 23 years as musicians for vaudeville acts and for silent films in movie houses before entering radio. They worked several years as freelance artists and then were hired at CBS along with Henry Gauthiere and George O'Donnell. They were the first staff sound effects personnel in the fledging radio industry. Ora Nichols directed the CBS Sound Effects Department for several years until Walt Pierson was hired in 1935 as director. This allowed Nichols to return to her preferred spot as a sound effects artist.

Arthur Nichols brought his skills as a craftsman to create props that were necessary to produce realistic sounds. According to Mott, Nichols spent nine months working on a piece of equipment that could produce a wide variety of sounds, from a small bird chirping to 500 gunshots a minute. Together with his wife, he created many of the props and tools that were used to create many of the sound effects for radio.

Orval White was the first African American sound effects artist. He started out in 1949 as an equipment man and worked his way up to be a talented and respected sound effects artist. White worked for 22 years on one of best-known programs on CBS, *Gangbusters*. He was, according to Mott, the first and only African American sound effects artist in radio, television, and film.

Jack Amrhein was considered by directors and peers to be one of the best sound effects artists in radio. His creativity and imagination were his strongest assets. Amrhein worked for the CBS radio network in New York as the sound effects artist for many shows, including *Mr. Keene, Tracer of Lost Persons; The Mysterious Traveler; Mr. Chameleon; The Fred Allen Show; Inner Sanctum; The Phillip Morris Playhouse;* and *The Robert Q. Lewis Show.*

With the rapid growth of radio, there was rapid growth among the cadre of artists who performed sound effects. CBS expanded its sound effects department from 8 artists to 40 in the early 1930s, despite the Depression, hiring engineers, music arrangers, film studio personnel, and other people with varied backgrounds who could contribute diverse skills to the growing department.

As the need for more complex sound effects expanded, four unofficial groups emerged: stars, artists, button pushers, and technicians. Though well known among the sound effects artists, this hierarchy was kept unofficial because CBS wanted to be able to randomly assign personnel to shows rather than having to accommodate requests for certain artists. The producers with shows that were rated highly did request certain personnel, but those requests were fulfilled only when the producers threatened to go to the advertisers. Sound effects personnel soon became specialized with different functions.

Lowest in the hierarchy were the technicians. They designed, built, and maintained the props. Occasionally a technician might fill when needed as a button pusher or for some other minimal sound effect production. The button pushers worked on shows with a limited need for sound effects and, according to Mott, literally pushed buttons to produce those sounds. These could be doorbells, buzzers on game shows, phone ringers, or oven timers on the soap operas. Mott noted that many sound effects artists hated the button pusher shows because they found the work boring, even though it was easy, the pay was the same, and the hours were better.

Next in the hierarchy were the sound effects artists, who made up the majority of the sound effects department at any radio network. These artists were used mostly for prime-time shows (both drama and comedy). Producers did not request them, and they were not paid as much as the artists who were considered stars.

Sound effects being created for *March of Time,* 1933
Courtesy CBS Photo Archive

The stars were those personnel at the top of the hierarchy. Mott wrote that these people had an uninhibited approach to performing sound effects. They didn't worry how they looked or what others thought of them. These artists were as successful in eliciting laughs from the studio audience as the comedians. Most often the producers allowed them to improvise, and the more freedom they had in their work, the more they performed for the audience. Some actors accused these sound effects artists of upstaging them, but audiences loved them. The stars demanded and received higher fees, overtime compensation, and a variety of other perks not given to most sound effects personnel.

Techniques for Creating Sound Effects

The tools used to produce sounds and noises varied greatly and often had no apparent connection to the sounds they were used to emulate. For example, in one program, Mott used a bowl of spaghetti to convince the audience a hungry worm was devouring people in their sleep. Other sound effects artists employed large open drums with BB shot to replicate the sound of waves at sea, a thunder screen to create claps of thunder, or a scratch box, a small wooden box with one side of tin. The tin was punched with nail holes, that, when swept with a wire brush, sounded like a steam engine train pulling out of the station.

A concept that became important to realism was layering. Layering sounds means using multiple sounds and mixing them to create a more realistic sound effect, an effort that requires the use of additional people (or recorded sounds). For example, to create the sound of a dinosaur for some of the early radio programs, sound effects artists mixed the sounds of real animals (a lion's roar, an elephant's trumpeting, and a tiger snarling), which when played at a slower speed became the industry standard for the sound of a dinosaur.

Sound effects personnel often had to make extraordinary efforts to create everyday sounds that sounded "right" on radio. For example, they used flash bulbs dropping into a glass for ice, a cork dipped in turpentine and then rubbed on a bottle to create the screech of a monkey or rat. They could squeeze a box of cornstarch to sound like footsteps in the snow or use their arms and elbows to hit a table to sound like a body falling. Splash tanks were important for sounds that required water, such as washing dishes on a soap opera or for creating storms at sea on the dramas.

Footsteps were the most commonly requested sound effect for radio. Most artists had special shoes, called "walking shoes," used only for creating the footsteps on air. The artists paid careful attention to the maintenance of the soles and heels of these shoes to be sure they would produce adequate sounds.

A variety of materials and techniques were used to create the illusion of walking on different surfaces. For example, plywood boards were most commonly used for floors and steps inside buildings such as offices or homes. Sound effects artists preferred a piece of plywood to a portable stair because they found the stairs too limiting; instead they perfected a technique of stepping on the board and then rubbing the sole over the end to replicate the sound of someone climbing or descending the stairs. When the script called for someone going up the stairs, they placed the weight of the step on the sole of the shoe, and for descending stairs, the weight was on the heel. To produce the sound of someone walking on a sidewalk, in the street, or indoors on a concrete floor, they used a slab of marble.

A large wooden box that could be filled with gravel, cornstarch (for footsteps in the snow), or other materials was used when needed, as were palm fronds or broom corn to imitate footfalls or movement through a jungle.

Recorded Sound Libraries

As the technology for recording sounds increased, recorded effects of those sounds not easily reproduced in the studio became commonplace. Initially, recording equipment was large and cumbersome, limiting artists' ability to go into the field and record needed sounds. Manufacturers responded by developing portable equipment that allowed artists the freedom to record realistic sounds to 78 rpm records and to alter the speed at which the records were played back, changing the nature of the sound. These recordings were used to provide general, background noises whereas the manually produced sounds were for specific actions such as someone walking down stairs.

Although recorded sound effects provided standardized sounds, the artists still had to be skilled to produce the correct volume, quality, and speed. According to Robert Turnbull (1951), sound effects artists had to be adept with the record to cross-arm, double-arm, segue, slip-speed, or spot-cue. By 1950, sound effects artists had access to over 15,000 recorded sounds. The major networks had extensive libraries of commercially recorded sounds for their sound effects artists. These were often complemented by a collection of specially recorded sounds that the artists recorded locally.

Well before the coming of television, most sound effects had been reduced to recordings, and stations as well as networks could stock hundreds of discs with all types of recorded sound carefully catalogued and indexed. These could be inserted or brought in under actors creating a precise and smoothly integrated sound—at a fraction of the cost of an extensive live sound effects operation. Virtually all modern television sound effects are achieved in this fashion.

MARGARET FINUCANE

See also Production for Radio

Further Reading

Creamer, Joseph, and William B. Hoffman, *Radio Sound Effects,* New York and Chicago: Ziff-Davis, 1945

Mott, Robert L., *Radio Sound Effects: Who Did It, and How, in the Era of Live Broadcasting,* Jefferson, North Carolina: McFarland, 1993

Mott, Robert L., *Radio Live! Television Live! Those Golden Days When Horses Were Coconuts,* Jefferson, North Carolina: McFarland, 2000

Nisbett, Alec, *The Technique of the Sound Studio,* New York: Hastings House, 1962; London: Focal Press, 1965

Turnbull, Robert B., *Radio and Television Sound Effects,* New York: Rinehart, 1951

Whetmore, Edward Jay, *The Magic Medium: An Introduction to Radio in America,* Belmont, California: Wadsworth, 1981

Soundprint

U.S. Documentary Series and Media Center

Soundprint is one of the few ongoing nationally broadcast, noncommercial radio documentary series in America. The series is produced by the Soundprint Media Center, Incorporated (SMCI), a nonprofit media production and training facility based in Laurel, Maryland, near Washington, D.C.

Since 1988 *Soundprint's* weekly broadcasts have featured documentaries by independent radio producers who explore a wide range of topics in half-hour segments. Most of the shows provide more context than one finds in deadline-driven, time-specific radio news stories. The series also provides an outlet for international voices through its documentary exchange program with producers in England, Canada, Australia, New Zealand, South Africa, Scotland, Hong Kong, the Netherlands, and Ireland.

In addition, *Soundprint's* parent company, SMCI, embraces cutting-edge media technologies. The center maintains state-of-the-art digital audio production facilities and offers world wide web hosting and other internet services.

Soundprint was created in 1986 by the meeting of the minds of some innovative public radio managers and producers. At the Johns Hopkins University station then known as WJHU in Baltimore, Maryland, station managers David Creagh and Dennis Kita hired William Siemering to be the executive producer of their proposed new national documentary series. Siemering was a visionary leader, having co-created National Public Radio's (NPR) flagship daily evening news program, *All Things Considered.* Also, at WHYY in Philadelphia, Pennsylvania, Siemering was a major force behind the creation of the station's celebrated daily national interview program *Fresh Air.* And multi-award winning producers Jay Allison and Larry Massett were two of *Soundprint's* first documentarians.

Although NPR, the largest of America's public radio networks, maintained its own vehicle for documentaries, *Soundprint* enabled producers outside of the NPR system to air programs dealing with extremely personal issues or with broader societal concerns. For example, one award-winning *Soundprint* show, "Mei Mei: A Daughter's Song" by producer Dmae Roberts (1989), provided an intimate view of a woman's difficult relationship with her mother. The Canadian production "Forever Changed" by Kelly Ryan (1999) examined how people from different walks of life in Nova Scotia dealt with the aftermath of the 1998 crash of Swiss Air Flight 111.

From the start, *Soundprint* encouraged producers to engage listeners aurally in such a way as to allow them to "see" a story through creative sound design—often a complex, multi-layered mix of ambient sound, music, narration, and sound bites from event participants. Moreover, *Soundprint* attempts to provide unique perspectives on social, political, cultural, or scientific issues. One of *Soundprint's* science programs, for example, entitled "A Plague of Plastic Soldiers" produced by Stephen Smith (1996), examined the technological impact of land mines in Southeast Asia—instruments of war that continue to maim and kill long after fighting has ceased. Another program, "Heavy Petting" by Gemma Hooley (1998), humorously chronicled America's infatuation with pets.

The series has always been produced in stereo—technically distinguishing itself from the monaural sound of most radio news programs. And as early as its first broadcast year, *Soundprint* employed the multitrack mixing format used by the music recording industry—enabling the series to better achieve its layered soundscape. In recognition of its technical and contextual daring, the series had won more than 50 major national and international awards through 2002.

Yet for all its success, *Soundprint* has had to wage an ongoing financial battle for survival. Documentary production can be expensive—requiring costly research, travel, sound gathering, editing, and mixing expenses. For the first five years of its life, the series was funded by grants from the Corporation for Public Broadcasting (CPB), the National Endowment for the Arts, and WJHU-FM funds. But as CPB funding (then *Soundprint's* largest funding source) ended in 1993, WJHU support also dwindled. *Soundprint* executive producer Moira Rankin and technical director Anna Maria de Freitas decided to incorporate the series into a nonprofit company called the Soundprint Media Center, Inc.

The newly formed company landed a major grant from the National Science Foundation (NSF) and rented space at American University's station WAMU-FM in Washington, D.C. But it was clear that the center could not survive on grants alone—especially grants exclusively linked to radio documentary production. So in an attempt to expand its support base, the SMCI began exploring the emerging world of the internet. Through this exploration, the company could also continue to satisfy its desire to utilize new media technology.

In 1994 the SMCI took its documentary series to high school students through the internet and developed some basic webpages for students and teachers. This work led to the center's development of a pioneering mini-network of public broadcasting stations on the internet in 1995. Participating stations included Norfolk, Virginia's WHRO; Athens, Ohio's WOUB; Boston, Massachusetts' WGBH; Minnesota Public

Radio; and the Louisiana Public Broadcasting system. Participants could share programs and compare programming strategies through SMCI's on-line network.

The SMCI later expanded its on-line efforts by providing the technical architecture, securing the funding, and project-managing "ArtsFest '97," one of the first public broadcasting internet arts festivals. Programming from about 15 noncommercial radio and television stations nationwide was featured almost 24 hours a day during a two-week period. Web users could hear arts programs as varied as cowboy poetry from KGNU-FM in Boulder, Colorado; zydeco music from WWOZ-FM in New Orleans, Louisiana; and radio theater from KCRW-FM in Santa Monica, California.

The SMCI is now continuing its internet innovations by offering database management along with website hosting, updating, and content development. In fact, the company's internet projects have brought in substantial funds from the NSF and the United States Department of Education, as well as contractual work with federal agencies and public radio and television research or production organizations.

SMCI's internet operations have financially bolstered the *Soundprint* radio series. Yet *Soundprint* continues to face challenges. During the mid-1990s, some independent producers boycotted the series until program ownership rights and compensation concerns were resolved. And although *Soundprint* can be heard on stations in several of the country's major markets, the series' station carriage numbers have dropped from a previous high of more than 100 stations during the early 1990s to about 50 stations in the year 2002. Since 1995, radio stations have had to pay for the series. It had been free to affiliate stations when *Soundprint* was distributed by the American Public Radio network (now known as Public Radio International) from 1988–93, and by NPR from 1993–95.

Soundprint produces roughly 45 new documentaries along with 53 reruns a year. Each week, two half-hour programs are fed via satellite to participating stations throughout the country. Current or archived programs can be heard at *Soundprint*'s website, located at www.soundprint.org.

SONJA WILLIAMS

See also Documentary Programs; Public Radio Since 1967

Hosts

Barbara Bogaev (2000–present), Lisa Simeone (1997–2000), Larry Massett (1988–1997), John Hockenberry (1988)

Creators

William Siemering, David Creagh, Dennis Kita, Larry Massett, Jay Allison

Producers

Moira Rankin, Anna Maria de Freitas

Programming History

American Public Radio/ Public Radio International	1988–93
National Public Radio	1993–95
Soundprint Media Center, Inc.	1995–Present

Further Reading

Coles, Robert, *Doing Documentary Work,* Oxford and New York: Oxford University Press, 1997

Conciatore, Jacqueline, "Wary of Losing Rights in New Media, Radio Producers Boycott 'Soundprint,'" *Current* (9 October 1995)

Conciatore, Jacqueline, "CPB Aims to Break Impasse in Talks on Indies Rights," *Current* (20 November 1995)

Conciatore, Jacqueline, "Negotiations Falter, Producers Continue 'Soundprint' Boycott," *Current* (29 January 1996)

"Independents End Boycott on 'Soundprint' Contracts," *Current* (20 January 1997)

Josephson, Larry, editor, *Telling the Story: The National Public Radio Guide to Radio Journalism,* Dubuque, Iowa: Kendall/Hunt, 1983; updated edition, as *Sound Reporting: The National Public Radio Guide to Radio Journalism and Production,* edited by Marcus Rosenbaum and John Dinges, Dubuque, Iowa: Kendall/Hunt, 1992

Latta, Judi Moore, "Wade in the Water—The Public Radio Series: The Effects of the Politics of Production on Sacred Music Representations," Ph.D. diss., University of Maryland, 1999

Miller, Doug, "Making Waves: 'Soundprint' Leaves Mark on Radio, Web," *The Laurel Leader* (3 September 1999)

South America

South America consists of 13 countries—ten with cultural roots tied to Spain or (in the case of Brazil) Portugal, and three smaller nations with non-Iberian backgrounds. Radio penetration is high and is an important part of the media environment, as many of the countries are poor and people rely on radio as an inexpensive means of communication, education, and entertainment.

Origins and Development

Radio developed during the 1920s and 1930s following the U.S. commercial model. Most governments allowed private development by large media companies owned by a few wealthy families. This relationship allowed the media companies to benefit from favorable decisions by the government. In addition, there was much influence by foreign media concerns, especially those in the United States. For example, in Argentina the big three U.S. networks—the American Broadcasting Company (ABC), the Columbia Broadcasting System (CBS), and the National Broadcasting Company (NBC)—were all present by the 1950s and influenced radio's structure and commercial development. Another influence was the political climate. Many countries of South America changed governments frequently, fluctuating between military dictatorships and ineffective democratic rule. Bolivia, the most extreme example, has had 200 governments (one every nine months) since gaining independence in 1825. Today many democratic reforms are influencing the political landscape but the history of repression is still felt, especially in media laws.

International radio broadcasts are used by the governments of South American countries, on shortwave and medium wave signals, to communicate political, economic, social, and cultural information to the world community. Many of these international services offer multilingual programming. For example, Radiodifusion Argentina al Exterior has English, German, Italian, Japanese, Portuguese, and Spanish services. International broadcasts are also popular with the local audiences. Los Medios y Mercados de Latino America, a marketing research company, reports that audiences in 1998 listened to Radio Nacional do Brasil, the Voice of America (VOA), the British Broadcasting Corporation (BBC), Radio Mexico International, Radiodifusion Argentina al Exterior, Radio France International, Deutsche Welle, and Radiotelevisione Italiana. The listeners to these broadcasts are usually more affluent, better educated, and of higher occupational status (managers and professionals) than average. The competition for these services are internet broadcasts that deliver clear signals across international borders. Of the 9.21 million people who listen to international radio broadcasts, 5 percent of them have internet access in their homes.

Despite poor economic conditions in many South American countries, new technology is present, such as cable television and web radio stations. Because of limited computer ownership, however, web access to radio stations in South America remains limited. Zonalatina, a comprehensive website covering Latin America, has links to 750 Latin American radio stations, including some that have live internet broadcasts. Most are from large cities such as São Paulo, and their audiences seem to be from outside the cities of origin, so the broadcasters can appeal to a larger audience over a wide area.

Community radio stations are owned, run, and controlled by the public they serve and provide "alternative" programming to commercial or state-run media. They generally have a very small range and operate on a small staff and budget. Although they are not profit oriented, these stations will sometimes accept advertising to survive. Since the 1950s, community radio has grown rapidly in Latin America and today numbers about 2000 stations. It currently operates in Argentina, Bolivia, Brazil, Columbia, Chile, Ecuador, Paraguay, Peru and Uruguay. One goal of the programming is to focus on local interests, rather than national or world interests caused by consolidation and globalization of the media. The World Association of Community Radio Broadcasters (AMARC; L'Association Mondiale des Radiodiffuseurs) helps with this goal. One example of globalization is the influence of international news agencies, primarily from Europe and the United States, on local radio news. To balance this influence, AMARC sponsors a news agency that provides community radio stations in Latin America with information useful for their listeners. Agencia Informativa Pulsar (referred to as Pulsar) began in March 1996 and produces daily bulletins and specialized services. One such service is "Nuqanchik," a daily news service in the Quechua language, which is spoken by almost 10 million people in the Andean region. Headquartered in Ecuador, services are distributed via the internet.

Although advertising and government support form the basis for much of South American radio, other means of funding are necessary for the small stations that operate in many economically poor areas. *Comunicados* (personal messages) are used as a primary means of communication between rural towns, government, businesses, and people. Listeners pay the local station to broadcast personal announcements, often at noontime when listenership is high. The price can be as low as 30 cents (U.S.) for three airings. Another source of income is song dedications, with listeners paying for the privilege. For as low as 20 cents (U.S.), the stations will play a record of one's choice and read an announcement such as "Happy Birthday" or "I love you."

Radio may follow formats (usually in urban areas) or block programming (usually in rural areas). The 1998 *Los Medios y Mercados de Latino America* (The Media and Markets in Latin America) listed the most popular radio formats for Spanish-speaking Latin America, grouping them into six general categories (percentages reflect the number surveyed who said they regularly listen to that format): popular music, which includes Spanish-language popular music (31 percent) and rock music (22 percent); information, including news (46 percent) and sports (24 percent); exotica, which includes classical music (10 percent) and jazz (4 percent); talk, including commentary (15 percent); relaxation, including tropical music (44 percent) and Spanish-language ballads (38 percent); didactic, which includes religious programs (5 percent); and *radionovelas,* or soap operas (2 percent).

Marketing research is used in larger cities. The results show similarities to other countries in listener preferences. Rock music is popular with males ages 12–29, advice/variety/family programs with older females, and sports with older men. The daypart patterns of listening are also very similar to other parts of the world. Radio listening is heaviest in the morning and lighter in the evenings and on weekends. The heaviest listening is on Monday through Friday from 6:00 A.M. to 10:00 A.M.

Radio in Selected Countries

Argentina

Argentina can claim the oldest radio station in the world. On 27 August 1920, ten weeks before KDKA in Pittsburgh signed on, Sociedad Radio Argentina began regular broadcasts from Buenos Aires. Until 1922, it was the only radio station on the air. It had no government license, as the Argentina government did not begin to issue them until 1923. By 1925, there were 12 stations in Buenos Aires and ten in the interior. Three networks also developed: Radio El Mundo, Radio Splendid, and Radio Belgrano.

Radio programs and music from the United States became popular in the early years of radio in Argentina. One of the most famous Argentine radio broadcasts was of the U.S. boxing match between Jack Dempsey and Argentine Luis Angel Firpo on 24 September 1923; the match increased the sale of radio receivers considerably. In the early years, U.S. dance music such as the fox trot, boogie woogie, and swing competed with the tango in Argentina. In 1934, Carlos Gardel, one of the most famous tango singers in Argentina, broadcast from NBC studios in New York to Buenos Aires. Orson Welles came to Buenos Aires in 1942 as a special guest of Radio El Mundo to talk about the impact of his famous broadcast, "War of the Worlds."

Broadcasting developed in complete freedom until 1943, when stations were seized by the fascist military government for propaganda purposes. All programs became scripted and no shortwave reception from any Allied power was permitted. Juan Perón, who was appointed head of the National Labor Department, was the first to use radio to communicate effectively to a large population. In 1945 when he was arrested, his wife Evita used Radio Belgrano to incite 200,000 workers to demand his release. (Evita had been working as one of the leading producers of *radionovelas* for Radio El Mundo and Radio Belgrano.) Perón eventually declared all radio licenses expired and passed ownership of them to his friends. When he was deposed in 1955, the number of both government and private stations increased. There was a virtual monopoly by wealthy media conglomerates who relied on advertising from U.S. companies doing business in Argentina. In the 1970s military coups repressed all media and put a former German Nazi radio propagandist in charge of all programming at government stations.

Today, Argentine radio broadcasts mostly music and news, with the Top 100 Hits format ranking among the most popular. Although formats are similar to U.S. programming, tango and tropical music is included. And The University of Buenos Aries broadcasts educational programs for credit over their radio station, UBA XXI.

Bolivia

Bolivia has a large number of radio stations because of the size of the country and its mountainous terrain. La Paz has about 40 stations that broadcast in Spanish and two native languages, Quechua and Aymara. The government operates its own stations and tries to maintain control over private stations, with little success due to the vague rules and regulations for telecommunications.

In Bolivia a small grassroots union movement became an important part of the radio system. Miners' radio began in 1946 as clandestine broadcasting to support the miners in the Siglo Viente mines, where conditions were abysmal. The movement was crushed by the government in 1949, but in 1952 when the mines were nationalized, the miners again began to broadcast. Included were news, folk music, education, information, and union news. Local events such as festivals were covered live. Some stations were supported by miners donating a day's pay every month. By 1956 there were 19 such stations. In 1959, fearing a communist takeover of the mines, the Roman Catholic church started Radio Pio Doce. Its broadcasts reached the entire country and spoke against the miners' movement, and twice the miners dynamited its transmitters. However, the competition improved the miners' broadcasts, and by 1963, when the government again closed the miners' radio stations and slashed their already low wages by 40 percent, the church took the side of the miners and Radio Pio Doce became their voice.

By 1974, after numerous coups in which the miners' stations were closed and reopened, the government distributed 5,000 TV sets in mining communities, attempting to get the miners away from the influence of radio. In 1980, the bloodiest coup of all took place and the miners' stations became the focus of resistance, staying on the air for 19 days following the coup. The Bolivian Air Force bombed at least one station, and no one knows how many miners and their families died during this struggle. In the 1980s Bolivia faced economic difficulties and many mines were closed. By 1993 the miners voted to support just one radio station in each of the three regions. Pio Doce also remains as a strong voice for the miners.

Brazil

Brazil is the largest country in South America and has an extensive communication system. There is a nationwide radio system and almost every household has at least one radio set. Radio is a very localized medium and most stations are independent. In major cities stations are usually affiliated with Radio Globo and Radio Bandeirantes.

Radio began in 1920. Most programming was live in the 1920s and 1930s; it included news, variety, and comedy. To support these early stations, radio clubs were formed, with donations given to the stations by the wealthy members. Often these stations were called "Radio Clube de . . ." or "Radio Sociedade . . ." There was also political influence for media owners who could provide coverage for politicians. One of the major networks was Diarios e Emissoras Associadas, which by 1938 owned 5 radio stations, 12 newspapers and 1 magazine led by Assis Chateaubriand. Other large media conglomerates include Radio Bandeirantes owned by Groupo Carvalho and Radio Globo owned by Roberto Marino. By the 1940s recorded music and soap operas were popular and commercial radio networks had been developed, primarily by newspaper chains. National and provincial governments also established networks, and the stations' main purpose, then as now, was to serve areas not economically viable for commercial interests, especially in the Amazon Basin.

Radio set a pattern in Brazil that television was to follow, which included a dominance of entertainment, advertiser supported stations, the importation of a considerable amount of programming in the early years, and, in later years, the use of much Brazilian material. Colgate was one sponsor of these early programs and used the same type of advertising aired in the United States. Variety shows developed a distinct local characteristic, the *show de auditorio* with a host and a live studio audience. This was one of the few formats in which the traditional Brazilian oral folk culture of story telling and song (including samba music, circus-type talk, and folk music) were brought to commercial media.

Today, radio offers diverse programming, market segmentation, and competition, especially in the large cities. Surveys indicate that radio is rated the most popular source of music, while television and newspapers are preferred for news. Entertainment varies by region, but radio dramas, soap operas, and variety shows are common throughout the country.

AM radio is more widely available; in cities, it focuses on music and formats that appeal to less affluent audiences, such as Brazilian country music, Brazilian popular music, sports, and talk. AM talk shows seem to appeal to a wider audience, especially during commuting hours. FM is primarily urban and plays more imported music as well as a great deal of Brazilian popular music. Large cities have 20 to 30 stations, which increases competition and audience segmentation. Many formats resemble U.S. radio.

Three government-sponsored programs must be carried by all stations. *Avoz do Brasil* is broadcast Monday through Friday at 7:00 P.M. and *Projecto Minerva*, sponsored by the Ministry of Education, is broadcast Monday through Friday from 8:00 to 8:30 P.M. Stations are also required to carry a one-hour evening newscast produced by the government, *Hora do Brazil*.

Chile

Chile is the only Latin American country classified as "developed" by the United Nations (UN). In the early 1920s, the Chilean congress considered setting up a BBC-type public broadcasting monopoly, but later decided to follow the U.S. model of private broadcasting. A solid radio industry was formed, and by the 1960s Chilean universities offered degrees in broadcasting and journalism. In 1971 when Salvador Allende nationalized the country's industry, advertising revenue dried up and many stations failed and were sold, mostly to political groups. During this period, about ten of 156 stations remained politically neutral; the others were owned by socialists (33), Christian democrats (29), and communists (28). Supporters of political agendas harassed the stations representing other viewpoints, by cutting their power lines, among other tactics.

After the coup that overthrew Allende, former staff members of Radio Magallanes moved to Moscow and were given a transmitter for special broadcasts to Chile. Other networks include the government-operated Radio Nacional (National Radio); Radio Chilean, run by the Roman Catholic church; Radio Mineria, which takes its name from the mining interests but is a source of reliable news; Radio Agricultura, which focuses on news and programs of interest to Chile's farming community; and Radio Tierra, established in 1983, which claims to be the first all-woman radio station in the Americas.

Colombia

Early on Colombia used radio for education to rural communities. Hundreds of thousands of transistor radios were given away by political groups and religious organizations. The radios were locked onto frequencies that broadcast housing repair, nutrition, health, history, and geography. Radio Sutatenza was a pioneer in educational radio. Begun in a small village in 1948, its purpose was to educate the rural adult compesino. A multimedia approach was used to eliminate illiteracy among the 8 million rural adults. Areas of education included health, reading, arithmetic, economy and work, and spirituality. A newspaper, *El Compesion,* was used to support the lessons. Also developed were rural libraries, extension schools, and institutes for farmers. Radio Sutatenza has served as a model for countries in Asia and Africa, as well as others in South America; however, due to financial difficulties it was sold to the Caracol network in 1990.

Today, government stations still provide this type of programming; private stations broadcast music, sports, and news. The National Institute of Radio and Television (Inravision) produces programs for the government-affiliated stations, but many are privately owned. Radio is viewed primarily as a source of entertainment or culture, rather than news. It has operated more freely than television and the two largest networks reflect opposing political viewpoints: liberal (Caracol) and conservative Radio Cadena Nacional (RCN). Caracol is the largest AM/FM network, with powerful transmitters that reach the entire country; the next largest network is RCN. Both have a large number of network-owned stations throughout the country that produce local news and information programs but also act as regional news bureaus. By law, they must broadcast some shortwave for Colombians out of the country, and the state imposes some guidelines to ensure equal time for political candidates. Caracol has an affiliate in Miami, Florida, that rebroadcasts from Bogota but also includes local news and information.

Other privately owned networks include Todelar, Super, and the evangelical Colmundo. Univalle Estereodio Station of the University of Valle from Cali broadcasts live on the internet 24 hours a day. Begun in 1995, it is concerned with news and culture and also broadcasts BBC and Radio France International.

Ecuador

Ecuador has over 260 commercial radio stations, including ten cultural and ten religious stations. The Voice of the Andes has operated for more than 50 years as a shortwave, evangelical Christian station, supported largely by contributions from the United States. Begun in 1931, it broadcasts in 16 languages and 22 dialects of the Quechua languages. In 1994, it began *America Latina via Satelite* (ALAS) as a joint project of HCJB World Radio and Trans World Radio to deliver quality gospel programs to more than 60 affiliate stations in 12 Latin American countries.

Media ownership has remained in the hands of a few large interests. By the late 1980s, all media were privately controlled, except for Radio Nacional which at the turn of the century was still operated by the National Communications Secretariat (SENAC-Secretaria Nacional de Communicaciones). The government controls the allocation of radio and television frequencies. The Febres Cordero government used the media in an effort to gain support for its free market economic policies and in the process infringed on press freedoms. In late 1984 the government temporarily closed five radio stations after they broadcast criticism of the government.

There is a well-developed infrastructure, so radio is better established and more professional than in some other South American countries. The Roman Catholic church has a large network run by the Franciscans. Programming usually includes Ecuadorian music and reciting of the rosary or inspirational thoughts for the day. The Bahai faith sponsors its only shortwave station here. There also is a trade union station owned and operated by the taxi drivers union in one province.

The largest commercial station, Radio Quito, was begun in 1940 and is owned by the largest newspaper, *El Comercio.* In the beginning, this station was affiliated with CBS and the BBC. In 1941 it carried the Joe Louis and Chilean Arturo Godoy boxing match and used loudspeakers in plazas for those who did not own radios. In 1949 Radio Quito did a takeoff on Orson Welles' "War of the Worlds" that created a riot in the city. The radio and newspaper offices were burned and 20 people died. Today, its programming is mostly news and sports.

Founded as a broadcast station in 1940, Acerca de CRE is now also transmitting on the internet with programs including information and sports events.

Paraguay

Although Paraguay was the first South American country to enjoy telegraph service, later communication systems developed slowly and the military maintained control over the media. In the 1980s, only one Paraguayan in 20 owned a radio. In every category of ownership, Paraguay ranked last in South America, well behind less developed countries such as Bolivia and Guyana. In the late 1980s there were 52 radio stations in Paraguay, only three of which were independent. One of these was Radio Caritas, sponsored by the Roman Catholic church. The most important independent radio station (until its closure in January 1987) was Radio Nanduti, which had a live phone-in program that frequently aired complaints about corruption and lack of democracy. In July 1983 station director Roberto Rubin was arrested several times, and in April and

May of 1986 the station was attacked by Colorado vigilantes. After months of jamming and other harassment, it was finally forced off the air.

Today, commercial and private stations are able to broadcast a range of views, including those of the opposition party. Radio Nanduti is back on the air. Programming includes news/talk, classical, jazz, pop, rock, and retro. A popular retro station, Radio Venus, can be heard on the internet at www.venus.com.py.

A popular form of folk music heard on many radio stations is harp music. Spanish and Guarani (the native language) are heard over the air. *Radio Oberdira,* located in southern Paraguay, is available on the internet. The station uses block programming, as various music styles are heard 20 hours per day, offering something for everyone.

Peru

Radio began in Peru in 1921, and military rulers maintained strict control over all forms of media. In 1971 a general telecommunications law gave the government a 25 percent ownership interest in all privately owned radio stations. A new telecommunications law in the 1980s, however, gave stations more freedom. In 1992 President Fujimori seized control of all media in the name of national security.

Today, Peru's media has much more freedom under President Alejandro Toledo, elected in 2001. Privately run radio stations are the most popular; state run media have small audiences. While broadcasters have regained some independence, the Fujimori-era scandals (Congress dismissed Fujimori on the grounds of "moral incapacity") have adversely affected public confidence in media.

Radio broadcasts consist mainly of music and news, with 30 percent of total time given to advertising. Regulations require that 65 percent of programming and 100 percent of advertising on radio be of domestic (Peruvian) origin. Small stations in rural areas are often not officially licensed and only broadcast part of the day, as electricity is often only available in the evening. Because of the country's geography, shortwave transmission is used. In the early years of radio, AM was in the cities and could not reach outlying areas. By the 1970s there were many shortwave stations, especially in the smaller towns in the rural north, which is either mountainous or jungle. Rural audiences prefer block programming and a popular folk music, *huayno.*

Many stations fail because competition is strong and costs are high. Stations often use kerosene generators for power, and at times there may not be enough money to buy the kerosene, as a station's income may only be 10 to 20 dollars per day. Due to economic factors, stations may be off the air for days, weeks, or months at a time. Radio Sol Armonia began on 1 January 1984. It is devoted to classical music and educational programs for the country. Although it is associated with commercial Radio El Sol, sponsors from all over the world provide money and material donations. International radio stations and broadcasting organizations that provide programs include Deutsche Welle, Radio France International, Nederland Radio, Radio Exterior Espana, Radio Canada International, and Swiss Radio. Programs are a mixture of lectures and conferences on cultural themes. The station publishes a monthly bulletin with information on specials and news and is available through the internet. Radio Sol Armonia also produced a cassette dedicated to the German-Peruvian composer Rudolfo Holzmann. In 1998 they released their first CD of Peruvian baroque music. The service has received several awards from Brazil, Chile, and the Peruvian congress for its cultural work. The station is owned by a non-profit organization, Asociacion Cultural Filarmonia.

Uruguay

The Uruguayan Broadcast Corporation has control over all communications media in Uruguay. Radio FEUU was a pirate station started by university students in Montevideo in 1995. They demanded a better budget and free radio stations as alternatives to the commercial system. However, the government eventually shut them down. All parts of the country receive at least one AM radio station, but 25 percent of its radio stations are found in the Montevideo area. Ten of the AM stations also broadcast on shortwave frequencies to reach a larger audience, both at home and abroad. All stations except for one government-owned transmitter are commercial and broadcast in Spanish.

Begun in 1924, Servicio Oficial de Difusion Radiotelevision y Espectaculos (SODRE) is the national government radio station. It primarily sponsors medium wave and FM transmissions. There are some international shortwave broadcasts for Uruguayans abroad. An internet station, Radio El Espectador, offers a mix of news, sports, information, and travel.

Venezuela

Venezuela's broadcast industry is a quasi-monopoly that was influenced during its development by foreign investment and personnel from the United States and Cuba. After the fall of dictator Perez Jimenez in the late 1950s, democratically elected leaders built a relationship with private media that allowed for largely unregulated development of commercial broadcasting. The government retains some control over content of an adult nature. The government operates the *Radio Nacional* network, which informs listeners (especially those in rural areas) about education, agriculture, and civic matters. Most stations are concentrated in Caracas, but transmitters are located throughout the country. There is also an international shortwave service.

There is very little religious broadcasting. Colombian radio can also be heard in western Venezuela.

FM is a relative newcomer. In the early 1950s, the government had a law that limited service to one radio station per 60,000 people, and by the time FM radio stations were developed in the 1970s, there was no room for them. The law made it impossible to operate FM stations except in the urban areas with sufficient populations to meet this requirement. In 1988 the government decided that the law applied only to AM stations, and officials began awarding numerous FM frequencies to political favorites. They also added Class D FM for lowpower FM stations in small towns. Some industry experts feel that the oversaturation of radio is a major crisis in Venezuelan broadcasting.

A unique type of financial support is used by Globo FM. Announcers contract with the station for hours of airtime and then are responsible for finding ads. Listener growth in FM is primarily among younger people. Radio Rumbos (anchor of the largest radio network) runs two AM stations and feeds newscasts to affiliates using a satellite link. Radio Mundal also has satellite uplinks and broadcasts news all over the country.

Other Nations

The non-Latin countries of South America are French Guiana, Guyana (a former British colony), and Suriname (a former Dutch colony). French Guiana operates a national radio broadcast, Radio-Télévision Française d'Outre-Mer (RFO), Radio FM from Cayenna, with several repeater stations throughout the country. Two radio stations are on the internet, one of which is part of a ten-station group of French-language stations providing coverage across three oceans.

In Guyana today, almost all people own a radio set. There is no clear telecommunications policy and the government controls what were once private stations. There are two AM stations and one FM station in the capital and one FM station in Lethem. None are on the internet. The government runs Radio Roraima and the Voice of Guyana.

Suriname has total radio saturation with one radio set per 1.5 persons. Reflecting its unusual past are broadcasts in Hindustani, Javanese, and Dutch. Four stations are available on the internet.

For most of South America, radio remains the most accessible and affordable mass medium. According to Bruce Girard, low production and distribution costs have made it possible for radio to focus on local issues, provide a local perspective on world issues, and to speak in local languages. For example, Quechua is almost absent from the television screens in Bolivia, Ecuador, and Peru, but in Peru it is estimated that 180 radio stations regularly offer programs in that language. Radio is also used in agricultural extension programs, education, and the preservation of local language and culture.

However, local and regional radio is threatened by national and international networks. For example, in Peru three satellite networks, broadcasting from the capital via repeater stations throughout the country, have more audience share than the 40 largest provincial stations put together. In Argentina and Brazil, national multi-media empires have built satellite radio networks that have converted hundreds of independent local radio stations into repeater stations that provide programs produced by the nation's capital. Technically superior, these networks have resulted in the loss of choice, local information, and alternative perspectives. Although the internet has the potential to develop a local role like radio's, internet access is very limited and requires an infrastructure that most countries in South America cannot afford. Currently, most radio programming remains local. The future of radio in South America could be one of consolidation and privatization, which will result in ownership and control in fewer hands and the loss of local and alternative voices.

MARY E. BEADLE

See also Brazil; Cuba; Developing Nations; Mexico

Further Reading

Cole, Richard R., editor, *Communication in Latin America: Journalism, Mass Media, and Society,* Wilmington, Delaware: Scholarly Resources, 1996

Fox, Elizabeth, *Latin American Broadcasting: From Tango to Telenovela,* Luton: University of Luton Press, 1997

Girard, Bruce, "Radio Broadcasting and the Internet: Converging for Development and Democracy," <www.communica.org/kl/girard.htm>

O'Connor, Alan, "Miner's Radio Stations in Bolivia," *Journal of Communication* (Winter 1990)

Patepluma Radio: International Radio Home Page, <www.swl.net/patepluma/>

Rufino, Padre Lobo, and Encinas, Padre Alfredo, *Radio y comunicacion popular en el Perú,* Lima: Centro Peruavno de Estudio Sociales, 1987

Strauhbaar, J., "Brazil," in *The World of International Electronic Media,* edited by Lynne Gross, New York: McGraw Hill, 1995

Ulanovsky, Carlos, Juan Jose Panno, and Gabriela Tijman, *Dias de radio: Historia de la Radio Argentina,* Buenos Aires: Espasa Calpe, 1995

Zago, Manrique, editor, "A Confluence of Cultures," in *Estados Unidos en la Argentina; The United States in Argentina* (bilingual Spanish-English edition), Buenos Aires: Manrique Zago Ediciones, 1997

Zona Latina, <www.zonalatina.com>

South Pacific Islands

Radio in the Pacific presents its own set of unique challenges to radio broadcasting that include remoteness, multiple languages, shortage of skills and resources, irregular transport, awkward low lying coral atolls, high humidity, salt spray, cyclones, electric thunderstorms, and less than certain power supplies. The South Pacific Ocean is home to a number of small island nations affected by the tyranny of distance. Vast oceans exist not only between states but also within them. The 84,000 citizens of Kiribati live on small islands dotted over 2,131,000 square kilometres. The state of Niue has fewer than 2,000 people. Only two countries other than Australia and New Zealand, treated separately, have populations in excess of a half million: Papua New Guinea (4.2 million) and Fiji (900,000). Of the 18 jurisdictions, nine microstates have independent sovereignty (Fiji, Kiribati, Nauru, Papua New Guinea, Samoa, Solomon Islands, Tonga, Tuvalu, and Vanuatu) and two have an independent association with New Zealand (Niue and Cook Islands). The remainder have some form of legislative arrangement with other countries (principally France, the United States, the United Kingdom, and New Zealand).

Papua New Guinea

Radio began in the South Pacific during the colonial period and continued to play a significant role as the region's states sought and gained independence. An early radio broadcast station was set up by Amalgamated Wireless in Papua New Guinea in October 1935. The station took advertising and provided a two-hour daily schedule except for Sundays when, in deference to the strong Christian traditions in the community, the station was silent. The station was not successful and closed in 1941.

In 1944 the U.S. military requested that Australia, a U.S. wartime ally and Papua New Guinea's neighbor, establish a radio station in Port Moresby for the United States and allied military. Following the war, Australia took administrative responsibility for Papua New Guinea and the Australian Broadcasting Corporation expanded from Port Moresby, developing a nationwide service that by 1963 could be heard over much of the territory. Programs were 40 percent indigenous and 30 percent expatriate; the rest were of general interest. During the 1960s the administration sought to develop a more proactive government broadcasting policy that could act as a barrier to the influence of broadcasts from the less stable Dutch and Indonesian northern neighbors. The Administration Broadcasting Service opened its first station in Rabul in 1961 and by the decade's end had a range of stations in densely populated areas with the aim of increasing awareness and acceptance of what the government was trying to do.

The two services combined following independence (in 1975) when the Papua New Guinea National Broadcasting Commission was established. It is the largest broadcasting organization in the Pacific islands with over 500 employees currently operating 19 provincial stations and three networks: the *Kundu* (provincial), the *Karai* (national), and Hits and Memories, FM100 (commercial) services. In 1994 private radio began in Port Moresby with the launch of Nau FM, a station locally owned but operated by Communications Fiji. A second station, Yumi FM, was added in 1997, broadcasting in local pidgin. Local stations have sprung up, including FM Central in Port Moresby and Lae. In addition a Christian radio network, supported by the Roman Catholic church, is being considered.

Fiji

Fiji, with a much smaller and more compact population, has been well served with radio coverage by the Fiji Broadcasting Commission, which was established with expatriate British Broadcasting Corporation (BBC) management playing a key role. In 1999 the Fiji Broadcasting Commission was corporatized, changing its name to the Fiji Broadcasting Corporation (FBC).

The FBC currently offers numerous radio services: two AM networks that reach most of the country's population centers—Radio Fiji One (in Fijian) and Radio Fiji Two (in Hindi)—and several FM stations including Bula 100 (in English), Bula 102 (in Fijian), and Bula 98 (in Hindi) and 2-Day FM.

Fiji was the first Pacific island state to permit private commercial radio broadcasting. Private broadcaster Communications Fiji, established in 1985, operates five networks and stations. Two are English language networks, FM96 broadcasting a contemporary Hit/Pop music format and Legend FM broadcasting a "Gold" format. It also broadcasts Radio Navtarang, a Hindi language network, and Viti FM, a Fijian language network. Both these programs are in an adult contemporary format. They also rebroadcast the World Service program of the BBC. Others include Radio Light, Radio Hope, and FM97, all independent Christian radio stations, as well as Z-FM, a private station operating in Lautoka and Nadi. Currently the radio market in Fiji is very "flexible," with Christian and other private stations coming and going depending on enthusiasm, finance, and opportunity.

Other Nations

Tuvalu also has a broadcasting system developed during colonial times by expatriate staff from the BBC. Local Tuvaluan staff even went to the United Kingdom for training, as did staff from other colonial broadcasting services. Radio Tuvalu, broadcasting on a medium wave station for about six hours each day, serves this nation of just 10,000 people.

Other small nations also maintain radio services. Niue (population 2,000) has Radio Sunshine, a part-time station providing local and overseas news, notices, advertisements, and light entertainment. Nauru (population 10,000) has Radio Nauru, a state-run service broadcasting 18 hours per day and providing local news coverage as well as rebroadcasts of news from Radio Australia. In Kiribati (84,000) the government-owned Broadcasting and Publications Authority operates Radio Kiribati. It has two transmitters to cover its far-flung atolls, and broadcasts (mainly in the local language) a morning and evening program each day.

The Cook Islands (20,000) has the Cook Islands Broadcasting and Newspaper Corporation (CIBNC), previously state owned but now privately owned and operated by religious broadcaster Elijah Communications. It operates on medium wave and shortwave broadcasts in English and also in Cook Islands Maori. It carries national and regional news bulletins and retransmits news from Radio Australia and Radio New Zealand International. A competing private FM radio station, KC FM, covers the main island of Rarotonga. The Kingdom of Tonga (99,000) has a state broadcasting system, the Tonga Broadcasting Commission, operating a medium wave station and an FM music station. A private Christian radio station, 93FM, also operates in Tonga. Two territories, Pitcairn Island (population 47) and Tokalau Islands (1,900) do not have broadcasting services.

In the Solomon Islands (404,000), the state-owned Solomon Islands Broadcasting Corporation operates the national AM service, Radio Happy Isles, as well as two AM stations on distant islands and a new youth-oriented FM service (Wan FM) broadcasting in both English and pidgin. In 1996 Z100, a locally owned private station, began broadcasting. In 1999 Paoa-FM ("Power FM" in pidgin) began operating nationally. This station is owned by a local newspaper *(The Solomon Star)*.

Vanuatu (population 178,000) has a state owned and operated Broadcasting and Television Corporation. This former French and British controlled nation recognizes three official languages: Bislama (local pidgin), French, and English. Radio Vanuatu broadcasts programs in all three. The corporation also operates a commercial FM station in the capital, Port Vila.

Samoa recognized 50 years of continuous radio broadcasting in 1997, as the state-owned radio station 2AP began operations in 1948. Its strong transmitter means that its programs are widely known through the Pacific, with the signal being received as far away as Fiji and the Cook Islands. An FM station began operation in 1992, and both stations operate commercially. Complementing state radio are two private FM radio stations, the fully commercial Magik FM and Radio Graceland, a Christian broadcasting station.

External Support

Many of the South Pacific island states rely on international aid for development and, in many cases, survival. A number of the region's state broadcasters have also had such support. Australia, New Zealand, and other countries provide direct financial support to island broadcasting. International organizations such as the United Nations Educational, Scientific, and Cultural Organization (UNESCO) and the Commonwealth Broadcasting Association have funded broadcasting activities in the islands. Two German agencies, Friedrich Ebert Stiftung and Deutsche Gesellschaft für Techniche Zusammenarbeit, have provided extensive funds to promote the development of broadcasting. Funds have been used to support regional training initiatives (PACBROAD), the establishment and operation of a regional independent news service (PACNEWS), a professional association of news organizations (Pacific Island News Association) and a professional association of broadcasting organizations (Pacific Islands Broadcasting Association). In 2003 the professional association merged with the news association to form one group.

Challenges

Three significant challenges face South Pacific islands radio broadcasting. First, aid support is neither guaranteed nor consistent. Thus organizations that are created and developed using such funds struggle to survive when funding is reduced or ceases altogether. All of the regional organizations mentioned above have had this experience. Second, independent journalism in the Pacific has always had an uncomfortable relationship with the region's conservative and cautious governments. Historically all of the broadcast media and much of the print media have been state controlled or influenced. Some governments object to criticism, reports that reflect badly on authorities, or any form of independent investigative reporting. Offending journalists have been treated harshly in the past, losing their jobs and even facing imprisonment. Pacific island states are slowly adapting to a more critical media environment but many see the form of freedoms and independence enjoyed by Western news media organizations as not appropriate nor in the best interests of their culture and politics.

Finally, and perhaps most significantly, many South Pacific island governments are following recent international trends and reducing expenditures. State media organizations are facing both budget cuts and requirements to provide a revenue stream. There is a move to privatize state media in some countries. Papua New Guinea is currently considering the privatization of several state-owned assets, including broadcasting. The

Cook Islands Broadcasting and Newspaper Corporation was privatized amid controversy in 1997, and others such as Fiji Broadcasting and the Samoa Broadcasting Services have incorporated their broadcasting organizations, requiring them to act as business entities, although they still receive significant state funding.

Commercial operators are a big threat to state broadcasters, which are often steeped in tradition, slow to react to market forces, and not responsive to audience wants. The independents are thus taking large audience shares. Navtarang, FM96, and Viti are in the first-, second-, and third-rated positions in the Fijian market, with nearly 70 percent of all listenership. In Papua New Guinea, Yumi FM and Nau FM are first and second in the market. Paoa FM and Z100 hold similar positions in the Solomon Islands (all 1999 figures). Meanwhile, despite the recent introduction of television (serving urban elites) in many countries, radio remains the only effective national communication system in most South Pacific island countries. But with competition from private radio and television, many national broadcasters have trouble servicing outer island communities, which thus risk increasing isolation and estrangement. However, as new technology is providing more efficient and cost-effective delivery systems, it is only a matter of time until even the smallest island nation will receive a commercial network feed from one or another of its larger neighbors.

BRIAN T. PAULING

See also Australia; New Zealand

Further Reading

Lent, John A., editor, *Broadcasting in Asia and the Pacific: A Continental Survey of Radio and Television,* Philadelphia, Pennsylvania: Temple University Press, 1978

Pacific Islands Communication Journal

Ryan, P., editor, *Encyclopaedia of Papua and New Guinea,* 3 vols., Melbourne: Melbourne University Press, 1972 (see article entitled "Broadcasting")

Seward, Robert, *Radio Happy Isles: Media and Politics at Play in the Pacific,* Honolulu: University of Hawaii Press, 1999

Soviet Union. *See* Russia/Soviet Union

Sportscasters

Sportscasters are the play-by-play and color announcers of sporting events and the hosts of shows that highlight games and athletic performances. From radio's earliest days, live sporting events have been surefire audience gatherers. Beginning in 1921 with Harold Arlin, announcer of the first professional baseball game over radio station KDKA, and lasting all the way to Bob Costas hosting the National Broadcasting Company's (NBC) 21st century's Olympic Games, sportscasters have been central to the popularity of electronic coverage of games, events, and sports-related celebrations. Indeed, a broadcast network once televised the first half of a National Football League (NFL) football game without any announcing merely to demonstrate to television audiences how crucial the announcers were to their entertainment and understanding of the game. Radio, of course, could not even attempt such a voiceless experiment.

Background

Although nearly all sportscasters began their careers as play-by-play announcers, a few came laterally from careers as players or coaches. Nonetheless, most sportscasters can point to years of routine play-by-play announcing of baseball, basketball, soccer, hockey, and football games, as well as tennis and golf competitions, swimming, wrestling, and track meets, at the local high-school and college levels before their moves into regional or national radio sports announcing of college and professional sports. At the regional and network levels, many began as spotters or statisticians before they were permitted on the air.

Although sportscasting has often been a profession of white men, African-American sportscasters began gaining small network on-air roles during the 1980s; brothers Greg and Bryant

Gumble, for example, moved into sportscasting, Greg permanently. Bryant, during his 15 years as host of *The Today Show* (1982–97), served as the main anchor for the 1988 Olympics. Greg hosted CBS's *The NFL Today,* NBC's *NFL Live,* and the 1994 Lillehammer Olympics, later becoming the host of HBO's *Real Sports.* Former Minnesota Viking Ahmad Rashad anchored the pregame *NBA on NBC, NBA Stuff,* and *Sports World* for NBC, while also serving as studio host and/or commentator for several Olympics (1988, 1992, 1996).

Women sportscasters first earned national visibility in the 1990s. Their voices quickly became part of radio as well as television sportscasting. Like their male counterparts, such nationally known sportscasters as Robin Roberts of ESPN, Leslie Visser of Columbia Broadcast System (CBS) and American Broadcasting Company (ABC), and Hannah Storm of NBC earned their network positions after years of local play-by-play announcing. Others, such as Mary Carillo, widely respected for her CBS and Olympic tennis reporting, came to announcing after playing and coaching.

Many luminaries of the television era began their careers in radio, including Jim McKay, Brent Musburger, and Curt Goudy. Moreover, these and such other famous television sportscasting personalities as Howard Cosell, Dick Vitale, Pat Summerall, and John Madden have regularly contributed to radio sports programs, generally as commentators within sports newscasts or sports talk programs. One of the most controversial of present-day sports commentators, Frank DeFord, contributes a weekly on-air essay to National Public Radio (NPR) and appears regularly on Home Box Office's (HBO) *Real Sports,* in addition to writing columns for *Sports Illustrated.*

Generalists and Specialists

Sportscasters come in two varieties: the generalists, who announce and host all kinds of sporting events and contests, and the specialists, who are experts in a single sport. Jim McKay and Bob Costas epitomize the generalists. Each has announced hundreds—if not thousands—of baseball and football games and hosted many Olympics. McKay was also the much respected host of the long-running *Wide World of Sports,* ABC's pioneer sports anthology TV program that covered numerous sporting events from around the world. Costas is especially known for his extraordinarily meticulous preparation before broadcasts, sometimes devoting months to thorough researching and writing before a major event. But in the very early days of radio, all announcers were expected to be able to move from sport to sport, just as newscasters reported on any newsworthy topic. The best generalists often migrate to the network level, and many big-name sportscasters of ABC, CBS, NBC, and Fox still call several different sports.

Announcers hired by teams (or in radio's early years, by advertisers) are usually specialists, either because they must be seen by fans as home team partisans or because they have special expertise. Specialists range from team baseball or basketball announcers (often doing play-by-play coverage of just one or two college and professional teams in a region) to professional auto racing or horse racing experts who bring a notable depth of knowledge to announcing their chosen sport. Professional football on radio is often the audio portion of the concomitant telecast with little added, but other sports with long moments of inaction (such as baseball and car racing) or very swift action (such as basketball and horse racing) require announcers who speak continually just for radio. These sportscasters need to be able to provide a wealth of background and statistical information for listening fans, who often can be highly critical of misinformation and outright errors.

Television versus Radio

Television has places for both the generalists and specialists, and at the network level it attracts the best of both groups. Each broadcast and cable sports network maintains a stable of on-camera house anchors, hosts, reporters, and commentators. They announce the big games and events on network television, especially all of NFL football, the top ratings generator, and the mega-events in golf and tennis, sports that attract the upscale executives sought by some advertisers. Radio, a more local or regional medium, provides jobs for hundreds of sometimes relatively anonymous announcers who call play-by-play for thousands of near-daily professional baseball and basketball games and college track and field, soccer, tennis, golf, and other sports for local radio over the course of a year. In each market, at least one commercial radio station pegs its image (its brand-name) on its status as the local provider of sports, and its coverage is often supplemented by even smaller college and/or high-school radio stations. As in most media, the large markets get the stars, and the small markets get the beginners.

Beginning in the 1950s and lasting well into the 1970s, sponsors controlled sportscasting to a degree hard to imagine today. Falstaff Beer and Gillette, for example, hired and fired their own announcers, and most commercials were read—and often acted out—by the announcers live on the air, lending whatever credibility they had to the product. The legendary Red Barber, for example, had to wear a gas attendant's cap during one set of commercials and dump Wheaties into a bowl in another. Being able to pour beer so that it made just the right noise for radio or just the right foaming head for television had enormous career implications in the era before audio cartridges and videotape. Big advertisers also intervened to avoid giving competitors even tiny on-air advantages. In one story, Chevrolet's objections led announcers to refer to the Southern Methodist University football team as the "Ponies" for the entire 1967 season, when the team was really the

"Mustangs," coincidentally the name of Ford's hot-selling car that targeted the collegiate market.

Recreations and Early Play-By-Play

During the 1920s, outside of Opening Day and the World Series, baseball fans had to follow teams through the local newspapers except in such large cities as Chicago, Boston, and Detroit (but not New York). Many owners were afraid that radio would hurt ticket sales. As late as 1936, only 13 teams in the major leagues consistently broadcast their games. Even then, few away games were broadcast because the team owners saw radio as a means of publicity and used it to raise stadium attendance. Well into the 1950s, outside of the largest markets, much of radio sportscasting consisted of summaries and recreations of regular season games. Recreations were dramatized studio versions of away games demanded by radio stations because announcers did not travel with the teams. The sportscasters usually had only the Western Union ticker from the field or a bad telephone line and had to make up action to account for the truncated facts coming from the field. Some stations wanted summaries of key action, sometimes a day or more after an event (before the invention of audio tape). Such recreations and summaries tended to resemble fantasies rather than the actual games. The announcers had far too little information and told long stories, invented plays, and concocted events in the grandstands to fill the airtime.

Because the audience at home could not see the action for itself, even during live events announced from the field, early sportscasters were free to make up events to keep games interesting or to account for game delays and crowd noise in the stadium. Bill Stern's reputation was as a premiere storyteller rather than an accurate reporter, for example, and he felt free to stick in lateral passes and fumbles to account for his mistakes about who was carrying the football. Other announcers routinely invented unbelievable catches or fan disputes in the stands to explain roaring crowds. As late as 1951, Gordon McLendon (the Old Scotchman) broadcast recreations as *The Game of the Day* over the Liberty Broadcasting System (LBS). These broadcasts were popular for their variety because, rather than sticking to one team, McLendon chose popular games for recreations from all over the country. Much of the concern for exceptional accuracy in announcing exhibited by Red Barber, for example, came from his distaste for recreations and summaries that had little to do with the reality of the games that were played. Barber was determined to report in detail what actually happened as it happened on the field. And by Barber's era, announcers had begun getting support from assistants, spotters, and statisticians who kept the facts accurate. In the 1920s and 1930s, radio sportscasters worked without scorecards, press guides, or a history of player or game statistics. A great deal had to be imagined.

While the networks have carried league championship and World Series games live since their inception, regular-season baseball and other sporting events got little or no national airtime, although baseball became a mainstay of local radio during the Depression of the 1930s. But during World War II, the Armed Forces Radio Service relayed shortwave broadcasts of daily baseball to service men overseas, and their immense popularity on wartime radio continued at home after the war ended. It was not until the postwar period that team or network announcers were allowed to travel with the home teams, and all games (at least those involving the most popular teams) began being delivered live to fans on radio. By the 1950s, sports had become a mainstay of the electronic media, with the clashes of football on the air in winter and the mellow sounds of baseball filling summer afternoons around the nation. Fifty years later, the pace of games has become faster, the variety of radio voices is much greater, and the competition from media coverage of other leisure activities splinters the audience. To compete, most radio stations and sports networks maintain interactive websites, and the internet provides up-to-the-minute sports scores, live play-by-play, ESPN's *SportsCenter,* as well as archived commentary, interactive games, and chat rooms devoted to fans of a sport or a particular team.

All-Sports Radio

One recent change in radio has been the rise of the all-sports format. Sport is especially successful on radio because the public has become so mobile, and radio lets fans stay in touch. In the largest markets, sufficient numbers tune in to successfully support one-format stations; in smaller markets, live sports coverage tends to be combined with syndicated talk and satellite music to keep costs down. Between 1988 and 2000, the number of stations claiming sports formats went from zero to more than 600. Such stations provide both live and taped play-by-play game and event coverage accompanied by constant updates on the day's game scores and sports news about individuals and the industry. For affiliated stations, part of the programming usually comes from the staffs and commentators of the network television sports divisions; in addition, syndicators, such as Sports ByLine USA, supply sporting events and a range of commentary to hundreds of subscribing stations, including the 400 stations of Armed Forces Radio. ESPN provides an audio feed of its celebrated *SportsCenter* that can fill the overnight time period or portions of daytime radio schedules. Mixing the voices of famous broadcast and cable sportscasters with local voices establishes an aura of credibility and provides a connection to big-time sports that listeners recognize, positively branding even the smallest station. Some stations specialize in an irreverent approach to sports, intended to appeal to younger fans; others take a more traditional, straightforward approach. Sports fans tend to be very loyal and listen for long periods of time to radio, and because listeners are largely male, stations carrying sports generally have strong appeal to advertisers. The first station to adopt an all-sports format, WFAN (AM) in New York was for years the top-billing radio station in the country. It combined a heavy load of play-by-play with a very local orientation.

Talking Personalities

Another big change from sportscasting's early days lies in the amount of time broadcasters have to talk on the air. With only a few commercials to insert, early sportscasters had to fill the time between innings and plays. They had the leisure to tell stories and discuss events on the field or dugout, in effect creating personalities for themselves and the players that resulted in long-term relationships with listeners. In contrast, present-day announcers can only talk during the action on the field, because commercials and cut-aways to studio announcers fill breaks in games. The upshot is that such beloved personalities as Mel Allen, Harry Caray, and Bill Stern would not be hired today. Management wants swift talkers who stick to a schedule, get in and out of commercials, and keep the games to the point—not storytellers or celebrities who generate controversy.

One current catchphrase is that fans don't tune in to hear broadcasters talk, but they did tune in to hear Red Barber, Mel Allen, Harry Caray, and Lindsey Nelson, as well as others of the great early radio and television sportscasters. For most fans, these much-beloved broadcasters were and are as much a part of the games as the players.

Major Sportscasters

More than a dozen announcers and commentators stand out in the history of sports radio; most of them have been elected to one of the halls of fame (Baseball Hall of Fame, Boxing Hall of Fame, Radio Hall of Fame, National Broadcasting Hall of Fame, etc.). Some of the younger individuals went on to careers in televised sports, but all made important contributions to the development of sports on radio.

Mel Allen (1913–96)

Idolized as "The Voice," Melvin Allen Israel was the mellifluous announcer for the New York Yankees for 25 years, including the six years in the 1950s when they were the annual World Series champions. He became the Yankee voice just when recreations of away games were dropped in favor of sending key play-by-play and color announcers to all games, and he benefited from this move, which allowed him to announce more games. Considered in his heyday the very best play-by-play man ever, Allen announced 20 World Series, did play-by-play for 24 All-Star Games, and broadcast more than 20 college bowl games. Allen's soft Alabama twang was instantly recognizable, and his trademark phrases ("How about that!" and "Going, going, gone!") have become legendary sportscasting calls. He received every award offered to a sportscaster, often many times over, demonstrating his peers' enormous respect and admiration for his sports knowledge, his communication abilities, and his love of baseball and football. Allen was voted the Sportscaster of the Year for an extraordinary 14 years in succession.

Harold Arlin (1895–1986)

Not to be forgotten, Harold Arlin was the first announcer of sports, broadcasting the first professional baseball game over KDKA in 1920, and in the next year, the first tennis match and the first football game (a college game between the University of Pittsburgh and the University of West Virginia). Arlin was not a career sportscaster, and he moved into corporate relations after about six years. But he left a unique legacy for radio sportscasting history.

Red Barber (1908–92)

A tally of sportscasting greats has to include patriarch Walter Lanier "Red" Barber, who was one of the early popularizers of daily radio play-by-play and one of the most respected and influential mentors of baseball announcers. The "Ol' Redhead" had one of the longest careers in broadcasting—about 63 years. He broadcast the Cincinnati Reds from 1934 to 1938, then became the voice of the Brooklyn Dodgers, the team with which he is most associated in sportscasting history, for more than two decades. However, he subsequently announced the New York Yankees and the New York Giants from 1954 to 1966 alongside such greats as Mel Allen, Ernie Harwell, and Vin Scully. Barber announced dozens of World Series, and after the Yankees fired him in 1966 (in part for commenting on the sparse attendance at a September home game), he ended his long career with a decade of Friday morning sports commentary on NPR. He anchored the first televised major league baseball game in 1953—to a minuscule audience. In contrast, the radio audience for the 1942 World Series, announced by Barber and Mel Allen, was estimated at 25 million. Known for his intense pre-game preparation and exceptional memory for and recognition of players' styles, on the air Barber charmed listeners with self-effacing humor and vivid homespun imagery. A fastidious and reserved man, he worked at a style that was precise and detached, reflecting integrity and professionalism. One of his greatest contributions was his quiet acceptance of Jackie Robinson in 1947, the first African-American player in major league baseball. Among many other honors, in 1978 Barber and Mel Allen were the first broadcasters to receive Ford C. Frick awards (Baseball Hall of Fame). Barber was installed in the National Sportscasters and Sportswriters Hall of Fame in 1973, inducted into the American Sportscaster Hall of Fame in 1984, and elected to the Radio Hall of Fame in 1995.

Jack Brickhouse (1916–98)
Before, during, and after World War II, Jack Brickhouse was the first daily announcer of the Chicago Cubs and White Sox, calling play-by-play games in his breezy style and imbedding his characteristic "Hey-Hey!" and "Back, back, back . . .That's it!" in the hearts of four decades of baseball fans. Atypical for local sportscasting, for 20 years he called the city's rival baseball teams on WGN radio and television: the White Sox from 1940 to 1967 and the Cubs from 1941 to 1981. Perhaps because his manner was warm and friendly, he was popular with both White Sox and Cubs fans, and like many generalists of the 1940s and 1950s, Brickhouse announced other local events in addition to baseball. He called Chicago Bears games for 24 years, and Chicago Bulls, Zephyrs, and Packers basketball, along with Notre Dame football, boxing, golf, and wrestling. As a national announcer for Mutual for one year in the 1940s, he called New York Giants baseball, and later for NBC he announced four World Series, five All-Star games, 12 NFL All-Star games, and many college bowl games. His style was avuncular and entertaining rather than abrasive or critical, and for 50 years he was a beloved part of Chicago history. Brickhouse died in 1998, just six months after his flamboyant Cubs successor, Harry Caray. He received a Ford C. Frick Award in 1983 (Baseball Hall of Fame), was installed in the National Sportscasters and Sportswriters Hall of Fame in 1983, and inducted into the American Sportscasters Hall of Fame in 1985 and the Radio Hall of Fame in 1996.

Jack Buck (1924–2002)
John Francis "Jack" Buck spent more than four decades broadcasting the St. Louis Cardinals, including the periods in the 1960s and 1980s when the team won pennants and the World Series. Alternating with baseball, Buck also had another career as a network football announcer, calling play-by-play for more Super Bowls than any other sportscaster. Beginning as a minor league baseball announcer after his military service and study at Ohio State University, Buck called the minors on radio for four years before moving to the big leagues at Busch Stadium in St. Louis. He started as a supporting announcer paired with the inimitable Harry Caray in 1954 and later worked with Joe Garagiola. His signature "That's a winner!" was a phrase beloved of generations of Cardinal baseball fans. After being crowded out at KMOX in 1959, Buck spent a year with ABC Television's baseball program *Game of the Week* and had the distinction of calling the first American Football League telecast. He returned to St. Louis to a warm welcome from Cardinal fans, and in 1969, when Caray went to Chicago, Buck stepped into the lead position as the Voice of the Cardinals. But Buck had a very different style: whereas Caray exhorted and stirred fans up, Buck was known for his silver voice, his satirical wit, his personal charm, and his low-key storytelling approach to sportscasting. Using no excess words, in a quietly emotion-laden voice, he called the game action as if it were a series of unfolding stories, causing listeners to hang on until each "story" concluded. Carried on KMOX, a 50-kilowatt station, the Cardinals could be heard all over the Midwest and built an enormous fan base across many states, creating enthusiastic devotees of the long-time St. Louis announcers (Caray and Buck), as well as of the team. In 1975 Buck once again took a stab at network sportscasting, this time as the host of the ill-fated *Grandstand* on NBC. Returning to St. Louis in 1977, Buck continued as the Cardinals' voice, but also called play-by-play for both baseball and football over CBS Radio, eventually becoming the radio Voice of the National Football League for many years. Buck was also widely recognized for his masterful hosting of ABC's *Monday Night Football,* and his technique influenced such familiar sports voices as Bob Costas, Tim McCarver, Dan Dierdorf, and his son, Joe Buck. His lasting presence remains a testimony to his greatness as a sportscaster. Buck received the prestigious Ford C. Frick Award in 1987 (Baseball Hall of Fame), and was inducted into the Broadcaster's Hall of Fame in 1990 and the Radio Hall of Fame in 1995.

Harry Caray (1920–98)
Born Harry Christopher Carabina, Harry Caray became a folk hero who stood larger than life. He was the gravelly-voiced, beer-swigging, beloved voice for Chicago Cubs fans on WGN for the 16 seasons between 1982 and his death. Much earlier, he was the darling of St. Louis Cardinal fans on KMOX for the 25 seasons between 1945 and 1969 and later for Chicago White Sox fans for the 11 seasons from 1971 to 1981. Caray is cherished in memory for his outspoken personality, his signature "Ho-o-o-ly Cow!," his home-run call of "It might be . . . It could be . . . It is!," and such diversions as participatory singing of "Take Me Out to the Ballgame" in the seventh inning stretch. Broadcasting more than 8,000 regular-season games over a long career, Caray made it fun to be a fan, even of a losing team. His raw exuberance and sarcasm over the air, combined with eternal optimism for the home team, set him apart from many of his contemporaries who tended toward the smoother styles of Barber and Allen. A showman with a can of Budweiser beer in one hand and a big wave in the other, Caray was part of the entertainment. He drew fans to the White Sox when the team had no stars, no marketing, and no winning record. He drew fans to Wrigley Field to cheer on the Cubs even when they hadn't won in decades. Successor to the kinder Jack Brickhouse, Caray loudly criticized the players and the management for plays on the field or decisions in the front office he didn't like, a practice that occasionally got him fired, while his energy and devotion to baseball got him hired again. Caray was a fan's fan: he praised, criticized, and rooted for the home team like fans do, and in several surveys, baseball fans chose him as the celebrity with whom they would most like to

share a beer. By the end of his career in Chicago, Caray was a bigger star than the team he announced, and he remains a giant figure among all sportscasters. For seven years in a row, he was named Baseball Announcer of the Year by *The Sporting News* for his work with the Cardinals. He received the Ford Frick Award in 1989 (Baseball Hall of Fame) and was inducted into the American Sportscasters Hall of Fame in that year; he was elected to the Radio Hall of Fame in 1990 and received a National Sportscasters and Sportswriters Hall of Fame award in 1994. When Caray died, he left behind two of the next wave of sportscasters, son Skip Caray and grandson Chip Caray. At Wrigley Field, fans built an oversized statue of him holding a beer can and waving to his audience.

Bob Costas (1965–)

Although best-known to present-day sports fans as the most celebrated of network television sports hosts for the NFL and the Olympics, Bob Costas, like so many other sportscasters, got his start in radio and continues to be heard over that medium. Costas is part of the generation that was able to study broadcasting in college, in his case, at Syracuse University. Relaxed in manner but passionate about baseball, with a seemingly bottomless knowledge of players and games, he attracted the networks when he did play-by-play of professional basketball on KMOX in the late 1970s. After a short time at CBS, he switched to NBC in 1980 and became one of the youngest ever NFL game announcers. Nonetheless, the slower pace and romantic past of baseball remained his passion, and after a stint as a play-by-play voice for NBC's backup game for the *Game of the Week,* he soon advanced to the status of NBC's centerpiece sportscaster. Rapidly recognized for his superb advance preparation, professionalism, and intense commitment, coupled with a wry wit that appeals to fans, Costas hosted NBC's 1992, 1996, and 2000 Olympics and received wide acclaim. He is frequently compared to the legendary Jim McKay, ABC's long-time Olympic host. In addition to superb play-by-play skills, Costas is widely believed to be the best interviewer and the best studio host in broadcasting. Host of a weekly radio show, *Costas Coast to Coast,* and of the late-night television talk show, *Later . . . with Bob Costas* early in his network career, in 2001 he launched a new sports magazine show on HBO, *On the Record.* Although primarily a big-event television announcer and host for NBC, his voice continues to be part of radio sports reporting and commentary. Able to mix anecdotes and history with perfect timing and delivery, Costas is the most recent of sportscasters to become an event within the events he covers. His exceptional vocabulary and wealth of poetic expression reaches across generational boundaries and sets an impossibly high standard for most aspiring broadcasters. He appeals to audiences because his great knowledge of baseball, football, and other sports is accompanied by a comfortable on-air style. Launched toward the peak of a luminous career as the 21st century commences with an extraordinary 11 Emmy awards, he has already been selected four times as the Sportscaster of the Year by the American Sportscasters Association. The much-respected Costas can most likely look forward to another decade of announcing professional football, baseball, and basketball games in addition to hosting NBC's Olympic coverage. While *Time* writer David Ellis called him "America's Host," *Newsweek*'s David Kaplan elevated him further, labeling Bob Costas "Anchor to the World," a fitting title for someone whose commitment to thoroughness and, above all, fairness stands out in the minds of millions of radio listeners and television viewers around the globe.

Dizzy Dean (1911–74)

Raised in rural Arkansas, Jay Hanna "Dizzy" Dean began his sports career as a pitcher for the St. Louis Cardinals in the 1930s, winning an astounding 120 games in his first five full seasons and achieving a career total of 150 games with 30 saves. After an injury and a few years with the Chicago Cubs, he turned to broadcasting the Cardinals and Browns games in 1941, and his distinctive folksy manner attracted enormous audiences on radio in the 1950s. His unique brand of upbeat, humorous color often appeared alongside the famed Mel Allen or paired with former player Pee Wee Reese. Dizzy Dean was a national voice of baseball in the period when baseball dominated the ratings. An eccentric and original with only a fourth-grade education, Dean had a total disregard for standard grammar and pronunciation, mangling the language to such a degree that colleagues used to joke that at least some of his words were English. In 1944 Judge Landis, the commissioner of Major League Baseball, refused to allow Dean to announce the World Series, calling his diction "unfit for a national broadcaster." His garbled syntax assaulted listeners' ears, and his long speeches often rambled on about something unrelated to the game at hand, but his exuberance, warmth, and humor endeared him to radio listeners. During the 1950s, he was one of radio's most widely recognized and most beloved broadcasters. Beginning in 1953, Dean provided color for Falstaff Beer's new *Game of the Week* on ABC, one of the events that helped make ABC a truly competitive national network. Moved to CBS in 1955 along with *Falstaff's Game of the Week,* Dean announced baseball on CBS Television until 1965, followed by three years at NBC (until 1968). In 1973, he had two popular guest appearances on ABC's *Monday Night Baseball.* He was selected for the National Sportscasters and Sportswriters Hall of Fame in 1976 for his decades of dedication to baseball.

Don Dunphy (1908–98)

Best known in the 1940s and 1950s, Don Dunphy announced the Friday night fights over the Mutual Broadcasting System (MBS) at the peak of boxing's popularity over network radio. Getting his start on WINS in New York, he began by interviewing fighters and then moved into event announcing, enduring for 24 years as the best-known voice of boxing. Dunphy was known for his eloquence and his ability to bring the drama and excitement of bouts to radio listeners. At one point, he and a colleague were hired by Gillette to cover the Giants and Yankees games over WINS in New York. During America's participation in World War II, there were no televised games, and radio reached enormous audiences. But Dunphy and his colleague Bill Slater were unsuccessful, with the huge audiences going to the Brooklyn Dodgers instead of the teams they announced. Dunphy stuck with boxing and continued to announce the big fights, including the classic matches of Joe Louis, until the boxing business declined in the 1960s. In 1964 he shifted largely to other sports but handled the national broadcasts of the *Fight of the Month* into the 1980s, and the inimitable Howard Cosell often provided color for him. Dunphy has broadcast track meets, bowling, wrestling, college football bowl games, and even some World Series, but he is remembered most for his clarity, skill, and drama as an announcer of boxing at the Polo Grounds in its heyday on radio. Dunphy was inducted into the American Sportscasters Hall of Fame in 1984 and into the National Sportscasters and Sportswriters Hall of Fame in 1986. He was elected to the Radio Hall of Fame in 1988 and to the International Boxing Hall of Fame in 1998.

Marty Glickman (1917–2001)

One of the first athletes to turn announcer, Marty Glickman popularized college and professional basketball games from Madison Square Garden. While still a college football player at Syracuse University, he began broadcasting and, after graduation in 1939, moved to WHN in New York, where he was one of a team doing pre- and post-game for the Brooklyn Dodgers. After military service during World War II, Glickman returned to WHN as sports director and became the New York Knickerbocker's first radio announcer in 1946. He remained the Knicks' voice for 21 more years, delivering the basketball play-by-play in his characteristic rapid-fire, staccato style. His widely recognized trademark call of "Swish" signaled a "nothing-but-net" Knicks' basket to decades of listeners. Beginning in 1948, Glickman did radio play-by-play of the New York Giants football games for 23 years, at a time when the New York television market was regularly blacked out, thus creating huge radio audiences for professional football. A versatile talent, Glickman also broadcast the Yonkers Raceway for 12 years and the New York Jets for 11 years on radio. In addition, he hosted pre- and post-game shows for the Dodgers and Yankees for 22 years, announced the first National Basketball Association (NBA) game carried on television, as well as many National Hockey League (NHL) games. On radio, he broadcast a plethora of wrestling meets, roller derbies, rodeos, and track and field events. He was also sports director for HBO Sports in its early years (the beginning of the 1970s) and a consultant for many years thereafter. Later in life, he became an executive of Manhattan Cable TV, which telecast all Madison Square Garden events to subscribers. He was particularly known for the enormous energy conveyed in his restrained voice that communicated excitement to the fans. Glickman provided an enduring model of restraint in early game announcing. For a time late in his 55-year career, he personally trained some of NBC's sports announcers, steering them away from the hyperbole of the first generation and toward concise, crisp, clear descriptions of events on the field. But it was his decades of nearly daily announcing that provided the strongest model for such later luminaries as Bob Costas and Marv Albert. In 1998 the United States Olympic Committee gave Glickman a plaque to make up for, in small measure, an Olympic medal he did not have a chance to win as a college runner. While a student at Syracuse, he had been chosen to run in the 400-meter relay in the now infamous 1936 Berlin Olympics, but the day before the race he and another runner were dropped from the team. Glickman always believed that Jesse Owen won the gold medal in his place. He maintained throughout his lifetime that he and his teammate were dropped because they were Jewish, a reasonable claim in the anti-Semitic climate of 1936. His 1996 autobiography recounts this bitter story, along with tales of many rewarding experiences in his long announcing career. Among his fellow broadcasters, Glickman was also known for his acidic criticism of colleagues' announcing (and decades later, he even had critical comments for Bob Costas after the 2000 Sydney Olympics). His reflexive analysis of the practice of broadcasting had the eventual benefit of improving sports coverage, and Glickman is admired for devising some of the enduring structures of sports documentary and top-quality event coverage. He was inducted into the American Sportscasters Hall of Fame in 1993 and the National Sportscasters and Sportswriters Hall of Fame in 1992. He died in Manhattan on 3 January 2001.

Curt Gowdy (1919–)

Known as the consummate professional to his colleagues and fans, after graduating from the University of Wyoming where he played basketball, Curt Gowdy began by announcing a variety of sports over local radio in Cheyenne during the early 1940s and soon moved to KOMA in Oklahoma City to do college football and basketball. In 1940 he was invited to New York to do radio play-by-play for the Yankees; there he worked under the celebrated Mel Allen until 1950, and also announced some college basketball, track, and boxing. From 1951 until 1965, he was the radio and television voice of the Boston Red Sox, and then became a network announcer for baseball's *Game of the Week*. Gowdy has been widely recognized as exceptionally versatile, equally competent at basketball, baseball, and football announcing, and as host of ABC's *The*

American Sportsman, endearing himself to hunting and fishing fans everywhere for over two decades. Younger football fans may know him best as a top television sportscaster for NFL football on NBC today, but he has also announced 15 All-Star games, 12 World Series, seven Olympics, and from 1966 to 1975, he called virtually all NBC network baseball games. In 1970 Gowdy was the first sportscaster to receive a Peabody Award for the highest achievement in radio and television. In the same year, he was honored as the Sportscaster of the Year for the fifth time. Indubitably the voice of NBC Sports in the 1960s and 1970s, in 1984 he received a Ford C. Frick Award (Baseball Hall of Fame), was included in the National Sportscasters and Sportswriters Hall of Fame in 1981, and was inducted into the American Sportscasters Hall of Fame in 1985.

Ernie Harwell (1918–)

Ernie Harwell is another voice from the earliest days of radio sports. He began announcing play-by-play for the minor league Atlanta Crackers in 1940 and is unique for being the only broadcaster ever traded for a player (Dodger catcher Cliff Dapper, so that Harwell could fill in as Brooklyn Dodger announcer for a hospitalized Red Barber). After working two years with a recovered Barber, Harwell joined Russ Hodges in 1948 with the New York Giants at the Polo Grounds. Beginning in 1954, he announced the Baltimore Orioles for six years, and then in 1969 Harwell went to Detroit as the voice of the Tigers, where he stayed for more than 40 years. Fired in 1991 (to the horror of fans and colleagues nationwide) after acrimonious verbal battles with the then-owner of the Tigers, he was rehired the next year by the new owner—after spending the intervening year with the California Angels. At the time he retired in 2002 at age 84, Harwell was said to be the most famous man in the state of Michigan. His name was synonymous with Motor City baseball. His peers say Harwell's lyrical announcing sounded as much like preaching and poetry as game calling; whatever he did, it was much admired by fans and colleagues. Harwell had two notable firsts in his long career: in 1942 he announced the first national broadcast of golf's Masters for CBS, and later he provided the first coast-to-coast television broadcast of a sporting event, the 1951 MLB playoff game when the Giants beat the Dodgers for the pennant. He has called three World Series, two All-Star games for CBS Radio, and numerous football games. He received the Ford C. Frick award in 1981 (Baseball Hall of Fame), was elected to the National Sportscasters and Sportswriters Hall of Fame in 1989, and was inducted into the American Sportscasters Hall of Fame in 1991 and the Radio Hall of Fame in 1998.

Graham McNamee (1888–1942)

Because he was the first network sportscaster, Graham McNamee virtually invented play-by-play and color commentary and, for many years, was the most prominent voice in America, enjoying celebrity status and a fan following. In the 1920s he announced the first World Series, the first Rose Bowl, and other major college football games as well as many tennis and boxing matches. Probably owing to his previous musical training as a singer, McNamee spoke with careful diction and used polished phrasing in his descriptions of the scene of a sporting event. Although his style was inventive and delivered in an enthralling voice, McNamee was not necessarily accurate or historically insightful about the players or the game. He was an unabashed generalist with little sports knowledge, convinced that he should broadcast sporting events much as he did news events. Nonetheless, McNamee significantly influenced the generation of sportscasters that followed by attracting many to this new profession and providing a model of how broadcasting a sporting event could be done. McNamee was installed in the National Sportscasters and Sportswriters Hall of Fame in 1964 and inducted into the American Sportscasters Hall of Fame in 1984.

Lindsey Nelson (1919–95)

A mellifluous charmer in the media spotlight and out of it, Lindsey Nelson is best known by fans for his trademark sports jackets—the loudest he could find—and his liquid voice. Near the beginning of his career, he determined that to stand out from crowds of sports journalists he needed some visual signature, so he adopted the practice of wearing the most outrageous plaid sports coats that he could buy (he eventually owned between 350 to 700 at any one time). Long after he was established at the top of the sports media world, he continued wearing such psychedelic jackets, fully understanding that they were truly horrible to behold but enjoying the joke. After army service during World War II, during which he became a war correspondent and friends with such giants as Dwight Eisenhower, Omar Bradley, and William Westmoreland, Nelson wrote for newspapers. Bored by the routines of reporting, he soon turned to baseball game recreations on Gordon McLendon's very successful Liberty radio network. Once in 1950, a very young Nelson served as game spotter for the legendary Bill Stern, but to his surprise, the following year he was asked to produce the NBC radio coverage of the U.S. Open golf tournament, with Stern and Dizzy Dean as his announcers. In 1952 Nelson went to NBC Sports in New York, now sharing office space with Stern, where he remained for the next decade. For many years, he was the all-purpose announcer for NBC, seemingly indefatigable and able to cover any sport with great accuracy, reliability, and humor. He broadcast major league baseball on radio and television from 1957 to 1961, and when the New York Mets were franchised in 1962, Nelson became their first voice. It was his view that many early network generalists burnt out from the travel and the stress of being the perfect announcers for a continuous stream of big events, whereas baseball announcers, employed by a team or sponsor, seemed to go on forever. At any rate, Nelson went to the Mets and stuck with them though all the crazy up and down years, eventually celebrating their 1969 World Series championship. In 1979, after 17 years, Nelson left the Mets to move to California, where for three seasons he served as the San Francisco Giants announcer. He then retired to teach at his alma mater, the University of Tennessee, and call some college football games for Turner Sports. Altogether, Nelson announced four Rose Bowls, 26 Cotton Bowls, two World Series, 19 years of National League Football, and five years of college basketball, but is best known for his 17 years as the voice of the Mets and 13 years with Notre Dame football. Among his special qualities were his exceptional storytelling ability, his great warmth and wit, and his reporting accuracy. He was elected to the Radio Hall of Fame in 1988 and beginning in 1959, was named the top sportscaster in the nation for four consecutive years by the National Sportscaster and Sportswriter Association; he was inducted into the NSSA's Hall of Fame in 1979 and into the American Sportscasters Hall of Fame in 1986. He received a Ford C. Frick Award in 1988 (Baseball Hall of Fame) and a Life Achievement Emmy Award in 1991.

Vin Scully (1928–)

Born in New York, after serving in the army Vin Scully attended Fordham University and then began his broadcasting career in 1950 with the Washington Senators on WTOP. After three years, he was lured to New York to join the Dodger team. Scully said the best thing that ever happened to him was learning the fine points of baseball announcing under the rigorous training of Red Barber at Ebbets Field. Replacing Ernie Harwell at Barber's side, Scully learned from a perfectionist, and he was later to assert that those paternal lessons were responsible for his own successful career. Succeeding Barber as the Dodger voice in 1954, four years later Scully moved with the Brooklyn Dodgers to California and became a crucial part of the Dodgers' popularity in Los Angeles. He brought with him the team's history and traditions and communicated them to southern California listeners who were establishing their first connection to major league baseball. In the poorly designed Los Angeles Memorial Coliseum, the initial home of the Dodgers, fans in the stadium had to listen to Scully on transistor radios to know what was going on, and that radio custom continues today. Beloved of California baseball fans, Scully was once voted the "most memorable personality" in Dodger history. In addition to carrying baseball tradition to the West Coast, he is known for his warm, resonant voice, his impeccable diction, and his intelligence and quiet erudition. Many of his peers have called Scully the best announcer ever (with the possible exception of Mel Allen), and many top sportscasters today exhibit the same smooth style, particularly Al Michaels and Dick Enberg. Besides announcing the Dodgers, Scully has called play-by-play for NFL football games and Professional Golf Association (PGA) Tour events on CBS-TV and Radio, and play-by-play baseball for *Game of the Week*, World Series, and All-Star Games on NBC. Altogether, he has announced 25 World Series and 12 All-Star Games. Widely recognized as the premiere baseball sportscaster, Scully's career has entered its sixth decade, and he has continuously broadcast one team longer than any other major league sportscaster. He was selected Outstanding Sportscaster of the Year four times, received a Fred C. Frick award in 1982 (Baseball Hall of Fame), and was inducted into the American Sportscasters Hall of Fame in 1992 and the Radio Hall of Fame in 1995. He has a Peabody Award and received a Lifetime Achievement Emmy Award in 1995.

Bill Stern (1907–71)

A major star of early radio sports talk and game hosting, Bill Stern was famous for his highly dramatic storytelling ability and his only passing concern for accuracy in game announcing. He was part of the first era of radio announcers along with Graham McNamee, sometimes called "the shouters" because they worked so hard to stir up excitement among the scattered listeners to the new medium of radio. He is especially remembered for his spellbinding stories, which got full rein in the sport talk programs he hosted and showed up often in the baseball and football games he called. One story told with affection about Stern is that he covered up his errors in early radio play-by-play by reporting imaginary lateral passes of the football to other players, an invention that has spawned decades of "lateraling" jokes among erring sportscasters. Operating in an era without extended pregame preparation and strong support from experienced assistants and statisticians, few announcers of the 1930s and 1940s managed an entire game without miscalling some action or some player's name. Stern started broadcasting as a teenager in Rochester, New York, and after college and a stint at the Radio City Music Hall, he assisted Graham McNamee with some football games and worked for Gordon McLendon's Liberty network doing recreations—a format at which he excelled. He was invited to become an NBC network sports announcer in 1937, and his rapid and dramatic calling of college football games set the pattern for the subsequent generation of football announcers. He was NBC's announcer for the first network telecast of a sporting event, a college baseball game in 1939. He hosted NBC's *Colgate Sports Newsreel* and other sports programs, and by 1941 was voted the most popular sportscaster in America in several polls. He broadcast for ABC in the mid-1950s, and after recovering from morphine addiction resulting from a gangrenous leg amputation, he resumed announcing for the Mutual network and hosted a television sports series. Like Mel Allen, Stern was wildly controversial and a bigger celebrity than the players he announced. He was selected for the National Sportscasters and Sportswriters Hall of Fame in

1974, inducted into the American Sportscasters Hall of Fame in 1984, and elected to the Radio Hall of Fame in 1988.

These legends of broadcasting inspired and enthralled generations of radio sports fans. There are others who could have been mentioned—one-time sportscaster turned sportswriter Grantland Rice; Tom Manning, Indians' announcer in the 1920s; Byrum Saam, voice of the Philadelphia Phillies for three decades; Russ Hodges of the New York Giants (who shouted "the Giants win the pennant! The Giants win the pennant!" in 1951); Milo Hamilton, voice of the Atlanta Braves and the Houston Astros; and on and on. All these radio sportscasters left their imprints on players and fans, and set the bar high for those who succeeded them.

SUSAN TYLER EASTMAN

See also Allen, Mel; Barber, Red; Crutchfield, Charles; Liberty Broadcasting System; McNamee, Graham; Radio Hall of Fame; Recreations; Rome, Jim; Sports on Radio

Further Reading

Albert, Marv, *Voices of Sport,* edited by Maury Allen, New York: Grosset and Dunlap, 1971

Creedon, Patricia J., *Women, Media, and Sport: Challenging Gender Values,* Thousand Oaks, California: Sage, 1994

De Simone, Anthony, *Going, Going, Gone! Music and Memories from Broadcast Baseball,* New York: Friedman/Fairfax, 1994

Edwards, Bob, *Fridays with Red: A Radio Friendship,* New York: Simon and Schuster, 1993

Poindexter, Ray, *Golden Throats and Silver Tongues: The Radio Announcers,* Conway, Arkansas: River Road Press, 1978

Smith, Curt, *Voices of the Game: The First Full-Scale Overview of Baseball Broadcasting, 1921 to the Present,* South Bend, Indiana: Diamond, 1987; revised edition, as *Voices of the Game: The Acclaimed Chronicle of Baseball Radio and Television Broadcasting—From 1921 to the Present,* New York: Simon and Schuster, 1992

Smith, Curt, *The Storytellers: From Mel Allen to Bob Costas: Sixty Years of Baseball Tales from the Broadcast Booth,* New York: Macmillan, 1995

Stone, Steve, and Barry Rozner, *Where's Harry? Steve Stone Remembers His Years with Harry Caray,* Dallas, Texas: Taylor, 1999

Sports on Radio

Sports have been a constant feature of radio literally since the beginning of the medium. When Guglielmo Marconi visited New York City in 1899 to demonstrate his wireless telegraphy equipment, he relayed the outcome of the America's Cup yacht race through the ether. Radio sports broadcasts provided some of the first mass-audience programming, enticed Americans to buy radio sets, helped national networks and local stations establish themselves as legitimate entities, and spurred technological and cultural innovations in attempts to capitalize on the appeal of sports. Radio broadcasts of sporting events have also been credited with, or occasionally blamed for, helping to create a sense of national identity, particularly during the 1920s and 1930s.

Origins

Sports—especially boxing, college football, and baseball—became common on some stations as early as the 1920s. The first boxing match broadcast, between heavyweight champion Jack Dempsey and challenger Georges Carpentier on 2 July 1921, was orchestrated by the Radio Corporation of America's (RCA) David Sarnoff and Major J. Andrew White. RCA applied for, and received, a one-day license to broadcast the event using a radio tower borrowed from the Lackawanna Railroad and equipment borrowed from the U.S. Navy. The improvised arrangement, fraught with problems, succeeded in its task. White's narration of Dempsey's victory, which traveled on a signal powerful enough to be received across much of the United States and even in Europe, is credited with sparking a surge in the construction of radio towers and the purchase of receivers. Radio broadcasts of other major bouts, particularly the battles between Dempsey and Gene Tunney in 1926 and 1927 and the Joe Louis–Max Schmeling fights in 1937 and 1938, became mass radio spectacles that attracted audiences of unprecedented size and helped make household names out of broadcasters such as Graham McNamee, Ted Husing, and Bill Stern.

The popularity of college football also grew rapidly in the first two decades of radio. During the 1920s, Notre Dame football became a regional and then a national phenomenon, thanks in part to its prominence on radio, and the annual Army-Navy contest, Ivy League games, and holiday bowl games became nationally recognized events. Bowl games in particular benefited from their appearance on radio: the Rose

Bowl, begun in 1927, was carried from the beginning by the National Broadcasting Company (NBC) and became the first coast-to-coast broadcast conducted by the fledgling network. The Sugar Bowl, Orange Bowl, and similar contests begun in the mid-1930s were hyped so vigorously by stations carrying them that radio can be credited for their continued existence. Ted Husing, who began announcing the Orange Bowl in 1937 for the Columbia Broadcasting System (CBS) and hyping the game on his weekly regular-season college broadcasts leading up to New Year's Day, almost single-handedly turned the little-known, sparsely attended event into a major spectacle. His success with the Orange Bowl convinced the industry that other bowl games could become similarly profitable.

Baseball, however, is the most remembered and romanticized radio sport. The inaugural baseball broadcast aired on 5 August 1921, when Harold Arlin called the Pittsburgh Pirates' 8-5 victory over the Philadelphia Phillies on KDKA in Pittsburgh; the first World Series broadcast also occurred that year. These early efforts were little publicized, however, and thus were not widely heard. The next year, however, New York station WJZ's broadcast of the 1922 World Series was heavily publicized by Westinghouse, General Electric, and RCA in an attempt to sell receivers, prompting other New York stations to sign off rather than interfere with the signal. The broadcast was a smashing success: it was heard clearly up to 800 miles away, and immense crowds gathered in front of radio stores to hear the game over loudspeakers.

Demand for baseball broadcasts, especially the World Series, also provoked some of the first experiments in networking. Two New York stations, WJZ and WEAF, were granted broadcast rights to the 1923 World Series; WEAF, owned by telephone giant American Telephone and Telegraph (AT&T), fed the broadcast to WMAF in Massachusetts and WCAP in Washington, D.C. Telephone lines linked seven stations in the Northeast, from Boston to Washington, to carry the Washington Senators' seven-game triumph over the New York Giants in 1924. The World Series quickly became one of radio's biggest and most popular events, with the distinctive voices of McNamee, Husing, Red Barber, Mel Allen, Jack Brickhouse, Jack Buck, Ernie Harwell, and Vin Scully becoming synonymous with baseball over the years for their work behind the microphone during both the regular season and the World Series.

When more regular schedules and commercial radio became established in the late 1920s and early 1930s, sports became more difficult for stations and networks to categorize. Sports resembled news programming in being live, unscripted, and spontaneous, but sports were rarely considered as weighty as other news items. Their mass appeal gave them nearly limitless profit potential, making them resemble entertainment programming, but their spontaneity, timeliness, and control by other entities (team owners, sports commissioners, universities, etc.) limited access and made them difficult to control—particularly because overtime or extra innings could disrupt schedules. In addition, the mixed feelings that outside entities had toward radio broadcasts were a particularly strong obstacle the radio industry had to overcome. Some, such as the Chicago Cubs and Chicago White Sox, viewed radio as a way to expand their sport's (or team's) fan base. Others—such as the New York Yankees, Brooklyn Dodgers, and New York Giants, which agreed to ban baseball broadcasts from 1932 through 1938—considered broadcasts that could be received free of charge a detriment to their box-office revenue. Still others, such as minor-league baseball clubs, believed radio undermined their own profitability while benefiting others in their industry (i.e., the major-league baseball clubs).

Developing Patterns in the 1930s

Reluctant to pay for broadcast rights early in the medium's history, stations and networks initially asked for, and often received, the same treatment as print news media in terms of access to sporting events. There was little exclusivity, and stations competed for listeners rather than for broadcast rights: for instance, during the early 1930s as many as five local stations in Chicago were broadcasting Cubs games from Wrigley Field simultaneously. Although most such broadcasts were noncommercial in character, this approach also allowed radio stations and networks to claim that they met federal responsibility/public service requirements by offering programming of substantial community interest. By the early 1930s, however, most stations accepted advertising revenue for sports programming, spending some of the money to obtain exclusive broadcasting rights. Cereal maker General Mills, Mobil Oil, Goodyear tires, and (after the repeal of Prohibition) numerous beer companies quickly became the major sponsors of major- and minor-league baseball, college football, boxing, horse racing, and, to a lesser degree, college basketball and professional football, in an attempt to reach male listeners.

To reduce costs, stations in the 1920s and 1930s rarely sent announcers on road trips with the local team. Instead, they relied on re-creations: broadcasts produced from skeletal Morse code descriptions of the contest relayed by Western Union to announcers in the radio studio. Baseball announcers became especially famous for their evocative accounts—and for their ingenuity when the wires failed, which announcers usually covered by inventing sudden storms, a ruckus in the stands, or innumerable foul balls. Re-creations slowly disappeared as technology became cheaper and the effects of the Depression were alleviated, and virtually none were heard after World War II except on the Armed Forces Radio Network and on Gordon McLendon's Liberty network.

The networks, oddly, eschewed exclusive sponsorship for sports programming longer than local stations did, for reasons best illustrated by examining the World Series. Network World

Series broadcasts from 1926 through 1934 were sustaining fare; NBC and CBS argued that the series was so important to the public that granting exclusive commercial rights to one network would prevent listeners from hearing this sacrosanct event. Because neither network could reach the entire United States through its affiliates, but taken together both could, they contended that nonexclusive rights were imperative. Moreover, the networks contended that commercializing the broadcasts would cheapen the great national pastime. This argument also enabled the networks to justify pre-empting commercial radio programming in order to air the sporting event: because the airtime was not given to another sponsor, an advertising agency that originally paid for the time slot had no basis for complaint.

When the Ford Motor Company acquired exclusive broadcast sponsorship rights to the Series from the commissioner of baseball from 1934 to 1937, the networks feared a backlash from the public, sponsors, and the federal government. Over Ford's objections, NBC and CBS persuaded Baseball Commissioner Judge Kenesaw Mountain Landis to make the broadcast available to all networks and unaffiliated stations; this provision was extended to the Mutual radio network in 1935. This allowed the networks to claim that they met the "public interest, convenience, and necessity" clause codified in the Communications Act of 1934, serving the entire nation in a way that no single station could and alleviating Federal Communications Commission (FCC) concerns. The print media and the public responded favorably to Ford's sponsorship. Jilted sponsors were livid about the arrangement, but they grudgingly agreed to accept compensation from the networks when the networks (legitimately) claimed they had not been privy to the deal. And even though Ford was upset at paying $350,000 for production costs on top of the $100,000 it had spent for exclusive rights, its sales for 1934 doubled, and the sponsorship was publicly hailed as a success.

The networks were ambivalent about this arrangement; although it served several of their needs, each network wanted exclusive rights to the event so it could turn a profit. CBS and NBC executives discussed forgoing their gentleman's agreement and competing for exclusive rights, but neither was able to obtain rights without a sponsor already lined up—and few sponsors were willing to advance the money without rights having been secured. Further, fears about FCC intervention made the networks reluctant to surrender their "public service" claim, and network executives fretted about how much the exclusive rights to all major sporting events would cost.

The commercial sponsorship model, however, developed quickly. In the wake of Ford's World Series deal, CBS acquired exclusive rights to the 1935 Kentucky Derby, and the major networks began to compete for rights to all major sporting events. College football bowl games, boxing matches, and other major events of national importance were all bought up quickly—except the World Series, the rights to which were already owned by Ford. After the automaker reneged on the final year of its contract, the series was broadcast on a sustaining basis by all major networks in 1937 and 1938. In 1939 Gillette began its long run of exclusive World Series advertising—and the upstart Mutual network acquired exclusive broadcast rights. The deal not only established Mutual's status as a competitive network rather than third banana to the older networks, but also led to FCC investigations of NBC and CBS when affiliates complained that they were pressured not to opt out of their affiliate contracts to carry Mutual's World Series broadcast.

Affiliates' local sports contracts also threatened to destabilize the networks. Although stations often wanted the high-quality programming, national appeal, and other benefits that network affiliation offered, they also wanted to appeal to a local audience—and many regularly pre-empted network programming to air local sports, particularly baseball. The NBC Blue network was particularly lax in policing its affiliates; after all, the high Crossley ratings earned by local baseball team broadcasts in cities such as Pittsburgh and San Francisco helped offset the lower ratings produced by its highbrow public-affairs programming. Both NBC and CBS used the "public interest" clause of the 1934 Communications Act to justify their affiliates' actions to national sponsors, but sponsors increasingly chafed at losing the exclusive access they paid for—particularly in the West, where broadcasts in prime time on the East Coast were often deferred in favor of sporting events in the late afternoon on the West Coast.

Since Television

The advent and diffusion of television throughout the United States in the 1940s and 1950s caused a slow but steady decline in national audiences for radio sports. Sports on American radio became an almost exclusively local or regional affair from the late 1940s through the late 1980s as radio networks declined in prominence and programming shifted from serial comedies and dramas to music, talk, and local-affairs formats. The primary innovation during these years came in the late 1940s, when networks began to cover several baseball or college football games simultaneously, airing different contests regionally while having a national anchor provide updates of distant contests—a practice since imitated by television. Though national broadcasts of regular-season baseball and professional football and of major events such as the World Series and the Super Bowl remain on network radio today, they have a much lower profile than in their heyday.

Radio sports broadcasts were marked by little change from the 1940s through the 1980s. The deregulation of radio and the proliferation of FM stations in the 1980s, however, siphoned listeners from AM radio, and the search for new, profitable formats resulted in a renewed emphasis on sports. The transformation of New York station WNBC into WFAN in 1987 marked the beginning of all-sports radio; other than a morning show hosted by Don Imus, the station devotes its entire programming day to sports talk and sports broadcasts. The debut of ESPN Radio in 1992, with 147 affiliates in 43 states carrying up to 16 hours of programming weekly, was another significant event in sports radio. Initially limited to sports news shows, update segments, and occasional features, the network has expanded to include seasonal baseball and football packages as well as talk shows featuring the likes of "The Fabulous Sports Babe" and Tony Kornheiser that can be acquired through affiliation or syndication, facilitating the development of the all-sports format.

Despite typically small audiences, often less than 3 percent of the market, the all-sports format's lucrative 25-to-54-year-old, mostly white, mostly affluent, male demographic has enabled WFAN to become the top-billing radio station in history, breaking the $50 million mark in 1997. Moreover, because sports talk shows are inexpensive to produce, because syndicated shows can be accepted in barter deals from networks like ESPN and CBS, and because local team broadcasts gain loyal audiences, the format offers a higher profit margin than other formats. The format appears to be proliferating rapidly: though estimates of the number of all-sports stations vary widely, *Broadcasting and Cable* magazine reported that there were more than 600 all-sports stations by 1998, with four dozen stations turning to an all-sports format in 1997 alone.

Many sports teams have also made their teams' radio broadcasts available over the internet, usually for a nominal fee, greatly extending their potential fan base without needing to rely on radio networks. The National Hockey League added access to live radio calls through its website, and the National Basketball Association has capitalized on basketball's popularity by allowing ESPN's website to carry its teams' radio broadcasts. Baseball and the National Football League have been slow to adopt webcasts, though some clubs, such as the Baltimore Orioles and the San Francisco Giants in baseball, have embraced this hybrid computer/radio transmission faster than others. This fusion, combined with the expansion of sports talk programming and the proven profitability of all-sports formats, may signal substantial future changes for sports radio.

DOUGLAS L. BATTEMA

See also Allen, Mel; Barber, Red; Crutchfield, Charles H.; McNamee, Graham; Rome, Jim; Sportscasters

Further Reading

Barber, Red, *The Broadcasters,* New York: Dial Press, 1970
Battema, Doug, "Baseball Meets the National Pastime," in *The Cooperstown Symposium on Baseball and American Culture,* edited by Alvin L. Hall and Peter M. Rutkoff, Jefferson, North Carolina, McFarland, 2000
Ghosh, Chandrani, "A Guy Thing," *Forbes* (22 February 1999)
Goldberg, David Theo, "Call and Response," *Journal of Sport and Social Issues* 22, no. 2 (May 1998)
Gorman, Jerry, Kirk Calhoun, and Skip Rozin, *The Name of the Game: The Business of Sports,* New York: Wiley, 1994
Haag, Pamela, "'The 50,000-Watt Sports Bar': Talk Radio and the Ethic of the Fan," *South Atlantic Quarterly* 95 (Spring 1996)
Halberstam, David J., *Sports on New York Radio: A Play-by-Play History,* Lincolnwood, Illinois: Masters Press, 1999
Hyman, Mark, "Do You Love the Orioles and Live in L.A.?" *Business Week* (12 May 1997)
McChesney, Robert W., "Media Made Sports," in *Media, Sports, and Society,* edited by Lawrence A. Wenner, Newbury Park, California: Sage, 1989
Smith, Curt, *Voices of the Game: The First Full-Scale Overview of Baseball Broadcasting, 1921 to the Present*, South Bend, Indiana: Diamond Communications, 1987; revised edition, as *Voices of the Game: The Acclaimed Chronicle of Baseball Radio and Television Broadcasting—From 1921 to the Present,* New York: Simon and Schuster, 1992

Stamberg, Susan 1938–

U.S. Newsmagazine Host

Public radio pioneer Susan Stamberg is probably best known as the first woman to anchor a nightly national news program, but she has had a far wider influence on the sound, style, and achievement of National Public Radio (NPR).

Stamberg came to NPR at its beginning in 1971, as Program Director Bill Siemering was developing NPR's first newsmagazine, *All Things Considered* (*ATC*). She began as a part-time tape editor and then moved up to reporter; after ten

Susan Stamberg
Courtesy National Public Radio

months Siemering and producer Jack Mitchell chose her to be the program's cohost. Siemering says he chose Stamberg "because I thought she expressed the most important quality for a host or a reporter—curiosity. I just assumed that the NPR listener was curious." Siemering says the fact that Stamberg was a woman wasn't the first consideration. "She was a bright, engaging person . . . and she had a good sense of radio." It was a revolutionary decision then to have a young woman anchor, report on the day's events, and interview world newsmakers about serious issues. There was opposition from some public radio station managers, who worried that a woman couldn't sound authoritative and wouldn't be taken seriously. Stamberg has said it was characteristic of Siemering's leadership that he didn't mention the station opposition to her until a dozen years later, preferring to let her on-air presence develop without that particular pressure. Noah Adams, who hosted *ATC* with Stamberg, says Siemering's confidence was rewarded: "I can't think of a tougher interviewer in a difficult situation."

Stamberg proved herself when the network was still a precarious operation, making do with scant funds, bad telephone lines, and the resourcefulness of its handful of reporters and cohosts. "In those days," says Adams, "a lot of the decisions about what went on the air were made at the microphone level." Stamberg's performance helped open doors for many other women in broadcasting. As *ATC* host Robert Siegel told the *Los Angeles Times*, "Once she established her presence on the air, it became unthinkable to have a broadcast with all male voices." Siegel also says the decision to put Stamberg on the air helped break down the barriers against regional accents. "The New Yorkisms in Susan's speech would have disqualified her" from the major networks, but Bill Siemering had pledged to let the American voice be heard in all its accents and dialects.

Stamberg's influence went further than her own example. She kept a "gender watch," urging producers to run stories by and about women and to use female analysts. Her answer to Freud's question "What does a woman want?" she says, is "to hear and see herself on the air."

Stamberg has said that she has mixed feelings about the role of public radio in bringing women into broadcasting, because some of it had to do with salaries that were too low to attract many talented men. As a married woman, it was assumed she could afford to accept less money, as she was only supplementing her husband's salary. She says that talented women came to NPR and put up with the low pay in order to do challenging work.

Stamberg cohosted *All Things Considered* for 14 years, developing an interview style that was fresh and sometimes startlingly down-to-earth. Toward the end of a long interview with orchestra conductor Kurt Masur, she asked whether his arms ever got tired. She brought warmth, curiosity, and a probing intelligence to encounters with Nancy Reagan, Annie Liebowitz, Rosa Parks, Dave Brubeck, James Baldwin, and thousands of others.

Stamberg's work helped to redefine the position of host on a newsmagazine. Robert Siegel says she came to the job with a background in radio, rather than reporting or newscasting. She didn't feel constrained by the old-fashioned stentorian style of radio delivery. She established that hosting was a job in its own right, making both guests and listeners feel at home in the program's world of ideas and experience. In the early years of *ATC*, Stamberg served as a managing editor of the program, suggesting and influencing story choices and decisions.

As a broadcaster, Stamberg sometimes revealed herself in a way that was unprecedented in broadcast journalism, sharing glimpses from her personal life when they touched on universal experiences, such as an essay on her son Josh's first day at kindergarten. In July 1986, Stamberg wept on the air after a farewell to commentator Kim Williams, who died from cancer just two weeks later. Just before conducting the interview, Stamberg had learned that she herself had breast cancer.

Stamberg left *ATC* to seek less stressful work and began planning a new NPR program, *Weekend Edition/Sunday*. The idea was to make a place for interesting radio on Sunday mornings, a traditional dumping ground where many stations aired the public-affairs programming that used to be required by the Federal Communications Commission. The current *Weekend Edition/Sunday* host, Liane Hansen, says Stamberg recognized that people listen to radio differently on Sunday, just as they look for different things in the Sunday *New York Times*. The result was a program that felt a bit like a Sunday paper, with a big "Arts and Leisure" section and even a puzzle.

Stamberg hosted *Weekend Edition/Sunday* from January 1987 through October 1989 and then became a special correspondent, covering cultural issues for all the NPR programs. Often she has reported on the visual arts. *ATC* host Robert Siegel says "she has almost a unique talent for describing pictures on the air, not just the images we all have in common, but an ability to convey images you haven't seen. It shows the quality of her writing." Stamberg also serves as a guest host on *Morning Edition, Weekend Edition/Saturday,* and *Weekly Edition.*

Stamberg was inducted into the Broadcasting Hall of Fame in 1994 and the Radio Hall of Fame in 1996. She has won almost every major award in broadcasting, including the Armstrong and duPont awards and the Edward R. Murrow Award from the Corporation for Public Broadcasting.

People who have worked with Stamberg say the essence of her personality, both on air and off, is her sincerity, curiosity, intelligence, and unrestrained exuberance. Noah Adams remembers that when Stamberg hosted *ATC,* the question most asked of reporters traveling around the country was "'what's Susan Stamberg really like?' . . . and the answer was always 'exactly what you hear, not any different.'"

COREY FLINTOFF

See also All Things Considered; Edwards, Bob; Morning Edition; National Public Radio; Siemering, William; Women in Radio

Susan Stamberg. Born in Newark, New Jersey, 7 September 1938. B.A. in English literature, Barnard College, 1959; editorial assistant, *Daedalus,* Cambridge, Massachusetts, 1960–62; editorial assistant, *New Republic,* 1962–63; host, producer, manager, program director, WAMU-FM, Washington, D.C., 1963–69; host and first woman to anchor national nightly news program, National Public Radio, *All Things Considered,* 1972–86; host, NPR's *Weekend Edition/Sunday,* 1987–89; NPR Special Correspondent, 1990– ; hosted various television series on the Public Broadcasting Service (PBS), and moderated three Fred Rogers television specials for adults in the early 1980s. Ohio State University Golden Anniversary Director's Award, 1977; Edward R. Murrow Award, Corporation for Public Broadcasting, 1980; honorary doctorates of humane letters from Gettysburg College, 1982 and Dartmouth College, 1984; Woman of the Year, Barnard College, 1984; Armstrong and duPont Awards; Distinguished Broadcaster award, American Women in Radio and Television; Jefferson Fellowship for Journalism, East-West Center in Hawaii; inducted into the Broadcasting Hall of Fame, 1994, and the Radio Hall of Fame, 1996. Serves on the board of the PEN/Faulkner Fiction Award Foundation and Northwestern University's Medill School National Arts Journalism program; Fellow of Silliman College, Yale University.

Radio Series

1972–86 *All Things Considered*
1987–89 *Weekend Edition/Sunday*
1990– *Morning Edition, Weekend Edition, Weekly Edition*

Television Series

Green Means, 1993

Films

The Siege, 1998

Selected Publications

Every Night at Five: Susan Stamberg's All Things Considered Book, 1982
The Wedding Cake in the Middle of the Road (coedited with George Garrett), 1992
Talk: NPR's Susan Stamberg Considers All Things, 1993

Further Reading

Adams, Noah, *Noah Adams on "All Things Considered": A Radio Journal,* New York, Norton, 1992
Collins, Mary, *National Public Radio: The Cast of Characters,* Washington, D.C.: Seven Locks Press, 1993
Looker, Thomas, *The Sound and the Story: NPR and the Art of Radio,* Boston: Houghton Mifflin, 1995
Siegel, Robert, editor, *The NPR Interviews, 1994,* Boston: Houghton Mifflin, 1994
Siegel, Robert, editor, *The NPR Interviews, 1995,* Boston: Houghton Mifflin, 1995

Stanton, Frank N. 1908–

U.S. Broadcast Executive and Pioneer in Radio Research

President of the Columbia Broadcasting System (CBS) for nearly three decades, Frank Stanton was an austere manager who, to a considerable degree, gave the network much of its "Tiffany" image. He joined the network as a scholarly audience researcher in the mid 1930s and rose rapidly in the network hierarchy, although he never lost his interest in or reliance upon research findings.

Early Years

Frank Stanton learned statistics and methodology while pursuing a graduate degree in psychology at Ohio State University. He sent a copy of his master's thesis, "A Critique of Present Methods and a New Plan for Studying Radio Listening Behavior," to Paul Kesten at CBS; Kesten was duly impressed. It laid the groundwork for his eventually joining CBS in October 1935.

In 1937 the Rockefeller Foundation funded the first large-scale research project on the nature and social effects of radio for the Office of Radio Research at Princeton. Hadley Cantrel and Frank Stanton recruited Paul Lazarsfeld to direct the project. One of its objectives was the study of attitudes and opinions of radio programs. Shortly after the project was completed, Stanton and Lazarsfeld developed media research's first in-process measure of audience response—the Lazarsfeld-Stanton Program Analyzer. The unit was a simple two-button device. Subjects were recruited to come to a small studio or theater to hear a test program. If a subject heard something that appealed to him or her, that subject pressed the green button—if something unappealing, the red button. The data were recorded on paper tape and summarized. In this fashion audience reaction could be matched with specific program content.

Stanton's relationships with the leading social scientists of the day were productive and influential. He had professional associations with both the Office of Radio Research at Princeton and Columbia University's Bureau of Applied Social Research, the two centers that were responsible for introducing and developing the empirical research tradition for mass communication research in the United States. Stanton coedited three volumes on communication research. Lazarsfeld and Stanton's chief editor for their *Communications Research 1948–1949* was Joseph Klapper, who would later become director of the Center for Social Research at CBS. Stanton and the other researchers of this era concluded that radio and other mass media did not have a direct effect on audience attitudes and behavior. Rather, the media's impact on attitudes, values, and beliefs was shown to operate among a host of other factors. Radio was seen as being used by listeners to meet their needs.

CBS Presidency

In 1945 Stanton was promoted to vice president and general manager of CBS to replace the ailing Paul Kesten. His leadership in research and management prompted this promotion to oversight of eight different network departments. Shortly thereafter, when Stanton was 37, he was named president of the network, when William S. Paley became chair of the board of directors. For the next 25 years, these two charted the course by which CBS became one of the most respected news and entertainment networks of its time.

A superb administrator, Stanton took on the day-to-day network management functions that no longer interested Paley. He was interested in technology and supported CBS engineer Peter Goldmark in the late 1940s in developing and promoting the long-playing record, as well as development of a partially mechanical system of color television that briefly enjoyed FCC acceptance as the national standard in the early 1950s (although it was eventually too expensive and unsuccessful). Stanton diversified CBS's holdings, reorganizing the network to give greater autonomy to the radio network and stations as well as television while decentralizing some decision-making. CBS invested in Broadway shows (most notably the highly successful *My Fair Lady* in 1956), built an impressive headquarters building, and bought the New York Yankees baseball team in addition to acquiring other information and entertainment businesses.

Stanton performed two vital roles for all of broadcasting in the 1950s and 1960s, and these formed his lasting legacy. He was a primary industry spokesperson before Congress and the Federal Communications Commission (FCC), where he evoked considerable respect and attention, even among industry critics. His measured views and words reflected well on CBS and broadcasting as a whole, and he was widely recognized as the thoughtful conscience of the industry. Stanton was also committed to the role and value of news and public affairs programming, although this led, ironically, to some notable clashes with the primary CBS news figure, Edward R. Murrow, as these two strong-minded men held differing views about the commercial role of radio and television journalism. Whereas Stanton held the business priorities of the network uppermost in his mind, Murrow had a more idealistic outlook.

Stanton is generally regarded as the prime force behind the U.S. Congress' suspension of Section 315 of the Communica-

Frank Stanton
Courtesy Radio Hall of Fame

tions Act in order to televise the Kennedy-Nixon presidential debates in the fall of 1960. Among his favorite stories are those recounting telephone calls from U.S. presidents trying to influence him and his fellow broadcasters to provide favorable coverage of events or complaining about news programs seen as unfair or critical.

On leaving CBS, Stanton accepted a presidential appointment to be the head of the American National Red Cross, serving two three-year terms. At the same time, he headed a panel of experts seeking ways to improve U.S. international information, education, and cultural relations. He served on the boards of directors for Pan American Airways, Atlantic Richfield (the petroleum company), and American Electric Power, among others. He was also elected as a Harvard University overseer for six years—the only non-alumnus to be so honored in the 20th century. He invested in a host of start-up firms in many lines of business, some of which did quite well.

The most noteworthy honors and awards received by Frank Stanton included a 1961 Peabody Award for his efforts leading to the Kennedy-Nixon presidential debates and a 1972 special Peabody Award for his response to a congressional contempt citation and his defense of "the people's right to know" that grew out of "The Selling of the Pentagon," a *CBS Reports* television documentary. The Radio and Television News Directors Association further honored his support of broadcast journalism in 1971 when he was given the Paul White Memorial Award for his advocacy on behalf of journalists' constitutional rights.

THOMAS A. MCCAIN

See also Columbia Broadcasting System; Kesten, Paul; Lazarsfeld, Paul; Murrow, Edward R.; Office of Radio Research; Paley, William S.; Peabody Awards

Frank Nicholas Stanton. Born in Muskegon, Michigan, 20 March 1908. Attended Ohio Wesleyan University, B.A. in zoology, 1930; M.A., 1932, Ohio State University, Ph.D. in psychology, 1935. Began working for CBS, New York City, 1935; promoted to director of research, 1938; vice president, 1945, president, 1946–71, vice chair, 1972–73, president emeritus, from 1973; founding member and chair, Center for Advanced Study in Behavioral Sciences, Stanford, California, 1953–60, trustee, 1953–71. Received Paul White Memorial Awards, Radio and TV News Director Association, 1957, 1971; George Foster Peabody Awards, 1959, 1960, 1961, 1964, 1972; International Directorate Award, National Academy of Television Arts and Sciences, 1980; named to TV Academy Hall of Fame, 1986.

Selected Publications

Radio Research 1941 (edited with Paul F. Lazarsfeld), 1941
Radio Research 1942–43 (edited with Paul F. Lazarsfeld), 1944
Communications Research 1948–1949 (edited with Paul F. Lazarsfeld), 1949

Further Reading

Boyer, Peter J. *Who Killed CBS?* New York: Random House, 1988
Metz, Robert, *CBS: Reflections in a Bloodshot Eye,* Chicago: Playboy Press, 1975
Paper, Lewis J., *Empire: William S. Paley and the Making of CBS,* New York: St. Martin's Press, 1987
Slater, Robert, *This . . . Is CBS: A Chronicle of 60 Years,* Englewood Cliffs, New Jersey: Prentice-Hall, 1988
Smith, Sally Bedell, *In All His Glory: The Life of William S. Paley: The Legendary Tycoon and His Brilliant Circle,* New York: Simon and Schuster, 1990

Star Wars

Public Radio Drama Series

The 1980 radio drama *Star Wars* was an adaptation of the groundbreaking motion picture of the same name. It can be seen as a radio landmark in several ways. Its production brought to bear the very best stereophonic, multitrack audio technology available. It was a collaboration between a major motion picture production company and a public radio network, with Lucasfilm, Limited supplying elaborate sound effects and music used in the original motion picture plus promotional and marketing practices hitherto thought beyond the scope of public radio. The series used six and a half hours of airtime to tell a story that, in the motion picture, was originally told in less than 30 minutes of dialogue, meaning that the characters could be treated in more depth and the story told in more detail. In addition, the series raised National Public Radio's (NPR) audience ratings spectacularly and brought a new awareness of the high quality of programs broadcast by public radio.

When the motion picture *Star Wars,* written and directed by George Lucas, opened in 1977, it was a great success, lauded for its music, special effects, and rip-roaring approach to telling an adventure story. It immediately became an icon of American culture, so embedded in the popular consciousness that, some time after the motion picture was released, a major shift in the defense policy of the Armed Forces of the United States was titled the Star Wars Strategic Defense Initiative.

A short while after the motion picture opened, Richard Toscan of the University of Southern California (USC) approached George Lucas about making a radio version of the film. This project was to be produced by the university's NPR affiliate, KUSC-FM. A USC alumnus, Lucas was fascinated by the idea of helping out his alma mater in such a novel way. Adaptations of film scripts were not new to radio—such programs as *Lux Radio Theater* were aired on the commercial networks in the 1940s, often with the stars of the motion picture re-creating their roles on radio as a means of advertising major films. But in the late 1970s, it was a given in the entertainment industry that radio drama, except for a few struggling exceptions, such as *Columbia Broadcasting System (CBS) Radio Mystery Theater* and NPR's *Earplay* and *Masterpiece Radio Theater,* was dead in the United States. The idea for a radio drama made in cooperation with a film company was communicated to NPR head Frank Mankiewicz, who was intrigued by it.

In July 1978 George Lucas' production company Lucasfilm, KUSC-FM, and NPR held initial meetings to get the project underway. In March 1979 an agreement was reached in which, for the price of $1, Lucasfilm subsidiary Black Falcon, Limited turned over to KUSC-FM the rights to write, produce, and broadcast a radio version of *Star Wars.* Lucasfilm also expressed interest in supplying technical help and in assisting to advertise the series. In April 1979 the project was publicly announced as an NPR/KUSC-FM coproduction with the British Broadcasting Corporation (BBC), in cooperation with Lucasfilm. The BBC was brought in by NPR in an effort to launch a cooperative venture with a well-respected colleague organization, experienced in the production of radio drama. But the BBC was uncomfortable with Lucasfilm's control over the script, which was being written by Brian Daley. Daley, who was not the first choice for writer, was familiar with the *Star Wars* stories, having written three novels based on *Star Wars* characters; however, for this undertaking, his scripts were subject to the approval of Lucasfilm's Carol Titelman. Eventually the BBC backed out of its agreement with NPR.

In May 1980 John Madden, who had directed for the National Theater in London and the BBC, as well as *American Playhouse,* was set as the director of the series. Tom Voegeli, who had worked on NPR's *Earplay* drama series, joined the production staff as sound mixer and supervisor of the postproduction period, the period when music and sound effects are added to the voice tracks already recorded by the actors.

Mark Hamill agreed to repeat his motion picture role as Luke Skywalker because he could not see anyone else in the part. Anthony Daniels, a veteran of British stage and radio, was eager to get back into a radio studio, so he, too, repeated his motion picture role as the robot C3PO. In late June 1980, the actors gathered at Westlake Audio Studios in Los Angeles for 13 weeks of recording. Many of the actors, more familiar with acting for screen than for radio, found the two media to be quite different in their demands. The actors had to learn to put the physicality of motion and facial expression into their lines because, in the recording studio, the voice had to do all the acting. With John Madden's help, actors and writer Brian Daley fine-tuned the scripts for the spoken word. Tom Voegeli engineered the stereophonic recording, using sensitive omnidirectional microphones; setting up a scene in which an actor moved across the room, which gave a sense of real spatial movement to the sound of the lines; and placing Anthony Daniels (C3PO) in a separate booth so that his voice could be processed to add a hollow, robotic sound.

The finished recordings of the actors were then taken to Minnesota, where Voegeli added John Williams' music, performed by the London Philharmonic, and sound effects, including the major characters R2D2 and Chewbacca, created by Ben Burtt, both from the original *Star Wars* motion picture.

A very small amount of dialogue had to be cut to accommodate the prerecorded music, but generally the actor tracks were left as they were originally recorded.

NPR distributed the finished series to its member stations via satellite in stereophonic sound for broadcast in 13 half-hour episodes beginning 2 March 1981. Audience response was overwhelming. In March 1981 NPR's special telephone number for the series received 40,000 calls. The network managed to answer more than 12,000 of the calls, some 7,000 of them from children. During the same period NPR received more than 10,000 letters from *Star Wars* listeners. Many of these people had never listened to public radio before. The network later calculated that its listening audience nearly doubled during the *Star Wars* broadcasts. Lucasfilm had insisted that the release of the radio series be scheduled to coincide with the release of its new motion picture, *The Empire Strikes Back,* and the rerelease of the original *Star Wars.* Perhaps not totally coincidentally, all this occurred during the annual public radio fund-raising drive, which saw an enormous increase in donations for that year. In May 1981 NPR's Frank Mankiewicz wrote to Lucasfilm expressing the positive impact of *Star Wars* on the network in terms of audience awareness, fund raising, and public perception of the quality of NPR's programming.

By June 1981 planning was underway for *The Empire Strikes Back,* the ten-episode radio sequel to *Star Wars.* Recording of the actors' tracks took place over ten days in June 1982, at A and R Studios on Seventh Street in New York City. Many of the *Star Wars* radio actors returned to continue the story. Mark Hamill and Anthony Daniels were joined by Billy Dee Williams, also from the original motion picture cast. Several noted theatrical names were added to the cast list in relatively small parts. *The Empire Strikes Back* was broadcast by NPR beginning 14 February 1982. Subsequently, planning began for a second sequel, *Return of the Jedi.* But all plans had to be laid aside when NPR found itself in a severe financial crisis. Over ten years passed before Highbridge Company, an affiliate of Minnesota Public Radio, managed to raise funds for a six-episode radio production of *Return of the Jedi,* with Tom Voegeli as producer, John Madden as director, and Brian Daley as writer. The actors were recorded in 1996 at Westlake Audio Studios in Los Angeles, and postproduction took place in Minnesota. The finished product was given to NPR, which broadcast it beginning 5 November 1996, bringing to a close the *Star Wars* radio trilogy. The satisfaction gained from this third successful production was tempered by the fact that Brian Daley, the writer of all three of the *Star Wars* radio series, who had become ill and was unable to attend the taping of the actor tracks, had died in February 1996, on the final day of recording.

FREDERICA P. KUSHNER

See also National Public Radio; Science Fiction Programs

Star Wars

Cast

Antilles	David Ackroyd
Fixer	Adam Arkin
Ben Kenobi	Bernard Behrens
Biggs	Kale Brown
Motti	David Clennon
Tion	John Considine
Grand Moff Tarkin	Keene Curtis
C3PO	Anthony Daniels
Prestor	Stephen Elliot
Aunt Beru	Anne Gerety
Luke Skywalker	Mark Hamill
Uncle Owen	Thomas Hill
Han Solo	Perry King
Darth Vader	Brock Peters
Princess Leia Organa	Ann Sachs
Heater	Joel Brooks
Rebel	John Dukakis
Customer #2	Phillip Kellard
Deak	David Paymer
Cammie	Stephanie Steele
Wedge	Don Scardino
Narrator	Ken Hiller
Various Roles	James Blendick, Clyde Burton, Bruce French, David Alan Grier, Jerry Hardin, John Harkins, Meschach Taylor, Marc Vahanlan, John Welsh, Kent Williams

Writer
Brian Daley

Directors
John Madden, Tom Voegeli

Producers
Carol Titelman, Richard Toscan

Programming History
NPR 2 March 1981–25 May 1981

Empire Strikes Back

Cast

Ben Kenobi	Bernard Behrens
Trooper	Brian Daley
C3PO	Anthony Daniels
Beta/Trooper	James Eckhouse
Deck Officer	Ron Frazier

Ozzel	Peter Michael Goetz
General Rieekan	Merwin Goldsmith
Veers	Gordon Gould
Renegade Four/ Second Rebel/Trooper	David Alan Grier
Luke Skywalker	Mark Hamill
Emperor	Paul Hecht
Narrator	Ken Hiller
Two-Onebee	Russell Horton
Needa	Nicholas Kepros
Han Solo	Perry King
P.A. Announcer	Michael Levett
Yoda	John Lithgow
Darth Vader	Brock Peters
Renegade Three/ First Rebel	John Pielmeier
Piett	David Rasche
Boba Fett	Alan Rosenburg
Princess Leia Organa	Ann Sachs
Wedge	Don Scardino
Lando Calrissian	Billy Dee Williams
RenegadeTwo/ CoordinatingDroid/ Zev/Crewman/Second Trooper/ Superintendent/Guard	Jerry Zaks
Dak	Peter Friedman
Controller	James Hurdle
Imperial Pilot	Jay Sanders
Various Roles	Sam McMurray, Steven Markle, Stephen D. Newman, Geoffrey Pierson

Producer
John Bos

Writer
Brian Daley

Directors
John Madden, Mel Sahr, Tom Voegeli

Programming History
NPR 14 February 1982–18 April 1982

Return of the Jedi

Cast

Jabba the Hutt	Edward Asner
Boba Fett	Ed Begley, Jr.
Anakin Skywalker	David Birney
C3PO	Anthony Daniels
Bib Fortuna	David Dukes
Luke Skywalker	Josh Fardon
General Madine	Peter Michael Goetz
Lando Calrissian	Arye Gross
Emperor Palpatine	Paul Hecht
Han Solo	Perry King
Yoda	John Lithgow
Darth Vader	Brock Peters
Princess Leia Organa	Ann Sachs
Arica	Samantha Bennett
Moff JerJerrod	Peter Dennis
Narrator	Ken Hiller
Barada	Martin Jarvis
Wedge Antilles	Jon Matthews
Mon Mothma	Natalia Nogulich
Admiral Ackbar	Mark Adair Rios
NineDeNine	Yeardley Smith
Major Derlin	Tom Virtue
Various Roles	Samantha Bennett, Ian Gomez, Rick Hall, Andrew Hawkes, Sherman Howard, Karl Johnson, John Kapelos, Ron LePaz, Joe Liss, Paul Mercier, Steven Petrarca, Jonathan Penner, Gil Segel, Nia Vardabs, Ron West

Writer
Brian Daley

Producer
John Bos, Julie Hartley, Tom Voegeli

Director
John Madden

Programming History
NPR 5 November 1996–10 December 1996

Further Reading

Brady, Frank, "A Journey from Outer Space to Inner Space on Public Radio: *Star Wars*," *Fantastic Films* (June 1981)
Collins, Glenn, "Metropolitan Diary: Interviews Anthony Daniels," *New York Times* (16 June 1982)
Daley, Brian, *Star Wars: The National Public Radio Dramatization,* New York: Ballantine Books, 1994
Daley, Brian, *Star Wars: The Empire Strikes Back: The National Public Radio Dramatization,* New York: Ballantine Books, 1995
Daley, Brian, *Star Wars: Return of the Jedi: The National Public Radio Dramatization,* New York: Ballantine Books, 1996

Klinger, Judson, "Radio: Interviews Mark Hamill, Perry King, and Anthony Daniels," *Playboy* (March 1981)

Lindsey, Robert, "Will *Star Wars* Lure Younger Listeners to Radio?" *New York Times* (8 March 1981)

Station Rep Firms

Representing Radio Stations to Advertisers

Station representative companies help local radio stations obtain national advertising. They have become known by many names—station reps, rep firms, media reps, sales reps, or simply "reps." Whatever the name, they exist to promote a station and its market and sometimes to assist client stations to improve their advertising appeal with changes in programming. For many years, a station rep firm did not represent competing stations in the same market, but that changed with consolidation of the industry in the late 1990s. From an industry once made up of several hundred companies, the radio station rep business has shrunk to a handful of major players.

Origins

As radio advertising became widespread in the late 1920s, a problem arose that had appeared decades earlier in the newspaper business: how could local stations successfully appeal to advertisers outside their immediate market area? The problem was, in part, a matter of communication, time, and efficiency. The station could not afford to have its own sales representatives in major cities, and advertisers and their agencies could not be troubled to contact dozens or even hundreds of individual stations across the country.

The first—and, as it turned out, temporary—solution was the rise of time brokering. A time broker represented no specific station or advertiser but rather sold (brokered) advertising time from many outlets to advertisers. A time broker might sell time on competing outlets in the same market. For example, around 1930 a broker named Scott Bowen began buying radio time for advertising agencies for a fee and then obtained a commission from the stations when he placed a schedule.

The Katz Agency

Emanuel Katz formed the Katz Agency in 1888 to represent newspapers. In 1931 the company sought to represent radio stations since several of the Katz newspaper clients had acquired radio licenses. Emanuel's son, Eugene Katz, the youngest member of the family and relatively new to the firm, was assigned the responsibility of selling time for the Oklahoma Publishing Company's new radio station, WKY in Oklahoma City. Hoping to organize a southwestern group of National Broadcasting Company (NBC) Radio affiliates, Eugene succeeded in gaining representation of KPRC in Houston, WFAA in Dallas, and WOAI in San Antonio.

Upon his return to New York, Eugene was told by his father to go back to Texas and call off the deal, because other Katz newspaper clients, including the *Houston Post* and *Dallas News*, had complained vigorously. Eugene was forced to withdraw the contracts and was unable to resume soliciting radio clients until a separate division was established in 1935. By that time Edward Petry, Paul Raymer, and the firm Free, John and Fields (later Peters Griffin Woodward) had all established themselves as radio reps, and Katz re-entered radio representation as a latecomer. The first non–newspaper-related radio clients at Katz were WGST, Atlanta; KRLA, Los Angeles; KRNT, Des Moines; and WMT, Cedar Rapids.

The Katz sales staffs for different media were separated after World War II. When television emerged in 1947, Eugene did not make the same mistake that his father had made with radio and moved quickly into television representation. He contracted with most of the big city television stations in the country. Ironically, other radio representatives were reluctant to enter the new medium, leaving Katz dominant until the major television groups such as Storer and Westinghouse formed their own in-house sales organizations. The Katz newspaper representation business continued its downward trend throughout the late 1960s, and the company ceased representing newspapers in 1973 to concentrate on electronic media.

In 1972 one of James Greenwald's first major steps as president of Katz Radio was to begin selling FM radio audience. Until then, most FM stations, if they were sold to national advertisers at all, were coupled with sister AM stations. Nearly all FM stations, except those that programmed classical music, simulcast programming with their AM counterparts, and Katz Radio was particularly steeped in the history of selling only large AM stations. Greenwald visited with the owners of the major Katz AM stations that also had FM stations and first convinced them to sell their fledgling FM stations in combination with their AM stations. In many instances the additional audience, which was essentially sold for the same price as the AM-only audience, resulted in higher rates and larger shares of budgets for the AM station. The Katz clients responded favorably.

When the Federal Communications Commission (FCC) in 1965 passed the rule limiting simulcasting to 50 percent of the program day, Greenwald formed an in-house programming consulting unit within the Radio Division that urged the owners of FM stations to program their FM properties independently. Greenwald foresaw a national sales market rapidly developing that was willing to spend large sums to reach the emerging FM audience. By 1976 national sales on Katz-represented FM stations had grown to represent over 20 percent of the company's total volume. By 1980 it had eclipsed 35 percent, and by 1990, 70 percent.

Eugene, the last member of the Katz family to be associated with the firm, helped to organize the company's sale to its employees in 1976 at the time he retired. Two years later, Katz had 450 employees and 17 sales offices, and the company represented 170 radio (and 108 television) stations with national spot billings of about $250 million. By 1980 the firm had grown to become the largest representative of radio and television stations in the nation. By the company's 100th anniversary in 1988 (two years after Greenwald retired as chairman), Katz had 1,400 employees in 22 offices and represented 1,440 radio (and 193 television) stations with total billings of $1.5 billion, two-thirds of that in television. By then it was the only rep firm still active in both radio and television.

In 1984 Katz took over the Henry Christal Company (which had specialized in high-power clear channel stations) and spun it off as a division along with RKO Radio Sales, which became the Republic Radio division. Katz took over the John Blair radio business in 1987, and it became the Banner Radio division. Katz also purchased the Jack Masla Company, Eastman Radio, and Metro Radio Sales. Katz set up a Hispanic radio division, and all the Katz divisions competed with one another nationally and in specific markets. By 2000 Katz Radio Group—Katz, Christal, and Clear Channel Radio Sales (set up in 2000, dedicated to the 1,200 stations in 48 states owned by Clear Channel Communications)—represented 2,000 stations in all. Emmanuel Katz died that same year.

Edward Petry

In 1932 Edward Petry established his own radio sales representative company—the first company devoted solely to radio. Petry was the first to develop the notion of "exclusivity," the idea that a station rep should handle only one station in any given market. He also developed a system of rates and standards to the spot broadcasting business that allowed it to grow and flourish. He was also the first rep to open a separate television division. Petry (as with several other rep firms) eventually left the radio business to focus on television.

Both Petry and Katz differed from time brokers in that they provided exclusive representation of client stations, never more than one in a given city. In this way the station rep could "sell" a market and the represented station as the best way to serve that market. By 1935 there were 26 such companies, and by 1937 at least 60 different rep firms were vying for radio station business.

The Industry Matures

Station rep firms increasingly competed with the national networks in the 1930s and 1940s, for the national chains usually represented not only their own stations, but also many of their affiliates. The National Association of Radio Station Reps (which became the Station Reps Association in the early 1950s) filed a complaint in 1947 with the FCC about networks representing non-owned-and-operated affiliate stations. After seven days of hearings in 1948–49 concerned with whether this was a violation of the chain broadcasting rules of the FCC, the commission took no action. American Broadcasting Companies (ABC) stopped representing affiliates in 1952. Only in 1959, by then alarmed at the degree to which television networks already dominated the advertising revenues of the relatively new medium, would the FCC ban networks from acting as reps for their affiliate stations.

With the demise of most programming on radio networks in the face of television competition after 1948, however, a new world opened for station reps. Not only did the networks rapidly fade from the competitive picture, but many more stations were going on the air each year, and each needed representation to reach national advertisers.

Following the Katz example, other radio rep firms were entering the television market. Starting in the late 1950s, a few stations (primarily those controlled by group owners) began again to represent themselves, and by the late 1970s these accounted for about a third of national spot billings for radio and television combined. By the 1970s there were some 230 rep firms, most of them regional, and they increasingly focused on radio or television, but not both.

Ralph Guild and Interep

The man who would change the face of the station rep business, Ralph Guild, began his radio career as an advertising salesman at KXOB in Stockton, California, in 1948. He moved to a similar post at a Sacramento station two years later and became manager of KROY in the state capital in 1955. He turned to the station representative business in 1957 when he joined McGavren-Quinn, then a San Francisco-based rep firm operated by Guild's college classmate Daren McGavren. Later that year, Guild moved to New York to open the company's first East Coast office. He became national sales manager in 1963 and moved up to become a partner of what became McGavren Guild in 1967. The firm was sold to employees in 1975 in an employee stock ownership plan.

In a break with station rep tradition, McGavren Guild began to represent more than one station in a given market. In 1981 Guild formed Interep as a holding company of separately managed and competing station rep firms. Over the next several years, several rep firms came under the Interep umbrella, including Major Market Radio in 1983 and Group W Radio Sales and Torbet Radio in 1987. Interep became the Interep Radio Store in 1988, all the while expanding its research and related services to both ad agencies and stations. By 1990 there were eight separate rep firms within Interep, which had become the largest radio rep organization. Billings rose from $60 million in sales in 1981 to $500 million by 1990 (half of the radio advertising in the largest 150 markets) and more than $1.25 billion a decade later. Through an initial public offering in December 1999, Interep became a publicly traded company.

With the consolidation of radio station ownership in the late 1990s, the rep companies' policy of exclusivity began to break down as stations changed hands. The huge merger between Clear Channel and AMFM in 1999, for example, caught the Katz and Interep firms in the middle. When Clear Channel gobbled up AMFM (which was the corporate parent of Katz), Guild promptly filed a $56 million lawsuit against Clear Channel for damages arising from Clear Channel's diversion of its business to Katz, and thus its alleged breach of the national sales representation agreement with Interep.

By the late 1990s, thanks in part to considerable ownership consolidation, the relatively small radio station rep firm was rapidly disappearing, unable to compete with the two dominant giant companies (Interep and Katz), each with hundreds of stations on their client list. Fewer than 40 radio station rep firms survived by 2000, and of those, 15 came under either the Interep or Katz umbrella. The largest independents were Lotus Hispanic Reps, Roslin Radio Sales, Savalli Radio and TV, and Howard C. Weiss Company.

Several smaller companies represented stations in specific parts of the country. Regional Reps, with offices in Cleveland, Cincinnati, Chicago, and Atlanta, was the largest. Others included Michigan Spot Sales, Midwest Radio, New England Spot Sales, and Western Regional Broadcast Sales. At the same time, radio station representative firms were venturing into internet sales with the rapid growth of internet radio.

GORDON H. HASTINGS, CHRISTOPHER H. STERLING, AND ED SHANE

See also Clear Channel Communications; Greenwald, James L.; Programming Strategies and Processes

Further Reading

"Katz 100th Anniversary," *Television-Radio Age* (November 1988)

Murphy, Jonne, *Handbook of Radio Advertising,* Radnor, Pennsylvania: Chilton, 1980

"Pulse Radio Executive of the Year: Ralph Guild," *The Pulse of Radio* (14 January 1991)

"Representatives," in *2000 Radio Business Report: Source Guide and Directory,* Springfield, Virginia: Radio Business Report, 2000

Schulberg, Bob, *Radio Advertising: The Authoritative Handbook,* Lincolnwood, Illinois: NTC Business Books, 1989; 2nd edition, by Schulberg and Pete Schulberg, 1996

United States Congress, House Committee on Interstate and Foreign Commerce, "Network Practices: Network Representation of Stations in National Spot Sales," in *Network Broadcasting,* Washington, D.C.: GPO, 1958

"U.S. Radio, TV Station, and Cable Representatives," *Broadcasting and Cable Yearbook* (1993–)

Stereo

Stereophonic sound, or "stereo" for short, is a system of sound reproduction in which separately placed microphones or loudspeakers enhance the realism of the reproduced sound. The effect of using multiple sound inputs and outputs in separated right and left audio channels is the creation of sound reproduction that is "three-dimensional," since aspects of right- and left-channel sound can be heard separately by persons with normal hearing.

Stereophonic sound is important to radio for two principal reasons. The popularity of stereophonic frequency modulation (FM) broadcasting in the second half of the 20th century contributed to public acceptance of that mode of radio transmission and reception. FM stereo gradually became the listening public's preferred medium for receiving music, which makes up the majority of entertainment programming for radio stations in industrialized nations. Second, the controversial method by which the Federal Communications Commission (FCC) authorized amplitude modulation (AM) stereo broadcasting in the United States in the early 1980s is thought by many to have contributed to its relative failure.

History of FM Stereo

Stereophonic radio broadcasting was invented in 1925, when WPAY radio in New Haven, Connecticut, experimented with two-station simulcasting. This early attempt featured the station's broadcast of the right channel of sound on one AM carrier frequency, while a second separate AM signal transmitted the left channel of sound. Despite experiments such as this, the real push for stereophonic broadcasting came in the 1950s, when the United States and British recording industries perfected "high-fidelity" sound reproduction, which included stereophonic recording technologies. The Record Industry Association of America (RIAA) adopted recording industry standards for stereo in January 1959. In the years preceding the adoption of the RIAA standards, interest in stereophonic radio broadcasting also increased. A variation of the technique used in the 1920s by WPAY was used experimentally in 1952 by station WQXR, owned by *The New York Times.* Like the WPAY system, this later variation featured a two-station approach, but with an AM signal for the right channel and an FM frequency for the left. In 1954, station WCRB in Boston began using this type of two-station stereo broadcasting for approximately four hours of programming per week, and for up to 40 hours per week by 1959. Nonetheless, there were problems with this type of stereophonic AM–FM broadcasting, mostly related to the wasted spectrum space of such two-station arrangements and to the fact that listeners needed two radios to get the full stereo effect, while listeners using only one radio receiver received just "half" of the intended sound. Further, the AM channel lacked the frequency response of the FM channel.

Such technical and practical limitations prompted both AM and FM broadcasters to push for the use of single-station, multiple-channel stereo broadcast authorization. Single-station stereo broadcast technology had become a reality with the FCC authorization of FM multiplexing in 1955. Multiplexing refers to the simultaneous transmission of two or more signals over the same radio channel. In FM broadcasting, a "carrier" frequency (the channel's center frequency) and its sidebands transmit the main electronic program information. However, additional electronic information can be transmitted using other frequencies within the station's designated channel, as long as the information generated and modulated on the sidebands does not interfere with the main carrier-frequency signal. This sideband frequency signal is called a "subcarrier," and the second-channel (right or left channel) audio information for FM stereo is carried in a subcarrier transmission. The technique was originally developed as a means to allow FM stations, which were financially struggling at that time, to pick up additional revenue by using the sidebands of their allocated frequencies to carry business background music or financial data information. The same technology that enabled this use of multiplexing for subsidiary communication authorization (SCA) broadcasts could be adapted so that the multiple portions of the signal would carry separate right and left audio channel information.

Developments in AM stereo broadcasting also moved forward. In 1959 AM stereo was successfully tested by the Radio Corporation of America (RCA) in conjunction with Belar Electronics Laboratory. That same year, Philco developed an AM stereo system that was tested on WABC in New York. Television industry engineers also developed their own adaptations of these stereophonic sound transmission techniques for the audio portion of TV transmission and reception systems.

In 1958 the FCC issued a *Notice of Inquiry* on further uses of FM radio, which included not only stereophonic broadcasting, but also other SCA services such as paging and calling services, traffic light–switching control, radio reading services for visually impaired persons, public utility load management, and specialized foreign language programming. A year later, the FCC separated the question of stereo from the more general SCA inquiry by issuing a *Further Notice of Inquiry.* A new industry testing group was set up in cooperation with the Electronic Industries Association (EIA) in order to sort out the features of the 17 proposed (and mutually incompatible) systems of FM stereo. This engineering test group was called the National Stereophonic Radio Committee (NSRC). Because of antitrust concerns, industry heavyweights RCA and the Columbia Broadcasting System (CBS) did not participate in the standardization testing process.

Boosters promoted FM stereo as "the one big thing" needed to ensure consumer acceptance of FM. Although several stations kept using the AM–FM two-station experimental procedure, most in the industry awaited permission to adopt single-station stereo. In the end, the FCC authorized *only* FM stereo in April 1961, accepting with modification the recommendations of the NSRC. In October of that year, the FCC denied petitions for AM stereo authorization, claiming that FM was "the ideal medium" for the development of high-quality stereo broadcasting and that the beneficial effects of AM stereo were *de minimis.* The FCC similarly denied two 1962 petitions to reconsider its negative AM stereo decision.

It appears that there were four overlapping reasons for the FCC to allow stereo FM while denying stereo AM radio or stereo television broadcasting. The most often-cited rationale is that the FCC recognized the need to give struggling FM stations a boost in order to allow them to compete economically with then-dominant AM radio stations. Second, the Commission recognized that FM, with its 200-kilohertz channel (20 times wider than the width of AM carrier frequencies) had the ability to faithfully reproduce a wider range of frequencies without suffering from fading, interference, or static. This made FM a technically superior medium for broadcasting with the use of sideband stereo technologies. Third, the FCC did not feel that it (or the industry) had adequate resources to introduce FM, AM, and TV stereo simultaneously. Finally, regarding stereo television, FCC engineers felt that "stereo sound mated with the small-screen pictures of a typical television set would be distracting and unsatisfying."

Although FM radio, with its full-frequency stereophonic sound, did gain consumer acceptance over the two decades that followed, the innovation diffusion period for FM stereo broadcasting was relatively protracted. Initially, the cost of stereo transmission equipment (estimated at $2,000 to $4,000—no small sum in the early 1960s) was considered prohibitive by many unprofitable FM broadcasters of the day, especially since there were too few stereo receivers in the consumer marketplace to make the investment pay off. Consequently, only about 25 percent of all FM stations in the United States were using stereophonic transmitters by 1965, and fewer than 50 percent were broadcasting in stereo by 1971. However, with the gradual growth of the FM industry in the 1970s, fueled by a turn from strictly upscale programming to more progressive rock music formats, a large majority of FM radio stations in the United States were broadcasting in stereo by 1975. In 1978, FM surpassed AM in terms of U.S. listenership.

Interest in AM Stereo Rekindled

The success of FM broadcasting, boosted in no small part by FM's ability to broadcast in stereo, was accompanied by a commensurate decline in AM listenership. By the late 1970s, once-dominant AM stations in several major markets expressed hope that AM stereo might be developed as part of a package of AM improvements that would enable them to compete more effectively with FM stereo.

A number of AM stereo proponents had continued to work on AM stereophonic transmission and reception throughout the 1960s and 1970s. In the early 1960s, CBS experimented with a modification of the AM stereo system developed by Philco and conducted transmission tests on its New York station, WCBS. AM stereo proponent Leonard Kahn, who in the late 1950s had introduced a "single-sideband" method of AM stereo, also refined his system and conducted stereo tests beginning in 1970 just south of San Diego at the 50,000-watt Tijuana, Mexico, AM station XETRA. The FCC granted permission for a six-month test of Kahn's system on WFBR, Baltimore, in 1974. In 1975 RCA demonstrated its AM stereo broadcasting system as its "big draw" at the National Association of Broadcasters (NAB) convention, and Motorola collaborated with a firm called Modulation Systems Laboratory to begin work on its C-QUAM system of AM stereo, which would eventually become the industry standard. Perhaps the greatest development was the 1975 united sponsorship of a new National AM Stereophonic Radio Committee (NAMSRC) under the auspices of the EIA, the NAB, the National Radio Broadcasters Association, and the Broadcasting Cable and Consumer Electronics Society of the Institute of Electrical and Electronics Engineers. Stations WGMS and WTOP in Washington, D.C., volunteered their facilities for on-air tests, and Charlotte, North Carolina's WBT was chosen as the site of the skywave tests. Although early proponent Leonard Kahn refused to participate in the joint-testing process, four companies did submit proposals to the NAMSRC in early 1976, and three systems were eventually tested.

The FCC adopted a *Notice of Inquiry* on AM stereo in 1977, but when no standard was announced by 1980, AM broadcasters grew restless. By that time a political and philosophical shift had taken place in Washington, and the consensus among the increasing number of economists at the FCC was to favor a "marketplace" option. Under this untested mechanism, the FCC would not pick one single AM stereo standard but would instead set only minimal technical standards that would enable any compliant system to be put on the air; theoretically, this system would allow the economics of the free marketplace to select its own de facto standard. Meanwhile, the FCC engineers continued to push for the more traditional single-standard outcome.

At first it appeared as if the traditionalists had prevailed when, in April 1980, an AM stereo standard decision favoring the Magnavox Corporation's proposal was announced. However, for a variety of reasons, this decision was reversed by the FCC shortly after it was announced, and in 1982 the commission adopted the "marketplace" option. This experiment was subsequently criticized as "technological Darwinism" and was widely blamed for the ultimate failure of the AM stereo technology to gain public acceptance. Although attractive in theory and certainly politically sensitive to the deregulatory impulse to create a less intrusive FCC, the marketplace experiment ultimately failed. However, it is impossible to sort out the exact reasons for the failure. Many felt that the lack of a single standard proved to be too economically unstable for broadcasters, who were in the position of trying to invest large sums of money in a transmission system that might not be adopted either by competing stations within its market or by portions of its listening audience. Likewise, the marketplace battle was also seen as too confusing for consumers trying to purchase home and auto receivers, which featured up to five different means of decoding AM stereo signals.

After a decade of uncertainty, during which time no AM stereo system emerged as the clear winner in the resulting innovation diffusion process, Congress stepped in and required the FCC to set an AM stereo standard. In 1993 the commission selected Motorola's C-QUAM system as the national AM stereo standard because, although it had not yet reached the level of acceptance that would make it the de facto standard, it nonetheless had the largest share of the AM stereo broadcast transmitter and receiver markets.

Television stereo, also approved in the 1980s, avoided this marketplace skirmish because the consumer electronics industry, through the formation of the Broadcast Television Systems Committee (BTSC) was able to agree on a preferred TV stereo transmission and reception system. As a result, in 1984 the FCC ruled unanimously that the Zenith-dbx TV stereo system's pilot subcarrier frequency would be "protected," without excluding the use of other competing systems. If other systems were to be used, a station would have to choose a subcarrier frequency different from that outlined in the Zenith-dbx standard. This was an unlikely outcome both because the Zenith system was the *only* system recommended to the FCC by the BTSC after exhaustive testing, and because Zenith had purposely used a stereo pilot subcarrier preferred by the television broadcast industry because of its compatibility with existing transmission systems.

The Future of Stereo Radio

In 1996 a system of high-capacity FM multiplex broadcasting called Data Radio Channel was proposed. This technology allows for additional text and graphics to be broadcast while maintaining compatibility with existing stereo-broadcasting technology by multiplexing digital signals at a higher frequency than the baseband FM stereo signals. The system was field-tested in NHK's Tokyo, Japan, FM station. However, such Radio Broadcast Data System technologies, which will be able to transmit text data such as artist information or station promotional graphics or text, have not yet caught on with consumers, and with only about 10 percent penetration in broadcast markets, they are compared to the unpopular AM stereo innovation.

Experiments in the late 1990s investigated the use of lasers as efficient high-speed subcarrier transmitters of stereo multiplexing. Although this method may be far off in the future of radio, digital audio broadcasting using S-band (2.3–2.6 gigahertz) or L-band (1.452–1.492 gigahertz) frequencies are very much on the horizon in the United States as part of the new digital audio radio services. In addition, five- and six-channel music-recording techniques are expected to revolutionize audio electronics and are being referred to as "beyond stereo" options.

MARK BRAUN

See also AM Radio; Dolby Noise Reduction; FM Radio; Radio Data System; Receivers; Recordings and the Radio Industry

Further Reading

Braun, Mark J., *AM Stereo and the FCC: Case Study of a Marketplace Shibboleth,* Norwood, New Jersey: Ablex, 1994

Johnson, Lawrence B., "Beyond Stereo: Sound Enters a Frontier of Many Dimensions," *New York Times* (10 September 1995)

Prentiss, Stan, *AM Stereo and TV Stereo: New Sound Dimensions,* Blue Ridge Summit, Pennsylvania: Tab, 1985

Smith, F. Leslie, Milan D. Meeske, and John W. Wright II, *Electronic Media and Government: The Regulation of Wireless and Wired Mass Communication in the United States,* White Plains, New York: Longman, 1995

Sterling, Christopher H., "Second Service: A History of Commercial FM Broadcasting to 1969," Ph.D. diss., University of Wisconsin, Madison, 1969

Sterling, Christopher H., "The New Technology: The FCC and Changing Technological Standards," *Journal of Communication* 32, no. 1 (Autumn 1982)

Sunier, John, *The Story of Stereo: 1881–*, New York: Gernsback Library, 1960

Warren, Rich, "RDS: What's the Story?" *Stereo Review* 62, no. 8 (August 1997)

Stereotypes on Radio

As with any other mass medium, early radio broadcasts made use of (some more recent critics might say "suffered") stereotypes in dramatic and other programming. Often, the use of clichés simplified groups by labeling them as "other" (that is, outside the mainstream of society) and emphasizing differences between outsiders and the core society. Such reductive portraits may not have promoted universal brotherhood, but they aided radio show popularity by relaxing audiences so that they would continue to listen and to buy the sponsor's products.

The focus here will be primarily on American radio's "Golden Age" (to about 1948) with its greater variety of programs and stereotypes, with a few comments about radio in the years since that time. That there is less stereotyping today is clearly owing to the stronger sense of political and social correctness now pervasive in society.

Precedents

The minstrel tradition began in the 1840s and produced two enduring stereotypes of African Americans: "Zip Coon" and "Jim Crow." The Zip Coon character was depicted as an individual who wore loud-colored clothes, used language inappropriately (malapropisms), and exhibited an air of self-importance. The Jim Crow character, on the other hand, was mentally slow and exhibited features that Caucasians associated with African-American field hands: speaking slowly and moving sluggishly, with thoughts that seemed to match both speech and movement. In addition to these two enduring stereotypes, other representations of African Americans included the trusted servant and maid. Thus, from the days of minstrelsy there were also such figures as Uncle Tom or Uncle Remus, Aunt Jemima or Mandy the maid, Preacher Brown and Deacon Jones, Rastus and Sambo and the old Mammy." These stereotypes persisted throughout the 19th century, became part of vaudeville, and later were transferred to radio.

Most Americans accepted these stereotypes as a real depiction of African Americans; they were, for the most part, unquestioned. Their comical nature became a defining feature of all of such stereotypes. They made Americans laugh and could be easily laughed at. Hence, racial stereotypes of African Americans served the interest of the status quo by articulating how African Americans would interact with white society, primarily as comedians and servants.

Stereotypes in Early Radio

Early radio often used stereotypes of other ethnic groups in addition to its portrayal of African Americans. For example, there were the *Cliquot Club Eskimos* and the *A&P Gypsies,* both programs featuring orchestras. Moreover, *The Goldbergs* also used heavy dialects and distinct accents, which had been part of the vaudeville and minstrel traditions.

Vaudeville programs that made heavy use of African-American stereotypes were also heard during the early years of radio. For example, the Columbia Broadcasting System (CBS) network broadcast a show featuring George Moran and Charlie Mack, cast as "The Two Black Crows," during the network's *Majestic Theater Hour.* New York radio station WEAF broadcast the *Gold Dust Twins* on Tuesday nights, another show that featured stereotypes of African Americans (played by two white men, Harvey Hindermeyer and Earl Tuckerman) in 1924.

Variety show formats often featured minstrel routines during the 1920s. *Dutch Masters Minstrels,* for example, was first broadcast by the National Broadcasting Company (NBC) in 1929. Moreover, daytime serials also featured caricatured stereotypes of African Americans. For instance, in 1929, NBC broadcast a serial based upon the Aunt Jemima trademark of the Quaker Oats Company, the show's sponsor. The focus of the program was the Aunt Jemima character and her family. All members of her family spoke with the heavy black dialect often heard in minstrel shows. (Significantly, and as was usual in this period, white actors played the parts of each character in this show.) Not to be outdone, the Cream of Wheat Company sponsored a program based upon its trademark African-American chef, Rastus. It featured musical selections performed by Rastus' imaginary animal friends and minstrel-type introductions to each song.

Sam 'n' Henry, created by Charles Correll and Freeman Gosden, made its radio debut on WGN, Chicago, in 1926. The program was based upon the minstrel tradition. Although this program never made network radio distribution, it served as the basis for *Amos 'n' Andy* which made its debut on 19 March 1928 on Chicago's WMAQ. The program changed stations and name because WGN refused Correll and Gosden a salary increase; as WGN owned the program/character names, a new name had to be chosen. Radio network NBC picked up the *Amos 'n' Andy* program a year later. As a network program, it soon became immensely popular—even among blacks—because it drew upon the minstrel tradition, made use of vaudeville ethnic humor, and offered sympathetic characters with whom the audience could identify.

Stereotypes of Foreigners, Women, and Children

Stereotypes of foreigners, women, and children appeared on dramatic, adventure, and comedy programs throughout radio's Golden Age. Scripts pictured foreigners, women, and children as predictable creatures who would not cause anxiety in listeners.

Foreigners

Historically, the number of immigrants to the United States between 1925 and 1950 barely equaled the number who entered in one important year—1907. Yet on radio, heavy accents and "ethnic" behavior routinely identified a large number of recent arrivals, nearly all of whose characters agreed to play by American rules. Radio boiled down the enormous Russian empire into Bert Gordon, the Mad Russian of *Eddie Cantor,* or Professor Kropotkin of *My Friend Irma*; all of Mexico's richness was diminished into Pedro, *Judy Canova*'s pal. On *Life with Luigi*, Luigi Basco told his "Mama mia" in Naples about America with the terminal vowels that placed him as one fresh from Ellis Island. Typically, he affirmed the values of his native-born listeners by studying English in night school, avoiding an old-world arranged marriage with Pasquale's daughter Rosa, and singing the ditty, "A-may-ree-kah, I love-a you, you like a papa to me." At least two other shows dealt more cautiously with Italian material: *Little Italy* and *The Great Merlini.*

Similarly, Englishmen, supposed masters of snobbishness, were neutered into stuffy blimps (Harry McNaughton, *It Pays to be Ignorant*; Count Benchley Botsford, *Judy Canova*); cool, work-obsessed police officials (*Scotland Yard's Inspector Burke*; *Hearthstone of the Death Squad*); or valets (*It's Higgins, Sir*). Even titled gentlemen were domesticated: Lord Bilgewater couldn't compete with *Al Pearce*, and Lord Henry Brinthrope catered to *Our Gal Sunday.* Untitled Britons such as Nicholas Lacey gratefully fit into *One Man's Family.*

French characters, too, lost touch with authentic identity. Jack Benny's violin teacher, Professor Le Blanc, suffered every time Jack produced a tortured "Love in Bloom" from his strings yet stayed because he'd still not been paid. He satisfied some comfortable expectation in the audience about starveling bohemians. *Alan Young* once disguised himself as "Pierre Eclair, decorator" in order to escape the rough treatment a real man might have expected at the hands of his girl's irked father. The supposed French connection to romance justified the character of Mademoiselle Fifi, the sultry flirt on *Eddie Cantor.* Similarly, a heavily-accented French teacher at Madison High School generated excitement in *Our Miss Brooks.*

The same Americans who slammed their geographic doors to genuine foreigners admitted through radio a surprising number of often vilified groups, especially Asian, Irish, Jewish, and African American. Some Asians on the air had been part of earlier tales in other media: *Fu Manchu; Charlie Chan;* Ming the Merciless, dread Emperor of Mongo, enemy of *Flash Gordon;* and *Mr. I. A. Moto.* Many "Easterners" were servants or, at best, sidekicks. *Bobby Benson* had a Chinese cook; so did *Little Orphan Annie* and *Tom Mix.* Ling Wee was a waiter in *Gasoline Alley* and, a little higher on the excitement ladder, Lai Choi San helped Terry against the pirates, Chula assisted on *Island Venture*, and Botak backed up *Green Lantern.* Asians often had simple two-syllable names such as Kato on *The Green Hornet* and Toku on *The Green Lama.*

Other linguistic clichés set Orientals apart from Caucasians. Gooey Fooey, laundryman on *Fibber McGee*, gibbered in a manic singsong; *Fred Allen*'s bumbling sleuth One Long Pan threatened crooks with his "lewoloweh." However, radio soothed listeners by implying that the ancient empires were eager to adopt Western ways. From 1938 to 1940, *This Day Is Ours* told how a dead missionary's daughter carried on his noble religious work, meeting small frustrations with grace because she had so much support from her adoring Chinese proselytes. The VJ episode of *The Charlotte Greenwood Show* (26 August 1945) featured the Chinese refugee Mrs. Lee who spoke, as the stage directions say, "definitely Oxford."

Irish characters used more recognizable words but expressed equally simplified personalities. Many real-life Irishmen had become police officers, so Mike Clancy aided *Mr. Keen*; Harrington helped *Mr. District Attorney*; Sergeant Velie supported *Ellery Queen*; Mullins abetted *Mr. and Mrs. North*; Sergeant O'Hara facilitated *The Fat Man*; and Happy McMann backed up *Martin Kane, Private Detective.* These Irish helpers loyally appreciated their more nimble-witted superiors. Such public servants softened a second Irish cliché, that of the bibulous blowhard. Best exemplified by Molly McGee's Uncle Dennis, this stereotype presented the Irish as ever thirsty and gregarious. *Duffy's Tavern* seemed the logical gathering place for them.

Jewish roles on radio exuded sentimentality. *The Goldbergs* led this saccharine parade, followed by Izzy Finklestein, the helpful foil on *Kaltenmeyer's Kindergarten.* Some characters, such as Papa David Solomon on *Life Can Be Beautiful,* became earth oracles in the pattern of Molly Berg. Others, such as the Levys of *Abie's Irish Rose*, radiated warm humor. Similarly, another Finklestein on *Houseboat Hannah* and *The House of Glass* series projected exuberant geniality. Mr. Kitzel, one of *Al Pearce and His Gang*, and his namesake on *Jack Benny*—the one who offered hotdogs having a "pickle in the middle, with the mustard on top"—glowed with the same lower-East-Side conviviality that made Pansy Nussbaum on *Fred Allen* so endearing.

Black characters best demonstrate how small a cookie cutter radio used to extract innocuous material from a complex culture. No George Washington Carvers or Marcus Garveys pushed their way to the front of radio's bus. The lethargic Lightning could never do more than run errands on *Amos 'n' Andy*; Molasses 'n' January (*Maxwell House Show Boat*) could only be minstrels; Cyclone could only be a ludicrous handyman for the equally silly *Hap Hazard.* The most independent, Birdy Lee Coggins, kept house for *The Great Gildersleeve,* and Geranium the maid chatted with *Judy Canova.* Even versatile African-American actresses such as Amanda and Lillian Randolph could only serve *Pepper Young's Family* and *Kitty Foyle*. Occasionally these characters bossed their bosses: Rochester van Jones twitted Jack Benny, and Beulah revealed a life outside the McGee household. Usually, like other outsiders, the characters portrayed by black actors merely augmented the lives of the characters they served, apparently content to live in the background and never rebel against middle-class expectations.

Women

Female characters on radio were squeezed into some confining aesthetic corsets. On soap operas they endured, suffered, and occasionally triumphed. Some women assisted male heroes on detective programs, either as compliant secretaries such as Effie Perine on *Sam Spade* or tagalong pals such as Margo Lane on *The Shadow*. Ironically, women were perhaps more fully represented on comedy programs. There they could stretch social molds and carry on at least a century's tradition of amusing, ironic, and flamboyant female speakers. Radio controlled the clichés so they would not discomfit audiences or sponsors. Robert J. Landry suggested in 1946 that the comedy programs (usually aired on Sundays and Tuesdays) repeated formulas because "American radio fans seem to be profoundly amused by the troublesomely imaginative adult and the juvenile equivalent, or brat" (in *This Fascinating Radio Business*). His typology can be expanded to include six major categories of funny females:

1. The brat
2. The teenager
3. The single working girl
4. The household servant (usually black)
5. The girlfriend or wife
6. The erratics: older spinsters, meddlers, society ladies, rebels.

Replicating Max und Moritz/Hans and Fritz models, brats relentlessly demanded attention or treats or information. Pipsqueak kids rose above gender so that the 10-year old boy on *Daddy and Rollo* couldn't claim much difference from the girlish Teeny who pestered *Fibber McGee*. Many of these characters incorporated the mannerisms of *Baby Snooks*.

Radio exploited the pre-World War I discovery of teenagers by unleashing a gaggle of adolescents. The females varied more than their dithery male counterparts. Admittedly there were the nonstop talkers, such as Gildersleeve's neighbor: by the time she pauses for breath, he has forgotten his message. (She had been commenting on what a quiet man Gildy was.) She belongs with chatty flirts such as Veronica on *Archie Andrews*.

A subdivision of teenage girls, the almost-mother, include Marjorie Forester, who managed much of the *Great Gildersleeve*'s household; Maudie, who kept *Maudie's Diary* with wry sensitivity; *Corliss Archer*; *My Best Girls*, who ran their widowed father's home near Chicago; Harriet Conklin, the mature daughter of *Our Miss Brooks's* school principal; Babs Riley, who assisted her mom in helping father Chester lead *The Life of Riley*; and Judy Foster, who did more than go out with Ooge Pringle on *A Date With Judy*. All of these buyers into adult responsibility helped to rectify the slur upon young women implied by the twit or coquette images.

The Single Working Girl stereotype offered more memorable characters than their accompanied or married sisters. These plucky females toiled in a world they did not create. Alone but not afraid, they confronted a commercial universe that insisted they were more bother than aid. Most radio singles were eager to remove themselves from the workplace to the sacred space of a kitchen. They lived according to Elizabeth Cushman's maxim, "No girl should remain in business more than five years" ("Office Women and Sex Antagonism," *Harper's Magazine*). *Maisie* pluckily endured low wages and unpromising boyfriends while dreaming of fulfillment.

The U.S. census for 1950 listed more than 1.6 million "stenographers, typists, and secretaries." However, these vital functionaries appeared on radio as airheads. In 1953 Lorelei Schmeerbaum, stalwart member of the club "Girls Who Say No But Mean Yes" and adviser to *My Friend Irma*, announced that Irma had won the money to go to England. Lorelei's group tells Irma to order everything new. She does, and then wastes the money by buying a ticket to *New* England.

Only a few women workers earned some validity as mature individuals. A predecessor of TV's *Moonlighting*, the 1941 *Miss Pinkerton* allowed one woman to enter a man's world. A pretty, bright, principled young woman who inherited a detective agency, she enlisted as her partner a brash, suggestive guy who both attracted and annoyed her. Likewise, *Penny Williamson*, a war widow with two children, coped poignantly with life in 1950 as a single parent by selling real estate in Middletown. Connie Brooks, the unsinkable English teacher at Madison High, and Miss Spaulding, who taught night school for immigrants on *Life with Luigi*, also managed to stay afloat in the workplace.

Household servants were predictable. One need only think of Beulah (*Fibber McGee and Molly*'s maid), or Geranium *(Judy Canova)* or Nightingale *(A Date with Judy)* to realize how automatically linked were the concepts of "house servant" and "woman of color." Here there exist traces of the wise woman archetype and a certain respect for people whom society often suppressed. Repeatedly, Birdy on *The Great Gildersleeve* moderated her portly employer's pomposity by reminding him of his own need to diet or to get closer to his ward Leroy.

The Girlfriend or Wife represented the grown-up female (as a group comprising nearly half of the total number of women in comedy). Whether she tried to teach *Slapsy Maxie*

Rosenbloom that there's more to life than boxing, or to soothe neighbors when *Lorenzo Jones*'s inventions made noise, or to moderate *Fibber McGee*'s bumptiousness, this helper civilized her man. Alice Faye took away Phil Harris' booze; Margaret Anderson sounded as wise as her husband on *Father Knows Best*; Mrs. Blandings altered her husband's schemes to build his dream home; Betty, *Alan Young*'s girl, encouraged him; and Judy Garland on *The Hardy Family* preserved Andy from embarrassment.

Erratics include the many censorious Mrs. Uppington/Mrs. Carstairs (*Fibber McGee and Molly*) types who corrected grammar and chastened mischief. Fussbudgets almost drowned out a small group of revolutionaries such as Lucy Arnaz or *Hogan's Daughter* or Jane Ace (*Easy Aces*). *Charlotte Greenwood* managed to be single, moral, and peppy. When Gracie Allen wandered onto other people's programs during 1937, apparently looking for her brother, she flummoxed normally self-possessed performers such as Walter Winchell, Fred Allen, Ben Bernie, and Singing Sam. The transgressions of erratics could be tolerated because everyone understood that it was temporary.

Children

Golden Age radio drew stereotypes of young characters from two deep wells of tradition. In public Americans looked up to the young. Citizens saw them as the lucky receptacles for their elders' accumulated wisdom and wealth; immigrants valued them because they could make a fresh start, learn to speak English well, and ascend socially. With luck and pluck, some admirable youths strove to succeed in adult-approved universes by helping their families like Horatio Alger heroes, or by comforting their elders with fey wisdom like that of *Pollyanna*, or by traveling so they could learn about grown-up activities like the jolly rovers of G. A. Henty and Edward Stratemeyer. Such characters might be called "collaborators."

The reverse of this optimistic view of children involved annoyance, helplessness, and embarrassment. Out of adult reach, past rational understanding, and immune to good advice, children were sometimes thought to have a life quite different from that of adults. This notion recognized that two forces contended in young people: the desire to belong and the bothersome urge to be an individual. Like the Katzenjammer Kids, spunky tykes discomfited adults. This second group of stereotypes may be called "confounders."

Radio judged, no doubt correctly, that abused, hungry, sexual, angry, homeless, or delinquent children would offend listeners. However, a medium that claimed to be immediate and realistic could not remain silent about young people, so it chose to present them nostalgically. Out of 55 programs that gave significant roles to young characters, 33 presented juniors who collaborated with adults. These collaborators worked to keep families intact. *Mrs. Wiggs of the Cabbage Patch* managed her modest household (during the Depression and on the wrong side of town) with the dependable aid of her little son Billy. The two Nolan kids, Francie and Neely, helped their similarly poor-but-proud family in *A Tree Grows in Brooklyn*. In 1936 *Wilderness Road* (an early version of TV's *The Waltons)* appeared, reporting how five Midwestern "younguns," the Westons, helped their folks homestead in the 1890s.

Even kids such as the orphans in 1942 Buffalo on *Miss Meade's Children*—who, in some real world, might exhibit anxiety or use unconventional language—merely frisked through one radio day after another. *Little Orphan Annie* defeated kidnappers and despair throughout the 1930s. *Mommie and the Men* had a level-headed mother managing four "children" in 1945: three kids and one infantile husband. They resembled *My Best Girls*, the three daughters of a widower who dealt amusingly with events in 1944 Chicago. Ethyl Barrymore and her daughter and son cooperated to keep the Thompson family intact on *Miss Hattie*.

Perhaps the most palatable form of the sugary category of home-centered helpers was the comedy program. Jack Barry's son and wife compensated for his flakiness on *It's the Barrys*. In *That's My Pop*, Hugh Herbert's son and daughter supported him (in 1945) because his last job had been peddling sunglasses during the eclipse of 1929. Niece Marjorie Forrester helped her aunt manage *The Great Gildersleeve*.

The kids who glued families together blended with a second subset of collaborators acting in non-residential settings. *Dick Cole* took time off from the Farr Military Academy to foil Nazi-type spies; *Jack Armstrong* skipped out of Hudson High to catch gamblers; and *Frank Merriwell*, no nearer shaving in 1946 than he was in the 1890s, found a huge underground reservoir of water that would enable farmers to make a profit.

Other compliant youths moved beyond home and school to work with adult mentors. Sixteen-year-old *Jimmy Allen* scurried about the 1930s-era Kansas City airport in order to teach 1946 listeners that a bright lad can rise if he keeps his eyes open for mechanics who might sabotage planes. Similarly, Jimmy Olsen and Beanie the office boy worked to keep *The Daily Planet* operating while *Superman* was on the road (or in the air). Junior interned with *Dick Tracy* and Pat Patton. Penny and Clipper aided *Sky King* so enthusiastically that audiences knew the maxims he spouted would inspire them to imitate his career as navy pilot, FBI agent, and rancher-detective. Jimmy, the heir of *Tom Mix*, resembled another apprentice, *Howie Wing*, who was learning to fly (as his name suggests) from Captain Harvey in 1938. Even 10-year old Barney Mallory helped his war-hero uncle Spencer Mallory during 1945 on *The Sparrow and the Hawk*.

A final group of collaborators performed noble deeds with little adult supervision, but still in harmony with adult aspirations. At one end of this spectrum of apparently individuated kids are Isabel and Billy, who hunted under the sea for misplaced toys in *Land of the Lost.* True, they were guided by a talking fish, but still they moved with relative autonomy. 1935's *Billy and Betty* scampered through perils, contacting adults only when they needed a policeman to take away the criminal they had collared. *Chick Carter* learned so much from his adoptive father that he could pursue criminals on his own or with his pal Sue.

In opposition to the goody-goodies, the confounders were an undisciplined parade of scamps who chipped away at adult composure. They were both male and female, with Red Skelton's "mean widdle kid" complementing Fanny Brice's *Baby Snooks.* For each pair of cooperators such as Tank Tinker who supported *Hop Harrigan*, there were opposites such as Archie and Jughead on *Archie Andrews* or Henry Aldrich and Homer Brown on *The Aldrich Family.* For caretaking niece Marjorie on *The Great Gildersleeve*, there was Leroy, the water commissioner's restive nephew; balancing dutiful daughter Babs was Junior, a true son of his fumbling father on *The Life of Riley.* In contrast to the attentive students of adult mentors (such as *Bobby Benson* and Tex Mason or Little Beaver and *Red Ryder),* there was Teeny, the exasperating kid who flummoxed *Fibber McGee.* Dinky added to the problems on *Today at the Duncans*, and teenagers such as those who dithered on *Junior Miss*, *Corliss Archer*, *A Date with Judy*, and *That Brewster Boy* did not exactly rebel, but their enthusiasms often torpedoed parental expectations.

Radio left each confounder's future in amiable doubt: would Harriet Conklin, sensible daughter of the high school principal, eventually marry Walter Denton, nemesis of authority but friend to *Our Miss Brooks*? Radio implied that this class of young people, like foreigners and women, might someday conform to the dictates of middle-class normalcy, but only after amusing tribulations. Darker visions of youth seldom surfaced. A few malevolent children appeared on science-fiction programs, but such characters were not typical in radio programs of the day.

Radio Stereotypes since the Advent of Television

After 1947 the radio industry was forced to change owing to the new competition for audiences from television and the subsequent loss of national advertisers, as well as the movement of radio stars and personalities to television. Of necessity, the kinds and types of radio programming changed.

Despite these changes in the medium, racial stereotypes of African Americans and others did not change quickly; as they had existed prior to radio's Golden Age, they persisted after it ended. In 1948 Joe Scribner developed *Sleepy Joe,* a children's show that used black dialect and "Uncle Tom" stereotypes in its broadcast. *Beulah* made its debut on network radio in 1947. This program made use of the "Mammy" stereotype with African-American actress Hattie McDaniel (of *Gone with the Wind* film fame) in the role of Beulah, after protests forced the network to replace a white man who had originally played the part. In addition to this program, several other network radio programs featured African-American women in stereotypical roles, often cast as maids and servants with flower names. (For example, Ruby Dandridge was cast as Geranium on the *Judy Canova Show.*)

Although the majority of stereotypes on network radio, even after the Golden Age, continued the negative portrayal of African Americans, other groups were also similarly depicted. Native Americans and immigrant ethnic groups were also stereotyped on network radio after radio's Golden Age. For example, *The Lone Ranger* used the Tonto character to denigrate Native Americans. Significantly, this Native American character referred to the *Lone Ranger* only as "Kemosabe," a word supposedly meaning "wise one" in an otherwise unidentified Indian language.

JAMES A. FREEMAN
("Stereotypes of Foreigners, Women, and Children")

GILBERT A. WILLIAMS
(opening and concluding sections)

See also, in addition to individual shows and people mentioned in this essay, Affirmative Action; African-Americans in Radio; Black Radio Networks; Black-Oriented Radio; Gay and Lesbian Radio; Hispanic Radio; Jewish Radio Programs in the U.S.; Native American Radio

Further Reading

Allport, Gordon W., *The Nature of Prejudice,* New York: Addison Wesley, 1954; abridged edition, New York: Anchor Doubleday, 1958

Bogle, Donald, *Toms, Coons, Mulattoes, Mammies, and Bucks: An Interpretive History of Blacks in American Films,* New York: Viking, 1973; 3rd edition, New York: Continuum, 1994

Daniels, Roger, *The Politics of Prejudice: The Anti-Japanese Movement in California, and the Struggle for Japanese Exclusion,* Berkeley: University of California Press, 1962

Dates, Janette L., and William Barlow, editors, *Split Image: African Americans in the Mass Media,* 2nd edition, Washington, D.C.: Howard University Press, 1993

Ely, Melvin Patrick, *The Adventures of Amos 'n' Andy: A Social History of an American Phenomenon,* New York: Free Press, 1991

Gossett, Thomas F., *Race: The History of an Idea in America,* New York: Schocken, 1965

Helmreich, William B., *The Things They Say behind Your Back,* Garden City, New York: Doubleday, 1982

Hilmes, Michele, *Radio Voices: American Broadcasting, 1922–1952,* Minneapolis: University of Minnesota Press, 1997
LaGumina, Salvatore J., editor, *"Wop": A Documentary History of Anti-Italian Discrimination in the United States,* San Francisco: Straight Arrow, 1973
Selzer, Michael, editor, *"Kike!": A Documentary History of Anti-Semitism in America,* New York: World, 1972
Steinberg, Stephen, *The Ethnic Myth: Race, Class, and Ethnicity in America,* New York: Atheneum, 1981; updated and expanded edition, Boston: Beacon Press, 1989
Wither, W. Tasker, *The Adolescent in the American Novel, 1920–1960,* New York: Ungar, 1964
Wu, Cheng-Tsu, editor, *"Chink!": A Documentary History of Anti-Chinese Prejudice in America,* New York: World, 1972

Stern, Howard 1954–

U.S. Radio Personality

Howard Stern is one of the best known and most controversial "shock jocks" on radio today. Stern's early interest in radio stemmed from his father's work as a radio engineer for WHOM (later WKTU). Stern attended Roosevelt High School in Long Island and went on to the Boston University School of Communication, where he first appeared on radio at the Boston University station. His first professional radio job was as a progressive rock disk jockey at WNTN AM in Boston. During the first 10 years of his professional career, he moved from station to station, working at WCCC, Hartford, Connecticut; DC-101 FM, Washington, D.C.; and WNBC, New York.

In 1985 Stern was fired from WNBC after pressure from upper management at the National Broadcasting Company (NBC). After several suspensions for refusing to follow station guidelines for on-air personnel, he was finally dismissed. Within a month, he was hired to work the afternoon drive shift on WXRK, a New York station owned by Infinity Broadcasting. This was the beginning of what was to be a long and profitable relationship for Stern and Infinity. After success in the afternoon shift, the show was moved to the morning drive slot.

The next year, Stern's show began to simulcast on Infinity's classic rock station in Philadelphia, WYSP. This was to serve as a test of the show's potential for simulcast on other stations around the country. It was also the beginning of problems arising from the content of Stern's program. By this time, Stern's reputation as a "shock jock" was beginning to attract attention from critics. His program combined profanity, sexual references, and an argumentative narrative style with a "nothing is sacred, no holds barred" approach to talk radio.

In 1986 the show became the target of a religious campaign led by the Reverend Donald Wildmon, who proposed that Stern's program be taken off the air because it was "indecent." Wildmon took his complaints, along with transcripts and tape recordings of Stern's show, to the Federal Communications Commission (FCC). He claimed that the content of the Stern show violated the indecency policy of the FCC.

The FCC had long held that indecency should not be broadcast during times when children were likely to be in the audience. The FCC developed a definition of indecency over time, initially focusing on indecent language or "dirty words." The policy was upheld by the Supreme Court in 1978 in the *Pacifica* case. A New York station challenged the FCC's indecency policy after complaints that it had aired George Carlin's "Seven Dirty Words" monologue during the afternoon hours. The FCC maintained that the station violated the indecency policy by airing the material when children were likely to be in the audience.

The *Pacifica* decision upheld the FCC's policy of channeling questionable content to a "safe harbor," a period of time during which stations could air indecent material. After several appeals, the safe harbor was settled at the period from 10 P.M. to 6 A.M. As of 1992, the FCC definition of indecency was "language or material that, in context, depicts or describes, in terms patently offensive as measured by contemporary community standards for the broadcast medium, sexual or excretory activities or organs."

Howard Stern's morning program on WXRK, New York, and WYSP, Philadelphia, was broadcast during a time that was outside the FCC safe harbor for indecency. So, in late 1986, after reviewing Reverend Wildmon's complaint, the FCC issued a *Notice of Apparent Liability* to Infinity Broadcasting. In it, the commission claimed that Stern's radio program was in violation of the indecency policy. Infinity was warned to bring Howard Stern under control or be fined.

Howard Stern, December 1993
Courtesy AP/Wide World Photos

In 1987 Stern held a free-speech rally in New York to protest the FCC indecency policy. Several thousand Stern fans showed up to support his right to free speech. Soon after the rally, the FCC broadened its indecency definition to include offensive references beyond the "seven dirty words" cited in the 1978 *Pacifica* case. Infinity and Stern claimed they would fight the FCC indecency policy in court.

In 1988 Stern added WJFK, Washington, D.C., to his program simulcast. In 1990 he signed a five-year contract with Infinity, which included the right to syndicate his show in other cities. Later that year, Infinity was fined $6,000, and the three stations carrying the Stern show were fined $2,000 each. Infinity and its stations refused to pay the fines.

Over the next five years, more FCC fines for indecency were issued. The fines were issued not only to Infinity and its own stations but also to independent stations that carried the Stern show. In 1992 KLSX, Los Angeles, was fined $105,000 for indecency on the Stern show, the first station not owned by Infinity to be fined. By 1995 FCC fines related to the Howard Stern show had reached a total of $1.7 million. Finally, the FCC and Infinity reached an agreement in which a "voluntary contribution" of $1.7 million was made to the U.S. Treasury by Infinity. It was rumored at the time that Infinity made the payment to clear the way for FCC approval of additional radio station purchases.

In 1996 CBS/Westinghouse purchased Infinity and Stern's radio show. Although the program has been highly rated in the 39 cities where it was sold as of 1997, expansion into other cities has been slow because of concerns about high cost and controversial content. In 1998 Stern took his radio show to television. The syndicated television program, *The Howard Stern Radio Show,* is distributed by Eyemark Entertainment, a division of the Columbia Broadcasting System (CBS). The program achieved high ratings initially but has been canceled in some cities because of controversial content. In spite of all the controversy, however, Howard Stern still proclaims himself "The King of All Media."

FREDERIC A. LEIGH

See also Censorship; Federal Communications Commission; Obscenity/ Indecency on Radio; Seven Dirty Words Case; Shock Jocks

Howard Stern. Born in Queens, New York, 12 January 1954. Attended Boston University, B.A., 1976; daytime host, WNTN-AM, Boston; disk jockey, WCCC, Hartford, Connecticut, DC-101FM, Washington, D.C., WWWW-radio, Detroit, Michigan, WNBC, New York City, 1976–1985; hired by Infinity Broadcasting's WXRK, New York City, 1985; simulcast on WYSP, Philadelphia, 1986; cited by the FCC for indecency, 1986; simulcast on WJFK AM/FM, Washington, D.C., 1988; billed as "radio's most notorious 'shock jock'" by *Time* magazine; first fine by FCC for indecency, 1990; first book, *Private Parts,* 1993; Libertarian candidate for governor of New York State, 1994; the movie, *Private Parts,* premiers, New York City, 1997; television version of *Howard Stern Radio Show* begins syndication, 1998.

Radio Series

1985– *Howard Stern Radio Show*

Television Series

The Howard Stern Interview, 1990–92; *The Howard Stern Show,* 1994; *The Howard Stern Radio Show,* 1998–

Films

Private Parts, 1997

Selected Publications

Private Parts, 1993
Miss America, 1995

Further Reading

Rathbun, Elizabeth A., "CBS Keeping Close Hold on Infinity Spin-Off," *Broadcasting and Cable* (23 November 1998)
Schlosser, Joe, "Howard Talks Dirty," *Broadcasting and Cable* (1 February 1998)
Schlosser, Joe, "Stern's Public Parts," *Broadcasting and Cable* (24 August 1998)
Trigoboff, Dan, "Two TV Stations Drop Stern," *Broadcasting and Cable* (7 September 1998)

Storer, George 1899–1975

U.S. Broadcast Executive

George B. Storer was one of radio's foremost entrepreneurial pioneers. Having entered radio at its inception in the 1920s, Storer became one of the medium's first important local group owners. From his early successes in radio, Storer went on to create a radio, television, and cable TV empire that was the nation's sixth-largest broadcast enterprise at the time of his death in 1975.

The product of a wealthy family in Toledo, Ohio, and raised and educated in Wyoming and Florida, Storer entered broadcasting by chance after having begun a career in the oil industry. What became Storer Broadcasting started as a Toledo-based service station chain that Storer, at the age of 25, formed in a partnership with his brother-in-law, J. Harold Ryan. After obtaining a franchise for Speedene gasoline products, Storer arranged advertising on Toledo's lone radio station to promote the Speedene brand. Impressed by the results, Storer then purchased this station in 1927. Originally known as WTAL, its call letters were changed to WSPD (to signify Speedene), and it would remain part of the Storer complex until 1979.

Storer's reputation for creating large value out of relatively small assets was demonstrated almost immediately at WSPD, initially a weak 50-watt outlet in a midsized city. Just months after the purchase, Storer extended overtures to Columbia Broadcasting System (CBS) founder William Paley, which resulted in WSPD's being named as the eighth affiliate of the year-old CBS radio network. By increasing the transmitting power of the facility, Storer transformed WSPD into a 5,000-watt station that attracted listeners and advertisers in many parts of the Midwest. With profits generated by the one Toledo station, Storer was able to launch a second outlet, CKLW in Windsor, Ontario, in 1932 and to purchase a third, WWVA in Wheeling, West Virginia, the following year.

Storer's most ambitious undertaking in radio was his unsuccessful attempt to form a fourth national radio network to compete with CBS and the Red and Blue networks of the National Broadcasting Company (NBC). This plan took shape in March 1934 when Storer acquired radio station WMCA in New York City. After designating this outlet as the key station, Storer formally chartered the network under the name American Broadcasting System. However, by this time, organizers of a competing venture, the Mutual network, had assembled a superior lineup of local affiliates. By 1937 Storer's hopes of entering network radio had collapsed.

It was largely because of his setback with the American Broadcasting System that Storer shifted his primary interests to the coming new medium of television, where his major achievements would unfold. At large premiums, Storer sold the stations in New York, Windsor, and Wheeling and used the proceeds in 1939 to initiate the first 50,000-watt station in Miami, with the call letters WGBS (for George Butler Storer). Then, in 1943, Storer was an indirect beneficiary when the U.S. Justice Department forced NBC to sell its Blue network. The buyer, Edward Noble, paid Storer $3 million for Storer's "American Broadcasting" moniker to begin the modern American Broadcasting Companies (ABC). This windfall cleared the way for Storer's formidable entry into TV.

Storer's most important accomplishment was licensing and launching three of the 108 prefreeze local television stations that were begun prior to the Federal Communication Commission's (FCC) 1948 "freeze." These stations were WSPD in Toledo, WJBK in Detroit, and WAGA in Atlanta. Shortly thereafter, Storer acquired a fourth station, WJW in Cleveland, and a fifth, KPTV in Portland, Oregon. In 1958, after the Portland station was sold, Storer acquired WITI in Milwaukee. In 1964 and 1965, Storer added two more outlets, WSBK in Boston and KCST in San Diego. Storer's company was the first to own seven television stations, the maximum originally allowed by the FCC. Among broadcasting companies, it trailed only NBC, CBS, ABC, Westinghouse, and Metromedia in annual revenues.

Radio, however, remained an important component of Storer's operations. Storer's entry into Cleveland in 1954 had included the acquisition of WJW radio. It was at this station, under a Storer-enlisted announcer named Alan Freed, that the rock and roll music craze was born. Storer's Detroit radio station, WJBK, likewise became a trendsetting rock and roll station. In 1960 Storer returned to New York and for $10 million, the largest amount that had then been paid for a radio station, acquired WINS. This outlet, too, was converted to rock music. Freed was transferred to WINS, and he, along with two other WINS personalities, Murray "The K" Kaufmann and Bruce "Cousin Brucie" Morrow, helped popularize the rock music trend.

Storer was featured in news reports in 1960 when FCC Chairman John Doerfer was forced out of office after accepting airplane flights and a six-day cruise on board the broadcaster's yacht. Congress did not pursue allegations, heard before a House Oversight Subcommittee, that Storer had given favors to Doerfer in order to expedite FCC licensing of a TV station in Miami that Storer had planned. The Miami station was abandoned. Two years earlier, Storer had been brought before the FCC to answer charges that his company was engaged in station "trafficking." These charges had stemmed from Storer's sale of KPTV and immediate purchase of WITI. Storer was cleared of wrongdoing following an FCC investigation.

Storer's business interests were not confined to broadcasting. Through the 1960s Storer owned Standard Tube and Nemir Industries, both plastics firms. He briefly controlled the Boston Gardens sports arena and in 1965 outbid Howard Hughes for the ownership of Northeast Airlines.

In the early 1970s, Storer passed control of his broadcast operations to his two sons, George B. Storer Jr. and Peter Storer. Between 1978 and 1980, the younger Storers sold all of the company's radio properties in order to expand holdings in cable TV. In 1993 the company broke up. Storer Cable, the country's third-largest multisystem operator, was absorbed by TCI, while several of the television stations were purchased by Rupert Murdoch and became cogs in Murdoch's Fox network.

The elder Storer died on 7 November 1975. Earlier that year, he had received broadcasting's top honor, the Distinguished Service Award of the National Association of Broadcasters. He remains the namesake of the Storer Foundation's philanthropic foundation.

CRAIG ALLEN

George B. Storer. Born in Champaign, Illinois, 10 November 1899. Established radio station WTAL (later WSPD), 1927; launched CKLW, Windsor, Ontario, 1932; WWVA, 1933; WGBS, Miami, Florida, 1939; received $3 million for "American Broadcasting" moniker, 1943; acquired WJW, Cleveland, Ohio, which launched rock and roll disc jockey Alan Freed's career, 1954; acquired WINS, New York City, to expand rock and roll format, 1960; owned seven television stations, 1965; rendered control holdings to sons, 1970s; recipient: Distinguished Service Award of the National

Association of Broadcasters. Died in Miami, Florida, 7 November 1975.

Further Reading

Allen, Craig M., *News Is People: The Rise of Local TV News and the Fall of News from New York,* Ames: Iowa State University Press, 2001
"A Combination Businessman, Engineer, Sportsman," *Broadcasting* (5 June 1961)
"George Butler Storer," *Broadcasting* (16 April 1951)
Van Tassel, David D., and John J. Grabowski, editors, *Cleveland: A Tradition of Reform,* Kent, Ohio: Kent State University Press, 1986

Storz, Todd 1924–1964

U.S. Station Owner and Music-and-News Innovator

Todd Storz was an entrepreneur who headed a chain of trend-setting AM radio stations in the Midwest and South from 1953 until 1964. He is best known for his role in the development of Top 40, a variant of the music-and-news management philosophy that evolved during the decade following World War II when television spread throughout the United States. The Top 40 formula popularized by Storz included three key elements: limited playlists, "giveaway" promotions requiring audience participation, and sensationalistic newscasts.

Origins

Robert Todd Storz was born into a prominent Nebraska family in 1924. His father, Robert H. Storz, was a wealthy Omaha civic leader who served as vice president of the Storz Brewing Company. Todd Storz was attracted to radio at an early age. He built a crystal set as a child and became a ham operator as a teenager. After an eastern college preparatory education at the Choate School in Wallingford, Connecticut, Storz attended the University of Nebraska in Lincoln for one year, followed by a three-year stint in the U.S. Army Signal Corps, an experience Storz referred to as the completion of his formal education. After his discharge, Storz attended a summer institute in radio sponsored by the National Broadcasting Company (NBC) and Northwestern University. He then worked briefly as a disc jockey and salesman at radio stations in Hutchinson, Kansas, and Omaha, Nebraska. In 1949 Storz and his father formed the Mid-Continent Broadcasting Company, which became the licensee of a marginal (500-watt, daytime-only) station in Omaha, KOWH.

Shortly after taking over KOWH, Storz replaced block programming with music and news throughout the broadcast day. This decision was far from innovative. As television emerged as the primary medium for expensive, nationally distributed programming, radio managers throughout the nation opted for low-cost schedules of recorded music hosted by local disc jockeys. Storz used the term *Top 40* as early as 1953, the year he acquired WTIX in New Orleans. At that time, disc jockey Bob Howard was using "The Top 20 at 1280" as his slogan at competitor WDSU. Storz and his New Orleans manager, George Armstrong, doubled it to "The Top 40 at 1450." Both the *Top 20* and *Top 40* terminology was an extension of the popular *Your Hit Parade* concept.

By 1953 music-and-news programming had pushed Storz's KOWH from the bottom to the top of the Omaha ratings, passing several full-time regional facilities and a 50,000-watt clear channel outlet. Storz began to promote the station aggressively in industry publications. He also began to acquire additional stations. After similar successes with music and news at WTIX in New Orleans and WHB in Kansas City, Storz began to attract both competitors from within his markets and imitators nationwide.

The Limited-Playlist Concept

Perhaps because of his untimely death at age 39, Storz become known in retrospect as "the man who invented Top 40 by watching people select songs on a jukebox." This legend is today part of radio's folklore. As the story goes, around 1955 Storz conceived the idea of repeatedly broadcasting only a few popular records after observing customers and waitresses at Omaha restaurants and bars play some records on the jukebox over and over while ignoring others. Most versions of the legend place Storz in the company of his associate, Bill Stewart. Although Storz did not deny that he had observed the jukebox phenomenon, he said he had first noticed it while in the army. Storz also considered the findings of a local research study suggesting that Omaha radio listeners preferred music to other forms of programming.

Regardless of origin, a limited playlist was implemented by Todd Storz and Bill Stewart at KOWH in early 1956 as a reaction to competition from KOIL, a full-time Omaha station that had begun airing a music-and-news format the preceding year. The limited playlist was an attempt to minimize dial switching to KOIL when KOWH played less popular songs.

Limiting all of the music played on a radio station to the most popular records of the week was counterintuitive to the conventional wisdom of programming variety, a vestige of the golden age of radio networks. During the early 1950s, disc jockeys at music-and-news stations selected from a wide array of recordings that included several renditions of particular songs and white "cover" versions of songs recorded by black artists.

After 1956 Storz began to remove much of the discretion his disc jockeys had previously exercised regarding which records were aired. He also began programming only the "best" version of a given song. It is important to note that because this occurred three years prior to the payola investigations, it represents a true programming innovation and not a reaction to the threat of regulatory scrutiny.

Audience-Participation "Giveaways"

In the summer of 1956, the four Storz stations were worth $2.5 million, and at age 32, Todd Storz was the fastest-rising figure in the radio broadcasting industry. *Time* magazine branded Storz the "giveaway king" and severely criticized his audience-participation promotional activities in Omaha, New Orleans, Kansas City, and Minneapolis–St. Paul. While drawing national attention to the Storz formula, reaction to the article in *Time* also might have influenced the members of the Federal Communications Commission (FCC) to entertain second thoughts about approving Storz's application for the license of WQAM in Miami. On 12 July 1956, the Mid-Continent Broadcasting Company received a letter from the FCC announcing that a hearing would be required on the pending WQAM purchase. Among their concerns, the commissioners cited "treasure hunts" in Omaha and Minneapolis/St. Paul, which indicated a "giveaway pattern" on the Storz stations designed to "buy" the listening audience.

One particular Storz giveaway received nationwide publicity. Criticized in the aforementioned *Time* article, the campaign landed a WTIX disc jockey in jail for stopping late-afternoon traffic in New Orleans. This giveaway was staged on a weekday in the middle of May 1956, during Bob "Robbin" Sticht's afternoon Top 40 program on WTIX. The promotion began with about 30 seconds of dead air at 5:05 P.M., after which a breathless announcer gasped, "I don't know what happened to Bob Robbin. He's gone. Oh, yes, we've just discovered that the cash box is missing. He apparently has skipped the station with the cash!"

By then, Sticht was atop a three-story building at the corner of Canal and Carondelet streets in downtown New Orleans. He was wearing a long raincoat in which he had stuffed 200 one-dollar bills. At about 5:15 P.M., after dropping a few bills to the street below, Sticht began shouting, "I've got this money and I'm going to give away this money. I hate money!" According to detailed instructions, several WTIX employees stationed at the four corners of Canal and Carondelet started to shout, "It must be Bob Robbin from WTIX!" People started fighting as he threw the rest of the bills over the side of the building. Some people almost fell through the plate glass window of a street-level clothier as they leaped into the air frantically trying to grab the money.

When the police arrived, Sticht was booked for disturbing the peace and inciting a riot. An announcement was broadcast on WTIX to the effect that Sticht had been arrested, the banks were closed, and the station did not have enough cash on hand for his bail. Listeners were asked to come to the police station and lend money to WTIX until the next morning so Sticht could be released. Sticht said that several hundred complied.

This type of "throwaway" promotion had been staged earlier in Omaha, when a Storz employee climbed a tree in a park and periodically tossed out money. Such promotions enabled the first two stations acquired by Storz to dominate their markets, even with limited facilities. For example, in 1956 WTIX operated full-time with 250 watts at 1450 kilohertz. Its signal barely covered the New Orleans city limits, especially at night. Yet in the April-June Hooperatings for 1956, the WTIX share of audience from 7 A.M. to 6 P.M. averaged 26.6. The next highest-rated station was 50,000-watt WWL, which averaged only 14.9. Storz later sold daytime-only KOWH in Omaha to William F. Buckley, Jr. and purchased a 5,000-watt regional station, WWEZ in New Orleans, to which he transferred the WTIX call letters after cleverly donating the original 250-watt facility for noncommercial use.

Storz's response to the FCC letter was to discontinue these types of giveaways on his stations in return for favorable FCC action on WQAM, which he received. The audience-participation contests later resumed and remained a programming staple throughout the Top 40 era.

Sensationalistic Newscasts

The third element of the Storz formula was designed to maintain the audience for his stations while fulfilling the news commitments of his licenses. This was accomplished by airing short newscasts during peak listening times and longer newscasts at other times. It also involved scheduling newscasts at five minutes before the hour so music could be played on the hour, the time most competitors aired news.

The newscasts on Storz's stations were delivered in a sensationalistic style designed to hook listeners and keep them listening. The procedures generally included program elements such as echo, shouted datelines, stories punctuated by the sound of a telegraph key à la Walter Winchell, and so on.

The Storz Disc Jockey Conventions

In 1958 the Storz organization sponsored "The First Annual Pop Music Disk Jockey Convention and Programming Seminar" in Kansas City. The objective was to improve the general perception of music-and-news stations by transferring an image of "professional" respectability to the Top 40 disc jockey, which was at best a low-prestige occupation.

The following year Storz sponsored a final gathering titled "The Second International Radio Programming Seminar and Pop Music Disk Jockey Convention," held in Miami Beach a few months before the first payola investigations. Both conventions were underwritten by record companies.

Impact

By the late 1950s Storz and his managers had perfected a fast-paced brand of music and news that came to be known as Top 40. The procedures changed slightly from time to time but basically involved airing records from the limited playlist in spaced repetition and sometime in a countdown order, skipping some records during certain time periods. News continued to air at five minutes before the hour, delivered in a sensationalistic manner. The rest of the time was filled with commercials, audience-participation giveaways, and various time, weather, public service, and call letter announcements.

By 1960 the Storz roster included WTIX, New Orleans (acquired in 1953, facility upgraded in 1958); WHB, Kansas City (acquired in 1954); WDGY, Minneapolis (acquired in 1955); WQAM, Miami (acquired in 1956); KOMA, Oklahoma City (acquired in 1958); and KXOK, St. Louis (acquired in 1960). When Todd Storz died less than a month before his 40th birthday, he had built a successful business by doing one thing well: Top 40 on AM. Top 40 would continue to be featured on AM stations until the early 1980s. Storz lived to see the programming formula he pioneered reach from the Midwest and South to all parts of the country, including major-market clear channel stations such as WABC in New York and WLS in Chicago. Within two years of Storz's death, RKO General's KHJ in Los Angeles became the most imitated radio station in North America when programming consultants Bill Drake and Gene Chenault achieved phenomenal ratings with a streamlined version of the basic Storz formula.

It would be inaccurate to infer that Storz's decision to program Top 40 on his stations was grounded in anything other than business interests. Storz said his mission in life was not to educate radio listeners. He maintained an objective stance regarding the program content of his stations and said he was ready to change if listener preference so warranted.

Moreover, it is doubtful that Storz anticipated the shift in listener preference from AM to FM. When he was acquiring stations, FM licenses were albatrosses. Storz therefore sold each FM station that he acquired in combination with an AM purchase. Although television licenses were hot properties in the early 1950s, Storz said he had no interest in that medium.

After Todd Storz died in 1964, the six Storz stations continued to operate as stand-alone AM facilities under the leadership of George W. Armstrong. By 1982 almost all former Top 40 AM stations had switched either to country or adult contemporary variants of the music-and-news format. Most of the Storz stations adopted country formats, but a news/talk format was eventually implemented at KXOK in St. Louis, a move that suggests adaptability on the part of Storz management to the rapidly changing market environment for AM radio in the early 1980s. Robert H. Storz nevertheless began to sell the stations when the pattern of audience migration to FM appeared irreversible. KOMA, WTIX, and WDGY were sold in 1984; WHB, KXOK, and WQAM were sold the following year. Sales prices averaged between $2 million and $3 million per station.

Todd Storz and Gordon McLendon

In many accounts, Todd Storz has been overshadowed by his contemporary, Gordon McLendon, who also made significant contributions to Top 40 and who outlived Storz by more than 20 years. Both Storz and McLendon were sons of the American heartland. They were about the same age, benefited from privileged upbringings, and became business associates of their successful fathers. In ways most directly related to their careers in radio, however, Storz and McLendon were quite different. McLendon possessed far more flair for creative programming (a floating pirate station, the first all-news station, the first beautiful music station, and legendary baseball recreations over his nationwide Liberty network). Storz stuck with Top 40. Perhaps most significant, McLendon and his representatives were more helpful to early scholars of the music-and-news era than were representatives of the Storz organization. As a result, far more has been written about McLendon than Storz.

The viability of commercial radio broadcasting during the early years of television sprang in large part from the efforts of young licensees such as Todd Storz who objectively conformed their management strategies to the changing structures within the radio industry. Storz received the National Association of Broadcasters Hall of Fame Award in 1987 for his contributions to radio broadcasting. In 1989 he entered the Nebraska Broadcasters Association Hall of Fame.

ROBERT M. OGLES

See also Contemporary Hit Radio Format/Top 40; McLendon, Gordon

Robert Todd Storz. Born in Omaha, Nebraska, 8 May 1924. Graduated from Choate School, Wallingford, Connecticut, 1942; attended University of Nebraska (one year); served in U.S. Army Signal Core three years; completed 12-week course on radio sponsored by NBC and Northwestern University; worked briefly as announcer at radio stations in Kansas and Nebraska; formed Mid-Continent Broadcasting Company with father, 1949, later named Storz Broadcasting; developed the "Top 40" format in early 1950s; became known for aggressive "giveaway" promotions; purchased WHB, Kansas City, Missouri, which became first 24 hour Top 40 station; moved company to Florida, 1961. Inducted posthumously into the Nebraska Broadcasters Hall of Fame, 1989. Died following a stroke in Miami Beach, Florida, 13 April 1964.

Further Reading

Fornatale, Peter, and Joshua E. Mills, *Radio in the Television Age,* Woodstock, New York: Overlook Press, 1980
"King of Giveaway," *Time* (4 June 1956)
Land, Herman, "The Storz Bombshell," *Television Magazine* (May 1957)
MacFarland, David T., "The Development of the Top 40 Radio Format," Ph.D. diss., University of Wisconsin, 1972
Ogles, Robert M., Steven O. Shields, and Herbert H. Howard, "Some Suggestions for Teaching the History of Contemporary Radio," *Feedback* 28, no. 3 (1987)
"Our Respects to Todd Storz," *Broadcasting* (19 September 1955)
Scherer, Steven Robert, "The Influence of the Limited Playlist at the Storz Broadcasting Company During the Payola Era," Master's thesis, Purdue University, 2002
"Throwaways: $1 Each," *Broadcasting* (21 May 1956)

Striker, Fran 1903–1962

U.S. Radio Scriptwriter

Fran Striker was a prolific writer of scripts for some of the most popular programs on network radio in the 1930s and 1940s, including *The Lone Ranger, The Green Hornet,* and *Sergeant Preston of the Yukon.* The character and adventures of *The Lone Ranger,* especially, have lived on as part of American popular culture, and the Ranger has continued in a wide variety of media forms, including books, comic strips, comic books, television, and film.

Fran Striker was working at WEBR radio in Buffalo, New York, when he was contacted in December 1932 by radio station WXYZ in Detroit, Michigan. The station had severed its relationship with the Columbia Broadcasting System (CBS) network, and one of the station owners, George W. Trendle, had determined that WXYZ could be more profitable if it developed its own programs to fill the air and attract both audience and advertisers. During an earlier conference, Trendle, James Jewell (a WXYZ dramatic director), and other station personnel had pooled their ideas in outlining the characteristics of a dramatic program that would be designed to appeal primarily to children. It was to feature a sort of modern Robin Hood, a hero who Trendle insisted should somehow symbolize justice. The scripts for the show should portray a character of the West who would right wrongs, who would accept no thanks, and who would be fulfilled simply by doing his job well. This hero was also to convey an aura of mystery and operate in a realm beyond the constraints imposed upon official lawmen. After trying various locally written scripts and finding them lacking, Trendle turned to Fran Striker.

Striker had attended the University of Buffalo from 1922 to 1925, but he left before graduating, finding work as a transmitter operator, announcer, program producer, and continuity editor, as well as employment in writing for radio. He had successfully scripted *Covered Wagon Days,* a show for WEBR in Buffalo, and *Warner Lester, Manhunter,* which was being aired on a station in Boston. His professional aspiration was to write scripts for programs that he could sell to stations around the country. He hoped this would provide a reasonable standard of living for himself and his family, even during the Depression years.

Striker accepted WXYZ's offer to write for *The Lone Ranger,* and his pilot scripts met with Trendle's approval. The character of the Ranger may have grown partly from Striker's *Covered Wagon Days,* although his initial idea that the Ranger should display a happy-go-lucky attitude and a strong sense of humor was soundly rejected by Trendle. Striker is generally credited with fashioning the character of Tonto, however, and both the use of silver bullets and the Lone Ranger's famous cry, "Hi Yo Silver, away," may also have been Striker's contribution. There is

Fran Striker, creator of *The Lone Ranger*
Courtesy AP/Wide World Photos

no doubt that his writing fleshed out the style and personality of *The Lone Ranger,* although strong differences of opinion as to exactly who was responsible for the character's initial creation continue today. Trendle consistently maintained that the Ranger was a result of his own conception, and he successfully convinced Striker to sign over all legal claims to having originated the Ranger. Striker gave up all legal rights to the ownership of current and future printed, published, or broadcast versions of the Ranger on 22 May 1934, although some historians believe he may have been inappropriately manipulated into doing so. On more than one occasion, Trendle vigorously defended his view of himself as the sole originator of the Ranger, and he moved quickly to counter references to anyone else as the creator. A modest, unassuming, friendly personality, Fran Striker did little to push for his own personal recognition as the Ranger's creator.

Striker continued to work for Trendle, however, as a paid employee, turning out three half-hour *Lone Ranger* scripts each week and for many years earning only $4.00 for each. Working in Buffalo, largely at night and sustained by huge quantities of coffee and cigarettes, Striker produced scripts for a variety of programs aired by more than 50 radio stations around the country, in addition to writing for WXYZ. Conforming to Trendle's demands and constraints, Striker developed a formula for scripting characters, setting, complications, and resolutions for *Lone Ranger* episodes, allowing him to maximize his efforts in turning out scripts for the other programs. He pounded heavily on the keys of his typewriter in order to produce the required eight carbon copies of each page of his scripts. Striker worked 14 hours a day, writing the equivalent in words of four Bibles a year, and found relaxation and inspiration in watching western movies.

WXYZ joined WGN Chicago, WOR New York, and WLW Cincinnati to share programs and sell commercial time in September 1934, offering *The Lone Ranger* as part of the package deal. The expenses and income were to be mutually shared; this agreement resulted in the creation of the Mutual Broadcasting System in 1934. Trendle persuaded Striker to move to Detroit later that year, and although Striker continued to write for other stations as well as for WXYZ, he was instrumental in the development of new shows for the station and network, including *The Green Hornet* and *Sergeant Preston of the Yukon,* serving as script writer for both. When Trendle incorporated *The Lone Ranger* in 1935, Striker was again left out of any actual ownership in the program.

By 1940, in addition to his many radio scriptwriting duties, Striker was writing for *The Lone Ranger* comic strip, children's books, and adult novels. A major challenge for Striker came in 1941, when Earl Graser, the actor who played the Lone Ranger, was killed in an auto accident. At Trendle's direction, Striker wrote scripts for programs in which Tonto nursed a wounded and silent Lone Ranger back to health. When Brace Beemer later replaced Graser in the role of the Ranger, few in the listening audience were aware of any difference in the Rangers.

Late in 1943, following an argument over salary, Trendle refused to increase Striker's pay, and Felix Holt was brought in to replace Striker as head of WXYZ's script department. Holt rejected several of Striker's scripts for *The Lone Ranger,* and Striker became depressed and disillusioned about his career and his future. His life seemed brighter by early 1944, when Trendle reinstated Striker as department head and chief writer for the Ranger. Striker continued in this role and later assisted in the selection of Clayton Moore to play the Ranger for television in 1949. He adapted many of the radio scripts for the television version of the Ranger and continued writing additional scripts for radio and television during the 1950s. In the early 1960s, Striker turned to teaching and was preparing to move his family from Arcade, New York, to Buffalo when he was killed in an auto accident near Buffalo on 4 September 1962.

B.R. Smith

See also Green Hornet; Lone Ranger; WXYZ

Francis H. Striker. Born in Buffalo, New York, 19 August 1903. Attended University of Buffalo, 1922–25; started career as announcer, WEBR, Buffalo, New York, 1926; script writer, radio program *Covered Wagon Days,* 1930; moved to Detroit, Michigan, 1933; key script writer for Detroit station WXYZ, companion shows, *Green Hornet* and *Challenge of the Yukon,* which eventually became *Sergeant Preston of the Yukon;* wrote and edited *Lone Ranger* material for novels, comic books, personal appearances and television; script writer, variety of radio and television shows during 1940s and 50s; creative writing teacher, University of Buffalo, 1959–61. Died (car accident) in Buffalo, New York, 4 September 1962.

Radio Series

1930s	*Covered Wagon Days; Warner Lester, Manhunter; Dr. Fang; Thrills of the Secret Service*
1933–54	*The Lone Ranger*
1936–52	*The Green Hornet*
1938–42	*Ned Jordan, Secret Agent*
1942–44	*Challenge of the Yukon,* aka *Sergeant Preston of the Yukon*
1946–48	*The Sea Hound*

Television Series

The Lone Ranger, 1949–57

Film

The Lone Ranger, 1939; *The Lone Ranger Rides Again,* 1939

Further Reading

Bickel, Mary E., *George W. Trendle, Creator and Producer of: The Lone Ranger, The Green Hornet, Sergeant Preston of the Yukon, The American Agent, and Other Successes,* New York: Exposition Press, 1971

Boemer, Marilyn Lawrence, *The Children's Hour: Radio Programs for Children, 1929–1956,* Metuchen, New Jersey: Scarecrow Press, 1989

Harmon, Jim, *Radio Mystery and Adventure and Its Appearances in Film, Television, and Other Media,* Jefferson, North Carolina: McFarland, 1992

Osgood, Dick, *WYXIE Wonderland: An Unauthorized 50-Year Diary of WXYZ, Detroit,* Bowling Green, Ohio: Bowling Green University Popular Press, 1981

Rothel, David, *Who Was That Masked Man? The Story of The Lone Ranger,* South Brunswick, New Jersey: Barnes, 1976; revised edition, San Diego, California: Barnes, and London: Tantivy Press, 1981

Striker, Fran, Jr., *His Typewriter Grew Spurs: A Biography of Fran Striker,* Lansdale, Pennsylvania: Questco, 1983

Studio Equipment. *See* Recording and Studio Equipment

Subsidiary Communications Authorization

Radio stations, both AM and FM, are permitted to generate programming in addition to their main programs. Subsidiary Communications Authorization (SCA) uses multiplexing techniques and transmits audio or data on a separate channel, but still as part of the modulated carrier. SCA services, called "subcarriers," are not receivable with a regular radio. A special receiver or adapter is required.

Origins

The principle of multiplexing (sending separate signals with one transmitter) was first demonstrated by FM system inventor Edwin Howard Armstrong in the mid 1930s. In 1948 the inventor returned to perfect the multiplex technology and announced it in 1953.Armstrong and his associates saw the system as a way to assist then hard-pressed FM outlets with an additional revenue stream by allowing them the ability to transmit—and sell—a secondary transmission different from the main broadcast signal. This derived from the fact that stations rented or sold the separate receivers needed to pick up the secondary (to the main broadcast channel) transmission, be it music (as most were) or some other format. Early in 1955 the FCC adopted rules for such a service, based in part on the experience from 20 experimental subsidiary multiplex operations already underway. The first new subsidiary communications authorizations were issued in October 1955 to WPEN-FM in Philadelphia and WWDC-FM in Washington, D.C.

In the years that followed, FM stations made wide use of SCAs. Some 30 subcarriers existed by 1958, and more than 600 by 1967. Most were used for "musicasting" (transmitting background music for stores and offices provided either by the station or a service leasing station facilities—the most common application), special news and information services, weather warnings, educational programming, and (by the 1970s) reading services for blind listeners. In the days before widespread FM listenership in the 1970s, operation of SCAs often made the difference between profit and loss (or prevented a larger loss) for stations.

Operations

SCA services are not allowed to disrupt or degrade the station's main programming or the programs of other broadcast stations. Permissible SCA uses fall into two categories: the first includes broadcast transmission of programs or data of interest to a limited audience. Examples include paging services, inventory distribution, bus dispatching, background music, traffic control signal switching, point-to-point or multipoint messages, foreign language programming, radio reading services for the blind, radio broadcast data systems (RBDS), storecasting, detailed weather forecasting, real-time stock market reports, utility load management, bilingual television audio, and special time signals. The second category includes transmission of signals that are directly related to the operation of the radio station. Examples include relaying broadcast material to other FM and AM stations, distribution of audio networks, remote cuing and order circuits, and remote control telemetry.

Many of the programming requirements for broadcast stations do not apply to SCA programming, including station identification, delayed recording, program logging, and sponsor identification announcements. For FM stations only, SCA operation may continue when regular FM programming is off the air. However, regular hourly station identification must continue. Noncommercial FM stations, usually located between 88.1 megahertz and 91.9 megahertz, may generate SCA programming for profit. But such stations are then required to provide another SCA channel for any radio reading services for the blind that may request such a channel. The station is limited to charging the radio reading service only for actual operating costs.

There are several technical restrictions for SCA services. SCA subcarriers must be frequency modulated (FM) and are restricted to the range of 20 to 75 kilohertz, unless the station is also broadcasting stereo, in which case the restriction is 53 to 75 kilohertz. This allows a subcarrier to be modulated at audio frequencies and prevents it from interfering with the main program, as listeners cannot hear modulation above 20 kilohertz. SCA use is secondary to the audio on the main channel and must not interfere with the main broadcast audio channel.

SCA programming is retrieved by a detector in a special receiver, in which a tuned circuit filters out all subcarrier signals except the desired one. A second detector retrieves the information that modulates the selected subcarrier. Generally, tunable subcarrier receivers are prohibited by the FCC.

DAVID SPICELAND

See also Armstrong, Edwin Howard; FM Radio; Licensing

Further Reading

Christiansen, Donald, editor, *Electronics Engineers' Handbook,* 4th edition, New York: McGraw Hill, 1997 (see "Subsidiary Communications Authorizations," "Frequency Modulation Broadcasting," and "FM Stereo and SCA Systems")

Federal Communications Commission: Audio Services Division, <www.fcc.gov/mmb/asd/subcarriers/sub.html>

Gibilisco, Stan, editor, *Encyclopedia of Electronics,* Blue Ridge Summit, Pennsylvania: TAB Professional and Reference Books, 1985; 2nd edition, 1990

Gibilisco, Stan, editor, *Illustrated Dictionary of Electronics,* 8th edition, New York: McGraw Hill, 2001

Parker, Lorne A., *SCA: A New Medium,* Madison: University of Wisconsin Extension, 1969

Suspense

Suspense Thriller

In 1941 there were 16 suspense programs on the radio networks; by the end of the war there were more than 40. The suspense-thriller was the fastest growing genre during the wartime period. The most famous and prestigious of these programs was *Suspense,* which debuted as a series in 1942 after a single episode premiered in the summer of 1940 (this first episode was directed by Alfred Hitchcock, his only direct connection to the show). *Suspense* set the artistic and thematic standard for the programs that followed.

Originally a sustaining program, *Suspense* was promoted as a prestige drama because of the talent of its creative team, its first-rate stars, and the high quality of its original scripts. Producer/director William Spier fine-tuned each episode, coordinating music, actors, and sound to maximum effect and earning himself the nickname of the "Hitchcock of the Air." Bernard Herrmann (famous for his musical scores for Hitchcock films) composed and conducted music for the series until 1948; his theme for the show was used throughout its 20-year run. *Suspense*'s popularity and effectiveness, however, were also due to its realism. Unlike previous thriller programs such as *Lights Out, Suspense* programs did not incorporate the supernatural but rather focused on the psychological and social horrors that could be visited on the lives of everyday people. Radio critics of the time saw the growth of the genre as a testament to the audience's need for "escape" during the war, but part of the impact of such programs lay in their ability both to capitalize on audience's wartime fears and to address some of the feelings of trauma the war produced. This is particularly obvious in the many programs that focused on mistrust between husbands and wives, which tapped into both men's wartime traumas and their fear of women's independence.

Suspense's popularity and influence can be traced to one particular episode during its second year, "Sorry, Wrong Number," which was broadcast on 25 May 1943. This half-hour program, written by *Suspense* regular Lucille Fletcher (Herrmann's wife) and starring Agnes Moorehead, was a watershed moment in the history of radio drama and became perhaps the most famous original radio play of all time. In "Sorry, Wrong Number," Moorehead plays an invalid who overhears a conversation on the telephone between two men who are planning to murder a woman in half an hour. Moorehead's character, known only as "Mrs. Elbert Stevenson," tries desperately to prevent the murder by calling on various public institutions for help—the police, the phone company, public hospitals—but they do nothing for her, and her frustration increasingly borders on hysteria. In the last few moments, she realizes that she is the intended victim, that her husband has paid to have her killed. She calls the police but she's too late, and the play ends with her desperate screams as she is stabbed to death.

The play touched a nerve, and the Columbia Broadcasting System (CBS) was flooded with calls commending the program's realism and Moorehead's performance. The program was repeated within a few months, and then seven more times within the next few years. Audiences identified with a character who ultimately has no control over her fate and whose cries for help are ignored by those in power; the insecurity this created in audiences was enhanced by the fact that the character's killers go unpunished (an exception to Spier's usual policy). The success of the play proved the popularity of the suspense genre and encouraged the proliferation of suspense-thriller programs. It also led to more programs with female leads and narrators (Moorehead herself would become *Suspense*'s most frequent star). Finally, it encouraged a focus on domestic tensions in the genre as a whole, in particular making the stalked wife and the killer husband staples of the genre for the next several years. Like film noir of the time, suspense programs seemed to mirror the frustration of many Americans faced with postwar social requirements, particularly the social conformity, suburban ideal, and standards of wealth (husbands frequently killed wives for money) expected of them during the Cold War period.

Although *Suspense* programs developed a stable of talented stars (most notably Nancy Kelly, Cathy Lewis, and Elliot Lewis), producers often called on Hollywood stars to fill the title roles. Frequently, the star's persona was tweaked to accommodate his or her role as a killer or psychotic, adding to both the thrill and the discomfort the program could cause. Stars enjoyed doing *Suspense* programs, in part because it gave them the opportunity to play against their Hollywood images. Ozzie and Harriet Nelson schemed to kill their elderly relative in "Too Little to Live On" (1947); Frank Sinatra played a murdering psychotic in "To Find Help" (1945); Robert Taylor shot and killed his crazed werewolf-wife in "The House in Cyprus Canyon" (1946); and Orson Welles dug into his own living son's skull in the gruesome "Donovan's Brain" (1944). Paul Henried, Joseph Cotten, Charles Laughton, and Lloyd Nolan all killed their wives or girlfriends on the show, and Eve Arden and Geraldine Fitzgerald killed their husbands or boyfriends.

Female stars could look forward to a particularly wide range of meaty roles, in which they had to use their smarts to outwit stalkers as well as to climb the corporate ladder. Some of the more memorable of these include Lucille Ball as a gold digger in "A Little Piece of Rope" (1948), Anne Baxter as a struggling career woman in "Always Room at the Top" (1947), and Ida Lupino as a businesswoman coping with her ex-convict husband in "The Bullet" (1949). *Suspense* also helped shape star personas, first casting Vincent Price as a murdering sophisticate in one of Fletcher's best stories, "Fugue in C Minor" (1944); in addition, Jimmy Stewart's turn as a paralyzed veteran who believes he is being stalked by his Japanese torturer in "Mission Completed" (1949) anticipates the actor's work for Hitchcock in the 1950s in films such as *Rear Window.*

In later years, the direction of the program shifted to Anton M. Leader and then to Elliot Lewis, but the high quality and star power of the programs continued until the mid-1950s. Comedians and musical stars continued doing interesting variations on their star personas, with Jack Benny as a bank thief in "Good and Faithful Servant" (1952), Red Skelton haunted by dreams in "The Search for Isabel" (1949), and Danny Kaye as a scapegoat for murder in "I Never Met a Dead Man" (1950). Past shows were frequently repeated using different stars. In 1949 *Suspense* made its television debut, and the two shows ran simultaneously until 1954, when Autolite dropped sponsorship of both. The television show ceased production, but the radio program continued until 1962 under multiple sponsorship, making it one of the longest-running programs in radio history. Fortunately, recordings of well over 900 of the program's 945 episodes are available commercially. Listening to them today, they not only provide thrills and chills but are also an invaluable historical record of their time.

ALLISON MCCRACKEN

Hosts
"The Man in Black" (1942–43), Joseph Kearns or Ted Osborne (1943–47), Robert Montgomery (1948)

Announcers
Truman Bradley, Ken Niles, and Frank Martin (1943–47), Bob Stevenson, Harlow Wilcox (1948–54), Larry Thor and Stu Metz

Actors
Cathy Lewis, Agnes Moorehead, Jeanette Nolan, Hans Conreid, Joseph Kearns, Elliott Lewis, Lurene Tuttle, Mary Jane Croft, Bill Johnstone, William Conrad, Lillian Buyeff, Paul Frees, Irene Tedrow

Producers/Directors
Charles Vanda (1942), William Spier (1942–48, 1949–50), William Robson (1948), Anton M. Leader (1948–49), Elliott Lewis (1950–54), Norman Macdonnell (1954), Antony Ellis (1954–56), William N. Robson (1956–59), Bruno Zirato Jr. (1959–62), Fred Hendrickson (1962)

Programming History
CBS 22 July 1940 (single episode); June 1942–September 1962

Further Reading
Grams, Martin, Jr., *Suspense: Twenty Years of Thrills and Chills,* Kearney, Nebraska: Morris, 1997

Kear, Lynn, *Agnes Moorehead: A Bio-Bibliography,* Westport, Connecticut: Greenwood Press, 1992

Krutnik, Frank, *In a Lonely Street: Film Noir, Genre, Masculinity,* New York: Routledge, 1991

Sustaining Programs

Sustaining programs are those not supported by advertising revenue; the cost of airtime is said to be *sustained* by the network or station. Sustaining programs may be of any format but most are usually (especially in recent years) of some public service variety.

Although never formally required by law, sustaining programs have been seen in the past as a key part of radio's responsibilities under the "public interest" portion of federal regulations concerning radio. The original regulatory theory held that only by providing programs on a sustaining basis could networks or stations offer the diverse points of view and coverage of public affairs that advertisers might not support. Deregulation has swept away most such thinking, and sustaining programs today are few and far between.

Origins

At first, virtually all radio time was provided on a sustaining basis; there was no commercial advertising on the air. Station operators sustained the entire cost of their broadcast activities.

This began to change in the early 1920s as various means of supporting the cost of radio broadcasting were discussed and tried and all proved unworkable—except for the sale of airtime for advertising.

After AT&T's New York City station WEAF first sold time in mid 1922, other stations slowly began to do the same. As the potential for revenue became clearer, more stations began the practice, so that by the end of the decade, time sold to advertisers was widely accepted as the standard for broadcasting.

Well into the 1930s, advertiser-supported time was typical only of the most popular programs on the air; substantial portions of the broadcast day were still sustaining. More than a third of U.S. radio network offerings were sustaining, even on the eve of World War II. Only the lack of print alternatives during paper-short World War II helped to fill most network time slots with paid advertising, reducing sustaining programs to a few public service offerings.

As the most popular (and highly rated) entertainment programs siphoned off advertising, such non-entertainment programs as religion, agriculture, children's shows, public affairs, discussion and talk, and news were offered on a sustaining basis to fill out station and network schedules with the broad program diversity sought by regulators. Many of these were broadcast in daytime hours when fewer people listened to radio. By 1940, "55 out of 59-1/2 daytime hours of sponsored programs per week [carried by the four national networks] were devoted to soap-operas. The broadcasting industry has thus permitted advertisers to destroy over-all program balance by concentrating on one type of program" (Warner, 1948). Similarly, stations in larger markets with more advertiser appeal were carrying fewer sustaining hours.

At the same time, many sustaining programs were providing important radio services. Some experimental drama work appeared, for example, on the Columbia Broadcasting System (CBS) sustaining program *The Columbia Workshop*, beginning in 1936. Stations in a number of larger markets provided sustaining time for programs concerning public affairs, agriculture, children's interests, and other important issues that usually appealed only to a minority of listeners. Although such programs rarely aired in the prime evening time of greatest interest to advertisers, the important fact is that they were provided and were also touted by broadcasters as evidence of their public service role.

The Public Interest

Government and industry perceptions of just what radio's "public interest" responsibilities were comprised a major factor in the long survival of sustaining programs. Prior to the Radio Act of 1927, there were no formal government-mandated rules for radio programs or advertising on the air. The 1927 law established the phrase "public interest, convenience or necessity" as the guiding principle for government licensing of radio stations. Regulators from the Federal Radio Commission (1927 to 1934) and the Federal Communications Commission (FCC) have attached various measures of importance to the provision of sustaining programs as a key part of meeting the public interest rubric.

The clearest statement of how the FCC saw sustaining programs as a key part of radio's fulfillment of its public interest responsibility came in its 1946 policy statement informally dubbed *The Blue Book*. In that high-water mark of pro-regulatory thinking—authored in part by former British Broadcasting Corporation (BBC) officials who thought only in terms of sustaining time—the commission's staff held that sustaining programs filled five essential functions: (1) to secure for the station or network a means by which, in the overall structure of its program service, it can achieve a balanced interpretation of public needs; (2) to provide programs that, by their very nature, may not be sponsored with propriety; (3) to provide programs for significant minority tastes and interests; (4) to provide programs devoted to the needs and purposes of nonprofit organizations; and (5) to provide a field for experimentation in new types of programs, free from the program restrictions dictated by an advertiser's interest in selling goods. Nearly half of the FCC report was devoted to a detailed discussion of these points and of actual industry practice statistics of sustaining programs during the war years (1940 to 1944).

Industry spokespersons responded by strongly arguing that they could fulfill all of their public service requirements without having broadcasters sustain program costs. Advertisers, they said, were more than willing to take up the slack. And industry economic realities—already evident in radio and soon to be so in television—made it impossible for the FCC to sustain its thinking about sustaining programs. As advertiser demand for radio time expanded after the war, stations sought and often found support for formerly sustaining programs.

When the FCC issued a new program policy statement in 1960, it clearly indicated its acceptance of industry arguments that "There is no public interest basis for distinguishing between sustaining and commercially sponsored programs in evaluating station performance. . . . Sponsorship of public affairs, and other similar programs may very well encourage broadcasters to greater efforts in these vital areas." Sustaining time was now merely that which had not been sold, so it no longer held a special interest for regulators.

When in the early 1980s the FCC removed radio license processing guidelines that called for at least minimal amounts of non-entertainment programming and strongly encouraged public service (i.e., sustaining) messages, another support mechanism for sustaining time disappeared. No longer was the commission interested in how much advertising time a station sold—the marketplace would set the standard.

The decline of the FCC Fairness Doctrine in 1987 took away another prop of sustaining program time. The commission no longer required careful station records demonstrating that various sides of public controversies were being aired even when that meant some had to be given sustaining airtime. On the other hand, expression of controversial points of view in paid time—another long-time industry taboo—was quickly accepted when it became clear that there were plenty of people and institutions eager to buy time in order to broadcast their views. The lapse of the Fairness Doctrine made such sales easier for stations, as they were no longer required to use expensive airtime to provide balancing points of view.

Religion

Religious programs offer a useful window of insight into the subsequent decline of sustaining time. As a category, these were once provided free of charge (i.e., on a sustaining basis) to established or mainline religious groups of the Protestant, Catholic, and Jewish faiths. The major networks offered such programs at least weekly, as did many other stations. Some of these programs ran for decades. For many years it was a proud radio industry policy (a boast?) that time was not to be sold for religious programs, as such programs were offered on a sustaining basis as part of radio's public interest responsibility. But some critics—especially those in smaller religious organizations not included in the mainstream programs—argued that the networks and stations were effectively censoring minority religious viewpoints with their refusal to give equal time to programs from other denominations.

Beginning on the fringes, with small stations in local markets in the late 1940s and early 1950s, evangelical and other generally conservative religious figures began to purchase time from financially-strapped outlets happy to make the sales and unconcerned with broader industry policies. Soon larger market stations were doing the same, while continuing to carry their traditional sustaining mainline religious programs. And gradually the mainline programs disappeared, religion on the air being effectively redefined to mean those denominations willing to pay for their time. Such paid programs often spent much of their time (or so it seemed) seeking donations to help purchase still more airtime.

At the beginning of the 21st century, some stations still provide sustaining time for community or other nonprofit organizations. And stations often provide time for their own special campaigns or public service benefits. But for the past four decades, sustaining time has been seen merely as time not (yet) sold, rather than as a special category in and of itself.

CHRISTOPHER H. STERLING

See also Advertising; Blue Book; Controversial Issues; Fairness Doctrine; Federal Communications Commission; Public Interest, Convenience or Necessity; Religion on Radio

Further Reading

Federal Communications Commission, "The Carrying of Sustaining Programs," in *Public Interest Responsibilities of Broadcast Licensees,* Washington, D.C.: GPO, 1946

Kahn, Frank, "The 1960 Programming Policy Statement," in *Documents of American Broadcasting,* edited by Kahn, New York: Appleton-Century-Crofts, 1968; 4th edition, Englewood Cliffs, New Jersey: Prentice Hall, 1984

Warner, Harry Paul, "Sustaining Programs," in *Radio and Television Law,* by Warner, Albany, New York: Bender, 1948

Swing, Raymond Gram 1887–1968

U.S. Radio News Commentator

With his earnest "Good Evening," Raymond Gram Swing introduced one of the most influential political and foreign affairs commentary programs on network and shortwave radio between 1930 and 1963. Within the United States, he developed a reputation as an uncompromising New Dealer and anti-isolationist, and to the largest international audience ever drawn by a newscaster he was considered a faithful friend to England, a virulent opponent of world fascism, and the "voice of America."

Swing was born in Cortland, New York, on 25 March 1887. In 1906, after spending only one year at Oberlin College, he launched his news career as a reporter for the *Cleveland Press* and a variety of other Midwestern newspapers. He became a foreign correspondent for the *Chicago Daily News* in

1912 and covered many of the principal events leading to World War I. In 1915 he was an eyewitness to the naval battle for Gallipoli. After the war he continued as a newspaperman, but in the 1920s he became fascinated with radio's ability to reach a vast audience and joined a number of other print journalists (Boake Carter, Edwin C. Hill, and others) who changed professional media.

Swing's first experience with broadcasting came in 1930, when he began a series of commentaries on U.S. affairs for the British Broadcasting Corporation (BBC). This series was soon followed by reports on the Geneva Disarmament Conference for the National Broadcasting Company (NBC) and on the London Naval Conference and the British election of 1931 for the Columbia Broadcasting System (CBS). In 1932 Swing helped make radio history when, from London, he participated in the first two-way transatlantic broadcast, with Cesar Saerchinger in New York.

By the end of 1934, Swing had developed a large following on both sides of the Atlantic. His *Things American* broadcasts were especially popular in England, where an estimated 30 percent of adults regularly tuned in. King George VI (who once requested an autographed photo) and Winston Churchill listened faithfully, and "Swing Clubs" were formed in parliament.

In 1935 Swing began a weekly "Behind the Week's Foreign News" feature for CBS's *American School of the Air.* Network executives were impressed by Swing's carefully measured delivery, precise timing, and meticulous and lengthy pre-broadcast preparation, but they felt his voice was uninspiring and his overall microphone presence pedantic and dry. When CBS Vice President Edward Klauber tried to entice him away from the studio with an offer of a position as European director of talks, Swing left the network (transferring the job to a relatively unknown Edward R. Murrow) and began a long association with the Mutual Broadcasting System (MBS). From 1936 to 1938, Swing broadcast every Friday night over MBS's flagship New York station WOR and occasionally spoke over the Canadian Broadcasting Corporation.

By the mid-1930s, Swing became increasingly conscious of the dangers of world fascism. In 1935 he published *Forerunners of American Fascism* and used his broadcasts to alert listeners to the demagogic abuses of Huey Long, Father Charles Coughlin, and Dr. Francis Townshend. In 1937 he urged Britain and France to intervene in the Spanish Civil War and defend the republican government against Franco's paramilitary Falange. In March 1938 Swing condemned the *Anschluss* (Germany's annexation of Austria). He reserved his sharpest criticism for the appeasement of Hitler at Munich in October 1938. Swing's firsthand accounts of the Czech situation for Mutual earned him a National Education Conference on Radio citation and an extra weekly slot in the network's broadcast schedule. At the end of the year, a radio editor's poll identified Swing as the third-most-popular commentator on

Raymond Gram Swing, 30 July 1941
Courtesy AP/Wide World Photos

the air. One of the 15 million listeners who now regularly tuned in to him remarked: "Everybody reads the foreign news these days, but very few can fit the myriad events and rumors into a coherent whole. Swing can. With deceptive ease he molds the hodgepodge of reports into a sharply-defined picture of the actual situation at the moment."

The Polish crisis of August 1939 gave Swing another opportunity to distinguish himself. When the Nazis and Soviets signed the non-aggression pact that would seal the fate of Poland, Swing was one of the first newscasters to comment on it. When Poland was invaded in September, Swing illuminated the military situation for listeners three times a day over Mutual's full 110-station network. His most memorable broadcasts were published at the end of the year in *How War Came.* In recognition of his growing appeal, the General Cigar Company offered Swing a contract with complete editorial freedom and limited commercial interruptions.

As the United States moved closer to war in 1940, Swing became an ardent supporter of the Roosevelt administration's interventionist policies. He made no effort, on the air or off, to conceal his anti-isolationist tendencies, and he became a prominent member of both the pro-war Council for Democracy and the Committee to Defend America by Aiding the Allies. In October 1940 he facilitated America's first peacetime draft by producing several broadcast appeals for the Selective Service. His commentaries in favor of Roosevelt's Destroyers for Bases agreement, Lend-Lease, and other measures to aid Britain against Hitler (even at the risk of the United States' own involvement) were delivered with an almost crusading zeal.

Although most of Swing's 1940–41 newscasts focused on the deteriorating European situation, he also supported Roosevelt's effort to halt Japanese expansion in the Far East through diplomatic and economic pressure. Roosevelt valued Swing's favorable treatment and allowed him to use government-owned shortwave facilities to broadcast daily Spanish and Portuguese translations of his analyses to Latin America. When combined with regular BBC relays of his programs to Australia and the British Commonwealth, this gave Swing the largest international audience of any newscaster (37 million).

The rapid pace of diplomatic and military events after the disaster at Pearl Harbor and the United States' official entry into the conflict ensured the need for serious commentators like Swing who could make sense of the daily flow of communiqués, speeches, and reports from distant battlefields. In 1942 Swing left Mutual and began a four-times-a-week program on NBC Blue for Standard Oil's Socony Vacuum. His 12 April 1945 commentary on Roosevelt's death is considered to be his finest. After the destruction of Hiroshima in August 1945, Swing began to devote considerable airtime to pressing the cause of global interdependence. By the end of the war, Swing was heard on 120 stations and enjoyed a Hooper rating of 14.5, a figure twice as high as the average newscaster and comparable to some of the more popular entertainment programs of the day. His wartime broadcasts and work on behalf of world peace earned him DuPont and Peabody awards and six honorary doctorates.

In 1947 Swing switched to the American Broadcasting Companies (ABC), but within a year declining health compelled him to turn over his thrice-weekly program to Elmer Davis and broadcast only on Sunday. He left the air completely in 1949 but returned briefly for WOR the following year when the United States became embroiled in Korea. In May 1951 Swing retired from commercial radio and became the first political commentator for the Voice of America (VOA). His nightly VOA broadcasts were shortwaved around the world in 36 languages. In 1953 he resigned to protest McCarthyism and served as a writer and editor for Edward R. Murrow's *This I Believe* series. He returned to VOA in 1959 and remained there until his final retirement in 1963. In 1964 he published his autobiography, *"Good Evening": A Professional Memoir.* He died of a heart attack in December 1968.

ROBERT J. BROWN

See also American School of the Air; Commentators

Raymond Gram Swing. Born in Cortland, New York, 25 March 1887. Son of Rev. Alfred Temple and Alice Mead Swing; attended Oberlin College, 1905; night reporter for *Cleveland Press;* court reporter for *Cleveland News;* editor of *Courier,* Orville, Ohio; reporter for *Item,* Richmond, Indiana; managing editor of *Indianapolis Sun,* 1906–11; reporter for *Cincinnati Times Star,* 1912; foreign correspondent for *Chicago Daily News,* 1912–17; War Labor Board examiner, 1918; Berlin correspondent for *New York Herald,* 1919–22; director of foreign service for *Wall Street Journal,* 1922–24; foreign correspondent for *Philadelphia Public Ledger* and *New York Evening Post,* 1924–32; member of board of editors of *The Nation,* 1934–36; commentator for various networks, 1930–53; writer for *This I Believe* (CBS). Died in Washington, D.C., 22 December 1968.

Radio Series

1934–36 *Things American*
1934–36 *American School of the Air*

Selected Publications

Forerunners of American Fascism, 1935
How War Came, 1939
Preview of History, 1943
Watchman, What of the Night?, 1945
In the Name of Sanity, 1946
"Good Evening": A Professional Memoir, 1964

Further Reading

Alexander, Jack, and Frank Odell, "Radio's Best Bedside Manner," *The Saturday Evening Post* (14 December 1940)
Culbert, David H., *News for Everyman: Radio and Foreign Affairs in Thirties America,* Westport, Connecticut: Greenwood, 1976
Fang, Irving E., *Those Radio Commentators!* Ames: Iowa State University Press, 1977
Smith, Robert R., "The Wartime Radio News Commentaries of Raymond Swing, 1939–1945," Ph.D. diss., Ohio State University, 1963

Syndication

Supplementing Local and Network Sources

Stations have three sources of programming: they can produce programming themselves, receive it from a network, or get it from an outside supplier, called a syndicator. For three decades, syndication provided dramatic series to stations; then, when radio formats changed, syndication provided music-oriented programming. Today, many stations rely on syndication for music and talk programming.

Origins

In the early days of radio, station owners relied on live talent and scratchy-sounding records for programming. The advent of the National Broadcasting Company (NBC) in 1926 and the Columbia Broadcasting System (CBS) in 1927 meant that stations could have access to the top talent in New York City for the production, writing, directing, and performing of programs. (Network programs were delivered to a permanent hookup of stations across the country, which were expected to broadcast the shows simultaneously as they were fed and which were compensated for carrying the programs and their network advertising.) But not all stations could be affiliated to the networks, and not all affiliates were satisfied with network programs and the compensation they received for carrying them.

To meet this demand, shows were distributed on records. The first syndicated radio program is credited to Freeman Gosden and Charles Correll, the creators of *Amos 'n' Andy.* When their series moved to WMAQ, which was owned by the *Chicago Daily News,* they acquired the right to record their program and sell it to other stations. In 1928 the newspaper mailed out the show to 30 stations.

Opposition

Although the idea of stations sharing programs on records seemed logical enough, problems arose. Stations, advertisers, and listeners were reluctant to accept an alternative to live programming. A major difficulty was that records were inferior in sound quality, partly because of the method of recording and partly because of the station's playback equipment. Development of electrical transcription in 1928 helped solve the problem, and stations slowly adopted the playback equipment.

Another problem was that some powerful organizations in the industry had a vested interest in live programming. The American Federation of Musicians insisted that major stations continue to employ staff musicians, even if recorded programming was used. If stations had to pay the salaries, they might as well use the employees rather than buying a recorded program. As transcriptions became more popular, the union stepped up its opposition. In 1942 the president of the musicians' union ordered his members to cease making transcriptions for broadcast use. The battle between syndicators and the union continued until television made the issue moot.

The radio networks also felt threatened by transcriptions. If a station could order a good-quality series through the mail, then who needed a network? Transcriptions also cut into the networks' revenue. Advertisers paid networks on the basis of the number of stations that aired or "cleared" their programs. If too many affiliates used transcriptions, then the networks would be hurt financially.

Network Option Time

To minimize the competition, NBC set up compensation rates to its affiliates in such a way that the use of recordings and the acceptance of national spot advertising were discouraged. These practices ended in 1941 after a Federal Communications Commission (FCC) investigation created new rules controlling network practices.

Both CBS and NBC required that their affiliates be willing to carry a certain amount of network programming at set times of the day. This practice of "optioning" portions of a station's schedule made the syndicator's job more difficult. Sponsors usually wanted their programs to run at the times when viewing levels were highest, but those were the hours the networks had claimed for themselves. If a station scheduled a syndicated program during those hours, the networks could demand that their own program replace it. Advertisers on syndicated programs had to settle for hours with fewer listeners or for nonaffiliated stations, which were usually less prestigious and had smaller audiences. The FCC tried to end the networks' optioning of their affiliates' time with its Chain Broadcasting Rules in 1941, but this move raised so much opposition that the Commission settled for reducing the amount of time the networks could stake out.

The failure to end option time was not the only reason syndicators complained about the Commission. The FCC insisted that announcers state after each record that the program had been transcribed. Syndicators claimed that the practice made their properties seem inferior to live programming. The rule had originated when recorded programs had poor quality, but despite improvement in recording techniques, the policy continued in effect until after World War II.

Advantages

Despite all of these problems, the syndication industry survived. By 1931, 20 percent of the stations in the United States could play electrical transcriptions, including many higher-powered stations. Syndicators claimed that their product was superior to network product because the listener received the best possible performance, without the mistakes that could happen in live situations. And stations could profit, because there was no network to keep commercial minutes and revenue for itself.

National, regional, and local advertisers found the use of syndicated transcriptions beneficial at times. The national advertiser could use syndication as a substitute for or a supplement to the networks. With a recorded program, a sponsor could reach and pay for only a certain region of the country, if desired. Or a seasonal product could be sold using different approaches in different parts of the country.

Stations usually charged less than the networks, and so the advertiser could get a bargain. For example, the Beech-Nut Packing Company syndicated *Chandu, the Magician* to 15 stations. Had syndication not been an option, the sponsor would have had to pay the networks for a cross-country lineup of stations that it didn't need. For local and regional advertisers, syndication was even more attractive. They could have big-name talent at local costs.

The Depression

By 1931, 75 commercially sponsored programs were available by syndication, an increase of 175 percent over the previous year. The most common syndicated shows were musical variety shows, dance bands, and programs that re-created news events.

The Lone Ranger was created on WXYZ in Detroit and syndicated to other stations. The program's success led to the founding of the Mutual Broadcasting Company, a cooperative network that began with the idea of stations sharing programs among themselves.

But as the Depression deepened, the industry faltered. (Sometimes syndicators would sign up some advertisers based on an audition disc but then go out of business if not enough sponsors were found.) During this period, smaller stations and regional advertisers were the main users of syndicated programs.

Two other organizations also became prominent in the syndication industry, NBC and CBS. Both networks maintained that live programs were better, but they recognized that recorded shows might be useful in some circumstances. NBC entered the field in 1935 with three services: a collection of musical numbers, series produced for syndication, and recorded series that had run live on the network. CBS followed in 1940.

Frederic Ziv

Perhaps the leading radio syndicator was Frederic Ziv, who began distributing radio programs in 1937 and later went on to syndicate television programs. He had owned an advertising agency in Cincinnati and wanted to create a product his bakery client could use. Once the program was produced, he started selling it to other stations. This first program was a series aimed at children, *The Freshest Thing in Town.*

World War II brought restrictions on the materials used to make electrical transcriptions, but after the war, radio syndication reached its height of popularity, as more stations went on the air and as the networks concentrated their efforts on television. Two of the more popular syndicated programs were produced by Ziv: *Boston Blackie* and *The Cisco Kid.* Stars such as actor Ronald Colman, singer Bing Crosby, and musician Kenny Baker were available on electrical transcriptions. With the advent of audiotape in the United States, syndicators had a new medium for distribution, and they used it mainly for features and short music programs.

By the mid-1950s, syndicators had to deal with the changes in the radio industry wrought by competition from television. Disc jockeys spinning records had replaced network serial programming. At first, stations desperate to maintain their traditional formats turned to the syndicators to provide the old-fashioned programs, but soon half-hour dramas and adventure shows became rare on radio.

Automated Formats

In the late 1950s, syndicators found another niche. They created hours of recorded music with announcements on audiotape and mailed the reels to stations around the country. Music format syndication offered several advantages to the station. It was cheaper and more reliable than the average live disc jockey. For unpopular time periods, late night and weekends, recorded programming could be the difference between profit and loss for a station. The supply of reels meant that even the smallest town could have an announcer with national appeal and a smoothly produced program aimed at a target audience. The main disadvantage was that the automated programming couldn't be localized. Some stations did use a local announcer during some of the programming in a practice called "live assist."

The FCC provided an impetus to the growth of automated radio by declaring in 1964 that FM and AM stations in the larger markets could no longer simulcast (i.e., carry the same programming at the same time). Because AM radio was the dominant medium at that time, stations scrambled to find programming to put on their FM channels. Automation was a cheap and easy solution.

Other Formats

Other types of programming were also syndicated. In 1968, 300 stations carried a special featuring mythical boxing matches between the all-time best heavyweights. By the mid-1970s, weekly programs were back in style. Successful formats included weekly musical specials, such as the *King Biscuit Flower Hour,* and musical countdowns hosted by Casey Kasem and Dick Clark. *The National Lampoon Radio Hour* and *Dr. Demento* were examples of successful comedies. These programs were carried mainly on weekends.

The mainstay of the industry remained automated programs. The most popular format was "beautiful music," standard songs done without lyrics and in lush arrangements, usually replete with strings. Other formats, though, such as rock and roll and country/western, were also syndicated. By 1977 almost 1,500 stations were fully automated, and another 1,000 relied at least some of the time on automation. The major syndicators were also expected to act as consultants, providing advice on technical matters, promotion, and advertising, as well as format.

Modern Syndication

In the 1980s, satellites made the delivery of programming cheaper and allowed the added quality of timeliness. Disc jockeys could comment on the day's events. A satellite could send a signal to a subscribing station's computer, which could be programmed with local news, commercials, and weather. Several channels of programming could be sent and received at the same time with no loss of quality. A station could carry some programming from one supplier and easily switch to another supplier on the satellite.

A new market developed as AM radio started to lose its audience to FM. Talk shows distributed by satellite began to be in demand. Call-in shows, such as those of Larry King and Tom Snyder, were popular late at night. In 1988 Rush Limbaugh's show was put into syndication for the daytime audience and became successful enough to attract imitators such as Howard Stern and Don Imus. Other types of talkers joined in the competition with topics such as sports, psychological advice, and business information.

Syndicators increased their use of satellites as the cost of equipment went down and as compression technology allowed them to squeeze more signals onto one transponder. The difference between them and networks blurred. Syndicators offered simultaneous delivery of signals to affiliates across the country via satellite, just as networks did, and the type of programming was the same. Traditional networks and syndicators became part of the same companies, as mergers and buyouts led to industry consolidation. Networks didn't always provide compensation, and both industries offered similar formats.

By 2000 the two most successful syndicated radio formats remain the talk show and automated programming delivered by satellite, tape, or disc. With the aid of a computer and the satellite, the syndicator can provide announcing, music, and cues for commercials, local news, and weather. This type of service had existed before but was delivered on tape to the station and was therefore not as versatile. For some markets, the national announcer records local announcements, which can be inserted smoothly into the program. Short features with medical advice, interviews with stars, and David Letterman's top 10 list are also syndicated.

Syndicated programming can be purchased outright, or the sale may involve barter or a combination of the two (cash plus barter). Some form of barter is the most common method: The producer sells some commercial minutes in the program but leaves time for the station to sell others. In other words, the station trades time for a free program or one with a reduced price. If cash is involved, the cost of the program will depend on the size of the market, its competitiveness, and the station's revenues.

Syndication should remain an important part of the industry. Because syndication of music formats is a cost-effective method of delivering programming, station group owners may rely more on it especially for their weaker outlets.

BARBARA MOORE

See also American Federation of Musicians; Amos 'n' Andy; Automation; Dr. Demento; King Biscuit Flower Hour; Limbaugh, Rush; Lone Ranger; Recording and Radio Industries; Simulcasting; Stern, Howard

Further Reading

Becker, Christine, "A Syndicated Show in a Network World: Frederic Ziv's *Favorite Story,*" *Journal of Radio Studies* 8, no. 1 (Summer 2001)

Eastman, Susan Tyler, Sydney W. Head, and Lewis Klein, *Broadcast Programming: Strategies for Winning Television and Radio Audiences,* Belmont, California: Wadsworth, 1981; 6th edition, as *Broadcast/Cable/Web Programming: Strategies and Practices,* by Eastman and Douglas A. Ferguson, 2002

Keith, Michael C., *Radio Programming: Consultancy and Formatics,* Boston and London: Focal Press, 1987

Keith, Michael C., and Joseph M. Krause, *The Radio Station,* Boston: Focal Press, 1986; 5th edition, by Keith, 2000

MacFarland, David T., *Contemporary Radio Programming Strategies,* Hillsdale, New Jersey: Erlbaum, 1990; 2nd edition, as *Future Radio Programming Strategies: Cultivating Listenership in the Digital Age,* Mahwah, New Jersey: Erlbaum, 1997

Rouse, Morleen Getz, "A History of the F.W. Ziv Radio and Television Syndication Companies: 1930–1960," Ph.D. diss., University of Michigan, 1976

Routt, Edd, James B. McGrath, and Fredric A. Weiss, *The Radio Format Conundrum,* New York: Hastings House, 1978

Sterling, Christopher H., and John M. Kittross, *Stay Tuned: A History of American Broadcasting,* 3rd edition, Mahwah, New Jersey: Lawrence Erlbaum, 2002

T

Taishoff, Sol 1904–1982

U.S. Editor and Publisher

For more than a half century, Sol Taishoff was a leader in radio's trade press as editor and publisher of *Broadcasting,* its most important weekly magazine. Based in Washington, D.C., the trade periodical reported important government (Federal Radio Commission, Federal Communications Commission, Congressional) decisions concerning radio. For 51 years, first as co-founder of the magazine in 1931, and then as the sole owner, Taishoff chronicled radio's spectacular growth. Through the 1940s, Taishoff also began reporting the early development of television and, to the chagrin of some radio station owners, renamed the magazine *Broadcasting-Telecasting* in 1950, before many Americans could even view the new medium. In the 1960s, cable television became part of Taishoff's beat, as did satellite distribution in the 1970s.

Through all of radio's growth and change, Taishoff became as much a part of the industry he covered as the network and station executives who bought and read the weekly *Broadcasting.* He even fit the mold of the typical broadcasting executive: he was personable, could tell a good story, and made friends easily. He walked a fine line: he was both a highly respected journalist and a beloved friend to hundreds in the American broadcasting community. In 1981 more than 1,200 people attended a banquet held to mark *Broadcasting*'s 50th anniversary, and then-President Ronald Reagan, a former WHO-AM sportscaster, lauded (via tape) Sol Taishoff's considerable talents and accomplishments.

Origins

Born in Minsk in Czarist Russia in 1904, Sol Taishoff immigrated to the United States when he was three years old. His family settled in Washington, D.C., where Taishoff lived and worked for the rest of his life. He dropped out of the city's Business High School when he was 16 to become a night-shift copy boy for the Associated Press. After leaving the Associated Press in 1926, Taishoff continued his journalism career by joining the original staff of David Lawrence's *United States Daily* (which became *U.S. News and World Report*).

Although he had occasionally written about radio while with the Associated Press, by the late 1920s radio had become Taishoff's entire beat. At that time, David Lawrence also owned a wire service called the Consolidated Press Association, and for his radio writings Taishoff used the pseudonym "Robert Mack." His predecessor as "Robert Mack" had been Martin Codel, who had moved on to work for a competing news wire. Taishoff and Codel had become friendly during their time with Lawrence and had often spoken of radio's need for a trade publication, its own version of the newspaper industry's *Editor and Publisher.* Taishoff and Codel raised barely enough capital to get their venture off the ground, and in October 1931 the Washington, D.C.–based *Broadcasting* was born as a biweekly publication. The FCC was established in 1934, forming one of Taishoff and Codel's key newsbeats.

Although 1931 was not viewed as an auspicious time to start any new business, Taishoff and Codel persevered and soon established their magazine as the bible of the fledgling industry. At first the battle was to stay afloat, so Taishoff and Codel met with some of their subscribers and advertisers and granted discounts to those who would pay in advance. In 1935 they issued the first *Broadcasting Yearbook,* a substantial directory of stations, networks, and related radio services. It continued to appear annually into the 21st century.

But in time, Taishoff and Codel disagreed about the magazine's direction, and after Codel went on leave in January 1943 to work for the Red Cross as director of information for the African Combat Area, he never returned to the magazine. Taishoff bought him out a year later and thereafter ran the magazine solely and completely. (Codel went on to create a rival publication, *Television Digest,* in 1945.)

So closely did Taishoff come to be associated with *Broadcasting* that network and station executives talked of placing ads in "Sol's magazine." In his many years at the helm of the magazine, Taishoff exhibited an editorial stance fundamentally in favor of free enterprise, opposing any proposal that might interfere with the free workings of the marketplace, including public broadcasting. He strongly advocated First Amendment rights for radio broadcasters, arguing consistently for freedom from government regulation and control.

But Taishoff knew that his magazine's strength lay in covering the Washington political scene, and he kept his subscribers well informed on any issue or proposal that could affect their interests. He made friends with FCC commissioners and politicians of both parties and became an unofficial adviser on communications policy to many of them, including several presidents. He was perhaps closest to Lyndon Johnson, whom he met when Johnson arrived in Washington as a congressional staffer in 1931. In the 1940s, when Johnson himself was in Congress, Lady Bird Johnson came into an inheritance. The story goes that Johnson was considering buying a newspaper, but Taishoff advised them to invest in an Austin, Texas, radio station. The future president protested that the station was not making any money, but Taishoff assured him radio was where large profits lay. He was, of course, right. And in time, the Texas Broadcasting Company, as it came to be known, owned, at least in part, nine radio and television stations and made Lyndon Johnson a multimillionaire (the licenses were always listed in Lady Bird's name).

At the end of his career, the broadcasting industry lauded Taishoff by presenting him with a George Foster Peabody Award and naming him the National Association of Broadcasters "Man of the Year." Though he slowed down during the 1970s, Taishoff never stopped working. Indeed, the last piece of copy to bear his name was written just weeks before his death in 1982.

The magazine lived on, first under Taishoff's son Larry, who sold the publication and its Washington headquarters in 1986 to Times Mirror, which in turn sold *Broadcasting* to Reed International in 1991. The magazine was part of the Cahners Business Information empire (as *Broadcasting and Cable*) and was published in New York, no longer from Washington, as the 21st century began.

DOUGLAS GOMERY AND CHUCK HOWELL

See also Trade Press

Sol Taishoff. Born in Minsk, Russia, 8 October 1904. Youngest of three sons of Rose Orderu and Joseph Taishoff; immigrated to the United States and settled with family in Washington, D.C., 1907; dropped out of business high school to become night-shift copy boy for Associated Press, 1920; joined *United States Daily,* which became *U.S. News and World Report,* 1926; created *Broadcasting* magazine with Martin Codel, 1931; bought out Codel, 1944, and ran magazine until death. Died in Washington, D.C., 15 August 1982.

Further Reading

"The 50th Anniversary Issue: On the Road to 2001," *Broadcasting* (12 October 1981)

The First 50 Years of Broadcasting: The Running Story of the Fifth Estate, Washington, D.C.: Broadcasting, 1982

"The First 60 Years," *Broadcasting* (9 December 1991) (a special supplement)

"Sol Taishoff, 1904–1982: The Best Friend the Industry Ever Had . . . ," *Broadcasting* (23 August 1982)

"Two Exciting Decades," *Broadcasting-Telecasting* (16 October 1950)

Talent Raids

In 1948 Columbia Broadcasting System (CBS) Chairman William S. Paley initiated a raid on the National Broadcasting Company's (NBC) top radio talent in order to compete with the better-fortified network. This bold move changed the balance of power in radio and affected key programming strategies in radio and later in television.

Background

Throughout the 1930s and much of the 1940s, NBC reigned as the dominant radio network. NBC had the financial backing of a wealthy corporate parent in the Radio Corporation of America (RCA), boasted a larger number of affiliated stations than

CBS, and had a popular roster of vaudeville-trained comedians. CBS initially tried to even the standings in the mid-1930s by capturing some of NBC's key affiliated stations. Paley also went after some of NBC's talent in 1936, luring such stars as Al Jolson and Major Edward Bowes over to the smaller network. NBC was angered by these maneuvers, particularly because the network thought it had an unwritten agreement with CBS not to participate in such raids. But given CBS's inferior position in the industry, Paley insisted it was the only way he could reasonably compete. NBC battled back heartily, winning back some of its stars and stations.

Thus, after World War II, CBS was still a distant number two. Dissatisfied that his previous coup attempts had failed, Paley began outlining a new strategy. Lew Wasserman and Taft Schreiber of MCA, the mammoth talent agency, ultimately helped lead Paley to a crafty solution in 1948, during a lunch date with CBS President Frank Stanton. MCA represented Freeman Gosden and Charles Correll, the stars of *Amos 'n' Andy,* and Wasserman offered the popular NBC actors to Paley along with a unique financial arrangement. Stars certainly earned a high salary at NBC (Jack Benny reportedly earned $12,000 a week), but this placed them within a very high income-tax bracket, and they were taxed at a rate as high as 77 percent. Wasserman and Paley thus devised a scheme wherein Gosden and Correll would incorporate, with CBS purchasing the resulting company and its assets, namely the characters and scripts for the shows. The money that CBS subsequently paid out to Gosden and Correll could thus be considered a capital gain, taxed at the considerably lower rate of 25 percent. Moreover, because CBS would now own the properties and names themselves, NBC would not be able to lure its talent back as easily as it did after Paley's first talent raids in the late 1930s.

The Raids

With Gosden and Correll signed, Paley next set his sights on Jack Benny, with Wasserman again brokering the deal. But because Benny and other NBC comedians went by their own names, rather than playing characters as did Gosden and Correll, it was initially unclear if they could legally incorporate their names for CBS to purchase. As a result, when Paley cemented a deal with Benny, the Internal Revenue Service challenged it in federal court. Though the Supreme Court did declare the maneuver legal in 1949, a more immediate obstacle came from Benny's sponsor, the American Tobacco Company, and his corresponding advertising agency, Batton, Barton, Durstine, and Osborn (BBD&O). Benny had a long-term contract with these companies, and they initially objected to the move to CBS, expecting a decline in Benny's ratings, if only because CBS had fewer affiliated stations. BBD&O thus vigorously complained to NBC about the decision to let go of Benny. NBC responded with a major counteroffer to Benny, totaling twice the value of CBS's offer. However, Lew Wasserman again intervened, obtained the NBC contract, changed every mention of NBC to CBS, and reoffered the deal to Benny, who then signed it. Reportedly, the personal attention given to Benny by CBS executives was enough to provide the deciding factor—Benny was continually insulted by the impersonal atmosphere of NBC and had reportedly never even met David Sarnoff, the head of the network's parent firm, RCA.

To counteract the sponsor's concerns about NBC's greater number of affiliates, Paley went to the unprecedented length of offering American Tobacco $3,000 for every rating point that Benny fell below his usual NBC total. Such a drastic move proved that CBS was not planning to merely buy out NBC's talent, but hoped to surpass NBC's success with this same talent. The deal was finally cemented in November 1948, despite the legal uncertainties at that point, and CBS bought Benny's company, Amusement Enterprises, for $2.26 million. Benny's CBS ratings were initially stellar, and despite a ratings decline shortly thereafter, Paley was pleased that he finally had a strategy in place to battle NBC. Bing Crosby, Red Skelton, Edgar Bergen, and George Burns and Gracie Allen were the next NBC stars to head to CBS.

Surprisingly, NBC and Sarnoff had little reaction to this continued upheaval. Some historians simply credit Sarnoff's arrogance for ignoring CBS's moves, and others highlight Sarnoff's belief that paying a performer so much money would set a dangerous precedent, resulting in a system that would give performers too much power over their network bosses. Whatever the reason for NBC's lack of a countermove, these events had the potential to devastate the network's industry standing. Indeed, by the end of 1949, CBS would tout 12 of the top 15 radio shows. But the emergence of television altered the playing field once again, and now both networks had to try out new strategies for the developing visual medium.

The success of its new talent did give CBS a profit infusion that helped launch the company into television. Lacking the benefit of a deep-pocketed corporate parent like RCA, CBS desperately needed such capital for its first steps into television. Additionally, CBS not only captured the radio ratings lead in 1949, but held onto that lead right into television and for the next 25 years. Finally, the talent raids related to a crucial industry strategy of developing and scheduling network-owned programming. Such direct connections with talent gave the network more control over their program decisions, rather than sponsors and advertising agencies making these decisions. This would prove to be a key difference between television and radio programming structures. In the end, an initial investment of less than $6 million brought huge benefits—the talent raid nearly eradicated NBC's top line-up of stars, brought CBS to equal status with NBC, and foretold of both networks' coming supremacy in television.

CHRISTINE BECKER

See also Amos 'n' Andy; Benny, Jack; Burns and Allen; Columbia Broadcasting System; Crosby, Bing; Edgar Bergen and Charlie McCarthy; Paley, William S.; Skelton, Red

Further Reading

Barnouw, Erik, *A History of Broadcasting in the United States,* 3 vols., New York: Oxford University Press, 1966–70; see especially vol. 2, *The Golden Web, 1933 to 1953,* 1968

Bergreen, Laurence, *Look Now, Pay Later: The Rise of Network Broadcasting,* Garden City, New York: Doubleday, 1980

McDougal, Dennis, *The Last Mogul: Lew Wasserman, MCA, and the Hidden History of Hollywood,* New York: Crown, 1998

Metz, Robert, *CBS: Reflections in a Bloodshot Eye,* Chicago: Playboy Press, 1975

Smith, Sally Bedell, *In All His Glory: The Life of William S. Paley,* New York: Simon and Schuster, 1990

Talent Shows

American radio in its earliest days was similar to an amateur hour, in that early performers were all volunteers, many of whom had great enthusiasm but minimal talent. As radio matured during the 1920s, more selective criteria for getting on the air were established, and that usually meant passing an audition. Local radio shows made a contest out of it—so-called *opportunity nights,* when those who envisioned themselves as tomorrow's radio stars could perform, and listeners voted by sending in postcards. Prizes were not very big, but the thrill of winning seemed to suffice.

Popular network talent shows had higher standards since performers would be heard by a national audience. The biggest shows held several rounds of auditions. To appear on *Roxy and His Gang,* would-be talent first auditioned for "Roxy" himself (Samuel Rothafel); he decided which amateurs would compete on the show. Some network programs also involved sponsors or advertising agencies in the decision-making process; because the money was coming from them, it seemed prudent to consider their input.

A successful show could be quite lucrative for its stars: in 1936, for example, Eddie Cantor was paid $10,000 a week by his sponsor, Texaco gasoline. George Burns and Gracie Allen made the same amount from their sponsor, Grape-Nuts Cereal. And topping the list was the $25,000 a week paid by Chrysler Motor Corporation to Major Edward Bowes. It is no wonder, with the media writing about these big salaries, that the average person dreamed of winning a talent show and becoming a network celebrity.

Radio talent shows became a national craze during the 1930s; with so many people out of work because of the Depression, the idea of striking it rich in radio was especially compelling. One network show that capitalized on this hope was *National Amateur Night,* which ran on the Columbia Broadcasting System (CBS) from December 1934 to December 1936. But thanks to its popular master of ceremonies, it was *Major Bowes' Original Amateur Hour* that would capture the largest audience. Major Edward Bowes (he had earned the rank of major in an obscure Reserve unit during World War I) had started out as a master of ceremonies for *Roxy and His Gang* at the Capitol Theatre in the mid-1920s. A few years later, while he was manager of New York radio station WHN, Bowes and two producers developed a new concept for a talent show: broadcast historian John Dunning (1998) explains that, unlike other shows, in which the master of ceremonies made fun of the contestants, "Bowes saw the amateur hour in terms of a prize fight. The amateurs were the combatants. . . . The bell between rounds [would be] utilized . . . to dismiss an amateur who wasn't making it. The gong was like sudden death, like the hook in the rough-and-tumble days of amateur nights in vaudeville. . . . its presence . . . add[ed] another element of suspense."

Major Bowes' Original Amateur Hour debuted on the National Broadcasting Company (NBC) radio network in late March of 1935, sponsored by Chase and Sanborn coffee. The show was an immediate sensation, both in its ratings and in the number of would-be participants. At its highest point, it was receiving more than 10,000 applications a week for the 20 available slots on the program. The show's opening lines—"the wheel of fortune goes round and round, and where she stops, nobody knows"—became an American catchphrase. But *Major Bowes' Original Amateur Hour* also attracted controversy, with critics questioning how honest the voting process was (people called in to vote for their favorite amateur, leading to charges that a sponsor or any amateur with a lot of friends could easily manipulate the totals) and whether Bowes decided on the winners in advance. And although the show did get huge ratings for a while, it resulted in the discovery of very few major stars—opera star Beverly Sills and crooner Frank Sinatra were the best known of the winners. (Some critics have also suggested that the idea for Major Bowes' program really came from comedian Fred Allen, who did a forerunner of the amateur hour in 1934 as a segment of his NBC show *Town Hall Tonight.*)

Another show that gave amateurs a chance at fame and fortune was *Arthur Godfrey's Talent Scouts,* which started in early July 1946 on CBS Radio and went on TV in 1948. Godfrey, who was also the master of ceremonies of a successful variety show, *Arthur Godfrey Time,* had a special reason for creating *Talent Scouts.* Unlike other talent shows, Godfrey did not give winners cash prizes or gong losers off the stage. Rather, he gave contestants the chance to graduate from *Talent Scouts* and become regular performers on *Arthur Godfrey Time,* a top-rated radio show that would be equally successful on television. Several *Talent Scouts* winners not only joined the "Little Godfreys" on *Arthur Godfrey Time,* but later had their own hit records on radio. Two of the best examples were the McGuire Sisters, whose song "Sincerely" went to number one on the U.S. pop charts in 1955, and the Chordettes, who had three top five songs during the mid-1950s. On the other hand, sometimes Godfrey got it wrong; among those who failed his audition was Elvis Presley.

One other network talent show with appeal was bandleader Horace Heidt's *Youth Opportunity Program,* heard on Sunday nights from 1947 to 1951. But by the early 1950s, most radio talent shows had moved to TV (where a few can still be found). And although the odds of becoming the next big celebrity are very small, the thrill of competing and the chance to get on the air still motivate people to come to auditions, hoping that this time the wheel of fortune will stop for them.

DONNA L. HALPER

See also Godfrey, Arthur

Further Reading

"Bowes Inc.," *Time* (22 June 1936)
Brindze, Ruth, *Not to Be Broadcast: The Truth about Radio,* New York: Vanguard Press, 1937
Maltin, Leonard, *The Great American Broadcast: A Celebration of Radio's Golden Age,* New York: Dutton, 1997

Talk of the Nation

Public Radio Call-In Program

Talk of the Nation is a two-hour radio show combining the news experience of National Public Radio (NPR) and the participation of call-in listeners from across the country. For the first four days of the week, the *Talk of the Nation* host (Neal Conan beginning late 2001) discusses a variety of national issues, while on Friday Ira Flatow hosts *Talk of the Nation Science Friday.* As stated by former host Ray Suarez, NPR designed the show to be "a news program that would bring in a caller who really wanted to understand better all the hanging questions out there." Although the show premiered in only nine markets, the midday voice of NPR News and NPR Talk, airing 2 P.M. eastern standard time (EST), boasted more than 2 million listeners from over 150 markets nationwide by 2001. The show has won several honors including the 1993 Corporation for Public Broadcasting Silver Award.

U.S. President Bill Clinton's presidential pardons in early 2001, U.S. international policies, environmental issues, school privatization, the economics of baseball, and international slavery are examples of the topics broached on *Talk of the Nation.* Jesse Jackson, Stephen King, Ralph Nader, Christopher Darden, Walter Mondale, and Yogi Berra all have graced *Talk of the Nation* as guests. As described by Suarez, "It's not a prissy, pointy-headed intellectual show, but we do give the audience a great deal of credit for being intelligent and literate, and they never let us down."

Originally hosted by John Hockenberry, *Talk of the Nation* began as a series of special call-in shows during the 1990–91 Gulf War and the Soviet coup. These shows received enough interest that they became a permanent addition to NPR on 4 November 1991. After approximately nine months with the show, Hockenberry departed to become an ABC news correspondent, and although Robert Siegel took over as interim host, it was Brooklyn-born Ray Suarez who took the torch from Hockenberry. Suarez remained the host for the next seven years and received two awards for journalistic excellence while watching the program's audience double. Specifically, during Suarez's tenure the program won the prestigious Alfred I. duPont-Columbia Silver Baton Award in 1994–95 for "The

Changing of the Guard: The Republican Revolution" and for NPR's coverage of the first 100 days of the 104th Congress; the program also won the 1993–94 duPont-Columbia Silver Baton Award for part of NPR's coverage of the South African elections. Suarez has been described by the Copley News Service as a highly intelligent navigator with "the compelling magic of intimacy spun from intense conversations while maintaining his professional distance." Suarez won honors from the *Los Angeles Times,* which listed him as one of "100 People to watch 1996."

By 2000 Suarez followed the path of the host he replaced when he too left *Talk of the Nation* for *The Newshour with Jim Lehrer.* At the end of February 2000, Emmy-award winning writer and reporter Juan Williams replaced Suarez. Williams, who has described radio as "the temperature of our times," and executive producer Greg Allen, took *Talk of the Nation* on the road as part of the continuing series "The Changing Face of America." The series, which debuted in February 2000 in Austin, Texas, travels cross-country on the final Thursday of every month and puts forth a nationally broadcast live town hall meeting. As described by the host, the goal is rather straightforward: "to paint a picture of America at the turn of the century." For example, this format took *Talk of the Nation* to Indiana to look at small-town life and to Los Angeles to look at spirituality. The attitude of the show and its host was described quite eloquently by Williams:

> Talk radio has been boiled down to the point where people tune in to have their prejudices confirmed. . . . The host drives the show by being intentionally provocative and taking very strong views intended to polarize the audience. You tune in to try to see what [outrageous thing] this guy is going to say today, or you tune in to hear him say what you believe but have never articulated. . . . I want people to tune in to *Talk of the Nation* because they want to know what's going on and want to have a full understanding of the arguments that people are making around the country and the different perspectives. . . . Hearing other people talk about the way they see American life today—that's energizing to me. And if I work as the host of the show, it will be energizing to the audience as well (*Minneapolis Star Tribune,* 1 March 2000).

Talk of the Nation has been described by one newspaper as "talk radio at its zenith, a rare quasar of civility and intelligence in the usually rabid world of boom box shriek and shout" (South Florida *Sun-Sentinel*), and the San Diego *Union-Tribune* stated, "In the argumentative archipelago of talk radio, *Talk of the Nation* is an island of civility"; however, the executive producer Allen may have said it best: "The main thing is we want a good talk show. . . . That's what we're there for."

JASON T. SIEGEL

See also National Public Radio; Talk Radio

Executive Producer

Leith Bishop

Hosts

John Hockenberry, November 1991–July 1992
Robert Siegel (interim host)
Ray Suarez, April 1993–February 2000
Juan Williams, February 2000–August 2001
Neal Conan, September 2001–present

Further Reading

Bauder, D., "Entertainment, Television, and Culture," *Copley News Service* (14 June 1999)
Becker, Dave, "Not Your Usual Call-In Show," *Wisconsin State Journal* (28 February 1997)
Craggs, Tommy, "Watch Out . . . NPR's 'Talk' Is Coming to Town," *Kansas City Star* (28 September 2000)
Feran, Tom, "Celevland Is Talk of the Nation As Public Radio Show Stops Here," Cleveland *Plain Dealer* (27 October 2000)
Holston, Noel, "Juan on Juan: New 'Talk of the Nation' Host Unloads," Minneapolis *Star Tribune* (1 March 2000)
Holston, Noel, "Mall of America Is Talk of the Nation," Minneapolis *Star Tribune* (17 May 1995)
Laurence, Robert P., "Suarez Is a Different Voice for Talk Radio," *San Diego Union-Tribune* (5 September 1994)
Walter, Tom, "Let's Talk about Race, Says NPR Host Williams, at Rhodes Tonight," *Commercial Appeal* (18 April 2000)

Talk Radio

Talk radio (sometimes referred to as All Talk) is a general term covering many closely related types of radio programs that do not focus on music or narrative comedy or drama. Most common are interview and call-in programs. Nearly all rely on a host, often a highly opinionated one. Some of the earliest talk programs in the 1920s and 1930s got station licensees in trouble for being one-sided. By the 1990s such polemical radio was common and widely popular. Talk programs today range from serious and balanced discussion of public affairs to scandal-studded rhetoric offering far more heat than light.

By the early 21st century, talk radio had become one of the most popular formats. More than 700 of the 12,000-plus radio stations in the United States identified their program format as talk. An additional 525 were in the hybrid category news/talk. Talk Radio tends to be found mostly on commercial AM stations, but it is starting to expand to FM outlets as well. The talk format is generally a mix of interviews and call-in programs frequently hosted by one or more personalities. Content ranges from information to politics to "shock" radio characterized by sexual innuendo and the flouting of social conventions. Many talk show hosts are as famous for the audiences they offend as for those they entertain and attract. News-talk became the generic radio industry term for all stations that carry both news and talk programming. The phrase *talk radio* includes the news-talk format as well as formats that rely on all-conversation programming.

Origins

Legendary Boston talk host Jerry Williams claimed he invented talk radio in 1950 when he worked at an obscure station in Camden, New Jersey, hosting a show called *What's on Your Mind?*

Actually, the form was developed—not as a radio format, but as individual programs—two decades before Williams' arrival. His may have been the first radio show to take calls, but he was hardly the first talker. As early as the 1930s, columnist Walter Winchell and Father Charles Coughlin were sowing the seeds of what would become talk radio. During the following decade, Arthur Godfrey mixed interviews and chat with his musical guests on *Arthur Godfrey Time,* and Don McNeill introduced regular talk segments on his *Breakfast Club* broadcasts from Chicago in the early 1950s.

Talk shows (including Williams' show in Camden) became familiar to listeners in the 1950s as music stations devoted time to discussion of local issues. Most were interview programs, although some stations devoted hours during evenings or overnight to call-in shows. Technology did not allow the protection of a delay system to delete objectionable phone calls. Many station operators avoided airing calls live, and often the host would repeat or paraphrase what a caller said. The motivation behind most early talk programming was to satisfy public affairs requirements as specified in licenses granted by the Federal Communications Commission (FCC).

KLIQ in Portland, Oregon, debuted a format described as All Talk in 1959. *Broadcasting* magazine headlined the story with the words "Talk, Talk, Talk" and called it "Top 30" because the station used 30 different announcers. The new format was a mix of news and talk. Twenty-five minutes of every hour were devoted to a "topic of the day" that listeners called in to discuss. Under special agreement, reporter-announcers read items from 15 national magazines from two to ten days prior to their publication. Taped interviews by British Broadcasting Corporation correspondents in England and Europe were sent daily by plane and aired in three-minute segments.

The term *talk station* came into being when KABC in Los Angeles discarded its music format in 1960 and filled its 24-hour day with talk shows. The station was originally promoted as "The Conversation Station." A four-hour news and conversation program was instituted from 5:00 A.M. until 9:00 A.M. using the title *News-Talk.* Not long afterward, KABC's sister station, KGO in San Francisco, adopted the phrase *News-Talk* as a positioning slogan because it carried news blocks in morning and afternoon drive-time periods with talk shows in between.

Listeners searching for something other than music could tune in Jean Shepherd and other hosts on WOR in the New York area. *Sponsor* magazine trumpeted the advertising and audience successes of WOR's talk radio formula early in 1964. With ratings among the highest in the New York metropolitan area, WOR was grossing $7 million annually. *Sponsor* cited WOR's success as reasons for shifts to talk by WNA Boston, KABC Los Angeles, KMOX St. Louis, and WNBC New York to WOR's success.

The success of early talk formats led to increased industry interest. In May 1965 the National Association of Broadcasters (NAB) featured talk radio at a Chicago programming clinic for program and management executives. The 125 slots quickly filled, and 75 broadcasters who wished to attend were turned away.

Pioneering Talk Hosts

One of the first stars of the new talk radio medium was Joe Pyne, who first appeared on another Los Angeles station that abandoned its music format in the early 1960s—KLAC. Pyne's reputation was built on his style of verbal bombast against almost everyone, guest and caller alike, leading pundits to call his brand of talk "insult radio" because his callers risked verbal skewering ("Look lady, every time you open your mouth to speak, nothing but garbage falls out!"). Pyne's listeners were generally delighted. Pyne had no philosophical leaning except to the contrary. He established the habit of hanging up the phone on callers he disagreed with, a habit later adopted by talk hosts seeking to create the reputation of firebrand.

Pyne achieved some national notoriety, primarily through a short-lived television show. His radio work was not syndicated because technology at that time was too expensive to link stations not already affiliated with one of the three major networks. During a stint at KABC, Pyne took controversy to new heights as a rabid and vocal hater of President John Kennedy. When the president was assassinated in 1963, Pyne was pulled off the air. His exile created opportunity for another KABC host, Bob Grant, who was assigned Pyne's slot during the hiatus.

In 1970 Grant moved to New York and created a name for himself by creating his own brand of controversy. Grant said what he felt, and his comments were often explicitly racist. He held forth for many years at WABC, New York, generally defended by station management as "misunderstood." When black activists picketed WABC after race-baiting by Grant, a WABC spokesman said, "He is extremely angry with rioters and criminals, period. If critics want to say that means blacks, that's their problem."

When the Walt Disney Company acquired WABC with the rest of the ABC Network properties, the tone of management changed. Grant's remarks after the death of Commerce Secretary Ron Brown in a plane crash proved too much for Disney, who yielded to public outcry and pulled Grant off the air. His absence was short-lived, since New York's WOR hastily made a place for Grant on their schedule and on their WOR Network.

Jerry Williams moved to Boston in 1957 where he sought out topics controversial enough to create talk about him and his show. He was one of very few media people to interview Malcolm X, and Williams took pride in announcing that the Boston TV stations and newspapers did not offer coverage to the black leader. For a brief period in the mid 1960s, Williams moved his show to Chicago and WBBM, where he would pit blacks against whites on the air to, in his words, "start a dialog." The result was often more of a fight. He returned to Boston in 1972, where he became a talk radio fixture until his death in 2003.

New York's Barry Gray was the longest running talk host on radio. He began as a celebrity interviewer in the mid-1940s and was still going strong when he died in 1996. For most of his career Gray was at WMCA in the overnight slot.

Evolution of the Host

Early talk radio was compared to broadcast journalism, but with the added pressures of live radio. As talk radio programmer Bruce Marr wrote:

> Management has to acknowledge that the members of the on-the-air team have their own biases and leave them free to express their stands on the issues being discussed. . . . This philosophy has grown slowly from the understanding that it is fruitless to ask on-air personalities to be unbiased. They are often investigators, sometimes advocates, and biases are doubtless part of their stock-in-trade.

There is structure in the talk show, but there is also enough fluidity for a host to change subjects and reflect the mood and interest of the audience. When an issue or news event is significant enough to capture the attention and interest of the listeners, the talk station changes subjects to respond. Thus a school shooting, a political scandal, or other "hot button" topics could prompt a talk station to abandon a pre-set list of guests or topics and take free-wheeling conversation about the topic at hand.

The most memorable talk hosts have been those whose political leanings are hardly the stuff of balanced journalism. By and large, talk radio is politically conservative. Those attempts at presenting liberal or left-leaning programming have been able to attract only a fraction of the audience that more conservative hosts can attract. The most notable experiment with a liberal bias on a talk station was at KFSO in San Francisco. The programming was dropped after a few months in favor of a lineup that included conservatives Pat Buchanan and Michael Reagan.

Former Texas Agriculture Commissioner Jim Hightower, a talk host syndicated at one time by ABC Radio, explains talk radio's conservatism this way:

> What happened is the progressive side forgot radio. My generation looked to television and mass demonstrations and other ways of communicating, whereas the conservatives—Ronald Reagan, Paul Harvey—hung in there and continued to build an audience. Now it's just follow the leader. People look across the street and say, "If that sucker is doing well with a conservative, that's what I need, too."

The rise of talk radio has been accompanied by an equally dramatic rise in harsh rhetoric under the guise of political opinion. The combination is often a breeding ground for what Peter Laufer called "hate, scapegoating and stereotyping" in his book, *Inside Talk Radio*. "The talk show demagogues are adept at manipulating anger and turning righteous resentment into fearful hatred of the oppressed," Laufer writes.

In *Hot Air*, Howard Kurtz says that the way to get attention in talk radio is "to shout, to polarize, to ridicule, to condemn, to corral the most outrageous or vilified guests." Kurtz points out that when White House chief of staff Leon Panetta wanted to attack House Speaker Newt Gingrich, he accused Gingrich of acting like "an out-of-control radio talk show host."

Scrappy and Intimate

In an era of blow-dried TV anchors, homogenized sitcoms and cookie-cutter Hollywood sequels, talk radio stands out as unpredictable. It is less politically correct than mass television. "It's more scrappy," said John Mainelli, the programmer who built WABC into a 1990s talk powerhouse. "It's also more intimate because it's radio," he said. That intimacy allowed Boston talk host David Brudnoy the permission to let his audience know that he was gay and that he had AIDS. There was a tremendous outpouring of affection from long time WBZ listeners who heard Brudnoy's late night program in 38 states. A number of listeners told Brudnoy that his confession had moved them to tell their own families that they, too, had the AIDS virus. As Brudnoy explained it, "Talk Radio is the last neighborhood in town. People know their talk hosts better than they know the person who lives next to them."

Technology and New Networks

In its early days, talk radio was expensive to produce. In addition to a skilled host, the station had to pay a producer, programmer, and engineer. A researcher was also a necessity. Much of talk radio consisted of interviews, easy enough to do over telephone lines, but long-distance charges could be substantial. To make talk radio a success, the stations needed the vast audience provided by large, urban areas. In the 1960s and 1970s it was nearly impossible to make talk a successful format in anything but a large market. It took technological innovation to make talk radio financially feasible.

Chief among the technical factors underlying talk radio as a mass format was the growing commercial success and expansion of FM radio in the 1970s. By 1977, a generation had grown up with their radios tuned as much to FM as to AM. Broadcasters who owned AM stations developed a strategy of counter-programming to retain audience. Some shifted their programming from music to the spoken word, as KABC and KGO had done a decade and a half earlier. The counter to music programming on FM gave rise to information programming.

Satellite technology gave syndicators the type of access to local stations that only networks had enjoyed previously. Initially, syndicated programs were sent via telephone lines at great expense. By the late 1970s geosynchronous satellites made it possible to distribute radio shows nationwide at a relatively reasonable cost, giving the radio program a national audience of simultaneous listeners. That gave rise to ABC Talkradio, featuring hosts from various ABC-owned talk stations. Host Michael Jackson's mellow Australian accent was networked from ABC's KABC in Los Angeles, as was Ira Fistell's daily program. From San Francisco's KGO came Dr. Dean Edell's medical advice and a general interest talk show with Ronn Owens.

The importance of the telephone caller led to improvements in on-air telephone systems. Nationally syndicated talk shows today use phone systems that handle more than 30 incoming lines. Special automatic gain-control devices compensate for the different levels and sound qualities of the callers' telephones. Specially designed computer software allows the producer and the show host to communicate with one another. Sitting in different rooms, the producer and host can both see on screen the caller's name and location and the topic of interest he or she called to discuss. More advanced versions can create a caller database that can include telephone number, address, regular topic of interest, occupation, birthday, and zip code. The information is available for the host the next time the listener calls. It can also be used for the radio station's marketing purposes.

Technology also fueled NBC's Talknet, where Bruce Williams answered financial and legal questions in an understanding, fatherly manner. If Williams was father, Sally Jesse Raphael was mother, offering personal advice and relationship counseling. New York personality Bernard Meltzer sounded like a loving grandfather presenting general interest topics. The family doctor was Harvey Rubin, who conducted a daily health program.

New York's WOR Radio, long a leader in talk programming in its own city, used satellite technology to deliver its talk shows to a national audience. Former libertarian candidate Gene Burns, money advisers Ken and Daria Dolan, psychologist Dr. Joy Brown, and others talked about issues that transcended New York. When WOR moved conservative host Bob Grant from rival WABC, he was added to the daily syndication schedule. Comedienne Joan Rivers joined the WOR Network in 1997.

Westwood One Radio Networks inherited the original talk programming of Mutual Radio in a series of mergers that ultimately put Westwood in the CBS family. Once the home of Larry King's overnight talk show, Westwood moved King to afternoons in order to increase his audience. Replacing King overnight was Jim Bohannon, who was the usual replacement when King was on vacation. King's move to afternoons proved unsuccessful, but Bohannon's installation in the overnight chair was a long-term proposition.

Westwood also carried the *Tom Leykis Show* from Los Angeles, the *G. Gordon Liddy Show* from Washington, and *Imus in the Morning* with long-time New York funnyman and curmudgeon Don Imus. Westwood's parent company, Infinity Broadcasting, a division of Viacom after its merger with CBS, also syndicated "shock jock" Howard Stern.

Stern is not a talk host per se, but his program abandoned music in favor of free-wheeling conversation often centered on

Stern and his private parts. Stern's program was seldom heard on talk stations; his affiliates were primarily rock music outlets. As one station manager described it, "Stern says what The Who used to sing." That statement reinforces the rock context of Stern's unpredictable broadcasts. In Los Angeles, Stern's program appeared on KSLX, one of a few talk stations on the FM band. In presenting Stern's and other hosts' shows to advertisers, however, KSLX account executives compared them to personalities on rock stations, not to traditional talk personalities such as Rush Limbaugh.

Other syndicators found success as consolidated radio companies attempted to cut expenses. A fledgling network called American View presented host Ken Hamblin. Known as "the Black Avenger," Hamblin was one of only two nationally known black conservative talk hosts in the late 1990s (the other was 2000 presidential candidate Alan Keyes). Cox Radio, owner of several news and talk stations, syndicated hosts Neal Boortz and Clark Howard from WSB in Atlanta for their own stations and others outside the Cox fold. Former WABC host Mike Gallagher created his own network and fed *The Mike Gallagher Show* to more than 200 stations. Jones Radio Networks, best known for long-form music programming, entered the talk field with several programs.

Exponential Growth

Talk radio's growth through the 1990s was fueled by Rush Limbaugh and the network built around the host by entrepreneur Edward F. McLaughlin. A former ABC Radio executive who had been Vice President and General Manager of KGO, McLaughlin rose to the presidency of the ABC Radio Networks in 1972, a position he held through 1986. In 1987, McLaughlin retired from the network and packaged Limbaugh, then a local Sacramento personality, and KGO's Edell into a fledgling syndication company. Officially called EFM Media (for McLaughlin's initials) the network was known publicly as the "EIB Network" because of Limbaugh's boastful catch phrase, "Excellence in Broadcasting." In 1997, EFM's assets, including *The Rush Limbaugh Show, The Dr. Dean Edell Show,* and the monthly publication *The Limbaugh Letter,* were sold to Jacor Communications and its Premiere Radio Networks. Both Jacor and Premiere were ultimately acquired by Clear Channel Communications.

Premiere began as a syndicator of short-form programming for music stations and developed through mergers and acquisitions into the leading provider of talk radio programming based on sheer numbers of shows. Premiere would ultimately be home to Limbaugh, Dr. Laura Schlessinger, UFO-chaser and expert in the paranormal Art Bell, sports personality Jim Rome, Michael Reagan, Edell, internet publisher Matt Drudge, and Los Angeles satirist Phil Hendrie. Because Premiere was acquired in 1999 by Clear Channel, that company was able to leverage key Premiere talents onto Clear Channel-owned news and talk stations.

Deregulation

If technology was one parent of talk radio, deregulation was the other. For 42 years, radio and television were ruled by the Fairness Doctrine, which required stations to broadcast opposing views on public issues. The Fairness Doctrine was born in 1949 in response to a court case involving the owner of powerful radio stations in Los Angeles, Detroit, and Cleveland, and an early organization of professional newspeople. The newspeople charged the licensee with slanting news on his radio stations.

The doctrine ordered stations to work in the public interest and guarantee equal time for disparate viewpoints. The FCC decided that, contrary to its stated purpose, the Fairness Doctrine failed to encourage the discussion of more controversial issues. There were also concerns that it was in violation of the free speech principles of the First Amendment. The FCC abolished the rule in 1987, leaving talk hosts unrestrained.

Congress tried to reinstate the Fairness Doctrine as law that same year, but President Ronald Reagan vetoed it. Later attempts failed even to pass Congress. Those failures have led many to believe that it will take an act of Congress to bring back the doctrine. In fact, however, the FCC has the power as an independent regulatory agency to re-impose the Fairness Doctrine without either congressional or executive action.

Talk radio as it developed through the 1990s could not exist in the Fairness Doctrine era. In 1987, the year the doctrine was discarded, there were 125 news-talk stations nationwide. By early 2003, there were a total of 1,785 news, talk, and information stations. In 1986, news, talk, and information stations captured 8.7 percent of the national listening audience, good enough for fifth place overall among all radio formats. The Fall 2002 Arbitron Format Trend Report showed news, talk, and information radio as the leading radio format, capturing 16.5 percent of the national radio audience.

Power Shift

The power of talk radio had been felt on a local level since the format's inception. Local stations traditionally staged debates and allowed unprecedented access to politicians. Deregulation merged journalism and populism. The public asked questions that had previously been the domain of reporters and gossip columnists.

Boston's Jerry Williams and listeners who heard his show on WRKO were credited with overturning Massachusetts' seat belt law in 1988. Williams and other show hosts claimed credit for public opposition to a congressional pay raise and for a boycott against Exxon in the wake of the Alaska oil spill.

At an organizational meeting of the National Association of Radio Talk Show Hosts (NARTSH) in 1989, Williams called talk radio "the greatest forum in history . . . the last bastion of freedom of speech for plain ordinary folks." The organization attempted to set a political agenda: "It's our government and we're going to take it back from those aristocrats," said Mike Siegel of Seattle's KING Radio on a broadcast from the NARTSH meeting. Siegel later had a short run in syndication on the Premiere Radio Network.

Talk radio created a new dynamic during the 1992 Presidential campaign. Most candidates, prompted by exposure given to independent Ross Perot, appeared on both radio and television talk shows to disseminate their views and corral support. The notable exception was incumbent President George H. Bush, who lost to a regular talk show guest, Bill Clinton. The proliferation of appearances by candidates on both television and radio talk shows prompted *Washington Post* television critic Tom Shales to label the 1992 election "the talk show campaign."

Aides in the Clinton campaign believed their candidate reversed his fortunes in New York with an early morning appearance with Don Imus, morning man on Sports station WFAN heard on a national network. As president, Clinton hosted talk personalities at the White House. California's Jerry Brown surprised political pundits by winning Connecticut's Democratic primary in 1992. Some observers felt his key move was an appearance on Michael Harrison's program then heard on WTIC in Hartford. Brown later hosted his own talk show.

Brown was not the only personality to join the talk host ranks from outside radio. G. Gordon Liddy, of Watergate infamy, had the highest profile because of his national network program. Conservative presidential candidate Pat Buchanan tried his hand for a short while at talk radio. On the local level, former Los Angeles police chief Daryl Gates, former New York Mayor Ed Koch, and former San Diego Mayor Roger Hedgecock gained new careers in their hometowns. Hedgecock became Rush Limbaugh's primary substitute host in 2002.

Audiences and Impact

For a while in the 1990s, talk radio and its effect on politics dominated media discussion. An entire issue of *CQ Researcher* in 1994 was focused on the question "Are call-in programs good for the political system?" A *Newsweek* cover story was headlined "The Power of Talk Radio." Other publications also expressed concern in articles entitled "Tower of Babble," "How to Keep Talk Radio from Deepening America's Divide" and "Talk Radio Lacking Real Dialogue." FCC Chairman Reed Hundt, speaking to members of the NAB, asked if talk radio created "such skepticism and disbelief that as a country we just can't get anything done?" The *New York Times* suggested that "modern politicians have become slaves to public opinion" in the "electronic din" of talk radio. Following the Oklahoma City bombing, there were suggestions that the rhetoric of talk radio might have fueled the bombers' discontent. *World Press Review,* in an article titled "A Bitter, Self-Doubting Nation," suggested that talk radio hosts who were cynical about the federal government and public officials more accurately represented the disillusionment of the U.S. voter than the mainstream press. Editorials and commentators pointed to talk radio's influence on the increasing incivility in U.S. society. Howard Stern continued to draw attention for his sexist and racist talk. Over the airwaves he regularly made fun of women, minorities, and disabled people. Toilet jokes and sex jokes continued to be regular fare on the Stern show in the 21st century.

Scholars began to study talk radio and its effects. In 1995 the Times Mirror Center for People and the Press challenged the view that talk radio listeners were ignorant and ill-informed. The research concluded that talk radio listeners paid close attention to the news and knew what was going on in the world. They were more likely to vote than the average American and were better educated, made more money, and were more focused on the issues than those who did not listen to talk radio. As an indication of its power, some supporters pointed to a three-way political race in Minnesota, where former wrestler turned radio talk host Jesse "The Body" Ventura was elected governor.

Randall Bloomquist, former news/talk editor of *Radio and Records* and a regular observer of the talk radio phenomenon, argues that its ability to influence political developments is heavily dependent on support from mainstream media. Talk show fans do not tune in to get their opinions changed. They listen to the hosts who affirm their personal and political beliefs. Bloomquist suggests that talk radio's power is not in its influence on its own listeners, but in its ability to trigger national media coverage of an issue. That national coverage is what truly affects issues.

Radio market research indicates that the average listener tunes in to talk radio for an hour or less at a time. In 1998, according to the Radio Advertising Bureau, 42 percent of listeners were doing so in their automobiles, 37 percent listened at home, and 21 percent tuned in at work. Most listened to radios while doing something else. In large metropolitan areas, more and more people are commuting greater and greater distances. Those commuters are the perfect captive audience for talk radio. In 1999 national research conducted for *Talkers* magazine, an industry trade journal, indicated that 52 percent of the audience was male, 72 percent had some college education, and nearly 60 percent had an annual household income higher than $50,000.

By the early 2000s, politics and "heavy" national issues were no longer the main focus of talk radio. Instead of the traditional older listener, radio stations were hustling to get younger (18 to 44) listeners. Sports, sex, and lifestyle were the "in" topics, and the goal was entertainment. Howard Stern imitators were as evident as Limbaugh clones. Hot talk ("rock and roll without the music," as one radio insider put it) was shouldering political talk out of the way. The new version of talk, "sports talk/guy talk," targets younger listeners with cruder, ruder talk.

Listeners were also discovering the internet, and their computers allowed access to stations too far away to tune in on radio. A number of websites provide listings of radio stations by format. Although some are web versions of broadcast stations, others are internet-only stations. Choices range from traditional talk show hosts to a variety of information and entertainment targeted at specific audiences such as computer technology specialists, lawyers, and entrepreneurs.

SANDRA L. ELLIS AND ED SHANE

See also All News Radio; All Night Radio; Canadian Talk Radio; Coughlin, Father Charles; Fairness Doctrine; Fresh Air; Godfrey, Arthur; Internet Radio; Jepko, Herb; King, Larry; Limbaugh, Rush; McNeill, Don; Pyne, Joe; Shepherd, Jean; Shock Jocks; Stern, Howard; Topless Radio; Williams, Jerry; Winchell, Walter; WOR

Further Reading

Bolce, Louis, Gerald DeMaio, and Douglas Muzzio, "Dial-in Democracy: Talk Radio and the 1994 Election," *Political Science Quarterly* (Fall 1996)

Bulkeley, William M., "Talkshow Hosts Agree on One Point: They're the Tops," *Wall Street Journal* (15 June 1989)

Heath, Rebecca Piirto, "Tuning in to Talk," *American Demographics* 20, no. 2 (February 1998)

Henabery, Bob, "Talk Radio: From Caterpillar to Butterfly," *Radio Business Report* (January 1994)

Hontz, Jenny, "Clinton Points Finger at Talk Radio," *Electronic Media* (1 May 1995)

Hoyt, Mike, "Talk Radio: Turning up the Volume," *Columbia Journalism Review* 31, no. 4 (November/December 1992)

Ivins, Molly, "Lyin' Bully," *Mother Jones* 20, no. 3 (May/June 1995)

James, Rollye, "The RUSH to Talk Continues," *Radio Ink* (17–30 January 1994)

Jost, Kenneth, "Talk Show Democracy," *The CQ Researcher* (19 April 1995)

Kurtz, Howard, *Hot Air: All Talk, All the Time,* New York: Times Books, 1996

Laufer, Peter, *Inside Talk Radio: America's Voice or Just Hot Air?* Secaucus, New Jersey: Carol, 1995

Levin, Murray, *Talk Radio and the American Dream,* Lexington, Massachusetts: Lexington Books, 1987

Marr, Bruce, "Talk Radio Programming," in *Broadcast Programming: Strategies for Winning Television and Radio Audiences,* by Susan Tyler Eastman, Sydney W. Head, and Lewis Klein, Belmont, California: Wadsworth, 1981

"Medium Is Message at Talk Radio Conference," *Broadcasting* (19 July 1989)

Michaels, Bob, *R&R Talk Radio Seminar 2003: Top Arbitron Performers in News/Talk,* Columbia, Maryland: The Arbitron Company, 2003

Munson, Wayne, *All Talk: The Talkshow in Media Culture,* Philadelphia, Pennsylvania: Temple University Press, 1993

"Programming for Profit with Satellite Talk," *Radio Business Report* (17 May 1993)

Rosenstiel, Thomas B., "The Talk-Show Phenomenon," *Houston Chronicle* (31 May 1992)

Seib, Philip M., *Rush Hour: Talk Radio, Politics, and the Rise of Rush Limbaugh,* Fort Worth, Texas: Summit, 1993

Shane, Ed, "The State of the Industry: Radio's Shifting Paradigm," *Journal of Radio Studies* 5, no. 2 (1998)

Shane, Ed, "Talk Radio," in *The Guide to United States Popular Culture,* edited by Ray Broadus Browne and Pat Browne, Bowling Green, Ohio: Bowling Green State University Popular Press, 2001

Shane, Ed, *Disconnected America: The Consequences of Mass Media in a Narcissistic World,* Armonk, New York: M.E. Sharpe, 2001

Taylor, Deems 1885–1966

U.S. Radio Commentator, Author, and Critic

Deems Taylor is perhaps best remembered as the narrator of Walt Disney's 1940 film *Fantasia,* the goal of which was to bring classical music to life by pairing it with animation to make it more accessible to the audience. When viewed in this light, Taylor was the perfect choice for the project. An unabashed populist, he was the composer of the first truly successful American opera; he was also a highly regarded radio commentator, author, and critic. Well known for his intermission talks during broadcasts of *The New York Philharmonic* on CBS, he was second only to Walter Damrosch as an advocate of music for all, not just an intellectual or economic elite.

Making a Name

A native of New York City, Taylor was born in 1885. Musically, he was largely self-taught, writing his first composition (a waltz) at the age of 10. He attended the Ethical Culture School of New York and went on to New York University (NYU), from which he graduated in 1906. While at NYU, Taylor honed his skills as a composer by writing Grand Opera parodies, as well as the music to four campus shows. The first of his works to garner serious attention was the prize-winning symphonic poem *The Siren Song* in 1913. *Witch Woman,* his first published work, appeared the following year.

True musical success being some years ahead, Taylor pursued an editorial career. A series of positions, including a stint with *Encyclopaedia Britannica,* led to his appointment as the Sunday Editor of *The New York Herald Tribune.* In 1916 he went overseas as a war correspondent for the paper, returning to the U.S. the following year. He later served as editor of *Musical America,* and was music critic of *The New York World, The New York American,* and *McCall's* magazine.

The war led, at least indirectly, to Taylor's first major success. Trying to transfer his wartime experiences into music, he proposed writing a series of "War Sketches." Finding the events still too fresh "to interpret into music," he turned his thoughts to a subject as far removed from the grim realities of war as possible. The result was a suite based on the works of a favorite author since childhood, Lewis Carroll. *Through the Looking Glass* is the most frequently performed of Taylor's works.

On the strength of this success, Taylor was approached by the Metropolitan Opera to compose a "successful American Opera." After convincing poet Edna St. Vincent Millay to write the libretto, Taylor set to work. The result was *The King's Henchman,* which opened at the Met in 1927 and played for the next three years—then a record for an American work.

Radio

It was the popularity of this opera that launched Taylor's radio career. *The King's Henchman* was chosen for the inaugural evening's broadcast on18 September 1927 of what was then called the Columbia Phonograph Broadcasting System (later CBS). Taylor provided the narration and commentary.

His second opera for the Met, *Peter Ibbetson,* debuted in 1931. An even larger success than his first, it brought him to the attention of the National Broadcasting Company (NBC), who engaged him to do commentary on its broadcasts of the Metropolitan Opera. His efforts were not well received, however. Still a radio neophyte, Taylor made the mistake of delivering his observations over the music. A storm of protest from opera purists led to his rapid ouster. He rebounded with his own program (briefly), and soon secured a spot on the popular *Kraft Music Hall,* then featuring Paul Whiteman and his Orchestra. Eventually a long-term assignment doing commentary for broadcasts of the New York Philharmonic on CBS brought him back to the classical arena. This time, he did not talk over the music.

In retrospect, the time was right for someone like Taylor. Both CBS and NBC's commitment to music was strong, and classical and light classical fare was an important part of the mix. The networks had their own symphony orchestras for most of the radio era, and big band, swing, sweet music, and jazz were all well represented on the dial. Taylor was an established composer with a strong journalism background, an acknowledged expert on things musical. He was articulate and possessed a dry wit. He looked the part of a musical scholar, and more importantly for radio, he sounded the part as well.

As a result, Taylor was in fairly constant demand throughout the peak years of network radio. He was a regular on many programs and a frequent guest on others. An inveterate replacement panelist on the erudite quiz show *Information, Please,* he appeared more than 40 times on the program during the 1940s. He even turned up on the comedy *Duffy's Tavern,* showing his willingness to poke fun at his "longhair" image in the process.

Taylor was the author or editor of a number of books, including three works of musical musings based largely upon his radio commentaries; *Of Men and Music, The Well Tempered Listener,* and *Music to My Ears.* Despite his accomplishments as an author and radio personality, however, he remained at heart a composer, viewing his other activities as merely a means to an end. "For many years," he said towards

Deems Taylor hosting a music program
Courtesy CBS Photo Archive

the end of his life, "I, the composer, have been supported by me, doing other things." He eventually wrote over 50 musical compositions.

A member of the Institute of Arts and Letters, Taylor was also a long-time Board member of the American Society of Composers, Authors and Publishers, serving as President from 1942–48. Each year since 1967, ASCAP has presented the "ASCAP-Deems Taylor Awards," honoring the best in print, broadcast, and now internet coverage of music. Deems Taylor died in 1966, at the age of 80.

CHUCK HOWELL AND DOUGLAS GOMERY

Joseph Deems Taylor. Born in New York City, 22 December 1885. Bachelor of Arts degree, New York University, 1906. Staff member, *Encyclopaedia Britannica,* 1907; his symphonic composition, *Siren Song,* won prize, National Federation of Music Clubs, 1913; first published work, *Witch Woman,* appeared in 1914; assistant editor, *New York Tribune* Sunday Magazine, 1916; war correspondent, *New York Tribune,* 1916–17; music critic, *New York World,* 1921–25; wrote music for Broadway, 1922–24; *Through the Looking Glass Suite* debut, 1922; commissioned by Metropolitan Opera to write a "successful American Opera," 1925; editor, *Musical America,* 1927–29; the opera, *The King's Henchman,* opened at the Metropolitan Opera, 1927; critic, *New York American,* 1931–32; made radio debut as commentator, 1931; noted for having more Metropolitan Opera performances than any other American composer; member, Board of Directors of ASCAP, 1933–66; narrator and musical advisor for the Walt Disney film *Fantasia,* 1940; president, American Society of Composers, Authors, and Publishers (ASCAP), 1942–48; honored by ASCAP with the establishment of the ASCAP-Deems Taylor Award for meritorious writing about music and musicians, 1967. Recipient: honorary doctorates, New York University, 1927; University of Rochester, 1939; Dartmouth College, 1939; Cincinnati Conservatory, 1941; Syracuse University, 1944. Died in New York City, 3 July 1966.

Radio Series

1931–32 *The Metropolitan Opera*
1933–35 *The Kraft Music Hall*
1934–35 *The Opera Guild*
1935–36 *Studio Party*
1936–43 *The New York Philharmonic Orchestra*
1937–39 *Paul Whiteman Band Remote*
1938–48 *Information Please*
1940–41 *Musical Americana*
1941–42 *America Preferred*
1941–44 *The Prudential Family Hour*
1943–46 *The Radio Hall of Fame*

Selected Operas

The King's Henchman, 1927; *Peter Ibbetson*, 1931

Selected Publications

Of Men and Music, 1937
Walt Disney's Fantasia, 1940
The Well Tempered Listener, 1940
Music to My Ears, 1949
Some Enchanted Evenings: The Story of Rodgers and Hammerstein, 1953

Further Reading

DeLong, Thomas A., *Radio Stars: An Illustrated Biographical Dictionary of 953 Performers, 1920 through 1960*, Jefferson, North Carolina: McFarland, 1996
Douglas, George H., *The Early Days of Radio Broadcasting*, Jefferson, North Carolina: McFarland, 1987
Dunning, John, *Tune in Yesterday: The Ultimate Encyclopedia of Old-Time Radio, 1925–1976*, Englewood Cliffs, New Jersey: Prentice-Hall, 1976; revised edition, as *On the Air: The Encyclopedia of Old-Time Radio*, New York: Oxford University Press, 1998

Taylor, Marlin R. 1935–

U.S. Music Programmer

Marlin Taylor once said, "My personal and professional goal was always to bring joy to others through what I could provide them in the way of musical entertainment." Taylor pioneered the "Good Music" format that was later popularized as "Beautiful Music."

At radio station WDVR in Philadelphia, Taylor built his Good Music format on his judgment of what songs and arrangements the station would play. Taylor said then, "The music selection is very much mine. No one really influences what I program." WDVR quickly began to achieve higher ratings and notice by the industry at large.

The sound was string-orchestra based and selections ranged from light classical melodies to the "Banjo Barrons Play Hawaiian Melodies." Taylor introduced heavy use of instrumental recordings of the most popular songs of the day, with artists such as Percy Faith, Mantovani, Bert Kaempfert, and Billy Vaughn. Very few vocals were incorporated into the music mix; only groups such as the Norman Luboff Choir and the Ray Conniff Singers were chosen. The concept of playing instrumental versions of popular songs was unique to the Taylor format and would be incorporated by programmers following him, including Jim Schulke. Taylor gives great credit to the use of instrumental records of popular songs for the format's ratings success, particularly among women.

The success of WDVR led Taylor to the Boston radio market, and in 1967 he was named program director of the Kaiser-owned station WJIB. At WJIB Taylor established a relationship with Bob Richer and Jim Schulke, who had formed a sales representative firm called Quality Media Inc. (QMI). The singular purpose of the company was to market Good Music stations to national advertisers. Schulke requested Taylor's help in setting the parameters for a new syndicated music service that QMI was establishing. Phil Stout, who later became the programmer for Schulke Radio Productions Beautiful Music format, was first hired by Taylor at WDVR and was recommended to Schulke and Richer as their programmer.

In 1969 Bonneville Broadcasting hired Taylor as the general manager of WRFM in New York City. In the summer of 1969 Taylor began programming music "bonus hours," which later became "total music hours." No public reference was ever made that the hours were commercial-free for fear of alienating advertisers. This concept was expanded to the Christmas season with the Christmas Festival of Music consisting of 36 hours of continuous holiday music on Christmas Eve and Christmas day.

Both in terms of ratings and profits, WRFM became a huge success. Because it was located in the nation's advertising capital, the station had an enormous pioneering impact on advertiser awareness of the success of the format. WRFM played a major role in overcoming a perception that Beautiful Music stations were playing "elevator" music that was not an attractive advertiser investment.

By the end of 1970, Bonneville recognized that Taylor could develop a company that could syndicate the WRFM format nationally. Bonneville Programming Services was formed, with its first client Bonneville-owned WCLR in Skokie (outside of Chicago), Illinois. The initial success of Beautiful Music led the company into several other formats, including Adult Contemporary and Bonneville Rock.

Bonneville Programming Services offered a unique twist to its Beautiful Music format, which was another Marlin Taylor innovation. The format was offered to broadcasters in two forms. The first was the traditional matched flow format, in which each quarter hour segment was packaged. A second option offered stations was a "random select" format based on single cuts that could be customized to individual markets and times of day. With the development of more sophisticated station automated systems, the random select service became much sought after, particularly in the most competitive markets. This concept was utilized by nearly all syndicated contemporary formats.

Bonneville Programming Services grew to serve stations in over 150 markets, and in the 1970s and 1980s was the chief head-to-head competitor with Schulke Radio Productions. Bonneville later bought Schulke Radio Productions from Cox Broadcasting. Taylor expanded Bonneville Programming Services to include marketing, research, and management consulting to its client stations.

After 18 years, Taylor left Bonneville at the end of 1987. He subsequently worked on a number of programs, wrote articles for industry publications, and packaged and distributed recordings of Beautiful Music under the Surry House Music label. In 2000, Taylor returned to the industry full time as a Program Director for XM Satellite Radio.

GORDON H. HASTINGS

See also Easy Listening/Beautiful Music Format

Marlin R. Taylor. Born in Abington, Pennsylvania, 26 August 1935. Worked at Armed Forces Radio in Thule, Greenland, June 1959–April 1960. Attended Army Information School, Ft. Slocum, New York, 1960. Developed Good Music (Beautiful Music) format at WDVR, Philadelphia; moved to WJIB, Boston, 1967; moved to WRFM, New York, 1969; began association with Bonneville Programming Services, 1970–87; Program Director for XM Satellite Radio, beginning in 2000.

Technical Organizations

Organizing Radio's Engineers

Four types of technical organizations have emerged during radio's development in the U.S. The first were professional organizations for engineers. The second helped to assemble amateur radio operators and were organized in the early years of radio. The third were company-specific employee organizations begun to counter independent unionization by their radio employees. And the fourth were independent labor unions for radio's engineers and related technicians.

Professional Organizations

Perhaps the most important organizations were those started by leading scientists, electrical engineers, and business leaders to advance the scientific study of radio. These have gone through several iterations but have consistently included the key figures and provided the most important publications.

American Institute of Electrical Engineers

On 13 May 1884, a group of scientists, inventors, and electrical engineers, which included Thomas Alva Edison, Alexander Graham Bell, and Norvin Green, president of the Western Union Telegraph Company, organized the American Institute of Electrical Engineers (AIEE). Spurred in part by an International Electrical Exhibition in 1884, the new AIEE included inventors, entrepreneurs, telegraph operators, and company managers. Technical publications (eventually *Electrical Engineering*), an annual meeting, and even a museum were contemplated. The AIEE changed as the industry became dominated by large manufacturing and service companies that employed "electricians." Creation of research laboratories added more members. AIEE also became centrally involved in the development of technical standards, allowing one company's device to work with another's. And it helped to promote professional standards. But the AIEE became increasingly focused on power engineering, leaving an opening for a new group focused on wireless and radio.

Society of Wireless Telegraph Engineers

The Society of Wireless Telegraph Engineers (SWTE) was started in 1907 by the head of the Stone Wireless Telegraph Company, John Stone Stone [sic]. In the first years, Stone limited membership in the SWTE to only the employees in his small Boston-based company, although he later permitted the rolls to be opened to the employees of Reginald Fessenden's National Electrical Signaling Company (NESCO) and other companies.

The Wireless Institute

The Wireless Institute (WI), a rival organization to the SWTE, began in 1909 when Robert Marriott organized about 100 wireless U.S. devotees in the United States. Three years later, however, WI membership dropped to 27 members and was teetering on collapse.

Institute of Radio Engineers

The SWTE and Marriott's WI merged on 5 April 1912 and agreed to name the new organization the Institute of Radio Engineers (IRE). The new organization became more international in scope, thus excluding the word "American" in its name. The IRE focused on extending the use of radio to protect and save lives at sea. Because of its ubiquitous nature, radio soon became a tool of war and many IRE members joined the effort as military needs meshed with rapid technical development. By 1916 there were 83 members in the IRE from 11 other countries besides the United States.

In addition to its prestigious monthly, *Proceedings of the IRE*, the IRE focused on improving technical standards as the international radio industry grew into a new economic sector. During the 1920s, the IRE synchronized its scientific and electronic research work with the National Electrical Manufacturers Association and the Radio Manufacturers Association. The IRE also assumed a leadership role as a trade representative for its members of U.S. radio manufacturers and broadcasters such as the Radio Corporation of America (RCA), American Telephone & Telegraph (ATT), and General Electric (GE).

Working for scientific progress and technological insight, the IRE lobbied for a major role in deciding how U.S. radio broadcasting would be regulated. The IRE was invited to participate in a series of National Radio Conferences between 1922 and 1925 that were organized by Secretary of Commerce Herbert Hoover to develop policies concerning station licensing and technology. These four meetings eventually led to Congress passing the Radio Act of 1927.

After World War II, the IRE focused on the developing technologies of television, FM radio, audio recording, and developing standards of excellence in engineering practice. The society delineated its membership along two specialties of interest: broadcast engineers and audio. With the rise of electronic technologies being used by the U.S. military during World War II, the IRE broadened its interest groups with radar, computers, television, solid-state electronics, and space exploration. By 1947, there were nearly 18,000 IRE members.

Institute of Electrical and Electronic Engineers

In January 1963, the AIEE and the IRE merged and formed the Institute of Electrical and Electronics Engineers (IEEE). As of the early 2000s, the IEEE remained the largest professional society in the world with more than a quarter of a million members.

The IEEE quickly expanded to include virtually all arms of electronics as shown by its steadily growing number of specialized technical journals, plus the general appeal *IEEE Spectrum*. Building on the IRE model, the IEEE developed dozens of professional divisions—called societies—each of which held their own meetings and issued their own publications. Among those of particular interest are the societies concerning broadcast technology, consumer electronics, and the society on social implications of technology. A history center was created in 1980.

Society of Broadcast Engineers

Not everyone was pleased with the AIEE-IRE merger, and the Society of Broadcast Engineers (SBE) was one dissident spin-off created at the same time to allow a focus on broadcasting. The SBE's international membership soon covered a broad scope of industry employees: studio and transmitter operators and technicians, supervisors, announcer-technicians, chief engineers of commercial and educational stations, engineering vice presidents, consultants, field and sales engineers, broadcast engineers from recording studios, schools, closed-circuit systems, cable TV, production houses, corporate audio-visual departments, and other facilities. The SBE began in 1977 to certify broadcast engineers in several categories, including the

Certified Broadcast Radio Engineer (CBRE) and Certified Senior Radio Engineer (CSRE). These certifications are renewed every five years. More than 100 local chapters meet regularly. As with other technical bodies, the SBE holds an annual meeting, issues publications, and holds training courses and workshops.

Amateur Radio Groups

To further the developing "ham" hobby, amateur radio clubs began to proliferate early in the 20th century.

The Wireless Club of America

The Wireless Club of America was begun in 1910 by Hugo Gernsback, a Luxembourg immigrant to the U.S. Sometimes referred to as the father of science fiction, Gernsback was an early wireless enthusiast who saw the potential in marketing wireless units to the general public while creating a market for the new technology of amateur radio. By 1912, the *New York Times* reported that there were some 12,000 amateur radio operators in the U.S. and some 122 radio clubs. Most of the meetings for these clubs took place over the air. Message handling, where one operator or groups of operators would relay messages and information to others, became a central feature of amateur wireless clubs. While these exchanges were most often for fun and fraternal reasons, they were sometimes needed to protect or inform others in public emergencies.

American Radio Relay League

In 1914, Hiram Percy Maxim, a Boston radio enthusiast, contacted others through his Hartford Radio Club and offered to set up a network of relay stations comprised of amateur radio operators. Calling the new network the American Radio Relay League (ARRL), Maxim tapped into the unrealized dreams of many in the field at the time. Within some four months, the League boasted 200 relay stations in the U.S. alone. The ARRL asked the Commerce Department to establish a special license for stations in order to make up a national relay network of stations. Maxim, the ARRL's visionary leader and an MIT graduate, set up a sophisticated system that would serve the nation in the event of war. When the U.S. entered World War I in 1917, amateur radio stations were ordered to close down. At the same time, the military began a full-blown recruitment effort to attract some 6,700 radio operators (many of whom began as amateurs) to the U.S. navy. Instead of being outside of the mainstream, amateur wireless operators became part of the system. After the war, the amateurs were intent upon getting back to where they left off when the war started. By 1921, the Commerce Department listed some 11,000 amateur radio operators in the U.S.

Amateurs were forced off the air again during World War II (1941–45), making ARRL an important means of keeping the hobby alive for what became strong postwar growth. Today ARRL thrives with meetings, the monthly *QST*, the annual amateur operators' handbook, and a host of other publications. In addition to the ARRL, more regional groups developed. Among them was the Radio Club of America (RCA—*not* to be confused with the former manufacturing company), a New York area group that enjoyed the participation of such key radio figures as Edwin Howard Armstrong, among others.

Technical Unions

The first attempts to organize radio technicians and engineers (often called the "below-the-line" employees to distinguish them from the "above-the-line" creative personnel) occurred in the 1920s. By the 1940s two national labor unions, one devoted to broadcasting, had between them organized most of radio's technical workers. They sought to overcome the relatively low pay and sometimes poor working conditions of workers. Broadcast employees often had to work split shifts, with no pay or time off for holidays, and no overtime pay. The typical wage was about $20.00 per week—$40.00 was big money in the 1930s. The same two unions—The International Brotherhood of Electrical Workers and the Association of Technical Employees—dominate the radio scene today.

International Brotherhood of Electrical Workers

The International Brotherhood of Electrical Workers (IBEW) was formed in 1891 out of several earlier attempts to organize those involved with electrical wiring and manufacturing. IBEW was behind what was probably the first labor strike in radio, in late 1925 at KMOX in St. Louis, Missouri. Concerned with wages and working conditions—the universal labor issues—the strike led to the union's recognition as the bargaining agent for KMOX technicians. In 1931, KMOX became a CBS owned-and-operated station that spread the IBEW idea to others at the network. IBEW also organized stations in Chicago and in Birmingham, Alabama. In 1939, the union hit the big-time when it successfully organized the technical employees at CBS. As it made plans to create more broadcast-centered locals on a national basis, however, IBEW increasingly came up against the National Association of Broadcast Engineers and Technicians (see below) in jurisdictional disagreements. In 1951 IBEW organized its broadcasting and recording members into a separate department, though they continued to represent but a tiny part of the larger union.

Association of Technical Employees

In 1933 some 300 employees at NBC formed the Association of Technical Employees (ATE) to represent themselves with the network. A year later they signed their first contract with NBC. This contract set a wage scale of $175.00 per month, rising to $260.00 after nine years' service. The work week was determined to be 48 hours. ATE was the first organization created exclusively to represent radio employees. By 1937 the ATE contract spelled out their jurisdiction, and the first independent (non-network) station joined the unit. A union shop clause came two years later.

National Association of Broadcast Engineers and Technicians

ATE became the National Association of Broadcast Engineers and Technicians (NABET) in 1940. Contracts negotiated the next year set the first eight-hour day. Some 23 small stations were also under NABET contracts. When NBC-Blue split off in 1943 (becoming ABC in 1945), NABET contracts carried over, giving the union two of the major radio networks. Seeking some organizational strength and thus organizing clout, NABET affiliated with the Congress of Industrial Organizations (CIO) in 1951, and changed its name to National Association of Broadcast *Employees* and Technicians, thus retaining the same acronym.

Recent Union Trends

The coming of television in the late 1940s, and later cable television, dramatically increased employment in the electronic media industries—and thus organizing opportunities as well as jurisdictional battles among the unions involved. One union, the International Alliance of Theatrical and State Employees (IATSE), never organized radio workers, building from its 1893 theater and later film bases to expand into television. But both the IBEW and NABET also grew into television, thus precipitating a host of jurisdictional disputes in the 1950s and 1960s, some resolved with strikes and others with arbitration.

The degree of unionization in radio varies greatly by market and region of the country. Union agreements are far more likely at the network level, in large markets, and in the Northeast and West Coast, as well as major Midwest cities. Smaller markets and stations tend not to be subject to union agreements. Recent mergers in the radio business, with huge numbers of stations coming under common control (including multiple stations in the same market) have not thus far had union implications. On the other hand, greater use of automation has trimmed employment ranks as has such FCC deregulation as no longer requiring licensed engineers to supervise radio transmitter operations.

In the constant search for new members and bargaining units, each union emphasizes what it has gained for its rank and file. NABET, for example, listed its accomplishments by the 1990s as the wide-spread acceptance of the union shop, a seven-hour day and 35-hour week, paid vacation and statutory holidays, lay-off and rehiring on a seniority basis, differential pay for night work, discharge only for just and sufficient cause, established grievance procedures (including arbitration), contract provisions covering increased automation, and pension plans. IBEW could make similar claims.

In 1993, NABET sought organizational support with the Communication Workers of America (CWA), heretofore a union of telephone workers, although it had also organized many cable television workers. After a one-year test run in 1993, the two merged into what became the NABET-CWA.

DENNIS W. MAZZOCCO AND CHRISTOPHER H. STERLING

See also American Federation of Musicians; American Federation of Television and Radio Artists; Ham Radio; KMOX; Trade Associations; WCFL

Further Reading

Brittain, James E., "The Evolution of Electrical and Electronics Engineering and the *Proceedings of the IRE," Proceedings of the IEEE,* 77 (June 1989; part 1) and 85 (May 1997; part 2)

"Fiftieth Anniversary of the AIEE," *Electrical Engineering* 53 (May 1934)

Finney, Robert, "Unions/Guilds," in *Encyclopedia of Television,* 3 vols., edited by Horace Newcomb, Chicago and London: Fitzroy Dearborn, 1997

Gardner, John, "The Below-the-Line Unions," *Television Magazine* 24 (December 1967)

Huntoon, John, *Fifty Years of ARRL,* Newington, Connecticut: American Radio Relay League, 1965

International Brotherhood of Electrical Workers website, <www.ibew.org>

Koenig, Allen E., "Labor Relations in the Broadcasting Industry: Periodical Literature, 1937–1964," *Journal of Broadcasting* 9, no. 4 (Fall 1965)

Koenig, Allen E., editor, *Broadcasting and Bargaining: Labor Relations in Radio and Television,* Madison: University of Wisconsin Press, 1970

McMahon, A. Michal, *The Making of a Profession: A Century of Electrical Engineering in America,* New York: Institute of Electrical and Electronics Engineers, 1984

Mermigas, Diane, "Unions under Fire," *Electronic Media* (4 and 11 September 1989)

National Association of Broadcast Employees and Technicians website, <www.nabetcwa.org/nabet>
Radio Club of America, *Seventy-Fifth Anniversary Diamond Jubilee Yearbook, 1909–1984*, Highland Park, New Jersey: Radio Club of America, 1984
Reader, William Joseph, *A History of the Institution of Electrical Engineers, 1871–1971*, London: Peregrinus, 1987
Whittemore, Laurens, "The Institute of Radio Engineers—Fifty Years of Service," *Proceedings of the IRE* 50 (May 1962)

Telecommunications Act of 1996

Changing Radio's Licensing and Ownership Rules

In February 1996 President Bill Clinton signed the Telecommunications Act into law, the result of two decades of industry and congressional effort to update government regulation of the industry. Cast as amendments to the benchmark Communications Act of 1934, the complex 1996 law wrought important changes in the radio industry as a fairly small part of a law primarily addressed to substantial policy change in the telephone business. Provisions of the law contributed to substantial consolidation of station ownership while at the same time furthering the aims of deregulation.

Origins

The 1934 Communications Act has often been amended in the decades since its passage. Important revisions were enacted every few years as regulated industries changed and expanded. Attempts to replace the law (especially a series of draft "rewrite" proposals from 1976 to 1981) failed to pass, however, because they tried to do too much at one time. The more focused public broadcasting act of 1967 and the cable acts of 1984 and 1992 are examples of more industry-specific legislation that did successfully amend the 1934 law.

Development of what became the 1996 amendments took several years. The issues involved were complex as Congress considered substantial deregulation of and other changes in traditional regulatory approaches to the telephone and electronic media industries. Growing digital convergence among all electronic communications services forced a rethinking of long-accepted regulatory assumptions. At the same time, largely defensive industry positions amidst rapid technical change were deeply entrenched, making compromise difficult (it is nearly always easier to stop legislative progress than to maintain its momentum). Moreover, a continuing trend toward less governmental and more marketplace control was changing the way Congress perceived the telecommunications sector.

For broadcast deregulation was not new. The Federal Communications Commission (FCC) had begun cautious moves in this direction in the mid–1970s, and they accelerated during the Reagan administration in the 1980s. To deregulate radio broadcasting, for example, the FCC lifted a number of "behavioral" regulations (such as guidelines encouraging at least minimal non-entertainment programming) in the early 1980s. Over the next decade, the commission also loosened its "structural" regulation by slowly increasing the number of radio stations any single entity could own from seven AM and seven FM stations (the longtime national limit) to 12 of each type in 1985, raised to 18 in 1992, and 20 by 1994. In an even more basic change, in 1992 the FCC removed a longtime restriction by allowing any entity to own more than one AM or FM station in the same market.

The first potential bills were considered by Congress, then under Democratic control, in the early 1990s, although none of them progressed far. Dramatic changes in telephone industry policy were at the core of each bill; electronic media provisions were relatively minor parts of the proposed legislation. Republican takeover of both houses of Congress in the 1994 elections delayed progress briefly as the longtime opposition party learned how to run things and trained a new cadre of staffers and members in the intricacies of the telecommunications field.

Finally, in the fall of 1995, both houses passed substantial telecommunications deregulation bills, albeit with the differences usually found in the legislative process in which each house acts independently of the other. A conference committee worked for many weeks and early in 1996 produced the compromise bill passed by both houses on the first day of February 1996. The president signed the bill a week later.

Radio Licenses

The new legislation had four important effects on the radio business, two concerning station ownership, and two focused on licensing. Reasoning that with nearly 13,000 radio stations on the air old restrictions (established when less than 10 percent of that number existed) could now be eliminated, Congress opted to free the marketplace. Provisions in the 1996 law (a) dropped limits on the number of radio stations that could be owned nationally; (b) increased the local market stations that could be owned by a single entity; (c) lengthened station licenses; and (d) eliminated competitive applications at license renewal time.

The license term change was relatively minor: Section 307 (c) of the 1934 law was changed to extend radio licenses from seven to eight years (until 1981 licenses for radio or television had run for only three years). But the end of potential competitive applications may have more far-reaching impact. Acting on a long-existing industry desire for license "renewal expectancy," the 1996 law added a new subsection (k) to Section 309 of the Communications Act to make clear that existing station licenses will be renewed unless the FCC finds important and continuing transgressions of its rules and regulations to have occurred. Even in such a situation, the commission retains the discretion to renew or deny a license. But the law also forbids the FCC from considering a rival application during a license renewal proceeding—until and unless it has first decided that the existing license must be terminated. Given the FCC's track record over seven decades of license renewals, such terminations are very unlikely. While only a tiny fraction of one percent of all stations were denied renewals under the old rules, the new law makes license renewal virtually automatic in the future.

Station Ownership

Statutory changes governing radio station ownership were more dramatic, as the initial years of station trading after passage of the 1996 law have shown. Stations may now be bought and sold or traded much as with any other business, although the FCC retains the right to approve each new licensee. The 1996 law directed the FCC to eliminate its national cap on radio station ownership (then standing at no more than 20 AM and 20 FM). Any company could now own as many stations as money would allow. Within a month of the law's passage, two radio station groups were approaching 50 stations each—by the end of the year the first group exceeded 100 stations. From there the pace of station buying and selling increased and station prices soared. By early in the 21st century, the largest radio group owner controlled more than 1,200 outlets, by which time 40 percent of all radio stations had changed ownership since 1996.

The market-level situation was more complex. The 1996 law allows a single owner to control up to eight stations in any of the largest markets (more than the *national* cap before 1985) as long as no one owner controls more than half the stations in a given market. The table provides the specific new limits, and the law allows for even these to be exceeded if such an action "will result in an increase in the number of radio broadcast stations in operation," although just how that might work in practice is not yet clear.

To further complicate matters, in August 1996 the U.S. Department of Justice's Antitrust Division announced that no single market radio owner would be allowed to control more than half of that market's radio advertising revenue. A few group owners had to divest themselves of one or more stations to comply.

In a market with this many radio stations:	*A single entity can control up to this many commercial stations:*
45 or more	Up to 8, no more than 5 in the same service (AM or FM)
30-44	Up to 7, no more than 4 in the same service
15-29	Up to 6, no more than 4 in the same service
14 or fewer	Up to 5, no more than 3 in the same service

Impact on Radio

Although many policymakers have cited its positive effects, such as eliminating outdated rules, encouraging innovation and development of new technology, and encouraging a more competitive environment, the 1996 law has had negative implications as well. For example, in some smaller markets, only two companies may end up owning all the available stations. The overall number of different radio station owners in the country declined by 30 percent in seven years, from the passage of the act to early 2003.

The Telecommunications Act has also allowed owners of multiple stations to operate multiple outlets with the same personnel (often with voice tracking technology), programming, and administration. Critics argue that such practices can also lead to more music homogenization and fewer different "voices" (points of view) being heard over the air. Major group owners now significantly control multiple local radio markets and can largely dictate the terms of advertising. CBS, for example, quickly expanded to control 40 percent of all radio revenue in Boston, Chicago, New York, and Philadelphia. At the same time, station ownership by ethnic minorities declined as radio station prices rose and groups expanded. Among FM stations, for example, minority-owned stations declined from 127 to 100 in just the first year after the 1996 law was passed.

When Congress passed the 1996 law, many proponents hailed it as an opportunity to create a pro-competitive, deregulatory national framework, as well as more industry employment. Then FCC Chairman William Kennard predicted that the 1996 Act would hasten "the transition to a competitive communications marketplace." Commissioner Susan Ness concluded that the main goals of the 1996 amendments were to promote competition, reduce regulation, and encourage rapid deployment of new telecommunication technologies.

On the other hand, commissioner Gloria Tristani later pointed out the dramatic and potentially negative impacts on radio station ownership and operations programming since passage of the Telecommunications Act. Tristani noted that group ownership reduces the number of different and competing voices and opinions heard on local radio stations. More recently, commissioner Michael Copps expressed concern that local radio markets had become oligopolies where programming originated outside the local station "far from listeners and their communities." Even FCC Chairman Michael Powell agreed that the consolidation of radio station ownership "concerned" him. The growing number of group-owned radio stations, of course, mirrors a similar trend in chain-ownership of newspapers, television stations, and cable systems. Many of the conglomerates buying radio stations also own other media.

Although proponents of the Telecommunications Act of 1996 still trumpet its successes, the law was subject to years of substantial litigation and has yet to achieve all of its goals. As with other deregulatory legislation, the 1996 amendments have tended to place business interests ahead of those of the listening public. A growing number of critics (some of them in Congress, others in the radio business) have argued that consolidation allowed by the 1996 law is at the heart of radio's declining audience appeal as more stations sound increasingly alike.

CHRISTOPHER H. STERLING

See also Clear Channel Communications; Communications Act of 1934; Deregulation of Radio; Federal Communications Commission; Licensing; Ownership, Mergers, and Acquisitions; U.S. Congress and Radio

Further Reading

Aufderheide, Patricia, *Communications Policy and the Public Interest: The Telecommunications Act of 1996*, New York: Guilford Press, 1999

Ferris, Charles D., Frank W. Lloyd, and Harold J. Symons, *Guidebook to the Telecommunications Act of 1996*, New York: Bender, 1996

Huber, Peter W., Michael K. Kellogg, and John Thorne, *The Telecommunications Act of 1996: Special Report*, Boston: Little Brown, 1996

Knauer, Leon T., Ronald K. Machtley, and Thomas M. Lynch, *Telecommunications Act Handbook: A Complete Reference for Business*, Rockville, Maryland: Government Institutes, 1996

Sterling, Christopher H., "Changing American Telecommunication Law: Assessing the 1996 Amendments," *Telecommunications and Space Journal* 3 (1996)

Sterling, Christopher H., "Radio and the Telecommunications Act of 1996: An Initial Assessment," *Journal of Radio Studies* 4 (1997)

Telecommunications Act of 1996, Public Law 104-104, 104th Congress, 2nd Session, February 8, 1996

Television Depictions of Radio

Fictional Portrayals in American Series

Television's eclipse of radio as the dominant mass medium of entertainment in the decade following World War II was propelled in large measure by the transformation of radio shows into video versions. More than 200 radio programs moved to television, including *The Adventures of Ozzie and Harriet*, *Truth or Consequences*, *The Lone Ranger*, *Your Hit Parade*, *Suspense*, *Arthur Godfrey's Talent Scouts*, and *Studio One*. Several comedians who had achieved enormous popularity on radio, such as Jack Benny, George Burns and Gracie Allen, Red Skelton, and Bob Hope, found similar success with their television series.

By the late 1950s, both the radio industry and American life had been profoundly altered by the rise of television. Big-budget network radio shows that appealed to the whole family had given way to cheaply produced local disc jockey programs catering to specialized audiences. Teenagers were an especially attractive market segment as the sale of transistor radios boomed and the age of rock and roll arrived.

The nostalgia many older Americans felt for the glory days of network radio in the 1930s and 1940s was inspiration for a March 1961 episode of *The Twilight Zone* entitled "Static." An elderly bachelor living in a boardinghouse retrieves his elegant radio console, circa 1935, from the basement. When he's alone in his room, he hears programs from the past. Fearing for his sanity, his former fiancée gives the radio to a junk dealer. Infuriated by her meddling, he gets the radio back and is relieved to find that it still works. When he calls the disbelieving woman to his room to hear for herself, she appears as she did in 1940 as his young sweetheart. He too has become a young man. The radio, a magic machine, sent them back in time and gave the couple a second chance.

But there was no return for the radio industry to its earlier splendor. It adapted to the modern era and itself became grist for TV's storytelling mill. Since the late 1960s a number of television series have had main characters who work in radio stations. *Good Morning, World,* for instance, was a 1967 Columbia Broadcasting System (CBS) situation comedy about a team of early-morning drive-time disc jockeys, Lewis and Clarke, who worked for an overbearing boss at a small station in Los Angeles.

WKRP in Cincinnati

A thoroughly realistic depiction of any occupation or workplace on television is limited by the narrative conventions of drama and comedy. But the spirit and flavor of a profession can be vividly conveyed. The series *WKRP in Cincinnati,* which ran on CBS for four seasons beginning in September 1978, was a show that earned high marks among radio industry insiders for its authentic ambience. WKRP creator, executive producer, and head writer Hugh Wilson received many letters complimenting the show's realism.

The show was, in fact, based on a real station—Atlanta's WQXI, a successful AM/FM combination with a rock format. In the early 1970s, Hugh Wilson, who was working in advertising in Georgia's capital city, met WQXI salesman Clark Brown at Harrison's, a bar that catered to the media crowd. Through Brown, Wilson was introduced to a number of people who worked in Atlanta radio. In 1977 these friendships proved valuable as a source of inspiration when the vice president for comedy development at CBS gave Wilson the go-ahead to write a pilot for a situation comedy about a radio station.

The premise of *WKRP in Cincinnati* was that the station, a ratings loser with an "elevator music" format, would change to rock and roll. A new young program director, Andy Travis, was brought on board to implement the switch, which alienated some longtime sponsors, such as Barry's Fashions for the Short and Portly.

The other employees of Arthur Carlson, the inept station manager whose mother owned WKRP, were Jennifer Marlowe, a brainy bombshell receptionist; Les Nessman, the naive, conspiracy theorist news director; Herb Tarlek, a salesman with a penchant for wearing white shoes and white belts; Bailey Quarters, Andy's shy assistant; and two disc jockeys—the burned-out hip cat, Dr. Johnny Fever, and the jive-talking sartorial sensation, Venus Flytrap.

Although the focus of *WKRP in Cincinnati* was character development, not the illumination of issues in the radio industry, viewers were introduced to the tribulations that came with the competitive territory. The show's theme song alluded to the uncertain lives of on-air talent and radio managers with a reference to "packing and unpacking up and down the dial." Throughout the 90 episodes, the WKRP staff was faced with many legal, ethical, and business matters that reflected the reality of local radio, including the dwindling length of playlists, the anxiety over the arrival of Arbitron ratings books, the use of programming consultants, the emergence of computer-operated radio stations, and the protests of disaffected listeners.

TV's Talk Radio

Another series of the late 1970s having a radio theme was *Hello, Larry,* a major disappointment for the National Broadcasting Company (NBC). It was hoped that the star power of McLean Stevenson, who had played Lt. Col. Henry Blake in *M*A*S*H* for three seasons, would ensure the success of the show. But viewers did not warm to the series about radio talk show host Larry Alder, who moved from Los Angeles to Portland, Oregon, after a divorce in which he gained custody of two teenaged daughters. Working with him on his phone-in show at KLOW were a female producer and an obese engineer. Even a crossover stunt with the popular lead-in series *Diff'rent Strokes,* in which Larry's old Army buddy Phillip Drummond buys the radio station, couldn't generate audience interest.

By the late 1980s, talk radio had become a growth format, and the hour-long dramatic series *Midnight Caller,* which debuted on NBC in October 1988, tapped into the trend. The lead character, Jack Killian, was a San Francisco police detective who had quit the force in despair after he accidentally killed his partner in a shoot-out. His new career as "The Nighthawk," host of an all-night call-in show, allowed him not only to offer advice, but also to become involved in the investigation of crimes and corruption.

The title character of the Fox comedy series *Martin,* which began in 1992, was also a talk show host. He worked for Detroit radio station WZUP until the end of the second season, when the station was sold to a large radio group, the format was changed to country, and Martin was fired by the new owner. Whether they realized it or not, viewers were getting a feel for the fruits of radio's deregulation.

Frasier

In September 1993, *Frasier,* an NBC spin-off of *Cheers,* introduced a radio-related character who would become one of the most popular in television history. Dr. Frasier Crane left his psychiatric practice in Boston, divorced his neurotic wife Lilith, and moved back to his hometown of Seattle, where his new job was hosting a radio advice show.

Unlike *WKRP in Cincinnati, Frasier* was not set principally in the workplace. Frasier's home—an ultramodern luxury apartment with a breathtaking view of the Seattle skyline—was just as often the scene of the action. But Frasier revels in his radio celebrity, and troubles at the station often overflow into his personal life. The heartless economics of radio in the 1990s created many complications for Frasier and his colleagues. Changes in management and a slavish adherence to the bottom line in station decision making are the only permanent features of their careers.

In a 1997 seminar at the Museum of Television and Radio, the executive producer of *Frasier,* Christopher Lloyd, acknowledged that faithful realism to the world of radio was not a consideration in the show's production: "We have people that we consult with that kind of help keep us in check as far as how legitimate things are that we do—you know, the buttons that they push and the carts they throw in and out are sort of like what would happen at a radio station. But beyond that, we don't hem ourselves in too much."

One memorable episode of *Frasier* that was based on an actual radio personality, however, took a swipe at Dr. Laura Schlessinger, whose syndicated daily talk show had become a phenomenon by the late 1990s. In the story, Dr. Nora joins the staff of KACL and begins to dispense harsh criticism and questionable advice to her troubled callers, such as calling a bisexual woman an equal-opportunity slut. But despite her rigidly moralistic approach, Dr. Nora in fact has a tarnished past—including two divorces, an affair with a married man, and estrangement from her mother—that renders her righteousness hollow.

Another 1999 episode of *Frasier* parodied the proliferation of crude shock jocks. KACL's new morning team, Carlos and the Chicken, sponsor a contest with a $1,000 prize to the listener who sends in "the best picture of Frasier Crane's humongous ass for our website." Though Frasier laments the success of "so-called humorists who rely on cruel pranks and scatological references," he's warned by his friends and family not to confront the duo or he'll continue to be fodder for their gags. As it turns out, Carlos and the Chicken become victims of their own pettiness and thin skins when an argument over who is the funnier of the two breaks up the team.

Alternate Formats

In 1995 two comedy series with a radio backdrop appeared in prime time. *The George Wendt Show* on CBS was based on a popular program on National Public Radio, *Car Talk,* hosted by brothers Tom and Ray Magliozzi. In the short-lived TV series, unmarried brothers George and Dan Coleman cohosted the radio call-in show *Points and Plugs* from the office of their auto repair shop in Madison, Wisconsin. The more successful entry was NBC's *NewsRadio,* which was set at WNYX, an all-news station in New York City. The domineering and abrupt station owner, Jimmy James, hires yet another in a long succession of news directors. The latest news director (played by Dave Foley), young and energetic, leaves Wisconsin for his big break in a big market. In over his head, he also has to contend with the idiosyncratic personalities of his staff, especially the huge ego of on-air anchor Bill McNeal (played by Phil Hartman).

The cable network American Movie Classics presented an original comedy series, *Remember WENN,* beginning in 1996. The show, set in Pittsburgh during the early years of World War II, soon developed a fanatically loyal audience. In each episode the cast and crew of station WENN struggled to create hours of ambitious daily programming on a shoestring budget—and as a result, viewers were well-schooled not only in the vintage art of sound effects and the logistics of microphone performance, but also in the radio genres of the era.

The heroine of *Remember WENN* is Betty Roberts, who came to the station as the winner of a writing contest with the prize of an unpaid internship. When the station's sole writer is overcome by alcoholism, Betty steps into his job and rises gloriously to the task. The show is evocative of the screwball comedies of the 1930s and 1940s but also weaves in elements of engrossing drama. The overarching theme of *Remember WENN*'s four seasons is the sheer romance and unbridled excitement of the medium at its zenith.

Other series that revolved around the radio industry include *The Lucie Arnaz Show* (CBS, 1985); *Knight and Daye* (NBC, 1989); *FM* (NBC, 1989); *Rhythm and Blues* (NBC, 1992); *Katie Joplin* (WB, 1999); and *Talk to Me* (ABC, 2000). Several made-for-television movies also depicted historical events and personalities in American radio, such as *The Night That Panicked America,* the story of the 1938 *War of the Worlds* broadcast, and biographies of Edward R. Murrow and Walter Winchell.

Radio in the Lives of Characters

In addition to television's bounty of direct portrayals of the radio industry, a vast amount of fictional TV programming has embedded in it a sense of the importance radio has always held in the daily lives of American listeners. Throughout the nine seasons of *The Waltons,* for instance, the family radio in the living room was part and parcel of their existence and even served as a key plot element in several episodes. In "The Inferno," aspiring journalist John-Boy travels to Lakehurst, New Jersey, in May 1937 to cover the landing of the German zeppelin *Hindenburg,* the world's largest airship. When he returns to Walton's Mountain after the traumatic incident, there's no need to explain why he's mired in depression. "We heard about it on the Blue Network," says his younger brother Jason. The next day, his little sister Elizabeth makes reference to announcer Herbert Morrison's famous eyewitness account. The night of the disaster, NBC had broken its rigid rules against the broadcast of recordings and aired the dramatic on-the-scene transcription. "Sure sounded gruesome on the radio," says Elizabeth. "The announcer was crying."

As the Depression years gave way to the war years, the Walton's tabletop radio continued to connect them with the world. In the episode "Day of Infamy," Christmastime 1941 is approaching. Oldest daughter Mary Ellen is planning to go to Hawaii to join her husband Curt, a doctor drafted into the U.S. Medical Corps, when, like millions of other stunned Americans, she learns from the radio that the Japanese had bombed Pearl Harbor.

Brooklyn Bridge, a 1991 series set in 1956, presents radio as an essential element in postwar popular culture. The lead character, 14-year-old Alan Silver, and his 9-year-old brother Nate follow the Dodgers baseball games on the radio with religious devotion. In a 1963 episode of *Leave It to Beaver,* Wally explains to his mother June the redeeming social value of the transistor radio. After every ten records they give a news report. "Heck," says Wally, "that's how Lumpy found out about Cuba."

Whatever changes technology will impose on the production and delivery of television programming in the decades ahead, stories of the American experience will continue to include radio as a key player and a rich source of plots and conflicts. In a world seemingly dominated by images, radio remains the most resilient and ubiquitous mass medium. Radio's pervasiveness in modern life cannot be overlooked by storytellers hoping to create characters and situations that, even if impressionistic, ring true.

MARY ANN WATSON

See also Film Depictions of Radio; Situation Comedy

Further Reading

Brooks, Tim, and Earle Marsh, *The Complete Directory to Prime Time Network and Cable TV Shows, 1946 –Present,* 7th edition, New York: Ballantine Books, 1999
Graham, Jefferson, *Frasier,* New York: Pocket Books, 1996
Kassel, Michael B., *America's Favorite Radio Station: WKRP in Cincinnati,* Bowling Green, Ohio: Bowling Green State University Popular Press, 1993
Mitz, Rick, *The Great Sitcom Book,* New York: Marek, 1980
Pease, Edward C., and Everette E. Dennis, editors, *Radio: The Forgotten Medium,* New Brunswick, New Jersey: Transaction Press, 1995
Stempel, Tom, *Storytellers to the Nation: A History of American Television Writing,* New York: Continuum, 1992
Zicree, Marc, *The Twilight Zone Companion,* New York: Bantam Books, 1982

Ten-Watt Stations

Educational FM Outlets

Ten-watt (Class D) FM stations were created in 1948 as an inexpensive way for noncommercial organizations to operate their own outlets and, at the same time, increase listener traffic on the slow-to-develop FM band. Although educational radio efforts had begun as early as 1930 with the formation of the National Committee on Education by Radio, early operations were limited to a few programs broadcast from commercial AM facilities. By 1936, however, more than three dozen stations licensed to educational entities had managed to get on the air and remain there. Even though these outlets were not operating on frequencies especially reserved for their use, they did provide a service to limited areas of the country and kept the dream of educational radio alive. Consequently, when the Federal Communications Commission (FCC) authorized full-scale FM broadcasting to begin in 1941, five of the original 40 channels (42–50 megahertz) were reserved for the use of noncommercial educational institutions. Seven school systems and universities were granted FM noncommercial licenses before wartime priorities brought most FM activity to a halt in early 1942.

At the war's end the FCC moved the FM band to a higher (88–108 megahertz) and larger band of 100 channels. The first 20 of these (88.1–91.9 megahertz) were again specifically reserved for noncommercial broadcasters. Educators moved quickly to take advantage of this greatly expanded allocation. Their sense of urgency was heightened by the realization that the commercial networks were abandoning their sustaining (unsponsored and often educational or cultural) programming in search of postwar profits.

The cost of building an FM station, however, remained an insurmountable barrier to many educational institutions. In 1948, at the urging of the National Association of Educational Broadcasters, the FCC approved a new low-power category of FM station. These Class D outlets could broadcast at an effective radiated power of just ten watts and were under less stringent operational and licensing requirements than were larger stations. The policy aim was to create a participatory broadcasting entity with low construction and operating costs that would make it affordable to educational institutions of all sizes.

Ten-watt stations were all to be allotted to the first and lowest channel (88.1 megahertz) on the FM band. Because the audio portion of television channel 6 ends at 88 megahertz, this meant that, by using the fine tuner on their new TV sets, consumers could pull in the signal from their neighborhood Class D radio stations without having to invest in FM radio receivers. If 88.1 had already been spoken for in its locale, a ten-watt applicant was free to request any higher available frequency within the educational band. Although Class D signals seldom carried more than four or five miles unless a more expensive high-gain antenna was deployed, school superintendents and college administrators saw these new audio vehicles as valuable community relations tools.

A typical example was WNAS in New Albany, Indiana, which went on the air in 1949. As chronicled in the high-school oriented *Senior Scholastic* magazine two years later, "Superintendent Henry Davidson wanted a 'voice' for his schools. He couldn't see a way to finance the only kind of station then possible—a high-power station costing from $50,000 to $100,000. He waited until the low-power FM station became a possibility for schools. Then he went into action. He found that a 10-watt FM station would cost about $3,000 to build and equip. (Actual final cost $3,500.)" Taking the air on 28 May 1949, WNAS was one of the pioneering Class Ds. But honors as the first such facility were claimed by De Pauw University's WGRE, which had fired up its Greencastle, Indiana, transmitter 33 days earlier.

At the time, the FCC required that stations in other classes have full-time first-class licensed engineers on site to perform required technical functions. In contrast, Class D stations were allowed to use non-technical third-class license holders to turn the transmitter on and off and operate the station. Technical servicing for a ten-watt outlet could be performed by a second-class operator available on call rather than an on-site first-class holder. Class Ds could also go on and off the air at will, a privilege denied higher power stations and one that meshed well with school and university calendars.

In the ensuing years, scores of Class D stations were built. Although the majority were constructed on college and university campuses, some of the most public-spirited were the licensees of independent school districts. The 20 such stations on the air in 1965 programmed a mix of in-school instructional lessons, general enrichment offerings (such as classical music and drama), and community-oriented services (ranging from school basketball games to school board meetings). The formats of college stations, on the other hand, often were more student-programmed and popular-music focused.

By the end of the 1960s, new realities began to threaten the Class D stations' continued existence. FM was now becoming widely popular—especially with younger listeners. As it rose to prominence, the medium's educational channels became increasingly occupied by high-power facilities whose professional staffs viewed ten-watt operations as inefficient amateurs clogging scarce spectrum space. That many of these now full-power outlets had originated as Class Ds was seldom mentioned. The raucous and undisciplined material aired by some under-supervised college-licensed Class Ds was used to indict all ten-watt outlets and undermine the case for their continued existence. The passage of the 1967 Public Broadcasting Act and its creation of the Corporation for Public Broadcasting (CPB) also worked against low-power stations. Expanded and more centralized government funding favored support for wide-coverage-area and professionally staffed public broadcast facilities, rather than limited-range student stations or volunteer-heavy ten-watt operations that were impractical to network and incapable of contributing quality programs for national or regional distribution. When National Public Radio (NPR) was founded in 1971, it brought further structure and substance to national noncommercial radio service, but it marginalized Class D stations even more.

In the early 1970s the public radio establishment introduced via NPR a series of gradually increasing facility, schedule, and personnel requirements for stations to remain NPR members. At the same time, CPB and other federal funding sources dovetailed their requirements for fiscal support with the standards for NPR membership. Although these moves made public radio much stronger, they walled off Class D and other small stations from most external funding. This happened despite the fact that by 1978, 426 outlets (almost half of U.S. noncommercial radio stations) were low-power operations. That same year, the FCC decreed that ten-watt facilities must make plans either to increase output to 100 watts or to assume secondary and preemptable status (meaning that a higher-power station could take their frequency or push them out of business from an adjoining one). Most Class Ds chose to increase power, although spectrum limitations forced some to relocate. By 1980, the ten-watt noncommercial station was functionally extinct.

Ironically, in 1999 the FCC proposed a new class of "micro radio" low-power FM outlets designed to better serve neighborhood needs and to advance the cause of minority ownership. These "secondary" stations could be commercial, and were projected to operate with outputs of as little as ten watts.

PETER B. ORLIK

See also College Radio; Community Radio; Corporation for Public Broadcasting; Educational Radio to 1967; Licensing; Low-Power Radio/Microradio; National Public Radio; Public Radio Since 1967

Further Reading

Martin, Howard, "Low Power Stations," *The Journal of College Radio* (November 1972)
McKown, Vernon, "Students Run WNAS," *Senior Scholastic* (4 April 1951)
Mead, James, "A Study of a Low-Power School Radio Station: WOAK," Ed.D. diss., Wayne State University, 1965
Orlik, Peter, *A Survey of Public School Radio in the United States,* Detroit: Wayne State University Mass Communications Center, 1966

Terkel, Louis "Studs" 1912–

U.S. Disc Jockey, Commentator, Interviewer, and Author

As a self-proclaimed "oral historian," interviewer Louis "Studs" Terkel practiced his craft on Chicago radio station WFMT-FM. During his career, which spanned 45 years, Studs interviewed the rich and famous as well as the average citizen and, in so doing, created a "bottom up" audio history of 20th-century America.

Origins

Louis Terkel was born to Russian-Jewish parents in New York City in 1912. In his youth, his family moved to Chicago, where his father was a tailor. During the 1920s, his family resided in a men's hotel in which transient workers of the day took up temporary residence and practiced their craft until they moved on to the next job in another city. It is there that young Louis spent hours in the main lobby listening to people's stories. In the 1920s, various political movements, such as the Communist and Socialist Workers' Parties, were rising in popularity in America, and many of the workers who lived in the hotel espoused those beliefs. Their stories not only helped to form Terkel's political views but also helped him to develop his unique style of interviewing.

After high school, Terkel attended the University of Chicago and graduated in 1932. He went on to study law at the same institution and received his law degree in 1934. It was in law school that he received his nickname "Studs." Louis was fond of the James T. Farrell *Studs Lonigan* novels. Since Terkel was often seen carrying those books around, the nickname "Studs" stuck.

After law school, Terkel found work producing radio shows in Chicago as part of the Federal Works Program. He also became involved with the Chicago Repertory Theater, acting as a producer and performer. At the outbreak of World War II, Terkel attempted to join the U.S. Army but was rejected because of a perforated eardrum. He joined the Red Cross, but was again unable to serve overseas. Later he learned that it was actually his left-wing political views that kept him from serving overseas in either capacity.

Terkel on Radio

During the 1940s, Studs Terkel became a familiar voice in Chicago radio as a news commentator and disc jockey. In 1949, he tried his hand at television interviewing when he was given the host's spot on *Studs' Place*, a series on the NBC station in Chicago. After one year, NBC canceled the program. Many believed that the cancellation of *Studs' Place* stemmed from an investigation of Terkel by Senator Joseph McCarthy and the House Un-American Activities Committee. For many years the impact of the investigation made it difficult for Terkel to work in broadcasting, but he found employment with the Chicago *Sunday Times* as a jazz columnist and working in plays around Chicago. Not until the mid-1950s was he offered a daily one-hour program on WFMT called *The Studs Terkel Show.* He stayed with the program for more than four decades until his last regularly scheduled show on 1 January 1998.

Toward the end of his radio career, Terkel began to flourish as an "oral historian." He had become famous for getting guests to open up about themselves. Even though he interviewed the famous, he found the average American to be his best subject. He had a reputation for being fair as an interviewer—eschewing the sensational or bombastic and always showing proper respect for his guests.

Perhaps his greatest asset was his unique interviewing style. While Terkel was known worldwide for talking about any subject, his listening skills made him such a good "oral historian." His subjects were allowed to think out their answers. While this often resulted in moments of "dead air," Terkel worked with the silence, allowing a person to think through a question and answer at his/her own pace.

Terkel's work evolved into many best-selling books and oral histories of the 20th century. More than 7,000 hours of interviews have been compiled by Terkel and WFMT and are housed at the Chicago Historical Society.

TIM POLLARD

Studs Terkel. Born Louis Terkel in New York City, 16 May 1912. Graduated from the University of Chicago, 1932, and the University of Chicago law school, 1934. Served in the American Red Cross, World War II. Employed in various radio producing jobs during the 1930s and 1940s. Moved to on-air disc jockey and commentator. Host of *Studs' Place* on WMAQ television, 1949–50. Jazz columnist for the *Chicago Sunday Times*. Host of *The Studs Terkel Show* on WFMT-FM Chicago, 1958–98.

Radio Series

1958–98 *The Studs Terkel Show*

Selected Publications

Division Street: America, 1967
Hard Times: An Oral History of the Great Depression in America, 1970
Working, 1974
Talking to Myself: A Memoir of My Times, 1977
The Good War: An Oral History of World War II, 1984
Race: How Blacks and Whites Think and Feel about the American Obsession, 1992
Coming of Age: The Story of Our Century by Those Who've Lived It, 1995
Voices of Our Time: Five Decades of Studs Terkel Interviews (cassettes), 1999
Will the Circle Be Unbroken? Reflections on Death, Rebirth, and Hunger for a Faith, 2001

Further Reading

Studs Terkel website, <www.studsterkel.org>

Tesla, Nikola 1856–1943

U.S. (Croatian-Born) Inventor and Radio Pioneer

Nikola Tesla, the man who made possible the control of electricity using alternating current (AC), also pioneered the wireless transmission of energy, the fundamental principle of radio. In 1943 the U.S. Supreme Court ruled that Tesla's radio patent for "System of Transmission of Electrical Energy," granted in 1900, in large part anticipated that of Guglielmo Marconi.

Early Years

Born of Serbian parents in 1856 in Smiljan, Croatia, Tesla demonstrated a phenomenal memory and performed brilliantly in school. He attended the Austrian Polytechnic School, but never completed a degree. Reportedly, Tesla then attended the University of Prague. He eventually found work in Budapest as a draftsman for the Hungarian government's newly established Central Telephone Exchange.

In 1884 Tesla immigrated to New York, where he worked briefly for Thomas Edison, the leading proponent of providing electricity by direct current (DC). As an independent inventor, Tesla developed the AC motor and eventually made the legendary decision to sell all of his 40 AC patents to George Westinghouse. Though his inventions could have made him a millionaire many times over, Tesla "had no business sense nor any real interest in commercializing his work" (Johnston, editor, *My Inventions*, by Tesla, 1982). Tesla received U.S. citizenship in 1891, an honor he cherished above all others.

Tesla's sensational experimentation and demonstrations with electricity made him a public celebrity and media attraction. Regardless of his showman persona, "Tesla's command of high frequency currents placed him at the forefront of late-nineteenth century research into x-rays, diathermy, discharge lighting, robotics, and wireless—his lectures on these and other subjects were stunning successes" (Johnston, 1982). Between 1886 and 1928, Tesla was awarded more than 100 U.S. patents.

Radio Years

Beginning in 1900, the visionary Tesla aimed to create a "World-System" of wireless transmission; the system would use a central transmitter to distribute virtually all forms of information, including voice, music, written communications, and photographs. In essence, it would be a universal wireless radio, fax, and telephone system, but financial difficulties prevented its realization. Tesla had already demonstrated his practical achievements in wireless in 1893 before the National Electric Light Association in St. Louis (Wagner, 1995), and in 1899, when he operated a radio-controlled boat in Chicago. Notably, Tesla's work influenced Lee de Forest, inventor of the Audion, which made wireless voice transmission possible: "*I aim at Tesla* . . . if I reach that I am a long way ahead," de Forest claimed (cited in Lewis, 1991).

Tesla's basic radio application, "System of Transmission of Electrical Energy," filed on 2 September 1887, was granted 20 March 1900 as patent number 645,576; a subsequent patent, number 649,621, "Apparatus for Transmission of Electrical Energy," was granted on 15 May 1900. Marconi filed for his first U.S. radio patent on 10 November 1900, but during the next three years, even after his historic transatlantic transmission, the U.S. Patent Office denied Marconi's patent because of the priority of Tesla, Sir Oliver Lodge, and German inventor Carl F. Braun.

In 1904 the U.S. Patent Office "suddenly and surprisingly" reversed its prior decisions and granted Marconi his patent for radio: "The reasons for this have never been fully explained, but there is little doubt that the decision was influenced by the powerful financial backing for Marconi in the United States" (Cheney and Uth, 1999). The editor of Tesla's autobiography, Ben Johnston, notes that Tesla "kept a toehold in wireless until World War I by licensing his potentially lucrative wireless patents, but his lack of either financing or corporate ties prevented his litigating the patents effectively" (Johnston, 1982).

Vindication

In 1943 the U.S. Supreme Court finally heard and decided a suit filed by Marconi against the U.S. government for its "infringement" during World War I, when the military took control of all U.S. wireless technology without paying patent royalties. As part of its decision in *Marconi Wireless Telegraph Company of America v United States,* the Court reviewed the radio patents of Tesla, Lodge, and John Stone. Specifically, it examined Marconi's patent concerning "the use of two high frequency circuits in the transmitter and two in the receiver, all four so adjusted as to be resonant to the same frequency or multiples of it."

Chief Justice Harlan F. Stone, in writing the Court's opinion, pointed out that Tesla's patent 645,576 "disclosed a four-circuit system, having two circuits each at transmitter and receiver and recommended that all four circuits be tuned to the same frequency." Although "devised primarily for the transmission of energy," Tesla also recognized that his apparatus "could, without change, be used for wireless communication, which is dependent upon the transmission of electrical energy." This included the transmission of an "intelligible message to great distances."

Regarding the specifications of Tesla's radio apparatus, Stone wrote that Tesla "anticipated the four circuit tuned combination of Marconi," save for one feature, the use of a "variable inductance as a means of adjusting the tuning [of] the antenna circuit of transmitter and receiver," which in actuality "was developed by Lodge after Tesla's patent but before the Marconi patent in suit." Based on its review of the patents in question, the Court concluded that "Marconi's patent involved no invention over Lodge, Tesla, and Stone." The events of World War II overshadowed the significance of the ruling; its impact diminished even further in that all the patentees and patents involved had expired.

Tesla died, bankrupt and alone, at age 86 in his New York hotel room on 7 January 1943, six months before the Supreme Court essentially rendered Marconi's patent invalid. More than half a century after his death, Tesla experts and advocates seek to rectify what they see as a historical bias against the unorthodox inventor, whose groundbreaking work in electricity and radio serves as the foundation for many of the devices modern society takes for granted.

ERIKA ENGSTROM

See also De Forest, Lee; German Wireless Pioneers; Lodge, Oliver; Marconi, Guglielmo

Nikola Tesla. Born in Smiljan, Lika, Croatia, 9 or 10 July 1856. Attended Austrian Polytechnic School and University of Prague; worked as draftsman at Central Telephone Exchange, Budapest, Hungary; Continental Edison Company, Paris, France, 1882; arrived in United States, 1884; worked for Thomas Edison briefly; sold patent rights to his system of alternating-current dynamos, transformers, and motors to George Westinghouse, 1885; became independent inventor, 1887; invented Tesla coil, 1891; U.S. citizenship granted 30 July 1891. Granted total of 112 U.S. patents between 1886 and 1928, including radio patents nos. 645,576 and 649,621 granted in 1900. Died in New York City, 7 January 1943.

Selected Publications

My Inventions: The Autobiography of Nikola Tesla, edited by Ben Johnston, 1982
The Complete Patents of Nikola Tesla, edited by Jim Glenn, 1994

Further Reading

Anderson, Leland I., "John Stone Stone on Nikola Tesla's Priority in Radio and Continuous-Wave Radiofrequency Apparatus," *AWA Review* 1 (1986)
Cheney, Margaret, and Robert Uth, *Tesla, Master of Lightning,* New York: Barnes and Noble Books, 1999
Lewis, Tom, *Empire of the Air: The Men Who Made Radio,* New York: E. Burlingame Books, 1991
Marconi Wireless Telegraph Co. of America v United States, 63 Supreme Court Reporter 369 (1943)
Martin, Thomas Commerford, editor, *The Inventions, Researches and Writings of Nikola Tesla,* New York: The Electrical Engineer, 1894
Seifer, Marc J., *The Life and Times of Nikola Tesla: Biography of a Genius,* Secaucus, New Jersey: Birch Lane, 1996
Wagner, John, "Nikola Tesla: The First Radio Amateur," *Amateur Radio Today* 73 (December 1995)

Theater Guild on the Air

Radio Drama Program

After World War II, the executive officers and board members of the U.S. Steel Corporation became willing converts to anthology drama showcasing company voice advertising. Their program *Theater Guild on the Air* (the *United States Steel Hour* after 1952) helped promote the corporation's public and government relations during reconversion, the postwar period that saw the lifting of price and wage controls and intertwined negotiation of agreements with the Truman administration and the United Steel Workers of America. *Theater Guild on the Air* featured distinct entertainment and educational components. By arrangement with New York's Theater Guild, the program presented adaptations of plays that had little bearing upon corporation "messages" (intermission talks) and messages with little connection to the plays. The separation of the program's dramatic and editorial control extended to U.S. Steel executives' admirable defense of their program in an era of rampant blacklisting, while the show simultaneously provided one of radio's last examples of corporate voice advertising read by corporation officers themselves.

Among large radio sponsors using the anthology format for institutional promotion, U.S. Steel was unique in contracting for program production and, effectively, dramatic control outside of its advertising agency, Batten, Barton, Durstine, and Osborn (BBD&O). Program production responsibility fell to the Theater Guild. Founded in 1918, the Guild aspired to the production of plays not then found in the commercial theater. The Guild championed the work of Bernard Shaw, Eugene O'Neill, Maxwell Anderson, Elmer Rice, Sidney Howard, William Saroyan, George Gershwin, Richard Rogers, and Lorenz Hart. By 1945 a back catalog of some 200 Guild productions provided a ready source of adaptable material for the *Theater Guild on the Air.* Under contract to U.S. Steel, the Guild supplied plays and casts, retaining artistic control under managing director Lawrence Langner, whose long career as a patent attorney representing inventor Charles F. Kettering and others enabled his easy circulation in the world of corporate affairs. The broadcasts were produced by BBD&O's George Kondolf, the former director of New York's Federal Theater Project, and directed by Homer Fickett, formerly the director of radio's *March of Time* and the *Cavalcade of America.*

The autonomy enjoyed by the Guild in the selection and casting of plays, and the confinement of corporation messages to two intermissions, conformed to the broad goals of public relations education and entertainment desired by the corporation. In addition to the *Theater Guild on the Air's* "commercial aspects," explained U.S. Steel public relations director J. Carlisle MacDonald, the program's "two main objectives were (1) To create a better understanding of the affairs of United States Steel through a series of weekly, informative messages explaining the corporation's policies and describing its widespread activities; (2) To provide the nation's vast listening audience with the finest in dramatic entertainment by bringing into millions of homes every Sunday evening the greatest plays in the legitimate theater." Exemplifying the rewards of such thinking, the first season's plays ("building bigger and bigger audiences for U.S. Steel") included *I*

Remember Mama, On Borrowed Time, and *The Front Page* and featured Lynne Fontaine and Alfred Lunt, Walter Huston, Katherine Hepburn, Ray Milland, Helen Hayes, Frederick March, Pat O'Brien, and Walter Pidgeon.

The hour-long *Theater Guild on the Air* featured two intermission talks prepared by BBD&O and read by announcer George Hicks, the "Voice of United States Steel." An American Broadcasting Companies (ABC) radio newsman who had broadcast the 1944 D-Day invasion in Normandy, Hicks brought his dispassionate reportorial style to the delivery of each week's talks. The first described the policies and objectives of the umbrella corporation, and the second described the activities of a subsidiary of "United States Steel—the industrial family that serves the nation." At intermission time, announcer Hicks served up veritable chestnuts of institutional promotion: paeans to the widespread ownership of U.S. Steel corporation stock among all classes of individuals and hospitals, schools, and charitable organizations; to the re-employment of veterans; and to the upgrading and training of personnel. An anthology of plays, including two intermission talks, published in 1947 suggests the program's aspiration to low-pressure salesmanship. A talk inserted between the acts of Sidney Howard's *They Knew What They Wanted,* for example, described the latent consequences of U.S. Steel's vast scale of production, namely, the employment of men and the movement of raw materials. Striving to convey the personal meaning of it all, the text concluded, "So, next time you use any product of steel from a can opener to an automobile, remember—you are benefitting from the skills and energies of literally millions of men who have helped to transfer the raw materials from the earth into the steel out of which come many things to make our lives more comfortable."

Again and again, the U.S. Steel board of directors expressed satisfaction with their radio program, renewing it on an annual basis from 1946 through the 1952 broadcast season. Chairman of the Board Irving S. Olds and President Benjamin F. Fairless remained sold on the bifurcated production arrangement, owing in part to the corporation's prestigious association with the Theater Guild and in part to the public platform that the program provided for Fairless, who personally took to the air to explain the corporation's position during the steel strike of 1949. Anticipating U.S. Steel's move to television, and with it the improved prospect of an agency-produced show, broadcast producer BBD&O successfully lobbied to change the program's title to the *United States Steel Hour* beginning with the fall 1952 broadcast season. The program's final radio season commenced with Joshua Logan's *Wisteria Trees,* starring Helen Hayes and Joseph Cotten, and concluded with Shakespeare's *Julius Caesar,* starring Maurice Evans and Basil Rathbone.

WILLIAM L. BIRD, JR.

Hosts

Lawrence Langer
Roger Pryor
Elliott Reid

Announcers

Norman Brokenshire; George Hicks

Programming History

CBS	6 December 1943–29 February 1944
ABC	9 September 1945–5 June 1949
NBC	11 September 1949–7 June 1953 (from 1952 as *United States Steel Hour*)

Further Reading

Bird, William L., Jr., *Better Living: Advertising, Media, and the New Vocabulary of Business Leadership, 1935–1955,* Evanston, Illinois: Northwestern University Press, 1999

Fitelson, H. William, editor, *Theatre Guild on the Air,* New York: Rinehart, 1947

Fones-Wolf, Elizabeth, *Selling Free Enterprise: The Business Assault on Labor and Liberalism, 1945–60,* Urbana: University of Illinois Press, 1994

Fones-Wolf, Elizabeth, "Creating a Favorable Business Climate: Corporations and Radio Broadcasting, 1934–1954," *Business History Review* 73 (Summer 1999)

MacDonald, J. Fred, *Don't Touch That Dial!: Radio Programming in American Life,* Chicago: Nelson-Hall, 1979

Nadel, Norman, *A Pictorial History of the Theatre Guild,* New York: Crown, 1969

O'Malley, Thomas, "Every Important Work of Art Has a Message," *Television Magazine* (11 October 1954)

"Steel Melts the Public," *Sponsor* (17 March 1950)

Theatre Guild, *The Theatre Guild Anthology,* New York: Random House, 1936

"Theatre Guild Show," *Tide* (15 October 1945)

This American Life

Public Radio Program

This American Life is an hour-long, weekly public radio program that stretches the boundaries of magazine-style entertainment radio with a poignant honesty and flair rivaling, and even exceeding, the journalistic efforts of groundbreaking programs such as *Morning Edition* and *All Things Considered.*

This American Life's website describes the show as follows: "It's a weekly show. It's an hour. Its mission is to document everyday life in this country. We sometimes think of it as a documentary show for people who normally hate documentaries. A public radio show for people who don't necessarily care for public radio."

The brainchild of former National Public Radio (NPR) reporter Ira Glass, *This American Life* is largely a collection of stories that endeavor to examine America from the inside out. Not unlike Charles Kuralt's video essays gathered "on the road," *This American Life* examines America by examining the lives and challenges faced by individual Americans. It does so with a writing style that is highly conversational and uses an "audio vérité" feel, with the frequent use of natural sound (ambient background sounds) and interview segments that are sometimes raw and unpolished.

Although a team of producers, regular contributors, and guests all provide pieces for the show, the soul of *This American Life* comes from host and producer Ira Glass. A Baltimore native who resisted his parents' idea of a medical career to pursue a career in media, Glass originally sold jokes to radio hosts and eventually found a job editing promotional spots for NPR as an intern at age 19.

During his internship at NPR, Glass immersed himself in all areas of news production and reporting, also filling in as a host on NPR programs *Talk of the Nation* and *Weekend All Things Considered.* As a reporter in NPR's Chicago bureau for six years, Glass emerged as an award-winning education reporter, receiving accolades from the National Education Association, Education Writers Association, and the Harvard Graduate School of Education. His best-known work was a longitudinal profile of several Chicago Public Schools students and the successes and failures of the changes imposed on their respective schools by education reforms.

Each hour-long program of *This American Life* is divided into acts—usually three—which communicate a general theme. Each segment is accompanied by a piece of music that supports the theme—either through its title or through the lyrics.

Among the segment titles *This American Life* aired in 2000 were: "Twenty-Four Hours at the Golden Apple" (a Chicago diner, where poignant stories are told by customers who wander in and out), "Election" (dealing with a high school class election, the production of negative presidential campaign ads, and the genealogy of then candidate George W. Bush, who, the piece claims, may be related to half of Americans), and "Immigration" (which studies the impact of immigration laws on individuals, including a deported legal alien whose country would not take him back). Contributors range from independent producers whose work is heard on other public radio programs to quirky characters such as "Dishwasher Pete," who share their wisdom and stories of real life.

Distributed by Public Radio International, *This American Life* is produced by WBEZ in Chicago, where it is mixed live each Friday evening for distribution to 370 public radio stations around the country. Glass introduces each segment in his matter-of-fact style, never afraid of a lengthy pause or unconventional voice inflections. While his delivery appears somewhat unpolished, the sound of the program is anything but. He and his producers spend a great deal of time editing pieces and copy to create the "relaxed" style of the program.

Glass is responsible for many of the stories told on *This American Life. Chicago Magazine* said this about his work: "Glass does stories that are casual and intimate in feeling, that seem almost to start in the middle of the story and are told in unfolding scenes. Sort of like a hipster version of Garrison Keillor." Marc Fisher, writing in the *American Journalism Review,* says that, "Glass is the boy wonder, a rumpled genius in the minuscule world of radio documentaries, a quizzical character who hides behind trademark oversized black plastic eyeglass frames and takes radio journalism to places it has not traveled before."

This American Life is a program with a clearly humanist slant, which is expressed through compelling stories and amusing send-ups. Its innovative approach and broad appeal helped land a $350,000, three-year production grant from the Corporation for Public Broadcasting's Program Development Fund in 1997.

As one of the most listened-to public radio programs, it also helps generate significant revenue for stations that carry it during pledge drives. Glass is among the most dedicated of public radio program producers in helping stations with their fundraising, producing a number of highly effective fundraising spots for station use and creating gimmicks such as a "decoder" ring as an incentive for listeners to contribute.

This American Life joins *All Things Considered, Morning Edition, A Prairie Home Companion, Michael Feldman's Whad'Ya Know?,* and *Car Talk* as one of the leaders in public radio's stable of national programs. The show was honored with a Peabody Award in 1996.

PETER WALLACE

See also National Public Radio

Host/Producer

Ira Glass

Production Staff

Senior Producer	Julie Snyder
Producers	Alex Blumberg, Diane Cook, Wendy Dorr, and Starlee Kine
Contributing Editors	Jack Hitt, Margy Rochlin, Alix Spiegel, Paul Tough, Nancy Updike, and Sarah Vowell
Writers/Contributors	David Sedaris, Joe Richman, Scott Carrier, Gay Talese, Tobias Wolff

Programming History

PRI 17 November 1995– (remained in production as of September 2003)

Further Reading

Barton, Julia, "It Takes Vision to Make Good Radio: Tales from *This American Life,*" *Salon Magazine* (23 July 1997)

Conciatore, Jacqueline, "*This American Life:* If You Love This Show, You Really Love It," *Current* (2 June 1997)

Fisher, Marc, "It's a Wonderful Life," *The American Journalism Review* (July–August 1999)

Mifflin, Margot, "*This American Life*: American Lives, Radio Journeys," *New York Times Sunday Magazine* (7 February 1999)

Sella, Marshall, "A Profile of Ira Glass and the Show: The Glow at the End of the Dial," *New York Times Sunday Magazine* (11 April 1999)

Snyder, Rachel Louise, "Lunch with Ira Glass," <www.salon.com/people/lunch/1999/07/16/glass/index.html>

This American Life website, <www.thislife.org>

Verhulst, Kari Jo, "Fearless Curiosity: The Irreverent Offerings of *This American Life,*" *Sojourners Magazine* 28, no. 5 (September–October 1999)

Wimsatt, William Upski, "Ira Glass: A Cure for the Common Radio," *Horizon Magazine* (1 September 1999)

Thomas, Lowell 1892–1981

U.S. Radio Newscaster and Author

Lowell Thomas was one of radio's best-known newsmen from 1930 into the 1970s, but he was also known as the voice of many movie newsreels, the author of more than 50 popular books, and a world traveler and lecturer. He is important in radio history as one of the first national radio newscasters. His evening network news program ran for 46 years.

Early Career

Thomas was born in 1892 in Ohio but spent his formative years growing up in the gold-mining towns of Cripple Creek and Victor, Colorado, where his father was a mining-town surgeon. In the second decade of the 20th century, he focused on developing a career in newspaper journalism while seeking an education, eventually earning several degrees from different schools.

He was sent by the federal government's Committee on Public Information (the Creel Committee) to Europe in 1917 with photographer Harry Chase to report on World War I battlefields in Italy and then in the Middle East. Many of the costs of the trip were underwritten by a number of Chicago businesses. During this period, Thomas met and reported on the activities of T.E. Lawrence, whom other reporters had generally ignored. Thomas helped to make him famous as Lawrence of Arabia with a series of highly popular illustrated lectures in Europe and the United States entitled "With Lawrence in Arabia and Allenby in Palestine." This highly romanticized version of the war in the Middle East played before some 6 million people over six months, helping to make Thomas and Lawrence household names. His 1924 book *With Lawrence in Arabia*, based on his experiences and the lectures, became a best seller.

Thomas spent the 1920s in travel, book writing, and promotion of air travel. He made two trips to the Arctic and a steady income as a popular lecturer, drawing on his extensive travels and adventures in remote areas of the world.

Radio Newscaster

Thomas' first radio broadcast was aired on Pittsburgh's KDKA in March 1925. He talked for an hour about the 1924 around-the-world flight sponsored by the U.S. Army in which he played a small part, about which he'd written a book and on which he was then lecturing. He continued his lucrative travel lectures for another five years. He was then approached by Columbia Broadcasting System (CBS) president William Paley (who had heard him speak in London) to consider anchoring (as we would term it today—the word was not used as such then) a news program on that radio network.

After several auditions, Thomas aired his first 15-minute newscast on CBS on 29 September 1930. In a unique arrangement, CBS carried *Lowell Thomas and the News* in the western United States and the National Broadcasting Company (NBC) carried it in the eastern states for the first year. In 1931 Sun Oil became the program's sponsor and NBC its home network. Over the next several years, Thomas became known for broadcasting from many places—not just a radio studio. He broadcast from an airplane circling above New York in 1930, from a coal mine in West Virginia shortly thereafter, and later from a submarine, a ship at sea, and (after World War II) a helicopter.

He quickly developed an audience who liked his conversational approach to news and his frequent use of anecdotes and human interest stories—perhaps a dozen stories in each newscast. He would make occasional personal asides or comments but avoided any on-air position or political commentary. Thomas wrote and edited his newscasts with the help of Prosper Buranelli (a feature writer for *New York World* and an editor at Simon and Schuster) and Louis Sherwin (a respected drama critic and columnist). His popularity with listeners and advertisers allowed him to negotiate with NBC for permission to broadcast from many remote sites as he continued to travel. He also frequently broadcast from a studio on his estate near Pawling, New York.

By 1936 Thomas' entertaining approach to news reached up to 20 million listeners each weeknight. By 1940, his audience had dropped to perhaps half that size, as many other newscasts had become available, and declined to about 8 million by 1947 when Procter and Gamble brought his weeknight news program back to CBS. In mid 1939 his radio news program was simulcast on W2XBS, the CBS experimental television station in New York, with Sun Oil as sponsor (this is said to have been the first regularly scheduled TV news broadcast). Thomas undertook regular television newscasts in 1943 and 1944, but always preferred radio, which allowed him to broadcast from a variety of places. His weeknight program ran 15 minutes for years, but by the 1970s, with the demise of most network programming, it had been reduced to six minutes.

Thomas helped cover the national political conventions of 1952, 1956, and 1960 for CBS. He retired with a final broadcast on 14 May 1976 at age 84 but continued to travel and write until his death five years later. Unique at the start of his career, he was outclassed and bypassed by a younger generation of war-trained radio correspondents during World War II. Still, he retained his popularity virtually to the end of his long radio (and television) career.

The prestigious personal Peabody Award given to Thomas in 1973 noted in its citation:

> To Lowell Thomas, twentieth century Marco Polo, a special, personal George Foster Peabody Award in recognition of his incredible 43 years of continuous daily broadcasts on CBS, often originating from every corner of the globe. During this record-setting series which has become the longest continuous run in network history, Lowell Thomas' voice has been heard by an estimated 70 billion persons. During his brilliant career, he has received 25 degrees from universities and colleges. For his authoritative voice and his friendly 'So Long . . .' he has become beloved by listeners of every age in every place. To Lowell Thomas, a Peabody Award in recognition.

Unfortunately, and somewhat strangely, Thomas' two-volume autobiography (*Good Evening, Everybody,* 1976; *So Long Until Tomorrow,* 1977) says little about his broadcasting career. Thomas was never a radio news heavyweight, providing a light touch in his newscasts rather than serious commentary. Although he traveled extensively, he was not a foreign correspondent as those are thought of today—his travels were reflected in his early lecture tours and his books, although not in his broadcasts, which focused on the events at hand. Only in a short-lived series (*High Adventure with Lowell Thomas*) on CBS television in the late 1950s did his programs focus on his travels.

In addition to his broadcast work, Thomas had narrated a host of motion pictures over the years, chiefly acting as the voice for the *Movietone News* series produced by the Fox Studios. In the 1950s, he also provided the voice (very well recognized by then) for early feature films shot in the wide-screen Cinerama process. Convinced that the new format was an important breakthrough, he was an investor and officer in one of the companies making the new movies. Thus he was able to film his commentary for *This Is Cinerama* (1952) in a specially built office (also used for his daily broadcasts) on his estate in Quaker Hill, New York. He also voiced *Cinerama Holiday* (1953) and several later titles, most of which paralleled his own adventurous traveling life.

CHRISTOPHER H. STERLING

Lowell Thomas
Courtesy CBS Photo Archive

Lowell Jackson Thomas. Born in Woodington, Ohio, 6 April 1892, first child of Harriet Wagner and Harry George Thomas, both teachers; his father later studied medicine and became a mining-town surgeon. Spent early years (1900–1907) in Cripple Creek and Victor, Colorado, gold-mining camps. Attended University of Northern Indiana (now Valparaiso) and graduated in two years (1911). Reporter, *Cripple Creek Times*; editor, *Victor Daily Record.* Earned second bachelor's degree and a master's degree from University of Denver while working on Denver newspapers. Reporter for *Chicago Journal* while attending Kent College of Law, 1912–14. Master's degree, Princeton University, 1916. World War I correspondent, 1917–1918. Travel and lectures, 1918–1930. CBS news division, 1930; NBC news division, 1931–47; CBS news division, 1947–76. Retired from radio, 1976. Received 30 honorary doctorates and innumerable awards, including the Peabody (1973) and the Medal of Freedom (1976). Died in Pawling, New York, 29 August 1981.

Selected Media Highlights

Radio
Lowell Thomas and the News (exact title varied), CBS and NBC, 1930; NBC, 1931–47; CBS, 1947–76 (weeknights)

Television
High Adventure with Lowell Thomas, CBS Television, 1957–59; rebroadcast June–September 1964

Motion Pictures
"Movietone News" (newsreels), Fox Studios, narrator, 1933–50
Cinerama films, narrator, 1951–55

Selected Publications
With Lawrence in Arabia, 1924
The First World Flight, 1925
European Skyways: The Story of a Tour of Europe by Airplane, 1927
Raiders of the Deep, 1928
The Untold Story of Exploration, 1935
Magic Dials: The Story of Radio and Television, 1939
History as You Heard It, 1957
Good Evening, Everybody: From Cripple Creek to Samarkand, 1976
So Long Until Tomorrow: From Quaker Hill to Kathmandu, 1977

Further Reading
Bowen, Norman R., editor, *Lowell Thomas: The Stranger Everyone Knows,* Garden City, New York: Doubleday, 1968
Crouse, R., "Yes, There Is a Lowell Thomas," *Reader's Digest* (April 1961)
Fang, Irving, "Lowell Thomas," in *Those Radio Commentators!* Ames: Iowa State University Press, 1977
Harris, Eleanor, "The Stranger Everyone Knows: Lowell Thomas," *Reader's Digest* (December 1948)
Levine, Brian H., *Lowell Thomas's Victor: The Man and the Town,* Colorado Springs, Colorado: Century One Press, 1982
Thomas, Lowell, "So Long Until Tomorrow," *Redbook Magazine* (August 1948)
Wecter, D., "Hearing Is Believing," *Atlantic Monthly* (July 1945)

Thomas, Rufus 1917–2001

U.S. Radio Personality

Rufus Thomas was one of the most colorful radio personalities and entertainers of the 20th century. Thomas' training on the American stage in the 1930s and 1940s as a dancer, singer, and comedian catapulted him into broadcasting in the early 1950s. By the turn of the 21st century he had become a highly respected broadcaster and legendary recording artist admired by fans on both sides of the Atlantic Ocean.

Thomas was born to Rachel and Rufus Thomas Sr. on 26 March 1917 in rural Cayce, Mississippi. Two years later the elder Thomas moved his family to Memphis and worked several jobs to support his wife and four children. At age six Thomas was inspired to pursue a career in show business after he danced on a Beale Street theater stage in the role of a jumping frog for his elementary school play. That experience led Thomas to tap dancing, and he set his goal of becoming the world's best. Later, many of his teachers at Booker T. Washington High School attempted to dissuade him, but Nat D. Williams, his history teacher and a Beale Street emcee, encouraged Thomas.

Williams became Thomas' mentor, but it was Thomas who taught Williams how to tap-dance.

In 1936, after graduating from high school and enduring one economically depressing academic term at Tennessee State University in Nashville, Thomas joined the famous Rabbit Foot Minstrels. He spent three seasons touring with the acclaimed troupe and danced with Johnny Dowdy as part of "Rufus and Johnny." Upon leaving the Minstrels around 1939, he returned to Memphis and began to forge a career on Beale Street. However, as the audience for tap dancing had begun to diminish, Thomas refocused his efforts to writing blues songs for nightclub singers. Thomas commenced his own singing career on Beale Street when a female blues singer did not appear for her show at the Elks Club. He filled in for her on the program and performed "Mr. Jelly Roll," a Lonnie Johnson tune: "She said, Mr. Jelly Roll baker let me be your slave. Be your good jelly I'll rise from my grave." The audience's enthusiasm surprised Thomas; spectators applauded and tossed coins, and Thomas was presented on stage with a $5 tip coupled with a kiss from a female patron.

By the early 1940s Thomas had made singing a permanent part of his repertoire. He performed with the Bill Ford Band, and worked with the Al Jackson Sr. Band on Beale Street. During the 1940s he rose through the ranks to become a complete entertainer. Always a natural comic, he expanded his skills by forming the comedy team of "Rufus and Bones" with Robert Couch. Under the influence of his mentor, Nat D. Williams, who hosted *Amateur Night from the Palace* on Wednesdays, Thomas perfected his own emceeing skills.

In 1951, three years after Nat D. Williams became a radio disc jockey and helped rescue WDIA from financial ruin, the station hired Thomas. Initially, he imitated a more stoic or "white announcers" style that did not project his personality. He was close to being terminated, but David James Mattis, WDIA's white program director, encouraged him to relax, have fun, and showcase his entertainment background and skills. Thomas credited Mattis with helping him to eschew the anger and hatred that had accrued in response to white-imposed segregation practices and enjoy the new WDIA environment. "After that," Thomas recalled, "I took off like a late freight!"

A relaxed Thomas unleashed his entertainment personality to combine rhythm and blues music with classic Beale Street patois: "I'm loose as a goose and full of juice, so what's the use?" His comedy, up-tempo pace, and rhyming style helped secure WDIA's historical significance as the nation's oldest full-time African-American–oriented broadcasting operation. Equally legendary is the teamwork Rufus Thomas and Nat D. Williams displayed Saturday afternoons on *Cool Train.* Williams hosted the first hour solo, followed by Thomas' own one-hour show. In the third hour they teamed up for pure original vaudeville jokes, laughter, and music (Cantor, 1992).

Thomas made his recording debut in 1943 with "I'll Be a Good Boy and I'm Worried" for Texas-based Star Talent Records, but sales were unimpressive. A decade later, at age 35, his recording of a Sam Phillips song, "Bear Cat," empowered the newly established Sun Records by hitting number 3 on *Billboard*'s Rhythm and Blues chart. His national acclaim quickly attenuated after rhythm and blues writers Jerry Lieber and Mike Stoller successfully sued Phillips for copyright infringement on their original "Hound Dog" composition. Thomas' second release on Sun was unsuccessful. Phillips released him, along with other black artists such as Little Milton and Junior Parker, and began recording whites, such as Elvis Presley, who sounded black. Thomas continued to work at WDIA and perform with his band on Beale Street and throughout the mid-South.

In 1959 Thomas wrote "Cause I Love You" and recorded it as a duet with his teenage daughter, Carla, for another newly established company, Satellite Records. The recording became a Southern hit, earned the nascent Memphis label a national distribution contract with Atlantic Records, and helped transform the company into the legendary Stax Records. Then, in 1962, while singing with his band, Thomas experienced a musical epiphany. He noticed a sexy female dancer engrossed in the spirit of a new dance called "The Dog" and performing directly in front him on the floor-level bandstand. In mid-performance he made a segue into free style rap, ad-libbed, and wrote the song "The Dog" on the spot.

By January 1963, Rufus Thomas' composition and recording of "The Dog" reached number 22 on the *Billboard* chart. His derivative mantra, "Walking the Dog" rose to number 5 on the *Billboard* chart in the fall of 1963. Thomas, then 46 years old, finally retired from his full-time day job as a boiler attendant at American Finishing Company, but not from WDIA. Between 1963 and 1971 he ran off a string of hits that included "(Do the) Push and Pull, Part 1," which went to number 1, and "Breakdown, Part 1," which rested at number 2. Overall, the popular Memphis disc jockey achieved 12 *Billboard* hits between 1953 and 1971.

By the mid–1970s, serious events had shaken Memphis and WDIA such as the civil rights movement, the James Meredith march and shooting, Martin Luther King Jr.'s assassination, and integration's impact on rhythm and blues music. Thomas was affected, too. New WDIA ownership and management did not fully appreciate him or the environment that Mattis had fostered at the station. Mattis left, and Thomas was eased off WDIA as an air personality and assigned to conducting tape delayed interviews. He took a position at cross-town rival WLOK radio for a brief period and then retired to concentrate on music and entertainment. His heart remained at WDIA.

In 1986 WDIA invited Thomas back to the station to make a Saturday morning guest appearance on the *All Blues Show,* in conjunction with the station's 38th anniversary celebration.

Original disc jockeys from the 1940s and 1950s were to be paired with WDIA's current on-air staff. Thomas was teamed with Jay Michael Davis, who had never heard Thomas on the air. Davis found himself brilliantly playing the straight man against Thomas' legendary personality and quick wit. The two men bonded instantly, and Davis asked Thomas back the next week and repeated the invitation until Thomas returned each week as de facto cohost. Thomas worked each Saturday morning call-in music show for a solid year without compensation. Said Thomas, "WDIA is more than a radio station, it is an institution!"

The Rufus and Jay Michaels combo went to number 1 in the market among 35 stations in Memphis, and Thomas was hired a second time by WDIA. The *All Blues Show* remained number 1 on Saturdays for nearly 15 years. At age 82, Thomas was still broadcasting with Davis, even from remote sites when he traveled around the world to perform as a famous recording artist.

In the mid 1990s, Porretta, Italy expanded Rufus Thomas Park and named it Rufus Thomas Camp in honor of his enormously successful annual appearances. In June of 1997 a star-studded gala with blues and rhythm and blues artists such as B.B. King and Millie Jackson saluted him in thanks and praise. The City of Memphis named a street in his honor, and The American Society of Composers, Authors, and Publishers (ASCAP) Foundation honored Thomas with its Lifetime Achievement Award.

Thomas died in December 2001. People lined the street and applauded as the hearse slowly carried his remains down Beale Street.

LAWRENCE N. REDD

Rufus Thomas. Born in Cayce, Mississippi, 26 March 1917; moved with his family to Memphis, Tennessee, 1919. Began performing at age six. Attended Booker T. Washington High School; graduated 1936. Performed as dancer, singer, and song-writer on Beale Street, from 1936. Began working at WDIA, Memphis, 1951; remained at WDIA off and on for the rest of his life. Helped found Stax Records, early 1960s. Died in Memphis, 15 December 2001.

Further Reading

Barlow, William, *Voice Over: The Making of Black Radio,* Philadelphia, Pennsylvania: Temple University Press, 1999

Cantor, Louis, *Wheelin' on Beale,* New York: Pharos Books, 1992

Thomas, Rufus, and Carla Thomas, *The Best of Rufus Thomas (Do the Funky Somethin'),* sound recording, produced by Steve Greenberg, Los Angeles: Rhino Records, 1996

"Tokyo Rose" (Ikuku Toguri D'Aquino) 1916–

Japanese (U.S.-Born) Wartime Radio Propagandist

No one person specifically identified as "Tokyo Rose" broadcast Japanese World War II propaganda. But after the war, one of several female propaganda broadcasters, U.S.-born Ikuku Toguri, was identified and achieved notoriety as *the* "Tokyo Rose." From 1943 to 1945, she had broadcast messages directed toward U.S. troops fighting Japan in the South Pacific for Radio Tokyo. The term *Tokyo Rose,* however, appears to have been a creation of those troops, as several studies found no trace of the name being used in the actual broadcasts.

Origins

Toguri was born in Los Angeles on the Fourth of July 1916. Her father had come to the United States from Japan in 1899 and married in 1907. His wife immigrated to the United States in 1913 and the family moved to Los Angeles. Toguri used the first name of "Iva" during her school years in Calexico and San Diego (where her father tried farming). She attended high school and junior college in Los Angeles, where her father had become a successful importer. Toguri then received a zoology degree from UCLA in January 1940 and continued with graduate work in pre-med for another six months.

On 5 July 1941, Toguri sailed for Japan from San Pedro, California; she later gave two reasons for her trip: to visit a sick aunt and to study medicine. That September she appeared before the U.S. vice consul in Japan to obtain a U.S. passport (she had only a birth certificate), stating that she wished to return to the United States for permanent residence there. Because she lacked a passport, her application was forwarded to Washington for consideration, but war intervened before the passport could be issued.

After the December 1941 Japanese attack on Pearl Harbor, Toguri applied for repatriation to the United States through the Swiss legation in Japan but later withdrew the application, indicating that she would voluntarily remain in Japan for the war's duration. Meanwhile she had enrolled in a Japanese language and culture school (having grown up in America, she spoke English far better than Japanese, which she barely understood). For about 15 months beginning in mid 1942, Toguri worked as a typist for the Domei News Agency in Tokyo. In August 1943 she obtained a second typing position with Radio Tokyo. This latter position led to the role for which she became famous.

"Tokyo Rose"

In November 1943, Toguri began her brief career (which would eventually result in her conviction for treason in the United States) as a broadcaster for Radio Tokyo. There were several on-air hostesses for the program known as the *Zero Hour* that became part of Japanese psychological warfare designed to lower the morale of U.S. soldiers in the Pacific theater. *Zero Hour* was broadcast daily (except Sundays), from 6:00 to 7:15 P.M. Tokyo time. Toguri was variously introduced as Orphan Ann, Orphan Annie, "Your favorite enemy Ann," or "Your favorite playmate and enemy, Ann," but never as Tokyo Rose. (She apparently adopted the "Ann" name from the abbreviation for "announcer" which appeared on her scripts.)

A typical program in October 1944 began, "Hello, boneheads. This is your favorite enemy, Ann. How are all you orphans of the Pacific? Are you enjoying yourselves while your wives and sweethearts are running around with the 4F's in the States? How do you feel now when all your ships have been sunk by the Japanese Navy? How will you get home? Here's another record to remind you of home." And with that, the band music that had begun the program (and which made it so popular with its soldier listeners) resumed. Toguri was on the air for about 20 minutes of each program, during which she made comments similar to those noted above and introduced popular records of the day. The rest of the program consisted of news items from the United States and general news commentaries by other members of the broadcasting staff.

It was not until early 1944 that Toguri became aware that U.S. troops had given her—and the other Japanese women broadcasting over Radio Tokyo—the "Tokyo Rose" title. She was the only U.S. citizen given that nickname; as far as is known, the others were all Japanese citizens (and thus were never tried after the war). Reportedly, Toguri was proud of the nickname. On 19 April 1945, she married Felipe D'Aquino, a Portuguese citizen of Japanese-Portuguese descent; their marriage was registered with the Portuguese consulate in Tokyo. The new Mrs. D'Aquino did not renounce her U.S. citizenship, nor did she discontinue her *Zero Hour* broadcasts despite apparent repeated warnings by her husband. (They were separated in the postwar confusion, though they remained in touch by letter, not divorcing until 1980.)

Postwar Trials

After Japan's surrender in August 1945, the U.S. Army arrested D'Aquino as a security risk, and she was held in various Japanese prisons until her release later that year. After further research, some ill-advised admissions by her to the press about her wartime role, and inflammatory stories by columnist Walter Winchell and others, she was again arrested in September 1948. She was then brought to the United States to stand trial for treason, "for adhering to, and giving aid and comfort to" Japan during the war.

The Federal Bureau of Investigation (FBI) took several years to probe D'Aquino's activities. Hundreds of former members of the U.S. military who had served in the South Pacific during World War II were interviewed; forgotten Japanese documents were unearthed; and six recordings of D'Aquino's broadcasts believed to have been destroyed were discovered. D'Aquino's trial began in San Francisco on 5 July 1949, ending 61 days later on 29 September, when the bitterly divided jury brought in a verdict of guilty on one of the counts against her, after four days of debate. The trial was said to have cost the federal government about a half million dollars and the transcript of the proceedings ran to more than a million words. Of the 46 government witnesses, 16 were brought from Japan, where they had been interviewed originally by the FBI; 26 witnesses appeared for the defense. On 6 October 1949, D'Aquino was sentenced to ten years of imprisonment, fined $10,000 for treason, and stripped of her U.S. citizenship. She had become only the seventh person in U.S. history to be convicted of treason.

On 28 January 1956, D'Aquino was released from the Federal Reformatory for Women (Alderson, West Virginia), where she had served six years and two months of her sentence. She successfully fought several government efforts to deport her and went to work for her father's store in the Chicago area. She later operated an oriental gift shop there. In 1971 a U.S. district judge held that she still had to pay the remaining $5,255 of her fine. In November 1976, D'Aquino filed a third petition seeking a presidential pardon (she had previously applied unsuccessfully in 1954 and 1968). This time, one of her supporters was the foreman of the jury that had convicted her in 1949. On 19 January 1977, President Gerald Ford issued her a full pardon.

CHRISTOPHER H. STERLING

See also Axis Sally; Lord Haw Haw; Propaganda by Radio; World War II and U.S. Radio.

Tokyo Rose (Ikuku Toguri D'Aquino). Born in Los Angeles, California, 4 July 1916. First of four children of Jun Toguri, an immigrant farmer and later successful importer, and Fumi Iimuro; grade school and high school in California; attended University of California, Los Angeles, 1936–40, graduated with zoology degree; six months of pre-med graduate work, UCLA, 1940; traveled to Japan, July 1941; employed as typist by Domei News Agency and Radio Tokyo, 1942–43; one of several hostesses of *Zero Hour* propaganda broadcasts on Radio Tokyo, 1943–45; married Felipe D'Aquino, 1945; arrested 1945, but freed; rearrested 1948 and tried for treason in San Francisco, California, 1949; convicted and imprisoned, 1950–56; employed as store clerk, Chicago, Illinois, 1956–1980s; granted presidential pardon, 1977.

Further Reading

De Mendelssohn, Peter, *Japan's Political Warfare,* London: Allen and Unwin, 1944; reprint, New York: Arno Press, 1972

Duus, Masayo, *Tokyo Rozu,* Tokyo: Simul Press, 1977; as *Tokyo Rose: Orphan of the Pacific,* translated by Peter Duus, Tokyo: Kodansha, 1979

Howe, Russell Warren, *The Hunt for "Tokyo Rose,"* Lanham, Maryland: Madison Books, 1990

Pierce, J. Kingston, "They Called Her Traitor," *American History* 37 no. 4 (October 2002)

Top 40. *See* Contemporary Hit Radio Format/Top 40

Topless Radio

Multiple sexual partners, methods of self-gratification and the pleasuring of others, odd sexual proclivities: though these may sound like some of the recurring topics of shock jocks like Howard Stern, they are actually examples of the hot topics discussed three decades ago on radio. The format of such programs became known as "topless radio."

Similar to much of today's "adult talk" radio and TV, topless radio was a format in which audience members called in to discuss graphically sexual issues with hosts who tried to titillate the audience by teasing every explicit detail out of a caller. Although a predecessor, and perhaps an ancestor, of today's "adult" radio, topless radio initially began as quite a different format and was certainly targeting an entirely different audience.

Origins

Topless radio's humble beginnings in the United States date back to the late 1960s, when some AM talk programs began to experiment with light, humorous discussions about relationships with female callers—aimed at younger female listeners. FM radio stations, with their higher-quality stereo signal, had begun replacing AM stations as the place of choice to listen to popular music. As traditional talk radio began to fill up the AM airwaves, female listeners tuned out. The new format was an attempt to bring younger female listeners to a format (talk) that attracted predominantly older listeners. Program hosts would ask female listeners to call in to have a candid discussion about "relationship issues." Up until then, radio had carefully avoided direct reference to sex—and innuendo was often dealt with swiftly by the Federal Communications Commission (FCC) with "cease and desist" orders. The medium, and the FCC's oversight of it, lagged behind television, print, and film of the era in terms of dealing with explicit subject matter.

The first topless radio programs required callers to phone the station the night before a program aired. Hosts would discuss topics with callers off the air and edit together a program for later broadcast. Compared to books, film, and even television of the time, the resulting programs were considered to be quite tame. Despite that, the format was considered somewhat risqué by the extremely conservative radio standards of the day. More important, producers felt the shows sounded "canned" and dry. So in 1971, KCBS in Los Angeles began experimenting with live discussions of sex by women callers that were aimed at female listeners. A male all-night disc jockey for the station, Bill Ballance, hosted the midday show, *Feminine Forum*.

Topless radio was an instant success and quickly spread across the nation. By 1973 there were 50 to 60 stations that allowed only women to call in and talk about the predetermined topic of the day. As the format became more popular and spread to other stations, the content became more explicit. Truly talented hosts were able to draw extremely detailed and explicit answers from their callers. Naturally, listenership grew dramatically.

Complaints to the FCC were also on the rise. As a result, the commission announced that it did not consider topless radio to be in the public interest, as prescribed by the Communications Act of 1934, and the FCC threatened to take action if the industry did not police itself. FCC Chairman Dean Burch considered the format "prurient trash" and did not feel that the format was broadcasting in the public interest, convenience, or necessity. Further, he did not feel that the First Amendment protected broadcasting discussions of this sort in such an easily accessible medium—a medium particularly available to children. Despite these warnings, topless radio programming did not change.

In 1973 the FCC announced its intention to fine WGLD-FM in Oak Park, Illinois, $2,000 based on two individual excerpts from a show called *Femme Forum*. This was the stiffest penalty then available under the Communications Act of 1934. The declaration did not go without dissent. Two organizations, the Illinois Citizens Committee for Broadcasting and the Illinois Division of the American Civil Liberties Union, along with one FCC member, complained that the ruling was outside the purview of the FCC and went against the organization's goal to maintain broadcasting in the public interest. They stated that the ruling would have a chilling effect on the discussion of important public issues and that, taken as a whole, the content of topless radio programming (specifically *Femme Forum*) was *not* patently offensive by community standards.

Hoping that this would be a test case of the FCC's ability to fine stations based on the commission's perceptions of the obscenity or indecency of the programming, the agency invited WGLD's parent company, Sonderling Broadcasting, to take the case to court. However, Sonderling, stating that they could not afford the cost of testing such broad constitutional issues in the legal arena, paid the fine instead, and the FCC was denied a judicial declaration of its ability to police radio decency. Despite the lack of a court ruling, the FCC achieved its goal. Not only did Sonderling pay the fine, they also canceled their sex-talk show. Indeed, such shows nationwide were canceled or drastically restructured after this event.

Topless radio was quickly banished. Thanks to the Sonderling fine and similar cases over the ensuing years—particularly the "Seven Dirty Words" case in 1978—the FCC managed to keep references to sex on radio primarily limited to risqué jokes and somewhat suggestive song lyrics. However, the FCC was not able to keep this format off the air for long. Not only did sex talk on the radio return, it evolved into a variety of forms, showed up in a number of parts of the day, and sought out multiple audiences. Particularly important were shifts in the regulatory focus of the FCC from behavioral regulation to allowing marketplace competition to "police" the actions of stations. In 1980, Dr. Ruth Westheimer began her serious but frank discussion of sex on local New York radio. In the early 21st century, Dr. Laura Schlessinger's nationally syndicated program dealt with moral and ethical discussions of relationships, sometimes resulting in discussions of sexual behavior and choices.

Another offshoot of topless radio is exemplified by Howard Stern—the self-proclaimed "King of All Media." In the mid-1980s, Stern and several other national and regional hosts stretched the limits of "patently offensive" to the breaking point—dealing with religion, politics, race, and naturally sex in a manner many consider particularly juvenile. Unlike earlier programming, shock radio sought out the lucrative male 18-to-49 demographic. These programs caught and held the attention of their audiences with guests from the porn industry, celebrity feuds, off-color phone pranks, stripping on the air, outlandish phone-in contests, and alternative dating games. Surprisingly, corporations backing this type of radio have managed to forestall significant FCC censure—in many cases simply paying massive fines after stalling the organization for a number of years. As the format cannot advance much further than it has, it appears to have simply spread into other parts of the day. Not only is this format aired at night, it has actually become most popular in evening and morning drive times.

Although the antics in this format have escalated since the early 1970s, topless radio may have helped usher in the new era of explicit radio discussions of sex.

PHILIP J. AUTER

See also Censorship; Controversial Issues; Federal Communications Commission; Licensing; Obscenity/Indecency on Radio; Seven Dirty Words Case; Shock Jocks; Stern Howard

Further Reading

Carlin, J.C., "The FCC versus 'Topless Radio,'" Master's thesis, University of Florida, 1974

Pierce, J. Kingston, "How Do Local Stations Boost Ratings? Give Seattle What It Wants—Sex in the Morning," *Seattle Weekly* (8–14 April 1999)

Pierce, J. Kingston, "Radio Raunch," <www.seattleweekly.com/features/9914/features-pierce.shtml>
Stern, Howard, *Private Parts,* New York: Pocket Star, 1994
"Touchiest Topic on Radio Now: Talk about Sex," *Broadcasting* 118 (19 March 1993)

Totenberg, Nina 1944–

U.S. Reporter and Legal Affairs Correspondent

Most of Nina Totenberg's radio reporting career has been focused on the Supreme Court and the other highest levels of America's legal system, including the investigations that have shaken presidents and the role of Congress in legal affairs. She may be best known for breaking stories that helped derail or disrupt the confirmations of Supreme Court nominees. In the process, Totenberg has made powerful enemies and won most of the top awards in broadcasting.

Born in New York City and reared in nearby Scarsdale, New York, Totenberg is the eldest of three daughters of concert violinist and music educator Roman Totenberg. She attended Boston University but left in 1965 to take various newspaper jobs until 1968, when she moved to Washington, D.C., and landed a job on the now-defunct *National Observer.*

While at the *Observer,* Totenberg wrote a profile of J. Edgar Hoover that so enraged the Federal Bureau of Investigation (FBI) director that he tried to have her fired. She recalls it as "the first time a credible news organization wrote a profile of Hoover that was neither a fan letter nor a hatchet job." At the *Observer,* Totenberg began to develop her interest in legal affairs, especially in the background of decisions at the Supreme Court.

In 1973 Totenberg moved on to *New Times,* an irreverent and short-lived national journal, where she created a stir on Capitol Hill with an article called "The Ten Dumbest Members of Congress." One of the men profiled, Senator William Scott of Virginia, compounded the publicity by holding a news conference in which he denied that he was the dumbest.

Totenberg went to work for National Public Radio (NPR) in 1974, learning the basics of radio production from colleagues as she perfected her legal research skills. Her persistent and aggressive reporting style won the admiration of many Washington news people but the ill will of those who were used to thinking of reporting from the Capital as an all-male club. "When I started," Totenberg recalls, "I was pretty much the only girl, and I thought the way to succeed was to be tough as nails. Over the years I've mellowed, but I'm also not the only girl anymore."

Linda Wertheimer, cohost of NPR's *All Things Considered,* ascribes Totenberg's success to hard work and persistence: "She'd do a tremendous amount of research on the whole Supreme Court docket before each session, so she'd go into all those cases knowing a lot about them." Wertheimer, who shared office space with Totenberg in the early years, also recalls that Totenberg was "dogged and tenacious when it came to following up leads. She just wouldn't take no for an answer."

Iran-Contra Special Prosecutor Lawrence Walsh says Totenberg has cultivated a wide network of sources over the years because of "her absolute honesty and trustworthiness" and because she is always imaginative in seeking out people to question.

Her imagination and persistence paid off in 1987, when Totenberg broke the story that Supreme Court nominee Douglas H. Ginsburg had openly smoked marijuana in the 1970s when he taught at Harvard Law School. Totenberg interviewed people who knew Ginsburg at the time, including former students and colleagues. "I was there before the FBI was," she recalls, "and I'm not sure they would have asked." The disclosure embarrassed the Reagan administration, which had promoted the federal appeals judge as a strict upholder of the law. Shortly after Totenberg's report, Ginsburg withdrew his name from consideration.

Totenberg's national fame stems from scoops such as the Ginsburg case and an even bigger one: uncovering sexual harassment allegations against Supreme Court nominee Clarence Thomas in 1991. Thomas' supporters were furious that the allegation had leaked. Senator Alan Simpson of Wyoming attacked Totenberg's integrity when the two appeared together on American Broadcasting Company's (ABC) *Nightline,* and the two had an angry exchange outside the studio that was widely reported. The *Wall Street Journal* ran an editorial accusing Totenberg of being fired from the *National Observer* for plagiarism 19 years earlier. In interviews with *Vanity Fair* and other journals, Totenberg has said that she did copy quotes for a story, calling it "a stupid mistake," but she maintains that she left the *Observer* because of sexual harassment from a supervisor. In the end, the Senate narrowly confirmed Thomas' nomination. Totenberg and Timothy Phelps were questioned by the Senate's general counsel but refused to reveal their sources, and Senate leaders declined to pursue contempt citations against them.

Although the more sensational stories have brought her fame, Totenberg has spent most of her career navigating the tamer twists and turns of Supreme Court arguments and congressional investigations. NPR editor Barbara Campbell says Totenberg's forte is summarizing the arguments in Supreme Court cases. "Her paraphrase can often tell you what's going on more succinctly than the speakers themselves, and she gives you the real flavor of the argument." Totenberg often recites portions of the dialogue among the justices and lawyers to show their thinking as it evolves. As she puts it, "The best thing I do is the everyday explanatory work of covering the law and making it interesting and understandable to people who might not otherwise listen, and at the same time, have lawyers say 'she got it right.'"

COREY FLINTOFF

See also All Things Considered; Morning Edition; National Public Radio; Wertheimer, Linda; Women in Radio

Nina Totenberg. Born in New York City, 14 January 1944, one of three daughters of violinist Roman Totenberg and Melanie (Shroder). Student at Boston University; reporter, *Boston Record American,* 1965, *Peabody Times,* 1967, *National Observer,* 1968–71, *New Times,* 1973; National Public Radio, 1974–present, *Inside Washington,* 1992, and American Broadcasting Company's (ABC) *Nightline,* 1993–present; covered Watergate Trials, Supreme Court nominations of G. Harrold Carswell, Douglas Ginsburg, Robert Bork, and Clarence Thomas, Iran-Contra and Whitewater investigations, and impeachment of President Clinton. Alfred I. duPont Award from Columbia University, George Foster Peabody Award, George Polk Award, American Judicature Society Tony House Award for outstanding legal coverage (first recipient), National Press Foundation Award for outstanding broadcast journalist of 1999, Joan Barone Award, Silver Gavel Award, Woman of Courage Award from American Women in Film, and Athena Award.

Further Reading

Bardach, Ann Louise, "Nina Totenberg: Queen of the Leaks," *Vanity Fair* 55, no. 1 (January 1992)

Collins, Mary, *National Public Radio: The Cast of Characters,* Washington, D.C.: Seven Locks Press, 1993

Looker, Thomas, *The Sound and the Story: NPR and the Art of Radio,* Boston: Houghton Mifflin, 1995

Mayer, Jane, and Jill Abramson, *Strange Justice: The Selling of Clarence Thomas,* Boston: Houghton Mifflin, 1994

Stamberg, Susan, *Every Night at Five: Susan Stamberg's All Things Considered Book,* New York: Pantheon, 1982

Wertheimer, Linda, editor, *Listening to America: Twenty-Five Years in the Life of a Nation, As Heard on National Public Radio,* Boston: Houghton Mifflin, 1995

Tracht, Doug "Greaseman" 1950–

U.S. Radio Personality

Doug Tracht's alter ego, The Greaseman, is rude, crude, and politically incorrect. He is also one of the funniest and most creative air personalities on the radio. Rooted in a boss jock routine based on the music-driven radio style of the 1960s, the persona of The Greaseman led Tracht into the new shock jock arena, using humor that is often sexist and offensive to many groups. The Greaseman character attracted a loyal fan base by telling elaborate stories and jokes with himself as the central character. He opines about people and events in the news, frequently taking his humor to rude, crude, and violent extremes. To avoid trouble with the Federal Communications Commission (FCC) and to go over children's heads, he created code words for body parts and bodily functions.

Origins

Doug Tracht was born and raised in the Bronx, New York, where he lived until graduating from DeWitt Clinton High School in 1968. Tracht, a tall, skinny kid, wanted to be macho, and he realized his ambition on radio. At Ithaca College, he worked at the student radio station, vomiting the first time he went on air. Noting that while other announcers "cooked," he told listeners he cooked with grease. Tracht began developing his unique style, preparing the macho character Nino "Greaseman" Mannelli. After graduating from Ithaca College in 1972 with a broadcasting degree, Tracht developed his Greaseman persona at a string of stations. He worked in New York in 1972 for both WTKO-AM 1470 in Ithaca and WENE-AM 1430 in Endicott. He worked at WAXC-AM 1460 in Rochester, New York, from 1972 to 1974 before moving to WPOP-AM 1410 in Hartford, Connecticut, where he worked from 1974 to 1975.

Washington, D.C.'s WRC-AM 980 hired Tracht to work evenings in 1975–76. Station management was not amused and asked Tracht to drop the Greaseman or leave. He chose to go to Jacksonville, Florida, where Greaseman hit his stride, rising to infamous heights at the Big Ape, WAPE-AM 690, from 1976–81. Tracht developed Greaseman into a "God-fearing, truck-driving redneck." Greaseman became known for his bits, parodies, running gags, ad libs, and his ability to play off of phone callers. For his shows, Tracht does no advance preparation, getting ideas from callers or newspapers while on the air. Tracht's characters include the Lawman, a career he would have chosen had he not found radio. In Jacksonville, Tracht worked as a reserve police officer at night, often doing his morning drive program in full uniform complete with a .44 Magnum pistol. He later volunteered as a deputy sheriff in Falls Church, Virginia.

After leaving WAPE, Tracht returned to Washington, D.C., and WWDC-FM 101.1, replacing Howard Stern in the morning drive slot after Stern's firing. Greaseman maintained Stern's number-one rating and increased the size of the audience. Stern's show, syndicated by Infinity Broadcasting, competed with Greaseman until Tracht joined the same company.

Greaseman prospered at WWDC from August 1982 until 22 January 1993, commanding more than 10 percent of Washington's morning drive listeners. He employed numerous publicity stunts, including a mock presidential run in 1984. In 1993 he turned down a $6.5 million renewal offer from WWDC to move to Los Angeles, where Infinity Broadcasting nationally syndicated *The Greaseman Show* until 1996.

In 1997 Tracht published *And They Ask Me Why I Drink?* a collection of Greaseman stories and anecdotes presented entirely "in character," with Tracht's real name mentioned only as the copyright holder. Greaseman joined Washington's WARW-FM 94.7 in May 1997, taking over the morning drive time. Tracht was suspended and subsequently fired on 24 February 1999, after playing an excerpt from a song by African-American Lauryn Hill and having "Greaseman" say, "No wonder people drag them behind trucks." His comment referred to the recent dragging death of a black man in Texas. Tracht appeared on TV and radio, including ABC's *Nightline*, *BET's Tonight with Tavis Smiley*, and MS-NBC's *Equal Time* to apologize. Although Tracht does not have a personal reputation as a racist or bigot, he had made a similar comment about Martin Luther King, Jr.'s birthday when it was made a national holiday in 1986. Greaseman suggested "killing four more and getting the rest of the week off." He was suspended but not fired after offering an apology over Washington's WRC-TV.

Tracht has moonlighted as an actor in TV movies and appeared in the play *The Last Session* in Los Angeles during the summer of 1996. Tracht portrayed a sleazy lawyer in a 7 December 1999 installment of *The FBI Files*, a Discovery Channel docudrama series. After being fired in Washington, Tracht hosted *Matchmaker.com*, a cable television dating show using the internet to make love matches. He also did some standup comedy at clubs in Washington, D.C., and Maryland.

The Greaseman returned to the airwaves via syndication on 5 March 2001, working from a studio at "The Grease Palace" (his name for his home). On 10 July 2002, Washington, D.C.'s WGOP-AM became the flagship station for The Grease Show, which is syndicated on stations in Maryland, New York, Pennsylvania, and Florida.

W.A. KELLY HUFF

Doug Tracht. Born in the Bronx, New York, 1 August 1950. Attended Ithaca College, New York, 1968–72, majored in broadcasting; worked in radio at WTKO-AM, Ithaca, New York, 1972; WENE-AM, Endicott, New York, 1972; WAXC-AM, Rochester, New York, 1972–74; WPOP-AM, Hartford, Connecticut, 1974–75; WRC-AM, Washington, D.C., 1975–76; WAPE-AM, Jacksonville, Florida, 1976–81; WWDC-FM, Washington, D.C., 1982–93; hosted nationally syndicated *Greaseman Show*, Los Angeles, 1993–96; worked at WARW-FM, Washington, D.C., 1997–99, and fired for racial comment; host of the syndicated *The Grease Show*, from 2001.

Television Movies
Jack Reed: Search For Justice, 1994; *Jack Reed: Death And Vengeance*, 1996

Stage
The Last Session, 1996

Selected Publications
And They Ask Me Why I Drink? 1997

Further Reading
Ahrens, Frank, "Radio's Shock in Trade," *Washington Post* (27 February 1999)

Ahrens, Frank, "The Silenced Greaseman: A Year After His Racist Slur the Deejay Remains an Outcast," *Washington Post* (9 March 2000)

Greaseman.com, <www.greaseman.org>
Greaseman/Kauzmo, <greaseman.kauzmo.net/>
The Reel Top 40 Radio Depository, <www.reelradio.com/gk/index.html> (a Greasman aircheck)

Trade Associations

The American radio business has organized a variety of trade associations to both lobby government and promote radio to the general public. While the largest and longest-lasting such group has been the National Association of Broadcasters (NAB), many others have focused on more specific concerns or groups, some of them lasting only a few years. This entry details a few such groups (many others, such as NAB, have their own entry) to illustrate the variety of their concerns.

Clear Channel Broadcasting Service

Organized in 1941, the Clear Channel Broadcasting Service (CCBS) was an association of a relative handful of large AM radio station owners whose goal was to operate maximum-powered (50,000 watt) AM radio stations on "clear channels" without being subject to co-channel skywave interference from other stations, which might reduce their nighttime coverage. Although its stated mission was to conduct "an educational and promotional campaign to acquaint Congress and the members of the public with the need for clear channel stations," the CCBS—the first special-interest trade organization in the broadcast industry—sought to use every political and legal means available to protect the frequencies its members occupied.

By 1934 numerous smaller stations were already asking the FCC for permission to operate on one of the 40 clear channel frequencies. To fight for the preservation of clear channels and repel this potential incursion, and to lobby for *superpower* license grants similar to the 1934–39 experimental permit obtained by WLW, Cincinnati, Ohio, Edwin Craig of WSM in Nashville, Tennessee, organized 13 independently owned (non-network) clear channel stations into the Clear Channel Group. (Stations owned by the National Broadcasting Company [NBC] and Columbia Broadcasting System [CBS] networks were not welcome to join, because it was felt that the networks had their own agendas and were not passionate about, or necessarily even in favor of, the clear channel movement.)

The CCBS was established on 4 February 1941 and eventually became a replacement for the predecessor Clear Channel Group, because the membership was essentially the same and Edwin Craig was the singular driving force behind both organizations. The CCBS, however, took a more aggressive stance than the Clear Channel Group had and, with the support of member station contributions, opened a Washington, D.C., office and employed Vic Sholis, a former public information official for the U.S. Department of Commerce, to lead the cause.

Following World War II, the Federal Communications Commission (FCC) launched a plan to break down a number of the clear channels (by letting other stations use them) in a proceeding labeled as Docket 6741. To fight the plan, the CCBS concentrated on building alliances with farm groups and others living in rural America who would support their claims that protecting the clear channel stations was vital to providing satisfactory service to these vast areas.

Although the CCBS was able to delay action, it was not able to stop the FCC from issuing its decision. This occurred on 13 September 1961, concluding that 13 of the clear channels would be duplicated, with the assignment of one fulltime Class II (regional) station to each, with the Commission designating in what area each could be located. CCBS spent the balance of the 1960s seeking congressional reversal of the FCC action while allowing superpower status for several stations. Although it received a degree of support, no bills passed into law. By the 1970s, members' interest in the issue was waning and the organization largely abandoned further legislative battles. The climax of the fight to retain major protected status for clear channel stations came on 20 June 1980, when the FCC released its decision in Docket 20642, declaring that the nation's population would be better served by allowing even more stations to operate on clear channel frequencies than had been permitted under the 1961 decision. Thus, all remaining clear channel stations were now subject to multiple Class II stations, which could operate on what had once been "cleared" channels.

The CCBS, which represented only a limited group of radio stations, was unique as it continued to champion that cause for nearly 50 years. Although CCBS became inactive after the 1980 decision, all of the AM frequencies established as "clear channels" in 1928 are still so designated, and stations originally licensed as Class I still receive a measure of interference protection over and above all other stations on the AM band.

National Association of Farm Broadcasters

By the 1930s, as farm-related broadcasters began meeting informally at various agricultural and broadcast industry gatherings, the need for an organization that focused on their specific needs became evident. On 4 May 1944 the National Association of Radio Farm Directors was officially formed to promote more and better programming directed to American farmers. As TV stations began operating (and, with them, farm-oriented TV reports), the group became the National Association of Television and Radio Farm Directors, adding more than 100 new members, for a total membership of approximately 500. The name was later shortened to the National Association of Farm Broadcasters (NAFB).

Members became aware that to survive, farm programming needed to produce revenue for stations. New emphasis was placed on sales and the acceptance of commercials. By the 1960s the NAFB had made good progress in becoming a business-oriented organization. They set out to tell about producing results for advertisers and proving there was an audience for farm programs.

Early in 1989 the NAFB employed its first fulltime executive director, Roger Olson, who served until 1996 and was followed by Steve Pierson. Under these two men, major focus has been placed on research and obtaining both qualitative and quantitative information about the farm market. By 1998 the NAFB had produced a farm broadcasting presentation on CD-ROM for use by media sales representatives in telling the story of how radio and television continue to provide the vital information needed by farmers as they labor to feed the nation.

The National Farm Broadcast Service was created in 1992 for the delivery of information via satellite, which also made possible the exchange of news stories and interviews by members.

Radio-Television News Directors Association

The Radio-Television News Directors Association (RTNDA) was founded in March 1946 under the name National Association of Radio News Editors for the purposes of setting standards for newsgathering and reporting, exchanging ideas, and "convinc[ing] news sources that broadcast reporters were legitimate members of the journalistic profession." Later that year the name was changed to the National Association of Radio News Directors. With the onset of TV news programming, the present name (including the word "Television") was put into place in 1952. Over the years, RTNDA focused on developing both ethical and operational standards for both radio and television news departments.

RTNDA is now a worldwide organization devoted to electronic journalism in all its formats. Thus it represents station and network news executives in radio, television, cable television, and other electronic media in more than 30 countries. By the turn of the century, membership in the RTNDA totaled more than 3,000 news directors, news reporters and editors, educators, and students.

RTNDA offers professional development programs as well as programs for students of journalism and young professionals, including scholarships and short-term paid Capitol Hill internships. RTNDA also produces an extensive lineup of publications and resources to support the work of electronic journalists, including the *Communicator,* a monthly magazine devoted to reporting on new technological advances, developments in reporting and newsroom management techniques, and topics vital to news facility managers.

Many attend the Association's annual international conference and exposition held in a different city each year. An independent affiliate of the RTNDA is the Radio and Television News Directors Foundation, which promotes "excellence in electronic journalism" through research, education, and training for news professionals and students.

National Association of College Broadcasters

The National Association of College Broadcasters (NACB) is a nonprofit trade organization for student-operated radio stations. NACB is fairly young, having been founded in 1988 by students at Brown University. In their efforts to establish a student TV station at the Providence, Rhode Island-based school, the Brown students realized the need for an entity that could assist student-operated radio and TV stations in their startup efforts and provide a conduit for exchanging programming, operational, and legal information with other student outlets.

NACB exists to provide students a ready resource for advice and information and a venue for exchanging ideas and innovative concepts. Through such events as its annual National Conference, the NACB functions as a link between the academic and professional worlds of the radio/TV industry. It presents an annual awards program, where individuals and station members from the United States and other countries honor the best work in student electronic media. In addition, the NACB devotes itself to encouraging and supporting student stations and individuals in reaching for and attaining high standards so as to enhance the communities they serve, to provide opportunities for individuals with an interest in media and communications, and to argue their position on pending laws or regulations that might affect student media.

Radio Music License Committee

The Radio Music License Committee has a very specific function. Made up of broadcasters who volunteer to represent their industry's interests, the committee negotiates with the two largest musical performing rights organizations, American Society of Composers, Authors, and Publishers (ASCAP) and

Broadcast Music Incorporated (BMI), to establish acceptable fees and terms for the performance of music by commercial radio stations in the United States.

Originally known as the All-Industry Music Committee, the committee took over this function from the NAB in the early 1940s. The NAB itself had been formed (in 1923) by a group of broadcasters who found ASCAP's fee requests to be unacceptable. In 1970 it was agreed that radio and television's performance of music raised different negotiating concerns requiring different approaches. Hence, the organization divided itself into two separate entities, and the Radio Music License Committee was born. It is an independent body, having no affiliation with the NAB.

Officially, the Committee negotiates on behalf of its member stations—those who voluntarily fund it—yet, for all practical purposes, it represents most of the radio industry, because the blanket and per-performance license fee structure agreed upon applies to all stations and producers. When unable to reach agreement with either ASCAP or BMI, the Committee more than once has commenced litigation under the federally imposed consent decrees, which require both music licensing organizations to set reasonable fees and terms for radio station licenses. Generally, agreements are renewed with the two entities for a four- or five-year term.

Broadcast Measurement Bureau

The Broadcast Measurement Bureau (BMB) was a brief experiment by the National Association of Broadcasters to provide a service comparable to that to the Audit Bureau of Circulation (ABC).

Beville (1988) describes the ABC as "a nonprofit, tripartite, self-regulatory, voluntary organization established in 1914 and supported by the entire print media and advertising industry." The ABC provides independent audits of circulation figures of newspapers, magazines, journals, and internet media. In its own words, "by creating an independent currency for measuring its value, the ABC makes the sale and purchase of print, exhibition, and internet media both easier and efficient." In the audits, individuals whom the various media assert are in the audiences of a publication or exhibition are independently contacted to assure that they are indeed readers, participants, or viewers as claimed. The governing board of ABC includes representatives of national and regional advertisers, newspapers, and other organizations. Advertisers, agencies, and publications pay an annual fee to receive publications reporting the independent audits.

In 1945—in an attempt to parallel ABC methods for radio—the broadcast industry organized the BMB to conduct radio station coverage surveys. Their method involved a survey form mailed to listeners, who would verify the stations they received and listened to. A private research firm that had conducted some methodological work along these lines was engaged to conduct a national survey. Broadcaster subscriptions supported this study.

Following the apparent success of an initial survey, a second was launched three years later. It did not meet its expenses because of a lack of subscriptions by radio stations. It appeared that some stations, pleased with the results of the first study, did not wish to risk less impressive results in a second survey, while other stations, dissatisfied with the first survey, saw no reason to support a survey likely to deliver bad news a second time.

Faced with this lack of support, BMB collapsed. The NAB paid $100,000 to the researchers so that the few broadcasters who had subscribed would receive complete reports. Beville theorizes that BMB was doomed to failure, if the objective was to impress advertisers. In his view broadcaster-supported research would always seem tainted to advertisers and their agencies.

International Radio and Television Society Foundation

The International Radio and Television Society (IRTS) Foundation is a New York City-based service organization whose goal is to "bring together the wisdom of yesterday's founders, the power of today's leaders, and the promise of tomorrow's young industry professionals." The emphasis here is more on education than lobbying.

The IRTS evolved from an organization founded in 1939 when a group of radio executives began meeting informally to discuss mutual interests. Because electronic media face continual change at every level, there is a constant need for development of training and information. IRTS seeks to provide education and on-going dialogue about important communication issues.

IRTS membership includes professionals across a variety of disciplines encompassing all modes of electronic content distribution, including radio, broadcast television, cable television, computer, direct broadcast satellite, and telephony.

Each year, IRTS presents approximately 45 programs, including monthly luncheons with a newsmaker as the guest speaker, seminars, and dinners, which help fund the organization's electronic media educational programs. These include a faculty/industry seminar, where university professors meet with industry leaders in New York for five days of intense sessions; the preparation of case studies to assist communications and business school professors in their teaching of media-related topics; minority career workshops; and an annual nine-week summer fellowship program.

Many other radio organizations have appeared and faded away over the years, including those representing radio DJs, sales personnel, financial managers, promotional personnel, stations interested in providing on-air editorials, and classical music stations.

MARLIN R. TAYLOR AND JAMES E. FLETCHER

See also American Federation of Musicians; American Federation of Television and Radio Artists; American Society of Composers, Authors, and Publishers; American Women in Radio and Television; Broadcast Music Inc.; Clear Channel Stations; Farm/Agricultural Radio; FM Trade Associations; Intercollegiate Broadcasting System; National Association of Broadcasters; National Association of Educational Broadcasters; National Federation of Community Broadcasters; National Religious Broadcasters; Promax; RADAR; Radio Advertising Bureau; Technical Organizations

Further Reading

Audit Bureau of Circulation: About ABC: The History of ABC, <www.accessabc.com/sub1/history.htm>

Beville, Hugh Malcolm, Jr., *Audience Ratings: Radio, Television, and Cable*, 2nd edition, Hillsdale, New Jersey: Lawrence Erlbaum Associates, 1988

Fletcher, J.E., editor, *Broadcast Research Definitions*, Washington, D.C.: National Association of Broadcasters, 1988

Foust, James C., *Big Voices of the Air: The Battle over Clear Channel Radio*, Ames: Iowa State University Press, 2000

Trade Press

Reporting Radio's Business

The radio or broadcast trade press consists primarily of weekly magazines and newsletters (some now available via internet delivery) that serve an audience of people working within the radio industry—including broadcasters, engineers, managers, program personnel, manufacturers, investors, and others. These publications report what is happening in the radio business and related fields, technological developments and trends, regulatory actions and decisions, and information about people in the industry. They often take a strongly pro-business editorial stance. Some trade periodicals are published as a function of professional organizations, others are published by manufacturing companies, but most are advertiser-supported commercial ventures. Most focus on the American scene, but some deal with comparative or international activities as well.

The trade press excludes publications directed at the general public, including fan magazines, hobby publications, scholarly journals, radio program guides, and the like. A selection of radio-specific titles (most of them American) are highlighted here. Some lasted for only a short time and either disappeared or merged with other periodicals, but others have continued on for decades.

Origins (to 1940)

The earliest related trade periodicals served the electrical and telegraph and telephone industries in the late 19th and early 20th centuries. Journals such as *The Electrical Engineer* (1882–99, weekly) and *Journal of the Telegraph* (1867–1914, monthly) included early reports on wireless telegraphy experiments and applications. The venerable show business papers *Billboard* (1894–present, weekly) and *Variety* (1905–present, weekly) created models of the entertainment trade paper genre that would blossom in the decades to come, and both regularly covered broadcasting from its inception. For a few years from the late 1930s into the early 1940s, *Variety* published an annual *Variety Radio Directory*.

Trade publications focusing on radio broadcasting emerged as regular broadcasting began in the 1920s. Some of these early publications included information of interest to amateurs, broadcasters, manufacturers, and even audience members, but they became more focused as the industry itself began to settle into a pattern. *Radio Broadcast* (1922–30, monthly) was the most important of these early titles, combining feature articles that at first appealed to both listeners and broadcasters. Technical information concerned building and operating receivers as well as station equipment, and the magazine carried more advertising than any of the other early radio journals. The monthly provided its readers with a broad cultural understanding of radio at first, but it became more technical and aimed at industry figures later in the decade, when the masthead noted that it was published "for the radio industry."

Focusing on the equipment side, as many early trade papers did, *The Radio Dealer* (1922–28, monthly) was a pioneer aimed at retailers of radio receivers. *Radio Retailing* (1925–39, monthly) took similar aim and was also supported by manufacturer advertising. Its editor, Orestes Caldwell, eventually served as a member of the Federal Radio Commission (most early editors were radio enthusiasts, and several others emerged as key figures in the development of the radio industry).

Perhaps the most influential of radio industry trade magazines is *Broadcasting* (1931–41, biweekly; 1941–present, weekly). The creation of Sol Taishoff and Martin Codel, this was primarily a newsmagazine from the start, and it built close relationships with the Washington industry and policy community. Taishoff and Codel's combination of journalistic experience with a wide network of contacts rapidly built the magazine into an industry staple, which added an annual yearbook directory number in 1935 (which is still published). Its advertising featured major stations, network and syndicated programs, and broadcast equipment.

As the industry grew in size and complexity, weekly publication was insufficient to follow all the developments and issues. This fact gave birth to *Radio Daily* (1936–50), which began to focus more on television and became *Radio Daily–Television Daily* (1950–62), and finally *Radio-Television Daily* (1962–66), "the national daily newspaper of commercial radio and television." The daily grew in size—from four pages in the 1940s to eight pages in tabloid format with photos. (It published a long-running *Radio Annual* beginning in 1938.) No publication devoted to radio has appeared on a daily basis since, though many broader publications (e.g., *Communications Daily,* which began in 1981) include radio issues and trends.

Radio Faces Television (1940–80)

The rise of a more complex broadcast industry during and after World War II gave rise to more specialized periodicals. *FM* (1940–48, monthly) underwent several title changes as it broadened its coverage to deal with shortwave and television in addition to frequency modulation broadcasting. *Frequency Modulation Business* (1946–48, monthly) was even more focused on management issues, perhaps explaining its shorter life, given FM's quick initial postwar peak and slow demise at the end of the 1940s. Because of wartime production shortages, *Radio Retailing Today* (1942–44, monthly) had a short life. *Radio Showmanship* (began monthly publication in 1939) offered stories and features for advertisers and station executives on program ideas and promotions for different types of products.

As radio began to face fierce competition from television, previously radio-only publications expanded their coverage. *Broadcasting,* for example, became *Broadcasting Telecasting* from 1949 to 1957, devoting increasing space to television beyond that point, and *Radio Daily* focused more and more on the visual medium. *Sponsor* (1946–68, monthly) increasingly focused on television, having begun with a devotion to radio advertising. On the other hand, a growing number of specialized radio-only publications began to appear. *Inside Radio* (1976–present, weekly) is a newsletter aimed at radio executives, programmers, and syndicators that is filled with competitive tips in a no-advertising format.

Development of a host of competitive radio management and programming journals demonstrated radio's post-television comeback and the growing competition among both FM and AM stations. *Radio and Records* (1973–present, weekly), touting itself as "the industry's newspaper," built on the symbiotic relationship between recorded music and broadcasting, including widely used music playlists. Claude Hall's *The International Radio Report* (1978–present, weekly) plays a somewhat similar role, claiming "the most accurate [music popularity] charts in the world." *Radio Only* (1978–present, monthly) calls itself "the management tool" and deals with all aspects of radio management, sales, and programming. *Inside Radio* (1976–present, weekly) is "the confidential newsweekly for radio executives, programmers, and syndicators," focusing on hot news, sales tips, and ratings news. With the dramatic post-1996 ownership changes in the radio business, *Inside Radio* began to issue *Who Owns What* (weekly), a newsletter listing the merger and acquisition activity of major radio group owners.

Company Publications. Several manufacturers published their own trade journals, which sometimes rose above being mere advertising vehicles. Perhaps the first was a product of the Marconi Company, which began publishing the monthly *Wireless Age* in 1913, well before broadcasting began, and continued it until 1925. *RCA Review* (1936–85, quarterly) and *Broadcast News* (1941–68, monthly), both published by the Radio Corporation of America (RCA), largely touted company products and applications but also provided useful information on studios and broadcast equipment. The *Review* offered research papers by RCA engineers. Archival copies remain a useful way to trace station technical development and design. Philips, General Electric, and several other firms issued house organs as well, many of which focused on radio technology.

Association Publications. Issued for members of organizations, such as engineers or broadcast journalists, or for broader trade associations, these often focused on radio. Over the years, for example, the National Association of Broadcasters (NAB) issued many (usually monthly) periodicals concerning radio, including *FMphasis* in the 1960s as the FM industry began to grow again, and *Radio Active,* a monthly, which became *Radio Week* (1988–present, weekly). All of these focused on Washington policy concerns, general industry trends, and NAB activities—and took a strongly pro-industry point of view. The monthly *RTNDA Bulletin* (1952–70) and *Communicator* (1971–88) helped to tie the nation's news directors together with reviews of common problems, though from the beginning the focus was on television rather than radio. With a focus on stations devoted to Christian programs and music, *Religious Broadcasting* (1969–present, monthly) is a publication of the National Religious Broadcasters.

Technical Journals. Engineering association publications are technical in nature or focus on radio production techniques. *The Proceedings of the Institute of Radio Engineers* (1913–62, monthly) was the vehicle for many important technical announcements, including Edwin Armstrong's pioneering FM paper in 1936. Aimed at electrical engineers, its contents were wholly technical, with advertising to match. It has been superseded by a host of publications from the Institute of Electrical and Electronic Engineers. *Audio* (1954–present, monthly) has long dealt with "The World of Sound." It began as *Pacific Radio News* (1917–21), then became *Radio* (1921–47), and later became *Audio Engineering* (1947–54). The *Journal of the Audio Engineering Society* (1953–present, quarterly, monthly, then semimonthly) began as the *Broadcast Equipment Exchange,* a tabloid newspaper dealing with both new and used station equipment. It took on its present title in 1980 and broadened its coverage to station engineering management as well as new technology. *Mix* (1977–present, monthly) is one of the recording industry journals that blur the line dividing radio and sound studio work. First a quarterly and then a monthly, *Mix* offers information on both new technology and its applications.

Non-U.S. Publications. Naturally, a thriving broadcast trade press exists in several other nations as well. There is space here to cite only a few English language titles. *Broadcast* (1973–present, weekly) began in 1959 as *Television Mail* and has become the key trade periodical for British broadcasting, covering all aspects of radio and television, including the British Broadcasting Corporation (BBC) and commercial services. *Broadcaster* (1942–present, monthly) does the same thing for all aspects of Canadian radio and television broadcasting. It provides directory issues listing stations and systems plus related firms and associations twice a year. *Asian Broadcasting* is a Hong Kong–based bimonthly reporting on programming and business aspects of radio and television; it also issues a technical overview covering the region from Egypt to Japan. *The Asian Broadcasting Technical Review* (1969–present, bimonthly) reports on technical developments and equipment trends among Asian nations.

Modern Era (Since 1980)

By the 1980s, specialized radio publications broadened their comeback, encouraged by the continued growth of the industry. *Radio Ink: Radio's Premier Management and Marketing Magazine* (1985–present, monthly) deals with all aspects of commercial radio operation, especially programming. *Digital Radio News* (1990–present, bimonthly) is a newsletter that first appeared just as serious thinking about digital radio began. Although regular digital audio broadcasting has been delayed in the United States, this publication has reported on related terrestrial and satellite developments.

On-Line Publications

The rise of the internet as a means of effective business communication is very evident in the radio trade press. Indeed, in time, the internet will probably totally transform the whole trade press business. Many radio magazines, including *Billboard, Broadcasting & Cable, M Street, Radio Business Report, Radio Ink,* and *Radio & Records,* offer extensive online versions of their print publications, some available only to subscribers, others to all comers. *Streaming: The Business of Internet Media* (formerly *eRadio,* it is published by Eric Rhodes, who also issues *Radio Ink*) first appeared as a monthly in May 2000 and offers (fittingly) on-line features. *Radio World* offers readers an email updating service. *FMQB* (*Friday Morning Quarterback*) began in 1968 and is now a 50-plus page glossy weekly, covering programming, management, music, promotion, marketing, imaging, and airplay for various rock and rhythm crossover formats—with an extensive internet presence as well. On-line versions of these titles often closely parallel the print editions, even to layout. But on-line editions often provide more stories and greater depth than the print version can.

Others, such as the *Radio and Internet Newsletter* (founded in 1999), are primarily on-line publications. Such "publication," of course, is vastly less expensive in that printing and distribution costs are non-existent. Further, on-line publication allows regular updating, even new daily releases. How many of the internet-only publications can survive without a strong advertising or subscriber base, however, remains to be seen.

CHRISTOPHER H. STERLING

See also Columnists; Critics; Fan Magazines; Taishoff, Sol

Further Reading

Brown, Michael, "Radio Magazines and the Development of Broadcasting: *Radio Broadcast* and *Radio News,* 1922–1930," *Journal of Radio Studies* 5, no. 1 (1998)

The First 50 Years of "Broadcasting": The Running Story of the Fifth Estate, Washington, D.C.: Broadcasting, 1982

Lyons, Floyd, "Publication Dates: Early U.S. Radio Magazines," *The Old Timer's Bulletin* 16, no. 2 (1975)

Slide, Anthony, editor, *International Film, Radio, and Television Journals,* Westport, Connecticut: Greenwood Press, 1985

Sova, Harry, and Patricia Sova, editors, *Communication Serials,* Virginia Beach, Virginia: Sovacom, 1992

Sterling, Christopher H., "Periodicals on Broadcasting, Cable, and Mass Media: A Selective Annotated Bibliography," *Broadcasting and Cable Yearbook* (1990–98)

Sterling, Christopher H., and George Shiers, "Periodicals," in *History of Telecommunications Technology: An Annotated Bibliography,* by Sterling and Shiers, Lanham, Maryland: Scarecrow Press, 2000

Topicator: Classified Article Guide to the Advertising/ Communications/Marketing Periodical Press (1965–)

Volek, Thomas, "Examining the Emergence of Broadcasting in the 1920s through Magazine Advertising," *Journal of Radio Studies* 3 (1993–94)

Transistor Radios

When the first transistor radio was introduced to the American market during the November 1954 holiday season, no one recognized it as the precursor to a technological revolution. Apart from electronics buffs, consumers appeared to greet the miniature radios with a collective yawn.

The initial development of transistors in the late 1930s was conducted by physicists working for Bell Laboratories, the research division of American Telephone and Telegraph (AT&T). They were trying to create an electronic device that could replace vacuum tubes, something much smaller that would consume significantly less electricity and generate less heat. World War II interrupted those efforts. In 1948 Bell Labs announced the development of the transistor. William Shockley, John Bardeen, and Walter H. Brattain would share the 1956 Nobel Prize for Physics for its invention.

Texas Instruments (TI) was a small seismic-survey company that built its own survey equipment. During World War II, there was little demand for oil exploration, so the company used its electronics capability to build systems for anti-submarine warfare. In 1952 TI was one of 20 companies that paid Western Electric (the manufacturing division of AT&T) $25,000 for the right to produce transistors.

In 1954 TI constructed a transistorized pocket radio. Concerned that it was introducing an unknown in the consumer electronics market, TI took its six-transistor design to radio manufacturers. The reception, according to marketing director S.T. Harris, was unimpressive. His phone calls, letters, and telegrams to every major radio manufacturer in the United States produced no response.

Finally, in June 1954 a small Indiana company, the Industrial Development Engineers Association (IDEA), agreed that its Regency division would produce and market the12-ounce, 3-inch by 5-inch radio. Production began in October just in time for Christmas sales in November. The TR-1 sold for $49.95, and the accompanying brochure extolled its pleasures: "in pocket or purse anytime, anywhere, you can be sure to hear that favorite program . . . be sure not to miss that vital installment of soap or horse or space opera . . . check scores, weather results, news . . . have music wherever you go at those times when music adds so much . . . and the Regency Radio can play for you alone without disturbing others around you or the whole group can share" (White, 1994).

Initially, consumers were slow to adopt transistor radios. After World War II, Americans wanted products that left behind the austerity of the war years. Automobiles and appliances were bigger and flashier. A tiny radio with an earphone resembling a hearing aid did not fit the American shopper's self-image as a prosperous trendsetter—and the relatively high price was an obstacle as well.

It was not until the end of the 1950s that the tiny portable radio would find its niche, but it would take the rock and roll revolution to fuel the increased demand for pint-size radios that teenagers could carry with them wherever they went. In 1955 Bill Haley and the Comets were the first to have a rock and roll record reach number one on the Billboard chart. "Rock Around the Clock" stayed in the number-one slot for eight weeks. A number of music historians mark the occasion as the birth of the rock and roll era.

Radio stations quickly realized that teenagers were a large part of the audience—and that they were clamoring to hear more Elvis, Paul Anka, the Everly Brothers, and others. Stations across the country quickly adopted the new Top 40 format. Parents were less than thrilled with rock music. "Turn that thing down," was the common refrain. Earplugs allowed teenagers to listen to transistor radios without antagonizing adults. The small size also permitted surreptitious listening while huddled under the covers at night after lights out. One of the inventors of the transistor is reported to have joked that he might have reconsidered the invention had he known it would allow kids across the United States ready access to rock and roll.

The head of International Business Machines (IBM), Thomas J. Watson, Jr., was another who recognized the importance of transistors. After reluctant engineers made little effort

The first commercial transistor radio, the Regency TR-1
Copyright 1998 John V. Terrey. Reprinted with permission of Antique Radio Classified

to incorporate transistor technology in computers they were designing, Watson bought several hundred Regency TR-1 radios. He ordered that no more computers be built using vacuum tubes after 1 June 1958. To those engineers who complained, he gave a TR-1 to make his point. Soon IBM was building computers using transistors from TI, $200 million worth by 1960.

American radio manufacturers started building transistor radios in the mid-1950s, with nearly 5 million radios produced by Admiral, Arvin, Emerson, General Electric, Raytheon, Radio Corporation of America (RCA), Westinghouse, and Zenith by 1957. That same year, the first Japanese transistor radio was sold in the United States. It was manufactured by Tokyo Telecommunications Engineering Company, a new corporation created immediately following World War II. Soon its name would be changed to something catchier for the international market: Sony. Other Japanese companies quickly followed Sony's lead. By 1959, 6 million radios were coming into the United States from Japan, a highly successful launch into the world market of consumer microelectronics.

In 1960 almost 10 million transistor radios were sold in the United States. That number would increase to 27 million radios in 1969. No longer were they owned exclusively by young people. Americans of all ages had discovered the convenience of a tiny radio that could go anywhere. The transistor radio had become ubiquitous in American life. Baseball fans kept up with the World Series; American soldiers in Vietnam listened to Armed Forces Radio broadcasts.

In the 1960s, innovators realized the transistor radio apparatus fit nicely into a variety of small cases. Model cars, soft drink bottles, and wristwatches were among the many novelty radios that needed only a company's name and message to make it a successful advertising tool. Novelty radios continue to be popular today, coming in every shape imaginable from a radio in a stuffed teddy bear to a fish containing a water-resistant shower radio.

SANDRA L. ELLIS

See also Bell Telephone Laboratories; Emerson Radio; General Electric; Radio Corporation of America; Receivers; Rock and Roll; Walkman; Westinghouse; Zenith

Further Reading

Handy, Roger, Maureen Erbe, and Aileen Antonier, *Made in Japan: Transistor Radios of the 1950s and 1960s,* San Francisco: Chronicle Books, 1993
Schiffer, Michael Brian, *The Portable Radio in American Life,* Tucson: University of Arizona Press, 1991
Smith, Norman R., *Transistor Radios: 1954–1968,* Atglen, Pennsylvania: Schiffer, 1998
Ward, Jack, "The Transistor's Early History from 1947 to the 1960s," *Antique Radio Classified* 15 (December 1998)
White, David, "With the Collectors: The Regency TR-1," *Antique Radio Classified* 11 (September 1994)

Tremayne, Les 1913–

U.S. Radio and Film Actor

With a long acting career marked by his first film role in 1917 as a child actor, initial radio role in 1931, an appearance on television in 1939, and re-entry into films in 1951, Les Tremayne was a frequently heard and seen actor in a variety of media. His role as a radio actor on soap operas and mysteries included *Adventures of the Thin Man* and *The First Nighter Program.*

Born in London, Tremayne moved with his family to America in 1930 while in his early teens. He was educated in classical Greek drama at Northwestern University, and later took courses in anthropology at both Columbia University and the University of California, Los Angeles. Tremayne first went on the stage in the early 1930s, where his distinguished approach and full deep voice served him well in a variety of roles.

He began his long radio career playing a suitor in *The Romance of Helen Trent* (1933), the first of what would become thousands of golden age broadcasts, notably as the star of the long-running anthology *The First Nighter Program,* where he played a variety of usually romantic leads (1936–43). At almost the same time he played regularly on the drama *Grand Hotel* (1934–40) and (more occasionally) the role of Bob Drake on *Betty and Bob* (1935–39). After the war he played the lead role of Nick Charles in *Adventures of the Thin Man* (1945–50) and Michael Waring in *The Falcon* (1946–49). He found time to star in a Broadway production in 1948–49. In 1949 he and his wife appeared on a noontime interview and chat program on New York's WOR.

With the decline of network radio acting opportunities, beginning in 1951 Tremayne turned primarily to playing roles as a film actor. His initial roles were as fairly serious characters (often military officers or scientists) in otherwise somewhat fantastic science fiction films, some of them distinctly "B" movies. In addition Tremayne showed up in several non-genre efforts, usually in small but substantial roles such as the auctioneer in Alfred Hitchcock's thriller *North by Northwest.*

Still later Tremayne became busy with television roles, including those as a commercial spokesman and voice-over artist. He appeared in, among other series, the prime-time TV version of radio's *One Man's Family* (1951); as Inspector Richard Queen in the reincarnation of the early 1950s *Ellery Queen* series (1958–59); and as the character Mentor on the Saturday morning weekly cartoon *Shazam!* (1974–77). He played various voice roles in a host of other animated television and film productions.

Tremayne was voted the best radio actor in 1938 and one of the three most distinctive voices in the United States in 1940. He was elected to the Radio Hall of Fame in 1995.

CHRISTOPHER H. STERLING

Lester (Les) Tremayne. Born 16 April 1913, in London to Walter Carl Christian and Dorothy Alice Gwilliam. Moved from England to the United States in 1930. Educated in Greek drama at Northwestern University (1937–39), and in anthropology at both Columbia (1949–50) and UCLA (1951–52), but no degrees earned. Many roles in radio dramas, 1933–49.

Major Radio Credits

1933	*The Romance of Helen Trent*
1934–40	*Grand Hotel*
1935–39	*Betty and Bob*
1936–43	*The First Nighter Program*
1945–50	*Adventures of the Thin Man*
1946–49	*The Falcon*
1949	*The Tremaynes*

Films

I Love Melvin, 1953; *The War of the Worlds,* 1953; *Susan Slept Here,* 1954; *A Man Called Peter,* 1955; *The Monolith Monsters,* 1957; *The Monster of Piedras Blancas,* 1958; *The Angry Red Planet,* 1959; *The Story of Ruth,* 1960; *The Slime People,* 1962; *The Fortune Cookie,* 1966; *The Phantom Tollbooth* (voice only), 1969; *Fangs,* 1973

Theater

Detective Story (Broadway, 1948–49)

Television

One Man's Family, 1951; *Ellery Queen,* 1958–59; *Shazam!* 1974–77

Further Reading

Lamparski, Richard, "Les Tremayne," in *Whatever Became Of . . .*, by Lamparski, New York: Crown, 1996

Trout, Robert 1908–2000

U.S. Broadcast Journalist and Radio Commentator

In a long and varied career, Robert Trout brought a professionalism and legitimacy to the fledgling news operations of commercial radio. Beginning with his work as announcer on President Franklin Roosevelt's "fireside chats," Trout also established himself as radio's premier commentator on the political process.

Born in 1909 in North Carolina, Trout became obsessed with radio, building a crystal set when still an adolescent. Like many young boys, he was amazed at tuning in to all parts of the country, and a wanderlust was also sparked that would define his adult personality. Wanting to become a novelist, he got into the radio business by accident. In 1931 he applied for a job for which he had no skills at WJSV, a small station in Mount Vernon Hills, Virginia, and was given an unpaid position of all-round handyman. When one announcer failed to show up, Trout was asked to substitute, and his deep, resonant voice impressed the station manager. Seeing himself as a "very poor man's Will Rogers," Trout performed a variety of tasks on the air: reporting local events; playing country records; giving folksy advice; and parodying the Ku Klux Klan, which he quickly learned was financially involved with the station.

In 1932 WJSV moved to Washington, D.C., and became affiliated with the Columbia Broadcasting System (CBS) network, changing its call letters to WTOP. Trout became a remote reporter for the network, covering such events as President Herbert Hoover's Christmas tree lighting and the 1932 election. With the inauguration of Franklin Roosevelt (FDR), Trout was appointed presidential announcer and was responsible for introducing FDR's radio broadcasts, which became known as "fireside chats." For the first talk he prepared two introductions: a formal one, underlining the dignity of the office, and a more familiar, folksier one. Roosevelt chose the informal opening, and this rapport with Trout helped to define the president's intimate style of communication and the importance of radio in his administration.

In 1935 Paul White, director of CBS's news and public-affairs unit, transferred Trout to the network's flagship station in New York. Trout joined a small staff that included Edward R. Murrow, who was head of talks. Trout's coverage of both 1936 political conventions underlined his facility with improvising under any circumstance and orchestrating complex proceedings. Reporting on lengthy, breaking events with an indefatigable stamina and verbal grace became Trout's forte, leading to his sobriquet, "the iron man of radio." In 1937 Trout accompanied Paul White to London and reported on the installation of George VI as king of England, the first broadcaster to cover a live coronation. The dapper Trout also made his first television appearance, as guest commentator on the British Broadcasting Corporation (BBC).

By the late 1930s, Trout had hosted several conventional public-affairs series, including *History Behind the Headlines* and *Headlines and Bylines*. But the developing European war in the late 1930s demanded a new form of news. On 13 March 1938, following the *Anschluss,* Trout anchored the first news roundup, with reports by Murrow and his team from the major overseas capitals. American radio now participated in the larger world, with Trout's sureness and expertise giving the new broadcast a credibility and urgency. As historian Susan Douglas notes, Trout used "language that helped listeners see and even hear what [the war] had been like, and related events to those that Americans might remember or have participated in."

Even as Nazism was on the rise in Germany, Trout continued to report on such domestic affairs as the opening of the 1939 World's Fair and the havoc wrought by Mississippi River floods. In October 1941 Trout replaced Murrow as the European news chief and began his coverage of the Blitz of London. He hosted a series, *Trans-Atlantic Call,* in which he interviewed ordinary Englishmen about their struggles.

Trout reported many of the most significant events of World War II. For more than seven consecutive hours on D-Day, he read the latest news bulletins about the Allied invasion of Normandy from a New York studio. *Broadcasting* cited his anchoring of these breaking reports, 35 times in a 24-hour stretch, as "a masterful orchestration of maintaining a running report of 'the greatest story ever told.'" Trout hosted the memorial service for President Roosevelt in April 1945, ad-libbing for almost 30 minutes with memories of the former leader. He covered the celebrations on V-E and V-J days, and the end of fighting inspired one of Trout's succinct wrap-ups: "This, ladies and gentlemen, is the end of the Second World War."

After the war, as CBS news experimented with television, Trout remained the reassuring voice of accuracy on radio. Managing a team of 22 correspondents, he anchored a weekday daily news broadcast, *The News Till Now.* After he was replaced by Murrow in 1948, Trout joined the National Broadcasting Company (NBC) and hosted two television series: *Who Said That?* a quiz show that took advantage of Trout's unflappability, and *Presidential Timber,* which featured talks with candidates for the White House. In 1952 Trout, the dean of political reporters, was wooed back to CBS to once again cover the conventions; "the iron man" proved his mettle, anchoring one convention broadcast on the radio for 15 consecutive hours. In retrospect, Trout considered his return to CBS a mistake because he was overshadowed by so many

Robert Trout
Courtesy CBS Photo Archive

younger personalities who were being groomed for television, including Walter Cronkite, Eric Sevareid, and Charles Collingwood. Throughout the 1950s, Trout continued to host convention coverage on radio and lent his voice to several television documentaries produced by Fred Friendly.

In 1964 Friendly, who had been promoted to president of CBS News, decided that television news was badly trailing the NBC team of Chet Huntley and David Brinkley, especially in ratings of convention coverage. For the Democratic National Convention, Friendly, in consultation with chairman William Paley, paired the elegant veteran Robert Trout with the up-and-comer Roger Mudd. This marked the first time that Trout anchored a convention on television, having reported on radio on every nominating gathering since 1936. As always, Trout received impressive reviews from the critics; Jack Gould of the *New York Times* cited two qualities that were forever associated with Trout—his "smooth and unruffled administration of the anchor desk" and his "effortless ad libs in periods of emergency." Unfortunately, the ratings did not improve dramatically, and Trout decided to go into semi-retirement in Spain while also serving as a roving reporter in Europe.

In 1974 Trout joined American Broadcasting Company (ABC) News and began his "third career." For ABC Radio he remained a fixture at all conventions, and for both the radio and television networks he reported on his new specialty, European culture. In 1982 he narrated a three-hour documentary on the centennial of Franklin Delano Roosevelt; for many, his rich voice evoked the entire era of the New Deal. He participated in the 50th anniversary of D day in 1994 with a series of reports from France, seeing firsthand the beaches of Normandy about which he had read news wires from his New York studio during the 1944 invasion. In the mid-1990s, he semi-retired again from the news profession that he had helped to shape beginning in the early 1930s. He sustained a long career with his ability to persevere at any cost and to ad-lib gracefully whenever the occasion arose. The essence of civility and objectivity, Trout helped to define the role of the anchorman in broadcasting.

RON SIMON

See also Commentators; Election Coverage; Friendly, Fred W.; Murrow, Edward R.; News; Politics and Radio; White, Paul; World War II and U.S. Radio

Robert Trout. Born Robert Albert Blondheim in Lake County, North Carolina, 15 October 1908. Son of Louis and Juliette (Mabee) Blondheim; graduated from Central High School, Washington, D.C.; worked at various odd jobs, including taxicab driver and bill collector; hired as handyman at WJSV, Mount Vernon Hills, Virginia, 1931; changed name to Robert Trout, 1932; wrote various pieces for station and promoted to announcer late in 1931; WJSV joined CBS network becoming WTOP and Trout became news reporter; selected as presidential announcer, 1933; transferred to CBS flagship station WABC, 1935; covered first conventions, which became his specialty, 1936; covered coronation of George VI, 1937; anchored premiere broadcast of *European News Roundup,* 1938; replaced Edward R. Murrow in London as European news chief; returned to United States and continued to anchor nightly news broadcast, 1943; joined NBC and served as moderator of television quiz show *Who Said That?* 1948; returned to CBS to cover political conventions for radio, 1952; anchored first television convention, 1964; became roving European correspondent for CBS, 1965; joined ABC News, 1974; commentary for National Public Radio, 1998. Died in New York City, 14 November 2000.

Radio Series

1937	*History Behind the Headlines*
1937–38	*Headlines and Bylines*
1938	*European News Roundup*
1946–47	*News Till Now*

Further Reading

Cloud, Stanley, and Lynne Olson, *The Murrow Boys: Pioneers on the Front Lines of Broadcast Journalism,* Boston: Houghton Mifflin, 1996

Douglas, Susan Jeanne, *Listening In: Radio and the American Imagination,* New York: Times Books, 1999

Gates, Gary Paul, *Air Time: The Inside Story of CBS News,* New York: Harper and Row, 1978

Trout, Robert, *Steven H. Scherer Collection at the Center for the Study of Popular Television: An Interview with Robert Trout by Ron Simon,* Syracuse, New York: Syracuse University, 1998

U

Underground Radio

Alternative and Free-Form Programming

The 1960s gave rise to one of radio's most unique programming genres. During its short existence (1966–72), this format became known variously as progressive, alternative, free-form, psychedelic, and acid and was ultimately dubbed underground radio because of its unorthodox and eclectic mix of music and features and disc jockeys who broke from the traditional delivery style embraced by other youth-oriented stations of the day.

Origin

FM provided the fertile soil from which commercial underground radio would grow. It was where experimentation was permitted, because there was so little to lose at the time. Until the mid-1960s, FM moved along in low gear. A nearly negligible listenership provided FM with little status and currency among the general public and industry. It was perceived by many as the province of the so-called eggheads and the terminally unhip—the place to tune for Mahler and fine-arts programming. Tuning to FM for most people was like choosing to attend a foreign film with subtitles when there was a new action-packed Audey Murphy movie just around the corner. Most 20-year-olds had never tuned between 88 and 108 megahertz, because the "in" music and "cool" disc jockeys were spinning the hits on AM.

Many social and cultural factors contributed to the rise of commercial underground stations. The repressive behavior and social conformity of the postwar years led to the volcanic eruption of the 1960s, particularly among youth. Political assassinations, racial upheaval, and an undeclared war in Asia, along with the growing use of mind-altering drugs by young people, contributed to the blossoming of what came to be called the counterculture.

Rock music began to more astutely and candidly reflect the troubles in American culture by incorporating thoughtful and challenging themes and more provocative and innovative scores and arrangements. The increasing popularity of rock albums among youth helped encourage FM stations to abandon their conventional fare and launched them on a quest for disenchanted and disenfranchised radio users—those who had rejected the 45 rpm–driven pop chart outlets. Also enhancing the enthusiasm for FM was its ability to broadcast in stereo—a process that recording companies had embraced for their best-selling groups and a feature that AM lacked.

Breaking the Mold

Several young programmers of the early 1960s had grown weary with the conventional sound of youth-oriented radio. Its frantic disc jockeys and two-and-a-half-minute doowop records left them wanting something more. The repetition and banality of Top 40 stations provided a primary impetus for movement in a very different direction. The pioneers of commercial underground radio, among them Tom Donahue, Larry Miller, Scott Muni, Thom O'Hair, Murray the K, Rosco, and Tom Gamache, took their lead from a couple of early 1960s noncommercial broadcasters and from a handful of innovators on the AM band in the 1950s, all of whom offered listeners a sound antithetical to the highly formulaic formats offered by mainstream stations.

Many stations claim to have debuted the new program genre, but two make the top of the list: WOR-FM in New York and KMPX-FM in San Francisco. The former went on the air in 1966 but changed format within a few months, and the latter was launched in 1967 and marked the beginning of a period in which the underground sound was sustained for several years. Within a year, KMPX lost its on-air staff during a strike to KSAN-FM, which grew to considerable prominence and won a special place in the underground firmament.

Although these stations are traditionally accorded landmark status, development of the underground format was foreshadowed by other stations as early as the 1950s. For example, WJR-AM in Detroit featured the "Buck Matthews Show," which mixed all kinds of music together in a fairly unrestricted, free-form way. Matthews employed a conversational, laid-back announcer style as well, which was atypical for disc jockeys of that day.

Other precursors to FM underground radio could be found on the AM band. For instance, Chicago's WCFL-AM offered a free-form mix of rock music in the 1960s. Soon Newton, Massachusetts, had progressive rock over WNTN-AM. Other low-power AM stations experimented with the "open" technique to music programming, despite the fact that the format was nearly the exclusive domain of FM.

A number of noncommercial stations also presaged the arrival of commercial underground radio. Perhaps most significant among them were WBAI-FM and WFMU-FM. At the former, young disc jockey Bob Fass worked the overnight slot airing a program called *Radio Unnamable.* Across the river in New Jersey, college station WFMU-FM's Larry Yurdin was doing much the same thing by offering a creative and innovative mix of sounds. Undoubtedly, like those mentioned above, others helped set the stage for the surfacing of commercial underground radio, which got under way at about the same time on both coasts.

Most radio historians point to WOR-FM in New York as the first commercial outlet to break from the "primary" or single-format approach to music programming. However, the station's free-form experiment lasted only a few months, and it was on to other things by the time KMPX-FM in San Francisco introduced Tom Donahue's version of the format in spring 1967. A few months after assuming the programming duties at KMPX-FM, Donahue took on its sister station, KPPC-FM in southern California, simultaneously working his format magic at both.

The underground radio programming genre, the "nonformat" format, as it has been described, was soon emulated by stations around the country. By 1968 dozens of stations around the United States were offering listeners their own brand of underground radio. Most large metropolitan areas (including Detroit, Cleveland, Chicago, and St. Louis) boasted what many were calling "flower power" stations. This was no longer an avant-garde form of radio restricted to the enclaves of the East and West Coasts.

By late 1968 there were over 60 commercial underground radio stations in operation around the country. By the summer of 1969, San Francisco alone could claim a half dozen, whereas New York had only three. One company (Metromedia) owned the two stations that *Billboard* magazine ranked the top underground stations in the country—KSAN (San Francisco) and WNEW (New York).

At this early stage in underground radio's evolution, these two stations were frequently held up as models of the genre. Both attracted listeners and advertisers. Though often compared, the stations had forged their own distinct personas, mainly by creating unique and distinctive sounds that reflected not only the times and the areas in which they broadcast but also the philosophies of their programmers.

The "Nonformat" Format

Programmer Tom ("Big Daddy") Donahue considered the underground radio sound the antidote to Top 40, and by declaring this, he wanted to make it amply clear to everyone that things were being done quite differently at his station. In fact, he even rejected the notion of the term *format,* believing it had little to do with his new brand of radio. In his eyes, underground radio, if anything, was the antithesis of standard format programming because it embraced the best of rock, folk, traditional and city blues, electronic music, reggae, jazz, and even classical selections as opposed to any single type of music. This musical ecumenism was evident at underground stations around the country.

Indeed, the way in which songs were presented by undergrounders was unlike that of any other contemporary radio station at the time. Interestingly, if not ironically, these new outlets did reflect an older adult format, which had been responsible for bringing the FM band to a larger audience in the 1960s. Its name was beautiful music or, as many called it, "elevator" music. It was the Muzak format of the radio world. The common ground between the two seemingly disparate forms of radio programming was the way in which they structured music into sweeps—that is, uninterrupted segments or blocks—typically of a quarter hour's duration. Evolving from the sweep approach was the idea of music sets, wherein a series of album cuts would establish a particular theme or motif.

Just as the approach to music programming in underground was antithetical to conventional AM radio, particularly Top 40, announcing styles were no less contrary to the long-standing norm. Since the medium's inception in the early 1920s, announcing techniques had undergone relatively subtle changes, never wandering too far from the affected "radioese" presentation style. The old-line announcing manner, characterized by its air of formality and self-consciousness, remained prevalent well into the second coming of the radio medium after the arrival of television.

The "stilts," as they have been called, found their way into the FM band as well, migrating to the beautiful music format and others. This announcing style was emphatically rejected by underground stations, which militated against its disingenuous affectations and mannerisms—the hype and histrionics. However, sounding "hip" was considered acceptable and even preferable, but not hip like the "screamers" on Top 40. Underground disc jockeys were intent on projecting a natural, friendly, and mild-mannered "grooviness" when they were on the air. In fact, the "stoned" announcer persona was often an integral part of this radio genre's repertoire. The idea was to be at one with the audience in every way possible. Staying "loose" was the underground disc jockey's mantra.

As with all formats, there are other programming ingredients besides the music and announcing that contribute to a station's general appeal, identity, and overall listenability. News and information broadcasts represent one of those elements. Despite the underground radio's dominant emphasis on album music designed for an under-30 crowd, it differed from other youth-oriented music outlets in that news and public-affairs features were frequently regarded as an integral part of what many of these stations sought to convey to their public. That is, they wished to be perceived as members of the caring and socially conscious community and not simply as record machine operators.

The Fate of the Nonformat Format

As the counterculture movement of the 1960s and early 1970s faded, so did commercial underground radio. Many members of the movement were embracing more mainstream and traditional goals and aspirations, if not values, while the anger and altruism inherent in rock music for nearly a decade bowed to the insipid patter and rhythms of disco and new-wave or "corporate" rock. Underground radio became a thing of the past as the baby boomers sought a less uncertain and chaotic future, taking refuge in that once unsavory realm known as the material world. A survey of published perspectives and conventional wisdom on the 1960s and 1970s and on the underground radio phenomenon itself reveals that numerous factors came into play that ultimately contributed to the nonformat's rather swift departure from the airwaves.

In addition to the changing cultural mores and attitudes, which diminished the relevance and appeal of the underground sound, the growing profitability of FM radio inspired a shift away from the nonformat programming approach to something with more advertiser appeal. Station owners sought greater control over program content to maximize bottom-line figures and profits. The role of disc jockeys in shaping the air product returned to the executive suit. FM was corporatized in the 1970s, and by the middle of the decade, commercial underground radio had been reconstituted in the form of album-oriented rock—a highly structured and formulaic offshoot of the former free-form sound.

MICHAEL C. KEITH

Further Reading

Anderson, Terry H., *The Movement and the Sixties,* New York: Oxford University Press, 1995
Fornatale, Peter, and Joshua E. Mills, *Radio in the Television Age,* Woodstock, New York: Overlook Press, 1980
Görg, Alan, *The Sixties: Biographies of the Love Generation,* Marina Del Ray, California: Media Associates, 1995
Keith, Michael C., *Voices in the Purple Haze: Underground Radio and the Sixties,* Westport, Connecticut: Praeger, 1997
Krieger, Susan, *Hip Capitalism,* Beverly Hills, California: Sage, 1979
Ladd, Jim, *Radio Waves: Life and Revolution on the FM Dial,* New York: St. Martin's Press, 1991

United Fruit Company

Early Wireless System Operator

Spurred by its need to communicate with its many banana plantations and with its shipping fleet, the United Fruit Company became a pioneer wireless operator in the early 20th century. Company operators developed many early radio techniques and helped to pioneer networking of stations and the use of crystal detector receivers in high-humidity conditions.

Origins

Through the merger of the Boston Fruit Company and other companies involved with the production of bananas in the Caribbean, the United Fruit Company was formed in 1899. The largest banana company in the world, the United Fruit Company was also the first transnational, corporate giant in the Americas. During the 20th century, the United Fruit Company would have a significant impact, both positive and negative, on many Central American nations, including Guatemala, Honduras, and Costa Rica. The company's influence in these countries was pervasive, extending into the political, transportation, and communication systems.

At its formation, the United Fruit Company owned banana plantations in seven countries throughout the Caribbean area.

The need for efficient means to harvest and transport the perishable banana product was of paramount concern, as errors in the timing of banana deliveries to United States markets often resulted in spoiled cargoes. As a result, the United Fruit Company took an aggressive role in developing railroad links to deliver the bananas to coastal ports for shipping. Because timely communications was essential to coordinate the harvesting and delivery of the bananas, the company also built and maintained telegraph and telephone lines between farms, local company headquarters, and Caribbean ports. Difficult terrain made the construction of land lines impractical in some areas, however, and the tropical weather conditions and political instabilities often rendered them unreliable when they were built. In addition, communication from shore to ship and from ship to ship required the use of radio. Faced with these obstacles, the United Fruit Company became one of the first companies, and the only major American company, to invest major capital and manpower into developing and adopting the new telecommunications mode of radio.

Implementing Wireless

During the summer of 1903, United Fruit Company employee and electrical engineer Mack Musgrave began investigating the possibilities of using radio for company business in Central America. United Fruit bought its first radio equipment from the American De Forest Wireless Company in 1904 and established its first two stations approximately 150 miles apart at Bocas del Toro, Panama, and Puerto Limon, Costa Rica. In 1906 two more stations were built at Bluefields and Rama, Nicaragua. The expenditures for these and future facilities were great, but the costs were deemed a necessity for company development. United Fruit faced many obstacles in its efforts to establish a radio communications network. One important obstacle was the comparative youth of the radio industry and the experimental nature of its equipment. Unlike the case today, transmission distances were relatively short and required the construction of relay stations throughout the Caribbean. Static, common to the tropics, interrupted transmissions and often made the point-to-point radio links unreliable. In addition, hurricanes destroyed island relay stations with alarming frequency.

By 1908 the United Fruit Company had built six more shore stations and had installed radio equipment on all of its ships. Later known as "The Great White Fleet," these ships delivered bananas but also transported passengers between the United States and the Caribbean. In addition to serving the company's needs in transporting its product, the radio network became an essential tool for ensuring the safety and convenience of the fleet's passengers. Recognizing the further commercial value of the network, United Fruit also made its communications system available to paying customers along the Caribbean coast and thus became Latin America's major commercial alternative to the European owned and operated radio systems of Marconi and Telefunken.

Now an independent provider of radio communications services, United Fruit purchased a controlling interest in the Wireless Specialty Apparatus Company in 1912. This acquisition provided United Fruit with ownership to important radio technology patents and thus the means to develop and manufacture advanced equipment. United Fruit would continue to operate the Wireless Specialty Apparatus Company until 1921 when, through a complicated process of corporate patent pooling, the newly formed Radio Corporation of America became owner of the patent rights.

In 1913 the United Fruit Company radio department was formally incorporated as a wholly owned subsidiary named the Tropical Radio Telegraph Company. The Tropical Radio Telegraph Company took over all company ship stations and most of the land stations. By 1920, the United Fruit Company had invested nearly $4 million in radio. The Tropical Radio Telegraph Company had established an extensive radio network including a radio-telephone communication system serving the general public and the banana trade and connecting many of the principal population centers of the Caribbean to the United States.

The United Fruit Company's remarkable adoption of the fledgling radio technology and its development of a large, privately owned radio communications system in Latin America is a success story. Other companies, including the U.S. Rubber Company subsidiary Amazon Wireless Telephone and Telegraph established in 1901, failed where United Fruit had succeeded. Numerous factors supporting the United Fruit Company's ultimate success in developing its radio system include the dominating U.S. influence in Central America, which assured minimal or no opposition to United Fruit's expansions; the demand for radio communication by passengers on the company ships and by shore customers complementing the banana business communication needs; and the fact that the United Fruit Company's product, bananas, was an economically stable product.

DOUGLAS K. PENISTEN

See also Crystal Receivers; Early Wireless

Further Reading

Douglas, Susan J., *Inventing American Broadcasting, 1899–1922*, Baltimore, Maryland: Johns Hopkins University Press, 1987

Mason, Roy, "The History of the Development of the United Fruit Company's Radio Telegraph System," *Radio Broadcast* (September 1922)

Melville, John H., *The Great White Fleet,* New York: Vantage Press, 1976
Schubert, Paul, *The Electric Word: The Rise of Radio,* New York: Macmillan, 1928; reprint, New York: Arno Press, 1971
Schwoch, James, *The American Radio Industry and Its Latin American Activities, 1900–1939,* Urbana: University of Illinois Press, 1990
Wilson, Charles Morrow, *Empire in Green and Gold: The Story of the American Banana Trade,* New York: Holt, 1947

United Nations Radio

United Nations Radio is a news and information service that has been provided by the United Nations Department of Public Information since 1946. Programs address United Nations (UN) initiatives and are produced in Arabic, Chinese, English, French, Russian, and Spanish, the official UN languages; many are also available in Bengali, Dutch, Papiamento, French Creole, Hindi, Indonesian, Kiswahili, Portuguese, and Urdu. The people of nearly every country are now within range of a UN Radio broadcast.

A primary method of broadcasting has been shortwave radio, although much UN Radio programming is distributed by mail, telephone, and the internet. Timely short-form programs are accessible around the clock through the UN Radio Information System by touch-tone telephone. Regular audio feeds from UN Headquarters (in New York City) include daily briefings, press conferences, news concerning peace-keeping missions, breaking news from UN member countries, and coverage of the activities of the Secretary-General, as well as open meetings of the General Assembly and the Security Council. Audio and written transcripts of long-form programs and archival material are provided to broadcasters by request. UN Radio facilitates access to historical recordings from the United Nations' audiovisual library as a service to media organizations, government agencies, and researchers. The library includes *UN-TV,* film documentaries, a photograph collection, and extensive archives.

Members of the UN Radio production staff in New York work with field reporters throughout the world and a production facility at the United Nations in Geneva to provide daily updates, weekly magazine shows, regional news briefs, and "round-ups" on a number of topics. These include events and issues of interest in the areas of health, human rights, drug trafficking, women's issues, solar energy, poverty, education, sustainable development, the rights of children, global treaties, the environment, refugees, natural disasters, and reports on the progress of the UN's humanitarian assistance.

UN Radio does not own or operate any broadcast facility, relying on other organizations and institutions to disseminate its programs. The internet has facilitated broad availability of its programming to anyone with computer access to the worldwide web. In its early years, France, Switzerland, the United States, and Britain were among the countries that leased transmission facilities to UN Radio. Its first worldwide broadcast, via the British Broadcasting Corporation (BBC) in 1946, proclaimed, "This is the United Nations Calling the People of the World," then soundly provided the entire proceedings of the United Nations Security Council.

The United Nations has never been a broadcast operation, per se. Debate on this issue ensued from 1946 to 1953, resulting in leasing agreements with shortwave and standard radio licensees throughout the world—including private, commercial, and state-run broadcasters—to transmit daily UN Radio programs. During the 1960s and 1970s, UN Radio expanded its coverage in Africa, Southeast Asia, the Middle East, and Latin America. By the mid 1980s (with program production growing rapidly and a lessening reliance on leasing shortwave facilities), UN Radio added Egypt, India, and China to the growing list of countries with radio operations that agreed to carry its programs.

From 1953 to 1985, the Voice of America broadcast UN Radio programs consistently throughout Asia and Africa. The 1990s saw more programs concentrating on Asia and the Caribbean, as well as an increase in distribution by tape mailings and telephony. The start of the 21st century marked an increased reliance on digital distribution methods and cable. In 2000, a daily live multilingual news program was developed, anchored at UN Headquarters in New York and featuring a mix of news, news analysis, and feature segments.

Issues related to population, water, indigenous peoples, health, and human rights are among the priorities for much of UN Radio's program content. Programs are available as event coverage, daily news feeds, radio documentaries, special reports, and regional sound recordings at virtually no cost to broadcasters. The primary production facility in New York is staffed by numerous editors and producers, providing thousands of hours of programming each year to broadcasters interested in advocating UN positions and providing quality world news coverage to their audiences. In addition, UN Radio is a mechanism by which the United Nations can inform the world population of its mission and garner public support.

JOSEPH R. PIASEK

See also BBC World Service; International Radio Broadcasting; Shortwave Radio; Voice of America

Further Reading

United Nations Radio, <www.un.org/av/radio/>

United States

This essay surveys highlights in American radio broadcasting's development, from the experimental era before 1920 to the rapidly changing industry eight decades later. That history is described using a series of specific periods that help to characterize key developments. Many of the topics touched on here are treated more extensively in their own entries; the reader should use the table of contents as well as the index for more in-depth discussions of specific topics.

Before 1920

Prior to the end of World War I, there was little sustained radio broadcasting, except for occasional experimental and amateur transmissions. This era was characterized by rapid wireless technical innovation over a three-decade period that made broadcasting possible.

After the theoretical foundations of wireless transmission of information were established by James Clerk Maxwell in the 1860s, and the theory was proven with Heinrich Hertz's experiments in the late 1880s, the stage was set for the key innovator, Guglielmo Marconi. Beginning in the mid-1890s he developed and improved the key elements for wireless telegraphy (code) transmission—transmitter, antenna, and receiver. By the end of the 19th century, Marconi was the head of a thriving British company introducing wireless transmission to merchant and navy ships and, with high-powered shore stations, long-distance competition to undersea telegraph cables. Other companies developed in France, Germany, Russia, and the United States, and all shared a common goal—perfecting point-to-point wireless telegraphy as the most lucrative future for wireless. Broadcasting was barely a glint in anyone's eye.

A few early wireless experimenters did stumble onto the key elements of radio broadcasting, beginning with Reginald Fessenden. On Christmas Eve of 1906 he transmitted voice and music in what many consider the first broadcast in the world. Another important inventor-innovator, Lee de Forest, offered occasional broadcasts in 1907 and 1908. Beginning the next year, Charles Herrold initiated a regular radio broadcast service in San Jose, California, as an adjunct to his radio school. He remained on the air (typically for a few hours per week) until World War I. Amateur experimenters in other cities also offered sporadic broadcasts for fellow amateurs to tune in, but these were seen largely as exceptions to the point-to-point focus of most people in the fledgling wireless business.

Recognizing the need for some order among the slowly increasing number of transmitters needing frequencies, Congress passed legislation in 1910 and again in 1912 to regulate the use of wireless at sea. These were followed by the more important Radio Act of 1912, which was to stand for 15 years as the basis for any government regulation of wireless transmission. Following industry advice and expectations, the Radio Act was predicated on wireless as a point-to-point means of communication. It made no mention of nor provision for radio broadcasting.

What grabbed the public's imagination early in the 20th century was the role of wireless in some spectacular maritime disasters. In 1909 the White Star liner *Republic* was rammed in the fog by an incoming vessel a few hours outside of New York. Wireless operator Jack Binns called for help (the other ship lacked wireless), and virtually everyone—some 1,500 people—was saved. Radio could not save as many in the horrific *Titanic* disaster of April 1912, when another White Star liner, on her glittering maiden voyage, hit an iceberg and sank, this time taking 1,500 lives with her. But Marconi operators Jack Phillips and Harold Bride stayed at their posts (Phillips died after the vessel sank) and, transmitting both the old "CQD" and newer "SOS" emergency signals, brought a rescue vessel to pick up the 700 survivors, who told the tale that fascinates people still.

Radio played a lesser role in World War I. Demands of reliability and secrecy and the relative immobility of trench warfare made telephone and telegraph service more common than wireless. But military and naval needs and orders helped increase the pace of technical development, and wireless manufacturers rapidly improved both transmitters and receivers. The U.S. Navy took over high-powered transmitting stations from private operators (most other transmitters, including those operated by amateurs, were closed down for the duration of the war) and trained thousands of radio operators to run them. Demand for the best equipment saw development of a Navy-sponsored pooling of patents of different companies to allow the latest developments to more rapidly reach battle fronts.

With the end of the war, Congress briefly considered making permanent the wartime system of government operation of both wired and wireless telecommunications (as was then the rule in most other countries). But despite urging by the Navy, Congress ordered that transmitters be returned to civilian control in 1919–20. The radio industry made plans to expand production and operations in peacetime with the thousands of trained radio technicians and amateurs. Some of them expressed interest in broadcast experiments.

1920–26

The initial period of regular radio broadcasting was one characterized by growth and excitement, little effective regulation, program experimentation, a lack of permanent networks, only limited advertiser interest, and almost no knowledge of its audience.

Among the pioneering American stations taking to the air in 1919–20 (others began at about the same time in Canada and in Europe) were WHA in Madison, Wisconsin (which had begun as a University of Wisconsin physics department transmitter 9XM, sending wireless telegraphy market reports to Wisconsin farmers before World War I); Charles Herrold's station, which became KQW in San Francisco; KDKA in Pittsburgh, Pennsylvania (which had begun as amateur 8XK station in 1916); and Detroit's WWJ, which grew out of amateur operation 8MK.

Each began under a different owner with a different purpose in mind. The University of Wisconsin's WHA, the first educational station on the air, was interested in spreading the university's courses and research to the boundaries of the state. KQW at first continued Herrold's interest in operating his radio school, but it soon passed into the hands of a local church and later a national network interested in commercial operations. KDKA was developed by Westinghouse, which in 1920 was interested in keeping its assembly lines of highly trained personnel together despite the loss of huge government contracts with the end of the war. By providing a radio music and talk service, the company figured it might attract people to buy receivers. Over the next two years, Westinghouse added other stations in different cities. General Electric and the Radio Corporation of America (RCA) entered the radio station business at about the same time. WWJ was the first of many stations to be owned by a local newspaper. Initially seen as another community service, operating a radio station would become a common means of hedging bets as to which medium would survive. Other stations went on the air as auxiliaries to the owner's primary business—a retailer or service firm.

The radio station business began slowly in 1920–21, but it exploded in 1922, when more than 600 stations went on the air. Many went off again in a matter of weeks or months, unable to find a means of supporting operations. Early stations were primitive operations, the studios of which were often merely hotel or office building rooms, with walls covered in burlap to deaden sound and equipped with a microphone and the near-ubiquitous piano. Transmitters were largely hand built, and many radiated with less power than a reading lamp. A few stations experimented with temporary hookups, using telephone lines to allow two or more stations to carry the same program at the same time (and perhaps share in its costs)—the germ of networking. But multiple station linkups were largely limited to special sports or political events, such as presidential speeches.

Strictly speaking, there was no "programming" at first—merely different times given over to talk, music, or comedy, and this was usually in an unplanned fashion. The paramount idea was to fill airtime, even for the very few hours that most stations were initially on the air. Ironically in light of more modern experience, few stations made use of phonograph records. Although records would have been an obvious and easy means of filling time, they were then of low acoustical quality, and their use was considered a poor application of radio's potential. As radio had no ready means of making money, stations could offer no payment to singers and performers, which further limited program options. It was a brief but golden age for amateurs of all types and ages, who would gladly come into the studio just for the chance to have an audience—even an unseen and unheard one. It sounded much like the vaudeville circuit from whence came many early radio stars.

More formal, preplanned schedules of programs, with clear formats, beginnings, and endings, developed only by the mid-1920s, and then first in the largest markets. One vaudeville pattern that carried over to radio was the "song and patter" team—usually two men with comedy and musical experience—who could easily expand or shorten their act as broadcast time permitted. About the only role for women in early radio was as singers. There was little or no drama or situation comedy, little play-by-play sports, and no regular newscasts or weather reports—all these would come later in the decade or in the 1930s.

Radio was a novelty for its audience as well—indeed, it became a huge fad, which many observers figured would not last, especially as multiple-station interference grew in the mid-1920s, raising listener frustration. But the excitement of hearing disembodied voices and music, often from a considerable distance, was enough to persuade more and more people to build or buy the available crude receivers, which required a fair bit of manual dexterity even to tune in to a nearby station. That ready-made receivers cost a lot was indicated in early radio programs, which included a substantial amount of classical music, aimed at discriminating (i.e., wealthy) ears. Countless radio books and magazines appeared to cater to the growing audience. There was no audience research in this initial period—stations determined who was listening merely by audience mail or requests for cards confirming programs (a spin-off on ham radio "QSL" cards).

Despite this audience interest, the biggest problem broadcasters faced in the 1920s was how to pay for their operations. Although advertising may now seem to have been preordained, there were then many strong arguments against allowing radio to sell airtime. Some observers suggested a tax to pay for programming, others called for voluntary contributions, and still others promoted an annual tax on receivers, a method adopted by many other countries. And a few states or cities operated stations as government services.

Interestingly, it was the American Telephone and Telegraph Company (AT&T) that brought advertising to radio, when their New York station WEAF first offered to sell airtime much as the company did for long-distance telephone calls, dubbing radio's version "toll broadcasting." A 15-minute real estate ad in August 1922 was probably the first radio commercial. Others slowly followed suit, encouraged by initial sales results. Yet these early ads almost never mentioned price—instead they promoted a kind of institutional or image advertising common on today's public radio. Another way advertisers made themselves known was to add their name to the program title or stars—thus the *Lucky Strike Hour* or the *A&P Gypsies*. But many in and out of broadcasting still resisted the notion of bringing business into the home (forgetting perhaps that newspapers and magazines had long done just that). The rapid expansion of advertising came only after 1926, when AT&T sold its stations and ended industry debates over whether the telephone company could force stations to pay a fee for the right to sell advertising.

That radio was developing as more of a business became increasingly evident in the industry's call for firm government licensing policies and related regulations. Secretary of Commerce Herbert Hoover called together four national radio conferences from 1922 to 1925, each of them larger and more strident in calling on Congress to reduce the interference resulting from trying to administer a growing broadcasting system under provisions of the 1912 legislation, which had not foreseen broadcasting. Acting largely on his own, Hoover did succeed in expanding the number of frequencies available for broadcasting from one (833 kilohertz) in 1920 to three in mid-1922 and to the beginnings of a "band" of frequencies in May 1923. More frequencies allowed more stations without the need to share time—but only if broadcasters cooperated, because the government had no enforcement power. Court decisions in 1926 took away what little authority Hoover had exercised and increased pressure on Congress to act.

1927–33

The 1927–33 period is one of the most important in the history of broadcasting. By the late 1920s American radio began to take on most of the characteristics of the system still recognizable today—definite program patterns, advertising support, reliable audience research, and far more effective regulation. The latter came with Congressional action early in 1927.

The Radio Act of 1927, though in force for only seven years, would set broadcast patterns that persist to the present. Stations were to be licensed for up to three years, on a specific frequency and with specified power that could only be changed on application to the new Federal Radio Commission (FRC), which had the legal power to enforce its decisions. No longer could stations shift frequency, increase power, or literally move their transmitters overnight, as had become common just before the act became law. The driving impetus of the new FRC and its rules and regulations (first codified in 1932) was to reduce interference on the air so that people could hear stations clearly. A key part of that process was to expand on Hoover's beginnings by allocating most of the modern AM radio "band" of frequencies and by classifying stations by power. A few stations were restricted to daytime-only transmission to reduce evening interference even further. The FRC often had to defend its expanding role in court. A series of four landmark cases in 1928–30 concerning program controls helped confirm the agency's authority and the constitutionality of its licensing decisions. Now subject to federal legislation and rules, it seemed clear radio was here to stay.

The other fundamental change in the business took place at nearly the same time—the development of permanent national networks. The National Broadcasting Company (NBC) was the first, built around the informal network of stations that WEAF had initially developed. After purchase of that station from AT&T late in 1926, RCA formed NBC as a subsidiary that in November 1926 began to operate a continuing network of entertainment and cultural programs.

In fact, NBC began—and would operate until 1943—as two networks, the Red and the Blue. The former was based on WEAF and its connected stations, and the Blue was built around RCA station WJZ (also in New York) and a parallel chain of affiliated stations. From the start, the Red network had the stronger stations, greater audience reach, and greater advertiser appeal. NBC's chief competitor, the Columbia Broadcasting System (CBS), had a far more complex birth and only became a stable competitor in 1928. Together, the two

New York–based networks soon contributed hugely to radio's expanding audience popularity and advertiser appeal.

Advertising became more widely accepted in this period—at least by the industry, if not by all of its listeners. Based on initial success stories, advertisers and ad agencies had begun to recognize the medium's potential, encouraged by the inception of regular audience research, which had been brought about by the new networks. The first books on radio advertising appeared in 1927. After 1929 the Depression pushed more direct or "hard sell" approaches to radio advertising, first at local stations and more slowly at the network level. Indeed, by 1931–32, major ad agencies had taken on the role of programmers for many networks and a few large station programs, providing casts and even finished productions. The first station representative (rep) firm, Edward Petry, was formed in 1932 to ease the buying of radio spot advertisements in markets across the country. That all of this was successful is indicated in radio's proportion of all advertising, which grew from only 2 percent in 1928 to nearly 11 percent just four years later. By 1932 an FRC survey found that 36 percent of all radio time had commercial sponsorship, meaning that for 64 percent, networks and stations sustained the costs of production.

With half or more of all broadcast time, music remained the most important kind of program on both networks and local stations, with variety (still drawn heavily from vaudeville) a close second.

Virtually all of these programs were broadcast live, because recorded programming was looked down upon by major broadcasters. Drama and comedy developed more slowly as writers and actors overcame the problem of an audience that could not see the action taking place. Though certainly racist by modern standards, *Amos 'n' Andy* became the first network comedy hit in 1929. The first westerns and thriller dramas began in 1929 and 1930, including such long-lasting hits as *The Shadow.*

Evening network news programs began with Lowell Thomas' weeknight 15-minute program on NBC in 1930. That even in its fledgling state, radio represented a threat to the press became increasingly obvious, and a short-lived press–radio "war" began in 1933, with newspaper and major news associations attempting—unsuccessfully as it would soon turn out—to limit the amount of news reports made available to radio. As it had since 1920, radio continued to cover national political conventions and election nights with the latest voter tallies and analysis.

Radio's audience continued to expand, thanks in part to better and easier-to-tune radio receivers. By 1928 plug-in receivers began rapidly replacing cumbersome battery-powered sets. Speakers improved, and users could tune radios using a single control rather than the former two or three. Radios also became fancier, virtually furniture in their own right, made by dozens of manufacturers. That more people were listening became obvious after 1929 and the inception of regular audience research. Archibald Crossley created the Cooperative Analysis of Broadcasting (CAB) to develop program ratings—how many people were tuning to CBS and NBC programs. The CAB relied on random telephone calls to households for its listening information.

1934–41

Viewed in retrospect, the golden age of network radio programming during the later Depression years saw a flowering of radio creativity in the last peacetime era that had no competition from television. More than 200 stations went on the air, including the first stations to serve many smaller communities. The growing number of stations forced many to utilize directional antennae to avoid interference with other outlets—fully a quarter of all stations used such antennae by 1941. Still, about a third of the nation's listeners got their only reliable evening radio service from one of 52 "clear channel" stations, all located in major cities.

The period's beginning is marked by the FRC's replacement in mid-1934 with the Federal Communications Commission (FCC), which had expanded regulatory responsibilities for wired as well as radio communications. Key provisions of the 1927 act, including most of those concerning radio, simply carried over into the new law.

By this time, success in broadcast station operation meant having an affiliation agreement with CBS or NBC Red or Blue—or, after mid-1934, with the Mutual Broadcasting System—and being able to carry their popular programs. A number of outlets also joined such regional chains as the Colonial or Yankee networks in New England, the Texas network, or the Don Lee network on the West Coast. The networks' efficient provision of programs and advertising truly dominated radio by the late 1930s. Networks also dominated their affiliates with one-sided contracts that bound the station for up to five years but bound the network for only one year at a time. This led to an FCC investigation of network operations from 1938 to 1941, which concluded that network control was too strong and that many practices needed change, chief among them the ending of NBC's operation of dual networks.

Most urban and some rural stations programmed 12 hours a day, some for as many as 18. An FCC survey of program patterns in 1938 found that 64 percent of programs were broadcast live (roughly half network and half local), with the remainder being some kind of recording. More than half of all programming was music; talks and dialogues accounted for 11 percent (including President Roosevelt's fireside chats); and 9 percent each were devoted to drama, variety programs, and news. Three new types of program soon dominated this period. Daytime hours were soon saturated with dozens of "soap operas," 15-minute domestic serial dramas, broadcast one after the other. By 1940 the national networks devoted no less than 75 hours a week to such programs, some of which would last into the late 1950s. Quiz and audience participation programs also became hugely popular. On the more serious side, news and news commentary programs greatly expanded with the rising world political crisis in Europe and the Far East and the beginning of World War II in 1939. From about 850 hours annually in 1937, network news and commentary programs grew to fill nearly 3,500 hours by 1941.

Perhaps the single most famous radio program, Orson Welles' dramatization of *The War of the Worlds,* was broadcast in October 1938. Realistic in its use of reporters breaking into music programs (as listeners had just heard during the Munich crisis a month before), the hour-long drama panicked millions and demonstrated both how radio had grown in importance in American life and how much it was trusted by its listeners to tell the truth. More than 90 percent of urban households owned a radio, as did 70 percent of rural homes, and half the country's homes now had at least two radios.

All the earlier talk about radio's educational potential (some 200 noncommercial stations took to the air in the 1920s) had dwindled to the activities of about 35 stations by 1941. These survivors, many licensed to universities, provided in-school enrichment and adult cultural and educational programs to small but loyal audiences. Several national organizations promoted conflicting notions of how radio might best serve educational needs, one urging cooperation with the commercial networks and stations, the other insisting on separation to promote purity of mission.

Another, the National Association of Educational Broadcasters, lasted from 1934 into the early 1980s. Frustrations of educators with an increasingly commercial system led to pressure on the FCC to set aside some channels specifically for noncommercial operations, which resulted in the first such set-aside when FM was approved for regular operation in 1941.

1941–45

In part because of wartime paper rationing, which limited newspaper reporting, radio journalism came of age in the war years, supported by high advertising income and popular entertainment programs. But the industry itself grew only a little during the war because of construction and material limitations. Full wartime restrictions remained in force until August 1945.

The proportion of stations affiliated with a network rose from 60 to 95 percent of all stations. Some stations in smaller markets held agreements with more than one network. The FCC's *Report on Chain Broadcasting* (1941) caused a huge controversy, in large part because it called for an end to NBC's operation of two networks, but also because it sought many other changes in the relationship between networks and their affiliate stations. Two years later the NBC Blue network was sold, becoming the basis for the American Broadcasting Company (ABC) in 1945. Radio prospered during the war as advertising spending rose sharply (in part so businesses could avoid simply paying profits out in federal taxes). Whereas a third of all stations reported financial losses in 1939, only 6 percent did by 1945.

Radio reporters became famous during the war, led by longtime newsman H.V. Kaltenborn on NBC and the CBS team that included Edward R. Murrow in London and William L. Shirer in Berlin. Covering cities under attack and forces fighting in Europe and the Pacific, radio brought the war to listeners at home while also providing propaganda broadcasts by enemy and Allied countries alike.

1945–52

Radio both expanded and contracted in this period—the number of stations more than tripled, but radio networks all but disappeared in the face of television competition. Most of radio's growth took place in smaller markets. What had been a largely AM-centered small industry grew in complexity to include FM and television outlets and networks. One measure of radio's declining glory was the release in March 1946 of the FCC *Blue Book,* which traced wartime profits and compared them unfavorably with radio's heavily commercial programming.

This weakening is most evident in the rapid decline of radio networks, which had dominated the business in 1945 but had all but disappeared by 1952. By then networks provided only a memory of their former service, with daytime soaps, some sustaining dramas, newscasts, and special events—including political year broadcasts. At the same time, their share of radio advertising dropped from nearly half in 1945 to just over a quarter by 1952. Large parts of the broadcast day once programmed by networks were now returned to stations to fill as best they could. One new network—the Liberty Broadcasting System—briefly thrived in 1949–52 based largely on Gordon McClendon's skillful recreations of professional baseball games based only on wire service reports and sound effects.

But radio's chief role in this brief but important period of transition was to provide the revenues that supported expansion of television service. Radio revenue increased each year, though it had to be divided among a vastly larger number of stations, which meant that many radio outlets operated in the red. Noncommercial radio expanded as well, thanks to reserved FM channels.

The most popular dramatic and comedy programs either began to "simulcast" on radio and television, or they moved over to television entirely. The networks began to offer cheaper music and quiz shows, of which *Stop the Music* was one example, to attract listeners by offering a chance at big-money prizes. ABC, the newest and smallest network, broke the long-lasting national taboo on recorded programs by using prerecorded transcriptions of many of its programs. Local stations paralleled that practice with "musical clock" programs that provided a local disc jockey with records, weather, news, occasional features—and constant time checks. Affiliates were soon providing more local than network programming—a throwback to the 1930s.

1952–80

This was an era of competition—with radio playing a distinct second fiddle to the country's fascination with television—and of creativity, as radio developed the many music formats that would give the medium a renewed lease on life. At the same time, the coming of stereo recording and then stereo FM gave that medium a huge boost.

The number of stations continued to expand—by about 100 AM outlets each year, and after 1958 by a revived FM business as well. The rising station population led to problems—where a third of AM stations operated only in the daytime in 1952, fully half were required to do so by 1960 in a continuing FCC attempt to reduce nighttime interference. More stations used directional antennae to reduce daytime and evening interference. Indeed, for much of the 1962–73 period, the FCC froze AM applications and tried—largely successfully—to steer radio growth to the FM band. The commission's decision to ban simulcasting by co-owned AM and FM stations in the late 1960s speeded the expansion of FM, which for the first time had a separate identity for most listeners.

Radio networks still demonstrated some program originality in the 1950s. NBC's *Monitor* created a weekend magazine program beginning in 1955, and the same network offered science fiction with *X-Minus-One;* CBS created *Gunsmoke* for radio before transferring the show to television. *Amos 'n' Andy* became a variety program. The last bastion of advertiser-supported network radio was the daytime soap opera, which only finally disappeared in 1960. The networks reverted to "news on the hour" for affiliates.

Freed of their network ties, most local stations at first stuck with middle-of-the-road (MOR) formats, trying to offer a bit of everything to everyone, including music, talk, variety, and features. Of more fundamental importance to radio's future was the slow mid-1950s development of formula or "Top 40" formats, which were dependent on the personality of a local disc jockey and on tightly formatted music and advertising. Tod Storz and Gordon McClendon are both credited with creating the program approach first used by about 20 stations in 1955 and by hundreds by 1960. Top 40 aimed primarily at teens with what became known as rock and roll music, which was based closely on African American rhythm and blues. The arrival of Elvis Presley in 1956 as the first rock superstar helped cement the new radio trend, though Congressional investigations of payola late in the decade cast doubts, certainly in adults' minds, about the wholesomeness of the format.

Educational radio received a huge boost with the establishment of National Public Radio (NPR) in 1968, the first national network for noncommercial stations. Though excluding hundreds of smaller FM outlets, the 300 to 400 large-market FM outlets that served as NPR "members" brought listeners the highly popular programs *All Things Considered* and *Morning Edition,* among others.

Since 1980

In one sense, the story of radio since 1980 is turned upside down from its history before that date, because beginning in 1979, radio was increasingly dominated by FM listening. By the late 1980s, FM stations attracted three-quarters of the national audience—and by 1995 nearly 60 percent of the stations. All radio outlets specialized in programming in their attempt to gain and hold a tiny sliver of the audience divided by the early 21st century across more than 13,000 stations.

Two technical developments with AM radio were failures. The first was an FCC attempt to narrow AM channels to 9 kilohertz (parallel to practice in most of the rest of the world) down from the 10-kilohertz channels standard since the 1920s. But industry opposition (based on fears of an influx of new competitors) stopped the idea cold in the early 1980s. The second attempted development, offering stereo on AM outlets, was seen by some as one means of slowing AM's decline. Competitors developed a half-dozen mutually incompatible means of delivering stereo in AM's narrow band. Though in 1980 the FCC selected one (by Magnavox) as it had done with prior new technologies, industry ridicule of the decision and the very close parameters of the competing systems led the commission to revisit the matter and in 1982 to allow any stereo system. This experiment in marketplace economics failed because nobody could agree on which system to select (or how best to make such a choice), and thus most stations ignored all of them. That most AM programs by then were talk and news formats less susceptible to the benefits of stereo contributed to the stillborn technology. Extension of the AM band up to 1705 kilohertz by 1990, the first change since 1952, allowed the FCC to reassign a number of stations to the new frequencies to reduce interference lower in the band.

A popular new format appearing by the late 1980s was nostalgia—"golden oldies"—stations, which allow former teens to relive their childhoods with the music of the 1950s through the 1970s. Another was religious stations. Although such outlets had existed from the early days of radio, by the 1980s hundreds of conservative and evangelical religious broadcasters were becoming a force in the business. By 2000 both AM and

FM radio were continuing to grow, with both high expectations and fears being expressed about the inception of digital audio broadcasting (DAB) in the new century. Further, many radio stations had expanded operations to the internet by means of audio streaming, thus greatly expanding their reach beyond their home markets.

Passage of the Telecommunications Act of 1996 along with related FCC rule changes triggered a wave of consolidation in radio station ownership lasting into the early 21st century. More than 40 percent of all stations changed hands and group owners with more than 100 stations became common; the largest owner controlled more than 1,200 outlets. Additionally, it became possible to own up to eight stations in the largest markets. Critics argued that this consolidation contributed strongly to programming that offered broad appeal but little specialization or inventiveness. One response was the launch of two satellite-delivered subscription digital radio services in the early 2000s, both of them providing 100 channels (with about 60 music and 40 talk formats), many without any advertising.

The FCC approved technical standards for digital terrestrial stations that over time would totally replace analog AM and FM outlets. Many stations offered their service over the internet (indeed, there were a growing number of internet-only audio services), although copyright controversies threatened their future.

In its ninth decade, radio in America was truly ubiquitous, ever-present yet often heard only in the background against a growing din of media services. The coming transition to an all-digital service promised finer-quality sound, though less change in program content.

CHRISTOPHER H. STERLING

Further Reading

Archer, Gleason Leonard, *History of Radio to 1926,* New York: American Historical Society, 1938
Archer, Gleason Leonard, *Big Business and Radio,* New York: American Historical, 1939
Barfield, Ray E., *Listening to Radio, 1920–1950,* Westport, Connecticut: Praeger, 1996
Barnouw, Erik, *A History of Broadcasting in the United States,* 3 vols., New York: Oxford University Press, 1966–70
Carothers, Diane Foxhill, *Radio Broadcasting from 1920 to 1990: An Annotated Bibliography,* New York: Garland, 1991
Douglas, Susan J., *Listening In: Radio and the American Imagination,* New York: Times Books, 1999
Fornatale, Peter, and Joshua E. Mills, *Radio in the Television Age,* Woodstock, New York: Overlook Press, 1980
Godfrey, Donald G., and Frederic A. Leigh, editors, *Historical Dictionary of American Radio,* Westport, Connecticut: Greenwood Press, 1998
Greenfield, Thomas Allen, *Radio: A Reference Guide,* New York: Greenwood Press, 1989
Hilliard, Robert L., and Michael C. Keith, *The Broadcast Century: A Biography of American Broadcasting,* Boston: Focal Press, 1992; 3rd edition, as *The Broadcast Century and Beyond,* 2001
Hilmes, Michele, *Radio Voices: American Broadcasting, 1922–1952,* Minneapolis: University of Minnesota Press, 1997
Hilmes, Michele, *Only Connect: A Cultural History of Broadcasting in the United States,* Belmont, California: Wadsworth, 2002
MacDonald, J. Fred, *Don't Touch That Dial: Radio Programming in American Life from 1920 to 1960,* Chicago: Nelson-Hall, 1979
Maclaurin, William Rupert, *Invention and Innovation in the Radio Industry,* New York: Macmillan, 1949
Nachman, Gerald, *Raised on Radio: In Quest of the Lone Ranger, Jack Benny . . . ,* New Pantheon Books, 1998
Settel, Irving, *A Pictorial History of Radio,* New York: Citadel Press, 1960
Smulyan, Susan, *Selling Radio: The Commercialization of American Broadcasting, 1920–1934,* Washington, D.C.: Smithsonian Institution Press, 1994
Sterling, Christopher H., *Electronic Media: A Guide to Trends in Broadcasting and Newer Technologies, 1920–1983,* New York: Praeger, 1984
Sterling, Christopher H., and John M. Kittross, *Stay Tuned: A History of American Broadcasting,* 3rd edition, Mahwah, New Jersey: Lawrence Erlbaum, 2002

United States Congress and Radio

Broadcasting of House and Senate Proceedings

Congress and radio broadcasting have been closely linked for more than 90 years. By virtue of the U.S. Constitution's "commerce clause," Congress has created and amended basic legislation to regulate the industry. It has also funded regulatory agencies and approved appointments to them. But only after years of debate was regular broadcast coverage of Congressional committee and floor activity finally allowed, thereby joining the long-time use of broadcasting media by individual members seeking reelection.

Legislative interest in radio began with the 1910 Wireless Ship Act that regulated maritime use of radio. The Radio Act of 1912 was the first comprehensive radio statute, but it was not designed to regulate radio broadcasting, which did not then exist. During and after World War I, Congress heard testimony from the military and from the Post Office urging federal control and operation of radio broadcasting. Turning away from that option, Congress was perplexed over what to do to regulate early radio broadcasting, and it did not pass the first law designed to regulate broadcasting until the Radio Act of 1927. The new law laid the regulatory foundation of U.S. broadcasting, and most of it is still in effect today. In mid-1934, acting on recommendations from President Roosevelt and its own extensive investigation of communication companies, Congress passed the comprehensive Communications Act of 1934, which is still in force in amended form nearly seven decades later. In the years since, Congress has regularly tinkered with the law, considering and sometimes adopting amendments (of which the Telecommunications Act of 1996 was by far the most extensive). When not actually legislating, Congress plays three other regulatory roles with radio: it provides annual budgets for operation of the Federal Communications Commission (FCC) and related agencies; the Senate approves (nearly always) presidential nominations of FCC commissioners; and commerce committees in both houses regularly conduct "oversight" hearings into industry and FCC activities and decisions.

Seeking to Broadcast Congress

From the inception of radio, some members of Congress perceived the new medium as the perfect way to carry the people's business to those unable to travel to Washington to visit the House and Senate in person. And such coverage would provide radio with just the kind of content that would help popularize the medium.

Efforts to initiate some form of congressional radio began with a resolution introduced by Representative Vincent Brennan (R-Michigan) in 1922 that was intended to allow for the "installation and operation of radiotelephone transmitting apparatus for the purpose of transmitting the proceedings and debates of the Senate and the House of Representatives." Though the Brennan resolution failed, its intent remained alive when Senator Robert Howell (R-Nebraska) introduced a resolution two years later directing that radio experts from the War and Navy Departments be appointed to study the feasibility of "broadcasting by radio of the proceedings of the Senate and the House of Representatives throughout the country, utilizing the radio stations of the War and Navy Departments." As the Senate discussed the Howell resolution, several arguments were made for and against broadcasting from the congressional chambers that would be repeated in years to come. On the one hand, listeners to congressional debate could hear their elected officials engage in the country's business—and in the process, that debate would be improved in order to give listeners a good impression of Congress. On the other hand, listeners' impressions might be detrimental to congressional members whose speaking skills were not up to radio standards or who were absent during chamber debate.

The report prepared by the War and Navy Departments finally materialized in 1927; the report concluded that broadcasting from the House and Senate chambers would be not only too costly, but also technically infeasible. This effectively ended any substantive efforts to implement congressional radio. Nonetheless, the idea remained alive. In fact, it was invigorated by radio's coverage of President Calvin Coolidge's 4 March 1925 inaugural address. Excitement over the president's radio remarks led the editor of *Radio Broadcast* to "hope that soon Congress will be forced to broadcast its activities." The editor's remark ironically draws attention to another argument that would be used in years to come by congressional broadcasting advocates: that of the imbalance in power that the president's effective use of radio caused with regard to clashes between presidential policy initiatives and congressional policy initiatives.

Efforts to persuade their congressional colleagues to allow broadcasting of Senate and House proceedings were continued by members of both bodies, most notably by Senators Clarence Dill (D-Washington) and Gerald Nye (R-North Dakota), into the 1940s. Meanwhile, parliamentary bodies in other countries experimented with broadcasting deliberative activities during the late 1920s. Germany and Japan were two countries where the idea seemed to be catching hold. The British House of Commons considered but rejected the idea of broadcasting its parliamentary sessions in 1926.

Congressional broadcasting of a very limited and decidedly clandestine nature did occur in December 1932, when the U.S. House voted on the repeal of the 18th Amendment. Radio network representatives had requested permission from House Speaker John Nance Garner (D-Texas) to cover debate on the matter from the House chamber. Undeterred by Speaker Garner's refusal to grant permission, the networks positioned microphones in the doorway of a library adjoining the House chamber and boosted the microphones' volume high enough to pick up the proceedings. Radio listeners had the rare privilege of hearing the repeal vote as it occurred. Afterwards, the *New York Times* took note of the event, saying that "broadcasters have taken a new hope that before long radio will invade Congress just as it has almost every other realm where people speak or sing."

Interest in congressional broadcasting emanated from other directions during the 1940s. For instance, a number of organizations, including the Congress of Industrial Organizations, the Writers War Board, and others, advocated allowing radio coverage of congressional proceedings. World War II certainly stimulated interest in the matter, but more important were the shortcomings of newspapers during the period in properly informing the public about congressional consideration and discussion of important wartime issues. A 1946 poll showed that the general public favored the idea of congressional broadcasting, and a poll of radio executives taken at roughly the same time showed that an overwhelming 70 percent of them also favored such broadcasts.

The first serious consideration of congressional broadcasting occurred during hearings conducted by the Joint Committee on the Organization of Congress in 1945. The committee, cochaired by Senator Robert M. LaFollette, Jr. (Progressive-Wisconsin) and Representative Mike Monroney (D-Oklahoma), was charged with exploring methods to improve the legislative process. Witnesses at hearings addressed many of the issues about congressional broadcasting that had been raised before. The issues dealt with both style (public reaction to poor or verbose speakers, to the Senate filibuster, or to the absent speaker) and substance (equal treatment of important issues during floor debate and the manner by which chamber proceedings from the House as well as the Senate might be broadcast simultaneously). Advocates for congressional broadcasting failed to win a great following, and in the end the Legislative Reorganization Act of 1946 that evolved from the work of the Joint Committee on the Organization of Congress ignored broadcasting.

Initial Efforts

Interest in congressional broadcasting changed direction by the late 1940s after television forced radio into a secondary role. Members of Congress had grown more at ease with broadcasting in general and thus were willing if not eager to allow television cameras to cover congressional hearings. Rules devised by each congressional committee soon emerged to regulate such coverage; these rules placed television and radio on an equal footing. Committee members had final approval as to whether television and radio could cover committee proceedings, but television and radio networks and stations that controlled the cameras and microphones determined how much of the proceedings would be broadcast. Several famous Senate investigations were televised in the 1950s, including the Kefauver crime hearings in 1951 and the Army-McCarthy hearings in 1954. Radio networks and local stations rarely chose to carry more than short excerpts from committee proceedings during newscasts. Only during the 1970s did National Public Radio (NPR) break tradition by airing some of the more notable congressional hearings virtually gavel to gavel, such as a hearing on a new Panama Canal treaty.

Efforts to regularly broadcast House and Senate chamber proceedings were reignited in the early 1970s, possibly owing to successful broadcasts of committee proceedings. The matter received special attention by the Joint Committee on Congressional Operations in 1973. The favorable view taken by the committee—that broadcasting from the House and Senate chambers could help Congress better communicate with the public—led directly to a decision by the House of Representatives to open itself to television cameras and radio microphones in March 1979. The Senate was slower to act but finally opened its own chamber to television and radio in June 1986.

Regular Live Coverage

Ironically, the Senate actually preceded the House in allowing live broadcast coverage to originate from the Senate chamber. A resolution introduced by Senator Robert Byrd (D-West Virginia) in 1977 called for the Senate to allow broadcast coverage of debate over ratification of the controversial Panama Canal Treaty, approval of which would return control of the Panama Canal from the United States to Panama. Senator Byrd's resolution would have allowed both television and radio coverage of the debate, which began on 8 February 1978, but unresolved technical questions forced the Senate to drop plans for television and to allow exclusive coverage to radio. The Columbia Broadcasting System (CBS) and the National Broadcasting Company (NBC) aired portions of the debate, and NPR aired the entire debate, with only brief interruptions to identify speakers.

Not to be outdone by the Senate, House Speaker Thomas "Tip" O'Neill (D-Massachusetts) announced that beginning on 12 June 1978, radio networks, stations, and news organizations would be able to pick up feeds from the House audio system for broadcast. The decision came in response to a request by Associated Press (AP) Radio. As it happened, AP Radio elected to carry only five minutes of live House debate on the day that radio was finally given a green light to cover chamber proceedings. This brief coverage, however, outdid other radio networks, which carried only brief taped excerpts of House debate during regular newscasts.

Just as rules exist for broadcast of congressional committee proceedings, so do similar rules exist for broadcast coverage of House and Senate floor debate. The Speaker of the House has final authority over all broadcast activity from that body, but House rules also stipulate that coverage of chamber proceedings must be unedited and must not be used for political purposes or commercial advertisements. The Senate Rules and Administration Committee has ultimate authority over the Senate broadcasting system, but the Senate sergeant at arms is authorized to act in the committee's behalf. The same rules on editing and on the political and commercial use of chamber broadcasts that exist in the House also govern the Senate.

Anything that emanates from the microphones in the House and Senate chambers may be broadcast by any radio network or station. As is the case with committee proceedings, neither networks nor stations generally choose to air more than brief excerpts of floor debate. A departure from that practice occurred during the House impeachment and Senate trial of President Bill Clinton. Radio coverage of these two events during late 1998 and early 1999 was extensive, and in the case of NPR the coverage was gavel to gavel.

Radio and Individual Members of Congress

The unsuccessful efforts to ignite interest in broadcasting chamber debates during the 1920s and 1930s might have stemmed from radio network policies that allowed U.S. senators and representatives free airtime. Records show, in fact, that between 1928 and 1940, CBS allowed some 700 U.S. senators and some 500 House members to speak on the network.

Possibly spurred by the interest shown in congressional radio during the 1945 Joint Committee on the Organization of Congress, in 1948 Robert Coor created a commercial operation known as the Joint Radio Information Facility, which provided a recording studio for individual members of Congress to record speeches and interviews for use by radio networks or local radio stations. Congress brought recording under its own control as part of the Legislative Appropriations Act in 1956, creating separate House and Senate recording studios. Studio operating costs were covered by charges to those members of Congress who actually used them. In later years, both the House and the Senate studios were upgraded for television and for live links with news-gathering organizations via satellite. And besides the congressional studios, individual senators and representatives by the 1990s had access to recording facilities operated by the Democratic and Republican parties.

Radio used purely as a campaign tool—a practice nearly as old as broadcasting itself—stands apart from the other uses of radio already noted. The first known use of radio for campaigning occurred in 1922, when Senator Harry S. New (R-Indiana) used the U.S. Navy's radio station in Washington, D.C., to address his constituents at home. The *New York Times* took note of the occasion, saying that "campaigning by radio soon might leave the field of novelty and become a practical everyday proposition during political fights." The small first step into campaign radio became a rush by 1924.

Radio was readily regarded as a valuable campaign tool, but fears arose over the possibility that those who owned radio stations might create unfair advantages by allowing candidates whom they supported easy access to the airwaves while refusing to allow use of their radio facilities to candidates whom they opposed. The issue was thoroughly discussed during the Washington Radio Conferences, with the resulting recommendation that Congress include some provision in the Radio Act of 1927 that would ensure not only that there would be equal opportunities for use of radio facilities by candidates for public office, but also that the content of any campaign message via radio would not be censored. The recommendation was approved by Congress and fashioned into Section 18 of the Radio Act, which was transferred intact as Section 315 of the Communications Act of 1934. Section 315 required that equal opportunities be extended to political candidates after their opponents had been granted use of a broadcasting facility. Broadcasters were still free to decide whether to extend initial use of their facilities for campaign purposes. That changed somewhat in 1971, when Congress amended the Communications Act to include language in Section 312 that required broadcasters to provide time for legally qualified candidates for federal elective office.

Radio and the Contemporary Congress

Radio has not only held its own as an important congressional medium, but its importance may be expanding. The "niche" medium that radio has become—meaning that radio stations now fine-tune their formats to reach specific demographic groups of listeners—means that members of Congress can design specific messages and utilize carefully chosen radio stations in their home states or districts to reach precisely the intended audience. This makes radio an especially valuable tool for campaigning. "Talk radio," one of the most popular formats in the 1990s, requires a steady flow of guests willing and able to discuss a range of issues, and the need of members of the U.S. Senate and House to have access to constituent listeners has created a near-perfect symbiosis between radio and members of Congress.

RONALD GARAY

See also Communications Act of 1934; Dill, Clarence; Equal Time Rule; Politics and Radio; Telecommunications Act of 1996; Wireless Acts of 1910 and 1912/Radio Acts of 1912 and 1927; White, Wallace

Further Reading

Archer, Gleason Leonard, *History of Radio to 1926*, New York: American Historical Society, 1938; reprint, New York: Arno Press, 1971
Chester, Edward W., *Radio, Television, and American Politics*, New York: Sheed and Ward, 1969
Garay, Ronald, *Congressional Television: A Legislative History*, Westport, Connecticut: Greenwood Press, 1984
Garay, Ronald, "Broadcasting of Congressional Proceedings," in *The Encyclopedia of the United States Congress*, edited by Donald C. Bacon, Roger H. Davidson, and Morton Keller, New York: Simon and Schuster, 1995
Jones, David, "Political Talk Radio: The Limbaugh Effect on Primary Voters," *Political Communication* 15 (1998)
Kahn, Frank J., editor, *Documents of American Broadcasting*, New York: Appleton-Century-Crofts, 1968
Katz, Jeffrey, "Studios Beam Members from Hill to Hometown," *Congressional Quarterly* 29 (November 1997)
Ostroff, David, "Equal Time: Origins of Section 18 of the Radio Act of 1927," *Journal of Broadcasting* 24 (Summer 1980)
Sarno, Edward, "The National Radio Conferences," *Journal of Broadcasting* 13 (Spring 1969)
Smith, F. Leslie, Milan D. Meeske, and John W. Wright II, *Electronic Media and Government: The Regulation of Wireless and Wired Mass Communication in the United States*, White Plains, New York: Longman, 1995

United States Influences on British Radio

Commercial Thinking and Public Service Systems

United States radio broadcasting has had varied impacts, both positive and negative, on British and European domestic radio systems over the years. Only relatively recently have most foreign radio systems begun to substantially parallel U.S. program and commercial approaches.

The attitudes of British radio broadcasters, managers, and listeners toward American radio have ranged from contempt, hostility, and condescension to craven admiration and imitation. The British system of a monopoly public-service broadcaster, largely financed by a compulsory license fee on radio receivers (which lasted for more than 50 years) was in part due to a reaction against the commercial U.S. system.

To World War II

In the 1920s and 1930s, the British establishment was appalled that advertisers might influence as well as finance programs on the new and powerful medium—as they were clearly doing in the U.S. The view of the British Broadcasting Corporation (BBC)—personified by its severe and highly religious first director-general, John Reith—was that radio should be used primarily as a means of educating and informing the public, ensuring that the highest and most sophisticated cultural forms are available to the humblest citizen. There was concern that British culture would be corrupted by "American-style" programming, and not just in "high culture." There was a belief among the British elites that U.S. popular music and comedy were inherently inferior and less intellectually demanding than the U.K. forms. United States radio, it was argued, was merely serving the "lowest common denominator," and any adoption of its forms and values would diminish and even corrupt the potential of the medium and the audience that it served. Many in the BBC and other parts of the establishment were especially dismissive of the production of drama in the United States in strict 30- or 60-minute slots, unlike the longer and often untidily timed dramas on the BBC (which of course did not have to worry about a matrix of programming to comply with numerous network/local "junctions"). British critics also inveighed against the preponderance of "soap operas" and thrillers, which were regarded as evidence of an intellectually limited use of the medium.

The idea that radio should have an entertainment and populist aim was the antithesis of Reith's driving philosophy; so long as the public did not have a choice of listening to alternative, lighter fare, then the "improving" nature of British radio would continue unchallenged. To Reith's dismay, however, in the early 1930s entrepreneurs aware of U.S. commercial stations realized there was a potentially lucrative gap in the market. Several commercial radio stations sprang up on the European continent (naturally outside the jurisdiction of British law), targeting British listeners with a diet of popular dance band music, comedy programs, and other forms of light entertainment. This was especially true on Sundays, which Reith insisted on dedicating to serious-minded programs, including religious services. These stations—with Radio Normandy and Radio Luxembourg having the greatest impact—copied many of the forms and financing of U.S. commercial radio of the time, including sponsored programs. The continental stations achieved audiences in the tens of millions and easily outstripped those tuned to the BBC services on Sundays and in the early-morning periods.

World War II and After

The European commercial stations closed down when Nazi Germany occupied most of Europe. More controversy was created in July 1943 when the American Forces Network (AFN) established a radio service carrying many American network programs for U.S. troops stationed in the United Kingdom using the BBC's emergency facilities. This was the only time in the half century from the inception of the BBC in 1922 to the introduction of authorized commercial radio in 1973 when the BBC's monopoly of British radio broadcasting was officially broken. There was much nervousness and even antagonism from many in the British establishment that listeners might find AFN's more informal and populist programming very attractive and demand something similar when the war ended. Indeed, many British servicemen heard AFN services in other parts of the globe and enjoyed the programming, in some cases more than their own Forces' Programme.

After World War II, the BBC made some concessions to popular fare by establishing the Light Programme, which shared its transmissions with the postwar Forces' Programme. However, for the most part, the BBC remained impervious to American style radio, and the policy of succeeding governments was to continue the BBC's radio monopoly—even after commercial television was introduced in 1955. Once again British listeners had to tune to Radio Luxembourg for a taste of U.S. style diet of record request programs and quiz shows.

It was the arrival of rock and roll, the attendant development of a distinctive youth culture, and television's growing domination of evening hours that presented the greatest challenge to the BBC's policies. The BBC seemed content to allow the slow death of radio as a mass medium. In particular its managers were appalled by rock and roll and only reluctantly allowed a few hours a week of such music even on the Light Programme. Only pirate stations and Radio Luxembourg provided an American-style alternative for British listeners.

Although the U.K. government forced most of the pirates off the air through an Act of Parliament that made it illegal for a U.K. citizen to supply or be employed by such a station, the BBC was ordered to produce a replacement, called Radio 1. Now the U.S. Top 40 station jingles were recycled yet again—"Wonderful Radio London" became "Wonderful Radio One," and the BBC employed many of the best-known pirate broadcasters—especially those from Radio London.

British Commercial Radio

Nevertheless, pressure for legalized commercial radio continued to grow, and in 1972 the Conservative government introduced an Act of Parliament that led to the establishment of a chain of independently owned commercial services. However, the Act specifically proscribed continuous music channels on the U.S. model: the U.K. stations were to be based on the public service model and to be "full service," including high levels of current affairs output and even drama and comedy programs, although many of these were modeled more on those common on U.S. network commercial radio in the 1930s and 1940s than on the BBC's output. Many U.K. broadcasters and managers who had hoped for something equivalent to the U.S. system were very disappointed by the strictures of the U.K. system. Some did their best to get around the regulations and produce as close a format to the U.S. stations as they could get away with. Piccadilly Radio in Manchester was perhaps the most notable example of this, which is not surprising, as its managing director, Philip Birch, had also been managing director of pirate Radio London. Several stations—including Radio Trent in Nottingham, Metro Radio in Newcastle-upon-Tyne, and Beacon Radio in Wolverhampton—imported U.S. jingles, although these had to be partly rerecorded in the United Kingdom to comply with Musicians' Union rules.

Nor were the attempts at imitation of U.S. radio confined to the Top 40 format. Many in the U.K. industry recognized that by the time commercial radio was legalized in the mid-1970s, this format, based on the sales of the 45-rpm single, was past its peak, with record sales now dominated by the 33 1/3-rpm album. Some managers and broadcasters also felt that the early vitality of Top 40 radio had now been dissipated and was now perceived as embarrassingly unsophisticated and "uncool." Consequently, stations such as Capital Radio in London (at least in its early days) and Beacon Radio opted for a U.S.-style FM format. Beacon in particular, which had an American managing director and a Canadian program director, attempted a U.S. West Coast FM-style format for its main daytime hours. This style was also much favored by Radio 1 disc jockey Johnnie Walker (formerly of Radio Caroline and Radio England) who, in ill-concealed protest at the "pap" he was forced to play, left Britain at the height of his popularity in 1976 to go to the United States, where he worked for a time at San Francisco's "free-form" FM station, KSAN.

A number of British stations continued to demonstrate a fascination with U.S. chart music: Beacon, for example, had a four-hour program on Sunday nights of the *Billboard American Hot 100,* and Casey Kasem's legendary *Countdown* program was transmitted on several U.K. stations in the 1980s and 1990s. BBC Radio 1—and later Radio 2—also featured a weekly U.S. chart countdown.

As the price of transatlantic air fares dropped in the 1970s, many British disc jockeys and programmers traveled to America for the first time. Although many had heard tapes of U.S. stations, the impact of hearing U.S. stations at first hand—in particular, the tightly defined format of most stations, their relative lack of public service elements, more aggressive presentation styles, and less reverential attitudes toward authority figures—greatly impressed some of the Britishers. Many returned from such vacations determined to radically alter their approach, sometimes with startling results. The use of U.S. topical humor services also became vogue among British disc jockeys. Contests, promotions, production techniques, technical innovations such as compression, the use of computer programs for music selection backed by intensive audience research, and, not least, the style of U.S. commercial copy provided much material for British stations.

Specialization

Nor was it only music radio formats that impressed U.K. station managers, investors, and broadcasters. The only commercial all-news service, LBC in London, was largely modeled on New York's WINS, and the informality of news presenters at WINS and their willingness to mix news with comment were enthusiastically adopted by the station. Some of this imitation in turn fed through to the BBC services nationally and locally.

Many local full-service commercial services included a late-night controversial phone-in show, some of which were clear imitations of the U.S. "shock jock" phenomenon—albeit considerably toned down in recognition of the different regulatory requirements, the absence in the United Kingdom of an equivalent to the U.S. Constitution's guarantee of free speech, and the necessity to maintain political balance and to conform to much stricter libel laws. But on both music and talk programs, the attitude of British presenters became noticeably less polite and deferential, especially toward figures in government and the royal family—although not all of this can be directly linked to U.S. radio's influence.

The imitation of the U.S. shock jock form reached its peak in the mid-1990s, when the last of the analogue national commercial stations, Talk Radio, adopted an all-phone-in format. However, the station produced disappointing audience figures, and after several changes of management and ownership, it changed format and name in January 2000 to Talk Sport.

In the meantime, there was continued political pressure for a more U.S.-based commercial model of licensing and regulation, which found favor under Conservative governments in the 1980s and 1990s. Most significantly, following the Broadcasting Act of 1990, there was a rapid end to the full-service or mixed British programming model—the middle-of-the-road format that had already been dropped by U.S. stations by the mid-1950s. Instead there was a trend toward a single format, usually either contemporary hit radio on FM or "Gold" on AM (most stations at that time were on both AM and FM; the ending of simulcasting was some 25 years behind the United States). Further, the large-scale switch to automation, networking, and conglomeration of radio companies in the United States has to some extent been replicated in the United Kingdom.

The manifest influence by the United States on British radio, however, seemed to be diminishing by the end of the 1990s. Radio 1 had restyled itself as a noncommercial, adventurous, youth-oriented station, less concerned with audience ratings than it had been in the past—in many ways the antithesis of most U.S. radio. Meanwhile, the commercial sector, now attaining at least 50 percent of all listening and with rapidly growing revenues, appeared to be less impressed by U.S. forms, and some managers even proclaimed that now the United States had more to learn from the United Kingdom than the other way around. The United Kingdom's adoption of the "European" digital radio standard (rather than the IBOC system developed in the United States), was another indication of a deliberate distinction between the two countries' radio services.

Nevertheless, much of the criticism of U.S. commercial radio—that it has become overcautious and that it increasingly lacks creativity and diversity—are also now leveled at radio in the United Kingdom. Some in the industry, along with many observers, believe that the swing from the peculiarly British, highly regulated public service commercial radio model to the laissez-faire United States style system has been too great. Some concede that Reith's belief that commercial broadcasting must inevitably lead to crass and undemanding broadcasting aimed at the "lowest common denominator" has been vindicated.

RICHARD RUDIN

See also British Broadcasting Corporation *and associated BBC essays*; British Commercial Radio; Capital Radio; London Broadcasting Company; Public Service Radio; Radio Authority

Further Reading

AFN History, <www.afneurope.army.mil/Program/history.htm>

Barnard, Stephen, *On the Radio: Music Radio in Britain,* Milton Keynes, Buckinghamshire, and Philadelphia, Pennsylvania: Open University Press, 1989

Chapman, Robert, *Selling the Sixties: The Pirates and Pop Music Radio,* London and New York: Routledge, 1992

Crisell, Andrew, *An Introductory History of British Broadcasting,* London and New York: Routledge, 1997

Crook, Tim, *International Radio Journalism: History, Theory, and Practice,* London and New York: Routledge, 1998
Douglas, Susan J., *Listening In: Radio and the American Imagination: From Amos 'n' Andy and Edward R. Murrow to Wolfman Jack and Howard Stern,* New York: Times Books, 1999
Fong-Torres, Ben, *The Hits Just Keep on Coming: The History of Top 40 Radio,* San Francisco: Miller Freeman Books, 1998
Garay, Ronald, *Gordon McLendon: The Maverick of Radio,* New York: Greenwood Press, 1992
Hilmes, Michele, *Radio Voices: American Broadcasting, 1922–1952,* Minneapolis: University of Minnesota Press, 1997
Keith, Michael C., *Voices in the Purple Haze: Underground Radio and the Sixties,* Westport, Connecticut: Praeger, 1997
Lister, David, *In the Best Possible Taste: The Crazy Life of Kenny Everett,* London: Bloomsbury, 1996

United States Navy and Radio

During the first two decades of the 20th century, the U.S. Navy served as a principal force in the development of radio communications in the United States. From the introduction of practical radio systems at the turn of the century to the beginning of U.S. radio broadcasting in the early 1920s, the Navy's influence on American radio was powerful and multifaceted. These significant influences included the application and expansion of radio for military and diplomatic purposes, the control of radio communications, technological developments, and the fostering of development in the radio industry.

Origins

At the beginning of the 20th century, electrical communication by telegraph and telephone provided Navy officials with rapid communications between most ports and naval stations worldwide. At sea, the situation was radically different. When a ship left port and disappeared over the horizon, it was isolated from shore communications. Free of new directives from Washington, the captain and fleet could act with complete autonomy.

The invention of radio communications and its application to maritime use brought both benefits and difficulties to the Navy. Some naval officials recognized very early the strategic and tactical benefits of being able to communicate between distant ships and from ship to shore, but the adoption and widespread use of radio on shore and ship installations was slow in coming. The early resistance to radio in the Navy can be attributed to several factors, including the traditional bound bureaucracy and the strong desire of many naval officers to preserve the independence a captain traditionally exercised at sea. To many naval officers, radio was viewed as the ultimate centralizing force from Washington. The Japanese Navy's success in using radio during the 1904–05 Russo-Japanese War aided in diminishing this resistance.

The Navy first tested radio apparatus in 1899, but serious attention and further testing did not occur again until 1901. Over the next decade, the navy tested and purchased various types of wireless equipment from both U.S. and foreign countries, often resulting in stations with unfavorable composite equipment. The Navy was an early and important potential client for new radio companies; it exerted a positive force in developing the U.S. radio industry, but it was also a force to be reckoned with. Inventors eager to gain valuable Navy contracts regularly found themselves frustrated by the Navy's hardball business tactics and lack of respect for patent rights. The Navy simply acquired equipment as needed and from whom they wanted, regularly ignoring patent restrictions. Complicating the Navy's relationship with U.S. radio equipment suppliers was the perception that the Navy often gave preference to foreign companies. A clear exception to this was its negative attitude toward the British Marconi Company.

The Navy's strong aversion to the Marconi Company and its equipment can be traced to concerns about British domination of the world cable system. Navy fears that the Marconi Company would grow to dominate radio communications, and therefore establish a British hegemony over worldwide communications, were very strong. As a result, Marconi equipment was rejected and the Navy began a two-decade-long effort to establish U.S. control of its own radio communications. Radio waves do not recognize national borders, and the need to dominate and so control the ether was a compelling security issue for Navy officials.

Beginning with the 1903 international radio conference in Berlin, the Navy took on the leadership role of representing U.S. radio interests worldwide. This leadership role had great influence over two decades of shaping the evolving international radio regulations. The Navy's influence in directing self-serving lobbying efforts was also ever-present on the Washington political scene. Congress recognized the importance of radio for the Navy and the national interest, placing the Navy in an influential and dominant position, especially for federal funding. Despite the Navy's efforts to support legislation that would wrest control of radio from commercial and amateur interests, however, access to the airwaves by non-military operators and interests prevailed.

World War I

U.S. entry into World War I on 6 April 1917 changed the entire radio communications scene. The following day, with the exception of army-operated transmitters, the Navy took control of all radio stations in the United States, acquiring 53 commercial stations and closing another 28 transmitters. In addition, all amateur radio stations were shut down "for the duration" of the war. In one swift move the Navy had taken control of the entire radio communications system in the United States, along with its existing international ship and port radio system.

Because of the sudden increase in radio communication traffic, Navy orders for equipment from private manufacturers increased dramatically. The Navy's centralized control over the equipping of ship and shore stations produced many improvements in equipment quality and, importantly, standardization in equipment design. The Navy also conducted significant wartime research and development in radio technology, greatly improving the consistency and quality of long-distance radio communications.

With the complete takeover of radio communications by the Navy, the question arose as to what would happen at the war's end. After the armistice on 11 November 1918, Secretary of the Navy Josephus Daniels began a strong lobbying effort for legislation that would leave radio permanently under Navy control. His efforts ultimately failed—Congress was in no mood to continue an activist federal establishment—and government control of the wartime-seized radio stations was eventually relinquished. On 11 July 1919 President Wilson ordered that all commercial stations be returned to their original owners by 1 March 1920. In addition, amateur radio stations were allowed to resume operation on 1 October 1919.

Radio Corporation of America

The Navy's postwar loss of the control of radio communications was a blow to its geopolitical aspirations, but its influence and power were still very significant. The drive to assure the U.S. a major role in global radio communications and to prevent the British Marconi Company from obtaining too strong a foothold in the U.S. communications market resulted in the Navy initiating and facilitating the formation of the Radio Corporation of America (RCA).

The specific technological innovation driving the navy's concern was the Alexanderson alternator, then the only effective means of achieving long-range radio communication. Developed from designs of Reginald Fessenden and greatly improved by General Electric (GE) engineer Ernst F.W. Alexanderson, rights to the device were controlled by GE. But GE was not interested in getting into the service side of radio; the company defined its role as a manufacturer and sought to sell the devices to recoup its extensive investment. Beginning during the war (1915), the most likely purchaser appeared to be Britain's Marconi Company which, however, sought full and exclusive rights to the alternator. GE was tempted at what would be a lucrative agreement. Navy officials, including assistant secretary Franklin Roosevelt, however, expressed strong concern about such an important device passing into the hands of even a friendly (but still foreign) country. Navy officials pressed GE to seek a domestic means of selling the alternator.

GE's concern paralleled another: the wartime pooling of radio patents so that manufacturers could perfect the best possible radio equipment for army and navy procurement contracts. The end of the war also ended the patent pools, and companies faced the need to develop a peacetime pool if radio was to continue its development. These related needs—GE's for the alternator, and the industry at large to develop a peacetime patent pool—led to the rise of RCA.

Through a series of complex company buyouts and new patent pooling agreements, General Electric established the Radio Corporation of America as a subsidiary in October 1919, and important technological patents (the alternator among them) were eventually pooled under RCA's corporate umbrella, thus preventing any possibility of British radio communications hegemony within the United States. Ownership in RCA was later spun off and it became the single most important firm in early radio broadcasting development.

DOUGLAS K. PENISTEN

See also Alexanderson, E.F.W.; Early Wireless; Fessenden, Reginald; General Electric; Marconi, Guglielmo; Radio Corporation of America; United States Congress and Radio

Further Reading

Aitken, Hugh G.J., *The Continuous Wave: Technology and American Radio, 1900–1932*, Princeton, New Jersey: Princeton University Press, 1985

Douglas, Susan J., *Inventing American Broadcasting, 1899–1922*, Baltimore, Maryland: Johns Hopkins University Press, 1987

Gebhard, Louis, *Evolution of Naval Radio-Electronics and Contributions of the Naval Research Laboratory,* Washington, D.C.: Naval Research Laboratory, 1976
Howeth, Linwood S., *History of Communications Electronics in the United States Navy,* Washington, D.C.: Government Printing Office, 1963
Hugill, Peter J., *Global Communications since 1844: Geopolitics and Technology,* Baltimore, Maryland: Johns Hopkins University Press, 1999
King, Randolph, and Prescott Palmer, editors, *Naval Engineering and American Seapower,* Baltimore, Maryland: Nautical and Aviation of America, 1989

United States Presidency and Radio

On 21 June 1923, when President Warren Harding stepped to the microphone to deliver a speech on the World Court from St. Louis as part of his tour of the western United States, he spoke not just to the citizens of St. Louis, but to those in Washington, D.C., and New York as well. This was the first time that a chain or network of radio stations had been assembled to carry a presidential message simultaneously to several parts of the nation. The speech was heard in St. Louis over KSA and in New York and Washington over American Telephone and Telegraph (AT&T) stations WEAF and WCAP, respectively. Perhaps 1 million Americans heard Harding speak, more than any president had reached before. No longer was a president bound by the flatness of daily newspaper coverage or the geographical limitations of single-station radio coverage; he now had the potential to speak to the entire electorate at once, a power that would enlarge the "bully pulpit" beyond any expectation of the day. With a single flip of the switch, broadcasters could help a president rise above his adversaries in Congress and go directly to the people.

A strong national broadcasting system contributed to a strong presidency. A politician whose voice commanded attention in every corner of the land simultaneously could build a strong national constituency. Conversely, a strong presidency contributed to a strong national broadcasting system. Presidential speeches created a demand and provided one of the few programs that could unite American interests. The American people were eager to hear their national leader over the fascinating new medium of radio, and broadcasters were pleased to be the purveyors.

Early Experiments in Networking

After the success of his first attempt at chain broadcasting in St. Louis, President Harding tried another major speech in Kansas City and had scheduled still another big speech on 31 July 1923. For the occasion, AT&T assembled its first coast-to-coast linkup stretching from San Francisco to New York. Radio had become such an important part of President Harding's western tour that he installed a powerful radio transmitter in his railroad car to give him a mobile broadcasting studio.

Harding died while on his western trip, and within a few days of his death the National Association of Broadcasters (NAB) was proposing to the new president, Calvin Coolidge, that he substitute radio addresses for public appearances to conserve his health and reach a wider audience. By the end of 1923, as the 1924 campaign approached, President Coolidge heeded the advice and took to the airwaves regularly. When Coolidge spoke to Congress at the end of the year, AT&T assembled a chain of seven stations to carry the speech. The president followed that speech with five additional nationally broadcast addresses. The 1924 party conventions reached an estimated 25 million listeners, and the subsequent campaign provided abundant opportunities for chain broadcasting.

A small group of broadcasters questioned whether politicians should be turned loose on radio. At the NAB's first convention in October 1923, John Shepard, III, of WNAC proposed that a political party applying for airtime be required to give comparable time to a speaker from the opposing party. The NAB accepted the measure and followed these "equal time" ground rules during the 1924 campaign. Presidential speeches, however, were not considered "political" except during campaigns and were not subjected to any "right of reply" mechanism until the campaign actually started. Still, voices were raised early to warn of the dangers of a one-sided political dialogue in which the party and congressional opposition had no standing.

By 1924 proponents of chain broadcasting had realized that politics was the perfect bait to lure America into a permanent national system of broadcasting. The presidential election campaign that year provided ample opportunities for demonstrating the virtues of chain broadcasting. AT&T was poised to erect a permanent network of stations and believed that political speeches were an excellent way to ensure frequent use of a system that could reach 78 percent of the nation's purchasing power through the top 24 markets.

The closer the fledgling broadcasting industry could bring itself to the presidency, the higher the status it could bring on itself. Thus, initially there was little concern for the newsworthiness of presidential addresses. Broadcasters saw presidential broadcasting as a means of providing a public service and basking in the prestige of the presidency. Given the open-ended invitation extended early on to presidents by broadcasters, it is no wonder that presidents ever since have regarded their access as a right of the office.

The *New York Times* reported in 1924, "It is a source of wonder to many listeners how a speaker can sit in the White House or stand before Congress and have his voice simultaneously enter the ether over Washington, New York and Providence." The inauguration of Calvin Coolidge on 4 March 1925 showed how far chain broadcasting had come in less than two years. On that day, President Coolidge reached at least 15 million Americans over a hookup of 21 stations from coast to coast. The transmission was so clear that people could hear rustling paper as the new president turned the pages of his text. The inaugural coverage was so successful that talk circulated about broadcasting sessions of Congress. Meanwhile, "Silent Cal" began speaking an average of 9,000 words a month over radio. Although his speeches lacked persuasive content, he was credited with being a strong and effective radio performer.

By the time Herbert Hoover took his presidential oath in 1929, two powerful broadcasting companies supporting three national networks were flourishing. President Hoover spoke on radio 10 times during 1929 and 27 times the following year. By the end of 1930, he had equaled the number of talks Coolidge gave during his entire administration. President Hoover's cabinet reinforced the administration's line by giving even more radio talks. In 1929 the National Broadcasting Company (NBC) devoted from 5 to 25 hours per week to presidential speeches, reports on national events, and addresses by public figures. By 1930 the government was using 450 hours of broadcasting time on NBC alone.

Despite his unmatched experience on the air, President Hoover was a reluctant and not particularly gifted participant in the broadcasting arena. Aides pushed him into making radio speeches. The president also had the unenviable task of selling an economic program that the American public did not want to hear in the depths of the Depression. Hoover's speeches progressively brought diminishing returns. Soon, everyone realized broadcasting was a two-edged sword; it could not only help elevate presidents but help bury them as well. Broadcasting worked its magic best when the potential for persuasion and good feeling were at a peak, a fact not lost on Hoover's successor.

A Radio Star Is Born

Until Franklin Roosevelt (FDR) became president, the networks were more captivated by the presidency than by any particular occupant of the office. The networks clung tightly to FDR's rising star and used his engaging personality on the airwaves to enhance their own status. Columbia Broadcasting System (CBS) commentator Frederic William Wile was one of the first at the networks to realize how high a priority Roosevelt placed on broadcasting. Wile emerged from a talk with the president-elect predicting that the new president would be "highly radio-minded," which would cause Washington to become more "radio conscious" and the American people to become more conscious of Washington.

Following his success in using radio to push a recalcitrant New York Legislature into action, Roosevelt told Wile that he expected to request time frequently. The new president's fascination with radio was surpassed only by the networks' fascination with him. Merlin Aylesworth, president of NBC, not wanting to miss a piece of presidential action, approached Roosevelt before he was inaugurated to offer him airtime on a regular basis. Although FDR was tempted by the alluring network offer, he feared overexposure and preferred to use the airwaves according to his own timing and his own priorities.

Starting with his first inaugural address, in which he told Americans the only thing they had to fear was "fear itself," the new president proved himself to be an exceptional communicator. The networks quickly realized that Roosevelt had different motivations for using radio than had his predecessors. To him, the medium was no longer a novelty for sending ceremonial greetings; it was a vital tool for persuasion. President Roosevelt used all the intimacy and directness radio could offer to rally support for his policies. Just eight days after being sworn in, FDR put radio to the test. Eschewing his fiery and strained campaign oratory, Roosevelt crafted a subdued, conversational style exclusively for radio. He spoke calmly, intimately, and above all, persuasively in what immediately became known as "Fireside Chats."

As more Fireside Chats poured out of the White House, the American people responded favorably. FDR became a friend and a neighbor who could captivate a nation and develop a truly national constituency. When FDR came to office, one employee could handle all the White House mail. By March 1933, a half-million letters sent the White House scrambling to hire additional staff.

Franklin Roosevelt presented only 28 Fireside Chats during his three terms (and a few months of his fourth), but they had an extraordinary impact on a nation seeking desperately to pull itself out of depression and to win a world war. He boosted his radio speeches' appeal by making them during the "primest" of prime time (between 9:00 P.M. and 11:00 P.M. EST) on weekdays when families were home together.

Roosevelt's political adviser Jim Farley said that radio could wash away the most harmful effects once "the reassuring voice of the President of the United States started coming through the ether into the living room."

FDR took advantage of his platform to persuade voters to support him and his programs. Later in the century, broadcasting executives would chafe at a president's using the airwaves solely to persuade voters or Congress to support a particular program, but there was no such resentment of Roosevelt. On the contrary, broadcasting executives delighted in the drama and excitement Roosevelt created. Merlin Aylesworth of NBC effusively wrote President Roosevelt that "I can honestly say that I have never known a public official to use the radio with such intelligence."

Behaviors were locked in during World War II that would greatly influence the way politicians communicated in the coming television age. Americans became even more dependent on the president as a national leader—accustomed to hearing the man who led them through the crisis of the Great Depression, they were prepared to hear him lead them confidently through another. More important, they expected direct communication from the commander in chief. Radio was no longer a novelty; it was a necessity. During the war, Roosevelt always had access to all four networks—CBS, Mutual Broadcasting System, and NBC Red and Blue—simultaneously. When Franklin Roosevelt spoke to the nation two days after the bombing of Pearl Harbor, he achieved the highest ratings of all time; 83 percent of American households that owned radios were listening to the president.

By the end of the Roosevelt administration, presidents had gained a de facto right of access to all radio networks simultaneously, creating a captive audience of millions. Perhaps most important, the president could enjoy this radio access without worrying about a direct rebuttal by the opposition party.

President Versus Candidate

President Roosevelt was a master at inching his presidential addresses closer and closer to election periods to avoid purchasing time from broadcasters. This tactic not only increased his exposure when the voters were starting to focus on the elections but also allowed him to speak without fear of opposition reply. Section 315 of the Communications Act of 1934 required broadcasters to offer "equal time" to opposing candidates during "candidate" uses of airtime. Roosevelt insisted that he was speaking as a "president" rather than as a "candidate" and that his speeches were not subject to the reply rule.

When the president held a Fireside Chat on 6 September 1936 about drought and unemployment, the Republicans charged (to no avail) that the speech was political. Roosevelt used the same successful strategy in the 1940 campaign to make as many free "presidential" rather than paid "candidate" speeches as possible. Roosevelt was particularly adamant about securing free network time for "fireside chats" because toward the end of the campaign, the Democrats were almost out of money. Earlier that year, the NAB had boosted the president's case by ruling that rival political candidates had to prove that FDR's speeches were political, something difficult for an opponent to do. The question of whether addresses were "presidential" or "political" continued into the Truman administration. The Republicans were outraged on 5 April 1947 when NBC, the American Broadcasting Company (ABC), and Mutual carried President Truman's Jefferson Day speech at a $100-per-plate Democratic fund-raising dinner. GOP National Chairman Carroll Reece said the networks' giving airtime constituted an illegal corporate campaign contribution. Reece also charged that "free radio time is a royal prerogative, something to be given without question whenever requested and without regard for the purpose to which it may be devoted."

When television ascended to preeminence in the early 1950s, radio was forgotten as a tool of presidential communication. During the Eisenhower, Kennedy, and Johnson administrations, radio found no place in the presidential arsenal—this despite the fact that Vice President Nixon had "won" the first "Great Debate" of the 1960 campaign on radio where people could not see his exhaustion. The televised presidential speech, broadcast simultaneously on all three major networks, had become the oratorical weapon of choice. Radio became a low-key form of communication that President Nixon used in 1973 to downplay the strident debate over highly controversial budget initiatives, choosing to deliver his State of the Union addresses as a series of radio speeches rather than making a formal, more visible presentation before Congress and a nationwide television audience.

Reagan on Radio

It wasn't until 1982 that the presidential radio speech made a comeback with the arrival of Ronald Reagan in the White House. The administration approached the networks about a series of radio speeches that would be broadcast each Saturday morning from the Oval Office. The Reagan administration saw radio as an opportunity to take its case directly to the people on a sustained basis in a way that would allow the White House total control of the broadcasts. At a time when the president's popularity was sagging, the radio addresses also presented an attractive complement to a broad multimedia offensive. Because the addresses would be broadcast on a traditionally "slow news" day, the White House expected abundant residual media exposure on Saturday network radio and television newscasts and Sunday newspapers. Finally, the White House staff realized how effective "the great communicator" could be on a medium with which he felt so confident.

With the radio speeches, the White House could achieve access in an entirely controlled way. There would be no editing of material, no filtering through reporters' minds, no distracting or nagging questions by the press. It would be a perfect opportunity to "let Reagan be Reagan." Even in the residual coverage, when reporters could edit as they wished, White House aides believed the Reagan momentum would still be present.

On Saturday, 4 April 1982, President Reagan began his radio initiative in the Oval Office with the first of ten five-minute speeches. The president said he was making the speeches to overcome "all the confusion and all the conflicting things that come out of Washington" by bringing "the facts to the people as simply as I can in five minutes." Aides said the speeches' brevity was an effort to prevent having Reagan's message "truncated" or "filtered" by the news media.

The White House staff said that Reagan wrote most of the first ten speeches himself, often rejecting drafts and writing the script in longhand shortly before airtime. Reagan set a folksy, conversational tone in the first address by beginning, "I'll be back every Saturday at this same time, same station, live. I hope you'll tune in." The president's personal input sometimes startled aides.

When President Reagan settled into his radio routine, it was obvious his targets were the Democrats in Congress. The president used six of his first ten speeches to defend his economic programs and to attack the Democrat-controlled House for not supporting him. Not surprisingly, these speeches gained the greatest press coverage: the *Washington Post* carried news about all six of them on the front page, and the broadcasting networks gave them prominent placement on the evening news. Although it is not difficult for the president to make news, it can be difficult for him to make news that he controls. Therefore, extensive regular coverage in Sunday newspapers and on Saturday network news about subjects that he initiated was a positive sign. Not surprisingly, when President Reagan signed off for the last of his ten speeches, he said, "I'll be back before too long."

After an 11-week hiatus, President Reagan continued his radio speeches on a regular basis. By the end of his term, they had become the longest-running regularly scheduled broadcast initiative ever taken by an American president, establishing Ronald Reagan as the person who resurrected radio as a persuasive tool of the presidency. These speeches also reinforced the value of radio as a campaign tool in non-election years. They brought the president a sustained, controlled forum for his views and significant residual media coverage. The fact that the Republican National Committee continued the radio speeches as paid political broadcasts gave clues to both their purpose and their effectiveness.

Opposition Response

Unlike televised speeches, in which the opposition got to reply on a hit-or-miss basis regulated by the networks, with radio speeches the Democrats in Congress were guaranteed automatic access. The automatic replies to President Reagan's radio speeches gave 28 senators, 41 congressmen, and 4 non-congressional Democrats an opportunity to go head-to-head with the president. Speaker Tip O'Neill and Senate Minority Leader Robert Byrd chose the spokesmen who wrote their own speeches, with guidance being made available from the leadership.

The opportunity to reply to one of President Reagan's radio speeches was sought by several rank-and-file members, especially younger, less visible congressmen. It was considered an honor and sign of approval to be asked by the Speaker to make the reply. Many members were more interested in the local audience in their districts rather than the nationwide audience. Some media-conscious congressmen heavily promoted the speeches in advance, advising their constituents to listen on Saturday.

The lure of a radio reply was not as appealing to senators and senior House members. The Saturday afternoon time posed an obstacle to recruiting the best and brightest of the party. Many senior members had important weekend commitments that they were not willing to change. Some senators resented the double standards the networks imposed on the Congress. The broadcasters followed President Reagan anywhere in the world for his Saturday speeches but made congressmen and senators come to a studio in downtown Washington, D.C. This gave President Reagan great flexibility—he gave less than one-third of his speeches from the White House—but posed a logistical problem for the Democrats in Congress. Furthermore, listening to the president, drafting a relevant reply, rehearsing it, and presenting it all in one hour did not strike some legislators as the most relaxing way to spend a Saturday afternoon. By the time President Reagan announced his bid for reelection in January 1984, the Democratic leadership and its members seemed ready to give up the Saturday replies. Still, the opportunity for "the loyal opposition" to have automatic, direct access to the president of the United States had been unprecedented, and the Democrats took full advantage of the opportunity.

President Reagan's success with getting residual print and broadcast exposure through his radio addresses caused his successors to carry on the tradition. Although few Americans listen to the Saturday radio speeches live, many see the aftermath of the speeches in the Sunday newspapers, on daily news broadcasts, or Sunday talk shows. The radio addresses have become a significant agenda-setting vehicle for television-age presidents. Despite the burden of having to make the addresses 52 Saturdays per year, Presidents George H.W. Bush, Bill Clinton, and George W. Bush embraced the radio speech format and found it to be beneficial to their presidencies. It is likely that the radio tradition resurrected by Reagan will continue.

JOE S. FOOTE

See also Election Coverage; Equal Time Rule; Fireside Chats; Politics and Radio

Further Reading

Balutis, A., "The Presidency and the Press: The Expanding Presidential Image," *Presidential Studies Quarterly* 7 (1977)

Barnouw, Erik, *A History of Broadcasting in the United States,* 3 vols., New York: Oxford University Press, 1966–70

Becker, S., "Presidential Power: The Influence of Broadcasting," *Quarterly Journal of Speech* 47 (February 1961)

Braden, W., and Brandenburg, E., "Roosevelt's Fireside Chats," *Speech Monographs* 22 (November 1955)

Brown, Charlene J., Trevor R. Brown, and William L. Rivers, *The Media and the People,* New York: Holt Rinehart and Winston, 1978

Chester, Edward W., *Radio, Television, and American Politics,* New York: Sheed and Ward, 1969

Cornwell, E., "Coolidge and Presidential Leadership," *Public Opinion Quarterly* 21 (Summer 1957)

Craig, Douglas B., *Fireside Politics: Radio and Political Culture in the United States, 1920–1940,* Baltimore, Maryland: Johns Hopkins University Press, 2000

Foote, Joe S., "Reagan on Radio," *Communication Yearbook* 8 (1984)

Foote, Joe S., *Television Access and Political Power: The Networks, the Presidency, and the "Loyal Opposition,"* New York: Praeger, 1990

Grossman, Michael Baruch, and Martha Joynt Kumar, "The White House and the News Media: The Phases of Their Relationship," *Political Science Quarterly* 94 (1979)

Grossman, Michael Baruch, and Martha Joynt Kumar, *Portraying the President: The White House and the News Media,* Baltimore, Maryland: Johns Hopkins University Press, 1981

Minow, Newton N., John Bartlow Martin, and Lee M. Mitchell, *Presidential Television,* New York: Basic Books, 1973

Paletz, David L., and Robert M. Entman, *Media Power Politics,* New York: Free Press, and London: Collier Macmillan, 1981

Rubin, R.L., "The Presidency in the Age of Television," In *The Power to Govern: Assessing Reform in the United States,* edited by Richard M. Pious, New York: Academy of Political Science, 1981

United States Supreme Court and Radio

The United States Supreme Court has played an important role in defining the relationship between government regulation and radio broadcasters, as well as what is permissible behavior by the broadcasters themselves. Numerous cases decided over six decades help to underpin and sometimes explain current regulatory practices.

Operation and Membership

Mandated by Article III of the U.S. Constitution, the Supreme Court stands at the apex of the judicial branch of the federal government. As its name implies, it is the final arbiter of legal issues involving the Constitution and governmental actions. It acts as an appellate court, meaning that it reviews decisions appealed from lower federal courts and from state supreme courts when cases involve questions of federal law—such as the First Amendment, a common ingredient in many media cases. Cases appealed from different states concerning the same federal issue may also end up in the Supreme Court. To be considered, the losing side in the lower court files a petition for *certiori* or review, along with supporting documents in an attempt to persuade the High Court of the importance of the case and of the wrong done to the litigant.

Unlike federal appeals courts, the Supreme Court need not accept all cases appealed to it. Indeed, the court selects a small portion (about 200 achieve the needed four votes for consideration) of the thousands of cases appealed to it annually. Most are denied (*certiori* or *cert.* denied) without any reason being given, leaving the lower court decision in place. Once a case is accepted, attorneys for each side file briefs (formal written arguments) and the court schedules an oral argument to highlight the key issues and give justices a chance to question counsel. A decision often appears months later, usually with an opinion representing the majority, and often with concurring or dissenting opinions as well.

Though its membership and operations have varied over time, for more than 100 years the court has operated with nine members, one of them designated by the president as Chief Justice of the United States. Appointments to the court, made by the president and subject to Senate approval, are for life-long terms. Vacancies, owing to retirement, death, or resignation (the latter is not common), thus occur at irregular times. President Johnson named the court's first minority justice, Thurgood Marshall (served 1967–1991), while President Reagan named the first woman, Sandra Day O'Conner (served 1981–present). The chief justice is usually a newly appointed justice; only three times in history (most recently in 1986 with William Rehnquist) has a sitting associate justice been so elevated.

Radio Rulings

On numerous occasions since the early 1930s, the court has issued decisions involving radio broadcasting. These have ranged from appeals of FCC decisions to specific sections of the Communications Act or other legislation. Most have concerned procedure and jurisdiction, while others have focused on permissible content. The majority of decisions have upheld commission actions.

The first important Supreme Court case to concern broadcasting was *Federal Radio Commission v Nelson Brothers Bond and Mortgage Company* (289 US 266, 1933), in which the court upheld the public interest statement in the 1934 Communications Act as being constitutional and not too vague for proper enforcement. The court also found that the commission had substantial discretion in applying the public interest standard to specific situations. Seven years later, in *Federal Communications Commission v Sanders Brothers Radio Station* (309 US 470, 1940), the court tackled the difficult question of how much the FCC had to be concerned with economic pressure on licensees caused by allowing additional stations on the air. In a decision that provided rhetorical meat for proponents of both sides of the question, the court determined that "economic injury to an existing station is not a separate and independent element to be taken into consideration by the Commission in determining whether it shall grant or withhold a license." Taken together, these first two court decisions served to strengthen the authority of the FCC to read the Communications Act with some discretion.

In *National Broadcasting Company v United States* (319 US 190, 1943), the court issued a landmark decision that further strengthened the FCC's discretionary power. Upholding the commission's network rules that, among other things, forced NBC to sell one of its two national radio networks (it became ABC in 1945), the court determined that the commission could regulate the business relationships between networks and their affiliates. A decision still widely cited in the legal literature, it included the key rationale for government's control of radio: "Freedom of utterance is abridged to many who wish to use the limited facilities of radio. Unlike other modes of expression, it is subject to governmental regulation. Because it cannot be used by all, some who wish to use it must be denied." The *NBC* case remains one of the most important media decisions of the Supreme Court.

On rare occasions, the FCC comes out on the short end of a Supreme Court decision. In *American Broadcasting Co. v Federal Communications Commission* (347 US 296, 1953), the court focused on commission procedures. The radio networks broadcast a variety of quiz programs in the late 1940s, and the commission determined that at least some of them violated the Criminal Code provision banning the broadcasting of lottery information. The Code and existing cases only partially defined what a lottery was, and the FCC had therefore issued a rule further defining what was illegal on the air. It was overturned in the lower court, and the Supreme Court agreed, holding that "the Commission has overstepped the boundaries of interpretation and hence has exceeded its rulemaking power."

More directly focused on program content was *Red Lion Broadcasting Co. v Federal Communications Commission* (395 US 367, 1969). A conservative religious radio station licensee in Red Lion, Pennsylvania had refused to allow free time to a man who had been attacked on the air during a syndicated religious program. After the station refused FCC orders to comply with its personal attack rules, the case went to court. At the same time, the Radio Television News Director's Association (RTNDA) appealed aspects of the FCC's "Fairness Doctrine." Lower courts found for the FCC in the Red Lion case and against the commission in the RTNDA case. Because they dealt with similar aspects of the law, they were combined when appealed to the Supreme Court. The court's ruling upheld the Fairness Doctrine, again justifying its decision because of the limited spectrum available for broadcasting that provided the groundwork for government regulation of the service. In its widely quoted line, the court concluded "It is the right of the viewers and listeners, not the right of the broadcasters, which is paramount."

Another radio content case came nearly a decade later in *Federal Communications Commission v Pacifica Foundation* (438 US 726, 1978) which effectively defined that which was indecent, and thus could be broadcast within certain conditions, as opposed to something obscene, which lacks First Amendment protection and may not be broadcast at all. Citing the "uniquely pervasive presence" of broadcasting (in this case a New York City noncommercial FM station) in the home as rationale for some limits on what could be broadcast, the court held (by a narrow 5-4 margin) that the FCC had been right to fine the station for broadcasting material (a satire on dirty words) in the early afternoon that might be permissible late at night when, presumably, children were not present in the radio audience.

In most matters dealing with media content, the Court has held in favor of the First Amendment and thus against government meddling or interference with media decisions. After a decade of legal wrangling, *Federal Communications Commission v WNCN Listeners Guild* (1981) resolved once and for all who would determine station formats—a government agency or the marketplace. After 1970, the Court of Appeals for the D.C. Circuit, in a series of cases, had held that the FCC did have to determine whether station format changes were in the public interest. In response, the FCC in 1977 issued a policy statement citing the *Sanders* case (that competition should "permit a licensee . . . to survive or succumb according to his ability to make his programs attractive to the public") and determining that the agency did *not* have to make public-interest determinations concerning format changes. An appeals court decision in 1979 once again held against the FCC position, and the agency appealed to the High Court.

In a 7-2 decision, the Supreme Court held that the lower court had made an "unreasonable interpretation of the act's public-interest standard" and reversed it, agreeing with the FCC finding in favor of marketplace determinations of station formats. The court determined that the FCC policy statement was both reasonable and consistent with the legislative history of the Act. Kahn notes of this decision that "[t]his document is a restatement of the notion that reviewing courts are to grant substantial deference to the discretion of the expert administrative agency Congress established to determine what serves the public interest in broadcasting" (Kahn, 1984).

In sum, the Supreme Court has been called on to determine whether FCC decisions meet Constitutional tests as to their substance or procedure, and in most cases (where lower court decisions have been accepted for review) has supported the federal agency. The court's rationale for upholding federal authority over radio has varied over time, but such rationale usually includes spectrum scarcity and thus the need to select among those who would broadcast, and the pervasive presence of the medium in the home. Combined, these considerations have led to a more limited First Amendment right for radio broadcasting compared to print media.

CHRISTOPHER H. STERLING AND
JOSEPH A. RUSSOMANNO

See also Censorship; Deregulation of Radio; First Amendment and Radio; Network Monopoly Probe; Obscenity/Indecency on Radio; Red Lion Case; Regulation; Seven Dirty Words Case

Further Reading

Campbell, Douglas S., *The Supreme Court and the Mass Media: Selected Cases, Summaries, and Analyses,* New York: Praeger, 1990

Carter, T. Barton, Marc A. Franklin, and Jay B. Wright, *The First Amendment and the Fifth Estate: Regulation of Electronic Mass Media,* Mineola, New York: Foundation Press, 1986; 5th edition, 1999

Creech, Kenneth C., *Electronic Media Law and Regulation,* Boston: Focal Press, 1993; 3rd edition, Oxford: Focal Press, 1999

Devol, Kenneth S., editor, *Mass Media and the Supreme Court,* New York: Hastings House, 1971; 4th edition: Mamaroneck, New York: Hastings House, 1990

Federal Communications Commission v Pacifica Foundation, 438 US 726 (1978)

Federal Communications Commission v Sanders Brothers Radio Station, 309 US 470 (1940)

Federal Communications Commission v WNCN Listeners Guild, 450 US 582 (1981)

Federal Radio Commission v Nelson Brothers Bond and Mortgage Company, 289 US 266 (1933)

Hindman, Elizabeth Blanks, *Rights vs Responsibilities: The Supreme Court and the Media,* Westport, Connecticut: Greenwood Press, 1997

Kahn, Frank J., editor, *Documents of American Broadcasting,* New York: Appleton-Century-Crofts, 1968; 4th edition, Englewood Cliffs, New Jersey: Prentice-Hall, 1984

Lipschultz, Jeremy Harris, *Broadcast Indecency: F.C.C. Regulation and the First Amendment,* Boston: Focal Press, 1997

National Broadcasting Co. v United States, 319 US 190 (1943)

Red Lion Broadcasting Co. v Federal Communications Commission, 395 US 367 (1969)

Ulloth, Dana Royal, *The Supreme Court: A Judicial Review of the Federal Communications Commission,* New York: Arno Press, 1979

United States v Edge Broadcasting Co., 509 US 418 (1993)

Urban Contemporary Format

Following the arrival of television, radio's new program specialization approach (formats for targeted audiences) provided a previously unavailable venue for a number of music genres, among them rhythm and blues, the "mother" of the urban contemporary radio format. Rhythm and blues, jazz, and black gospel music were developed from the lifestyles and experiences of African-American musicians and artists. Rhythm and blues provided the optimum entertainment and storytelling experiences about life, love, and pain in the African-American community. Artists such as Little Richard, LaVerne Baker, Ruth Brown, Jerry Butler, Jackie Wilson, James Brown, The Platters, The Coasters, The Drifters, Fats Domino, Ray Charles, Chuck Berry, The Spaniels, Faye Adams, Little Anthony and the Imperials, The Moonglows, The Flamingos, The Five Satins, Oscar Brown, Buster Brown, Frankie Lymon and the Teenagers, Big Jay McNeely, and a host of others sang, wrote, and performed their music with a unique flair that captivated listeners throughout the world. These artists provided an excellent source of radio programming material for the development of the rhythm and blues radio format.

In 1946 WDIA-AM, Memphis, Tennessee, became the first radio station to air a complete rhythm and blues radio format. Its 50,000-watt signal introduced new African-American musicians and disk jockeys to the airwaves, including B.B. King and Rufus Thomas, who would go on to legendary careers in the music industry. The powerful AM signal provided coverage throughout the Southeastern and Midwestern states. The station was so successful with the rhythm and blues format that its owners proclaimed that there was "gold in the cotton fields of the South." WDIA became the "Mother Radio Station" of the rhythm and blues format, thus laying the historic foundation for this broadcast style and its future offspring, urban contemporary.

Rhythm and blues radio stations grew in number during the 1950s to include such stations as WOOK, Washington, D.C.; WERD, WAOK, Atlanta, Georgia; WRAP, Norfolk, Virginia; WEBB, Baltimore, Maryland; WANN, Annapolis, Maryland; WDAS, WHAT, Philadelphia, Pennsylvania; WLIB, WWRL, New York, New York; KNOLL, San Francisco, California; KGFJ, Los Angeles, California; WVON, Chicago, Illinois; WYLD, New Orleans, Louisiana; WCHB, Detroit, Michigan; WSRC, Durham, North Carolina; WANT, Richmond, Virginia; WILD, Boston, Massachusetts; WJMO, Cleveland, Ohio; KATZ, St. Louis, Missouri; and WAAA, Winston-Salem, North Carolina.

Broadcast groups such as Rollins, Rousanville, Sonderling, Speidel, and United Broadcasting established rhythm and blues stations in large and medium markets throughout the country. Rollins was important because it operated stations in major markets, including WNJR, New York, New York/Newark, New Jersey; WBEE, Chicago, Illinois; KDAY, Los Angeles, California; WGEE, Indianapolis, Indiana; and WRAP, Norfolk, Virginia. The Sonderling group laid the original foundation for the "new" sound of rhythm and blues radio, soon to be called urban contemporary. Jerry Boulding, program director of WWRL, New York, during the 1960s, was the architect of this new sound of rhythm and blues radio. This air sound, which was very smooth, became the sound of soul on WOL, Washington, D.C.; WWRL, New York; WDIA, Memphis; WBMX, Chicago; and KDIA, San Francisco/Oakland.

The major change that transformed rhythm and blues radio into urban contemporary was the shift of the format from AM stations to FM outlets in the early 1970s. Four radio stations, WBLS-FM, New York; WDAS-FM, Philadelphia; and WHUR-FM, WKYS-FM, Washington, D.C., were major players in this transformation. WBLS-FM in New York received programming directions from veteran disc jockey Frankie Crocker. His efforts transformed WBLS-FM into the number one station with an urban contemporary format in New York City. This success also was responsible for the early ratings erosion of the then-number-one-rated Top 40, 50,000-watt WABC-AM. The format of WBLS-FM was a mix of rhythm and blues, jazz, Latin, and gospel. WDAS-FM, Philadelphia, with disc jockey personality Hy Lit and Doctor Perri Johnson, also pioneered in urban contemporary radio. This format was a mix of jazz, blues, rhythm and blues, reggae, Latin, and urban rhythms. The format soon moved the station to high ratings.

Howard University's WHUR-FM in Washington, D.C., first aired in December 1971 with an urban contemporary format. The format was quite similar to the WBLS-FM urban sound, but with an added emphasis of news, educational features, and community programs. Just about every urban contemporary radio station in the United States has adopted one of its programs, *The Quiet Storm*. This program, which features love ballads and slow tunes, was originally aired under the *Quiet Storm* title at WHUR-FM and has become an integral part of the urban contemporary format.

WKYS-FM played a major role in giving the urban contemporary format a mainstream audience and the popular nickname "Kiss." In July 1975 the National Broadcasting Company (NBC) station switched from "beautiful music" to a disco format, a variation of urban contemporary. This helped the station move from a number 17 Arbitron rating to number three in Washington, D.C.

WKYS-FM, Washington, D.C., made format adjustments and moved urban contemporary radio to new programming heights. Veteran radio programmers Donnie Simpson, Bill Bailey, Eddie Edwards, Melvin Lindsay, Ed McGee, Rick

Wright, and Jack Harris laid the original foundation. In 2000 WKYS-FM was owned by Radio One, the largest African-American–owned radio broadcasting company in the world.

Urban contemporary is a radio format designed to attract an urban but demographically diverse audience. The overall air sound is that of music performed mainly by African-American artists, with the announcers presenting material such as commercials, features, public service announcements, station breaks, announcements, jingles, and news in various styles ranging from smooth and mellow to wild and zany to authoritative and informational.

Urban contemporary radio stations have become major electronic media entertainment, informational, promotional, and marketing tools. In Washington, D.C., the top three stations, WHUR-FM, WPGC-FM, and WKYS-FM, are all urban contemporary, but syndication, technological innovations, and station competition have brought major changes to urban contemporary radio. Most stations have decreased the use of local disc jockeys and moved toward the use of nationally syndicated programs.

One major syndicated urban contemporary service is "The Touch," which is delivered by satellite and downlinked to station affiliates. This service allows staff on-air announcer Tim Garrison to air his *Love Zone* show, with its mellow urban sounds, to audiences at night throughout the world.

The highly popular Tom "Flyjock" Joyner and the Doug Banks Show are two other examples of successful syndicated morning drive-time urban radio programs heard on many stations. The American Broadcasting Company (ABC) radio networks both syndicate radio shows and program a mix of music, news, comedy, contests, political information, features, and interviews with celebrities and people in the news. Both originate from studios in Dallas but also broadcast from various locations around the country, thereby attempting to localize the programming content and focus while still reaching a national audience.

The sound characteristics of urban contemporary have changed from its rhythm and blues roots. This change has been caused by competition from other contemporary formats, such as contemporary hit radio and its variation, called CHURBAN. The latter combines trendy popular musical hits from Top 40 and urban charts and presents it in a tight, fast-paced manner.

To maintain its popularity, urban contemporary radio must seek a balance between the hot, hip-hop, contemporary music choices and the basic staples of radio programming. These staples include the use of local disk jockey personalities, on-going musical experimentation, creative use of technology, news programs, and programming research for discovery of new audience demographics and format designs.

The power of urban contemporary radio is best defined and reflected in its rich history of strong community service (fundraising efforts, crime-fighting campaigns, and drug-awareness programs foremost among them) and the solid presentation of news, education, and entertainment.

ROOSEVELT "RICK" WRIGHT, JR.

See also Blues Format

Further Reading

Barlow, William, *Voice Over: The Making of Black Radio,* Philadelphia, Pennsylvania: Temple University Press, 1999

Williams, Gilbert Anthony, *Legendary Pioneers of Black Radio,* Westport, Connecticut: Praeger, 1998

V

Vallee, Rudy 1901–1986

U.S. Radio and Film Star

Radio personality Rudy Vallee is associated with many radio landmarks: he was the first national network star created by radio; he was the first crooning idol, predating both Bing Crosby and Frank Sinatra; and he inaugurated the most influential radio variety show in the history of the medium. Vallee was a product of both the traditional middle-class culture of his New England youth and the sophisticated urban values of 1920s New York City. An emotional, even volatile performer credited with giving the radio voice sex appeal, he was also an exceptionally hard worker who strove for and demanded the highest standards of performance and who took pride in his responsiveness to fans.

Hubert Prior Vallee was born in Island Pont, Vermont, on 28 July 1901, to middle-class parents with French and Irish roots. His family moved to Westbrook, Maine, when he was very young, and he prided himself on his New England heritage. His father was a pharmacist, and it was expected that Vallee would follow in his father's footsteps. Instead, young Vallee developed an early interest in music and participated in amateur bands in high school as a singer and drummer. After graduation, he became passionately devoted to the saxophone (a relatively unknown instrument at the time) and began a correspondence and friendship with Rudy Wiedoeft, a saxophonist whose talent so impressed Vallee that he changed his first name to Rudy.

Vallee became one of the most famous collegians to emerge on the national scene in the 1920s, a time when the college culture had an unprecedented influence on popular culture. A student first at the University of Maine and then at Yale, where he transferred in 1922, Vallee was at the center of this new youth culture because of his work as a "sweet" jazz musician. By this time, Vallee was not only playing the saxophone but was also singing through a makeshift megaphone, a necessary tool of amplification that would become his trademark.

After taking his degree at Yale in 1927 (he took time off to perform in London in 1924–25), Vallee worked with several bands in the New York and Boston areas, most prominently Vincent Lopez's orchestra. He was soon given the opportunity to form and lead his own band; in December 1927, Vallee and his Yale Collegians (soon to be renamed the Connecticut Yankees) opened a posh new nightclub in New York called "The Heigh-Ho Club," with Vallee as leader, saxophonist, and singer. It was while at the Heigh-Ho that Vallee became famous, largely because some of his performances were broadcast over WABC. Vallee proved so immensely and immediately popular with radio listeners that other stations signed him up within days of his first broadcast. Within a year, the National Broadcasting Company (NBC) signed Vallee to an exclusive contract and began broadcasting his band nationally. In October 1929, just two days after the stock market crash, Vallee debuted his own weekly program, sponsored by Fleischmann's Yeast. The program consisted of Vallee announcing, singing, and playing music with his band for an hour, with interruptions for commercial announcements. The show was a huge hit, generating thousands of fan letters daily.

In many ways, Vallee's approach to his music made him ideally suited to radio broadcasting. He made several innovations in the way he presented music that allowed radio audiences to become easily involved in the program and that kept them from switching the dial. Because he wanted his music to be accessible to people, Vallee typically cut out the verses of a song and played only the chorus. Where other musicians focused on the music and treated the singer's voice as just another instrument, Vallee emphasized a song's lyrics in order to encourage listener identification and emotional involvement, and he employed more vocal solos than any other bandleader. Although Vallee played a variety of music, he also understood the value of repetition in fostering audience

Rudy Vallee
Courtesy Rudy Vallee Collection, American Library of Radio and Television, Thousand Oaks Library

familiarity and anticipation, and he developed several "signature" tunes that clearly identified him to listeners. He was also the first bandleader to do his own announcing; his friendly, conversational speaking style (marked by his famous greeting "Heigh-Ho Everybody") and the anecdotes he told about each number endeared him to his audience and created an identity that made him seem like a familiar friend to his listeners.

Most important, Vallee gave the radio voice sex appeal. He was the first radio performer to take full advantage of the intimate singing that radio microphones and amplification made possible, a style of singing that soon became widely known as "crooning." During his early years on radio, Vallee sang his love songs in a soft, yearning voice that made him the idol of millions of female fans. By romancing women within their homes, Vallee broke down established boundaries between public and private and helped to inaugurate an era of unprecedented mass media availability in domestic space. Vallee's popularity proved both radio's potential as a starmaker and the power of its domestic audience, resulting in airwaves full of crooners by the early 1930s.

Crooning's popularity did not come without controversy. Because male crooners amplified their voices and sang so emotionally, critics and moral watchdogs complained that they were effeminate and worried about their tremendous influence on popular culture. Vallee was singled out for criticism, and his handlers at NBC worked hard to change his "slushy" image. The network and the sponsor, J. Walter Thompson, restructured Vallee's program in 1932–33 to focus more on variety, bringing in more guest stars to balance Vallee's singing. They also reworked Vallee's persona into more of a "master of ceremonies" than a star performer.

This format proved to be enormously successful, and *The Rudy Vallee Hour,* which aired from 1929 to 1939, became one of the most influential early radio programs and is considered by many to be the best variety program in the history of radio. Based in New York, Vallee was able to take full advantage of access to Broadway and vaudeville stars, as well as visiting international performers. The hour-long show combined musical pieces with dramatic and comic sketches from recent shows, films, and vaudeville acts. Many already famous performers were given their first radio exposure on Vallee's program, among them Fred and Adele Astaire, George Burns and Gracie Allen, and Fannie Brice; the show also launched the careers of many others, such as Alice Faye, Edgar Bergen and Charlie McCarthy, Bob Hope, Frances Langford, Red Skelton, and Milton Berle.

Vallee's later attempts at radio were never as successful as his first show, but he became quite popular as the quintessential "stuffed shirt" figure in films of the 1940s, most notably those of director Preston Sturges. He also had a career-reviving turn in the Broadway show *How to Succeed in Business without Really Trying* in 1961. Although today Vallee is best known for his film roles, his groundbreaking broadcasting work as both a crooner and a showman remains his most significant legacy to American entertainment. Vallee died at his home in Beverly Hills on 3 July 1986.

ALLISON MCCRACKEN

See also Singers on Radio; Variety Shows

Rudy Vallee. Born Hubert Prior Vallee in Island Pond, Vermont, 28 July 1901. Attended Yale University, degree in philosophy, 1927; tried to serve in U.S. Navy during World War I, 1917, but was underage; United States Coast Guard, lieutenant second grade, bandmaster, 1942–45; became saxophonist and bandleader, various bands, 1922–27; leader of Yale Collegians, which eventually became Connecticut Yankees, 1927–39; radio debut, WABC, New York, 1928; own radio series, NBC, 1936–39; debuted Edgar Bergen and Charlie McCarthy; first film appearance, *Rudy Vallee and His Connecticut Yankees* (short), 1929; discovered Victor Borge, 1940s; noted for appearances in *How to Succeed in Business without Really Trying* on Broadway, 1961–64, and in film, 1967. Received New York Critics Award, 1962. Died in Beverly Hills, California, 3 July 1986.

Radio Series

1928, 1940–43, 1946–47, 1950–51, 1955	*The Rudy Vallee Show*
1929–36	*The Fleischmann Hour* (aka *The Rudy Vallee Hour*)
1936–39	*The Royal Gelatin Hour* (aka *The Rudy Vallee Hour*)
1944–46	*Villa Vallee*
1950–52	*The Big Show*

Television Series

On Broadway Tonight, 1964; *Preston Sturges Porträt eines Hollywood Regisseurs,* 1970; *The Perfect Woman,* 1978

Films

Rudy Vallee and His Connecticut Yankees (short), 1929; *Radio Rhythm* (short), 1929; *The Vagabond Lover,* 1929; *Glorifying the American Girl,* 1929; *Campus Sweethearts,* 1929; *The Stein Song* (animated short), 1930; *Betty Co-Ed* (animated short), 1931; *Kitty from Kansas City* (animated short), 1931; *Musical Justice* (short), 1931; *Knowmore College* (short), 1932; *Musical Doctor* (short), 1932; *Rudy Vallee Melodies* (short), 1932; *International House,* 1933; *A Trip thru a Hollywood Studio,* 1934; *George White's Scandals,* 1934; *Sweet Music,* 1935; *Golddiggers in Paris,* 1938; *Second Fiddle,* 1939; *Time Out for Rhythm,* 1941; *Too Many Blondes,* 1941;

Hedda Hopper's Hollywood No. 6, 1942; *The Palm Beach Story,* 1942; *Happy Go Lucky,* 1943; *Rudy Vallee and His Coast Guard Band,* 1944; *It's in the Bag,* 1945; *Man Alive,* 1945; *People Are Funny,* 1946; *The Fabulous Suzanne,* 1946; *The Sin of Harold Diddlebock,* 1947 aka *Mad Wednesday,* 1950; *The Bachelor and the Bobby Soxer,* 1947; *I Remember Mama,* 1948; *So This Is New York,* 1948; *Unfaithfully Yours,* 1948; *My Dear Secretary,* 1948; *Mother Is a Freshman,* 1949; *The Beautiful Blonde from Bashful Bend,* 1949; *Father Was a Fullback,* 1949; *The Admiral Was a Lady,* 1950; *Ricochet Romance,* 1954; *Gentlemen Marry Brunettes,* 1955; *The Helen Morgan Story,* 1957; *How to Succeed in Business without Really Trying,* 1967; *Live a Little, Love a Little,* 1968; *The Night They Raided Minskys,* 1968; *The Phynx,* 1970; *Sunburst* aka *Slashed Dreams,* 1974; *Won Ton Ton, the Dog Who Saved Hollywood,* 1976

Stage
The George White Scandals, 1931–32, 1933–34; *How to Succeed in Business Without Really Trying,* 1961–64

Publications
Vagabond Dreams Come True, 1930
My Time Is Your Time (with Gil McKean), 1962
Let the Chips Fall, 1975

Further Reading
Douglas, Susan J., *Listening In: Radio and the American Imagination,* New York: Times Books, 1999
Hilmes, Michele, *Radio Voices: American Broadcasting, 1922–1952,* Minneapolis: University of Minnesota Press, 1997
Kiner, Larry F., *The Rudy Vallee Discography,* Westport, Connecticut: Greenwood Press, 1985
McCracken, Allison, "God's Gift to Us Girls: Crooning, Gender, and the Re-Creation of American Popular Song, 1928–1933," *American Music* vol. 17 (Winter 1999)
Pitts, Michael, and Frank Hoffmann, *The Rise of the Crooners,* Lanham, Maryland, Scarecrow Press, 2002
Vallee, Eleanor, and Jill Amadio, *My Vagabond Lover: An Intimate Biography of Rudy Vallee,* Dallas, Texas: Taylor, 1996

Variety Shows

Music and Comedy Formats

Inspired by live stage vaudeville, variety programs on network radio through the 1930s and 1940s—whether oriented to music or comedy—offered significant examples of core network radio programming. Variety programs on network radio proved a powerful lure, but except for the popularity of Arthur Godfrey, they declined and disappeared as radio programming changed in the 1950s.

Origins

Vaudeville started in the late 19th century, and by the early years of the 20th century, it was a staple of mass entertainment in U.S. cities. Jugglers, comics, singers, tumblers, and indeed any talent that could operate in 10- to 20-minute units toured the United States, honing their acts in front of live audiences. This infrastructure of well-practiced talent was in place when radio appeared as a mass medium starting in the 1920s.

Just as radio began to make individual vaudeville stars famous, the new medium slowed and then killed the live vaudeville circuit. Along with the Great Depression, the coming of radio and the arrival of movie sound in the late 1920s ended live stage vaudeville, because live shows could not amortize their costs over huge audiences as did radio and the movies. Thus, vaudeville talent either went west to Hollywood or shifted to radio, particularly the comics and singers whose voices represented the core of vaudeville's appeal.

The National Broadcasting Company (NBC) and the Columbia Broadcasting System (CBS) sought what they called "variety talent" to host shows in prime time. Some long-time vaudeville stars such as Eddie Cantor and Ed Wynn had grown up on the boards and made a successful transition to radio. Others, such as Buster Keaton and Harry Lauder, did not. Indeed, most did not, but Jack Benny and the Marx Brothers smoothly moved to radio (and the movies), bringing their considerable talents to listeners across the United States. Even ventriloquists such as Edgar Bergen thrived in radio, where the skill of throwing one's voice was not necessary, but Bergen's comedy overwhelmed any consideration of the radio inappropriateness of his act.

And the radio variety shows with guest stars thrived. Comics such as Eddie Cantor on Sunday night drew audiences measured in the millions and inspired millions more to purchase radio sets. Others from vaudeville adapted to radio. Will Rogers, for example, who talked on stage as he did his rope tricks, gave up the rope to just tell stories on radio. There was a constant need for new talent, and so amateur hour programs began in 1934 with the introduction of *Major Bowes' Original Amateur Hour.* Scouts roamed the United States looking for skilled entertainers who might win the big prize and then make the next step into radio variety show stardom. According to one report, at the height of his popularity in the late 1930s, Major Bowes received more than 10,000 applications per week, and when local Hoboken, New Jersey, singer Frank Sinatra won, the legend of discovery became firmly established as part of radio's myth.

These radio variety shows seemed safe, because they booked stars who appealed to those in what surveys determined was a group and/or family listening demographic. As was the practice of the day, advertising agencies developed most such programs on behalf of their sponsor clients. Rudy Vallee, for example, was an employee of the J. Walter Thompson advertising agency as a client for Standard Brands. He was not employed by NBC, his network, although most fans surely thought so. But this practice would change as the star system—adapted from the movies and live stage vaudeville—developed and soon dominated.

By the early 1940s, radio stopped inheriting stars from other sources, such as the live vaudeville stage or Hollywood. Instead, radio began to make its own stars. Both Kate Smith and Arthur Godfrey started in radio and then moved to other media to exploit their radio fame.

The Variety Show Schedule

From the beginning—as early as January 1927—listings of network schedules included numerous prime-time variety radio programs. Through the late 1920s, dozens of network prime-time radio variety shows aired on NBC Blue and Red, and later also on CBS. Most programs from this early stage were either music or comedy oriented. Musical variety shows made up a third of all prime-time radio at that early point. These shows were headlined by bands or groups of musicians whose very names incorporated references to their sponsors' products: A&P Gypsies, Goodrich Zippers, the Cliquot Club Eskimos, and the Hires (Rootbeer) Harvesters. These were mostly 30-minute shows, with few running an hour.

With the rise of CBS in 1928 came a doubling of variety programming, and although sponsorship still dominated, a new trend arose, because CBS chairman William S. Paley immediately embraced the star system. The first such star was probably *Roxy and His Gang* on NBC Blue, hosted by the noted movie theater impresario, and Roxy's name—not the sponsor's—was above the title. But gradually CBS was to lead the way in exploiting the star system. Such stars would come to equal sponsorship, as Hollywood and then radio itself created the larger-than-life figures the public sought out in their daily listening.

In the early 1930s, with dozens of variety shows on the air, most still featured the names of their sponsors—with new attractions including the Atwater Kent (a radio set manufacturer) Dance Orchestra, the Happy Wonder Bakers, and the Palmolive Hour. But as the 1930s progressed, the stars began to headline in a variation of the star system that Hollywood, and vaudeville before that, had used since the late 19th century. There were variety shows with such names as *The Ben Bernie Orchestra, The Guy Lombardo Orchestra,* and *The Paul Whiteman Orchestra.*

Many combinations of variety talent were tried as NBC and CBS tried to win the ratings war and thus to be able to charge more for their increasingly precious airtime. The search was intense for new stars. Some still came over from the New York stage, none more successfully as a variety host and talent than Eddie Cantor, who gained initial stardom as a member of the annual Follies of Florence Ziegfeld. Cantor adapted his talents neatly to radio, and from 1931 through 1949 his NBC variety show was a highly rated network fixture. Jimmy Durante was surely Cantor's biggest rival, but radio-made comic talent such as Bob Hope would take over this subgenre of the variety program.

For example, Fred Waring headed a relatively unknown big band in the 1920s. When he started his radio variety show in 1933 on CBS, he quickly rose to stardom. He moved to NBC in 1939 with a show that was titled *Chesterfield Time* until 1945, when he had become such a big star that his name went into the title. The show became *The Fred Waring Show* and lasted until 1950, spanning the whole of the variety show era.

Rudy Vallee was a nightclub and vaudeville star when he came to radio on NBC in 1929, first in *The Fleischmann Hour,* and then under his own name, with Vallee telling jokes and singing but also delivering guest talent. Vallee first introduced Eddie Cantor, Noel Coward, Beatrice Lillie, Alice Faye, Edgar Bergen, and Red Skelton to radio fans—all of whom would go on to host or star on radio through the 1930s and 1940s. But it was Vallee's name that carried this radio variety show.

Ed Sullivan would become a television variety show legend, but he never claimed that he did any more than simply to offer New York–centric entertainment gossip and to bring as guests the top music, comic, and other talent. Supposedly, comic Fred Allen once remarked that Sullivan would last as long as someone else had talent that radio could showcase.

Some radio-made stars came through the ranks of local stations. For example, bandleader Guy Lombardo and his newly minted Royal Canadians made their radio debut on Chicago's WBBM-AM in 1927, and within a couple of years Lombardo was appearing on the CBS network, of which WBBM was a long-time affiliate.

Robert Ripley proved that stardom could be transferred to a radio variety show—even if it seemed improbable. Ripley had first introduced his *Believe It or Not* newspaper cartoons in 1918; then, using vivid descriptions and sounds, he was able to develop a variety act for NBC, CBS, and the Mutual Broadcasting System. He was a good host as well and invited his guests to share in his glory.

By 1940 the movies and radio were inexorably linked. Long-retired silent film star Mary Pickford—with third husband, bandleader Buddy Rogers—hosted her own variety shows, first titled simply *Mary Pickford and Buddy Rogers* and later titled *Parties at Pickfair,* referring to their posh and noted Hollywood estate. Here was a star play pure and simple. And once performers such as Bob Hope or Bing Crosby became big in radio, they could also go to Hollywood and then smoothly back to radio. Indeed, both CBS and NBC set up major studios in Los Angeles in the late 1930s to take advantage of this growing Hollywood-radio connection.

Program Strategies

Variety shows almost always revolved around music and/or comedy, because the dancing, acrobatics, magic tricks, and other staples of the vaudeville stage had no appeal on radio. There were general variety shows, music-oriented variety shows, and comedy-oriented variety shows. Subgenres included the amateur hours, ethnic music comics (such as Jewish comics who had crafted their comedy on the Catskills circuit), and subcategories of music (such as the barn dance programs, which would later evolve into the country music format radio).

The distinction between programs was in how much of a story was told. *The Jack Benny Show* offered a variety of musical acts with Jack and his gang of comics, but the show was more the story of the Benny group. On the other hand, Al Jolson's various radio variety shows featured the singing of Jolson, with only an occasional sketch and comic guest star.

Sometimes the appeal for radio made little sense. Ben Bernie was a vaudeville star as the genre was dying, killed by the movies. He started on a local New York City radio station in 1923, and thanks to a long-running feud with gossip columnist Walter Winchell, he became famous and was a radio fixture from 1931 until his death in 1943. Like many famous variety talents, he had one strength—in his case, music—but could offer enough talent in comedy, simple patter, and an ability to play master of ceremonies that he could become a star in variety.

Stars were often made by first appearing on a variety show and then moving to their own programs. For example, Dinah Shore was discovered in the late 1930s and early 1940s by Ben Bernie and Eddie Cantor, and by 1940 she was voted top new star in a Scripps Howard newspaper chain national radio poll. A year later, NBC had her on its schedule with her own show.

By the late 1930s, variety, whether categorized as comedy driven, music driven, or general, offered the most common program type on network radio. Musical comedy always trailed musical variety, but sometimes it was hard to tell the difference. Indeed, Harrison Summers categorized the *Al Jolson Program* on CBS during the 1937–38 season as "comedy variety." Even programs centered on individual stars frequently filled out their 60-minute time slots with guest stars and so should be thought of as variety shows, such as, for example, *The Edgar Bergen and Charlie McCarthy Program.*

Orchestras dominated, because this was the big band era. There were Horace Heidt, Morton Gould, Russ Morgan, Sammy Kaye, Benny Goodman, and Tommy Dorsey, as well as the long-running Paul Whiteman, Wayne King, and Rudy Vallee.

In the 1940s, bands made their singers into individual stars in their own right. Doris Day, Bing Crosby, and Frank Sinatra could be backed by any band. The fans wanted to hear them as singers, not as appendages of notable big bands. Radio indeed sparked the sales of phonograph records, as stars introduced their new tunes and stylings via radio and then fans flocked to purchase copies of the discs of their favorites.

But there were other forms of music than big bands, notably hillbilly, which was later called country music. On Saturday nights, NBC broadcast both the *National Barn Dance* and the *Grand Ole Opry,* complete with singers, dancers, and comics. But these variety shows borrowed from another, alternative musical tradition and aimed exclusively at a rural audience. Variety shows most often were fed by the New York City–based Tin Pan Alley tradition and then later melded this with the music and comedy coming from Hollywood.

When in 1943 NBC spun off what became the American Broadcasting Company (ABC), the new company began with the *Alan Young Show,* the *Mary Small Revue,* the *Paul Whiteman Radio Hall of Fame,* and the *Woody Herman Band Show.* This is a partial listing, but surely ABC embraced all the strategies and scheduling opportunities that NBC and CBS had pioneered. ABC hit an apex in 1946 when Bing Cosby moved to the network for reasons of convenience and technical change. During the summer of 1946, Crosby left NBC after fulfilling his obligation with long-time sponsor Kraft and signed with Philco (maker of radio and later television sets) for a weekly salary of $7,500. Philco and ABC would permit Crosby to prerecord his *Philco Radio Time* on newly developed audiotape. He did not need to be in the studio when his show debuted (on 16 October 1946), nor as it ran on ABC until 1 June 1949. With the transcription ban broken, Bing Crosby then took advantage of the talent raids by CBS and so switched to a new sponsor—Chesterfield cigarettes—and to a new network—CBS—on which *The Bing Crosby Show* debuted 21 September 1949 and ran until the end of the 1951–52 radio season.

CBS and NBC built studios in both New York and Los Angeles to house these variety productions. Sometimes they simply adapted old vaudeville or movie palaces, but more often through the late 1930s and before building restrictions imposed by World War II, they built original, art deco–style studio spaces made for radio. The early 1930s NBC studio at Rockefeller Center—Radio City—best represented these sizable commitments to variety show's popularity.

Demise

The war years were the final hurrah for the radio variety show. National defense bond rallies often functioned as all-star radio variety shows, meant to outdo all other radio extravaganzas. Programs such as *Music for Millions, Treasury Star Parade,* and *Millions for Defense* not only drew needed bond sales, but also were beamed across the ocean or recorded for later playback for the troops fighting in Europe and the Pacific. The stars of radio—led by Bob Hope and Bing Crosby—toured for the United Service Organizations (USO) and went abroad to entertain the soldiers near the fronts. Indeed, radio star and big band leader Glenn Miller was killed in a plane crash while traveling from one show to another.

In the 1940s there was no more popular genre of radio program than variety. But late in the decade, both CBS and NBC shifted their variety stars to their television networks. Music for radio started with the combination of one of network radio's subgenres—hillbilly turned into country music—and another that never appeared on network radio—rhythm and blues. This amalgamation became rock and roll, and television maintained the Tin Pan Alley variety format through the 1950s and 1960s until rock proved so powerful that the variety tradition all but disappeared from television, as it had from radio two decades earlier.

But radio did have a last hurrah during the late 1940s, when CBS's William S. Paley—in alliance with Hollywood agent Lew Wasserman—incorporated stars such as Jack Benny, Red Skelton, Bing Crosby, Groucho Marx, and Edgar Bergen and then had their corporations sell their shows to the networks and sponsors. This incorporation meant vast savings in income taxes for the stars, but it led to the creation of a stable of CBS-housed talent that then transferred—by the middle 1950s—to the dominant CBS television network. By 1949 CBS was winning the radio rating wars with all but four of the top 20 radio shows; a decade later, the ratio would be about the same for CBS television in its battle with NBC.

It was not that participants in the radio variety show tradition gave up easily. There were sizable institutions that had a vested interest in keeping the live variety show going. First, there were the performing music societies: the American Society of Composers, Authors, and Publishers (ASCAP) and the new Broadcast Music Incorporated (BMI). Even more concerned was the union of musicians, the American Federation of Musicians, who tried to slow the innovative recording techniques—tape and discs—that union leaders and members correctly felt would lessen the demand for their live services.

NBC tried to keep the radio variety show alive with *The Big Show* on Sunday nights in 1951, but even this splashy revue could not keep the variety format from switching over to TV. A better symbol of the change was Paul Whiteman, who had hosted many golden age radio variety shows but who in 1947 became an early disc jockey on ABC, symbolizing the transformation to a new form of musical presentation on radio.

One man did keep radio variety alive until 1960 by doing variety in both media—Arthur Godfrey. His morning show was in the variety tradition, and his prime-time hit *Talent Scouts* was able to continue the form through the 1950s, with simulcast over radio until 1956. Godfrey's gift for gab was so popular on radio that he could book whomever he wanted, and fans tuned in. But with the cancellation of first *Talent Scouts* and later his morning show, variety on radio reverted to pure and simple nostalgia. These shows—more than any other form—offer in their preserved form a record of the top variety talent of the first two-thirds of the 20th century.

DOUGLAS GOMERY

See also Benny, Jack; Cantor, Eddie; Crosby, Bing; Durante, Jimmy; Godfrey, Arthur; Hollywood and Radio; National Barn Dance; Singers on Radio; Talent Raids; Vaudeville and Radio; Your Hit Parade; Wynn, Ed

Further Reading

Bilby, Kenneth W., *The General: David Sarnoff and the Rise of the Communications Industry,* New York: Harper and Row, 1986

Buxton, Frank, and William Hugh Owen, *The Big Broadcast, 1920–1950,* New York: Viking Press, 1972

Crosby, Bing, and Pete Martin, *Call Me Lucky,* New York: Simon and Schuster, 1953

DeLong, Thomas A., *The Mighty Music Box: The Golden Age of Musical Radio,* Los Angeles: Amber Crest Books, 1980

Hickerson, Jay, *The Ultimate History of Network Radio Programming and Guide to All Circulating Shows,* Hamden, Connecticut: Hickerson, 1992; 3rd edition, as *The New, Revised, Ultimate History of Network Radio Programming and Guide to All Circulating Shows,* 1996

Hilmes, Michele, *Hollywood and Broadcasting,* Urbana: University of Illinois Press, 1990

Joyner, David Lee, *American Popular Music,* Madison, Wisconsin: Brown and Benchmark, 1993

Lombardo, Guy, and Jack Altshul, *Auld Acquaintance,* Garden City, New York: Doubleday, 1975
MacDonald, J. Fred, *Don't Touch That Dial! Radio Programming in American Life, 1920–1960,* Chicago: Nelson-Hall, 1979
Paley, William S., *As It Happened: A Memoir,* Garden City, New York: Doubleday, 1979
Rhoads, B. Eric, *Blast from the Past: A Pictorial History of Radio's First 75 Years,* West Palm Beach, Florida: Streamline Press, 1996
Summers, Harrison Boyd, editor, *A Thirty-Year History of Programs Carried on National Radio Networks in the United States, 1926–1956,* Columbus: Ohio State University, 1958

Vatican Radio

The Vatican's need for an effective means of communicating with the outside world can be traced back at least to 1870, when Pope Pius IX was restricted to Vatican City by the Italian Army. Only after the Lateran Pact of 1929—four decades and five popes later—was Vatican City recognized as a sovereign and independent state. But as tensions began to grow between Pius XI and Italy's fascist dictator, Benito Mussolini, the implicit threat of isolation lingered.

Cardinal Eugenio Pacelli (then Vatican secretary of state) was determined to avoid any possibility of future papal isolation. He suggested that the Holy See integrate a new medium—radio—into religious dissemination, thereby making geographic and political borders virtually meaningless when challenged by the "airwaves" of broadcast technology. After gaining papal support, Pacelli began negotiations with inventor Guglielmo Marconi to create a powerful shortwave radio system as well as an efficient telephone operation for Vatican use. Within months, Marconi's plans became a reality, and on 12 February 1931 Pope Pius XI sent his first message via "Vatican Radio" to the world.

During its first seven years of operation, Vatican Radio, under Jesuit management, became a significant force in the propagation of church views, programming portions of its content to diverse audiences in seven languages. In 1938 Pius XI increased his use of the medium by instituting a "Catholic Information Service," created solely to attack the atheistic propaganda coming from Germany, Italy, Japan, and Russia. With added broadcast power (now transmitting in ten languages on both shortwave and medium wave), radio had now become the primary medium for the Pontiff's anticommunist message. This "new" propaganda campaign continued through the last days of Pius XI's life, as well as during the reign of his successor, Cardinal Pacelli, as Pope Pius XII.

Pius XII quickly confirmed the importance of radio as a message disseminator to vast territories: he reached far greater numbers of people within a shorter period of time than any of his predecessors. Further, he realized that radio often communicated more emotionally, dramatically, and persuasively than other informational sources—a characteristic that figured prominently when open aggression became more severe and global battle seemed to be inevitable.

Against this backdrop, personnel at Vatican Radio were asked to do many things, involving covert activities as well as humanitarian assistance. Some of the most courageous broadcasts aired by the transnational service at this time were those that unveiled the horrors of the Nazi Holocaust. News reports featured stories on concentration camps, torture, and the ghettos, based on eye-witness testimony. From 1940 to 1946, Vatican Radio also ran an Information Office, transmitting almost 1.25 million shortwave messages to locate prisoners of war and other missing persons. Later the radio system combined its information services with the International Refugee Organization, forming a team "Tracing Service" to reunite war-torn families and friends.

After World War II, Vatican Radio returned to its prewar programming schedule, broadcasting in 19 languages throughout the world and competing for transnational listenership with such networks as the Voice of America, the British Broadcasting Corporation (BBC), Radio Moscow, and Radio Peking. However, the onset of the Cold War (and Stalin's growing number of "godless communism" messages) renewed Vatican Radio's commitment to boost the church's religious message.

In 1950 Vatican Radio officials asked for contributions from the faithful to expand the transnational system's facilities. With this money, Pius XII proposed to use Vatican Radio vigilantly, broadcasting 24 hours each day, in at least 28 languages. The "free world" responded to this announcement enthusiastically, contributing almost $2.5 million to the cause. The new facilities took almost six years to build, but in 1957, Radio Vaticana introduced its newly finished, high-powered station to the world. The renovated system now sent its signal via two new 10-kilowatt shortwave transmitters (to add to the

Pope John XXIII records radio and TV speech from the Vatican, 11 September 1962
Courtesy AP/Wide World Photos

shortwave transmission power of its 100-kw transmitter), one 250-kw medium wave transmitter (augmenting the earlier 100-kw medium wave transmitter), a 328-foot multidirectional antenna for medium wave broadcasts to Southern, Central, and Eastern Europe, and 21 additional antennas. In addition, the new transmission power of Vatican Radio reached new areas in North and South America and much of Asia. Coupled with the establishment of a Pontifical Commission for Cinematography, Radio, and Television, the Holy See clearly established its commitment to transnational media propagation of the faith.

In 1958 Vatican Radio had the unhappy task of broadcasting the last days of Pius XII's life. The next pope, John XXIII, was modern, liberal, and extremely media-literate. Immediately after his election, the new Pontiff proclaimed plans for a Second Vatican Council; much of the progress of the council was conveyed through Vatican Radio.

On 12 February 1961, the Pope celebrated Vatican Radio's 30th anniversary by imparting his apostolic blessing to use radio "to overcome the barriers of nationality, of race, [and] of social class." By 1962 the world's oldest transnational broadcast system communicated in 30 languages, 17 of which were specifically intended for nations behind the Iron Curtain. The Pope used the airwaves in an attempt to mediate the Soviet-American standoff over missiles found in Cuba, broadcasting pleas to both Khrushchev and Kennedy in 1962.

After John XXIII's death in 1963, however, Pope Paul VI was not as supportive of Vatican Radio as his predecessor had been. Some analysts attribute the new Pontiff's indifference to his general mistrust of the secular media. Whatever the reason, station personnel suffered from low morale through the 1960s and 1970s. Their program schedule was singularly noncreative, filling valuable broadcast time with organ recitals, sacred choral presentations, and detailed announcements of minor papal appointments. Fortunately, however, the dark era at the Catholic station was temporary and would later be reversed by an increase in professional station management and creative programmatic direction, as well as by a more media-wise pope.

In 1995 Vatican Radio began broadcasts on its new satellite network. Combined with newly developed cooperative programming efforts on AM and FM stations throughout the world and an internet homepage, one of the oldest transnational broadcast services continues its work as a major voice of the Roman Catholic church.

In 1998 the Italian government issued stringent restrictions on radio emissions (to make them three times lower than most other European countries), and thus began a long struggle with Vatican Radio over the amount of power used by the international service. Some people living near the antennas (on Vatican land and thus not directly under Italian jurisdiction) claimed various health hazards from the power emissions. Early in 2001, things went so far as to witness the environmental minister threatening to cut off the electricity used to power the antennas, an action overridden by the prime minister. Instead Vatican Radio officials agreed to begin to limit the amount of transmitter power used.

Vatican Radio in 2002 programmed a 24-hour news day, using more than 200 correspondents from 61 countries. The content included classical, jazz, and popular music, as well as news/commentary programs and daily church services. Much of this programming was being relayed to millions of Catholics in 40 languages, by a staff of more than 425 employees.

MARILYN J. MATELSKI

See also Italy; Marconi, Guglielmo; Propaganda by Radio; Religion on Radio

Further Reading

Bea, Fernando, *Mezzo secolo della radio del Papa: Qui Radio Vaticana,* Vatican City: Edizioni Radio Vaticana, 1981

Graham, Robert A., *Vatican Diplomacy: A Study of Church and State on the International Plane,* Princeton, New Jersey: Princeton University Press, 1959

Hanson, Eric O., *The Catholic Church in World Politics,* Princeton, New Jersey: Princeton University Press, 1987

Hofmann, Paul, *O Vatican! A Slightly Wicked View of the Holy See,* New York: Congdon and Weed, 1984

Jolly, W.P., *Marconi,* New York: Stein and Day, 1972

Marconi, Degna, *My Father, Marconi,* New York: McGraw-Hill, 1962

Matelski, Marilyn J., *Vatican Radio: Propagation by the Airwaves,* Westport, Connecticut: Praeger, 1995

Radio Vaticana, 1931–1971, Vatican City: Tipografia Poliglotta Vaticana, 1971

Vaudeville

In a world where vaudeville is all but forgotten, it is difficult to imagine what a great impact this art form had on people's lives. For some immigrants who had talent, it was a way out of poverty. It gave others comfort and cheer when they were lonely. It was catharsis. It was amusement. In some ways, it was remarkably egalitarian. To truly understand the development of radio, one has to understand the vaudeville circuit that nurtured most of radio's early stars and provided the model for many early radio programs.

Origins

The origin of the word "vaudeville" is uncertain; some reference books say it comes from the French drinking songs called "chansons du Val de Vire." Others say it comes from the phrase "voix de ville," meaning "voice of the city" (or "voice of the people"). In Europe the term came to mean comic entertainment, comprised of farce and satire, often in song or skit. In America vaudeville developed gradually, emerging from the burlesque shows performed in frontier towns and mining camps. As America became more urban, a growing middle class wanted entertainment, and that potential audience included women and children. Vaudeville historian Frank Cullen credits Tony Pastor with giving this form of entertainment some much-needed refinement. Pastor, Cullen writes, used a variety format that would become the standard for vaudeville: his first performances featured "a concert singer, a popular balladeer, a lady who played a [number of] instruments, an Irish act . . . and a comedian who did only clean material."

Vaudeville theaters multiplied; by 1910, there were 2000 of them across the United States. In a world where radio broadcasting did not yet exist, movies were silent, and phonograph records were still fairly new, an enjoyable way to pass the time was to take the family to a vaudeville performance, where there was something for everyone. Some of the comedians used old jokes that the audience knew, and the fans would say the punchline along with the performer. People also sang along if they knew a song, and requests were sometimes taken. Song pluggers, with no radio to promote their potential hits, would persuade performers to do a certain song and then hire people to stand up and request it again or to cheer loudly at the end; the plugger would then stand outside the theater to sell the song on record or sheet music. On the other hand, if spectators did not like a performance or found a routine boring, they might express their disapproval in a chorus of boos. The vaudeville stage was no place for a person with a fragile ego.

The lives of vaudeville performers must have seemed exciting and glamorous to the audience, but the reality of being a performer was endless touring, sometimes playing several theaters a night, staying in cheap hotels, and sharing cramped dressing rooms with others on the same bill. Many impresarios—such as E.F. Albee, Samuel Rothafel, the Shuberts, and Florence Ziegfield—booked the major theaters and acts, though they were not always the kindest people with whom to deal. There was also a hierarchy of vaudeville theaters: the unknowns played in smaller houses and rural towns, while the biggest and the best got booked for New York City's Palace Theatre.

Until an entertainer developed a following and became a star, there was not much luxury, and sometimes there wasn't much money. But having to pay some dues did not dissuade the hopefuls: for those who dreamed of being famous, vaudeville was their best chance. In that era before talking pictures and broadcasting, there was a constant need for new and interesting live performers at the many vaudeville houses; some theaters had nine or ten acts on the bill, and performers tried their hardest to be memorable, or at least unique in some way. A few performers who were moderately successful hired their own press agents to get them even more visibility, and, they hoped, better bookings. Nellie Revell, one of the few women press agents, not only worked with vaudeville stars and for several theaters, but also wrote about vaudeville for such publications as *Variety.*

Vaudeville helped the sons and daughters of immigrants become successful in America. Al Jolson and Eddie Cantor are two examples of young men from impoverished immigrant backgrounds whose careers took off thanks to their time on the vaudeville stage. In the North talented black performers often performed as well. (In the South, there was a separate circuit for blacks only.) In New York the legendary black comedian Bert Williams offered his amazing routines consisting of song and dance interfused with comedy; Williams became one of the highest paid black performers of his day and earned the respect of his white colleagues. Earning respect was not easy for a minority performer, especially when vaudeville reinforced every stereotypic representation in society—the greedy Jew, the cheap Scotsman, the drunken Irishman, the unintelligent black man. Williams was able to bring dignity to even those skits where he was supposed to play a bewildered Negro. There was no "political correctness" on the stage—immigrants who could not speak proper English, the nouveau riche who did not know how to behave—any foibles of any group could become the butt of jokes. At the same time, numerous foreign language theaters featured performers who poked fun at life in mainstream America.

Vaudeville on Radio

When radio developed, many of the performers who had made a name for themselves on the stage ignored the new medium. *Variety,* the "bible" of show business, said radio was a fad that would not last, and besides, most stations had no money to pay the big name stars from Broadway. But by 1922 it was obvious that radio was winning new friends every day. Some of the vaudevillians decided that making an appearance on radio might be useful after all. At first, most performers and all of the impresarios had been opposed to going on the air; a 3 March 1922 cover story in *Variety* headlined, "Vaudeville and musicians declare against Radiophone," with several major impresarios expressing the belief that radio would only encourage people to stay home and not come to the theater anymore. But the novelty and the chance for publicity had already attracted a few entertainers, and more would follow.

One of the first big vaudeville names to do so was comedian Ed Wynn. In February of 1922, he performed the first live play, "The Perfect Fool," on WJZ in Newark, New Jersey. The legend is that Wynn, ill at ease about performing in a silent studio with only an engineer to watch him, gathered up whoever was still in the building, including the cleaning crew and even a few people on the street, and invited them to watch his routine, thus creating the first studio audience. Their natural reactions to his humor greatly aided his timing. Radio made most vaudevillians uncomfortable because in the early days, studios were usually located in factories or on top of a roof, and there was no audience with which to interact. But Ed Wynn's innovation soon changed that, and as stations began building nicer studios (or moving into hotels that had ballrooms) it became acceptable to allow the public to watch performances.

Paying the big names was still a problem, but several of the impresarios decided to expand their use of radio and began putting their stars and theater acts on the air. Samuel Rothafel (better known as "Roxy") and Charles Carrell were two who did very well with this initiative. Roxy broadcast from the

Capitol Theater over WEAF in New York City as early as 1923; he would be heard on the NBC network starting in 1927. Carrell, who ran numerous theaters in the Midwest in the mid 1920s, had a novel way of attracting attention to his vaudeville shows: he brought a "portable" station into towns where people had few opportunities to see a performance. His traveling companies then put on a show and the entire event was broadcast on his own station, which often encouraged listeners to make a trip to Chicago (where one of his theaters was located) to see a show in person. Of course, once the networks began operations, the problem of paying talent was solved. Early advertisers hired and paid the most famous stars.

Most of radio's best loved early entertainers got their start in vaudeville: Eddie Cantor, Al Jolson, Ed Wynn, comedienne Fanny Brice, singer Ruth Etting, comedians Jack Benny, George Burns and Gracie Allen, singer Sophie Tucker (who billed herself as the "Last of the Red Hot Mamas"), and many more. Nellie Revell (whose clients had included Jolson) ended up offering her show business gossip column over the air on NBC, under the name "Neighbor Nell." She also invited the stars to her show for interviews.

The variety show on radio operated somewhat as the vaudeville show had—a number of acts in which each performer's job was to win over the audience both in the studio and listening at home. The mid 1930s talent show craze (characterized by Major Bowes' program) also harkened back to vaudeville days when managers offered an "opportunity night" for new performers to try out. If the public liked them, they might win a small prize (Eddie Cantor won $5); the real prize was the chance to come back and perform again, and ultimately to be hired.

Some critics later accused radio of killing vaudeville, but the truth is that the genre had begun to decline before the radio craze really took hold. Perhaps radio hastened its demise, but then, vaudeville really didn't die: it became a part of radio, and later a part of television. Thanks to vaudeville, a generation of entertainers perfected their craft and brought it to the airwaves, where a national audience could appreciate it all over again.

DONNA L. HALPER

See also Allen, Fred; Benny, Jack; Brice, Fanny; Cantor, Eddie; Comedy; George Burns and Gracie Allen Show; Jewish Radio Programs in the United States; Stereotypes on Radio; Talent Shows; Variety Shows; Wynn, Ed

Further Reading

Csida, Joseph, and June Bundy Csida, *American Entertainment: A Unique History of Popular Show Business,* New York: Watson-Guptill, 1978

Green, Abel, and Joe Laurie, Jr., *Show Biz from Vaude to Video,* New York: Holt, 1951

Nye, Russel B., *The Unembarrassed Muse: The Popular Arts in America,* New York: Dial Press, 1970

Sobel, Bernard, *A Pictorial History of Vaudeville,* New York: Citadel Press, 1961

Vic and Sade

U.S. Radio Comedy Serial

For 13 years, from 1932 until 1945, devoted radio listeners tuned in daily to "smile again with radio's home folks," Vic and Sade Gook. Over the course of more than 3,500 scripts, writer and creator Paul Rhymer produced an intimate, idyllic, and eccentric portrait of small-town life in Depression-era America.

Neither its serial format (*Vic and Sade* appeared alongside dozens of daytime soap operas) nor its subject matter (*The Aldrich Family* and *One Man's Family* also featured accounts of white, middle-class American life) made *Vic and Sade* unique. What distinguished Rhymer's radio tales of life in "the small house half-way up on the next block" were its odd placement on the network schedule, the inimitable perspective of its creator, and its creative use of the aural medium. James Thurber (1948) wrote that amidst the tears and tragedy of daytime soap operas, *Vic and Sade* "brought comedy to the humorless daytime air." Indeed, *Vic and Sade* was one of the earliest and most enduring radio comedies about middle-class families in the American Midwest. Because of its unwavering focus on the humor of domestic life and its large fan following, *Vic and Sade* influenced the shape and form of situation comedies on both radio and television. Using only three (and later four) voices, a microphone, and a vivid imagination, Paul Rhymer collected everyday conversations, trivial events, and

Bernardine Flynn as "Sade" and Art Van Harvey as "Vic"
Courtesy CBS Photo Archive

mundane details and wove them into fantastic vignettes of small-town life. Few other radio programs capitalized so successfully on the intimacy and imaginative potential of radio as did *Vic and Sade.*

Rhymer wrote all the scripts during the serial's run. His experiences as a young boy growing up in Bloomington, Illinois, served as a model for the Midwestern town life chronicled in *Vic and Sade.* After attending Illinois Wesleyan University, he was hired by the National Broadcasting Company (NBC) in Chicago to write continuity for music programs. As a special assignment, Rhymer was asked by NBC's program director, Clarence Menser, to develop a skit for an up-and-coming client, Procter and Gamble. Although Procter and Gamble did not recognize the early promise of this serial, Menser put the program on the air on a sustaining basis on 29 June 1932. (It was briefly sponsored in 1933 by Jelke and Ironized Yeast.) But not until November 1934 did Procter and Gamble realize its mistake; for the remainder of its run, *Vic and Sade* was sponsored by Procter and Gamble's Crisco. Paul Rhymer quit his position at NBC and devoted himself full-time to writing the serial. Although Rhymer was once fired as a journalist for writing stories about people he had not yet interviewed, his talent for creating stories about eccentric townsfolk was rewarded on radio.

Many contemporaries sought to distinguish *Vic and Sade* from the serials surrounding it on the network schedule. In addition to sharing the format of daytime soap operas (running five times a week), *Vic and Sade* also shared the serials' focus on family life and interpersonal relationships. *Vic and Sade* was set in the Gook household on Virginia Street in Crooper, Illinois. The serial focused on Victor Rodney Gook (Art Van Harvey), a bookkeeper for Plant No. 14 of the Consolidated Kitchenware Company; his wife, Sade (Bernardine Flynn), a not-so-brilliant housewife who talked in mixed metaphors and malapropisms; their adopted son, Rush (Billy Idelson); and absent-minded Uncle Fletcher (Clarence Hartzell). Each episode focused on the conversations of the Gook family about household events and the daily happenings of their small town—Rush's stomachache; Vic's failure to notice that Sade has cut her hair; or Sade's fight with her best friend, Ruthie Stembottom. As Fred E.H. Schroeder (1978) observed, there is never a scene in *Vic and Sade* "that goes farther than the front porch, attic, or cellar." What distinguished *Vic and Sade* from other daytime serials were its noncontinuous story lines and quirky account of life in Crooper, Illinois. Unlike most radio serials, each episode of *Vic and Sade* was self-contained; the conflict introduced in each episode was often resolved by the end of the 15-minute program.

Through references and conversations of the main characters, a whole town came to life on the air. Listeners knew that Sade was a member of the ladies' Thimble Club, loyally attended the washrag sale at Yamilton's Department Store with Ruthie Stembottom, and specialized in making "beef punkle" ice cream for her family. Vic, the Exalted Big Dipper of his lodge, the Sacred Stars of the Milky Way (Drowsy Venus Chapter), was both antagonized by and devoted to his family. Young Rush often went to the Bijou theater with his friends Smelly Clark, Bluetooth Johnson, and Freeman Scuder to catch a feature film such as *You Are My Moonlit Dream of Love* or *Apprentice Able-Bodied Seaman McFish* when he wasn't playing rummy with Vic and Uncle Fletcher. Listeners knew the places in Crooper, Illinois, visited by Vic and Sade—the Butler House; the Bright Kentucky Hotel; and the Tiny Petite Pheasant Feather Tea Shoppe, which served scalded cucumbers and rutabaga shortcake—and the friends and neighbors who inhabited Vic and Sade's world—Reverend Kidney Slide; Chuck and Dottie Brainfeeble; Jake Gumpox; Rishigan Fishigan of Shishigan, Michigan (who married Jane Bayne from Paine, Maine); and Robert and Slobert Hink, who had brothers named Bertie and Dirtie and sisters named Bessie and Messie. When the eccentric Uncle Fletcher became a permanent character in 1940, the absurdity of the serial reached new heights. Uncle Fletcher entertained audiences with his rambling tales about Vetha Joiner, who went daffy after reading dime novels, or Ollie Hasher, whose friend painted his table every day so he wouldn't have to dust it.

Vic and Sade soon became one of the most popular serials on radio. In fact, in early 1935 *Radio Stars* reported that business in one South Dakota town literally halted each day so people could listen to the program. That same year, the Women's National Radio Committee named *Vic and Sade* one of the few daytime programs worth listening to. Just four short years after the program's introduction, a promotional offer featured on the show prompted an extraordinary 700,000 requests. By 1938 nearly 7 million listeners tuned in daily. Over 600 radio editors polled named *Vic and Sade* the best radio serial. *Vic and Sade* was also admired by contemporary writers and humorists, including Jean Shepard, James Thurber, John O'Hara, Sherwood Anderson, Edgar Lee Masters, Ogden Nash, and Ray Bradbury, for its whimsical and humorous look at small-town America.

In its focus on three or four central characters, *Vic and Sade* remained largely unchanged until the last years of the serial. Art Van Harvey (Vic) was temporarily written out of the show while recuperating from a heart attack, as was Billy Idelson's character (Rush) when the actor enlisted in the navy during World War II. In this period, many of the characters previously only described by the Gooks (such as Orville Wheeney and Mayor Geetcham) were given voice as supporting characters in the serial. Although the program ended its continuous run in 1944, it was briefly revived by the Columbia Broadcasting System (CBS) in mid-1945 as a variety show and in 1946 by the Mutual network as a 30-minute sitcom. Two unsuccessful attempts were made to bring the program to television—to

NBC's *Colgate Theater* in 1949 and to a local station, Chicago's WNBQ, in 1957. But the aural magic and the serial's peculiarity did not translate easily to a new visual medium (it was not perhaps until the debut of the *Andy Griffith Show* that eccentric small-town life successfully appeared on television). Although lack of storage space led Procter and Gamble to destroy original recordings of more than 3,000 episodes of the show, some remaining scripts were preserved in two edited volumes and in the archives at the State Historical Society of Wisconsin. Fan clubs such as the Vic and Sadists and the Friends of Vic and Sade and websites such as "Stephen M. Lawson's *Vic and Sade* Fan Page" and "Rick's Old-Time Radio *Vic and Sade* Page" have emerged since the serial's demise to share information, scripts, and the few recordings of the program that remain.

JENNIFER HYLAND WANG

Cast

Victor Gook	Art Van Harvey
Sade Gook	Bernardine Flynn
Rush Meadows	Billy Idelson, Johnny Coons, Sid Koss
Uncle Fletcher	Clarence Hartzell (1940–46), Merrill Mael (briefly in 1943)

Announcers

Mel Allen, Charles Irving, Bob Brown, Roger Krupp, Ralph Edwards, Vincent Pelletier, Jack Fuller, Glenn Riggs, Clarence Hartzell, Ed Roberts, and Ed Herlihy

Creator/Writer

Paul Rhymer

Directors

Caldwell Cline, Clarence Menser, Earl Ebi, Paul Rhymer, Homer Heck, Charles Rinehardt, Ted MacMurray, and Roy Winsor

Programming History

NBC Blue/NBC Red/NBC	1932–44
CBS	May 1938–November 1938; 1941–43; August 1945–December 1945
Mutual	March 1941–September 1941; June 1946–September 1946

Further Reading

"Meet Vic and Sade," *Radio Stars* (March 1935)
Rhymer, Paul, *The Small House Half-Way Up in the Next Block: Paul Rhymer's Vic and Sade,* New York: McGraw-Hill, 1972
Rhymer, Paul, *Vic and Sade: The Best Radio Plays of Paul Rhymer,* New York: Seabury, 1976
Schroeder, Fred, E.H., "Radio's Home Folks, Vic and Sade: A Study in Aural Artistry," *Journal of Popular Culture* 12, no. 2 (Fall 1978)
Thurber, James, *The Beast in Me and Other Animals,* New York: Harcourt Brace, 1948

Violence and Radio

Among the issues associated with any medium is a concern about violent and antisocial content. Many of the concerns about television and violence in contemporary society have antecedents in the history of radio. Violence, for the purposes of this essay, is defined as physical aggression toward humans by other humans, or the threat of such aggression. Radio has been associated with both "real" and fictional violence since its inception.

War and Radio's Beginnings

Radio has been related to violence since its inception in the late 19th century. Indeed, radio's applications for purposes of war and defense nurtured its early development. Guglielmo Marconi persuaded Great Britain to utilize his invention for military and commercial ships before the turn of the century. Radios were installed in military ships to enable communication between the sea and the shore, radically changing the nature of sea warfare. The ability to communicate with shore and with other ships greatly enhanced the ship as a weapon.

In the years between 1907 and 1912 in the United States, amateur radio grew steadily, to the agitation of the military. The U.S. Navy became concerned when official messages were undeliverable because of East Coast amateur chatter. President Taft signed the first general radio licensing law in 1912 in part as a response to military concerns. A clause of this law stated that "in time of war or public peril or disaster," the President might seize or shut down any radio station (Public Law No. 264, 62nd Congress, Sec. 2). In 1915 the Navy, acting on a tip from an amateur monitor, took control of a Telefunken-owned station on Long Island, New York, that had been transmitting radio messages regarding movements of neutral ships, presumably to German submarines. On 6 April 1917, as the United States declared war on Germany, all amateur radio operations in the United States were ordered to be shut down. On 7 April all commercial wireless stations were taken over by the Navy. Amateur radio enthusiasts protested the new regulatory atmosphere, but to no avail.

Radio coverage of violence and warfare during World War II ensured radio's place at the pinnacle of journalism. Live reports from Europe enabled audiences to learn about the aggression of Hitler's Germany. Radio contributed to the United States' move out of isolation and into World War II. Journalists such as Edward R. Murrow had broadcast vivid reports of the European conflict that helped convince the populace of the United States that it would be in the best interest of the nation to enter the war. After the bombing of Pearl Harbor, radio brought news of war from both Europe and the Pacific. The American public became adjusted to radio's providing details of violence during the worst war of human history.

Contemporary War and Radio

Radio has played important roles in contemporary violent conflicts. Among the unique uses of radio in wartime occurred during the Gulf War in 1990–91. Israelis tuned in to radio stations for warnings of incoming SCUD missile attacks from Iraq. The Israeli Broadcasting Authority and the radio station of the Israel Defense Forces unified to form the Joint Channel in an effort to keep the Israeli populace informed. The Joint Channel became immensely important to Israeli civilians. An interesting variation was the Quiet Channel. Reacting to audience suggestions, the Joint Channel broadcast silence on one frequency with the exception of missile warning alarms. This permitted Israelis unable to sleep with music or other radio programs an innovative alternative that allowed them to receive missile alarms.

Christine L. Kellow and H. Leslie Steeves (1998) documented the role of radio in the Rwandan genocide of 1994. The government-controlled Radio-Television Libre des Mille Collines (RTLM) served as the propaganda mouthpiece of the Hutu government. Messages intended to incite violence by Hutus against Tutsis were broadcast over Radio Rwanda, the official government station, immediately following the mysterious downing of a plane carrying the presidents of Rwanda and Burundi. With radio as the prominent source of information, RTLM was able to have a greater impact on the Rwandan populaces, both Tutsi and Hutu, in terms of both attitude and behavior. The Hutu-controlled RTLM used reversal techniques, emphasizing Tutsi hatred of the Hutus in order to encourage Hutu hatred of the Tutsis. In this fashion, Rwandan radio audiences were manipulated by messages of violence.

Terrorism

Terrorism has a particular and controversial relationship to media, including radio. To be successful, terrorism relies upon the public reporting of violence. A violent act itself gains nothing without public knowledge of the event. Thus, terrorism is inherently linked to the propaganda value of violence enacted. Terrorist acts are often consciously planned to gain media attention, making media's role even more controversial. The ethics of reporting terrorism are complex. Media organizations face the difficult task of determining what are newsworthy events without encouraging terrorism. At times, radio has inadvertently served the needs of terrorists by reporting police and military activity around the event, placing victims in more danger. Also of concern is violence directed against radio personnel or stations. Dozens of journalists, some working in radio, are killed every year. Government radio stations are often initial targets in military coups.

Violence and Children

Much of the controversy in recent years concerning television's violent content parallels charges raised against radio in the 1930s and 1940s. Concern over violent programming may be seen in the production process undergone in the making of *Jack Armstrong, the All-American Boy,* broadcast on several radio networks from 1933–51. This serial cliff-hanger adventure series featured Jack, a high school athlete, who had many adventures, each and every one of which was examined by child psychologist Martin Reymert before production to ensure that the program was not excessively violent. Still, complaints were heard from critics and some listeners, largely because the program targeted younger, more impressionable listeners.

Mrs. George Ernst of Scarsdale, New York, organized a 1933 campaign against the "Ether Bogeyman" of radio, whose characters were said to be causing nightmares among children. Her group analyzed 40 popular children's radio programs and found 35 unacceptable, including *Little Orphan Annie* and *Betty Boop,* both for violent and suspenseful content. In a 1934 symposium held in New York City, members of the Ethical Cultural Society, the Columbia Broadcasting System (CBS), the National Broadcasting Company (NBC), the Child Study Association of America, and other groups recommended the formation of a clearinghouse to offer a mechanism by which advertisers, the public, and broadcasters might make more informed decisions concerning the content of radio programs, especially those aimed largely at children.

Few studies examined psychological effects of radio listening on children during this period; most focused instead on what children wanted to hear. Paul Dennis (1998) characterized the experts who examined the psychological effects of radio listening during this time as falling into two camps. The dominant view held that dramatic and violent programming was cathartic for children, providing an outlet for tendencies of aggression. A smaller number of researchers argued that violent dramatic content promoted violent behavior, delinquency, and negative emotions.

Proponents of the dominant cathartic model relied on the idea that radio's programming functioned as pragmatic fantasy. In 1924, well before the children's programming boom, Mansel Keith claimed that a 41 percent reduction in juvenile court cases was attributable to radio's provision of adventure and romance for youth audiences. Jersild stated in 1938 that the vicarious enjoyment of excitement was a right of children, for "a cold, intellectual diet does not fill all of their needs." Ricciuti's comprehensive 1951 study found there was no scientific evidence that thriller programs contributed to fears or daydreaming in child audiences. Comparative studies frequently found no difference between audiences of violent radio content and nonlisteners.

Those arguing for scrutiny of violent radio content claimed negative impacts of exposure. The early 1940s witnessed a rise in juvenile delinquency, which was frequently attributed to violence in radio programming and comic books. Herzog (1941) found that 72 percent of fourth to sixth graders who dreamed about radio said their dreams were unpleasant. Several other researchers claimed to have found evidence that dramatic radio program content had a significant influence on children's reality expectations.

Dramatic radio programming sometimes had very tangible and obvious effects. An understanding of both the penetration and the potential impact of radio in terms of perception of violence may be gleaned from an examination of audience reactions to CBS's Halloween broadcast of a dramatic adaptation of H.G. Wells' *War of the Worlds,* on 30 October 1938. The program simulated a news program, with announcers interrupting seemingly standard-formatted programming for reports of aliens invading New Jersey. The American Institute of Public Opinion poll estimated that 1.7 million people believed the program was a newscast, and that 1.2 million people were at least excited by the news, approximately one-sixth of the total audience. Hundreds of people left their homes in fear. Such a mass reaction to dramatized violence in a radio program indicated the potential impact of radio and of American reliance upon the medium as a primary source of information at that time.

By the 1960s social learning theory had gained credence and was used as a foil for the cathartic model. Most of the research community turned away from the cathartic theory for lack of any evidence that such a positive impact existed. Many researchers in the 1960s, generally examining television and not radio, used social learning theory to explain how and why violent content leads to aggressive behavior in audience members.

Research findings concerning violent content of radio remain inconclusive. Many studies indicate that violent content is a cause of social violence, whereas many other studies conclude that violent content is not a cause of social violence. Many social scientists agree that violent media content can be a contributing factor to social violence, but there is disagreement concerning the magnitude of this relationship and the role and extent of other factors.

In 1929 the National Association of Broadcasters (NAB) established self-regulatory practices for program content and advertising. These codes of practices were non-binding. Often revised, by 1967 the NAB Radio Code reflected the nation's concern for violent material and responsibility toward children. Specifically, the 1967 code stated: "They (radio programs) should present such subjects as violence and sex without undue emphasis and only as required by plot development or character delineation. Crime should not be presented as attractive or as a solution to human problems, and the inevitable retribution should be made clear." While modified slightly from year to year, the Code wording did not have much impact on radio programming at any time. The Code was eliminated in the early 1980s.

Radio Music and Violence

The violent content in the lyrics of some popular songs broadcast over radio has caused controversy for years. There has been concern about lyrics of music since the first broadcasts of rock and roll records, especially with regard to lyrics dealing with sex, drugs, and violence. In the 1980s and 1990s, the wording of heavy metal, rap, and alternative songs was scrutinized by and became targets of advocacy groups and congressional inquiries. The Parents' Music Resource Center was founded in 1985 by Tipper Gore and Susan Baker, wives of powerful political figures in Washington, to advocate labeling music that dealt with violence, drug usage, suicide, sexuality, or the occult. In 1985 the Recording Industry Association of America agreed to use a uniform warning phrase, "Parental Advisory Explicit Lyrics," for such content.

In 1990, 8.5 percent of all music sold in the United States was classified as rap music, with lyrics often thematically linked to urban lifestyles and hip-hop culture. What became known as "gangsta rap" used themes of crime, violence, anti-police sentiments, and gang activities. Some of the notable figures affiliated with gangsta rap as it became popular in the late 1980s and early 1990s were Ice-T, Easy-E, Dr. Dre, Tupac Shakur, Biggie Smalls, and Ice Cube. Radio was a vital link in the distribution of such music, since after hearing songs on the air listeners then sought out their own copies.

Several politicians and public figures, including Senator Joseph Lieberman (D-Connecticut), Senator Sam Nunn (D-Georgia), and former Education Secretary William Bennett, held a series of news conferences in 1995 to protest lyrics in the music industry that were considered pro-drug, degrading to women, or advocating violence against the police. Senator Bob Dole (R-Kansas) praised the effort, calling for an outright ban on gangsta rap during his presidential campaign of 1996. The advocacy group targeted Time Warner because of its 50 percent ownership of Interscope Records, which served as distributor for such artists as Dr. Dre and the band Nine Inch Nails. The group accused Interscope of distributing music of artists who promoted drug use as well as the rape, torture, and murder of women. Time Warner sold its stake in Interscope records later in 1995, insisting the sale was not precipitated by pressure from the senators' advocacy group.

There have been several efforts by the radio industry to self-censor violent lyrical content. In 1993 WBLS (FM) of New York and KACE (FM) of Los Angeles publicly pledged not to give airtime to the most offensive and violent rap songs. In December 1993 Inner City Broadcasting Corporation, one of the largest African-American-owned broadcasting firms in the United States, announced it had banned lyrics it considered derogatory, sexually explicit, or violent. Entercom Communications, owner of many radio stations around the country, announced in July 1999 that it had adopted a policy to reject any music or advertising with violent content or that condoned violence.

The historical controversies concerning radio's violent dramatic programming and potential impacts on its audience reflect current debates concerning violent content of television. The controversy of the 1980s and 1990s concerning violent song lyrics; the contemporary concern with hate groups' use of radio; and the penetration of radio in developing nations, greatly exceeding that of other mass media, ensure that the topic of violence and radio will have great importance into the 21st century.

D'ARCY JOHN OAKS AND THOMAS A. MCCAIN

See also Obscenity/Indecency on Radio

Further Reading

Cantril, Hadley, *The Invasion from Mars: A Study in the Psychology of Panic,* Princeton, New Jersey: Princeton University Press, 1940

Dennis, P.M., "Chills and Thrills: Does Radio Harm Our Children? The Controversy Over Program Violence during the Age of Radio," *Journal of the History of the Behavioral Sciences* 34, no. 1 (1998)

Herzog, Herta, *Children and Their Leisure Time Listening to the Radio: A Survey of the Literature in the Field,* New York: Office of Radio Research, Columbia University, 1941

Jersild, A.T., "Children's Radio Programs," *Talks* 3 (1938)

Kellow, C.L., and H.L. Steeves, "The Role of Radio in the Rwandan Genocide," *Journal of Communication* 48 (1998)

Klapper, Joseph T., *The Effects of Mass Communication,* New York: Free Press, 1960

Skornia, Harry Jay, and Jack William Kitson, editors, *Problems and Controversies in Television and Radio,* Palo Alto, California: Pacific Books, 1968

Virtual Radio

Though a trademarked phrase, Virtual Radio has become a generic term in the radio industry for the practice of using voice tracks to produce a radio station's programming, usually at a location other than at the station. At the most basic level, voice tracking is the integration of prerecorded tracks into music programming, with the intention of giving small market stations access to air talent that only larger markets could afford.

Origin

The true origins of Virtual Radio were tape syndication companies that began to provide services to stations in the 1960s. First was International Good Music, or IGM, which prerecorded classical music with announcer tracks and distributed the programming on tape to automated stations.

In the 1970s, the firm Drake Chenault did the same with Top 40 programming based on the "Boss Radio" concept founder Bill Drake had pioneered at KHJ in Los Angeles. TM Productions—later known as TM Century after a merger—became the largest supplier of syndicated programming for automated radio. Satellite radio formats were never classified as Virtual Radio because they were delivered in real time and not in disassembled form to be reassembled by the radio station.

As a generic name, Virtual Radio is also applied to internet audio streaming services, which deliver radio-like content without transmitters. Some of the content was produced for internet entities, and some was originated by terrestrial radio and repackaged for internet use.

Further Definition

The phrase was given specific definition in 1997 by The Research Group, a Seattle–based consumer research firm serving the radio industry. As the company expanded beyond audience research studies, it added voice tracking as a service for their client radio stations using announcers from the Seattle area. The new service was termed Virtual Radio, and The Research Group sought trademark registration.

The concept was both praised and derided in the industry—praised by station operators who saw voice tracking from distant studios as a way to reduce costs and increase efficiencies, and derided by local disc jockeys displaced by the systems. Internet chat rooms were filled with postings from air talents complaining about "corporate radio" and the lack of localism that resulted from most virtual radio operations. "How does someone in Seattle know what's going on in Fayetteville, Arkansas?" asked a typical posting on broadcast.net. The answer to that question was in the original plan for voice tracking systems: the local station was expected to provide content information to the voice-tracked disc jockeys so their performances could contain references to events, landmarks and personalities in the town.

Writing in the *BP Newsletter*, Klem Daniels of Broadcast Programming, Inc., stressed that voice tracking "is not an exercise in mediocrity, but a chance to achieve perfection." Daniels said, "Many feel that this is a blow to the creativity that has made great jocks and entertaining radio for years. Actually, only the most talented and creative personalities can make voice tracking a success."

Broadcast Programming (later known as Jones Radio Networks) provided a voice tracking service called Total Radio, and its accompanying *Total Radio Users Guide* asked local stations to provide the following information on a long questionnaire:

> Call letters, dial position and station name (i.e., "Mix 96").
> A list of personalities and disc jockeys on the station.
> What sets your city apart from other area cities? Manufacturers, football teams, universities, landmarks?
> Names of dignitaries and famous residents.
> The target audience: Married? Kids? Income level? Hobbies?

A Boon to Cash Flow

The efficiencies and cost savings of Virtual Radio were so lucrative that Capstar Broadcasting Partners built its business plan around computer and internet links among its stations. The first of the links was the Austin, Texas–based Star System, which was first used in Capstar's Gulf Star division, made up of Capstar stations in Texas, Louisiana, and the Southwest. In a 1998 analysis of Capstar's business by Credit Suisse/First Boston financial group, technology, including the Star System, prompted a "buy" rating for shares of Capstar stock:

> Using T-1 lines and other Intranet like systems, Capstar is able to have a DJ in Austin doing a live show in Waco or Tyler. Morning shows are primarily kept local, but other dayparts are done remotely. If the need arises, a local manager can tell a remote DJ what is happening in town over a computer screen, and the DJ can then report. So, if there is a large police chase in Tyler, a DJ in Austin can report on it as if he or she is actually there.
>
> By linking all of its stations to a central location in each "Star" region, interviews of large celebrities can be made to sound local. News and weather are localized, yet music playlists are selected for the individual markets. Thus, local listeners get the researched programming they demand, and Capstar is able to save money on talent by getting more productivity out of each member of its on-air staff.

Credit Suisse/First Boston pointed to Capstar's Baton Rouge stations as good examples of the results of using the Star System: in 1997 revenues were up 18 percent and broadcast cash flow was up 45 percent because of the savings on talent. The Star System digital voice-tracking network became part of AMFM, Inc., in a merger with Capstar. AMFM added a second voice track studio location in Fort Lauderdale, Florida. At its height, Star System fed 400 shows a day from the two operations, employing about 50 full-time air talents in each city.

After yet another merger that absorbed AMFM, Inc., into Clear Channel Communications, Clear Channel elected to close the Star System operations in 2001 in favor of its "hub and spoke" voice tracking concept. Hub and spoke is also virtual radio with voice tracks recorded at major market stations (the hub) and fed to nearby smaller markets in the same region (the spokes). Thus, a Clear Channel station in Columbus, Ohio, would feed voice tracks to sister stations in nearby Ohio towns. At about the same time Clear Channel closed the Star System, Jones Radio ended their marketing efforts for Total Radio because client demand for the service had dwindled to half a dozen stations. The Research Group had folded Virtual Radio in 1999 when Jacor Communications acquired the company before Jacor's merger with Clear Channel.

Local Input

Virtual Radio remained as a local operation, either in the hub-and-spoke style of Clear Channel or with individual station clusters using in-house talent to feed multiple stations in the same market. It would not be unusual to hear the same voice using one name on a Top 40 station and then later that same day using another name on the co-owned Country station. In some situations, voice-tracked disc jockeys even competed with themselves in the same time slot.

The advantage of voice tracking within the local station or cluster is the effective use of time and talent. Instead of waiting for songs to end before performing, disc jockeys and other air talent can spend time producing commercials, making local appearances, or selling advertising. Also, by keeping the talents local, it is less likely that they will mispronounce an important local celebrity's name or miss a reference to a local event.

ED SHANE

See also Audio Streaming; Automation; Clear Channel Communications; Drake, Bill; Internet Radio

Further Reading

Daniels, Klem, "Voice Tracking: Doing It Right," *BP Newsletter* 12, no. 10 (October 1998)

Voice of America

International Radio Service

The legacy of the Voice of America (VOA), which has offered nearly six decades of broadcasting in multiple languages, began on 11 February 1942 with this opening for the first broadcast to Germany:

> OPENING MUSIC: The Battle Hymn of the Republic
> ROLAND WINTER: Attention! This is the Voice of America!
> WILLIAM HARLAN HALE: The Voice of America at War.
> PETER KAPPEL: Our voices come to you from New York across the Atlantic Ocean.
> STEFAN SCHNABEL: America is today in its sixty-sixth day of the war.
> HALE: Today and every day from now on we shall be speaking to you about America and the War. Here in America we receive news from all over the world. This news may be favorable or unfavorable. Every day we shall bring you this news THE TRUTH.

Later, Fred Waring's orchestral rendition of *The Battle Hymn of the Republic* was replaced with *Yankee Doodle* when it was discovered that the German marching song *Laura, Laura* had the same tune.

Origins

From that first broadcast during World War II to the present, the VOA grew to be a dominant presence in international broadcasting. Reaching the world through shortwave radio, the service was initiated to counter propaganda broadcasts from Germany and Japan. When Secretary of War Henry L. Stimson warned that Germany was undermining the American institution of free speech, Robert E. Sherwood, a noted playwright and President Franklin Roosevelt's speechwriter, observed the national reaction to the president's Fireside Chats and responded with a plan. With presidential authorization, Sherwood created the Foreign Information Service (FIS) in 1941.

After the FIS became the Overseas Branch of War Information (OWI) under Executive Order 9182, President Roosevelt authorized the VOA to become part of OWI in December 1942. Sherwood persuaded John Houseman to serve as VOA's first director. According to Houseman, Sherwood instructed him to consider all VOA transmissions as continuations of Roosevelt's speeches.

Houseman had an extensive background in radio, which included being the producer of the 1938 "War of the Worlds" broadcast for the *Mercury Theatre of the Air.* Drawing on his training in drama, Houseman decided to avoid the single-voice format used by the British Broadcasting Corporation (BBC). Houseman gave credit to Norman Corwin's radio styling for the idea of using several voices to maximize variety and energy. Moreover, he also hoped to counteract jamming with the diversity of pitch and rhythm.

Once launched, the VOA was praised for becoming a credible source of news for the rest of the world. In his book *Front and Center,* Houseman quoted a letter from Cannes, France, showing VOA's impact on the Resistance in France:

> You in America cannot imagine how even a few minutes of news from America, heard by a Frenchman, is spread around. An hour after it is heard hundreds, thousands know the truth. (Houseman, 1979)

Since the goal was to reach as many people worldwide as possible, VOA transmitted via shortwave, which allows low-powered, high-frequency transmissions to be received thousands of miles from the point of origin. VOA also deemed it an advantage that shortwave transmission was not considered an effective delivery system by the commercial American radio industry. Moreover, though this type of transmission posed no threat to commercial broadcasters in the United States, all the other major foreign powers were sending international communications via shortwave. Through VOA, the United States joined the international broadcasting ranks—especially when all American shortwave transmitters were placed under government control eight months after the bombing of Pearl Harbor.

Postwar Change

The postwar period brought many changes for the VOA. In late August 1945, President Harry Truman abolished the OWI. The VOA remained on the air, and the State Department took over operations under the Interim International Information Service. Archibald MacLeish, assistant secretary of state for public and cultural affairs, coordinated VOA until he was replaced by William B. Benton, who cut expenditures to the bone, trimmed VOA programming, and terminated many members of the staff. Soon Congress decreased appropriations. VOA was in danger of ceasing operations until Truman and Secretary of State George C. Marshall intervened with the advent of the Cold War in 1948.

The threatening aspects of the Berlin Blockade heightened feelings that an American radio voice was important. In January 1948 Congress passed the Smith-Mundt Act, which authorized the VOA to continue its service to "disseminate abroad information about the United States, its people and policies promulgated by the Congress, the President, the Secretary of State and other responsible officials of government having to do with matters affecting foreign affairs."

However, in 1953 Senator Joseph McCarthy attacked VOA's programming practices and its propaganda implications with charges of subversive activity. Although never proved, widespread dismissals and resignations, and a sharp drop in agency morale, followed.

Just as the administration of VOA was changing, so also was the radio format for newscasts. Speculating on this new style of reporting, Shulman comments:

> The shift from multiple-voice plays to single-voice news stories that were based on outside sources emphasized the reliability and objectivity. . . . The Voice of America spoke in terms that were increasingly concrete and factual. In an atmosphere in which "propaganda" . . . was discredited both abroad and at home, precise information sounded neutral. . . . News stories replete with official quotations and a remote authorial voice established distance and command. Increasingly, Voice writers used the passive voice or impersonal form to indicate a weighty authority. These overall changes . . . . emerged from discussions. . . . Was it artistic, dramatic radio, or was it clear informational journalism? (Shulman, 1990)

This movement to increase the prestige of the Voice was reflected both in VOA's move to Washington, D.C., in 1954, and in President Gerald Ford's signing of the VOA Charter (Public Law 94-350, FY 1977 Foreign Relations Authorization Act) on 12 July 1976. The Charter's purpose was to protect the integrity of VOA programming and define the organization's mission:

> The long-range interests of the United States are served by communicating with the peoples of the world by radio. To be effective, the Voice of America (the broadcasting service of the United States) must win the attention and respect of listeners. These principles will therefore govern Voice of America (VOA) Broadcasts.
>
> 1. VOA will serve as a consistently reliable and authoritative source of news. VOA news will be accurate, objective and comprehensive.
> 2. VOA will represent America, not any single segment of American society, and will therefore present a balanced and comprehensive projection of significant American thought and institutions.
> 3. VOA will present the policies of the United States clearly and effectively, and will also present responsible discussion and opinion on these policies.

VOA kept the world informed about developments in the United States. For example, in 1969 when Neil Armstrong took those steps for mankind on the moon, nearly 800 million people were tuned to VOA or to one of the hundreds of stations around the world that were relaying VOA's live coverage. Outpacing its counterparts throughout the world, in 1977 VOA became the first international broadcast service to use a full-time satellite circuit to deliver programming from its own studios to an overseas relay station, which was located on the Greek island of Rhodes. Because of its commitment to present factual information to the world, VOA dramatically enhanced its credibility through its candid reporting of two events that traumatized the nation—the 1965–72 war in Vietnam and the 1972–74 Watergate scandal.

Challenges Since 1980

The 1980s began with promise. In 1983 VOA launched a $1.3 billion program to rebuild and modernize programming facilities. In 1985 Radio Martí, which was affiliated with VOA, began daily broadcasts to Cuba. In 1988 Congress enlarged the VOA Charter to include WORLDNET Film and Television Service. VOA Mandarin and Cantonese broadcasts were increased in 1989.

On 30 April 1994, President Bill Clinton signed the International Broadcasting Act of 1994 (Public Law 103-236), establishing the International Broadcasting Bureau (IBB) within the United States Information Agency (USIA), which consolidated all civilian U.S. government broadcasting, including VOA, WORLDNET, and Radio and TV Martí, under a Broadcasting Board of Governors (BBG), and funding a new surrogate Asian Democracy Radio Service (later called Radio Free Asia [RFA]). This extended the BBG's oversight to government grantee organizations Radio Free Europe/Radio Liberty (RFE/RL) and to RFA. The bipartisan Board includes the Director of the USIA (ex officio) and eight other members who are appointed by the president and confirmed by the Senate. The first Board of Governors was sworn in on 11 August 1995.

The Board oversees the operation of VOA, WORLDNET Television Service, and Radio and TV Martí to Cuba, as well as two other international broadcast services—RFE/RL and RFA, which were shifted to its jurisdiction under the 1994 legislation. VOA, WORLDNET, and Radio and TV Martí are part of the USIA and are U.S. government entities; RFE/RL and RFA were nonprofit, grantee organizations that received annual grants of congressionally appropriated funds from the BBG.

In 1994 VOA began distributing its newswire and selected newscast and program audio files in 19 languages, VOA frequency and satellite information, and other general material, via the internet. The Office of Business Development was established in 1994 to investigate the possible privatization of VOA language services, procurement of corporate underwriting for broadcasts, coproductions with major broadcast networks, and fund raising from various foundations. From 1994 through 1996, the office raised $4 million. VOA became the first international broadcast service to launch a webpage on the internet in May 1996.

On 25 October 1996, VOA broadcast its first simulcast radio and TV program—an hour-long Farsi broadcast—from a new TV studio at VOA's Washington, D.C., headquarters. Programs in Arabic, English, Mandarin, Russian, Serbian, and Spanish followed.

In 1997, an agreement was signed between the IBB and the Asia Satellite Telecommunications Company (AsiaSat) giving VOA, WORLDNET Film and Television Service, RFA, and RFE/RL access to AsiaSat 2 with a footprint reaching more than 60 percent of the world's population.

In October 1998, President Clinton signed the Foreign Affairs Reform and Restructuring Act. This abolished the USIA effective 1 October 1999 and integrated all the agency's elements, except the IBB, into the Department of State.

For 60 years, the VOA has earned the reputation of providing current, accurate, and balanced news, features, and music to its international audience. VOA continues to attempt to reach its announced goal of providing listeners from every walk of life, race, and religion with reliable, comprehensive news of events from around the world. Occasionally, VOA also offers practical information about how to maintain new democracies and free-market economies.

Fourteen relay stations worldwide transmit VOA's programs 24 hours a day to international audiences via satellite, shortwave, and medium wave. Most "affiliates" now receive their programming via one of the 37 satellite circuits that deliver VOA broadcasts to virtually every corner of the globe. According to VOA figures, the service is successfully competing with nearly 125 similar broadcast services worldwide and remains one of the top international broadcasters in today's vast global media market. Each week 86 million listeners around the world tune in to VOA programs broadcast in 53 languages via direct medium wave (AM) and shortwave broadcasts. Millions more listen to VOA programs placed on local AM and FM stations around the world and give VOA a vast global reach that is unequaled by any of the other international broadcasting services.

VOA credits its extensive programming to the more than 80 writers/editors in its newsroom and to the 40 correspondents at 22 news bureaus in the United States and throughout the world. These broadcast journalists write and report an average of 200 news stories each day. VOA's original programming totals almost 700 hours each week.

MARY KAY SWITZER

See also Board for International Broadcasting; Broadcasting Board of Governors; Cold War Radio; International Radio Broadcasting; Jamming; Office of War Information; Radio Martí; Radio Sawa/Middle East Radio Network; Shortwave Radio

Further Reading

Alexandre, Laurien, *The Voice of America: From Detente to the Reagan Doctrine,* Norwood, New Jersey: Ablex, 1988

Brown, John Mason, *The Ordeal of a Playwright: Robert E. Sherwood and the Challenge of War,* New York: Harper and Row, 1970

Browne, Donald R., *The Voice of America: Policies and Problems,* Lexington, Kentucky: Association for Education in Journalism, 1976 (Journalism Monographs, no. 43)

Culbert, David Holbrook, *News for Everyman: Radio and Foreign Affairs in Thirties America,* Westport, Connecticut: Greenwood Press, 1976
Fitzgerald, Merni Ingrassia, *The Voice of America,* New York: Dodd Mead, 1987
Grandin, Thomas, *The Political Use of the Radio,* Geneva: Geneva Research Center, 1939; reprint, New York: Arno Press, 1972
Halberstam, David, *The Powers That Be,* New York: Knopf, 1979
Houseman, John, *Run-Through: A Memoir,* New York: Simon and Schuster, 1972
Houseman, John, *Front and Center,* New York: Simon and Schuster, 1979
Irons, Peter H., *Justice at War,* New York: Oxford University Press, 1983
Jurey, Philomena, *A Basement Seat to History: Tales of Covering Presidents Nixon, Ford, Carter, and Reagan for the Voice of America,* Washington, D.C.: Linus Press, 1995
Kolko, Gabriel, *The Politics of War: The World and United States Foreign Policy, 1943–1945,* New York: Random House, 1968
Krugler, David F., *The Voice of America and Domestic Propaganda Battles, 1945–53,* Columbia: University of Missouri press, 2000
Pirsein, Robert William, *The Voice of America: A History of the International Broadcasting Activities of the United States Government, 1940–1962,* New York: Arno, 1979
Shulman, Holly Cowan, *The Voice of America: Propaganda and Democracy, 1941–1945,* Madison: University of Wisconsin Press, 1990

Voice of the Listener and Viewer

British Audience Pressure Group

The Voice of the Listener and Viewer (VLV) has become an influential audience pressure group, describing itself as the "citizens' voice in broadcasting in Britain." It was founded by a group of mainly professional and middle class radio listening enthusiasts who had heard of proposed changes to one of the British Broadcasting Company's (BBC) national radio networks.

Origins

In 1982, during the war in the South Atlantic between British and Argentinean forces for control of the Falkland Islands, the BBC had used its Radio 4 network to carry live news of the conflict. Radio 4, although it carried more news than any of the BBC's other networks, was a mixed service, and it included regular radio drama, a daily "soap" (*The Archers*), culture and arts, religious, and magazine programs, documentaries, and much else. But the news coverage of the war was deemed a success, and some senior BBC executives thought that the network could in future become a mainly news network, equipped to carry breaking news, as well as analysis, background, and comment.

The protestors, all very keen Radio 4 listeners, saw the possible disappearance, or at best the serious downgrading, of many of the programs they most enjoyed. Although the BBC is a publicly owned corporation, they felt powerless to make the broadcast planners listen. As license payers, they believed that the BBC had an obligation to listen to them. They met with Geoffrey Cannon, the media correspondent of the *Sunday Times.* Cannon wrote about their concerns in his paper, and many more people added their support. They wanted to create a voice for the listener, and hence the name of the new group was born. In late 1983 the Voice of the Listener and Viewer came into existence at a public meeting in London. By this time the BBC plan to change Radio 4 had been shelved, but the momentum for a consumer group for radio (and later for television) was unstoppable.

It is significant that changes to public radio rather than television had stimulated the formation of the group. In Britain and elsewhere, radio stations and networks often evoke greater feelings of ownership and consumer concern than television stations and networks. Listeners often develop stronger attachments to radio than they do to television.

The VLV is widely recognized as a significant and important consumer voice in British broadcasting policy making. It makes representations to the government, to the BBC and commercial broadcasters, and to the broadcasting regulatory authorities. It holds regular meetings on current topics, arranges visits to broadcasting stations, and issues a quarterly newsletter. Through its charitable arm, the Voice of the Listener Trust, it also promotes public education about all aspects of broadcasting. Finally, the VLV makes annual awards to programs and broadcasters, both in radio and television that, unlike most others, are chosen by its members: ordinary listeners and viewers.

The main thrust of the organization's efforts is in support of quality in broadcasting. As the deregulation of both radio and television in Britain has progressed, concerns have been expressed by many at what they see as reduction in the number of high-quality programs being made, as public and private networks compete for audiences. It makes a strong defense of the principle of public funding for the BBC through the compulsory license fee for all television set owners, unless and until an alternative, non-commercial means of funding can be found. It says that the license fee gives it the freedom to concentrate on delivering quality programs to audiences, while their commercial competitors have to deliver audiences to the advertisers who fund them. The VLV thus sees the BBC as the standard setter of British broadcasting.

Another major concern of the VLV has been to maintain and even increase the quality and range of programs for children. And in the increasingly commercial world of broadcasting in the U.K., it seeks to ensure that older listeners and viewers are not neglected as a result of being of less interest to commercial sponsors and advertisers.

The VLV has developed links with similar bodies in other parts of the world, especially those with related origins and philosophies, such as the Friends of New Zealand Radio, the Friends of the ABC in Australia, and the Friends of Canadian Broadcasting. It has organized several successful international conferences on wider global issues. At its third international conference in 1996, it joined with other groups in Europe to form the European Alliance of Listeners' and Viewers' Associations (EURALVA). This group seeks to represent the interests of listeners and viewers of broadcasting services in Europe and maintain the principle of public service in broadcasting. The member groups come from Denmark, Finland, France, Portugal, Spain, and the U.K.

The VLV has its critics, who often say it represents a middle class, well-educated, and privileged minority who has dominated the style and content of British radio broadcasting for too long. But it has gained respect and support from many in the industry, and some well-known broadcasting names are to be found among its membership.

GRAHAM MYTTON

See also British Broadcasting Corporation; British Commercial Radio; British Radio Journalism; Public Service Broadcasting

Further Reading

Voice of the Listener and Viewer website, <www.vlv.org.uk/>

Vox Pop

Radio Interview Program

The radio program *Vox Pop* (from the Latin for "voice of the people"), one of the first "man on the street" interview shows, was also one of the earliest quiz programs. Later it became a popular human-interest program and one of the biggest home-front morale boosters of World War II. It was also probably the best-traveled program in broadcasting history.

Origins

The show began in 1932 at station KTRH in Houston, Texas. Someone had the idea of dangling a microphone on a very long cord out the window of the hotel from which KTRH broadcast so that passersby could be interviewed. Station ad man Parks Johnson and station manager Jerry Belcher took on the task of talking to the man or woman on the street. They started out by asking about current events, then segued into lighter topics. The results were alternately fascinating and hilarious.

Once in those early days, after a large storm had swept through Houston, the hosts found themselves facing an empty street. They had no one to interview. Necessity being the mother of invention, Johnson quickly relieved the program's crew of all their money, emptied his pockets as well, and had it all changed into dollar bills. He collared an usher from a nearby theater and proceeded to ask him questions, giving him a dollar for every correct answer. Soon the street was mobbed, and the "quiz show" was born.

This device worked very well during the depths of the Depression, and it began to alter the focus of the show. The opinions of the people, their voices, were downplayed as the quiz element gained in popularity. Current events questions were used as a warm-up to the *real* questions, the ones worth a dollar. Johnson often asked questions that originated from everyday life and was known for carrying a notebook with him and jotting down new topics for questions. Participants were asked questions that tested their knowledge of the Bible ("What did Pharaoh's daughter find?") or their vocabulary ("Can you ad-lib?") or questions meant to elicit a humorous response ("Why can't a cat be called 'Fido' or 'Rover'?"). Questions regarding the so-called war between the sexes were practically a regular feature. Such queries as "What makes a person fall in love?" and "What is a woman's place?" were sure to trip up the guest.

Network Popularity

Vox Pop was almost totally unrehearsed, and this spontaneity proved to be a hit with listeners. The show was broadcast from the streets of Houston for more than two years, but it also attracted attention outside of Texas. On 7 July 1935, *Vox Pop* began appearing on the National Broadcasting Company (NBC), broadcasting from the sidewalk at New York's Columbus Circle. The following week, the show took its microphones to the waiting room of Grand Central Station, moving the show around New York City for the rest of the summer. Brief interviews with contestants before the quizzing began became part of the proceedings, and soon *Vox Pop* was dropping in on events as varied as a Hollywood movie premiere and the "Days of '76" celebration in Deadwood, South Dakota.

Into early 1940, the program was still giving away dollar bills at a furious pace, but as the nation edged closer to war, the focus of *Vox Pop* began to change once again. Johnson, a World War I veteran, threw the program wholeheartedly into the war effort at least 17 months before the bombing of Pearl Harbor. The quiz show structure was phased out, and the program changed to a focus on human interest, going into a community or attending an event and deciding beforehand which guests to interview.

The show traveled up to 1,000 miles per week throughout the United States, visiting military bases, military schools, and factories and showcasing different communities that were helping on the home front. Themes such as "Lumber at War," "Food at War," and "Dogs for Defense" were typical of this period; later in the war years, the show's visits included military hospitals. The show broadcast from 45 states as well as to Canada, Mexico, Puerto Rico, and Cuba. At this time, Parks Johnson's wife Louise began helping the show by buying gifts for the interviewed guests. "Mrs. Santa Claus," as she became known, eventually had a budget of $1,000 per week and was quite adept at locating hard-to-find items during the war years. These were halcyon days for *Vox Pop*. The show seems to have connected with the country in a very real way, and its ratings climbed steadily during this period, reaching a respectable 15.3 during the last two years of the war.

Parks Johnson was the guiding force throughout the life of the radio program. Jerry Belcher left the show in 1936 and was replaced by Wally Butterworth, a well-known radio announcer. Butterworth hosted the show with Johnson from 1936 until 1941. Neil O'Malley filled in briefly but was replaced by Warren Hull in 1942. Hull, an actor and announcer who had played the Green Hornet in the movies and was later master of ceremonies for the popular TV program *Strike It Rich,* stayed with the program until *Vox Pop* left the air in 1948.

Advertising Squabble

The show had several sponsors through the years. Deals with Kentucky Club Tobacco and Bromo-Seltzer lasted longest, but it was the sponsor *Vox Pop* had for the least amount of time that made the biggest impact. Lipton Tea sponsored the show in a Tuesday night slot starting in 1946, but the relationship between program and sponsor quickly soured. T.J. Lipton, Inc. thought its products did not get enough attention on the program and insisted that each guest (as many as six per show) be presented with a box of Lipton products before receiving his or her personalized gift. These presentations were in fact commercials, commercials that "must do a hard selling job," according to a memo from Lipton.

Johnson, as sole owner of the show, did not like the new requirement, and after negotiations failed, he took the unusual step of firing his sponsor. Many a sponsor had canceled a program for low ratings, but never had a performer canceled a sponsor! Newspapers around the country picked up the story. Johnson was hailed as a "radio knight," a man of high moral principles who refused to compromise. This favorable publicity meant that *Vox Pop* had little trouble finding a new sponsor, and the show continued until 1948 for American Express. By this time, however, radio was changing, and Johnson was weary of traveling. The last show aired on 19 May 1948, and Johnson retired to his ranch in Wimberly, Texas, where after a second career of civic boosterism he died in 1970.

CHUCK HOWELL

Hosts

Parks Johnson, Jerry Belcher, Wally Butterworth, Neil O'Malley, Warren Hull

Announcers

Graham McNamee, Ford Bond, Milton Cross, Ben Grauer, Ernest Chappell, Dick Joy, Tony Marvin, and Roger Krupp

Producers/Directors

Arthur Struck, John Becker, Rogers Brackett, Thomas Ahrens, Don Archer, and Glenn Wilson

Programming History

KTRH (Houston)	1932–35
NBC	1935–39
CBS	1939–47
ABC	1947–48

Further Reading

DeLong, Thomas A., *Quiz Craze: America's Infatuation with Game Shows*, NewYork: Praeger, 1991

Manning, Jerald, "Ready Wally? Ready Parks? How the *Vox Pop* Boys Run Their Show, and Some Questions for You," *Radio Stars* (November 1938)

Sammis, Fred, "The Program on Which YOU Are the Star!" *Radio Mirror* (October 1935)

Sullivan, Ed, "Radio Award" *Modern Screen* (November 1946)

W

WABC

New York City Station

One of the most powerful New York City stations (in terms of both transmission and ratings strength), WABC has been successful first as a Top 40 station and more recently as a talk-radio outlet. The flagship station of Disney/American Broadcasting Companies (ABC), WABC traces its history back to 1921. (It should not be confused with another WABC, also in New York City, that served as the flagship for the CBS network and became WCBS in 1946.)

Origins

WABC began broadcasting in 1921 in Newark, New Jersey, with the call letters WJZ, a 3,000-watt station owned by the Westinghouse Broadcasting Company transmitting at 833 kilohertz. Two years later it moved into New York City and was purchased by the Radio Corporation of America (RCA), which in 1926 created the National Broadcasting Company (NBC). NBC, in turn, established its Blue network with WJZ as the flagship station. The station increased its power to 50,000 watts in 1935 and changed its frequency several times, finally settling on 770 kilohertz in 1941.

In 1943 the Blue network and WJZ were sold to Edward J. Noble and Associates, which in 1945 changed its licensee name to the American Broadcasting Companies. Station call letters were altered to WABC in 1953 to reflect the transition. The ABC networks and WABC were taken over by United Paramount Theaters in 1953. Decades later the station and network were sold (1996) to the Disney Corporation.

Programs and Promotion

In the late 1950s WABC was lagging behind New York's leading popular music stations, WMCA and WINS, and was struggling to find its own niche in the highly competitive market. By 1964 WABC was not only the top station in New York, it was also the most-listened-to radio station in America. With weekly audiences of between 5 and 6 million from the mid-1960s to the mid-1970s, WABC could be heard in 38 states and Canada. Its Saturday night *Dance Party* with Bruce Morrow (universally known as "Cousin Brucie") reached 25 percent of the total radio audience in the New York metropolitan area.

The phenomenal change in WABC's fortunes is largely credited to the vision of Rick Sklar, station program director from 1963 to 1977. Sklar helped WABC stand out from its competition with a number of approaches. He limited the music playlist while at the same time trying to bridge the generation gap of rock and non-rock listeners. Sklar held weekly music meetings with his staff, assiduously choosing appropriate selections for diverse groups and tapping into the rapidly changing music of the era. The repeated airing of music by the Beatles (and on-air interviews with the members of the group) led to the hyping of the station as "W A Beatle C."

WABC's on-air lineup was built to create recognizable personalities, among them radio legends Harry Harrison, Dan Ingram, and Ron Lundy, as well as one of the best-known disc jockeys in the country, "Cousin Brucie" Morrow. These men had mellifluous voices, but more important, they established a rapport with their devoted listeners and hosted programs that were fun to listen to.

To establish loyalty and to garner attention, the station conducted unusual and ultimately enormously effective promotions. One such gimmick allowed people to vote for the School Principal of the Year. Listeners of all ages could vote as often as they wanted; the idea was to get entire families and school faculties tuned to the station. By the second year of the promotion in 1963, the program was so successful that the station received 176 million ballots—many of them from well beyond the tristate (New York, New Jersey, and Connecticut) standard audience area.

In 1964 WABC sponsored a contest honoring the best and the worst copy of the *Mona Lisa,* which was being flown from Paris to New York for exhibition at the Metropolitan Museum of Art. With Nat King Cole singing the promotional jingle and surrealist painter Salvador Dali serving as the judge, WABC received more than 30,000 entries.

Jingles and slogans were a key aspect of the station and were broadcast constantly. There were special jingles promoting the station's place on the dial, its news and sports reports, its disc jockeys and their time slots, and—above all—its ranking of songs and attention to music. The station's sounds became so distinct that even listeners tuning in during advertisements knew they were hearing WABC

News and Talk Radio

The demise of all-music WABC came on 10 May 1982–dubbed by some as "the day the music died"—when "MusicRadio 77" gave way to "NewsTalk Radio 77." A range of factors contributed to the need for the switch, among them the development of FM radio with its superior sound quality and the transfer of most music formats (and their audiences) to FM. AM stations like WABC needed to develop viable talk-based formats where sound quality was less crucial. By switching to an all-talk format, WABC was following the trend of many AM stations across the country.

By 2000 WABC was a powerhouse in personality-driven talk radio. As in years past, its promotions emphasize the station's approach: "If you're talkin' about it, we're talkin' about it." The station's two daytime hosts, Dr. Laura Schlessinger (9:00–11:45 A.M.) and Rush Limbaugh (12:00–3:00 P.M.) were the two top-rated hosts in the county; Limbaugh's Excellence in Broadcasting production operation shares studio space in the same building as WABC. Others in WABC's program lineup include the locally based conservative host Sean Hannity, liberal Lynn Samuels, former New York mayor Rudolph Giuliani, and former Guardian Angel Curtis Sliwa.

The station is also the broadcast home of the New York Yankees and New York Jets, and scheduled programs are preempted to broadcast the baseball and football games live. The only indication that WABC was once the premier music station in the nation is a three-hour program on Saturday nights devoted to Frank Sinatra.

RUTH BAYARD SMITH

See also American Broadcasting Company; Limbaugh, Rush; Morrow, Cousin Brucie; Promotion on Radio; Sklar, Rick

Further Reading

Battaglio, Stephen, "When AM Ruled Music and WABC Was King, *New York Times* (10 March 2002)

Freedom Forum Media Studies Center, *Radio, the Forgotten Medium,* New York: The Center, 1993

Jaker, Bill, Frank Sulek, and Peter Kanze, "WABC-II," "WJZ," in *The Airwaves of New York,* by Jaker, Sulek, and Kanze, Jefferson, North Carolina: McFarland, 1998

Musicradio 77 WABC, <www.musicradio77.com>

Passman, Arnold, *The Deejays,* New York: Macmillan, 1971

Sklar, Rick, *Rocking America: An Insider's Story: How the All-Hit Radio Stations Took Over,* New York: St. Martin's Press, 1984

WABC website, <www.wabcradio.com>

Walkman

Portable Audio

Begun in 1979 as a risky consumer electronics experiment by Sony and soon after almost a generic term (it was added to the Oxford English Dictionary in 1986), the Walkman was the first personal stereo, a tiny portable cassette tape playback device with a lightweight headset. Perhaps the most important consumer audio product since the development of the transistor radio in the mid 1950s and of stereo FM in 1961, the Walkman and its many imitators became one of the biggest world-wide consumer successes of the 1980s and 1990s. Begun as a tape-playing device, the Walkman changed over the years to add recording capability, radio reception, and eventually digital formats including CDs. The trade name became almost synonymous with any portable radio or disc player.

Origins

The Walkman began with an earnest request from engineer (and Sony co-founder) Masaru Ibuka, (1908–97) in late 1978 for a device on which he could listen to music while on long international flights from Japan. The first prototype was developed in Sony's tape recorder division in just a few days early in 1979, based on the existing "Pressman," a tape recorder designed for reporters. Given the short time for development, the new device made use of a tape transport and stereo circuits from existing Sony products.

As it was not a sophisticated piece of electronics that provided something technically new, many Sony engineers and some managers were not interested in the device. But Sony cofounder Akio Morita (1921–99), with his strong sense of product appeal and marketing, soon became a strong proponent of the innovation he saw as a potential best-seller. At his urging, engineers added dual headphone connections (the "his and hers" option) to make the device less off-putting, and an orange button mute to reduce volume so that the listener could hear outside sounds without having to remove the headphones (the "hotline function").

Perhaps the most important part of the package was the new lightweight headphones. Sony would emphasize the stylish nature of the tiny headphones in comparison to what people then used—large earmuff-size devices. This attempt to create a headphone culture was risky in a society with a phobia about deafness or other physical impairment. Also risky was the potential market value of a machine that could play cassette tapes but could not record. Batteries could operate the tiny machine for up to eight hours (two decades later batteries could last for 60 hours). Finally, the machine offered quality sound reproduction despite its tiny size—a key selling point.

Sensing an untapped market, Morita wanted to focus sales efforts on youngsters during the summer of 1979. There was no advance market testing, in part for lack of time. A relatively low sales price (projected to be about $125) would mean thousands would have to be sold to break even. Morita suggested an initial batch of 30,000—easily twice what their most popular tape recorder sold in a year—and again horrified his colleagues. The first product announcement and demonstration took place in June. Reporters were invited to an outdoor park in Tokyo to demonstrate the many ways the "Walkman," as it was now named, could be used while performing other activities.

Phenomenon

The first Walkmans went on the Japanese market 17 July 1979 (for $200), three weeks later than Morita had hoped. The delay made Sony even more concerned when initial sales were very slow, especially given its huge stock of the devices. By mid-August, however, word of mouth began to propel the Walkman to widespread popularity and sales took off. The first 30,000 were sold in a month, largely to the teenage buyers Morita had projected would be most interested. Sony had to constantly increase production to keep up with demand. Indeed, the first foreign sales had to be postponed for months—despite already running advertising campaigns—just to keep up with burgeoning Japanese demand.

The Walkman device was first sold as the "Soundabout" in the U.S. and Europe only in February 1980. Initially the plan was to sell the device under various names, depending on the country's language, but Sony quickly decided to stick with the Walkman label everywhere, which made advertising campaigns easier and helped the product build a global image.

Some later models added radio reception and by the 1990s were increasingly digital, built around CDs (as the "Discman"). Walkman had become a generic (though still trademarked) term for all the many small, portable CD or tape players and recorders as well as small radios. A variety of Sony models were made available—by 1990 more than 80 different models; by 1999, 180 in Japan alone and more than 600 worldwide. More than 250 million had been sold worldwide by late 1998 at prices ranging from as little as $25 up to $500. By the mid-1990s, the "portable audio product" category of consumer electronics devices was selling more than 25 million units a year. The Walkman principal was extended to television, with the eventual development of the Watchman (tiny portable TVs).

Perhaps the major social impact of the Walkman was to create a "personal sound" space for its users. Now one could listen to a recording or broadcast in a crowd—as in public transportation or even an elevator—without invading others' space. The very unobtrusiveness of the light, tiny device allowed users to carry sound with them almost everywhere—even on strenuous workouts or hikes. The Walkman represented ultimate portability even two decades after its introduction.

CHRISTOPHER H. STERLING

See also Receivers

Further Reading

Du Gay, Paul, editor, *Doing Cultural Studies: The Story of the Sony Walkman,* Thousand Oaks, California: Sage, 1997

Hooper, Judith, and Dick Tersi, *Would the Buddha Wear a Walkman? A Catalogue of Revolutionary Tools for Higher Consciousness,* New York: Simon and Schuster, 1990

Joudry, Patricia, *Sound Therapy for the Walk Man,* St. Denis, Saskatchewan: Steele and Steele, 1984

Morita, Akio, Edwin M. Reingold, and Mitsuko Shimomura, *Made in Japan: Akio Morita and Sony,* New York: Dutton, 1986

Nathan, John, *Sony: The Private Life,* Boston: Houghton Mifflin, and London: HarperCollinsBusiness, 1999

Patton, Phil, "Humming Off Key for Two Decades," *New York Times* (29 July 1999)

Sanger, David, "Stalking the Next Walkman," *New York Times* (23 February 1992)

War of the Worlds

Radio Drama

Broadcast as part of the *Mercury Theater of the Air* in October 1938, the style and format of *War of the Worlds,* an hour-long live drama, coming just a month after radio's news bulletins about Europe's Munich crisis, pushed some of the U.S. listening audience into a panic—and clearly demonstrated the growing trust in and power of radio broadcasting.

The Script

In 1898 English author and social thinker H.G. Wells (1866–1946) published *War of the Worlds,* a novella concerning an attack on Earth by creatures from Mars. Drawing on fears of a rearming Germany and concerns about the impact of such modern technologies as the telegraph and improved modes of transport, Wells' invasion story took place in England at the end of the 19th century.

Four decades later, in July 1938, the Columbia Broadcasting System (CBS) radio network began to offer weekly radio adaptations of literary works on the *Mercury Theater of the Air* on Monday evenings at 9:00 P.M. (changing to Sunday in September). The program featured the cast and producers of Broadway's successful and creative Mercury Theater troupe, headed up by the 22-year-old *wunderkind* Orson Welles (1915–85) and his producer John Houseman (1902–88). For each weekly broadcast, the cast would have only a few days to develop, rehearse, and finally broadcast each script. Several weeks into the radio season, Welles came upon H.G. Wells' story, which he felt would be perfect for the Halloween eve broadcast of his series.

By this point, Howard Koch (1901–95), undertaking his first professional scriptwriting job, had joined Houseman (who had written all program scripts to that point). In what became his third script for the series, Koch modified the original Wells story to take place in the present time and in tiny Grovers Mill, New Jersey (picked at random), not far from Princeton. Most important, in light of what would happen, Koch applied Welles' idea of a radio news bulletin approach (rather than straight narration) for the first portion of the drama. Listeners would hear what appeared to be news bulletins breaking into ongoing network musical programming. But Koch had to create a modern radio script from the 40-year-old story in less than a week.

He wrote the 60-page script over four days, with about 20 pages at a time being turned over to the producers for their comments and changes. With Houseman's aid, the script was finally finished on Wednesday so that the cast (save Welles, who was committed to Broadway activities) could rehearse on Thursday. The recorded results, however, sounded stilted and dull to all involved, and the script was revised again. Houseman and Koch added bits and pieces to enhance the eyewitness sound of the story with real places and some realistic government voices at key moments. When CBS received a copy on Friday, the network censor also asked for numerous script changes to make the story's fictional basis more obvious. This usually involved the changing of place names or organizations. Whereas the network seemed concerned that listeners might think the story too realistic, Mercury people feared listeners would not stay tuned in or would find the story too fantastic.

Busy with Broadway commitments, Orson Welles did not even see the script until Sunday morning, 30 October—mere hours before it would be broadcast live. At that point he took charge, making further script changes at the opening of the drama to increase the reality of the news bulletins breaking into other programs.

The Broadcast

Sunday evening network radio listening was dominated by Edgar Bergen and Charlie McCarthy on National Broadcasting Company (NBC) Red, the only commercial program in the time slot. On NBC Blue, listeners heard *Out of the West,* narrated dance music from San Francisco presented on a sustaining (noncommercial) basis, while Mutual stations were carrying the WOR symphony orchestra, though many carried a speech by Father Charles Coughlin, as it was sponsored (paid) time. Not all CBS stations carried the Mercury broadcasts—the Boston outlet, for example, was carrying a local program.

At 8 P.M. in CBS Studio 1 on the 20th floor of CBS headquarters at 485 Madison Avenue in New York, the cast and sound effects people were in place. Besides Welles playing Professor Richard Pierson, Dan Seymour played the New York studio announcer, Kenneth Delmar played several roles (he became famous several years later on the Fred Allen show), and Ray Collins also performed several different roles. The music background was by Bernard Hermann.

Relatively few listeners heard the program opening—a fairly standard announcement of what was to come followed by Welles setting the stage for the actual inception of the story. Some simply tuned in late, but many were listening to Bergen and McCarthy on NBC and would tune over to CBS after the play had begun. These patterns were a critical factor in the panic created by the program, for when listeners tuned in they

Orson Welles broadcasts H.G. Wells' *War of the Worlds,* 30 October 1938
Courtesy AP/Wide World Photos

heard what seemed like a normal weather forecast and then a cut-away to a hotel orchestra in downtown New York. The music was announced and was just getting under way when an announcer cut in with the first bulletin—about "reports observing several explosions of incandescent gas, occurring at regular intervals on the planet Mars." More music, then another bulletin, followed by a brief interview with a Princeton astronomer played by Welles. Then a bit of piano music, and still another news bulletin about those Mars explosions, but this time adding that "a huge, flaming object, believed to be a meteorite" had fallen on a farm "in the neighborhood of Grovers Mill, New Jersey." After about 20 seconds of a hotel swing band, another announcer broke in—and indeed, the next half hour was a series of increasingly exciting live on-the-scene reports from different reporters at various points in New Jersey. After the hostile nature of the "invasion" became clearer, about 25 minutes into the program, an announcer said,

> Ladies and gentlemen, I have a grave announcement to make. Incredible as it may seem, both the observations of science and the evidence of our eyes lead to the inescapable assumption that those strange beings who landed in the Jersey farmlands tonight are the vanguard of an invading army from the planet Mars.

This was followed by several more reports—from an army detachment and then a bombing aircraft—all of them giving way to the seeming invincibility of the invaders. By 8:40 P.M., clouds of poisonous gas were reported to have covered Manhattan, and a lone radio operator elsewhere was heard calling out with no response.

At about this point, a CBS announcer made a brief station break, noting that the program in progress was a drama, and then the final portion of the program began. Little remembered today, this consisted of Welles playing the Princeton astronomer, Richard Pierson, fearing he may be the last person on earth. He puts down his thoughts in a diary; then he runs across another survivor and, at the end of the hour, the remains of the invading Martians, killed by earthly organisms and bacteria. This segment ran 20 minutes.

At the end, and "out of character" as he put it, Welles spoke briefly to "assure you that the *War of the Worlds* has no further significance than as the holiday offering it was intended to be." A network announcer wrapped up with a hint of next week's broadcast and a Bulova Watch advertisement for the 9:00 P.M. hour.

The Effect

The first the program's cast knew of the commotion being caused came as the broadcast ended and police entered the studio, confiscating copies of the script and questioning actors on how much they knew about what was going on outside. Network telephone lines were flooded with calls of concern. Three more times during the evening, CBS announcers made clear the broadcast had been merely a drama and not a real news event. More than half of the stations that carried the play also made their own announcements.

But these cautionary announcements came too late for many listeners. Thousands of them, especially those in the seeming New Jersey and New York "target" zone of the Martian attack, had heard more than enough well before the program was over and were trying to flee the scene. If they heard one of the characters note that people were fleeing, they would, as later reported to researchers, look out the window. If streets were busy, they would assume people were fleeing. If streets were empty, the conclusion often reached was that others had already fled. Amazingly, few of those who panicked thought to tune another radio station to check or even to look at newspaper listings of programs to see what CBS was supposed to be broadcasting at that time. Few even telephoned others. If radio said we were being invaded, then it must be true.

Hundreds of calls were placed to newspapers (the *New York Times* alone received more than 800), radio stations, and police. The Associated Press put out a bulletin to its member papers explaining what had happened. Only slowly on the evening of the broadcast was widespread panic reduced. Though there were many accidents on crowded roadways, luckily no one was killed.

Over the next several days, there were widespread press reports about what had happened. Ironically, the program's impact that night helped it to gain commercial sponsorship, and it became the *Campbell Playhouse* in December and lasted in modified form until mid-1941.

Fascinated by the reaction to the program, the newly formed Office of Radio Research at Princeton University undertook a research study, under sociologist Hadley Cantril, to determine why so many had been driven to panic. The results, published as a book two years later, were largely based on interviews with 135 listeners. Researchers learned that people had grown accustomed to radio breaking in with important news during the Munich crisis of a month before and that such bulletins were assumed to be true. Of the roughly 6 million who tuned in, 1.7 million reportedly believed what they heard.

The Aftermath

A number of lawsuits were filed against CBS, all of which were eventually settled out of court. The Federal Communications Commission (FCC) launched a brief investigation out of which came an industry ban on program interruptions for fake news bulletins.

Rebroadcasts or sequels appeared on the anniversary of the original broadcast. In one case, a 1949 Spanish language sequel in Quito, Ecuador, led to enraged listeners' burning down the station, with several lives being lost. The story became a popular 1953 movie, with the setting shifted to Los Angeles. The first commercial recording of the original broadcast was issued in 1955 by Audio Rarities. "The Night America Trembled," presented on CBS's *Studio One* in September 1957, provided television viewers with a dramatic portrayal of the radio program. Buffalo station WKBS offered an updated version on the 30th anniversary of the original broadcast (30 October 1968), with radio in a Top 40 format style using station disc jockeys and newspeople. It was replayed a year later. Scriptwriter Koch provided a brief 1970 book with his own version of what had happened. For the 50th anniversary (30 October 1988), National Public Radio (NPR) offered a program based on a modified script by Howard Koch. A long documentary of the whole story appeared on the Discovery Channel in October 1998.

No other single radio program has had such a long-lasting impact as *War of the Worlds*. Both as a highlight of creative radio writing, and as an inadvertent measure of radio's growing place in society, the 1938 drama stands alone.

CHRISTOPHER H. STERLING

See also Hoaxes on Radio; Mercury Theatre of the Air

Programming History

CBS Mercury Theater of the Air, Sunday, 30 October 1938, 8–9 P.M.

Further Reading

Cantril, Hadley, *The Invasion from Mars: A Study in the Psychology of Panic*, Princeton, New Jersey: Princeton University Press, 1940
"Grim Fantasy in Ecuador," *New York Herald Tribune* (15 February 1949)
Homsten, Brian, and Alex Lubertozzi, editors, *The Complete War of the Worlds: Mars' Invasion of Earth from H.G. Wells to Orson Welles*, Naperville, Illinois: Sourcebooks MediaFusion, 2001
Houseman, John, "The Men from Mars," *Harper's* (December 1948)
Klass, Phillip, "Wells, Welles, and the Martians," *New York Times* (30 October 1988)
Koch, Howard, *The Panic Broadcast: Portrait of an Event*, Boston: Little Brown, 1970
Noble, Peter, "Orson Scares America," in *The Fabulous Orson Welles*, by Noble, London: Hutchinson, 1956
"Radio Listeners in Panic, Taking War Drama As Fact," *New York Times* (31 October 1938)
Wolff, G. Joseph, "War of the Worlds and the Editors," *Journalism Quarterly* 57, no. 1 (Spring 1980)

WBAI

New York City Station

From its inception, WBAI (99.5 FM) has broadcast alternative and often controversial programming to the greater New York City metropolitan area. WBAI has become one of the largest stations in the community radio network, with an operating budget of about $3 million per year.

Origins and Free Speech Heritage

The station became the third in the educational and noncommercial Pacifica network when then-owner and philanthropist Louis Schweitzer donated it to the Pacifica Foundation in the midst of a contentious city newspaper strike in 1959 because he believed that media should be used for the public interest.

WBAI has had a rich and sometimes combative broadcast history and is renowned for programs raising issues related to the First Amendment. In 1968 WBAI incited controversy in a widely publicized incident when a guest on writer Julius Lester's program read an anti-Semitic poem written by one of his students. The airing of the poem, "Anti-Semitism," which vividly described the African-American teenager's views toward the largely Jewish population of schoolteachers, again raised issues of what could and could not be broadcast. Although the United Federation of Teachers union filed a complaint with the Federal Communications Commission (FCC), that agency ruled in favor of the station, asserting that it allowed for "reasonable opportunity for the presentation of conflicting viewpoints."

The most prominent case involved the 1973 broadcast of "Seven Words You Can't Say on Television" by comedian George Carlin. The FCC cited WBAI's owner, Pacifica, for indecency for airing the 12-minute monologue in the early afternoon when children might be in the audience. The case eventually reached the Supreme Court, which ruled in 1978 that the government could regulate "indecent" broadcast speech from the airwaves but could not ban it.

In 1987 WBAI addressed free speech concerns again when it sent the FCC a list of questionable words and phrases without acknowledging that they were from James Joyce's *Ulysses*. Though originally the FCC responded that it could not judge the material until it was broadcast, the agency later ruled that context was critical and allowed the text to be aired as part of a program on Bloomsbury writers.

Programs

WBAI is best known for its extensive, thorough, and often subjective reporting. Beginning in the 1960s the station sent reporters into the South to report on the civil rights movement, including coverage of the murders of three young volunteer workers (James Cheney, Mickey Schwerner, and Andrew Goodman) in 1962. In 1965, WBAI participated with the other two Pacifica stations in holding Vietnam Day "teach-ins," providing nontraditional discussion-type broadcasts.

WBAI pushed the limits even further that same year when program director Chris Koch traveled to North Vietnam to report on the war. His reports were attacked by supporters of the Vietnam War and even by many on the Pacifica Board who were conflicted about the coverage. After attempts were made to have Koch delete parts of his reports, he and five others resigned. Nearly one-third of station member subscribers also canceled their subscriptions in protest.

WBAI's exhaustive foreign coverage has continued through the years, and long after the mainstream media have left the scene, WBAI has reported from battlefronts in countries such as Iraq, Haiti, and East Timor. Over the years the station has applied the same scrutiny to domestic conditions, examining the practices of the Federal Bureau of Investigation (FBI), federal and state prisons, and city homeless shelters, in addition to other problems.

While news is clearly a mainstay of WBAI's orientation, just as significant are its music and innovative programming. The station is known for introducing free form radio, specifically in *Radio Unnameable,* a program developed in 1963 by WBAI's Bob Fass, which still runs today. The broadcasts—precursors of much subsequent radio—feature a blending of sounds from many different sources. Musicians who would later become popular entertainers appeared regularly on *Radio Unnameable,* among them Judy Collins, Bob Dylan, Jose Feliciano, and Arlo Guthrie (who debuted "Alice's Restaurant" on the show).

It continues its hard-hitting news reporting and also broadcasts an eclectic range of programming, including shows on alternative lifestyles, health, labor, technology—and, of course, music. In the noncommercial "free radio" tradition, WBAI continues to inform, provoke, and rankle its listeners. It has also been honored for its coverage: in 1999 Amy Goodman, the morning co-host of *Wake Up Call,* a local call-in show, won a George M. Polk reporting award for a piece about two unarmed environmentalists who were killed in Nigeria as a result of the Chevron Corporation's involvement with the military there. More recently, the station won a Rodger N. Baldwin Award for outstanding contributions to the cause of civil liberty. As the plaque read, "From the armies converging on Iraq to the march for women's lives in Washington, from the killing fields of East Timor to the mean streets of Manhattan's homeless, WBAI covers the local, national and international scene with a depth and integrity not even conceived of by commercial broadcasting."

Staff Protests

WBAI has regularly been subject to various protests within its own staff. As an operation making heavy use of volunteers, and catering to a broad range of political opinion, heated disagreements are not surprising.

In 1977 a bitter strike over issues of race, staff authority, and finances kept the station silent for seven weeks and still remains one of the most contentious episodes in New York radio history. The strike came about because the staff protested management's reorganization of the station, which they perceived as taking away their power and softening the station programming. Very little was resolved as a result of the strike; it ended when the staff voted (by a narrow margin) to obey a court order requiring them to leave the transmitter room they had taken over.

An even longer and fiercer battle broke out in 1999. Pacifica Foundation management moved to make WBAI sound more professional and mainstream in its programming as they sought larger audiences and additional sources of funding. The move immediately ran into opposition from staffers wedded to their own sometimes controversial programs, often with strong listener support. Many staffers saw the move as an attempt to muzzle unpopular points of view. The host of *Democracy Now!* got into a battle with station management over the direction of her program. By late 2000 a 20-year veteran (including a decade as station manager) had been fired and the staff was in an uproar, talking to newspapers and filing various legal grievances. Soon some staffers were being locked out of the station (somewhat ironically housed on Wall Street) by foundation management. The battle—which had even featured street protests by listeners and disaffected staff—was finally settled in December 2001. WBAI retained the right to work with its own board of directors, only loosely controlled by the foundation in California. After a six-month search, a new station manager was appointed late in 2002.

RUTH BAYARD SMITH

See also Community Radio; Educational Radio to 1967; Free Form Format; Pacifica Foundation; Public Radio Since 1967; "Seven Dirty Words" Case

Further Reading

Engelman, Ralph, *Public Radio and Television in America: A Political History,* Thousand Oaks, California: Sage, 1996
Lester, Julius, *Lovesong: Becoming a Jew,* New York: Holt, 1988
Post, Steve, *Playing in the FM Band: A Personal Account of Free Radio,* New York: Viking Press, 1974
WBAI website, <www.wbai.org>

WBAP

Fort Worth, Texas Station

Clear channel WBAP in Fort Worth, Texas, has been a dominant force in Texas and Southwest broadcasting since the very dawn of the radio-television era.

Like many infant stations in the early 1920s, WBAP was an extension of a powerful newspaper, the *Fort Worth Star-Telegram,* published by Amon G. Carter Sr. and Carter Publications. *Star-Telegram* circulation manager Harold Hough convinced a skeptical Carter to put WBAP on the air. Carter grudgingly approved the expenditure of $300 to put the 10-watt station on the air on 2 May 1922. "But when that $300 is gone, we're out of the radio business," Carter admonished Hough. Of course, it soon became obvious that radio was more than a $300 experiment. Hough became WBAP's most popular on-the-air personality in the early days, going by the moniker "The Hired Hand."

Dual Channels

For more than 40 years, WBAP shared airspace with WFAA, owned by the *Dallas Morning News,* first on a single frequency, and later, uniquely, on two separate frequencies. It is the only known case of stations sharing more than one frequency.

WFAA had been saddled with a frequency-sharing deal not to its liking with KRLD, owned by the *Morning News'* bitter rival, the *Dallas Times Herald.* The stations were assigned to share 1040 kilohertz, also a clear channel. *Dallas Morning News* publisher George B. Dealey apparently preferred to cooperate with Carter's station, across 30 miles of then open prairie, rather than with the station of his intra-city rival. WBAP had been paired with KTHS in Hot Springs, Arkansas. The station found the awkwardness of sharing airtime with such a distant station untenable. So WBAP and WFAA petitioned the Federal Radio Commission (FRC) to allow them to share the 800 kilohertz channel (820 kilohertz beginning in 1941), and the FRC granted the change effective 1 May 1929. At the same time, the FRC authorized WBAP and WFAA to broadcast at 50,000 watts on 800 kilohertz, an increase of 100-fold from their previous wattage.

In 1935 the resourceful Carter acquired KGKO in Wichita Falls, Texas, broadcasting on 570 kilohertz, setting the stage for the unique dual-channel time-sharing arrangement. Carter had the station's license transferred to Fort Worth, and WBAP began broadcasting on KGKO's channel in 1938. Soon WFAA bought into KGKO, and in 1940 the two stations began sharing time on two separate frequencies.

Don Harris enjoyed a career of 33 years on the air at WBAP starting in 1965, in the final years of the dual-frequency days. "There was a 10-second changeover [for each station to sign off and the other to sign on]," Harris recalls. "Sometimes, those WFAA guys would cheat us a couple of seconds, and we would have two or three seconds of dead air [as WBAP waited for WFAA to complete its sign-off]." Harris admitted that sometimes it happened the other way around.

Fort Worth is also known as "Cowtown," and WBAP became famous for a cowbell station identification. The cowbell was almost certainly the innovation of the ubiquitous Hough, the station's most identifiable air personality in the 1920s and 1930s. WBAP claimed that its cowbell jangle was U.S. broadcasting's first "memory signal." The cowbell would be heard when WBAP identified itself before and after exchanging frequencies with WFAA.

WBAP prided itself (then and now) on its farm and ranch programming. A brief 1947 history of the station commented, "The station had a definite field to serve—the vast area of Texas (and West Texas in particular), where distance is most unusual, and which was, and is, a ranching and farming area." Until radio came to Texas, farmers and ranchers relied only on their limited powers of observation to prepare for rough weather. "More than once, ranchers were saved millions of dollars in livestock losses by WBAP's warning of approaching blizzards and storms. National recognition has been given the station for that service," a 1949 *Star-Telegram* article reported.

In the 1940s WBAP broadcast a night-time program from the state prison in Huntsville called *Thirty Minutes behind the Walls,* which was written, performed, and produced by prisoners. WBAP claimed in 1943 that the program received nearly 200,000 fan letters in one month from 45 states, Canada, and Mexico.

As television emerged in the 1950s, the WBAP and WFAA formats gradually evolved toward middle-of-the-road music. The decision by WBAP to hire the legendary Bill Mack to play country music on 820 kilohertz from midnight to 7:30 A.M. in 1969 marked the first clear break between the two stations. Harris credits general manager James A. Byron, who had been the station's news director, for having the foresight to make the change to country music.

In April 1970 the Federal Communications Commission (FCC) approved the sale of the Belo Corp's share of the 820 channel to Carter Publications for $3.5 million. Belo and WFAA received WBAP's interest in the 570 channel facilities. The end of the partnership came on 1 May 1970, when WFAA went to 570 kilohertz full-time and WBAP took over 820 kilohertz. It was the 41st anniversary of the time-sharing agreement. WFAA radio left the air in 1983.

After the split, WBAP initiated a country music format full-time. Other stations were playing country music, but Harris says WBAP immediately shot right past them and everyone else. "Six months after the change, we were number 1 in the market," he said, "It was an overnight success, just phenomenal."

In 1973 Capital Cities Communications bought WBAP and WBAP-FM (now KSCS) from Carter Publications for $80 million. In 1996 WBAP and KSCS became part of the merger of Capital Cities/American Broadcasting Companies (ABC) and the Walt Disney Company. In 1993 WBAP switched from country music to news/talk.

J. M. DEMPSEY

Further Reading

Dempsey, John Mark, "WBAP and WFAA: A Unique Partnership in the Era of Channel Sharing: 1929–1970," *Journal of Radio Studies* 7, no. 1 (Spring 2000)

Glick, Edwin, "WBAP/WFAA: Till Money Did Them Part," *Journal of Broadcasting* 21 (Fall 1977)

"Radio History Made by WFAA and WBAP in New Joint Service," *Dallas Morning News* (8 September 1940)

Schroeder, Richard, *Texas Signs On: The Early Days of Radio and Television,* College Station: Texas A&M University Press, 1998

"Station WFAA Assigned New Air Channel," *Dallas Morning News* (1 May 1929)

WBAP News/Talk 820: A Brief History of WBAP, <wbap.com/aboutwbap.asp>

"'We'll Spend but $300' and WBAP Was Begun: Hoover Named Station," *Fort Worth Star-Telegram* (30 October 1943)

"WFAA, WBAP to End Dual Frequency Use," *Dallas Morning News* (25 April 1970)

WBBM

Chicago, Illinois Station

This AM radio station, the long-time Columbia Broadcasting System (CBS) affiliate in Chicago, has been linked to the history of CBS as a radio (and later television) network. Chicago has long been famous as a pioneer in radio, but WBBM-AM never set in motion any significant shows. It simply aired what CBS sent along, and for radio in the late 1940s and network television through the 1970s, this usually meant the top-rated shows being broadcast. There is only one sense in which WBBM was a pioneer, and that was as an early FM station, beginning its broadcasts on 7 December 1941.

With WBBM standing for "World's Best Broadcast Medium," WBBM went on the air on 14 November 1923. WBBM has spent three-quarters of a century on Chicago's AM dial, beginning its life in the basement of the Atlass family home, then moving to the Broadmore Hotel, and then to its long-time home in the Wrigley Building on Chicago's north side. Within three years of its debut, the station bragged that it had aired the first dance music and the first church service, carried regular remote broadcasts, and acquired the leading number of local advertisers.

In 1933 CBS purchased WBBM, and station cofounder H. Leslie Atlass was made its general manager on a lifetime basis as part of the sale. By the late 1930s WBBM transmitted with 50,000 watts, and in 1936, the station often ranked number one in Chicago with a significant set of locally produced shows, such as *Piano from Warehouse 39, Sunday Night Party, One Quarter of an Hour of Romance, Dugout Dope, Women in the Headlines, Radio Gossip Club, Man on the Street,* and *Sports Huddle.* But none of these—unlike many other Chicago-based programs—ever made a national splash. From its 410 North Michigan Avenue studio, the station could air from the WBBM Air Theatre, which seated 300 persons. Through most of the 1930s and 1940s, it was a clear channel powerhouse at 780 kilohertz, able to be picked up throughout the upper Midwest and beyond. This meant that WBBM broadcast CBS news into Chicago and served as the Midwest base for the Columbia News Service set up in 1933.

WBBM-FM went on the air on 7 December 1941, but the FM station long merely duplicated the AM broadcast. In recent years, as FM has ascended in popularity, it has tried a number of different music formats. In 2000, WBBM-FM (96.3 FM) had the format of "dance hits," but the station had the same studio as WBBM-AM at 630 North McClurg Court. It was not until well into the 1960s that WBBM-FM did more than duplicate WBBM-AM.

With the rise of television, WBBM-AM needed to reinvent itself. The new format grew out of CBS chairman William S. Paley's desire to come up with a format that was on one hand prestigious (and not simply another rock variation) while also making money. WBBM found this with all-news radio in 1968, and the network reinvented itself as the link among the CBS all-news radio stations. WBBM-AM has provided all-news radio broadcasting in the nation's second city for more than a third of a century, with classic drive-time formatting. WBBM-FM has not been so lucky and has tried a number of formats that might work—with "dance hits" the format by 2000. Both AM and FM remained at studios at 630 North McClurg Court, though there was talk in 1999 of moving to another part of downtown. WBBM's AM antenna is located on the west side of the region, in Elk Grove Village, and the FM antenna is located atop the landmark John Hancock Building.

DOUGLAS GOMERY

See also All News Format; Columbia Broadcasting System

Further Reading

Boyer, Peter J., *Who Killed CBS? The Undoing of America's Number One News Network,* New York: St. Martin's Press, 1989
Broadcasting in Chicago, 1921–1989, <www.richsamuels.com>
Paley, William S., *As It Happened: A Memoir,* Garden City, New York: Doubleday, 1979
WBBM Newsradio 780, <www.wbbm780.com>

WBT

Charlotte, North Carolina Station

Dubbed the "Colossus of the Carolinas," WBT was the first commercially licensed station in North Carolina and, with WSB in Atlanta, was one of the first two such licensees in the southeastern region of the United States. From its debut in 1922, the station has also been one of the region's most powerful. In 1929 William S. Paley purchased WBT, establishing it as a key link in the distribution of the Columbia Broadcasting System (CBS) programming.

WBT's roots stretch back to 1920, when Fred M. Laxton (a former General Electric employee), Fred Bunker, and Earle Gluck began transmitting experimental station 4XD from Laxton's home in Charlotte. Laxton and his friends had constructed the transmitter in a chicken coop behind Laxton's house; they broadcast music from phonograph records to the few people who owned receivers at the time. On 10 April 1922, Laxton, Bunker, and Gluck—operating as the Southern Radio Corporation—received a license from the federal government to broadcast with 100 watts of power as WBT. Initially, the station aired from 10:00 to 11:45 in the mornings and from 7:30 to 9:45 in the evenings.

Southern Radio Corporation sold WBT in 1926 to C.C. Coddington, a local Buick automobile dealer, and over the following three years, the station grew to be a regional power. During the years following Coddington's purchase of WBT, the station acquired permission to broadcast at 5,000 watts and affiliated with the National Broadcasting Company (NBC).

In the late 1920s William S. Paley of CBS, in fierce competition with NBC to garner affiliates, had begun purchasing established stations in key regions to ensure the expansion of the CBS network. In late 1928 the broadcasting titan bought WABC (later WCBS) in New York, and in 1929, he added to his collection WCCO in Minneapolis, Minnesota, and, with an eye toward reaching the southeastern United States, WBT.

Soon after acquiring WBT, CBS persuaded the Federal Radio Commission to allow the station to broadcast with 25,000 watts of power. Four years later, in 1933, the station's power rose to 50,000 watts. Listeners from Maine to Florida could clearly receive WBT's signal.

WBT would be a CBS-owned-and-operated station until 1945, when the Federal Communications Commission's (FCC) network rulings forced CBS to divest itself of certain high-power stations. Reluctantly, CBS sold its gold mine in the Southeast to Jefferson Standard Life Insurance Company, the enterprise (known today as Jefferson Pilot) that currently owns the station. WBT would maintain its CBS affiliation and continue to be a powerful outlet for the network's programs.

Just as WBT proved to be an important cog in the distribution of CBS network programming, by the late 1920s the station had also become an important source of local news, information, and entertainment for listeners in the Charlotte region. Led by program director (and later station manager) Charles H. Crutchfield and personalities such as Grady Cole and Kurt Webster, the station broadcast tobacco auctions, professional baseball games, old Confederate veterans reeling off their "rebel yells," country music performances, and other events of local flavor. Many of WBT's local productions were picked up by the CBS network for national and regional distribution.

Numerous entertainers would receive valuable exposure via their appearances on WBT productions. Before achieving fame as *Amos 'n' Andy* over the NBC network, Freeman Gosden and Charles Correll performed on the station in the 1920s. In addition, saxophonist Hal Kemp, who would become a well-known big band leader, had performed over WBT in the 1930s, as had bandleader Kay Kyser. Country music performers, such as the Briarhoppers and Arthur Smith and the Crackerjacks, watched their reputations grow on regularly scheduled WBT programs such as the *Carolina Hayride,* a weekly barn dance program picked up by the CBS network in the 1940s. Perhaps the best-known of the traditional music performers who emerged from WBT were the Johnson Family Singers, a gospel-singing sextet who, as a result of their success on WBT in the 1940s, broadcast regularly over the CBS network and attracted recording contracts from Columbia and, later, Radio Corporation of America (RCA) Victor. WBT also claimed a role in grooming newsmen for CBS: Charles Kuralt and Nelson Benton worked at the station before going on to the network.

WBT helped pioneer radio broadcasting in the Southeast and became a widely disseminated source of regional news, information, and entertainment. The station would also spawn North Carolinian entertainers and news personalities who would go on to national fame in broadcasting. WBT will be most remembered for its role in building the CBS network.

MICHAEL STREISSGUTH

See also Columbia Broadcasting System

Further Reading

Eberly, Philip K., *Music in the Air: America's Changing Tastes in Popular Music, 1920–1980,* New York: Hastings, 1982

Johnson, Kenneth M., *The Johnson Family Singers: We Sang for Our Supper,* Jackson: University Press of Mississippi, 1997

Wallace, Wesley Herndon, "The Development of Broadcasting in North Carolina, 1922–1948," Ph.D. diss., Duke University, 1962

WBZ

Boston, Massachusetts Station

One of the Westinghouse group of stations, WBZ Radio went on the air on 19 September 1921, with studios at the Westinghouse Electric plant in East Springfield, Massachusetts, approximately 85 miles west of Boston. WBZ's first broadcast was a remote from the Eastern States Exposition, a large New England county fair; among the speakers helping to dedicate the station were the governors of Connecticut and Massachusetts. WBZ operated on a frequency of 800 kilohertz, with a power of 100 watts. Within several years, the station would move to 900 kilohertz, and by 1928 it was placed at 990 kilohertz by the Federal Radio Commission.

By early 1922, WBZ had moved its studio to Springfield's Hotel Kimball. Station programming was typical of radio in those pioneering days: an occasional star but mostly eager amateurs willing to perform free. By early 1924, it was already becoming difficult to get good free talent, as more stations competed for performers. WBZ decided to open a Boston studio in conjunction with the *Boston Herald* and *Boston Traveler* newspapers, from which they got their news at that time; the new station was known as WBZA. The studios were first in Boston's Hotel Brunswick; they moved to the Statler Hotel in 1927 and to the Bradford in 1931. Meanwhile, the Boston station grew in importance, and its ability to attract major talent was a big plus for Westinghouse. WBZ was the first Boston station to broadcast Boston Bruins hockey games (featuring *Herald* sportswriter Frank Ryan doing play-by-play reports), and when not doing sports, the station provided a regular schedule of dance bands, well-known singers, political talks, a storyteller for kids, and a staff of announcers who became very popular in their own right. By March of 1931, it was decided that the WBZ call letters should belong to Boston and the WBZA letters should go to the Springfield station.

In 1926 WBZ began broadcasts of the Boston Symphony Orchestra, and in 1927, the station became one of five original affiliates of the National Broadcasting Company (NBC) Blue radio network. (As time went on, many of the WBZ announcers would be hired by the network.) When WBZ had its tenth anniversary celebration in September 1931, the NBC Blue network carried it. But in April 1932 WBZ got a rather dubious bit of publicity, which showed the perils of live radio. A supposedly trained circus lion (he was trained to roar on cue) was brought to the studio, and for some reason, he broke away and went rampaging through the studios, destroying equipment, terrifying spectators, and injuring seven people before the police arrived and had to shoot him.

By the mid-1930s, WBZ was using 50,000 watts; in 1941, the station was moved to 1030 kilohertz, a dial position it still has today. WBZ began doing a morning show featuring country vocalist Bradley Kincaid during the late 1930s. In 1942 Carl DeSuze joined the station; he would go on to a long career as the morning show host. The programming in the 1930s and 1940s included a daily women's show (radio homemakers were very popular, and WBZ had Mildred Carlson and later Marjorie Mills), as well as an increasing emphasis on news. The WBZ news staff included some of the best-known reporters, many of whom would later go on to join WBZ-TV.

Experiments with FM were taking place in the late 1930s and early 1940s; WBZ used its FM station (W1XK) to broadcast the Boston Symphony Orchestra in 1941, which was probably the first time the orchestra had been heard via the new FM technology. In June 1948 WBZ radio was joined by WBZ-TV, channel 4, and the stations moved into their own new facility, which had been specially designed for both radio and TV.

In the mid-1950s, WBZ moved away from its previously "middle-of-the-road" programming, dropping NBC and abandoning big bands for a more popular, hit-oriented sound. The station hired five well-known announcers and called them "The Live Five" to let the audience know that WBZ was now locally programmed. By the 1960s, WBZ had moved to a soft Top 40 format (detractors called it "chicken rock"—rock without any loud songs), but it still offered a nightly talk show and a heavy news commitment. WBZ's disc jockeys were personalities, and the younger audience found them very entertaining. Disc jockeys such as Bruce Bradley, Dick Summer, and Dave Maynard did more than just play the hits—they also interacted with their fans at numerous remote broadcasts and events: it was the era of the "record hop," and WBZ announcers were masters of ceremonies at dances all over Massachusetts.

By the 1970s, WBZ was moving away from Top 40 to a more Adult Contemporary sound and increasing its news. The station had long been known for public service, raising money for worthy charities; this activity was also increased, such that by the 1980s music was gradually being phased out in favor of longer news blocks, more sports, and more talk shows. By December 1985, WBZ had an all-news and information format during afternoon drive time, and by September 1992 the station completed the transition to being an all-news station.

Ironically, the city where it all began for WBZ—Springfield—was no longer a part of the WBZ game plan after 1962. WBZA was shut down by the parent company in order to buy another station in a larger market. But as a tribute to the station's beginnings, when WBZ celebrated its 50th anniversary in 1971, festivities were held both in Boston and at the Eastern States Exposition in Springfield.

WBZ is one of the few stations that still has its original set of call letters, and it has experienced minimal staff turnover on air: for example, several current members of the sports department have worked there since the 1960s, the news staff includes men and women with 10 to 15 years of service, and announcers such as Carl DeSuze and Dave Maynard retired after more than 40 years on air. This is a tribute to WBZ's stability: although it is no longer called a Westinghouse station (Westinghouse purchased the Columbia Broadcasting System (CBS) in 1995, and by the end of 1996 the parent company became known as CBS rather than Westinghouse), it is still known for news, public service, sports, and night-time talk shows. The annual telethon/radiothon for Children's Hospital consistently brings in large sums of money; in fact, WBZ Radio has won numerous awards for public service and excellence in broadcasting, including the National Association of Broadcasters' Crystal Award and the Marconi Award. WBZ has consistently earned number one ratings in Boston since the station changed over to an all-news format. And with a signal that can reach 38 states at night, WBZ continues to be among the most respected and most listened-to AM radio stations.

DONNA L. HALPER

See also Columbia Broadcasting System; Group W; Westinghouse

Further Reading

"WBZ News Radio 1030 Celebrating 75 Years: 1921–1996," Boston: WBZ, 1996

WCBS

New York City Station

The flagship station of the Columbia Broadcasting System (CBS) radio network, WCBS traces its lineage back to WAHG, which first aired in 1924. It was long the radio home of Jack Sterling and other local radio figures before becoming an all-news outlet in 1967.

Origins

WCBS got its start as part of a radio manufacturer's business. Alfred H. Grebe began to make radio receivers in 1922, and two years later he placed WAHG ("Wait and Hear Grebe") on the air on 24 October 1924 at 920 kilohertz with 500 watts of power. As was common at the time, WAHG soon shifted frequencies to 950 kilohertz and had to share time with another New York City station. By 1926 power had been increased tenfold. That same year, Grebe organized the Atlantic Broadcasting Company and changed the call letters to WABC, moving the studios from Queens into Manhattan.

Grebe had plans to develop a network, following the example set by the National Broadcasting Company (NBC) beginning in 1926. His outlet became one of two New York stations for the new CBS network, because as stations commonly shared time, more than one was necessary for the new network to provide a full week's coverage. Instead, late in 1928 Grebe sold his station to William S. Paley, head of CBS.

Network Station

CBS moved the new flagship station and its network headquarters into 485 Madison Avenue, where both would remain for several decades. The station moved to 860 kilohertz and by late 1929 was up to 50,000 watts of power. The final shift, to the present 880 kilohertz, came in 1941, part of the readjustment of American stations because of the North American Regional Broadcasting Agreement treaty.

WABC carried virtually all of the developing network's programming, plus some New York–only shows. With the creation of the new American Broadcasting Companies (ABC) in 1945, however, CBS was broadcasting from a station with a competitor's initials, and in November 1946, WABC became WCBS (after involved negotiations, as the new call letters had been used by a small station in Illinois since 1927). At the same time, it began to offer *This Is New York,* a combination of interviews and features on the city that would run for 17 years, first in the morning and finally as an evening program. Another long-lasting program was *Music Til Dawn,* which began in 1953 and lasted until 1970 as an inexpensive way of filling the late night and early morning hours, sponsored by American Airlines (then still headquartered in New York).

With the demise of evening network daytime programs in the early 1950s, WCBS tried a variety of programs with a "middle of the road" approach. Among them was *The Jack Sterling Show,* a morning music DJ show beginning in 1948 that ran for years. Not finding success, WCBS converted to an all-news format (for most of its schedule) in August 1967. Ironically, the premier of the new service was delayed when an airplane hit the station's transmitter tower. But, as often happens in such disasters, WCBS was able to use other station facilities while getting its own back on the air full-time. In the early 1970s the remaining non-news programming was terminated, making the station all news, all the time.

In recent years advertised as "WCBS 880," the station is formally owned and operated by Infinity Broadcasting Corporation, a publicly traded subsidiary of CBS Corporation. Offices and studios are located in the CBS Broadcast Center on West 57th Street in Manhattan.

CHRISTOPHER H. STERLING

See also Columbia Broadcasting System; Infinity Broadcasting; North American Regional Broadcasting Agreement

Further Reading

Jaker, Bill, Frank Sulek, and Peter Kanze, "WABC," "WAHG," and "WCBS," in *The Airwaves of New York: Illustrated Histories of 156 AM Stations in the Metropolitan Area, 1921–1996*, by Jaker, Sulek, and Kanze, Jefferson, North Carolina: McFarland, 1998

WCBS, <www.newsradio88.com/main/home/index>

WCCO

Minneapolis, Minnesota Station

The Twin Cities' "Good Neighbor to the Northwest," a 50,000-watt, clear channel AM station, is said to have dominated local audience ratings like no other station in the country. Legend has it that radio executives from across the United States would travel to Minnesota and sit in their hotel rooms listening to WCCO in order to "research" its format. They could only shake their heads in wonder. What they heard was a homespun blend of often insipid comedy featuring Scandinavian humor; remote-location broadcasts of amateur glee clubs and ethnic musicians; a schedule heavy on local news, farm markets, and weather; and ubiquitous live chats with political figures such as Hubert Humphrey. It was a sound that could not be duplicated elsewhere, called "Just Folks Radio" by *The Wall Street Journal.* At the height of its popularity, surveys showed WCCO to be the favorite among listeners in 128 counties by a 19-to-1 margin. WCCO was as much a part of the fabric of the Northwest as were snow days in January.

The station's predecessor, WLAG, went on the air in 1922 as a marketing tool for a Minneapolis radio manufacturer. In September 1924 the failed station was resurrected when a promoter convinced executives of the Washburn Crosby Company to buy it (hence the call letters "WCCO"). The company, known today as General Mills, used its new "Gold Medal Station" to promote its many brands of flour and cereal products. The first day's log boasts two home service programs performed by "Betty Crocker," and on Christmas Eve 1926, WCCO aired radio's first singing commercial, "Have you tried Wheaties?"

The station was one of 21 original affiliates of the National Broadcasting Company (NBC) Red Network, but within two years it switched to the fledgling Columbia Broadcasting System (CBS). In 1929 CBS chief William Paley bought one-third of WCCO for $150,000, with an option to buy the rest in three years for $300,000. The station's first general manager, Henry Bellows, was a University of Minnesota rhetoric professor who later served on the Federal Radio Commission and is said to have helped WCCO gain its favorable 50,000-watt clear channel designation in the early 1930s. Bellows and his successor at WCCO, Earl Gammons, eventually became CBS vice presidents. Although CBS network programming came to fill about three-quarters of WCCO's broadcast day, Gammons built an immensely loyal audience for local programming in the 1930s. Gammons' trademark was hiring air personalities with a warm, neighborly style of delivery that contrasted with the more sophisticated sound of most New York–based network entertainers. The sound was called "personality radio."

In 1934 the resonant voice of Cedric Adams took over the WCCO evening news, and, bolstered by a nighttime signal that spanned half the country, WCCO and Adams together became a U.S. radio phenomenon. He was perhaps the most widely known and highest paid local radio personality for over a quarter of a century. At the height of his popularity, he did a daily five-minute program on CBS and occasionally substituted for network star Arthur Godfrey. WCCO took *Cedric Adams' Open House* on the road, annually logging over 15,000 miles with 49 weekend road trips. At one time his weekend traveling troupe consisted of an accordion player, a magician, three singing sisters, a comedy act, two baton twirlers, an eight-year-old girl who yodeled, and the Minneapolis Aquatennial Queen. CBS executives in New York may have snickered at the hokey lineup, but they never failed to notice the bottom line. Cedric Adams' 12:30 P.M. newscast had the largest Hooper rating of any non-network radio show in the country.

In 1938 the station moved from its cramped studios in the Nicollet Hotel to a spacious headquarters in the Minneapolis Elks Club building. The new location had the high ceilings favored by CBS architects and was quickly transformed into an art deco landmark with cream-and-Columbia-blue interiors. It was remodeled in the classic CBS design that had been used in other network projects in Boston, Saint Louis, and Chicago, featuring "floating studios"—in which the floors, walls, and ceilings were separated from the main building structure. Best of all, the new location had a fourth-floor auditorium that seated 700 people. Although unpretentious, with its tiny stage and its stage right main-floor engineering booth, the room became the source throughout the 1940s for some of the country's best live radio. At its peak, the WCCO auditorium was home to seven consecutive individually sponsored live-audience programs on Friday evenings.

The growth of television caused a change in WCCO ownership in the 1950s. Because CBS could own only a minority interest in a Twin Cities television station (it already owned five television stations in other markets), CBS merged WCCO with Mid-Continent Radio and Television in 1952. The firm became Midwest Radio-Television. The Ridder and the Murphy newspaper families held a 53 percent majority, and CBS owned the remaining 47 percent. Two years later Federal Communications Commission (FCC) rules (requiring that minority interests be counted) forced CBS to sell its minority share of the company. The *Minneapolis Star and Tribune* bought this minority interest for $4 million. Under the reorganization, WCCO Radio was set up as an independent entity within the company, controlled by neither the parent company's television nor the newspaper's interests, which now included both the *St. Paul Dispatch–Pioneer Press* and the *Minneapolis Star and Tribune.* This independence probably helped WCCO Radio compete and even thrive in the post–television era and to maintain its market dominance well into the decades that followed. The "variety" programming of the 1960s and 1970s gave way to a more uniform "news-talk" format in the 1980s and 1990s. As late as the mid–1980s, Arbitron showed that WCCO-AM was still the number one rated major-market station in the entire country, although this unusual market dominance was later moderated by local FM competition.

In 1992, 40 years after WCCO was sold to local owners, CBS reacquired the stations, purchasing the assets of Midwest Communications in a deal worth over $200 million. In addition to WCCO Radio and WCCO-TV, the acquisition included WLTE-FM, Minneapolis (formerly WCCO-FM); three smaller television stations in Wisconsin and Minnesota; and the MSC regional sports cable TV channel.

MARK BRAUN

See also Columbia Broadcasting System

Further Reading

Haeg, Lawrence P., *Sixty Years Strong: The Story of One of America's Great Radio Stations, 1924–1984: 50,000 Watts Clear Channel, Minneapolis-St. Paul,* Minneapolis, Minnesota: WCCO Radio, 1984
Sarjeant, Charles F., editor, *The First Forty: The Story of WCCO Radio,* Minneapolis, Minnesota: Denison, 1964
WCCO, 1924–1949: For 25 Years Good Neighbor to the Northwest: The CBS Station in Minneapolis and St. Paul, Minneapolis, Minnesota: n.p., 1949
Williams, Bob, and Chuck Hartley, *Good Neighbor to the Northwest, 1924–1974,* Minneapolis, Minnesota: n.p., 1974

WCFL

Chicago, Illinois Station

The Chicago Federation of Labor's (CFL) 1926 plan to open its own radio station initiated one of the most notorious controversies in early broadcasting. As planned by the CFL, the proposal would create the only radio station in the country specifically devoted to the interests of organized labor. The CFL planned to call its new station WCFL and decided to use the same frequency as KGW in Portland, Oregon, and WEAF, the popular American Telephone and Telegraph (AT&T) station in New York City. These plans were drawn without first consulting the U.S. Department of Commerce, which was then charged with station licensing.

When the Commerce Department was approached for a license, the CFL was told that none would be forthcoming. Acting Secretary of Commerce Stephen Davis explained that there was no room in the overcrowded broadcast frequency spectrum to accommodate the station. In response, the CFL declared on 17 May 1926 that the organization fully intended to broadcast with or without a license. The CFL intended to transmit primarily educational programming, including labor-related information, public affairs, and discussions of economic issues, in addition to entertainment features.

CFL secretary Edward N. Nockels protested the commerce department's preferential treatment of corporate giant AT&T. Nockels expressed shock that no room could be found on the airwaves for WCFL, while AT&T's WEAF was able to occupy a clear channel frequency. Nockels wrote Davis asking if it might be possible for WCFL's signal to occupy half of the territory reached by WEAF.

The Department of Commerce intimated that if the CFL would broadcast illegally, the department would bring the matter before a federal court to test the legality of the Radio Act of 1912. Nockels said in reply that the CFL was willing to allow the courts to settle the matter. He expressed confidence that the courts would support the CFL's plans. The CFL began construction of its new station, WCFL, without official approval.

Ultimately the CFL was able to secure an inspection for the station from the Department of Commerce, and WCFL received its license. After a week of experimental transmissions, the station officially opened on 22 July 1926 with a special two-hour inaugural broadcast from 6:00 to 8:00 P.M. on 491.5 meters, "just a shade away" from WEAF. WCFL broadcast from transmitter facilities on Chicago's Municipal (Navy) Pier, which was linked to studios elsewhere in the city. WCFL was supported by the contributions of labor unions and quickly established itself as "the Voice of Labor."

The station quickly became the target of attacks because of its hindrance of long-distance ("DX") reception of AT&T's WEAF in New York. *Radio Broadcast,* for example, suggested that WCFL had chosen its frequency as a direct attack on AT&T, out of indignation that WEAF could control an exclusive frequency. *Radio Broadcast* engaged in something of a crusade against WCFL, at one point putting the CFL in the same class as the Ku Klux Klan, since both were special-interest broadcasters.

By October 1926, the station was on the air from 6:00 P.M. until midnight six days a week. The 6:00 to 7:00 slot each evening was devoted to labor issues discussed in a program called *CFL Talks and Bulletins.* The program's length was quickly cut in half, however, because of a lack of material. WCFL filled the remainder of the evening with entertainment features, predominantly music. The station later also offered educational lessons, farm market reports, weather reports, and religious services.

Throughout the history of the CFL's involvement in the station, most WCFL programming was not directly union-oriented, a fact that occasionally generated complaints from union officials but that did not stop the Federal Radio Commission (FRC) from labeling WCFL a "propaganda station." To some extent, WCFL became a victim of General Order 40, issued by the commission in 1928. Severe limitations were placed on the station's activities. WCFL was made to share its frequency, and its power was drastically reduced before the commission finally insisted that the station could not operate in the evening, so as not to interfere with other stations.

After the CFL found that the FRC's stance could not be shifted, the organization sought support for the station from Congress. As a result of congressional pressure, the FRC moved WCFL in 1929 to a new frequency, on which the station could again operate on a full-time basis. When Nockels found that reception of the station was poor on that frequency, the commission allowed the station to return to 970 kilohertz and to broadcast for an additional four hours in the evening. By 1932 WCFL and the National Broadcasting Company (NBC) had entered into an agreement whereby WCFL would broadcast at 970 kilohertz full-time and at increased power (5,000 watts), despite the NBC station KJR in Seattle, which shared the frequency. The proposal quickly found approval with the FRC.

WCFL faced financial woes in its early days, which were to some extent alleviated by the acceptance of advertising, a practice that began in 1927. The station ushered in a lucrative rock and roll format in 1965 and subsequently became known as the home of some of radio's legendary disc jockeys, including Dick Biondi and Larry Lujack. The station's format change, however, prompted controversy among the membership of organized labor, with AFL-CIO president George Meany asking at one point what rock and roll had to do with the labor movement. By the early 1970s, WCFL had become Chicago's leading rock station.

Emphasis on entertainment, coupled with the competitive pressures of the Chicago commercial radio market, resulted in WCFL's abandoning an ever-increasing portion of its remaining public service programming. In spite of criticisms that the station was no longer significantly engaged in public service, WCFL survived challenges to its 1975 license renewal. Because the station eventually lost its ability to compete adequately in the Top 40 market, however, WCFL switched to an easy listening format in 1976. The station increasingly became a burden on the CFL without financial reward, until WCFL was finally sold for $12 million to the Mutual Broadcasting System in 1979.

Unable to make the station financially viable as an all-news operation (which would have cost twice the then $4 million annual operating budget), Mutual attempted by 1980 to turn WCFL into an adult contemporary music station. However, that failed as well, because most music listening by then was to FM, and AM outlets focused more on talk. Three years later, Mutual sold the station to Statewide Broadcasting, which turned WCFL into a religious music operation. Consistently last in city-wide ratings, WCFL was finally merged with Heftel Broadcasting's WLUP-FM, and in April 1987 the long-time voice of labor became WLUP-AM, an adult rock station.

STEVEN PHIPPS

Further Reading

"Chicago Labor Unions Plan to Build or Buy Station," *New York Times* (21 February 1926)

Godfried, Nathan, "The Origins of Labor Radio: WCFL, the 'Voice of Labor,' 1925–1928," *The Historical Journal of Film, Radio, and Television* 7, no. 2 (1987)

Godfried, Nathan, *WCFL, Chicago's Voice of Labor, 1926–78*, Urbana: University of Illinois Press, 1997

McChesney, Robert W., "Labor and the Marketplace of Ideas: WCFL and the Battle for Labor Radio Broadcasting, 1927–1937," *Journalism Monographs* 134 (August 1992)

WDIA

Memphis, Tennessee Station

Known as the "Mother Station of the Negroes," WDIA in 1949 became the first black-oriented radio station in the United States. Radio had featured black talent and black-appeal programs since the 1920s, but never before had a station directed its entire broadcast schedule to the black audience. This novel and long-overdue development in radio programming that WDIA pioneered spawned dramatic growth in the number of radio stations tailored exclusively for the black market.

The station, whose original studios were at 2074 Union Avenue in Memphis, had signed on the air in June 1947 with a white-oriented format that featured country and western music; classical music; and a smattering of news, sports, religion, and children's programs. This format garnered pale ratings, so in the fall of 1948 WDIA's founders, John Pepper and Bert Ferguson, began to experiment with black-appeal programming. It proved to be a profitable decision for the white owners and one that would open opportunities in broadcasting to black men and women. Initially, WDIA featured only a handful of black-oriented programs, but by 1949 the station had made a complete conversion.

To spearhead the new format, Pepper and Ferguson hired Nat D. Williams, a leader, educator, and impresario in the black community of Memphis. Williams's established popularity and ebullient on-air style virtually guaranteed the success of the programs he hosted; he and his shows became the core around which WDIA built the remainder of its black-appeal format.

In establishing the first black-appeal radio station in the United States, Pepper and Ferguson capitalized on the large black population in the Memphis area and its growing postwar prosperity. The ratings and revenue improvements for WDIA were startling and almost immediate. Among the tens of thousands of black Memphians, WDIA was number one. Soon the station was reporting that it attracted 33 percent more daytime listeners on weekdays than any other Memphis station did. Naturally, with more listeners came more advertisers, and although many advertisers, fearing racist backlash, avoided WDIA, they soon shed their fears and latched on to the very real earning potential that the station offered. Ford, Kellogg's, Sealtest, Lipton's Tea, and other companies with deep advertising budgets became regular sponsors of WDIA programs. Confirming that their venture into black-oriented programming had succeeded, Pepper and Ferguson reported in 1951 that their local and national advertising sales had increased 75.4 percent and 80 percent, respectively, in 1950 over 1949.

In 1954, over protest from radio station WMPS in Memphis, the Federal Communications Commission (FCC) granted WDIA the authority to operate with 50,000 watts daytime and with 5,000 watts nighttime with different antenna patterns both day and night. As a result, the station expanded its influence well beyond Memphis into parts of Mississippi, Arkansas, Texas, Louisiana, and Illinois. WDIA claimed to have access to 1,466,618 blacks, or ten percent of the black population in the United States. By 1957 WDIA had the highest ratings and advertising income among radio stations in the mid-South.

Fueled by the example WDIA set and by the general decline in radio audiences, other radio stations sought to bolster their bottom line by becoming exclusively black-oriented. In WDIA's wake followed WMRY in New Orleans, WEFC in Miami, WCIN in Cincinnati, WJNR in Newark, and others.

WDIA employed various tactics to build loyalty among the black population it reached. Noting the success in 1948 of Nat D. Williams—first on his afternoon *Tan Town Jamboree* and then on his morning *Tan Town Coffee Club*—management quickly set about hiring more black talent until virtually the entire on-air staff was black. (Despite this groundbreaking opportunity for black on-air talent, it is important to note that Jim Crow still hovered about WDIA in the 1940s and 1950s: blacks were barred from sales, management, and engineering positions.) Joining Nat D. Williams were disc jockeys Martha Jean Steinberg, Maurice "Hot Rod" Hulbert, B.B. King, and Rufus Thomas, all of whom featured blues, rhythm and blues, and other popular music of the day. (B.B. King and Rufus Thomas would achieve fame as blues and rhythm and blues performers.) Ford Nelson, Theo "Bless My Bones" Wade, and

Rev. Arnold Dwight "Gatemouth" Moore presented gospel music programs, and the popular Willa Monroe hosted the *Tan Town Homemaker's Show,* which was geared to women in the radio audience. Blacks who listened to WDIA heard their own voices, which nurtured loyalty to the station as well as pride in the status that blacks had achieved on Memphis radio.

WDIA also built and maintained its listenership with comprehensive community relations efforts. The station's public service pervaded the black community: it traced missing persons, pleaded for blood donors, sponsored baseball teams, and collected food and clothing for needy Memphians. The centerpieces of the station's community focus were its roundtable broadcasts, which often dealt with racism, and the *Goodwill Revue* and the *Starlite Revue,* live concerts that raised money for a school for handicapped black children. Throughout the 1960s, WDIA's programs played an important role in the civil rights movement in Memphis, appealing for black equality and for calm when violence threatened to erupt.

A price was put on WDIA's success when, in 1957, John Pepper and Bert Ferguson sold their station to the Sonderling Broadcast Corporation of Chicago for $1 million. Although Bert Ferguson stayed on as executive vice president and general manager until 1970, many complained that absentee ownership resulted in WDIA's losing its local focus. The entertainment conglomerate Viacom took control of the station in 1980.

In 1983 Ragan Henry of Philadelphia added WDIA to his chain of stations and became the station's first black owner. Henry's ownership capped a period of growing black influence on the station's operations and management. Since the 1960s, blacks had been working in all facets of the station, and in 1972, WDIA welcomed its first black general manager, Chuck Scruggs. In 1996 Ragan Henry sold WDIA to Clear Channel Communications, the company that currently owns and operates the station.

MICHAEL STREISSGUTH

See also Black-Oriented Radio; Blues Format; Gospel Music Format; Williams, Nat D.

Further Reading

Cantor, Louis, *Wheelin' on Beale: How WDIA-Memphis Became the Nation's First All-Black Radio Station and Created the Sound That Changed America,* New York: Pharos Books, 1992

Dates, Jannette Lake, and William Barlow, editors, *Split Image: African Americans in the Mass Media,* Washington, D.C.: Howard University Press, 1990; 2nd edition, 1993

King, B.B., and David Ritz, *Blues All around Me: The Autobiography of B.B. King,* New York: Avon Books, and London: Sceptre, 1996

Newman, Mark, *Entrepreneurs of Profit and Pride: From Black-Appeal to Radio Soul,* New York: Praeger, 1988; London: Praeger, 1989

Streissguth, Michael, "WDIA and the Rise of Blues and Rhythm and Blues Music," Master's thesis, Purdue University, 1990

WEAF

New York City Station

WEAF is significant in radio history as the first station to broadcast a paid commercial and as the originating point for American Telephone and Telegraph's (AT&T) pioneering network experiments. AT&T owned the station for only four years, after which it became the flagship of the developing National Broadcasting Company (NBC) network, becoming WNBC after World War II.

Origins

During the 1920s, AT&T was actively involved in the development of radio. The corporation was a member of the radio patent pool (along with the Radio Corporation of America [RCA], Westinghouse, and other companies), owned radio stations, and experimented with networking. It also owned the highest-quality intercity connection circuits in the United States and so was crucial to the efforts of other radio broadcasters to interconnect stations. WEAF actually began as short-lived radio station WBAY. Construction of its antenna was begun in March 1922, and it broadcast its first program on 25 July 1922. WBAY was a technological failure, its signal barely audible. After reconfiguration and location elsewhere in the city, WBAY gave way to WEAF on 16 August.

WEAF was the first station in the United States to broadcast a paid commercial, engaging in what AT&T called "toll broadcasting" at the time. AT&T saw the possibilities for treating radio the same way it treated its long-distance telephone service: it would provide the facilities, and others would provide the content. It would lease its facilities to those who had a message to deliver to the public, just as it provided long-distance service to customers who paid for the time they used its lines. At the same time, it saw toll broadcasting as a means of destroying its most significant rival in the provision of long-distance services, RCA. Although RCA did not own long-distance telephone lines, it had been established in 1919 to provide point-to-point wireless communication services, thus making it a potential rival to AT&T. RCA was also involved in radio broadcasting but planned to provide radio services on a public service model, subsidizing program production and broadcasting through the profits from the sale of radio receivers. AT&T, whose long-distance services provided massive profits, saw the potential of toll broadcasting to accomplish the same thing, which would give it an unassailable financial base from which to battle RCA.

WEAF broadcast its first commercial on 28 August 1922 at 5:00 P.M. It was a commercial for the Queensboro Corporation that lasted ten minutes and promoted the sale of apartments in Jackson Heights, New York City. The cost was $50. Queensboro Corporation quickly became a repeat customer, leasing time from WEAF on five additional occasions over the ensuing few weeks. Its initial broadcast was shortly followed by others for Atwater Kent (a radio receiver manufacturer), Tidewater Oil, and the American Express company.

WEAF did not permit these early advertisements to make direct pitches to the public. It was concerned to preserve the dignity of broadcasting—and of advertising—so it did not allow prices to be mentioned, and the advertisements provided little in the way of graphic descriptions of products. Queensboro's initial advertisement, for instance, told the audience that its apartments had been named to honor Nathaniel Hawthorne, "the greatest of the American fictionists," and invited people to visit the development "right at the boundaries of God's great outdoors." The commercial claimed that just this sort of residential environment had influenced Hawthorne, and it "enjoined" the listeners to "get away from the solid masses of brick" with meager access to the sun. But even this indirect sort of appeal was enough to change the philosophy of broadcasting in America.

WEAF's second significant impact on American radio stemmed from its role as the central point for AT&T's network experiments. The first networked broadcast in the United States occurred on 4 January 1923, when WEAF was connected with WNAC in Boston, although the broadcast lasted only five minutes. In June of that year, WEAF was networked with WGY (Schenectady, New York); KDKA (Pittsburgh); and KYW (Chicago) for a single program broadcast, and the following month WEAF and WMAF (Portsmouth, New Hampshire) were permanently networked, allowing WMAF to take a three- to four-hour-a-day feed from WEAF. By the middle of 1925 WEAF had become the flagship of the AT&T radio network, which connected 20 stations. All of these early experiments consisted of other stations' carrying programming that originated at WEAF.

National Broadcasting Company Flagship

On 15 November 1926, the true era of network radio arrived when a permanent arrangement was concluded whereby programming could originate at multiple points and be carried by multiple stations. On that date, WEAF originated remote programming from the Waldorf Astoria Hotel ballroom in New York City, and stations in both Chicago and Kansas City originated programs as well. These programs were carried on 21 network affiliate stations and 4 other stations. This broadcast was the result of complicated industry negotiations.

By the mid-1920s the cross-licensing agreement that formed the basis of the radio patent pool was in jeopardy. RCA and AT&T were part of arbitration hearings that began in early 1924, but before the final report was issued, the legal basis for the cross-licensing agreement itself was called into question, and the arbitrator's report, released in November, essentially affirmed RCA's position, which, among other assertions, denied AT&T's claim that it had an exclusive right to toll broadcasting. On 11 May 1926, AT&T announced that it was forming a separate company to conduct its broadcasting business, and on 7 July the final agreement between AT&T and RCA gave RCA an option to purchase AT&T's broadcasting interests, including WEAF—an option that RCA exercised on 21 July. On 1 November RCA paid the telephone company $1 million for WEAF, which became the flagship station of NBC's "Red" network.

A year later, NBC moved the station's transmitter from downtown Manhattan out to Bellmore, Long Island, increasing its power tenfold, to 50,000 watts. Studios moved from what had been AT&T space to a new headquarters at 711 Fifth Avenue, expanding from several rooms to five full floors of operating space. In 1933 the station shifted once more, this time to the new Radio City complex of Rockefeller Center in midtown Manhattan. Nearly 30 studios fed signals to the NBC Red and Blue networks, as well as providing resources for WEAF.

The station's programs were effectively those of the NBC Red network. Only relatively late in its history (during and after World War II) did the outlet develop a local sound for part of the day, adding morning and other talk shows appealing specifically to New Yorkers. In late 1946 the WEAF call letters were dropped, as the station became WNBC.

ROBERT S. FORTNER

See also American Telephone and Telegraph; National Broadcasting Company; Radio City; WNBC

Further Reading

Archer, Gleason L., Big Business and Radio New York: American Historical Company, 1939; reprint, 1971
Banning, William Peck, *Commercial Broadcasting Pioneer: The WEAF Experiment, 1922–1926,* Cambridge, Massachusetts: Harvard University Press, 1946
Jaker, Bill, Frank Sulek, and Peter Kanze, "WEAF" in *The Airwaves of New York,* Jefferson, North Carolina: McFarland, 1998

Weaver, Sylvester (Pat) 1908–2002

U.S. Broadcast Network Producer and Executive

Sylvester (Pat) Weaver is remembered as one of television's most innovative and farsighted executives. Although the changes he instituted as president of the National Broadcasting Company (NBC) were responsible for the network's assuming control of its television programming, his tenure at the network during the 1950s lasted less than seven years. For 20 years before, Weaver's programming philosophy had been shaped through his work as a radio producer for local stations and advertisers. During his years as a radio executive, Weaver conceived many of his revolutionary ideas about broadcasting, including the magazine concept of advertising and the educational potential of the electronic media.

Early Radio Work

After graduating *magna cum laude* from Dartmouth College with a degree in philosophy and the classics, Weaver worked as a direct mail advertiser and in 1932 joined Don Lee's regional network of eight West Coast radio stations. Because the country was not yet entirely hooked up for coast-to-coast transmission, Lee's stations, although affiliated with the Columbia Broadcasting System (CBS) network, developed their own programming. Weaver began at the Los Angeles station KHJJ as a comedy writer on *The Merrymakers,* a variety show starring Raymond Paige and his orchestra. Weaver used his marketing expertise and began to create specific programs targeted to potential sponsors. For example, for the Western Auto Supply Company, which wanted to underwrite an uplifting show during the Depression, Weaver concocted the first of his many educational series, *America Victorious,* which dramatized examples of the country overcoming severe obstacles.

In 1934 Weaver was put in charge of news, sales, and programming at Lee's San Francisco station KFRC. Weaver supervised a weekly two-hour extravaganza, *The Blue Monday Jamboree,* which featured the music of Meredith Wilson, who would later create *The Music Man.* Believing from the beginning that "comedy will always be the key ingredient of successful programming," he also produced a daily humor show, *Happy Go Lucky,* spotlighting the antics of Morey Amsterdam, who would star in Weaver's first television hit, *Broadway Open House.* Weaver's successes gained attention, and he was hired in New York to create and package musical and variety series for Bourjois Toiletries (*Evening in Paris Roof*) and for United-Whelen stores (*Evening Serenade*).

Realizing the importance of advertising agencies in producing network radio shows, Weaver joined Young and Rubicam in 1935. He was put in charge of one of radio's most popular shows, *Town Hall Tonight,* starring the acerbic comedian Fred Allen. Two years later, Weaver was managing all of Young and Rubicam's radio shows, and he was still intrigued by how radio could market ideas and products during this so-called golden age of the medium. Using his extensive knowledge as producer and executive, Weaver balanced the needs of sponsors with those of Young and Rubicam talent, whose roster included Jack Benny, Kate Smith, and George Burns and Gracie Allen.

In 1938 a sponsor, the American Tobacco Company, impressed with Weaver's accomplishments, recruited him to reposition its flagging Lucky Strike brand. He worked for the legendary champion of the hard sell, George Washington Hill, who made LSMFT ("Lucky Strike Means Fine Tobacco") his incessant advertising mantra. Weaver found an affluent, educated audience for the cigarette by sponsoring the popular quiz series *Information Please;* he also lured the Jack Benny program from the General Foods account at Young and Rubicam to secure mainstream exposure for Lucky Strike. Weaver's judicious placement of advertising money, actually using fewer dollars than had been allocated when he arrived, reinvigorated the brand, and Lucky Strike became the best-selling cigarette again.

During World War II, Weaver took a leave of absence from the tobacco battles and organized antifascist broadcasts to South America as part of the Office of the Coordination of Inter-American Affairs. He also joined the navy and produced the popular radio series *Command Performance* for the armed forces overseas.

Despite his success in and out of the corporate world, Weaver was passed over for a senior management position at American Tobacco after Hill's death in 1946. Thoroughly embarrassed, he not only returned to Young and Rubicam as vice president in charge of radio, television, and movies, but also went cold turkey on his own four-packs-a-day smoking habit. Instead of creating programs, Weaver began to advertise on such radio series as *Arthur Godfrey's Talent Scouts* and *My Favorite Husband,* which were owned by the networks. The advertising industry was in flux, grappling with a whole new environment: a peacetime economy and the advent of the most potent selling tool in history, television. Weaver was convinced that the old system would not remain the same for the new medium: the cost of television production would be too prohibitive for any one agency to create and own the programs.

NBC Years

To implement his vision for the new technology, Weaver joined NBC as vice president in charge of television. He maintained complete control over the network's schedule and refashioned its production philosophy, inviting multiple sponsors to invest in new series and specials. With his structural changes, Weaver shifted the power of production and scheduling from the advertising agencies to the networks. In the process, he also envisioned programming concepts that had not been tried in agency-controlled radio: early and late-night programs (*Today* and *The Tonight Show*), big-budget spectaculars (*Peter Pan* and *Amahl and the Night Visitors*), and ambitious public-affairs and cultural series (*Wisdom* and *Wide Wide World*).

Weaver was renowned as "thinker-in-chief" of NBC, inspiring his company to create "an aristocracy of the people . . . to make the average man the uncommon man." He was appointed president of the NBC network in 1953 and chairman two years later. Weaver talked and wrote extensively about the new role of radio, no longer "the dominant element in American leisure time activities." He conceived of building programs for both radio and television simultaneously; radio could replay in a variety of formats the best of *Today* or *Tonight* with new introductions. In addition to radio programs for a mass audience, Weaver foresaw a specialized use for the medium. He dreamed of shows for special interest groups, notably the business community and the book-reading public. These new concepts were planned to get leaders and opinion makers involved in radio as both listeners and advertisers, opening up NBC to greater influence on the national stage. In 1955 Weaver introduced his defining radio project, *Monitor,* a 48-hour weekend radio spectacular of news and special events.

By the mid-1950s, David Sarnoff, chairman of NBC, found most of Pat Weaver's ideas financially extravagant. Weaver's vision of an uplifting medium of live programming had given television an artistic legitimacy earlier in the decade, but now NBC's task was to make the business economically viable with a predictable schedule of mostly film programs. Sarnoff gave the reins of NBC power to his son Robert, and Weaver left the company in September 1956. After that, Weaver kept his eye on the future of telecommunications. In the early 1960s, he ran the first major pay television operation. He later became a consultant to cable and videocassette ventures. Throughout his career, Pat Weaver was among the more unorthodox of broadcasting executives, trying to keep a simultaneous focus on both the bottom line and the big picture.

RON SIMON

See also Monitor; National Broadcasting System

Pat Weaver. Born Sylvester Laflin Weaver, Jr., in Los Angeles, California, 21 December 1908. Educated at Dartmouth College, B.A. magna cum laude and Phi Beta Kappa, 1930. Writer and salesman for Young and MacCallister Print Company; writer, producer, announcer, director, and salesman, KHJ, Los Angeles, 1932; program manager, KFRC, San Francisco, 1934; produced radio programs for United Cigar Stores and Whelan Drug Stores, 1935; joined Young & Rubicam advertising agency, 1935; supervisor of programs, Young & Rubicam's radio division, 1937; advertising manager, American Tobacco Company, 1938–46; associate director of communications, Office of the Coordination of Inter-American Affairs, 1941; served in the U.S. Navy, 1942–45; vice president in charge of radio, television, and motion pictures for Young & Rubicam, 1947–49; vice president, vice chairman, president, then chairman, NBC, 1949–56; chairman, McCann, Erickson, 1958–63; president, Subscription TV, Los Angeles, 1963–66; member, board of directors, Muscular Dystrophy Association, from 1967. Recipient of Peabody Award, 1956, Emmy Award, 1967; named to Television Hall of Fame, 1985. Died in Santa Barbara, California, 15 March 2002.

Selected Publications

The Best Seat in the House: The Golden Years in Radio and Television, 1994

Further Reading

Bergreen, Laurence, *Look Now, Pay Later: The Rise of Network Broadcasting,* Garden City, New York: Doubleday, 1980

Boddy, William, "Operation Frontal Lobes versus the Living Room Toy," *Media, Culture, Society* 9 (1987)

Kepley, Vance, Jr., "The Weaver Years at NBC," *Wide Angle* 12 (1990)

Welles, Orson. *See* Mercury Theater of the Air; War of the Worlds

Wertheimer, Linda 1943–

U.S. Commentator and Host

The distinctively elegant voice of Linda Wertheimer has been heard on National Public Radio's (NPR) afternoon newsmagazine, *All Things Considered,* since 1971. Wertheimer remains today one of the best-known figures on National Public Radio.

Linda Cozby was born in Carlsbad, New Mexico, on 19 March 1943, the first of two daughters of Miller and June Cozby. After graduating from Carlsbad High School in 1961, Cozby received a scholarship to Wellesley College in Massachusetts, where in 1965 she earned a Bachelor of Arts degree in English literature. Her childhood heroine was Pauline Frederick, a pioneer female radio and later television reporter for National Broadcasting Company (NBC) News.

Cozby began her broadcasting career as an intern at the British Broadcasting Corporation (BBC) in London, serving as a production secretary responsible for writing radio copy and audio production. Two years later, in 1967, she accepted a job as a researcher at the all-news radio station WCBS in New York. Lou Adler, a well-known journalist at WCBS, worked closely with Cozby to help her polish her skills as a writer. The mentoring proved fruitful, and she was eventually allowed to produce features and write radio copy. During this time she was the only woman in the building who was not a secretary, aside from the vice president of advertising. In 1969 Cozby married Fred Wertheimer.

From 1969 to 1971, Wertheimer took time off from her broadcasting career until she heard about a new radio network that was being created as a result of the Public Broadcasting Act of 1967. The first NPR broadcast of *All Things Considered* was on 3 May 1971, and Wertheimer worked as a director of the afternoon program that broadcast in-depth news reports and offbeat features. Wertheimer states, "I have never forgotten the terror of directing the early programs . . . . I remember mistakes, dead air, missing tapes, and moments of panic." Before 1971 concluded, her dream of becoming an on-air reporter came to fruition when she became NPR's congressional correspondent. By 1976 she was promoted to political correspondent. In 1989, Wertheimer was promoted to cohost of the afternoon program with Bob Siegel and Noah Adams. Wertheimer became part of a collaborative team that decided what would be aired during the two-hour program; the job included writing news copy, introducing reports, and conducting interviews.

Wertheimer has covered a number of significant news events, ranging from Watergate to Iran-Contra to all presidential campaigns since 1976. Moreover, she has reported on every major congressional news story since Watergate as well as major elections and national politics. Wertheimer anchored NPR's coverage of the Iran-Contra hearings in 1987 and provided summaries of each day's testimony. She covered the 1976, 1980, 1984, and 1988 presidential campaigns full-time. She hosted *All Things Considered* after that and only traveled part-time during the 1992, 1996, and 2000 presidential campaigns. While covering the presidential campaigns, she followed major candidates as they campaigned across the country during both state primaries and national conventions. Wertheimer has also anchored NPR's live coverage of nominating conventions and presidential debates.

Wertheimer's tenure at NPR has afforded her the opportunity and knowledge to edit a book titled *Listening to America,* the purpose of which was to mark the 25th anniversary of National Public Radio. It included NPR's coverage of major stories as well as features from each of the years NPR had been on the air. Wertheimer placed the NPR features in historical context.

Wertheimer has become one of America's most experienced and highly regarded broadcast journalists as host of NPR's *All*

Things Considered, with a radio audience of 15 million listeners. She also appears as a contributor on Columbia Broadcasting System (CBS) Television's *Face the Nation* and as a political analyst on Cable News Network (CNN). She has paved the way for women to enter the broadcasting industry as on-air talent or broadcast journalists, as opposed to starting out as researchers and working their way into broadcast journalist positions.

JOHN ALLEN HENDRICKS

Linda Wertheimer. Born Linda Cozby in Carlsbad, New Mexico, 19 March 1943. B.A. in English literature, Wellesley College, Massachusetts, 1965; intern at the British Broadcasting Corporation (BBC), London, 1965–67; researcher, news writer, and producer for WCBS, New York, 1967–69; married Fred Wertheimer, 1969; director, congressional correspondent, political correspondent, and host of *All Things Considered,* National Public Radio (NPR), Washington, D.C., 1971–present. American Women in Radio/TV award for "Illegal Abortion," 1992; Corporation for Public Broadcasting award for "The Iran-Contra Affair: A Special Report," 1988; Alfred I. duPont-Columbia University Award, 1978; Distinguished Alumnae Achievement Award from Wellesley College, 1985; honorary degrees from Colby College in Maine (1990), Wheaton College in Massachusetts (1992), and an honorary doctor of humane letters degree for her lifetime achievements in journalism from Wesleyan University in Illinois (1999).

Publications

Listening to America: Twenty-Five Years in the Life of a Nation, as Heard on National Public Radio, 1995

Further Reading

Dreifus, Claudia, "Cokie Roberts, Nina Totenberg and Linda Wertheimer," *The New York Times Magazine* (2 January 1994)

Lycan, Gary, "Wertheimer Remembers NPR from Day 1," *The Orange County Register* (4 June 1995)

Westerns

Radio Drama Format

"Hi-yo, Silver, away!," "You betchum, Red Ryder!," and "Happy trails to you" are just a few of the expressions that have entered America's lexicon from the majestic Old West as presented on network programming during the golden age of radio. From the 1930s through the 1950s, westerns entertained millions of listeners—and sold lots of sponsors' products. While westerns varied in their narrative styles and tone, they shared the customary motifs that defined the genre: six-guns and horses, heroes and villains, cowboys and Indians, dusty trails and mountain passes. Radio westerns constitute four overlapping sub-genres: anthology series, singing cowboy shows, juvenile adventures, and so-called adult westerns.

Origins

One of the earliest western series was *Death Valley Days,* broadcast on NBC-Blue from 1930 to 1941 and on CBS from 1941 to 1945. As an anthology series, *Death Valley Days* featured new characters each week, although all stories were set in 19th-century California. Stories were reportedly based on fact as collected and dramatized by producer Ruth Cornwall Smith, one of the few women radio producers of the era.

Anthology series not specifically dedicated to the western genre sometimes featured western dramas as well. For example, *Lux Radio Theatre,* which presented weekly radio versions of famous movies, occasionally broadcast western stories such as *The Plainsman* (1937). Similarly, anthology programs featuring original radio dramas, such as *Suspense,* broadcast western radio stories from time to time.

Most radio westerns, however, were episodic series or serials with recurring characters, many of which cultivated fanatically loyal audiences—particularly among children. As such, these programs were a boon to advertisers whose products were carefully associated with the values embodied in and verbalized by mythic, straight-shooting, clean-cut cowboys.

Song-filled programs with a western flavor and recurrent characters included *Grapevine Rancho* (CBS, 1943), *The Hollywood Barn Dance* (CBS, 1943–47), *The National Barn Dance* (various networks, 1924–1950), and *The Grand Ole Opry,* which began in 1925 and continues to this day. While these musical programs lack some of the generic qualities of the classic western—shootouts, for example—their emphasis on country music and folksy dialog delved heavily into romanticized imagery of the mythic West. In contrast, *Hawk Larabee* (CBS, 1946–47) also featured songs but leaned in the opposite direction; its primary emphasis was on adventures of the series' protagonist, and songs linked narrative segments.

Children's Series

Juvenile adventure westerns were more ubiquitous than their musical counterparts. The most famous was *The Lone Ranger,* which was developed at Detroit station WXYZ in 1933. The story of the famous western hero quickly became a local sensation. Soon, it functioned as the programming glue that connected stations in the fledgling Mutual radio network. In 1942, the series moved to NBC-Blue (later ABC) where it remained until 1955. Throughout its production, the series featured voices of a well-honed stock company who also were heard in other WXYZ programs including *The Green Hornet* (various networks, 1938–52) and *Sergeant Preston of the Yukon* (ABC, 1947–50; Mutual, 1950–55)—the former a Lone Ranger spin-off set in contemporary America (the Green Hornet was the Lone Ranger's great-nephew) and the latter a Canadian Mountie, aided by his famous dog, Yukon King.

The Lone Ranger was warmly received, not only by avid young listeners, but also by the national press as a successful and laudable role model. This was particularly true during World War II when *The Lone Ranger* was peppered with patriotic dialog intended to boost the domestic war effort and overseas troop morale. Also, despite its simplistic story lines and two-dimensional characters, the Lone Ranger was a highly polished program penned by a one-man script mill, Fran Striker, who reportedly wrote some 60,000 words of radio dialog per week. The program soon was spun off into highly profitable movie series, a comic strip, and numerous toys.

The tremendous success of *The Lone Ranger* led to many copycat series for young listeners, especially boys. The Lone Ranger's major ratings competitor was *Red Ryder,* heard on various networks from 1942 until the 1950s. Other western heroes were featured in series such as *Hopalong Cassidy* (Mutual, 1950; CBS, 1950–52) and *Wild Bill Hickok* (Mutual, 1951–56). *Tom Mix* (various networks, 1933–50), *Bobby Benson* (various networks, 1932–55), and *Sky King* (various networks, 1946–54) were similar programs set in the modern West.

Among these series' most memorable features was their unrelenting push for sponsors' products, often transforming young listeners into a potent sales force. For instance, in 1939, *The Lone Ranger* mixed hero worship with salesmanship by offering a Lone Ranger badge to listeners who convinced three of their neighbors to buy the sponsor's product. A reported 2 million such badges were distributed within the first year of the campaign.

Like the singing cowboy programs, juvenile adventure westerns were highly formulaic, once a successful pattern had been established. *The Lone Ranger* episodes, for instance, were formed around recurring narrative elements, beginning with a character and his/her seemingly hopeless problem and moving through the Ranger's appearance and involvement, a trap set for the Ranger, the Ranger's overcoming the trap, his solution of the problem, and resolution of the narrative. Fistfights, gunplay, and chases on horseback were predictable narrative devices. Despite the fact that these programs were intended for young audiences, bloodshed and death were dolled out on a regular basis. In one episode of *The Lone Ranger* from 1939, for example, no less than twelve people were killed within the first three minutes of the program.

Western series with a strong emphasis on cowboy songs and campfire chat included *Gene Autry's Melody Ranch,* which ran on CBS from 1940 to 1956, sponsored by Wrigley's Gum. Episodes of this series consisted of jokes and stories told by Autry and his pals, dramatized tales, and plenty of songs, including the classic closing theme "Back in the Saddle Again." Gene Autry's major ratings competitor was *The Roy Rogers Show,* a program with a similar format that ran from 1937 to 1955. Stars of both series were also featured in popular films of the era.

Contrary to their musical counterparts, heroes of juvenile adventure series tended to live outside the bounds of their communities, even while working to preserve and help expand white society throughout the West. They roamed the frontier looking for adventure and people in need of help, but they never tarried long. (An exception was Red Ryder, who lived with his aunt.) Radio cowboy heroes' combination of placelessness and service to strangers is typical of the western mythos prevalent in other forms of popular culture, such as dime novels and western films, with roots reaching back to James Fennimore Cooper's *Leatherstocking Tales.*

Indeed, the western outsider/hero motif has a deep cultural resonance, and given their immense popularity, juvenile western adventures are worth considering as important cultural artifacts. For example, while heroes such as the Lone Ranger deliberately exemplified values such as honesty and fair play, closer reading also suggests that these programs served to affirm white and male forms of cultural dominance and westward expansion. The Lone Ranger, for instance, may be indebted to Tonto for certain "native" skills such as tracking and healing; nevertheless, power in the relationship is clearly skewed toward the Ranger, as Tonto acts as intermediary between settlers and "hostile" Indians, always under the Ranger's guidance. Similarly, Red Ryder was aided by his Indian protege, Little Beaver. Other sidekicks who helped their white friends tame the West included a Latino character named

"Pablo" on *Dr. Sixgun* (NBC, 1954–55) and the Asian Heyboy of *Have Gun, Will Travel* (CBS, 1958–60). A variant of this pattern was *Straight Arrow* (Mutual, 1949–51), whose white hero, Steve Adams, had been raised by Comanches. (One exception to the pattern was *The Cisco Kid*, which was broadcast in the 1940s; in this series, Cisco was a vaguely Latino adventurer who had a humorous sidekick named Pancho.)

Similarly, in juvenile western adventures, female characters tend to be victims in need of heroic rescue, rather than strong, independent forces in their own right. The mythic West of juvenile western adventures, in other words, was rhetorically constructed as a domain to be settled and supervised by white men. While this was never explicitly spelled out as such, recurring rhetorical tropes in thousands of episodes heard on a regular basis (in combination with accompanying costumes, decoder rings, and other accessories) endowed juvenile western adventures with a ritualized aura, reaffirming dominant values of their era.

Adult Westerns

While juvenile westerns ran well into the 1950s, by 1953 a new breed of western drama appeared. These "adult westerns" were grittier, more realistic, and clearly intended for an older audience. Adult westerns were less the descendants of their juvenile predecessors than they were cousins of western films such as *Shane* and *High Noon*, which were produced at about the same time. These postwar western films (as well as detective movies—with their own radio counterparts) reflected national feelings of postwar disillusion as the U.S. plunged into the Cold War and communists were thought to infiltrate all aspects of American society. In both film and radio, good and evil were far more difficult to distinguish than in juvenile fare. Both heroes and villains were finely shaded, unstable mixes of virtue and vice, suggesting a re-working and reinterpretation of dominant ideological assumptions.

The most notable reluctant hero of the adult radio westerns was Matt Dillon in *Gunsmoke*, played on CBS by gravelly-voiced radio veteran William Conrad from 1952–61. Matt Dillon was the Marshall of Dodge City, a frontier town that attracted the worst sorts of outlaw scum ever heard on radio. Dillon's sidekick was his dimwitted deputy Chester Proudfoot, played by Parley Baer. Georgia Ellis played Kitty, and Howard McNear played Doc. While Kitty's job was never spelled out, *Gunsmoke* creator Norman McDonnell described her as "just someone Matt has to visit every once in a while. . . . We never say it, but Kitty is a prostitute, plain and simple."

Kitty, in fact, represents a complete revolution in portrayals of women in radio westerns; a shrewd, outspoken professional, Kitty is a far cry from the innocent school marm and rancher's daughter common among juvenile western programs. Whereas women on other western programs personified dominant values of white America (specifically values regarding family, religion, and education), Kitty simultaneously represents challenges to the moral status quo and the gutsy determination of independent women.

McDonnell's relish in describing Kitty's profession exemplified the realistic presentation of the West that *Gunsmoke* strove to create. Matt Dillon clearly hates his job, and his decisions are never easy or clear-cut. *Gunsmoke* is also notable for its intense emphasis on sound effects, ranging from the clinking of spurs to ever-present background sounds such as twittering birds and far-off barking dogs. Having established himself as a force in adult radio westerns, McDonnell went on to produce *Fort Laramie*, a series about a Wyoming cavalry outpost broadcast on CBS in 1956.

Other adult westerns of note included *Frontier Gentleman* (CBS, 1958), a well-written, highly polished drama concerning an English newspaper reporter who wandered the Old West. And *Luke Slaughter of Tombstone* (CBS, 1958) was set amidst cattle drives in 19th-century Arizona.

The shift to adult fare among radio westerns can partly be explained by the Cold War milieu within which they were produced. The Red Scare hit the radio industry hard, as many top performers and other creative personnel were blacklisted, while many more were silenced by intimidation. Westerns were removed in time and place from the reality of urban America. As such, they could be arenas in which writers expressed dangerous political concepts and dramatized social issues such as racism, albeit couched in the past and far from contemporary society. (Similarly, adult science fiction—notably *X Minus One* [NBC, 1955–58]—played upon the same themes, but was set in the future instead.)

Beyond sociopolitical conditions of the times, the development of adult westerns can also be explained by analysis of the economic state of radio in the 1950s. The popular and trade presses had long since announced the death knell of radio drama's golden age, correctly emphasizing television's immediate and immense popularity. As audiences moved toward the newer medium, radio broadcasters needed fresh, new products to hold onto their advertising base. Also, network affiliates were going independent, having discovered that local programs were more lucrative than network fare. Networks, in turn, scrambled to cancel most of their long-running daytime series and develop evening and weekend programming with ambitious new concepts in order to retain their rapidly dwindling number of affiliates. *Gunsmoke* was clearly such a product. Other new series featuring major film stars were created to keep dwindling audiences tuned to radio. These programs included *Tales of the Texas Rangers* (NBC, 1950), a modern-day western starring Joel McCrea, and *The Six Shooter* (NBC, 1953), starring Jimmy Stewart as a well-meaning, though intensely lethal, gunfighter. Both series were high-quality productions, although of the two, *The Six*

Shooter was more in the adult-western mold, given its finely nuanced protagonist, compelling stories, and close attention to historical detail.

Third, the development of adult westerns resulted from changes in the producer/network/sponsor relationship in radio. Prior to the popularity of television, most radio programs were produced by advertising agencies on behalf of sponsors. Censorship was strict, because neither agencies nor sponsors wanted their programs or products associated with controversy. As sponsors moved into television, radio dramas were increasingly produced by networks in hopes of selling commercial time. If left unsponsored, these "sustaining" programs provided a public-service function for networks. Sustaining programs were touted as experimental venues whereby the networks gave something back to the people for the use of public airwaves. Nevertheless, commercial time on series such as *Gunsmoke* would be sold when possible. Norman McDonnell noted the dilemma of sponsorship versus sustaining programming, quipping "I'd feel great if someone did buy it, but there would be problems. We'd have to clean the show up. Kitty would have to be living with her parents on a sweet little ranch." Whether sponsored or sustaining, network-produced radio dramas, including westerns, were less burdened by direct censorship from the advertising industry.

Decline

Despite the effort put into their development, westerns eventually met the same fate as other forms of network radio theater. The final network radio western was *Have Gun, Will Travel*, last heard on CBS in 1960. *Have Gun, Will Travel* was something of an oddity in radio drama. Whereas most radio westerns, such as *The Lone Ranger* and *Gunsmoke*, moved to television after (or even during the time) they were on radio, *Have Gun, Will Travel* began on television and then was developed for a dual radio/TV presence. As such, the series exemplified not only the end of radio westerns, but also the undeniable ascendancy of television.

Reruns of original western radio dramas, now often termed "old-time" radio, are still heard on some stations, where they have a loyal following. Also, westerns from radio's golden age are readily available on cassette tapes. Radio Spirits of Schiller Park, Illinois, for example, markets *The Lone Ranger*, *Gunsmoke*, *Frontier Gentleman*, and other series. The Smithsonian Institution, in association with Radio Spirits, similarly offers a package of classic western series for sale on cassette and compact disk. These series are marketed not just as nostalgia but also as first-rate forms of entertainment, which, given their fairly high costs, suggests that many radio westerns retain their vitality decades after their original production.

Although new radio westerns are rarely produced apart from a handful of special-event programs, a descendent of the radio western can be found in books on tape which feature a variety of western novels read by famous actors. Also, a well-produced series of dramas based on the novels of Louis L'Amour is available, often for sale in interstate highway gas stations—strongly appealing, no doubt, to truck drivers and other lonesome travelers eager to pass the time on long hauls under western skies.

WARREN BAREISS

See also Autry, Gene; Country Music Format; Grand Ole Opry; Gunsmoke; Lone Ranger

Further Reading

Allen, Chadwick, "Hero with Two Faces: The Lone Ranger As Treaty Discourse," *American Literature* 68 (1996)
Barabas, SuzAnne, and Gabor Barabas, *Gunsmoke: A Complete History and Analysis of the Legendary Broadcast Series with a Comprehensive Episode-by-Episode Guide to Both the Radio and Television Programs,* Jefferson, North Carolina: McFarland, 1990
Bryan, J., "Hi-Yo Silver!" *Saturday Evening Post* (14 October 1939)
Harmon, Jim, *The Great Radio Heroes,* Garden City, New York: Doubleday, 1967
Holland, Dave, *From Out of the Past: A Pictorial History of The Lone Ranger,* Granada Hills, California: Holland House, 1988
"Is Network Radio Dead?" *Newsweek* (20 August 1956)
Nachman, Gerald, *Raised on Radio: In Quest of the Lone Ranger, Jack Benny . . . ,* New York: Pantheon Books, 1998
"Network Drama," *Time* (12 January 1959)
"Weeks of Prestige," *Time* (23 March 1950)
Wylie, Max, editor, *Best Broadcasts of 1939–40,* New York: Whittlesey House, 1940

Westinghouse

Electrical and Radio Manufacturer

The Westinghouse Electric Company was founded in 1886 and became well known for both industrial and consumer products, such as washing machines and refrigerators, as well as many consumer electronic products. For many years its widely known advertising slogan was "You can be sure if it's Westinghouse."

Origins

American engineer George Westinghouse (1846–1914) would eventually receive more than 360 patents including the invaluable one in 1869 for the air brake used by railroads. Although he formed nearly 60 companies, the largest and longest lasting, Westinghouse Electric, was created in 1886, with headquarters in Pittsburgh, Pennsylvania. Westinghouse received numerous honors in the U.S. and abroad. Perhaps his finest tribute came from inventor Nikola Tesla, whose patents for one system of alternating current and the induction motor were acquired by Westinghouse in 1888 and gave the company its early leadership in electric power developments. Westinghouse used Tesla's system to light the World's Columbian Exposition at Chicago in 1893.

By the turn of the 20th century, Westinghouse had become one of the two or three largest electrical manufacturers in America, employing more than 50,000 workers. The company was active in electric power generation, electric traction (trolleys) and railway equipment, and various industrial applications of electricity. A research department was formed in 1904, becoming a division two years later.

In the financial panic of 1907, Westinghouse lost control of the several companies he had founded and then still headed. In 1910, he founded his last firm to exploit the invention of a compressed air spring for absorbing the shock of riding in automobiles. By 1911, however, he had severed all ties with his former companies. He had shown signs of a heart ailment by 1913 and was ordered to rest by doctors. Not long after deteriorating health confined him to a wheelchair, Westinghouse died in March 1914. His last patent was granted four years later.

As with some other electrical companies, Westinghouse Electric began to investigate the manufacturing potential of wireless. During World War I, Westinghouse held huge government contracts to manufacture wireless equipment for the army and navy. Those contracts were canceled with the end of the war in November 1918, leaving the company with a trained cadre of workers and a considerable investment in production equipment but not enough work to keep them busy.

Radio Receivers

To better its manufacturing position for a possible civilian radio market, in 1920 Westinghouse purchased the International Radio Telegraph Co. to obtain important Fessenden heterodyne circuit patents. A few months later, seeking to head off developing competition from the new Radio Corporation of America (RCA), Westinghouse also purchased Edwin Armstrong's regeneration and superheterodyne tuning circuit patents.

At about the same time, a chance event helped pull all the pieces together. Westinghouse engineer Frank Conrad had been experimenting with wireless since 1912. In 1919–20 he operated amateur ("ham") station 8XK, playing recorded music one or two nights a week for the amusement of fellow hams. While hobbyists preferred to build their own equipment, others wanted to tune in as well. A September 1920 newspaper advertisement by a Pittsburgh department store seeking to sell receivers to those who wanted to hear Conrad's broadcasts caught the eye of Harry Phillips Davis, a Westinghouse vice president in charge of radio work. Davis perceived that making receivers for the public to receive a possible new radio broadcasting service could be the answer to Westinghouse's canceled-contract predicament.

Within a year or so, a very basic Westinghouse crystal receiver marketed as the "Aeriola Jr." could be purchased for about $25 in department stores. A more sophisticated tube set, sold as the "Aeriola Sr.," offered better reception and cost about $60. Westinghouse was soon turning out thousands of radios from its factories, primarily those in Springfield, Massachusetts. Unlike many other set makers, Westinghouse had the advantage of owning several radio stations, which were useful as a means of promoting company products. In addition to radio receivers, Westinghouse also built transmitters for its own and some other stations. The company made increasingly powerful transmitters throughout the 1920s: whereas a typical station of that time transmitted with 100 watts, in the mid-1920s Westinghouse station KDKA was broadcasting with 10,000 watts.

Westinghouse's control of the Fessenden, and especially the Armstrong patents, gave the company the clout it needed to enter into agreements with its main competitors, General Electric (GE), RCA, and the Western Electric manufacturing arm of American Telephone and Telegraph. The "patent pool" allowed each company to license the patents of the others, and established a division of the equipment market. The pool had a flaw, however, for it concerned only the international or "point-to-point" radio market, not broadcasting which was then still small and seemingly insignificant. According to the company's version of the story, Westinghouse saw this alliance as a way to get into the international market, but media historians have suggested that Westinghouse felt it could not remain independent and still succeed in outselling the others, especially RCA, which was already solidifying its power and could easily have shut out any meaningful competition. Since Westinghouse couldn't beat its competitors, it decided to join them.

After 1922—and until an anti-trust case was filed against RCA in 1930—Westinghouse radios were sold under the RCA trade label, usually as "Radiola" receivers. The patent pool agreement reserved 40 percent of the business for Westinghouse while the larger General Electric took 60 percent.

Expanding Consumer Electronics

Westinghouse began to expand its consumer electronics business early on. Television pioneer Vladimir Zworykin undertook his early development work for the television camera tube (dubbed the iconoscope) at Westinghouse before transferring to RCA in 1930. Nearly four decades later, a Westinghouse television camera accompanied the first men to land on the moon.

In 1945, and again fattened by large wartime government contracts (including many for radar, which remained a company specialty), what had long been Westinghouse Electric and Manufacturing Company reverted to an older and simpler name, becoming known simply as the Westinghouse Electric Corporation.

Postwar Westinghouse products included a full range of radios and phonographs, and soon, television sets. In 1954 Westinghouse formed a Westinghouse Credit Corporation subsidiary that assisted consumers in making major appliances over time. The 1950s were a boom period for the company, whose public face was the ubiquitous spokesperson Betty Furness, who touted refrigerators and other Westinghouse appliances on prime-time television.

But as was true for all American manufacturers of consumer electronics, competition was fierce, and by the late 1950s inexpensive foreign imports began to cut into sales of radios and related products. Soon Westinghouse-labeled models were being manufactured overseas and American plants were closing down.

Demise

Westinghouse's manufacturing divisions began to show some signs of financial trouble and seemed to lose their way. By the 1970s the firm was beginning to sell off some of its units—including the once-lucrative major appliances division in 1974. The electric lighting business was sold in 1983. On the other hand, Westinghouse also got into the cable business, paying $646 million to acquire the Teleprompter Corporation (renamed Group W Cable) in 1981. When it proved extremely expensive to upgrade the cable systems, they were sold at a profit in 1985. Throughout this period, Westinghouse made some bad real estate loans and made a number of risky acquisitions—all in an attempt to diversify the industrial firm more into the services sector. One important part of the company's service continued to be the broadcast station-owning subsidiary, known as Group W. The East Pittsburgh operations, once the birthplace of commercial radio broadcasting, were closed in 1987. The power transmission and distribution unit was sold two years later.

Major change continued into the 1990s for what had once been one of the country's most stable manufacturing concerns. Layoffs were widespread as a series of restructurings took place. The motor manufacturing divisions were spun off in 1995 to become an independent firm. A new chairman and chief executive officer, Michael Jordan (who had formerly been at PepsiCo), was brought in to redirect Westinghouse.

In 1994 Jordan allied the Group W broadcast stations with CBS in a joint venture. Just a year later came a more dramatic move as Westinghouse agreed to purchase CBS for $5.4 billion. After the Telecommunications Act of 1996 lifted limits on radio station ownership, Westinghouse acquired a much larger group of radio stations, Infinity Broadcasting. Defense electronics manufacturing was sold off in 1996.

But what was supposed to make a hero of Jordan and save Westinghouse had ironic results. Infinity's chairman, Mel Karmazin, became the largest shareholder in Westinghouse, and as financial columnist Steve Massey (1998) observed, he "had no sentimental attachment to the old Westinghouse." He also did not get along with Jordan. The plan had been for Westinghouse Electric to survive and for Jordan to be in charge. In the end, Jordan was forced out (early retirement, the story went), Karmazin took over, and the Westinghouse name officially vanished from the broadcasting division in December 1997 as the huge entity became known as CBS/Infinity. The nuclear power manufacturing parts of the company were sold in 1999, and with them the Westinghouse Electric Company name, and became a part of the Nuclear Utilities Business Group of British Nuclear Fuels.

DONNA L. HALPER AND CHRISTOPHER H. STERLING

See also Armstrong, Edwin Howard; Columbia Broadcasting System; Fessenden, Reginald; General Electric; Group W; Infinity Broadcasting; Karmazin, Mel; Radio Corporation of America; Receivers; Tesla, Nikola

Further Reading

Douglas, Alan, "Westinghouse," in *Radio Manufacturers of the 1920s, Volume 3*. Vestal, New York: Vestal Press, 1991

Massey, Steve, "Who Killed Westinghouse," *Pittsburgh Post-Gazette* (1–7 March 1998); also at <www.post-gazette.com/westinghouse/default.asp>
Passer, Harold C., *The Electrical Manufacturers, 1875–1900*, Cambridge, Massachusetts: Harvard University Press, 1953
Prout, Henry G., *A Life Of George Westinghouse*, New York: The American Society of Mechanical Engineers, 1921
Schatz, Ronald W., *The Electrical Workers: A History of Labor at General Electric & Westinghouse, 1923–60*, Champaign: University of Illinois Press, 1983
Schiffer, Michael Bryan, *The Portable Radio in American Life*, Tucson: University of Arizona Press, 1991
Westinghouse history and timeline, <www.westinghouse.com/A1.asp>
Woodbury, David O., *Battlefronts of Industry: Westinghouse in World War Two*, New York: Wiley, 1948

Westwood One

Radio Program Service

Westwood One is one of the largest radio service companies in the United States, producing and distributing entertainment, news, sports, talk, and traffic programming to more than 7,500 stations. It runs the nation's largest radio network, which includes National Broadcasting Company (NBC), Columbia Broadcasting System (CBS), Cable News Network (CNN) and Fox radio programs, and maintains an international radio programming service called Westwood One International. It owns local programming subsidiaries Shadow Broadcast Services and Metro Networks, which dominate the industry in localized traffic reports; and it has an ownership stake in an internet radio company called WebRadio.com. With its emphasis on music, personalities, and large-scale events, Westwood One has made a significant contribution to the return of entertainment to network radio.

Origins

Westwood One was founded by Norman J. Pattiz in 1974 after he lost his job as a sales manager at KCOP-TV in Los Angeles. Pattiz had heard a weekend-long Motown music program on local rhythm and blues radio station KGFJ and had the idea of producing such programs for national syndication, as was common in the television industry. Although he had no radio production experience, Pattiz convinced the KGFJ station manager to let him put together another Motown show for national distribution. After nine months of production and lining up advertisers, *The Sound of Motown* aired on about 250 stations and grossed several hundred thousand dollars. The program was provided to the stations at no cost. Local stations were allowed to sell local ad time while Pattiz and KGFJ collected the revenue from national advertisers.

From there Pattiz went on to produce more programs and incorporated his own company, Westwood One, late in 1974. Its name is derived from the Westwood neighborhood of Los Angeles where Pattiz started his new company in a one-room office. In the next nine years, Westwood One would continue to produce programs for syndication and to package the programs with national advertisers, eventually offering 52-week vehicles for those advertisers.

In 1984 Westwood One became a public company. The infusion of capital allowed the company to change quickly from a small syndicator wholly dependent on the success of each program to a national radio network with multiple resources. In 1985 Westwood One acquired the struggling and aged Mutual Broadcasting System from Amway for $39 million. Pattiz consolidated operations and reprogrammed the network, turning the debt-ridden Mutual into a profit source. Mutual, like other traditional radio networks, had become primarily a distributor rather than a producer of radio programming, although it still retained its news division. By acquiring Mutual, Westwood One was able to offer and market sports, news, and talk radio, as well as its original entertainment programming, becoming a more full-service network. Two years later Westwood One bought the NBC Radio Network from General Electric for $50 million, and in 1989 it launched the Westwood One News and Entertainment Network.

Sharpening the Focus

After several years of expansion, Westwood One found that its debt burden was becoming a problem. NBC Radio was losing $11 million a year when it was purchased by Westwood, but unlike Mutual, its employees were heavily unionized, making cost-cutting more difficult. To make matters worse the radio industry went into a recession shortly after the purchase and advertising revenue declined by nearly 15 percent. In 1988 the company made a secondary stock offering, but after posting continued losses, its stock plummeted. By the early 1990s it became clear that cutbacks had to be made. Pattiz sold off his three radio stations and his trade publication *Radio and Records*. He also consolidated news operations, merged four networks into three, and reduced compensation payments to stations.

By 1994 the company had reduced its debt significantly and could concentrate on its programming. That same year, in an effort to secure station contracts and expand its network capacity, Westwood One made another very significant move. It traded 25 percent of its ownership to Infinity Broadcasting, the largest owner of radio stations in the United States, in exchange for Infinity's Unistar networks. Infinity was hired to take over management and Pattiz gave up the position of CEO while retaining his position as chairman and executive producer of all programming. (He also remained a major stockholder.) Mel Karmazin, then CEO of Infinity, became the new CEO of Westwood One. After merging with Unistar, the company formed two new divisions: Westwood One Radio Networks and Westwood One Entertainment. The networks division managed the six networks, news, and 24-hour formats, and the entertainment division produced programming and live concerts.

In order to tap into the local advertising revenue stream, Westwood One purchased Shadow Broadcast Services in 1996. Shadow provided localized traffic reports as well as local news, sports, weather, and entertainment. Three years later, Westwood bought the number one traffic news service, Metro Networks, in a $900 million stock deal and merged it with Shadow Broadcast Services. In 2000 it added SmartRoute Systems to this lineup, providing wireless and internet services as well. By providing localized products in exchange for ad time, these services account for a substantial percentage of Westwood's revenue.

Shortly after Westwood picked up Shadow Broadcast Services, Karmazin attempted to buy CBS's owned-and-operated radio stations. When CBS CEO Michael H. Jordan declined, Karmazin offered to sell Infinity to CBS in return for allowing Karmazin to run the radio group. Jordan accepted the offer and in December 1996, CBS bought Infinity for $4.9 billion in stock. By the following March, management of the CBS Radio Networks division was spun off to Westwood One. The deal provided that Westwood One would represent CBS Radio Networks, managing its sales, marketing, and promotion, and that CBS would continue to produce and control the programming. CBS, by virtue of owning Infinity Broadcasting, now owned a stake in Westwood One.

After becoming president of CBS in 1998, Karmazin decided to move out of the CEO position at Westwood One, making way for a new full-time president and CEO, Joel Hollander, formerly head of New York station WFAN. Hollander led the company through the Metro acquisition and continued to secure deals for college and professional sports programs, new entertainment venues, and web-based radio programming. In 1999 Westwood made the decision to shut down its Mutual News division, ending the network's 65-year broadcasting legacy. The following year, Westwood bought a six percent stake in WebRadio.com, an internet broadcasting company, and secured an equity stake in Fanball.com, a sports-fantasy multimedia company. In May 2000 Viacom merged with CBS, making Viacom the new parent company of Westwood One. Karmazin became president and chief operating officer of Viacom at that time.

Programs and Operations

Westwood One made a name for itself by focusing on entertainment programming for radio and particularly on large-scale entertainment events. Its advertising slogan became "Westwood One, for the biggest events in radio." In the 1970s, when Westwood first started, traditional radio networks provided mostly news, sports, and talk, and very little (if any) musical or dramatic entertainment. Starting with *The Sound of Motown*, Westwood went on to produce dozens of radio specials. In the early 1980s, it produced a live broadcast of the US Festival, a four-day rock concert in Riverside, California. Although individual stations such as KRLA had sponsored concerts in the 1960s, no radio network or syndicator had ever produced such a large event for live broadcasting. At that time Westwood also decided to hire its own sound crew and build a state-of-the-art mobile recording studio for concert production. It then produced the first live stereo radio broadcast from Japan with its concert "Asia from Japan."

When Westwood One began to acquire traditional networks, its youth-oriented entertainment emphasis clashed with the staid culture of the older networks. Although these networks were brought in specifically to provide a wider variety of programming, Pattiz set about to restructure and reprogram them to reflect the energetic style of Westwood One. When Karmazin came on board with the purchase by Infinity Broadcasting, he pushed for expansion into new markets such as traffic and satellite formats. However, as executive producer of programming, Pattiz continued to pursue the exclusive superstar concert broadcasts for which Westwood was known, such as the Rolling Stones, the Eagles, and Barbra Streisand.

Westwood One's network division currently provides programming in several categories. In news it carries Fox News Radio, CNN Radio, NBC Radio Network, CBS Radio News and CBS MarketWatch.com. Both CNN Radio and CBS

MarketWatch.com air 24-hour services. The Associated Press (AP) is the primary provider of news content to Westwood's syndicated and network programs. In sports Westwood features broadcasts for the National Football League, the National Hockey League, the National Collegiate Athletics Association, the Olympics, championship boxing, and professional golf. Westwood's talk personalities include Jim Bohannon, Larry King, G. Gordon Liddy, and Tom Leykis. Dozens of music formats are available, including a number of full-time and international services.

Shadow Broadcasting Services and Metro Networks provide local traffic reports, localized news, weather, and sports to hundreds of local radio markets across the country. Each program is customized for the individual station and made to sound as though it is being delivered by local talent. Traffic reports make up the bulk of affiliate contracts, but the other areas, particularly news, are growing with increased demand. The Shadow division has added short-form entertainment and health news reports to its lineup of services. SmartRoute Systems provides traffic, news, sports, and weather information to wireless and web services.

Westwood One also provides prep services such as the MTV Morning Facts and The CBS Morning Resource, which supply stations with interviews, celebrity and entertainment industry facts, sound bites, news, and gossip. Program directors integrate this material into their local programming, using local talent.

CHRISTINA S. DRALE

See also Columbia Broadcasting System; Infinity Broadcasting; Karmazin, Mel; Mutual Broadcasting System; National Broadcasting Company

Further Reading

Borzillo, Carrie, "New Challenges Afoot for Network 'King' Westwood One," *Billboard* (8 July 1995)
Dunphy, Laura, "The Advantage of Getting Fired," *Los Angeles Business Journal* (27 December 1999)
Grossman, Lawrence K., "The Death of Radio Reporting," *Columbia Journalism Review* 37, no. 3 (September/October 1998)
Petrozzello, Donna, "Traffic Services Dish Out News," *Broadcasting and Cable* (9 June 1997)
Viles, Peter, "'Leaner, Meaner' Westwood Cuts Losses," *Broadcasting and Cable* (19 July 1993)
Westwood One website, <www.westwoodone.com>

WEVD

New York City Station

WEVD-AM was established in 1927 as a memorial to socialist leader Eugene Victor Debs. A famous unionist and five-time presidential candidate on the Socialist ticket, Debs died in 1926, and the Socialist Party was moved to erect a monument in his honor. The Party raised enough money to buy Long Island radio station WSOM in 1927 and subsequently changed the station's call letters to WEVD. The station signed on at a frequency of 1220 kilohertz on 20 October 1927, the anniversary of Debs' death, and immediately became an electronic voice for the ideas and causes that Debs had championed.

On a typical day, WEVD would broadcast a mixture of poetry, music, and speeches reflecting the ideals of labor and socialism. The station presented shows for New York's minority population, including a *Jewish Hour*, a *Negro Art Group* program, and shows on African-American literature, music, and history. Debates on topics ranging from foreign policy in Nicaragua to general labor conditions were common. It was station policy to provide free access to labor unions, including, among others, the Teacher's Union, the Union of Technical Men, the Office Workers, the Garment Workers, and the Neckwear Workers. WEVD's fiscal health was dependent on donations from these and other unions. Consequently, the station was plagued with financial difficulties from the start, and plans to increase power and form a network were suspended.

The first task confronting the Federal Radio Commission (FRC) in 1927 was to resolve the problem of congestion on the airwaves. The commission concluded that the solution was to eliminate at least 100 radio stations. Toward that end, the FRC issued its famous General Order Number 32, which asked 164 stations, including WEVD, to show cause why their licenses should not be revoked.

The commission held hearings for two weeks in July 1928 in Washington, D.C., and the WEVD case was the first to be heard. The burden was placed on WEVD and 163 other stations to demonstrate why license renewal would be in the public interest. The FRC insisted that WEVD was placed on the list because of complaints of interference and technical violations. Station officials believed WEVD was singled out because of the unpopular and controversial doctrines expressed in its programming. Station officials invoked the First Amendment, arguing for the right of dissident minorities to free speech and arguing that labor, socialist, and other forms of unpopular rhetoric are in the public interest.

Approximately one month after the hearing, WEVD's license was renewed. Of the 164 radio stations called before the commission, only 81 (including WEVD) escaped adverse action by the FRC. In 1929 WEVD moved to 1300 AM, a frequency it would share with three other stations. By 1930 continued financial problems, the Great Depression, and the weakening of the Socialist Party would bring WEVD into a second confrontation with the FRC. The commission accused WEVD of several operational violations, but station officials again suspected they were being singled out for their unpopular programming.

WEVD's license was renewed in early 1931, but in a highly unusual decision, the FRC changed its mind three days later and revoked the license. The revocation was apparently prompted by a competing applicant, the Paramount Broadcasting Corporation of New York, which vowed to provide better public service. WEVD officials pledged to continue operating the station in defiance of the commission's decision. The FRC decided to grant temporary license extensions to WEVD, during which time the commission would conduct hearings on WEVD's renewal application. One hearing before the commission in 1931 lasted an entire week, which was the longest hearing before the FRC to that time.

In a narrow decision, the FRC finally renewed the license of WEVD in October 1931. However, the delays had involved WEVD in a long and costly battle, and the station's financial problems became acute. The *Jewish Daily Forward* rescued WEVD with a commitment of $250,000, but this and later contributions from the daily newspaper gave that organization effective decision-making power. Within a year, the station's staff was reorganized, and the studios moved to the Times Square area. The Debs Memorial Radio Fund later merged with the Forward Association to run WEVD as a commercial outlet. Six years later the station moved again, to occupy its own building on West 46th Street.

Throughout this period, WEVD shared the broadcast day with other New York stations. Gradually those other operations either sold out, moved, or were taken over by WEVD, which was on the air nearly 90 hours a week by 1938. In the 1980s, the station underwent a host of ownership and frequency changes, selling the AM outlet in 1981 and operating only with FM for much of the decade. In a complex exchange, WEVD sold the now-valuable FM station for $30 million in 1988, and as part of the deal it took over the 1050 kilohertz AM frequency that had been WSKQ. By the mid-1990s, the 50,000-watt station specialized in syndicated talk shows and news programming.

PAUL F. GULLIFOR

Further Reading

Brindze, Ruth, *Not to Be Broadcast: The Truth about Radio,* New York: Vanguard Press, 1937

Friendly, Fred W., *The Good Guys, the Bad Guys, and the First Amendment: Free Speech vs. Fairness in Broadcasting,* New York: Random House, 1976

Godfried, N., "Legitimizing the Mass Media Structure: The Socialists and American Broadcasting, 1926–1932," in *Culture, Gender, Race, and U.S. Labor History,* edited by Ronald Kent et al., Westport, Connecticut: Greenwood Press, 1993

Jaker, Bill, Frank Sulek, and Peter Kanze, *The Airwaves of New York: Illustrated Histories of 156 AM Stations in the Metropolitan Area, 1921–1996,* Jefferson, North Carolina: McFarland, 1998

McChesney, Robert Waterman, *Telecommunications, Mass Media, and Democracy: The Battle for the Control of U.S Broadcasting, 1928–1935,* New York: Oxford University Press, 1993

WGI

Boston, Massachusetts Station

Although WGI never had expensive studios, it did have a woman engineer/announcer, a morning exercise program, on-air college courses, and the best-known children's show in town. Furthermore, it may have been the first station to run paid commercials, and several well-known performers got their start in those not very opulent studios. Yet today, few people know that the Boston station ever existed.

Origins

Harold J. Power fell in love with "wireless" when he was nine years old, and by the time he attended Tufts College, at Medford Hillside (about five miles from Boston), he was already an experienced ham radio operator who enjoyed building his own receiving equipment. After graduating in 1914, he and several fellow hams decided to start their own station, along with a company to manufacture receivers. They named this new venture the American Radio and Research Company; most people knew it as AMRAD. As for the new station, because radio broadcasting was still considered experimental by the Department of Commerce, it received the call letters 1XE. But even six years later, when the station was assigned the commercial call letters "WGI," listeners still thought of it as either "the AMRAD station" or "the Medford Hillside station."

Harold Power became the president of AMRAD, and he soon began to air wireless concerts of phonograph records to promote the new company. In early 1916 this was so unusual that the *Boston Globe* wrote an article about the amazing music programs being heard by the ships at sea. At first, AMRAD targeted the ham radio audience (since there was no commercial broadcasting yet), but at some point in 1920, everything began to change.

We may never know who was really first to broadcast commercially. Scholars have debated endlessly whether KDKA, WWJ, or any of several other stations were the first, but there is evidence that 1XE was in the elite group of stations on the air in the fall of 1920. Unfortunately, most of AMRAD's and 1XE/WGI's files were long ago destroyed in a fire, but based on the radio columns of several Boston newspaper reporters of the early 1920s and an interesting interview from the manager of a competing Boston radio station, the consensus is that 1XE was on the air at around the same time as the much better known KDKA.

By the spring of 1921, 1XE was on the air every day with a regular schedule. The station's air staff included a popular woman announcer, Eunice Randall. She read the nightly police reports, gave children their bedtime story several nights a week, sang when a guest didn't show up, and worked as one of the station's engineers—all highly unusual for a woman in those days. Randall also had a show that may have been sponsored—although as with many such arrangements in the early days, it may have been a barter arrangement. Randall's bedtime stories were presented by *Little Folks Magazine,* which may have provided free copies as prizes rather than paying for sponsorship.

The need for revenue was a constant problem for small stations of the early 1920s. Stations owned by individuals or small companies like AMRAD ran into financial trouble when lightning struck their towers or when equipment broke and was expensive to repair. At first, entertainers performed free because radio was a novelty, but eventually the bigger names wanted compensation; small stations had to depend on eager volunteers or up-and-coming performers who would still work in exchange for exposure.

AMRAD had originally been backed by financier J.P. Morgan's son Jack, who had known Harold Power ever since Power worked for him while still in high school. Morgan had been persuaded by Power's dreams of success in the radio business and was a silent partner in AMRAD for its first few years. But although Power had big dreams and good intentions, running a company was problematic for him. AMRAD became famous for good concepts but poor implementation: equipment was often delivered to suppliers late, and AMRAD was slow to react to new trends. Morgan gradually phased out his support of AMRAD, leaving the company to deal with its own financial problems by 1923.

Decline

In spite of financial and technical problems, 1XE gained fans all over the eastern United States (the station's 100-watt signal was even heard in England one night). Guest speakers, from politicians to professors, and even the famous economist Roger Babson, gave talks from Medford Hillside. By late 1921, WBZ was on the air (although it did not yet put a good signal into Boston), but musicians and celebrities continued to appear at the AMRAD station, because it was so close to the theaters and clubs of Boston, whereas WBZ was 80 miles away in Springfield. 1XE became WGI in February 1922: Power later said that he never saw a good reason to get a commercial license until then; he believed his experimental license was sufficient. By March of that year, it was airing Boston's first radio newscasts, courtesy of the *Boston Traveler* newspaper. WGI consistently provided good entertainment throughout the early

1920s: musicians such as *Hum and Strum* and Joe Rines went on to successful careers on the networks and on records; the popular children's show the *Big Brother Club* enjoyed a 45-year run on radio and then TV; and the famous poet Amy Lowell and the African-American actor Charles Gilpin were heard first over WGI, as was Harry Levi, the "Radio Rabbi."

Harold Power's attempt to bring in some revenue at WGI got the station in trouble with the Department of Commerce in the spring of 1922. In early April, Power had accepted money from a car dealer and an advertising agency to air some commercial announcements. Evidently, somebody notified the department (direct advertising was frowned upon, and Secretary of Commerce Herbert Hoover wanted to keep it that way). A series of "cease and desist" letters were sent by Radio Inspector Charles Kolster, beginning on 18 April and continuing into May. His correspondence to the station suggests that WGI aired commercials in April (well before WEAF's first commercial in August) and that the department had to warn WGI management several times before these commercials stopped.

By the fall of 1922, a new station, WNAC, was on the air, owned by the Shepard Department Stores. With substantial financial backing, it paid its talent and soon enjoyed handsome studios as well. In the meantime, AMRAD's financial worries increased, and some of WGI's best performers (and several announcers) accepted jobs at WNAC. Over the next two years, WGI continued to win praise from listeners, magazine editors, and radio columnists, but as more Boston stations came on the air—all with more powerful signals, better equipment, and money to pay the performers—it was only a matter of time before WGI could no longer compete. AMRAD went into bankruptcy in late April 1925, and as a result one of America's pioneer stations came to an end. When a buyer was not found, WGI left the air.

DONNA L. HALPER

See also WEAF

Further Reading

Halper, Donna L., "The Rise and Fall of WGI," *Popular Communications* (June 1999)

WGN

Chicago, Illinois Station

Named for the "World's Greatest Newspaper," this Chicago radio (and later TV) station was not actually started by its long-time owner, the Tribune Company, but was purchased by the newspaper giant soon after it went on the air in June 1924.

Origins

By no means the first Chicago broadcaster (which was KYW), the *Chicago Tribune* entered radio by providing that pioneering outlet with news and market reports. At least four other stations soon followed KYW on the air. The first station located in the handsome Tribune Tower building was WDAP, which aired from May 1922 until a July tornado destroyed its antenna. The station moved and continued operating with a single (very busy) employee, one Ralph Shugart. When station mail piled up, the second employee, Myrtle Stahl (who would stay in radio until 1960) joined him. In 1923, the Chicago Board of Trade purchased WDAP. By March 1924, however, the *Tribune* had purchased enough air time to take control of the station, and changed its call letters to WGN. Full page ads in the newspaper announced the "new" station on 28 March 1924, and WGN's inaugural broadcast came the next evening from studios in the Edgewater Beach Hotel.

WGN pioneered a significant number of radio firsts. Probably the best-known special news radio broadcast of the middle 1920s was WGN's broadcasts from the Scopes "monkey" trial from Dayton, Tennessee, which cost the station $1,000 a day in personal expenses and in telephone lines to send the signals back to Chicago. Freeman F. Gosden and Charles J. Correll, who started together in vaudeville, created *Sam 'n' Henry* for WGN, but in 1929 they were tempted by the National Broadcasting Company's (NBC) greater offer; they then moved to WMAQ and came up with a new form of the same act, *Amos 'n' Andy,* national radio's first great hit. Soap opera pioneer Irna Phillips created *Painted Dreams* on WGN as one of the early soap operas.

Tribune executives tried to take early advantage of radio and newspaper common ownership. In the immediate wake of the Saint Valentine's Day Massacre in 1929, at company expense, WGN installed radio receivers in all 40 of the light blue touring cars driven by members of the Chicago Police Detective Squad. Detectives were instructed to listen to

WGN—and only WGN—throughout their shifts. When word of a crime, either in progress or recently completed, reached police headquarters, a dispatcher was instructed to telephone WGN and pass along whatever details were available to the announcer on duty, who interrupted programming and broadcast the information in the form of a bulletin. The nearest squad car would hear the bulletin and rush to the crime scene (but so might others who heard the same bulletin). WGN and the *Tribune* boasted that this experiment was a success, but it lasted only a few years, proving how difficult what would be labeled "synergy" two generations later was to accomplish.

Robert McCormick, long-time *Tribune* editor and publisher, saw radio as an ally, not as an adversary. His most important innovation in radio was to assist in forming the Mutual radio network in 1934. From its studios at 435 North Michigan Avenue, WGN aired Mutual with its 50,000 watts and poured the network's offerings into homes all across the upper Midwest.

Through the 1930s and 1940s, WGN functioned as Chicago's link to the Mutual radio network. As a founding station of Mutual—along with WOR (New York), WXYZ (Detroit), and WLW (Cincinnati)—through this period WGN and WOR functioned as the only clear channel stations not affiliated with either NBC or the Columbia Broadcasting System (CBS). On 1 March 1941, W59C (later WGNB) aired as the first WGN-owned FM radio outlet. Just a year before WGN had initiated 24-hour operation, a fairly rare service at the time.

Decline of Mutual

In reaction to TV's growing popularity after 1948, WGN-AM had to lessen its dependence on the Mutual radio network. WGNB, WGN's FM sister, was donated to Chicago Educational TV (WTTW) in the early 1950s and became WFMT. Receiving fewer network programs as Mutual declined, WGN had to resort to developing its own programming. Station manager Ward Quaal developed local talk stars as many of the radio studios were converted for use by WGN television. In October 1956 Quaal hired Wally Phillips away from another powerhouse clear channel station, WLW in Cincinnati. Along with Bob Bell, who later became WGN-TV's Bozo the Clown, the duo did comedy and talk to entertain Chicago's growing mass of daily commuters. Phillips proved so popular that he also appeared on WGN-TV in *Midnight Ticker,* but it was his solo drive-time show, which debuted in 1959, that lasted until 1986.

In 1961 WGN radio and television moved to a new building on Chicago's north side. At the time the radio station adopted a new slogan: First in sound, first in service, first in sports. By the turn of the century, WGN had long operated as a high-class major market talk station. Its broadcast day started at 5:00 A.M. with *The Bob Collins Show,* then switched at 9:00 A.M. to *The Kathy and Judy Show,* followed by 20 minutes of news starting at 11:55 A.M., and then offered talk with John Williams and Spike O'Dell all afternoon. The evening was allocated mostly to sports broadcasting, and the station, at 720 kilohertz, broadcast the baseball games of the Chicago Cubs, which was owned by the Tribune Company.

DOUGLAS GOMERY

See also Mutual Broadcasting System; Quaal, Ward

Further Reading

Broadcasting in Chicago, 1921–1989, <www.mcs.net/~richsam/home.html>

Fink, John, and Francis Coughlin, *WGN: A Pictorial History,* Chicago: WGN, 1961

Linton, Bruce A., "A History of Chicago Radio Station Programming, 1921–1931, with Emphasis on Stations WMAQ and WGN," Ph.D. diss., Northwestern University, 1953

WGN Continental Broadcasting Company, *The Wonderful World of WGN,* Chicago: Bassett, 1966

WGN Radio 720 Online: Chicago's News and Talk, <www.wgnradio.com>

WGN-TV Online: WGN: Welcome to Chicago's Very Own, <www.wgntv.com>

WHA and Wisconsin Public Radio

WHA (originally 9XM) is significant for three reasons: (1) it has a disputed claim to the title "oldest station in the nation"; (2) it has a strong, and less disputed, claim to having provided educational radio in the U.S. with its guiding philosophy and operating model; (3) it undisputedly pioneered the nation's first statewide FM network, which was later complemented by a second statewide network and the concept of dual program services. Today that network, Wisconsin Public Radio, remains the country's largest institutionally based public radio operation and a significant provider of national programming.

"The Oldest Station in the Nation"

A historical marker on the University of Wisconsin campus proclaims 9XM/WHA to be "the oldest station in the nation." Other stations dispute that claim, of course, but if nothing else, the early history of WHA demonstrates the importance of land grant universities in the early development of radio, particularly in the period before anyone recognized the commercial potential of the medium. The University of Wisconsin's story is emblematic of similar, if smaller-scale, efforts at more than 200 colleges and universities in the second and third decades of the 20th century.

Physicists and engineers began the University of Wisconsin's activities in radio. First Professor Edward Bennet (Engineering) and then Professor Earl Terry (Physics) experimented with radio apparatus in the first two decades of the 20th century. The Commerce Department issued a license for experimental wireless telegraphy station 9XM to Professor Bennet in 1912. The license later was transferred to the Regents of the University, who have been licensed to use radio since 1916, perhaps Wisconsin's strongest claim to the title of "oldest."

In December 1916, 9XM joined stations at the University of North Dakota, the University of Nebraska, and Nebraska Wesleyan University in regularly scheduled daily noontime wireless telegraphy broadcasts of weather and agricultural markets in cooperation with the U.S. Department of Agriculture. With these reports, the four institutions began "broadcasting" in the sense that they sought to serve a dispersed audience on a regular basis. Of course, only those familiar with Morse code could understand the broadcasts, severely limiting the effectiveness of the service. For the most part, local offices in scattered communities received the messages and posted the information for farmers and other citizens to read. By early 1917, however, Wisconsin had added voice to the telegraphic broadcasts, making them accessible to the few ordinary listeners who built or owned receivers. Hence, the historical marker cites 1917 as the beginning of "broadcasting" on station 9XM.

Wisconsin's claim to "oldest" status draws on its exemption from the government order closing down all private radio apparatus during World War I. Rather than cease operations, 9XM formed a partnership with the U.S. Navy to help in developing a cadre to radio operators at the Great Lakes Naval Training Station. That wartime activity allowed the proponents of WHA to trace its continuous operation back to 1916–17, even though the "broadcasts" during the war years could not be heard legally by anyone outside the U.S. Navy.

Broadcasting to the public resumed on a regular basis in 1919, the second date cited on the historical marker as the beginning of continuous regularly scheduled broadcasting. That continuity was not unbroken, however. 9XM stopped broadcasting for six months in late 1920 in order to build a new, larger transmitter. During those six months of silence, KDKA in Pittsburgh, Pennsylvania, broadcast its famous coverage of the November 1920 election results, an event often cited as the beginning of radio broadcasting. When it returned to the air on 3 January 1921, two months after KDKA's "birth of broadcasting" event, 9XM could no longer be dismissed as an engineering experiment. It had a program director, speech professor William Lighty. It published its program listings in the local paper. It broadcast voice and music. The call letters changed from experimental 9XM to WHA a year later in 1922.

The Wisconsin Idea

The debate over who broadcast *first* obscures the true significance of WHA. More than any other educational station, WHA enunciated a clear mission of public service and implemented it on a scale far beyond any other. WHA developed as a unique broadcasting institution at a unique time. The combination of a progressive state government working closely with a service-oriented university became known as "The Wisconsin Idea," and out of that idea came a unique commitment to serve the state with radio from the campus in Madison, combining education with broader public-service goals.

In its educational role, WHA's programming initially emphasized the practical, particularly agricultural information and "home economics." In this, the radio station complemented the work of the university's network of county agents who assisted farmers and their families throughout the state. The station offered, in addition, music-appreciation series and talks and dramatizations written in conjunction with faculty members on a wide range of historical, literary, and contemporary topics. In 1932 the charismatic director of WHA for almost 40 years, H.B. (Mac) McCarty, and his meticulous and diplomatic deputy, Harold Engel, gave the umbrella title "College of the Air" to these diverse offerings, which they said would make a college experience available to those whose circumstances had denied them the opportunity. When portable tape recorders liberated radio from the studio in the 1940s, the College of the Air literally moved into the lecture halls of the University of Wisconsin. At its height, lectures from university courses filled three hours each day on the WHA schedule. Faculty lecturers became statewide celebrities.

A year earlier, in 1931, McCarty and Engel had begun a separate series of programs called "School of the Air" for rural "one room school houses," of which Wisconsin still had many. *Let's Draw, Let's Sing,* and *Ranger Mac* provided the art, music, and science lessons otherwise unavailable in small schools without specialist teachers. More than anything else, the School of the Air justified continued state investments in educational radio, particularly in the lean Depression years, and introduced hundreds of thousands of children and their families to the state radio service.

Narrowly defined education constituted only part of the programming mix, however. Variety shows, radio drama, folk, and popular music found their way to the airwaves. The station also pioneered in covering public issues. Radio provided a means for people to learn about issues and to consider alternative directions. In 1931, for example, WHA offered advocates of different farm policies the opportunity to present their views. This led, in turn, to offers of free airtime to all candidates for statewide office, including those of the minor parties, in the 1932 elections and in every subsequent election for the next 40 years. The 1932 "Political Education Forum" marked the first time any American radio station had used its facilities as a forum for public debate. *The New York Times* praised this initiative, and Professor Bennet made it his theme when he testified before Congress in an unsuccessful effort to reserve 25 percent of radio frequencies for noncommercial stations in the Communications Act of 1934.

The State Radio Network

The dramatic boom in commercial radio in the late 1920s and early 1930s forced most educational radio stations off the air, their frequencies taken over by commercial stations, either through purchase or reassignment by the Federal Radio Commission. The handful of educational stations that survived found themselves relegated to inferior "regional" channels and usually only during daytime hours. Whereas in its earlier years WHA transmitted broadly outside Wisconsin, it ended up with a daytime regional frequency in Madison that covered only a portion of the state by day and none of the state by night, when most people listened to the radio. The addition of a second regional daytime station (WLBL) in the central part of the state helped, but it still left WHA with inadequate facilities.

To solve its coverage problems, in the early 1940s WHA proposed to take over the frequency of the National Broadcasting Company (NBC)'s Chicago station, WMAQ, arguing that Chicago had five clear channel stations and Wisconsin had none. Not surprisingly, NBC was able to beat the challenge, as did Atlanta's WSB when WHA proposed to share its clear channel. Thwarted in two attempts to improve its AM facility, Wisconsin looked to the new technology of FM, which had been authorized by the FCC in 1941 but had remained mostly undeveloped in the war years. In 1944, the Wisconsin Legislature created a State Radio Council separate from the university to develop a network of FM stations to carry WHA programming throughout the state. Although legally separate from the University of Wisconsin, the State Radio Council effectively operated as an adjunct to the university broadcasting operation. McCarty headed broadcasting for both the university and the State Radio Council, and WHA was the sole source of programming for the network. In March 1947 the State Radio Council activated WHA-FM in Madison, the first of nine FM stations that would go on the air at the rate of one per year until the network achieved statewide coverage in the mid-1950s. Because McCarty and Engel saw the role of these stations as broader than narrowly defined education, they named the system the Wisconsin State Radio Network or the State Stations rather than the university network or the educational network. Already the oldest and largest educational broadcaster in the country, the Wisconsin operation soared past any potential rival with the state's commitment to build and operate those nine FM stations. On the eve of the Public Broadcasting Act of 1967, the budget of the Wisconsin State Stations tripled that of any other educational radio operation.

Wisconsin Public Radio

McCarty and Engel chose 1967 to retire, just as Congress enacted the Public Broadcasting Act, which gave a new name and a somewhat different concept to "educational" broadcasting. In a sense, the national legislation vindicated Wisconsin's vision of radio as a public service broader than university-level education. Indeed, the principle author of the vision for the new enterprise that would be called National Public Radio (NPR) was Bill Siemering, a former WHA staff member who often acknowledged his debt to McCarty and the Wisconsin Idea.

Without McCarty to hold his creation together, the university and the Educational Communications Board (successor to the State Radio Council) struggled for control of the state stations. At first, the advocates of a narrow view of educational radio seemed to prevail. Indeed, the Educational Communications Board changed the call letters of WHA-FM to WERN, for Wisconsin Educational Radio Network, its new designation for the State Radio Network. That name, however, ran directly counter to what was happening to the stations. School of the Air programming essentially moved over to television, and programming from NPR replaced some of the College of the Air programming. The State Network was becoming less "educational" at precisely the time it added "educational" to its name.

In 1978 the university and the Educational Communications Board accepted this reality and agreed to designate their joint enterprise "Wisconsin Public Radio." They appointed Jack Mitchell, formerly of NPR, as director of radio for both organizations. They put in place a long-range strategy to provide two formatted services in most parts of the state. One of the services—headed by WERN (the former WHA-FM)—would build a format around music and arts. The other service—headed by WHA (AM)—would feature news and information. The music and news division became a common pattern among public radio organizations that controlled two stations in one community. In 1989 WERN's music and arts service evolved into the "NPR News and Classical Music Network," while WHA's information service narrowed its focus to emphasize unique Wisconsin talk programming, particularly statewide call-in shows on a range of informational, educational, and public-affairs topics. As residents across the state talked with academics, authors, officials, advocates, and one another, this service echoed the traditional educational and public-service purposes of the Wisconsin Idea, as did the name of the new service, "The Ideas Network of Wisconsin Public Radio."

As the "Ideas Network" focused inward on the state, Wisconsin Public Radio exported several programs nationally. For a decade in the late 1970s and early 1980s, WHA served as home to public radio's national drama project *Earplay,* under the direction of Karl Schmidt. Although *Earplay* won critical acclaim, it fell victim to the budget cuts of the Reagan administration. Since 1985 WHA has been the home of Michael Feldman's *Whad'Ya Know?*—a comedy quiz program carried by more than 200 public radio stations to a weekly audience of more than 1 million people. The station also distributes two national advice programs and *To the Best of Our Knowledge,* a weekly two-hour interview magazine covering the world of ideas. Often considered the "brainiest" program on public radio, *To the Best of Our Knowledge* is a fitting product of the radio station that gave America the concept of educational radio.

JACK MITCHELL

See also Educational Radio to 1967; Farm/Agricultural Radio; Minnesota Public Radio; National Public Radio; Public Radio Since 1967; Siemering, William

Further Reading

Baudino, Joseph, and John M. Kittross, "Broadcasting's Oldest Stations: An Examination of Four Claimants," *Journal of Broadcasting* 21 (Winter 1977)

Engel, Harold A., "Wisconsin's Radio Pattern," *The National Association of Educational Broadcasters Journal* 24 (January 1958)

Engel, Harold A., "The Oldest Station in the Nation," *The National Association of Educational Broadcasters Journal* 25 (February 1959)

The First 50 Years of University of Wisconsin Broadcasting: WHA, 1919–1969, and a Look Ahead to the Next 50 Years, Madison: University of Wisconsin, 1969

Frost, S.E., Jr., *Education's Own Stations: The History of Broadcast Licenses Issued to Educational Institutions,* Chicago: University of Chicago Press, 1937

McCarty, H.B., "WHA, Wisconsin's Radio Pioneer: Twenty Years of Public Service Broadcasting," *Wisconsin Blue Book* (1937)

McCarty, H.B., "Educational Radio's Role," *The National Association of Educational Broadcasters Journal* 25 (October 1959)

Penn, John Stanley, "The Origin and Development of Radio Broadcasting at the University of Wisconsin to 1940," Ph.D. diss., University of Wisconsin, 1958

Penn, John Stanley, "Earl Melvin Terry, Father of Educational Radio," *Wisconsin Magazine of History* 50 (Summer 1961)

Smith, R. Franklin, "Oldest Station in the Nation?" *Journal of Broadcasting* 4 (Winter 1959–60)

Witherspoon, John, Roselle Kovitz, Robert K. Avery, and Alan G. Stavitsky, *A History of Public Broadcasting,* Washington, D.C.: Current, 2000

WHER

Memphis, Tennessee Station

When WHER-AM 1430 broadcast for the first time on 29 October 1955, it was staffed almost entirely by women, a phenomenon never before seen in U.S. radio. An experiment in novelty during a period of declining radio audiences and revenues, WHER would demonstrate and confirm women's competencies in the radio industry and inspire women to pursue careers in the industry.

The brainchild of Memphis record producer Sam Phillips (who first recorded Elvis Presley, Johnny Cash, and other important figures in American music on his Sun Records label), WHER came to life with a $25,000 investment from Kemmons Wilson, the founder of the Holiday Inn hotel chain. Together, Phillips and Wilson formed Tri-State Broadcasting Service, Inc. It was a difficult time for radio in 1955 when the station began operations; television was stealing radio listeners and gobbling up advertising dollars. To remain profitable, station owners sought new ways to reach audiences and looked to audiences that radio had traditionally ignored.

After hiring station manager Dotty Abbott, a veteran radio manager from Phoenix, Phillips selected seven other women (Teresa Kilgore, Marion Keisker, Dot Fisher, Pat McGee, Denise Howard, Barbara Gurley, and Laura Yeargain) to run the station. Phillips' wife Becky, a prominent on-air personality in Southern radio, also joined the team. WHER set up its first studios in several rooms provided by Wilson at the Memphis Holiday Inn (only the third Holiday Inn in existence at the time). Painted in pink and purple pastels, the radio studio featured distinctively feminine decor, somewhat resembling a dollhouse.

Despite skepticism and shock when Keisker's first broadcast aired in Memphis with little warning on the morning of 29 October 1955, the station was an immediate success. Audiences enjoyed tuning in to hear female voices and perspectives on radio. The "girls" lived up to their slogan, "A smile on your face puts a smile in your voice." Their broadcasts were energetic and fun, and they were not afraid to laugh at themselves and their mistakes. Initially the station catered primarily to female homemakers and featured love ballads, jazz, and light content. But as the station evolved, the format matured and diversified. The station fielded competent news and sports staffs, and by the 1960s it programmed one of Memphis' early call-in talk shows, *Open Mike*, hosted by Marge Thrasher. *Open Mike* and the station's other news programs addressed issues of importance in Memphis, such as the city's festering racial tensions that led to the 1968 sanitation workers' strike, the backdrop of Martin Luther King Jr.'s assassination.

Memphis radio had already seen such experimentation in 1949 when radio station WDIA began offering an all-black format, and although WHER's format was never strictly all-female-oriented, the station's all-female staff did represent the lengths radio entrepreneurs were going to in order to distinguish themselves in the ever-tightening radio market.

Although Sam Phillips and Kemmons Wilson's endeavor bordered on outright gimmickry, the station remained viable for almost 20 years and became a landmark in the history of women's broadcasting. During a period when female reporters were still banned from the National Press Club in Washington, D.C. and most radio stations rarely hired more than one female air personality, WHER boasted a staff of forty women who held positions ranging from general manager to program director to disc jockey. (Among the few stations that adopted WHER's all-girl approach was WSDM, Chicago, which employed "Hush Puppy" Linda Smith, later known as Linda Ellerbee.)

The social upheaval that accompanied the late 1960s would ultimately contribute to WHER's demise. The notion of an "all-girl" radio station that traded on stereotypical feminine qualities seemed out of step with the resurging women's rights movement (hosts of music programs were known as "jockettes," and one of WHER's taglines was "One thousand beautiful watts"). As a result, the station added more men to its staff and sought to broaden its appeal. In 1971 (after 16 years on the air) WHER was changed to WWEE, a talk radio station that later became a gospel station. In 1988, Phillips and Wilson sold the station first known as WHER. Currently a WHER broadcasts from Hattiesburg, Mississippi, but the outlet has no connection to the original WHER.

In October 1999, to commemorate WHER's inception, Davia Nelson and Nikki Silva produced a documentary highlighting the history of WHER for National Public Radio's "Lost and Found Sound" series on *All Things Considered*. Also featured on local and national news programs, the women were brought together by Sam Phillips in New York for a reunion.

WHER's female orientation represented radio's general effort to find stable ground in the burgeoning television age of the 1950s, but the station will be best remembered as having played an important role in the development of opportunities for female radio professionals in all areas of radio operations. Many women working in radio today owe a tremendous debt to Sam Phillips and the more than 50 women who worked at WHER from 1955 to 1971.

MICHAEL STREISSGUTH AND
ALEXANDRA HENDRIKS

See also Women in Radio

Further Reading

Aherns, Frank, "Memphis's Other Music Revolution," *Washington Post* (29 October 1999)
Blumenthal, Ralph, "Spinning a Little History in a Studio Painted Pink," *New York Times* (30 October 1999.)
Burch, Peggy, "When Radio Did It Her Way," *Memphis Commercial Appeal* (29 October 1999)
Chin, Paula, and Barbara Sandler, "A League of Their Own," *People Weekly* (29 November 1999)

White, Paul 1902–1956

U.S. Executive, Editor, Creator of CBS News

Because Paul White built one of the first network news operations and then led it through its glory days of covering World War II, he can be said to have established precedents for broadcast journalism that are still followed and principles that are still admired.

Early Years

The son of Paul W. White, a stone contractor, and Anna Pickard, White was born in Pittsburg, Kansas, in 1902. His interest in journalism began early. In high school, he reported for the *Pittsburg Headlight*. After he graduated, rather than go to business school as his parents intended, White ran away from home and became a reporter for the Salina, Kansas, newspaper.

He majored in journalism at the University of Kansas for a year in 1920 while working as telegraph editor at the *Kansas City Journal*. He then earned a bachelor's degree in literature at Columbia University in New York City and in 1924 a master's degree from Columbia's School of Journalism. During his years at Columbia, White contributed articles to the *New York Evening Bulletin* and the *New York Sunday World*. After graduation he was hired as a reporter for the United Press in 1924 and later was promoted to features editor.

Developing CBS News

White came to the Columbia Broadcasting System (CBS) in 1930 with the title of news editor, but he was in the publicity department. Under him were only three people—an assistant, a secretary, and an announcer—and their main job was to cover special events rather than to gather news. CBS's journalism at that time consisted of weekly news commentaries and occasional bulletins.

Like other broadcasters, CBS depended on wire services as the major source of information, but the newspapers controlled them and felt that broadcasting was becoming too strong a competitor. When the wire services cut off the flow of information to radio, CBS was forced to go out and collect its own stories. The Columbia News Service was organized by White with news bureaus in major U.S. cities and stringers from around the world.

The results were so successful that the print industry applied more pressure by threatening to charge for the listing of radio programs and by refusing to do publicity on sponsors of radio news. Broadcasters signed the "Biltmore" agreement (much to White's dismay), severely limiting their ability to gather and report the news. They had to dismantle their newsgathering efforts (including the Columbia News Service) and be satisfied with short bulletins provided by the wire services. In 1935, though, the more independent wire services resumed supplying broadcasters with the news, and the agreement was ignored by everyone soon. The result of this intermedia conflict was that the National Broadcasting Company (NBC) and CBS were forced to hire more news people and expand their newsgathering process.

Radio journalism entered its maturity with the coming of war in Europe. At first, White was reluctant to allow his special-events employees, Edward R. Murrow and William Shirer, to go beyond their assigned duties of arranging talks by European leaders, but Germany's 1938 takeover of Austria led to a temporary change in CBS's policy. For a while, news became the norm rather than a special event.

To cover the crisis, White developed the first *CBS News Roundup*, which allowed the audience to hear live reports via shortwave from each of the trouble spots in Europe with an anchor to provide continuity—much the pattern for today's nightly TV newscasts.

Then, when Germany threatened to invade Czechoslovakia in 1938, the network made a full commitment to international coverage under White's leadership. His team covered the crisis with hours of live newscasts. When World War II broke out a year later, CBS became the leader in news both because of the quality of its correspondents on the scene, headed up by Murrow in London, and because of White's leadership. At the end of the war, the network received the Peabody award for its outstanding coverage.

White was an actively involved news director. He talked frequently via shortwave from his desk to correspondents around the world. It is no surprise that he and his crew of broadcast journalists sometimes clashed. The main cause of conflict was the CBS policy stipulating that reporters and commentators analyze the news without offering their own opinions. His staff maintained that pure objectivity was difficult to achieve and did not serve the audience.

After the war, Murrow returned to New York City and was promoted to network vice president for news and public affairs. White resigned soon afterward in 1946.

Later Years

From 1939 to 1946, White also taught at his alma mater, Columbia University. His textbook on the practices and techniques of broadcast journalism was published in 1947 and

Paul White
Courtesy CBS Photo Archive

became a standard for many years. He also taught at the University of Iowa for a short time.

The Associated Press in 1947 hired White as a consultant to improve the radio wire service. In 1948 he became associate editor of the *San Diego Journal.* When it went out of business in 1951, White became executive news director of KFMB AM and TV in San Diego, where his main duty was delivering editorials. (He explained that he had changed his mind and now thought under some circumstances, opinions *should* be allowed on the air.) He also helped the American Broadcasting Companies (ABC) with its coverage of the 1952 Republican and Democratic national conventions. White died in 1955.

BARBARA MOORE

See also Columbia Broadcasting System; Murrow, Edward R.; News; Press-Radio War; World War II and U.S. Radio

Paul White. Born in Pittsburg, Kansas, 9 June 1902. Son of Paul W. White and Anna Pickard. Reporter for the *Pittsburg Sunlight* (Kansas), 1918, and for the *Salina Journal* (Kansas), 1919; telegraph editor, *Kansas City Journal,* 1920. Attended Columbia University, New York City, B.Litt., 1923, M.S., 1924. Reporter for United Press, 1924–30; news editor, vice president, and general manager of the Columbia News Service, and director of public affairs for CBS, 1930–46; journalism faculty member, Columbia University, 1939–46; author of the textbook *News on the Air,* 1947; associate editor, *San Diego Journal,* 1948–51; executive news editor, KFMB AM-TV, San Diego, California, 1951–55. Died in San Diego, 9 July 1955.

Publication
News on the Air, 1947

Further Reading
Bliss, Edward, Jr., *Now the News: The Story of Broadcast Journalism,* New York: Columbia University Press, 1991

Cloud, Stanley, and Lynne Olson, *The Murrow Boys: Pioneers on the Front Lines of Broadcast Journalism,* Boston: Houghton Mifflin, 1996

White, Wallace H. 1887–1952

U.S. Legislator of Radio

Wallace H. White Jr. was a member of the U.S. House of Representatives from 1917 to 1931 and the Republican senator from Maine from 1931 to 1949. He was the radio authority in Congress—no one knew more about radio legislation than White. He was a major influence in drafting the Radio Act of 1927 and the Communications Act of 1934. As the radio act wound its way through Congress, it was known as the White Radio Bill. After the passage of the 1927 Radio Act, it was White who, without the flair of oratory, acquired the support of Secretary of Commerce Herbert Hoover and the signature of President Coolidge. He had at first proposed control of radio within the Department of Commerce, but by 1928 he was an outspoken proponent of the Federal Radio Commission and an ardent supporter of the "public interest" provisions of the law.

It was White's work in the Merchant Marines and Fisheries Committee and the Committee on Commerce that placed him in early association with Secretary of Commerce Hoover and the challenge of radio legislation. White's first attempt at radio legislation was in 1919. The bill, H.R. 10831, directed control of radio to the president of the United States and the secretary of the navy. Authored by White, it was referred directly to the Merchant Marine and Fisheries Committee. The bill and its predecessor, H.R. 11779, introduced 15 January 1920, were written "only to fill the gap until a general [radio] bill [could] be gotten into shape." H.R. 10831 was the first to designate control of radio with the secretary of commerce (White Papers). The exigency of World War I and the rapidly growing technical complexities of radio deflected the interest of legislation, and control at the time was turned over to the navy for the duration of the war.

Following the war, when the navy relinquished control of radio, White again began work at drafting legislation for the new industry. After Secretary of Commerce Herbert Hoover called the First Radio Conference, 27 February 1922, Representative White proceeded to draft proposed legislation from the recommendations of the conferees. The first radio bills fell on an uninterested House of Representatives, but White hammered away at passage until 1927. During each of the Radio Conferences called by Hoover, White provided leadership and created legislative action. By the late 1920s, he was known in the House for his "perennial radio bills" (Archer,

Wallace H. White
Courtesy AP/Wide World Photos

1938). During the Third Radio Conference, White participated in the Coordinating Committee, which Hoover chaired, and White chaired the Committee on Problems with Marine Communication. During the Fourth Radio Conference, he chaired the Committee on Copyright Relations to Broadcasting. White's perennial bills fell upon deaf ears, but each succeeding bill grew with the refinements and complexities of the technology.

Most historians associate the White Radio Bills with placing licensing control in the hands of the secretary of commerce. However, it was White's work that specified the concept of public service as the foundation of radio legislation. He was an outspoken advocate of the public's rights in electronic media. In a speech before the National Association of Broadcasters on 26 October 1931, White extolled the public-interest virtues of the law. It was the public-interest provision that, according to White, "gave it [the law] the virtue of flexibility" for a growing and dynamic new industry (White Papers).

White was born in Lewiston, Maine, on 6 August 1877. He graduated from Bowdoin College in Brunswick, Maine, in 1899 and studied law at Columbia University. He was admitted to the bar in the District of Columbia in 1902 and began his career in Lewiston in 1903 at his father's firm, White and Carter. He was first elected to the House of Representatives in 1916, as Maine's second district congressman. His first service was in the extra session called by President Woodrow Wilson in April 1917. White was one of the few men in Congress who served through World War I. Many of his colleagues, including his radio colleague, Washington's Representative Clarence C. Dill, had their careers cut short because of their votes against U.S. entry into World War I.

The State of Maine's coastal interests in fishing and marine industries dictated White's initial legislative agendas. As the grandson of the former Senator William P. Fry of Maine, White had served for a time as his grandfather's personal secretary while Fry was chair of the Committee on the Merchant Marine and Fisheries and the Committee on Commerce. This experience turned out to be an important foundation, as White later occupied his grandfather's positions.

White's career was not marked with spectacular notoriety, but he was nevertheless an effective legislator. He was not a press hound or even a legislative debater. In his legislative exchanges with his colleagues, his voice could barely be heard in the audience galleries. *Newsweek* described him as "short, slight and sad-eyed; amiable, but reticent" ("GOP: Top Ten," 13 January 1947). White was earnest, knowledgeable, and persuasive. *U.S. News* reported that he was "gentle and uncontentious, soft spoken and retiring. Mr. White was more valuable . . . as a quiet backstage negotiator rather than as a debater" ("Republicans Who Take Over Leadership . . . ," 3 January 1947).

White's achievements were found in the services he rendered. Primarily known for his work on radio, White nonetheless devoted the bulk of his work to marine and fishery issues. He was the author of 550 bills, resolutions, and amendments; 216 congressional reports; and 385 addresses. He was a trusted confidant of both Presidents Coolidge and Hoover, serving under both Republican presidents. According to the *Marine Journal,* his crowning achievement was the passage of the Jones-White Shipping Bill, which was passed into law as the Merchant Marine Act of 1928. During the last years of his tenure, he was described by *Newsweek* as one of the top ten senators in the nation. White served for 30 years—an outspoken proponent of people's rights to enjoy radio as means of communication and of his state's interest in maritime law. He was a practical politician with a humanistic approach to the issue of public concern.

DONALD G. GODFREY

See also Communications Act of 1934; Dill, Clarence C.; Federal Radio Commission; Hoover, Herbert; Wireless Acts of 1910 and 1912/Radio Acts of 1912 and 1927; United States Congress and Radio

Wallace Humphrey White. Born in Lewiston, Maine, 6 August 1877. Attended public schools in Lewiston, graduated from Bowdoin College, Brunswick, Maine, 1899. Assisted his grandfather, William Pierce Frye, chair of the Committee on Commerce, 1899–1903; admitted to the bar in 1902 and began his legal practice. Served in the U.S. House of Representatives, 1917–31 and the Senate, 1931–49; work included appointments as chairman to the Department of Justice's Committee on Expenditures, Committee on Woman Suffrage, Committee on Merchant Marine and Fisheries, and Committee on Interstate and Foreign Commerce. Minority leader in the Senate, 1944–47, and majority leader, 1947–49. Retired from politics in 1949. Died in Auburn, Maine, 31 March 1952.

Further Reading

Archer, Gleason Leonard, *History of Radio, to 1926,* New York: American Historical Society, 1938; reprint, New York: Arno Press, 1971

Bensman, Marvin R., *The Beginning of Broadcast Regulation in the Twentieth Century,* Jefferson, North Carolina: McFarland, 2000

Godfrey, Donald G., "The 1927 Radio Act: People and Politics," *Journalism History* 4, no. 3 (Autumn 1977)

Godfrey, Donald G., and Val Limburg, "The Rogue Elephant of Radio Legislation: Senator William E. Borah," *Journalism Quarterly* 67, no. 1 (Spring 1990)

Godfrey, Donald G., and Louise M. Benjamin, "Radio Legislation's Quiet Backstage Negotiator: Wallace H. White, Jr.," Journal of Radio Studies 10, no. 1 (Summer 2003)

"GOP: Top Ten," *Newsweek* (13 January 1947)

McChesney, Robert Waterman, *Telecommunications, Mass Media, and Democracy: The Battle for the Control of U.S. Broadcasting, 1928–1935,* New York: Oxford University Press, 1993

Paglin, Max D., editor, *A Legislative History of the Communications Act of 1934,* New York: Oxford University Press, 1989

"Republicans Who Take Over Leadership in the New Congress," *U.S. News* (3 January 1947)

Tucker, Ray Thomas, and Frederick R. Barkley, *Sons of the Wild Jackass,* Boston: Page, 1932

"Wallace H. White: Congressman from Maine," *Marine Journal* (2 January 1930)

Williams, Bruce 1932–

U.S. Radio Talk Show Host

Bruce Williams began dispensing financial advice on the air in 1975, when he first hosted *At Your Service* on WCTC, a small New Brunswick, New Jersey, station. At the beginning of the 21st century, he was broadcasting for three hours every weeknight on about 400 stations nationwide, offering the same homespun philosophies and common-sense wisdom to a radio audience of about 8 million.

Origins

Born in 1932, Williams sees himself first and foremost as a "regular guy" who has been an entrepreneur since his junior high school days in East Orange, New Jersey. At the age of 11 he took advantage of the toy shortage during World War II by melting down lead pipes in the coal furnace of his basement, casting toy soldiers, and selling them to friends, thus establishing the first of his many enterprises. Over the years his ventures have included floral concessions in hospitals, barbershops, insurance agencies, preschools, nightclubs, and radio stations in several cities.

Before entering the world of broadcasting, Williams served in the U.S. Air Force during the Korean War and attended Newark State College—now Kean College—in New Jersey. In the 1960s he was a city councilman in Franklin Township, New Jersey, where he later served as deputy mayor and then mayor.

Radio Career

After his defeat in a race for the state's General Assembly, by his own account Williams bombarded New Brunswick station WCTC with letters and phone calls. Several years later he moved to WMCA in New York, where he hosted a finance show six days a week. In 1981 the National Broadcasting Company (NBC) network put together a nightly package of advice-oriented talk shows, pairing Bruce Williams with former WMCA colleague Sally Jessy Raphael. The format for *Talk-Net* placed Williams' focus on finance and Raphael's on sex and relationships; *Talk-Net* is now part of the Westwood One Radio network.

What sets Williams apart from other broadcasters is his avuncular approach. With a decidedly folksy style, Williams can be both soothing and blunt in his broadcasts from his suburban home in New Port Richey, Florida, just north of Tampa. During his program—which generally runs from 7:00 to 10:00 P.M. Eastern time—he shares his own life experiences, setting the stage for others to do the same. Williams talks to callers about their lives and their dreams for the future. Although his advice is centered mostly on financially oriented issues of credit, investments, and decision making, he sprinkles in old-fashioned bits of philosophy on tenacity, common sense, and the importance of maintaining good physical and emotional health. He establishes a rapport with his listeners, referring to the women as "honey" or "sweetheart" and ending calls with his trademark, "I wish you well, my friend."

Each year Williams is cited by *TALKERS Magazine: The Bible of Talk Radio and the New Talk Media* as one of the 100 most important radio talk show hosts in America; the magazine describes him as a "solid nighttime advisor for everyday business situations." In 1994 he was honored as the host of the year by the National Association of Radio Talk Show Hosts.

Although Williams' broadcasts are not outwardly political, he is quick to attack certain business practices that he perceives as unfair to consumers. He also encourages his listeners and readers to let their views be known to candidates for public office as well as to those already elected. For example, his website features a link to his strong stance against airline pricing; in it he states that he is opposed to the "tyranny inflicted on the public by airlines with regards to certain ticketing practices."

Williams eschews what he perceives as elitists who dismiss the appeal of programs like his. After hearing the assertion that "nobody" listens to talk radio, he urged his listeners to prove otherwise. Soon, Williams' legions of listeners, the "nobodies," responded with tens of thousands of postcards and letters; in turn he displayed the correspondence prominently at the National Association of Broadcasters annual convention.

In 1995 Bruce Williams told the Freedom Forum First Amendment Center, "We're not in the news business; we're in the entertainment business." As he is quick to point out, the information and wisdom he instills to his listeners night after night comes from firsthand experience and is meant to interest and inform. His audience responds in kind; they know his father had owned a profitable shoe salon in New York City but lost money in the Great Depression, and as a result, his listeners are open to his thoughts on how to protect their financial futures. Many have heard of Williams' successes and failures in a variety of different fields and are phoning to learn what he has to say about which risks to take and which to avoid. Although they may share some of Williams' frustrations with

Bruce Williams
Courtesy Radio Hall of Fame

bureaucracy and politics, they are after practical information, not speeches or news analyses.

Other Activities

In addition to his nightly broadcasts, Williams writes "Smart Money," a column syndicated by United Features that appears three times a week in about 600 papers nationwide. He has written several books, including *In Business for Yourself, HouseSmart, CreditSmart,* and *America Asks Bruce.* He has also produced the tapes *The Road Map to Financial Security* and *The Bruce Williams One-Hour Crash Course in Getting the Job.* Williams is also popular on the public-speaking circuit, presenting "An Evening with Bruce Williams" to diverse groups of audiences.

A man of many passions, Williams has been a small-plane enthusiast for 50 years, even though in the mid-1980s he nearly died when his plane hit a tree while he attempted to abort a landing. He is currently part of Young Eagle Flights, a program that pairs pilots with students ages 8–17 to take them flying. Pilots such as Williams donate their time and use of their aircraft as part of the larger goal of promoting aviation as a career or as recreation.

RUTH BAYARD SMITH

See also Talk Radio; Westwood One

Bruce Williams. Born in 1932. Served in the U.S. Air Force during Korean War. Hosted *At Your Service* on WCTC beginning in 1975; from 1981, hosted talk show syndicated by NBC and then Westwood One.

Selected Publications

In Business for Yourself, 1991
HouseSmart, 2000
CreditSmart
America Asks Bruce

Further Reading

Borton, Petrini, and Conron, LLP's, Bakersfield Business Conference: 1999 Conference: Bruce Williams, Syndicated Radio Talk Show Host, <www.bpcbakbusconf.com/99-williams.htm>

Bruce Williams Official Home Page, <www.brucewilliams.com/test/home.html>

Laufer, Peter, *Inside Talk Radio: America's Voice or Just Hot Air?* Secaucus, New Jersey: Carol, 1995

Williams, Jerry 1923–2003

U.S. Talk Radio Host

In over half a century on the air, more than any other broadcaster Jerry Williams has played a crucial role in the invention and development of nearly every feature of the modern talk radio format. Known nationally as the "Dean of Talk Radio," Williams owes his success to his ability to adapt to the changing technological and political terrain of the medium. Despite a career based almost exclusively in local, rather than national, broadcasts, Williams gained a national reputation by helping to invent talk radio as a forum for political advocacy and dissent, rather than just another form of audience participation programming. Moving from angry populism to friendly chatter to old-fashioned shtick, Williams connected with audiences and earned consistently high ratings in cities all over the nation. Callers to Williams' program trusted him with their stories of economic ruin and the most intimate details of their sex lives; stories like these helped boost his ratings into the stratosphere.

Williams began his career in radio fresh out of military service in World War II, at WCYB in Bristol, Virginia, where the Brooklyn, New York, native hosted a country music program and a variety show entitled *Farm and Fun Time at Noontime* and read the news. From there he bounced around several stations in Pennsylvania including WKAP in Allentown, where he served as program director and as disk jockey on a hit parade program called *A Date with Jerry.* Inspired by the radio comedy of Henry Morgan, Williams helped to invent a style of talk that combined humor, listener interaction, and attention to the issues of the day.

Throughout the postwar years Williams sought new ways to engage listeners in his broadcasts and, in the process, invented some of the first programs of the talk-back radio genre. In 1948, while hosting a noontime program on WKDN, a Philadelphia, Pennsylvania, station, Williams invited listeners to call in and voice their opinions on a variety of local topics. Lacking the technological wherewithal to place callers' voices on a tape delay, Williams, who was also the program director at WKDN, took calls and then repeated the callers' comments to great comic effect. Around this same time, broadcasters in various markets began experimenting with live, impromptu interview programs—in many cases, from restaurants frequented by celebrities. Barry Gray at WXXX and Jack ("I'm at the Copa, where are you?") Eigen pioneered this form. Williams mastered this interactive format as well, hosting his own interview show from a table in a Philadelphia restaurant for two years.

In 1957 Williams, now at WMEX, Boston, Massachusetts, made use of tape delay equipment to field "live" phone calls about Boston politics on the air. His high ratings in Boston—one month he earned a phenomenal 46 share—were so tough on competitor WEEI that its parent company, Columbia Broadcasting System, lured him away from Boston to its Chicago, Illinois, affiliate, WBBM. In 1968 Williams returned to Boston for an eight-year run on WBZ, a clear channel station that could be heard in 38 states and six provinces of Canada. Despite his popularity, Williams was still subject to the peripatetic nature of the business and so his stint in Boston was followed by briefer sojourns at stations in New York, Philadelphia, and Miami, Florida.

With his return to Boston in 1981, Williams became a fixture on that city's 50,000-watt station, WRKO. The power of his program to evoke the spirit of populist discontent, political cynicism, and economic despair was analyzed in Murray Levin's *Talk Radio and the American Dream,* a study of two Boston-area talk radio programs in the post-Watergate, post-Vietnam era of liberal disillusionment and resurgent conservatism. Levin categorized Williams as "liberal" and, compared with the conservative Avi Nelson, whose program Levin also studied, he surely was, but the central themes of Williams' program in this period tapped into a broader populist distrust of government, big business, and the mass media.

In 1986 Williams served as the key rallying point for widespread opposition to a proposed seat-belt law in Massachusetts. Energized by the host's attention to the issue, impassioned callers decried the governmental intrusion into personal liberties. Williams' consistent attention to the matter helped to galvanize opposition to the bill, which took the form of letter-writing and phone-call campaigns to state lawmakers. The issue took on a surprising ferocity and, when the bill was soundly defeated, many gave Williams much of the credit.

In 1989 Williams demonstrated the grassroots power of talk radio on a national level when he led a campaign to halt congressional pay raises. Working with Ralph Nader and fellow talk radio hosts around the country, Williams helped galvanize powerful nationwide opposition to the bill. This incident focused national attention on the political and cultural power of talk radio, particularly on its ability to reach a vital demographic: politically aware citizens who felt alienated from the political process and traditional forms of news media. Williams helped to consolidate talk radio's national influence by founding the National Association of Radio Talk Show Hosts (NARTSH), an organization that conferred a degree of legitimacy and political clout on the upstart medium. This newfound influence was most clearly in evidence during the presidential election of 1992, when the candidates accorded talk radio hosts in general, and NARTSH in particular, a new level of respect.

In addition to his passionate interest in politics, Jerry Williams also became notorious during this era for his semi-annual "Sex Survey." Inspired by *Cosmopolitan* magazine, Williams first tried the survey in 1976 at WWDB in Philadelphia, but it became an enduring institution on Boston's WRKO. Williams set aside a week of his program to conduct a surprisingly frank survey of the sexual habits, pleasures, and disappointments of his female listeners. Williams' remarkable success in soliciting willing female callers, their accounts often graphic, was most likely because of his deadpan delivery and his reputation for avoiding exploitative or nasty forms of talk show humor.

Williams was inducted into the Radio Hall of Fame in Chicago in 1996, the same year that Congress passed legislation weakening most of the ownership limitations on broadcasters. Williams credits this latest round of deregulation with the declining quality of talk radio nationwide, because corporate-owned stations increasingly lack the local touch, and respect for callers and listeners is replaced by a focus on the bottom line. In 1998 the Dean of Talk Radio himself became a victim of corporate consolidation in the industry. Days after Entercom purchased WRKO, Williams was summarily fired. In 2000 Williams tried for a comeback attempt on the Boston airwaves, as host of WMEX interview show to be broadcast from local restaurants, harking back to the days of Jack Eigen, Barry Gray, and the first stirrings of talk radio. Never inclined to retire, he finished his career at Boston's WROL, as host of a one-hour call-in program even as he battled the early stages of Parkinson's disease, kidney disease, and other ailments. He suffered a stroke in March 2003 and died 29 April of that year.

JASON LOVIGLIO

See also Talk Radio

Jerry Williams. Born in 1923. Founder, National Association of Radio Talk Show Hosts (NAARTSH), 1987. Inducted into Museum of Broadcast Communications' Radio Hall of Fame, 1996; Paul Harvey Lifetime Achievement Award, 1996. Died 29 April 2003

Further Reading

Douglas, Susan J., *Listening In: Radio and the American Imagination: From Amos 'n' Andy and Edward R. Murrow to Wolfman Jack and Howard Stern,* New York: Times Books, 1999

Levin, Murray Burton, *Talk Radio and the American Dream,* Lexington, Massachusetts: Lexington Books, 1987

Williams, Nat D. 1907–1983

U.S. Radio Personality

In 1948 Nat D. Williams became the first black personality featured on the historically important station WDIA in Memphis, Tennessee. WDIA would become the nation's first all-black-oriented radio station in 1949.

Williams, also the first black air personality in Memphis and one of the first in the South, was hired by WDIA's owners, John Pepper and Bert Ferguson, in the fall of 1948 after they had decided to feature black-oriented programs. Williams' *Tan Town Jamboree,* which featured jazz and blues music, debuted in the afternoon of 25 October 1948 and met immediate and overwhelming success. The popularity of the new disc jockey among black Memphians convinced Pepper and Ferguson that black-oriented programming could be profitable. They soon plugged Williams into additional time slots, including their keystone morning show, *Tan Town Coffee Club,* and set about building the station's all-black format around him. By 1949, with Nat D. Williams as its primary ambassador to the Memphis market, WDIA had completed the conversion to an all-black-oriented format, featuring an all-black on-air staff and programming tailored exclusively to the black audience. Within a year, WDIA boasted the highest radio ratings in Memphis' black market as well as the city's highest overall ratings.

The wisdom of hiring Williams and then featuring him heavily in the format became increasingly evident as WDIA's popularity and profits grew with zephyr-like speed. His well-established prominence in the Memphis community and his exuberant on-air style virtually guaranteed that black listeners in Memphis would flock to WDIA. For more than 15 years before joining WDIA and continuing throughout his tenure at the station, Williams taught history at Booker T. Washington High School, which produced many of Memphis' black elite, and he wrote a weekly column in the *Memphis World,* a black newspaper. Williams was also well known in Memphis entertainment circles as the originator and host of *Amateur Night on Beale,* a weekly black talent show staged at the Palace Theater. There is no question that by 1948 much of the black community in Memphis was quite familiar with Nat D. Williams; they willingly followed him to WDIA.

WDIA co-owner Bert Ferguson had known Williams since 1937, when the station Ferguson worked for at the time, WHBQ, began airing *Amateur Night on Beale.* WHBQ aired the program only for a short time, but ten years later Pepper still remembered Williams' skills behind a microphone and brought him to WDIA. (Williams' appearance on the WHBQ broadcast of his show in the late 1930s places him among the first black announcers to appear on radio in the South.)

Williams had an on-air ebullience that caught the ear and brought a smile. Williams told author Mark Newman, "When [the engineer] pointed his finger at me I forgot everything I was supposed to say. So I just did what became typical of me. I laid out for dead. I just started laughing 'cause I was laughin' my way away. And the man said, 'the people seem to like that thing' and they told me to make it standard. . . . So ever since then . . . Nat has started his program laughin' and closed his program laughin'" (see Newman, 1988). Williams' style—and his established prominence in the Memphis community—helped him and WDIA maintain a loyal listenership.

In the wake of Williams' hiring and WDIA's conversion to an all-black-oriented format came a host of black disc jockeys and announcers, many of whom had been Williams' students at Booker T. Washington or performers on *Amateur Night on Beale.* Many were recruited by Williams or drawn to WDIA because of the respectability that Williams' name and presence brought to the station. He became the godfather of Memphis radio for blacks, attracting talented employees, recommending new employees for hire, and, of course, shepherding listeners to the station. As WDIA gained national prominence in the 1950s for its pioneering and complete march into black-oriented programming, Nat D. Williams, too, became nationally prominent. His name and face were frequently featured in national stories about black radio and in WDIA's national trade advertisements. Williams became one of the deans of black disc jockeys in America, inspiring many blacks to enter professions in radio and demonstrating that blacks could make important contributions to the radio industry. Nat D. Williams broke ground for the legion of black radio personalities who would enter the profession in the 1950s and 1960s, just as the nation's first black announcer and disc jockey, Jack L. Cooper, had broken ground for the handful of blacks hosting radio shows in the late 1940s.

Nicknamed "The Professor" by many, Williams continued to host musical programs and public-affairs programs on WDIA until 1972, when a stroke forced him to retire. Until the end, his programs continued to post high ratings. A series of strokes would bring on his death in 1983.

MICHAEL STREISSGUTH

See also African Americans in Radio; Black-Oriented Radio; WDIA

Nathaniel D. Williams. Born in Memphis, Tennessee, 19 October 1907. Son of Albert and Hattie Williams. Attended Tennessee Agricultural and Industrial School (now Tennessee State University), Nashville, B.S. in English, 1928, M.S. in secondary education, 1956. Reporter for various newspapers in New York City, 1928–30; teacher, Booker T. Washington High School, Memphis, 1930–72; wrote for various black newspapers in Memphis, 1931–72; disc jockey and announcer, WDIA, Memphis, 1948–72. Died in Memphis, 27 October 1983.

Further Reading

Cantor, Louis, *Wheelin' on Beale: How WDIA Memphis Became the Nation's First All-Black Radio Station and Created the Sound That Changed America,* New York: Pharos Books, 1992

Dates, Jannette Lake, and William Barlow, editors, *Split Image: African Americans in the Mass Media,* Washington, D.C.: Howard University Press, 1990; 2nd edition, 1993

Newman, Mark, *Entrepreneurs of Profit and Pride: From Black-Appeal to Radio Soul,* New York: Praeger, 1988; London: Praeger, 1989

Wilson, Don 1900–1982

U.S. Radio Announcer

Beginning in 1934 and for more than 40 years, Don Wilson was known primarily as the heavy-set announcer with a hearty laugh who acted as a foil to Jack Benny's comedy. Wilson began as Benny's radio announcer and made a smooth transition to being the announcer for Benny's television program. He was strongly identified as the announcer of Benny's radio sponsor, Jell-O (at a time when most programs were sponsored by one advertiser). Like most announcers of that time, he was well educated and a credible actor, and he performed in many radio and television sketches with Benny.

Born in Lincoln, Nebraska, on 1 September 1900, Wilson grew up in Colorado after his parents moved there when he was two years old. He was a star on the University of Colorado football team, graduated in 1923, and began his professional life as a salesman. Shortly thereafter, he began his show business career as a member of a vocal trio that toured the western United States. In 1927 in San Francisco, an advertiser heard the group perform and put them on the air on KFRC. When the contract ended after more than a year, Wilson and another member of the trio went to Los Angeles, where they worked on various radio shows.

Finally, in 1929 Wilson abandoned his singing career and became an announcer on KFI in Los Angeles and soon moved up to chief announcer. He covered the Rose Bowl from 1930 to 1933. He was a very popular sports announcer, and as a result, in 1933 the vice president of the National Broadcasting Company (NBC) invited him to move to New York to cover sports for NBC. At the same time, he was also working on many of the high-rated NBC shows as an announcer. In 1934 he began working on Jack Benny's Sunday night broadcasts; the first program was *The General Tire Revue,* which aired on 20 July 1934.

When Jack Benny offered him the announcing job for the Benny show, Wilson joined regulars such as Dennis Day, Eddie Anderson (Rochester), Phil Harris, Mary Livingston, and Mel Blanc both as a character in skits and as a straight man and announcer. Wilson helped to develop the job that became known as the radio announcer. During the golden years of radio in the 1930s, Wilson established himself as a strong personality, as important as that of the star of the show. Benny recognized this, and when Benny took the show on the road, he made an effort to bring Wilson along.

His large size, 6 feet 2 inches tall and 220 pounds, made him the butt of many of Benny's jokes. The constant joking made it seem that Wilson was actually larger than he was, and Benny once said that he made Wilson "the biggest man in radio." On one program, Wilson began by teasing Benny for not having an overcoat for a New York visit. Benny replied, "Listen, Don, are you selling clothing or Jell-O?" Wilson replied, "Jell-O." Benny's response was, "Stick to that or I'll fatten up [announcer] Graham McNamee for your job!" Wilson's association with Jell-O was so strong that one of the most memorable aspects of the program was Wilson's weekly selling of the "the six delicious flavors" and "a treat without equal." Wilson claimed that one of the most attractive gifts was a "big shimmering mold of Jell-O." For the 1942–43 radio season, Benny's sponsorship was changed by General Foods to Grape-Nuts. Wilson's reaction to the change of sponsors was panic, and on the first program of the new season, he claimed, "I won't do it, I tell ya, I won't do it!"

On another program, Wilson introduced Benny with a poem honoring their location in Palm Springs: "And there, out by the pool / far from strife and toil / is our blue-eyed star / selling suntan oil." Benny reacted by calling Wilson "Henry Wadsworth Fatfellow." Wilson once tried to get back at Benny by saying that Fred Allen was his favorite comedian. Another running gag featured Wilson trying to get the commercial on the air with Benny trying to stop him. Wilson would sometimes have the Sportsman Quartet sneak in the commercial. He would tell Benny there was just one more chorus for the quartet to sing, and then they would sing the commercial.

From 1950 to 1965, Wilson was part of Benny's television show. He continued his role as announcer and appeared in many television skits. In one parody of Art Linkletter's *Kids Say the Darndest Things,* he dressed in a child's sailor suit and was interviewed by Linkletter, along with Benny and Rochester, dressed in similarly outrageous costumes.

From 1937 to 1944 he was voted the most popular announcer by the press and the audience. When *The Jack Benny Show* ended as a weekly program, Wilson spent a year as the announcer for the *Kraft Music Hall* and *The Tommy Riggs and Betty Lou Show,* both on radio. Through 1973, he continued in his role as announcer on Jack Benny television specials.

Wilson's wife, Lois Corbet, appeared on the Jack Benny radio show, and when they moved to Palm Springs in 1967, both Wilson and his wife worked in radio and television. He hosted *Town Talk* until 1975, and in 1975 they both appeared on a television show that lasted six months, *The Don and Lois Wilson Show.* He died 25 April 1982.

MARY E. BEADLE

See also Benny, Jack

Donald Harlow Wilson. Born in Lincoln, Nebraska, 1 September 1900. Graduated from University of Colorado, 1923. After a short stint as a salesman, toured the West as a member of a singing trio; abandoned singing career and became an announcer on KFI, Los Angeles, 1929; joined NBC as sports announcer, 1933; served as announcer and cast member for the Jack Benny radio and television programs, 1934–65; announced for radio programs *The Kraft Music Hall* and *The Tommy Riggs and Betty Lou Show;* hosted *Town Talk* with wife in Palm Springs, California, 1967–75; hosted *The Don and Lois Wilson Show* in Palm Springs for six months, 1975. Voted by press and listeners most popular radio announcer, 1937–44. Died in Cathedral City, California, 25 April 1982.

Radio Series

1934–55 *The Jack Benny Program*
1967–75 *Town Talk*
1975 *The Don and Lois Wilson Show*

Television Series

The Jack Benny Program, 1950–64 (CBS); *The Jack Benny Program,* 1964–65 (NBC); *The Don and Lois Wilson Show,* 1975

Further Reading

Godfrey, Donald G., and Frederic A. Leigh, editors, *Historical Dictionary of American Radio,* Westport, Connecticut: Greenwood Press, 1998
Maltin, Leonard, *The Great American Broadcast: A Celebration of Radio's Golden Age,* New York: Dutton, 1997
Museum of Television and Radio, *Jack Benny: The Radio and Television Work,* New York: HarperPerennial, 1991

Winchell, Walter 1897–1972

U.S. Radio Commentator

One of the most popular American radio gossip journalists in the 1940s, Walter Winchell reached a huge audience with his newspaper column and radio program mixture of gossip and news tips, and he became one of the first celebrity journalists to build a career partially in the electronic media.

Origins

Winchell was born in Harlem and had a difficult childhood of poverty and parental discord. He sold newspapers and early on practiced to perform in the vaudeville circuit. He left school in the sixth grade. Working with George Jessel and Jack Weiner as the "Imperial Trio," he began to entertain while still a youngster. He worked through a number of vaudeville troupes until 1918, when he served for five months as an admiral's aide in the navy.

In February 1920 *Billboard* published the first of his columns, "Stage Whispers," which combined gossip and one-liners about Broadway and vaudeville. The column was credited not to the unknown Winchell but to "the busybody." He then began contributing to the new *Vaudeville News* a column called "Newssence." At the same time, he and spouse Rita Greene continued on the vaudeville circuit. He soon moved off the stage and into full-time journalism with "On Broadway" for the New York *Evening Graphic* (1924–29) and later for the *Daily Mirror* (1929–63). The column would become syndicated in nearly 1,000 papers by the 1930s and 1940s.

Radio Career

Winchell moved into radio for the first time with several guest appearances on various programs and, in May 1930, a program on the Columbia Broadcasting System (CBS) called *Before Dinner—Walter Winchell,* which provided an audio version of what he was writing in his daily column. Shortly thereafter he moved to the National Broadcasting Company (NBC) with a gossip news program three times a week. He was soon tapped to be a part of NBC Red's *Lucky Strike Hour,* having attracted the attention of advertiser George Washington Hill. By 1932 Winchell was being touted on billboards across the country.

Winchell's chief radio vehicle, *Jergens Journal,* began in December 1932 on NBC Blue with a Sunday evening 15-minute broadcast that was briefly carried on CBS as well. The program began with Hollywood and Broadway gossip but moved into political news over time. The program continued on NBC Blue and later on the American Broadcasting Companies (ABC) networks until 1948. His opening line—"Good evening Mr. and Mrs. North America and all the ships at sea—let's go to press!"—and his tapping telegraph key became nationally recognized. He tapped the key himself (it was mere background noise, as he did not know Morse code) while chattering away at some 200 words per minute. The program was a mixture of rumor and "fact"—and was often wrong. It mixed minor Broadway or Hollywood gossip with more important items. As his fame increased, Winchell basically edited what others often wrote (including many press agents eager for a good mention of their client). He constantly sought—and received—favors from those he covered. Along the way he continued a pattern begun years before with created words and phrases that listeners loved and that often entered the language: "making whoopie" (having fun) and "phfft" (getting divorced) are but two examples.

Throughout this 16-year period sponsored by the lotion company, Winchell was essentially liberal in his views, supporting President Roosevelt, but at the same time cozying up to Federal Bureau of Investigation chief J. Edgar Hoover. He was strongly antifascist in tone. He strode mightily across the New York social scene, holding court at table 50 of the swank Stork Club almost nightly. His was a voice to be reckoned with by Hollywood, Broadway, and even politicians.

After the war this began to change, as he slowly swung to the right and became a critic of President Harry Truman. As complaints rose, Jergens dropped the program. Other sponsors quickly picked up Winchell, who continued on ABC on Sunday evenings until 1955. But the final years of his weekly broadcast took on a darker tone, with constant attacks on the left and support for Senator Joseph McCarthy and others on the far right. His coverage of show business seemed to give way to a sharper political tone, and his once huge audience began to drop off. In 1955 Winchell walked away from a lifetime ABC contract and continued his weekly Sunday program on the Mutual network for two more years. A thinly veiled portrayal of him in the movie *The Sweet Smell of Success* (1957) showed the gossip columnist in the most negative fashion.

Winchell made several attempts in the 1950s to carry his persona to television, but the programs were all short-lived, as his intense approach did not work well on the small screen, seeming dated to younger viewers. In his one television success, he reverted to an audio role as the fast-talking narrator of ABC's *The Untouchables* on ABC from 1959 to 1963. His press column was a shadow of itself by the time the New York *World Journal Tribune* closed in 1967, ending his newspaper outlet in New York after more than four decades.

CHRISTOPHER H. STERLING

See also Commentators; Hill, George Washington; News

Walter Winchell. Born in New York City, 7 April 1897, older son of Jacob Winschel, a salesman and Janette Bakst. (The "s" in Winschel was dropped when he was young; the extra "l" was added on a theater marquee early in his vaudeville career.) Left school at age 13. Vaudeville circuit in New York, 1910–1917. Admiral's aide, 1918. Reviewer, *Vaudeville News*, then *Billboard*, 1922–24. Columnist, New York *Evening Graphic*, 1924–29. Columnist, New York *World*, 1929–63. *Jergen's Journal* on NBC-Blue/ABC, 1932–49. News program on ABC, 1949–55; continued on Mutual 1955–57. Several short television program series, 1950s. Narrator ABC television's *The Untouchables*, 1959–1963. Died in Los Angeles, 20 February 1972.

Further Reading

Fang, Irving E., "Walter Winchell," in *Those Radio Commentators!* by Fang, Ames: Iowa State University Press, 1977

Gabler, Neal, *Winchell: Gossip, Power, and the Culture of Celebrity*, New York: Knopf, 1994; London: Picador, 1995

Herr, Michael, *Walter Winchell: A Novel*, New York: Knopf, and London: Chatto and Windus, 1990

Klurfeld, Herman, *Winchell: His Life and Times*, New York: Praeger, 1976

McKelway, St. Clair, *Gossip: The Life and Times of Walter Winchell*, New York: Viking Press, 1940

Mosedale, John, *The Men Who Invented Broadway: Damon Runyon, Walter Winchell, and Their World*, New York: Marek, 1981

Stuart, Lyle, *The Secret Life of Walter Winchell*, New York: Boar's Head Books, 1953

Thomas, Bob, *Winchell*, Garden City, New York: Doubleday, 1971

Weiner, Edward Horace, *Let's Go to Press: A Biography of Walter Winchell*, New York: Putnam, 1955

Winchell, Walter, *Winchell Exclusive*, Englewood Cliffs, New Jersey: Prentice-Hall, 1975

WINS

New York City Station

WINS, important as an all-news radio pioneer in the nation's largest market, traces its history back to the 1920s.

Origins

WINS-AM emerged from pioneering WGBS-AM, part of the Gimbels Brothers department store empire. In 1922 Gimbels had put WIP-AM in its Philadelphia store, and thus it was logical that in 1924 Gimbels would follow with a New York City broadcasting station from its landmark 33rd Street and 6th Avenue location. On opening night, Eddie Cantor was the master of ceremonies, with guests George Gershwin, Rube Goldberg, and the Vincent Lopez Orchestra. In November 1928 Gimbels reorganized its radio operations as the General Broadcasting System and announced plans for national expansion. But the General Broadcasting System failed, and on 10 October 1931, as the Depression deepened, Gimbels sold out to William Randolph Hearst, which changed the call letters to WINS (the *INS* stood for Hearst's International News Service).

In July 1932 WINS moved its studios to Park Avenue and 58th Street. Through the 1930s Hearst tried unsuccessfully to use its newspapers, the *New York Journal-American* and the *New York Daily Mirror,* to make WINS a success by carrying feature stories on station programs and stars, including full listings of the daily schedule. WINS moved to 1010 on the AM dial and to new news studios at 28 West 44th Street in the heart of Times Square. But when these changes brought no higher ratings, in 1945 Hearst sold WINS to Crosley Broadcasting for a reported $2 million, at the time a record amount paid for a single radio station.

Postwar Changes

The deal was consummated in July 1946, and WINS began to carry programming from WLW-AM Cincinnati, including news broadcasts by Gilbert Kingsbury, the Cincinnati Symphony Orchestra, and *Top o' the Morning.* Crosley also tried *Going to Town, Morning Matinee,* and the *Three Corner Club.* New York City was not the Midwest, however, and the station's only ratings winner was New York Yankees baseball.

Crosley owned WINS for seven years before selling the station to a consortium headed by J. Elroy McCaw (a radio station owner on the West Coast), Charles F. Skouras (a movie-theater exhibitor), and Jack Keating (a Honolulu and Portland, Oregon, radio station owner) in 1953. Operating as Gotham Broadcasting, WINS was reformatted as a disc jockey station centered on the talents of Mel Allen, Johnny Clark, Jack Eigen, and Jack Lacy. A pivotal moment came when WINS's union contract with musicians expired in the summer of 1954, and McCaw said that WINS would no longer air any live music. The American Federation of Musicians protested, but to no avail.

Disc jockeys Murray "the K" Kaufman, Paul Sherman "the Clown Prince of Rock and Roll," Sam Z. Burns, and Herb Sheldin boosted WINS to the top tier of New York radio. Murray Kaufman became so popular that when the Atlantic record label issued Bobby Darin's "Splish, Splash," composer Darin assigned half the publishing rights to Murray Kaufman's mother so that Murray would plug the song on WINS. This transformation had formally begun in the fall of 1954 when McCaw hired Alan Freed from Cleveland. However, Freed proved so controversial that, despite his popularity, he was fired in May 1958. As the "Fifth Beatle," Murray the K would prove a far more lasting figure.

All News

In July 1962 McCaw and his Gotham group cashed in and sold WINS to Westinghouse Broadcasting Company for a reported $10 million. Westinghouse at first continued to seek a musical format to top market leader WABC-AM, but it never succeeded. So on Monday, 19 April 1965, WINS went to an all-news format, becoming one of the first stations to make what was then considered a radical format transition. Importantly, WINS pioneered the all-news format in the largest media market in the United States. The station donated its massive music library to Fordham University, and a radio era was over.

WINS-AM became the station where New Yorkers tuned to learn about breaking news. In November 1965, when a major blackout darkened northeastern cities, WINS kept millions informed during the crisis as they listened on battery-powered portable radios. That same year WINS-AM became New York City's first all-computerized news operation, but advanced automation sometimes caused problems. For example, in December 1973 the station falsely reported that New York's Governor Nelson Rockefeller had been stabbed during a visit to Atlanta. Better for the station's image was the March 1974 event when Joseph Yacovelli, wanted for nearly two years by New York City police in connection with the shooting death of underworld leader Joseph Gallo, turned himself in at WINS studios. Yacovelli's attorney said that the purpose of broadcasting the surrender was to protect against police "manufacture" of evidence against his client. Credit for arranging the public surrender went to WINS newsman Paul Sherman, longtime friend of Yacovelli's attorney.

By 2000 WINS was still part of Westinghouse, which was in turn part of media conglomerate Viacom. WINS-AM was simply one profitable AM radio station within a vast group of radio stations, allied with Viacom's other interest in television and film. To New Yorkers, WINS-AM was still "all news, all the time."

DOUGLAS GOMERY AND CHUCK HOWELL

See also All-News Format; Murray the K; Westinghouse

Further Reading

Jackson, John A., *Big Beat Heat: Alan Freed and the Early Years of Rock and Roll,* New York: Schirmer Books, 1991

Jaker, Bill, Frank Sulek, and Peter Kanze, *The Airwaves of New York: Illustrated Histories of 156 AM Stations in the Metropolitan Area, 1921–1996,* Jefferson, North Carolina: McFarland, 1998

Rhodes, B. Eric, *Blast from the Past: A Pictorial History of Radio's First 75 Years,* West Palm Beach, Florida: Streamline, 1996

Ward, Ed, Geoffrey Stokes, and Ken Tucker, *Rock of Ages,* New York: Rolling Stone Press, 1986; London: Penguin, 1987

WINS website, <www.1010wins.com>

Wireless Acts of 1910 and 1912; Radio Acts of 1912 and 1927

Pioneering U.S. Legislation

Four acts of Congress concerning radio preceded the definitive Communications Act of 1934. The first two focused on maritime wireless telegraphy and are mostly important as initial precedents. The Radio Act of 1912 was to stand for 15 years and was thus in force for the first seven years of regular broadcasting. Its many defects led to the more complex Radio Act of 1927, many provisions of which remain in force three-quarters of a century later.

Origins

The history of congressional action in the field of communications reaches well back into the 19th century with the Post Roads Act of 1866, in which Congress sought "to aid in the construction of telegraph lines and to secure to the government the use of the same for postal, military, and other purposes."

As experimental wireless activity increased after 1903, attempts were made to regulate the "wireless telegraph" industry. Between 1902 and 1912, some 28 bills were introduced in the U.S. Congress to deal with the problem of interference. In 1903 Germany called the First International Convention on radio, and the U.S. Navy made its first attempt to regulate wireless transmission. On 12 July 1904, President Theodore Roosevelt formed an inter-departmental radio advisory board consisting of the departments of commerce and labor, navy, war, and agriculture. Its recommendations constitute the first well-defined radio policy of the U.S. government. One of its recommendations was the necessity for legislation to prevent the control of radio telegraphy by monopolies or trusts by placing supervision in the Department of Commerce and Labor.

In 1906 the German government called the Second International Radio-Telegraph Conference, which was attended by 27 nations. The U.S. Senate's failure to ratify the resulting Berlin Convention caused all government departments concerned with radio to intensify their efforts to obtain legislation for federal supervision of radio usage. The navy department led these efforts, because the commercial wireless companies and amateur interests were opposed to any legislation that would affect their interests.

The marine disaster on 23 January 1909—when the liner *Republic,* with 440 passengers, collided with the Italian *SS Florida,* crowded with 830 immigrants, virtually all of whom were saved thanks to wireless distress messages—focused the public's attention on the safety applications of wireless. Within days there was considerable editorial comment on the role wireless had played in limiting the loss of life, creating such a favorable impression that radio, like life preservers, came to be considered a necessity by individual sea voyagers.

Wireless Ship Acts of 1910 and 1912

On 8 February 1909, President Theodore Roosevelt sent a special message to Congress recommending the immediate passage of legislation requiring, within reasonable limits, oceangoing vessels to be fitted with efficient radio equipment. The Wireless Ship Act of 1910 contained in just one page nearly all that was called for in the 1906 Berlin protocol. The first step to carrying out the provisions of the law was the creation of a radio inspection service.

On the night of 14 April 1912, two months before the Third Wireless Conference was to be held in London, the liner *Titanic,* on its maiden voyage, struck an iceberg 800 miles off the coast of Nova Scotia. The *Titanic* disaster, in which some 1,500 people lost their lives, is often cited as the reason for amending the Wireless Ship Act. However, the subcommittee of the Senate commerce committee had completed its work and the bill had been reported out prior to the *Titanic* disaster. It had become apparent that the United States would have to ratify the 1906 Berlin Convention in order to be invited to a forthcoming London Conference. The new bill became the Wireless Act of 1912. The *Titanic* disaster, however, had awakened congressional concerns for such legislation and ensured its final enactment.

Radio Act of 1912

The Radio Act of 13 August 1912 was a totally new piece of legislation to provide for the licensing of terrestrial (not maritime) radio operators and transmitting stations. For nearly 15 years, radio operators and radio stations were licensed under this law, which was in effect until 1927.

The law was designed to serve two purposes: (1) to promote safety of life and property at sea and to promote commerce by facilitating the dispatch of ships; and (2) to secure the fullest use of radio communication by means of federal regulation, which was made necessary by the fact that in the state of the art at that time, unregulated use and resulting interference would impair or prevent almost all use. To give effect to these two purposes, both based on the Commerce Clause of the Constitution (Article I, Section 8), Congress provided for licensing and entrusted the administration of the system to the Department of Commerce and Labor's Bureau of Navigation, which previously had inspected ships leaving the United States' harbors for proper wireless apparatus.

The term *radio communication* instead of *radio telegraphy* was used throughout the bill so that its provisions would cover the possibility of the commercial development of radio telephony (or the use of radio waves to carry voice and other noncode signals). One feature of the Radio Act of 1912 that had far-reaching consequences was the fact that 19 specific regulations were embodied in the law, and thus no discretion to make further regulations was allowed to the Secretary of Commerce and Labor. It was the judgment of some members of Congress that doing so would be a surrender by Congress of its powers and would, to all intents and purposes, bestow discretionary legislative power on administrative officers.

Secretary of Commerce and Labor Charles Nagel soon attempted to deny a license to a station that was a subsidiary of certain German interests. Germany did not allow American-owned or -controlled stations to operate in that country, and the secretary wished to apply pressure until a reciprocal arrangement could be made with Germany that would allow U.S. capital the right of investing in and controlling corporations organized under German laws. The secretary of commerce asked for an opinion from the U.S. attorney general concerning his licensing power, and on 22 November 1912, he was advised that Congress had not intended to repose any discretion in the secretary. Although the 1912 opinion clearly restricted the secretary of commerce, he used as a lever a clause that directed him to license for the "least possible interference." However, in reality no one had been given the authority to meet the new problems that were to arise with the rapid development of radio telephony. The fact that a "normal" wavelength might be written on the face of the license did not give the Department of Commerce (as it was after 1913) actual power of wavelength assignment.

Nevertheless, the assumption of controls was a step forward, and no serious problems arose for some eight years, until the era of regular broadcasting began in 1920. The development of radio broadcasting was delayed first by the threat of government ownership and then by renewal of the patent wars of the radio manufacturing industry. Bills on behalf of the Navy Department, which desired to take over wireless, were presented in January 1917 and again late in 1918. On both occasions, Congress, reluctant to establish outright governmental control, tabled the proposals. The cross-licensing agreements of General Electric, Western Electric, and Westinghouse in 1920–21, involving some 1,200 radio patents, ended the long patent war in radio.

Early Broadcast Regulation

After 1920, the biggest problems the Department of Commerce faced were the licensing of broadcast stations and the control of interference. Twenty bills were placed before the 67th Congress (1921–23); 13 proposed laws were submitted to the 68th Congress (1923–25); and 18 bills were introduced to the 69th Congress (1925–27)—all to regulate radio communication. Of these 51 bills, only one was to pass both houses of Congress—the Radio Act of 1927. Of importance is what took place while these various bills were being debated.

The first step to controlling interference was the closing down of amateur radio transmission. Another reason for stopping amateur work was the fact that many amateurs were attempting "broadcasting." To prevent this, the Department of Commerce began stipulating on all licenses the material that particular classes of licenses could transmit, restricting music, weather, market reports, speeches, news, and so forth to the "limited commercial" or broadcasting stations.

Radio conferences were held to obtain industry support. At the First National Radio Telephony Conference, held at Secretary of Commerce Herbert Hoover's invitation in Washington in 1922, priorities were assigned to stations according to the services they rendered, with toll stations (i.e., those that sold their airtime) being last on the list. The concepts that the wavelengths being used were public property, that broadcasting should be performed by private enterprise, that there should be no monopoly, and that there must be regulation by the government were presented at these meetings.

The numbers of stations continued to increase, and time sharing became increasingly difficult. The Department of Commerce made available another wavelength for a special class of stations, called Class B. These stations were to be the higher-grade stations in terms of both equipment and programming. Congestion still grew on all three available frequencies. Time sharing of Class B stations began to be required before 1923. The new station classification system with the extra frequency did not solve the problems of interference or time sharing.

The Department of Commerce then decided that a complete band of frequencies was necessary, with specific wavelengths assigned to cities for use by broadcasting stations in those localities. Upon consultation with the navy, which agreed to relinquish its control of the 600- to 1600-meter wavelengths to obtain new equipment, the band of 500 to 1500 kilohertz was made available exclusively for broadcasting on 15 May 1923.

The Second National Radio Conference in Washington then formally recognized and supported the Department of Commerce's classification system. However, with the increasing number of stations, all of which had great difficulty maintaining a constant frequency with accuracy, the interference and time-sharing problems continued.

A third conference was called by Hoover in an effort to deal with interference. Power increases continued, and experimentation was conducted to reduce the kilocycle separation between adjacent stations. The very success and popularity of broadcasting gave rise to its principal difficulty, which came to a head in 1925. The frequencies of broadcasting stations, which were in 10-kilohertz bands, provided 89 channels or frequencies. Since there were about 578 stations operating in 1925, not every station could have an exclusive frequency, and most of the stations had to share time with one or more stations. This duplicate assignment of frequencies required that stations alternate in the use of the frequencies, for instance, by transmitting on alternate evenings. This was generally recognized as undesirable. In spite of this, the building of stations and the applications for broadcasting licenses increased. Because all of the channels for broadcasting were already completely filled, the Department of Commerce could see no way of complying with more applications.

The discussions at the Fourth (and final) National Radio Conference in 1925 clearly brought out the fact that broadcasting would be harmed unless a severe check were put on the numbers of stations being authorized. The Department of Commerce in 1926 refused to license any further radio stations, leading to increased pressure from those individuals and corporations who wished to enter this growing field. Scarcity inevitably brought about an increase in the practice of renting the airtime of broadcasting stations to parties who wished to reach the public.

The Zenith Radio Corporation, which had obtained a license in 1925 by promising to restrict their schedule to two hours of operation a week, pressed for either increases in broadcast time or another frequency. Without permission, Eugene F. McDonald placed the company's station, WJAZ, in operation on a Canadian frequency. The Department of Commerce initiated court action, and on 16 April 1926, the court found that the Department of Commerce had no right to make regulations other than those prescribed in the Act of 1912 and could not, therefore, limit a license as to frequency.

Secretary of Commerce Herbert Hoover, instead of appealing the court decision, forced the issue by requesting an opinion on his powers from the attorney general of the United States. This opinion supported the WJAZ–Zenith Company decision and restricted the Department of Commerce's powers to the issuance of licenses to any and all applicants. Because of this decision, at least 100 stations changed frequencies, and over 200 were issued licenses when they applied for them. This created chaos. The main difficulty in achieving legislation was resolving the question of which agency should control radio regulation. Secretary Hoover compromised his stand of keeping the control within the Department of Commerce and agreed to leave some control in the radio service and regulatory and licensing control in an independent commission.

Radio Act of 1927

The Department of Commerce's involvement with broadcast regulation diminished but did not cease. The Act of 1927 was an experiment in the field of administrative legislation, as it combined a semi-independent agency with the Department of Commerce's newly formed Radio Division.

First, the Act created a Federal Radio Commission of five members, appointed by the president with the advice and consent of the Senate. The commission was given broad administrative and quasi-judicial powers to classify radio stations, prescribe the nature of their service, assign frequencies and wavelengths, determine locations for classes of stations, regulate the apparatus used, prevent interference through regulation, hold hearings, and summon witnesses.

The Radio Division of the Department of Commerce retained the power to accept applications for station licenses, renewals, or changes, but these were to be referred to the commission for definite actions. The secretary of commerce might refer to the commission any matter upon which he desired its judgment. An appeal could be made to the commission from any decision or regulation that the Radio Division, through the secretary of commerce, made.

Second, certain purely administrative powers were left in the hands of the secretary of commerce. He was to receive all applications, although he could not act on them. He was to license and fix the qualifications of station operators and suspend such licenses for cause. He was to inspect, through the Radio Division, transmitting equipment; designate call letters; and conduct investigations designed to uncover violations of the act or the terms of the licenses.

Third, this division of labor was to continue for one year only. The secretary of commerce was then to take over all the powers and duties of the Federal Radio Commission except its power to revoke licenses and its appellate powers, and the commission itself was to become merely an appellate body. However, at the end of one year it was apparent that only the worst cases of radio interference had been eliminated, and for the next two years Congress continued the year-by-year status of the commission.

In December 1929 the Federal Radio Commission was made a permanent agency of government. The Radio Division of the Department of Commerce became the field staff of the Federal Radio Commission on 20 July 1932.

The Department of Commerce radio actions influenced almost all of the provisions of the law that was passed in 1927. Through trial and error, the essential ingredients of the regulatory scheme embodied in the Radio Act of 1927 had been developed and refined. The Radio Act of 1927 codified that (1) the radio waves or channels belong to the public; (2) broadcasting is a unique service; (3) not everyone is eligible to use a channel; (4) radio broadcasting is a form of expression protected by the First Amendment; (5) the government has discretionary regulatory powers; and (6) the government's powers are not absolute. Perhaps most importantly, the 1927 law created the principal under which radio was to be regulated—"the public interest, convenience, or necessity," a phrase not defined in the act. Carried over into the definitive 1934 Communications Act, and still in force, the words have been varyingly defined over the years by the courts, adding to a considerable degree to the Act's flexibility.

MARVIN BENSMAN

See also Communications Act of 1934; Federal Radio Commission; Hoover, Herbert; Public Interest, Convenience, or Necessity

Further Reading

Aitken, Hugh G.J., "Allocating the Spectrum: The Origins of Radio Regulation," *Technology and Culture* 35 (1994)

Benjamin, Louise M., "Working It Out Together: Radio Policy from Hoover to the Radio Act of 1927," *Journal of Broadcasting and Electronic Media* 42, no. 2 (Spring 1998)

Bensman, Marvin R., The Beginning of Broadcast Regulation in the Twentieth Century, Jefferson, North Carolina: McFarland, 2000

Garvey, Daniel E., "Secretary Hoover and the Quest for Broadcast Regulation," *Journalism History* 3, no. 3 (Autumn 1976)

Godfrey, Donald G., and Val E. Limburg, "The Rogue Elephant of Radio Legislation: Senator William E. Borah," *Journalism Quarterly* 67 (1990)

Hazlett, Thomas W., "The Rationality of U.S. Regulation of the Broadcast Spectrum," *Journal of Law and Economics* 33, no. 1, (April 1990)

Holt, Darrell, "The Origin of 'Public Interest' in Broadcasting," *Educational Broadcasting Review* 1 (October 1967)

Streeter, Thomas, "Selling the Air: Property and the Politics of U.S. Commercial Broadcasting, " *Media, Culture, and Society* 16 (1994)

United States Department of Commerce, Bureau of Navigation, *Radio Communication Laws of the United States,* Washington, D.C.: GPO, 1914

Webster, E.M., "The Interdepartmental Radio Advisory Committee," *Proceedings of the I.R.E.* 33, no. 8 (August 1945)

Wire Recording

Early Means of Radio Recording

Wire recording technology, used briefly in the 1940s, was an interim approach to recording of radio programs. It helped mark the transition from electrical transcriptions in the 1930s to the soon-to-be-developed plastic tape recording process introduced in the late 1940s and widespread by the 1950s.

Origins

The phonograph and later the wire recorder were developed as dictation devices for stenographers, not for entertainment value. After the 1877 invention of the phonograph, Oberlin Smith, an American mechanical engineer, suggested in the September 1888 issue of *Electrical World* that a thread or ribbon of magnetizable material could record and play sound electromagnetically.

Building on Smith's suggestion, Danish inventor Valdemar Poulsen built the first magnetic recorder in 1893 using a steel piano wire. Poulsen's "telegraphone," patented in 1898, was designed as a dictation device for office use and as an alternative to the phonograph. A working model was demonstrated at the 1900 Paris Exposition and was apparently received well, winning the *Grand Prix*. It was capable of recording for 30 minutes with the wire traveling seven feet per second. Following the device's success, Poulsen and others searched for financial backing but were not successful.

Early wire recording had inherent technical problems. The first wire recorders utilized acoustical (mechanical) technology as opposed to electronic recording. This frequency response was limited; the dynamic recording range did not exceed about 20 decibels, and there were high noise levels and low acoustical output in comparison with other mechanical systems then in place. In time some of these technical problems were overcome. Wire recorders recorded crosswise in a perpendicular direction as opposed to longitudinal magnetization. The discovery of AC or high frequency bias technique and better recording head design aided the process.

Most wire recording improvements occurred in Europe. In 1930 several movies were completed by Ludwig Blattner in England using a sound track recorded on synchronized steel tape. The British Marconi Company acquired Blattner's company and improved the recorder (called the "Blattnerphone") to produce the "Marconi-Stille" machine used by the British Broadcasting Corporation (BBC). But this machine weighed almost a ton and thus was not portable.

In Europe the "Dailygraph" was a wire recorder used for dictation and telephone recording by the Echophone Company. One interesting feature was the device's cartridge loading capability. In 1933 the C. Lorenz Company sold the "Textophone," an improved version of the Dailygraph that featured wire or steel bands. Both machines were used throughout Europe to provide a central station telephone answering service or for centralized office dictation systems. The machine was eventually used by Nazi party officials in 1933.

About this time various versions of tape recording entered the marketplace. One example was the "magnetophone" exhibited at the 1935 Radio Exposition in Berlin. The medium used was a 6.5-millimeter-wide plastic tape. Another machine, the German "Lorenz Stahltonmachine," used steel tape. Other steel tape machines followed.

Government-operated radio systems were another catalyst for the development of wire recording in Europe, as a method of storing programming and broadcasting programs across various time zones was needed. Military interest in the United States also spurred wire recording. Immediately prior to the U.S. entry into World War II, wire recorders built by Armour Research Foundation were used for research in submarine detection, language classes, and music recording. Wire recorders also were used increasingly by the U.S. military in the war effort. Manufacturers included Peirce Wire Recorder, the Armour Research Foundation, and Minifon.

An early American pioneer in the development of wire recording was Marvin Camras (1916–95). In 1938 he helped develop a consumer wire recorder, "Model 50," from the Armour Research Foundation. The U.S. military used these wire recorders to train pilots. They were also used to record battle sounds and then play them back, amplified, in places where the D-Day invasion would not take place, thus deceiving the German military. Brush Development (a company in Cleveland, Ohio) also designed and built wire recorders for the U.S. military. One model had a magazine that totally enclosed the spools of wire, level winders, heads, and indicators; it also featured bronze wire. (The first use of stainless steel in 1943 was an important development for wire recorders. It was magnetically superior to previously used carbon steel and even to chromium and tungsten alloy magnet steels; it did not rust or corrode.)

Active Use

After World War II, the production of consumer wire recorders increased until it peaked around 1948. Webster-Chicago and Sears, Roebuck and Company began large-scale wire recorder production. Wire recorders were an alternative to high-priced dictation equipment as they featured advanced electronics and were erasable, a feature not feasible with wax or vinyl. A typical wire machine would use a wire gauge from .004 to .0036 inches. At a speed of 24 inches per second, typical recording time would be 15 minutes, 30 minutes, or one hour. Magnecord produced a wire machine with a frequency response from 35 to 15,000 hertz and flutter below 0.1 percent. (School systems had been among the first to buy wire recorders; a typical model would cost about $150, an affordable price for many school systems. However, the wire would easily snarl and the devices were too difficult for young children to use.)

One of the most popular models was the Webster-Chicago (Webcor) Model 80 because of its low cost, portability, and relative reliability. A similar model was built by Crescent Industries of Chicago. A compact automobile recorder was built by WiRecorder Corporation of Detroit. However, it saw only limited production. Utah Radio Products introduced its "Magic Wire" machine in 1945, touting its ability to record up to 66 minutes of talk or music on one spool. Advertisements emphasized the point that the portable recorder-reproducer had been originally developed for the military.

The value of wire recorded audio segments became clear in postwar radio news departments, several of which were using the device by 1946 and 1947. In January 1948 Mutual's Washington bureau reported that the chairs of all congressional committees had agreed to wire recorder use in committee sessions. By this time some reporters were using wire recorders to create "cut-in" recordings that allowed the broadcast of speech or interview highlights within newscasts.

Decline

The rise of a vastly improved competing device—plastic tape for recording—soon spelled the end of wire recording as a mainline technology. The formation of Ampex in 1946 (and the financial support of its research by singer Bing Crosby) helped to focus work on seeking an effective means of recording popular weekly shows to avoid live rebroadcasts for different time zones. Wire recording continued to be used for many years by military personnel, who appreciated the ruggedness of the equipment and had less need for high fidelity recordings.

DAVID SPICELAND

See also Audiotape; Recording and Studio Equipment

Further Reading

Camras, Marvin, *Magnetic Recording Handbook,* New York: Van Nostrand Reinhold, 1988
Lowman, Charles E., *Magnetic Recording,* New York: McGraw-Hill, 1972
"Magnetic Wire Recorder," *Life* (1 November 1943)
Morton, David, *Off the Record: The Technology and Culture of Sound Recording in America,* New Brunswick, New Jersey: Rutgers University Press, 2000

WJR

Detroit, Michigan Station

Perhaps best known in the latter half of the 20th century for its award-winning news, documentaries, and sports coverage, WJR is a 50,000-watt clear channel station with a long and colorful broadcast history.

Origins

When newspaper rival *The Detroit News* put its station WWJ on the air in 1920, *Detroit Free Press* owner and publisher E.D. Stair felt compelled to begin his own station. So WCX, WJR's precursor, began broadcasting on 4 May 1922 from a studio located on the ninth floor of the Free Press Building. Operating at 580 kilohertz, WCX became known as "The Call of the Motor City." One of its popular programs was a variety show called the *Red Apple Club,* named after WCX's first manager, C.D. Tomy, who offered a "nice red apple" to the first person to call in with the name of the next singing guest. Tomy later became "Uncle Neal" to two generations of listeners who grew up with his children's programs on WJR.

On 16 August 1925, the Jewett Radio and Phonographic Company of Pontiac bought into WCX and moved it to the Book-Cadillac Hotel. The station became WCX/WJR (the "JR" stood for Jewett Radio). Power was increased from 517 watts to 5,000 watts, making it the second "super power" station in the country (Cincinnati's WLW was the first). For more than two years the *Free Press*'s WCX broadcast news, sports and *The Red Apple Club* while WJR aired commercial programs on their shared frequency.

On 20 December 1926 WCX's call letters were changed to WJR and it moved from 580 to 680 kilohertz. Jewett had hired Leo J. Fitzpatrick, a popular personality from Kansas, as program director in hopes of increasing sales of his radios, but Jewett went out of business. It soon became clear to most radio manufacturers who owned radio facilities that stations could not survive on radio set sales alone.

In 1926 WJR offered several religious programs on a commercial basis, and Sundays were especially profitable. Now station manager, Fitzpatrick persuaded Father Charles Coughlin of the Shrine of the Little Flower in Royal Oak to experiment in using radio for fund-raising. The program was a success and the controversial priest soon became known over the Columbia Broadcasting System (CBS) network for his vitriolic political views.

George A. Richards (president of Pontiac automobiles of southern Michigan) and the Richards-Oakland Motor Car Company bought the *Detroit Free Press*'s interest in WJR in 1927 and constructed a street-level studio for it in a showroom in the General Motors Building. It was perhaps the only station in the country to operate a ground-floor studio, and entertainers could be seen by passersby. Its new slogan was "The Goodwill Station." In April 1927 WJR/WCX became affiliated with the Blue Network of the National Broadcasting Company (NBC), and in 1928 the WJR orchestra was formed, beginning almost 40 years of music performed live by various WJR staff.

After the Federal Radio Commission revised the entire broadcasting band in 1928, WCX/WJR moved to 750 kilohertz. In December 1928 WJR physically separated from WCX. The station installed studios in the new Fisher Building for a token rental fee and regular on-air mentions of its location. On 17 April 1929 WJR bought all of WCX's equipment and WCX left the air. In July 1929 WJR, the Good Will Station, was formed.

In 1932 WJR increased its power to 10,000 watts and became "In the Golden Tower." Programs during the 1930s included *Detroit Police Drama,* based on actual crimes, and *The Seven-Day Trial of Vivienne Ware,* featuring Judge John Brennan overseeing a trial and then basing his verdict and sentence on listener votes.

WJR switched from the NBC network to CBS in 1935 and constructed a 50,000-watt transmitter. On 29 March 1941 WJR moved again from 750 to 760 kilohertz, where it remains today. In 1942 it began operating 24 hours a day. Future Federal Communications Commissioner (FCC) James H. Quello was hired in 1947 as publicity and promotions director, rising to become vice president and station manager of WJR 22 years later.

In the 1940s WJR created hundreds of special programs devoted to the war effort. In 1944 owner George Richards began *Victory F.O.B.*, a series that featured a businessman supposedly discussing postwar problems. By this time Richards owned two other 50,000-watt stations. It had become clear that he had a strong political agenda and wished to use his stations to further his causes. According to broadcast historian Erik Barnouw, Richards wanted to use *F.O.B.* to influence the 1944 and 1948 elections. He had a history of encouraging both anti-Semitism and comments against President Franklin D. Roosevelt on his stations. A petition in March 1948 to the FCC from the Radio News Club of Southern California accused Richards of instructing his newsmen to slant, distort, and falsify news. It led to a struggle by Richards to keep his station licenses. Richards died in 1951 during the extensive hearings, but the station licenses were renewed by his family.

The Modern Station

By the late 1950s WJR had developed an intensive news schedule, producing eight five-minute daily newscasts as well as five-minute network news summaries throughout the day. A variety of programs aired in the 1950s, including symphony concerts, opera, and sports. Rock and roll was excluded, according to then executive vice president Worth Kramer, because it was "music to steal hubcaps by." Popular host J.P. McCarthy began at WJR in 1956 as a staff announcer and took over *Music Hall,* a morning music show, in 1958. He went on to dominate the Detroit market for many years with his blend of music and talk on WJR. In 1959 the station left the CBS network to commit itself to more local programming.

CBS and WJR joined forces again on 30 December 1962. CBS agreed to let the popular WJR censor any network advertisements and programs. It was the only CBS station in the country that didn't broadcast Arthur Godfrey's show live, as it aired at the same time as WJR's showcase program, *Adventures in Good Music* (hosted by Karl Haas, the station's director of fine arts). One of the most celebrated educational programs during the 1960s was the award-winning *Kaleidoscope*, a blend of recorded music and dramatic narrative on a particular topic, hosted by Mike Whorf. WJR was sold to Capital Cities Broadcasting Corporation on 9 September 1964. Its on-air slogan became "The Great Voice of the Great Lakes," and it changed to a middle-of-the-road talk and variety format.

On 1 January 1976 WJR dropped its CBS affiliation and joined the NBC radio network. In early 1983 WJR "Radio 76" began C-QUAM stereo broadcasting. When Capital Cities Broadcasting merged with the American Broadcasting Companies (ABC) in the spring of 1985, WJR's NBC affiliation was dropped for the ABC Information Network.

In 1990 WJR changed to an adult contemporary music/news/talk format but dropped the music in 1993. On 9 February 1996 Capital Cities/ABC, including WJR, was purchased by the Walt Disney Company, and its licensee name was shortened to ABC. WJR continued into the 21st century as "The Great Voice of the Great Lakes" with its award-winning news and sports coverage.

LYNN SPANGLER

See also Adventures in Good Music; Columbia Broadcasting System; Coughlin, Father Charles; National Broadcasting Company

Further Reading

On the Air: History of Michigan Broadcasting, <http://www.sos.state.mi.us/history/museum/explore/museums/hismus/special/ontheair/index.html>

WJR website, <www.wjr.net>

WLAC

Nashville, Tennessee Station

WLAC is the powerful radio station in Nashville, Tennessee, that played an influential role in the national diffusion of music recorded by African-American artists from the late 1940s through the mid-1960s. The call letters represent the Life and Casualty Insurance Company of Tennessee, which established the station in 1926. In its early years WLAC was known as the "Thrift Station," reflecting the inscription "Thrift—the Cornerstone" chiseled on the Life and Casualty headquarters building in downtown Nashville.

After operating the station for about a decade, in 1935 Life and Casualty sold WLAC to a company executive, J. Truman Ward. During Ward's tenure as licensee, WLAC enjoyed a close relationship with the Columbia Broadcasting System (CBS). Ward's station manager, F.C. Sowell, served as head of the CBS affiliate group, and the network carried a few of WLAC's local programs coast to coast.

WLAC began operation with 1,000 watts of power and increased to 5,000 watts within a few years, but in 1941 Ward won a class 1-B clear channel assignment. For full-time operation with 50,000 watts at 1510 kilohertz, WLAC engineers designed a new transmitter facility about seven miles north of Nashville. The antenna system accentuated night-time skywave radiation, resulting in a reliable signal that carried the CBS prime-time schedule into 28 states and parts of Canada.

Following World War II, WLAC developed a lucrative niche advertising market by promoting the products of the many small, independent record companies of the period that specialized in "race music" or "sepia and swing." Sometime around 1946 or 1947 WLAC announcer Gene Nobles began to generate a large volume of mail by playing African-American artists on a late-night disc jockey program sponsored by a middle-Tennessee record store. WLAC's management soon recognized the revenue potential of selling access to its widespread night-time audience on a per-inquiry (PI) basis. As network programming dwindled, PI programming on WLAC was gradually extended into the earlier evening hours. During the height of WLAC's PI years, the station maintained a large mail room to handle the orders for products advertised nightly such as 45-rpm recordings, pomade, petroleum jelly, and live baby chicks. These orders were delivered daily to the station by the postal service in large canvas bags. WLAC also sold blocks of time at night for paid religious broadcasts.

In addition to Gene Nobles, the night-time announcing staff included Herman Grizzard, John Richbourg ("John R"), Bill Allen ("The Hossman"), Hugh Jarrett ("Huey Baby"), and newscaster Don Whitehead. The music varied according to the products being pitched. On-air descriptions included "the sweet and the beat," "rock and roll," "rhythm and blues," "spirituals," and "gospel." Several African-American artists who became major recording stars after Motown brought rhythm and blues music into the cultural mainstream have attributed their early success to WLAC airplay in general and in particular to Bill Allen and John Richbourg, both of whom maintained close ties to the recording industry.

Truman Ward sold WLAC back to Life and Casualty in 1953 during the early years of the music-and-news era. His son, Jim Ward, later served as the station's general manager during Life and Casualty's second period as licensee. Although well positioned for a television license, Life and Casualty chose to merge interests with two competing local applicants in order to facilitate the construction of Nashville's third television station, WLAC-TV, in 1954. From 1964 to 1967, WLAC-TV aired a local weekend variety program, *Night Train*, which featured many of the African-American artists heard nightly on WLAC "blues" radio.

Life and Casualty also operated a successful separately programmed FM station during the 1960s and 1970s. From studios on the observation deck of the company's Nashville skyscraper—the tallest building in the southeastern United States when built in 1957—WLAC-FM featured an easy listening format that introduced many middle Tennesseans to FM.

It is important to recognize that during the day, when its signal covered a 125-mile radius of Nashville, WLAC sounded like an entirely different radio station from the one listeners in distant places received via night-time skip signals. Daytime listeners in middle Tennessee, southern Kentucky, and northern Alabama heard a full array of mass-appeal adult programming that included CBS news and features, local news and interview programs, editorials by F.C. Sowell ("The South's Foremost Radio Commentator"), helicopter traffic reports, and upbeat middle-of-the-road music.

In 1971 WLAC introduced Nashville's first news-talk format, but the next year the station dropped its CBS affiliation and switched to a Top 40 music format. WLAC enjoyed several years of good audience ratings with Top 40. It was during those years that the station gradually retreated from its legendary night-time schedule. John Richbourg retired in 1973 rather than play Top 40 music. Bill Allen and Gene Nobles continued to host a few late-night PI shows for long-time sponsors, but by 1977 Top 40 music aired until midnight. The PI and paid religious programming that had distinguished WLAC from other clear channel stations was relegated to overnights.

In 1968 the Life and Casualty Insurance Company of Tennessee was acquired by American General, the Houston-based insurance conglomerate. Life and Casualty's 50 percent interest in WLAC-TV was sold in 1975, and in 1977 WLAC AM-FM were sold to the publishers of *Billboard* magazine.

By 1980 historians of popular culture had begun to recognize Gene Nobles as the first person in the United States to play "race music" on a "power" station. WLAC's impact as a principal conduit of rhythm and blues music throughout the eastern half of North America during the 1950s and 1960s has made it one of America's most frequently cited radio stations in historical treatments of the early rock and roll era. In 2003 WLAC was again a CBS affiliate and featured an award-winning all-news-talk format.

ROBERT M. OGLES

See also Black-Oriented Radio; WSM

Further Reading

Cohodas, Nadine, *Spinning Blues into Gold: The Chess Brothers and the Legendary Chess Records*, New York: St. Martin's Press, and Maidenhead, Berkshire: Melia, 2000

Cooper, Daniel, "Boogie-Woogie White Boy: The Redemption of 'Hossman' Allen," *Nashville Scene* (4 March 1993)

Egerton, John, *Nashville: The Faces of Two Centuries, 1780–1980*, Nashville, Tennessee: PlusMedia, 1979

Fong-Torres, Ben, *The Hits Just Keep on Coming: The History of Top 40 Radio*, San Francisco: Miller Freeman Books, 1998

Smith, Wes, *The Pied Pipers of Rock 'n' Roll: Radio Deejays of the 50s and 60s*, Marietta, Georgia: Longstreet Press, 1989

WLS

Chicago, Illinois Station

Chicago's WLS-AM started as WJR, but in April 1924 the Sears-Roebuck company took over the station and renamed it—in one of the most famous of radio's early logos—for the "World's Largest Store." First under Sears, then after 1928 under a new owner—the *Prairie Farmer* magazine—and finally later in the 20th century under the American Broadcasting Companies (ABC), WLS-AM has long helped define radio broadcasting in Chicago and the upper Midwest. Clear channel status, granted in the early 1930s, made WLS-AM a fixture in homes from Minnesota to Ohio, from the Upper Peninsula of Michigan to the cotton fields of Arkansas and Mississippi.

WLS-AM ought to be remembered as the home base for one of radio's most popular programs of its golden age. Although the *Grand Ole Opry* survived longer, the WLS *National Barn Dance* before World War II pulled in a far larger listenership and ranked as America's most popular country music program. After being picked up by the National Broadcasting Company (NBC) network in 1933, the *National*

Barn Dance expanded to all Saturday night once *Prairie Farmer* took charge. Renamed simply *The WLS National Barn Dance,* the show became an NBC Saturday night fixture. Indeed, it was only seven years later, in 1940, that NBC paired the *WLS National Barn Dance* with the *Grand Ole Opry* on the network. The two programs battled evenly during World War II, but the postwar years saw the *Opry* surpass its predecessor in ratings. In time, country format radio, playing recorded music, would replace all barn dance live radio shows—except for the *Opry.*

Still, WLS-AM offered more than the *Barn Dance. Prairie Farmer* sold not only its magazines, but also membership in the WLS–Prairie Farmer Protective Union, which promised farmers that it would keep thieves away from listeners' farms. Resident announcers, such as Hal O'Halloran, Martha Crane, Al Rice, Jack Holden, Margaret McKay, and Bill Cline became household names. Ralph Waldo Emerson, organist for WLS, offered an alternative to string hillbilly music. Bill Vickland, the voice of the *Book Shop,* produced live drama. The Sunday School Singers Trio was heard each week on the *Cross Roads Sunday School.* The WLS Staff Orchestra, under Herman Felber's direction, played Tin Pan Alley standards and the latest big band hits. The Chicago Gospel Tabernacle offered another alternative. And Jim Poole offered the necessary reports from the Chicago Live Stock Exchange.

WLS-AM became part of the ABC radio network in 1960. By then, in response to the ascendance of television, WLS-AM needed to reinvent itself. In May 1960 it joined the Top 40 ranks and helped bring mainstream rock sound to the Chicago area through the 1980s. New management hired and promoted many local disc jockey stars—none hotter or more famous than Larry Lujack, who, based at WLS-AM, became one of the symbols of the Top 40 era.

In the 1990s, WLS-AM became news plus talk radio 890, adapting to the new ownership by the Walt Disney company in 1995. The program lineup by 2000 included Dr. Laura Schlessinger, Rush Limbaugh, and Electronic Town Hall meetings over radio.

DOUGLAS GOMERY

See also Biondi, Dick; Clear Channel Stations; Country Music Format; Farm/Agricultural Radio; National Barn Dance

Further Reading

Baker, John Chester, *Farm Broadcasting: The First Sixty Years,* Ames: Iowa State University Press, 1981

Broadcasting in Chicago, 1921–1989, <www.mcs.net/!richsam/home.html>

Evans, James F., *Prairie Farmer and WLS: The Burridge D. Butler Years,* Urbana: University of Illinois Press, 1969

Hickerson, Jay, *The Ultimate History of Network Radio Programming and Guide to All Circulating Shows,* Hamden, Connecticut: Hickerson, 1992; 3rd edition, as *The New, Revised, Ultimate History of Network Radio Programming and Guide to All Circulating Shows,* 1996

Stand By (12 April 1949) (special 25th anniversary issue of WLS's magazine)

WLS Newstalk 890 AM website, <www.wlsam.com>

WLW

Cincinnati, Ohio Station

"The Nation's Station," founded in Cincinnati in 1922 by Powel Crosley, Jr., was for many years the United States' most powerful radio station, not only in wattage but in the wide geographical range of its audience. From 1935 to 1939, it used 10 times the wattage allowed today. WLW produced and broadcast popular local programming over a large area in Ohio, Indiana, and Kentucky, and also had listeners in Michigan, West Virginia, Illinois, and Tennessee—and in many more states at night. Numerous stars began their careers at WLW (the talent roster includes Fats Waller and Rod Serling), and the station developed programs that were later carried on the national networks. After nearly 80 years WLW remains the most popular radio station in Cincinnati and a large surrounding area, aided by its favorable combination of power and frequency—50,000 watts at 700 MHz.

Origins

Crosley, a manufacturer in Cincinnati, Ohio, became interested in radio in 1921 when his son asked for a "radio toy" as a birthday present. Crosley's interest in radio grew to fascination as he built a set for his son from an instruction booklet and an assortment of loose parts. That same year he started an amateur radio station in his home, licensed as 8XAA, and within about a year the Crosley Manufacturing Corporation was the world's largest manufacturer of radio sets and parts. In March 1922 Crosley's station was assigned (at random, as was the practice) the letters WLW as a call sign for a "land radio station" of 50 watts on 360 meters. Such a station could have had a range of about 100 miles, but it shared that frequency with hundreds of other stations.

Because he was manufacturing small, inexpensive, and therefore less sensitive radio receivers, Crosley had a more compelling interest in higher power than did many other radio broadcasters. The station's power was increased to 500 watts in April 1923, then to 1,000 watts a year later, and the Commerce Department announced it might use five kilowatts on a "strictly experimental" basis.

At first WLW was operated mainly to provide programming for purchasers of Crosley radio sets "as a medium of advertising and publicity." The company's weekly magazine, distributed to radio retailers and purchasers of new sets, asked listeners to fill out a questionnaire to vote for the type of programs they would most like to hear, out of a choice of music, talk, and drama categories.

The station's earliest programs were musical variety shows featuring amateur talent, talks, and soloists. In August 1922 Crosley hired Fred Smith as station director; he was WLW's first employee. Smith, who later conceived of the news digest program that became *The March of Time,* inaugurated a regular daytime schedule of market reports, financial news, weather, and recorded music. For the evening hours he arranged musical variety shows and live music remotes. Smith also wrote original radio dramas, the first of which aired on 22 December 1922. On 3 April 1923, WLW broadcast *When Love Wakens,* an original play that Smith wrote especially for radio—probably the first in radio history—and for one station in particular (note the title's initials).

Fred Smith tried many other formats at the station, too, to provide entertainment and information, such as programs for children, lectures to teach swimming, and re-creating a boxing match based on telephone reports from a station staff member at the arena. He also read news items interspersed with musical selections played by an organist in the studio; Smith later revised this news digest program as the basis for the news drama program *The March of Time.*

The U.S. Commerce Department designated WLW as a class B station in June 1923, making the Cincinnati station one of only about 39 (of 500 total stations) in the United States that could use higher power. When the Federal Radio Commission (FRC) set about "cleaning up the broadcast situation" and announced frequency assignment for 694 stations beginning 1 June 1927, WLW was assigned 700 kilocycles (kHz). Soon it was the only station in the United States using that frequency, and thus became a "clear channel" station. It was one of only ten stations using five kilowatts of power, and it grew more powerful after 25 May 1928, when the FRC authorized WLW to begin construction of facilities for 50 kilowatts. At that time there were probably only four other stations with that output, but WLW's staff kept producing original programming that (in Crosley's view, anyway) warranted further expansion.

By the late 1920s, WLW was affiliated with both the NBC Red and Blue networks, but the station originated more expensive and high-quality local programming than most other stations in the country. Only the network-owned stations in New York, Los Angeles, and Chicago had staffs as large as WLW's. In June 1929, a hookup was arranged connecting WLS Chicago, WOR New York, and WLW. The Quality Radio Group, as the coalition came to be called, carried each other's programs and made available to advertisers a huge audience in most of the northeastern United States. In 1934 WLW got together with WOR New York, WGN Chicago, and WZYZ Detroit (with its prize show *The Lone Ranger*) to establish the Mutual Broadcasting System (MBS).

Superpower Experiment

In June 1932 the FRC authorized WLW to construct a 500-kilowatt (500,000-watt) experimental station and to conduct tests from 1 A.M. to 6 A.M. On 2 May 1934, the station began using this "super power" at all hours after a dedication ceremony in which President Franklin Roosevelt activated the new water-cooled transmitter by pushing a gold key on his desk at the White House. This power increase allowed WLW to transmit with ten times more power than any other AM station—then or now—until 1 March 1939, when the Federal Communications Commission (FCC) refused to renew the "experimental" license for higher power. After a Canadian station in Toronto complained of interference with its signal on 690 kHz, WLW was briefly limited to 50 kilowatts after local sunset. Soon a directional signal was arranged so that no more than 50 kilowatts was transmitted northwest of Buffalo, New York.

The decision to end the 500-kilowatt experiment was controversial and complex for political and technical reasons, but it was primarily the result of complaints by other stations resentful of WLW's enormous economic advantage. A "sense of the Senate" resolution had been passed that directed the FCC to limit stations to 50 kilowatts. The resolution stated that an increase of superpower stations in the United States (at least 15 other clear-channel licensees had applied to the FCC for superpower status) would deprive other local and regional stations of valuable network affiliations and national advertising revenues.

Although its 500-kilowatt experiment was discontinued, WLW had proved that higher power was possible, that it did not cause more than normal interference with adjacent stations, and that it did not blanket out other stations for nearby listeners. Many other AM stations around the world soon began using this much and even more power. However, it was because of its superior programming, not its powerful wattage, that WLW achieved much higher ratings than even the hometown station in many cities in its vast coverage area, which extended in a circle around Cincinnati stretching nearly to Chicago, Detroit, Pittsburgh, and Nashville. (As an indication of WLW's reach, one fundraising appeal during floods on the Ohio River brought donations from 48 states.) In 1936, as part of an FCC study of channel allocations and utilization, a survey returned by more than 32,000 rural listeners (of 100,000 surveys mailed) showed that WLW was by far the favorite station and the listeners' first choice in 13 states and second in six more, from Michigan to Florida and Texas. While its powerful signal made WLW available in homes out of the reach of many other stations at night, its popularity derived from its use of the most popular shows from four networks, a large staff of local talent and specific programming—especially in the early-morning hours aimed at farm audiences—and its geographical position near the center of the agricultural Midwest.

WLW also began a shortwave service and relayed the station's programs around the world until the facility was taken over during World War II for government wartime propaganda broadcasts. In 1942 six new transmitters installed at that facility became the largest installation for Voice of America.

During the 1930s WLW's staff numbered about 350, of whom about 200 worked in the programming division. The station carried programs from both NBC networks (Red and Blue), CBS, and MBS. The station called itself "The Cradle of the Stars," in reference to those who worked there early in their careers, including Virginia Payne (who created the serial heroine Ma Perkins), the Mills Brothers, Andy Williams, writer Rod Serling, Betty and Rosemary Clooney (their younger brother Nick was later a news anchor at WLW's television affiliate), the McQuire Sisters, actor Frank Lovejoy, Red Skelton, Durward Kirby, Eddie Albert, Thomas W. "Fats" Waller, Red Barber, the Ink Spots, Norman Corwin (who quit after only a few weeks when the station refused to broadcast news about labor strikes), and Erik Barnouw. Many performers soon moved on to stations in New York, Chicago, and Los Angeles, and some joked that the call letters WLW stood for World's Lowest Wages.

In the 1930s about 50 percent of all WLW programming was local, 40 percent in evening hours. The station's staff originated programming carried on NBC, especially variety and hillbilly (later called country) music, and many Mutual programs. It also produced original drama—some destined for the networks included *Ma Perkins* (soap opera) and *Mr. District Attorney* (crime). In the late 1930s, station executives added a great deal of agriculture programming, and WLW even started its own experimental farm in an attempt to retain the 500-kilowatt superpower transmitter by providing more unique and rural programming.

The station produced popular early morning hillbilly variety programs and, in cooperation with Ohio State University and the state's department of education, an educational series for in-school listening. WLW was a "regional" service, nearly a network unto itself. The same could be said of other major clear-channel stations such as WGN, WCCO, and WSM, but none matched WLW's quantity of original local programming.

Postwar Change

From 1949 to April 1953, as the station sought to duplicate the huge WLW AM coverage in the new medium of FM (frequency modulation), WLW's programming was broadcast simultaneously on FM stations in Cincinnati, Dayton, and Columbus, Ohio, but the growth of the FM audience was slow. The company concentrated on television. Many WLW radio programs were first simulcast on the television outlet, then later were shown only on TV. In 1955 WLW was the first radio station to provide weather forecasts based on radar (which it shared with the TV station), and in 1958 it became one of the first with helicopter traffic reports. In the 1960s WLW often ranked only third or fourth in Cincinnati ratings (losing out to Top 40 formats), but it had the largest total audience of any station in the city because of its larger coverage area. For example, in 1961 Nielsen reported WLW as having listeners in 184 counties in four states.

As network programming declined in the last half of the 1950s and into the 1960s, WLW became what was often called a Middle of the Road (MOR) station, producing "magazine" programs of news, information, talk, and limited amounts of recorded music during the morning and afternoon hours. Other types of music, including classical all night long, filled most of the rest of the hours. The station's format in the 1970s and 1980s was "adult contemporary." In the 1990s and into the new century, by which time there were many competing stations with a variety of popular music formats (most derived from the earlier Top 40 formula), WLW was usually the station with the highest audience share in the Cincinnati market. The station consistently maintains the largest total audience of any Cincinnati station because it reaches audiences over a wide area in Ohio, Indiana, and Kentucky.

Modern WLW

The programming on WLW is now what is generally described as "full service," being a combination of news, talk, and sports with well-known local personalities—a format that is usually found only on a few stations (mostly 50-kilowatt stations) similar to WLW in the largest markets with very big revenues. More news and information are offered during peak listening times in the mornings and afternoons. Evenings are generally call-in talk, including sports. Typically WLW has slightly higher overall ratings during the summer and spring, when Cincinnati Reds play-by-play baseball is broadcast.

At the turn of the century, WLW was estimated to have revenues of about $21 million, which would place it in the top 20 of all-talk, news/talk and so-called full service radio stations. All the other full service stations earning more money were in markets larger than Cincinnati (ranked 26th in radio), mostly in New York, Los Angeles, and Chicago, the three largest markets. It is regularly rated among the top 20 stations in other markets in a radius of 35 to 135 miles from its transmitter in Mason, Ohio, such as Dayton, Columbus, Lima, Lexington, and Fort Wayne. After several changes of ownership and mergers, WLW is now owned by Clear Channel Communications, which operates seven other facilities in the market, and owns more radio stations than any other single company.

LAWRENCE W. LICHTY

See also Clear Channel Communications; Clear Channel Stations; Crosley, Powel; Middle of the Road Format; Mutual Network

Further Reading

Lichty, Lawrence W., "'The Nation's Station': A History of Radio Station WLW," PhD. diss., The Ohio State University, 1964
Lichty, Lawrence W., "Radio Drama: The Early Years," *National Association of Educational Broadcasters Journal* (July–August 1966)
Perry, Dick, *Not Just a Sound: The Story of WLW,* Englewood Cliffs, New Jersey: Prentice-Hall, 1971
WLW website, <www.700WLW.com>

WMAQ

Chicago, Illinois Station

The origins of WMAQ-AM go back to the beginnings of radio when, in the spring of 1922, the *Chicago Daily News* put WGU on the air as a 500-watt forerunner of what the National Broadcasting Company (NBC) would turn into a mighty 50,000-watt broadcast giant. In 1923 the station acquired formal studios in the LaSalle Hotel; in 1929 the station moved to the new *Daily News* building. Two years later, on 1 November 1931, WMAQ-AM was purchased by the Radio Corporation of America's (RCA) NBC, which moved the studios to the Merchandise Mart and made WMAQ a flagship station.

WMAQ-AM remained a key NBC-owned and -operated station until General Electric (which purchased RCA, NBC's parent, in 1985) sold it to Group W in 1988. (There was also a WMAQ-FM, which was sold in the early 1970s.) The related television operation—also named WMAQ—was also an NBC-owned-and -operated station in Chicago.

NBC's acquisition of WMAQ-AM gave the network primary Chicago outlets for both its Red and Blue network programming, because WENR-AM had been purchased by NBC earlier. WMAQ-AM was a programming pioneer. Freeman Gosden and Charles Correll started *Sam 'n' Henry* for WGN-AM, but they moved to NBC and WMAQ to become *Amos 'n' Andy* in 1929. Another one of WMAQ's biggest achievements was the discovery of a new act by Marian and Jim Jordan, later widely popular on the show *Fibber McGee and Molly,* who made their first appearance on radio in February 1931. Other first-timers were Count Ilya Tolstoy, son of the noted author; Cyrens Van Gordon, opera star; Lorado Taft, the sculptor; Rosa Raisa, the famed soprano; George Arliss; Ben Hecht; Otis Skinner; Ruth Chatterton; and Jane Addams.

Indeed, the station was within a day of being six months old when it presented the first music appreciation program—on 12 October 1922. Mr. and Mrs. Max E. Oberndorfer began a series of broadcasts with an analysis of the opening program of the Chicago Symphony Orchestra that year. On the following day, WMAQ-AM led the radio industry into the field of children's programs with Mrs. Oberndorfer's *Hearing America First* series. This phase of broadcasting was expanded on 16 October of the same year when Georgene Faulkner, the "Story Lady," gave the first of her *Mother Goose* broadcasts. On 28 November 1922, the first educational broadcast was presented by WMAQ when Professor Forest Ray Moulton, head of the astronomy department at the University of Chicago, gave a lecture on "The Evening Sky." It was the first in a series of broadcasts by University of Chicago professors and was the forerunner of the *University of Chicago Round Table.*

WMAQ-AM also led the field in sports and news broadcasting. It presented one of the first daily play-by-play descriptions of major-league baseball on 20 April 1925 and one of the first play-by-play descriptions of a football game on 3 October 1925. For news, WMAQ-AM ran the first transatlantic news broadcast in history, on 4 December 1928, which consisted of a telephone conversation between John Gunther, then *Chicago Daily News* correspondent in London, and Hal O'Flaherty, then foreign news editor of the *News,* regarding the condition of King George V, who was seriously ill. WMAQ-AM was also the only Chicago station to broadcast the first presidential inaugural ever put on the air, that of Calvin Coolidge, on 4 March 1925.

That the best in radio entertainment was continually on WMAQ-AM is obvious after a glance at a list of stars, both past and present, who made their radio debuts over Chicago's oldest station. Wayne King, for instance, made his first broadcast anywhere over WMAQ-AM on 28 January 1928. Fred Waring and his Pennsylvanians made one of their earliest broadcasts on WMAQ in August 1922; Ed Wynn made his initial radio broadcast in October 1922; Vincent Lopez made his radio debut in September 1924; and so on.

Because of the success of WMAQ-TV, WMAQ-AM, as a NBC network mainstay, had to reinvent itself as TV forced radio's redefinition. It was never very successful, and so few were surprised when WMAQ-AM was acquired by Westinghouse and became a news/talk station, still based in the NBC Tower on North Columbus Drive.

DOUGLAS GOMERY

See also Amos 'n' Andy; Clear Channel Stations; National Broadcasting Company

Further Reading

Caton, Chester, "WMAQ: Her Independent Years," Ph.D. diss., Northwestern University, 1950

Linton, Bruce A., "A History of Chicago Radio Station Programming, 1921–1931, with Emphasis on Stations WMAQ and WGN," Ph.D. diss., Northwestern University, 1953

WNBC

New York City Station

Begun as WEAF by AT&T in 1922 and becoming the flagship of the new National Broadcasting Company (NBC) network in September 1926, this network-owned-and-operated station's call letters changed to WNBC in November 1946 as part of a process of "rationalizing" network station call letters in the city. Operating with 50,000 watts on 660 kilohertz, WNBC was both the network flagship and a local market station.

With the decline in network radio in the early 1950s, WNBC took on a middle-of-the-road format like many other big-city stations, offering a variety of programs aimed at different tastes during the day, trying to provide something for everyone. Boston radio humorists Bob and Ray joined the station in 1951 with their parade of characters and radio takeoffs. Public service features such as daily pollen counts and traffic reports became important program elements by mid decade. From 1954 to 1960, WNBC became WRCA to better promote the initials of owner Radio Corporation of America (the NBC call letters went to a network-owned UHF television station in Connecticut, so the network did not lose control of the identity).

WNBC, as with most stations, had for its entire history signed off the air after its late-evening programs concluded. Beginning in 1952 it began to provide 24-hour service with the *Music through the Night* mixture of easy talk and both classics and light classical music. In 1959 daytime programming featured "wall-to-wall" music, a combination of easy listening and pop tunes and talk. For a time the program could be heard in a kind of stereo with one channel broadcast on the network's AM outlet and the other on the co-owned FM station.

But even in New York City, radio was in decline. By 1962 the station's facilities had shrunk from studios on five floors of the RCA building in Rockefeller Center to just two. A longtime transmitter location on Long Island was closed down and NBC radio shared a transmitter with WCBS, also on Long Island. By the 1960s WNBC was trading more on its traditions and history than its current listenership, and ratings were far from market leaders. From 1964 to 1970, WNBC programmed a talk format, including some of New York's first call-in programs.

By 1970 music had once again taken over, although the format concentration on "66 NBC" (emphasizing the station's location on the AM dial) varied. The music was all contemporary, sometimes current hits and sometimes a mixture of new and older top songs. A major change came with the 1972 hiring of a Cleveland disc jockey, Don Imus, for the morning drive time period, and the beginning of shock jocks in the New York market. A decade later Howard Stern moved into the afternoon drive time period. WNBC soon added former TV entertainers Soupy Sales and Joey Reynolds. Although their comments were embarrassing at times, the jocks and their music helped the station climb in audience ratings.

But the ratings turnaround could not save the station. In 1986 parent company RCA was sold to General Electric, which soon decided not to continue radio operations. The NBC network was sold to Westwood One and the New York AM and FM outlets were sold to Emmis Broadcasting. WNBC's last broadcast was heard on 7 October 1988. The next day, its frequency was taken over by WFAN, the city's first all-sports station.

CRAIG ALLEN AND CHRISTOPHER H. STERLING

See also Imus, Don; Shock Jocks; Stern, Howard; Talk Radio; WEAF

Further Reading

Jaker, Bill, Frank Sulek, and Peter Kanze, *The Airwaves of New York: Illustrated Histories of 156 AM Stations in the Metropolitan Area, 1921–1996,* Jefferson, North Carolina: McFarland, 1998

Lieberman, Philip A., *Radio's Morning Show Personalities: Early Hour Broadcasters and Deejays from the 1920s to the 1990s,* Jefferson, North Carolina: McFarland, 1996

The 66 WNBC Tribute Pages, <www.imonthe.net/66wnbc>

Tracy, Kathleen, *Imus: America's Cowboy,* New York: Carroll and Graf, 1999

WNEW

New York City Station

A comparative novelty during network radio's pre–World War II heyday, independently programmed WNEW generated respectable ratings and profit without ties to the likes of the National Broadcasting Company (NBC), the Columbia Broadcasting System (CBS), or Mutual. By the late 1940s, television's widening influence cut deeply into revenues of radio network affiliates that continued to embrace soap operas, sitcoms, and drama. Meanwhile, WNEW's pop music format proved relatively impervious to video competition. Curious broadcast insiders were drawn to WNEW. They wanted to see how the locally programmed AM outlet could succeed with just a handful of disc jockeys, some phonograph records, and a bit of news.

The WNEW story begins in New Jersey, where the station was formed primarily from remnants of comedian Ed Wynn's failed network venture. The performer had tried competing against CBS and NBC with his Amalgamated Broadcasting System, but the effort succumbed to fiscal woes. The Wynn connection is mentioned owing to its legendary nature, but was largely a province of studio facilities acquired from the comic's then-defunct broadcast ownership foray. WNEW more accurately originated from the amalgamation of New Jersey stations WAAM at Jersey City and WODA in Paterson, which timeshared 1250 kilohertz in the metropolitan region. (Another outlet, WHBI of Newark, occupied the 1250 dial position on Sundays and Monday nights.)

Watch manufacturer Arde Bulova, advertising man Milton Biow, and WODA's owner Richard O'Dea were principals in the station consolidation. Because—at least for a time—their broadcast enterprise would qualify as the New York area's newest, WNEW seemed a perfect call sign choice. President Roosevelt was selected to inaugurate the station, and on 13 February 1934, he pushed a button in the White House that was wired to a light in WNEW's Carlstadt, New Jersey, transmitter building. The button signaled a singer to belt out the "Star Spangled Banner," and the facility's 1250-kilohertz channel carried it with 1,000 watts. In 1939, its frequency got reassigned to 1280 kilohertz while the wattage was raised to 5,000.

For two decades, the station's real power came from a metropolitan socialite blessed with an intuitive knack for programming. Bernice Judis had no previous radio industry experience when a friend (wife of co-owner Milton Biow) suggested that she become part of the WNEW staff. Consequently, she thought more like a listener than a detached, by-the-book executive. This, as well as a limited budget, caused Judis to fill much of WNEW's schedule with a pleasant output of pop music records and smooth-voiced announcers skilled at describing the music's performers. Audiences soon equated the station with an endearing kind of companionship that gave them a sound track for their daily routine. Ratings rose, prompting Bulova and Biow to name Judis general manager of what had evolved from Newark, New Jersey–based studios to an exclusively downtown New York operation. Judis felt that a city that never slept needed a station that stayed on all night. In 1936 with an early-morning broadcast dubbed *Milkman's Matinee,* WNEW became the first to do so.

The thought of their records being freely spun, even at 3:00 A.M., irritated some musicians, who would rather perform live and be paid. In a resulting lawsuit, the plaintiff artists were disappointed to learn that because WNEW had purchased the recordings, it could use the transcriptions for any desired purpose. This decision prompted independents in other U.S. communities to proudly play records (a practice "big radio" had considered laughably low class in the 1920s and 1930s). Columnist Walter Winchell so enjoyed the way Judis' announcers deftly rode ad-libs around their discs that he coined their professional title—disc jockey.

In a 1941 frequency swap with co-owned WOV, WNEW was moved from 1280 to 1130 kilohertz, and power was increased to 10,000 watts. This shift also rid WNEW of the time-share arrangement with WHBI. Programs included news, public service, celebrity interviews, and even a series about good grooming. Broadcast schedule diversity included actor James Earl Jones' WNEW debut, circa 1945, in an American Negro Theater radio production. The next year, radio's first two-man morning show hit WNEW's air. This wake-up session garnered a following as loyal as that of the station's cornerstone, *Make Believe Ballroom,* a show on which a recording star might just drop in while disc jockey Martin Block happened to be playing his or her latest release.

A group including general manager Judis purchased WNEW in 1950. The station saw revenues jump due in part to a 1949 quintupling of its transmitter power (to 50,000 watts). The group sold to Dick Buckley and associates in 1954. Subsequently, Buckley's principals merged with remnants of the DuMont Television Network, before selling in 1957 to what became John Kluge's prominent Metromedia.

As radio "flagship" for this growing media conglomerate, WNEW received what was arguably the largest and most skilled news department of any independent outlet. By 1961 New York Giants (football) play-by-play broadcasts brought listeners to the 1130 dial position. And the list of excellent air personalities (William B. Williams, Ted Brown, Gene Klavan, Jim Lowe, Bob Landers, Gene Rayburn, and others) made for a very secure-sounding product. That's why, whenever they were in Manhattan, America's most sophisticated pop music icons (such as Dean Martin, Jack Jones, Peggy Lee, Steve Lawrence, and Eydie Gorme) eagerly sampled WNEW. Toting their just-released 33 1/3–rpm album, they would often drop by the Fifth Avenue studios for a surprise on-air appearance.

But by the late 1960s, most advertisers had become more interested in baby boomers and focused their advertising budgets on stations playing records kids demanded. Metromedia watched its venerable AM property slip, while WNEW-FM (established in 1958 as the remaining supply of vacant 50,000-watt New York City–area FM allocations had dwindled to three) achieved high ratings among the young demographic by dumping its "all-girl" announcers and easy listening music format in favor of progressive rock.

The old WNEW tried attracting new audiences by trying various incarnations of adult contemporary records and sports talk shows. A variety of air-personality and other changes never drained WNEW of broadcasters who could make a shampoo commercial entertaining, but such classic announcing talent was a poor fit with disco-era chart toppers. Fortunately for WNEW, advertising priorities had changed by the late 1970s. Advertising agencies were suddenly interested in America's comfortably affluent 40-plus population, giving the original "good" music, personality, and news station an economic reason to return to the most familiar shelves in its (1930s through 1960s non-rock hits) record library. Targeted listeners gratefully responded. For most, the change represented a surprise homecoming, lifting the station into a second heyday circa 1979. Other AM operators in what was by then an FM world studied this round of WNEW success. Arguably, all music stations that transmit disc jockey patter and recorded music can trace some of their roots to WNEW's original programming practices.

Rupert Murdoch's 1986 Fox acquisition of the Metromedia television properties orphaned WNEW radio. The FM station was bought by Westinghouse (in 1989), and WNEW-AM eventually ended up with the Bloomberg organization for a 1992 rechristening as the 24-hour, business-formatted (Bloomberg Business Radio) WBBR. So great was WNEW's personality-delivered big band/standards void that the *New York Times* came to the rescue. The paper's perennially classical WQXR-AM switched call letters to a mnemonic WQEW (Westinghouse didn't wish to relinquish the WNEW name) and accurately reconstructed, in full AM stereo, WNEW's programming spirit there. That legacy ended in late 1998, when Disney contracted with the *Times* to shift WQEW's programming to a juvenile focus.

PETER E. HUNN

See also All Night Radio; Block, Martin; Disk Jockeys; Judis, Bernice; Metromedia; Radio Disney; WQXR; Wynn, Ed

Further Reading

Eberly, Philip K., *Music in the Air: America's Changing Tastes in Popular Music, 1920–1980,* New York: Hastings House, 1982

Gordon, Nightingale, *WNEW: Where the Melody Lingers On,* New York: Nightingale Gordon, 1984

Jaker, Bill, et al., "WNEW" in *The Airwaves of New York: Illustrated Histories of 156 AM Stations in the Metropolitan Area, 1921–1996,* Jefferson, North Carolina: McFarland, 1998

Passman, Arnold, *The DeeJays,* New York: Macmillan, 1971

Rhoads, B. Eric, *Blast from the Past: A Pictorial History of Radio's First 75 Years,* West Palm Beach, Florida: Streamline, 1996

WNYC

New York City Station

For more than 75 years, WNYC, the nation's "most listened to" public radio station (93.9 FM and 820 AM), was funded by New York City municipal tax dollars. The station has weathered a long history of stormy relations as a result of its control at the hands of a line of New York City mayors extending from Mayor John F. Hylan in the 1920s to Mayor Rudolph W. Giuliani in 1995. This mayoral prerogative ended when ownership of the AM and FM stations was sold by the Giuliani administration to the WNYC Foundation, a nonprofit entity, in 1995.

Origins

In 1922 city official Grover A. Whalen, then commissioner of the Department of Plant and Structures, convinced Mayor John F. Hylan to appoint a committee to study the feasibility of operating a city radio station. The committee recommended the establishment of a municipally owned station, despite opposition from city Republicans and various big-business interests. When Western Electric was the sole bidder for the construction of the station and charged "exorbitant prices" for the use of its wires for remote broadcasts, the city searched for another alternative. In March 1924, a solution appeared in the form of a slightly used 1,000-watt broadcasting plant, which Westinghouse sold to the city after removing the equipment from its site at a Brazilian Centennial Exposition in Rio de Janeiro. During construction, an experimental license was granted to "station 2XHB." The station premiered on 8 July 1924, broadcasting from the top of the Municipal Building using the new official call letters WNYC. The opening evening featured crooners and instrumentals by the Police Band and the Vincent Lopez Orchestra. Mayor Hylan and city officials provided stirring orations, solemnized by blessings by clergy from three major faiths. The premier broadcast interfered with broadcasts of ships at sea and annoyed WEAF listeners when the transmission interrupted a broadcast from the Democratic National Convention.

Early broadcast schedules were erratic, beginning in the evenings at approximately 6:00 P.M. and concluding at around 11:00 P.M. Live performances from musical artists promoting the sale of sheet music and a series of one-hour foreign language lessons were among the staples of the evening broadcasts. Ad hoc longer broadcast hours accommodated visiting dignitaries and events ranging from band concerts to visiting monarchs, record-breaking aviators, and channel swimmers. H.V. Kaltenborn was an early contributor to WNYC and organized a radio quiz, and *WNYC's Air College* offered scholarly discussion on a variety of topics.

The station's service to the city consisted mostly of a nightly broadcast of missing persons. When Mayor Hylan sought to air a report of progress to the board of aldermen over the air in 1925 while running for his third term of office, the Citizen's Union, a watchdog advocacy group, brought legal action to prevent the mayor from using the station as a tool of propaganda. It was the beginning of a long history of contentious relationships between the mayor or board of city aldermen and the station. In the late 1920s, the station was criticized by a "Freethinker" for broadcasting religious services promoting the Catholic and Jewish faiths. To avoid such criticism, management codified a statement of its mission in 1930; the station would feature music, concerts, and entertainment; talks on current affairs; meetings of civic bodies, associations, and societies; lectures and addresses; and the reception of distinguished visitors.

Upon his election, Depression-era Mayor Fiorello H. La Guardia planned to sell the station as a cost-cutting move. A report by La Guardia aide Walter Chambers argued eloquently for the retention of the station as a public relations arm of the mayor's office. La Guardia detested the station at first, reportedly shouting upon spotting the old-fashioned carbon microphones of WNYC at a speech at the Commodore Hotel in 1934, "Get that damn peanut whistle out of here!" WNYC weathered the Great Depression partially through staffing provided by the Works Progress Administration (WPA) from government relief rolls, and the station embarked on a "Radio Project" to increase cultural programs, vary content, and find new sources of information and entertainment. In 1935 WNYC received a $30,000 WPA grant, and its transmitter was moved to the Greenpoint section of Brooklyn.

Controversies

Almost from its inception, the station was embroiled with federal regulators, fighting to protect its frequency assignment or to maintain broadcast hours. Under the Radio Act of 1927, the station was assigned 570 kilocycles and would be forced to share its frequency with radio station WMCA, broadcasting on alternate days at a reduced power of 500 watts. The city fought the assignment, and the Federal Radio Commission agreed that WNYC could swap frequencies with another station also owned by the management of WMCA, at 810 kilocycles. The swap allowed WNYC to operate during the daytime on clear channel until sunset at Minneapolis, when the 50,000-watt WCCO assigned to the same frequency came on the air.

In 1937 city councilmen seeking to embarrass Mayor La Guardia accused the station of anti-Semitism and racial hatred following a broadcast discussion of the Arab position on Palestine. The city council attempted to engineer another frequency swap that would have given the 810 AM frequency to the Paulist Fathers and forced WNYC to operate on a half-time basis from frequency 1130. La Guardia successfully fought the move. He was then targeted by political opponents as a supporter of communism for appointing the former secretary to the American Labor Party, Morris S. Novik, as director of radio communications for the city. WNYC was again accused of communist sympathies when the producer George Brandt's National Travel Club program on Soviet Russia on 27 February 1938 expressed admiration for Soviet accomplishments, but station management successfully defended itself against the charges. Concurrently, WNYC began broadcasting city council meetings, and the proceedings quickly became a source of public amusement. The *New York Times* commented that whereas the former board of aldermen had taken 135 years to make fools of themselves, the council had accomplished the same in only 2 years of radio broadcasts. Despite WNYC political broadcasts, the *New York Post* noted in 1938 that the station also provided more live unrecorded music than any other station in the city.

By 1940 La Guardia was a WNYC enthusiast, using the station to promulgate his political agenda. In 1944 the mayor began a weekly radio show, most memorably reading the Sunday comics to New York City children during a newspaper strike in July 1945. Comedian Fred Allen quipped to the *New York Times:*

> The Mayor's . . . program is a happy blend of *Mary Margaret McBride, Information Please,* and *Gang Busters.* One week the Mayor will tell you how to make French-fried potatoes with artichoke roots. The next week he gives you the name of the bookmakers and hurdy-gurdy owners he has chased out of the city." (cited in Scher, 1966)

Mayor William O'Dwyer, a former district attorney and judge, was the last of the line of early New York City mayors to routinely review the merits of WNYC and contemplate its sale immediately following election. O'Dwyer eventually was convinced to maintain the station and its noncommercial status.

During World War II, the station requested additional hours of broadcast for wartime information, and in 1942 the Federal Communications Commission granted a Special Service Authorization to allow WNYC to broadcast from 6:00 P.M. until 10:00 P.M., despite the conflict with Minnesota's clear channel frequency. The additional time slot was extended after retired Mayor La Guardia lobbied the Columbia Broadcasting System (CBS), owner of the Minneapolis clear channel station, to permit the Special Service Authorization to continue, and the hours were made permanent in 1955. In 1943 New York City launched a companion FM station that would allow WNYC to operate around the clock. Initially, the two stations carried the same programming until sign-off at 10:00 P.M. for the AM band. In 1953 WNYC also received a permit to construct an ultrahigh-frequency (UHF) band television station, which was renamed WNYC-TV in 1962. In 1987 WNYC moved to 820 kilohertz and boosted power tenfold—to 10,000 watts.

WNYC has built its reputation over the years on its ability to craft a mix of local, national, and world politics; culture; education; the arts; and classical music. The station claims it was the first to broadcast the Japanese attack on Pearl Harbor, and it has broadcast more Senate hearings, conferences, and conventions, including the presidential conventions of smaller political parties, than any other station. In the 1960s, the station offered an outlet for foreign programs and the views of various national news such as the *French Press Review* and the *Review of British Weeklies.* Its broadcast of public hearings on the increase of the New York City subway fare was called "one of the greatest mass civic lessons in the history of radio" by *Variety* magazine. Bob Dylan made his radio debut in 1961 on Oscar Brand's *Folk Song Festival,* and experimental and modern music found an outlet on WNYC.

In 1994 Mayor Giuliani announced that the city would sell the station, and a deal was struck whereby the AM and FM stations would be sold at a reduced price of $20 million to the WNYC Foundation, and the television station would be sold to commercial interests. For the first time, WNYC would be a public- and grant-supported station, and its president would no longer be a political appointee of the mayor but rather hired by the board of the WNYC Foundation. The board appointed Laura Walker, a former producer of Children's Television Workshop, as president. Her management team succeeded in doubling public contributions to the station and raising $1.4 million in June 1995 in a four-day campaign while garnering support from corporations and other high-profile supporters. WNYC 93.9 FM is a member station of National Public Radio (NPR) and of Public Radio International (PRI), and it features NPR and WNYC news and cultural programming along with classical music. WNYC Radio New York at 820 kilohertz broadcasts news, talk, and public-affairs programs. In 1998 the station launched a website, wnyc.org, which also allows web users to receive radio programming, both live and in archived segments.

In 2001, the station lost its FM transmitter and antenna in the September 11th attacks on the World Trade Center and temporarily used borrowed transmitters and antenna to simulcast the news talk programming of its AM station over both AM and FM outlets. The subsequent soar in ratings and fundraising was instructive. Within a few months of restoring the FM tower, the station announced it was eliminating five hours of classical music programming each day on the FM channel in favor of more news/talk programming, to the consternation of many listeners and performers who saw WNYC as one of the few outlets for both original and recorded classical programming.

L. CLARE BRATTEN

See also Educational Radio to 1967; Public Radio since 1967

Further Reading

Gladstone, Valerie, "Having Important (and Generous) Friends Helps," *New York Times* (21 November 1999)
Hinckley, David, "WNYC and Public Radio Become More of a Turn-On," *New York Daily News* (26 August 1997)
Jaker, Bill, Frank Sulek, and Peter Kanze, *The Airwaves of New York: Illustrated Histories of 156 AM Stations in the Metropolitan Area, 1921–1996*, Jefferson, North Carolina: McFarland, 1998
Lewine, Edward, "Making Radio Waves," *New York Times* (12 January 1997)
Scher, Saul Nathaniel, "Voice of the City: The History of WNYC, New York City's Municipal Radio Station, 1924–1962," Ph.D. diss., New York University, 1966
Van Gelder, Lawrence, "Morris S. Novik, 93, Early Director of WNYC," *New York Times* (12 November 1996)
Van Gelder, Lawrence, "Footlights," *New York Times* (7 April 1999)
"WNYC-FM to Cut Back Classical Music," *New York Times* (8 March 2002)

Wolfman Jack (Robert Smith) 1938–1995

U.S. Radio Personality

Robert Weston (Bob) Smith, born in Brooklyn, New York, on 21 January 1938, was mesmerized as a youth by Jocko Henderson at WDAS in Philadelphia, John R. (John Richbourg) at WLAC in Nashville, and Alan Freed at WINS in New York. Smith's career started modestly, as a gofer for WNJR in Newark, New Jersey. When his parents discovered he was at WNJR and missing school, they banished him from home. Smith left for Hollywood but made it only to his sister's Alexandria, Virginia, home, where he entertained her children as "Wolfman." Smith recalled, "The 'Jack' part got added on because that was a hipster's style of speaking, to end your sentences with the name Jack."

Smith sold encyclopedias and Fuller brushes until attending Washington, D.C.'s National Academy of Broadcasting. Upon graduation in 1960, Smith worked at WYOU-AM in Newport News, Virginia, a station programmed for African-Americans. Smith (a.k.a. "Daddy Jules") did everything from sales to cleaning. WYOU became WTID and changed formats to beautiful music. Smith was retained by the new management but changed his air name from "Daddy Jules" to the more sedate "Roger Gordon." In his autobiography, Smith wrote: "[I]t wasn't where my heart was. To this day, I feel a lot worse about cracking open a mike and announcing, 'Good morning. This is Roger Gordon bringing you Music in Good Taste,' than

Wolfman Jack
Courtesy Radio Hall of Fame

I do about the fact that I was transporting weed and brokering mattress action on the side."

Mo Burton, WTID sales manager, bought KCIJ-AM in Shreveport, Louisiana, in 1962 and hired "Big Smith with the Records" to play country music. By 1963 KCIJ merged with KREB in Marshall, Texas. KREB could not boost its power as long as KCIJ was on the air, so the owners worked out an unprecedented deal with the FCC. The owner of KCIJ took its station off the air in exchange for half ownership in KREB, which was then allowed to increase its power. A weak station, then, went dark so a stronger one could grow. (Such an arrangement has now become commonplace.) Big Smith was now on KREB, but he had bigger aspirations.

All his life, Smith had listened to the most powerful commercial radio station on earth. XERF was a 250,000-watt blowtorch on the border near Del Rio, Texas, in Ciudad Acuña, Mexico. XERF was notorious for transmitting, as Smith explained, "old-time, prayer-cloth-selling preachers, fervent gospel singers, hillbilly mountain music, all kinds of stuff that was just a little too weird, a little too dangerous for American stations." The station "sold gain-weight plans, lose-weight plans, baby chicks, pep pills, sex-drive-boosting pills, songbooks, records, hair dye, anything that could be stuffed into a box or corked up in a bottle and sent off in the mail for 'cash, check, or money order.'" Smith initially got in on the action by selling KCIJ airtime to about half the preachers who bought slots on XERF. The preachers taped their shows and paid XERF about $1,500 to air 15-minute programs five days a week. Smith gave them a cut rate to air the programs over KCIJ, too. Meanwhile, he continued to dream of developing his own show for XERF. Smith recalled: "I'd do exactly like the preachers do, make my money on mail order. Except there wouldn't be any preaching. Instead, I'd stir people up with the power of the music, and a crazy DJ character who had the same blend of coolness and hipness as all the great ones that had zapped my imagination when I was a strange little kid transfixed by what was coming through the radio."

Smith secretly developed his Wolfman Jack character. In December 1963 he took a demo tape to XERF but instead ended up operating the station. For eight months, he aired *The Wolfman Jack Show,* featuring a "superjock" who relied on personality as much as music. Wolfman Jack played rock and blues, had an unconventional delivery, used strange language, and howled like a wolf.

Smith returned to KCIJ but continued to tape *The Wolfman Jack Show* for XERF. Few people knew that Smith was Wolfman Jack. In character he recorded a live singing performance. *Wolfman Jack—Live at the Peppermint Lounge*, an embarrassing yet profitable recording, was sold over XERF, which Smith continued to control. Burton bought KUXL-AM in Minneapolis in 1964, and Smith ran the station. XERF continued airing tapes of *The Wolfman Jack Show,* but Wolfman Jack was not seen for years. The show soon aired on other border stations, XEG and XERB. In 1965 Smith moved to Los Angeles to run XERB as though it were local.

Smith had always kept the Wolfman Jack character separate from his real identity. To almost everyone, he was radio businessman Bob Smith. Wolfman Jack was something Bob Smith did for fun, but as the popularity and profits rose, people wanted to see Wolfman Jack and they paid big money to see him. For Wolfman Jack to appear in public, Smith's wife designed the look, and a makeup artist added a wig, dark makeup, a big nose, fangs, false fingernails, and facial hair. In 1971, Smith's empire crashed when the Mexican government banned evangelical religious programming, costing XERB 80 percent of its income. As XERB went out of business, Smith turned completely to Wolfman Jack.

In 1972 Wolfman Jack was hired for the 7 P.M. to midnight shift at Los Angeles' KDAY for one-tenth of his XERB salary. He put together tapes of old XERB broadcasts for syndication and for Armed Forces Radio. Wolfman Jack played himself in 1973's *American Graffiti,* where he was seen widely for the first time. He surprised fans who assumed he was black. In 1973 he started a nine-year stint as host of NBC TV's concert show *The Midnight Special* (which ran from 2 February 1973 to 1 May 1981). The program was network TV's first foray into late-late-night hours. The same year, New York's WNBC hired Wolfman Jack to compete with WABC's Cousin Brucie. WNBC wanted Wolfman Jack to be crazy but constantly complained about things he said and did. He made it to number one doing a shock jock routine that influenced Howard Stern. At one point, Wolfman Jack, Don Imus, and Stern worked simultaneously at WNBC. The Guess Who, who did the 1974 song *Clap for the Wolfman,* asked Wolfman Jack to join them on tour that same year. To get out of his contract, he convinced WNBC to hire Cousin Brucie. He continued to fly to Los Angeles twice monthly to tape *The Midnight Special.*

Wolfman Jack recorded his second album in 1974, called *Fun and Romance.* In the 1980s, he appeared at oldies concerts and car shows. He hosted a short-lived oldies revival on the Nashville network (TNN) in 1989. In 1995 he hosted a live weekly radio show on WXTR from Washington, D.C.'s *Planet Hollywood.* On 1 July 1995, he died at home in Belvidere, North Carolina, of a heart attack after a 20-city promotional tour for his autobiography. He was inducted posthumously into the Museum of Broadcast Communications Radio Hall of Fame on 27 October 1996 and into the National Association of Broadcasters Hall of Fame on 20 April 1999.

W.A. KELLY HUFF

Wolfman Jack. Born Robert Weston Smith in Brooklyn, New York, 21 January 1938. Began radio career as a gofer at WNJR, Newark, New Jersey; graduated from National Academy of

Broadcasting, Washington, D.C., 1960; on-air and sales, WYOU-AM, Newport News, Virginia, 1960; on-air at KCIJ-AM, Shreveport, Louisiana, 1961; operated XERF, Mexico, 1963–64; *The Wolfman Jack Show,* XERF, 1964; operated KUXL-AM, Minneapolis, Minnesota, 1964; operated XERB, Los Angeles, California, 1965–71; permanently became Wolfman Jack, 1971; on-air at KDAY, Los Angeles, 1972; syndicated XERB broadcasts, 1972; appeared in *American Graffiti,* 1973; hosted NBC's *The Midnight Special,* 1973–81; on-air at WNBC, New York, 1973–74; on-air at WXTR, Washington, D.C., 1995; inducted into Museum of Broadcast Communications Radio Hall of Fame, 1996, and NAB Hall of Fame, 1999. Died in Belvidere, North Carolina, 1 July 1995.

Radio Series
1970–86 *The Wolfman Jack Show*

Television Series
The Midnight Special, 1973–81 (NBC); *Superstars of Rock,* 1973 (syndicated); *The Wolfman Jack Show,* 1974 (syndicated)

Films
The Seven Minutes, 1971; *American Graffiti,* 1973; *Murder at the Mardi Gras* (TV Movie), 1977; *Deadman's Curve* (TV Movie), 1978; *Sgt. Pepper's Lonely Hearts Club Band,* 1978; *Hanging on a Star,* 1978; *More American Graffiti,* 1979; *Motel Hell,* 1980; *Conquest of the Earth* (TV Movie), 1980; *The Midnight Hour* (TV Movie), 1985; *Garfield in Paradise,* 1986; *The Return of Bruno,* 1988; *Mortuary Academy,* 1988

Stage
I Saw Radio, 1975

Publication
Have Mercy! Confessions of the Original Rock 'n' Roll Animal (with Byron Laursen), 1995

Further Reading
Smith, Wes, *The Pied Pipers of Rock 'n' Roll: Radio Deejays of the 50s and 60s,* Marietta, Georgia: Longstreet Press, 1989

Women in Radio

Why is it necessary to have a separate entry on the subject of women and radio? Haven't women—as producers, stars, industry personnel, and audiences—been so deeply interwoven in radio's history, development, innovation, marketing, and reception as to make them an integral part of a work such as this? Won't women's contributions and concerns turn up as a natural component of the general run of entries, without need for special consideration? The answer to the former question is undoubtedly "yes"—women have been deeply and centrally involved in radio's development from the 1920s to the present. Yet their near-exclusion from most standard histories of the medium, along with special characteristics of the ways that U.S. radio has handled the presence and contributions of women, underline the need for additional consideration.

The basic structures, programming, marketing strategies, and audience definitions of U.S. radio have been deeply affected by gender. From radio's earliest days, women's interests were segregated in a separate "women's ghetto" of daytime programs, while the prime-time hours—those most often written about by historians and enshrined in popular memory—carefully relegated women to ancillary roles. The industry very early on devised an understanding of its public in which daytime audiences were seen as fundamentally female (despite a sizable male minority) and night-time audiences were addressed as primarily male (despite a sizable feminine majority). These understandings carried over into television and have only begun to change since the 1980s.

Yet this kind of gender segregation—conforming to a held-over "separate spheres" philosophy from Victorian times—also worked to create a space on radio that was dedicated to female interests and points of view. That this space was commercially driven (women have long been understood as the primary consumers for the household) marks out its limitations but doesn't take away from its unique contributions to radio's development. Women became powerful independent producers, writing and marketing programs that employed hundreds of men and women and creating genres that, 70 years later, form the backbone of popular television. Daytime programs addressed topics of direct relevance to women's lives, created dramas in which women controlled events and saw their experiences played out, and encouraged talk programs that discussed women's interests and concerns. Meanwhile, night-time radio, and later television, carefully contained female voices and faces within secondary roles, situation comedies, and family programs; excluded them from "serious" genres such as news, discussion, and documentaries; and marginalized women's contributions to prime-time drama. But gradually daytime conventions crept into the night-time hours, and with them a more extensive female presence, especially during the World War II years.

On post-television radio, a similar route was traced as female disc jockeys and artists slowly developed a more extensive presence on the airwaves. This overview of how women have operated within the cultural structures of the radio broadcasting field should provide a contextual framework for the separate entries on some of the individuals who appear elsewhere in this encyclopedia. Their achievements—and limitations—took place within a highly gendered social and industrial system that guided not only these women's lives and practices but the way they have been treated by historians, critics, and regulators.

Early Experimenters, Amateurs, and Managers

Although it has become a cliché of history to think of the period of radio amateurs—roughly 1915 to 1926—as the domain of "small boys," in fact wireless experimentation offered a fairly welcoming venue for women. The American Radio Relay League (ARRL), one of the earliest and largest of the radio amateur organizations, began to recognize women members in 1916 and predicted a flood of female membership after World War I, because the army had recruited hundreds of young women as wireless instructors. The ARRL's journal, *QST,* offers a number of profiles of female amateur operators during the early years, including Emma Candler of station 8NH in St. Mary's, Ohio, one of the key operators in the ARRL's cross-country relay network, and amateur M. Adaire Garmhausen, who was a regular contributor to the journal. Other early experimenters—selected for their presence in the historical record more than for any difference from hundreds of others—were Eunice Randall of 1XE, Boston, an engineer with the American Radio and Research Corporation (AMRAD) company who became chief announcer for WGI from 1922 until 1925; Marie Zimmerman, who with her husband Bob broadcast from WIAE in Vinton, Iowa, from 1920 to 1923; Eleanor Poehler, station manager of WLAG, Minneapolis, from 1922 to 1924 and later music director at WCCO, Minneapolis; and Ida McNeil, who ran station KGFX in Pierre, South Dakota, from 1922 until the mid-1940s.

Announcers and Network Executives

Many more women came on board fledgling commercial stations as they began to make their debut in the mid-1920s, and radio shifted from the province of amateurs toward becoming an industry. Gwen Wagner of station WPO in Memphis, Tennessee, began as the station's one on-air employee, announcing, planning the broadcast schedule, inviting the guests, and reporting the news. Both Bertha Brainard and Judith Waller started out in such jill-of-all-trades on-air positions. Brainard began as a volunteer program producer on WJZ, "Broadcasting Broadway" over the Radio Corporation of America (RCA) lead station. She then went on to become first program director, then station manager, and then director of commercial programming for the newly organized National Broadcasting Company (NBC). Similarly, Waller started out on Chicago's WMAQ as an announcer and station manager. When NBC purchased the station in 1932, Waller became the network's director of public service programming for the central region, where she innovated with such respected shows as *American School of the Air* and *University of Chicago Round Table.*

A number of female pioneers, whose voices were heard on the air during the early 1920s, also filled secretarial roles and later went into audience relations (via their duties of answering audience mail); still more, like most of those above, were directed toward women's, children's, and public service programming. A turning point occurred in 1924, when Jennie Irene Mix, a columnist for *Radio Broadcast,* one of the most influential of the new radio magazines, opened up a discussion on the suitability of women's voices for radio transmission. Though many station managers—mostly male—dismissed the notion that there was any significant difference between male and female broadcast capabilities, others argued that women voices were monotonous, were overly emotional, lacked personality, had too much personality, didn't transmit well over radio because they employed higher frequencies (a frequently repeated error, only applicable to the highest soprano notes and not to the spoken voice), or "lacked body." The opinion that women's voices were untransmittable, however specious, had a great effect on the standardization of the broadcast schedule that would take effect over the next five years, and on the exclusion of women from many aspects of broadcasting.

Program Producers and Innovators

Radio, as a new field whose barriers for women were only slowly developing, attracted female pioneers through a variety of routes. Gertrude Berg developed her popular and long-running radio serial *The Rise of the Goldbergs* in 1928, writing, acting, and producing throughout the show's radio decades and into its run on television. The show ran on NBC's nightly schedule with high ratings until 1936, when it moved to daytime. College sorority sisters Louise Starkey, Isobel Carothers, and Helen King created their heavily improvised comic serial *Clara, Lu, and Em* for WGN in Chicago in 1930, about three neighbors in a Chicago apartment building and their humorous travails; it became a prototypical soap opera when it transferred to daytime in 1932, and it ran there until Carothers' death in 1936. Myrtle Vail created the dramatic serial *Myrt and Marge* with her daughter Donna Damerel in 1931; it kept audiences tuning in for its tales of a mother-and-daughter theatrical team on Broadway until Damerel's untimely death in 1941. Also a night-time original, it went to daytime in 1937. Although *Amos 'n' Andy*, created in 1926, must be counted as the first radio program to develop the genre of serial comedy/drama, these programs took the form in a feminine direction and led not only to the daytime serial format but to the later genre of situation comedy.

As the history of these programs shows, the debate over women's suitability on the air led to a major change in programming practices in the early to mid-1930s. As radio's commercial support grew larger, and as networks began to turn over much of their programming function to the radio departments of advertising agencies, a new way of understanding the radio audience arose. Research conducted by first the Crossley and later the Hooper companies, as well as in-house network and agency data, showed that the daytime audience consisted of about 70 percent women and that the night-time audience had a slightly smaller feminine majority, about 55–60 percent. Women were known to purchase roughly 85 percent of all household goods, making this an extremely desirable market to approach. Yet radio's regulatory status mandated a restriction of its purely commercial aims, requiring public service standards to be applied to its program offerings.

Broadcasters resolved these two contradictory claims on their attention by separating daytime and night-time schedules by gender. Women's programs—those featuring primarily female characters, usually produced by women, and selling products to women—were placed on the daytime schedule, even though many had had a fairly large mixed audience at their former night-time hours. Here the networks could sell unabashedly to their primary audience, indulging in program types and subject matter that were not seen as wholly respectable and that often provoked controversy, as long as they attracted the target audience of women. Night-time programs, although still commercially supported, were increasingly seen as more prestigious and of higher quality; the night-time schedule became a place for sponsors and networks to display their big names and big-budget productions for an audience increasingly defined as masculine despite its majority female composition. Advertising was held to a minimum and was often worked into the content of the show to keep it as unobtrusive as possible. This "separate spheres" approach would at once spark a lively wave of creative activity addressed to and created by women in the daytime and create serious barriers for women's ability to compete on the night-time schedule.

Ladies of the Daytime

In the space allotted to them, the ladies of radio's daytime sphere developed programs that not only appealed to women but also tended to feature women in central roles in a way that created a woman-centered discourse that few other media genres cared to touch. Its closest precedent was probably the field of women's magazines, whose serialized fiction; interview pieces; discussion of the "private" realm of home, family, health, and relationships; and interaction with their readership in many ways influenced the development of daytime genres. Irna Phillips is often credited with being "the mother of soap operas," and indeed, her pioneering serial *Painted Dreams,* begun on WGN in 1930, represents one of the earliest examples of a woman-centered drama created specifically for the daytime schedule. This program mutated into *Today's Children* on rival station WMAQ, and Phillips went on to become one of the most prolific individual producers of daytime soaps. Perhaps more important, she discovered that retaining ownership of her intellectual property and incorporating as her own producer made far more sense—and gave her far more creative control—than working as an agency or network employee. Her most famous serials include *Women in White* and *The Right to Happiness,* along with the still-running *Guiding Light.* Production control was a lesson most female innovators needed to learn. Jane Crusinberry, author of the extremely popular serial *Story of Mary Marlin*, learned the hard way that her own property could be taken from her: after a series of escalating disagreements with the ad agency that produced the show after the war—mostly about Mary's "redomestication" after serving as a U.S. senator—Crusinberry was fired from her own soap, and she never wrote another.

Anne Hummert, in partnership with her husband Frank, owned one of the biggest radio production houses in the country, at one point producing almost half of the soaps on the air. These included *Ma Perkins, The Romance of Helen Trent,* and *Mary Noble, Backstage Wife.* Elaine Carrington produced a number of soaps that enjoyed a relatively high critical reputation (*Pepper Young's Family, When a Girl Marries,* and *Rosemary*), as did Sandra Michael (*Against the Storm*—the only soap ever to win a Peabody award—*Lone Journey,* and *Open Door*). Many other popular serials were written by women who, not having become producers, labored in obscurity for agencies and other producers: Helen Walpole of *Our Gal Sunday* and *Stella Dallas;* Elizabeth Todd of *Young Widder Brown* and *Amanda of Honeymoon Hill* (and later of the night-time *Fred Allen Show*); and Addy Richton and Lynn Stone, who wrote under the joint pen name Adelaide Marston for *Valiant Lady, Hilltop House,* and *This Life Is Mine,* just to name a few.

Besides the ever-popular and ever-expanding soaps, a staple of daytime was the talk show. This form evolved from the "household chat" format of advice aimed at modern busy housewives. Early innovators of this genre include Mrs. Julian Heath, Ida Bailey Allen, Ruth Crane, the various Betty Crockers (a made-up identity filled by a series of female broadcasters), and, perhaps most famous, Mary Margaret McBride. In the mid-1930s this format began to make the transition from solely domestic topics to a wider range of material, including news and interviews. It also went from a highly scripted format to one that emphasized informal, ad-libbed discussion. In both of these trends, McBride was a leader. A former journalist and feature article writer, McBride brought her reportorial and interviewing skills to a freely ad-libbed format that also involved highly effective product pitches. Starting out as *Martha Deane,* a Betty Crocker–like fictional character, she soon broke out on her own. By 1940 she had two 45-minute shows daily, one on NBC and one on the Columbia Broadcasting System (CBS), with over 6 million regular listeners.

Others took note, and by the mid-1940s a variety of "breakfast shows" abounded on the airwaves, usually featuring a husband-and-wife team who bantered back and forth on a wide range of topics, bringing in a guest or two and providing light news and weather updates. Innovators here include Pegeen Fitzgerald, whose *Pegeen Prefers* was modeled after McBride's show; husband Ed Fitzgerald also had a similar program called *Small Talk* on WOR. Their combined efforts produced *The Fitzgeralds* over WOR from 1940 to 1945, which then ran on the American Broadcasting Company (ABC) network from WJZ from 1945 to 1947. WOR replaced them with the breakfast team of Dorothy Kilgallen and Richard Kollmar in *Breakfast with Dorothy and Dick.* A third team graced the airwaves from 1946 until 1948, *Tex and Jinx,* featuring married duo Tex McCrary and Jinx Falkenberg. From here, the *Today Show* was only a short step in the future, although the early television version would "defeminize" itself, relying on all-male hosts until Barbara Walters' debut in 1964.

The breakfast show also marks one of the few venues for women to report the news. News formats remained largely closed to women on radio unless, as with print journalism, they could adapt their reportorial skills especially for the "female" audience and air in the daytime. Some early broadcasters of "news for women" include Ann Hard; Kathryn Cravens, whose *News through a Woman's Eyes* aired on CBS from 1936 to 1938; and Marian Young, who took over the *Martha Deane* show after McBride left and who began to include more outright news content. The years before and during World War II brought increased opportunities for women, as network news operations began to build up, and a number of journalists found their way to radio, including Helen Hiett, who had a morning news spot daily on NBC Blue from 1941 to 1942, and Mary Marvin Breckinridge, who was recruited by Edward R. Murrow as the first CBS female staff broadcaster from Europe in 1939. Dorothy Thompson, the world-famous journalist, had held a news commentator spot on NBC in the mid-1930s until her strong anti-Hitler stance led the sponsor to dismiss her. But she returned in the middle of the war years after a brief stint at Mutual Broadcasting Company, providing news reports from various locations around the world. Janet Flanner, too, had established a reputation as a columnist at the *New Yorker* magazine before turning to radio in 1944 with a series on NBC Blue.

Pauline Frederick represents a breakthrough for women in broadcast news, as the first to build a career in the medium. Beginning as a print journalist for the North American Newspaper Alliance, she appeared on NBC starting in 1938 for occasional pieces and covered the later years of the war in Europe. In 1946 she became the first woman to cover "hard" political news on the ABC network, and she began to appear on television. By 1953 she had her own NBC interview show, *Pauline Frederick Reporting,* a daily 15-minute program aired in the afternoon, as well as a program called *Listen to the Witness* on Sunday afternoons. She later joined National Public Radio (NPR) as foreign-affairs commentator.

Another important innovator was journalist Martha Rountree, whose *Meet the Press* debuted in 1945 with herself as coproducer and cohost. Although she sold her interest in the program to her partner, Lawrence Spivak, in 1953, the show continues on the air today, widely recognized as one of the most successful panel formats on television. Rountree's previous radio effort, *Leave It to the Girls,* was a witty, fast-paced all-female panel show that ran on Mutual from 1945 to 1948.

Women also featured prominently in the field of children's programming. One of the best known is Dorothy Gordon, who began on station WEAF in 1924 with children's story hours and continued into the *Children's Corner* on CBS in the mid-1930s. She also served as music director for *American School of the Air.* The male/female duo format made a children's hit in the *Roy Rogers Show* with Dale Evans from 1944 well into the television years. Many local stations included children's shows in their schedules, and many of these featured female hosts or were produced by women.

Ladies of the Night

Although night-time radio remained a less hospitable spot for woman-controlled productions, many talented performers lent their names to the worlds of variety, comedy, and musical performance that made up the prime-time hours. Kate Smith remained for many years the only female host of a prime-time variety show, the *Kate Smith Hour,* which ran on CBS from 1931 to 1947 under a number of titles and sponsors but always featured Smith's singing interspersed with guest acts and sketches. Not until 1943, when Joan Davis took over Rudy Vallee's spot for Fleischmann's Yeast, did radio hear the second woman to single-handedly headline a major night-time vehicle. Fanny Brice would make a third female headliner after 1940, when her Baby Snooks routine slowly took over the *Maxwell House Coffee Time* and gradually metamorphosed into the *Baby Snooks Show.*

Until the early 1940s, almost all of night-time's female stars appeared either only occasionally, as guests, or as part of the traditional male/female duo act carried over from vaudeville. Usually the male/female duo involved a male straight man and a female "dumb Dora," a successful format illustrated not only by the teams of Jack Benny and Mary Livingston, and Fred Allen and Portland Hoffa, but most famously by George Burns and Gracie Allen. All of these famous radio duos were married in real life, too. So were Jim and Marian Jordan, playing *Fibber McGee and Molly,* although this time with the husband in the dumb Dora role.

In the late 1930s, a number of shows spun off from variety sketches also featured such teams, as in the *Bickersons, Easy Aces,* and *Duffy's Tavern.* However, the absence of available male talent, as well as the larger domestic audience of women and better audience research techniques, led to a breakthrough for women during the war years. Former Hollywood second-tier stars such as Lucille Ball (*My Favorite Husband*), Ann Sothern (*Maisie*), Eve Arden (*Our Miss Brooks*), Hattie McDaniel (*Beulah*), and Marie Wilson (*My Friend Irma*) took the opportunity to move into night-time radio, and most of them obeyed the rule that Irna Phillips had set down years before: form your own production company and retain production control. Not only would these women's efforts transform the night-time comedy scene, making important innovations on the emerging situation comedy format, but their shows would dominate early television as well.

Radio in the Age of Television

After television's widespread debut in the early 1950s, radio was transformed into a local, music-oriented medium whose new stars were the disc jockeys who introduced songs and provided chatter in between. Though a few women moved into this sort of work, the remnants of the "separate spheres" system mandated that they mostly stay in the daytime, speaking to an audience still conceived of as women in the home. Until the early 1960s, a few network daytime serials and talk shows still remained, and in the newly disorganized world of local radio, many different experiments could be tried. Martha Jean "the Queen" Steinberg became well known on the new black-format station WDIA Memphis in the 1950s. She cohosted an evening swing and rhythm and blues show with famed disc jockey Nat Williams; soon, she had her own show, called *Nite Spot,* and she had also replaced an earlier female talk host, Willa Monroe, on a daytime program called *Tan Town Homemaker.* She introduced her own rhythm and blues show, *Premium Stuff,* a little later and soon became known nationwide, eventually moving to Detroit station WCHB. Other African-American women rose to prominence in the postwar years of black radio: Vivian Carter on WGRY in Gary, Indiana; Louise "Louisville Lou" Saxon on WLOU in Louisville, Kentucky; "Chattie" Hattie Leeper on WGIV in Charlotte, North Carolina; and Irene Johnson on WGOC, Mobile, Alabama. These disc jockeys and their male counterparts became important sources of community inspiration and information as the civil rights struggle heated up in the late 1950s and early 1960s.

Other women served as announcers and disc jockeys on stations of all kinds in the 1950s and 1960s, but they rarely surface in the historical record. Dottie Miller was featured on WBBQ Atlanta as "the miss with the hits" in the mid-1950s. Pola Chasman anchored a classical music program on WQXR in New York. In 1955 Memphis recording entrepreneur Sam Phillips experimented with an "all-female" format on station WHER, whose slogan was "1,000 Beautiful Watts." To differentiate himself in a market that was already becoming crowded, Phillips determined that this would be a station run by women, for women. Specializing in "pleasant, light music" and the usual mix of news, weather, and talk, this time all the voices heard were female, as was the station's management. The station was successful enough to last until 1966, and it spawned several imitators. As the climate of the country started to change in the late 1960s, underground radio brought its style of female broadcasting onto the airwaves. Rachael Donohue and Dusty Street were important early alternative broadcasters. WNEW in New York put out a nationwide call for female disc jockeys in 1966, and at least one stayed on: Alison Steel, known as the "Nightbird," whose "sultry voice and iron will" in her late-night slot made her one of the few women in rock radio throughout the 1970s and 1980s.

National Public Radio

In 1967 the United States finally caught up with many other countries by creating its first public broadcasting system; in 1970 National Public Radio (NPR) was founded. Its first major offering, the daily afternoon newsmagazine *All Things Considered* (*ATC*), introduced a new breed of serious female journalists to the still deeply gender-divided world of radio. With Susan Stamberg in the cohost position from the program's inception in 1971 and soon adding talented reporters and correspondents such as Linda Wertheimer, Cokie Roberts, and Nina Totenberg, NPR provided a different take on the news, broadening its mission to include those issues that affect the lives of ordinary people—including women. By 1984 *ATC*'s staff was more than 50 percent female, which may have reflected the lower salaries women journalists were obliged to take (NPR's salaries were a third lower than the industry average); nevertheless, women had an enormous effect on the news content provided. Not only everyday news but also politics, law, international affairs, science, sports, and culture became the proper domain of women in this female-anchored medium for a general audience of the entire national public.

NPR's ability and mission to cover stories not just as breaking news but in depth and over time helped to build these women's careers into national reputations, and they are now frequently seen on television. Terry Gross developed one of NPR's top talk programs, *Fresh Air,* in 1975; it is still going strong. NPR's two additional newsmagazines, *Morning Edition* and *Weekend Edition,* continue the emphasis on women's voices as equals. Public radio remains a hospitable forum for female broadcasters on both national and local levels, although its lower pay scales and frequent use of volunteer talent may not do much to dispel the wage discrimination prevalent in society at large. Yet as talk radio developed into a forum for conservative white men in the 1980s and 1990s, public radio's feminine alternative occupies an increasingly important position in the U.S. public sphere.

Talk Radio

With the development of music formats on the FM band, AM became a radio graveyard, largely empty of all except news and sports and struggling local mavericks. Then talk radio came along in the late 1970s, made interactive by the new 800 number telephone technology that allowed listeners to call in from all over. Just as format radio had begun to go nationally syndicated, so too could popular local talk show hosts begin to attract national audiences.

One of the earliest to discover radio's potential in the field of talk and advice was the well-known psychologist Dr. Joyce Brothers. In the late 1960s, she took her advice to radio with syndicated shows on both the ABC and NBC networks, and she was later featured on WMCA in New York. Her shows, usually built around questions called in by listeners, were among the first to bridge the public/private divide, bringing personal problems into the light of publicity, including issues of sexuality, menopause, intimacy, and mental illness. Following in Brothers' footsteps came two more divas of the airwaves, Sally Jesse Raphael and Dr. Laura Schlessinger. Raphael got her start in Miami, with both a morning television show and an afternoon radio interview program. In 1976 she won a morning slot on New York's WMCA, and in 1979 hers was one of the original programs gathered together in NBC's syndicated Talknet package of night-time radio call-in shows. Schlessinger got her start at KABC in Los Angeles in 1975 as a frequent guest on another show, as an expert on human sexuality issues. This format had been introduced to the Los Angeles market by Dr. Toni Grant over KABC that same year. In 1976 Schlessinger began her own show on KWIZ Santa Ana, later moving back to Los Angeles at KMPC. Her major vehicle, *The Dr. Laura Schlessinger Show,* debuted in 1990 over KFI as a three-hour daily call-in program focused around personal advice and went into national syndication in 1994. In the late 1990s, Schlessinger's show surpassed talk king Rush Limbaugh's as the most popular talk program on the air, although her conservative views brought controversy and declining status by 2002. Another broadcaster famous for her sexual and medical advice is Dr. Ruth Westheimer, whose show *Sexually Speaking* debuted in 1980 on WYNY-FM; by 1983 it was the top-rated radio show in the New York City area. In 1984 NBC began syndicating her show, which was renamed the *Dr. Ruth Show,* and Westheimer also ventured into television on the Lifetime network.

Into the Future

As the discussion of popular female radio talk show hosts above indicates, a peculiar kind of "separate spheres" philosophy still operates in the radio industry today. Most nationally recognized female broadcasters, with the exception of those on NPR, operate in the world of personal and private issues, giving advice in the areas of psychology, health, lifestyle, and culture, whereas political talk radio is largely the province of men. In commercial music formats, the "morning team" program, epitomized nationally by such "shock jocks" as Howard Stern and Don Imus, typically features a male host with a female sidekick in a markedly subordinate position (sometimes called "the giggle box"). Though a few women achieve prominence in the highly rated morning slots, their numbers are small, and it is usually still as half of a male/female team.

Meanwhile, behind the scenes and out of the public eye, more and more of radio's program, music, promotion, and marketing directors are women, and women make up the largest slice of the commercial audience. There are a few women station owners and syndicated producers, but the vast bulk of management remains male. Things have changed, but in many ways they have remained the same. The "women's daytime ghetto" may be gone, but certain occupational ghettoes remain. However, turn on your radio and the chance that you'll hear a female voice announcing the upcoming hits, reporting the news, or delivering a commercial has increased enormously over the last 20 years. The controversy over women's voices on the air would appear to be settled, though now we must wait for industry control to catch up.

MICHELE HILMES

See also Association for Women in Communication; Brice, Fanny; Female Radio Personalities and Disk Jockeys; Frederick, Pauline; Fresh Air; Goldbergs; Phillips, Irna; Hummert, Frank and Anne; McBride, Mary Margaret; Our Miss Brooks; Serials; Smith, Kate; Soap Opera; Stamberg, Susan; Totenberg, Nina; Wertheimer, Linda; WHER

Further Reading

Barlow, William, *Voice Over: The Making of Black Radio,* Philadelphia, Pennsylvania: Temple University Press, 1999

Halper, Donna, *Invisible Stars: A Social History of Women in American Broadcasting,* Armonk, New York: Sharpe, 2001
Hilmes, Michele, *Radio Voices: American Broadcasting, 1922–1952,* Minneapolis: University of Minnesota Press, 1997
Isber, Caroline, and Muriel Cantor, *Report of the Task Force on Women in Public Broadcasting,* Washington, D.C.: Corporation for Public Broadcasting, 1975
Keith, Michael C., *Voices in the Purple Haze: Underground Radio and the Sixties,* Westport, Connecticut: Praeger, 1997
Knight, Ruth Adams, *Stand By for the Ladies! The Distaff Side of Radio,* New York: Coward-McCann, 1939
Lacey, Kate, *Feminine Frequencies: Gender, German Radio, and the Public Sphere, 1923–1945,* Ann Arbor: University of Michigan Press, 1996
Signorielli, Nancy, editor, *Women in Communication: A Biographical Sourcebook,* Westport, Connecticut: Greenwood, 1996
Stamberg, Susan, *Talk: NPR's Susan Stamberg Considers All Things,* New York: Turtle Bay Books, 1993

WOR

New York City Station

WOR-AM (710), now a leading talk/news format radio station in New York City that is heard in 35 states, began in 1922 as the New York metropolitan area's second radio station. WOR was constructed as a marketing promotion for the L. Bamberger and Company Department Store, and it originally served the New York area from Newark, New Jersey. Although its transmitter remains in New Jersey, its studios have been located in New York City for more than 70 years.

After New York station WJZ (later WABC) was launched in October 1921, Edgar Bamberger decided to start a radio station to promote both the crystal radio sets sold at the L. Bamberger and Company store and the store itself. Jacob R. Poppele, a former wireless operator for the U.S. Army who worked in the store's radio department, strung up antenna wires and launched station WOR on 22 February 1922 as its chief engineer. Initially, WJZ and WOR alternated daylight and evening hours. WOR operated at 833 kilohertz with 250 watts of power and moved to 740 kilohertz when it increased its power to 500 watts. When the American Society of Composers, Authors, and Publishers (ASCAP) sued WOR for copyright fees in 1923, WOR argued that it was providing a cultural service rather than broadcasting for profit. However, the courts decided that repeated emphasis of the station's sponsorship by L. Bamberger and Company as "one of America's Great stores" excluded it from the realm of charitable enterprises.

Early schedules alternated vocal and instrumental numbers with talk and various programs aimed at self-improvement. One of the most enduring of these early shows was a morning exercise program begun in 1925 featuring Bernarr Macfadden, a physical culture enthusiast and publisher of the magazines *Physical Culture* and *True Story.* Macfadden paid WOR for the privilege of hosting the program. When Macfadden came down with laryngitis, John B. Gambling filled in and eventually took over the morning exercise show. That show ended in 1934, but John Gambling stayed on as a morning fixture. Two succeeding generations of Gamblings have held the morning-show host position since—John A. until 1991 and John R. Gambling for the final decade, 1991–2001.

In 1927 WOR moved to the 710-kilohertz clear channel frequency. That same year, WOR joined with a group of radio stations enlisted by Arthur Judson and George A. Coats to form the United Independent Broadcasters (UIB), which required that participating stations sell ten hours of station time each week to UIB. WOR was the key station for the UIB group, which merged with Columbia Phonograph Broadcasting System and later became the Columbia Broadcasting System (CBS). However, WOR separated from the CBS chain in 1928 when WABC joined and became the key transmitter station for the chain. Competing department store chain R.H. Macy and Company acquired a controlling interest in L. Bamberger and Company in 1929, and the licensee name for WOR was changed to the Bamberger Broadcasting Service. By 1935 a new 50,000-watt transmitter and directional antenna sent WOR airwaves over a "bean-shaped area" up and down the eastern seaboard, garnering it a huge audience.

WOR joined three other major-market stations in 1934 to form the Mutual network, and it was the anchor station for Mutual for many years before leaving the network in 1959. Mary Margaret McBride got her start as *Martha Deane* on WOR from 1934 to 1940. *Uncle Don,* played by Donald Carney, a popular figure with young listeners, told stories and provided avuncular advice. In 1940 Ed Fitzgerald and his wife Pegeen created *The Fitzgeralds,* a breakfast show broadcast from the couple's 16th-floor apartment overlooking Central Park that achieved an audience of 2 million listeners as they discussed the day's upcoming social events and the morning news. The station featured another long-lived radio family—the McCanns. Alfred W. McCann, a muckraking journalist, used *The McCann Pure Food Hour* to expose the practices of the food industry in the late 1920s. After his death in 1931, son Alfred W. McCann Jr. took over and was later joined by his wife, Dora, to broadcast *The McCanns at Home* from their house in Yonkers. Daughter Patricia McCann continued the family tradition with *The Patricia McCann Magazine* until 1983.

In 1940 J.R. Poppele oversaw the creation of another WOR entity—New York's first commercial FM station, carrying WOR programs. During the 1940s, the *American Forum of the Air* staged debates on various topics. The station also originated dramatic series, such as *Nick Carter, Master Detective;* game shows; and soap operas. It attracted talent such as Henry Morgan, Cab Calloway, Arlene Francis, and theater critic Jack O'Brian. In the mid-1950s, Jean Shepherd joined WOR and stayed for 21 years as a personality and raconteur. In 1948 Bamberger Broadcasting System became a part of General Teleradio, and in 1952 Don Lee Broadcasting System purchased WOR. In 1955 the station was acquired by RKO Pictures (later RKO General).

WOR-FM developed a new format described as "ahead of its time," featuring "alternative" music such as the Beatles and Rolling Stones album cuts in 1966. It switched to an oldies format called the "Big Town Sound" in November 1967. When, in 1972, it dropped its oldies format and lost half its audience, WOR-FM became WXLO and later WRKS.

Attempting to update its programming in the 1970s, WOR-AM moved increasingly toward a talk format. A 1978 New School of Social Research survey reported that independent stations such as WOR produced more public-affairs programming than network radio competitors. From 1973 and for the next 20 years, "folksy foggy-voiced" radio personality Bernard Meltzer hosted the programs *What's Your Problem?* and *Guidance for Living,* which combined advice on finances, real estate, taxes, and personal problems. During the 1980s, both WOR and its archrival WABC adopted the talk radio format and struggled with revamping their programming to attract a younger audience. In 1988 the Federal Communications Commission (FCC) approved the sale of WOR-AM to S/G Communications, which in turn sold it a year later to Buckley Broadcasting Corporation for $25.5 million, after RKO General lost its licenses because of fraudulent billing practices in the mid-1970s.

Along with its mellow programming personalities, WOR also had its share of controversial figures. Irwin H. Sonny Bloch, a financial talk show host, was aired by WOR until Bloch admitted to swindling his listeners and was jailed. Eric Braverman, a doctor with a weekly show on alternative medicine, lost his medical license in 1996. WOR also hired controversial talk show host Bob Grant from rival WABC-AM in 1996 after Grant was fired for racial slurs against deceased Commerce Secretary Ron Brown.

WOR-AM talk format includes *Rambling With Gambling; Dr. Joy Brown,* a radio psychologist; *The Bob Grant Show;* financial advice from Ken and Daria Dolan; *The Joan Rivers Show; Health Talk* with Dr. Ronald Hoffman; and a food program by *Daily News* critic Arthur Schwartz. The station also carries sports coverage of the New Jersey Nets and Rutgers University.

WOR-AM talk format includes journalist Ed Walsh on *The Morning Show, Dr. Joy Brown,* a radio psychologist; *The Bob Grant Show,* financial advice from Ken and Daria Dolan, and Joan Hamburg, a talk show host wildly popular with women over 30 and whose style is reminiscent of WOR's earlier successes such as *The Fitzgeralds. Health Talk* with Dr. Ronald Hoffman, a food program by *Daily News* critic Arthur Schwartz, and an evening talk show host Tom Marr, brought in from Baltimore to replace Joan Rivers, were on the 2003 roster. Joey Reynolds, a former shock jock who is billed by the station as the "Mr. Nice Guy of Night Radio," holds down the night time hours; his show is distributed by satellite to 35 stations nationwide. WOR does use some outside programming—Fox-syndicated commentator Bill O'Reilly's program *The Radio Factor* is featured during late afternoon hours.

L. CLARE BRATTEN

See also Gambling, John; Mutual Broadcasting System; WABC

Further Reading

Jaker, Bill, Frank Sulek, and Peter Kanze, *The Airwaves of New York: Illustrated Histories of 156 AM Stations in the Metropolitan Area, 1921–1996,* Jefferson, North Carolina: McFarland, 1998

Wilson, George, "With Bob Grant at the Mike, WOR Means 'White Only Radio,'" *New York Amsterdam News* (11 May 1996)

World Radiocommunication Conference

The World Radiocommunication Conference (WRC) is an important global forum in the planning for and utilization of radio spectrum. Though only created in 1992, it builds on more than a century of international cooperation in telecommunications.

In response to unprecedented communication and technological convergence, the International Telecommunication Union (ITU) was restructured in 1992, and a new sector, the Radiocommunication Sector (ITU-R), was created. Under the 1992 ITU Constitution and Convention, the ITU-R replaces the World Administrative Radio Conference (WARC), which met every few years on matters related to radio spectrum.

The ITU-R is responsible for all ITU radiocommunication activities. Its mission includes ensuring worldwide rational, equitable, efficient, and economical use of frequency spectrum by all radiocommunication services and users around the world. WRCs are organized under the auspices of the ITU-R and serve as its governing body.

Functions

As the main legislative and policy-making body of the ITU-R, the WRC coordinates allocation of spectrum bands for national and international radio communication services among the 189 member countries. Unlike its WARC predecessor (which convened on an ad hoc basis to discuss specific matters related to international regulation of the radio spectrum), the WRC meets regularly (every two years) to discuss and adopt regulations governing the use of the radio-frequency spectrum. The WRC is designed to maintain a fair and efficient allocation of the radio spectrum for multiple uses by avoiding or at least reducing international disagreement and technical interference. The rationale for WRC's action is that the electromagnetic spectrum is a limited resource that must be managed for the benefit of all nations.

The WRC serves as an ITU legislative organ, and its resolutions, recommendations, and regulations have the status of treaties and conventions. However, ITU member countries are not bound by all WRC agreements. Although WRC resolutions and decisions are adopted by consensus, member countries are allowed to issue declarations reserving their right to take actions necessary to safeguard their national interests. For example, the United States has taken a number of these so-called exceptions to WRC or ITU decisions. Various national declarations and reservations notwithstanding, however, most nations participate in the technical standardization of worldwide radio communication to facilitate communications as well as world trade in radio equipment.

Nongovernmental Organization Participation

The ITU has over 600 nongovernmental organization (NGO) members representing a wide range of interests in telecommunications. These private-sector NGOs include major telecommunication and internet service providers, equipment manufacturers, broadcasters, and network and radio infrastructure designers, as well as regional and international organizations. These all play an increasingly important role in the WRC. They serve as information technology developers, exhibitors, consultants, and advisers, among other roles. Despite the active participation of the private sector in WRCs, NGOs do not have the rights and obligations generally appertaining to ITU member states. Only member countries or their accredited representatives are allowed to vote on WRC resolutions and conventions.

The WRC in the Digital Age

WRC meetings may revise radio regulations and associated frequency assignment and may address any radiocommunication matter of international importance. The WRC is concerned with the entire range of radio services: both AM and FM broadcasting, satellite broadcast services, mobile and amateur broadcasting, and a host of nonbroadcast services. In the last ten years, WRCs have allocated frequencies and set standards for wideband high-definition television (HDTV), mobile satellite services, terrestrial public communication, aeronautic services to provide telephone communication for passengers in commercial aircraft, satellite or terrestrial digital audio broadcasting, and maritime distress communications. The most significant recent WRC resolutions and agreements include treaties governing the use of the geostationary-satellite orbit, HDTV, low earth-orbiting satellites, and high-frequency national and international radio services (including amateur and government-sponsored international broadcasting). WRC meetings are usually forward-looking conferences held to consider and update regulations on specific terrestrial and space radiocommunication services in order to keep up with technological changes. For example, the WRC-2000 conference, which took place in Istanbul, Turkey, allocated additional radio spectrum for third-generation international mobile telecommunication; agreed on conditions under which a new wave of non-geostationary satellites will operate; and assigned new broadcasting satellite orbits to Europe, Africa, and the Asia-Pacific region.

WRCs and International Politics

Because all countries negotiate WRC agreements in the light of their national interests, WRC meetings have not been free of controversy. For example, in the WRC that charted the method to be followed in planning fixed satellite services ("Space WARC 1985"), regional and national interests were very evident. Developing countries, which did not have the financial or technological resources to launch their own satellites, feared that all satellite orbital slots would be occupied by industrial nations before poorer countries could develop means of launching their own satellites. The larger number of poorer countries therefore demanded that slots be reserved for their future use as a means of creating equitable access to the geostationary orbit for all countries. The United States and other developed countries preferred an evolutionary system in which countries would be assigned bands on a first-come, first-served basis. The same kinds of pressures were evident in more recent HF (high-frequency or shortwave radio) WRC conferences.

Technological developments in the field of radiocommunication and the convergence of telecommunications, voice, video, and information technology have tendered the role of the WRC more crucial than ever before. Without WRCs' forward planning, resolutions, and treaties, radiocommunication around the world would be chaotic at best and impossible at worst.

LYOMBE EKO

See also Frequency Allocation; International Telecommunication Union

Further Reading

Anselmo, Joseph, "GPS Threat Averted, LEOs Win Big at WRC," *Aviation Week and Space Technology* 147 (1997)

Butler, Richard, "World Telecommunications Conference 95: News from the Front," *Telecommunications* 30 (1996)

Clegg, Andrew, "Protecting Our Turf," *Sky and Telescope* 90 (1995)

De Vroey, Vincent, "Aviation Spectrum under Attack," *Interavia Business and Technology* 55 (2000)

Doyle, Stephen, "Space Law and the Geostationary Orbit: The ITU's WARC-ORB 85-88 Concluded," *Journal of Space Law* 17 (Winter 1989)

ITU Radiocommunication Sector webpage, <http://www.itu.int/ITU-R/>

Jessell, Harry, "U.S. Declares WARC a Success," *Broadcasting* (9 March 1992)

Liching, Sung, "WARC-92: Setting the Agenda for the Future," *Telecommunication Policy* 16 (1992)

Robinson, Glen, "Regulating International Airwaves: The 1979 WARC," *Virginia Journal of International Law* 21 (1980)

Salameh, A., "Key Wireless Issues to Be Resolved at WARC-'92," *Telecommunications* 26 (May 1992)

Smith, Milton, "Space WARC 1985: The Quest for Equitable Access," *Boston University International Law Journal* 3 (1985)

Steiglitz, Martin, and Christine Blanchard, "Frequency Allocations Accommodate New Commercial Applications," *Microwave Journal* 35 (1992)

World War II and U.S. Radio

A brief radio bulletin brought the reality of another world war to American listeners on 7 December 1941. At 2:26 P.M. EST, the Mutual Broadcasting System (MBS) interrupted coverage of a Sunday afternoon football game to announce that Japanese warplanes had bombed U.S. forces at Pearl Harbor, Hawaii. The other networks quickly followed suit, John Daley of CBS mispronouncing the name of Oahu in his excitement, for example. America and its Allies soon were engaged in a monumental struggle, every aspect of which was in one way or another connected with radio. U.S. radio's prominence during World War II proved to be the medium's journalistic high-water mark. In a sense, radio was the perfect medium to undertake the wartime task of informing, entertaining, and boosting the morale of the American public. In doing so, radio adapted itself to serve three program settings—international, domestic, and military.

International Radio

Such a purpose for radio came not from the United States but rather from the Axis nations of Germany, Italy, and Japan. As early as 1933, German shortwave broadcasts were flooding the airwaves with programming designed initially to provide Germans living abroad a link to the "Fatherland." Such programs gradually became conduits for coded messages meant for German agents living in the United States. And once the United

States' intention to ally with Great Britain became obvious, the German broadcasts began filling with propaganda messages, of which the purpose, according to Adolf Hitler, was "psychological decomposition of the masses."

The international or "external" voice that radio provided the Axis powers soon spread worldwide in some 30 languages. Shortwave listening nonetheless was limited to a U.S. audience of between 3 and 7 million, according to a 1941 survey by the American Institute of Public Opinion. The survey also found that persons listening specifically to German shortwave programming numbered only about 150,000 on any particular day. Those in charge of Germany's North American Service were well aware of the small audience but still found success with the manner in which loyal listeners (many of whom had formed into "listening groups") were able to spread radio's propaganda messages to non-listeners by word of mouth.

In order to counter the prewar propaganda available to American listeners via shortwave radio, President Franklin D. Roosevelt encouraged several members of Congress to introduce legislation that would create a government-owned shortwave radio station to aim programming toward Axis countries. When several bills to this effect failed, Roosevelt encouraged the existing privately owned U.S. shortwave stations not only to assist the government's propaganda efforts, but also to air propaganda programs produced by government agencies.

Several U.S. companies had constructed shortwave radio stations during the 1930s in order to experiment with high-powered, high-frequency broadcasting. The experimental nature of these stations changed in 1936 when the Federal Communications Commission (FCC) reclassified them as "international broadcast stations." Six companies or organizations—National Broadcasting Company (NBC), Columbia Broadcasting System (CBS), Westinghouse, General Electric, Crosley Corporation, and World Wide Broadcasting Foundation—transmitted shortwave signals to Europe, Latin America, and Asia via 14 transmitters just prior to America's entry into World War II. The number of transmitters jumped to 38 as the war began. Similarly, the number of languages in which international programming was aired jumped from six in 1939 to nearly two dozen by 1942. Privately owned U.S. shortwave stations were not openly cooperative with the government's efforts to use their facilities. Nonetheless, the Office of the Coordinator of Inter-American Affairs (OIAA) began supplying news items in August 1941 for broadcast to Latin American countries.

Matters changed substantially when the United States entered the war in December 1941. War emergency provisions of the Communications Act of 1934 empowered the president to order any U.S. radio station, domestic or international, either to cease operation or to operate under government control. Neither of these options was applied to domestic stations (with the exception that all amateur radio operators were ordered off the air), but such was not the case for international stations. The government arranged to "lease" the transmitters of these stations and to place their control under the Office of War Information (OWI) in June 1942. Assurances were given all international broadcast station licensees that stations would be returned to them following the war. Station staffs were retained by the government, partly because of their experience and partly as reassurance to station owners that government supervision was only temporary. The broadcast service of the now-combined international stations was called the Voice of America (VOA). Seventy-five percent of VOA programs consisted of news, with the remainder of program time devoted to music and features.

The broad objectives of the VOA were to engage in psychological warfare against the enemy and to provide news and morale-building programs for U.S. Allies and armed forces. Programs of the former type were supplied by the Office of Strategic Services (OSS). Much of the OSS effort was aimed at formulating quick responses to Nazi propaganda broadcasts that had been intercepted and translated by the newly created Foreign Broadcast Monitoring Service (later called the Foreign Broadcast Intelligence Service). News programs were typically 15 minutes in length and were designed to provide a true account of whatever events were reported.

VOA entertainment programming was meant to reflect life in the United States. Programs ranged in length from 15 to 30 minutes and were produced primarily in the New York studios of CBS and NBC. By September 1943 the VOA was transmitting approximately 2,600 shortwave programs per week. A majority of these programs were beamed to Europe, but two West Coast shortwave stations also beamed VOA programming to the Far East in Japanese and a number of Chinese and Filipino dialects. The VOA was assisted by Australian radio stations that picked up and rebroadcast its programming. The same was true in Europe, where the British Broadcasting Corporation (BBC) relayed VOA programming into occupied countries via BBC standard-band radio stations. The European operation actually began in May 1944, just prior to the D-Day invasion, and was called the American Broadcasting Station in Europe (ABSIE). ABSIE was programmed much like the BBC, with music and news. The news was broadcast in seven languages and often carried comments from exiled leaders; nearly a third of the ABSIE broadcast day was aimed into Germany. ABSIE disbanded in July 1945, after reaching a peak of nearly 80 percent of listeners in occupied Europe.

Domestic Radio

Nearly 82 percent of U.S. households and 30 percent of U.S. automobiles were equipped with at least one radio receiver as World War II began, and more than 900 U.S. radio stations, 76 percent of which were affiliated with one of the four major networks—NBC Red, NBC Blue, CBS, and MBS—served listeners before war's end. A demonstration of radio's power to reach listeners occurred on 8 December 1941 when President Roosevelt delivered his noontime address asking that Congress declare war on Japan. A daytime record of 66 percent of radio listeners tuned in to the address. The night-time record was broken the next evening when 83 percent of the audience heard another address by the president.

As the United States entered the war, radio, especially radio news, had achieved stature as a major unifying force. Radio network and station executives who had been reluctant to program anything that appeared to breach the nation's declared neutrality now were able to act in concert. As a result, practically all of the radio programs that listeners were accustomed to hearing—from drama to comedy and from variety show to soap opera—incorporated patriotic, morale-boosting themes.

Radio news during World War II was, for all intents and purposes, NBC and CBS News. CBS News became a force in journalism in a sudden burst of improvised reporting activity at the very outset of the war. On 11 March 1938, when the German army occupied Austria, Edward R. Murrow and William L. Shirer were the only two foreign correspondents employed by CBS Radio. Both, however, were resourceful enough to arrange to cover the *Anschluss*, and at 8 P.M. on 13 March 1938, Murrow and Shirer, along with several newspaper reporters scattered in various European cities, went on the air with live reports of reaction to the German occupation. The CBS *World News Roundup*, as it came to be called, was successful enough to be made a permanent wartime fixture. Radio from then until war's end was not only the chief information source for most Americans, but also the information source that Americans held in highest esteem.

Radio performers such as Jack Benny mixed regular program fare with wartime requests to conserve scarce items or to grow "Victory Gardens." Networks also produced special programs built around stars to make particular mobilization appeals. One such program with Kate Smith, airing on CBS in February 1944, was successful in raising some $112 million in war bond sales. Comedian Bob Hope began a tradition that lasted until the Vietnam War era of taking his show on the road, entertaining radio listeners and military personnel alike at various military locations at home and abroad. The music programs (most of which were in the popular or "pop" category) that filled nearly a third of the network schedules also joined the war effort with the likes of Frank Sinatra or Glenn Miller and his Orchestra performing tunes such as "Praise the Lord and Pass the Ammunition" and "This Is the Army, Mr. Jones."

New dramatic programs such as *Counter Spy* and *Alias John Freedom* incorporated wartime themes exclusively, as did documentary series such as *To the Young, Report to the Nation,* and *This Is War.* All of these were meant to stimulate morale as well as to inform listeners about U.S. war policy. Youthful radio listeners were introduced to Americanism and patriotism via the heroic deeds of such characters as Jack Armstrong, Superman, Tom Mix, and Captain Midnight. These and similar characters battled the treacherous deeds of assorted Axis villains. Messages imploring children to collect paper or to buy war stamps were also inserted into programming especially produced for that age group. And many of the nearly 50 morning and afternoon serials or "soap operas" such as *Young Dr. Malone, Backstage Wife,* and *Front Page Farrell* adapted their plots and characters to wartime situations.

Special programs that demonstrated the unified spirit of the American entertainment industry occasionally appeared throughout the war. One of these, "We Hold These Truths," was produced by the U.S. government and had been intended as a commemoration of the 150th anniversary of America's Bill of Rights. It aired live simultaneously in prime time on all four U.S. radio networks only eight days after the Japanese attack on Pearl Harbor. Created by noted writer Norman Corwin, "We Hold These Truths" reached a radio audience estimated at the time to have been the largest ever to hear a dramatic performance. President Roosevelt, along with a cast of some of America's most famous performers, including Lionel Barrymore, Walter Huston, Marjorie Main, Edward G. Robinson, Orson Welles, and James Stewart, all lent their voices and talent to a program in which the original intentions of celebration gave way to a more somber call to patriotism.

The U.S. government played other prominent roles in radio programming during the war. For example, *The Army Hour,* financed by NBC but produced by the U.S. Army, began its hourly Sunday afternoon broadcasts in April 1942. *The Army Hour* was produced not only for domestic listeners, but also for military personnel around the globe. The show gave listeners a blend of entertainment and information about the army's mission and featured well-known celebrities and army personnel from privates to generals. For instance, it was here that listeners heard General Jimmy Doolittle tell of Army Air Corps bombing runs over Tokyo. *The Army Hour,* originating from military bases around the globe, remained popular throughout the war.

The U.S. War Department found radio of use in other ways. Noted writers such as Norman Corwin and William Robson were commissioned to prepare scripts for programs meant to raise public perception regarding women serving in the armed forces and regarding the fighting abilities of African-American troops. And when morale problems crept into the numerous military training bases around the country, the War Department requested that networks provide some entertainment relief by originating programs from the bases themselves. The networks complied, but before the originations began, the networks had to promise to abide by another request—to begin each program with a disclaimer for endorsements by either the army or War Department of commercial products represented by the programs' sponsors.

A number of government agencies produced radio programs that were available to networks and local radio stations wishing to air them. All such programs had to be approved by the OWI, and many of the agencies producing them were assisted by such civilian groups as the Writers' War Board, the War Advertising Council, the Council for Democracy, and the War Activities Committee of the Motion Picture Industry. Members of these wartime organizations devoted time, talent, and expertise to the cause. *Treasury Star Parade* was one such product of this civilian/government collaboration. The 15-minute series was produced by the U.S. Treasury Department and was meant to persuade listeners to buy war bonds. Leading performers of the day donated their dramatic and musical talents to *Treasury Star Parade*. The program, reproduced on electrical transcriptions (called "ETs") that resembled the contemporary long-playing record, was distributed free upon request to over 800 radio stations nationwide.

The government played other key roles during World War II that affected the radio industry. For instance, in April 1942 the FCC discontinued issuing permits to construct both AM and FM stations until war's end. The commission's rationale for its decision was the shortage of construction material resulting from wartime needs and the shortage of trained personnel necessary to operate radio stations. Even before the war had begun, in September 1940, Roosevelt had created the Defense Communications Board (later renamed the Board of War Communications), which was made up of government representatives and civilian representatives of the radio industry and was charged with deciding how best to utilize American radio during the war. The Board's most significant contribution was allowing radio networks and stations to operate as usual with only a modicum of extraordinary government oversight.

Practically all of that oversight came via the Office of Censorship (see following) and units of the FCC. The FCC's Radio Intelligence Division had been created prior to the war to search for unlicensed radio stations. That job became more critical during the war with the threat of clandestine stations broadcasting subversive messages. The FCC's Foreign Broadcast Intelligence Service also existed prior to the war to monitor and transcribe foreign broadcasts and to make their transcriptions available to the appropriate government agencies.

Voluntary censorship was assisted by an Office of Censorship (OC), which was created by the president's executive order only two days after the Pearl Harbor attack. One of the OC's first objectives was to prepare a code for broadcasters that would include advice on what should or should not be said on the air and how certain kinds of information might be handled.

The OC issued the first of five editions of its *Code of Wartime Practices for American Broadcasters* in January 1942. Because of ways in which the enemy might make use of otherwise innocent kinds of programming, radio station licensees were cautioned to abstain from broadcasting, among other things, weather reports, information about military troop locations or deployments, identification and location of naval ships or military aircraft, military base locations, and casualty counts as a result of military action. Licensees were also cautioned to avoid musical request programs or interview programs that relied on extemporaneous comments and to avoid dramatic content that portrayed the horrors of battle. The OC staff spot-checked network programs for any content that might prove problematic. The staff also reviewed program scripts that networks or stations submitted voluntarily. With few reported breaches of security, the radio industry's ability to look after its own house during the war worked exceptionally well.

The one new FCC unit was the War Problems Division, which the commission established in its law department in 1942 to monitor the nation's 169 radio stations that broadcast all or a major portion of their programming in foreign languages. Because German and Italian were the predominant languages of these programs, there were fears that agents of the Axis powers could easily use the programs to plant subversive messages. The stations could have been ordered off the air, of course, but the FCC realized that they served a significant number of listeners (estimated at 14 million foreign-born and first-generation Americans) who might otherwise listen to shortwave stations originating in Axis countries if they were not served by American stations. In order to police themselves, the stations created the Foreign Language Radio Wartime Control, which adopted a code requiring strict oversight of program content and station employee loyalty. This code eventually was incorporated into the *Code of Wartime Practices for American Broadcasters*.

The abundance of programming for World War II radio was matched by the abundance of advertising. In fact, only one year after the war began, the U.S. radio industry billed a record $255 million for commercial time. And things simply got better after that. Two factors contributed to radio's economic well-being. One was the shortage of newsprint, which curtailed the amount of space many newspapers could allow for advertising. Businesses naturally turned to radio, where no such space constraints existed. As a result, radio's total share of the advertising dollar climbed steadily throughout the war years. The second economic factor favoring radio derived from a tax levied by the government on excess profits as a means of curtailing wartime profiteering. The 90 percent tax on profits exceeding what a company would normally be expected to make was exempted when these profits were spent for advertising. Instead of keeping what amounted to only ten cents for every dollar, companies dumped much of their excess profits into radio advertising.

Radio advertising in the form of program sponsorship also benefited from government largesse. Companies under contract to manufacture war-related products were allowed to count advertising along with other manufacturing costs and to pass along to the government all the bills for doing business. This formula allowed a number of companies to dedicate large amounts of money to sponsoring radio programs. Such sponsorship amounted to paying the full production costs of programs, with the results that a number of network programs either remained in production or were created as a result of funding that came indirectly from the government.

Wartime conditions changed radio advertising content in a number of ways. For one, a number of consumer products either were not available or were in short supply and thus rationed. Companies whose products were thus affected often turned to institutional advertising just to keep their name or brand in the listener's mind. This kind of advertising was particularly beneficial to the fine arts when, for example, a corporation such as General Motors that had no cars to sell nonetheless gave its name to sponsor the NBC Symphony Orchestra. Other companies tied their name or product to the war effort by the messages they chose for their radio commercials. A famous western hat company, for example, mixed its product name with a plea to guard against spreading rumors by admonishing Americans to "Keep it under your Stetson."

Finally, domestic radio stations underwent some wartime changes. Many stations located in markets that were also the homes of important war industries adjusted their broadcast day in order to provide industry and plant workers something to listen to as they worked. Radio was also considered an essential industry by the government, and many key radio station personnel were granted deferments for military duty by the Selective Service System. Of course, although some decisions were made to take advantage of these deferments, many broadcasters still joined the military when duty called. Broadcast engineers in particular provided technical expertise that was useful to the military. They contributed considerably in developing America's wartime electronics capability. And their skills advanced in such ways that the radio engineers who entered World War II left it in 1945 to form the advance troops in postwar development of the television industry.

Military Radio

U.S. military personnel were among the most avid listeners of the shortwave international stations that had been transformed by the OWI into the Voice of America. Programming produced especially for the "G.I." began appearing in early 1942 as a result of the U.S. Army's Bureau of Public Relations (BPR). The BPR's first and perhaps best-known program was *Command Performance.* The one-hour program aired weekly on Sundays and carried a blend of comedy, sports summaries, popular music, and celebrity appearances. The *Command Performance* title derived from the program's key element of entertainers responding to military personnel requests such as singing a special song. The program was produced from several takes, allowing deletion of questionable material or material deemed censorable by the military, and then recorded finally onto wax disks for the Sunday airing. *Command Performance* originated in New York but was moved to Los Angeles shortly thereafter because of the Hollywood talent pool. After producing 44 of the programs, the BPR relinquished control of *Command Performance* to the Armed Forces Radio Service (AFRS).

The AFRS was created in 1942 as a unit within the War Department's Information and Education Division and was headed by Col. Thomas H.A. Lewis. The AFRS was headquartered in Los Angeles and run by staff members from the army and navy as well as civilians. The service was closely connected to the entertainment industry and thus attracted top talent for AFRS programs. By the end of 1943 AFRS had 306 outlets and stations in 47 countries. Each outlet received more than 50 hours of recorded programming per week, delivered by plane, as well as special shortwave programming (e.g., news and ball games) and programs that were produced in the field or at the stations themselves. All programs were reproduced on unbreakable 12-inch vinylite disks, some 83,000 of which were shipped every month from Los Angeles to AFRS outlets. Reuse of commercial programs required cooperation of network and transcription companies as well as concession of rights of entertainers, sponsors, advertising agencies, musicians, copyright claimants, and publishers.

AFRS programs either were produced by the AFRS itself or were network series that had been "denatured." Denaturing meant deleting commercials that normally appeared in the network programs from the transcribed versions heard by the G.I. so as not to give unfair advantage to particular advertisers. There also was fear that soldiers far from home and living under less than ideal conditions would be depressed to hear commercials for food and drink products or goods meant to provide comfort and relaxation. But when soldiers complained of missing the familiar sound of commercials, the army began inserting gag commercials that spoofed the real ones. And since many commercial network programs used the name of the program sponsor in the show's title, completely new openings and closings had to be created by the AFRS and substituted for the actual title. Thus, *Camel Caravan* became *Comedy Caravan* and *Chase and Sandborn Hour* became *Charlie McCarthy.* AFRS also combined dramatic episodes of several programs into anthology-like series with titles such as *Front Line Theater, Globe Theater,* and *Mystery Theater.*

Besides entertainment programs, educational and documentary programs were carried on the AFRS network. *Heard at Home* was another combination program that pulled segments from such shows as *The Chicago Roundtable, America's Town Meeting of the Air,* and *People's Platform.* The AFRS also produced its own educational programs, such as *Know Your Enemy* and *Know Your Ally. Mail Call,* a musical variety program that began in August 1942, was the first AFRS-produced program. For commercially produced shows, several different AFRS offices were responsible for examining program content for security violations, technical quality, authenticity, and compliance with current policy, such as race relations policy. AFRS could not supply all programming needs, and besides, troops preferred the familiarity of U.S. commercial network programs.

Most of the AFRS military radio stations operating in the field were called "American expeditionary stations." Their prototype was a station in Kodiak, Alaska. Soldiers stationed there soon after the war began were unhappy with the unreliable shortwave service in that remote area. To improve matters, the soldiers built and operated their own low-powered transmitter and a makeshift station that broadcast to troops stationed nearby. Programs consisted mainly of recorded music supplemented by occasional news reports as items were received via shortwave.

A more significant makeshift radio operation began in Casablanca shortly after U.S. forces invaded North Africa in November 1942. Present with the troops was Major Andre Baruch, who had been a CBS radio announcer prior to the war and who asked permission of General George H. Patton to construct a radio station for his comrades. The general approved, and Baruch set about molding spare parts into what became the U.S. Army's first expeditionary radio station on 15 December 1942. Major Baruch entertained troops with "platter-chatter" and music from his own limited collection of records. Radio receivers were not plentiful and either had to be purchased from French Moroccan radio shops or built from spare parts. A short time later, soldiers at Anzio were creative enough to build a simple receiving device called "Foxhole Radio" that was similar to the crystal sets of the 1920s.

The idea that radio was an important element in keeping troop morale high led to General Dwight D. Eisenhower's ordering the creation of additional stations that either were stationary or traveled with troops. Transmitters for the 50-watt mobile stations were carried in five portable cases and accompanied the Fifth Army as it advanced into Italy. At key locations along the route, permanent stations were established by Major Baruch. By March 1943 these stations were joined to form the Mediterranean Army network. A similar network of stations, called the Armed Forces network, was built to serve American troops stationed in England and began broadcasting in July 1943. And in due course, an equivalent South Pacific operation called the "Mosquito" network was created with stations built on islands scattered throughout the area. All of these stations, of course, carried the AFRS programming supplied on disk, but they also provided listeners with localized programming produced at the stations themselves.

Radio at War's End

Reporting about military matters became more sophisticated as the war progressed. Radio reporters who accompanied troops were assisted in their reporting efforts by field recording techniques using portable wire recorders. During the latter part of the war, these reporters were indirectly assisted by German technical advances that had replaced the difficult-to-edit wire recorders with plastic tape recorders. Captured recorders of this new variety allowed reporters to cover a variety of dramatic wartime events and to quickly get their eyewitness accounts back to their respective networks. Thus, ABC's George Hix provided listeners with vivid accounts of the 6 June 1944 D-Day invasion from his observation point aboard a navy ship. And listeners heard the recorded reactions to nearly incomprehensible horrors on 13 August 1944, when a Mutual reporter first entered Maidanek extermination camp, which had been liberated only a few days earlier by the Russian army.

At least three months of fighting remained in the Pacific when Germany surrendered in May 1945. There was much V-E Day celebrating over the triumph of the Allied forces, and radio was in the thick of it. On 8 May 1945, CBS aired *On a Note of Triumph,* written by Norman Corwin to commemorate the event. *On a Note of Triumph* proved so popular that CBS rebroadcast it live on 14 May. The program eventually was released as a commercial record album, and Simon and Schuster put out a book version of the script that quickly made the best-seller list. The program in a symbolic sense brought radio full circle. Radio had made a commitment of so many of its resources to join in the collective experience that World War II became. Radio and its listeners—wherever they might have been and in whatever endeavor they might have been engaged—had formed a community during the war and had effectively woven themselves into the very fabric of that event.

RONALD GARAY

See also American Broadcasting Station in Europe; Armed Forces Radio Service; British Forces Broadcasting Service; Churchill, Winston S.; Commentators; Corwin, Norman; Davis, Elmer; International Radio Broadcasting; Murrow, Edward R.; News; Office of War Information; Propaganda; Shirer, William L.; Voice of America

Further Reading

Bannerman, R. LeRoy, *Norman Corwin and Radio: The Golden Years,* Tuscaloosa: University of Alabama Press, 1986

Barnouw, Erik, *A History of Broadcasting in the United States,* 3 vols., New York: Oxford University Press, 1966–70; see especially vol. 2, *The Golden Web, 1933 to 1953,* 1968

Browne, Donald R., *International Radio Broadcasting: The Limits of the Limitless Medium,* New York: Praeger, 1982

Brylawski, Samuel, "Armed Forces Radio Service: The Invisible Highway Abroad," *The Quarterly Journal of the Library of Congress* 37 (Summer-Fall 1980)

Davis, Elmer Holmes, and Byron Price, *War Information and Censorship,* Washington, D.C.: American Council on Public Affairs, 1957

Dryer, Sherman Herman, *Radio in Wartime,* New York: Greenberg, 1942

Garay, Ronald, "Guarding the Airwaves: Government Regulation of World War II American Radio," *Journal of Radio Studies* 3 (1995–96)

Kirby, Edward Montague, and Jack W. Harris, *Star-Spangled Radio,* Chicago: Ziff-Davis, 1948

Landry, Robert John, *This Fascinating Radio Business,* Indianapolis, Indiana: Bobbs-Merrill, 1946

MacDonald, J. Fred, *Don't Touch That Dial! Radio Programming in American Life, 1920–1960,* Chicago: Nelson-Hall, 1979

Redding, Jerry, "American Private International Broadcasting: What Went Wrong—and Why," Ph.D. diss., Ohio State University, 1977

Rolo, Charles James, *Radio Goes to War: The "Fourth Front,"* New York: Putnam, 1942

Rose, Ernest, "How the U.S. Heard about Pearl Harbor," *Journal of Broadcasting* 5 (Fall 1961)

Siepmann, Charles Arthur, *Radio in Wartime,* London and New York: Oxford University Press, 1942

Summers, Robert Edward, *Wartime Censorship of Press and Radio,* New York: Wilson, 1942

Willey, George, "The Soap Operas and the War," *Journal of Broadcasting* 7 (Fall 1963)

Winkler, Allan M, *The Politics of Propaganda: The Office of War Information, 1942–1945,* New Haven, Connecticut: Yale University Press, 1978

WQXR

New York City Station

One of the very few radio stations to begin as part of an experimental TV operation, WQXR (as it became in 1936) and its later FM associate station became in 1944 the voice of the *New York Times* in New York City, renowned for classical music and arts programming. Elliott Sanger, both father and son, played key roles in the station's creation and operation for decades.

Origins

This New York classical music station had a unique start—as the sound channel for a very early experimental *TV* transmitter. In 1929 radio inventor J.V.L. Hogan received a Federal Radio Commission (FRC) license for W2XR, based in Long Island City, New York, to develop a mechanically scanned TV and facsimile service. Whereas the pictures were sent out on 2100 kHz, as of 1933 a second, sound, channel at 1550 kHz was added. In accord with FRC rules of the time, this was one of three 20 kHz channels (each twice the width of existing AM radio channels). Using this new channel, Hogan used classical music recordings as his sound background for the crude pictures. Audience interest in the music (most could not receive the pictures) prompted the idea of a commercial station providing such content. Indeed, the eventual failure of mechanical approaches to television led to an increased focus on the sound channels.

Working with engineer Al Barber, Hogan focused on providing high fidelity sound by using special broadcast-only electrical transcription recordings and various filters to improve the music transmitted. Equipment was the best available—or was made especially for W2XR operation. By 1934, using 250 watts, the station was on the air four hours daily and was providing a program log for listeners.

The big change came two years later when, teamed with advertising and public relations expert Elliott M. Sanger, the experimental W2XR became commercial station WQXR with 1,000 watts of power, still on 1550 kHz. But this was a different kind of commercial operation, because the station (with all of six employees) picked and chose its advertisers and their messages carefully to match the high tone of the programming.

WQXR would not accept advertising for products that were in poor taste or represented a bad value—and the method of presentation of those that were accepted "must be in keeping with the quality of the broadcast programs." Prophetically, in light of what happened eight years later when the *Times* bought the station, the original license application stated that the new station would seek to emulate the values of the *New York Times*. Many radios could not even tune in to the station, which was just above what were then the uppermost frequencies on the AM band.

A focus on top-quality sound continued and in 1938 WQXR broadcast the first tape-recorded program (part of the opera *Carmen*), although the method of recording used was short-lived. Power rose to 10,000 watts (on 1560 kHz) in 1941. Hogan and Sanger's Interstate Broadcasting Co. became an FM licensee, with New York's first FM station, W2XQR (with a transmitter loaned by FM inventor Edwin Armstrong) on 26 November 1939, which became W59NY (45.9 MHz) when the FCC approved commercial operation in 1941. When FM call letter rules changed in 1943, the station became WQXQ, and with the service's frequency change it shifted operations up to 97.7 MHz after World War II. Both stations provided the same programming and were now operating from a Fifth Avenue address.

Voices of the *Times*

In mid-1944, the *New York Times* purchased the stations from Sanger and Hogan for $1 million. The program emphasis on fine music continued, as did the extensive schedule of live studio performances. Six years later, the studios moved to the *Times* building. In October 1952 the stations began a two-station transmission of stereo, with AM providing the right channel and FM the left. These were replaced in 1961 with the inception of FM multiplex transmission. The AM outlet's power received a final power boost, to 50,000 watts, in 1956.

In a decade-long experiment, the New York-based classical programming was shared with more than a dozen "affiliate" stations in the WQXR network. Beginning in 1953 stations from Buffalo, New York, to Boston, Massachusetts, to Washington, D.C., formed the "WQXR Network." It closed 10 years later because of insufficient income and the desire of the other stations to have more local say in their program offerings. December 1963 saw the end of the station's monthly program guide, which had first appeared in 1936 (the *Times* carried the program listings in its pages).

The stations won a Peabody Award in 1959, which noted "no station anywhere has devoted more time or more intelligent presentation to *good* music than has WQXR." In 1967, in response to FCC rules not allowing AM-FM outlets to duplicate their programming, WQXR began to focus on lighter classics, show tunes, and jazz. Several other approaches were tried, but all angered listeners and did not please advertisers. The *Times* briefly considered selling the outlets in 1971, but instead obtained a nonduplication exemption from the FCC and went back to both stations providing the standard classical repertoire that had done so well for so long.

In December 1992 the AM outlet (which over the years had lost much of its audience to its higher fidelity FM twin) became WQEW, with much of the popular music programming that had been on now-defunct WNEW. At the end of 1998 WQEW entered a new stage when the *Times* leased the outlet to Disney for an eight-year period and the format was directed toward children (with a back-up of New York Jets game broadcasts), delivered by satellite and carried by about 40 other stations across the country. WQXR (FM) continued to provide classical music and in January 1997 began an internet website.

CHRISTOPHER H. STERLING

See also Armstrong, Edwin Howard; Classical Music Format; Hogan, John V.L.

Further Reading

New York Radio Guide, <www.nyradioguide.com>

Sanger, Elliott M., *Rebel in Radio: The Story of WQXR*, New York: Hastings House, 1973; as *Rebel in Radio: The Story of the New York Times "Commercial" Radio Station*, London: Focal Press, 1973

"WQXR," in *The Airwaves of New York: Illustrated Histories of 156 AM Stations in the Metropolitan Area, 1921–1996*, by Bill Jaker, Frank Sulek, and Peter Kanze, Jefferson, North Carolina: McFarland, 1998

WQXR website, <www.wqxr.com>

Wright, Early 1915–1999

U.S. Disc Jockey

Early Wright was one of the earliest and longest-running black disc jockeys in the South. In 1947 he began as a disc jockey at WROX in Clarksdale, Mississippi. For nearly 50 years Wright broadcast blues and gospel to the people on the northern Mississippi Delta. Many internationally known blues artists began their careers on his show. He is credited with popularizing the Delta blues sound.

Wright claimed he was born in 1915 on a farm in Jackson, Mississippi. In 1937, after moving to Clarksdale, he became an auto mechanic. During his spare time, he was the manager for the Four Star Quartet Singers, a local gospel group. The group appeared on KFFA's *King Biscuit Time* program, which originated across the Mississippi River in Helena, Arkansas. Named after the sponsor's product, King Biscuit Flour, the show was one of the earliest (1941) to feature live blues artists.

When WROX started broadcasting from Clarksdale in 1944, the Four Star Quartet appeared there. After hearing Early Wright deliver Sunday morning announcements for the quartet, WROX's station manager, Buck Hinman, offered him a job. Two weeks later, after consulting with a preacher to be sure there was nothing sinful about playing blues records on the radio, Wright accepted the job.

From 1947 Early Wright's radio career continued until his heart surgery in 1997. His radio fame is interwoven with the rich blues tradition surrounding Clarksdale and Coahoma County. W.C. Handy, "Father of the Blues," lived there, and the area has been a gathering place for blues musicians since the 1920s.

In the early years Wright hosted live broadcasts of blues artists he knew from the *King Biscuit Time* show. Many, such as Sonny Boy Williamson, Robert Nighthawk, and Joe Willie "Pinetop" Perkins, achieved worldwide fame. Wright's programs were broadcast from the WROX studio at Clarksdale's Alcazar Hotel. The program reached listeners across the northern Delta and sections of eastern Arkansas.

During the 1940s Clarksdale's Issaqueena Avenue was the hub of blues entertainment. Early Wright's program on WROX set the pace for the festivities. Roaring bands led by emerging musicians such as Ike Turner played all night. The Riverfront Hotel was another hot spot for blues music. Artists performed there while rooming at the hotel and traveling through Clarksdale. Before becoming a hotel, the Riverfront had been a black hospital; Bessie Smith, "Empress of the Blues," died there after a car accident in 1937.

At WROX Wright opened the door for Ike Turner in the early 1950s. Turner first appeared on Early Wright's program with the "Kings of Rhythm." Later Turner did his own 30-minute show before moving to WHBQ and Sun Records in Memphis. According to Wright, Ike Turner first recorded "Rocker 88," considered by many to be the first rock and roll tune, at WROX.

Other well-known blues musicians, such as Muddy Waters and B.B. King, gave live performances hosted by Wright. In 1989 the *New York Times* reported that a little-known Mississippi country boy named Elvis Presley had made a radio appearance on the Early Wright show. Impressed with Presley's politeness and showmanship, Wright said "He always had a motion, you know" (Applebome, 1989).

In time the format of his program involved two segments. Usually beginning in the evening around 7:00 P.M., he was the "Soul Man," playing blues music. Later in the evening he was "Brother Early," playing gospel songs.

His delivery, which did not change during his career, was folksy, country, and deep-South sounding. *Living Blues Magazine* wrote of the time Wright delivered a warning about snakes in the black neighborhood. "I want to let you know that some snakes has been seen in the Round-yard neighborhood. The grass has grown up around the sidewalks and snakes has been seen, looking for water. And a man told me the other day, he saw a snake in the street. That's right—snakes has been seen in the street; it's been dry and they are looking for water. . . . Tell the children to be cautious in the street, too, and look out for cars" (McWilliams, 1988).

His down-home style engaged listeners, many of whom were white. Even during the heated 1960s racial disturbances, Wright held a sizable white audience, about a 50-50 split with black listeners. Responding to phone callers, he would say, "Soul Man speaking." On occasion, he left the microphone on, and the audience could hear him talking over the music: "and you wanna dedicate it to who?"

On a typical program, as reported by Peter Applebome of the *New York Times,* he introduced "an extinguished guest," then played the blues of Bobby Rush or the gospel tunes of the "Mighty Sons of Glory," and then reported on who was about to be "funeralized."

He would deliver commercials with the same folksy style. "Now I want to tell you about the Meat House. It's on Highway 61 South, owned and operated by Mr. Askew, the nicest man who ever sliced a piece of meat. . . . They've got two female meat cutters. They're A's instead of B's. The cashier's so nice, it makes the meat taste even better."

Early Wright is a radio legend who spoke the language of his listeners. He died in a Memphis hospital in December 1999. The Delta Blues Museum in Clarksdale has exhibits on the history of the blues and Early Wright. In his honor the city named the street where he lived Early Wright Drive.

FRANK J. CHORBA

See also Black-Oriented Radio; Blues Format; Disk Jockeys; Gospel Format

Early Wright. Born in Jackson, Mississippi, 1917. Began disk jockey career at WROX, Clarksdale, Mississippi, 1947; continued broadcasting and popularizing the Delta Blues on WROX for 50 years. Died in Memphis, Tennessee, 10 December 1999

Further Reading

Applebome, Peter, "Early Wright: An Institution on Mississippi Radio," *New York Times Biographical Service* (28 June 1989)

Barlow, William, *Voice Over: The Making of Black Radio*, Philadelphia, Pennsylvania: Temple University Press, 1999

The Delta Blues Museum, <www.deltabluesmuseum.org>

Martin, Douglas, "Early Wright, 84, Disc Jockey Who Made the Delta Blues, Dies," *New York Times* (17 December 1999)

McWilliams, Andy, "The Night Time Is the Wright Time," *Living Blues Magazine* (September/October 1988)

Oakley, Giles, *The Devil's Music: A History of the Blues*, London: British Broadcasting Company, 1976; 2nd edition, updated, New York: DaCapo Press, 1997

Palmer, Robert, *Deep Blue*, New York: Viking Press, 1981

WRR

Dallas, Texas Station

WRR-FM has the distinction of being one of the few municipally owned commercial FM stations in the United States. WRR-AM began broadcasting in 1920; the FM station began broadcasting in 1948. Rising station values convinced the Dallas city government to sell the AM station in 1978, but the city retained the FM service. Although rising station values have again prompted the city to consider selling WRR-FM, the city has so far elected to retain the station and the distinct fine-arts and classical programming contribution the station makes to the city of Dallas.

A Radio Emergency Service

WRR-AM is one of the pioneer radio stations in the United States. In March 1920 WRR was issued a limited commercial license by the Department of Commerce to begin a specialized wireless service. While WRR claims to be the first station west of the Mississippi River, its original specialized function excludes it from being considered for many of the "firsts" in radio history. WRR was conceived as a two-way communication system to send and receive emergency calls from the city's fire and police departments. Thus its services were not intended for reception by the general public, nor did the station provide general-interest programs in its early years. When not in use, the transmitter's carrier frequency would remain on the air. In the event of a fire or rescue emergency, the silence of the carrier frequency was interrupted with an emergency announcement for firefighters. What would today be termed "programming" was added during the interludes between emergency calls to enable listeners to know that their receivers were set on the correct station frequency.

As with radio in other parts of the United States, the presence of a station and at least limited programming service encouraged listeners to tune in to receive the miracle of wireless transmission. As the radio system developed, the central Dallas fire station served as the base of transmission. Fire Department personnel and dispatchers soon developed an eclectic broadcasting schedule, typical of many early radio stations, consisting of weather reports, joke telling, reading the newspaper on the air, announcing birthdays, and playing music. Such general-interest programming served to encourage more citizens to buy receivers, which in turn increased listener demand for more and improved programming. Eventually, the station was airing not only recorded music but broadcasts of local amateur musicians as well.

The success of WRR led its operation to shift from the fire department to being a municipally operated station. The station began accepting advertising in 1927. By 1939 the station's trailblazer operations led to the construction of a radio building and transmitter tower at the Fair Grounds Park near downtown Dallas. In 1940 the station affiliated with the Mutual Broadcasting System and the Texas State network. Much of the station's format consisted of block programming.

By the 1970s WRR-AM was operating with 5,000 watts of power on 1310 kilohertz. Rising station values led the city of Dallas to sell WRR-AM for $1.9 million in 1978 to Bonneville Broadcasting. Later, Bonneville sold the station to Susquehanna Broadcasting. The station is still on the air using the call letters KTCK and is a profitable all-sports station.

WRR-FM Begins Operation

WRR-FM began broadcasting in 1948, a time when few members of the public owned an FM receiver, but when hopes for the new radio service were high. As with many other FM outlets, WRR-FM began by programming classical music.

WRR-FM has survived a series of city government debates regarding its ownership and operation. Some members of the Dallas City Council have characterized ownership of a commercial radio station as socialism and have viewed the station as being in direct competition with commercial station owners. WRR-FM has consistently maintained a classical music format with a strong emphasis on local news. It is perhaps the classical format that has enabled the station to continue its status as a municipally owned station. Few commercial station operators would be likely to continue the commercial classical music format if the station were sold. The Dallas–Fort Worth market is also served by a classical and news affiliate of National Public Radio (NPR).

WRR-FM Extends Its Coverage

WRR increased its transmitter power and relocated the transmitter and antenna in 1986. The station currently broadcasts in stereo with 98,000 watts on 101.1 megahertz. The station's primary coverage radius extends more than 65 miles from its transmitter location, enabling the station to reach listeners in Fort Worth and the suburbs in north Dallas.

Though a commercial radio station, WRR has adopted a tactic of many noncommercial classical-formatted stations. The group "Friends of WRR" is a nonprofit organization whose mission is to support the station's classical programming. WRR also generates revenue by selling time for religious broadcasts, and it sells a weekday half-hour segment at 6:00 P.M. to a local television station to air the station's local newscast.

WRR programming includes symphonic and opera broadcasts, local arts and cultural programming, and children's programming; the station also broadcasts the bimonthly meetings of the Dallas City Council. WRR-FM continues to be a self-sustaining and profitable radio station. The station is typically one of the 20 most-listened-to stations in the Dallas–Fort Worth radio market, one of the ten largest radio markets in the United States.

GREGORY G. PITTS

Further Reading

Schroeder, Richard, *Texas Signs On: The Early Days of Radio and Television,* College Station, Texas: Texas A&M University Press, 1998

WRR website, <www.wrr101.com>

WSB

Atlanta, Georgia Station

Atlanta's WSB radio is the oldest radio station in the South, born of a competition between two Atlanta newspapers. A clear-channel station, WSB signed on the air 15 March 1922.

Origins

As early as 1921, ham radio operators in Atlanta were asking for a station. Since the sign-on of KDKA in 1920, radio was becoming a big amateur venture, and Atlanta—and surrounding Georgia—jumped on the bandwagon. A former Navy wireless operator named Walter Tison approached Major John S. Cohen, editor and publisher of *The Atlanta Journal,* to discuss the feasibility of a local radio station. Tison persuaded Cohen, and plans went into action for the new station; Tison would become the first federally licensed operator. George A. Iler, an engineer with the Georgia Power Company, was hired to be the first station director.

The Atlanta Journal and *The Atlanta Constitution,* then rival newspapers, were fighting to get the first Atlanta radio station on the air. On 15 March 1922, *The Journal* won when it received this telegram, signed by acting secretary of commerce C. H. Huston:

> *The Atlanta Journal* is authorized to temporarily broadcast weather reports on the wavelength of four hundred eighty five meters pending action on the formal application for a radio license. Station must use radio call letters WSB repeat WSB and employ commercial second class or higher radio operator licensed by this department.

The telegram ended with these words: "If you desire to broadcast news entertainment and such matter this is permitted on wavelength of three hundred sixty meters only."

According to the station's history, by nightfall the station was on the air with these words, "Good evening: This is the Radiophone Broadcasting Station of *The Atlanta Journal.*" That first broadcast came from *The Atlanta Journal* office on Forsyth Street; the station broadcast at a mere 100 watts of power. *The Journal* wrote the next day, "Atlanta is on the radio map of the world today." Three months later, according to the station history, that map grew as the station increased to 500 watts.

WSB Radio remained a noncommercial station located at *The Journal*'s office for its first three years but moved in 1925 to the Biltmore Hotel. On 9 January 1927, according to the station's official history, WSB became a charter affiliate of the National Broadcasting Company and began to sell advertising. These moves paralleled the station's power boost to 1,000 watts.

WSB was a station of firsts. In addition to being the first station in the South, it was the first in the country to have a slogan. According to the station's golden anniversary history, a listener coined the phrase "The Voice of the South," writing, "Because of its remarkable powers of transmission, penetrating alike into lonely cottages in isolated sections and palatial residences in distant cities, WSB has truly become 'The Voice of the South'." A listener contest gave meaning to the call letters WSB—"Welcome South, Brother." Lambdin Key, the first full-time general manager, led programming.

According to the station's official history, WSB was also the first station in the nation to present an entire church service—on Easter Sunday in 1922. The station claims many other firsts, including the first radio fan club, the WSB Radiowls.

By 1933, the station's wattage was increased to 50,000, and WSB had become a permanent and national fixture at 750 kHz. The station made its name by offering first-class news coverage as well as entertainment options.

In December 1939, Governor James M. Cox of Ohio bought the station and *The Atlanta Journal,* adding them to his company that would become the Cox Broadcasting Company. He put J. Leonard Reinsch in charge of the station as general manager. In 1944, Governor Cox made Reinsch managing director of all of his broadcasting properties. Reinsch and former vice president and general manager Elmo Ellis moved the radio station through the initial days of television in Atlanta, a time that saw a national decline in radio usage. Both men saw the station through the turbulent 1960s and into the 1970s. Ellis retired from the station in 1982.

According to the station's history, WSB achieved another first on 16 November 1944, when it put Georgia's first ever Frequency Modulation (FM) station on the air. That station, now known as B 98.5, has changed music format over the years, but it remains a leader and soft rock alternative in the vast Atlanta market.

WSB radio made its last big location move on 28 December 1955 joining its sister television station at Cox's new broadcasting facility known as White Columns, and both remain at that location.

WSB-AM at the turn of the century led its market as an all-news radio station, a format it adopted in the 1980s. It remained a clear-channel station at 50,000 watts and was the national broadcasting center for the Atlanta Braves.

GINGER RUDESEAL CARTER

Further Reading

Welcome South, Brother: Fifty Years of Broadcasting at WSB, Atlanta, Georgia, Atlanta: Cox Broadcasting, 1974

WSM

Nashville, Tennessee Station

WSM, "The Air Castle of the South," is the powerful radio station in Nashville, Tennessee, largely responsible for that city's emergence as a major entertainment and media center. The call letters WSM represent "We Shield Millions," slogan of the National Life and Accident Insurance Company of Nashville, the station's founding licensee.

Other Tennessee radio stations had already begun broadcasting in Memphis, Knoxville, and Nashville by the time WSM signed on the air in 1925. Although most of the inaugural WSM schedule consisted of programs characteristic of the mid-1920s (i.e., light classical and dance music), some of the station's programs were tailored for the farmers and wage earners who constituted National Life's main business, the sale of weekly premium insurance. Very early in WSM's history, a program of rural-flavored music was modeled after the WLS (Chicago) *Barn Dance*. Shortly after its start, WSM's program became known as the *Grand Ole Opry*. The content of the *Grand Ole Opry* engendered some distaste in certain sections of Nashville, but the "*Grand Ole Opry* Insurance Company" prospered from its affiliation with the broadcast.

A "W S *Em*pire" emerged from National Life's 1931 authorization to operate WSM as an unduplicated class 1-A clear channel (650 kilohertz), with 50,000 watts of power full-time and a nondirectional radiation contour. When National Life won the clear channel for WSM, it erected North America's tallest radio tower near Brentwood, about 15 miles south of downtown Nashville. At the dawn of radio's golden age, WSM thus became one of about 25 stations in the United States that could be received reliably at night throughout much of North America.

Although barn-dance programs aired on other clear channel stations during the 1930s and 1940s, Nashville's *Grand Ole Opry* emerged as the leader of that genre for three principal reasons. First, the National Broadcasting Company (NBC) Red network carried a portion of the *Grand Ole Opry* coast to coast for several years starting in 1939, thereby extending the reach of WSM entertainers beyond the range of "clear channel six-fifty." Second, WSM's association with a rural style of music diffused by word of mouth during World War II, as people from various parts of the country served together in the armed forces. Third, touring *Grand Ole Opry* musicians and personalities who entertained World War II troops fostered the image of Nashville as a center for "hillbilly" or "country and western" music.

National Life also was an early commercial FM licensee but abandoned FM broadcasting in favor of television in 1950. The present WSM-FM at 95.5 megahertz (formerly WLWM) was purchased in 1968. In 1950 National Life established Nashville's first television station, WSM-TV, which operated without competition for three years. This head start in television did much to solidify WSM's position of dominance in the Nashville radio and television market, which continued at least through the 1970s.

WSM remained a basic affiliate of the NBC radio network during radio's music-and-news era. Only after television displaced much of the evening radio audience in the 1950s did WSM begin to devote large amounts of its weeknight schedule to recorded country and western music. During the day, however, WSM remained a full-service, mass-appeal station. Middle-of-the-road and, later, adult contemporary recorded music filled the time between network features, local news, and business and agricultural reports. The station maintained a staff of musicians who performed pop standards during a live studio morning show well into the 1970s.

By 1968 National Life had become Nashville's largest corporation, the nation's sixth largest stock life insurance company, and the principal subsidiary of NLT Corporation. In 1972 NLT capitalized on WSM's *Grand Ole Opry* as the theme for Opryland USA, a $40 million entertainment complex followed in 1977 by a $26 million hotel.

The decade of the 1970s probably captures the peak of WSM's influence in Nashville and middle Tennessee. As the radio and television industries began to undergo tremendous change during the late 1970s, NLT broadcasting executives led by Tom Griscom envisioned a way to extend the company's advertising base beyond the reach of WSM, WSM-FM, and WSM-TV. They foresaw the inevitable decline in the audience for AM stations in general and in particular the erosion of the clear channels, as federal regulators increasingly viewed the extensive protection of night-time signals such as WSM's as a vestige of the radio age. They also recognized the widespread market potential of cable television. In response to these structural changes in telecommunications, NLT executives advanced a plan that would fully utilize Nashville's extensive talent pool and NLT's large investment in television production while maintaining *Grand Ole Opry*'s core audience.

That plan eventually took form as the Nashville Network. Seed money for the Nashville Network was obtained from the 1981 sale of WSM-TV to George N. Gillette. An even more significant change at WSM occurred that same year, when NLT was acquired by American General, the Houston financial services giant. Executives at American General made it clear that their interest was limited to NLT's insurance business and that the NLT broadcasting and entertainment operations were to be sold. For a time, there was concern that a single buyer would not be found for WSM, WSM-FM, the Opryland complex, and the Nashville Network. Speculation abounded that splitting the NLT broadcasting and entertainment division through sales to multiple buyers would have the effect of ending the historic *Grand Ole Opry,* which by then had been recognized as the world's longest-running live radio program.

Gaylord Broadcasting Company, a family-owned firm itself wholly owned by the Oklahoma Publishing Company (publisher of the *Daily Oklahoman*), stepped in and purchased the NLT broadcasting and entertainment properties as a group in 1983, and the *WSM Grand Ole Opry* continued uninterrupted. In 2000 WSM featured a fine local news department and a country music format carried live on Sirius Satellite Radio.

ROBERT M. OGLES

See also Clear Channel Stations; Country Music Format; Grand Ole Opry; WLAC

Further Reading

Crabb, Alfred Leland, *Nashville: Personality of a City,* Indianapolis, Indiana: Bobbs-Merrill, 1960
Egerton, John, *Nashville: The Faces of Two Centuries, 1780–1980,* Nashville, Tennessee: PlusMedia, 1979
Emery, Ralph, and Tom Carter, *Memories: The Autobiography of Ralph Emery,* New York: Macmillan, and Toronto, Ontario: Macmillan Canada, 1991
Hoobler, James A., *Nashville Memories,* Knoxville: University of Tennessee Press, 1983
Ogles, Robert M., and Herbert H. Howard, "The Nashville Network," in *The Cable Network Handbook,* edited by Robert G. Picard, Riverside, California: Carpelan, 1993
Streissguth, Michael, *Eddy Arnold, Pioneer of the Nashville Sound,* New York: Schirmer Books, and London: Prentice-Hall International, 1997
Zibart, Carl F., *Yesterday's Nashville,* Miami, Florida: Seemann, 1976

WTOP

Washington, D.C. Station

"The spot at the top of your dial," WTOP-AM, at 1500 kilohertz, has been a fixture in the Washington, D.C., market (with various owners, frequencies, and call letters) since 1927. Still an affiliate of the Columbia Broadcasting System (CBS), WTOP was owned by the network from 1932 to 1949 and then by the *Washington Post* before undergoing a series of ownership changes in the 1980s and 1990s. The station's Washington location and the reach of its signal made WTOP a key originating station for CBS, and it launched important careers, including Arthur Godfrey's. In 1969 WTOP became a pioneer of the all-news AM format, and in the 1990s it began providing its signal over the world wide web.

Origins

Although long based in the Washington, D.C. area, WTOP began under different call letters more than 200 miles to the north. A Republican Party political club placed station WTRC on the air in Brooklyn in September 1926. It offered music and some talk programs. Just a year later the station's equipment was sold to John S. Vance, a Virginia publisher with Ku Klux Klan affiliations. Vance's station, first as WTFF and then carrying his initials as WJSV, began broadcasting from Mount Vernon Hills, Virginia, as a self-styled "independent voice from the heart of the nation." CBS purchased WJSV in 1932, took it off the air for three months, then returned to broadcasting from a facility beside the Potomac River in Alexandria, Virginia, using a submarine telephone cable to communicate with a second studio in Washington's Shoreham Hotel. An increasing number of live location broadcasts motivated a full move into Washington in 1933, with a new studio in the Earle Theatre Building. Station engineers set up for Franklin Roosevelt's fireside chats, concerts from Constitution Hall, the visit of George VI and Queen Consort Elizabeth in 1938, and other live broadcasts throughout the 1930s and 1940s.

A complete tape exists of WJSV's broadcast day on 21 September 1939, an invaluable snapshot of radio's golden era. Signing on at 5:58 A.M., the station broadcast *Sundial with Arthur Godfrey* at 6:30, CBS serials (including *The Goldbergs* at noon), a live Roosevelt address to Congress (repeated later in the evening), a Washington Senators baseball game, *Major Bowes' Original Amateur Hour,* and Louis Prima's orchestra at midnight, prior to a 1:00 A.M. signoff. During World War II, because its night-time signal reached the entire East Coast, the station was designated by the Federal Communications Commission (FCC) as a conduit for alerts, and its studios were staffed 24 hours a day even when not broadcasting.

After several frequency changes, the station in 1941 settled at 1500 kilohertz, at that time the highest AM frequency. This move in turn motivated new call letters, WTOP, adopted in 1943 along with the "top of your dial" slogan. A 50,000-watt transmitter had been established in a new international-style building in Wheaton, Maryland, in 1940, where it remains today. (In the 1960s, Sam Donaldson hosted *Music Till Dawn* from the Wheaton transmitter.) The Washington Post Company purchased 55 percent of WTOP in 1949 and assumed full ownership in 1954. Under *Post* control, the station adopted an all-news format in 1969. For 25 years, WTOP-AM shared "Broadcast House" studios and some personnel with WTOP television, channel 9, which was also a Post-Newsweek station and CBS affiliate. The connection ended in 1978 with the sale of the AM station to the Outlet Company of Providence, Rhode Island; WTOP radio moved its studio to an office building next door to Broadcast House. WTOP was acquired in 1997 by Bonneville International Corporation, which also owns classical, contemporary, and country music stations in the Washington market.

From 1947 to 1966, WTOP simulcast its programming on FM. At that point, FCC regulations intervened to stipulate that eight hours daily be separately programmed for FM. In the uncertain regulatory environment, the Post-Newsweek Company eventually sold WTOP-FM for one dollar to Howard University, where it became WHUR. In an attempt to improve its signal in Virginia, another attempt at FM began in 1997, with a signal purchased, upgraded, and moved to 107.7 megahertz (near the top of the FM frequency band). In 1998 WTOP launched its website, wtopnews.com. Web newscasts quickly became popular in Washington's government offices (where broadcast signals were sometimes weak), and webcasting figures significantly in WTOP's plans for the future.

Like other AM stations, WTOP had lost audience in the 1960s, and in response the station experimented with a variety of formats before adopting an "all-news" format in 1969. (Originally "all news, all the time," the station soon went from 24 hours a day to a 2 A.M. sign-off; WTOP returned to around-the-clock broadcasting, initially with a talk program, after its purchase by the Outlet Company.) News was a smart choice given the station's history, its location in the political capital, and its affiliations with the *Washington Post* and CBS. Through three subsequent decades of ownership changes and technical developments, WTOP has maintained its highly successful version of all news. One or two anchors coordinate a mix of live reporting and feeds from CBS. There is a strong emphasis on local reporting, not necessarily focused on the federal government. Local leaders are featured on regularly scheduled call-in shows. Well-coordinated local news resources culminated in WTOP's award-winning live coverage of the July 1998 shootings inside the U.S. Capitol.

In its successful news mix, WTOP's sports coverage emphasizes local professional teams, and "traffic and weather together" appear every ten minutes. WTOP has not been immune to trends that are only marginally related to traditional "news." The long-running *Call for Action* consumer feature was joined in the late 1990s by a series of *Place for the Kids* fund-raising activities for boys clubs and girls clubs. Although on-air personnel still emphasized professionalism over personality, a *WTOP's Man about Town* was created through sponsorship by a luxury automobile. Activities stretching the all-news focus at WTOP had a long-standing precedent: the station for many years broadcast the full season of baseball games by the Baltimore Orioles. By 2000, three technologies—AM, FM, and the world wide web—were delivering WTOP's "all-news" content.

GLEN M. JOHNSON

Further Reading

"WTRC," in *The Airwaves of New York,* by Bill Jaker, Frank Sulek, and Peter Kanze, Jefferson, North Carolina: McFarland, 1998

WWJ

Detroit, Michigan Station

WWJ Radio was the first radio station in the world to be started by a newspaper, the *Detroit News,* in 1920. By 1924 the radio station was to achieve a number of other first-time events in the history of radio.

Origins

At 8:15 P.M. on 20 August 1920, WWJ radio was born when eight automobile batteries powered what was then called 8MK. The equipment for the station came from a local electrical retail store. The station put out a 20-watt signal on that first broadcast. Listeners (mainly ham operators) were asked to call in if they could hear the broadcast.

According to newspaper accounts of the historic event, the first words on the station were spoken by a 17-year-old Canadian named Elton Plant: "This is 8MK calling." A Windsor, Ontario, native, Plant had worked his way up to cub reporter at the *Detroit News* when the managing editor approached him with the idea of going on the air. Plant agreed to do it. He noted later, "It didn't mean a thing as far as I was concerned . . . because I didn't know what it all was."

The name of the station was changed from 8MK to WBL when it received its radio license in October 1921. In March 1922 the call letters were changed again to WWJ, reportedly because listeners kept getting the call letters wrong. Over the years, a dispute developed over whether WWJ in Detroit, KDKA in Pittsburgh or WHA in Madison, Wisconsin, was the oldest radio station in the United States. Those who favor WWJ argue that it was the first station to actually get on the air when wireless restrictions were lifted after World War I.

Initial Programs

Between 1920 and 1924, WWJ aired the first news program, the first election returns (a Michigan race in the summer of 1920), the first complete symphony broadcast, the first regularly scheduled religious broadcast, and the first sports broadcast. The early sports broadcasting duties for WWJ were handled by Ty Tyson, later credited with being the world's first radio sports broadcaster. He got his job at WWJ through an orchestra leader whose band was invited to play on WWJ. Tyson was hired to do weather reports but went on to do sports and live interviews with celebrities such as Charles Lindbergh and Will Rogers. In 1924 he broadcast the first college football game and the Gold Cup powerboat races. On 19 April 1927 Tyson became the first radio sports broadcaster to do a regular season major league baseball game. He was also known for his ability to communicate the essence of the game even when he was not physically present. When the Detroit Tigers played out of town before direct radio lines became common, a telegraph operator in the opponent's park would tap out coded play-by-play messages to Tyson back in Detroit. Tyson would decode the taps and broadcast the plays as if he were seeing them himself.

Later Years

By the mid-1970s, WWJ radio was moving to an all-news format. In 1978 the *Washington Post* acquired WWJ-TV from the Evening News Association. In exchange the Evening News Association acquired WTOP, a Washington, D.C. television station owned by the *Washington Post*. The station trade followed speculation about forthcoming FCC rules banning local market cross-ownership. The Evening News Association kept WWJ radio. In 1985 the Gannett Company announced that it would sell five broadcast properties to satisfy FCC rules affecting its proposed $717 million purchase of the Evening News Association in Detroit. (FCC rules at the time prohibited companies from owning newspapers and broadcast properties in the same city.) The broadcast properties in Detroit that were to be sold included WWJ radio. WWJ was purchased by Federal Broadcasting, but in March of 1989 the Columbia Broadcasting System (CBS) announced that it had acquired WWJ and its sister station WJOI from Federal Broadcasting for $58 million. WWJ and WJOI continue to be a part of the CBS news family, even though the network has been through several ownership changes, and WWJ remains the only commercial all-news radio station in Michigan.

RICK SYKES

Further Reading

Baudino, Joseph E., and John M. Kittross, "Broadcasting's Oldest Stations: An Examination of Four Claimants," *Journal of Broadcasting* 21 (Winter 1977)

Baulch, Vivian M., "The Stars Who Turned Detroiters into Couch Potatoes," <detnews.com/history/tvhist/tvhist.htm>

Bradford, Doug, "Elton M. Plant, Pioneer Voice of Fledgling Radio Station WWJ," *Detroit News* (5 January 1992)

Eden, David, "New Owner Takes over WWJ-TV with 'Sign-on,'" *Detroit News* (26 June 1978)

McFarlin, Jim, "Radio's First Voice Gets a Hero's Welcome," *Detroit News* (21 August 1980)

Plant, Elton M., *Radio's First Broadcaster: An Autobiography*, N.p.: Plant, 1989

"Ty Tyson, the World's First Sports Broadcaster," <www.detroitnews.com/history/tyson/tyson.htm>

WWL

New Orleans, Louisiana Station

The first radio station in the lower Mississippi Valley, WWL has long provided a 50,000-watt clear channel voice to one of America's most culturally distinctive cities, serving as a window on New Orleans for much of the nation. This powerful commercial radio station was owned and operated through most of its history by a Jesuit university.

Origins

From the opening of Loyola University of New Orleans in 1914, its physics department offered courses in "wireless telegraphy." By 1922, after several years of accumulating radio equipment to support its curriculum and two years after the U.S. debut of regular broadcasting, Loyola established WWL radio. Although the goals outlined for the station included some educational, cultural, and public service programs (the latter in the form of weather and agricultural reports for farmers), WWL was intended first and foremost to serve as a fund-raising tool for the private university. The station's first broadcast, on 31 March 1922, was a direct appeal for funds by the university president. Not only were listeners urged to contribute to a $1.5 million campaign for the construction of six new classroom buildings, they were urged to spread word of the university's financial needs to those not fortunate enough to own radio receiving sets.

WWL's first two years saw it off the air more than on. Once the initial enthusiasm wore off, live, original programs became burdensome to produce. The fund-raising appeals were not having the desired effect; the station cost more to operate than it was bringing in. The original 100 watts of operating power was reduced to 10 watts in an effort to cut costs. This only served to diminish further the station's broadcast range and fund-raising potential.

The station seemed doomed in 1924, when new physics faculty decided to attempt resuscitation. Committing to a reliable if extremely modest broadcast schedule of one hour per week, WWL's power was increased in increments over the next few years to 500 watts. In an effort to find a safe haven from the rampant interference problems of the era, the station moved up the radio dial, from 833.3 kc, to 1070 kc, to 1090 kc, and finally up to 1220 kc.

The Federal Radio Commission permitted WWL to increase its power to 5,000 watts, but the Commission moved the station back down the dial to 850 kc, a frequency that it was forced to share with KWKH of Shreveport, Louisiana, a city 350 miles northwest of New Orleans—the stations were too close to both be on the air at the same time. KWKH was owned by W.K. Henderson, a social activist who used his station to broadcast his political views. For the next several years, WWL's growing program schedule had to be squeezed into a complicated time-sharing arrangement with Henderson. Depending on the time of day or day of the week, radio listeners tuning in to 850 kc would hear either WWL's classical music and lectures or Henderson's political harangues.

In 1929 the administration at Loyola made a pivotal decision: the university's goals would be best served if WWL operated as a money-making commercial enterprise, thus providing a continuing endowment to the university. Educational and religious programming, with few exceptions (such as the long-running *Mass from Holy Name Church*), gave way to popular entertainment. However, for WWL to realize its full revenue-producing potential, it needed the reliable source of quality programs that only affiliation with one of the major networks could provide. Although WWL was the most powerful radio station in New Orleans (reaching 10,000 watts in 1932), the networks were not interested in a part-time affiliate. As long as it was saddled with the KWKH time-sharing arrangement, WWL would have to continue producing most of its own programming. To this end, Loyola moved the station from the university campus to new studios in New Orleans' Roosevelt Hotel in 1932.

In 1934, KWKH moved to 1100 kc, and WWL achieved full-time status. With this hurdle cleared, negotiations regarding network affiliation could begin in earnest. WWL joined the Columbia Broadcasting System (CBS) on 2 November 1935 with a live, one-hour network show entitled *A City of Contrasts,* which dramatized events from New Orleans history.

The year 1937 saw the debut of *Dawnbusters,* a local morning show featuring live music and comedy. The program would be one of the most popular in WWL's history, running until 1959. In 1938 WWL reached the maximum permissible power of 50,000 watts, permitting its signal to cover much of eastern North America at nighttime. President Franklin Roosevelt sent the station a congratulatory telegram, noting that the station's far-reaching signal "should be a source of great satisfaction to the Jesuit Fathers who have worked so assiduously building up the station from a small beginning. I trust that its future will be one of great usefulness in the service of God, of Home and of Country." In 1941, to comply with a treaty seeking to reduce interference throughout North America, WWL changed frequencies one final time to its current 870 kilohertz.

Postwar Change

After World War II, WWL, like all of the nation's AM radio broadcasters, faced a new set of options: expand into television and/or FM or stay the course with AM. WWL's initial decision was to move into FM and leave television to others. WWL's FM station, WWLH, went on the air in 1946. A halfhearted effort from the outset, WWLH mainly simulcast WWL programming. The public failed to buy FM receivers in sufficient quantities to make the enterprise viable, and WWLH went off the air in 1951.

Loyola now decided that television was the more valuable path, but between the Federal Communication Commission's (FCC) four-year freeze on new television licenses and a post-freeze battle with competing interests for the few channel allocations available to the New Orleans market, WWL-TV (channel 4) didn't become a reality until 1957.

With the advent of television and the decline of traditional radio network programming in the late 1950s, WWL's schedule became increasingly centered around local talk shows. WWL's present-day programming is dominated by live, local news programs; "topic of the day" phone-in shows; sporting events; and sports talk programs.

In the late 1980s, the Loyola administration decided to abandon broadcasting. WWL-AM and WWL-TV were sold to different companies, and, through a series of sales and mergers, the AM station has since changed owners several times. Its current ownership, Entercom Communications, holds seven other stations in New Orleans, including WSMB, a long-time National Broadcasting Company (NBC) radio network affiliate that, along with WWL and WDSU, once defined broadcasting in "the city that care forgot."

RICHARD WARD

See also KWKH

Further Reading

Pusateri, C. Joseph, *Enterprise in Radio: WWL and the Business of Broadcasting in America,* Washington, D.C.: University Press of America, 1980

WWVA

Wheeling, West Virginia Station

WWVA is the oldest station in West Virginia, and from early in its history it proved to be an important factor in the popularization of country music.

WWVA first aired on 13 December 1926. Founder John Stroebel, a physics teacher and for years an experimenter with crystal sets and wireless telephone, transmitted from his basement in Wheeling. The 50-watt station was licensed to broadcast on 860 kilohertz.

Offering a menu of local information and entertainment, the station received permission to broadcast at 500 watts in 1927. WWVA evolved quickly over the next five years. Stroebel sold the station to Fidelity Investments Associates in 1928, and the Federal Radio Commission raised the station's power to 5,000 watts in 1929. "The Friendly Voice from out of the Hills of West Virginia" affiliated with the Columbia Broadcasting System (CBS) in 1931, which helped to fill out the station's program schedule. In 1931 Fidelity Investments sold WWVA to George B. Storer, whose Fort Industry Company (later renamed Storer Broadcasting Company) would hold the license for three decades.

At the time, Storer was collecting a wide array of radio stations, which would make him one of the largest chain owners of radio stations. Under Storer's ownership, WWVA would see its most dramatic developments: the introduction in 1933 of the *Jamboree,* a regular Saturday night broadcast of country music, and the increase of the station's power to 50,000 watts in 1941, just as the station moved to 1170 kilohertz.

The station had featured country music almost since its inception, but the debut of the *Jamboree* on 7 January 1933 would secure its prominent role in the popularization of country music. The show became an important stage for regional artists, whose reputations grew with every appearance, and when the station began broadcasting with 50,000 watts, the *Jamboree* became a force in the dissemination of country music in the Ohio Valley and far beyond. WWVA sent the *Jamboree* and country music surging into 18 eastern states and 6 Canadian provinces during the night-time hours on Saturdays. Along with the *Grand Ole Opry* on WSM (Nashville), the *National Barn Dance* on WLS (Chicago), and the *Louisiana Hayride* on KWKH (Shreveport), the *WWVA Jamboree* was one of a number of widely heard barn dance radio shows that carried country music to large audiences around the nation.

Country music scholar Bill C. Malone (1985) has noted that although many of the barn dance radio programs were important in expanding country music's audience, the *Jamboree* did the most to carry country music to Northeast audiences and to help create new audiences in the Northeast and Canada for country music. The program also helped propel to national prominence the careers of a number of country and bluegrass artists. Artists who prospered from their exposure on the *Jamboree* included Grandpa Jones, Hank Snow, Hawkshaw Hawkins, Wilma Lee and Stoney Cooper, and Reno and Smiley. For a period in the mid-1950s, the CBS network carried the *Jamboree,* giving the program an even wider sphere of influence.

The *Jamboree* wasn't the only conduit for country music on WWVA. Starting in the 1950s, the station featured a popular overnight disc jockey program (hosted for many years by performer Lee Allen, "The Coffee Drinking Nighthawk") that covered the station's wide listening area, and in the 1960s, the station was among the first to adopt a "modern country" format, which featured the lush musical stylings popularized by performers such as Eddy Arnold and Jim Reeves.

WWVA, owned today by AM-FM Incorporated, dropped its all-country programming in 1997 in favor of a news/talk format. But the station's Saturday night country show survives; it is the second-longest-running country music stage show on radio, behind only WSM's *Grand Ole Opry. Jamboree U.S.A.,* as the barn dance is known today, still airs from Wheeling's Capitol Musical Hall, which first hosted the show in 1933 (although the show changed venues a number of times after its debut before returning to the Capitol in 1969).

MICHAEL STREISSGUTH

See also Country Music Format; Storer, George

Further Reading

Kingsbury, Paul, editor, *The Encyclopedia of Country Music,* New York: Oxford University Press, 1998

Malone, Bill C., *Country Music U.S.A.,* Austin: University of Texas Press, 1968; revised edition, 1985

Snow, Hank, Jack Ownbey, and Bob Burris, *The Hank Snow Story,* Urbana: University of Illinois Press, 1994

Tribe, Ivan M., *Mountaineer Jamboree: Country Music in West Virginia,* Lexington: University Press of Kentucky, 1984

"WWVA-Wheeling," *West Virginia Review* 23, no. 2 (Nov. 1945)

WXYZ

Detroit, Michigan Station

Despite its comparatively remote Midwest location, WXYZ pioneered network radio and radio drama. The station and its *Lone Ranger* series were instrumental in the founding of the Mutual Broadcasting System.

Early Years

WXYZ began life on 10 October 1925 as WGHP, a Class B (medium power) station at 1270 kilohertz. This frequency was far enough removed from other Detroit stations to avoid serious interference. The WGHP call letters signified founding owner George Harrison Phelps, who had directed automobile advertising for Dodge since 1914. The station secured an affiliation with the fledgling Columbia Broadcasting System (CBS) in 1927 and moved its studios from an alley garage to the 15th floor penthouse of the Maccabees Building, near the Detroit Institute of Arts, and the new Detroit Public Library main branch. The station was a money-loser, however, and when Dodge was purchased by Chrysler Corporation in 1928, Phelps' agency lost the Dodge account. To lessen the cash crunch, Phelps peddled WGHP to J. Harold Ryan and his brother-in-law George Storer, owners of profitable WSPD in Toledo, Ohio. In October 1928 these two oil and steel magnates purchased WGHP for $40,000.

WGHP's airtime was now aggressively marketed in a variety of commercial lengths. Just 18 months later as the Depression was deepening, the Storer group sold the station for $250,000 to John Kunsky and George Trendle, owner and manager, respectively, of the Kunsky Theatres movie chain. In July 1930 the new owners unleashed a movie-business-like promotional campaign for their outlet, whose call sign they also changed to WXYZ—call letters Trendle had persuaded the U.S. Army and Navy (separately) to relinquish. Programming now included *Carl Rupp and His Orchestra,* the first network show to originate from Detroit.

Trendle's background as a newsboy and lawyer and his grasp of management, promotion, and show business would soon propel the station onto the national stage as he stocked WXYZ with exceptional executives, writers, and on-air talent. All of this potential was nearly discarded, however, when station management decided they were being forced to give up too much time—and therefore advertising revenues—to the Columbia Broadcasting System (CBS). At the end of 1931, WXYZ abruptly canceled its network affiliation and suddenly faced vast amounts of empty airtime for which programming quickly had to be found. Two studios now became four, rehearsals for one show cleared the studio only moments before the cast for another arrived to take the air, and announcers, actors, and musicians jostled each other in the narrow corridors.

Creating National Programs

The first network show to emerge from this creative chaos was *The Lone Ranger,* which debuted in January 1933 with scripts developed by Buffalo syndicated writer Fran Striker and a concept refined by George W. Trendle himself. Only days after the program's first broadcast, the Michigan Radio Network began linking stations in the state's major cities with WXYZ as the key outlet and *The Lone Ranger* as a centerpiece offering. By November the show was also airing over Chicago's WGN, and New York's WOR was added in early 1934. A few weeks later, the series was made available via transcription to stations in seven southern states under the sponsorship of American Bakeries. WXYZ's sales manager H. Allen Campbell then persuaded the general managers of WGN and WOR to expand their relationship into a program-sharing network that would feed multiple programs among their stations as well as to new partners in Cincinnati, Pittsburgh, St. Louis, and Washington, D.C. On 19 September 1934, the seven outlets were linked with telephone lines and became the Mutual Broadcasting System.

In the next year, WXYZ's partner stations began adding far more outlets to Mutual than Trendle thought wise. So while WXYZ continued to feed *The Lone Ranger* to Mutual, it joined the National Broadcasting Company (NBC) Blue Network to secure its own source of programming. In a successful attempt to repeat *The Lone Ranger*'s success, Trendle and Striker debuted *The Green Hornet* in early 1936.

WXYZ's success with these and other programs was recognized in 1937 by *Variety's* award of its Citation for Showmanship in Program Origination, given each year to the station judged best in new show production. Three years later, to ensure continued access to the talent that made such distinctions possible, WXYZ became the first Detroit station to sign a contract with the American Federation of Radio Artists (AFRA). Future television news stars getting early experience on the station at this time included Douglas Edwards, Hugh Downs, and Myron Wallace, whose first name was changed by WXYZ executives to "Mike."

Postwar Transition

To enlarge its production space, the station moved in 1944 to the Mendelssohn mansion in suburban Grosse Pointe. WXYZ's continued success made it a desirable purchase for the former NBC Blue Network, which had become the separately owned American Broadcasting Company (ABC). ABC badly needed to upgrade its owned-and-operated station holdings, and WXYZ was among the most desirable of its affiliates. In April 1946 the sale was consummated. Key programs such as *The Lone Ranger* and *The Green Hornet* remained the property of George Trendle and longtime station sales executive H. Allen Campbell. The other former station owners, John Kunsky and Howard Pierce, cashed out of the business entirely. The sale of WXYZ marked the end for the Michigan Radio Network stations, which now received program feeds directly from ABC.

Despite wooing by rival WJR, WXYZ sales manager Jim Riddell accepted ABC's Trendle-brokered offer to stay on as the new general manager. A separately programmed WXYZ-FM went on the air on 1 January 1948 at 101.1 megahertz, but before the end of the year it reverted to simulcasting the AM station signal. Meanwhile, WXYZ-TV had taken to the air from the Maccabees Building, under the direction of the radio operation's former wire recording technician John Pival. Pival lured some of the radio outlet's top personalities to the television side and the shift of dominance began. As happened around the country, the number of network radio shows withered, to be replaced by local disc jockeys. Chief among them on WXYZ were Paul Winter, Jack Surrell, Ed McKenzie, and Fred Wolf. *The Green Hornet* went off the air in 1952. *The Lone Ranger* hung on, but the last live broadcast was in 1954.

The radio station moved to a caretaker's cottage near the transmitter in 1955 while Fred Wolf's converted house trailer, the "Wandering Wigloo," became a vehicle for a hugely successful remote program; radio executives came from around the country to study the show. Unfortunately for Wolf and other program hosts, however, WXYZ embraced the Top 40 concept in 1958 and tight music formatting now overshadowed individual air personalities. In 1959 the WXYZ television and radio properties moved to the newly constructed suburban Broadcast House—in which radio was relegated to an obscure corner. Four years later, Charles Fritz, former manager of Blair Radio's Detroit office (the firm representing the station to national advertisers), became WXYZ's general manager, but the station's prominence continued to wane. Even though disc jockey Lee Alan's record hops were proving tremendously popular, the station subsequently lost its Detroit market dominance to WKNR (programmed by Mike Joseph). In the years that followed, a string of competitors would continue to beat both the AM and FM (which became the harder-rocking and separately owned WRIF in the 1980s) at the music game.

Fritz bought the AM from ABC in 1984, changed its call letters to WXYT, and thereby launched Detroit's first all-talk outlet. The WXYZ designation thus disappeared from radio but survived as the call letters of the formerly co-owned television station that was purchased by the Scripps Howard News Service in 1986. WRIF (the old WXYZ-FM) continued its mainstream rock format and ultimately was purchased by Greater Media. The AM facility that began it all was acquired by Infinity Broadcasting in 1994. Ironically, when Infinity subsequently merged with CBS, the station found itself owned by the same entity its first owners had unceremoniously jettisoned in 1931.

PETER B. ORLIK

See also Clear Channel Stations; Green Hornet; Lone Ranger; Mutual Broadcasting System; Striker, Fran

Further Reading

Harmon, Jim, *The Great Radio Heroes,* Garden City, New York: Doubleday, 1967; 2nd edition, Jefferson, North Carolina: McFarland, 2001

Lackman, Ronald W., *Remember Radio,* New York: Putnam, 1970

Osgood, Dick, *WYXIE Wonderland: An Unauthorized 50-Year Diary of WXYZ Detroit,* Bowling Green, Ohio: Bowling Green University Popular Press, 1981

Zier, Julie, "Fritz Selling Its Detroit AM-FM," *Broadcasting and Cable* (3 January 1994)

Wynn, Ed 1886–1966

U.S. Radio Comedian

Ed Wynn, arguably the first all-out clown superstar of radio, began in the medium reluctantly. His career would cover radio's golden years, extended into television, and included several other media as well.

Origins

A "Broadway Baby" if ever there was one, Wynn (born Isaiah Edwin Leopold) ran away from home at age 15 to join a theater company. After that stint ended with the company's bankruptcy, Wynn returned home, only to run off again, this time to New York. By the age of 19, he was a headliner on the vaudeville stage, and by 1914 he was working for Ziegfeld. By this time, Wynn had perfected his comic persona, "The Perfect Fool," a moniker he took from his stage show of the same name.

Wynn's act—old-school probably even at that time—was pun-heavy with shopworn, groan-inducing lines and relied heavily (though with great success) on exaggerated shoes, over-the-top costumes, and funny hats. (In some ways Wynn was carrying on the family tradition; his father had been a successful hat manufacturer.) But, nevertheless, Wynn's schtick, his rubbery face, and his all-out, eager-to-please style were huge crowd-pleasers.

Radio Years

Wynn made his first radio appearance in a 1922 in-studio broadcast of his show *The Perfect Fool.* It was not a complete success, certainly not for Wynn. Accustomed to the stage, Wynn was used to playing to an audience, not just to a lone microphone. Legend has it that on learning he would have no people to perform in front of, he quickly rustled up an impromptu audience at the station that consisted of cleaning women, stage hands, and technicians. After that appearance, Wynn returned to the stage.

Despite additional offers, Wynn resisted the new radio craze, thinking (perhaps rightly) that his physical clowning and dependence on props and costumes wouldn't translate well to a nonvisual medium. But when promised a weekly salary of $5,000 by Texaco to star in their program *The Fire Chief,* Wynn saw it as what it was: an offer too good to refuse. Wynn's *The Fire Chief* debuted on 28 April 1932, broadcast live and, at Wynn's insistence, in front of an audience.

Ed Wynn
Courtesy CBS Photo Archive

At the time, doing a radio show in front of a crowd seemed foolhardy, but Wynn's instinct would prove prophetic for the entire radio industry. Later, almost all radio comedy programs played to packed theaters while being broadcast. In its own way, this approach was the precursor to Lucille Ball's and Desi Arnaz's revolutionary "live on film" recording technique pioneered for their *I Love Lucy* television show.

To further blur the line between radio show and live theater, Wynn continued to perform in full makeup and in outrageous costumes. He even went so far as to change clothes several times during a performance. Each of Wynn's entrances would be greeted with applause and laughter, which obviously left radio listeners totally in the dark about just what was going on; in turn, it gave listeners the feeling they were not so much hearing something *meant for them* as much as they were eavesdropping onto a stage show already in progress. Nevertheless, Wynn persisted with his clothes and his silent, goofy antics, believing that listeners would assume something funny was going on and enjoy the merriment as if by proxy. (It should be noted, though, that radio audiences were surprisingly tolerant of "visual" comedy: Eddie Cantor, a comic of the Wynn variety, thrived for a time, as did Edgar Bergen's ventriloquism, where listeners just had to assume that his dummy was really there.)

Ed Wynn's on-stage, on-microphone comic persona was an interesting one. In describing it, writers and critics over the years have used a long list of colorful adjectives: "giggling," "befuddled," "frantic," and "fey." Wynn also used on the air a lispy, high-pitched voice (supposedly originally evoked by him out of "mike fright," but considered funny enough for him to retain it). Had Wynn entered the popular culture not in the 1930s but today, when the media is anxious to label, to "out," actors and characters, Wynn would no doubt have found himself labeled as "gay." But, of course, the subject of homosexuality was completely closeted at that time, and despite whatever signals Wynn's on-air personality sent out to the public, he was granted an entertainer's license to exist, in fact flourish, without being co-opted by or associated with any political or social agenda.

Wynn, married three times and a father, was not gay in real life, and despite the stereotypically "feminine" qualities of his show-business character, his comic self was not really a "sexual personae": indeed his character (similar in some ways to Jerry Lewis' boy–man persona in his early work or even to Chaplin's Little Tramp) was more asexual, practically genderless, possessing a childlike innocence that belied Wynn's real age.

But whatever Wynn was or represented, he was certainly popular at least for a time. As soon as *The Fire Chief* premiered, it was a hit, one of the largest in radio up to that time. But his on-air success was short-lived, declining each month the show was on, making it something of a flash in the pan. Only three years after its debut, Wynn's *Fire* went out.

By this time comics like Jack Benny and Bob Hope had premiered with their shows, which were more attuned to the limits and possibilities of radio and were, when compared with Wynn's vaudeville style, downright sophisticated and urbane. Wynn, by contrast, soon found himself something of a dinosaur in the medium he had conquered only a few years before.

Television and Film

Wynn tried other radio vehicles (*Gulliver, Happy Island*) with mixed artistic and little popular success. Luckily though, by that time, the age of television had dawned. Wynn was one of the first big-name stars to enter television, having realized it would be better suited to his brand of sight-gag humor. Though he didn't quite cotton to all the technical possibilities in the way that Ernie Kovacs later would, Wynn's clowning was well appreciated, at least briefly, by kids, until they, like the generation that preceded them in radio, gradually grew bored with Wynn's clown-at-all-costs identity.

By about 1950 Wynn, trapped by his own show-biz invention of himself, had nothing left to do but reinvent his persona and career. With the help of his talented son, actor Keenan Wynn, Wynn developed into an acclaimed character actor, especially good at playing saddened clowns or washed-up old men. Reining in his trademark voice and gestures, Wynn delivered acclaimed performances in television's landmark *Requiem for a Heavyweight* (1956) and later in the film *The Diary of Anne Frank* (1959), among other productions.

Wynn had also by this time developed into a singular presence in the canon of popular culture, as television and radio commercials, impressionists, and others all mimicked his voice and mannerisms to sell products or gain immediate, positive recognition from an audience. Wynn's place as a permanent fixture in the lexicon of Americana was further solidified with his long professional association with the Walt Disney Company. Disney's farcical, far-fetched stories, whether live-acted or drawn, seemed perfect for Wynn's gifts—he was always a sort of human cartoon anyway. The company paid him perhaps the highest compliment when they used his voice (and his visage) for the role of the Mad Hatter in their animated *Alice in Wonderland* (1951). After all, who besides Wynn could be more believable in that role—more delightfully clowny, wildly irrepressible, or completely madcap?

CARY O'DELL

See also Amalgamated Broadcasting System; Cantor, Eddie; Comedy; Edgar Bergen and Charlie McCarthy; Skelton, Red

Ed Wynn. Born Isaiah Edwin Leopold in Philadelphia, Pennsylvania, 9 November 1886. Appeared in theater and vaudeville beginning in 1901; made Broadway debut in 1910 (*The Deacon and the Lady*); worked for Ziegfeld beginning in

1914; appeared in *The Perfect Fool* stage show, 1921; made radio debut, 1922; began radio series career, 1932; began television series career, 1949; appeared in many films. Died in Beverly Hills, California, 19 June 1966.

Radio Series

1932–35	*The Texaco Fire Chief*
1936	*Gulliver*
1936	*Ed Wynn's Grab Bag*
1936–37	*The Perfect Fool*
1944–45	*Happy Island*

Television

The Ed Wynn Show, 1949–50; *All-Star Revue,* 1950–52; *The Ed Wynn Show,* 1958–59; *Requiem for a Heavyweight,* 1956; *Meet Me in St. Louis,* 1959; *For the Love of Willadean,* 1964

Films

Rubber Heels, 1927; *Follow the Leader,* 1930; *The Hollywood Parade,* 1932; *The Chief,* 1933; *Stage Door Canteen,* 1943; *Alice in Wonderland* (voice only), 1951; *The Great Man,* 1956; *Marjorie Morningstar,* 1958; *The Diary of Anne Frank,* 1959; *Cinderfella,* 1960; *The Absent-Minded Professor,* 1961; *Babes in Toyland,* 1961; *Golden Horseshoe Revue,* 1962; *Son of Flubber,* 1963; *The Patsy* (cameo), 1964; *Mary Poppins,* 1964; *Those Calloways,* 1965; *Dear Brigitte,* 1965; *The Greatest Story Ever Told,* 1965; *That Darn Cat!,* 1965; *The Daydreamer,* 1966; *The Gnome-Mobile,* 1967

Stage

The Deacon and The Lady, 1910; *The Follies of 1914,* 1914; *The Zeigfeld Follies of 1915,* 1915; *The Passing Show of 1916,* 1916; *Doing Our Bit,* 1916; *Over the Top,* 1918; *Sometime,* 1918; *The Shubert Gaieties of 1919,* 1919; *Ed Wynn's Carnival,* 1920; *The Perfect Fool,* 1921; *The Grab Bag,* 1924; *Manhattan Mary,* 1927; *Simple Simon,* 1930; *The Laugh Parade,* 1931; *Alice Takat,* 1936; *Hooray for What!,* 1937; *Boys and Girls Together,* 1940; *Laugh, Town, Laugh,* 1942

Further Reading

Maltin, Leonard, *The Great American Broadcast: A Celebration of Radio's Golden Age,* New York: Dutton, 1997

Wertheim, Arthur Frank, "The Fire Chief," in *Radio Comedy,* by Wertheim, New York: Oxford University Press, 1979

Wynn, Keenan, *Ed Wynn's Son,* Garden City, New York: Doubleday, 1959

Y

Yankee Network

New England Regional Network

The Yankee network was one of several regional radio networks from the 1930s into the 1950s that linked stations to share programs and advertising.

Although Boston broadcaster John Shepard III knew little about engineering, he knew enough to hire good people who did understand the technical side of the radio business. In early 1923 he encouraged them to experiment with networking (WNAC linked up briefly with New York's WEAF). It was not long before WNAC in Boston and WEAN in Providence, Rhode Island, were frequently sharing programming, connected by a telephone line. But Shepard wanted to expand: he had begun paying salaries to talented musicians so they would appear on his stations (early radio was still mainly volunteer, so being able to pay was a major plus in getting the big names to appear), and he felt confident he could offer good programs. When the National Broadcasting Company (NBC) and the Columbia Broadcasting System (CBS) were formed in 1926–27, Shepard was convinced that a local network that emphasized New England news, sports, and music would be well received. He called it the Yankee network, and by early in 1930 he had begun signing up a number of stations in New England. The first affiliates were WLBZ in Bangor, Maine; WNBH in New Bedford, Massachusetts; and WORC in Worcester, Massachusetts. In August 1932, *Broadcasting* magazine published a tribute to Shepard, noting that he now had eight affiliates, with number nine soon to go on the air. In 1939, Shepard would put the first experimental FM station in Massachusetts (W1XOJ) on the air, and it too would carry Yankee network programming. By then, the network had its own house orchestra, a music director, staff vocalists, and a large number of talented performers who could offer the affiliates everything from a radio drama to an evening of hit songs. Always innovative, Shepard sometimes ran synagogue services on the Yankee network, as he would also run church services and sermons by well-known priests and ministers.

Perhaps his biggest innovation was with radio news: in March 1934, thanks in large part to the hard work of editor in chief Leland Bickford, the Yankee News network went on the air. In a jab at newspapers, the network used the slogan "News while it IS news; the Yankee Network is on the air!" A former Boston newspaper reporter, Dick Grant, was hired to run the news department at a time when relationships between radio and newspapers were becoming more contentious. Local newspaper reporters were not amused and tried to bar the network's reporters from getting press passes and covering city hall. But Shepard and his team persisted, and gradually radio reporters gained credibility and came to be accepted as journalists. The Yankee News network made "radio news reporter" a career choice: in radio's first decade, what little news radio stations offered came mainly from newspapers, many of which had agreements with a local station that allowed a reporter to go on the air several times a day with headlines and top stories. But for radio journalists to cover news and generate their own stories (the network even established a news bureau in Washington, D.C.) was something new, and it made the Yankee News network unique in New England.

One popular news program the Yankee network offered was *Names in the News* (late 1930s through early 1940s), in which local heroes and newsmakers were invited to talk about their achievements against a backdrop of Yankee network performers dramatizing the important events that made the guests famous. This was similar to the famous CBS program *The March of Time,* but with a New England emphasis. Such radio newsmagazines were very popular and helped make the news more interesting to the average person.

The more benefits the Yankee network offered, the more New England stations wanted to affiliate. By retaining their affiliation with a national network—which provided the major music, drama, and comedy programs—as well as the regional link, smaller stations benefited from the best of both worlds: professional-sounding local news coverage and access to the best-known national radio stars. By the early 1940s, the Yankee network had 19 affiliates. Shepard was becoming more involved with FM and was also active in Mutual Broadcasting. In late 1937 Bostonians had been shocked when he sold the Shepard Store in downtown Boston; the store in Providence still remained under Shepard family control, however. Shepard invested in technological improvements for his Boston stations—in early 1942, six new studios (for Shepard's AM stations, his FM stations, the Yankee network, and the Yankee News bureau) were dedicated; he was also attempting to organize a national FM network (this venture was not successful; his interest in FM was ahead of its time).

In late 1942 Shepard, rumored to have health problems that led him to sell off various assets, agreed to sell the Yankee network to the General Tire and Rubber Company, although he stayed on as a board member and general manager. (Later, in 1958, long after Shepard's death, the corporate ownership's name would change to RKO [Radio-Keith-Orpheum] General.) Shepard's poor health forced him to retire altogether from radio in 1948; he died two years later. The Yankee network acquired more affiliates and remained a major player in New England through the 1950s. But radio was changing: the youth market wanted Top 40, and news was not as important to that demographic. Affiliates began programming for the younger audience, and gradually they dropped the Yankee network to "play the hits." Although a few stations did remain faithful to the older audience, allowing the Yankee network to survive into the 1960s, in early 1968, without much fanfare, RKO General disbanded the network, ending its 38 years of distinguished service.

DONNA L. HALPER

See also Don Lee Network; FM Radio; Mutual Broadcasting System; Shepard, John

Further Reading

Bickford, Leland, *News While It Is News: The Real Story of the Radio News,* with Walter Fogg, Boston: G.C. Manthorne, 1935

"News While It Is News: Yankee Network Sets up Own 24 Hour Service," *Broadcasting* (15 March 1934)

You Bet Your Life

Comedy Quiz Show

Reruns have made Groucho Marx's *You Bet Your Life* familiar to generations of television viewers. Few fans realize that they are also "watching" radio. For 6 seasons of its 14-year run, the program was recorded simultaneously for both media. The versions were then edited separately and broadcast on successive nights. Its circumstances of production, editing, and broadcast were just one aspect of the show's distinctiveness. A comedy show masquerading as a quiz, *You Bet Your Life* was "postmodern" before the term was invented.

The origin of *You Bet Your Life* was the appearance by Groucho Marx on a radio variety show in April 1947, when an ad-lib by Groucho led to a verbal duel with Bob Hope that made the segment run many minutes over. Producer John Guedel (who had made Art Linkletter a radio success) immediately went backstage and suggested to Marx a quiz show with an emphasis on ad-libs. Groucho replied, "I've flopped four times on radio before. . . . I might as well compete with refrigerators. I'll give it a try." *You Bet Your Life* premiered Monday, 27 October 1947, on the American Broadcasting Companies (ABC), then moved to Wednesday night as lead-in to Bing Crosby's popular variety show.

A success by any measure, *You Bet Your Life* secured for Groucho the career he sought apart from the Marx Brothers. It also made a celebrity of its announcer and Groucho's comic foil, George Fenneman. The first season sold out the entire stock of its sponsor, Elgin Watches. Groucho received a 1949 Peabody Award as best radio entertainer, with cover stories in *Newsweek* and *Time.* Guedel and Marx moved *You Bet Your Life* in 1949 to the larger Columbia Broadcasting System (CBS) network for a longer, 45-week season. (The show continued to precede Bing Crosby's.) A year later, the National Broadcasting Company (NBC) won a bidding war to begin a televised version. The radio broadcasts moved to NBC in October 1950, where they remained in the Wednesday 9 P.M. time slot. The televised version, recorded simultaneously but edited separately, aired a day later, on Thursdays. At the

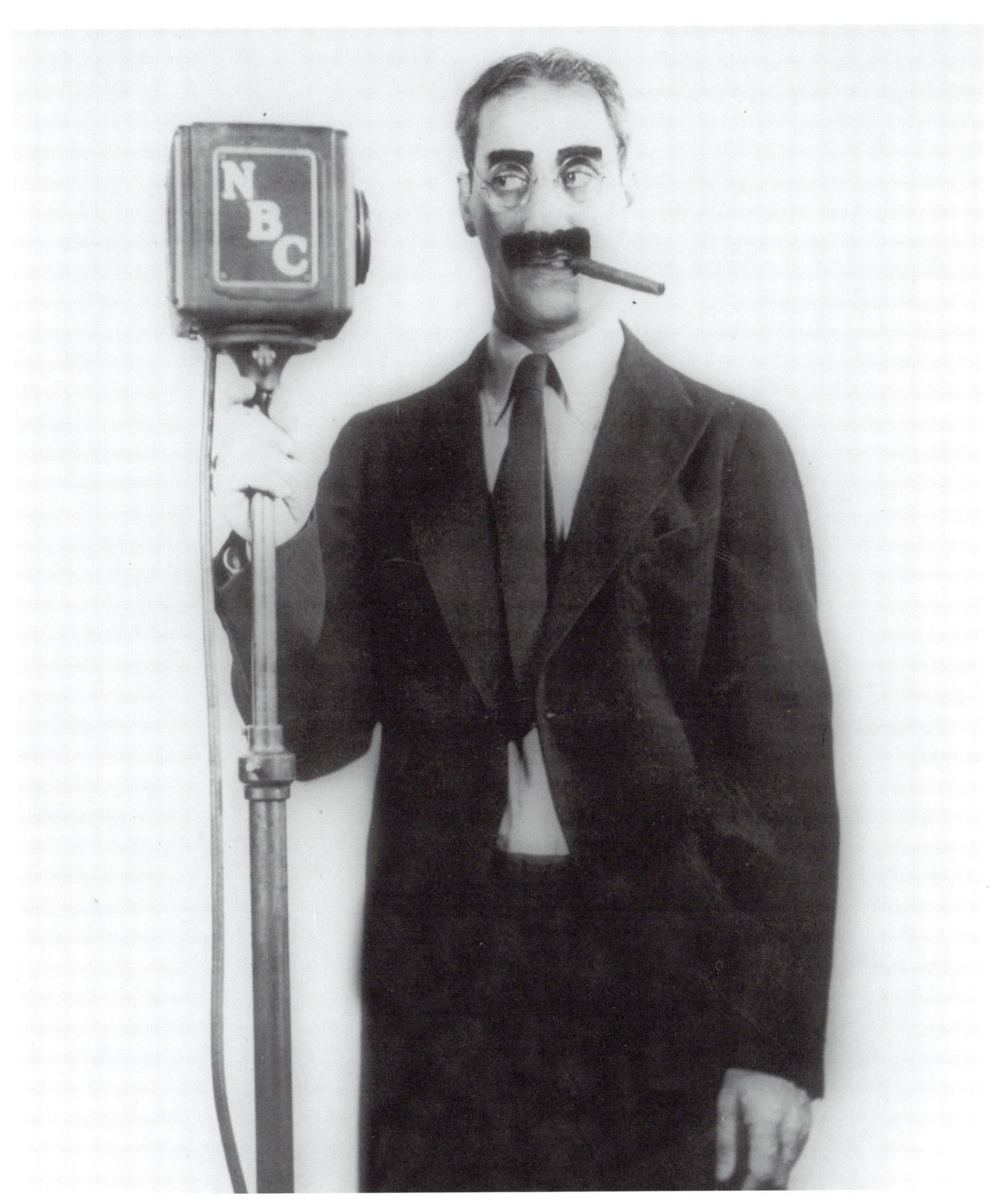

Groucho Marx, *You Bet Your Life*
Courtesy of family of Groucho Marx

show's peak, in 1955, the broadcasts drew a combined audience of 35 million. The radio version folded late in 1956, with *You Bet Your Life* continuing on television until 1961.

Though its content was decidedly low-tech—Groucho interviewed contestants and asked simple quiz questions—*You Bet Your Life* was innovative in its production and delivery. *You Bet Your Life* pioneered a version of what became the "live on tape" approach adopted by television talk shows in the 1950s. Guedel's original intention to broadcast live was scrapped, apparently at the last minute, because of concerns about Groucho's ad-libs. The producers then procured acetate disks, of the kind used by Armed Forces Radio, which had the advantage of allowing content to be minimally edited. Later, the program was a pioneer in the use of magnetic tape. For a standard program, one hour of tape was edited down to 26 minutes. When the American Federation of Musicians in 1948 changed its policy to allow network radio shows to be prerecorded, *Daily Variety* attributed the "cry and hullabaloo for tape" to the influence of a single program, *You Bet Your Life*. With the debut of the televised version, the producers recognized the need for separate postproduction for different media. The Wednesday radio program and the Thursday television version were often quite different, to the extent of presenting different contestants because of time shifts during editing.

Ironically, neither version was the spontaneous fest of ad-libs originally conceived by Guedel and Marx. On the contrary, pre-production was as crucial to *You Bet Your Life* as its postproduction editing was. Groucho's writers were disguised in the program's credits, but most of his repartee was scripted. Room was left for spontaneity: for example, Marx declined to meet contestants beforehand, but his writers did extensive pre-interviews with them. Genuine ad-libs were always a prospect: director Bernie Smith commented, "At his peak you could never write for this man." Nevertheless, the key was Groucho's ability to deliver scripted lines as if they were ad-libbed. Thus, to a tree surgeon: "Have you ever fallen out of a patient?" To a cartoonist: "If you want to see a comic strip, you should see me in the shower." To a fat woman: "I bet you're a lot of fun at a party. . . . In fact, you *are* a party." The remarks often had a cruel edge, but as writer Howard Harris observed, "If they weren't insulted, they were insulted."

Scripted ad-libs and edited "live" content were aspects of what might now be called the "postmodern" approach of this quiz show. *You Bet Your Life* was almost pure process, inverting the conventions of its ostensible genre. For example, introductory interviews with contestants, ordinarily perfunctory on quiz shows, occupied half the running time of Groucho's program. Contestants usually appeared in male-female couples—carefully paired to create possibilities for comic repartee—yet despite Groucho's standard compliment to "an attractive couple," they rarely knew each other. The quiz portion of the program was played straight: contestants began with $20 (later $100) and bet on four questions in a set category; the couple with the highest total for each program got a chance at a jackpot question for $1,000 (increased by $500 per week if nobody won). Nevertheless, prizes were never very important; in an era of big-money quiz shows (and scandal), *You Bet Your Life* awarded an average of $333 to 2,100 contestants over a decade. Besides, it was impossible *not* to win: if contestants blew the standard quiz, Groucho would ask a variation of the most famous of all quiz show questions: "Who's buried in Grant's Tomb?" Contestants could also win money accidentally, by speaking the previously announced "secret word."

Periodic journalistic exposés, such as *TV Guide*'s 1954 "The Truth about Groucho's Ad Libs," had no effect whatever on his program's popularity. The audience knew they were listening to a comedy program in quiz show guise. If Groucho was funny, nobody cared that his quips were scripted and edited. *You Bet Your Life* presented a perfect match of star and vehicle. Later attempts to duplicate its success on television with Buddy Hackett and Bill Cosby failed. The show's opening audience-response formula turned out to be literally accurate: "Here he is—the one, the only . . . GROUCHO!"

GLEN M. JOHNSON

Host
Groucho Marx

Announcer
George Fenneman, Jack Slattery

Producer/Directors
John Guedel, Bernie Smith, Bob Dwan

Writers
Ed Tyler, Hy Freedman, and Howard Harris

Programming History

ABC	October 1947–May 1949
CBS	October 1949–June 1950
NBC	October 1950–September 1956

Further Reading

Arce, Hector, *Groucho*, New York: Putnam, 1979
Marx, Groucho, and Hector Arce, *The Secret Word Is Groucho*, New York: Putnam, 1976
Tyson, Peter, *Groucho Marx*, New York: Chelsea House, 1995

Your Hit Parade

Musical Variety Program

Your Hit Parade reflected popular music trends of its era, especially the big band sound; the program also helped sell millions of Lucky Strike cigarettes. Yet despite its status as a Saturday night radio staple, *Your Hit Parade* underwent many changes over its long history, notably its continual shifting of length, its scheduled time slot, and even its network.

Your Hit Parade emerged in early 1935 as the National Broadcasting Company (NBC) looked to fill its Saturday night schedule. The Rogers and Hart ballad "Soon" ranked as the number-one hit. This alliance between big band hits and Lucky Strike cigarettes would continue to define the style and shape of *Your Hit Parade*: as a generation of executives for the American Tobacco Company's Lucky Strike division correctly figured, the public would tune in to the cover versions of hit songs by unknowns, and so show costs would be low while retaining a broad-based appeal.

Although its Saturday night venue never changed, its line-up of announcers, orchestras, and singers surely did. For

W.C. Fields and soundman Al Span, *Your Hit Parade*
Courtesy CBS Radio Archive

example, in 1937, when *Your Hit Parade* shifted to the Columbia Broadcasting System (CBS), out went the old talent, and in came the Lanny Ross Orchestra, with Barry Wood and Bonnie Baker as the leading vocalists. But in 1939, out went Ross, and in came the Mark Warnow Orchestra. Such shifts were frequent for the program. Its time slot and length also varied over the years. Starting times of 8:00 P.M., 9:00 P.M., and 10:00 P.M. were tried and retried, as running time fell from one hour, to 45 minutes, down to 30 minutes.

Generally the names of the *Your Hit Parade* vocalists and orchestras have been forgotten, with a few exceptions such as Dinah Shore, Frank Sinatra, and Doris Day. Sometimes, to boost ratings, American Tobacco brought in guest stars, including most notably W.C. Fields and Fred Astaire. Indeed, *Your Hit Parade* reached its peak during the World War II years when executives—in a rare spending spree—hired Frank Sinatra, and thus the CBS Radio Theater at Broadway and 53rd Street became the focus of young female fan attention. The theater, which held 1,200, filled with teenagers who roared as Sinatra rendered hits such as "Paper Doll," "You'll Never Know," "Long Ago and Far Away," and "I'll Be Seeing You." In January 1945 Sinatra's contract expired, and rather than pay a higher wage to this budding star, American Tobacco reverted to its low-cost approach. Sinatra would return in September 1946, bringing as his costar former Les Brown Orchestra star Doris Day, but only temporarily.

If there was an omen of the impending end of radio's *Your Hit Parade,* it was surely when Mark Warnow, the show's longest orchestra leader, died in 1949, immediately after completing his 493rd *Your Hit Parade* broadcast. He was replaced by Raymond Scott, and it was Scott who led the show to television by hiring and developing Snooky Lanson, Dorothy Collins, and Russell Arms. Scott did not change the programming formula.

The constant was that both the radio and television versions featured relative unknowns reprising the most popular pop songs of the week as determined by a national "survey" of record and sheet music sales. (The methodology of this survey was never revealed, but it could hardly have been scientific, as it probably never went beyond calls to a few major city record stores and to the leading publishers of sheet music.) Repeated chart toppers were simply played again and again, with slight variations. "Race" music from and for African Americans and "Hillbilly" music from and for rural and small-town whites was wholly ignored unless a version "crossed over" and was covered by a mainstream crooner or band. So although Texas western swing band Bob Wills and the Texas Playboys composed, created, and initially recorded "The New San Antonio Rose," it would be Bing Crosby's version that would make it onto *Your Hit Parade.* This Tin Pan Alley focus and inability to deal with the synthesis of Race and Hillbilly music that eventually led to rock and roll signaled the end of the formula and of *Your Hit Parade*—on both radio and television.

DOUGLAS GOMERY

See also Recordings and the Radio Industry; Singers on Radio

Cast

Vocalists (partial list)	Buddy Clark, Frank Sinatra, Joan Edwards, Freda Gibbson (later Georgia Gibbs), Lawrence Tibbett, Barry Wood, Jeff Clark, Eileen Wilson, Doris Day, Bonnie Baker, and Andy Russell
Announcers (partial list)	Martin Block, Del Sharbutt, Andre Baruch, Kenny Delmar, and Basil Ruysdael

Programming History

NBC	Spring 1935–Fall 1937
CBS	Fall 1937–Fall 1947
NBC	Fall 1947–Winter 1953

Further Reading

Buxton, Frank, and William Hugh Owen, *The Big Broadcast, 1920–1950,* New York: Viking Press, 1972
DeLong, Thomas A., *The Mighty Music Box: The Golden Age of Musical Radio,* Los Angeles: Amber Crest, 1980
Williams, John R., *This Was Your Hit Parade,* Camden, Maine: n.p., 1973

Yours Truly, Johnny Dollar

Drama Program

Yours Truly, Johnny Dollar was the last surviving network dramatic show after the inception of television. From 1949 to its demise in 1962, Johnny Dollar entertained those detective fans who had not yet been seduced by "the tube." *Yours Truly, Johnny Dollar* and *Suspense* were the last two original radio dramatic series produced for the Columbia Broadcasting System (CBS), and they ended their run on 30 September 1962.

The radio series recounted the detective cases of Johnny Dollar, "America's fabulous freelance insurance investigator." He would often receive his assignments from Pat McCracken of the Universal Adjustment Bureau, a clearinghouse for several insurance firms. Hartford, Connecticut, the headquarters for many major insurance companies, was his home base, but his assignments took him all over the world. His investigations of such matters as stolen jewels, paintings, or furs; missing persons; and insurance fraud of various types would inevitably lead to a murder investigation and an encounter with the criminal element. However, Johnny Dollar could take care of himself; he could be as hard-boiled as the toughest detective. His wisecracking betrayed a cynical attitude, and his encounters with women certainly resulted in some suggestive language.

Johnny Dollar was a confirmed bachelor, although he did have a girlfriend, Betty Lewis, who appeared occasionally. He was basically a loner, and each story was told from his first-person point of view. "Dollar" was a metaphor for the detective's interest in money. Described as the detective "with the action-packed expense account," he tallied each and every expenditure, no matter how small. Each show concluded with the revelation of his total expenses, as if dictating a memorandum to his employer, before he signed off with "yours truly, Johnny Dollar."

The series premiered on 11 February 1949 with a 30-minute episode entitled "The Parakoff Policy," in which the insured was being held for the murder of Mr. Parakoff. Johnny Dollar's encounter with Parakoff's widow allowed for some suggestive dialogue. Paul Dudley and Gil Doud wrote the pilot script for the series, and actor Dick Powell auditioned for the title role on 8 December 1948, but he went on to star in *Richard Diamond, Private Detective* instead. Charles Russell was the first of six radio actors to play Johnny Dollar on the air. Russell played the role as the stereotypical hard-boiled investigator with his own little quirks, such as flipping silver dollars to hotel bellboys.

Russell played the role of Johnny Dollar for one year, through 34 half-hour episodes. Edmond O'Brien assumed the role in February 1950, starring in 103 episodes until September 1952, and John Lund continued in the role for the next two years, starring in 92 episodes through September 1954, when the show was canceled, probably because of a lack of sponsorship. Most often the shows were broadcast on a sustaining basis. Wrigley's gum had the longest continuous sponsorship, from 10 March 1953 to 10 August 1954.

Yours Truly, Johnny Dollar returned to the air in October 1955 with a new format, star, and producer/writer/director. Instead of the 30-minute series format, the show moved to a serial format, with five 15-minute episodes per week. The listeners seemed to like this format because it allowed more time for story and character development—75 minutes each week, including commercials or other promotional material, of course. They liked the new star as well. Bob Bailey played Johnny Dollar as a more caring and less cynical and hard-boiled investigator. Bailey's portrayal made the hero seem more human, but nevertheless a tough and smart detective. Gerald Mohr, who had played the lead in *The Adventures of Philip Marlowe,* made an audition tape on 29 August 1955, but it never aired. Jack Johnstone, who was responsible for the new directions in the program, began producing and directing the show at this time, and he contributed several scripts before the series ended in 1962.

Bailey played in 55 of these weekly serials before November 1956, when CBS reverted back to the original 30-minute, once-a-week format. Continuing until 27 November 1960, Bailey played in 203 episodes, more than any other star of the series. At that time, the show was moved from Hollywood, where it had been produced from its beginning, to New York City. Robert Readick assumed the role on 4 December 1960 and played in 28 episodes until 11 June 1961, when Mandel Kramer took the part. Kramer played Johnny Dollar for 69 episodes until the series ended with the last case, "The Tip-Off Matter," on 30 September 1962.

PHILIP J. LANE

Cast

Johnny Dollar — Charles Russell (1949–50), Edmond O'Brien (1950–52), John Lund (1952–54), Bob Bailey (1955–60), Robert Readick (1960–61), Mandel Kramer (1961–62)

Directors

Richard Sanville, Norman Macdonnell, Gordon Hughes, Jaime del Valle, Jack Johnstone, Bruno Zirato, Jr., and Fred Hendrickson

Robert Bailey in *Yours Truly, Johnny Dollar*
Courtesy CBS Photo Archive

Writers

Gil Doud, Paul Dudley, David Ellis, John M. Hayes, E. Jack Neuman, Les Crutchfield, Blake Edwards, Morton Fine, David Friedkin, Sidney Marshall, Joel Murcott, John Dawson, Jack Johnstone, and Robert Ryf

Programming History

CBS February 1949–September 1962

Further Reading

Dunning, John, *Tune in Yesterday: The Ultimate Encyclopedia of Old-Time Radio, 1925–1976,* Englewood Cliffs, New Jersey: Prentice-Hall, 1976; revised edition, as *On the Air: The Encyclopedia of Old-Time Radio,* New York: Oxford University Press, 1998

Maltin, Leonard, *The Great American Broadcast: A Celebration of Radio's Golden Age,* New York: Dutton, 1997

Widner, James F., "Yours Truly, Johnny Dollar," <www.otrsite.com/articles/widner002.html>

Wright, Stewart, "Johnny Dollar," <www.thrillingdetective.com/dollar_johnny.html>

Z

Zenith Radio Corporation

Radio and Electronics Manufacturer

Zenith Radio Corporation, now Zenith Electronics Corporation, was the longest-surviving American-owned consumer electronics corporation. Founded in 1919 as Chicago Radio Laboratory, Zenith manufactured a wide range of electronic products for 80 years and continues to be one of the most respected and widely known American names in consumer electronics. The United States–based company has been a wholly owned subsidiary of Korean electronics giant LG Electronics since late 1999.

Under the guidance of founding genius "Commander" Eugene F. McDonald, Jr., and innovative financial manager Hugh Robertson, Zenith grew from its beginnings on a kitchen table on Chicago's North Side to a leadership position in radio and, along with archrival Radio Corporation of America (RCA), to continued dominance in the postwar television boom. Along the way, Zenith and McDonald made significant contributions to the very form of the consumer electronics and broadcasting industries. Zenith is best known for the high quality and reliability of its products and its innovative concepts in product development and marketing.

Origins

The founders of what was to become Zenith Radio Corporation were two radio amateurs, Ralph H.G. Matthews and Karl Hassel. Matthews built his first amateur station in Chicago in 1912. In 1913 and 1914 he perfected a distinctive aluminum sawtooth rotary spark gap disk that later became the company's first product. Matthews also became heavily involved in the newly formed Amateur Radio Relay League (ARRL), and in 1917 his radio call sign was changed to 9ZN. While serving as a radioman at the Great Lakes Naval Training Station at the end of World War I, he met a radio code instructor, Karl Hassel. Upon release from the navy, the two entered into a partnership producing first the aluminum spark gap transmitting disk and then other amateur equipment. They were soon producing complete receivers and transmitters of their own design. Operating as the Chicago Radio Laboratory, the two quickly outgrew their manufacturing space at 1316 Carmen Avenue (actually Matthew's house) and moved into half of a garage on Sheridan Road, on the lakeside grounds of the Edgewater Beach Hotel. The other half of the garage served as the home of 9ZN, one of the best-known amateur stations in the United States. Because their equipment was built for the radio amateur, the earliest advertising was placed in *QST,* the magazine of the American Radio Relay League. By late 1921 *QST* advertisements listed the 9ZN call followed by a small "ith," the origin of the trade name Z-Nith.

In 1921 Eugene F. McDonald became involved with the Chicago Radio Laboratory. McDonald was a savvy businessman who was looking for a business investment when he discovered Matthews, Hassel, and radio. McDonald offered to become a financial partner in their undertaking, and a partnership was formed, with McDonald as the general manager. A period of rapid growth followed.

As demand for the product increased in the spring of 1922, McDonald engaged his friend Tom Pletcher, a well-known figure in the music industry and president of the QRS Music Company, to take over the sales and manufacturing of CRL receivers in his large (and partially empty) new factory. By July, production had reached 15 sets per day.

Because the Armstrong receiver circuit patent was licensed to the Chicago Radio Laboratory, which produced Z-Nith products, McDonald formed Zenith Radio Corporation to become the marketing arm for the Z-Nith radios. The corporation was founded on 30 June 1923 with capital of $500,000 derived from common stock sold at $10 per share.

McDonald's Zenith

In 1923 McDonald built one of Chicago's pioneer radio stations, WJAZ, to stay in contact with the 1923–24 MacMillan

Arctic Expedition, which was carrying Zenith radio equipment. The experiment was successful, allowing the expedition to be the first to maintain contact with civilization during the long polar night and generating considerable publicity for the small radio company.

McDonald also equipped the 1925 MacMillan Arctic Expedition with Zenith shortwave equipment. That expedition was the first to use shortwave in the Arctic and the first to fly heavier-than-air craft in the Arctic; it was also Richard Byrd's first introduction to the polar regions. McDonald accompanied the expedition as second-in-command, and the Zenith equipment performed flawlessly. Experimental shortwave communications from the expedition in North Greenland to U.S. Navy vessels in New Zealand played a seminal role in the adoption of shortwave radio for long-distance communications.

McDonald's work with WJAZ highlighted for him emerging problems with the American Society of Composers, Authors, and Publishers (ASCAP) over royalties for performers whose music was played on the radio. Dissatisfied with the arbitrary nature of ASCAP's rate schedules, McDonald organized a meeting of a small group of broadcasters in Chicago in early 1923 to oppose ASCAP; this organization was to become the National Association of Broadcasters, with McDonald serving as its first president.

By the end of 1924, the production rate at the QRS factory could not keep up with increasing demand, and Zenith resumed manufacturing its own products in a new four-story plant on Iron Street in Chicago.

In 1925, Zenith introduced the grandest Zenith radio models the company had ever manufactured, the ten-tube Deluxe receivers. There were five cabinet styles, each handmade: the Colonial, the English, the Italian, the Chinese, and the Spanish. The price for these models ranged from $650 to $2,000 ($5,800 to $21,400 in 2000 dollars) and were the most expensive radios being manufactured at that time. They illustrated the company's commitment to building the very best equipment, regardless of cost.

McDonald became embroiled in another broadcasting battle in 1926, when his WJAZ shifted ("jumped") to another frequency, seeking a less-congested channel, but also challenging the authority of the Secretary of Commerce to assign radio frequencies. On 16 April 1926, the case was decided in federal court in McDonald's favor (*United States v Zenith Radio Corporation*, 1926), proving finally that the existing frequency allocation laws, dating to 1912, were unenforceable and that the secretary lacked authority.

Zenith Radio Corporation was first listed on the Chicago Stock Exchange in March 1928 and on the New York Stock Exchange in July 1929. Stockholders increased from 250 in April 1928 to 2,750 in April 1929. Fiscal year 1929 earnings exceeded $1 million.

When the stock market crashed in 1929, Zenith found itself with a large inventory of materials to build new sets, but not many finished sets, primarily because of an innovative inventory control plan. That, and the selling of 100,000 shares of stock (worth $1 million) just before the market collapsed, enabled Zenith, with proper management, to ride out the rough times without missing loan payments, borrowing money, or releasing large numbers of employees. The company continued to manufacture high-quality, high-priced radios during the Depression but also added a less expensive line, the Zenette series, to appeal to the average buyer.

Recovery for Zenith began in 1933 when deficits, which had been running at about $500,000 a year, were converted into a $50,000 profit for the fiscal year ending 30 April 1934. At the beginning of 1934, Zenith was the lowest-priced radio stock quoted on the New York Stock Exchange; by the end of 1934, it was the highest. The Depression recovery assumed spectacular proportions for Zenith in 1935, when net earnings returned to the pre-Depression high of just over $1 million. Zenith also undertook major efforts to maintain its distributors' profit margins during the rough times, and Zenith emerged from the Depression with a fiercely loyal band of distributors who would serve the corporation admirably for many decades in the boom ahead.

In 1937 Zenith supplemented its factory space with the addition of the 400,000-square-foot West Dickens Avenue facility. In 1937 the radio industry as a whole showed a 15 percent drop, but Zenith's sales rose. New developments prior to World War II included the chairside radio-phonograph; a "Radio Nurse" baby monitor; and a line of portable radios, including the venerable Zenith Trans-Oceanic radio, which would go on to become the longest lasting radio brand in radio history. By 1938 most Zenith radios contained the Zenith-patented "Wavemagnet" antenna.

The ensuing years were marked by steady progress. The Zenith experimental television station, W9XZV, began operating in black and white in February 1939 and began color transmissions in 1941.

Because of the war, all domestic production stopped on 1 April 1942. Zenith's war efforts centered on development and production of sophisticated frequency meters, work on the V-T proximity fuse, and military-grade radio communications devices. It was through Zenith's efforts that most manufacturers, except RCA, granted the government free license under all patents covering war work. Zenith was awarded the Army-Navy "E" in November 1942, the first of five it would receive. Zenith was given special permission to manufacture only one civilian product during the war years, an inexpensive hearing aid that allowed the hard-of-hearing to be gainfully employed in war work.

Postwar Radio

Zenith planned for the resumption of civilian production in the closing years of the war and was among the earliest to attain volume production after the war. In 1945 Zenith began production of many of its own components, such as loudspeakers, record changers, and coils. The company was also an important early manufacturer of FM receivers. In 1947 Zenith introduced the "Cobra" phonograph arm. In 1948 the company introduced turret tuning for television, allowing the expansion of the tuner for future UHF reception. Zenith acquired television tube manufacturer Rauland Corporation in 1948 and in 1949 introduced the first "black tube" television sets, which quickly became the industry standard. In 1950 Zenith stopped manufacturing automobile radios, in spite of excellent sales, to provide space for the rapidly expanding television business. The continuously variable speed (10- to 85-rpm) Cobra-Matic record changer was introduced in 1950.

Major expansion of manufacturing occurred again in 1950–51, when a large facility in Chicago was acquired for television production and for Korean War military contracts. The removal of the television station "freeze" in 1952 greatly stimulated the company's television business, and the Zenith turret tuner made Zenith the only sets in production that could be easily converted to UHF. In the fall of 1953 Zenith introduced a three-transistor hearing aid, the first of many solid-state models to follow. By 1954 Zenith was selling more hearing aids than all other companies combined, and their dominance of the industry continued through the 1970s. Zenith entered the high-fidelity market in 1953 and was among the first to provide high-fidelity sound for television receivers. The ultrasonic Zenith "Space Command" remote TV control was introduced in 1956. Zenith's founder, Eugene F. McDonald, Jr., died in 1958.

By the mid-1970s, because of increasing competitive pressures from offshore (mostly Asian) manufacturers of radio and television, Zenith established its own manufacturing operations in Mexico and Taiwan, while forging an alliance with LG Electronics to build Zenith-brand clock radios in Korea. While growing its television business and venturing into new areas such as VCRs and cable set-top boxes, Zenith continued to play a role in transistor and portable radios and in component and console stereos until the company phased out its radio and audio products business in 1982 to concentrate on television and other video-related products. That year, the last of the legendary Trans-Oceanic radios was produced, marking the completion of the four-decade reign of that famous series of multiband shortwave radios.

HAROLD N. CONES AND JOHN H. BRYANT

See also McDonald, Eugene F.; National Association of Broadcasters

Further Reading

Bensman, Marvin, "The Zenith-WJAZ Case and the Chaos of 1926–27," *Journal of Boradcasting* 14, no. 4 (Fall 1970)

Bryant, John H., and Harold N. Cones, *The Zenith Trans-Oceanic: The Royalty of Radios,* Atglen, Pennsylvania: Schiffer, 1995

Bryant, John H., and Harold N. Cones, *Dangerous Crossings: The First Modern Polar Expedition, 1925,* Annapolis, Maryland: Naval Institute Press, 2000

Cones, Harold N., and John H. Bryant, *Eugene F. McDonald, Jr.: Communications Pioneer Lost to History,* Record of Proceedings, 3rd International Symposium on Telecommunications History, Washington, D.C.: Independent Telephone Historical Foundation, 1995

Cones, Harold N., and John H. Bryant, *Zenith Radio, The Early Years: 1919–1935,* Atglen, Pennsylvania: Schiffer, 1997

Cones, Harold N., and John H. Bryant, "The Car Salesman and the Accordion Designer: Contributions of Eugene F. McDonald, Jr., and Robert Davol Budlong to Radio," *Journal of Radio Studies* 8, no. 1 (Spring 2001)

Douglas, Alex, "Zenith," in *Radio Manufacturers of the 1920s,* by Douglas, vol. 3, Vestal, New York: Vestal Press, 1991

Zenith Radio Corporation, *The Zenith Story: A History from 1919,* Chicago: Zenith Radio Corporation, 1955

CONTRIBUTOR AFFILIATIONS

Michael H. Adams. San Jose State University.
Essays: Children's Novels and Radio; Crystal Receivers; De Forest, Lee; Herrold, Charles D.; KCBS/KQW; Low-Power Radio/Microradio; Milam, Lorenzo

Alan B. Albarran. University of North Texas.
Essays: Clear Channel Communications; Hicks, Tom

Pierre Albert. Université Panthéon-Assas, Paris II
Essay: France

Craig Allen. Arizona State University
Essays: Italy; Storer, George; WNBC

Steven D. Anderson. James Madison University
Essays: Broadcast Education Association; Internet Radio; KOA

Larry Appelbaum. Library of Congress
Essay: KPFA

Edd Applegate. Middle Tennessee State University
Essays: Conrad, William; Howe, Quincy

Sousan Arafeh. American Institutes for Research
Essays: *American School of the Air;* National Telecommunications and Information Administration

John S. Armstrong. University of Utah
Essay: Gabel, Martin

Philip J. Auter. University of Louisiana at Lafayette
Essays: Science Fiction Programs; Situation Comedy; Topless Radio

Robert K. Avery. University of Utah
Essays: Corporation for Public Broadcasting; National Association of Educational Broadcasters; Public Broadcasting Act of 1967; Public Service Radio

Glenda R. Balas. University of New Mexico
Essay: KOB

Mary Christine Banwart. University of Kansas
Essay: Election Coverage

Warren Bareiss. University of Scranton
Essays: Free Form Format; Westerns

Ray Barfield. Clemson University
Essays: *Edgar Bergen and Charlie McCarthy Show;* Fadiman, Clifton; Premiums; *Shadow*

Kyle S. Barnett. University of Texas, Austin
Essay: Heavy Metal/Active Rock Format

Douglas L. Battema. Western New England College
Essay: Sports on Radio

Mary E. Beadle. John Carroll University
Essays: Australian Aboriginal Radio; Clandestine Radio; Collingwood, Charles; Group W; South America; Wilson, Don

Christine Becker. University of Notre Dame
Essays: Ameche, Don; Dr. Demento; *Lux Radio Theater;* Talent Raids

Johnny Beerling. RDS Forum
Essay: Radio Data System

Alan Bell. Grand Valley State University
Essay: Comedy

Louise Benjamin. University of Georgia
Essays: Brinkley, John R.; General Electric; Peabody Awards

ElDean Bennett (deceased). Arizona State University
Essay: Mormon Tabernacle Choir

Marvin Bensman. University of Memphis
Essays: Audiotape; Hoover, Herbert; Nostalgia Radio; Wireless Acts of 1910 and 1912/Radio Acts of 1912 and 1927

Jerome S. Berg. Author, *On the Shortwaves*
Essays: DXers/DXing; Shortwave Radio

Rosemary Bergeron. National Archives of Canada
Essays: Canadian Radio Archives; Canadian Talk Radio

William L. Bird, Jr. Smithsonian Institution
Essays: *American Family Robinson; Cavalcade of America; Theater Guild on the Air*

Howard Blue. Independent Scholar
Essay: Playwrights on Radio

A. Joseph Borrell. Shippensburg University
Essays: KYW; Simon, Scott

Douglas A. Boyd. University of Kentucky
Essay: Arab World Radio

John Bradford. The Radio Academy
Essay: Radio Academy

L. Clare Bratten. Middle Tennessee State University
Essays: WNYC; WOR

Mark Braun. Gustavus Adolphus College
Essays: American Federation of Television and Radio Artists; Keillor, Garrison; Marketplace; Stereo; WCCO

Jack Brown. Author, *Please Stand By: A History of Radio*
Essay: Armed Forces Radio Service

Michael Brown. University of Wyoming
Essay: Fan Magazines

Robert J. Brown. Syracuse University
Essays: Axis Sally; Churchill, Winston S.; Swing, Raymond Gram

Donald R. Browne. University of Minnesota
Essays: International Radio Broadcasting; Kling, William; Radio in the American Sector (Berlin); Radio Free Europe/Radio Liberty

John H. Bryant. Oklahoma State University
Essays: McDonald, Eugene F., Jr.; National Association of Broadcasters; Zenith

Joseph G. Buchman. Indiana University
Essay: Jepko, Herb

Karen S. Buzzard. Southwest Missouri State University
Essays: Arbitron; Cooperative Analysis of Broadcasting; Hooperatings

Paul Brian Campbell. Le Moyne College
Essay: Religion on Radio

Dom Caristi. Ball State University
Essays: Copyright; Emergency Broadcasting System; Federal Communications Commission; Federal Radio Commission; First Amendment and Radio

Ginger Rudeseal Carter. Georgia College and State University
Essays: Ellis, Elmo; WSB

Dixon H. Chandler II. Florida State University
Essay: Horror Programs

Frank J. Chorba. Washburn University
Essays: Rogers, Will; Wright, Early

Lynn A. Christian. Independent Scholar and Radio Consultant
Essay: FM Trade Associations

Claudia Clark. University of Alaska, Fairbanks
Essay: Psychographics

Kathleen Collins. Independent Scholar
Essay: Critics

Jerry Condra. State University of New York, Oswego
Essay: KGO

Harold N. Cones. Christopher Newport University
Essays: McDonald, Eugene F., Jr.; National Association of Broadcasters; Zenith

Bryan Cornell. Library of Congress
Essays: American Broadcasting Station in Europe; *Easy Aces*

Elizabeth Cox. Writer and Columnist
Essays: *Adventures in Good Music*; Intercollegiate Broadcasting System

Steve Craig. University of North Texas
Essay: Farm/Agricultural Radio

Tim Crook. Goldsmiths College, University of London
Essays: British Broadcasting Corporation; BBC Broadcasting House; BBC Local Radio; BBC Radio Programming; British Radio Journalism; Cooke, Alistair; Cooper, Giles; Drama, U.S.; Drama, Worldwide; Gillard, Frank

Marie Cusson. Concordia University
Essay: Canadian Radio Satire

Keri Davies. British Broadcasting Corporation
Essay: *Archers*

E. Alvin Davis. E. Alvin Davis and Associates
Essay: Oldies Format

J.M. Dempsey. University of North Texas
Essays: Commercial Load; Recreations [of events]; Singers on Radio; WBAP

Corley Dennison. Marshall University
Essay: Digital Satellite Radio

Neil Denslow. Independent Scholar
Essay: *Mercury Theater of the Air*

Steven Dick. Southern Illinois University
Essay: Audio Streaming

John D.H. Downing. University of Texas, Austin
Essays: British Pirate Radio; Social and Political Movements

Pamela K. Doyle. University of Alabama
Essay: National Association of Educational Broadcasters Tape Network

Christina S. Drale. Southwest Missouri State University
Essays: Association for Women in Communications; KRLA; Westwood One

Susan Tyler Eastman. Indiana University
Essays: Allen, Mel; Kuralt, Charles; McNamee, Graham; Promax; Sportscasters

Bob Edwards. National Public Radio
Essay: Red Barber

Kathryn Smoot Egan. Brigham Young University
Essay: Programming Research

Lyombe Eko. University of Maine
Essays: Africa No. 1; World Radiocommunication Conference

Sandra L. Ellis. University of Wisconsin, River Falls
Essays: All News Format; Philco Radio; Polk, George; Prairie Home Companion; Talk Radio; Transistor Radios

Ralph Engelman. Long Island University, Brooklyn Campus
Essay: Friendly, Fred

Erika Engstrom. University of Nevada, Las Vegas
Essays: Adult Contemporary Format; Consultants; Hottelet, Richard; Middle of the Road Format; Soft Rock Format; Tesla, Nikola

Stuart L. Esrock. University of Louisville
Essay: Motorola

Charles Feldman. Monmouth College
Essay: Shock Jocks

Michel Filion. Université du Québec en Outaouais
Essay: Canadian Radio Policy

Howard Fink. Concordia University
Essay: Canadian Radio Drama (English-language section)

Seth Finn. Robert Morris University
Essay: Office of Radio Research

Robert G. Finney. California State University, Long Beach
Essays: Jazz Format; Netherlands; Radio Corporation of America

Margaret Finucane. John Carroll University
Essays: Freed, Paul; Murray, Lyn; Sound Effects

James E. Fletcher. University of Georgia
Essays: Audience Research Methods; Commercial Tests; Demographics; Pulse Inc.; Trade Associations

Corey Flintoff. National Public Radio
Essays: Edwards, Bob; *Fresh Air; Morning Edition;* National Public Radio; Stamberg, Susan; Totenberg, Nina

Joe S. Foote. Southern Illinois University
Essay: United States Presidency and Radio

Robert C. Fordan. Central Washington University
Essays: Donahue, Tom; Pyne, Joe

Robert S. Fortner. Calvin College
Essays: Board for International Broadcasting; Cold War Radio; Far East Broadcasting Company; Frequency Allocation; Inter-

national Telecommunication Union; Propaganda by Radio; Radio Free Asia; Radio Monte Carlo; Radio Moscow; WEAF

James C. Foust. Bowling Green State University
Essays: Ground Wave; Quaal, Ward L.

Ralph Frasca. Marymount University
Essays: Dunbar, Jim; Lord, Phillips

James A. Freeman. University of Massachusetts
Essays: *Aldrich Family; Fibber McGee and Molly; Great Gildersleeve;* Stereotypes on Radio

Elfriede Fürsich. Boston College
Essays: European Broadcasting Union; Germany

Charles F. Ganzert. Northern Michigan University
Essays: Block, Martin; Freed, Alan; Narrowcasting

Ronald Garay. Louisiana State University
Essays: Liberty Broadcasting System; Long, Huey; McLendon, Gordon; United States Congress and Radio; World War II and U.S. Radio

Philipp Gassert. University of Heidelberg
Essay: Fritzsche, Hans

Judith Gerber. Freelance Writer
Essays: Morgan, Robert W.; Museums and Archives of Radio

Norman Gilliland. Wisconsin Public Radio
Essay: *Earplay*

Donald G. Godfrey. Arizona State University
Essays: Capehart; Dill, Clarence Cleveland; White, Wallace H.

Douglas Gomery. University of Maryland
Essays: American Broadcasting Company; Border Radio; Clark, Dick; Country Music Format; Don Lee Broadcasting System; Godfrey, Arthur; *Grand Ole Opry;* Karmazin, Mel; Lewis, William B.; Local Marketing Agreements; Metromedia; Music; *National Barn Dance;* Ownership, Mergers, and Acquisitions; Taishoff, Sol; Variety Shows; WBBM; WGN; WINS; WLS; WMAQ; *Your Hit Parade*

Jim Grubbs. University of Illinois, Springfield
Essays: Biondi, Dick; Cable Radio; Citizen's Band Radio; Ham Radio; Jamming

Joanne Gula. Hawaii Pacific University
Essay: Emergencies, Radio's Role in

Paul F. Gullifor. Bradley University
Essays: Radio Martí; WEVD

Linwood A. Hagin. North Greenville College
Essay: Contemporary Christian Music Format

Donna L. Halper. Donna Halper & Associates
Essays: Cantor, Eddie; Columnists; Jewish Radio; Portable Radio Stations; Shepard, John; Silent Nights; Talent Shows; Vaudeville; WBZ; Westinghouse; WGI; Yankee Network

Tona J. Hangen. Harvard University
Essays: Evangelists/Evangelical Radio; McPherson, Aimee Semple

Margot Hardenbergh. Fordham University
Essays: Hummert, Frank and Anne; Roosevelt, Eleanor

Jeffrey D. Harman. Muskingum College
Essays: Audio Processing; Digital Recording; Production for Radio; Recording and Studio Equipment

Dorinda Hartmann. University of Wisconsin, Madison
Essays: *Duffy's Tavern; Lum 'n Abner*

Gordon H. Hastings. Broadcasters Foundation
Essays: Greenwald, James L.; McLaughlin, Edward F.; Schulke, James; Station Rep Firms; Taylor, Marlin R.

Joy Elizabeth Hayes. University of Iowa
Essay: Mexico

John Allen Hendricks. Southeastern Oklahoma State University
Essay: Wertheimer, Linda

Alexandra Hendriks. University of Arizona
Essay: WHER

Ariana Hernandez-Reguant. Tulane University
Essay: Cuba

Robert L. Hilliard. Emerson College
Essay: Hate Radio

Jim Hilliker. Independent Scholar
Essays: KFI; KNX

Michele Hilmes. University of Wisconsin, Madison
Essays: Brice, Fanny; Hollywood and Radio; Women in Radio

John Hochheimer. Ithaca College
Essay: Public Affairs Programming

Jack Holgate. Francis Marion University
Essay: Columbia Broadcasting System

Herbert H. Howard. University of Tennessee, Knoxville
Essays: Blue Network; Cross, Milton; Kaltenborn, H.V.

Chuck Howell. University of Maryland
Essays: Kirby, Edward M.; McGannon, Don; Office of War Information; Taylor, Deems; *Vox Pop*

Kevin Howley. Northeastern University
Essays: *Flywheel, Shyster, and Flywheel;* National Federation of Community Broadcasters

W.A. Kelly Huff. Truett-McConnell College
Essays: Benny, Jack; Dees, Rick; Digital Audio Broadcasting; *George Burns and Gracie Allen Show;* Morrow, Bruce "Cousin Brucie"; Owens, Gary; Sarnoff, David; Tracht, Doug "Greaseman"; Wolfman Jack

Peter E. Hunn. State University of New York, Oswego
Essays: CONELRAD; Drake, Bill; Gambling, John; Judis, Bernice; WNEW

John D. Jackson. Concordia University
Essay: Canadian Radio and Multiculturalism

Randy Jacobs. University of Hartford
Essay: Advertising Agencies

Glen M. Johnson. Catholic University of America
Essays: WTOP; *You Bet Your Life*

Phylis Johnson. Southern Illinois University
Essays: Hoaxes on Radio; Gay and Lesbian Radio; Female Radio Personalities and Disk Jockeys

Sara Jones. Sheffield Hallam University
Essays: Reith, John C.W.

Lynda Lee Kaid. University of Oklahoma
Essays: Election Coverage; Politics and Radio

Stephen A. Kallis, Jr. Independent Scholar
Essay: *Captain Midnight*

Steve Kang. St. Hugh's College, Oxford University
Essay: Fairness Doctrine

Michael C. Keith. Boston College
Essays: All Night Radio; Corwin, Norman; Disk Jockeys; Hate Radio; Morning Programs; Native American Radio; Retro Formats; Underground Radio

Ari Kelman. New York University
Essays: Kasem, Casey; Non-English-Language Radio in the United States

Colum Kenny. Independent Scholar
Essay: Ireland

John Michael Kittross. Emerson College
Essay: Educational Radio to 1967

Frederica P. Kushner. Library of Congress
Essay: *Star Wars*

Philip J. Lane. California State University, Fresno
Essay: *Yours Truly, Johnny Dollar*

Matthew Lasar. University of California, Santa Cruz
Essays: Hill, Lewis; KPFA; Pacifica Foundation

Laurie Thomas Lee. University of Nebraska, Lincoln
Essays: Bell Telephone Laboratories; Contemporary Hit Radio/Top 40 Format; Formats

Renée Legris. University of Montreal, Quebec
Essay: Canadian Radio Drama (French-language section)

Frederic A. Leigh. Arizona State University
Essays: Classic Rock Format; *Mayflower* Decision; Stern, Howard

Lawrence W. Lichty. Northwestern University
Essays: Audimeter; Crosley, Powel; Diary; Documentary Programs on U.S. Radio; *Gunsmoke; Hindenburg* Disaster; Imus, Don; *March of Time;* WLW

Lucy A. Liggett. Eastern Michigan University
Essay: Maxwell, James Clerk

Val E. Limburg. Washington State University
Essays: KSL; Murrow, Edward R.; Obscenity/Indecency on Radio

Robert Henry Lochte. Murray State University
Essays: Bob and Ray; Early Wireless; *King Biscuit Flower Hour*

Jason Loviglio. University of Maryland, Baltimore County
Essay: Williams, Jerry

Gregory Ferrell Lowe. Finnish Broadcasting Company
Essays: Digital Audio Broadcasting; Scandinavia

Christopher Lucas. University of Texas, Austin
Essay: Social Class and Radio

Mike Mashon. Library of Congress
Essay: National Broadcasting Company

Marilyn J. Matelski. Boston College
Essays: Soap Opera; Vatican Radio

Peter E. Mayeux. University of Nebraska, Lincoln
Essays: Fleming, John Ambrose; Hertz, Heinrich

Dennis W. Mazzocco. Hofstra University
Essay: Daly, John Charles

Thomas A. McCain. Ohio State University
Essays: Ireland; Stanton, Frank; Violence and Radio

Jeffrey M. McCall. DePauw University
Essay: Press-Radio War

David McCartney. Independent Scholar
Essays: Automobile Radios; *Little Orphan Annie; Ma Perkins; Monitor;* Skelton, Red

Tom McCourt. University of Illinois at Springfield
Essays: Community Radio; Minnesota Public Radio; Public Radio International

Brad McCoy. Library of Congress
Essays: Dolby Noise Reduction; Poetry and Radio

Allison McCracken. Temple University
Essays: Crosby, Bing; *Inner Sanctum Mysteries; Lights Out; Our Miss Brooks;* Smith, Kate; *Suspense;* Vallee, Rudy

Drew O. McDaniel. Ohio University
Essay: Asia

Michael A. McGregor. Indiana University
Essays: Armstrong, Edwin Howard; Licensing; Red Lion Case

Robert McKenzie. East Stroudsburg University of Pennsylvania
Essay: News

Elizabeth McLeod. Freelance Journalist
Essays: Amalgamated Broadcasting System; *Amos 'n' Andy;* Jehovah's Witnesses

Mike Meeske. University of Central Florida
Essay: Combo

Fritz Messere. State University of New York, Oswego
Essays: American Telephone and Telegraph; American Top 40; Goldsmith, Alfred; High Fidelity; Regulation

Colin Miller. Freelance Journalist
Essay: Canadian Broadcasting Corporation

Toby Miller. New York University
Essay: Australia

Bruce Mims. Southeast Missouri State University
Essays: Automation; Easy Listening/Beautiful Music Format

Jack Minkow. Broadcasting Asset Management Corporation
Essay: Brokerage in Radio

Jack Mitchell. University of Wisconsin, Madison
Essays: *All Things Considered;* Mankiewicz, Frank; Public Radio Since 1967; Siemering, William; WHA and Wisconsin Public Radio

Jason Mittell. Middlebury College
Essay: Quiz and Audience Participation Programs

Barbara Moore. University of Tennessee, Knoxville
Essays: Davis, Elmer; Schechter, A.A.; Syndication; White, Paul

Matthew Murray. University of Illinois, Chicago
Essay: Censorship

Graham Mytton. BBC World Service
Essays: Africa; Digital Audio Broadcasting; Radio Authority; Voice of the Listener and Viewer

Gregory D. Newton. Ohio University
Essays: College Radio; Localism in Radio

Greg Nielsen. Concordia University
Essay: Canadian Radio Satire

D'Arcy John Oaks. Independent Scholar
Essays: Violence and Radio

William F. O'Connor. Asia University, Tokyo
Essay: Lord Haw-Haw

Cary O'Dell. Discovery Channel
Essays: *Jack Armstrong, the All-American Boy;* Wynn, Ed

Robert M. Ogles. Purdue University
Essays: KHJ; Storz, Todd; WLAC; WSM

Ryota Ono. Aichi University
Essay: Japan

Peter B. Orlik. Central Michigan University
Essays: American Society of Composers, Authors and Publishers; Broadcast Music Incorporated; Coughlin, Father Charles; Ten-Watt Stations; WXYZ

Pierre-C. Pagé. University of Montreal, Quebec
Essays: Canadian Radio Programming (French-language sections); CKAC

Brian T. Pauling. New Zealand Broadcasting School, Christchurch Polytechnic
Essays: New Zealand; South Pacific Islands

Manjunath Pendakur. University of Western Ontario
Essay: All India Radio

Douglas K. Penisten. Northeastern State University, Oklahoma
Essays: United Fruit Company; United States Navy and Radio

Stephen D. Perry. Illinois State University
Essays: Autry, Gene; Marconi, Guglielmo

Patricia Phalen. George Washington University
Essays: American Women in Radio and Television; Beville, Hugh Malcolm; *Family Theater*

Steven Phipps. University of Missouri, St. Louis
Essay: WCFL

Joseph R. Piasek. Quinnipiac College
Essays: Album-Oriented Rock Format; Alternative Format; Progressive Rock Format; Radio Disney; United Nations Radio

Gregory G. Pitts. Bradley University
Essays: Advertising; Radio Advertising Bureau; Simulcasting; WRR

Mark Poindexter. Central Michigan University
Essays: KCMO; KMOX

Tim Pollard. Ball State University
Essay: Terkel, Studs

Robert F. Potter. Indiana University
Essays: Music Testing; Promotion on Radio

Alf Pratte. Brigham Young University
Essays: Hope, Bob; Limbaugh, Rush

Patricia Joyner Priest. University of Georgia
Essays: Brown, Himan; Isay, David

Dennis Randolph. Calhoun County Community Development
Essay: RADAR

Lawrence N. Redd. Michigan State University
Essays: African-Americans in Radio; Blues Format; Murray the K; Thomas, Rufus

David E. Reese. John Carroll University
Essays: Call Letters; Control Board/Audio Mixer; Harvey, Paul

Patton B. Reighard. Appalachian State University
Essay: Crutchfield, Charles H.

Edward A. Riedinger. Ohio State University
Essays: Brazil; Landell de Moura, Roberto

Terry A. Robertson. University of South Dakota
Essay: Politics and Radio

Melinda B. Robins. Emerson College
Essay: Developing Nations

América Rodríguez. University of Texas, Austin
Essay: Hispanic Radio

Eric W. Rothenbuhler. The New School, New York
Essay: Rock and Roll Format

Richard Rudin. Liverpool John Moores University
Essays: British Disk Jockeys; British Forces Broadcasting Service; Capital Radio; Digital Audio Broadcasting; Everett, Kenny; Jingles; London Broadcasting Company; Radio Luxembourg; United States Influences on British Radio

Joseph A. Russomanno. Arizona State University
Essay: Rome, Jim

Anne Sanderlin. Freelance Writer
Essays: Durante, Jimmy; Gordon, Gale

Erica Scharrer. University of Massachusetts, Amherst
Essay: Children's Programs

Steven R. Scherer. Purdue University
Essay: Payola

Karl Schmid. University of North Carolina, Chapel Hill
Essay: Infinity Broadcasting

Clair Schulz. Museum of Broadcast Communications
Essay: Morgan, Henry

Ed Shane. Shane Media Services
Essays: Blore, Chuck; Bose, Amar G.; Chenault, Gene; Drew, Paul; Freberg, Stan; KTRH; Market; Programming Strategies and Processes; Shaw, Allen; Station Rep Firms; Talk Radio; Virtual Radio

Pam Shane. Shane Media Services
Essay: *Car Talk*

Mitchell Shapiro. University of Miami
Essays: Allen, Fred; *Gangbusters*

Jason T. Siegel. University of Arizona
Essay: *Talk of the Nation*

Ron Simon. Museum of Television and Radio
Essays: King, Larry; Markle, Fletcher; Oboler, Arch; *One Man's Family; Red Channels;* "Seven Dirty Words" Case; Trout, Robert; Weaver, Sylvester (Pat)

B.R. Smith. Central Michigan University
Essays: Conrad, Frank; Fireside Chats; *Lone Ranger;* Paley, William S.; Striker, Fran

Ruth Bayard Smith (deceased). Montclair State University
Essays: WABC; WBAI; Williams, Bruce

Lynn Spangler. State University of New York, New Paltz
Essays: CKLW; WJR

David R. Spencer. University of Western Ontario
Essay: Canadian News and Sports Broadcasting; Canadian Radio Programming (English-language sections)

David Spiceland. Appalachian State University
Essays: Subsidiary Communications Authorization; Wire Recording

Laurie R. Squire. Freelance Radio Producer
Essay: Shepherd, Jean

Michael Stamm. University of Chicago
Essay: Delmar, Kenneth

Christopher H. Sterling. George Washington University
Essays: A.C. Nielsen Company; AM Radio; American Federation of Musicians; Antenna; Asia; Audience; Awards and Prizes; Barnouw, Eric; *Beulah Show;* Blue Book; Broadcasting Board of Governors; *Can You Top This?;* Classical Music Format; Clear Channel Stations; Commentators; Controversial Issues; Deregulation of Radio; Disk Jockeys; Dunlap, Orrin E.; Editorializing; Education about Radio; Edwards, Ralph; Emerson Radio; "Equal Time" Rule; Faulk, John Henry; Fessenden, Reginald; FM Radio; Foreign Broadcast Information Service; Frederick, Pauline; German Wireless Pioneers; *Green Hornet; Hear It Now;* Hill, George Washington; Hogan, John V.L. "Jack"; *I Love a Mystery;* Kent, A. Atwater; Kesten, Paul; Klauber, Ed; Kyser, Kay; Lazarsfeld, Paul F.; Lodge, Oliver J.; McBride, Mary Margaret; McNeill, Don; Metropolitan Opera Broadcasts; Morning Programs; National Radio Systems Committee; Network Monopoly Probe; News Agencies; North American Regional Broadcasting Agreement; Pay Radio; Popov, Alexander; Radio City; Radio Hall of Fame; Radio Sawa/Middle East Radio Network; Receivers; Recordings and the Radio Industry; Sevareid, Eric; Shirer, William L.; Siepmann, Charles A.; Sklar, Rick; Station Rep Firms; Sustaining Programs; Technical Organizations; Telecommunications Act of 1996; Thomas, Lowell; Tokyo Rose; Trade Press; Tremayne, Les; United States; United States Supreme Court and Radio; Walkman; *War of the Worlds;* WCBS; Winchell, Walter; WQXR

Will Straw. McGill University
Essay Canadian Radio and the Music Industry

Michael Streissguth. Le Moyne College
Essays: *Big D Jamboree;* Cooper, Jack L.; Drake, Galen; Hulbert, Maurice "Hot Rod"; KFFA; King, Nelson; KWKH; Perryman, Tom; *Renfro Valley Barn Dance;* WBT; WDIA; Williams, Nat D.; WWVA

Mary Kay Switzer. California State Polytechnic University, Pomona
Essays: Duhamel, Helen; *Let's Pretend;* Osgood, Charles; *Voice of America*

Rick Sykes. Central Michigan University
Essay: WWJ

Marlin R. Taylor. XM Satellite Radio
Essays: Smulyan, Jeffrey; Trade Associations

Matt Taylor. Rutgers University
Essays: Birch Scarborough Research; Media Rating Council

Herbert A. Terry. Indiana University
Essays: Communications Act of 1934; Public Interest, Convenience, or Necessity; Russia and Soviet Union

Richard Tiner. Belmont University
Essays: British Broadcasting Corporation Orchestras; Israel; Promenade Concerts

Regis Tucci. Mesa State College
Essay: KDKA

David E. Tucker. University of Toledo
Essay: Blacklisting

Don Rodney Vaughan. Mississippi State University
Essay: Gospel Music Format

Mary Vipond. Concordia University
Essay: Canada

Randall Vogt. Shaw University
Essays: Alexanderson, E.F.W.; Cowan, Louis

Ira Wagman. McGill University
Essays: CFCF; CHED; CHUM

Andrew Walker. Independent Scholar
Essays: British Broadcasting Corporation: World Service; British Broadcasting Corporation: Monitoring

Peter Wallace. Wisconsin Public Radio
Essays: Auditorium Testing; *This American Life*

Jennifer Hyland Wang. University of Wisconsin, Madison
Essays: *America's Town Meeting of the Air; Goldbergs;* Phillips, Irna; *Vic and Sade*

Richard Ward. University of South Alabama
Essays: Mutual Broadcasting System; WWL

Mary Ann Watson. Eastern Michigan University
Essay: Television Depictions of Radio

Brian West. Independent Scholar
Essay: British Commercial Radio

Gilbert A. Williams. Michigan State University
Essays: Joyner, Tom; Black Oriented Radio; Black Radio Networks

Sonja Williams. Howard University
Essays: Durham, Richard; *Soundprint*

Wenmouth Williams, Jr. Ithaca College
Essay: Affirmative Action

Roger Wilmut. Author, *The Goon Show Companion*
Essay: *Goon Show*

Stephen M. Winzenburg. Grand View College
Essay: National Religious Broadcasters

Richard Wolff. Dowling College
Essay: Film Depictions of Radio

Roosevelt "Rick" Wright, Jr. Syracuse University
Essay: Urban Contemporary Format

Edgar B. Wycoff. University of Central Florida
Essay: Nightingale, Earl

Thimios Zaharopoulos. Washburn University
Essay: Greece

INDEX

Page numbers in **boldface** indicate subjects with their own entries; page numbers in *italics* indicate photographs.

Abbot, Waldo, 530
Abbott, Dotty, 1514
Abbott and Costello Show, 361
Abie's Irish Rose, 1338
Abott, Roger, 287
Abrams, Benjamin, 549
Abrams, Lee, 382, 466, 1115
Abrams, Lewis, 549
Abrams, Max, 549
Abramsky, Jenny, 210, 242
Abramson, Stacy, 764
ABSIE. *see* American Broadcasting Station in Europe
A.C. Nielsen Company, **1–2**
Academy Award Theater, 718
Access Radio, 1160
account executives (AEs), 7
Accuweather, 1020
Ace, Goodman, 521–23
Ace, Jane, 521–23
Ackerman, Harry, 1049
acoustical phase, definition, 120
Active Rock Format. *see* Heavy Metal/Active Rock Format
Adams, Bill, 320, 857
Adams, Cedric, 806, 1489
Adams, Franklin Pierce, 561, 1147
Adams, Noah, 44, 45
Adaptive Delta Modulation, 478
Adarand v Pena (1995), 14, 16
Adelaide Hawley Homemaking, 957
Adie, Kate, 209
Adorno, Theodor, 855, 1040
Adrian, Rhys, 498
Adult Contemporary Format, **3–4,** 1105
Adventure, 349
Adventures, 1046
Adventures in Good Music, **4–5,** 339, 1535
Adventures of Dick Tracy, 493
Adventures of Ellery Queen, 962
Adventures of Helen and Mary, 320, 857
Adventures of Jimmy Dale, 262
Adventures of Maisie, 492
Adventures of Nero Wolfe, 492
Adventures of Philip Marlowe, 492, 962
Adventures of Sam Spade Detective, 492
Adventures of the Thin Man, 492, 1415
Advertising, **6–10**
  blacklisting and, 168
  commercials, 7, 1100
  elections, 541–42
  growth of, 1427
  production of commercials, 1100
  start on radio, 1426
  during WWII, 1562–63
Advertising Agencies, **10–13**
Advertising Hall of Fame, 971
Advice to the Housewife, 173
AEF Jukebox, 95
Affirmative Action, **14–17**
Affluenza, 1270
Afghanistan, 102
Africa, **17–22,** 499
African-Americans in Radio, **25–29**
  Amos n' Andy and, 79–82
  Black-Oriented Radio, 169–72
  Black Radio Networks, 173–74
  comedy, 360–61
  radio drama and, 491–92
  Rock and Roll Format and, 1219
  stereotypes, 1337
  WDIA and, 1492
Africa No. 1, **22–24,** 751
Afri-Star, 467
Afropop, 1004
AFTRA. *see* American Federation of Television and Radio Artists
Against the Storm, 1552
Agencia Informativa Pulsar, 1303
Agresti, Karen, 369
Agricultural Radio. *see* Farm/Agricultural Radio
aided recall measurement, 114
AIDS Coalition to Unlease Power (ACT UP), 646
Aird, John, 259, 263
Alan Burke Show, 1142
Alan Freed's Rock and Roll Party, 623
Alan Young Show, 56, 287, 1338, 1340, 1454
Album-Oriented Rock Format, 30
Aldana, Carlos, 431
Alder, Cyrus, 784
Aldrich Family, **30–32,** 1341
Alexanderson, E.F.W., **32–34,** *33,* 586, 649, 1438
Alexanderson alternator, 1438
Algeria, 1293–94
Alias John Freedom, 1561
Alicoate, Jack, 356
Al Jolson Program, 1454
Alka-Seltzer, 320, 994
All-American Alarm Clock, 322
All American Boy, 55
Allan, Andrew, 29–83, 283, 284, 286
All Blues Show, 1399, 1400
All Canada News Radio, 291
Allen, Bill, 1536
Allen, Fred, **46–49,** *47,* 57, 363, 1554
Allen, Frederick Lewis, 417
Allen, Gracie, *651, 1061*
  CBS contract with, 353
  comedic style, 360, 1554
  Gracie Awards, 75
  income, 1366
Allen, Greg, 1368
Allen, Ida Bailey, 1552
Allen, Lee, 1577
Allen, Leslie, 415
Allen, Mel, **50–52,** *51,* 1314
Allen, Peter, 939

Allen, Steve, 835
Allgemeine Elektricitäts Gesellschaft, 125
Allied Products, 1078
All India Radio, **35–37**, 500
Allison, Jay, 1301
All Negro Hour, 26, 169, 397
All News Format, **37–40**
All Night Radio, **41–42**
All-Radio Methodology Study, 1141, 1156
All Talk Format. *see* Talk Radio
All Things Considered, **42–46**
Bob Edwards and, 537, 955
concept, 632, 1005, 1122, 1133
female journalists in, 1554
premiere, 993
ranking, 1130
Siemering and, 1265
Stamberg and, 1323–24
time slot, 1002
Wertheimer on, 1497
Alltid Nyheter (Always News) (Norway), 40
Al Pearce and His Gang, 1338
Alphabet Soup, 322
Alternative Format, **52**
Alternative Radio, 1124
Alvear, Clemente Serna, 942
Amalgamated Broadcasting System, **52–54**
Amalgamated Wireless Company of Australia (AWA), 128
Amanda of Honeymoon Hill, 1552
Amari, Carl, 1031
Amateur Night from the Palace, 1399
Amateur Night on Beale, 1524
Amateur Radio Relay League (ARRL), 1593
amateur talent programs, 46
Amazon Country, 646
Amberg, Alan, 647
Ameche, Don, **54–56**, 526
America and the World, 733
America Latina via Satelite, 1306
American Association for the Advancement of Science, 321
American Association of Public Opinion Research (AAPOR), 161
American Bakeries, 1578
American Bandstand, 336
American Broadcasting Company, Federal Communications Commission v (1954), 1151
American Broadcasting Company (ABC), **56–60**. *see also* American Contemporary Networks
beginning of, 1428
FM stations, 606
foundation of, 181
Network Monopoly Probe, 1014–15
newscasts, 38
Radio Disney and, 1055, 1167
Watermark purchase by, 920
American Broadcasting Company v Federal Communications Commission (1953), 1444
American Broadcasting Station in Europe, **61–62**, 1560
American Broadcasting System, 1345
American Business Consultants, 167
American Civil Liberties Union, 1038
American Contemporary Networks, 58, 59
American Country Countdown, 921
American De Forest Wireless Company, 1422
American Entertainment Network, 59
American Express Company, 1494
American Family Robinson, **62–64**
American Federation of Labor (AFL), 64, 66
American Federation of Musicians, **64–66**, 1194
American Federation of Radio Artists (AFRA), 66, 294
American Federation of Television and Radio Artists, **66–67**
American FM Network, 59
American Forces Korea Network, 109
American Forces Network, 1435
American Forces Radio, 68
American Forum of the Air, 1121, 1557
American Gold, 60
American Graffiti, 593
American Half Hour, 394
American Hot Wax, 592
American Independent Radio, 315
American Indian Movement (AIM), 1011
American Indian Radio on Satellite (AIROS), 1012
American Institute of Electrical Engineers, 1378–79
American Library of Radio and Television, 972
American Marconi Company, 1163
American Museum of Radio, 974
American Newspaper Publishers Association, 1095
American Oil Company, 1274
American Pilgrimage, 1078
American Public Radio. *see also* Public Radio International
concept, 1134
creation, 808, 948
drama on, 495
Kling and, 830
programming, 1004
American Public Radio Associates, 1130
American Radio Company, 1092
American Radio Company of the Air, 808
American Radio Relay League (ARRL), 567, 689, 1380, 1551
American Radio Warblers, 320
American Research Bureau. *see* Arbitron
American Safety Razor, 57
American School of the Air, **67–68**, 351
concept, 474, 957
Dorothy Gordon and, 1553
Judith Waller and, 1551
poetry on, 1079
Raymond Swing on, 1357
Americans for Radio Diversity, 15
American Society of Composers, Authors, and Publishers, **69–70**
contracts with African-Americans, 26
Internet payment models, 124
NAB and, 986
rights controlled by, 245
royalty collection, 401
WOR lawsuit, 1556
American Sportsman, 1318
American Telephone and Telegraph, **71–73**
advertising, 1426
cost of land lines, 58
NESCO purchase by, 587
RCA and, 996
stations, 1439
WEAF operation by, 6
American Tobacco Company, 848, 1365, 1495, 1588
American Top 40, **73–74**, 320, 801, 921
American Urban News (AUN)/SBN, 174
American Urban Radio Networks (AURN), 29, 174. *see also* National Black Network; Sheridan Broadcasting Network
American View network, 1372
American Week, 1246
American Women in Radio and Television, **75–76**
America's Negro Soldiers, 170
America's Town Meeting of the Air, 27, **76–77**, 182, 294, 1121, 1564
America Victorious, 1495
Ameri-Star, 467
AMFM Incorporated, 341, 702, 703, 1333, 1467
Amici, Dominic. *see* Ameche, Don
Amos, Deborah, 1004
Amos 'n' Andy, **78–82**
on Blue Network, 181
Canadian audiences, 269
CBS raid of, 353
and minstrel tradition, 361
popularity, 1026
on Red Network, 996
stereotypes, 26, 170, 1337, 1338
TV series, 1277
on WMAQ, 1541
Amos 'n' Andy Music Hall, 81
Ampex Corporation, 126, 1097
Ampex Model 100, 126
amplitude, definition, 120
amplitude processors, 120
AMRAD (American Radio and Research Company), 1508
AM Radio, **83–85**
antennas, 86
in Australia, 130, 133
in Britain, 229
in Canada, 265, 270
frequency allocation, 631
ground wave, 681–82

receivers, 1184–85
simulcasting, 1271–72
stereo, 1334–36
technical developments, 1429
in the USSR, 1228–29
Amrhein, Jack, 1298
Amusement Enterprises, 1365
Anacin, 504, 522
An American in England, 475
An American Moment, 845
Anderson, Arthur, 857
Anderson, Betty Baxter, 317
Anderson, Eddie, 156, 170, 361
Anderson, Jon, 466
Anderson, Marion, 27
Anderson, Marjorie, 1251
Anderson, Sam, 183
Anderson, Stanley, 416
André, Pierre, 319, 868
Andres, Robert Hardy, 771
Andrews, Charles, 737
Andrews, Robert D., 494
Andrews, Stanley, 868
Andrews Sisters, 56
Angel's Curse, 499
Angle, Jim, 907
Angola, 1294
Anheuser Busch, 57
M.L. Annenberg, 568
announcements, classes of, 7
"Another One" (Geer), 515
Another World, 1073, 1288
Antenna, **86–87**
AntenneX, 569
Anthem, 1005
Anthology, 1080
Anthony, Allan C., 844
Anthony, Earle C., 815
Anthony, John, 890
anti-aliasing, 463
Antony, William E., 846
Anzac Tattoo: From the Enemy to the Enemy, 146
AP Broadcast Wire, 1022
A&P Gypsies, 1337, 1426, 1453
Aplon, Boris, 868
Apostolides, Penelope, 1122
Apple, QuickTime, 123
Appleton, Victor, 316
AP Radio Network, 39, 1022
À propos, 273
Aptheker, Herbert, 1058
AQHs (average quarter hours), 8–9
Arabsat, 467
Arab World Radio, **87–89,** 1182–83
Arbitron, **90–92**
audience measurement, 115
data on children's listening, 322
definition of audience, 1291
diaries, 455
foundation, 119
Hooperatings and, 723
Nielsen litigation, 91
on overnight time slots, 41–42
Pulse and, 1141
Arcand, Pierre, 311
Archer, John, 1251
Archers, **92–94,** 218, 497
Archibald MacLeishTribute, 1081
Archie Andrews, 31, 319–20, 1339, 1341
Archives. *see* Museums and Archives of Radio; Nostalgia Radio
Arch Oboler's Plays, 1033
Arco, George von, 653
Arden, Eve, 1049–51, 1554
Area of Dominant Influence (ADI), 91, 1101
Argentina, 1304
Arlen, Harold, 1196
Arlin, Harold W., 805, 1314, 1321
Arlott, John, 220
Armed Forces Network, 61, 1313
Armed Forces Radio and Television Service (Saigon, Vietnam), 97
Armed Forces Radio Service (AFRS), 27, **94–97,** 726, 774, 1563
Armed Forces Vietnam Network (AFVN), 97
Armen, Kay, 57
Armour Research Foundation, 125, 1533
Armstrong, Ben, 1007
Armstrong, Edwin Howard, **98–101,** 440, 704, 1234, 1502
Armstrong, Garner Ted, 559
Armstrong, George W., 1348
Armstrong, Marion, 99, 101
Armstrong Awards, 143–44
Armstrong of the SBI, 772
Army Hour, 27, 826, 1561
Army-McCarthy hearings, 1432
Arnheim, Rudolf, 1040, 1288
Arnold, Elizabeth, 1004
Art Baker's Notebook, 816
Arthur C. Nielsen Jr. Research Center, 971
Arthur Godfrey and His Friends, 661, 662
Arthur Godfrey's Talent Scouts, 353, 661, 1367, 1455, 1496
ratings, 56
Arthur Godfrey Time, 661, 662, 1367, 1369
Arthur Kudner agency, 152
ASCAP. *see* American Society of Composers, Authors, and Publishers
Asch, Moses, 785
Ashenhurst, Anne, 250
Asia, **101–10,** 499
Asian Broadcasting, 1412
Asian Broadcasting Technical Review, 1412
Asia-Pacific Broadcasting Union, 499
Asia Satellite Telecommunications Company, 1469
Asia-Star, 467
As It Happens, 270, 272, 273, 274, 291
Askey, Arthur, 221
Ask-It-Basket, 1147
As Others See Us, 968
Aspel, Michael, 297
Associated Press (AP), 39, 1021, 1022, 1095
Association for Professional Broadcasting Education (APBE), 243, 244, 530
Association for Women in Communications, **110–11**
Association of German Broadcasters (ARD), 655
Association of Independent Radio Contractors (AIRC), 228
Association of National Advertisers, 114, 399
Association of Public Radio Stations, 404, 1133
Association of Radio News Analysts, 366, 734
Association of Technical Employees, 1381
As The World Turns, 1073, 1288
ASTRA satellite, 467, 1176
Athens 98.4, 677
Atlanta Journal, 1569, 1570
Atlantic 252, 759
Atlas: The Magazine of the World Press, 734
Atwater Kent company, 811, 1453, 1494
Atwater Kent Hour, 811
At Your Service, 1520
Audience, **112–13**
all-night radio, 41
applied studies of, 1039–40
Arbitron measurement, 90–92
auditorium testing, 127–28
Beville in research on, 160–61
Commercial tests, 370–72
measurement, 1–2
programming research, 1101–2
psychographics, 1118–19
Pulse, Inc, 1140–41
quantifying, 8
RADAR, 1153–54
sampling methods, 115–16
satellite digital audio broadcasting, 468
talk radio, 1373–74
turnover, 9
Voice of the Listener and Viewer, 1471–72
Audience 88, 1136
Audience Building Task Force, 1135
Audience Participation Programs. *see* Quiz and Audience Participation Programs
Audience Ratings: Radio, Television, Cable, 161
Audience Research Methods, **113–18,** 454–456
Audimax, 121
Audimeter, 1, 90, 115, **118–19,** 454
Audio Engineering magazine, 1412
Audio magazine, 1412
Audio Mixer. *see* Control Board/Audio Mixer
Audion, 71, 98, 440
Audio Processing, **119–23,** 1192
Audio Streaming, **123–24**
Audiotape, **125–26,** 1097
audio-video (AV) content, 123
Auditorium Testing, 117, **127–28,** 980, 1102
Aunt Jemima, 1337

AUSSAT (Australia), 135, 478
Australia, **128–34**
Australian Aboriginal Radio, **134–36**
Australian Broadcasting Authority (ABA), 132
Australian Broadcasting Company, 128
Australian Broadcasting Control Board (ABCB), 129, 130
Australian Broadcasting Corporation (ABC), 131, 132, 134–35
Australian Broadcasting Tribunal, 130
Autolite, 1354
Automation, **137–37**
Automobile Radios, **138–39**, 1069
Autry, Gene, **139–43**, *141*, 995, 1274
Ave Maria Hour, 1215
Avoz do Brasil, 1305
Awards and Prizes, **143–45**
AWARE, Inc., 574
Awaye, 135
AWRT. *see* American Women in Radio and Television
Axis Sally (Mildred Gillars), 95, **145–48**, 1115
AX.25 packet protocol, 690
N.W. Ayer agency, 11, 416, 1140
Aylesworth, Merlin H., 996, 1441
Azcárraga, Emilio, 713, 941
Azcárraga, Luis, 940
Azcárraga, Raul, 940

Bab-O, 250, 1094
Babson, Roger, 1508
Baby, 1034
Baby Snooks Show, 196, 1339, 1341, 1553
Bacall, Lauren, 96
Backstage Wife, 1561
Back to God Hour, 1214
Back to the Bible Hour, 556, 1214
Baer, Parley, 685, 1500
Bagley, William C., 68
al-Bahri, Yunus, 89
Bailey, Deford, 25
Bailey, Jack, 481
Bailey, Robert, 1589, *1590*
Baillie, Hugh, 1022
Baker, Ed, 26
Baker, Susan, 1465
Baker's Dozen, 908
Bakker, Jim, 1008
Balaguer, Ramón, 431
Baldwin, Alec, 1251
Baldwin, Ben, 844
Ball, Lucille, 1554
Ball, Zoe, 234
Ballad Hunter, 474
Ballance, Bill, 1403
Bamberger, Edgar, 1556
Bamberger Broadcasting System, 1557
Bandstand, 472
Band Wagon, 221
Bangladesh, 103
Banneker Radio Club, 25
"banner ads," 757
Bannerman, Richard, 219
Banner Radio, 1332
Barbary Coasts Nights, 1046
Barber, Al, 1565
Barber, Red, **149–51**, *150*
  Bob Edwards and, 537
  Hear It Now, 696
  Mel Allen and, 50
  Morning Edition, 956
  sportscasting by, 1314
  Wheaties commercial, 1312
Bardeen, John, 154, 1413
Barker, Brad, 868
Barkley, Roger, 816
Barnett, Bill, 322
Barnouw, Erik, **151–53**, 494
Barrymore, John, 719
Barsamian, David, 1124
Bartlett, Hugh, 272
Bartlett, Vernon, 240
Bartley, Dick, 60
Bartley, Robert, 608
Baruch, Andre, 1564
Baruck, Allan, 868
Baseley, Godfrey, 92
Batavia Radio Vereniging (Jakarta, Indonesia), 106
Batman and Robin, 801
Baton Broadcasting, 272
Batten, Barton, Durstine, and Osborn (BBD&O)
  Cavalcade of America, 302–3
  Jack Benny and, 1365
  March of Time, 11
  Office of War Information and, 1043
  security officers at, 1198
  Theater Guild of the Air, 1392
Battle of the Warsaw Ghetto, 494
Baudry, Eduard, 287
Bauer, Robert, 61
Bauersfeld, Eric, 839
Baukhage, Hilmar Robert, 366, 621
Bawdry, Jack, 286
Beacon Radio, 1435
Beagle, Shyster, and Beagle, Attorneys at Law, 601
Beal St. Caravan, 182
Bean, Orson, 168, 574
Beasley Broadcast Group, 1253
Beatles, 964
Beat the Band, 1148
Beaubien, Philippe de Gaspé, 332
Beaudin, Ralph, 58
Beaulne, Guy, 281
Beech-Nut Packing Company, 1360
Before Dinner—Walter Winchell, 1526
Begin, Karin, 584
Belar Electronics Laboratory, 1334
Belcher, Jerry, 1472
Believe It or Not, 1454
Bell, Alexander Graham, 518
Bell, Art, 41
Bell, Bill, 1288
Bell, Bob, 1510
Bell, Haley, 28
Bell, Shirley, 868
Bell, William J., 1073
Bellows, Henry Adams, *1209*, 1489
Bell Telephone Hour, 338
Bell Telephone Laboratories, 72, **154–55**
Bell Telephone System, 71
Ben Bernie Orchestra, 1453
Benedaret, Bea, 589
Benèt, Stephen Vincent, 1075
Ben Hunter's Nite Owls, 816
Benjamin, Walter, 498
Bennet, Doug, 45, 1003
Bennet, Edward, 1511
Bennett, William, 1465
Benny, Jack, **155–59**, *157*, *1061*
  comedic style, 361
  Don Wilson and, 1525
  sponsors of, 1365
  transition to TV, 1277
Benson, Al, 26, 183
Bentine, Michael, 668
Bentley, Julian, 356
Benton, Nelson, 1486
Benton, William B., 1469
Berelson, Bernard, 1039
Berg, Alan, 836
Berg, Gertrude, 250, *665*, *785*, 1551
Berg, Louis, 1072
Bergen, Edgar, 11, 353, *527*, 1452
Berkeley, Reginald, 203, 217
Berland, Pierre, 311
Berle, Milton, 363, 793
Berlin, Irving, 69
Berlin Alexanderplatz, 498
Berlin Diary (Shirer), 1259
Berliner Rundfunk (Berlin Broadcasting), 657
Berlin International Wireless Telegraph Convention, 255
Berlusconi, Silvio, 767
Berman, Doug, 302
Bernays, Edward, 706, 827
Bernie, Ben, 1454
Bernier, Jovette, 268
Bernstein, Dennis, 839
Berry, Chuck, 978
Berwin, Bernice, 1046
Best Food Boys, 360
Best Ideas You'll Hear Tonight, 273
"Better Things for Better Living," 302–3
Betty and Bob, 11, 26, 55, 737, 1415
Betty Boop, 1464
Between the Bookends, 1077, 1078, 1080
Beulah Show, 26, **159–60**, 170, 360, 1341, 1554
Beveridge Committee, 12
Beville, Hugh Malcolm, **160–61**, 855
Bewitched, 1034

Bhutan Broadcasting Service (BBS), 104
Biblical Dramas, 474
Bickel, Theodore, 786
Bickersons, 55, 1554
Bickford, Leland, 1583
Biden, Joseph, 1169
Bierly, Kenneth, 1198
Big Broadcast, 592, 726, 1030, 1069, 1282
Big Brother Club, 1509
Big D Jamboree, **162–63**
Big Picture, 826
Big Question, 349
Big Show, 363, 1455
Big Sister, 639
Billboard American Hot 100, 1436
Billboard magazine
 AT40 and, 73
 charts, 3, 4
 coverage of radio, 356, 357
 methodology, 74
 model used by, 1410
 online version, 1412
Bill Stern's Sports Reports, 57
Billy and Betty, 1341
Biltmore Agreement, 1018, 1095, 1096
Bing Crosby Enterprises, 126
Bing Crosby Show, 126
Binns, Jack, 355, 1424
Biography in Sound, 476
Biondi, Dick, **163–64,** 473, 714
Biow, Milton, 53, 1543
Birch, Philip, 1435
Birch, Thomas, 90, 165
Birch Scarborough Research, **165–66**
Birthday Club, 824
Black, Don, 695
Black, Harold, 704
Black and White Minstrel Show (BBC), 221
Black Castle, 730
Black Chapel, 729
Black College Radio, 29
Black Efforts in Soul Television, 28
Blackett, Hill, 737
Blackett-Sample-Hummert (BSH) agency, 11, 522, 718, 737
Black Falcon, Limited, 1328
Black Liberation Radio, 886
Blacklisting, **166–69**
Black-Oriented Radio, 25–29, **169–72**
Black Radio Networks, 25–29, **173–74**
Blackwell, Noble, 28
Blair, Frank, 1019
Blair, John, 1332
Blake, Eubie, 25, 356
Blattner, Ludwig, 1533
Blatz Brewing Company, 504
Blayton, J.B., 28, 171
Blazer, Phil, 786
Bliss, Ed, 537
Bloch, Irwin H. Sonny, 1557
Block, Joel, 175
Block, Martin, *175,* **175–76**
 Bernice Judice and, 792
 Make Believe Ballroom, 57, 471, 1274, 1544
Bloomberg Business News Network, 40
Bloomquist, Randall, 1373
Blore, Chuck, **176–78,** *177*
Blue, Howard, 494
Blue, Ira, 695
Blue Book, **178–80,** 391, 416, 532, 1205, 1267, 1355, 1428
Blue Coal, 250
Blue Coal Review, 1248
Blue Hills, 130
Blue Jam, 221–22
Blue Monday Jamboree, 1495
Blue Network, 72, **180–82**
Blues Format, **182–84**
Blue Theatre Players, 181
BMI. *see* Broadcast Music Incorporated
Board for International Broadcasting, **185–86**
Boardman, True, 567
Bob and Ray, **186–88,** *187,* 363
Bobby Benson's Adventures, 152, 1338
Bob Collins Show, 1510
Bob Hope Pepsodent Show, 726
Bob Hope Show, 56, 728
Bogart, Humphrey, 96
Bogart, Leo, 160
Bohannon, Jim, 1371
Bolivia, 1304–5
Bongo, Omar, 24
Bonneville Broadcasting Services, 1377, 1378
Bonteque, Elinor, 178, 179
Book at Bedtime (BBC), 219
Bookbeat, 1081
Booker T. Washington in Atlanta, 491
Book Shop, 1358
Book Show, 1081
Booktalk, 1081
Bookworm, 1081
Boole, Emma, 1121
Boortz, Neal, 1372
Booth, Shirley, 503
Borach, Fannie. *see* Brice, Fanny
Border Radio, **189–91**
Bork, Robert, 1059
Bornstein, Ronald, 1003, 1133, 1135
Bose, Amar G., **192–93**
Boss Radio, 315, 486, 952
Bostic, Joe, 26
Boston Blackie, 1360
Boston Globe, 355
Boston Post, 355
Boston Radio Exposition, 25
Boston Traveler, 354
Boulding, Jerry, 1446
Bourdieu, Pierre, 1290
Bourjois Toiletries, 1495
Bouvard, Pierre, 9
Bowes, Edward, Major, 46, 1147, 1366
Bowman, Robert. *see* Far East Broadcasting Company
Boyle, Andrew, 241
Boyle, Harry, 284
Brackbill, Charles, 557
Bradford, John, 1155
Bradley, Bruce, 1487
Bradley, Ed, 38
Bradley, Preston, 1072
Bradshaw, Thornton, 1165
Brailsford, Glynn, 1109
Brainard, Bertha, 355, 1551
Brain's Trust (BBC), 219
Brancaccio, David, 906, 907
Brand, Oscar, 1081, 1546
Brandt, George, 1546
Branly "coherer" device, 874
Brasilsat, 467
Brattain, Walter H., 154, 1413
Braun, Ferdinand, 653–54
BraveNew Waves, 274
Braverman, Eric, 1557
Brazil, **193–95,** 1305
Breakfast Brigade, 1214
Breakfast Club, 57, 928, 957, 1369
Breakfast with Dorothy and Dick, 719, 1553
Brecht, Bertolt, 498
Breckenridge,Gerald, 316
Breckinridge, Mary Marvin, 1553
Breeden, Frank, 385
Breedon, Robin, 584
Bremen Sender, 145
Brennan, Vincent, 1431
Brennan, William, 1247
Brenner, Zev, 786
Briarhoppers, 424, 425
Brice, Fanny, **195–98,** *197,* 1553
Brickhouse, Jack, 1315
Bride, Harold, 1424
Bride and Groom, 1150
Bridson, D.G., 491
Briggs, Asa, 201, 206
Brighter Day, 1072
Brinkley, John R., 189, **198–200,** 581
Bristol-Myers, 504
British Broadcasting Corporation (BBC), **200–205**
 ABSIE transmitters built by, 61
 Arabic service, 749–50
 Asian Network, 210
 BBC Local Radio, **207–11**
 BBC Monitoring, **211–14**
 BBC Orchestras, **214–15**
 BBC Radio Programming, **216–22**
 BBC World Service, **222–27,** 751, 1117
 Broadcasting House, **206–7**
 censorship of radio, 306
 Churchill and, 325
 conditions of license, 12
 digital audio broadcasting and, 458–59
 Empire Service, 88, 1263
 establishment, 1137

FM stations, 607
influence from U.S. radio, 1435
John C.W. Reith at, 1208–11
lobby by advertising agencies, 10
pay radio, 1064
Radio Leicester, 208
British Commercial Radio, **227–32**
British Cultural Studies approach, 1290
British Disk Jockeys, **232–35**
British Disregard for American Rights, 498
British Forces Broadcasting Service, **235–36**
British Forces Network (BFN), 235
British Malaya Broadcasting Service (Singapore), 105
British Marconi Company, 1437
British Pacific Programme, 235
British Pirate Radio, **237–38**
British Radio Journalism, **239–43**
British West Africa, 18
BRMB (Birmingham, England), 234, 298
Broadcast Advertising Bureau, 1156
broadcast cartridge recorder/players, 1191
Broadcast Data Systems (BDS), 70
Broadcast Education Association, **243–44,** 531
Broadcast Equipment Exchange, 1412
Broadcaster magazine, 1412
Broadcasting: A Study of the Case for and against Commercial Broadcasting under State Control in the United Kingdom, 12
Broadcasting Act of 1958 (Canada), 263
Broadcasting Act of 1968 (Canada), 264
Broadcasting Act of 1972 (United Kingdom), 13
Broadcasting Act of 1990 (United Kingdom), 1160, 1436
Broadcasting Board of Governors, **246–47**
Broadcasting & Cable, 1412
Broadcasting Corporation of China (BCC; Formosa), 108
Broadcasting House. *see* British Broadcasting Corporation: Broadcasting House
Broadcasting in Remote Aboriginal communities (BRACS), 135
Broadcasting magazine, 38, 1363, 1411
Broadcasting Rating Council. *see* Media Rating Council
Broadcasting's Better Mousetrap, 608
Broadcasting Telecasting magazine, 1411
Broadcasting Yearbook, 1363
Broadcast magazine, 1412
Broadcast Measurement Bureau, 1409
Broadcast Music Incorporated, 69–70, 124, **245–46,** 1274
Broadcast News, 1411
Broadcast over Britain, 201
Broadcast Pioneers Library, 970
Broadcast Promotion and Marketing Executives, 1109
Broderick, Gertrude, 534
Broger, John, 570
Brokerage in Radio, **247–49**
Bromo-Seltzer, 1473
Bronx Marriage Bureau, 250
Brooklyn Bridge, 1386
Brooklyn Daily Eagle, 355
Brooks, Garth, 412
Brothers, Joyce, 1555
Brown, Bob, 1078
Brown, Cecil, 366
Brown, Dick, 57
Brown, F.J., 201
Brown, Harry, 272
Brown, Helen Gurley, 819
Brown, Himan, **250–52,** *251,* 494, 730, 746, 1198
Brown, James, 28
Brown, Joy, 1371
Brown, Maurice, 857
Brown, R.R., 1211
Brown, Willet, 315
Brownmiller, Susan, 1123
Brudnoy, David, 1371
Brunei, 106
Brunson, Dorothy, 28
Brush Development Company, 125
Bryan, Arthur Q., 589
Bryan, Ed, 1078
Bryson, Lyman, 416
bubble jamming, 774
Buchanan, Bill, 357
Buck, Jack, 1315
Buckley, Dick, 1544
Buckley, William F., 1142
Buckley Broadcasting Corporation, 1557
Buck Matthews Show, 1420
Buck Rogers in the Twenty-fifth Century, 318, 493, 1243
Budd, Barbara, 273
"Bugbomb," 334
Buggins, 217
Bull Durham tobacco, 1221
Bullis, Harry A., 62
Bulova, Arde, 1543
Bulova Watch, 1480
Bundy, McGeorge, 1123
Bunker, Fred, 1485
Buranelli, Prosper, 1396
Bureau of Navigation, 255
Burke, Dan, 59
Burke, Georgia, 26
Burke, J.R. Frank, 840
Burkhart, Kent, 382
Burlingame, Elmer, 440
Burma, 104–5
Burma Broadcasting Service, 105
Burnett, Lester "Smiley," 140
Burns, George, *651, 1061. see also George Burns and Gracie Allen Show*
CBS contract, 353
comedic style, 360, 1554
income, 1366
Burns, George A., 1105
Burns, Pat, 292
Burns, Sam Z., 1528
Burrows, Abe, 696
Burrows, Ed, 1132
Burton, John, 816
Burton, Mo, 1549
Burton, Ralph, 803
Burtt, Ben, 1328
Burtt, Robert, 299
Bush, George H.W., 1085, 1169, 1442
Bush, George W., 1442
Bush House, 223
business news, 40
Business Radio Network, 1023
Business Sense, 274
Bustany, Don, 73, 74, 801
Butler, Daws, 618
Butsch, Richard, 1292
Butterworth, Wally, 1473
But We Were Born Free, 437
Buxton, Frank, 1030
Bwana Devil, 1034
By Kathleen Norris, 884
Byrd, Robert, 1432, 1442
Byron, James A., 1484

Cable Radio, 230, **253–56**
Cable Television Consumer Protection and Competition Act of 1992, 14
Cadburys Chocolate, 204
Cadena Oriental de Radio, 429
Cadman, S. Parkes, 1212
Caldwell, Orestes, 1410
California, Miller v (1973), 1036
California Prune Advisory Board, 620
California Theatre Concert Orchestra, 834
Calkins, Ernest Elmo, 905
Call for Action, 1573
call-in music tests, 980
call-in shows, 1361
Call Letters, **254–56**
call-out and call-in, 116–17
callout research, 980
Calloway, Cab, 29, 173
Calvert, Reginald, 238
Cambodia, 107–8
Camel Caravan, 509, 1563
Camel Quarter Hour, 152, 1078
Cameron, James, 241
Campbell, H. Allen, 1578
Campbell, Julie, 317
Campbell, M.G., 332
Campbell Playhouse, 936, 1480
Campbell's Soup, 935, 1074
Camras, Marvin, 1533
Canada, **257–61,** 498–99. *see also under* Canadian entries
Canada Dry Program, 156
Canada in the World, 274
Canada Today, 290
Canadian Association of Broadcasters (CAB), 263, 276

Canadian Broadcasting Corporation, 261–62, 263, **271–75**, 278
Canadian Broadcast Standards Council (CBSC), 276
Canadian Charter of Rights and Freedoms, 275
Canadian Human Rights Act of 1978, 275
Canadian National Railway (CNR) network, 269–70, 283, 311
Canadian News and Sports Broadcasting, **289–91**
Canadian Radio and Multiculturalism, **275–77**
Canadian Radio and Television Commission, 278
Canadian Radio and the Music Industry, **278–79**
Canadian Radio Archives, **261–62**
Canadian Radio Broadcasting Commission (CRBC), 259
Canadian Radio Drama, **280–85**
Canadian Radio Policy, **263–66**
Canadian Radio Programming, **266–71**
Canadian Radio Satire, **285–89**
Canadian Radio—Television and Telecommunications Commission (CRTC), 260, 264, 272
Canadian Talk Radio, **292–93**
Canary Pet Shows, 320
Cannon, 380
Cannon, Geoffrey, 1471
Cantor, Charlie, 503
Cantor, Eddie, **294–96**, *295*
 background, 1459
 CBS raid of, 351
 comedic style, 360
 Fred Allen and, 49
 income, 1366
 popularity, 1453
Cantril, Hadley, 610, 854, 1039, 1480
Can You Top This?, **257**, 1149
Capehart, Homer Earl, 296
Capehart Corporation, **296–97**
Capital Cities/ABC, 59
Capital Cities Broadcasting, 59
Capital Radio, 228, **297–99**, 559
Capitol Cloakroom, 1246
Caplan, Rupert, 283, 284
Capnews, 1023
Capstar Broadcasting Partners, 1467
Captain Midnight, **299–300**, 318, 319, 1094
Captain Tim Healy's Stamp Club of the Air, 1094
Caray, Harry, 1315–16
Cardiff, David, 1291
Cardin, Joseph, 262
Carillo, Mary, 1312
Carlin, George, 309, 1036, 1058, 1246–48, 1481. *see also* "Seven Dirty Words" Case
Carlin, Phillips, 927
Carl Laemmle Hour, 718
Carl Rupp and His Orchestra, 1578
Carlton, Leonard, 416
Carmichael, Ralph, 384
Carney, Art, 951
Carney, Donald, 1556
Carolina Hayride, 1486
Carothers, Isobel, 1551
Carpenter, Ken, 816
Carpenter, Thelma, 27
Carpenter, William S., 610
Carpentier, Georges, 1320
Car Radio. *see* Automobile Radio
Carrell, Charles, 1459–60
Carrington, Elaine, 494, 1287, 1552
Carrington, Roland, 25
Carroll, Gene, 589
Carrus, Gerald, 745
Carson, Johnny, 48
Carson, Saul, 416
Carstensen, Svend, 1236
Car Talk, **301–2**, 364, 1004, 1136
Carte Blanche, 288, 289
Carter, Amon G., 1483
Carter, Andrew "Skip," 28
Carter, Angela, 218
Carter, Boake, 307, 366
Carter, Vivian, 1554
Carter Family, 190, 977
Cartier, Jacques-Narcisse, 267, 331
Casey at the Mike, 799
Casey's Coast to Coast, 801
Casey's Top 40, 74, 801
Cash, Dave, 298, 559
Castleberry, Ed, 173
Catholic Academy for Communication Arts Professionals, 1008
Catholic Hour, 1212, 1215
Catto, Henry, 242
Caul, Dorothy, 26
Cavalcade of America, **302–5**, *304*
 format, 474, 1074
 playwrights for, 152, 1075, 1076
Cavalieri, Grace, 1081
Cavendish Laboratory, 911
CBA (Sackville, New Brunswick), 271
CBC. *see* Canadian Broadcasting Corporation
CBC Overnight, 273
CBC Wednesday Night, 270, 271, 284
CBF (Verchères, Quebec), 271, 280
CBI Roundup, 634
CBK (Watrous, Saskatchewan), 271
CBL (Hornby, Ontario), 271
CBS Evening News with Dan Rather, 1048
CBS Evening News with Walter Cronkite, 1246
CBS Is There, 476
CBS Morning Show, 732
CBS Mystery Theater, 495, 1244
CBS Radio Mystery Theater, 250, 353, 730
CBS Radio News, 40
CBS Radio Workshop, 353
CBS Reports, 636
CBS Reviews the Press, 968
CBS Sunday Morning, 845
CBS Sunday Night News, 1048
CBS This Morning, 1048
CBS World News Roundup, 352, 1259, 1515, 1561
CDQ emergency signals, 545
CD Satellite Radio Service. *see* Sirius Corporation
Cedric Adams Open House, 1489
Celler, Emmanuel, 1121
Cellular I, 1020
Censorship, **306–10**, 1562
Centennial Broadcasting, 1253
Central Australian Aboriginal Media Association (CAAMA), 131, 135
Central Broadcasting Station (Nanking, China), 108
Central Intelligence Agency, 1116, 1171
Centre Radio (Leicester), 228
Cerf, Bennett, 562
Ce Soir en France, 335
C'est la vie, 273
Ceylon, 104
CFCA (Toronto, Ontario), 267
CFCF (Montreal, Quebec), 258, 270, **310–11**, 331
CFL Talks and Bulletins, 1491
CFRB (Toronto, Ontario), 262, 270
CFTR (Toronto, Ontario), 291
CFWE (Alberta, Canada), 277
Chabin, Jim, 1109
chain broadcasting, 71, 997
Chain Broadcasting Regulations, 1205
Chalmers, Tom, 18
Chamber Music Society of Lower Basin Street, 181
Chancellor Media Corporation, 1054
Chandler, Emma, 1551
Chandler, Jeff, 1051
Chandu, the Magician, 319, 1360
Chapin, Katherine Garrison, 1080
Chaplain Jim, 414
Chapman, Alan, 316
Charbonneau, Danielle, 274
Charlie and Harrigan Show, 559
Charlie Chan, 493, 1338
Charlie McCarthy Show, 527, 1563. *see also Edgar Bergen and Charlie McCarthy Show*
Charlot, Harry Engman, 1248
Charlotte Greenwood Show, 1338, 1340
Charren, Peggy, 1168
Charrier, Reine, 292
Chase, Harry, 1395
Chase, Ron, 312
Chase and Sanborn coffee, 1366
Chase and Sanborn Hour
 AFRS transmission, 1563
 Arch Oboler and, 1033
 Eddie Cantor and, 294

Edgar Bergen on, 526
Fannie Brice on, 196
format, 718–19
NBC and, 996
production, 11
Chasman, Pola, 1554
Chatterji, P.C., 500
Chauncey, Tom, 142
CHCB (Marconi, Ontario), 267
CHCY (Edmonton, Alberta), 780
Checkpoint, 242
CHED (Edmonton, Alberta), **312–13**
Cheerio, 1077, 1078
Chenault, Gene, **313–15,** 819
Chernobyl disaster, 1172
Cherrio Exchange, 1078
Cherry, William, 792
Chess, Leonard, 1220
Chesterfield cigarettes, 1455
Chesterfield Sound Off Time, 49, 1453
Chesterfield Supper Club, 1277
Chevrolet, 138, 1312
Chevrolet Program, 156
Chevron Corporation, 1482
Chevron Texaco, 939
Chez Miville, 268, 288
Chiarito, Americo, 839
Chibás, Eduardo, 429
Chicago Broadcasters Association, 1268
Chicago Civic Opera, 938
Chicago Daily News, 1541
Chicago Defender, 491, 511
Chicago Radio Laboratory, 916, 1593
Chicago Roundtable, 1564
Chicago Tribune, 355, 1509
Chick Carter, 1341
Children's Corner, 1553
Children's Hour, 318
Children's Novels and Radio, **316–18**
Children's Programs, **318–23**
Children's Radio Network, 1168
Children's Radio Theater, 321, 1081
Children's Satellite network, 322
Children's Song Bag, 320
Child's Garden of Freberg, 618
Chile, 1305
China, 108
China Radio International, 108
Chinatown Tales, 1046
CHIR-FM (Toronto, Ontario), 277
Choi Jae-Do, 499
CHOM-FM (Montreal, Quebec), 276
Choquette, Robert, 287
Cho Won-Suk, 499
CHRC (Quebec City, Quebec), 268
Christian, Fred, 834
Christian Science Monitor, 415, 416
Christmas Story, 1257
Chronicle, 349
Chrysalis Radio, 230, 877
Chrysler Motor Corporation, 138, 1222, 1366
CHSR-FM (New Brunswick, Canada), 277
Chubu Nihon Hoso, 776
Chuck Blore Creative Services, 176
Chuck Shaden Radio Collection, 972
CHUC (Saskatoon, Saskatchewan), 780
CHUM (Toronto, Ontario), 293, **324–25,** 965
Chun King Chinese foods, 620
CHUO (Ottawa, Canada), 277
Churchill, Caryl, 497
Churchill, Winston Spencer, 224, **325–28**
Church of the Air, 351, 352
CIBL-FM (Montreal, Quebec), 277
CILQ-FM (Toronto, Ontario), 276
Cinerama, 813
CIQC (Verdun, Quebec), 311
Circuito Nacional Cubano, 429
Cisco Kid, 1360, 1500
CISM (Montreal, Quebec), 277
Cities Service Concerts, 338
Citizens Band Radio, **329–30**
Citizens Committee to Save WEFM v Federal Communications Commission (1976), 15
Citizens Communications Center v Federal Communications Commission (1974), 15
City of Contrasts, 1575
CJCA (Edmonton, Canada), 267
CJCE (Vancouver, Canada), 267
CJCG (Winnipeg, Canada), 267
CJMS (Montreal, Quebec), 293
CJOR (Vancouver, Canada), 292
CJSC, 267
CJSO (Sorel, Quebec), 262
CJSW-FM (Calgary, Canada), 277
CKAC (Montreal, Quebec), 199, 258, 267, 268, 280–81, **331–32**
CKCX (Toronto, Ontario), 780
CKGW (Toronto, Ontario), 270
CKLW (Windsor, Ontario), **332–33,** 1345
CKNW (Vancouver, Canada), 292
CKOK (Windsor, Ontario), 332
CKON-FM (Akwesasne, Canada), 277
CKRK-FM (Kahnawake, Quebec), 277
CKVL (Quebec, Canada), 280
CKWR-FM (Waterloo, Ontario), 277
Clandestine Radio, **334–36**
Clap for the Wolfman, 1549
Clara, Lu, and Em, 1094, 1551
Clarence Jones and His Wonder Orchestra, 25
Clarion Awards, 144
Clark, Dick, 60, 324, **336–37,** *337,* 472
Clark, Montgomery, 114
Clark, R.T., 240
C.P. Clark, Inc, 825
Clark–Hooper, Inc., 114, 721
Classical Music Format, **338–39**
Classic Rock Format, **340**
Clay, Tom, 1063
Clayton, Lou, 509
Clear Channel Broadcasting Service, 1145, 1333, 1407
Clear Channel Communications, Inc., **341–42**
acquisitions, 1054–55
hub and spoke system, 1106
mergers, 1372
studio automation and, 137
Tom Hicks and, 703
virtual radio and, 1467
Clear Channel Stations, **342–44**
Cleary, W.J., 129
Cleveland Bandstand, 799
Cleveland Plain Dealer, 357
Cline, Patsy, 662
Cline, William Hamilton, 355
Clinton, Bill, 1085, 1469
Clio Awards, 144
Cliquot Club Eskimos, 10, 996, 1337, 1453
Close, Upton, 366–67
Close Up, 291
Close-Ups of the News, 366
Club Alabam, 25
CMJ New Music Report, 346
CMQ (Havana, Cuba), 428, 429
CNBC Business Radio, 40
CNRV Players, 283
CNRV (Vancouver, Canada), 283
Coatman, John, 240
Coats, George A., 1556
cochanneling, 774
Cocoanut Grove Orchestra, 834
Cocoon, 55
Coddington, C.C., 1485
Codel, Martin, 1363, 1411
Code of Federal Regulations, 255
Code of Wartime Practices for American Broadcasters, 1562
Coffin, Charles, 648
Cognizant Corporation, 1
Cohen, Barbara, 955
Cohen, John S., 1569
Cohen, Lizabeth, 1291
Cohen, Phil, 61
Cold War Radio, **344–45**
BBC Monitoring Service, 212–13
international broadcasting, 750–51
Radio in the American Sector (Berlin), 1158–60
Radio Moscow and, 1181
radio propaganda, 1115–16
Cole, Alonzo Deen, 729, 884
Cole, Nat King, 170
Coleman, Jim, 290
Colgate, 1305
Colgate Comedy Hour, 49, 363, 509
Colgate-Palmolive, 429, 941
Colgate Palmolive Peet Company, 1286
Colgate Sports Newsreel, 1319
Colgate Theater, 1463
College of the Air, 1511
College Radio, **346–47**

Collier Hour, 729
Collingwood, Charles, 168, **347–49,** *348,* 352, 574
Collins, Carr P., 190
Collins, Ray, 1478
Collins, Ted, 1282
Collison, Perce, 415
Colman, Ronald, 1360
Colombia, 1306
colonialism, radio and, 87
Colonna, Jerry, 726
Columbia Broadcasting System (CBS), **350–54**
- advertising effectiveness and, 11
- Audimax, 121
- Canadian affiliates, 278, 282
- creation of, 72, 1426
- Infinity Broadcasting merger with, 745–46, 1054, 1055
- Karmazin and, 798
- morning programs, 957
- music on, 976
- Network Monopoly Probe, 1014–15
- news stations, 38–39
- *Quiz Kids,* 320
- self-regulation, 308
- Volumax, 121

Columbia Broadcasting System v Democratic National Committee (1973), 597, 598
Columbia Broadcasting System v Federal Communications Commission (1981), 597
Columbia News Service, 1095
Columbia Presents Corwin, 639, 962
Columbia Workshop
- CBS and, 352
- Fletcher Markle and, 908
- format, 405
- launch of, 1074
- public service role, 1355
- sound effect use by, 492
- techniques, 282
- William Lewis and, 859

Columnists, **354–58**
Combo, **358–59**
Comedy, **359–64**
Comedy Caravan, 1563
Comedy Theatre, 1034
Come unto These Yellow Sands (Carter), 218
Comhairle Radio Eireann, 760
Coming Soon, 173
Comissão Técnica de Rádio (Radio Technical Commission), 194
Command Audio, 459
Command Performance, 95, 826, 1496, 1563
Commentators, **365–69**
Commercial Load, **369–70**
Commercial Radio Australia (CRA), 132
Commercial Recording Corporation, 788
Commercials. *see* Advertising
Commercial Tests, **370–72**
Committee for Economic and Industrial Research (CEIR), 92
Communications Act of 1934, **372–74**
- censorship and, 1433
- Corporation for Public Broadcasting, 403
- equality of service provision, 870
- equal time rule, 550–52
- Fairness Doctrine, 579
- FCC powers and, 1201
- licensing and, 689, 1128
- payola and, 70
- public interest standard, 862
- war emergency provisions, 1560

Communications Daily, 1411
Communications Decency Act, 1038
Communicator, 1411
Communist Party, 166–69
Community Broadcasting Association of Australia (CBAA), 132
Community Media Association (CMA), 461
Community Radio, **375–76,** 1058–159, 1303
Community Radio News, 999
Como, Perry, 1277
compact discs, 464, 705, 1191–92, 1195
Compagnia Marconi, 767. *see also* British Marconi
Compagnie générale de TSF, 614
Compagnie Luxembourgeoise de Radiodiffusion, 1175
Company of Wolves (Carter), 218
Compton Agency, 539
Comrade in America, 908
Conan, Neal, 1004, 1367
Condon, Richard, 61
CONELRAD, **375–78,** 546, 548
confidence intervals, 116
Congress Cigar Company, 350
Conley, Robert, 43, 1001
Conquest, 1246
Conrad, Frank, **378–79,** *379*
- Group W and, 682
- KDKA radio and, 805
- shortwave broadcasts by, 1262, 1290, 1502
- telephony experiments, 83

Conrad, William, **380–81,** 685, *686,* 1500
Conseil Supérieur de l'Audiovisuel (CSA), 616
consoles, 1189–90
Constandurous, Mabel, 217
construction permits, 862
Consultants, **381–84**
Contadina tomato products, 620
Contemporary Christian Music Format, **384–85**
Contemporary Hit Radio/Top 40 Format, 59, 163, **386–87,** 1105
contests, 1112, 1281
Continental Radio News Service, 1023
Control Board, **387–89**
Control Board/Audio Mixer, 1188
Control Data Corporation, 1141
Controversial Issues, Broadcasting of, **390–92**
Conversation, 562
Cook, Alton, 355, 416
Cook, Fred J., 1200
Cook, Peter, 221–22
Cook, Ray, 1011
Cook, Roger, 242
Cooke, Alistair, 221, 242, **392–95,** *393*
Cooke, Donald, 840
Cooke, Jack Kent, 840
Cook Islands, 1310
Coolidge, Calvin
- campaigning, 540
- inaugural, 1431, 1440, 1542
- use of radio, 1083, 1439

Cool Train, 1399
Cooper, Alex, 471
Cooper, Giles, **395–97,** 498
Cooper, Jack L., 26, 169, **397–98**
Cooper, Roger, 92
Cooper, Wyllis, 729, 730, 864, 1033
Cooperative Analysis of Broadcasting (CAB), 114, 160, **398–400,** 721
Coopman, Ted, 885–86
Coops, Michael, 1384
Coor, Robert, 1433
Coordinator of Inter-American Affairs (CIAA), 942
Cope, F.W., 842
Cope, John, 842
Copyright, **400–403,** 1030
Copyright Act of 1976, 401
Copyright Office, U.S., 70
Copyright Tribunal, 230
Corday, Ted, 1073
Cordell, Lucky, 28
Cordic, Rege, 805
Corliss Archer, 1339, 1341
Corporation for Public Broadcasting, **403–4,** 871, 1388
NPR and, 991
Correll, Charles J., 79
- *Amos 'n' Andy,* 78–82, 998
- MCA and, 1365
- minstrel tradition, 361
- move to ABC, 1509
- *Sam 'n' Henry,* 1337
- syndication, 1359
- on WBT, 1486

Cortleigh, Steve, 1251
Corwin, Norman, **405–8**
- blacklisting of, 1198
- CBS programs, 492
- documentary programs, 475–76
- Martin Gabel and, 639
- *On a Note of Triumph,* 352, 1564
- poetry by, 1077, 1079
- *We Hold These Truths,* 498
- WWII efforts, 1075

Cosell, Howard, 433, 1312, 1317
Costas, Bob, 1312, 1316
Costas Coast to Coast, 1316

Costen, Dean, 1132
Cost Per Thousand, 8, 9
Cotten, Joseph, 1069
Cotton, James, 814
Cotton, Paul, 416
Couch, Robert, 1399
Coughlin, Father Charles, **408–10**, *409*
anti-Semitic comments, 307, 694
fundraising on radio by, 1535
Jewish objections to, 786
political use of radio, 1083
talk radio and, 1369
Coulter, Douglas, 639
Counterattack (newsletter), 167
Counter Spy, 884, 1561
Country Crystals, 840
Country Journal, 352
Country Music Association, 411, 1067
Country Music Disc Jockey Association, 1067
Country Music Format, **411–13**, 1105
Court of Missing Heirs, 1094
Covered Wagon Days, 1349
Cowan, Louis, **413–14**, 826, 1148
Cowan, Maurice, 784
Cowles, Gardner "Mike," Jr., 1041
Cox, Eugene E., 1015
Cox, James M., 1570
Cox, Lester E., 804
Cox, Sara, 234
Cox Radio, 1055–56, 1372
Coyle, Jim, 506
Coyle and Sharpe, 818
C-QUAM, 1335, 1336
Crackerjacks, 1486
Craig, Edwin, 825, 1407
Crane, Bob, 835
Crane, Martha, 1358
Crane, Ruth, 1552
Cranston, Alan, 1027
Cravens, Kathryn, 1553
Crazy Water Crystals, 190
Creagh, David, 1301
Cream of Wheat, 857, 1337
Crenna, Richard, 1051
Crescent Communications, 1253
Crescent Industries, 1534
Crisco, 1462
Critics, **414–17**
Crocker, Frankie, 1446
Croft, Mary Jane, 1051
Cronkite, Walter, 804, 844
Crosby, Bing, **418–20**
Bob Hope and, 726
CBS contract with, 353
Kraft Music Hall and, 11
move to ABC, 56, 1454
Personal Album, 95
recordings, 977, 1194
sponsorship of, 1069
style, 1273
syndication of, 1360
Crosby, John, 416
Crosley, Powel, **420–22**, *421*
Crosley, Powel, Jr., 1078, 1538
Crosley Broadcasting Corporation, 1145, 1528
Cross, Milton J., 181, **423–24**, 938, 939
Cross Country Checkup, 272, 292–93
Crossley, Archibald M., 114, 398–99, 721
Crossley, Inc., 398, 399
Crossroads, 782
Cross Roads Sunday School, 1358
Crown FM, 876
Crusade in Europe, 898
Crusellas, 429
Crusinberry, Jane, 494, 1552
Crutchfield, Charles H., **424–26**, *425*, 1486
Crutchfield, Les, 687
Crystal Radio Awards, 144
Crystal Receivers, **426–28**, *427*
Cuba, **428–32**, 1177–79
Cuban Missile Crisis, 1171
Cullen, Hugh Roy, 861
Culligan, Matthew J., 138
Cummings, Irving, 892
Cumulus Media, 1055–56
Curb, Mike, 801
Curbside Carnival, 321
Curran, Don, 818
Current Affairs, 216
Curwood, Steve, 1123
Cutforth, Rene, 241
CYB (Mexico City, Mexico), 940
CYFC (Vancouver, British Columbia), 780

Daddy and Rollo, 1339
Dagget, John S., 355
Dagmar, Doris, 890
Dahl, Christopher, 1168
Daily Mail, 239
Daley, Brian, 1328
Dall, Curtis B., 53
Daly, John Charles, 352, **433–35**
Damerel, Donna, 1551
Damrosch, Walter, 338
Dance Party, 1475
Dancer Fitzgerald company, 718
Daniels, Anthony, 1328
Daniels, Josephus, 1438
Daniel Starch organization, 721
Danmarks Radio, 1236
D'Arcy MacManus Masius agency, 12
Dark Fantasy, 729
Darrow, Mike, 324
Darsa, Josh, 43
data collection methods, 114–17
Data Radio Channel, 1336
Date with Judy, 31, 1339, 1341
Daunais, Lionel, 267
Davenport, Ronald R., 29, 173
David Harum, 250, 1094
Davis, Benjamin J., 167
Davis, Elmer, 366, **435–38**, *436*, 828, 1041
Davis, Harry P., 378, 682, 805, 1502
Davis, Jay Michael, 1400
Davis, Joan, 1553
Davis, Moishe, 1215
Davis, Stephen, 530
Davis, Thomas, 647
Davis Amendment, 870, 1203
Davis Baking Powder Company, 114
Davisson, Clinton J., 154
Dawnbusters, 1575
Dawson, Tom, 1109
Day, Doris, 1454
Days of Our Lives, 1073, 1288
Dayton Hudson, 948
dbx noise reduction processing, 121
DC/MD/DE Broadcasters Association v Federal Communications Commission (2001), 14
DC-100 (Washington, D.C.), 1342
Dean, Dizzy, 1316
Death Valley Days, 1498
Deaville, Frank, 286
"Deborah" commercial, 177
Debs, Eugene Victor, 1506
decibels, definition, 120
decoder pins, 299, 319
Dee, Ruby, 173
Deeb, Gary, 357
Deep in the Heart, 575
Dees, Rick, **438–39**
De Forest, Lee, **440–43**, *441*, 938, 1272, 1424
De Forest Wireless Telegraph Company, 440
de Freitas, Anna Maria, 1301
Delafield, Ann, 731
DeLeath, Vaughn, 355
Delmar, Kenneth, 48, 360, **443–45**, 1478
Delphi, 468
Delta Melodies, 735
Delta Rhythm Boys, 27
De Lue, Willard, 355
DeMars, Paul, 1254
Demento. *see* Dr. Demento
DeMille, Cecil B., 718, 890, *891*
Democracy Now!, 1124, 1482
Democracy USA, 511
Democratic National Committee, Columbia Broadcasting System v (1973), 597, 598
Democratic People's Republic of North Korea, 109
Demographics, 8, **445–47**
DeMott, Benjamin, 1290
Dempsey, Jack, 1320
Denison, Merrill, 283
Denkitsushinshou, 776
Denmark, 1236
Dennis, Paul, 1464
Denny, George V. Jr., 76, 1121
Denny, James, 673
Depression, 6, 26
Deregulation of Radio, **448–50**

Der Tog, 1026
Desert Island Discs, **450–51**
Desert Storm, 97
Designated Market Areas, 91
Desmond, Connie, 149
Destination Freedom, 170, 491, 511, 513, 1076
DeSuze, Carl, 1487
Detective Story Hour, 1248
Detroit Free Press, 1534
Detroit News, 354, 1534, 1573–74
Detroit Police Drama, 1535
Deutsche Stunde (German Hour), 565
Deutsche Welle (German Wave), 467, 658, 751, 1159
Deutschlandfunk, 1160
Deutschlandradio, 657
Deutschlandsender (Germany's Station), 657
Developing Nations, **452–54**
Devil's Roost, 729
Devil's Scrapbook, 729
DeVoto, Bernard, 417
Dewey, Thomas, 541, 1084
Dialogue, 1123
Diamond Lane, 499
Diane Rehm Show, 1005, 1123
Diarios e Emissoras Associadas, 1305
Diary, 90, 115, **454–56**
Dick, Elsie, 556
Dick Barton—Special Agent!, 92–93, 218
Dick Biondi Road Show, 163
Dick Clark's National Music Survey, 74
Dick Cole, 1340
Dick Tracy, 250, 319, 1340
Digital Audio Broadcasting, **456–62**
  audience demographics and, 133
  Bell Telephone Laboratories patents, 155
  Capital Radio and, 298
  control board, 389
  developments, 1430
  Eureka 147 standard, 231
  National Radio Systems Committee and, 1006
  pay radio and, 1065
  United Kingdom, 230–31
  World Administrative Radio Conference and, 630
Digital Audio Radio Services (DARS), 869
Digital Millennium Copyright Act (DMCA), 254
Digital Music Express (DMX), 254
Digital One, 458
Digital Radio Express (DRE), 461
Digital Radio Mondiale, 22, 457
Digital Radio News, 1412
Digital Recording, **463–66**, 1191–92
Digital Satellite Radio, **466–69**
  American Indian Radio on Satellite (AIROS), 1012
  antennas, 87
  ASTRA satellite, 1176
  AUSSAT, 135
  BBC World Service and, 226
  Children's Satellite network, 322
  digital audio broadcasting and, 456, 462
  distribution, 1134
  international broadcasting, 751
  NPR and, 895
  OSCAR, 690
  pay radio, 1064–65
  Radio Authority licenses, 230
  SATNET, 97
  Sirius, 182
  SKY TV, 1176
  Vatican Radio, 1458
  XM service, 182
Digital Studio to Transmitter Lines (DSTLs), 478
Dill, Clarence Cleveland, **469–70**, 987, 1202, 1431
Dillard, Everett, 609
Dimbleby, Richard, 240, 241, 242
Dimension, 1019
Dimensione Suano (Rome, Italy), 767
Dimension X, 1243
Direct to Sailors (DTS) service, 97
DirecTV, 779
Disagreeable Oyster, 395
Disc Jockeys (DJs or DeeJays), **471–73**
Discriminate Audio Processor, 121
Disney Corporation, 59–60, 1055, 1475. *see also* Radio Disney
DisneyRadio.com, 720
Dispatches, 274
diversity, 14–17
Dixon, Franklin W., 317
DKDAV (Lubbock, Texas), 411
Dobson, James, 1007, 1215
Doctor Fights, 1076
Documentary Programs on U.S. Radio, **474–77**
Dodge, 1221
Dolan, Daria, 1371
Dolan, Ken, 1371
Dolbear, Amos, 518
Dolby, Ray, 477
Dolby Noise Reduction, 121, **477–79**
Dole, Robert, 1466
Donahue, Tom, 472, **479–80**, 1107, 1253, 1420
Donald, Peter, 48
Donaldson, Sam, 1573
Don Ameche's Real Life Stories, 55
Don and Lois Wilson Show, 1525
Don Lee Broadcasting System, 72, 351, **481–82**, 1557
Don McNeal's Breakfast Club, 181, 818
Donnelley, Thorne, 987
Donohue, Rachael, 1554
Don't Knock the Rock, 592
Don't Touch That Dial, 1288
Doordarshan, 36
Dorrance, Dick, 608
Dorrough Electronics, 121
Do the Right Thing (film), 593
Double or Nothing, 1150
Doud, Gil, 1589
Doug Banks Show, 1447
Douglas, Don, 730
Douglas, Doug, 355
Douglas, Susan, 1290
Douglas, Van, 26
Douglas, William O., 1122
Douglas-Home, William, 660
Dove, Rita, 1081
Downs, Hugh, 1019, 1578
Doyle Bulletin, 262
Dr. Dean Edell Show, 921, 1372
Dr. Demento, 482, **482–84**, 1361
Dr. I.Q., 844, 1147
Dr. Joy Brown, 1557
Dr. Laura Schlessinger Show, 1555
Dr. Ruth Show, 1555
Dr. Sixgun, 1500
Dracula, 730
Dragonette, Jessica, 338
Drahtloser Dienst AG (Wireless Services, Inc), 656
Drake, Bill, **484–88**
  Chenault and, 137, 315
  at KHJ, 369, 472, 819
  oldies format, 1218
  Top 40 format, 386
Drake, Galen, **488–90**, 489
Drake, O. Burtch, 370
Drake Chenault agency, 137, 1466
Drama on U.S. Radio, **490–96**
Drama Worldwide, **496–501**
Dreft Star Playhouse, 718
Drene Time Show, 55
Drew, Paul, 333, **501–3**
Drewry, John E., 1065
Driscoll, Marian, 589
Driscoll, Tom, 801
Drivetime, 233
Drudge, Matt, 60
Drum, 1124
Dubois, Cynthia, 281
Ducretet, Eugene, 1090
Dudley, Paul, 1589
Duffy's Tavern, **503–5**, 1375, 1554
Duhamel, Helen, **505**
Duke University Library Advertising History Archive, 973
Dumm, William, 840
DuMont, Bruce, 971
Dunbar, Jim, **506–7**, *507*
Duncan, Robert, 1057
Duncan's American Radio, 7, 369
D'une certaine manière, 268
Dunham, Franklin, 534
Dunifer, Stephen, 885–86
Dunlap, Orrin E., 355, **508**
Dunphy, Don, 1316–17
Dunphy, Eamon, 761
Dunwich Horror, 730

Dupont, J.-Arthur, 267, 331
DuPont chemical company, 474
E.I. du Pont de Nemours and Company, 302
Duquesne, Albert, 267
Durante, Jimmy, **509–11**, *510*, 1453
Durham, Richard, 170, 491, **511–14**, *512*, 1076
Dutch Masters Minstrels, 1337
Dutt, Kamal, 500
DVDs (digital versatile discs), 464
DXers/DXing, 454, **514–16**
Dynamical Theory of Electromagnetic Field (Maxwell), 911
dynamic range, definition, 120

Earl, Guy C., 834
Earl Carroll Vanities, 156
Early, Steve, 594
Early Wireless, **517–20**
Earplay, 495, **520–21**, 1003, 1513
Earth and Sky, 175
Eastland, James, 1058
Eastman Kodak, 1274
Eastman Radio, 1332
Easy Aces, **521–23**, 1554
Easy Listening/Beautiful Music Format, **523–26**
Ebony/Jet Celebrity Showcase, 790
Ebony magazine, 511
Echo Moskvy (Moscow's Echo), 1230
Eckersley, Peter, 1210
Economic Opportunity Act of 1965, 1011
Economics in a Changing World, 1120–21
Economics of the Radio Industry (Jome), 530
Economic World Today, 1120
Ecuador, 1306
Eddie Cantor Show, 26, 719, 1338
Eddy, Nelson, 526
Edell, Dean, 921, 1371, 1372
Eden, Anthony, 225
Edgar Bergen and Charlie McCarthy Show, 55, 361, **526–28**, 1454
Edge of Night, 1288
Edison, Thomas, 518
Edison Media Research, 7
editing process, 1099–1100
Editorializing, **528–29**
Edna Fischer Show, 26
Ed Sullivan Show, 509
Education about Radio, **530–31**
Educational Radio to 1967, **532–36**
Educational Television Facilities Act of 1962, 1126
Educational Television Stations, 1126
Edward R. Murrow Awards, 144
Edwards, Bob, 44, 45, **537–38**, *538*, 955
Edwards, Charles, 290
Edwards, Douglas, 1578
Edwards, Ralph, **539–40**, 1149
Edward Stratemeyer Syndicate, 316
Ed Wynn, The Fire Chief, 592
EFM Media, 1372
Egleston, Charles, 896
Egypt, 88
Ehret, Cornelius, 603
EIB Network, 1372
Eigen, Jack, 1522
8XK (Wilkinsburg, Pennsylvania), 378, 689, 805
Eikerenkoetter, Frederick J., II, 1215
Einstein, Albert, 701, 709
Einstein, Jack, 1056
Eisenhower, Dwight D., 542
Eisenhower, Milton, 185
Eisner, Michael, 59, 1055
El Buen Tono, 940
Elder, Robert, 1, 115, 118
El Drecho de nacer (Right to be Born), 942
Election Coverage, **540–43**
Electrical Engineer, 1410
Electrolytic Detector, 427
Electronic Industries Alliance, 86
Electronic Media Rating Council. *see* Media Rating Council
Electronics World, 568
Elgin Watches, 1584
Elijah Communications, 1310
Elinor, Carli, 834
Ellery Queen, 1338, 1415
Ellington, Duke, 25, 27
Elliot, Bob. *see* Bob and Ray
Ellis, Bobby, 31
Ellis, Elmo, **544–45**, 1570
Ellis, Georgia, 685, 1500
El Programa de Ramón (Ramon's Show), 431
Emergencies, Radio's Role in, **545–47**, 1568–69
Emergency Alert System (EAS), 547
Emergency Broadcast System, 377, 546, **547–49**
Emerson, Ralph, *888*
Emerson, Ralph Waldo, 1358
Emerson, Victor Hugo, 549
Emerson Radio, **549–50**, 972, 1173
Emery, Bob, 355
Emmis Broadcasting Corporation, 1284
Empire Builders, 54
Empire Service. *see* British Broadcasting Corporation: BBC World Service
Empire Strikes Back, 1328
Empower MediaMarketing of Cincinnati, 369
Empresa Brasileira de Radiodifusão (RADIOBRÁS), 195
enhanced other networks (EON) features, 1166
Enjoyment of Poetry, 1081
Ennis, Skinnay, 726
Entercom Communications, 1575
Entwistle, Guy, 354
Epp, Theodore, 556, 1214
Equal Employment Opportunity (EEO) program, 14
equalizers, 120
"Equal Time" Rule, **550–52**
eRadio, 1412
Eritrean People's Liberation Front, 335
Erwin Wasey agency, 152
Escale, 281
Escape, 1243
ESPN Radio Network, 40, 59, 1323
Es spricht Hans Fritzsche (This is Hans Fritzsche), 637
Eternal Light, 786, 1076, 1215
Ethiopia, 19, 20
Ethnikon Idryma Radiophonias (ETR), 676
EUREKA 147, 231, 456, 457
Euroclassic-Notturno, 553, 554
Euro Jazz, 779
Europäische Fremdsprachendienste (European Foreign Language Service), 145, 146
European Broadcasting Union, **552–55**, 607
European Song Contest, 554
Euroradio Control Centre, 553
Eutelsat array, 467
Evangelical Council for Financial Accountability, 1008
Evangelical Organization, 1013
Evangelists/Evangelical Radio, **555–58**
Evans, Chris, 235
Evans, Leonard, 173
Evans, Richard L., 954
Evans, Thomas L., 804
Evans, W. Leonard, 29
Evans, Walter, 683
Evans, Dale, 1553
Evening in Paris Roof, 1495
Evening Serenade, 1495
Evening Theater, 322
Eventide Ultra-Harmonizer, 122
Eveready Hour, 11, 180, 491, 1221
Eveready House, 491
Everett, Kenny, 204, 233, 297–98, **558–60**
Everett, Peter, 219
Evergreen, 1054
Everybody Wins, 1151
Everyman's Theatre, 1034
Everything for the Boys, 1034
Everything in the Garden, 395
Everywhere Show, 953
Ewing, John D., 847
EWTN network, 1216
Exacting Ear, 1122
exclusive cume listeners, 9
Eyewitness to History, 349, 845
e-zines, 569

Facts and Figures of Commercial Broadcasting, 12
Fadiman, Clifton, **561–63**, *562*, 1147
Fairless, Benjamin F., 1392
Fairness Doctrine, **563–66**
  dropped, 1129
  public interest standard, 1128
  Reagan veto of, 1146

Red Lion Case, 14, 564, 1200–1201, 1444
Faith in Our Times, 1215
Falcon, 1415
Falkenberg, Jinx, 1553
Fall of the City, 492, 859, 1074
Falstaff Brewing Company, 861, 1312, 1316
Falstaff's Game of the Week, 1316
Falwell, Jerry, 1215
Family Favourites, 232, 235
Family Theater, **566–67**
Fanball.com, 1505
Fan Magazines, **567–70**
Fanon, Frantz, 1293–94
Farber, Erica, 1109
Far East Broadcasting Company, 570, **570–71**
Farm/Agricultural Radio, **571–74**
Farm and Fun Time at Noontime, 1522
Farm Question Box, 836
Farnsworth, Philo T., 1069
Farnsworth corporation, 296
Farrar, Eugenia, 1272
Fass, Bob, 627, 1420, 1482
Fat Man, 492, 1338
Faulk, John Henry, 168–69, **574–76,** 1199
FCC Record, 578
Feather, Leonard, 779
Featherbee, Harry, 238
Feature Broadcasting Company, 173
Feature Series Plan, 1043
FEBC (Karuhatan, Philippines), 570, 571, 751
Fecan, Ivan, 272
Federal Broadcasting, 1574
Federal Bureau of Investigation, 1198
Federal Communications Commission
American Broadcasting Company v (1953), 1444
Citizens Committee to Save WEFM v (1976), 15
Citizens Communications Center v (1974), 15
Columbia Broadcasting System v (1981), 597
DC/MD/DE Broadcasters Association v (2001), 14
Lamprecht v, 15
Lutheran Church v (1998), 14
Metro v, 15
Red Lion Broadcasting Company v (1969), 14, 564, 1200–1201, 1444
TV 9 Inc. v (1973), 15
United Church of Christ v (1966), 14
Federal Communications Commission (FCC), **576–79**
air traffic control interference and, 775
AM-FM Program Nonduplication Rule, 136
AM radio and, 84
Blue Book, 178–80, 391, 416, 532, 1205, 1267, 1355, 1428
call letters and, 255
censorship and, 307
Citizens Band Radio and, 330
Clear Channel Stations and, 342–43
College radio and, 346
Communications Act of 1934, 372–74
creation of, 1009
Deregulation of Radio, 448–50
Digital Audio Broadcasting and, 460, 461, 1065
diversity of information and, 14
enhancement credits, 171
"Equal Time" Rule, 551
Fairness Doctrine, 563–66, 1128, 1129, 1148, 1444
Femme Forum case, 1403
FM radio and, 604
Foreign Languages Division, 1027
frequency allocation, 629
hoaxes and, 714
Howard Stern and, 1342
licensing, 248–49, 862–64
Local Marketing Agreements, 872–73
mandate, 1203–4
In the Matter of Editorializing by Broadcast Licensees, 563
Mayflower Decision, 528–29, 912–13
microradio and, 885
modulation limits, 119–20
Network Monopoly Probe, 1014–15
network probe, 983
pay radio, 1065
on portable stations, 1091
Pot o' Gold issue, 1148
profanity and, 1038
Programming Policy Statement (1960), 1205–6
public interest in licensing, 1035
Radio Act of 1927, 199
"radio quacks" and, 199
Report on Chain Broadcasting, 997, 1428
Satellite Digital Audio Broadcasting Systems and, 467–68
self-regulation and, 308
"Seven Dirty Words" Case, 1246–48
stereo broadcasting and, 1334–35
syndication and, 1359
ten-watt stations, 1388–89
United Paramount takeover of ABC, 57
Volunteer Examiner Program, 690
War Problems Division, 1562
Federal Communications Commission versus
American Broadcasting Company (1954), 1151
League of Women Voters of California (1984), 564, 598
Pacifica Foundation (1978), 1444
Sanders Brothers Radio Station (1940), 1444
WNCN Listeners Guild (1981), 1445
Federal Council of Churches, 1007
Federal Power Commission, NAACP v (1976), 14
Federal Radio Commission, 532, **580–83,** 1444
General Order No. 40, 342, 1203
power of enforcement, 1426
Federal Radio Commission v Nelson Brothers Bond and Mortgage Company (1933), 1444
Federal Radio Education Committee, 533
Federal Register, 578
Federation of Australian Radio Broadcasters (FARB), 130
Feder, Robert, 417
Feedback, 243–44
"Feed Back," 282
Feinburg, Abraham, 785
Felber, Herman, 1358
Feldman, Marty, 221, 1513
Fellows, James A., 991
Female Radio Personalities and Disk Jockeys, **583–85**
Feminine Forum, 1403
Femme Forum, 1261, 1403
Fennelly, Parker, 48
Fenneman, George, 1584
Ferguson, Bert, 27, 1492, 1523
Ferguson, Don, 287
Ferguson, Max, 286–87
Ferlinghetti, Lawrence, 1057
Fessenden, Reginald, **585–88**
Alexanderson influenced by, 32, 1438
first voice broadcast, 258, 1272, 1424
Liquid Barretter of, 427
telephony experiments, 83
Fessenden Wireless Company of Canada, 587
Fibber McGee and Molly, **589–91**
Beulah Show spinoff, **159–60**
concept, 362
films, 592
Gale Gordon and, 670
popularity, 1278
stereotypes, 1338, 1340, 1341, 1554
on WMAQ, 1541
Fickett, Homer, 1392
Fidenas Investments, 550
Fidelity Investments Associates, 1576
Fidler, Jimmy, 719
Field, Charles K., 1078
Fielden, Lionel, 35, 240
Fields, W.C., 527, 1587
Fifth Estate, 291
Fight of the Month, 1317
Fiji, 1309
File on Four, 242
Film Depictions of Radio, **592–93**
Financial News Network (FNN), 649
Finkel, Bernie, 786
Finkelstein, Louis, 786
Finland, 1237–38
Finlay, Mary Lou, 273
Fire Chief, 1579

Fireside Chats, **594–96**, 1440, 1441
Firing Line, 1142
First Amendment and Radio, **596–99**, 1200–1201, 1247
First in the Air, 1075
First Nighter, 54–55, 1415
Firth, John, 587
Fisher, Cyrus, 416
Fistell, Ira, 1371
Fitz-Allen, Adelaide, 729
Fitzgerald, Pegeen, 719, 1557
Fitzgerald, Ed, 1557
Fitzgeralds, 719, 1553, 1557
Fitzmaurice, Frank, 955
Fitzpatrick, Howard, 355
Fitzpatrick, Leo J., 355, 1535
Flair Reports, 1048
Flanner, Janet, 1553
Flash Gordon, 1243, 1338
Flatow, Ira, 1367
Fleetwood, Harry, 1080
Fleischmann Hour, 140, 196, 1273, 1453
Fleischmann's Yeast, 1449
Fleming, John Ambrose, **599–601**, 701
Fleming Valve, 440
Fletcher, C. Scott, 1126
Fletcher, Harvey, 704
Fletcher, James, 906
Fletcher, Lucille, 1074, 1353
Florence, John. *see* Allen, Fred
Floyd, J.Q., 814
Fly, James Lawrence, 577, 1015
Flynn, Bernardine, *1461*, 1462
Flynn, Charles, 772
Flywheel, Shyster, and Flywheel, **601–2**
FM (film), 592
FMphasis, 1411
FMQB magazine, 1412
FM Radio, **602–8**
  advertising sales and, 6
  in Africa, 18–19
  AM-FM Program Nonduplication Rule, 136
  AM radio and, 85
  antennas, 86–87
  Arab broadcasting, 88
  in Australia, 130–32, 133
  in Britain, 229
  British Forces Radio, 236
  Canada, 265, 270
  in France, 616
  frequency allocation, 631
  Group W and, 683
  invention of frequency modulation, 99–100
  low-power, 309
  receivers, 1185
  simulcasting, 1271–72
  stereo, 1334–35
  Ten-Watt Stations, 1387–89
  in the USSR, 1229
FM Trade Associations, **608–10**
focus groups, 117, 447, 1102
Focus on the Family, 1007, 1215
Folger's Coffee, 712
Folk Festival USA, 1002
Folk Song Festival, 1546
Fondation Hirondelle (Swallow Foundation), Switzerland, 20
Foote, Cone and Belding, 13
Footprints in the Sands of Time, 634
Forbes, Murray McIntyre "Jerry," 312, 896
Ford, Gerald, 1469
Ford Motor Company, 138, 909, 1069, 1322, 1492
Ford Theater, 909, 962
Foreign Affairs Reform and Restructuring Act (Public Law 105-277), 247
Foreign Broadcast Information Service, **610–12**
Foreign News Roundup, 1018
Foreign Service Broadcast Information Service, 211, 212
Forgotten Footsteps, 283
Formats, **612–14**
Formosa, 108–9
Forsythe Saga, 493
Fort Laramie, 1500
Forverts, 785
Foster, Harry E., 262
Foster, Solomon, 784
Foundation of Music, 219
Four Girls and a Radio (Anderson), 317
Fowler, Mark, 1206
Fox, Carolyn, 584
Fox All Access Countdown, 322
Fox FM (Oxford, England), 298
Fox Kids Countdown, 322
Fox News Radio, 40
Foy, Fred, 319
France, 40, 607, **614–17**, 1115, 1186
Francis, Dick, 1155
Frank, J.L., 140
Franklin, John Thomas, 814
Frank Merriwell, 1340
Frank Seaman Advertising, 114
Frasier, 1386
Fraylekher Kabtsen, 786
FRC, Great Lakes Broadcasting Company et al v, 1204
Freberg, Limited agency, 620
Freberg, Stan, **618–20**, 1031
Fred Allen Show, 48–49, 443, 1338
Frederick, Pauline, **621–23**, 1553
Fred Waring Show, 1453
Free, John, and Fields agency, 1331
Free Company, 1075
Freed, Alan, 472, **623–25**, 1063, 1220, 1345
Freed, Paul, **625–27**
Freedom Foundation National Awards, 144
Freedom's People, 27, 170
Free Form Format, **627–28**
Freeman, Gosden, 1337
Free Voice of Iran, 1116
Freiheitssender 904 (Freedom Station 904), 335, 657
French Agency Press, 1020
French Congo, 18
French Guiana, 1308
French Press Review, 1546
Frequency Allocation, **629–31**
Frequency Modulation Business, 1411
Frequency Registration Board, 630
Fresh Air, **632–33**, 1004, 1122, 1131, 1265, 1555
Fresh Air (Canada), 274
Freshest Thing in Town, 1360
Fridays with Red, 537
Frigidaire, 663
Friendly, Fred, **634–36**, 697
  Hear It Now, 476, 696, 968
  Robert Trout and, 1418
Friendly Five Footnotes, 1094
Friends of Old Time Radio, 970
Fries, Gary, 1157
Friml, Rudolph, 69
Frischnecht, Lee, 1002
Frith, Simon, 1291
Fritsch, Babe, 844
Fritz, Charles, 1578
Fritzsche, Hans, **637–38**
From Our Own Correspondent, 221, 241
Frontier Gentleman, 1500, 1501
Frontline Family, 218
Front Line Theater, 1563
Front Page Challenge, 291
Front Page Farrell, 1561
Front Row, 219
Frost, Robert, 1080
Frost, Wesley, 290
Frum, Barbara, 272–73
Fuller, Barbara, 1046
Fuller, Charles, 557
Fu Manchu, 493, 1338

Gabbert, James, 609
Gabel, Martin, **639–40**
Gable, Bill, 333
Gabriel Awards, 144
Gale, Don, 529
Galen Drake Show, 489
Gallagher, Mike, 1372
Galling, Ben, 786
Gallup, George, 721
Galvin, Paul, 138, 960
Galvin Manufacturing Corporation, 961
Gambling, John A., **641–43**
Gambling, John B., **641–43**, 1556
Gambling, John R., **641–43**, 1556
Game of the Day, 1313
Game of the Week, 1316, 1319
Gammons, Earl, 1489
Gangbusters, **643–45**, 644, 883
GapKids, 323
Garay, Ronald, 1197
Garde, Betty, 951

Garden Gate, 352
Gardner, Ed, 503
Garmhausen, M. Adaire, 1551
Garner, Michele, 787
Garrison, Garnet, 454, 455
Garrison, Tom, 1447
Gaslight Revue, 524
Gates Radio Company, 136
Gatti-Casazza, Giulio, 938
Gaudet, Hazel, 1039
Gauthiere, Henry, 1298
Gavin Report, 176
Gay and Lesbian Radio, **645–48**
Gaylord Broadcasting Company, 1572
Geddes, Norman Bel, 549
Geer, Charlotte, 515
Gélinas, Gatien, 287
Gene and Glen, 957
Gene Autry Program, 140
Gene Autry's Melody Ranch, 141, 142, 1499
General Cigar Company, 1357
General Electric, **648–50**
  alternator, 32–34, *33*
  Fessenden and, 587, 1163
  FM transmitters, 603
  radio station business, 1425
  RCA and, 1165
  receivers, 1184
  sports sponsorship, 1321
General Foods, 728, 1525
General Mills, 57, 319, 351, 828, 1321
General Mills Adventure Theater, 251
General Mills Hour, 1072
General Mobile Radio Service (GMRS), 329
General Motors, 620, 1563
General Motors' NBC Symphony Orchestra, 11
General Strike of 1926, 325
General Telephone and Electronics, 1069
General Teleradio, 983, 1557
General Tire and Rubber Company, 482, 1584
General Tire Revue, 156, 1525
George, Vic, 311
George Burns and Gracie Allen Show, **650–53**, 719
George Carlin Again, 1247
George Clark Radioana Collection, 972
George Polk Awards, 144
George Wendt Show, 1386
GE Radio News, 835
German Freedom Station 904, 335, 657
German General Electric Company (AEG), 653
German Wireless Pioneers, **653–55**
Germany, **655–59**. *see also* Nazi Germany
  FM stations, 607
  propaganda on radio, 1114
  radio drama in, 498
  table radios, 1186
Gernsback, Hugo, 356, 568, 1380
Gersback, Sidney, 355
Get Rich Quick, 1151
Ghana, 18, 21
Ghana Broadcasting Corporation, 21
Ghetto Life, 764
Gibb, Bart, 312
Gibbons, Floyd, 367
Gibson, Mel, 321
Gibson, Walter B., 1249
Gibson Family, 26
Gielgud, Val, 208, 217
Giellerup, Frank, 114
Gifford, Walter S., 71
Gill, Eric, 206
Gillard, Frank, **659–61**
Gillars, Mildred, 145–48, *147*
Gillette, 1014, 1312, 1317, 1322
Gillette, Don Carle, 356
Gillette, George N., 1571
Gilliam, Laurence, 219
Gilliland, Ezra, 518–19
Gillmore, Jack, 283
Gilman, Page, 1046
Gilmour, Clyde, 274
Gilmour's Albums, 274
Gimbel's Department Store, 1121, 1528
Ginny Gordon and the Broadcast Mystery (Campbell), 317
Ginsburg, Allen, 645
Ginsburg, Douglas H., 1404
Ginzburg v United States (1966), 1036
Giuliani, Rudolph, 1476
Give and Take, 1150
GLAMA (Gay/Lesbian American Music Awards), 647
Glaser, Rob, 123
Glass, Ira, 1394
Glasser, Gerald J., 1153
Glickman, Marty, 1317
Global Positioning System (GPS), 755
Globe Theater, 1563
Glorious Monster in the Bell of the Horn, 495
Gluck, Earle, 1485
Gluck, Louise, 1081
G-Men, 643
Goddard, Ralph W., 837, 838
Goddes, Jeff, 274
Godfrey, Arthur, **661–64**, *663*
  Fred Allen and, 48–49
  popularity, 957
  talent shows and, 1367, 1455
Godin, Claude, 281
Goebbels, Joseph, 656, 1114
Goff, Norris "Tuffy," 361, 887, *888*
Going to Town, 1528
Go Johnny, Go, 592
Goldberg, Melvin A., 935
Goldbergs, **664–66**
  ethnic characters on, 785
  Himan Brown and, 250
  stereotypes on, 1337, 1338
Gold Coast. *see* Ghana
Gold Dust Twins, 1337
Golden Hour of the Little Flower, 409
Golden Memories, 836
Golden Mike Awards, 144
Goldenson, Leonard H., 57–59
Goldin, J. David, 970, 1030
Goldmark, Peter C., 812, 1194, 1326
Gold Medal Awards, 144
Gold Medal Flour, 11
Goldsmith, Alfred Norton, 532, **667–69**
Goldsmith, Clifford, 31
Golenpaul, Dan, 561, 1147
Golor Rossii (Voice of Russia), 1230
Gompers, Samuel, 64
Gonzáles, Ismael, 431
Gonzáles, Pedro, 712
Goode, Bob, 127
Good Gulf Show, 1221, 1222
Goodman, Amy, 1482
Goodman's Matzos, 250
Good Morning America, 693
Good Morning Vietnam, 593
Good Morning World, 1385
Goodness Gracious Me, 499
Good News Broadcaster, 557
Good News of 1938, 196
Goodrich Company, 10
Goodrich Zippers, 1453
Goodyear tires, 1321
Goon Show, 204, 221, **668–69**
Gorbachev, Mikhail, 225, 1116, 1173, 1229
Gordon, Bert, 360, 1338
Gordon, Dorothy, 1553
Gordon, Gale, 589, **670–71**, 1049
Gordon, Greg, 647
Gordon, John, 332
Gordon, Larry, 609
G. Gordon Liddy Show, 1371
Gore, Tipper, 1465
Gorham, Maurice, 760
Gosden, Freeman F. Sr., 79
  Amos 'n' Andy, 78–82, 998
  MCA and, 1365
  minstrel tradition, 361
  move to ABC, 1509
  and stereotypes, 1337
  syndication, 1359
  on WBT, 1486
Gospel Music Format, **671–72**
Goss, Jim, 772
Gosteleradio, 1229
Gotham Broadcasting, 1528
Goudy, Curt, 1312
Gould, Jack, 416
Gould, Morton, 561
Gould, Sandra, 503
Goulding, Ray. *see* Bob and Ray
Goulet, Charles, 267
Gowdy, Curt, 1317–18
Gracie Awards, 75
Graf Spee, 1240
Graham, Billy, 425, 555, 557, *1213*, 1215

Gramsci, Antonio, 1290
Grand Central Station, 250
Grand Hotel, 1033, 1415
Grandin, Thomas, 610
Grand Ole Opry, 25, 412, **672–74**, 1067, 1454
Granik, Theodore, 1121
Grant, Amy, 385
Grant, Bob, 695, 1370, 1557
Grant, Dick, 1583, 1584
Grant, Heber J., 842
Grant, Toni, 1555
Grape-Nuts Cereal, 1366, 1525
Grapevine Rancho, 1498
Grauer, Ben, 1198
Graves, Harold N., 610
Gray, Barry, 41, 1370, 1522
Grayson, Mitchell, 1076
Great American Broadcast, 592
Greatest Story Ever Told, 1214
Great Gildersleeve, **674–76**
  films, 592
  popularity, 1278
  stereotypes on, 1338, 1339, 1340, 1341
Great Lakes Broadcasting Company et al v FRC, 1204
Great Lakes Review, 156
Great Merlini, 1338
Great Moments in History, 474, 816
Great Play Series (BBC), 218
Great Temptations, 156
Greb, Gordon, 803
Grebe, Alfred H., 1488
Greece, **676–78**
Greely, Walter, 278
Green, Abel, 356
Green, Eddie, 503
Green, H.R., 71
Greene, Lloyd, 355
Greene, Rosaline, 491
Greene, Vernon, 1251
Green Hornet, 319, 493, **678–79**, 1338, 1350, 1499, 1578
acquisition by WXYZ-AM, 56
Green Lama, 1338
Green Lantern, 1338
Greenspring Company, 948
Green Valley, 250
Greenwald, James L., **679–81**, 1331–32
Greenwich Village Follies, 25
Gregg, United States v, 580
grid rate cards, 9
Griffin, Robert, 1132
Grisby-Grunow Company, 67
Griscom, Tom, 1571
Grizzard, Herman, 1536
Gross, Ben, 53, 416
Gross, Terry, 632, 1555
Gross Impressions (GIs), 8
Ground Wave, **681–82**
groundwave jamming, 774
Group W, **682–85**
Group W Cable, 1503
Group W Radio Sales, 1333
Gude, Jap, 696
Guedel, John, 1584
Guest, Edgar, 1079
Guevara, Che, 429
Guglielmo Marconi Foundation, 974
Guidance for Living, 1557
Guiding Light
  Irna Phillips and, 493, 1070, 1072, 1073
  length of run, 1287
  theme, 1288
Guild, Ralph, 1332–33
Gulf Oil Company, 1221
Gulf War, 89, 212–13, 1464
Gulf War: Special Edition, 1123
Gumble, Bryant, 1311–12
Gumble, Greg, 1311–12
"Gumps," 78
Gunsmoke, 380, **685–87**, 1429, 1500, 1501
Gunther, John, 1542
Gus Arnheim's Orchestra, 418
Gustavson, E. Brandt, 1008
Guterma, Alexander, 983
Guthrie, Tyrone, 217, 270, 283
Guyana, 1308
Guy Lombardo Orchestra, 1453
GWR Group, 231
Gyri, Ota, 54

Haas, Bobby, 703
Haas, Karl, 4–5, 5, 339, 1535
Haciendo Radio (Doing Radio), 431
Hadden, Briton, 898
Haggin, B.H., 417
Haley, William, 224
Hall, Alan, 219
Hall, Claude, 357
Hall, David G., 816
Hall, Juanita, 173
Hall, Lee, 219
Hallmark Cards, 1078
Hallmark Playhouse, 962
Hall of Fantasy, 730
Halop, Florence, 503
Halper, Donna, 382
Hal Roach Studios, 983
Halton, Matthew, 261
Ham, Al, 1218
Hamblin, Ken, 1372
Hamill, Mark, 1328
Ham Radio, **689–91**
Ham Radio Online, 569
Hancock, Harrie Irving, 316
Handbook of Broadcasting (Abbot), 530
Handbook of Radio Writing, 152
Handie-Talkie, 961
Handley, Tommy, 221
Handy, W.C., 25
Hanfmann, George, 61
Hannity, Sean, 1476
Hansen, Barry, 482, 483
Hansen, Liane, 1325
Hap Hazard, 1338
Happiness Boys, 359
Happy Gang, 272
Happy Go Lucky, 1495
Hard, Ann, 1553
Hardcastle, William, 241
Harding, Warren G., 1082, 1120, 1439
Hardy, Bob, 833
Hardy, Miles, 25
Hardy Boys and the Short Wave Mystery (Dixon), 317
Hardy Family, 1340
Hare, Ernie, 359, 788
Hargis, Billy James, 1200
Harley, William G., 990
Harnett, Vince, 1199
Harper, Johnny, 162
Harris, Bass, 26
Harris, Howard, 1586
Harris, Nick, 816
Harris, Oren, 934
Harrison, Harry, 1475
Harrold College of Wireless and Engineering, 802
Harron, Don, 287
Harry Potter (Rowling), 219
Harry Salter Orchestra, 57
Hart, Fred, 802
Hartley, Harold, 704
Hartnett, Vincent, 167, 168
Hartzell, Clarence, 1462
Hartz Mountain Canary Hour, 320
Harud (Autumn), 500
Harvey, Paul, 192, **691–94**, 692
Harvey, Paul, Jr., 693
Harvey, Lynne, 691, 693, 971
Harwell, Ernie, 1318
Hassel, Karl, 916, 1593
Hastings, T. Mitchell, Jr., 609
Hate Radio, **694–95**
Hauntings Hour, 730
Have Gun, Will Travel, 1500, 1501
Havel, Vaclav, 1117, 1173
Haverlin, Carl, 245
Hawkins, Reginald, 441
Hawk Larabee, 1499
Hay, George D., 672
Hayes, Albert, 890
Haynes, Dick, 471
HCJB (Quito, Ecuador), 557, 558, 749, 751, 1263
Headline Hunter, 367
Headliners, 262
Headlines and Bylines, 1416
Health Talk, 1557
Hearing America First, 1541
hearing limits, 120
Hear It Now, 353, 476, 636, **696–97**, 969, 1122
Hearst, William Randolph, Jr., 1121
Heart at Home, 1564

Heartbeat Theater, 489
Hearthstone of the Death Squad, 1338
"Heart of George Cotton," 491, 513
Heath, Mrs. Julian, 1552
Heatter, Gabriel, 366, 367
Heavenly Days, 592
Heavy Metal/Active Rock Format, **698–99**
Hedges, William S., 970
Hee-Haw, 575
Heidt, Horace, 1367
Heinl, Robert D., 355
Hellenic Radio Hour, 1122
Hellenic Radio-Television S. A. (ERT), 676
Hello, Larry, 1385
Hello Again, 970
Helmholtz, Hermann von, 701
Helms, Jesse, 1177
Henderson, Doug "Jocko," 28
Henderson, Fletcher, 25
Henderson, William Kennon, 846, 881, 1575
Hendrie, Phil, 816
Hennacy, Ammon, 1122
Hennock, Frieda, 75, 577
Henry, Joseph, 517
Henry, Pat, 779
Henry, Ragan, 1493
Henry Christal Company, 1332
Henry Morgan Show, 950–51
Henry Woods Promenade Concerts, 214, 219
Henshaw, Don, 283
Herald Tribune Forum, 992
Herbert, Ira, 792
Herbert, Victor, 69
Herbert v Shanley, 69
Here, There and Everywhere, 1079
Here Comes Tomorrow, 511
Here's Morgan, 362, 950
Here We Go Again, 592
Hermit's Cave, 729
Hernandez, Juano, 26
Heroines in Bronze, 27
Herpe, Robert, 609
Herrera, Nibaldo, 429
Herrmann, Bernard, 908, 936, 1353, 1478
Herrold, Charles D., 83, 532, **699–700**, 1424
Hertz, Heinrich Rudolph, 653, **700–702**, 874
Herzog, Herta, 855, 1040
Hewitt, Don, 636
Hewitt, Foster, 272
Hewitt's Bookstore, 1077
Hibberd, Stuart, 239
Hickerson, Jay, 970, 1030
Hicks, George, 37, 1019
Hicks, Johnny, 162
Hicks, Muse, Tate, and Furst, 702
Hicks, Tom, **702–3**
Hicks Muse, 1054
Hidden Medium; Educational Radio, 991
Hiett, Helen, 1553
Higgy, Robert, 989
High Fidelity, **704–5**
Hight, Jean, 858
Hightower, Jim, 1370
Hill, Edwin C., 367
Hill, Francis, 183
Hill, George Washington, 11, **706–8**, *707*
Hill, Holliday, 369
Hill, Lewis, **708–9**, 838, 1057
Hillbilly Jamboree, 824
Hilltop House, 1552
Hilmes, Michele, 1291
Hinckley, David, 417
Hindemith, Paul, 498
Hindenburg Disaster, 546, **710–11**
Hinman, Buck, 1567
Hinojosa, Maria, 1123
Hinshaw, Ed, 529
Hires (Rootbeer) Harvesters, 1453
Hispanic Radio, **712–14**
History Behind the Headlines, 1416
History of Rock n Roll, 316
Hitchcock, Alfred, 719
Hitchhiker's Guide to the Galaxy, 218, 496, 1244, 1278
Hite, Katherine, 687
Hit Parade, 592
Hix, George, 1564
Hoaxes, **714–16**
Hockenberry, John, 1004, 1367
Hoffa, Portland, 360, 1554
Hoffman, Hallock, 1058
Hofheintz, Roy, 609
Hogan, John V.L. "Jack," **716–17**, 1565
Hogan, Michael, 217
Hogan and Sanger's Interstate Broadcasting Company, 1566
Hogan's Daughter, 1340
Holden, Jack, 1358
Hole, Tahu, 241
Holiday, Pat, 333
Hollenbeck, Don, 367, 696
Holloway, Sally, 241
Hollywood and Radio, **717–20**
Hollywood Barn Dance, 835
Hollywood Hotel, 495, 719, 819
Hollywood Melody Shop, 835
Hollywood Premiere, 718
Hollywood Review of 1929, 156
Hollywood Star Preview, 495
Hollywood Startime, 718
Holmes, Art, 271
Holmes, Oliver Wendell, 69
Holzmann, Rudolfo, 1037
Home News from Britain, 225
Home Sweet Home, 146
Honesty, Eddie, 26
Hong Kong, 109
Hong Kong Commercial Broadcasting Company Limited, 109
Hood, Raymond, 1161
Hooks, Benjamin L., 28, 577
Hooley, Gemma, 1301
Hooper, C.E., 118, 400, 454, 721–23
Hooperatings, 118, 399–400, **721–23**, 1291
Hoover, Herbert, 541, **723–26**, *725*, 883, 1083, 1440
Hoover, J. Edgar, 1058
Hopalong Cassidy, 1499
Hope, Bob, 363, **726–28**, *727*, 835, 1453
  sponsorship, 11
  WWII radio, 1561
Hop Harrigan, 319, 1341
Hopkins, Arthur, 53
Hopper, Hedda, 719
Hora do Brazil (Hour of Brazil), 194
Horizon FM (Ouagadougou, Burkina Faso), 19
Horlicks, 12
Horn Blows at Midnight, 158
Horror Programs, **729–31**
Hõrspiel, 495, 498
Hottelet, Richard C., **731–33**
Hough, Harold, 1483
Hour of Decision, 557, 1215
Hour of Faith, 1215
Hour of Smiles, 46
Houseboat Hannah, 1338
House by the Side of the Road, 1078
House (Canada), 272
House in the World, 857
Houseman, John
  background, 1074
  Mystery Theater of the Air, 492, 859, 936
  Voice of America and, 1041, 1468
  War of the Worlds, 1478–81
House of Glass, 1338
House of Myths, 1046
House Un-American Activities Committee, 167, 1198
Housewives' Choice, 232
Housewives Protective League, 488, 835
Howard, Clark, 1372
Howard, Eddie, 1079
Howard, Margaret, 241
Howard Stern Show, 276, 293, 745, 1344
Howe, Quincy, 366, **733–35**, *734*
Howell, Robert, 1431
Howie Wing, 1340
How to Write for Radio, 494
How War Came, 1357
Hoyos, Rodolfo, 712
Huckle, Paul. *see* Allen, Fred
Hugenberg, Alfred, 637
Hugh, Mrs. David, 1077
Hughes, Cathy, 28
Hughes, David, 518
Hughes, Floy, 868
Hughes, Howard, 690
Hughes, Langston, 491, 1075
Hulbert, Maurice "Hot Rod," Jr., 27, **735–36**
Hull, Warren, 1473
Hulsen, Albert, 404, 1132
Hum and Strum, 1509
Human Rights Radio, 886
Human Side of the News, 367
Hummert, Frank and Anne, **736–39**, *738*

Betty and Bob and, 55
formats, 1286, 1287
Just Plain Jane, 11
program themes, 896
soap operas and, 493, 1552
Hundt, Reed, 15
Hungarian Uprising, 1159
Hunt, Frazier, 835
Hunt, Marsha, 527
Hunt, Reed, 577
Huntley, Chet, 816
Hurd, Douglas, 229
Hurd, Volney, 415
Hurt, Marlin, 159, 360, 589
Husing, Ted, 1321
Hussey, Marmaduke, 201
Hyde, Douglas, 759
Hyland, Robert F., Jr, 832–33

I, Libertine, 1257
Ibuka, Masaru, 1476
I Can Hear It Now, 634, 968
Iceland, 1237
Icelandic Radio, 1237
Ideas, 273
Idelson, Billy, 1462
IG Farben, 125
I Hear America Listening, 394
Iler, George A., 1569
I Live on Air (Schechter), 1240
I Love a Mystery, 493, 730, **741–42,** 1046
I Love Lucy, 363
Imislund, Clancy, 486
imru, 646
Imus, Don, **742–44,** 743
MSNBC and, 958
style, 1037, 1261
WNBC and, 1543
Imus in the Morning, 1371
in-band, on-channel (IBOC) systems, 155, 460, 468
indecency, 308
Independent Broadcasting Authority (IBA), 227–28, 229
Independent Federation of Free Radios (FIEL), 429
Independent Radio and Television Commission, 760
Independent Television Authority (ITA). *see* Independent Broadcasting Authority
Independent Television Network (ITN), 242
India, 500. *see also* All India Radio
Indian Broadcasting Company, 35
Indian State Broadcasting Service (ISBS), 35
Indian Telegraph Act of 1885, 35
Indicatif présent, 288
Indochina, 106–8
Indonesia, 106
Industrial Development Engineers Association, 1413
Infinity Broadcasting, **745–46**
CBS Radio in, 350
CBS/Westinghouse merger, 684, 1054, 1055
FCC and, 1037, 1261
Howard Stern at, 1342
Information Please
Deems Taylor on, 1375
Clifton Fadiman on, 561
format, 1147, 1148
sponsorship, 708, 1495
Ingersoll, Charles, 970, 1030
Ingram, Dan, 1475
Inner Broadcasting, Inc., 28
Inner Sanctum, 1103
opening signature, 250
Inner Sanctum Mysteries, 729, **746–47**
Inside Radio magazine, 1411
Inside Talk Radio (Laufer), 1370
Inside the News, 816
Inside Track, 273
Institute of Electrical and Electronic Engineers, 1379
Institute of Practitioners in Advertising (IPA), 12
Institute of Radio Engineers, 716, 1379
Inter-American Association of Radiobroadcasters (AIR), 941
Intercity Radio, 580
Intercollegiate Broadcasting System, **748–49**
International Amateur Radio Union (IARU), 689
International Bible Students Association, 270
International Broadcasting Act (Public Law 103-236), 246, 1173, 1178
International Broadcasting Awards, 144
International Broadcasting Bureau (IBB), 246, 1173, 1469
International Broadcasting Company (IBC), 12
International Brotherhood of Electrical Workers, 1380
International Business Machines, 1413
International Evangelism, 626
International Frequency Registration Board (IFRB), 754
International Good Music, 1466
International News Service (INS), 1021, 1095
International Press Service (IPS), 1096
International Radio and Television Society Foundation, 1409
International Radio Broadcasting, **749–53**
International Radio Consultative Committee, 754
International Radio Report, 1411
International Radiotelegraph Convention, 754
International Red Cross, 226
International Silver, 1078
International Telecommunication Union, 457, 629, 631, **753–55**
International Telegraph Union, 753
International Telephone and Telegraph Committee (CCITT), 754
International Telephone Consultative Committee, 754
Internet Radio, **756–58**
audio streaming, 123–24
censorship and, 309
fan magazines, 569
online radio, 1027–28
poetry webcast, 1080
virtual radio and, 1466–68
Interrep, 1332–33
Interrep Radio Store, 1333
Interstate Grocer Company, 814
interviews, research, 114
Interwoven Pair, 360
In the Native State (Stoppard), 218–19
In the Shadow of the Swastika, 498
Into the Night Starring Rick Dees, 438
In Town Tonight, 221
inventory management, 7, 9
Invitation to Reading, 1080
Ionized Yeast, 1462
Ipsos-ASI, 127
Ireland, **758–63**
I Remember, 846
Irving, Larry, 869
Isaacs, Edith, 417
Isaacs, Godfrey, 201
Isay, David, 477, **763–65**
Iskowitz, Isidor. *see* Cantor, Eddie
Islamic Republic of Iran Broadcasting, 102
Island Adventure, 1338
Israel, **765–67**
Israel Today Radio, 1215
Italy, 607, **767–69,** 1294–95
ITAR-Tass, 1020
ITMA (It's That Man Again), 221
It Pays to Be Ignorant, 1149, 1338
It's Higgins, Sir, 1338
It's the Navy, 1075
It's What's Happening, Baby, 964
It's Your Nickel, 1142
Ives, Raymond, 31
Ivory soap, 1094, 1149
Ivory Tower, 1034
I was a Communist for the FBI, 494

Jack Armstrong, the All-American Boy, **771–73**
Ameche on, 55
on NBC, 181
premiums, 1094
stereotypes, 1340
theme, 318, 319
violence, 1464
Jack Benny Program, 170, 1454. *see also* Benny, Jack
advertising on, 11
cost, 708
feud with Fred Allen, 48
films, 719

format, 360, 362
"Rochester" on, 27
TV and, 158
Jack Benny Show, 170, 1454
Jack L. Cooper Presentations, 397
Jack Masla Company, 1332
Jackpot, 1149
Jack R. Howard Awards, 144
Jackson, Eddie, 509
Jackson, Eugene D., 173
Jackson, Hal, 26
Jackson, Jack, 233
Jackson, Michael, 1371
Jack Sterling Show, 1488
Jaco, Charles, 833
Jacobellis v Ohio (1964), 1036
Jacobs, David, 232
Jacobs, George, 774
Jacobs, Ron, 73, 486, 801, 819
Jacor Communications, 341, 1372, 1467
Jake and the Fat Man, 380
JAM, 789
Jamboree, 1576
James, Bill, *686*
Jameson, House, 31
Jam for Supper, 824
Jamming, **773–75**
Jammin Live, 163
Jane Ace, Disc Jockey, 523
Janet Hardy in Radio City (Wheeler), 317
"Janitor" commercial, 177
Janssen, Werner, 526
Japan, 499, **775–78,** 1186
Japan Radio Network (JRN), 777
Jarrett, Hugh, 1536
Jarvis, Al, 471
Jaxon, Frankie "Half Pint," 25
Jazz Alive, 779, 1002
Jazz Format, **779–80**
Jean Shepherd Show, 1256
Jefferson-Pilot Broadcasting Company, 425, 1486
Jeffrey, R.E., 217
Jehovah's Witnesses and Radio, 270, **780–81**
Jelke, 1462
Jell-O, 11, 31
Jell-O Program Starring Jack Benny, 156
Jeno's Pizza Rolls, 620
Jepko, Herb, **782–84,** *783*
Jergens Journal, 1526
Je vais et je viens entre les mots, 281
Jewell, James, 878, 1349
Jewett, Frank B., 154
Jewett Radio and Phonographic Company, 1534
Jewish Community Hour, 786
Jewish Giant, 764
Jewish Hour, 786, 1506
Jewish Radio Programs in the United States, **784–87**
Jimmy Allen, 1340
Jimmy Durante-Garry Moore Show, 509
Jimmy Durante Show, 55
Jimmy Got His gun, 1034
Jimmy Wakely Trio, 142
Jim Rome Show, 1223
Jingles, **788–89**
JM in the AM, 786
Joan Rivers Show, 1557
Joe Penner Show, 361
Joe Pyne Show, 1142
John Henry, Black River Boat Giant, 26, 170
John Henry Faulk Show, 574
Johnny Home, 272
Johnny Mann Singers, 176
Johnny's Front Porch, 574
Johns-Manville Company, 1226
Johnson, Bruce, 315
Johnson, Dave, 324
Johnson, Edward, 939
Johnson, Irene, 1554
Johnson, Johanna, 858
Johnson, Lawrence, 167, 168
Johnson, Lyndon B., 1126
Johnson, Nicholas, 28
Johnson, Parks, 1472
Johnson, Perri, 1446
Johnson, Raymond Edward, 747
Johnson's Wax, 591, 1078
John's Other Wife, 250
Johnston, Brian, 220
Johnstone, Bill, 1251
Johnstone, Jack, 1589
John XXIII, Pope, *1457*
Joint Radio Information Facility, 1433
Jolly Bill and Jane, 957
Jolson, Al, 351, 353, 1459
Jome, Hiram, 530
Jones, Billy, 359, 788
Jones, Candy, 783
Jones, Clarence, 557
Jones, Dickie, 31
Jones, Jesse, III, 844
Jones, Jesse H., 843
Jones, John T., 844
Jones, LeAlan, 764
Jones, Quincy, 466
Jones Radio Networks, 316, 1372
Jordan, Jim, 362, 589, 1286
Jordan, Marian, 362, 1286, 1554
Jordan, Michael, 1503
Jory, Victor, 1251
Joseph, Mike, 381, 386
Josephson, Larry, 627, 628
Journal of Broadcasting, 243
Journal of Broadcasting and Electronic Media, 243, 244
Journal of Radio Studies, 243, 244, 531
Journal of the Audio Engineering Society, 1412
Journal of the Telegraph, 1410
Joyce, William. *see* Lord Haw-Haw
Joyner, Tom, 29, **790–91,** *791,* 1447
Judge for Yourself, 49
Judis, Bernice, 175, **792–93,** 1543
Judson, Arthur, 72, 1556
Judy Canova Show, 1338, 1341
Jugendradio DT64 (Youth Radio DT64), 657
Jugoslavenska Radiotelevizija (JRT), 306
Julian, Joseph, 494
Junior Miss, 31, 1341
Just Concerts, 274
Just Plain Bill, 11, 737, 1287
Juvenile Jury, 1122

KABC (Hollywood/Los Angeles, California), 58, 255, 786, 1369
KABL (San Francisco, California), 524, 924–25
KACE (Los Angeles, California), 1466
KADS (Los Angeles, California), 925
Kael, Pauline, 1057
Kahn, Leonard, 1335
KAJA (San Antonio, Texas), 341
KAKC (Tulsa, Oklahoma), 315
Kaleidoscope, 1535
KALE (Portland, Oregon), 481
KALI (Salt Lake City, Utah), 730
Kallinger, Paul, 190
Kaltenbach, Frederick Wilhelm, 498
Kaltenborn, H.V., **795–97,** *796*
on Churchill, 326
commentary by, 355, 365, 1018, 1120
NYC's Air College and, 1545
Kaltenmeyer's Kindergarten, 1338
Kamarak, Rudolph, 549
Kamin, Jeff, 1001
Kaneo Naru Oka (Hill without a Bell), 777
Kantako, Mbanna, 886
Karloff, Boris, 747
Karmazin, Mel, 745, **798–99,** 1054, 1503
Karpoff, David, 632
Kasem, Casey, 73, 74, **799–802,** *800,* 921
Kate Smith and Her Swanee Music, 1282
Kate Smith A&P Bandwagon, 1282
Kate Smith Hour, 718, 1282, 1553
Kate Smith Sings, 1282
Kate Smith Speaks, 1282
Kate Smith's Swanee Review, 1282
Kathy and Judy Show, 1510
Katie Joplin, 1386
KATL (Miles City, Montana), 256
KATO (Reno, Nevada), 481
Katz, Elihu, 855
Katz, Emanuel, 1331
Katz, Eugene, 1331
Katz, Mickey, 786
Katz Agency, 679, 1331
Katz AID Plan, 679
Katz Communications, 679
Katz Media Corporation, 679
Katz Radio Probe Research System, 679
KATZ (St. Louis, Missouri), 1446
Kauff, Peter, 820
KAXE-FM (Grand Rapids, Minnesota), 830
Kay, Lambdin, 1065

Kaye, Sidney M., 245
Kay Kyser's College of Musical Knowledge, 413, 848, 1148
Kay Kyser's Kampus Class, 848
Kazakhstan, 103
Kazakh State Television and Radio Broadcasting Company (Kazakhstan), 103
KBIG (San Francisco, California), 584
KBOO (Portland, Oregon), 375
KBOY (Medford, Oregon), 256
KCAT (Pine Bluff, Arkansas), 256
KCBS/KQW (San Jose, California), 177, **802–3**, 1403, 1425
KCHU (Dallas, Texas), 375
KCIJ (Shreveport, Louisiana), 1549
KCMO (Kansas City, Missouri), **804**
KCNA (Tucson, Arizona), 481
KCRO (Omaha, Nebraska), 1284
KCSB (Santa Barbara, California), 1223
KCST (San Diego, California), 1345
KDAY (Los Angeles, California), 1446, 1549
KDEO (El Cajon, California), 801
KDIA (San Francisco/Oakland, California), 1446
KDKA (Pittsburgh, Pennsylvania), **804–6**
- AT&T chain, 71
- baseball on, 1321
- beginnings, 683, 1425
- election returns, 1920, 37, 540
- first broadcast, 1196
- Frank Conrad and, 378
- Lowell Thomas on, 1396
- news on, 568
- poetry on, 1077
- religious programs, 1211
- status as oldest station, 803
- Symphonium Serenaders, 25
- Vatican reception of, 114
- WEAF and, 1494
- Will Rogers on, 1221

KDNA (St. Louis, Missouri), 375
KDRU (Dinuba, California), 1214
KDRY (Alamo Heights, Texas), 256
KDWB (Minneapolis, Minnesota), 176
KDYL (Salt Lake City, Utah), 782
Keating, Jack, 1528
KECA-AM. *see* KABC-AM (Hollywood, CA)
Keefe, Anne, 833
Keenan, John G., 1198
Keep 'Em Rolling, 562
Keep Happy Club, 824
KEEZ (San Antonio, Texas), 341
Kefauver crime hearings, 1432
KEGL (Fort Worth, Texas), 256
Keighley, William, 892
Keillor, Garrison, **806–9**, *807*
- on KSJR, 947
- move to Denmark, 45
- poetry readings, 1081
- *Prairie Home Companion,* 1092–93
- style, 364

Keith, Mansel, 1465
Keith, Michael C., 1104
Kelk, Jackie, 31
Keller, Arthur C, 704
Kellogg, 13, 319, 1198, 1243, 1492
Kellow, Christine L., 1464
Kelly, Joe, 320
Kelly, Shawn, 584
KELP (El Paso, Texas), 176
KELW (Burbank, California), 712
Kemp, Hal, 844, 848, 1486
KEND (Roswell, New Mexico), 256
Kennard, William, 29, 577, 886, 1384
Kennedy, John, 1084, 1441
Kennedy, Paul, 273
Kennelly–Heavyside layer, 681
Kenny and Cash Show, 559
Kent, A. Atwater, **809–11**, *810*
Kentucky Club Tobacco, 1473
Kentucky Minstrels (BBC), 221
Kenya, 20
Kenyon, Peter, 1004
Kenyon & Eckhardt agency, 1043
Kernis, Jay, 537
Kershaw, Andy, 219
Kesten, Paul, **812–13**
KEWB (San Francisco, California), 176, 801, 952, 1052
Key, Lambdin, 1570
KFAI-FM (Minneapolis/St. Paul, Minnesota), 830
KFAX (San Francisco, California), 38, 1019
KFBI (Abilene, Kansas), 691
KFBK (Sacramento, California), 921
KFCC (Bay City, Texas), 578
KFC (Seattle, Washington), 25
KFFA (Helena, Arkansas), 27, 183, **814–15**, 1567
KFI (Los Angeles, California), **815–17**, 1274
KFKB (Kansas), 199, 581
KFKX (Hastings, Nebraska), 805
KFOG (San Francisco, California), 256
KFON (Long Beach, California), 1077
KFOX (Long Beach, California), 488
KFPT. *see* KSL (Salt Lake City, Utah)
KFRC (San Francisco, California), 26, 481
KFRE (Fresno, California), 315
KFSG (Los Angeles, California), 556, 931, 933, 1212
KFSO (San Francisco, California), 1370
KFUO (St. Louis, Missouri), 556, 1211
KFWB (Los Angeles, California), 176, 684, 718, 1052
KFWM (Oakland, California), 780
KFXX (Hastings, Nebraska), 683
KGAY (Denver, Colorado), 647
KGB (San Diego, California), 315, 481, 485, 819
KGBS (Los Angeles, California), 920
KGC (Los Angeles, California), 834
KGEE (Bakersfield, California), 136
KGEI (San Francisco, California), 94, 571
KGFJ (Los Angeles, California), 1446
KGKO (Wichita, Kansas), 1483
KGO Players, 817
KGO (San Francisco, California), 58, 506, **817–18**, 1046, 1369
KGRI (Henderson, Texas), 1067
KHJ History of Rock and Roll, 953
KHJ (Los Angeles, California), 315, 438, 472, 481, 485, **818–20**, 931, 952
Khmer Republic, 107–8
KHMO (Hannibal, Missouri), 178
KHON (Honolulu, Hawaii), 481
KHOW-AM (Tampa, Florida), 937
KICK (Palmyra, Missouri), 256
KICY (Nome, Alaska), 256
Kid Millions, 294
Kids America, 1168
Kid's Choice Broadcasting network, 321
Kid's Corner, 320
KidStar Radio, 323
Kids Voting USA, 321
Kiernan, John, 561, 1147
Kierulff, C.R., 818
KIEV (Geldale, California), 178
KIIS (Los Angeles, California), 177, 438
KIKX-AM (Tucson, Arizona), 15
Kilgallen, Dorothy, 1553
KILI-FM (Porcupine, South Dakota), 375, 1011
KILT (Houston, Texas), 1052
Kiminonawa (What's Your Name?), 777
KIMN (Denver, Colorado), 1052
Kinard, J. Spencer, 954
8 KIN (Australia), 134
Kincaid, Bradley, 995, 1487
Kinetic City Super Crew, 321
King, B.B., 814
King, Brian, 219
King, Helen, 1551
King, Larry, 41, **821–23**, *822*, 1361
King, Loyal, 840
King, Mackenzie, 259
King, Martin Luther, Jr., 28, 171
King, Nelson, **824–25**
King, Wayne, 1542
King Biscuit Flower Hour, **820–21**, 1361
King Biscuit Time, 27, 183, 814, 820, 1567
Kingdom of Tonga, 1310
KING (Seattle, Washington), 26
King's Henchman, 1375
Kingsley, Bob, 921
Kintner, Robert, 57
KIOI (San Francisco, California), 256
KIQQ-FM (Los Angeles, California), 316, 487
Kirby, Edward M., 413, **825–27**
Kiribati, 1310
Kirkpatrick, Theodore, 1198
Kiss FM, 1239
Kita, Dennis, 1301
KITT (San Diego, California), 609
Kitty Foyle, 1339

KIXL (Dallas, Texas), 524
KJAZ (San Francisco, California), 779
KJOI (Stockton, California), 742
KJQY-FM (San Diego, California), 684
KJUL-AM (Las Vegas, Nevada), 1253
KKDA-AM (Dallas, Texas), 790
KKFI-FM (Kansas City, Missouri), 647
KKGO (Los Angeles, California), 779
KKLA-FM (Los Angeles, California), 933
KKOB. *see* KOB (Las Cruces/Albuquerque, New Mexico)
KKTK (Waco, Texas), 256
KLAC-AM (Los Angeles, California), 471, 937, 1053, 1142, 1369
Klapper, Joseph, 855, 1040, 1326
Klauber, Edward A., 351, 409, 812, **827–29**, *828*
KLIF (Dallas, Texas)
  baseball on, 860, 922, 1274
  jingles on, 788
  McLendon and, 924
  Top 40 format on, 1197
Kling, William, **829–31**, 947, 1004, 1130, 1133
KLON-FM (Long Beach, California), 779
Klose, Kevin, 1005
KLSX (Los Angeles, California), 1343
Kluge, John W., 745, 937, 1230
Klugh, Paul B., 987
KMAK (Fresno, California), 314, 485, 952
KMAN (Manhattan, Kansas), 256
KMA (Shenandoah, Iowa), 1052
KMBY-AM (Monterey, California), 952
KMEO-AM (Phoenix, Arizona), 684
KMET-FM (Los Angeles, California), 482
KMLE (Chandler, Arizona), 256
KMOO (Mineola, Texas), 256
KMOX (St. Louis, Missouri), **832–34**, 1380
KMPC (Beverly Hills/Hollywood/Los Angeles, California), 142, 175, 380, 570, 1052
KMPX-FM (San Francisco, California), 472, 479, 480, 627, 1253
  format change, 1419
KNBC. *see* KPO (San Francisco, California)
KNBR. *see* KPO (San Francisco, California)
KNET (Palestine, Texas), 922
KNEW (Oakland, California), 931
KNEW (Spokane, Washington), 481
Knight, Raymond, 415
Knight and Daye, 1386
KNOB-FM (Los Angeles, California), 779
KNOLL (San Francisco, California), 1446
Know Your Ally, 1564
Know Your Enemy, 1564
KNX (Los Angeles, California), 488, **834–35**
KOA (Denver, Colorado), **835–36**
KOA Staff Orchestra, 836
KOAX-FM (Dallas, Texas), 684
Kobak, Edgar, 983
KOBH (Rapid City, Iowa), 505
KOB (Las Cruces/Albuquerque, New Mexico), **837–38**
Koch, Chris, 476, 1482
Koch, Howard, 492, 1074, 1478
KODA-FM (Houston, Texas), 684
KODK (Kodiak, Alaska), 94
Koehler, Ted, 1196
Kohn, Melvin, 1290
KOHW (Omaha, Nebraska), 1274
KOIL (Omaha, Nebraska), 1052, 1347
Kollmar, Richard, 1553
Kol Yisrael (Voice of Israel), 765
Kolynos toothpaste, 11
KOMA (Oklahoma City, Oklahoma), 691, 1317, 1348
Kondolf, George, 1392
KOOL (Phoenix, Arizona), 142, 481, 1044
Korea, 109, 499–500
Korean Broadcasting System (KBS), 109
Korean War, 97, 1198
Kornheiser, Tony, 60
KORN (Mitchell, South Dakota), 1052
KOSI-FM (Denver, Colorado), 684
"Kosmos" radio, 677
Kostelanetz, Andre, 976
KOTA (Rapid City, Iowa), 505
KOTO (Telluride, Colorado), 375
KOWH (Omaha, Nebraska), 924, 1346
KPAL (Little Rock, Arkansas), 1168
KPCC-FM (Pasadena, California), 831, 948
KPEN (San Bernardino, California), 485
KPFA (Berkley, California), 627, 645, 708, **838–40**, 946, 1057, 1059, 1080
KPFK-FM (Los Angeles, California), 627, 645, 1057, 1058
KPFT (Houston, Texas), 1057, 1058
KPHO-AM (Phoenix, Arizona), 142
KPIG (Freedom, California), 256, 757
KPO (San Francisco, California), 817
KPOW (Powell, Wyoming), 256
KPPC (Los Angeles, California), 480, 482, 1420
KPRC (Houston, Texas), 1331
KPRS (Kansas City, Missouri), 28
KPRZ (Los Angeles, California), 1053
KQV-AM (Pittsburgh, Pennsylvania), 58
KQW. *see* KCBS/KQW (San Jose, California)
KQXT-FM (San Antonio, Texas), 684
KQZY-FM (Dallas, Texas), 684
KRAB Nebula, 535
KRAB (Seattle, Washington), 375
Kraft Foods, 56, 1454
Kraft Music Hall
  Bing Crosby and, 418
  Deems Taylor on, 1375
  Don Wilson on, 1525
  film stars on, 719
  live shows, 126
  production, 11
KRAK (Hesperia, California), 256
Kramer, Mandel, 1589
KRAY (Sitka, Alaska). *see* KRB (Sitka, Alaska)
KRB (Sitka, Alaska), 94
KREB, 1549
Kreer, Henry B. "Pete," 330
Kreymborg, Alfred, 1079
Kristiania Broadcasting, 1237
KRKD (Los Angeles, California), 933
KRLA (Los Angeles, California), 163, 164, 176, 801, **840–41**
KRLD (Dallas, Texas), 162
Kroll, Woodrow, 557
KROQ (Los Angeles, California), 715
KROW (Oakland, California), 539
KROY-AM (Sacramento, California), 952
KRSI-AM (St. Louis Park, Minnesota), 830
KRTH (Los Angeles, California), 487, 1044
Krud Radio, 569
KRWG (Las Cruces, New Mexico), 838
KRZA (Alamosa, New Mexico), 375
KSAN (San Francisco, California), 480, 584, 937, 1124, 1420
KSA (St. Louis, Missouri), 1439
KSCA-FM (Los Angeles, California), 142
KSD (St. Louis, Missouri), 308
KSFO (San Francisco, California), 488
KSHE (St. Louis, Missouri), 548, 715
KSIJ (Texas), 1067
KSJN (St. Paul, Minnesota), 806, 808
KSJR (Collegeville, Minnesota), 806, 829–30, 947
KSKY-AM (Dallas, Texas), 190
KSL (Salt Lake City, Utah), 256, 529, 782, **842–43**
KSLX-FM (Scottsdale, Arizona), 715
KSLX (Los Angeles, California), 1372
KSPC (Boise, Idaho), 256
KSTP (St. Paul, Minnesota), 256
KTAB (Oakland, California), 539
KTAR (Phoenix, Arizona), 1024
KTCK (Dallas, Texas), 1569
KTEE-AM (Carmel, California), 952
KTHS (Hot Springs, Arkansas), 888, 1483
KTIM (San Rafael, California), 920
KTKK (Salt Lake City, Utah), 783
KTLA (Los Angeles, California), 142, 1241
KTMS (Santa Barbara, California), 1223
KTRH (Houston, Texas), **843–45**, 1472
KTSA (San Antonio, Texas), 1052
KTTV (Los Angeles, California), 1142
KTWV (Houston, Texas), 684
Kubelsky, Benjamin. *see* Benny, Jack
Kunsky, John, 1577
KUOM (Minneapolis/St. Paul, Minnesota), 829
Kuralt, Charles, **845–46**, 1486
Kureishi, Hanif, 497
KUSC (Los Angeles, California), 1134, 1328
KVOO (Tulsa, Oklahoma), 140, 691, 782
KWE-FM (Belleville, Ontario), 277
KWK-AM (St. Louis, Missouri), 790

KWKH (Shreveport, Louisiana), 671, **846–47**, 881, 1067
KWY (Chicago, Illinois), 718
KXIV (Phoenix, Arizona), 782
KXLA (Pasadena, California), 411
KXOA (Sacramento, California), 742
KXOK (St. Louis, Missouri), 691, 1348
KXXL (Denver, Colorado), 1045
KYA (San Francisco, California), 479, 485
KYLD (San Francisco, California), 1253
KYNO (Fresno, California), 314, 485, 819
Kyser, Kay, **848–49**, *849*, 1148, 1486
KYSO (Alexandria, Louisiana), 163
KYW-AM-TV (Cleveland, Ohio), 918
KYW (Chicago, Cleveland, and Philadelphia), 39, 338, 471, 683, **850–51**, 987, 1019
  Clarence Jones and His Wonder Orchestra, 25
  news format, 38
  "Shuffle Along," 25
  WEAF and, 1494
KZAS (Manila, Philippines), 570
KZJZ (St. Louis, Missouri), 779
KZN. *see* KSL (Salt Lake City, Utah)
KZZZ (Bullhead City, Arizona), 255

L. Bamberger and Company Department Store, 1556
L.A. Radio Guide, 569
Laboe, Art, 840
Labor for Victory, 1075, 1076
LaCurto, James, 1248, *1250*
La Demi-heure théâtrale du Docteur J.O. Lambert, 280
Laemmle, Carl, 718
La Fabuliste La Fontaine à Montréal, 281
La Feuillaison, 281
La Historia de Quien Soy (Story of Who I Am), 567
Laine, Bob, 324
Lalonde, Louis-Philippe, 331
Lambda Report, 647
Lamour, Dorothy, 526, *527*
Lampell, Millard, 1075
Lamprecht v FCC, 15
Landell de Moura, Father Roberto, 193, 519, **853–54**
Land of the Lost, 1341
Landon, Alfred, 1084
Landrith, John, 92
Lane, Minetta, 1046
Langmuir, Irving, 649
Langner, Lawrence, 1392
Languirand, Jacques, 282
Lanny Ross Orchestra, 1588
Lansman, Jeremy, 375, 1215
Laos, Peoples Republic of, 107
La Palina cigars, 1060
La Palina Smoker, 350
Lardner, Ring, 416
Lareux, Mike, 1242
La Rosa, Julius, 662
Larry King Show, 822
Larsen, Rob Edward, 1196
Lasker, Albert, 11, 706
Lasswell, Harold, 413
Last Word, 1223
Late Junction, 219
Latimer, Lewis H., 25
Latino USA, 1123
Latouche, John, 475
Lauck, Chester, 361, 887, *888*
Laufer, Peter, 1370
Laughter, Victor, 25
Laurents, Arthur, 1076
Laval, Pierre, 1179
La voix du Vietnam (Voice of Vietnam), 107
Lavoris mouthwash, 522
La Voz de América Latina (voice of Latin America), 713
Law, Bob, 174
Lawley, Sue, 450
Law of Radio Communication (Davis), 530
Lawrence, T.E., 1395
Lawton, Sherman, 530
Laxton, Fred M., 1485
layering, 1299
Lazareff, Pierre, 61
Lazarsfeld, Paul F., **854–56**, 1039, 1140, 1326
Lazarsfeld–Stanton Program Analyzer, 1040
LBJS Broadcasting Company, 255
Lea Act, 65
League of Political Education, 76
League of Women Voters, 1083
League of Women Voters of California, Federal Communications Commission v (1984), 564, 598
Lear, William, 138
Leary, Chris, 322
Leave It to Beaver, 1386
Leave It to the Girls, 1122, 1553
Leaving(s) Project, 495
Lebanon, 88, 97
Lecous, Tex, 288–89
Lee, Don, 481–82
Lee, Robert E., 577
Leeper, Hattie "Chattie," 1554
Le festival de l'humour, 288
Légaré, Olivier, 287
Legislative Appropriations Act of 1956, 1433
LeGoff, Jack, 1063
LeGrand, Dick, 589
Lemke, William, 410, 1084
Lenin, Vladimir, 1227–28
Lennon, Florence Becker, 1081
Lennon and Mitchell agency, 812, 827
Le Petit théâtre de poche, 281
Le Radiothéâtre de Radio-Canada, 281
Les Amours de Ti-Jos et les mémoires de Max Potvin, 287
Lesbian Radio. *see* Gay and Lesbian Radio
Les Crane Life from the Hungry I, 818
Les insolences d'un téléphone, 288
Leslie, Phil, 589
Les nouvelles de chez-nous (Local News), 267, 331
Lesueur, Larry, 352
Le Théâtre de chez-nous, 280
Letondal, Henri, 280, 331
Let's Draw, 1511
Let's Pretend, 320, **857–58**
Let's Sing, 1511
Letter From America, 221, 242, 392
Letters to Laugh-In, 1053
Letter to Daniel, 242
Levant, Oscar, 561, 1147
Lever Brothers, 13, 728
Levi, Harry, 784, 785, 1509
Lewis, Al, 1049
Lewis, Delano, 45, 1004, 1131
Lewis, Draper, 1080
Lewis, Elliot, 362
Lewis, Fulton, Jr., 367, 694
Lewis, Jerry, 49
Lewis, Sheldon, 173
Lewis, Thomas H., 95
Lewis, Tom, 866
Lewis, William B., 405, 474–75, 492, **859–60**, 935
L'heure catholique, 331
L'heure provinciale, 267, 280, 331
L'heure universitaire, 331
Liasson, Mara, 1004
Liberty at the Crossroads, 1084
Liberty Broadcasting System, 72, **860–61**, 922, 1313, 1428
Library of American Broadcasting, 970–71
Library of Congress, 1030
Licensing, **862–62**, 1383
  FCC decisions, 176
Lieberman, Joseph, 1465
Life and Casualty Insurance Company of Tennessee, 1536, 1537
Life and Music of George Gershwin, 394
Life Begins at 80, 1122
Life Can Be Beautiful, 1338
Life of Riley, 1339, 1341
Lifetime Reading Plan, 562
Life with Luigi, 1338, 1339
Light and Life Hour, 1214
Lightning Action, 498
Lights Out, **864–65**
  creation, 729
  Oboler and, 1033, 1075
  revival, 730
  themes, 1243
Lighty, William, 1511
Limbaugh, Rush, **866–67**, *867*
  EFM and, 1372
  McLaughlin and, 921
  politics and, 1085
  style, 694
  syndication, 1361
  WABC and, 1476
Limbaugh Letter, 1372

Lindsay, James Bowman, 517
Linit Bath Club Review, 46
Linkletter, Art, 1149
T.J. Lipton, Inc, 1473
Lipton's Tea, 663, 1492
Liquid Barretter, 427
listener demographics, 8
listeners per dollar, 9
listening mentions, 116
Listen to the Witness, 1553
Lit, Hy, 1446
Literary Digest, 37
Little, Mary, 416
Little Folks Magazine, 1508
Little Italy, 250, 1338
Little Orphan Annie, 319, 493, **868–69,** 1094, 1340, 1464
Little Red Schoolhouse, 534
Livek, William, 165
Living on Earth, 1123
Livingston, Mary, 360, *1061,* 1554
Ljungh, Esse W., 283, 284
Lloyd, Christopher, 1386
Localism in Radio, **869–72**
Local Marketing Agreements, **872–73**
Lockwood, Robert, Jr., 27, 183
Lodge, Oliver J., **874–76**
Lodge, Tom, 238
"A Logic Named Joe," 1243
Lohman, Al, 816
Lomax, Alan, 474
Lomax, John, 474
London Broadcasting Company (LBC), 228, **876–77**
London International Radiotelegraphic Convention, 255
London Letter, 392
London News Radio (LNR), 876–77
London Press Exchange, 1175
London Talkback Radio, 876
Lone Journey, 1552
Lonely Women, 493, 1072
Lone Ranger, **878–79**
  acquisition by WXYZ-AM, 56
  development, 1499
  Fran Striker and, 1349, 1350
  sponsorship, 57
  stereotypes on, 1341
  syndication, 1360, 1578
  TV show, 1501
Lone Star Jamboree, 162
Long, Huey, 410, **879–81,** *880*
Longines Symphonette, 338
Long Lines, 73
Look Who's Laughing, 592
Loomis, Henry, 1002
Loomis, Mahlon, 517
Lopez, Vincent, 1542
Lord, Phillips H., 643, **883–85**
Lord and Thomas agency, 11, 706, 1043, 1070
Lord Haw-Haw (William Joyce), 334, **882–83,** 1114–15, 1176
Lorenz, George "Hound Dog," 163
C. Lorenz Company, 1533
Lorenzo Jones, 1340
Lorenz Stahltonmachine, 1533
Lorre, Peter, 730
Los Angeles Times, 355
Los Madrugadores (Early Risers), 712
Los Medios y mercados de Latino America (Media and Markets in Latin America), 1303, 1304
Lost and Found Sound, 1514
loudness, definition, 120
Louis, Joe, 1320
Louisiana Hayride, 847, 1067
Love, Gordon, 290
Lovecraft, H.P., 730
Love in Bloom, 156
Love Is a Many Splendored Thing, 1288
Love of Life, 1288
Love Story Drama, 1249
Love Zone, 1447
Lowell Thomas and the News, 37, 1396
Low-Power Radio/Microradio, 309, **885–87,** 946
"LSMFT" (Lucky Strike Means Fine Tobacco), 706, 1495
Lucas, George, 894
Lucas, Rupert, 283, 284
Lucasfilm, Limited, 1328
Luce, Henry R., 898
Lucent Digital Radio (LDR), 461
Lucent Technology, 1, 154
Lucie Arnaz Show, 1386
Lucky Strike, 1495, 1587
Lucky Strike Dance Orchestra, 11
Lucky Strike Hit Parade, 835
Lucky Strike Hour, 706, 1426, 1526
Lujack, Larry, 58, 1358
Luke Slaughter of Tombstone, 1500
Lum 'n' Abner, 361, **887–90,** 1277
Lunchpail, 1247
Lund, John, 1589
Lund, Robert L., 62
Lundy, Ron, 1475
Lurtsema, Robert, 339
Lust for Life, 407
Lutheran Church v Federal Communications Commission (1998), 14
Lutheran Hour, 556, 557, 1212, 1214
Lutton, Dorothy, 269
Lux Radio Theater, 55, 129, **890–92,** 1498
Lyman, Henry, 1081
Lyn, Mary, 789
Lynne Harvey Radio Center, 971
Lyon, Peter, 1075
Lyric FM (Ireland), 760
Lyric Ohio, 1080

Macfadden Bernarr, 1556
MacDonald, J. Fred, 360, 1288
MacDonald, Trevor, 242
Macdonnell, Norman, 685
MacDougall, Georgina, 110
MacDougall, Ranald, 1075
Macer-Wright, Philip, 240
Macfarlane, W.E., 982
MacGrath, Raymond, 206
MacGregor, Art, 286
MacGregor, C.P., 94
Macgregor, Sue, 241
Machines, 203
MacInnis, Marion. *see* Armstrong, Marion
Mack, Bill, 1484
Mack, Nila, 320, 857
Mackenzie, Compton, 220
MacLeish, Archibald
  Fall of the City, 492, 859, 1074, 1079
  tribute, 1081
  verse plays, 1077
  Voice of America and, 1041, 1469
MacMillan Arctic Expedition, 1593
MacNeice, Louis, 218
R.H. Macy and company, 1556
Madden, John, 1328
Maddy, Joseph E., 65
Madsen, Arch, 782
Magazines. *see* Fan Magazines; Trade Press
Magliozzi, Ray, *301,* 364. *see also Car Talk*
Magliozzi, Tom, *301,* 364. *see also Car Talk*
Magnavox Corporation, 1241, 1335
Magnetophone, 125
Mahaweli Radio (Sri Lanka), 104
Maier, Walter A., 557, 1211, 1212, 1214
Mail Call, 1564
Maill, Leonard, 241
Mainelli, John, 1371
Mainly about Manhattan, 394
Main Street, 250
Maisie, 719, 1554
Maison de la Radio (House of Radio), 616
Majestic Theater Hour, 1337
Major Bowes' Original Amateur Hour, 351, 1366, 1453
Major Electronics Corporation, 550
Major League Baseball Notebook, 173
Major Market Radio, 1333
Make Believe Ballroom. *see* Block, Martin
Malawi, 18
Malaya/Malaysia, 105
Maldives, 104
Malesky, Bob, 1002
Malone, Ted, 1077, 1078
Man Behind the Gun, 352, 1075
Man Born to be King (Sayers), 218
Mandel, William, 1058
Man in the Street, 824
Mankiewicz, Frank, 537, **893–95,** *894,* 1002, 1134
Mann, Johnny, 176
Man Who Went to War, 491
Ma Perkins, 11, **896–97,** 1287, 1540, 1552
Maple Leaf Mailbag, 274

March of Time, **897–900**
  format of, 474
  Fred Smith and, 1539
  Kenneth Delmar and, 443
  Peter Lyons and, 1075
  ratings, 56
  recreation of events, 1196
  sound effects, 1299
  sponsorship, 11
  start of, 351
Marconi, Guglielmo, **900–904**, *901*
Marconi Company, 71
Marconi Radio Awards, 144
Marconi Wireless and Telegraph Company, 1233
Marconi Wireless Telegraph Company of America v United States (1943), 1391
Marian Theater, 567
Marie, the Little French Princess, 250
Marine Offences Act, 238
Market, **904–6**, 1303
market censorship, 307
Marketplace, **906–8**, 1131
Market Place (Canada), 291
market research, 1–2, 7, 127–28. *see also* Audience
Market Statistics, Inc, 160
Markle, Fletcher, **908–10**
Markle Foundation, 1168
Marks, Sadie, 156
Mark Warnow Orchestra, 1588
Marr, Tom, 1557
Marshall, George C., 1469
Marshall, Herbert, 719
Marshall, Pluria, 28
Martha Deane, 1553, 1556
Martin, 1385
Martin, Dean, 49
Martin, Halloween, 471, 583
Martin Kane, Private Eye, 493, 1338
Marx, Chico, 601–2
Marx, Groucho, 353, 601–2, 1149, *1585*
  You Bet Your Life, 1584–86
Marx, Karl, 1290
Marx, Steve, 9
Marx of Time, 602
Mary Noble, Backstage Wife, 1552
Mary Pickford and Buddy Rogers, 1454
Mary Small Review, 1454
Mason, Edward J., 93
Masquerade, 1072
Massachusetts, Memoirs v (1966), 1036
Mass Communication Research, 855
Massett, Larry, 1301
Massey Commission, 259
Massey Report, 281
Masterpiece Radio Theater, 1003
Mather and Crowther Limited, 12
Matheson, Hilda, 240
Mathews, R.H.G., 916
Matrix (magazine), 110–11
Matthew, Brian, 233
Matthews, Cleve, 43
Matthews, Grace, 1251
Matthews, Ralph H.G., 1593
Mattola, Tony, 1080
Maudie's Diary, 31, 1339
"Maurice Cole Quarter of an Hour Show," 558
Maxim, Hiram Percy, 689, 1380
Maxwell, James Clerk, **910–12**
Maxwell House Coffee Time, 652, 1553
Maxwell House Showboat, 718, 996
Mayak (beacon), 1229
Mayak jamming, 774
Mayer, G. Val, 206
Mayflower Decision, 528–29, **912–13**, 1214
Maynard, Dave, 1487
Mays, L. Lowry, 341
McAdorey, Bob, 324
McArthur, Daniel, 290
McArthur, J.D., 1242
McBride, Mary Margaret, **913–15**, *914*
  career start, 1556
  poetry and, 1080
  talk format, 719, 1552–53
McCain, John (R-AZ), 16
McCall, C.W., 330
McCambridge, Mercedes, 908, 1047
McCann, Alfred W., 1557
McCann-Erickson agency, 601
McCann family, 1557
McCann Pure Food Hour, 1557
McCanns at Home, 1557
McCarthy, Charles, 803
McCarthy, Charlie, 11, 527
McCarthy, Eugene, 1085
McCarthy, Joseph, 166–69, 437, 694, 1469
McCarthy, J.P., 1535
McCarthy-Ferguson Group, 983
McCarty, H.B. "Mac," 1511
McCaw, J. Elroy, 1528
McClendon, Rose, 26
McCombs, B.J. "Red," 341
McCord, Bob, 312
McCord, Charles, 743
McCormick, Robert, 1510
McCosker, Alfred, 409
McCoy, Sid, 173
McCrary, Tex, 1553
McCrutcheon, B.R., 610
McDaniel, Hattie, 26, 159, 170, 1341, 1554
McDonald, Bob, 273
McDonald, Eugene F., Jr., **916–18**, *917*, 1593
McDonald, Slim, 162
McDonnell, Norman, 1500, 1501
McDowell, Jack, 26
McFadden, Louis, 782
McGannon, Don, 683, **918–20**, *919*
McGarth, Paul, 747
McGavin, Bob, 312
McGavren, Daren, 1333
McGavren Guild, 1333
McGavren-Quinn, 1332–33
McGregor, Byron, 333
McGuire Sisters, 1367
McInnes-Rae, Rick, 274
McJunkin Advertising Agency, 812
McKay, Don, 312
McKay, Jim, 1312
McKay, Margaret, 1358
McKenzie, Ed, 1062
McKinney, Eleanor, 1122
McLaughlin, Bob, 471
McLaughlin, Edward F., **920–21**, 1372
McLean, Ross, 291
McLean, Stuart, 274
McLemore, Ed, 162
McLendon, Gordon, **922–26**, *923*
  all-news format and, 1019
  KLIF and, 471, 1274
  Liberty Broadcasting and, 860
  recreations, 1196–97, 1313
  Storz and, 1348
  XETRA and, 38, 190–91
McLuhan, Marshall, 991
McMillan, Gloria, 1051
McNally, Edward, 1163
McNamara, Francis A., 69
McNamee, Graham, 360, **926–28**, *927*, 1318
McNear, Howard, 685, 1500
McNeil, Ida, 1551
McNeill, Don, **928–31**, *929*, 957, 1369
McPherson, Aimee Semple, **931–33**, *932*, 1211
McWinnie, Donald, 396
Mediabase 24/7, 74
MediaCorp Radio (Singapore), 105
MediaPlayer software, 123
Media Rating Council, 160, **934–35**
Media Zone, 274
Medical Question Box, 199
Medical Reports, 146
Meed Corliss Archer, 31
Meehan, Eileen, 1291
Meeks, Bill, 788
Meetings of the Foreign Policy Association, 1120
Meetings of the Government Club, 1120
Meet Mr. Morgan, 950
Meet the Press, 1122, 1553
Meiklejohn, Alexander, 1122
Meisler, Beatrice, 1077
Mellon Bank, 620
Meloche, Line, 281
Melody and Rhyme, 1078
Melody Mike's Music Shop, 270
Melody Puzzles, 1148
Melody Ranch, 141, 835
Melody Roundup, 95
Melomanía, 431
Melrose Quartet, 25
Meltzer, Bernard, 1371
Memoirs v Massachusetts (1966), 1036
Memos to a New Millenium, 407
Men in Scarlet, 262

Menjou, Adolph, 719
Mennonite Hour, 1214
Men o' War, 27
Menser, Clarence, 1462
Mercury Theater of the Air, 492, 639, 859, **935–37**, 1243
Meredith Corporation, 564–65
Mergers. *see* Ownership, Mergers, and Acquisitions
Merriman, Tom, 788
Merrow, John, 1123
Merrow Report, 1123
Merrymakers, 1495
Merton, Robert K., 855, 1040
Merwin, W.S., 1081
Meserand, Edyth, 75
Message of Israel, 785, 1212
Messianic Vision, 787
Messter, Charles, 1090
Meston, John, 685
Mestre, Abel, 429
Mestre, Goar, 429
Metcalfe, Jean, 235
Metromedia, **937–38**, 1142, 1544
Metro Morning, 270
Metro Networks, 40, 1505
Metropolitan Opera Auditions of the Air, 939
Metropolitan Opera Broadcasts, **938–40**
Metropolitan Television Company (MTC), 835
Metro Radio, 1435
Metro Radio Sales, 1332
MetroSource, 1021
Metro v Federal Communications Commission, 15
Metzger, Gale, 1153
Mexico, 38, 189–91, **940–44**
Meyrowitz, Bob, 820
MGM Theater of the Air, 719
Miami Herald, Tornillo v (1974), 564
Michael, Sandra, 1552
Michaels, Jay, 1400
Michelmore, Cliff, 236
Michelson, Charles, 1030
Mickey Mouse Club, 1168
Microphone, 356
microphones, 1188
Microradio. *see* Low-Power Radio/ Microradio
Microsoft MediaPlayer, 123
Microsoft Windows Media, 756
microwave radio, 87, 631
Middle East. *see* Arab World Radio; Israel
Middle East Radio Network. *see* Radio Sawa/ Middle East Radio Network
Middle of the Road Format, **944–45**
Midge at the Mike, 146
MIDI (Musical Instrument Digital Interface) time code, 1192
Midnight Caller, 1385
Midnight Special, 1549
Midnight Ticker, 1510
Mighty Allen Art Players, 48
Mighty Carson Art Players, 48
Mike Gallagher Show, 1372
Milam, Lorenzo, 375, 535, **946–47**, 1215
Mile High Farmer, 836
Miles Laboratories, 994, 1046
Milkman's Matinee, 41, 471, 792, 1544
Miller, Arthur, 1076
Miller, Danny, 632
Miller, Dottie, 1554
Miller, Howard, 684
Miller, Larry, 472
Miller, Rice, 814
Miller v California (1973), 1036
Millett, Art, 890
Millie and Lizzy, 286
Milligan, Spike, 668
Millington, Caroline, 1155
Millions for Defense, 977, 1455
Mills, Florence, 25
Mills Brothers, 27
Milne, Alisdair, 201
Milwaukee Sentinel, 416
Mineralava, 10
Minnesota Mining and Manufacturing (3M), 126, 984
Minnesota Public Radio, 829–31, **947–48**, 1004, 1092, 1133
Minor, Dale, 476
Minow, Newton N., 577
minstrelsy, 26–27
Mintz, Leo, 1220
Miranda, Sandy, 839
Misica (Milan, Italy), 767
Miss Hattie, 1340
Miss Meade's Children, 1340
Missouri Honor Awards, 144
Miss Pinkerton, 1339
Mister Rock and Roll, 592
Mitani, Koki, 499
Mitchell, Bobby, 479
Mitchell, Elvis, 1269
Mitchell, Jack, 43, 1512
Mitchell, Shirley, 589
Mitsubishi, 1161
Mix, Jennie Irene, 356, 1551
Mobil Oil, 1321
Moby Dick, 1297
Modern Electrics, 568
Modern Times with Larry Josephson, 1123
Modulation Systems Laboratory, 1335
Moffatt, Lloyd, 312
Mommie and the Men, 1340
Money Smarts, 173
Mongolia, 109–10
Mongol Yaridz Radio (Mongolia), 109
Monitor, 943, **949–50**, 1019, 1429, 1496
Monitorradio, 1131
monitors, 1192–93
Monroe, Harriet, 1079
Monroe, Willa, 1554
Montaigne, Rene, 45
Montgomery, Ralph, 385
Montgomery Ward, 1072
Montiegal, Robert, 1003
Montpetit, Édouard, 331
Moods by Maurice, 735
Moody, Dwight, 556
Moog synthesizer, 788
Moondog Show, 623
Mooney, Francis, 410
Moon River, 1078
Moore, Garry, 509
Moore, Max, 183, 814
Moore, Wilfred, 299
Moorehead, Agnes, 730, 1249, 1353
Moores, Shaun, 1291
More You Watch the Less You Know, 1123
Morgan, Brewster, 61, 859
Morgan, Claudia, 492
Morgan, Edward P., 367
Morgan, Henry, 362, **950–51**, 1198
Morgan, Jane, 1051
Morgan, Robert W., 315, 316, **952–54**
Morgan, Rus, 1081
Morishige Audio Drama Contest, 499
Morita, Akio, 1477
Mormon Tabernacle Choir, **954–55**
Morning (Canadian satire), 287
Morning Edition, **955–56**
  Bob Edwards and, 537
  format, 958
  Red Barber on, 151
  success, 1002
Morning Ireland, 761
Morning Matinee, 1528
Morning Musical Clock, 957
Morning Programs, **957–58**
Morning Pro Musica, 339
Morning Show, 349
Morningside, 270
Morris, Chris, 221–22
Morris, Willie, 1197
Morrison, Herbert, 546, 710, 1018
Morrow, Bruce "Cousin Brucie," 959, **959–60**, 1345, 1475
Morse, Carleton E., 741, *741*, 1045–46
Morse, Samuel F.B., 517
Mother Goose, 1541
Mother's Midday Meal, 221
Motor Boat Club and the Wireless; or The Don, Dash and Dare Cruise (Hancock), 316
Motorola, 138, **960–62**, 1335, 1336
Mott, Robert L., 1297
Moulin Rouge Orchestra, 25
Moulton, Forest Ray, 1541
Mountain Stage, 1131
Mount Royal Broadcasting, 311
Mourning Dove, 499
Movie Man, 522
Movies. *see* Film Depictions of Radio; Hollywood and Radio
Mozambique, 17, 18, 1294

MP3.com, 124
MP3 compression, 123, 465, 756, 1030
MPEG-I layer 3. *see* MP3
mr. ace and JANE, 523
Mr. and Mrs. North, 1249, 1338
Mr. District Attorney, 884, 1338, 1540
Mr. I.A. Moto, 1338
Mr. Keen, 1338
Mrs. Dale's Diary, 218
Mrs. Wiggs of the Cabbage Patch, 1340
MSNBC, 743
M Street Radio Directory, 1105, 1412
MTV, 720
Mudd, Roger, 1418
Muir, Jean, 168
Muirhead, Alexander, 874, 875
Mukou Sangen Ryoudonari (My Neighbors), 777
Mullally, Don, 1003
Muller, Erich "Mancow," 1261
Mulligan, John M., 329
Mullin, John T., 125
Multiculturalism Act of 1971 (Canada), 275
Multiple Access, 311
multiplexing, 1351
mults, 72
Munhwa Broadcasting Corporation (MBC), 109
Murdoch, Richard, 221
Murdoch, Rupert, 937
Murphy, Rex, 272
Murphy, Thomas, 59
Murray, Lyn, **962–63**
Murray, Pete, 232
Murray, Ralph, 240
Murray the K (Murray Kaufman), 959, **963–65**, *964*, 1345, 1528
Murrow, Edward R., 697, **966–69**, *967*
  on blacklisting, 169
  CBS and, 352
  on Churchill, 326
  Foreign News Roundup and, 1019
  Hear It Now! and, 476, 696
  William Lewis' influence on, 859
  WWII reporting by, 1561
Musburger, Brent, 1312
Museum of Broadcast Communications, 971, 1173
Museum of Radio and Technology, 975
Museum of Television and Radio, 972
Museums and Archives of Radio, **970–75**
Musgrave, Mack, 1422
Musical Clock, 641
Musical News, 898
Music and the Spoken Word, 842, 954, 1212
Music Appreciation Hour, 338
Music Choice, 254
Music for A While, 274
Music for Millions, 977, 1455
Music from Studio X, 524, 642
Music Hall, 1535
MusicMaster software, 1105
Music of the World, 839
Music of Your Life, 1053
Music on Radio, **976–79**
Music Testing, 127, **979–81**
Music Through the Night, 1542
Music Til Dawn, 1488, 1573
Music to My Ears, 1375
Music While You Work (BBC), 220
Mutual Black Network (MBN), 29, 173
Mutual Broadcasting System, **981–84**
  creating of, 1578
  foundation of, 1360
  Lone Ranger and, 1350
  Network Monopoly Probe, 1014–15
  newscasts, 38
  and Orson Welles, 492
  religious programming, 1212
  WCFL and, 1491
  Westwood One and, 1504
Mutual of Omaha, 57
MXR consortia, 459
Myanmar, 104–5
My Best Girls, 1339
Myers, Paul, 570
My Favorite Husband, 363, 670, 1496, 1554
My Friend Irma, 353, 592, 1338, 1339, 1554
My Little Margie, 1278
Mylo Ryan Phonarchive, 1030
Myrt and Marge, 493, 1287, 1551
Mysterious Traveler, 730
Mystery in the Air, 730
Mystery Serial, 1046
Mystery Theater, 1563

NAACP v Federal Power Commission (1976), 14
Nabors, Ted, 844
Names in the News, 1583
Nantz, Jim, 844
Napster, 124
Narrowcasting, **985–86**
Nation, 561
National Alternative Radio Konference (NARK), 999
National Amateur Night, 1366
National Archives, 973, 1030
National Association for the Advancement of Colored People (NAACP), 81, 170
National Association of Black-Owned Broadcasters, 15, 28
National Association of Broadcast Engineers and Technicians, 1381
National Association of Broadcasters, 28, 41, 69, 243, 308, **986–89**, 1274, 1465
National Association of College Broadcasters, 1408
National Association of Disc Jockeys, 471
National Association of Educational Broadcasters, 534, **989–92**
National Association of Educational Broadcasters Tape Network, **992–93**
National Association of Evangelicals (NAE), 1007
National Association of Farm Broadcasters, 573, 1408
National Association of FM Broadcasters, Incorporated, 608, 609, 1241
National Association of Manufacturers (NAM), 62
National Association of Radio Station Reps, 1332
National Association of Radio Talk Show Hosts, 1373
National Barn Dance, 140, **994–95**, 1274, 1357–58, 1454
National Black Media Coalition, 15, 28
National Black Network (NBN), 29, 173–74. *see also* American Urban Radio Networks (AURN)
National Broadcast Editorial Association, 529
National Broadcasting Company et al v United States (1943), 596, 997, 1205, 1444
National Broadcasting Company (NBC), **996–98**
  ABC spinoff, 56
  advertising, 11, 1426
  Blue Network, 72, 180–82
  Canadian affiliates, 278, 282
  creation of, 72
  formation, 1163
  General Electric and, 649–50
  Group W and, 683
  market, 904
  morning programs, 957
  music on radio, 976
  NBC Radio Network distribution, 40
  Network Monopoly Probe, 1014–15
  newscasts, 38
  news format stations, 39
  Quiz Kids, 320
  Red Network, 72
  self-regulation, 308
National Council of Churches, 1007
National Disc Jockey Association, 472
National Educational Radio (NER) Network, 991
National Electric Signaling Company, 586–87
National Emergency of 1975 (India), 36
National Farm and Home Hour, 181, 572
National Federation of Community Broadcasters, 375, **999–1000**
National Headliner Awards, 144
National Industrial Recovery Act (NRA), 62
National Iranian Radio and Television (NIRTV), 102
National Jazz, Rhythm-and-Blues Disc Jockey Association, 28. *see also* National Association of Radio and Television Announcers (NATRA)
National Lampoon Radio Hour, 1361

National Lesbian and Gay Country Music Association, 647
National Life and Accident Insurance Company, 572, 1571
National Music Camp, Interlochen, MI, 65
National Negro Network (NNN), 173, 174
National Public Broadcasting Archives, 971
National Public Radio, **1000–1007**
  All Things Considered, 42–46
  cable radio study, 253–54
  children's programs, 320–21
  David Isay and, 763
  digital audio broadcasting and, 460–61
  drama on, 495
  education and, 991
  election coverage and, 542
  FM stations, 606
  foundation of, 1388
  Frank Mankiewicz and, 893–95
  Morning Edition, 537
  National Federation of Community Broadcasters and, 999
  poetry on, 1081
  satellite radio and, 895
  start of, 1429
National Radio Awards, 144
National Radio Broadcasters Association, 609
National Radio Conferences, 1202
National Radio Network (NRN; Japan), 77
National Radio Plan (NRP; Australia), 133
National Radio Pulpit, 1212
National Radio Systems Committee, 84, **1006–7**
National Religious Broadcasters, **1007–8**
National Spot Plan, 1043
National Stereophonic Radio Committee, 605
National Telecommunications and Information Administration, **1009–10**, 1011
National Union Electric, 550
National Union for Social Justice, 409
National Voice of Iran, 335, 1116
Native American Radio, **1011–12**
Native American Restoration Act, 1011
Nauru, 1310
Naylor, Brian, 1004
Nazaire and Barnabie, 287
Nazi Arabic Service, 89
Nazi Germany, 61, 89, 145–48
NBA on NBC, 1312
NBA Stuff, 1312
NBC Symphony Orchestra, 181, 338, 996
Nealk, Hal, 58
Neble, Long John, 783
Nederlands Indische Radio Omroep Maatschappij (Indonesia), 106
Negroes in Today's World, 430
Negro in the War, 27
Neil, Ronald, 209
Nelson, Davia, 1514
Nelson, "Jungle" Jay, 324
Nelson, Lindsey, 860, 1197, 1318–19
Nelson Brothers Bond and Mortgage Company, Federal Radio Commission v (1933), 1444
Nepal, 104
Netherlands, **1012–13**
Netherlands Broadcasting Foundation, 1013
Netherlands Radio Union (NRU), 1012
NET Radio (New Hellenic Television), 677
Network Allocation Plan, 1042
Network Monopoly Probe, **1014–15**
Neue Hörspiel, 498
Never Get Out!, 395
New, Harry S., 1433
New Adventures of Sherlock Holmes, 730
New British Broadcasting, 334
Newby, Ray, 699
New City Communications, 1055
New Day Highway, 321
Newell, Lloyd, 954
New Herald Tribune, 416
New Letters on the Air, 1081
Newman, Fred, 1168
Newman, Gerald, 284
Newman, Lloyd, 764
Newman, Robert, 1111
New Music Canada, 274
New Republic, 417
New Research Group, 127
News, **1017–21**
News about Britain, 225
NewsActing, 898
News Agencies, **1021–23**
News and Comment, 691
News and Information Service (NIS), 38
News at Midnight (BBC), 221
News Briefing, 221
NewsCasting, 898, 1196
NewsDirect (London, United Kingdom), 40
Newshour, 241, 1123
New Sounds, 1019
NewsRadio (NBC), 39
News-Talk format, 1105
News through a Woman's Eyes, 1553
News Till Now, 1416
New Waves, 1168
New World a'Comin', 170, 1076
New York, Redrup v (1967), 1036
New Yorker, 416
New York Festivals Awards, 145
New York Philharmonic, 338
New York Pulse, 1140
New York Times, 416, 508, 1565–66
New York Tribune, 355
New York World-Telegram, 355
New York Yankees, 50
New Zealand, **1016–17**
New Zealand Broadcasting Board, 1016
NFL Live, 1312
NFL Playbook, 173
NFL Today, 1312
Nichols, Arthur, 1297, 1298
Nichols, Ora, 492, 1297, 1298
Nick Carter, Master Detective, 493, 1557
Nickell, Paul, 909
Nielsen. *see* A.C. Nielsen Company
Nielsen, Arthur C., Jr, 1, 2, 115, 722
  Arbitron litigation, 91
  end of radio measurement, 119
Nielsen Radio Index (NRI), 1, 115, 118, 445
Nielsen Station Index (NSI), 2, 118
Nielsen Television Index, 90, 118
Nigeria, 18
Nightingale, Anne, 234
Nightingale, Earl Clifford, **1024–25**, *1025*
Night People, 363
Night Talk (film), 593
Night that Panicked America, 1386
Night Train, 1357
Night Waves (BBC), 219
Nihon Television (NTV), 77
Niles, Chuck, 779
Nine O'clock News (BBC), 221
9XM (Madison, Wisconsin), 346
Nippon Hoso Kyokai (NHK), 776, 777–78
Nitecap Radio Network, 782
Nitecaps, 782
Nitecaps International Association (NIA), 783
Nite Spot, 1554
Niue, 1310
Niven, Harold, 244
Nixen, John, 241
Nixon, Agnes, 1073, 1288
Nixon, Richard, 1084, 1085, 1441
Nobel Broadcast Group, 1223
Noble, Daniel, 961
Noble, Edward J., 56, 57, 997, 1345
Nobles, Gene, 183, 1536
Nockels, Edward N., 1491
Nodojiman Shirouto Ongakukai (Amateur Song Contest), 776, 777
nonemanating outputs (NEMO), 72
Non-English-Language Radio in the United States, **1026–28**
Nonn, Mary, 521
Norberg, Eric, 871
Nordine, Ken, 1080
Norman, Gene, 471
North American Philips Corporation, 1069
North American Radio Archive, 1030
North American Regional Broadcasting Agreement, 343, **1028–29**
North by Northwest, 1415
Northern Electric Hour, 262
Northern Ireland, 306–7
Northern Rhodesia. *see* Zambia
North Toward Home (Morris), 1197
Norway, 40, 1237
Norwegian Broadcasting Company (Kringkastingselskapet AS), 1237
No School Today, 320
Nostalgia Radio, **1029–31**, 1030

Nouveautés dramatiques, 281
Novik, Morris S., 1546
NPR Playhouse, 495, 520
NPR Worldwide, 1004
NRK (Norway), 40
Nugent, Ted, 466
Nunn, Sam, 1465
NVNU, NV, 1
NYAB (Bhutan), 104
Nye, Gerald, 1431

Oak Knoll Broadcasting, 840
Oberndorfer, Max E., 1541
Oboler, Arch, **1033–35**
Lights Out, 729, 730, 864
NBC and, 492
Plays for Americans, 1075
O'Brien, Edmond, 1589
Obscenity and Indecency on Radio, 308, **1035–39**
Ocean Wireless Boys, 316
O'Connell, Kathy, 321
O'Dea, Richard, 1543
O'Dell, Spike, 1510
O'Donnell, George, 1298
Ofcom, 1161
Office of Censorship, 309, 1562
Office of Radio Research, **1039–40**
Office of War Information, **1041–44**
abolition of, 1469
ABSIE and, 61–62
censorship and, 309
Elmer Davis and, 437
Psychological Warfare Branch, 1060
School of the Air of the Americas, 68
war emergency provisions, 1560
Official Languages Act of 1969 (Canada), 275
O'Flaherty, Hal, 1542
Of Men and Music, 1375
OFRH, 1214
Ogilvy and Mather, 13
O Globo, 194
O'Hair, Madelyn Murray, 1215
O'Halloran, Hal, 1358
O'Hehir, Michael, 759, 761
Ohio, Jacobellis v (1964), 1036
Ohio School of the Air, 534
Ohls, Erik, 500
Old Fashioned Revival Hour, 556, 557, 1212
Old Gold Hour, 151
Oldies Format, **1044–45**
Olds, Irving S., 1392
Old-Time Radio. *see* Nostalgia Radio
O'Leary, Gratton, 290
Olsen, Harry, 704
Olson, Roger, 1408
O'Malley, Neil, 1473
On a Note of Triumph, 352, 406, 639, 1564
onda tropical (tropical wave), 195
One-a-Day vitamins, 320
O'Neill, Jimmy, 840
O'Neill, Thomas "Tip," 1432, 1442
One Man's Family, 493, **1045–47**, 1094, 1338, 1415
One World Flight, 407
Only A Game, 1005
On Physical Lines of Force (Maxwell), 910, 911
On the Record, 1316
On the Road, 845
Open Door, 1552
Open End, 1142
Open Letter on Race Hatred, 27, 170
Open Mike, 1514
open-reel audiotape recorders, 1190–91
Opportunities in Broadcast Careers (Ellis), 545
Opportunity Knocks, 1176
Opry Star Spotlight, 1067
Optimistic Doughnut Hour, 170
Optimod-FM 8000, 121
Optimum Effective Scheduling (OES), 9, 446
O'Rahilly, Ronan, 237
Orange Network, 180
Orban Associates, 121
Orbit Satellite TV and Radio Network, 467
O'Reilly, Bill, 1557
Original Amateur Hour, 46, 1147
Ormandy, Eugene, 338
Ory, Kid, 25
Osborne, Ted, 729
OSCAR (Orbiting Satellite Carrying Amateur Radio), 690
Osgood, Charles, 38, **1048–49**
Osgood Files, 1048, 1081
O'Shea, Daniel T., 167, 169
Ossenbrink, Luther, 995
Ossie Davis and Ruby Dee Story Hour, 174
Otis, James, 316
O'Toole, Darian, 584
Ottley, Roy, 170
Ouimet, Marcel, 261
Our America, 764
Our Changing World, 1024
Our Gal Sunday, 27, 413, 737, 1338, 1552
Our Government, 1120
Our Miss Brooks, **1049–51**
duration, 1278
Gale Gordon and, 670
stereotypes, 1338, 1339, 1341
Our Radio, 335
Our Radio into Turkey, 1116
Our Secret Weapon, 352
Outlet Company of Providence Rhode Island, 1573
Outline of Radio (Hogan), 716
Outlook, 241, 1123
Out of the Blue, 220
Out of the West, 1478
Outskirts of Hope, 476
Ovaltine
children's programs sponsored by, 299, 319, 493, 868
League of Ovaltineys, 1175
premiums, 1094
Ovaltiney's Concert Party, 1175
Overseas Press Club Awards, 145
Owens, Gary, 1030, **1052–53**
Owens, Ron, 1371
Ownership, Mergers, and Acquisitions, **1053–56**, 1383
Oxydol, 11, 665, 737

Pabst Blue Ribbon Town, 602
Pacelli, Eugenio, 1456
Pacifica Foundation, 309, 476, 495, 645–46, **1057–60**, 1132
Lewis Hill and, 708–10
WBAI and, 1481–83
Pacifica Foundation, Federal Communications Commission v (1978), 1036, 1444
Pacific Radio News, 1412
Packard Ballad Hour, 815
Packard Fiesta, 816
Packard Radio Club, 815
Packard Six Orchestra, 815
Page, Benjamin, 1292
Painted Dreams, 1070, 1286, 1509, 1552
Painting, Norman, 93
Pakistan, 103
Pakistan Broadcasting Corporation (PBC), 103
Pakistan Broadcasting Service, 103
Pakistan Television (PTV), 103
Palestine Broadcasting Service (Jerusalem), 88, 765
Paley, William S., **1060–61**
aid for ABSIE, 61
CBS and, 1095
Crosby's acquisition by, 56
Godfrey and, 663
Hear It Now and, 696
Murrow and, 968
Palmer, Bill, 125
Palmolive Hour, 1453
Paltridge, J.G., 38
Panama Coast Artillary Command (PCAC), 94
Pan American coffee Association, 1224
Panamsat, 467
Papua New Guinea, 1309
Parade of Stars, 173
Paraguay, 1306–7
Paramount mergers, 720
Paramount Pictures, United States v (1948), 596
Parents' Music Resource Center, 1465
Parks, Bert, 57
Par le trou de la serrure, 287
Parsons, Louella, 719, 819
Parties at Pickfair, 1454
Pastor, Tony, 1458
Patricia McCann Magazine, 1557
Patrick, Dan, 60

Patterson, Richard, 76
Pattiz, Norman J., 1182, 1504
Pauline Frederick Reporting, 621, 1553
Pauline Frederick's Guest Book, 621
Paul Whiteman Orchestra, 1453
Paul Whiteman Radio Hall of Fame, 1454
Paulynne Productions, 691
Pavek Museum of Broadcasting, 973–74
Payne, Jack, 220
Payne, Sonny, 182, 814
Payne, Virginia, 896
Payola, 171, 245–46, **1062–64,** 1220
Pay Radio, **1064–65**
Peabody, George Foster, 1065
Peabody Awards and Archive, **1065–66**
Peacock Committee, 13
Peale, Norman Vincent, 1215
Peale, Ruth, 1215
Pear Orchard, Texas, 575
Pearson, Drew, 56, 367
Peary, Hal, 674
Pederson, Wayne, 1008
Peel, John, 233, 234
Peer, Ralph, 411
Peerce, Jan, 785
Pegeen Prefers, 719, 1553
Pena, Adarand v (1995), 16
Penn Tobacco, 1046
Penny Williamson, 1339
Penzias, Arno, 154
People Are Funny, 1149
People Meter system, 160
Peoples Choice, 855
Peoples Court, 540
People's Platform, 27, 1564
Pepper, John, 27, 184, 1492, 1523
Pepperday, T.M., 837
Pepper Pot, 930
Pepper Young's Family, 494, 1287, 1339, 1552
Pepsi Cola Cousin Brucie Saturday Night Party, 960
Pepsodent, 665, 1094
Performance Today, 1004, 1005
Performing Rights Society (PRS), 230
Performing Rights Tribunal, 230
Perowne, Lesley, 451
perpetual audio coding (PAC), 155
Perrin, Nat, 602
Perry, John, 954
Perryman, Tom, **1067–68**
Persia/Iran, 102
Persian Service, 224
Personal Album, 95
Personal Influence, 855
personal music tests, 980
Person to Person, 349
Peru, 1307
Peter Ibbetson, 1375
Peter Pan Peanut Butter, 319
Peters Griffin Woodword agency, 1331
Petrillo, James Caesar, 64, 65
Petry, Edward, 1331, 1332, 1427
Peyton, Patrick, 566
Phelps, George Harrison, 1577
Philadelphia Battery Storage Company, 1068
Philadelphia Symphony Orchestra, 338
Phil Blazer Show, 786
Philco Hour, 196, 1068
Philco Radio, **1068–70,** 1334, 1335
Philco Radio Playhouse, 1069
Philco Radio Time, 56, 419, 977, 1097, 1454
Phil Harris and Alice Faye Show, 362, 670
Philippines, 106
Phillips, Dewey, 1220
Phillips, Irna, 493, **1070–74,** 1288, 1509, 1552
Phillips, Jack, 1424
Phillips, Sam, 583, 1220, 1513, 1554
Phillips, Wally, 1510
Phillips H. Lord, Inc., 1199
Phoenix, 357
Phonographic Performance Limited (PPL), 230
phonographs, 705
Photophone, 518
Picard, Laurent, 272
Piccadilly Radio, 1435
Pierce, Webb, 190
Piercy, Marge, 1081
Pierson, Richard, 1478
Pierson, Steve, 1408
Pierson, Walter, 857, 1298
Pilgrimage of Poetry, 1078
Pilgrim's Hour, 1214
Pillsbury, 663
Pinsky, Robert, 1080, 1081
Pioneer Radio Society, 25
Pious, Minerva, 360, 951
pirate radio. *see* British Pirate Radio
Pival, John, 1578
Place for the Kids, 1573
Plainsman, 1498
Planet Hollywood, 1549
Plant, Elton, 1574
Plantation Club (Los Angeles, California), 25
Plantation Club (New York City), 25
Plantation Nights, 1274
Plays for Americans, 1075
Playwrights on Radio, **1074–77**
Pleikys, Rimantas, 773
Pletcher, Thomas, 987, 1593
Plomley, Roy, 450
Plot to Overthrow Christmas, 405, 492
Plowright, Piers, 219
Plugge, Leonard, 12
PM, 216, 241
Poehler, Eleanor, 583, 1551
Poems, 1077
Poet and the Poem from the Library of Congress, 1080, 1081
Poetic License, 1079
Poetry Hour, 1079
Poetry of Lawrence Ferlinghetti, 1080
Poetry of Our Time, 1080
Poetry on Radio, **1077–82**
Poet's Gold, 1077, 1078, 1079
Poet Speaks, 1079, 1080, 1081
Police Cruiser, 961
Policy Statement on Comparative Broadcast Hearings, 15
Politics and Radio, **1082–86**
Politische Zeitungsschau (Political Newspaper Digest), 637
Polk, George, **1086–88**
Pollyanna, 1340
Pond, Wayne, 1081
Ponds, 12
Pond's Radio Program Speeches, 1224
Poniatov, Alexander, 126
Poole, Jim, 1358
Popov, Alexander, **1089–90,** 1227
Poppele, Jacob R., 1556, 1557
Pop Question Game, 1147
Popular Radio, 356, 568
Portable People Meter system, 92
Portable Radio Stations, **1090–91**
Porter, Katherine Anne, 1080
Porter, Shirley, 876
Portrait, 349
Portraits in Blue, 182
Post, Steve, 627
Post-Newsweek Company, 1573
Pot o' Gold, 1148
Potter, Charles, 495
Pouillot, Jean Adelard, 311
Poulsen, Vlademar, 125, 1533
Pound Ezra, 1115
Powell, Dick, 1589
Powell, Lewis, 1247
Powell, Michael, 29, 1384
Powell, William, 719
Power, Harold J., 1508
Power, Ralph, 355
PowerGold software, 1105
Prairie Farmer, 1357, 1358
Prairie Home Companion, **1092–93**
Keillor and, 806–9
marketing of, 831
popularity, 364, 948, 1004, 1130
style, 1134
PRA-2 (Rio de Janeiro, Brazil), 193
Prasar Bharati Act of 1990, 36
Preece, William, 518, 902
Premiere Radio Network, 74, 1223
Premières, 281
Premiums, 319, **1093–94,** 1243
Premium Stuff, 1554
Presidential Timber, 1416
Presley, Elvis, 847, 1567
Press News Ltd., 268
Press-Radio Bureau, 1023
Press-Radio War, **1095–96**
Pretesting Company, 92
Priessnitz, Horst, 498
Prism Fund, 16

private censorship, 307
Private Line Services, 72
Private Parts, 593
Prizes. *see* Awards and Prizes
Proceedings of the Institute of Radio Engineers, 667, 1412
Proctor and Gamble
advertising in Mexico, 941
Goldbergs sponsorship, 665
Lowell Thomas and, 1396
Ma Perkins sponsorship, 11
soap opera sponsorship, 1072, 1286, 1287
Truth or Consequences, 539
Vic and Sade sponsorship, 1462
Production Advertising Marketing Services (PAMS), 788
production directors, 1097, 1099
Production for Radio, 122, **1096–1101**
profanity, definition, 1038
Professor Quiz, 1147
Programming Research, **1101–2**
Programming Strategies and Processes, **1102–6**
Progressive Rock Format, **1107–8**
Projecto Minerva, 1305
Promax, **1108–10**
Promenade Concerts, 214, 219, **1110–11**
Promenades en Nouvelle-France, 281
Promotion on Radio, **1111–13**
Propaganda by Radio, **1113–18**, 1400–1402, 1560
Psuan Munwha Broadcasting Station (Korea), 109
Psychographics, **1118–19**
PTL Club, 1008
Public Affairs Programming, **1120–25**
Public Broadcasting Act of 1967, 536, 1001, **1125–26**, 1388
Public Broadcasting Service, 991, 1133
Public Interest, 1005
"Public Interest, Convenience, or Necessity," **1127–30**
Public Radio International, 808, **1130–31**. *see also* American Public Radio
Public Radio Program Awards, 145
Public Radio Since 1967, **1132–37**
Public Service Radio, **1137–40**
Public Service Responsibility of Broadcast Licensees. *see* Blue Book
Public Telecommunications Review, 991
Publishers' National Radio Committee, 1095
Puffed Wheat Sparkies, 868
Puhan, Alfred, 61
Pulse, Inc., The, **1140–41**
Pumarejo, Gaspar, 429
Pure Oil Company, 797
Pursuit of Happiness, 475, 492, 1274
Purtan, Dick, 333
Puss in Boots (Carter), 218
Putting Your Best Face Forward, 221
PWX (Havana, Cuba), 428
Pyne, Joe, 695, **1142–43**, 1369–70
Pyramid Radio (Kyrgyzstan), 103

Q'part, Jack, 324
QSLs, 515
QST, 356, 567, 1593
Quaal, Ward L., **1145–46**, 1510
Quaker Oats, 1094, 1337
Quality Group, 982
Quality Media Incorporated (QMI), 1242
Quayle, Donald, 1001, 1133
Queen for a Day, 481, 1150
Queensboro Corporation, 1494
Quelles nouvelles, 268
Quello, James, 577
questionnaires, 117
Question of Place, 1003
Quetzal/Chase Capital Partners, 16
Quick and the Dead, 634, 636
Quick as a Flash, 1251
QuickTime software, 123
Quiet Please, 730
Quiet Storm, 1446
Quillan, Joe, 1049
Quinn, Don, 589
Quirks and Quarks, 273, 274
Quivers, Robin, 584, 593
Quiz and Audience Participation Programs, **1146–52**
Quiz Kids, 181, 320, 413, 1148
Q'zine, 646

Rabbit Ears Radio, 321
Rabbit Foot Minstrels, 1399
Rabell, Fred, 609
RADAR, 115, **1153–54**
Rader, Paul, 1211, 1214
Radio, 356
Radio 1212, 334, 1115
Radio, Inc., 28
Radio, Telegraph and Telephone Company, 442
Radio Aahs, 322, 1168
Radio Academy, **1154–55**
Radio Active, 1411
Radio Act of 1912
call letters and, 255
no-censorship clause, 307
passage of, 580, 724, 1431
shortwave band allocation, 1262
Radio Act of 1927
advertising-supported radio and, 1138
ham radio and, 689
impact of, 1426, 1433
local zones assigned by, 199
local zones established by, 870
passage, 724, 862
public interest standard and, 550
role of, 1201, 1202–3
Wallace H. White and, 1517
Radio Acts. *see* Wireless Acts of 1910 and 1912; Radio Acts of 1912 and 1927
Radio Advertising Bureau, 8, 13, 369, **1156–57**
Radio Afghanistan, 102
Radio Age, 568
Radio Alger, 615
Radio Alice (Bologna, Italy), 1295
Radio Almaty (Kazakhstan), 103
Radio Amateur News, 356
Radio and Internet Newsletter, 1412
Radio and Records, 1411, 1505
Radio and the Printed Page (Lazarsfeld), 855, 1039
Radio Annual, 356, 1411
Radio Ashkhabad (Turkmenistan), 103
Radio Australia, 751
Radio Authority, 230, 231, **1160–61**
Radio Azad, 1173
Radio Bandeirantes, 1305
Radio Bangladesh, 103
Radio Bantu, 18
Radio Bari, 88, 749, 1114
Radio Base Popolare (Venice, Italy), 767
Radio Berlin International, 657
Radio Betar (Bangladesh), 103
Radio Bible Hour, 556, 557, 1214
Radio Bishkek (Kyrgyzstan), 103
Radio Boys as Soldiers of Fortune, 316
Radio Boys (Breckenridge), 316
Radio Boys (Chapman), 316
Radio Brazzaville, 615
Radio Broadcast, 356, 415, 568, 1410, 1491
radio broadcast data systems (RDS), 1006
Radio Broadcasting Act (Canada), 259
Radio Broadcasting News, 568
Radio Broadcast of Will Rogers, 1222
Radio Bucharest, 1116
Radio Budapest, 1116
Radio Business Report, 1412
Radio Cadena Nacional, 942, 1306
Radio Cambodge (Phnom Penh, Cambodia), 107
Radio Canada International, 751
Radio-Carabin, 288
Radio Caritas, 1306
Radio Caroline, 233, 237–38
Radio Ceylon, 104
Radio Chapel Service, 1211
Radio Church of the Air, 1211
Radio City, **1161–63**
Radio City Music Hall of the Air, 785
Radio City Playhouse, 1243
Radio Ciudad de La Habana (Havana City Radio), 431
Rádio Clube Português, 1294
Radio Club of Burma (Rangoon), 104
Radio Collectors of America, 1030
Radio-Collège, 281
Radio Corporation of America, **1163–65**
ABC spinoff, 56
ABSIE transmitters built by, 61
Alfred Goldsmith and, 667
AM stereo, 1334

AT&T holdings in, 71
automation equipment from, 136
creation of, 34
election coverage, 540
General Electric and, 648
NBC and, 996
patent rights acquisition, 1422
radio station business, 1425
recording discs, 1194
Sarnoff and, 1234–35
superheterodyne circuit, 1184
U.S. Navy and, 1437
Radio Corporation of the Philippines, 106
Radio Daily, 356, 1411
Radio Daily–Television Daily, 1411
Radio Data System, **1166–67,** 1187
Radio Days (film), 593
Radio DDR, 657
Radio Dealer, 1410
Radio de la Fidelité, 335
Radio des Mille Collines (Radio of a Thousand Hills; Rwanda), 19
Radio Dial, 970
Radiodiffusion Nationale, 615
Radio Diffusion Nationale Khmère (Khmer Republic), 107
Radiodiffusion Nationale Lao (Laos), 107
Radio Digest, 356, 490
Radio Disney, 59–60, 322, **1167–68.** see also Disney Corporation
Radio-Divertissement Molson, 287
Radio Drama in Action, 494
Radio Drama Network, 251
Radio Eireann, 758
Radio El Espectador, 1307
Radio El Sol, 1037
Radio Enciclopedia, 430
Radio España Independiente, 335
Radio Expeditions, 1005
Radio Factor, 1557
Radio Farda, 1173
Radio FEUU, 1307
Radiofictions en direct, 282
Radio Five Live, 204, 242
Radio Folio, 908
Radio France, 616, 617
Radio Free Afghanistan, 1117, 1173
Radio Free Alcatraz, 1011
Radio Free Asia, 750, 1116, **1168–09,** 1173, 1469
Radio Free Cuba, 430
Radio Free Dixie, 430
Radio Free Europe/Radio Liberty, **1170–73**
Board for International Broadcasting and, 185
creation of, 750
Cuba and, 430
funding, 344
governance, 247
jamming of, 774
mission, 345
radio propaganda, 1116
Radio Free Iraq, 1173
Radio Free Luxembourg, 1115
Radio Free Portugal, 335
Radio Free Spain, 334
Radio Frunze (Kyrgyzstan), 103
Radio Ghana, 751
Radio Girls, 316
Radio Globo, 194, 1305
Radio Group, 181
Radio Guide, 356, 569
Radio Guild, 181, 282, 283
Radio Gune Yi (Senegal), 21
Radio Habana Cuba–Cadena Azul, 429, 430
Radio Hall of Fame, 971, 972, **1173–75**
Radio Hanoi (Vietnam), 107
Radio Hauraki, 1016
Radio Havana Cuba, 750
Radio Historical Society of America, 970, 1030
Radio History Society's Radio-Television Museum, 975
Radio Imagery: Strategies in Station Positioning (Burns), 1105
Radio Ink, 1412
Radio in the American Sector (Berlin), 1116, **1158–60**
Radio in the Home, 356, 568
Radio Invicta, 238
Radio Iran, 102
Radiojänst AB (Svergies Radio), 1237
Radio Japan, 751
Radio-Keith-Orpheum (RKO)
Movie Studios, 1163
RKO General Corporation, 485–87, 502, 1557
RKO Radio, 91, 952
Radio Kiev, 345
Radio Kotmali (Sri Lanka), 104
Radiola, 1030
Radioland, 356
Radiola station, 615
Radio Liberation. *see* Radio Free Europe/ Radio Liberty
Radio Liberation from Bolshevism, 750
Radio Liberty. *see* Radio Free Europe/Radio Liberty
Radiolog, 356
Radio London, 233, 237, 238, 559, 789
Radio Londres, 615
Radio Luxembourg, 204, 617, 749, **1175–77,** 1435
Radiomafia, 1238
Radio Maldives (Maldives), 104
Radio Mambí, 431
Radio Manual, 508
Radio Marathon. *see* Radio Martí
Radio Marketing Guide and Factbook for Advertisers, 1156
Radio Marketing Research, 90, 165
Radio Martí (Marathon, Florida), 334, 430, 502, 774, 1029, 1173, **1177–79**
radio propaganda, 1116
VOA and, 1469
Radio Mayak (Radio beacon), 1230
Radio Méditerranée Internationale, 751
Radio Mercur, 237
Radio-Mercury Awards, 145
Radio Mirror, 356, 569
Radio Monte Carlo, 617, **1179–80**
Radio Monte Carlo Middle East, 751
Radio Moscow, 425, 615, 749, **1181–82**
radio propaganda, 1116
Radio Mundal, 1308
Radio Music License Committee (RMLC), 70, 246, 1408–9
Radion, 13
Radio Nacional (Rio de Janeiro, Brazil), 194
Radio Nanduti, 1306
Radio Nepal, 104
Radio Netherlands, 751
radio networking, 71–72, 90
Radio News, 356, 568
Radio News Association, 1023
Radio Newsreel, 224
Radionews (Rome, Italy), 767
Radio Nord, 924
Radio Normandie, 12
Radio Nostalgie, 21
Radio Nova, 129, 760
Radio Oberdira, 1037
Radio of the Sudanese People's Liberation Army, 335
Radio Only, 1411
Radio Orfey (Radio Orpheus), 1230
Radio Pakistan, 103
Radio Peace and Progress, 345
Radio Peking, 750
Radioperedacha, 1228
Radiophone, 518
Radio Planet FM, 767
Radio Popolare (Milan, Italy), 1295
Radio Prague, 1116
Radio Program Awards, 145
Radio Programs of Mexico, 941
Radio Progreso, 429, 430
Radio Publicity Limited, 12
Radio Pyongyang, 750
Radio Quince de Septiembre, 335, 1116
Radio Quinto, 1306
Radio Rating, 1410
Radio Reader's Digest, 962
Radio Rebelde, 429, 431
Radio & Records magazine, 386, 1412
Radio Red, 942, 943
Radio Reloj, 429
Rádio Renascença, 1294
Radio Republik Indonesia (RRI), 106
Radio Research, 855, 1027
Radio Retailing Today, 1411
Radio Roma, 767
Radio Rossii (Radio Russia), 1230
Radio Rumbos, 1307
Radio Sagarmatha (Nepal), 104

Radio Sawa/Middle East Radio Network, 247, 767, **1182–83**
Radio Service Bulletin, 532
Radio's Golden Age, 1030
Radio Showmanship magazine, 1411
Rádio Sociedade do Rio de Janeiro (Radio Society of Rio de Janeiro), 193
Radio Society of Great Britain, 203
Radio Sofia, 1116
Radio Sol Armonia, 1037
Radio Sonic, 274
Radio Sottens, 615
Radio Speech (Lawton), 530
Radio's Second Chance (Siepmann), 1267
Radio Stars, 356
Radio Station Kilroy (Taegu, Korea), 799
Radio Station Management (Reinsch and Ellis), 545
Radio Sunshine, 767
Radio Suomi, 1237
Radio Sutatenza, 1306
Radio Sutch, 238
Radio Swan, 334, 430, 1116
Radio Taiso, 776
Radio Tara, 759
Radio Teheran (Tehran, Iran), 102
Radio Telefis Eireann, 758, 760
Radio-telegraph Act of 1913 (Canada), 263
Radio Television and Society (Siepmann), 1267
Radio-Television News Directors Association, 1408
Radio Television News Directors Association, United States v, 1200
Radio Thailand, 105
Radio-théâtre canadien, 281
Radio Theatre Guild, 283
Radio The Voice of Vietnam (Hanoi, Vietnam), 107
Radio Times (magazine), 203, 208
Radio Tokyo, 750
Radio Trent, 1435
Radio Unnameable, 1482
Radio Vaticana, 749
Radio Venceremos, 334, 335
Radio Venus, 1307
Radio Veritas (Philippines), 106
Radio Veronica, 237
Radio Voice of Shari'a (Afghanistan), 102
Radio Warsaw, 750
Radio Wayne Awards, 145
Radio Week, 1411
Radio World, 356
RadioYesteryear, 970
Radio Yunost (Radio Youth), 1230
Raffetto, Michael, 1046
Raina, Shankar, 500
RAI (Radiotelevisione Italiana), 767
Raleigh Cigarettes, 1279
Ralston-Purina, 319, 1094
Rambling with Gambling, 641–43, 1557
Randall, Eunice, 583, 1508, 1551
Randolph, A. Phillip, 26
Randolph, Amanda, 1277
Randolph, Isabel, 589
Ranger Mac, 1511
Rankin, Moira, 1301
Rankin, William H., 10
Rankin advertising agency, 10
Raphael, Sally Jesse, 584, 1371, 1520, 1555
Rashad, Ahmad, 1312
Raskin, Rhona, 292
rate cards, 9
Rathe, Steve, 1002
Rather, Dan, 844
ratings, meaning of, 114
Rauland Corporation, 1595
Rawhide, 286–87
Rawls, Lou, 173
Raymer, Paul, 1331
Raytheon, 550
RCA. *see* Radio Corporation of America
RCA Photophone, 1163
RCA Radiotron Company, 1163
RCA Review, 1411
RCA Victor, 1163
RClub (Naples, Italy), 767
R-DAT, 464
REACT program, 330
Reader's Digest, 76
Readick, Frank, Jr., 1248
Readick, Robert, 1589
Reading I've Liked, 562
Reagan, Ronald, 1085, 1196, 1441–42
RealAudio, 123
RealOne, 123, 756
Real Sports, 1312
Rear Bumpers, 1256
Reber, John U., 718
recall surveys, 114, 399
Receivers, **1183–87,** 1502–3
Recollections at 30, 476
Recording and Studio Equipment, **1187–93**
Recording Industry Association of America (RIAA), 124
recording process, 1099
Recordings and the Radio Industry, **1194–95**
Re-Creations of Events, **1196–98**
Red Adams, 494
Red Among Us, 494
Red Apple Club, 1534, 1535
Red Channels, 167–68, 169, 951, 962, **1198–99**
Red Davis, 494
Red Dragon (Wales, Great Britain), 298
Red Foley Show, 1279
Red Ladies, 494
Red Lion Broadcasting Company, 1200–1201, 1444
Red Lion Case, 14, 564, **1200–1201,** 1444
Red Network, 72, 180, 181
Red Power Movement, 1011
Redrup v New York (1967), 1036
Red Ryder, 1499
Red Skelton Show, 1279
Red Waves, 494
Reece, Carroll, 1441
Reeves, Jim, 190
Reeves, John, 284
Regulation, **1201–7**
Reichsministerium für Volksaufklärung und Propaganda (Reich Ministry of Popular Enlightenment and Propaganda), 637
Reichsrundfunkgesselschaft (German Broadcasting Corporation), 125, 637
Reid, Charlotte, 930
Reinsch, J. Leonard, 1570
Reis, Irving, 859
Reith, John C.W., **1208–11,** *1209*
  BBC local radio and, 208
  BBC World Service and, 222–23
  Broadcasting House and, 206
  Broadcast over Britain, 201
  debut, 894–95
  impact of U.S. radio on, 1434–35
  radio journalism and, 239
  South African broadcasting and, 17
  women's voices on, 1555
Religion on Radio, **1211–16**
Religious Broadcasting, 1412
Reller, Betty, 55
Remarkable Mouth commercial, 177
Remember WENN, 1386
Remington-Rand, 899
Remley, Frankie, 362
Remorse, 764
Renaud, Emiliano, 267
Renfroe, Everett, 25
Renfro Valley Barn Dance, **1217–18**
Report from Washington, 991
Report of the National Advisory Commission on Civil Disorders, 14
Report to the Nation, 1561
Republic ocean liner, 1424
Republic Radio, 1332
request programs, 979–80
Resnick, Max, 786
Rest of the Story, 691
Retro Formats, **1218–19**
Return of the Jedi, 1328
Reuters, 1020, 1021, 1023
Revell, Nellie, 356, 1459
Review of the British Weeklies, 1546
Rewind, 364
Rexroth, Kenneth, 709, 1057
Reymert, Martin, 772, 1464
R.J. Reynolds, 1078
Reynolds, Joey, 1543, 1557
Rhapsody of the Rockies, 836
Rhodes, Ray, 38
Rhodesia. *see* Zimbabwe
Rhymer, Paul, 361, 1460
Rhymes and Cadences, 1079
Rhythm and Blues, 1386
Ribbon of Song, 1081

Rice, Al, 1358
Richards, George A., 1535
Richards-Oakland Motor Car Company, 1535
Richardson, Bill, 273
Richardson, T.W., 53
Richardson's Roundup, 273
Richbourg, John, 1220, 1536
Richfield gasoline, 816
Richman, Don, 176
Richter, Bob, 1377
Richton, Addy, 1552
Rick Dees' Weekly Top 40, 74, 320, 438
Riddell, Jim, 1578
Riegel, O.W., 610
Righi, Augusto, 900
Right to Happiness, 493, 1072, 1552
Rinehart, Nolan "Cowboy Slim," 190
Rinker, Al, 418
Rin Tin Tin, 54
Rio, Rosa, 1078
Ripley, Robert, 1454
Rise and Fall of the Third Reich (Shirer), 1260
Rise of the Goldbergs, 250, 1551
Ritchie, Daniel L., 684
Rivera, Henry, 577
Rivers, Joan, 1371
Rivertown Trading Company, 948
RKO. *see* Radio-Keith-Orphenum
Roadhouse Nights, 509
Road to LIfe, 493, 1072, 1073
Robbins, Marty, 673
Robert Burns Panatella Program, 652
Roberts, Cokie, 45, 956, 1004, 1134, 1554
Roberts, Dmae, 1301
Roberts, John, 324324
Roberts, Ken, 1251
Roberts, Oral, 1008
Roberts, Robin, 1312
Roberts, William J., 570
Robertson, Hugh, 916
Robertson, Pat, 1008, 1215
Robeson, Paul, 475, 1274
Robinson, Claude, 118
Robinson, Earl, 475
Robinson, Jackie, 149
Robinsons (BBC). *see Frontline Family* (BBC)
Robson, William, 1198
"Rochester," 27
Rock, Rock, Rock, 592
Rock and Roll Format, 1105, **1219–21**
Rock Around the Clock, 592
Rockefeller, John D., 1161
Rock et Belles oreilles, 288
Rocket Ship Show, 28
Rodriguez, Jose, 816
Rodzinski, Artur, 338
Rogers, Roy, 95, 1274
Rogers, Shelagh, 273
Rogers, Will, 351, 362, **1221–22,** 1453
Rogers Communications (Vancouver, Canada), 40
Román, Enrique, 431
Romance of Canada, 270, 283
Romance of Helen Trent, 27, 414, 1287, 1415, 1552
Rome, Jim, **1223–24**
Ronald, William, 272
Roosevelt, Eleanor, 1121, **1224–27,** *1225*
Roosevelt, Franklin D., 541, 594–96, 1083, 1121. *see also* Fireside Chats
Ropin' in the Wind, 412
Rose, Billy, 195
Rosemary, 1552
Rosen, George, 416
Rosenblatt, Josef (Yossele), 785
Roslow, Sydney, 1140, 1141
Ross, David, 1078, 1080
Ross, Helen, 110
Ross, Les, 234
Ross, Sam, 290
Rosten, Norman, 1075
roster interviews, 114
Roth, Edward, 760
Roth, Sid, 787
Rothafel, Samuel L., 717–18, 1459
Rothenberg, Jerome, 1081
Roth v United States (1957), 1036
Rounds, Tom, 73, 801
Round the Horn, 221
Rountree, Martha, 1553
Rousseau, Alfred, 287
Rowlands, Sarah, 219
Roxy and His Gang, 718, 1366, 1453
Roy, James, 499
Royal Canadian Air Farce, 272, 287, 288
Roy Rogers Show, 1499, 1553
RTNDA Bulletin, 1411
Rubin, Harvey, 1371
Rubin, Roberto, 1306
Rudy Vallee Show, 26, 196, 718, 1279, 1451
Runyan, Brad, 492
Rusher, William, 1122
Rush Limbaugh Show, 1372
Russell, Audrey, 203–4, 241
Russell, Charles, 1589
Russell, James, 906
Russell, Tim, 907
Russia and Soviet Union, **1227–31**
  Cold War Radio, 344–45
  Radio Moscow, 1181–82
  radio propaganda, 1116
Rutherford, Joseph Franklin, 780
Rwanda, 19, 1464
Ryan, Frank, 1487
Ryan, J. Harold, 1577
Ryan, Kelly, 1301
Ryan, Meg, 321
Ryder, Alfred, 665
Ryman Auditorium, 673

Saatchi and Saatchi agency, 12
Sabates, 429
Sabbagh, Isa, 89
"safe harbor" concept, 1037
SAFRA Radio (Singapore), 105
Saga FM, 234
Sahayon, Robert Lewis, 416
Saint, Eva Marie, 1047
Salad Bowl Review, 46
Sales, Soupy, 1543
sales managers, 7
Sal Hepatica Review, 46
Salisbury, Cora, 156
Salyer, Steven, 1131
Samara, Noah, 21
samizdat radio stations, 306
Sam 'n' Henry, 78, 1337, 1509, 1541
Samoa, 1310
Sample, John Glen, 11, 737
sampling error, 116
sampling methods, 115–16
Sam Spade, 57, 1339
Samuels, Lynn, 1476
Sanders Brothers Radio Station, Federal Communications Commission v (1940), 1444
Sandler, Jerrold, 1126
Sanford, Eugene, 1065
Sanger, Elliot M., 1565
Sapoznik, Henry, 764, 787
Sarnoff, David, **1233–35,** *1234*
  cultural programming and, 1078
  Dempsey fight and, 1320
  E.H. Armstrong and, 98, 99, 100
  NBC and, 338, 918, 996
  RCA and, 1163
  reaction to talent raids, 1365
Satellite Radio. *see* Digital Satellite Radio
SATMUSIC, 554
SATNET programming, 97
Saturday Club, 233
Saturday Evening at Seth Parker's, 884
Saturday Night Beechnut Show, 336
Saturday Night Sewing Club, 26
Saudek, Robert, 61
Saulnier, Jean-Pierre, 281, 282
Savage, Barbara, 77
Savile, Jimmy, 232
Saxe, Henry, 868
Saxon, Louise, 1554
Scandinavia, 460, **1236–39**
Schafer, Paul, 136
Schaffner, Franklin, 909
Schechter, A.A., 1095, **1240–41**
Schechter, Danny, 1123
Schlesinger, Arthur, Jr., 1123
Schlessinger, Laura, Dr., 584, 1403, 1476, 1555
Schmeling, Max, 1320
Schmidt, Karl, 520, 1513
Schorr, Daniel, 368, 1269
Schramm, Wilbur, 417
Schreiber, Taft, 1365
Schuler, Robert, 581
Schulke, James, 524, 609, **1241–42,** 1377

Schulke Radio Productions (SRP), 1242
Schuster, Frank, 286
Schwartzkopf, Norman, 644
Schweitzer, Louis, 645
Science Fiction Programs, **1243–44**
Scooby Doo, Where Are You?, 801
Scooler, Svee, 786
Scotland Yard's Inspector Burke, 1338
Scott, Jane, 357
Scott, Raymond, 1588
H.H. Scott brand, 550
Screen Actors Guild (SAG), 66–67
Screen Directors' Playhouse, 718
Screen Guild Theater, 718
Scrimgeour, Colin, 1016
SCR-70 radio receiver, 805
Scruggs, Chuck, 1493
Scully, Vin, 149, 1319
Seagall, Lee, 524
Sealed Book, 730
Sealtest, 1492
Search for Tomorrow, 1288
Searching Paradise, 499
Sears, Roebuck and Company, 572, 994, 1357, 1533
Sears Radio Theater, 909
Secombe, Harry, 668
Second Honeymoon, 1150
Secret Storm, 1288
Seeing Ear Theater, 1244
See It Now, 353, 969
Seems Radio Is Here to Stay, 492
Segal, Nachum, 786
Seiler, James, 90, 91, 115, 455
Seldes, Gilbert, 415, 417
Selector (software), 1105
Self-Regulation for Broadcasters? (ABT), 130
Seligman, Olivia, 450
Selinger, Henry, 1070
Sellers, Peter, 668
Selvage, James P., 63
Send for Paul Temple, 218
Senegal, 21
Sepia Swing Club, 735
Sergeant Preston of the Yukon, 1094, 1350, 1499
Serguera, Jorge "Papito," 429
"Serious Literary, Artistic, Political or Scientific" value rule, 1036
Sermons in Song, 1214
Services Kinema Corporation (SKC), 236
Services Sound and Vision Corporation (SSVC), 236
Sevareid, Eric, 326, 352, **1244–46**, *1245*
Seven Day Trial of Vivienne Ware, 1535
"Seven Dirty Words" Case, 597, 1036, 1058, **1246–48**, 1481
Seven Lively Arts, 417
Sexually Speaking, 1555
Seymour, Dan, 1478
Shaden, Chuck, 1031
Shadow, 443, 493, 729, 730, 1030, 1103, **1248–52**, 1339
Shadow Broadcast Services, 40, 1505
Shadow Traffic, 1020
Shafer, Dave, 333
Shakespeare Festival (Ontario, Canada), 284
Shalit, Gene, 1019
Shane, Ed, 446
Shane Media Services, 1104–5
Shanley, Herbert v, 69
Shannon, Tom, 332, 333
Sharma, Reotic Sharan, 500
Sharpe, Mel, 506
Shaw, Allen, **1252–54**
Shaw, George Bernard, 217, 1121
Shaw, Stan, 471
Shayon, Robert Lewis, 476
Shebang, 801
Sheekman, Alan, 602
Sheen, Fulton, 1215
Sheldin, Herb, 1528
Shell Chateau, 718
Shepard, John, 608, **1254–56**, 1439, 1583
Shepard Department Stores, 1509
Shepherd, Jean, 363, **1256–58**, *1257*, 1369
Sheridan Broadcasting Network, 29, 173
Sherlock Holmes, 181, 493
Sherman, Paul, 1528
Sherwin, Louis, 1396
Sherwood, Robert, 61, 1041
Shields, Del, 28
Shin Nihon Hoso, 776
Shirer, William L., **1258–60**, *1259*
  blacklisting of, 1198, 1199
  CBS and, 352
  WWII reporting by, 1561
Shock Jocks, 309, 1037, **1261–62**
Shockley, William, 154, 1413
Sholis, Vic, 1407
Shore, Dinah, 977, 1454
Shortwave Radio, **1262–64**
  in Africa, 18–19
  antennas, 87
  Arab broadcasting, 89
  8 KIN (Australia), 134
  Westinghouse experimentation, 683
  World War II, 85
  Zenith Radio and, 1593
Showboat, 26, 170
Show Business Week, 696
Showcase, 173
Showers of Blessings, 1214
"Shuffle Along," 25
Shugart, Ralph, 1509
Shuman, Adrian, 274
Shuster, Frank, 272
Shutty-McGregor, Jo Jo, 333
Siam, 105
Sibbald, Hugh, 312
Sidewalk Backtalk, 634
Siegel, Mike, 1373
Siegel, Robert, 44, 45, 632, 1324
Siegel, Seymour, 535, 992
Siemering, William, **1264–66**, 1512
  NPR and, 42–43, 1001, 1132
  Sound Print and, 1301
  Susan Stamberg and, 1323, 1324
Siepmann, Charles A., 178, 208, 416, **1266–68**
Sieveking, Lance, 220
Sigma Delta Chi Distinguished Service Awards, 145
Sigmon, Lloyd, 95
Silber, Roslyn, 665
Silent Nights, **1268–69**
Silk Stockings, 55
Sills, Beverly, 1367
Silva, Nikki, 1514
Silver, Douglas, 63
Silver Baton Awards, 145
Silverblatt, Michael, 1081
Silver Theater, 835
Silvey, Robert, 204
Simmons Mattress company, 1224
Simon, Cadi, 1004
Simon, Scott, 1003, 1004, **1269–70**
Simon and Schuster, 561
Simplement Maria (Simply Maria), 942
Sims, Monica, 450
Simulcasting, **1271–72**
Sinatra, Frank, 353, 1367, 1454, 1588
Sinfonietta, 338
Singapore, 105
Singers on Radio, **1272–75**
Singing Story Lady, 320
Sing It Again, 1151
single sideband (SSB), 689
Sioussat, Helen, 883
Sirius Satellite Radio Network, 182, 254, 466–69, 779, 869, 1065
Sissle, Noble, 25, 356
Situation Comedy, **1275–78**
Six Hits and a Miss, 726
Six Shooter, 1500
$64,000 Question, 414, 1150
Skelly Oil, 299–300, 1094
Skelton, Red, 353, 363, **1279–80**
Sketches in Melody, 836
Skiffle Club, 233
Sklar, Rick, **1280–81**, 1475
Sklar Communications, 1281
Skornia, Harry, 417
Skouras, Charles F., 1528
Sky Blazers, 884
SkyFi radio, 468
Skyful of Lies, 226
Sky King, 318, 1024, 1340, 1499
SKY TV, 1176
Slaby, Adolph, 653
Slager Radio, 1284
Slaight, Allan, 324
"SLAPS rule," 1036
Slapsy Maxie Rosenbloom, 1339–40
Sleepy Joe, 1341

Sliwa, Curtins, 1476
Sloane, Allan, 1075
Small, Sidney, 174
Small Talk, 1553
Small Things Considered, 1168
Small Webcaster Settlement Act of 2002, 756
SmartRoute Systems, 1505
Smedley, Oliver, 238
Smiley, Tavis, 790
Smith, Alfred E., 541, 1083, 1120
Smith, Bernie, 1586
Smith, Bessie, 25, 170, 182
Smith, Fred, 474, 897, 1196, 1539
Smith, Howard K., 352, 368, 1198
Smith, J. Harold, 557
Smith, James, 440
Smith, Jim, 383
Smith, Kate, 583, 977, **1282–84,** 1553, 1561
Smith, Linda, 1514
Smith, Malcolm, 983
Smith, Oberlin, 1533
Smith, Ralph Lewis, 415
Smith, Ruth Cornwall, 1498
Smith, Stephen, 1301
Smith, William Wiley, 518
Smith, Willoughby, 518
Smith Family, 1286
Smith-Mundt Act of 1948, 345, 1116
Smithsonian Institute, 972
SMPTE (Society for Motion Picture and Television Engineers) time code, 1192
Smulyan, Jeffrey H., **1284–86**
Smythe, J. Anthony, 1046
Snow, Anthony, 957
Snowden, Mrs. Phillip, 1210
Snyder, Tom, 1361
Soap Opera, 11, 27, 493, **1286–89**
Social and Political Movements, **1293–95**
Social Class and Radio, **1289–93**
Société de Financement de Radiodiffusion (SOFIRAD), 22
Société de Radio-Canada, 259, 260, 272, 285
Société de Radio-Diffusion de la France d'Outre-Mer (Society for Radio Transmission to French Overseas Territories, SORAFOM), 18
Society of Broadcast Engineers, 1379–80
Society of Composers, Authors, and Publishers, 1274
Society of Wireless Telegraph Engineers, 1379
Society to Preserve and Encourage Radio Drama, Variety, and Comedy, 975, 1030
Socony Vacuum, 1358
Soft Rock Format, **1296–97**
Sokolsky, George, 167
Soldatensender, 750
Soldatensender 935 (Soldiers' Station 935), 657
Solid Gold, 438
Solomon Islands, 1310
Somalia, 19
Sonderling Broadcast Corporation, 1493
Song Hits, 356
Sonovox, 788
Sony, 1413, 1477
Sorry, Wrong Number, 1074
SOS emergency signals, 545
So This Is Radio, 639
sound, characteristics of, 120
Sound Effects, 1097, 1100, **1297–1300**
sound envelope, 120
Sound of Motown, 1504
Sound Portraits Productions, 764
Soundprint, 1266, **1301–2**
Soundprint Media Center, Incorporated, 1301
sound recorders, 115
Sounds Like Canada, 273, 274
Sounds Like Science, 1005
South Africa, 17
South African Broadcasting Corporation (SABC), 17
South America, **1303–8**
South East Asian Command Radio (SEAC), 104, 105
Southern, Ann, 719, 1554
Southern Baptist Hour, 1214
Southern California Broadcasters Association, 38
Southern Radio corporation, 1485
South Pacific Islands, **1309–11**
Soviet Union. *see* Russia and Soviet Union
Sowell, F.C., 1536
So You Think You Know Music, 1148
Space Patrol, 1243
Span, Al, *1587*
Sparrow and the Hawk, 1340
Spaulding, Norman, 173
Speaking Out, 135
Special Broadcasting Service (SBS; Australia), 132
Special of the Week, 993
Speiker, Carl, 1115
Spencer, June, 93
Spicer, Jaci, 1057
Spier, William, 1353
splicing blocks, 1100
Split-Second Tales, 1046
Sponsor magazine, 38, 1411
Spoonface Steinberg (Hall), 219
Spoon River Anthology, 1079, 1081
Sporting Chance, 220
Sporting News, 1316
Sports ByLiine USA, 1313
Sportscasters, **1311–20**
SportsCenter, 1313
Sports on Radio, 40, 72, 95, **1320–23**
Sports Report (BBC), 220
Sports World, 1312
"Spot" campaigns, 6
Spotlight, 274
Springbok Radio, 17
E.R. Squibb and Sons, 1221
Squier, Owen, 524
Sri Lanka, 104
Sri Lanka Broadcasting Corporation, 104
St. James, Fred. *see* Allen, Fred
St. John, Robert, 1198
St.-Laurent, Bernard, 273
Stage: Canada's National Theatre on the Air, 284, 286
Stahl, Myrtle, 1509
Stamberg, Susan, **1323–25,** *1324*
  All Things Considered and, 44–45, 1001–2, 1554
  Bob Edwards and, 537
  Weekend Edition Sunday, 45
Standard Brands, 1094, 1453
Standard-Cahill Corporation, 53
Standard Oil Company, 534, 601, 1358
Standard School Broadcasts, 534
Standby, 356
Stan Freberg Here, 620
Stan Freberg Show, 619
Stang, Arnold, 951
Staniar, Burton B., 684
Stanley, William, 648
Stanton, Frank N., **1326–27,** *1327*
  audience research by, 118, 854
  CBS and, 352, 812, 1039
Starch, Daniel, 112, 114
Starkey, Louise, 1551
Starlite Revue, 1493
Starr, Henry, 26
Star-Spangled Radio, 826
Stars Salute, 250
Star System, 1467
Star Theater, 495
Start the Week, 216
Star Wars, 495, 1003, **1328–31**
Stassen, Harold, 542, 1084
State Radio and Television Company (Dushanbe, Tajikistan), 103
Station Announcement Plan, 1043
Station Live Plan, 1043
Station Rep Firms, **1331–33**
Station Reps Association, 1332
Station Resources Group, 1135
Station Transcription Plan, 1043
station turnover *(T/O),* 9
Statistical Research Incorporated, 90
Stax Records, 1399
Steele, Alison, 583, 1554
Steele, Bob, 958
Steele, Don, 316
Steeves, H. Leslie, 1464
Stein, "Sleepy," 779
Steinberg, Martha Jean "The Queen," 583, 1554
Steinmentz, Charles, 648
Stella Dallas, 1552
Stenman-Rotstein, Eva, 499
Stephens, Evan, 954
Stephens, Larry, 668
Stepmother, 1288–89

Stereo, **1333–36**
Stereo Radio Productions, 1242
Stereotypes on Radio, **1337–42**
Stern, Bill, 1313, 1319–20
Stern, Howard, **1342–44**, *1343*
FCC and, 745
style, 363–64, 646, 1261, 1403
syndication, 276, 1371
Sternberg, George, 1141
Stevens, John Paul, 1247
Stevens, Shadoe, 74
Stewart, Bill, 614, 922
Stewart, Rex, 779
Stimme der DDR (Voice of German Democratic Republic), 657
Sting, 466
Stokowski, Leopold, 338
Stolen Husband, 737
Stone, Christopher, 220
Stone, Ezra, 31
Stone, Harlan F., 1391
Stone, John, 1379
Stone, Lynn, 1552
Stonewall Rebellion, 646
Stoppard, Tom, 218–19
Stop the Music!, 57, 1151
Storer, George, **1344–46**
Storer Broadcasting Company, 1576
Stories and Poems, Read by Uncle Ed over KDKA, 1077
Storm, Hanna, 1312
Stormfront, 695
Story-Corps, 764
Story of Alexander Graham Bell, 55
Story of a Wireless Telegraph Boy (Trowbridge), 316
Story of Mary Marlin, 495, 1552
Story of Ruby Valentine, 29, 173
Storz, Todd, 314, 472, 922, 1274, **1346–49**
Stouffer, Samuel, 1040
Stout, Phil, 524, 1242, 1377
Straight Arrow, 1500
Strange, Cy, 272
Strange Dr. Weird, 730
Strange Holiday, 1034
Strangest Secret, 1024
Streaming: Business of Internet Media, 1412
Street, Dusty, 1554
Street & Smith Detective Story Magazine Hour, 729
Stricht, Bob "Robbin," 1347
Strike It Rich, 1150
Striker, Fran, 878, **1349–51**
Stroebel, John, 1576
STRZ Entertainment network, 174
Stubblefield, Nathan, 519
Studebaker, John W., 68
Studio d'essai, 281
Studio Equipment. *see* Recording and Studio Equipment
Studio One, 908, 1481
studios, 136, 1193
Studs' Place, 1389
Studs Terkel Show, 1389
Study of Media and Markets, 934
Sturgeon, Theodore, 1257
Sturtevant, Emilie, 583
Suarez, Ray, 1367
Subsidiary Communications Authorization, 605, **1351–52**
Sudan, 19
Sudbrink, Woody, 1242
Sue and Irene, 1070
Suleman, Farid, 746
Sullivan, A.M. (Aloysius Michael), 1079
Sullivan, Ed, 156, 719, 1453
Sullivan, Elizabeth, 355
Summer, Dick, 1487
Summers, Harrison, 1454
Summit Communications, 1253
Sun Company (Shanghai), 108
Sunday, Billy, 556
Sunday Morning, 1048
Sunday Show, 1002
Sunday Vespers, 1211
Sun Oil, 1396
Sunrise Salute, 488
Sunshine Hotel, 764
Sunshine Orchestra, 25
Sunshine Radio, 760
Superman, 1243, 1340
Super Suds, 1094
Surf (detergent), 13
Suriname, 1308
Sur toutes les scènes du monde, 268, 281
surveys, 117
Suskind, David, 1142
Suspense, 730, 1074, **1353–54**
Sustaining Programs, **1354–56**
Sutherland, Sid, 816
Sutton, Percy, 28
Swaggert, Jimmy, 1008
Sweden, 499, 1237
"sweeps" periods, 91
Sweetland, Sally, 1080
Sweets, William, 1199
Swift, Hughie, 25
Swing, Raymond Gram, 168, 326, 366, **1356–58**, *1357*
Sykes, Eric, 668
Sykes, Eugene O., 580
Sykes, John, 746
Sykes Committee on Broadcasting, 12
Symphonium Serenaders, 25
Syndication, 137, **1359–62**, 1466–68
syntony, 874

Taishoff, Sol, 53, **1363–64**, 1411
Taiwan, 108–9
Take a Chance, 55
Take It or Leave It, 1150
Take Your Pick, 1176
Talent Raids, **1364–66**
Talent Shows, **1366–67**
Tales of the Texas Rangers, 1500
Talk2, 1223
Talkline with Zev Brenner, 786
Talk of the Nation, 1004, 1081, 1122, **1367–68**
Talk Radio, 41, 59, **1369–74**
Talk Radio (film), 593
Talk to Me, 1124, 1386
Tamm, Edward A., 1247
Tanganyika. *see* Tanzania
Tannenbaum, Jason, 1292
Tanneyhill, Ann, 27
Tan Town Coffee Club, 1492, 1523
Tan Town Homemaker's Show, 1493, 1554
Tan Town Jamboree, 1492, 1523
Tanzania, 18, 20
Tanzi, Callisto, 767
tape recorders, 1190–91
Tapex News, 290
Tapscan, 445
Tarplin, Maurice, 730
Tarrant, Chris, 298
Tashkent, Uzbekistan, 103
tax certificates, 15
Taylor, Davidson, 908
Taylor, Deems, 939, **1375–77**, *1376*
Taylor, Marlin R., 524, **1377–78**
Taystee Loafers, 360
Technical Organizations, **1378–82**
Technical Radio Laboratory, 580
technological developments, 66
Ted Mack's Original Amateur Hour, 56
Tedro, Henrietta, 868
Teenage Diaries, 320
Teilhet, Darwin, 416
Teishinshou, 775
Telecommunications Act of 1934, 469, 1382, 1427, 1517
Telecommunications Act of 1996, **1382–84**
deregulation and, 449, 1054, 1201
licensing and, 577
ownership rules, 28, 137, 171, 1129
radio station sales and, 249
on rebroadcasting signals, 253
Telefunken company, 653, 654
Telegram and Telephone Law of 1924 (Turkey), 101
Telegraphone, 125
telegraphy system, 874
Telekom-TELEFUNKEN Multicast, 457
telephone call-out and call-in, 116–17
Teleprompter Corporation, 1503
Tele-Que, 90, 455
Telescope, 909
Televerket, 1237
Televisa, 713
Television Depictions of Radio, **1384–87**
Television Mail, 1412
television receive-only (TVRO) antennas, 87
TelSat satellites, 467
Tender Leaf Tea, 1094
Tenney Committee (California), 167

Ten-Watt Stations, **1387–89**
Terkel, Louis "Studs," **1389–90**
terminal block, 72
terrorism, 1464
Terry, Earl, 1511
Terry and the Pirates, 250, 319
Tesla, Nikola, 519, **1390–92**
Test Match Special, 220
Texaco, 938, 939, 1366, 1579
Texaco Star Theater, 48, 719
Tex and Jinx, 719, 1553
Texas Barn Dance, 162
Texas Instruments, 1413
Thailand, 105
That Brewster Boy, 31, 1341
That's My Pop, 1340
Theater Guild on the Air, 152, **1392–93**
Théâtre de chez-nous, 331
Theocaris, John, 219
Theodore Bickel at Home, 786
These Are My Children, 1073
Theusen, Len, 312, 313
They Call Me, Joe, 27
They Fly Through the Air with the Greatest of Ease, 405, 492
They Live Forever, 352
Thieves Rush In, 395
Things American, 1357
13 by Corwin, 407
Thirty Minutes behind the Walls, 1484
This American Life, **1394–95**
This Day Is Ours, 1338
This Is New York, 1488
This Is Show Business, 562
This Is War, 405, 859, 1561
This Is Your Life, 539
This Life Is Mine, 1552
This Little Light, 476
This Morning, 270, 273, 274
This Way Out, 647
This Way Please, 592
This Week in Baseball, 50
Thomas, Clarence, 1404
Thomas, Dylan, 218
Thomas, Lowell, 1018, **1395–98,** *1397*
Thomas, Norman, 1121
Thomas, Parnell, 1198
Thomas, Rufus, **1398–1400**
Thompson, Bill, 589
Thompson, Dorothy, 368, 1553
Thompson, Elihu, 518
Thompson, Elsa Knight, 839, 1058
Thompson, Hank, 190
Thompson, Roy, 290
Thompson, Virgil, 696
J. Walter Thompson agency
  Chase and Sanborn Hour and, 11
  Eleanor Roosevelt and, 1226
  I Love a Mystery and, 1046
  Kellogg and, 13
  Kraft Music Hall production, 11
  Office of War Information and, 1043
  promotional recording, 12
  Radio Luxembourg and, 1175
  Rudy Vallee and, 1451, 1453
Thompson S.A. of France, 1163
Thorne, Richard, 730
Those Were the Days, 972
Thought for a Day, 1070
Thrasher, Marge, 1514
Three Corner Club, 1528
Three Days Lost, 494
Through the Looking Glass, 1375
Thunder Rock, 908
Thurber, James, 1072
timbre, definition, 120
Time for Beany, 618
Time for Reason—About Radio, 416
Time magazine, 11
time processors, 120–21
time spent listening (TSL), 447, 468
Tin Pan Alley, 976–77
Tisch, Lawrence, 684
Tison, Walter, 1569
Titanic, 546, 1424, 1530
Titleman, Carol, 1328
Titterton, Lewis (L.H.), 63, 1033
Today (U.K.), 242
Today at the Duncans, 1341
Today FM (Ireland), 761
Today's Children, 493, 1072, 1552
Today Show, 1241, 1312
Todd, Elizabeth, 1552
Tokar, Norman, 31
"Tokyo Rose" (Ikuku Toguri D'Aquino), 95, 1115, **1400–1402**
Tokyo Telecommunications Engineering Company, 1413
toll-free 800 numbers, 41
Tom Corbett, Space Cadet, 1243
Tom Joyner Morning Show, 790
Tom Joyner Movin' on Weekend Show, 790
Tom Leykis Show, 1371
Tom Mix, 318, 1094, 1340
Tommy Rigs and Betty Lou Show, 1525
Tom Swift and His Wireless Message (Appleton), 316
Tonchi Kyoushitsu (Wit Class), 777
Toni products, 663
Tony's Scrapbook, 957, 1077, 1078
Took, Barry, 221
Too Long America, 27
Top 40. *see* Contemporary Hit Radio/Top 40 Format
Topless Radio, **1402–4**
Top of the Pops, 233
Top o' the Morning, 1528
Torbet Radio, 1333
Torin, "Symphony Sid," 779
Torme, Mel, 868
Tornillo v Miami Herald (1974), 564
Toscan, Richard, 1328
Toscanini, Arturo, 338
Total Radio Users Guide, 1467
total service area (TSA), 91, 115
Total Telephone Frame listing, 165
Totenberg, Nina, 45, 1004, 1134, **1404–5,** 1554
To the Best of Our Knowledge, 1513
To the Young, 1561
Touch Radio (Wales, Great Britain), 298
Tovrov, Orin, 896
Town Hall Tonight, 46, 48, 719, 1495
To Your Health, 845
Tracht, Doug "Greaseman," **1405–7**
Tracy, Spencer, *891*
Trade Associations, **1407–10**
Trade Press, **1410–13**
traffic message channels (TMCs), 1166
traffic reports, 38
Trans-Atlantic Call, 1416
transcription discs, 1097, 1190
Transistor Radios, 154, 1185, **1413–14**
transmitters, 71, 226
Trans-Oceanic Radios, 1595
Transradio Press Service, 1018, 1023, 1095
Trans World Radio, 625, 626
Treasury Agent, 884
Treasury Star Parade, 977, 1034, 1043, 1455, 1562
Tree Grows in Brooklyn, 1340
Tremayne, Les, **1415**
Trendle, George W., 53, 1577
Tristani, Gloria, 1384
Trombley, Rosalie, 333
Tropical Radio Telegraph Company, 1422
Trotter, John Scott, 418
Trout, Robert, 352, **1416–18,** *1417*
Trowbridge, John, 316, 518, 519
True, Harold, 878
True or False, 1147
True Story, 998
Truman, Harry, 541, 1084, 1469
Truth or Consequences, 539, 1149
Tumbare Ghum Mere Hain (Your Woes are Mine), 500
tuning, single dial, 717
Tunney, Gene, 1320
Turkey, 101–2
Turkish Radio and Television Corporation (TRT), 102
Turkish Wireless Telephone Co., 101–2
Turner, Al, 162
Turner, Ike, 1567
Turow, Joseph, 905
Tuvalu, 1309–10
TV 9 Inc. v Federal Communications Commission (1973), 15
TV Radio Mirror, 569
Twende na Wakati (Let Us Go with the Times), Tanzania, 21
Twenty-Six by Corwin, 492
Twilight Zone, 1385
2 Live Crew, 1037
2XN (New York City), 532

Ulett, John, 548, 715
Ullswater Committee, 12
Ultimate History of Network Radio Programming, 1030
Ultra Violet, 1243
Uncle Don, 1556
Uncle Jim's Question Bee, 1147
Uncle Sam, 1043
Uncle Sam Speaks, 1027
Uncommon Sense: The Radio News Essays of Charles Morgan, 1124
Undecided Molecule, 602
Undecided Moment, 492
Underground Radio, **1419–21**
Under Milk Wood (Thomas), 218
Understanding Media, 991
Under the Loofah Tree, 395
Unger, Sanford, 45
Union Radio, 429
Unistar, 1023
United Church of Christ, 14, 448
United Church of Christ v Federal Communications Commission (1966), 14
United Fruit Company, **1421–23**
United Independent Broadcasters, 72, 1556
United Kingdom. *see also* British Broadcasting Corporation (BBC); British entries
  all-news radio, 40
  BBC Empire Service, 88, 1263
  censorship of radio, 306
  commercial radio in, 10, 12–13
  digital radio, 230–31
  FM stations, 607
  radio drama in, 497–98
  receivers, 1186
  Voice of the Listener and Viewer, 1471–72
United Nations Radio, 23, 754, **1423–24**
United Paramount, 57–59
United Press International (UPI), 1021, 1022–23, 1095, 1096
  purchasers of, 40
United Press Radio News Style Book, 1022
United States, **1424–30**. *see also under* American entries
  censorship of radio, 307–8
  Cold War Radio, 344–45
  digital audio broadcasting in, 460–61
  propaganda on radio, 1114
United States, Ginzburg v (1966), 1036
United States, Marconi Wireless Telegraph Company of America v (1943), 1391
United States, National Broadcasting Company v (1943), 596, 997, 1205, 1444
United States, Roth v (1957), 1036
United States Congress and Radio, **1431–34**
United States Department of Agriculture (USDA), 572
United States Influences on British Radio, **1434–37**
United States Library of Congress, 973
United States Navy and Radio, **1437–39**
United States Presidency and Radio, **1439–43**
United States Supreme Court and Radio, **1443–45**
United States v Gregg, 580
United States v Paramount Pictures (1948), 596
United States v Radio Television News Directors Association, 1200
United States v Zenith Radio Corporation (1926), 469, 1202, 1203, 1594
United-Whelen stores, 1495
University Association for Professional Radio Education (UAPRE), 243, 244, 530
University of Chicago Round Table, 76, 1094, 1121, 1541, 1551
University of Memphis Radio Archive, 974
Upson, Dean, 847
Up to the Minute, 1048
Urban Contemporary Format, **1446–47**
Urban Public Affairs network (UPAN), 174
URI (Union Radio Italiana), 767
Uruguay, 1307
U.S. Information Agency (USIA), 246
U.S. National Marconi Museum, 974
U.S. Steel Corporation, 1393
USA Digital Radio, 458, 460
USA Radio Network, 1023
Utah Radio Products, 1534
Utvarp HF, 1237

Vail, Myrtle, 1551
Valentine, Lewis, 644
Valera, Eamon de, 758, 759
Valiant Lady, 1552
Vallee, Rudy, 526, 1273, **1449–52**, *1450*
Vampirella (Carter), 218
Vance, John S., 1572
Vancouver Theatre, 284
Van Harvey, Art, *1461*, 1462
van Horne, Harriet, 416
Vanuatu, 1310
Van Voorhis, Westbrook, 897, 899
Vargas, Getúlio, 194
Variety, 355, 1410
Variety Radio Directory, 1410
Variety Shows, **1452–56**
Vatican Radio, 767, **1456–58**
Vaudeville, 11, 27–28, 359–60, **1458–60**
Vaughan, Guy, 312
VE9DR shortwave station, 310
Vendig, Irving, 1288
Venezuela, 1307–8
Vennat, Raoul, 267
Venture, 291
Veronica Omroep Organisatie (VOO), 1013
"V for Victory" campaign, 223–24
VHF radio. *see* FM radio
V-I: Story of the Robot Bomb, 908
Viacom, Inc., 350, 720, 798, 799, 1493
Vic and Sade, 361, **1460–63**
Vickland, Bill, 1358
Victor Talking Machine Company, 504, 1272–73
Victory F.O.B., 1535
Videodex, 90
Vietnam War, 97, 107
Views on the News, 822
Vignettes of Melody, 332
Village Voice, 357
Vinyl Cafe, 274
Violence and Radio, **1463–66**
Virgin Radio, 298
Virtual Radio, **1466–68**
Virtual Safari, 323
Vision of Invasion, 146
Visser, Leslie, 1312
Vitale, Dick, 1312
Vitale, Tom, 1081
Voegeli, Don, 43
Voegeli, Tom, 1328
Voice of America (VOA), **1468–71**
  ABSIE transmission of, 61
  Cold War and, 1115
  creation of, 750, 1041
  expansion of, 751
  jamming of, 774
  John Daly at, 433, 435
  programming, 1560
  Raymond Swing on, 1358
Voice of Firestone, 338
Voice of Freedom (Vietnam), 107
Voice of Greece, 677
Voice of Liberation, 334
Voice of Mongolia, 109
Voice of Prophecy, 1214
Voice of Taipei (Taipei, China), 109
Voice of Tangiers, 626
Voice of the Andes, 1306
Voice of the Broad Masses of Iritrea, 335
Voice of the Listener and Viewer, **1471–72**
Voice of the National United Front of Kampuchea, 334
Voice of the Poet, 1081
Voice of the Sarbedaran, 335
Voice of the Turkish Communist Party, 335
Voice of the United Front of Kampuchea, 1116
Voice of Truth, 335
Voices and Events, 696
Voices in the Purple Haze, 1253
Voices in the Wind, 1002, 1081
voice tracking, 1467
Volksempfänger (peoples receiver), 656
Volumax, 121
Voluntary Listener Sponsorship (Hill), 709
Volunteer Examiner Program, 690
Von Zell, Harry, 360
Vorhees, Donald, 338
Voron, Abe, 609
Vox Pop, 1147, **1472–74**
VU meter, 72

W1XAY (Boston, Massachusetts), 1255
W1XAZ (East Springfield, Massachusetts), 683
W1XER (Mt. Washington, New Hampshire), 1255
W2XMN (Alpine, New Jersey), 100, 603
W2XR (New York City), 716
W3XE, 1069
W8XO (Cincinnati, Ohio), 342
W9XEN, 917
W9XZV, 917, 1594
W47NV (Nashville, Tennessee), 604
WAAA (Winston-Salem, North Carolina), 255, 1446
WAAB (Boston, Massachusetts), 528, 912–13
WAAF (Chicago, Illinois), 65
WAAM (Jersey City, New Jersey), 1543
WABC (New York City), **1475–76**
  Alan Freed at, 623
  as Disney outlet, 59
  jingles used by, 789
  profitability, 57–58
  programming at, 57
  purchase price for, 350
  reverb on disk jockey voices, 121
  Rick Sklar at, 1281
  Rush Limbaugh on, 866–67, 921
  talk radio on, 1370
  Top 40 format, 59, 1103
Wace, Marjorie, 242
WACO (Waco, Texas), 256
Wade, Ernestine, 1277
Wade, Jimmy, 25
Wade in the Water: African-American Sacred Music Traditions, 1005
Wagner, Gwen, 1551
Wagner-Hatfield Amendment, 532
Wagontrain Enterprises, 316
Wahl, John, 816
WAHR (Miami, Florida), 821
WAIF (Cincinnati, Ohio), 375
Wait, Wait...Don't Tell Me, 364, 1005
WAIT-AM (Chicago, Illinois), 680
WAKE (Atlanta, Georgia), 484, 502
Wakelam, H.B.T., 220
Wake Up Call, 1482
WAKR (Akron, Ohio), 623
WALE (Rhode Island), 715
Walker, Johnnie, 233, 234, 1435
Walker, Laura, 1547
Walkie-Talkie, 961
Walkman, **1476–77**
Wallace, Henry A., 542
Wallace, John, 415, 416
Wallace, Mike, 319, 433
Wallace, Myron, 1578
WALL-AM-FM (Middletown, New York), 960
Wallenstein, Alfred, 338
Waller, Judith, 1551
Wallington, Jimmy, 360
Walpole, Helen, 1552
Walter, Eugene, 490
Walter Damrosch Music Appreciation Hour, 181
Walters, Mike, 44
Walter Winchell, 57
Waltons, 1386–87
WAMF (Amherst, Massachusetts), 1081
WAMO AM/FM (Pittsburgh, Pennsylvania), 173
WAMU-FM (Washington, D.C.), 1301
Wang, Deborah, 1004
WANN (Annapolis, Maryland), 1446
WANT (Richmond, Virginia), 1446
WAOK (Atlanta, Georgia), 1446
WAPE-AM (Jacksonville, Florida), 1406
warble jamming, 774
Ward, J. Truman, 1536
Ward, Jim, 1536
Ward, William Henry, 517–18
Ward L Quaal Company, 1145. *see also* Quaal, Ward L.
WARE (Ware, Massachusetts), 256
Waring, Fred, 1453, 1542
WARL (Arlington, Virginia), 411
Warner, Gertrude, 1251
Warner Lester, Manhunter, 878, 1349
Warner mergers, 720
Warnock, Tom, 1003
Warnow, Mark, 1588
War of the Worlds, **1478–81**
  impact of, 112, 492, 1074, 1243, 1428
  presentation, 443
  Welles in, 714
War Report (BBC), 221, 241
Warriors (film), 593
WARW-FM (Washington, D.C.), 1406
Washburn Crosby Company, 1489
WASH (Grand Rapids, Michigan), 1079
Washington, George Dewey, 25
Washington Post, 355, 1574
Wasserman, Lew, 1365
Watashiwa Daredeshou (Who Am I?), 777
Watch Tower Society, 780
Waterman, Willard, 674
Watermark, Inc., 73, 74, 920, 965
Watermark Productions, 801
Waters, Allan, 324
Waters, Ethel, 29, 173, 1274
Waters, James, 665
Waters, Mike, 1002
Watson, Thomas J., 1413
Watt, John, 240
WATT (portable station), 1090
Watts, Alan, 1057
WAVA (Arlington, Virginia), 38
Wavering, Elmer, 138
WAVN-FM (Memphis, Tennessee), 182
WAXC-AM (Rochester, New York), 1406
Way Down East, 250
Wayne, Johnny, 272, 286
Wayne and Shuster, 272
WBAI (New York City), **1481–83**
  Dick Biondi at, 164
  format, 627, 628, 1420
  Jerry Williams at, 1522
  Lunchpail, 1247
  Moulin Rouge Orchestra, 25
  Pacifica Foundation and, 645, 646, 1057, 1058
  programming, 495, 496
WBAL (Baltimore, Maryland), 178, 783
WBAP (Fort Worth, Texas), **1483–84**
WBAY (New York City), 71. *see also* WEAF (New York City)
WBBM (Chicago, Illinois), **1484–85**
  CBS and, 351, 352
  Democracy USA on, 511
  Halloween Martin at, 471
WBBR, 781, 782
WBBZ (portable station), 1091
WBEE (Chicago, Illinois), 1446
WBEM (Buffalo, New York), 1140
WBEO (Marquette, Michigan), 1145
WBEZ (Chicago, Illinois), 1394
WBFC (Stanton, Kentucky), 255
WBFO (Buffalo, New York), 632, 1001, 1265
WBGO-FM (Newark, New Jersey), 779
WBGR/WEBB (Baltimore, Maryland), 736
WBKB-TV (Chicago, Illinois), 58
WBKR (Barrow, Alaska), 1124
WBLI (Long Island, New York), 369
WBLS (New York City), 1446, 1466
WBMX (Chicago, Illinois), 1446
WBNX (New York City), 53
WBNY (Buffalo, New York), 801
WBOE (Cleveland, Ohio), 535
WBPI (New York City), 718
WBT (Charlotte, North Carolina), 424, 1335, **1485–86**
WBUF-FM (Buffalo, New York), 609
WBUR (Boston, Massachusetts), 301, 1124
WBZ (Boston, Massachusetts), 683, 684, 850, 1079, **1486–88**
WBZA (Springfield, Massachusetts), 684
WCAG (New York City), 959
WCAL (Northfield, Minnesota), 829
WCAP (Asbury Park, New Jersey), 53, 397, 1439
WCAP (Washington, D.C.), 26, 71, 169, 1321
WCAR (Detroit, Michigan), 176, 256
WCAU (Baltimore, Maryland), 409, 957
WCAU (Philadelphia, Pennsylvania), 962, 1044, 1060
WCBM (Baltimore, Maryland), 53
WCBR-FM (Chicago, Illinois), 647
WCBRJ (portable station), 1091
WCBS (New York City), **1488–89**
  CBS acquisition of, 352
  Faulk on, 574
  Galen Drake at, 488, 489
  news format, 38

oldies format at, 1044
poetry on, 1081
purchase price, 350
stereo broadcasting by, 1335
WCCC-AM/FM (Hartford, Connecticut), 715, 1342
WCCO (Minneapolis, Minnesota), **1489–90**
WCDA (New York City), 53
WCFL (Chicago, Illinois), **1490–92**
Allen Shaw at, 1253
Dick Biondi at, 153–64
format, 1420
Halloween Martin at, 471
jingles, 176–77
non-English language programs on, 1026
ownership, 1292
WCHB (Detroit, Michigan), 28, 1446
WCHD-FM (Detroit, Michigan), 28
WCHI (Batavia, Illinois), 780
WCIN (Cincinnati, Ohio), 1492
WCKY (Cincinnati, Ohio), 824
WCLR (Skokie, Illinois), 1378
WCLU (Cincinnati, Ohio), 825
WCLV (Cleveland, Ohio), 4
WCMZ (Miami, Florida), 84
WCNN (New York City), 26
WCNW (Brooklyn, New York), 53
WCOG (Greensboro, North Carolina), 438
WCOP (Boston, Massachusetts), 320
WCOW (Sparta, Wisconsin), 256
WCPO (Cincinnati, Ohio), 824
WCRB (Boston, Massachusetts), 339, 1334
WCRW (Santa Monica, California), 1081
WCX (Detroit, Michigan), 1534
WCYB (Bristol, Virginia), 1522
WDAF (Kansas City, Missouri), 41
WDAP (Chicago, Illinois), 987
WDAS (Philadelphia, Pennsylvania), 174, 1446
WDBS (Champaign-Urbana, Illinois), 253
WDEL (Wilmington, Delaware), 53, 1142
WDGY (Minneapolis, Minnesota), 1348
WDIA (Memphis, Tennessee), **1492–93**
African-American DJs on, 27
creation of, 184
format, 1446
Martha Steinberg at, 583
Maurice Hulbert at, 735
Nat D. Williams at, 1523
programming, 170, 171
Rufus Thomas at, 1399
WDM (Washington, D.C.), 1211
WDOG (Allendale, South Carolina), 256
WDSU (New Orleans, Louisiana), 506
WDT (New York City), 25, 356
WDVR (Philadelphia, Pennsylvania), 606, 1377
We, the People, 1147
WEAF (New York City), **1493–95**. *see also* WNBC (New York City)
advertising, 6, 10, 1426
ASCAP test case, 69
AT&T and, 71, 180
Atwater Kent Hour, 811
beginning of, 1426
ecumenism at, 784
Eveready Hour, 491
NBC and, 72
political programming, 1022, 1120
President Harding on, 1439
RCA and, 1163
sale of time, 1355
"Shuffle Along" cast, 25
singing on radio, 1272–73
sports on, 926, 1321
Will Rogers on, 1221
WEAN (Providence, Rhode Island), 634, 1254, 1583
WEAT-FM (Palm Beach, Florida), 1242
weather information, 1540
Weaver, Henry, 558
Weaver, Sylvester (Pat), 819, 949, **1495–97**
Webb, Dick, 92–93
WEBB (Baltimore, Maryland), 1446
WEBC (Duluth, Minnesota), 950
WebRadio.com, 1505
WEBR (Buffalo, New York), 1349
Webster, Jack, 292
Webster, Lance, 1109
Webster-Chicago, 1533, 1534
WEEI (portable station), 1091
Week Around the World, 696
Weekend Edition, 1555
Weekend Edition Saturday, 1269, 1270
Weekend Edition Sunday, 45, 1004
Susan Stamberg and, 1324–25
Weekending (BBC), 221
Week in Westminster, 221, 242
Weekly Top 40, 74
WEFC (Miami, Florida), 1492
WEFM (Chicago, Illinois), 339
We Hold These Truths, 492, 498
Wehrnalp, Erwin Barth von, 498
Weill, Kurt, 498
Weiner, Lazar, 785
Weiner, Michael A., 745
Weinstein, Richard, 165
Weir, Austin, 290
Weird Tales, 729
Weiss, Michael, 905
Welch, Elizabeth, 203
Welcome Back Mr. McDonald, 499
Welles, Orson, 48, 1249, 1479. *see also* *Mercury Theater of the Air; War of the Worlds*
Wells, H.G., 1478–81
Well Tempered Listener, 1375
Weltrundfunk-sender, 749
WENE-AM (Endicot, New York), 1406
WENR-AM (Chicago, Illinois), 693, 729
WENS (Shelbyville, Indiana), 1284
Wente, E.C., 704
Wentworth, Martha, 729
WERD (Atlanta, Georgia), 28, 171, 1446
Wertheimer, Linda, 44, 1134, **1497–98**, 1554
All Things Considered and, 45
West, Mae, 526
West, Ray, 834
Westaway, Jennifer, 273
Western Auto Supply Company, 1495
Western Electric, 71
Western Public Radio, 1000
Westerns, **1498–1501**
Westheimer, Dr. Ruth, 1403, 1555
Westinghouse, **1502–4**
Group W, 682–85
KDKA development, 1425
KYW and, 850–51
news format stations, 39
receivers, 1184
sports programming, 1321
WINS and, 38, 919
Westinghouse Broadcasting Company (WBC), 683, 1528
Westinghouse Radio Stations (WRS), 683
Westwood One, 40, 350, 1107, 1371, **1504–6**
WETA (Washington, D.C.), 1080
We the People, 474, 884
WEVD (New York City), 785, 786, 1026, 1081, **1506–7**
WFAA (Dallas, Texas), 162, 340, 1331, 1483
WFAN (New York City), 255, 743, 1284, 1314, 1322–23
WFAS (White Plains, New York), 53
WFBR (Baltimore, Maryland), 1335
WFCR (Amherst, Massachusetts), 1081
WFHB (Bloomington, Indiana), 999
WFIL (Philadelphia, Pennsylvania), 534
WFIL Studio Schoolhouse, 534
WFLN (Philadelphia, Pennsylvania), 339
WFMR (Milwaukee, Wisconsin), 339
WFMT (Chicago, Illinois), 253, 339, 477, 1080, 1389
WFMU (Jersey City, New Jersey), 786, 1420
WFNX-FM (Boston, Massachusetts), 647
WFOR (Hattiesburg, Mississippi), 671
WFOX (Atlanta, Georgia), 256, 1045
WFRG (Utica, New York), 256
WFYR-FM (Chicago, Illinois), 1044
WGAY-FM (Silver Spring, Maryland), 937
WGBG (Greensboro, North Carolina), 438
WGBH (Boston, Massachusetts), 993, 1134
WGBS (Miami, Florida), 1345. *see also* WINS (New York City)
WGEE (Indianapolis, Indiana), 1446
WGES (Chicago, Illinois), 26
WGI (Boston, Massachusetts), **1508–9**
WGLD-FM (Oak Park, Illinois), 1261, 1403
WGL (Fort Wayne, Indiana), 683
WGL (Philadelphia, Pennsylvania), 780
WGMS (Washington, D.C.), 339, 1048
WGN Barn Dance, 995
WGN (Chicago, Illinois), **1509–10**
Amos 'n' Andy rejection, 78, 1337
call letters, 255

children's programs, 320
Earl Nightingale at, 1024
Little Orphan Annie, 868
Quality Group and, 982
Scopes trial, 1018
Ward Quaal at, 1145
WGOD (Charlotte Amalie, Virgin Islands), 255
WGR (Buffalo, New York), 256
WGRE (Greencastle, Indiana), 1388
WGRL (Noblesville, Indiana), 256
WGST (Atlanta, Georgia), 680
WGTO-AM (Hanes City, Florida), 1252
WGUC (Cincinnati, Ohio), 1134
WGY (Schenectady, New York), 71, 490, 649, 1494
WHA and Wisconsin Public Radio, 346, 520, 1265, 1425, **1510–13**
Whad'Ya Know?, 364, 1131, 1151, 1513
WHAK (Rogers City, Michigan), 256
WHAM (Rochester, New York), 71, 256, 704, 1178
WHAS (Louisville, Kentucky), 782, 1217
WHAT (Philadelphia, Pennsylvania), 256, 736, 1446
What's My Line?, 49, 363, 433
What's New?, 55
What's on Your Mind?, 1369
What's Your Problem?, 1557
WHAW-AM (Hanes City, Florida), 1252
WHB (Kansas City, Missouri), 1064, 1346, 1348
WHBM (portable station), 1091
WHBQ (Memphis, Tennessee), 438, 1524
WHCT (Hartford, Connecticut), 1048
WHDH (Boston, Massachusetts), 53, 339
Wheaties, 319, 772, 1094, 1312, 1489
Wheaties Quartet, 976
Wheaton, Glenn, 95
Wheeler, Dan, 666
Wheeler, Ruthe S., 317
WHEL-AM (New Albany, Indiana), 537
When A Girl Marries, 1552
When Love Awakens, 1539
When Radio Was, 620
WHEN (Syracuse, New York), 256
WHER (Heidelberg, Mississippi), 256, 583
WHER (Memphis, Tennessee), **1513–14**
Whetstone, Walter, 53
White, Abraham, 440
White, Byron, 1200
White, J. Andrew, 1320
White, Jim, 833
White, Llewellyn, 417
White, Orval, 1298
White, Paul, 796, **1515–17**, *1516*
CBS and, 351, 365, 828, 1095
Robert Trout and, 1416
White, Steve, 1080
White, Wallace H., 582, 1202, **1517–19**, *1518*
White City, 207
Whitehead, Don, 1536
Whiteman, Paul, 418
Whitfield, Anne, 1047
Whitman, Ernest, 26
Whitton, John R., 610
WHKC (Columbus, Ohio), 563
WHN (New York City), 25, 785, 1317
WHO (Des Moines, Iowa), 256, 1196
Who Owns What, 1411
Whorf, Mike, 1535
Who Said That?, 634, 1416
WHOT-AM (Youngstown, Ohio), 163
WHTZ (Newark, New Jersey), 255
WHUR (Washington, D.C.), 28, 1446, 1447
WHYY (Philadelphia, Pennsylvania), 256, 632
WIBG (Philadelphia, Pennsylvania), 479
Wicker, Irene, 167, 1198
Wicker, Walter, 1072
Wide World of Sports, 1312
Wiersbe, Warren, 557
Wilcox, Harlow, 591
WILD-AM (Boston, Massachusetts), 173, 1446
Wild Bill Hickok, 1499
Wilderness Road, 1340
Wildmon, Donald, 1342
Wile, Frederic William, 1440
Wiley, Fletcher, 488
Wiley, Richard, 448
Wilkerson, Jim, 833
Williams, A.C. "Mooha," 27
Williams, Albert N., 416
Williams, Bert, 294, 1459
Williams, Billy Dee, 1328
Williams, Bruce, 1371, **1520–21**, *1521*
Williams, Jerry, 1369, 1370, 1372, **1522–23**
Williams, John, 1510
Williams, Juan, 1368
Williams, Kenneth, 221
Williams, Kim, 1325
Williams, Nat D., 184, 1399, 1492, **1523–24**
Williams, Robert F., Jr., 430
Williamson, Sonny Boy, 27, 183
Willis, J. Frank, 284
WILM (Wilmington, Delaware), 1142
Wilson, Don, 360, 816, **1525–26**
Wilson, Earl, 719
Wilson, Hugh, 1385
Wilson, Kathleen, 1046
Wilson, Kemmons, 1513
Wilson, Marie, 1554
Wilson, Robert, 154
Wilson, Woodrow, 1082, 1114, 1120
WIL (St. Louis, Missouri), 1052
Winchell, Walter, 694, 719, 792, 1369, **1526–27**
WIND (Chicago, Illinois), 256, 684, 953, 1044
WINR (Binghamton, New York), 163
WINS (New York City), **1528–29**
all-news format, 38, 919, 1019
call letters, 255
DJs, 623, 684
Dunphy on, 1317
Kaufman at, 963
Morrow at, 959
Rick Sklar at, 1280
Storer and, 1345
Winsor, Roy, 1288
Winter, Jonathan, 974
WINZ (Miami, Florida), 960
WIOD (Miami, Florida), 821
WIP-AM (Philadelphia, Pennsylvania), 937
WiRecorder Corporation, 1534
Wireless Acts of 1910 and 1912; Radio Acts of 1912 and 1927, **1529–32**. *see also* Radio Act of 1912; Radio Act of 1927
Wireless Age, 1411
Wireless Club of America, 1380
Wireless Institute, 1379
Wireless Ship Act of 1910 (PL 262, 61st Cong.), 1201, 1431
Wireless Specialty Apparatus Company, 1422
Wireless Station at Silver Fox Farm (Otis), 316
Wireless Telegraphy Act of 1905 (Canada), 263
"Wireless Workshop," 560
Wire Recording, **1533–34**
Wisconsin School of the Air, 534
Wise, Stephen S., 784, 786
Wishengrad, Morton, 494, 1076
Witches' Tale, 729
Witching Hour, 729
WITH (Baltimore, Maryland), 735
WITI (Milwaukee, Wisconsin), 1345
Witness, 291
WIXY (Cleveland, Ohio), 357
WJAR (Providence, Rhode Island), 634
WJAY (Cleveland, Ohio), 1026
WJAZ (Chicago, Illinois), 916
WJBK (Detroit, Michigan), 26, 799, 1063, 1345
WJBT (Chicago, Illinois), 1211, 1214
WJFK (Washington, D.C.), 1343
WJHU (Baltimore, Maryland), 1301
WJIB (Boston, Massachusetts), 1377
WJJD (Chicago, Illinois), 511
WJMK (Chicago, Illinois), 163, 164
WJMO (Cleveland, Ohio), 1446
WJNR (Newark, New Jersey), 1492
WJOB (Hammond, Indiana), 26
WJPC-AM (Chicago, Illinois), 790
WJR (Detroit, Michigan), 729, 1145, 1420, **1534–36**
WJSS (Rosswell, Georgia), 255
WJSV (Mount Vernon Hills, Virginia), 1416. *see also* WTOP (Washington, D.C.)
WJW (Cleveland, Ohio), 623, 799, 982, 1220, 1345
WJZ (Newark, New Jersey), 683, 850
WJZ (New York City). *see also* WABC (New York City)
beginning of, 1163, 1426

Cross at, 423
educational programming, 489, 534
Henry Morgan at, 950
Melrose Quartet, 25
NBC and, 72
"Shuffle Along" cast, 25
sports on, 1321
WKAP (Allertown, Pennsylvania), 1522
WKAT (Miami, Florida), 821
WKBS (Buffalo, New York), 1481
WKBW (Buffalo, New York), 163
WKDN (Philadelphia, Pennsylvania), 1522
WKHX (Atlanta, Georgia), 177
WKIX (Raleigh, North Carolina), 438
WKOZ (Kosciusko, Mississippi), 671
WKRP in Cincinnati, 1385
WKYC (Cincinnati, Ohio), 176
WKY (Oklahoma City, Oklahoma), 729, 1331
WKYS (Washington, D.C.), 1446, 1447
WKZ (New York City), 784
WKZO (Kalamazoo, Michigan), 691
WLAC (Nashville, Tennessee), 27, 183–84, 1220, **1536–37**
WLAK (Chicago, Illinois), 680, 1242
WLBW (Miami, Florida), 822
WLBZ (Bangor, Maine), 1583
WLIB (New York City), 1446
WLIF (Baltimore, Maryland), 1242
WLIR (Long Island, New York), 786
WLOK-AM (Memphis, Tennessee), 790, 1399
WLS (Chicago, Illinois), **1537–38**. *see also* *National Barn Dance*
country music, 1217
creation of, 572
Dick Biondi at, 163
educational programming on, 534
Hindenburg Disaster, 710
National Barn Dance, 140, 994, 1274
ownership, 58
programming on, 573
Stand By, 356
talk format, 59
WLW (Cincinnati, Ohio), **1538–41**
country music, 1217
Crossley and, 421, 422
drama on, 491
power, 342
Red Foley Show, 1279
WMAC (Chicago, Illinois), 718
WMAC (New York City), 26, 950
WMAF (South Dartmouth, Missouri), 71
WMAL (Washington, D.C.), 955, 957
WMAQ (Chicago, Illinois), **1541–42**
Amos n' Andy and, 78
drama on, 491
Irna Phillips at, 1072
Lum 'n' Abner, 888
ownership, 409
purchase of, 684
Richard Durham on, 513
WMAX-AM (Grand Rapids, Michigan), 381
WMBI (Chicago, Illinois), 556, 1212
WMCA (New York City), 963, 965, 1345, 1370, 1520
WMC (Memphis, Tennessee), 25, 182
WMET-FM (Chicago, Illinois), 937
WMEX (Boston, Massachusetts), 1522
WMGM (New York City), 186, 951, 963, 1281
WMGR (Bainbridge, Georgia), 484
WMMN (Fairmont, West Virginia), 711
WMMS (Cleveland, Ohio), 357
WMPR-FM (Jackson, Mississippi), 182
WMPS (Memphis, Tennessee), 438
WMRA (Montgomery, Alabama), 790
WMRY (New Orleans, Louisiana), 1492
WMSG (New York City), 53
WMTS (Murfeesboro, Tennessee), 1067
WMXJ-FM (Miami, Florida), 1045
WNAC (Boston, Massachusetts), 784, 1254, 1439, 1494, 1583
WNAS (New Albany, Indiana), 1388
WNAX (Yankton, South Dakota), 1077
WNBC (New York City), 742, 1162, 1322–23, 1342, **1542–43**. *see also* WEAF (New York City); WFAN (New York City)
WNBH (New Bedford, Massachusetts), 1583
WNCN Listeners Guild, Federal Communications Commission v (1981), 1445
WNDO (Auburn, New York), 163
WNET (New York City), 1001
WNEW (New York City), 53–54, **1543–45**
African-Americans and, 175
all night radio, 41
DJs on, 471, 583
format, 614, 1420
Group W and, 684
jingles on, 798
Karmazin at, 937
Metromedia and, 788
singers on radio, 1274
WNJR (New York City; Newark, New Jersey), 1446
WNOE (New Orleans, Louisiana), 460, 1052
WNOR-FM (Norfolk, Virginia), 715
WNTN (Newton, Massachusetts), 1342, 1420
WNTS (Indianapolis, Indiana), 1284
WNUS (Chicago, Illinois), 38, 925
WNYC (New York City), **1545–47**
APR and, 1134
educational programming, 536, 990
Rick Sklar at, 1280
Small Things Considered, 1168
WNYC's Air College, 1545
WNYE (New York City), 536
WOAI (San Antonio, Texas), 783, 1026, 1331
wobbler jamming, 774
WOC (Davenport, Iowa), 1196
WODA (Paterson, New Jersey), 1543
Wolf, 490
Wolfe, Edwin, 896
Wolfe, Miriam, 729
Wolfert, Jonathan, 789
Wolfman Jack—Live at the Peppermint Lounge, 1549
Wolfman Jack (Robert Smith), 190, 714, 960, **1547–50**, *1548*
Wolfman Jack Show, 1549
Wolters, Larry, 355
WOL (Washington, D.C.), 28, 53, 1446
Woman in White, 493, 1072, 1552
Woman of America, 494
Woman's Day, 415
Woman's Hour, 221
Women in Communications, Incorporated (WICI). *see* Association for Women in Communications
Women in Radio, **1550–56**
Women of Courage, 957
WOMN-AM (Connecticut), 583
WOMP (Bellaire, Ohio), 256
Wonder Bakers, 1453
Wons, Tony, 1078
Wood, Henry, 1110–11
Woodbury Soap Show, 726
WOOD-FM (Grand Rapids, Michigan), 1242
Woodhouse and Hawkins, 286
Woodruff, Louis, 1, 115, 118
Woods, Granville, 519
Woods, Lesley, 1251
Woods, Randy, 1220
Woody Herman Band Show, 1454
WOOK (Washington, D.C.), 26, 1446
WOR (Newark, New Jersey), 71, 641
WOR (New York City), **1556–57**
format, 524, 1419
Galen Drake at, 489
Henry Morgan at, 950
Jean Shepherd Show, 1256–57
Kaufman at, 965
McBride at, 913
Newscasting, 898
programming, 729
Quality Group and, 982
talk radio on, 1369
WORC (Worcester, Massachusetts), 256, 1583
WORD (Batavia, Illinois), 780
Word Jazz, 1080
Words in the Night, 1080
Words without Music, 475, 492, 859, 1079
Work, George, 1078
World, 1123, 1131
World Administrative Radio Conference (WARC), 630
World Association of Community Radio Broadcasters, 1303
World at One (BBC), 221, 241
World at Six (Canada), 272, 274
WORLDNET, 1173, 1469

World Radiocommunication Conference, **1558–59**
WorldSpace, 21–22, 457, 467, 751
World Today, 1123
World Tonight, 216, 241
World War I
  impact on musicians, 64
  radio in, 1424–25
  U.S. Navy and radio, 1437
  violence and radio, 1463–64
World War II
  African-American radio during, 27
  African broadcasting during, 18
  all-night radio programming, 41
  American Broadcasting Station in Europe, 61
  American Forces Radio Services, 94–95
  AM radio and, 84, 85
  Arab radio and, 88–89
  Axis Sally, 145–48
  BBC and, 204
  BBC Monitoring Service, 211–14
  Canadian broadcasting during, 271
  censorship during, 309
  Edward R. Murrow and, 966, 968
  fireside chats, 594–95
  Foreign Broadcast Intelligence Service, 611
  France during, 615
  impact on musicians, 65
  international broadcasting, 750–51
  Lord Haw-Haw, 882–83
  Mexican radio and, 941–42
  news radio and, 38
  radio advertising and, 6
  radio documentaries, 474–75
  radio journalism and, 240–41
  radio propaganda, 1114–15
  radio sponsorship during, 11
  rationed newsprint, 130
  shortwave radio and, 85
  Soviet radio during, 1228
  violence and radio, 1464
  Winston Churchill, 325–28
  Zenith products, 1594
World War II and U.S. Radio, **1559–65**
WORM (Savannah, Tennessee), 256
Wormys, Rudolf, 334
WORT (Madison, Wisconsin), 375
WOSU (Columbus, Ohio), 534
WOV (New York City), 574, 1026
WOWO (Fort Wayne, Indiana), 683
WOW (Omaha, Nebraska), 256, 1211
WPAC (Patchogue, New York), 1280
WPAP (New York City), 718
WPAT (Paterson, New Jersey), 524, 574
WPAY (New Haven, Connecticut), 1334
WPAY (Portsmouth, Ohio), 824
WPCH (Atlanta, Georgia), 1242
WPEN (Philadelphia, Pennsylvania), 53, 1352
WPFK (Los Angeles, California), 1124
WPFW (Washington, D.C.), 321, 1057, 1058
WPGC-FM (Washington, D.C.), 1447
WPOP-AM (Hartford, Connecticut), 1406
WPOW, 782
WPRO (Providence, Rhode Island), 256, 714, 958
WPTR (Albany, New York), 1253
WQAM (Miami, Florida), 1347
WQSR-FM (Baltimore, Maryland), 1045
WQXI (Atlanta, Georgia), 502, 1385
WQXR (New York City), **1565–66**
  Corwin on, 405
  Dolby broadcasts, 478
  FM broadcasts, 605
  high fidelity broadcasts, 704
  music format, 339, 716
  poetry on, 1079
  Quincy Howe at, 733, 734
  stereo broadcasts, 1334
WRAP (Norfolk, Virginia), 1446
Wray, Fay, *891*
WRC-AM (Washington, D.C.), 1406
Wren, Jack, 167
WRFM (New York City), 524, 1377
Wright, Bill, 28
Wright, Early, 27, **1567–68**
Wright, Frank, 1008
Wrigley's gum, 142, 1499, 1589
Writer's Almanac, 1081
WRKO (Boston, Massachusetts), 1372, 1522, 1523
WRKS (New York City), 1557
WRNG (Atlanta, Georgia), 680
WRNR (Washington, D.C.), 1056
WRNY (New York City), 784
WROX (Clarksdale, Mississippi), 27, 1567
WRR (Dallas, Texas), **1568–69**
WRUF (University of Florida), 149
WSAI (Cincinnati, Ohio), 1256
WSAR (Fall River, Massachusetts), 53
WSAZ (Huntington, West Virginia), 824
WSB (Atlanta, Georgia), 71, 256, 544, 1065, **1569–70**
WSBC (Chicago, Illinois), 25, 26, 169
WSDZ (St. Louis, Missouri), 320
WSGN (Birmingham, Alabama), 438
WSIA (Staten Island, New York), 787
WSM (Nashville, Tennessee), **1571–72**
  call letters, 572
  Deford Bailey on, 25
  Grand Ole Opry, 573, 672, 673, 1274
  marketing through, 825
  Perryman at, 1067
WSNO (Barre, Vermont), 256
WSNY (Schenectady, New York), 178
WSPA (Spartenburg, South Carolina), 424
WSPD (Toledo, Ohio), 1344–45
WSRC (Durham, North Carolina), 1446
WSUN (Tampa, Florida), 256
WSXK (Pittsburgh, Pennsylvania), 683
WTAE (Pittsburgh, Pennsylvania), 177
WTAL (Toledo, Ohio), 1344
WTAT (portable station), 1090
WTIC-AM (Hartford, Connecticut), 958
WTIP (Charleston, West Virginia), 479
WTIX (New Orleans, Louisiana), 1346, 1348
WTKO-AM (Ithaca, New York), 1406
WTMJ (Milwaukee, Wisconsin), 529
WTOB (Winston-Salem, North Carolina), 438
WTOL (Toledo, Ohio), 178
WTOP's Man about Town, 1573
WTOP (Washington, D.C.), 256, 1416, **1572–73**
WTVM (Syracuse, New York), 564
WUFO-AM (Buffalo, New York), 173
WUHY (Philadelphia, Pennsylvania), 632
WUNC (Chapel Hill, North Carolina), 845
WUNR (Brookline, Massachusetts), 786
Wurlitzer Company, 296
WVLS (Monterey, Virginia), 375
WVOK (Birmingham, Alabama), 671
WVOL (Nashville, Tennessee), 28
WVOM-FM (Bangor, Maine), 546
WVON (Chicago, Illinois), 28, 1446
WVPO (Stroudsburg, Pennsylvania), 208, 660
WWBA-FM (Tampa, Florida), 937
WWBX-FM (Bangor, Maine), 546
WWDB (Philadelphia, Pennsylvania), 1523
WWDC (Washington, D.C.), 339, 1352, 1406
WWEE (McMinniville, Tennessee), 256
WWET (Valdosta, Georgia), 256
WWEZ (New Orleans, Louisiana), 1347
WWIN (Baltimore, Maryland), 736
WWJ (Detroit, Michigan), 1425, **1573–74**
WWKX (Providence, Rhode Island), 478
WWL (New Orleans, Louisiana), 529, **1575–76**
WWNO (New Orleans, Louisiana), 461
WWNS (Statesboro, Georgia), 484
WWRL (New York City), 736, 1446
WWSW (Pittsburgh, Pennsylvania), 1045
WWTC (Minneapolis, Minnesota), 1168
WWVA (Wheeling, West Virginia), 1345, **1576–77**
WWVZ (Washington, D.C.), 369
WWWW (Detroit, Michigan), 925
WXEL-TV (Cleveland, Ohio), 623
WXLO (New York City), 1557
WXPN (Philadelphia, Pennsylvania), 320, 321, 646
WXRK (New York City), 1342
WXVI (Montgomery, Alabama), 256
WXXX (Philadelphia, Pennsylvania), 1522
WXYZ (Detroit, Michigan), **1577–78**
  ABC purchase of, 56
  foundation of, 1360
  Fran Striker at, 1349
  Green Hornet, 678
  Kasem at, 799
  Lone Ranger, 878, 1499
  ownership, 58

payola scandel and, 1062
Quality Group and, 982
WYIS (Philadelphia, Pennsylvania), 1215
WYLD (New Orleans, Louisiana), 1446
Wynn, Ed, 52–53, 1459, 1542, **1579–82**, *1580*
WYNY (New York City), 789
WYPL, 1081
WYSP (Philadelphia, Pennsylvania), 1342

XEAW-AM (Reynosa, Mexico), 189
XEB (Mexico City, Mexico), 940
XED-AM (Reynosa, Mexico), 190
XEEP, 942
XEFD (Tijuana, Mexico), 175
XEOY-Radio Mil, 941
XEPN-AM (Piedras Negras, Mexico), 190, 199
XER/XERA-AM (Villa Acuna, Mexico), 189, 190, 199
XERF (Ciudad Acuna, Mexico), 190, 557, 1549
XETRA-AM (Tijuana, Mexico), 38, 190–91, 1335
XEW (Mexico City, Mexico), 713, 941
XM Corporation, 466–69, 1065
X Minus One, 1243, 1429
XM Satellite Radio Service, 182, 254, 779, 869
XREB (Los Angeles, California), 1549
Xtra-AM (Birmingham, England), 298
XTRA (Los Angeles, California), 925
XTRA (San Diego, California), 1223
XWA (Montreal, Quebec), 258, 269, 310

Yahraes, George, 1253
Yankee Network, 912–13, 1254, **1583–84**
Yankee News Service, 1023
Yankovic, "Weird Al," 483
Yarborough, Barton, 1046
Yarrow, Peter, 1168
Ydstie, John, 1004
Year 2000 Plus, 1243
Yeltsin, Boris, 1173, 1229
Yentob, Alan, 450
Yiddish Melodies in Swing, 785
Yiddish Radio Project, 764, 787
Yiddish Voice, 786
YMCA radio classes, 25
Yore, J.J., 906
You Are There, 476, 968
You Bet Your Life, 602, 1149, **1584–86**
Young, Ella, 1122
Young, Jimmy, 232
Young, Marian, 1553
Young, Owen D., 176, 649, 1163
Young, Scott, 290
Young Ambassador, 557
Young and Rubicam agency, 11, 12, 718, 1241, 1495
Youngbloods of Beaver Bend, 283
Young Dr. Malone, 1561
Young Widder Brown, 1552
Your Hit Parade, 56, 708, 1273, **1587–88**
Your Hollywood Parade, 726
Your Lucky Strike, 55
Yours Truly, Johnny Dollar, 270, **1589–91**
Youth Opportunity Program, 1367
Youth Takes a Stand, 349
Yugoslavia, 306
Yurdin, Larry, 1420
"*Yuseishou,*" 777

Zack and Zoey's Survival Guide, 323
Zambia, 18, 20–21
Zapple, Nicholas, 564
Zapple Rule, 564
ZBM (Hamilton, Bermuda), 959
Zenith-dbx TV stereo system, 1336
Zenith Radio Corporation, 580, 916, 1185, **1593–95**
Zenith Radio Corporation, United States v (1926), 469, 1202, 1203, 1594
Zero Hour, 1401
Ziegfeld Follies, 195, 196
Ziegfeld's Follies of the Air, 1221
Zimbabwe, 18
Zimmerman, Marie, 583, 1551
Ziowe Advertising Agency, 798
Zip codes, 905
Z-Nith products, 1593
Zola Levitt Presents, 787
Zworykin, Vladimir, 1163, 1227, 1503